TRANSACTIONS

OF THE

INTERNATIONAL

ASTRONOMICAL UNION

VOLUME XXIIIB - PROCEEDINGS

INTERNATIONAL ASTRONOMICAL UNION

98bis Bd Arago 75014 PARIS France

Tel. 33 1 43 25 83 58 - Fax 33 1 43 25 26 16

e-mail: iau@iap.fr - www: http://www.iau.org

INTERNATIONAL ASTRONOMICAL UNION
UNION ASTRONOMIQUE INTERNATIONALE

TRANSACTIONS

OF THE

INTERNATIONAL ASTRONOMICAL UNION

VOLUME XXIIIB

PROCEEDINGS OF THE TWENTY-THIRD

GENERAL ASSEMBLY

KYOTO 1997

Edited by

JOHANNES ANDERSEN
General Secretary of the Union

KLUWER ACADEMIC PUBLISHERS
DORDRECHT / BOSTON / LONDON

A C.I.P. Catalogue record for this book is available from the Library of Congress.

ISBN 0-7923-5588-1

*Published on behalf of
the International Astronomical Union
by
Kluwer Academic Publishers, P.O. Box 17, 3300 AA Dordrecht, The Netherlands.*

*Sold and distributed in North, Central and South America
by Kluwer Academic Publishers,
101 Philip Drive, Norwell, MA 02061, U.S.A.*

*In all other countries, sold and distributed
by Kluwer Academic Publishers,
P.O. Box 322, 3300 AH Dordrecht, The Netherlands.*

Printed on acid-free paper

Printed in the Netherlands

Président de l'Union Astronomique Internationale

LODEWIJK WOLTJER

PRESIDENT OF THE INTERNATIONAL ASTRONOMICAL UNION

1994-1997

CONTENTS

Preface xiii

CHAPTER I - INAUGURAL CEREMONY 1

Address by Prof. Dr. H. Yoshikawa, President of the Science Council of Japan 1
Address by Prof. Dr. H. Okuda, President of the Astronomical Society of Japan 2
Address by Prof. Dr. L. Woltjer, President of the International Astronomical Union 3
Okotoba (Address) by His Majesty the Emperor 4
Message from Mr. R. Hashimoto, Prime Minister of Japan 6

CHAPTER II - TWENTY THIRD GENERAL ASSEMBLY 7

First Session 7

1. Opening of the Session 7
2. Appointment of Official Interpreters 7
3. Appointment of Official Tellers 7
4. Report of the Executive Committee 1994-1997 *(summary)* 7
5. Report on the Work of the Special Nominating Committee 8
6. Official Representatives of Adhering Organisations at the General Assembly and on the Nominating Committee 8
7. Acting Presidents of Divisions and Commissions 9
8. Appointment of the Finance Committee and Corresponding Votes 10
9. Revision of the Statutes and Bye-Laws 11
10. Resolutions Proposed by Divisions 11

Second Session 13

11. Financial matters 13
    Report of the Finance Committee 13
    Treasurer's Report 15
    Approval of budget for 1998-2000 15
    Appointment of a Standing Finance Sub-Committee 16
12. Appointment of Division Presidents, Commission Presidents and Vice-Presidents, and Presidents of EC Working Groups 16
13. Resolutions Proposed by the Executive Committee and by Working Groups of the EC 18
14. Appointment of the Resolutions Committee 1997-2000 18
15. Appointment of the Special Nominating Committee 1997-2000 18
16. National Membership of the Union 18
17. Individual Membership of the Union 18
18. IAU Representatives to Other International Organizations 19
19. Places and Dates of the XXIVth and XXVth General Assemblies 19
20. Election to the Union of a President, a President-Elect, six Vice-Presidents, a General Secretary and an Assistant General Secretary 20

CHAPTER III - CLOSING CEREMONY 21

Address by the President 1994-1997, Prof. L. Woltjer 21
Address by the President 1997-2000, Prof. R.P. Kraft 21
Address Dr. Robert J. Rubin, on behalf of the Registered Guests 22
Address by Dr. G. Cayrel de Strobel on behalf of the participants 22
Address by the retiring General Secretary, Prof. I. Appenzeller 23
Address by the incoming General Secretary, Dr. J. Andersen 24

CHAPTER IV - RESOLUTONS OF THE GENERAL ASSEMBLY 25

Report of the Resolutions Committee 25
Resolutions A1 - A4 30
Resolutions B1 - B7 36

CHAPTER V - REPORT OF THE EXECUTIVE COMMITTEE 1994-1997 53

Executive Committee 1994-1997 53

1. **Administrative Matters** 53
Executive Committee Meetings 53
Officers' Meetings 53
IAU Secretariat 53
IAU Archives 53
Revision of Statutes, Bye-Laws, and Working Rules 54
Adhering Countries 54
Individual Members and Consultants 54
Divisions 55
Commissions 55
Working Groups of the Executive Committee 56
Representation at other International Organizations 56

2. **Scientific Matters** 58
Scientific Meetings 1994-1997 58
IAU Publications 60

3. **Financial Matters** 63
Budget 1994-1997 63
Statement of Income and Expenditure 1994-1996 66
Proposed Budget 1998-2000 69

CHAPTER VI - REPORTS OF DIVISIONS, COMMISSIONS, AND WORKING GROUPS 73

DIVISIONS AND THEIR COMMISSIONS

I Fundamental Astronomy 75
4 Ephemerides 77
7 Celestial Mechanics and Dynamical Astronomy 79
8 Positional Astronomy **and** 85
24 Photographic Astrometry
19 Rotation of the Earth 101
31 Time 107

II Sun and Heliosphere 111
10 Solar Activity **and** 113
12 Solar Radiation and Structure **and**
49 Interplanetary Plasma and Heliosphere

III	**Planetary System Sciences**	117
15	Physical Study of Comets, Minor Planets and Meteorites	119
16	Physical Study of Planets and Satellites	123
20	Position and Motions of Minor Planets, Comets and Satellites	NA
21	Light of the Night Sky	127
22	Meteors and Interplanetary Dust	129
51	Bioastronomy: Search for Extraterrestrial Life	131
IV	**Stars**	133
26	Double and Multiple Stars	135
29	Stellar Spectra	NA
35	Stellar Constitution	139
36	Theory of Stellar Atmospheres	141
45	Stellar Classification	143
V	**Variable Stars**	147
27	Variable Stars	149
42	Close Binary Stars	153
VI	**Interstellar Matter**	157
34	Interstellar Matter	159
VII	**Galactic System**	163
33	Structure and Dynamics of the Galactic System	NA
37	Star Clusters and Associations	167
VIII	**Galaxies and the Universe**	169
28	Galaxies	171
47	Cosmology	173
IX	**Optical Techniques**	177
9	Instruments	179
25	Stellar Photometry and Polarimetry	181
30	Radial Velocities	185
X	**Radio Astronomy**	189
40	Radio Astronomy	191
XI	**Space and High Energy Astrophysics**	193
44	Space and High Energy Astrophysics	195

COMMISSIONS OF THE EXECUTIVE COMMITTEE

5	Documentation and Astronomical Data	201
6	Astronomical Telegrams	207
14	Atomic and Molecular Data	209
38	Exchange of Astronomers	211
46	Teaching of Astronomy	215
50	Protection of Existing and Potential Observatory Sites	221

COMMISSION NOT ATTACHED TO A DIVISION

41	History of Astronomy	223

WORKING GROUPS OF THE EXECUTIVE COMMITTEE

Working Group on Encouraging the International Development of Antarctic Astronomy 229
Working Group on Planetary System Nomenclature 231
List of newly approved names in the Solar System 232
Working Group on the World Wide Development of Astronomy 253
Working Group on Future Large Scale Facilities 255

CHAPTER VII - STATUTES, BYE-LAWS AND WORKING RULES 257

English texts: Statutes 257
Bye-Laws 261
Working Rules 265

French texts: Statuts 271
Règlement 275
Directives 279

CHAPTER VIII – RULES FOR SCIENTIFIC MEETINGS 285

CHAPTER IX – MEMBERSHIP 299

List of Adhering Organizations 299

Commission Membership 309

List of Division & Commission Working Groups 405

Individual Membership from Adhering and *Associate* Countries 405

ALGERIA 411
ARGENTINA 411
ARMENIA 411
AUSTRALIA 412
AUSTRIA 413
AZERBAJAN 413
BELGIUM 413
BOLIVIA 414
BRAZIL 414
BULGARIA 415
CANADA 415
CHILE 417
CHINA Nanjing 417
CHINA Taipei 419
CROATIA 420
CZECH Rep 420
DENMARK 420
EGYPT 421
ESTONIA 421
FINLAND 421
FRANCE 421
GEORGIA 425
GERMANY 425
GREECE 428
HUNGARY 429
ICELAND 429
INDIA 430
INDONESIA 431
IRAN 431
IRELAND 431
ISRAEL 432
ITALY 432
JAPAN 434
KOREA Rep 437
LATVIA 438
LITHUANIA 438
MALAYSIA 438
MEXICO 438

NETHERLANDS 439
NEW ZEALAND 440
NORWAY 440
PERU 440
POLAND 440
PORTUGAL 441
ROMANIA 441
RUSSIA 442
SAUDI ARABIA 434
SLOVAK Rep 444
SOUTH AFRICA 444
SPAIN 444
SWEDEN 446
SWITZERLAND 446
TAJIKISTAN 447
TURKEY 447
UK 447
UKRAINE 451
URUGUAY 452
USA 452
VATICAN CITY STATE 466
VENEZUELA 466

Individual Membership from Non-Adhering Countries 467

ALBANIA 467
BAHRAIN 467
COLOMBIA 467
CUBA 467
HONDURAS 467
IRAQ 467
JORDAN 467
KAZAKHSTAN 467
KOREA Dem Rep 468
LEBANON 468
MAURITIUS 468
MOROCCO 468
NIGERIA 468
PAKISTAN 468
PARAGUAY 468
PHILIPPINES 468
SINGAPORE 468
SLOVENIA 469
SRI LANKA 469
THAILAND 469
UZBEKISTAN 469
VIETNAM 469
YUGOSLAVIA Fed Rep 469

Alphabetical List of Members ..... M 1-M 267

PREFACE

The XXIIIrd General Assembly of the International Astronomical Union was held August 17 to 30, 1997, in the Kyoto International Conference Hall in Kyoto, Japan, at the invitation of the Science Council of Japan and the Astronomical Society of Japan. The National Organising Committee, chaired by Prof. Daiichi Sugimoto, and the 40-strong Local Organising Committee chaired by Dr. Toshio Fukushima, had arranged a very well-organised and pleasant meeting, which was much appreciated by the nearly 2000 IAU Members and Invited Participants from 59 countries, and nearly 200 Registered Guests. Substantial financial support for the participants was received from NASA, ESA, and ESO, and through the generosity of many individual members of the Astronomical Society of Japan, and is gratefully acknowledged here.

The General Assembly featured an exceptionally rich scientific programme, organised by Prof. Immo Appenzeller, IAU General Secretary 1994-1997. Between three Invited Discourses, six Symposia, 23 Joint Discussions, three Special Scientific Sessions, and well over 100 Commission and Working Group meetings, over 800 oral and 1100 poster papers were presented. Proceedings of the Symposia will appear in the regular *IAU Symposium Series* as IAU Symposia Nos. 183 - 188, and condensed proceedings or summaries of the other scientific events of the General Assembly are in press in the series *Highlights of Astronomy*, Vol. 11A&B.

The present Vol. XXIIIB of the *Transactions of the International Astronomical Union* records the organizational and administrative business of the XXIIIrd General Assembly. Chapter I contains the addresses given at the Inaugural Ceremony on August 20, which was honoured by the presence of the their Majesties, the Emperor and Empress of Japan, and representatives of the two host organisations. Chapter II records the results and decisions of the two Working Sessions of the General Assembly itself on August 20 and 27, while Chapter III contains the addresses given at the Closing Ceremony on August 27.

Chapter IV presents the Report of the Resolutions Committee and the Resolutions approved by the General Assembly. The Reports of the Executive Committee and of the Divisions, Commissions, and Working Groups are the subject of Chapters V and VI, while Chapter VII contains the Statutes, Bye-Laws and Working Rules as modified and approved by the General Assembly and the Executive Committee. In Chapter VIII, a revised edition of the *Rules for Scientific Meetings* is published as a guide to potential organisers of future scientific meetings sponsored or co-sponsored by the Union. Finally, Chapter IX provides details on the Membership: The Adhering Bodies representing our Full and Associate Member Countries, and the names of the Individual Members listed by Commission membership and member country as well as in alphabetical order. The list of addresses of Individual Members is also printed in a separate volume, and is also maintained accessible on-line from the IAU WWW page (*www.iau.org*).

I wish to acknowledge the cooperation of the Division and Commission Presidents for providing their contributions in standardised or camera-ready format, and of the Resolutions Committee under the Chairmanship of Prof. Jean-Claude Pecker and the official translators, Drs. Janet Rountree and Thierry Courvoisier, for ensuring the correct preparation of the Resolutions of the General Assembly. I also sincerely thank our Secretariat staff, Mss. Monique Léger-Orine, Julie Saucedo, and Jodi Greenberg, for carrying a major share of the work of editing this volume; I regret that a number of human and computer health problems have delayed the completion of this volume despite their efforts. Finally, I wish to acknowledge the IAU Publishers 1967-1997, Kluwer Academic Publishers (formerly D. Reidel) for their pleasant cooperation in the preparation of this and numerous previous publications of the IAU.

Paris and Copenhagen, September 1998

Johannes Andersen
General Secretary

CHAPTER I

TWENTY THIRD GENERAL ASSEMBLY

INAUGURAL CEREMONY

August 20, 1997, 14.00
Kyoto International Conference Hall, Main Hall

Address by Prof. Dr. H. Yoshikawa, President of the Science Council of Japan

Your Majesties the Emperor and Empress, Distinguished Guests, Ladies and Gentlemen.

It is my great pleasure to speak at the Opening Ceremony of the 23rd General Assembly of the International Astronomical Union, in the presence of Their Majesties the Emperor and Empress, on behalf of the Science Council of Japan, The Prime Minister's Office.

The Science Council of Japan is honored to co-host, on behalf of the Government of Japan, with the Astronomical Society of Japan the largest international conference in astronomy held every three years by the International Astronomical Union, which is the union of astronomers from all over the world.

Inter-disciplinary researches and international collaborations have become more and more important in all fields of science. Since its establishment in 1949, The Science Council of Japan has been active in both domestic and foreign affairs to represent Japanese scientists of various fields in order to contribute to the promotion of science, in cooperation with the scientists of all over the world. The Science Council of Japan is making efforts to have better research structure to meet the demand of scientists with the aim of promoting further development of science toward the 21st Century. We have also been promoting international research in collaboration with corresponding organizations in other countries.

Many Japanese participants in this General Assembly are members of the Astronomical Society of Japan, the Physical Society of Japan, and the Japan Society for Planetary Sciences. The representatives of these Societies belong to The Japan National Committees for Astronomy, Physics, Space Research and Planetary Sciences, which are among 180 National Committees under the umbrella of the Science Council of Japan. We recognize the recent progress of astronomy and its important role in the human culture through these National Committees.

We have here many participants from overseas. Astronomers from all over the world meet here together to make research presentations and active discussion on the latest topics in the field. Many people still have memory of the beautiful comet Hale-Bopp. It is quite timely, in terms of both progress and spread of astronomy, to have this General Assembly and Symposia now.

I conclude with my sincere hope that this conference is successful and that participants from overseas promote friendship with Japanese scientists and enjoy their stay in Japan, especially, in Kyoto, the city of traditional Japanese culture.

J. Andersen (ed.), Transactions of the International Astronomical Union Volume XXIIIB, 1–6.

Address by Prof. Dr. H. Okuda, President of the Astronomical Sociey of Japan

Your Majesties the Emperor and Empress, Distinguished Guests, All Participants

It is a great pleasure and honor for the Astronomical Society of Japan to perform the Opening Ceremony of the 23rd General Assembly of the International Astronomical Union. We are particularly honored to hold it in the presence of Their Majesties the Emperor and Empress.

Since the first meeting in Rome in 1922, the IAU General Assemblies have been held every three years excluding a short period during the last War and this is the 23rd General Assembly. Since the last time, we have improved the arrangements by holding the scientific symposia concurrent with the General Assembly. This is indeed the biggest international conference of astronomy where thousands of astronomers gather from all over the world and report their latest scientific results and discuss future research.

In the long history of the IAU General Assembly of 75 years, this is the first time that the General Assembly has been held in our country and the second time in Asia following the conference held in Delhi India in 1985. It is a great pleasure and honor for the Astronomical Society of Japan to host this conference together with the Science Council of Japan, having many participants from abroad.

Astronomy is one of the oldest sciences in human history. But, at the same time, it is also one of the newest sciences.

The invention of the telescope in the 17th century dramatically expanded our visible world and revealed the new faces of the universe. Now the space telescope is flying and many large telescopes with apertures of 8 to 10 m are newly built or under construction. They will see almost to the edge of the universe.

In the latter half of the 20th century, we have had new experiences of seeing invisible worlds by expanding observable wavelength to the full range of electromagnetic waves from radio, to infrared, ultraviolet, X-ray and gamma rays. In addition, observations with high energy particles started with cosmic rays, have been extended to neutrinos and even detection of gravitational waves is expected in the near future. Observations by these new techniques have succeeded in revealing unimaginably diverse morphologies of the universe and the creation of the universe and its global structures.

In the coming 21st century, observations in space will be accelerated together with building up giant observational facilities on the ground, which will give us new dreams in astronomy.

In our country, people have been attracted by astronomical phenomena from the earliest times and we can find many records of astronomical events in the old books beginning with "Kojiki" or "Nihon-shoki" compiled in the 8th century. These have also been drawn in many pictures and modeled in various works of art. In our country, however, modern astronomy as a science started only about a hundred years ago. Fortunately, since then, we have made rapid progress and are making important contributions in fields such as X-ray astronomy and radio astronomy. We are now constructing a large optical/infrared telescope on Mauna Kea in Hawaii and are expecting its completion next year. Now new satellite missions are under development and a large submillimeter radio interferometer is under planning, by which we hope to make much more contribution to the future astronomy.

Needless to say, we cannot talk about modern astronomy without international collaborations. In this conference, many results obtained by such collaborations will be reported. Based on the progress brought by all these efforts, together with the new techniques developed in the 20th century, we can only anticipate the flourishing astronomy of the coming 21st century. I really hope that this conference will be a most fruitful meeting for future research. It is still very hot and humid in Kyoto, but this city is the old capital of our country with a vast cultural heritage and with its unique institutions and culture. I hope, all participants, particularly those from abroad, will have the chance to visit and see some of the treasures of Kyoto and that you will enjoy your stay in Kyoto. I also hope this will be a good chance to renew old friendships and develop new friendships among the participants.

Thank you very much for your attention.

Address by Prof. Dr. L. Woltjer, President of the International Astronomical Union

Your Majesties, Distinguished Guests, Members and Friends of the Union,

It is a pleasure to be here for the XXIII General Assembly of the IAU in Japan, a country in which astronomical science has flourished for so long. We are particularly honoured that this opening ceremony is being held in the presence of Their Majesties the Emperor and Empress.

The oldest astronomical records in Japan go back to the 7th century. Particularly important astronomical observations were made of a guest star which appeared at the beginning of this millennium. In the Meigetsuki, "Diary of the Full Moon" written by Fujiwara-no Sadaie in the thirteenth century, the following report appears: "Third year of the Kanko reign, period of Ichijo In, 4th month, 2nd day, kuei-yu. After nightfall within Ch'i-kuan (part of the constellation Lupus) there was a large guest star." This was the famous supernova of 1006, the brightest exploding star of the last millennium, so bright that objects were clearly visible by its light. It was observed also in China, in the Arab world and in Europe. Already then astronomy was an international science. In the "Gonki" written by Fujiwara-no Yukinari soon after the event, we are informed that on several occasions reports of this "guest star" were read to the Emperor.

This "guest star" of 1006 has left a mass of rapidly expanding matter composed of heavy elements and very energetic particles. Its astonishing characteristics have recently been elucidated by the ASCA satellite launched by the Institute of Space and Astronautical Science in Japan. It is an honour to report to Your Majesties on the most recent developments concerning an object that so interested your illustrious ancestor so long ago.

Of course, there is more to Japanese astronomy than this. Other space crafts study the sun, the environment of the earth, and radio and X-ray sources, while with the Kamiokande neutrino detector, with the new 8-m telescope still under construction and with the radio astronomical instruments at Nobeyama Japan has clearly signalled its ambition to undertake world research at the highest level. Many researchers from around the world are participating in the elaboration of results of your satellites and many Japanese scientists participate in research elsewhere. In this way science advances and at the same time astronomers make their contribution to the harmony between nations. It is the principal function of the International Astronomical Union to facilitate such international collaboration.

The IAU has also many specific tasks. We discuss here questions relating to time, one of the most basic issues in modern society. We discuss information systems, standardisation, the naming of objects and of features on the planets. Such issues are important if effective communication of scientific data and results are to remain possible notwithstanding forever increasing specialisation.

We discuss the need for preserving an environment suitable for astronomical observation. There are many threats to the purity of the space environment, to the freedom from interference at essential radio frequencies, and to the darkness of the night sky which is as much part of the heritage of mankind as other natural or archeological monuments on earth. The IAU is the organization which brings such threats to the attention of the appropriate governmental and intergovernmental organizations and suggests remedies.

Astronomical research has become very expensive. There is, therefore, a need for international joint ventures and for avoiding duplication of effort. The IAU provides a forum for discussing such issues.

All these efforts are necessary. But the principal reason most of us come to this General Assembly and its associated symposia is to learn about the most recent scientific progress, to make plans for the future or just to renew old friendships. To make all of this possible requires a substantial organisational effort and a suitable environment. We are much indebted to the Japanese scientists, to the Science Council of Japan, to the Astronomical Society of Japan, and to the Local Organizing Committee for the General Assembly for the immense amount of effort they have devoted to hosting this meeting in Japan's cultural capital.

Thank you very much. The General Assembly is now in session.

Okotoba (Address) by His Majesty the Emperor

天皇陛下おことば

平成九年八月二十日（水）国立京都国際会館
第二十三回国際天文学連合総会開会式

第二十三回国際天文学連合総会が、世界の各地から多数の天文学の専門家を迎え、ここ京都の地において開催されることを誠に喜ばしく思います。開会式に当たり、総会の開催に尽力された関係者に対し、深く敬意を表します。

古来人々は、天体に深い関心を寄せてきました。世界の各地において、天体に関する物語が生まれ、占星術が行われ、また、暦が作られました。我が国で暦が使われ始めたのは、百済から暦博士が来日したという記録のある、六世紀半ばのころと考えられます。以来、我が国では中国の暦が用いられてきましたが、九世紀末遣唐使が廃されてからは、中国の新しい暦を使うことなく、八百年もの間改暦は行われませんでした。十七世紀に至り、この長年使われた暦の欠陥が指摘されるようになり、初めて渋川春海により日本の地に適した暦が作られています。その後更に正確な暦を作るべく、ヨーロッパの科学に対する関心が深まり、鎖国下の我が国で禁止されていた、科学に関する漢訳洋書の輸入が許されることになりました。後には、オランダ語からの科学書の翻訳も行われ、コペルニクスの地動説も、発表より二百年余を経て我が国に紹介されております。鎖国の厳しい状況下で、ヨーロッパの科学を学び、我が国の科学を育てた先人の努力を思うとき、深い感慨を覚えます。

我が国が諸外国と国交を開いた十九世紀半ば以降、我が国も国際共同観測網の中に組み入れられ、ここに、日本の天文学界は海外の天文学から大きな刺激を受けることになりました。万国緯度観測事業における今世紀始めの木村栄博士のＺ項の発見は、こうした時代を背景にして生まれた日本天文学界の誇りとする成果であったといえると思います。

我が国のこうした天文学の歩みを振り返るとき、学問の発達における国境を越えた交流、協力の重要性を深く感じるものであります。

最近の天文学は、地上の望遠鏡ばかりでなく人工衛星を利用するなど、高度な技術を駆使して、宇宙に生起する様々な現象を、あらゆる角度から探求する先端的学問に成長しています。宇宙の始まりから、物質世界の進化の過程を解き明かし、同時に多くの天体現象を究明してきた天文学は、人類の文化に重要な地位を占めており、多くの人々から幅広い関心を持たれています。

今回のこの総会が実り多いものとなり、その成果が参加者の今後の研究や活動にいかされるとともに、世界の天文学の進歩と普及に大きく貢献することを衷心より願い、開会式に寄せる言葉といたします。

Okotoba (Address) by His Majesty the Emperor

English translation of the Japanese original:

It is truly a great pleasure for me to see the 23rd General Assembly of the International Astronomical Union being held here in Kyoto with the participation of many astronomers from all over the world. On this occasion of the opening ceremony, I would like to express my deep appreciation to all concerned for their unsparing efforts in organizing this General Assembly.

From time immemorial, people have shown great interest in the heavenly bodies. In all parts of the world, stories concerning the heavenly bodies came into being, astrology was practiced and calendars were drawn up. The first use of a calendar in Japan may be dated to about the middle of the 6th century when, as extant records have it, a scholar of the calendar came to our country from Paekche. From that time on the Chinese calendar was used in Japan but, after the dispatch of diplomatic missions to Tang Dynasty China was suspended in the late 9th century, new Chinese reforms of the calendar were not put into use, and for some 800 years the calendar remained unrevised in Japan. In the 17th century, upon indication of deficiencies in the calendar used down all those long years, for the first time a calendar better suited to Japan's specific location was worked out by Shibukawa Harumi. Later, in an effort to produce a more accurate calendar, Japanese took a deepening interest in European science. The import of Chinese translations of western books on science, which were forbidden under Sakoku (a policy of seclusion) was eventually allowed. Later on, scientific books were also translated directly from the Dutch language, and Copernicus' heliocentric theory of the universe came to be introduced to Japan a little more than 200 years after its initial announcement. I am deeply moved when I think of the great endeavors of our forebears, who, in spite of their severely restricted circumstances under Sakoku, nevertheless managed to learn European science and nurture the early growth of Japanese science.

From the middle of the 19th century when Japan established diplomatic relations with various foreign countries, our country too was incorporated into the joint international observation network and the Japanese astronomical world came to receive a great stimulus from overseas astronomy. Dr. Kimura Hisashi's discovery of the "Z Term", in connection with the International Latitude Service at the beginning of this century, may be said to be the proud accomplishment of Japanese astronomy against the background of such a time in history.

When I look back over the road trodden by astronomy in Japan, I feel deeply the importance of exchanges and cooperation across national boundaries for the advancement of knowledge.

Modern astronomy, using highly advanced technology, not only ground-based telescopes but also artificial satellites, to explore every aspect of the various phenomena occurring in the universe, has developed into a most advanced science. Simultaneously investigating the multiple phenomena of the heavenly bodies, probing the origins of the universe and elucidating the evolution process of matter, astronomy occupies a position of great importance in human culture, and attracts the wide-ranging interest of a great many people.

In concluding my address at this opening ceremony, I sincerely hope that the present General Assembly will bear much fruit and that its results may be used to advantage by the participants in their researches and activities, thus contributing greatly to the further progress and dissemination of astronomical science.

Telegram message from Mr. R. Hashimoto, Prime Minister of Japan

Pleased to extend a hearty welcome to all participants from all over the world on this occasion of the Opening Ceremony of the XXIIIrd General Assembly of the International Astronomical Union in Kyoto, in the presence of Their Majesties the Emperor and Empress, under the joint-sponsorship of the Science Council of Japan and the Astronomical Society of Japan.

I wish a great success of this General Assembly for advancement in the field of astronomy.

Ryutaro Hashimoo, Prime Minister

CHAPTER II

TWENTY THIRD GENERAL ASSEMBLY

First Session

August 20, 1997, 15.15
Kyoto International Conference Hall, Main Hall
Prof. L. Woltjer, President, in the Chair

1. Opening of the Session

Following the Inaugural Ceremony, the President opened the meeting and welcomed Members to the first working session of the General Assembly.

2. Appointment of Official Interpreters

Drs. Thierry Courvoisier (English-French) and Janet Rountree (French-English) had agreed to serve as Official Interpreters. Their election was approved by acclamation.

3. Appointment of Official Tellers

Drs. Martin Ward, Jean-Pierre Swings and Patricia Whitelock had agreed to serve as Official Tellers. Their election was approved by acclamation.

4. Report of the Executive Committee 1994-1997

The General Secretary briefly presented the Report of the Executive Committee 1994-1997. The main issues and decisions are summarised as follows:

- The implementation of the new Division structure of the IAU following the proposal adopted by the XXIInd General Assembly in The Hague;
- Revision of the IAU Statutes, Bye-Laws and Working Rules to accommodate the Division structure;
- Modification of the Statutes and Bye-Laws to accommodate the individual membership of astronomers not represented by a national adhering organisation;
- Introduction of the Teaching of Astronomy Development (TAD) Programme;
- Establishment of a new Working Group for Future Large Scale Facilities;
- Conclusion of a new IAU Publishing Contract for the years 1998-2003;
- The use of the World Wide Web as a new major means of communication within the Union.

The full version of the Report is published as Chapter V of this volume. An abbreviated version of the Report appeared in IB 80, p. 24-36 (June 1997).

J. Andersen (ed.), Transactions of the International Astronomical Union Volume XXIIIB, 7–11.

5. Report on the Work of the Special Nominating Committee

The President reported that the Special Nominating Committee proposed the following slate of IAU members for Officers and of the Executive Committee for the period 1997-2000:

R.P.	Kraft	President	USA
F.	Pacini	President-Elect	Italy
J.	Andersen	General Secetary	Denmark
H.	Rickman	Assistant General Secretary	Sweden
C.	Anguita	Vice-President (1994-2000)	Chile
B.	Hidayat	Vice-President (1994-2000)	Indonesia
V.	Trimble	Vice-President (1994-2000)	USA
C.	Cesarsky	Vice-President (1997-2003)	France
N.	Kaifu	Vice-President (1997-2003)	Japan
N.	Kardashev	Vice-President (1997-2003)	Russia

6. Official Representatives of Adhering Organisations at the General Assembly and on the Nominating Committee

	General Assembly	*Nominating Committee*
Algeria	-	
Argentina	H. Levato	H. Levato
Armenia	-	-
Australia	J. Mould	J. Mould
Austria	M. Breger	M. Breger
Azerbajan	-	-
Belgium	-	-
Brazil	R.C. De Carvalho	R.C. De Carvalho
Bulgaria	M. Tsvetkov	M. Tsvetkov
Canada	J. Landstreet	J. Landstreet
Chile	-	-
China Nanjing	Li Qi-bin	Yang Fu-min
China Taipei	C. Chi-kang	C. Chi-kang
Croatia	V. Ruzdjak	V. Ruzdjak
Czech Rep.	P. Harmanec	P. Harmanec
Denmark	P.E. Nissen	L.K. Kristensen
Egypt	M.I. Wanas	M.I. Wanas
Estonia	J. Einasto	J. Einasto
Finland	I. Tuominen	E. Valtaoja
France	J.P. Zahn	R. Courtin
Georgia	J. Lominadze	J. Lominadze
Germany	M. Grewing	G. Hasinger
Greece	E. Kontizas	E. Kontizas
Hungary	K. Olah	M. Paparo
Iceland	-	-
India	J.V. Narlikar	T.N. Rengarajan
Indonesia	-	-
Iran	Y. Sobouti	Y. Sobouti
Ireland	R.M. Redfern	R.M. Redfern
Israel	-	-
Italy	L. Padrielli	L. Padrielli
Japan	Y. Osaki	M. Iye
Korea Rep.	H.M. Lee	H.M. Lee
Latvia	-	-
Lithuania	-	-
Malaysia	-	-
Mexico	M. Peimbert	E. Recillas

Netherlands	H.J. Habing	E.P.J. Van den Heuvel
New Zealand	P. Cottrell	P. Cottrell
Norway	K. Aksnes	K. Aksnes
Peru	-	-
Poland	K. Stepién	K. Stepién
Portugal	J.P. Osório	J.P. Osório
Romania	-	-
Russia	A.A. Boyarchuk	A.A. Boyarchuk
Saudi Arabia	-	-
Slovak Rep.	J. Zverko	J. Zverko
South Africa	B. Warner	R.S. Stobie
Spain	-	-
Sweden	D. Dravins	D. Dravins
Switzerland	A.O. Benz	A.O. Benz
Tajikistan	P.B. Babadzhanov	P.B. Babadzhanov
Turkey	M.E. Ozel	M.E. Ozel
UK	M. Longair	D. McNally
Ukraine	V.V. Tel'nyuk	V.V. Tel'nyuk
Uruguay	-	-
USA	V.C. Rubin	A.U. Landolt
Vatican City State	C. Corbally	C. Corbally
Venezuela	G. Bruzual	G. Bruzual

7. Acting Presidents of Divisions and Commissions

Division	*Acting President*
I	J. Kovalevsky
II	O. Engvold
III	M. A'Hearn
IV	D.L. Lambert
V	Y. Kondo
VI	M. Dopita
VII	J.J. Binney
VIII	R.B. Partridge
IX	C.D. Scarfe
X	J.B. Whiteoak
XI	G.G. Fazio

Commission	*Acting President*
4	H. Kinoshita
5	B. Hauck
6	R.M. West
7	S. Ferraz-Mello
8	T.E. Corbin
9	G. Lelièvre
10	O. Engvold
12	F.-L. Deubner
14	F. Rostas
15	M.F. A'Hearn
16	Y. Marov
19	J. Vondrak
20	D.K. Yeomans
21	C. Leinert
22	I.P. Williams
24	C. Turon
25	J.D. Landstreet
26	E.C. Worley

27	M. Jerzykiewicz
28	V.L. Trimble
29	M.S. Bessell
30	C.D. Scarfe
31	H.F. Fliegel
33	J.J. Binney
34	M.A. Dopita
35	J.P. Zahn
36	L.W. Cram
37	G. Da Costa
38	H.E. Jørgensen
40	J.B. Whiteoak
41	S.M.R. Ansari
42	M. Rodono
44	G.G. Fazio
45	O.H. Levato
46	J.R. Percy
47	J.V. Narlikar
49	H.W. Ripken
50	S. Isobe
51	J.C. Tarter

8. Appointment of the Finance Committee and Corresponding Votes

FULL MEMBER/ *ASSOCIATE MEMBER*	*Category*	*Votes* (a*)	*Votes* (b*)	*Representative(s)*
Algeria	I	-	-	-
Argentina	III	1	4	H. Levato
Armenia	I	-	-	-
Australia	IV	1	5	B. Boyle
Austria	I	1	1	M. Breger
Azerbajan	I	-	-	-
Belgium	IV	1	5	-
Brazil	II	1	3	R.C. De Carvalho
Bulgaria	I	1	2	M. Tsvetkov
Canada	VI	1	7	D. Morton
Chile	I	1	2	-
China Nanjing	V	1	6	Zhang He-qi
China Taipei	I	1	2	Chou Chi-kang
Croatia	I	1	2	V. Ruzdjak
Czech Rep.	II	1	3	J. Vondrak
Denmark	III	1	4	L.K. Kristensen
Egypt	III	1	4	M. Wanas
Estonia	I	1	2	J. Einasto
Finland	II	1	3	I. Tuominen
France	VII	1	8	R. Courtin/J. Kovalevsky
Georgia	I	1	2	-
Germany	VII	1	8	M. Grewing
Greece	III	-	-	E. Kontizas
Hungary	II	1	3	K. Petrovay
Iceland	I	1	2	-
India	III	1	4	N.K. Rao
Indonesia	I	1	2	-
Iran	I	1	2	Y. Sobouti
Ireland	I	1	2	M. Redfern
Israel	II	1	3	-
Italy	V	1	6	L. Padrielli

Japan	VII	1	8	S. Ikeuchi
Korea Rep.	I	1	2	H.S. Park
Latvia	I	1	2	-
Lithuania	I	-	-	-
Malaysia	*I*	*1*	*NA*	-
Mexico	II	1	3	E. Carrasco
Netherlands	I	-	-	J.M.E. Kuijpers
New Zealand	IV	1	5	P. Cotrell
Norway	I	1	2	K. Aksnes
Peru	I	-	-	-
Poland	III	1	4	M. Sarna
Portugal	II	1	3	J.P. Osório
Romania	*I*	*1*	*NA*	-
Russia	V	-	-	A.A. Boyarchuk
Saudi Arabia	I	1	2	-
Slovak Rep.	I	1	2	J. Zverko
South Africa	III	1	4	P.A. Whitelock
Spain	IV	-	-	-
Sweden	III	1	4	A. Ardeberg
Switzerland	III	1	4	A.O. Benz
Tajikistan	*I*	*1*	*NA*	P.B. Babadzhanov
Turkey	I	-	-	M.E. Ozel
UK	VII	1	8	A. Boksenburg
Ukraine	II	1	3	V.V. Tel'nyuk
Uruguay	I	-	-	-
USA	VIII 1/2	1	10	R.W. Milkey
Vatican City State	I	1	2	C. Corbally
Venezuela	I	1	2	G. Bruzual

*(a) On questions concerning the administration of the union, not involving its budget, voting at the General Assembly is by Adhering Country, each country having one vote. Adhering Countries which have not paid their annual contributions up to 31 December of the year preceding the General Assembly may not participate in the voting.

*(b) On questions involving the budget of the Union, voting is similarly by Adhering Country, under the same conditions and with the same reservations as in (a), the number of votes for each Adhering Country being one greater than the number of its category.

9. Revision of the Statutes and Bye-Laws

The revised Statutes and Bye-Laws, having been submitted to the Adhering Organizations in due time, were unanimously approved (See Chapter VII of these Transactions).

10. Resolutions Proposed by Divisions

The Chairman of the Resolutions Committee, Prof. J.-C. Pecker, presented the Report of the Committee concerning the proposed Resolutions of Type B, which had been submitted by Divisions in due time before the General Assembly and had been posted before the session. The full text of the Report appears in Chapter IV of this volume. The proposed resolutions were unanimously approved by the General Assembly (Resolutions B1 - B7, Chapter IV).

Second Session

August 27, 1997, 14.00
KYOTO International Conference Hall, Main Hall
Prof. L. Woltjer, President, in the Chair

11. Financial matters

REPORT OF THE FINANCE COMMITTEE OF THE IAU, BY D.C. MORTON, CHAIR

The Finance Committee had appointed a subcommittee to examine the accounts and proposed budget of the IAU and prepare a draft report for presentation to the General Assembly. The subcommittee was composed of D. C. Morton (Canada), Chair, A. Boksenberg (U.K.), G. Bruzual (Venezuela), S. Ikeuchi (Japan), and R. Milkey (USA). The report by the Finance Sub-Committee, as approved by the full Finance Committee, is reproduced in the following:

Prior Years: The Finance Committee reviewed the accounts for 1994, 1995, and 1996 as well as the audit reports for these years, and is pleased to report that the funds were used responsibly in accordance with the budget and policies adopted by the 22nd General Assembly.

For the current year, 1997, the Committee found that both the income and expenditures are on track to meet the budget approved by the 22nd General Assembly, and that no revisions are indicated.

Budget for the Triennium 1998 - 2000: The Finance Committee examined the assumptions used to construct the budget for the period 1998 through 2000, as published in IAU Bulletin No. 80, and found that these are reasonable and relatively conservative. We note that a substantial deficit is planned for this period, and we offer some suggestions to assist in controlling this deficit. In particular, we are concerned that in the proposed budget the reserves at the end of 2000 will only be at a level of approximately 85% of the expected operating budget for the following year. In the following sections, we address the specific items of the budget.

A. Income: The assumption for annual inflation rates for the Unit of Contribution is appropriate and consistent with the best information currently available. These rates, quoted below, are intended to result in a Unit that is constant in real terms.

	1998	1999	2000
Inflation Rate (from prior year)	2%	2.5%	3%
Unit of Contribution (CHF)	2880	2950	3040

We caution, however, that when future budgets are constructed the IAU must consider that the science budgets in many countries are not growing at a rate to compensate for inflation and therefore the increases in the contribution may come at the expense of other scientific programs.

1. *Adhering Organizations:* The budget is based on the current categories of adherence by members and the assumption that all members will pay dues as assessed. This is a reasonable approach to budgeting; however the General Secretary must continue to track any shortfall and be prepared to take compensating action on the expense side of the ledger, if required.

2. *Special Contributions:* The budget assumes no income from special contributions. This is a very conservative approach, but reasonable in view of the fact that there is no guarantee of such income. We do

J. Andersen (ed.), Transactions of the International Astronomical Union Volume XXIIIB, 13–20.

note, however that contributions at a level of 176,300 CHF were received in 1991-93 and 208,730 CHF in 1994-96. The IAU should make every effort to maintain the special contributions at a level consistent with past experience.

3. *Publications:* Income from royalties on IAU publications has been reduced in the new publishing contract, consistently with the goal of reducing the cost of IAU publications to libraries. While it would be unwise to depend on a growth in circulation to offset the reductions in income per volume, it should certainly be the goal to increase the circulation with more modestly priced volumes, and the publisher should be encouraged to market these aggressively. Any additional income so realized should be used to reduce the deficit.

B. Expenses: The specific categories of expense are discussed below.

1. *Scientific Activities:* The last triennium saw an increase in expenditures for scientific activities by approximately 27% over the period 1991-93. This was a response to a recommendation of the Finance Committee and clearly had a beneficial effect for the IAU as a whole. The budget for the next triennium forecasts a growth of yet another 10%, primarily for travel grants for Symposia and Colloquia. While continued growth of these activities will have a positive effect on the conduct of astronomical research, it must be moderated after consideration of the forecast of available income. In particular, we recommend that the Executive Committee consider the financial aspects when determining how many and which Symposia and Colloquia to support, and apply the highest scientific standards in the selection of these and the level of support to be provided.

2. *Dues to Unions and Organizations*: The amount forecast in the budget appears adequate to meet the obligations of the IAU to such organizations.

3. *Executive Committee:* The amount budgeted is reasonable. However, the Committee and the General Secretary should continue to minimize the costs of this activity, consistently with accomplishing the necessary business of the IAU, and any funds so realized should be applied toward reduction of the deficit.

4. *Publication Costs:* The forecast budget reflects reduced costs resulting from the change of publishers and is judged to be adequate to continue the Information Bulletins and free distribution programmes at an appropriate level. The General Secretary is encouraged to investigate the options for the use of electronic distribution to reduce the mailing costs and to accelerate the flow of information to the membership. The option for mail distribution must be retained to assure that we do not cut off those members who do not have reliable electronic access.

5. *Administration/Secretariat:* There is minimal growth in expenditures in this area and we wish to express our satisfaction with the efficiency of operations of this office. It has been possible to provide salary increases consistent with inflation, without having a similar overall growth in expense in this category.

C. Reserve Funds: The IAU policy has been to maintain a reserve equivalent to one-year's operating budget. We anticipate ending 1997 with a reserve that is approximately 110% of projected budget for 1998, but with the level of deficit in this budget we will end 2000 with reserve funds equal to only about 85% of the budget extrapolated for 2001. We feel that it is necessary for the General Secretary to adopt an aggressive approach to attempt to reduce this deficit by about 100,000 CHF and to restore the reserves as desired by the policy of the Executive Committee. Special contributions could be an important element in achieving the desired programme content without unduly depleting the reserves.

Although the interpretation of the reserve policy as requiring a balance equivalent to the operating level of a year in which a General Assembly does not occur is acceptable for this budget, we would be more comfortable with a level equal to the average of a triennium's expenses. We note that due to variations in timing of receipts from adhering organizations and payment of expenses, the cash flow during any given year may temporarily draw the bank accounts well below the balance at the beginning of the year, so that the General Secretary must assure that the IAU has the resources to deal with such situations in a flexible manner.

D. Budgeting for the Period 2001 - 2003: It is essential that the IAU return to a balanced budget for this period and restore a stable reserve equivalent to an annual operating budget, averaged over three years. This may require either a reduction of expenditures, increase in revenues, or a combination of both.

Special Support: The Finance Committee notes with gratitude the special contributions by ESA, ESO, and NASA in 1997 for travel grants for the 23rd General Assembly. The Finance Committee also recognizes the important indirect financial contributions of the Institut National des Sciences de l'Univers of CNRS by providing office space in Paris for the Secretariat, and of Landessternwarte Heidelberg-Königsstuhl and Copenhagen University for the time which the General Secretary and the Assistant General Secretary, respectively, spend on IAU affairs.

Contribution Levels: The IAU differs from other international unions in its substantial support of scientific meetings, publications, and related activities which directly benefit its individual members worldwide. It is important for the membership to understand that for these reasons the IAU requests significantly higher contribution levels from the adhering agencies than do other unions.

Treasurer's Report and Financial Advice: The Finance Committee received a useful report (see following paragraph) from the Treasurer of the IAU, a position which the Executive Committee created at the recommendation of a previous Finance Committee. It is appropriate for the Executive Committee to assess the need for continuing this position and to define the duties and appointment procedure or eliminate the position.

The Finance Committee recommends that the Executive Committee adopt procedures in which a standing Finance Sub-Committee is consulted formally in the preparation of the budget for each triennium and that there is some continuity with the membership of the next Finance Subcommittee.

TREASURER'S REPORT, BY MORTON S. ROBERTS, JULY 30, 1997

The annual expenditures of the Union over the six year period 1991-1996 averaged $8.0x10^5$ CHF, the annual income over this same period averaged $8.1x10^5$ CHF, i.e., close to steady state. The projection for the four year period 1997-2000 is less encouraging.

Average Annual Expenditures	$9.4x10^5$ CHF/year
Average Annual Income	$8.4x10^5$ CHF/year

Note however that the figure for Income does not include "Special Contributions" which by their nature cannot be forecast. In the period 1991-1996 they averaged $0.6x10^5$ CHF/year. If such special contributions continue into the future, the projected average annual deficit will be significantly reduced. That this may well happen is indicated by the generous subventions made in 1997 by ESA, ESO, and NASA for travel grants to the Kyoto General Assembly.

The reserves of the Union are $1x10^6$ CHF so that this projected deficit can be handled. However, as the Finance Committee has noted in the past, a million Swiss Francs is a prudent reserve. It is an amount close to one year's operation and should be maintained near this value. Some 88 percent of the Union's income originates from contributions by Adhering Organizations. This average percentage includes a modest increase of two to three percent per year for inflation.

Close to 60 percent of the Union's expenditures are for Scientific Activities. These are detailed in the budget document and include the General Assembly, support for Symposia, Colloquia, and Regional Meetings. Various working groups and commission activities are also supported as is the Exchange of Astronomers programme.

Dues to other Unions and Organizations are more than compensated by the annual ICSU contribution. The remaining third of the budget is attributable to (1) expenses incurred by officers of the Union, (2) publications, and (3) the Administration/Secretariat office.

The Union is growing in membership and in the activities it supports. It is doing this with a modest increase in budgetary outlay. In one specific instance it has purposefully reduced its expected income by lessening the royalties from the sale of Symposia volumes to lower their cost to libraries and to individuals.

The proposed budget for the coming triennium is well thought out and looks to a successful entry for the IAU into the next millennium. Our General Secretary, Immo Appenzeller, is to be complimented on a task well done.

The President thanked the Chairman of the Finance Sub-Committee and the Treasurer for their reports and asked the National Representatives to vote on the proposed budget for 1998-2000.

APPROVAL OF BUDGET FOR 1998-2000

The proposed budget was approved.

APPOINTMENT OF A STANDING FINANCE SUB-COMMITTEE

Following the recommendation of the Finance Committee, the General Assembly appointed a Standing Finance Sub-Committee for the period 1997-2000, composed of:

A. Boksenberg	UK	
G. Bruzual	Venezuela	
S. Ikeuchi	Japan	
R. Milkey	USA	
D.C. Morton	Canada	Chair

12. Appointment of Division Presidents, Commission Presidents and Vice-Presidents, and Presidents of EC Working Groups

DIVISION/COMMISSION(S)

I	***Fundamental Astronomy***	*President*: P.K. Seidelmann
	Commissions	*President / Vice-President*
4	Ephemerides	E.M. Standish / J. Chapront
7	Celestial Mechanics and Dynamical Astronomy	C. Froeschlé / J.D. Hadjidemetriou
8	Positional Astronomy	H. Schwan / Jin Wenjing
19	Rotation of the Earth	D.D. McCarthy / N. Capitaine
24	Photographic Astrometry	E. Schilbach / Jin Wenjing
31	Time	T. Fukushima / G. Petit
II	***Sun and Heliosphere***	*President*: P.V. Foukal
	Commissions	*Presidents / Vice-President*
10	Solar Activity	Ai Guoxiang / A. Benz
12	Solar Radiation and Structure	P.V. Foukal / S. Solanki
49	Interplanetary Plasma and Heliophere	F. Verheest / M. Vandas
III	***Planetary System Sciences***	*President*: M.F. A'Hearn
	Commissions	*President / Vice-President*
15	Physical Study of Comets, Minor Planets and Meteorites	V. Zappala / H.U. Keller
16	Physical Study of Planets and Satellites	C. de Bergh / D. Cruikshank
20	Position and Motions of Minor Planets, Comets and Satellites	H. Rickman / E.L.G Bowell
21	Light of the Night Sky	S. Bowyer / Ph. Lamy
22	Meteors and Interplanetary Dust	W. Baggaley / V. Porubcan
51	Bioastronomy: Search for Extraterrestrial life	F.R. Colomb / S. Bowyer
IV	***Stars***	*President*: L. Cram
	Commissions	*President / Vice-President*
26	Double and Multiple Stars	H. Zinnecker / C. Scarfe
29	Stellar Spectra	B. Barbuy / G. Mathys
35	Stellar Constitution	J.-P. Zahn / D. VandenBerg
36	Theory of Stellar Atmospheres	R.O. Pallavicini / D. Dravins
45	Stellar Classification	M. Gerbaldi / T.L. Evans
V	***Variable Stars***	*President*: M. Jerzykiewicz
	Commissions	*President / Vice-President*
27	Variable Stars	D.W. Kurtz / J. Christensen-Dalsgaard
42	Close Binary Stars	E.F. Guinan / P. Szkody

VI	***Interstellar Matter***	*President*: M.A. Dopita
	Commission	*President / Vice-President*
34	Interstellar Matter	M.A. Dopita / B. Reipurth
VII	***Galactic System***	*President*: K.C. Freeman
	Commissions	*President / Vice-President*
33	Structure and Dynamics of the Galactic System	K.C. Freeman / D.N. Spergel
37	Star Clusters and Associations	G.S. Da Costa / G. Meylan
VIII	***Galaxies and the Universe***	*President*: P. Shaver
	Commissions	*President / Vice-President*
28	Galaxies	F. Bertola / S. Okamura
47	Cosmology	A. Szalay / J. Peacock
IX	***Optical techniques***	*President*: Ch.L. Sterken
	Commissions	*President / Vice-President*
9	Instruments	I. McLean / Su Ding-qiang
25	Stellar Photometry and Polarimetry	Ch.L. Sterken / A. Landolt
30	Radial Velocities	J.B. Hearnshaw / A. Tokovinin
X	***Radio Astronomy***	*President*: J.M. Moran
	Commission	*President / Vice-President*
40	Radio Astronomy	J.M. Moran / L. Padrielli
XI	***Space and High Energy Astrophysics***	*President*: W. Wamsteker
	Commission	*President / Vice-President*
44	Space and High Energy Astrophysics	G. Srinivasan / H. Okuda

COMMISSION NOT ATTACHED TO A DIVISION

		President / Vice-President
41	History of Astronomy	Dick S. J. / F.R. Stephenson

COMMISSIONS OF THE EXECUTIVE COMMITTEE

		President / Vice-President
5	Documentation and Astronomical Data	O.B. Dluzhnevskaya / F. Genova
6	Astronomical Telegrams	R.M. West / B.G. Marsden
14	Atomic and Molecular Data	Rostas F. / P.L. Smith
38	Exchange of Astronomers	M. Roberts / R.M. West
46	Teaching of Astronomy	J. Fierro / S. Isobe
50	Protection of Existing and Potential Observatory Sites	W.J. Sullivan III / J. Cohen

WORKING GROUPS OF THE EXECUTIVE COMMITTEE

	President
Working Group for Encouraging the International Development of Antarctic Astronomy	M. Dopita
Working Group for Planetary System Nomenclature	K. Aksnes
Working Group for the World Wide Development of Astronomy	A.H. Batten
Working Group for Future Large Scale Facilities	R.D. Ekers

13. Resolutions Proposed by the Executive Committee and by Working Groups of the EC

The Chairman of the Resolutions Committee, Prof. J.-C. Pecker, summarised the recommendations of the Committee concerning the resolutions which had been proposed by the Executive Committee and its Working Groups. Four of the proposals were submitted to the vote of the General Assembly and unanimously approved as Resolutions A1-A4, as recommended. The full text of the Report and of the Resolutions is reproduced in Chapter IV of this volume.

14. Appointment of the Resolutions Committee 1997-2000

By acclamation, the General Assembly appointed the following Resolutions Committee for the triennium 1997-2000 (Working Rules 37):

J. Kovalevsky	France	
Y. Kozai	Japan	
D. McNally	UK	Chair
V. Trimble	USA	ex officio
P. Wayman	Ireland	

15. Appointment of the Special Nominating Committee 1997-2000 (SNC)

Following the vote of the Division Presidents, the following composition of the Special Nominating Committee for the period 1997-2000 was unanimously approved by the General Assembly:

B. Barbuy	Brazil	
R. Kraft	USA	ex officio, Chair
M.S. Longair	UK	
D.C. Morton	Canada	
D. Sugimoto	Japan	
P.A. Whitelock	South Africa	
L. Woltjer	Netherlands	ex officio

16. National Membership of the Union

By vote of the National Representatives, the General Assembly admitted the following new Associate Members of the Union:

- Bolivia, the Academy of Science being the Adhering Organisation;
- The Central American Assembly of Astronomers, representing the astronomers in Costa-Rica, El Salvador, Guatemala, Honduras, Nicaragua and Panama, and conforming to the same rules as the Adhering Organisations of normal Associate Members.

An application by the The Macedonian Astronomical Society was found not to be in order, but the EC was authorised to accept this application as soon as a proper official membership application has been received from the corresponding Academy of Sciences.

17. Individual Membership of the Union

By acclamation, the General Assembly welcomed the 748 new IAU members who had been admitted on the advice of the Nominating Committee. This brings the total number of IAU Members to 8328, effective as of the day of the session.

The General Assembly stood in silence while the General Secretary read the names of the 136 Members whose death had been reported to the Secretariat since the XXIInd General Assembly. The full list of these names is given in the Report of the Executive Committee (Chapter V) in these Transactions.

18. IAU Representatives to Other International Organisations, 1997-2000

By acclamation, the General Assembly approved the following proposal by the Executive Committee for representatives of the IAU in ICSU and other international organisations for the period 1997-2000:

Acronym	*Organisation*	*Representative(s)*
ICSU	International Council of Scientific Unions General Committee	*J. Andersen*
BIPM	Bureau International des Poids et Mesures	*J. Kovalevsky*
CCDS	International Consultative Committee for the Definition of the Second	*S. Débarbat*
CIE	Compagnie Internationale de l'Eclairage	*S. Isobe*
CODATA	Committee on Data for Science and Technology	*E. Raimond*
COPUOS	Committee on the Peaceful Uses of Outer Space Scientific and Technical SubCommittee	*J. Andersen*
COSPAR	Committee on Space Research	*J. Andersen*
	COSPAR SC B	*C. de Bergh*
	COSPAR SC D	*F. Verheest*
	COSPAR SC E	*W. Wamsteker*
	COSPAR Sub. Committee E1	*R. Sunyaev*
	COSPAR Sub. Committee E2	*O. Engvold*
COSTED	Committee on Science and Technology in Developing Countries	*J. Andersen*
FAGS	Federation of Astronomical and Geophysical Services	*P. Pâquet/ E. Tandberg-Hanssen*
IAF	International Astronautical Federation	*Y. Kondo*
IERS	International Earth Rotation Service	*B. Kolaczek*
IGBP	International Geosphere-Ionosphere Programme	*J. Eddy*
IUCAF	Inter-Union Commission on Frequency Allocation for Radio Astronomy and Space Science	*B.A. Doubinsky I Kawaguchi/ S. Ananthakrishnan/ A.R. Thompson*
IUPAP	International Union of Pure and Applied Physics	*J. Andersen*
	IUPAP C4 Commission on Cosmic Rays	*C. Cesarsky*
ITU	International Telecommunication Union	
	IUT-BR Radiotelecommunication Bureau	*W. Klepczynski*
	IUT-R Radiocommunication Bureau	*J. Whiteoak/ A.R. Thompson*
IUWDS	International Ursigram and World Day Service	*H. Coffey*
QBSA	Quarterly Bulletin on Solar Activity	*P. Lantos*
SCOPE	Scientific Committee on Problems of Environment	*D. McNally*
SCOSTEP	Scientific Committee on Solar-Terrestrial Physics	*B. Schmieder*
URSI	Union Radio-Scientifique Internationale	*J. Moran*
WMO	World Meteorological Union	*L. Woltjer*

19. Places and Dates of the XXIVth and XXVth General Assemblies

On behalf of the Royal Society, Dr. Malcolm Longair invited the Union to hold its XXIVth General Assembly in Manchester, UK, from August 7 to 19, 2000. The invitation was accepted by acclamation.

An invitation from the Australian Academy of Sciences, presented by Dr. Jeremy Mould, to hold the XXVth General Assembly in Sydney in the year 2003, probably in August, was also accepted by acclamation.

20. Election to the Union of a President, a President-elect, six Vice-Presidents, a General Secretary and an Assistant General Secretary

As proposed by the Special Nominating Committee, the General Assembly elected:

R.P. Kraft	as	President
F. Pacini	as	President-Elect
J. Andersen	as	General Secretary
H. Rickman	as	Assistant General Secretary
C. Cesarsky	as	Vice-President (1997-2003)
N. Kaifu	as	Vice-President (1997-2003)
N. Kardashev	as	Vice-President (1997-2003)

Accordingly, the composition of the Executive Committee for the period 1997-2000 is as follows:

R.P. Kraft	President
F. Pacini	President-Elect
J. Andersen	General Secretary
H. Rickman	Assistant General Secretary
C. Anguita	Vice-President (1994-2000)
C. Cesarsky	Vice-President (1997-2003)
B. Hidayat	Vice-President (1994-2000)
N. Kaifu	Vice-President (1997-2003)
N. Kardashev	Vice-President (1997-2003)
V. Trimble	Vice-President (1994-2000)
L. Woltjer	Adviser
I. Appenzeller.	Adviser

This being the last item of the business session, the President invited the newly elected members of the Executive Committee to the podium, and the General Assembly proceeded to the Closing Ceremony (Chapter III).

CHAPTER III

TWENTY THIRD GENERAL ASSEMBLY

CLOSING CEREMONY

August 20, 1997, 14.00
Kyoto International Conference Hall, Main Hall

Address by the President 1994-1997, Prof. L. Woltjer

Looking back over my six years on the Executive Committee, there are two sources of satisfaction: The divisional structure of the Union has become firmly implanted with its potential for greater effectiveness and representativity, and the integration of the Symposia into the General Assembly has greatly increased the scientific interest in this triennial event. The result is visible: At Buenos Aires there were a thousand attendees, here at Kyoto the double.

On this occasion, I would like to thank the General Secretary and Assistant General Secretary for a period characterized by cordial cooperation. The General Secretary has a very heavy task and Immo Appenzeller has executed his function with calm, judgement and effectiveness. Also the Paris Office with Monique Orine and Julie Saucedo has contributed much. It is not always easy to run a rather lonely office for an organization with nearly ten thousand members; it has been done well, with administrative expenses going down while the number of members went up.

This General Assembly has been resounding success thanks to the unceasing efforts of Daiichiro Sugimoto and Toshio Fukushima and their numerous collaborators. Our former President Yoshihide Kozai did much to raise funds, and I also wish to acknowledge the members of the Astronomical Society of Japan who made a substantial personal contribution.

Address by the President 1997-2000, Prof. R.P. Kraft

First of all, on behalf of all IAU members, spouses and friends, I want to express our sincere appreciation to those persons and organizations who made possible this especially productive General Assembly: to the members of the Advisory Board of the Host Organizations and its Chairperson, Dr. Yoshio Fujita, President of the Japan Academy, to the members of the National Organizing Committee and its Chairperson, Dr. Daiichiro Sugimoto, and to members of the Local Organizing Committee and its Chairperson, Dr. Toshio Fukushima. Their hard work and attention to many details have been the major factors in making this GA so exciting socially, culturally and scientifically.

As we approach the new millenium, our Union appears strong and, in my view, headed in the right direction. On the adminstrative level, the new Divisional structure admits of a close connection of the Division Presidents with the Executive Committee. This, for example, will greatly assist the EC in the process of rank-ordering proposals for future Symposia and Colloquia. On the scientific front, the integration of several timely Symposia with the GA yields an important new dividend: in what other venue could one explore, in the same assembly, an organized exposition of topics ranging from solar physics to cosmology? At the same time, the rule which insures a wide national representation on SOCs helps to maintain the truly international, and to a great extent unique, flavor of IAU sponsored Symposia and Colloquia.

J. Andersen (ed.), Transactions of the International Astronomical Union Volume XXIIIB, 21–24.

Returning to affairs of this GA, it was very impressive to hear the many excellent papers presented by our Japanese colleagues, and to recognize the increasing importance of the work of Japanese astronomers on the world astronomical scene. We are all aware of significant advances made here in space sciences, mm wave astronomy, and nuclear astrophysics. The completion of the Subaru Telescope on Mauna Kea and the recent launch of HALCA will provide new and virtually unparalleled opportunities for Japanese observational astronomers. We wish them good luck and "happy hunting" with the new facilities.

Address by Robert J. Rubin, on behalf of the Registered Guests

We, the Registered Guests and Accompanying Persons wish to express our appreciation and thanks to Professor Fujita's Advisory Board, to the National Organizing Committee, to the Local Organizing Committee, their staffs and all the volunteers who have created and participated in our excellent program. Particularly for those of us from abroad, we have learned much about the history and culture of Kyoto and Japan.

We especially note the marvellous organization of the tour to Nara, the strategic placement of Japanese guides with cold drinking water (in plastic bags), their directions to and from the chartered trains, as well as the placement of guides in Nara Park. It all made for a memorable day.

Kyoto is one of the greatest cities in the world in which to be Registered Guest or Accompanying Person (and in which to a General Assembly).

Again, we thank you all. Domo arigato-o gozai masu.

Address by Dr G. Cayrel de Strobel, on behalf of the Participants

President Woltjer, General Secretary Appenzeller, Assistant General Secretary Andersen, the IAU Secretariat with Monique and Julie, Prof. Fujita, Prof. Sugimoto, Dr Fukushima, all Members of the Executive Committee of the IAU, of the Japanese National Organizing Committee and of the Japanese Local Organizing Committee, on behalf of all Participants, I have the great honor of expressing our deep gratitude for the wonderful organisation of the XXIIIth General Assembly of the IAU.

Everything was perfect: from the breathtaking KICH Building to the choice of: 6 Symposia, 24 Joint Discussions, and 3 Special Sessions. All these scientific events have occurred smoothly, in an orderly manner. What we have learnt in the KICH Building will come back to us, once back in our respective countries.

In 1988, during the XXth IAU General Assembly in Baltimore, we had the feeling that something unique would happen in Astronomy soon. Nine years later the Space Telescope is distributing to all of us its marvellous scientific results.

The Invited Discourse of Prof. Williams has illustrated how far the eye of the Space Telescope can penetrate in the deep Universe, opening new puzzles, hopefully to be solved by its successor the Next Generation Space Telescope (NGST).

Prof. Warner in his Invited Discourse has given an impressive view of one of the most capricious astronomical objects: the well named Cataclysmic Variables.

Prof. Novikov has with great courage, in less than one hour, introduced us, with beautiful viewgraphs, in the most dramatic astronomical situation: that of the Black Holes.

I cannot finish this talk without expressing my deep gratitude to two Astronomers with whom I collaborated for many years: Jun Jugaku and Nobuo Arimoto. Arriving in Kyoto I heard that, surprisingly, they have changed activity: one is now Newspaper Editor, and everyone of us has enjoyed the sense of humour present in the "Sidereal Time", and one became Managing Director of the Local Finances of the IAU. I hope, however, that soon they will return to their original job: our dear Astronomy.

Again a very warm THANK YOU to all the Organizers of the XXIII IAU General Assembly.

Address by the retiring General Secretary, Prof. I. Appenzeller

Members of the Union, Dear Friends,

When I accepted the office of General Secretary three years ago I expected a fair amount of work, the pleasure of interacting with colleagues in many different countries, and the satisfaction of rendering a valuable service to the international astronomical community.

Looking back today I am pleased to state that all these expectations turned out to be correct.

There was, in fact, plenty of work to do. When I recently brought the part of my IAU correspondence which I had kept in Heidelberg to our archives in Paris I found that I had accumulated during the three years about 4000 letters, faxes and e-mails. But although there has been a lot of work - and although I am definitely not unhappy to hand over today this office to my successor - the last three years were also pleasant and satisfying.

Among the reasons why this office remained a pleasure and did not become a burden I would like to mention first the excellent cooperation with our President, with the Assistant General Secretary and with the other members of our Executive Committee. Therefore, many thanks to all of you here on the stage. Of course, much of the work in such an organisation goes on back-stage, behind the scenes. In the case of the IAU this means in our office in Paris. Hence I would like to express my special thanks to our Paris staff, Monique Orine and Julie Saucedo, for their hard and dedicated work and for the pleasant and cheerful atmosphere at the Paris office which I will surely miss in the future.

Another factor which made this work easy and pleasant was the very good cooperation which I received from all sections of our Union. This cooperative spirit is certainly one of the characteristics of our astronomical community, perhaps reflecting the fact that astronomers are particularly attached to their science and fascinated by the work which they are doing.

My term of office ends here in Kyoto during the first IAU General Assembly taking place in Japan. I am sure that this meeting will be remembered for a long time not only for its beautiful venue, for attracting a record number of participants, but also for an exceptionally efficient and smooth local organisation. Thus, I would like to conclude this address with a few personal words to our Japanese hosts:

今回、日本で開催される第一回国際天文学連合総会の組織委員に参加させて頂くことができ、大変光栄に思います。多くの方はご存知かもしれませんが、私が日本の天文学界に関わるようになったのは、２５年以上も前に宇野わさぶろう、上条ふみお両教授に東京大学に招かれのがきっかけでした。それ以来、私は日本における天文学に多大な関心を寄せ続け、そのすばらしい進歩にはただただ敬服するばかりです。今日、天文学の分野においては日本は重要な役割を果たしており、その意味でもIAU総会がここ日本で開催されるのは当然かと思われます。期待した通り、杉本だいいちろう教授のNational Organizing Committeeと福島としお教授のLocal Organizing Committee の働きによって、総会の準備も整い、すばらしく組織されたものになりました。

お二人をはじめメンバーのみなさまのすばらしい働きに感謝いたします。

Address by the incoming General Secretary, Dr. J. Andersen

Fellow members of the IAU, dear colleagues and friends,

It is with considerable trepidation that I venture to fill the last place of the century in the long line of IAU General Secretaries, which includes so many of the most distinguished astronomers of their time. Yet, like in astronomical research, there are useful functions for people of many kinds. So, with the guidance of the Executive Committee, perhaps even someone like I might be of the useful kind. Et à mes amis français: Je ferai mon mieux pour utiliser les deux langues officielles de l'Union selon les circonstances de la situation!

I suppose my predecessors have also been asked the question, "What attracts a person to the job of IAU General Secretary?". Apart from the opportunity to help our science flourish by international cooperation, an attractive challenge in itself, a quality of the IAU which I find particularly charming is what might be called "constructive disobedience". Maybe because astronomy is such a wholly peaceful, inherently apolitical, and generally non-lucrative enterprise, the IAU has had the opportunity, from time to time, to circumvent political conventions to some extent and help penetrate or undermine the political walls that still separate some of our communities. As a recent example I was delighted to see, at the Asian-Pacific Regional Meeting in Korea last year, how astronomers from both parts of China were happily discussing the results of their ongoing joint research projects. And the very successful International School for Young Astronomers in Iran which Don Wentzel and his team have just held, is another small, but significant step in the same direction: It is good to remember that walls are torn down the same way they are built, stone by stone.

In practical terms, I often think of the IAU as a ship: On the bridge, the Officers (the Executive Committee) set out the course, and down below the Chief Engineer (the General Secretary) and the crew try to maneuver the machinery to make the ship actually move in that direction. To someone on the shore the ship may seem to move very slowly, but below deck a lot of activity is going on.

Similarly, to some of you the IAU may seem to move frustratingly slowly on your favourite issues. Yet, despite the fact that we are running on very limited supplies of fuel (=money), there is indeed "evolution on human time scales", as the working title of one Joint Discussion was at one time: In The Hague, the new format of the General Assemblies was introduced, and while some fine tuning may remain, this is essentially a great and undisputed improvement. The pooling of Symposium and General Assembly travel funds that has allowed such a large attendance at this General Assembly is one of the ways in which we try to maximise the scientific returns of our funds.

Moreover, under my predecessor, the new Division structure was introduced on an trial basis that has led to its formal acceptance at this General Assembly. I look forward to working with the Division Presidents to turn the Division structure into a real "turbo charger" for the IAU machinery. And finally, Saturday's Joint Discussion showed that the IAU is finding a useful place as a forum for early, informal discussions of possible future large-scale facilities much sooner than I would have dared hope even a year ago.

You will be able to judge the progress in these matters for yourself when we meet again in Manchester in the year 2000. I look forward to working with our British friends to make the XXIVth General Assembly another memorable event. Our Japanese hosts at this General Assembly have set the standards by which we shall be judged extremely high, but we shall take their example as our best inspiration.

Finally, I look forward to working, on the one hand with the new Executive Committee and with you, the Members, and on the other hand with our small, but very capable staff, Monique Orine and Julie Saucedo. Many of you have already expressed your appreciation for their always kind and efficient help, as I shall do again three years from now.

My last words must go to my predecessor, Immo Appenzeller, whose kind and diplomatic instruction over the last three years to the intricacies of the job of General Secretary of the IAU will be my compass over the next three years.

CHAPTER IV

RESOLUTIONS OF THE GENERAL ASSEMBLY

Rapport du Comité des Résolutions/Report of the Resolutions Committee

The Report was presented by the Chairman of the Committee, Prof. J.-C. Pecker.

PREMIERE SESSION DE L'ASSEMBLEE GENERALE

Monsieur le Président, Chères et Chers Collègues,

Contrairement à la procédure suivie jadis, nous discuterons, dès la première session de l'Assemblée Générale, d'un certain nombre de résolutions, celles dont le Comité des Résolutions, en fonction depuis la précédente Assemblée Générale, a eu la possibilité de débattre déjà, par correspondance, avec l'aide des traducteurs officiels élus alors.

Je rappellerai tout d'abord que les propositions de résolutions sont issues, comme indiqué dans l'article VIII.36, § a des Directives :

(A) du Comité Exécutif (ainsi que, implicitement, de Comités du Comité Exécutif, ou de groupes de travail désignés par le Comité Exécutif), et des Organisations Nationales adhérentes,

(B) des Divisions, des commissions ne dépendant pas d'une Division, (et, implicitement, des Discussions communes, ou des Symposiums, organisés par diverses Divisions ou Commissions).

RESOLUTIONS DE TYPE A

Nous n'avons été saisis d'aucune résolution en provenance d'une Organisation Adhérente, et allons donc maintenant débattre des résolutions de type B.

RESOLUTIONS DE TYPE B

Il s'agit des résolutions proposées par des Divisions et des Groupes de travail mis sur pieds par l'UAI, seule, ou en commun avec l'UGGI. Ces résolutions (lues en séance, dans les deux langues de l'Union, et présentées ci-après, légèrement amendées par rapport au texte initial qui nous a été soumis, et avec leurs appendices le cas échéant) portent les numéros B1 à B7. Elles sont d'une grande importance pour le travail astronomique. Sans répercussion financière pour l'UAI, elles ont paru assez importantes pour être discutées et votées séparément.

DISCUSSION ET VOTE

Les résolutions de type A soumises par le Comité Exécutif (ou par des Comités du Comité Exécutif) seront discutées et votées lors de la seconde séance de l'Assemblée Générale, ainsi que les résolutions de type B éventuellement proposées à la suite des débats spécialisés qui auront lieu entre les deux sessions. La suite de ce rapport sera présentée lors de la seconde session de l'Assemblée Générale, comme indiqué sur l'ordre du jour.

J. Andersen (ed.), Transactions of the International Astronomical Union Volume XXIIIB, 25–51.

Je voudrais, Monsieur le Président, Chères et Chers Collègues, insister auprès des Présidents de Division ou de Commission pour que les nouvelles résolutions qui seront discutées lors de la seconde session, si elles devaient avoir une implication financière directe, soient examinées directement par le Comité Exécutif, dans le cadre de la politique budgétaire de l'Union, sans avis du Comité des Résolutions.

Je souhaiterais aussi préciser que si les résolutions soumises par des Divisions ou des Commissions devaient contredire des résolutions antérieures, il n'est pas certain que le Comité des Résolutions puisse en prendre conscience dans un bref laps de temps. Dans ce cas, dès lors qu'un doute serait émis par qui que ce soit, à ce sujet, au cours de la seconde session de l'Assemblée Générale, la résolution concernée devra être ajournée, après avis du Comité des Résolutions sur la validité du doute exprimé.

Je terminerai ce rapport lors de la seconde session de l'Assemblée Générale, comme prévu par l'ordre du jour.

Monsieur le Président, Chères et Chers Collègues, je vous remercie de votre attention.

SECONDE SESSION DE L'ASSEMBLÉE GÉNÉRALE (Suite)

Monsieur le Président, Chères et Chers Collègues,

Nous avons aujourd'hui à discuter et à voter un certain nombre de résolutions, quelques-unes soumises depuis la première session de l'Assemblée Générale, et que nous avons considéré comme recevables. Je lirai les résolutions qu'il nous reste à examiner dans la langue dans laquelle elles ont été écrites et soumises. Elles seront projetées par rétroprojection dans l'autre langue de l'Union (le texte du rapport a été prononcé tantôt en anglais, tantôt en français; nous donnons ici les deux versions intégrales de ce rapport).

Conformément à l'ordre du jour, nous commencerons par les

RESOLUTIONS DE TYPE A

La résolution A1, soumise par le Comité Exécutif, est particulièrement importante. Elle vise à donner au Comité Exécutif de l'UAI l'autorité que confère un vote de tous les Membres de l'Union pour lui permettre de s'adresser directement, avec une efficacité optimale, à diverses autorités politiques responsables, en vue de la protection du ciel nocturne.

La commission 5, qui est un Comité du Comité Exécutif, a soumis 6 résolutions. Trois d'entre elles sont maintenant soumises au vote de l'Assemblée Générale (Les résolutions A2, A3, A4, sont discutées, et approuvées par vote ; leur texte figure ci-après dans la liste des résolutions approuvées).

La commission 5 a proposé trois autres résolutions.

- L'une concerne **l'accessibilité des bases de données astronomiques**. Il nous a semblé que l'action légitime demandée par la Commission est du ressort du Comité Exécutif, plus que de l'Union en général. Elle est soumise directement à l'examen du Comité Exécutif.
- La seconde concerne **le soutien nécessaire pour la publication de la bibliographie Astronomy and Astrophysics Abstracts**. Cette résolution exige aussi une action de la part du Comité Exécutif auprès de certaines autorités administratives; elle est renvoyée directement au Comité Exécutif, sans vote de l'Assemblée Générale.
- Enfin, la commission 5 a soumis une résolution assez générale demandant **un soutien en vue de définir un protocole standard d'accès aux données d'observation individuelles, et un format unique pour les décrire**. Nous considérons cette importante résolution comme exigeant une action de la Commission 5 elle-même. Elle sera diffusée par la commission 5 dans ses rapports ou publications, comme résolution de type C.

Aucune résolution n'ayant été proposée par le Comité des Résolutions, nous en arrivons aux

RESOLUTIONS DE TYPE C

Il s'agit des résolutions proposées par les Divisions, les Commissions ou les Groupes de travail ayant proposé des résolutions depuis la première session de la XXIIIème Assemblée Générale.

La Commission 50 a soumis trois résolutions de catégorie C.

- Le première concerne **la pollution lumineuse** du ciel: elle constitue une mise en oeuvre partielle des buts que se propose la résolution A1, déjà adoptée.
- La seconde concerne **le bruit radioélectrique**; elle reprend essentiellement des résolutions antérieures de l'UAI et les développe.
- La troisième concerne **les engins spatiaux et les débris de l'espace**. Elle demande une action de la part d'organismes extérieurs à l'UAI.

Par suite, bien que nous soyons en accord avec l'esprit de ces trois résolutions, nous proposons de les soumettre directement, sans vote, au Comité Exécutif.

La JD 22 a soumis une résolution sur **la protection d'un site éventuel d'observatoire sur la face cachée de la Lune, pour l'avenir de la radioastronomie à haute sensibilité**. Nous sommes très conscients de l'importance de cette résolution. Mais nous considérons qu'elle devrait être rédigée à nouveau par les commissions impliquées dans la JD 22, et proposée ultérieurement à un vote de l'Assemblée Générale. Elle devrait à notre avis donner lieu à un échange préliminaire entre les Divisions et Commissions impliquées, d'une part, et le Comité Exécutif d'autrepart, en vue de l'interaction nécessaire avec des organismes autres que ceux dépendant directement de la seule UAI.

Le Groupe de Travail **sur la nomenclature des planètes du système solaire** a proposé une résolution importante ; elle nous semble devoir être mise en oeuvre par les spécialistes du domaine concerné, c'est-à-dire s'adresser au groupe de Travail lui-même, et aux Commissions concernées, comme résolution de type C.

Les commissions 25 et 36 ont soumis une résolution **sur le point zéro de l'échelle des magnitudes bolométriques**. Cette résolution est aussi très importante. Elle sera considérée comme résolution de type C, relevant des deux Commissions concernées.

Les résolutions ci-dessus mentionnées, essentiellement de type C, ne sont donc pas soumises au vote de l'Assemblée Générale. Soit elles sont soumises directement au Comité Exécutif, soit elles sont du seul ressort des Divisions, Commissions, ou Groupes de travail qui les ont proposées. Elles devront être diffusées par ces Divisions, Commissions ou Groupes de travail, dans leurs publications et rapports.

Monsieur le Président, chères et chers Collègues, il est temps de conclure ce point de l'ordre du jour. Nous avons eu beaucoup de mal à concilier divers impératifs : le caractère durable de nombreuses résolutions, le besoin de les savoir préparées le mieux possible avant le vote, la nécessité du multilinguisme de l'Union, et - last but not least ! - le fait que le temps nous ait été mesuré.

J'adresserai en temps utile à mon successeur, et au Comité Exécutif, des propositions en vue d'un meilleur fonctionnement du système qui régit la rédaction, l'examen, et le vote des résolutions.

Je voudrais enfin exprimer ici mes remerciements sincères et cordiaux à mes collègues, Membres pendant le triennium passé, et pour deux Assemblées Générales consécutives, du Comité des Résolutions, les Professeurs Martin McCarthy, Jorge Sahade, Joseph Smak, Patrick Wayman, Bernard Yallop, et à nos deux traducteurs officiels, les Drs Roger Cayrel et Janet Rountree. Notre petit groupe a travaillé entre les deux Assemblées Générales par correspondance et s'est réuni quatre fois au cours de la présente Assemblée Générale. Ce fut toujours dans une atmosphère de travail et d'amicale collaboration. J'exprime aussi la reconnaissance du Comité des Résolutions au Secrétariat de l'UAI et spécialement à Madame Monique Léger-Orine, pour l'aide précieuse qui nous a été dispensée, au Professeur Jun Jugaku, Rédacteur en Chef du "Sidereal Times", qui a si bien maîtrisé certains de nos problèmes, ainsi qu'au Comité Local d'Organisation, présidé si efficacement par le Professeur Toshio Fukushima, avec l'aide de nombreux collègues japonais. Grâce à tous, nous avons bénéficié d'excellentes conditions de travail.

Monsieur le Président, Chères et Chers Collègues, je vous remercie de votre attention.

Jean-Claude Pecker
Président (1994-1997) du Comité des Résolutions.

FIRST SESSION OF THE GENERAL ASSEMBLY

Mr President, Dear Colleagues,

Contrarily to the previous procedure, we shall debate, already at the first session of the General Assembly, of some resolutions, those of which the Resolutions Committee, in function since the preceding General Assembly, has had the possibility of discussing already by correspondence, with the help of the official translators then elected.

I would like first to remind the audience that draft resolutions are submitted, as indicated by article VIII.36 § a of the Working Rules:

(A) by the Executive Committee (as well as, implicitly, by the Committees of the Executive Committee, or by Working Groups set up by the Executive Committee), and by Adhering bodies

(B) by Divisions, and Commissions not attached to a Division (and implicitly by Joint Discussions, or Symposiums, organised by various divisions or commissions).

RESOLUTIONS OF TYPE A

We have not received any resolution submitted by an Adhering Body. Point 9 of the agenda is therefore dealt with, and we pass to point 10 of the agenda.

RESOLUTIONS OF TYPE B

We shall now open the debate concerning resolutions of type B proposed by Divisions or Working Groups set up by the IAU alone, or in common with IUGG. These resolutions (read in both official languages of the Union, and printed hereafter, slightly amended with respect to the draft resolutions submitted to us, and with their appendices in the appropriate cases) are numbered B1 to B7. They are of a great importance for astronomical daily work. Without any financial implications for the IAU, they seemed to us important enough to be discussed and voted upon separately

VOTE AND DISCUSSION

Resolutions of type A, submitted by the Executive Committee (or by Committees of the Executive Committee) will be discussed and voted upon at the second session of the General Assembly, as well as resolutions of type B that might be submitted as follow-ups to specialised debates having taken place between the two sessions. The second part of the report will be presented during during the second session of the General Assembly, as indicated in the Agenda.

I would like, Mr President, Dear Colleagues, to insist upon the Presidents of Divisions or Commissions so that new draft resolutions to be discussed during the second session of the General Assembly, in the case they would have direct financial implications, be examined directly by the Executive Committee, in the framework of the budgetary policy of the Union, without an advice of the Resolutions Committee.

I would like also make clear that if resolutions submitted by Divisions or by Commissions would contradict previous resolutions, it is not certain that the Resolutions Committee could be conscious of that in a short time. In that case, as soon as a doubt would be expressed by anyone on that matter, during the second session of the General Assembly, the concerned resolution will have to be adjourned, after advice of the Resolutions Committee upon the validity of the expressed doubts.

I shall terminate my report during the second session of the General Assembly, as indicated in the agenda.

Mr President, Dear Colleagues, I thank you for your attention..

Second part of the Report of the Resolutions Committee/Second session of the General Assembly

Mr President, Dear Colleagues,

We have got to debate today, and to vote upon, several resolutions, some of them submitted since the first session of the General Assembly, and which we have taken into consideration. I shall read the resolutions which we have still to examine in the language in which they have been submitted. They will be projected by overhead projection in the other language of the Union (the text of the report has been pronounced partly in English, partly in French; we give here both versions of the full report).

As indicated in the agenda, we shall start with

RESOLUTIONS OF TYPE A

Resolution A1, submitted by the Executive Committee is of particular importance. It aims at giving the Executive Committee of the IAU the authority given by a vote of all the Members of the Union in order to allow it to address directly with an optimum efficiency to various responsible political authorities, in view of ***the protection of the night sky***. (Resolution A1 is discussed and approved by vote; its text is printed hereafter in the list of resolutions approved by the General Assembly of the Union).

Commission 5, which is a Committee of the Executive Committee, submitted 6 resolutions. Three of them are now put to the vote of the General Assembly. (Resolutions A2, A3, A4 are discussed and approved by vote; their text is printed hereafter in the list of resolutions approved by the General Assembly of the Union).

Commission 5 has proposed three other resolutions

- *The first one concerns* ***the accessibility of astronomical data bases***. *It seemed to us that the legitimate action required by the Commission is the responsibility of the Executive Committee, more than that of the Union itself. It is therefore submitted directly to the Executive Committee.*
- *The second one concerns* ***the needed support for the publication of the bibliography "Astronomy and Astrophysics Abstracts"***. *This resolution requires also an action of the Executive Committee towards some administrative authorities, and it is referred directly to the Executive Committee.*
- *Finally, Commission 5 has submitted another resolution, rather general. It asks for* ***the support to define a standard protocol to access individual observatory data, and a unique format to describe individual observations***. *We consider that this important resolution requires action by Commission 5 alone. It will be circulated by Commission 5, in its reports or publications, as a resolution of type C.*

No resolution having been proposed by the Resolutions Committee, I am coming now to the

RESOLUTIONS TYPE C

Those are resolutions submitted by Divisions, Commissions, or Working since the first session of the XXIII-rd General Assembly

Commission 50 has submitted three resolutions, of category C..

- *The first one concerns light pollution of the sky. It essentially implements in part the resolution A1, already adopted.*
- *The second one concerns the radio noise; it implements and develops past resolutions of the IAU.*
- *The third one, on space craft and space debris, requires an action on the part of bodies outside IAU.*

Therefore, we have agreed with the spirit of these three important resolutions, but we propose to refer them directly, without a vote, to the Executive Committee.

Joint Discussion 22 has submitted a resolution concerning the protection of a potential observatory site on the far-side of the Moon for the future of high sensitivity radio astronomy. We are quite conscious of the importance of this resolution. But we consider that it should be rewritten by the relevant commissions and proposed to the IAU later. It should perhaps give place to a preliminary exchange between the Divisions and Commissions involved, on one side, and the Executive Committee, on the other side, in view of the necessary interaction with bodies other than the IAU.

The Working Group on Planetary System Nomenclature has submitted an important resolution concerning solar system planets nomenclature. In our opinion, it is to be implemented by the specialists in the field, i.e. to stay internal to the WG itself.

Commission 25 and 36 have submitted a resolution on the zero point of bolometric magnitudes scales. This resolution is also very important. As a resolution of type C, it will be considered as addressing the two commissions involved.

The six above mentioned resolutions will not be submitted to the vote of the General Assembly, being essentially of type C. Either they are directly referred to the Executive Committee, or their implementation is of the responsibility of the Divisions, Commissions or Working Groups of their origin. They shall be circulated by these Divisions, Commissions, or Working groups, in their publications and reports.

It is time to conclude this point of the agenda. We had some trouble in satisfying various requirements: the sustainable importance of several resolutions, the need to ensure they have been drafted and are feasible before the vote, the necessity of the Union multilinguism, and (last, but not least!) the fact that time was rather limited.

In order to improve the procedure for writing, reviewing, and voting on resolutions, I shall make some suggestions to my successor and to the Executive Committee, at the appropriate time.

I would like finally to express my sincere and cordial thanks to my colleagues, Members of the Resolutions Committee during the past triennium, and during two consecutive General Assemblies, namely Professors Martin McCarthy, Jorge Sahade, Joseph Smak, Patrick Wayman, Bernard Yallop, and to our two official translators, Drs Janet Rountree and Roger Cayrel. Our small group has worked between the two General Assemblies, by correspondence, and met five times during the present General Assembly. It has always been in a fruitful and friendly atmosphere. I express also the gratitude of the Resolutions Committee to the Secretariat of the IAU, in particular to Ms Monique Léger-Orine, for the precious help dispensed to us, to Professor Jun Jugaku, Chief Editor of the "Sidereal Times", who handled so well some of our problems, and finally to the Local Organising Committee, so efficiently chaired over by Professor Toshio Fukushima, with the help of several Japanese colleagues. All of them contributed greatly to the excellent working conditions of which we have benefited.

Mr. President, Dear Colleagues, I thank you for your attention.

Jean-Claude Pecker
President (1994-1997) of the Resolutions Committee.

TEXT OF APPROVED RESOLUTIONS/TEXTE DES RESOLUTIONS APPROUVEES

RESOLUTION A 1

PROTECTION OF THE NIGHT SKY

The XXIIIrd International Astronomical Union General Assembly,

Considering that

proposals have been made repeatedly to place luminous objects in orbit around the earth to carry messages of various kind and that the implementation of such proposals would have deleterious effects on astronomical observations,

and that

the night sky is the heritage of all humanity, which should therefore be preserved untouched,

Requests the President

to take steps with the appropriate authorities to ensure that the night sky receive no less protection than has been given to the world heritage sites on earth.

RESOLUTION A 1

PROTECTION DU CIEL NOCTURNE

La XXIIIe Assemblée Générale de l'Union Astronomique Internationale.

Considérant

que des propositions ont été faites de façon répétée afin de mettre sur orbite terrestre des objets lumineux véhiculant des messages de diverse nature et que la réalisation de tels projets aurait des effets nuisibles sur les observations astronomiques,

et que

le ciel nocturne est un heritage de l'Humanité qui doit, en tant que tel, demeurer intact

demande à son Président

de prendre toutes mesures avec les autorités compétentes de façon que le ciel nocturne ne soit pas moins protégé que les sites du Patrimoine Commun de l'Humanité.

RESOLUTION A2

PROPOSALS FOR REGISTERING A NEW ACRONYM

The XXIIIrd International Astronomical Union General Assembly,

Recognizing

the many benefits that would follow from the clear and unambiguous identification of all astronomical objects outside the solar system to which reference is made in astronomical journals and other sources of data,

and noting

that the "Memorandum on Designations" (which accompanied Resolution C3 - New Delhi) presented the basic FORM for designations, namely:

acronym sequence (e.g. NGC 6334, PSR J1302-6350)

that since the "Memorandum on Designations" was issued in 1985, much progress has been made which includes:

the latest version, IAU Recommendations for Nomenclature, on the World Wide Web (WWW) with URL: http://cdsweb.u-strasbg.fr/iau-spec.html

and the on-line "Second Reference Dictionary of the Nomenclature of Celestial Objects" on the WWW with URL: http://astro.u-strasbg.fr/cgi-bin/Dic

and realizing

that much confusion still exists with duplicate acronyms and non-conforming designations appearing in the literature,

Acknowledges

the need for a voluntary registry of new acronyms where the entries are reviewed by the Task Group on Designations before publication to facilitate the discovery and elimination of potentially confusing and inadvertently non-conforming designations **BEFORE** they appear in print or in data archives,

that registering an acronym would be especially advantageous for large, on-going surveys where images and source lists may be produced in stages and/or may be published in electronic form **BEFORE** the final printed catalogue,

that registering the acronym ensures the availability of a suitable, unique acronym for the survey and that the proposed designation conforms to the IAU Recommendations,

Endorses

the continued development by members of the Task Group on Designations of the Experimental Acronym Registry which is now part of the on-line "Second Reference Dictionary",

and supports

the efforts of the Task Group to encourage authors, referees, and editors to use this new tool to help guarantee that designations in future papers conform to IAU recommendations.

RESOLUTION A2

PROPOSITIONS POUR L'ENREGISTREMENT DE NOUVEAUX ACRONYMES

La XXIIIe Assemblée Générale de l'Union Astronomique Internationale.

Reconnaissant

les nombreux avantages qui découleraient d'une identification claire et sans ambiguité de tous les objets astronomiques extérieurs au système solaire, nommés dans les revues astronomiques et autres sources de données,

et considérant

que le "Memorandum sur les désignations" (accompagnant la résolution C 3 prise à Delhi en 1985) a présenté la forme type pour les désignations, à savoir

séquence acronyme (par exemple NGC 6334, PSR J1302-6350),

que depuis que le "Memorandum sur les désignations" a été publié en 1985, beaucoup de progrès ont été réalisés, incluant:

la dernière version des recommandations de l'UAI pour la Nomenclature, sur le "World Wide Web" (WWW), à l'URL: http://cdsweb.u-strasbg.fr/iau-spec.html

et le "Deuxième Dictionnaire de référence sur la nomenclature des objets astronomiques" sur le WWW, à l'URL: http://astro.u-strasbg.fr/cgi-bin/Dic,

et se rendant compte

qu'il existe encore une certaine confusion avec des acronymes en double et des désignations non conformes dans la littérature,

reconnaît

le besoin pour un enregistrement spontané des nouveaux acronymes, avec un examen des entrées nouvelles par le groupe de travail sur les Désignations avant publication, pour faciliter la découverte et l'élimination des désignations non conformes ou potentiellement ambigues **AVANT** qu'elles ne paraissent, imprimées ou dans des bases de données,

que l'enregistrement d'un acronyme serait particulièment avantageux pour les grands relevés en cours, où les images et les listes d'objets peuvent être produits par étapes et/ou peuvent être publiés sous forme électronique **AVANT** le catalogue final imprimé,

que l'enregistrement de l'acronyme garantit la disponibilité d'un acronyme unique et adéquat pour le relevé et que la désignation proposée est conforme aux recommandations de l'UAI,

appuie

la poursuite de l'activité de l'enregistrement expérimental des acronymes par le groupe de travail sur les Désignations, dont le résultat est maintenant une partie du "Second Dictionnaire de Référence", accessible électroniquement,

et soutient

les efforts du Groupe de Travail pour encourager les auteurs, Comités de lecture et éditeurs à garantir que les désignations dans les publications à venir soient conformes aux recommandations de l'UAI.

RESOLUTION A3

ON THE NEED FOR ARCHIVING ASTRONOMICAL DATA

The XXIIIrd International Astronomical Union General Assembly,

Considering

the continuing important role of astronomical data from the past, including bibliographical information,

Considering

the phenomenal increase in these data,

Considering

the importance of their safeguarding and of their accessibility to the entire astronomical community,

Recommends

that the archiving of these data be an integral part of all major research projects and be taken into account by the editors of journals. The IAU recommends that astronomy archives be coded in the FITS format,

Supports

the continued maintenance of the Data Centers whose role in the distribution of information is of prime importance for astronomy, and supports their collaboration.

RESOLUTION A3

SUR LA NECESSITE D'ARCHIVER LES DONNEES ASTRONOMIQUES

La XXIIIe Assemblée Générale de l'Union Astronomique Internationale.

Considérant

l'importance permanente du rôle des données astronomiques du passé, y compris les informations bibliographiques,

Considérant

l'accroissement phénoménal du volume des données,

Considérant

l'importance de leur sauvegarde et de leur mise à disposition de la communauté astronomique,

Recommande

que l'archivage de ces données soit partie intégrale de tout projet de recherche majeur, et soit pris en considération par les éditeurs de revues,

et que les archives astronomiques soient codées en format FITS,

Soutient

la permanence du fonctionnement des Centres de Données, dont le rôle dans la distribution de l'information est de première importance pour l'astronomie ainsi que leur collaboration mutuelle.

RESOLUTION A4

ON THE MODIFICATION OF DATE VALUES ON FITS SOFTWARE

The XXIIIrd International Astronomical Union General Assembly,

Recognizing

that the two-digit year numbers in the date values of keywords such as DATE-OBS='31/12/99' in FITS files will become ambiguous on the day 2000-01-01,

and noting

that the IAU FITS Working Group has adopted new rules for DATExxxx value strings which specify that the previous convention applies only to dates in the range 1900-1999 and that the new convention DATE-OBS='1999-12-31' is to be used in data interchange and in data archives beginning 1998-01-01,

Urges all IAU members

to ensure that their FITS writing and reading software is modified before 1998-01-01 to support both the new convention and the old convention, in accordance with the rules specified by the IAU FITS Working Group.

RESOLUTION A4

SUR LA MODIFICATION DES DATES EN FORMAT FITS DANS LES LOGICIELS

La XXIIIe Assemblée Générale de l'Union Astronomique Internationale,

Constatant

que les nombres à deux chiffres dans la valeur de la date des mots clefs tels que DATE-OBS='31/12/99' dans les fichiers FITS vont devenir ambigus à partir de la date 2000-01-01,

et notant

que le groupe de travail IAU FITS a adopté de nouvelles règles pour les chaînes de valeur DATExxxx, qui spécifient que la précédente convention s'applique seulement aux dates dans l'intervalle 1900-1999 et que la nouvelle convention DATE-OBS='1999-12-31' doit être utilisée dans l'échange de données et dans les archives à commencer en 1998-01-01,

Demande à tous ses membres

de s'assurer que leur logiciel d'écriture et de lecture FITS est modifié avant 1998-01-01 pour accepter concouremment la nouvelle et l'ancienne conventions, en accord avec les règles spécifié par le groupe de travail IAU FITS.

RESOLUTION B1

ON THE USE OF JULIAN DATES

The XXIIIrd International Astronomical Union General Assembly,

recognizing

a. the need for a system of continuous dating for the purpose of analyzing time-varying astronomical data, and

b. that both Julian Dates and Modified Julian Dates have been employed for this purpose in astronomy, geodesy, and geophysics.

recommends

a. that Julian Date (as defined in the appendix) be used to record the instants of the occurrences of astronomical phenomena,

b. that for those cases where it is convenient to employ a day beginning at midnight, the Modified Julian Date (equivalent to the Julian Date minus 2 400 000.5) be used, and

c. that where there is any possibility of doubt regarding the usage of Modified Julian Date, care be exercised to state its definition specifically,

d. that, in all languages, Julian Date be abbreviated by "JD" and Modified Julian Date be abbreviated by "MJD".

APPENDIX. PROPOSED DEFINITIONS

The following definitions are recommended

1. Julian day number (JDN)

The Julian day number associated with the solar day is the number assigned to a day in a continuous count of days beginning with the Julian day number 0 assigned to the day starting at Greenwich mean noon on 1 January 4713 BC, Julian proleptic calendar -4712.

2. Julian Date (JD)

The Julian Date (JD) of any instant is the Julian day number for the preceding noon plus the fraction of the day since that instant. A Julian Date begins at 12h 0m 0s UT and is composed of 86400 seconds. To determine time intervals in a uniform time system it is necessary to express the JD in a uniform time scale. For that purpose it is recommended that JD be specified as SI seconds in Terrestrial Time (TT) where the length of day is 86,400 SI seconds.

In some cases it may be necessary to specify Julian Date using a different time scale. (See Seidelmann, 1992, for an explanation of the various time scales in use). The time scale used should be indicated when required such as JD(UT1). It should be noted that time intervals calculated from differences of Julian Dates specified in non-uniform time scales, such as UTC, may need to be corrected for changes in time scales (e.g. leap seconds).

An instant in time known in UTC can be converted to Terrestrial Time if such precision is required. Values of TT-UT are available using tables in McCarthy and Babcock (1986) and Stephenson and Morrison (1984, 1995). Table 1 provides the difference between TAI and UTC from 1961 through 1 January 1996. The difference between TT and UTC can be calculated knowing that TT = TAI + 32.184s. The Annual Reports of the International Earth Rotation Service should be consulted for dates after 1996. The data of Table 1 are also available electronically at

http://hpiers.obspm.fr or ftp:hpiers.obspm.fr/iers/bal/bulc/TC-TAI

or at http://maia.usno.navy.mil or at ftp://maia.usno.navy.mil/ser7/tai-utc.dat.

TABLE 1

Difference between the TAI and UTC time scales.
TT-UTC can be calculated by adding 32.184s to TAI-UTC.

1961 Jan 1	=	JD 2 437 300.5	TAI-UTC	=	1.4228180s + (MJD - 37300.) x 0.001296s
1961 Aug 1	=	JD 2 437 512.5	TAI-UTC	=	1.3728180s + (MJD - 37300.) x 0.001296s
1962 Jan 1	=	JD 2 437 665.5	TAI-UTC	=	1.8458580s + (MJD - 37665.) x 0.0011232s
1963 Nov 1	=	JD 2 438 334.5	TAI-UTC	=	1.9458580s + (MJD - 37665.) x 0.0011232s
1964 Jan 1	=	JD 2 438 395.5	TAI-UTC	=	3.2401300s + (MJD - 38761.) x 0.001296s
1964 Apr 1	=	JD 2 438 486.5	TAI-UTC	=	3.3401300s + (MJD - 38761.) x 0.001296s
1964 Sep 1	=	JD 2 438 639.5	TAI-UTC	=	3.4401300s + (MJD - 38761.) x 0.001296s
1965 Jan 1	=	JD 2 438 761.5	TAI-UTC	=	3.5401300s + (MJD - 38761.) x 0.001296s
1965 Mar 1	=	JD 2 438 820.5	TAI-UTC	=	3.6401300s + (MJD - 38761.) x 0.001296s
1965 Jul 1	=	JD 2 438 942.5	TAI-UTC	=	3.7401300s + (MJD - 38761.) x 0.001296s
1965 Sep 1	=	JD 2 439 004.5	TAI-UTC	=	3.8401300s + (MJD - 38761.) x 0.001296s
1966 Jan 1	=	JD 2 439 126.5	TAI-UTC	=	4.3131700s + (MJD - 39126.) x 0.002592s
1968 Feb 1	=	JD 2 439 887.5	TAI-UTC	=	4.2131700s + (MJD - 39126.) x 0.002592s
1972 Jan 1	=	JD 2 441 317.5	TAI-UTC	=	10.0s
1972 Jul 1	=	JD 2 441 499.5	TAI-UTC	=	11.0s
1973 Jan 1	=	JD 2 441 683.5	TAI-UTC	=	12.0s
1974 Jan 1	=	JD 2 442 048.5	TAI-UTC	=	13.0s
1975 Jan 1	=	JD 2 442 413.5	TAI-UTC	=	14.0s
1976 Jan 1	=	JD 2 442 778.5	TAI-UTC	=	15.0s
1977 Jan 1	=	JD 2 443 144.5	TAI-UTC	=	16.0s
1978 Jan 1	=	JD 2 443 509.5	TAI-UTC	=	17.0s
1979 Jan 1	=	JD 2 443 874.5	TAI-UTC	=	18.0s
1980 Jan 1	=	JD 2 444 239.5	TAI-UTC	=	19.0s
1981 Jul 1	=	JD 2 444 786.5	TAI-UTC	=	20.0s
1982 Jul 1	=	JD 2 445 151.5	TAI-UTC	=	21.0s
1983 Jul 1	=	JD 2 445 516.5	TAI-UTC	=	22.0s
1985 Jul 1	=	JD 2 446 247.5	TAI-UTC	=	23.0s
1988 Jan 1	=	JD 2 447 161.5	TAI-UTC	=	24.0s
1990 Jan 1	=	JD 2 447 892.5	TAI-UTC	=	25.0s
1991 Jan 1	=	JD 2 448 257.5	TAI-UTC	=	26.0s
1992 Jul 1	=	JD 2 448 804.5	TAI-UTC	=	27.0s
1993 Jul 1	=	JD 2 449 169.5	TAI-UTC	=	28.0s
1994 Jul 1	=	JD 2 449 534.5	TAI-UTC	=	29.0s
1996 Jan 1	=	JD 2 450 083.5	TAI-UTC	=	30.0s

RÉSOLUTION B1

SUR L'UTILISATION DES DATES JULIENNES

La XXIIIème Générale Assemblée de l'Union Astronomique Internationale

reconnaissant

a. le besoin d'un système de datation continue pour l'analyse des données astronomiques dépendant du Temps,

b. que la date julienne et la date julienne modifiée ont, toutes deux, été employées dans ce dessein, en astronomie, en géodésie et en géophysique

recommande

a. que la date julienne (telle que définie dans l'appendice) soit utilisée pour enregisrer les moments où se produisent les phénoménes astronomiques

b. que dans les cas où il est commode d'employer des jours commençant à 0h, on emploie la date julienne modifiée (équivalente à la date julienne moins 2 400 000,5 jours).

c. que, dans le cas où il pourrait y avoir doute sur l'emploi de la date julienne modifiée, on prenne soin d'en rappeler la définition.

d. que, quelle que soit la langue utilisée, l'abréviation pour date julienne soit "JD" et l'abréviation pour date julienne modifiée soit "MJD".

APPENDICE, DEFINITIONS PROPOSEES

1. Numéro du jour Julien (JDN)

Le numéro du jour Julien est le numéro assigné au jour considéré, dans une numérotation continue partant du jour Julien zéro, affecté au jour commençant à midi moyen à Greenwich, le 1er janvier 4713 avant Jésus-Christ, calendrier julien proleptique (-4712).

2. Date julienne (JD)

La date julienne (JD) à un instant quelconque est le numéro du jour Julien du midi précédent plus la fraction de jour depuis cet instant. La date julienne commence à 12h 0m 0s et est composée de 86400 secondes. Pour déterminer des intervalles de temps dans un système de temps uniforme, il est nécessaire d'exprimer la date julienne dans une échelle de temps uniforme. Il est recommandé pour cela de spécifier la JD en temps terrestre (TT), où la longueur du jour est de 86400s SI.

Dans certains cas il est nécessaire de spécifier la date julienne à partir d'autres échelles de Temps (cf. Seidelmann 1992, pour des explications sur les différentes échelles de temps en usage). L'échelle de temps utilisé devrait être indiquée, en cas de besoin, sous la forme JD (UT1). A noter que les intervalles de temps, calculés par différence de deux dates juliennes spécifiées dans des échelles de temps non uniformes, comme UTC, peuvent nécessiter des corrections pour des changements dans les échelles de temps (par exemple des secondes intercalaires).

Un instant connu dans l'échelle de temps UTC peut être converti en temps terrestre si une telle précision est requise. Les valeurs de TT-UTC sont disponibles dans les tables de McCarthy et Babcok (1986) et Stephenson et Morrison (1984, 1995). La table 1 donne les différences TAI (temps atomique international) et UTC de 1961 jusqu'au 1er janvier 1996. La différence entre TT et UTC peut être calculée sachant que TT = TAI + 32,184s. Les rapports annuels du Service International de la Rotation de la Terre pourront être consultés pour les dates postérieures à 1996. Les données de la table 1 sont aussi disponibles électroniquement à:

http://hpiers.obspm.fr or ftp:hpiers.obspm.fr/iers/bal/bulc/TC-TAI

ou à http://maia.usno.navy.mil ou à ftp://maia.usno.navy.mil/ser7/tai-utc.dat.

TABLE 1

Différence entre les échelles de temps TAI et UTC.
La valeur de TT-UTC s'obtient en ajoutant 32.184s to TAI-UTC.
Voir table de la version anglaise

RESOLUTION B2

ON THE INTERNATIONAL CELESTIAL REFERENCE SYSTEM (ICRS)

The XXIIIrd International Astronomical Union General Assembly

Considering

a. That Recommendation VII of Resolution A4 of the 21st General Assembly specifies the coordinate system for the new celestial reference frame and, in particular, its continuity with the FK5 system at J2000.0;

b. That Resolution B5 of the 22nd General Assembly specifies a list of extragalactic sources for consideration as candidates for the realization of the new celestial reference frame;

c. That the IAU Working Group on Reference Frames has in 1995 finalized the positions of these candidate extragalactic sources in a coordinate frame aligned to that of the FK5 to within the tolerance of the errors in the latter (see note 1);

d. That the Hipparcos Catalogue was finalized in 1996 and that its coordinate frame is aligned to that of the frame of the extragalactic sources in (c) with one sigma uncertainties of ± 0.6 milliarcseconds (mas) at epoch J1991.25 and ± 0.25 mas per year in rotation rate;

Noting

That all the conditions in the IAU Resolutions have now been met;

Resolves

a. That, as from 1 January 1998, the IAU celestial reference system shall be the International Celestial Reference System (ICRS) as specified in the 1991 IAU Resolution on reference frames and as defined by the International Earth Rotation Service (IERS) (see note 2);

b. That the corresponding fundamental reference frame shall be the International Celestial Reference Frame (ICRF) constructed by the IAU Working Group on Reference Frames;

c. That the Hipparcos Catalogue shall be the primary realization of the ICRS at optical wavelengths;

d. That IERS should take appropriate measures, in conjunction with the IAU Working Group on reference frames, to maintain the ICRF and its ties to the reference frames at other wavelengths.

Note 1: IERS 1995 Report, Observatoire de Paris, p.II-19 (1996).

Note 2: "The extragalactic reference system of the International Earth Rotation Service (ICRS)", Arias, E.F. et al. A & A 303, 604 (1995).

RESOLUTION B2

SUR LE SYSTEME CELESTE INTERNATIONAL DE REFERENCE (ICRS)

La XXIIIème Assemblée Générale de l'Union Astronomique Internationale

Considérant

a. que la recommandation VII de la Résolution A4 de la 21ème Assemblée Générale spécifie le système de coordonnées pour le nouveau repère de référence céleste et, en particulier, sa continuité avec le système du FK5 à J2000,0;

b. que la Résolution B5 de la 22ème Assemblée Générale donne une liste de sources extragalactiques candidates potentielles pour l'établissement du nouveau repère de référence céleste;

c. que le groupe de travail de l'UAI sur les repères de référence a fixé les positions de ces sources extragalactiques potentielles dans un système de coordonnées aligné sur celui du FK5 à l'intérieur d'une zone de tolérance compatible avec les erreurs de ce dernier (voir note 1);

d. que le catalogue Hipparcos a été achevé en 1996 et que ses axes de coordonnées sont alignés avec ceux du système des sources extragalactiques mentionnées en (c) avec une incertitude quadratique moyenne de ± 0.6 millièmes de seconde de degré (mas) pour l'époque J1991,25 et ± 0.25 mas par an pour la vitesse de rotation.

Notant

que, ainsi, toutes les conditions imposées par les résolutions de l'UAI sont à présent remplies,

Décide

a. qu'à compter du 1er janvier 1998, le système céleste de référence sera le Système Céleste International de Référence (ICRS) tel qu'il est décrit par la Résolution de l'UAI de 1991 sur les repères de référence et tel qu'il est défini par le Service International de la Rotation de la Terre (IERS) (voir note 2);

b. que le repère fondamental correspondant sera le repère céleste international de référence (ICRF) construit par le Groupe de travail de l'UAI sur les repères de référence;

c. que le catalogue Hipparcos sera la réalisation primaire de l'ICRF pour les longueurs d'onde optiques;

d. que l'IERS devrait prendre des mesures appropriées, conjointement avec le Groupe de Travail de l'UAI sur les repères de référence, pour la maintenance de l'ICRF et de ses liens avec les autres repères de référence à d'autres longueurs d'ondes.

Note 1: IERS 1995 Report, Observatoire de Paris, p.II-19 (1996).

Note 2: "The extragalactic reference system of the International Earth Rotation Service (ICRS)", Arias, E.F. et al. A & A 303, 604 (1995).

RESOLUTION B3

ON THE ESTABLISHMENT OF A RELATIVISTIC COHERENT FRAMEWORK

The XXIII General Assembly of the IAU,

considering that

- the IAU Resolution A4 (1991) has set up a general relativistic framework to define reference systems centered at the barycenter of the solar system and at the geocenter,
- the Sub Working Group on Relativity in Celestial Mechanics and Astrometry, established by IAU Resolution C6 (1994), reports that relativity has to be taken into account for all astronomical and geodynamical observations but that the framework of IAU Resolution A4 (1991) is not sufficient for some applications, and that the current terminology should be changed to be consistent in the general relativistic framework,

a consistent system of notations is desirable and should be used in all fields of astronomy, geodesy and metrology that deal with space-time references,

noting that

- work on these matters is also being carried out in several other organizations of different types; in the BIPM (an intergovernmental organization), in the IAG (an international association of scientific unions), in the IERS (a service of IAU and IUGG),
- it is of utmost importance that all interested parties adopt consistent definitions and conventions in a coherent general relativistic framework,
- the BIPM has proposed a collaboration with the IAU to realize this goal,

recommends that

- a Joint Committee of the BIPM and the IAU be formed, its tasks being to establish definitions and conventions, to provide a coherent relativistic frame for all activities in space-time references and metrology at a sufficient level of uncertainty, to establish a uniform system of notations for quantities and units, and to develop the adopted definitions and conventions for practical application by the user,
- the IUGG be invited to participate in this Joint Committee to ensure that a coherent system is agreed by the scientific community,
- the organizations taking part in the Joint Committee adopt Resolutions or Recommendations, each following its own procedures, with the aim of having identical definitions, conventions and notations based on the conclusions of the Committee.

BIPM: Bureau International des Poids et Mesures
IAG: International Association of Geodesy
IERS: International Earth Rotation Service
IUGG: International Union for Geodesy and Geophysics

RESOLUTION B3

SUR UN SYSTEME RELATIVISTE COHERENT DE REFERENCE

La XXIIIe Assemblée Générale de l'UAI,

Considérant

- que la Résolution A4 (1991) de l'UAI a établi les bases dans le cadre de la Relativité Générale permettant de définir les systèmes de Référence centrés au barycentre de l'ensemble du système solaire ou au centre de la Terre,
- que le sous-groupe de travail sur la Relativité en Mécanique Céleste et Astronomie, créé par la Résolution C6 (1994) de l'UAI a conclu que la théorie de la Relativité Générale doit être prise en compte dans toutes les observations astronomiques et géodynamiques, mais il a noté que la Résolution A4 (1991) de l'UAI ne suffit pas pour certaines applications et que la terminologie courante devrait être changée pour être en accord avec la Relativité Générale
- qu'un système cohérent de notations est souhaitable, qui devrait être utilisé dans tous les domaines de l'Astronomie, de la Géodésie et de la Métrologie se rapportant à des références spatio-temporelles,

notant

- que des travaux relatifs à ces problèmes sont également effectués dans plusieurs organisations de statuts différents: au BIPM (Organisation intergouvernementale), à l'IAG (Association d'une union scientifique), à l'IERS (Service de l'UAI et de l'UGGI),
- qu'il est de la plus grande importance que toutes les communautés intéressées adopent uniformément des définitions et des conventions dans un cadre cohérent,
- que le BIPM a proposé une collaboration avec l'UAI pour atteindre ce but,

recommande

- qu'un Comité mixte BIPM-UAI soit mis en place, ses tâches étant d'établir des définitions et des conventions pour toutes les activités dans le domaine des références spatio-temporelles et la Métrologie et ce, à un niveau d'incertitude suffisant, d'établir un système uniforme de notations pour les quantités et les unités, et de décrire les conséquences de ces définitions et conventions pour les applications pratiques par les utilisateurs,
- que l'UGGI soit invitée à participer à ce comité mixte afin de s'assurer que ce système sera cohérent et aura l'accord de la communauté scientifique
- que les organisations qui prendront part aux travaux de ce comité devraient adopter des résolutions ou des recommandations selon leurs procédures propres ayant pour objectif d'avoir des définitions, des conventions et des notations identiques, basées sur les conclusions de ce comité.

BIPM: Bureau International des Poids et Mesures
AIG: Association Internationale de Géodésie
IERS: Service International de la Rotation de la Terre
UGGI: Union Géodésique et Géophysique Internationale

RESOLUTION B4

ON NON-RIGID EARTH NUTATION THEORY

The XXIIIrd International Astronomical Union General Assembly

Recognizing

that the International Astronomical Union and the International Union of Geodesy and Geophysics Working Group (IAU-IUGG WG) on Non-rigid Earth Nutation Theory has met its goal by identifying the remaining geophysical and astronomical phenomena that must be modeled before an accurate theory of nutation for a non-rigid Earth can be adopted, and

that, as instructed by IAU Recommendation C1 in 1994, the International Earth Rotation Service (IERS) has published in the IERS Conventions (1996) an interim precession-nutation model that matches the observations with an uncertainty of ± 1 milliarcsecond (mas),

endorses

the conclusions of the IAU-IUGG WG on Non-rigid Earth Nutation Theory given in the appendix,

requests

the IAU-IUGG WG on Non-rigid Earth Nutation Theory to present a detailed report to the next IUGG General Assembly (August 1999), at which time the WG will be discontinued,

and urges

the scientific community to address the following questions in the future:

- completion of a new rigid Earth nutation series with the additional terms necessary for the theory to be complete to within ± 5 microarcseconds, and
- completion of a new non-rigid Earth transfer function for an Earth initially in non-hydrostatic equilibrium, incorporating mantle inelasticity and a Free Core Nutation period in agreement with the observations, and taking into account better modeling of the fluid parts of the planet, including dissipation.

APPENDIX

The WG on Non-rigid Earth Nutation Theory has quantified the problems in the nutation series adopted by the IAU in 1980 by noting:

1. that there is a difference in the precession rate of about -3.0 milliarcseconds per year (mas/year) between the value observed by Very Long Baseline Interferometry (VLBI) and Lunar Laser Ranging (LLR) and the adopted value,
2. that the obliquity has been observed (by VLBI and LLR) to change at a rate of about -0.24 mas/year, although there is no such change implied by the 1980 precession-nutation theory,
3. that, in addition to these trends, there are observable peak-to-peak differences of up to 20 milliarcseconds (mas) between the nutation observed by VLBI and LLR and the nutation adopted by the IAU in 1980,
4. that these differences correspond to spectral amplitudes of up to several mas, and
5. that the differences between observation and theory are well beyond the present observational accuracy.

The WG has recognized the improvements made in the modeling of these quantities, and recommends, in order to derive a more precise nutation model, at the mas level in spectral amplitudes and at a few mas level in the peak to peak analysis, the use of models:

1. based on a new non-rigid Earth transfer function for an Earth initially in non-hydrostatic equilibrium, incorporating mantle inelasticity, a core-mantle-boundary flattening giving a Free Core Nutation (FCN) period in agreement with the observed value, and a global Earth dynamical flattening in agreement with the observed precession, and
2. based on a new rigid Earth nutation series which takes into account the following perturbing effects:
    1. in lunisolar ephemerides: indirect planetary effects, lunar inequality, J_2-tilt, planetary-tilt, secular variations of the amplitudes, effects of precession and nutation,
    2. in the perturbing bodies to be considered: in addition to the Moon and the Sun, the direct planetary effects of Venus, Jupiter, Mars, and Saturn, should be included,
    3. in the order of the external potential to be considered: J_3 and J_4 effects for the Moon, and
    4. in the theory itself: effects of the tri-axiality of the Earth, relativistic effects and second order effects.

The WG recognizes that this new generation of models still has some imperfections, the principal one being poor modeling of the dissipation in the core and of certain effects of the ocean and the atmosphere, and urges the scientific community to address these questions in the future.

The WG recognizes that, due to the remaining imperfections of the present theoretical nutation models, the nutation series published in the IERS Conventions (1996), following 1994 IAU recommendation C1, still provides the users with the best nutation series. This IERS model being based on observations of the celestial pole offset, the WG supports the recommendation that the scientific community continue VLBI and LLR observations to provide accurate estimations of nutation, precession and rate of change in obliquity.

RESOLUTION B4

SUR LA THEORIE DE LA NUTATION D'UNE TERRE NON RIGIDE

La XXIIIème Assemblée Générale de l'Union Astronomique Internationale

reconnaissant

que le Groupe de Travail de l'Union Astronomique Internationale (UAI) et de l'Union Géodésie et de Géophysique Internationale (UGGI) sur la Théorie de la Nutation d'une Terre non-rigide a atteint son but en identifiant les phénomènes géophysiques et astronomiques qui doivent encore être modélisés avant qu'une théorie précise de la nutation d'une Terre non-rigide ne puisse être adoptée, et

que le Service International de la Rotation de la Terre (IERS), comme il en a été chargé par la recommandation C1 de l'UAI en 1994, a publié dans les Conventions du IERS (1996) un modèle provisoire de précession-nutation en accord avec les observations avec une incertitude de ±1 milliseconde de degré (mas),

approuve

les conclusions du Groupe de Travail de l'UAI et de l'UGGI sur la Théorie de la Nutation d'une Terre non-rigide, données en annexe,

invite

le Groupe de Travail sur la Théorie de la Nutation d'une Terre non-rigide à présenter un rapport détaillé à la prochaine Assemblée Générale de l'UGGI (Août 1999), moment où le Groupe de Travail cessera ses activités, *et encourage*

la communauté scientifique à étudier dans le futur les questions suivantes:

- la mise au point d'une nouvelle série de nutations d'une Terre rigide comprenant les termes additionnels nécessaires pour une théorie complète à ±5 microsecondes de degré, et
- la mise au point d'une nouvelle fonction de transfert d'une Terre non-rigide pour une Terre initialement en équilibre non-hydrostatique, avec un manteau inélastique, incorporant une période de la nutation libre du noyau (FCN) en accord avec les observations, et qui tienne compte d'une meilleure modélisation des parties fluides de la planète en incluant de la dissipation.

APPENDICE

Le Groupe de Travail sur la Théorie de la Nutation d'une Terre non-rigide a quantifié les problèmes dans la série de nutations adoptée par l'UAI en 1980 en notant :

1. qu'il y a une différence dans le taux de précession d'environ -3.0 millisecondes de degré par an (mas/an) entre la valeur observée par interférométrie à très longue base (VLBI) et par Tirs au Laser sur la Lune (LLR) et la valeur adoptée,
2. qu'un taux d'accroissement de l'obliquité d'environ -0.24 mas/an est observé (par VLBI et LLR) alors qu'il n'existe pas dans la théorie de précession/nutation de l'UAI 1980,
3. qu'en plus de ces tendances, il y a des différences observables (pic à pic) dans le domaine temporel allant jusqu'à 20 mas (milliseconde de degré) entre la nutation observée par VLBI et LLR et la nutation adoptée par l'UAI en 1980,
4. que ces différences correspondent dans le domaine des fréquences à quelques mas, et
5. que les différences entre observation et théorie sont bien au-dessus de la précision actuelle des observations.

Le Groupe de Travail a reconnu les améliorations apportées dans la modélisation de ces quantités et recommande, en vue d'obtenir un modèle de nutation plus précis, au niveau du mas dans le domaine des fréquences et au niveau de quelques mas dans le domaine du temps, d'utiliser des modèles :

1. basés sur une nouvelle fonction de transfert pour une Terre non rigide initialement en équilibre non-hydrostatique, incorporant l'inélasticité du manteau, un aplatissement de la frontière entre le noyau et le manteau donnant une période pour la Nutation Libre du Noyau (FCN) en accord avec celle observée, et un aplatissement dynamique global en accord avec la précession observée, et

2. fondés sur une nouvelle série de nutations pour une Terre rigide qui tient compte des effets perturbateurs suivants:
    1. au niveau des éphémérides lunisolaires: effets planétaires indirects, inégalité lunaire, basculement del'orbite de la Lune dû à la forme de la Terre (J_2-tilt), basculement de l'orbite de la Lune dû à la présence des planètes (planetary-tilt), variations séculaires des amplitudes, effets de la précession et de la nutation,
    2. au niveau des corps perturbateurs à considérer: outre la Lune et le Soleil, il convient d'inclure les effets planétaires directs de Venus, Jupiter, Mars et Saturne,
    3. au niveau de l'ordre du potentiel extérieur à considérer: les effets provenant de J_3 et J_4 pour la Lune, et
    4. au niveau de la théorie elle-même: les effets de la tri-axialité de la Terre, les effets relativistes et les effets de second ordre.

Le Groupe de Travail reconnaît que cette nouvelle génération de modèles a encore quelques imperfections, les plus importantes étant une mauvaise modélisation de la dissipation dans le noyau et de certains effets de l'océan et de l'atmosphère, et invite la communauté scientifique à étudier ces questions dans le futur.

Le Groupe de Travail reconnaît qu'à la suite des imperfections résiduelles dans les modèles actuels de nutation théorique, la nutation publiée dans les Conventions du IERS de 1996 suivant la recommandation C1 de l'UAI de 1994 fournit encore aux utilisateurs la meilleure série de nutations.

Comme ce modèle adopté par l'IERS est basé sur les observations des écarts du pôle céleste, le Groupe de Travail approuve la recommandation stipulant que la communauté scientifique poursuive les observations VLBI et LLR pour fournir des estimations précises de la nutation, de la précession et du taux d'accroissement de l'obliquité.

RESOLUTION B5

ON THE INTERNATIONAL CELESTIAL REFERENCE SYSTEM (ICRS)

AND THE HIPPARCOS CATALOGUE

The XXIIIrd International Astronomical Union General Assembly

considering

1. that the International Astronomical Union (IAU) has adopted an International Celestial Reference System (ICRS) in which the axes are fixed relative to the distant background as implied by observations of extragalactic sources,
2. that the realization of the ICRS is based on observations made from the Earth, the axes of which precess and nutate relative to the ICRS,
3. that there are significant differences between the nutation adopted by the IAU in 1980 and astronomical observations,
4. that a rate of variation of the obliquity is observed, which is not predicted by the 1980 IAU precession-nutation theory,
5. that there is a difference in the precession rate of about -3.0 milliarcseconds per year (mas/year) between the observed and adopted values,

recommends

1. that Division I form a new Working Group to report to the IAU General Assembly in 2000 which will
   a. examine and clarify the effects on astrometric computations, of changes such as the adoption of the International Celestial Reference System, the availability of the Hipparcos catalogue, and the change expected in the conventional precession-nutation model, and
   b. make recommendations regarding the algorithms to be used,
2. that this Working Group study these questions jointly with the International Earth Rotation Service (IERS) and maintain a close connection with the IAU Working Group on Reference Frames, the IAU Working Group on Astronomical Constants, and the IAU-IUGG Working Group on Non-rigid Earth Nutation Theory (up to its discontinuation at the 1999 IUGG General Assembly), through exchange of representatives.

RESOLUTION B5

SUR LE NOUVEAU SYSTEME CELESTE INTERNATIONAL DE REFERENCE (ICRS) ET LE CATALOGUE HIPPARCOS

La XXIIIème Assemblée Générale de l'Union Astronomique Internationale

considérant

1. que l'UAI a adopté un Système Céleste International de Référence (ICRS) dans lequel les axes sont fixés par rapport au ciel lointain comme l'impliquent les observations de sources extragalactiques,
2. que la réalisation du ICRS est basée sur des observations effectuées à partir de la Terre dont les axes présentent un mouvement de précession et des mouvements de nutation par rapport au ICRS,
3. qu'il y a des différences significatives entre les nutations adoptées en 1980 par l'UAI et les observations astronomiques,
4. qu'un taux de variation de l'obliquité est observé, qui n'est pas prédit par la théorie de précession/nutation de l'UAI 1980,
5. que l'on note une différence du taux de précession de l'ordre de -3,0 millisecondes de degré par an (mas/an) entre la valeur observée et la valeur adoptée,

recommande

1. que la Division I crée un nouveau Groupe de Travail qui fera rapport à l'Assemblée générale de l'UAI en 2000 et qui
   a. examinera et clarifiera les effets sur les calculs astrométriques des changements tels que l'adoption du Système Céleste International de Référence, la mise à disposition du catalogue Hipparcos, et le changement à venir du modèle conventionnel de précession-nutation et,
   b. présentera des recommandations au sujet des algorithmes à utiliser,
2. que ce Groupe de Travail étudie ces questions conjointement avec le Service International de la Rotation de la Terre (IERS) et soit en relation étroite avec le Groupe de Travail de l'UAI sur les Systèmes de Référence, et celui sur les Constantes Astronomiques ainsi qu'avec le Groupe de Travail de l'UAI et de l'UGGI sur la Théorie de la Nutation d'une Terre non-rigide (jusqu'à sa dissolution à la prochaine Assemblée Générale de l'UGGI en 1999), par échange de représentants.

RESOLUTION B6

ON RELATIVITY IN CELESTIAL MECHANICS AND IN ASTROMETRY

The XXIIIrd General Assembly of the International Astronomical Union

considering that

- a relativistic solar system barycentric four-dimensional coordinate system with its coordinate time scale TCB was defined by International Astronomical Union (IAU) Resolution A4 (1991),
- a relativistic geocentric four-dimensional coordinate system with its coordinate time scale TCG was defined by IAU Resolution A4 (1991) and International Union of Geophysics and Geodesy (IUGG) Resolution 2 (1991), and
- the basic physical units of space-time in all coordinate systems were recommended by IAU Resolution A4 (1991) to be the SI second for proper time and the SI meter for proper length,

noting that

- practical realization of barycentric and geocentric coordinate systems in many groups (see International Earth Rotation Service (IERS) Standards, 1992) is based on time scales TDB and TT instead of TCB and TCG, respectively, and involves the scaling factors $1-L_B$ and $1-L_G$ for the spatial coordinates and mass factors GM in barycentric and geocentric systems, respectively, L_B and L_G being given in IAU Resolution A4 (1991),
- even more complicated scaling factors are introduced in the VLBI (Very Long Baseline Interferometry) model of IERS Conventions (1996), and
- astronomical constants and currently employed definitions of fundamental astronomy concepts are based on Newtonian mechanics with its absolute space and absolute time leading to ambiguities in dealing with relativistic effects,

recommends that

- the spatial coordinates of the Barycentric and Geocentric Reference Systems as defined by the IAU (1991) resolutions be used for celestial and terrestrial reference frames, respectively, without any scaling factors,
- the final practical realizations of the coordinate systems for use in astronomy and geodesy be implementations of the systems defined by IAU-IUGG (1991) resolutions,
- the use of TT for convenience of observational data analysis not be accompanied by scaling of the spatial geocentric coordinates,
- algorithms for astronomical constant determination and definitions of fundamental astronomy concepts be explicitly given within the basic reference systems envisaged by IAU-IUGG (1991) resolutions, and
- the IAU Working Group on Astronomical Standards (WGAS) continue the consideration of relativistic aspects of the concepts, algorithms and the constants of fundamental astronomy.

RESOLUTION B6

SUR LA RELATIVITE EN MECANIQUE CELESTE ET EN ASTROMETRIE

La XXIIIème Assemblée Générale de l'Union Astronomique Internationale

considérant

- qu'un système de coordonnées relativiste, à quatre dimensions, dont l'origine est au barycentre du système solaire et avec l'échelle de temps TCB comme coordonnée-temps, a été défini par la Résolution A4 de l'UAI (1991),
- qu'un système de coordonnées relativiste, à quatre dimensions, géocentrique et avec l'échelle de temps TCG, a été défini par la Résolution A4 de l'UAI (1991) et la Résolution 2 de l'Union Géophysique et Géodésique Internationale (UGGI) (1991), et
- que les unités physiques de base de l'espace-temps dans tous les systèmes de coordonnées sont, comme recommandé par la Résolution A4 de l'UAI (1991), la seconde du Système International d'unités (SI) pour le temps propre et le mètre du SI pour la longueur propre,

notant

- que la réalisation pratique des systèmes de coordonnées barycentrique et géocentrique est basée, dans beaucoup de groupes (voir les Standards du Service International de la Rotation de la Terre (IERS), 1992), sur les échelles de temps TDB et TT à la place de TCB et TCG respectivement, et comporte les facteurs d'échelle $1\text{-}L_B$ et $1\text{-}L_G$ pour les coordonnées spatiales et des facteurs de masse GM dans les systèmes barycentrique et géocentrique, L_B et L_G étant donnés par la Résolution A4 de l'UAI (1991),
- que des facteurs d'échelle encore plus compliqués ont été introduits par le modèle VLBI (interférométrie à très longue base) donné dans les Conventions du IERS (1996), et
- que des constantes astronomiques et des définitions de concepts de l'astronomie fondamentale employées couramment sont basées sur la mécanique Newtonienne qui impose un espace absolu et un temps absolu conduisant à des ambiguïtés liées aux effets relativistes,

recommande

- que les coordonnées spatiales des systèmes de référence barycentrique et géocentrique telles que définies par les résolutions de l'UAI (1991), soient utilisées pour les repères de référence céleste et terrestre, respectivement, sans facteur d'échelle,
- que les systèmes de coordonnées finaux, à utiliser en pratique en astronomie et géodésie réalisent les systèmes définis par les résolutions de l'UAI et de l'UGGI (1991),
- que l'utilisation du TT pour la commodité de l'analyse des observations ne soit pas accompagnée de facteur d'échelle dans les coordonnées spatiales géocentriques,
- que, d'une part, des algorithmes pour la détermination des constantes astronomiques et, d'autre part, des définitions des concepts de l'astronomie fondamentale soient donnés explicitement dans le cadre des systèmes de référence de base envisagés par les résolutions de l'UAI et de l'UGGI (1991), et
- que le Groupe de Travail de l'UAI sur les Standards Astronomiques continue à prendre en considération les aspects relativistes des concepts, des algorithmes et des constantes de l'astronomie fondamentale.

RESOLUTION B7

ENCOURAGING VLBI AND LLR OBSERVATIONS

The XXIIIrd International Astronomical Union General Assembly

noting

1. resolution B4
2. resolution B5

considering

1. that regular observation by Very Long Baseline radio Interferometry (VLBI) is the only way to maintain the International Celestial Reference Frame (ICRF),
2. that observation by Lunar Laser Ranging (LLR) is important to connect the solar system reference system with the ICRF, and
3. that VLBI and LLR are the basic observational techniques for determination of the precession and nutation of the Earth,

recommends

that high-precision astronomical observing programs be organized in such a way that

1. astronomical reference systems can be maintained at the highest possible accuracy for both northern and southern hemispheres, and
2. high accuracy observations of precession-nutation will be made available for comparison with geophysical models and for astronomical and geodetic applications.

RESOLUTION B7

ENCOURAGEANT LES OBSERVATIONS VLBI ET LLR

La XXIIIème Assemblée Générale de l'Union Astronomique Internationale

notant

1. la Résolution B4 et
2. la Résolution B5

considérant

1. que l'observation régulière par interférométrie à très longue base (VLBI) est le seul moyen d'assurer la maintenance du Repère Céleste International de Référence (ICRF),
2. que l'observation par Tirs au Laser sur la Lune (LLR) est importante pour lier le système de référence du système solaire au ICRF, et
3. que le VLBI et le LLR sont les techniques de base d'observation pour déterminer la précession et la nutation de la Terre,

recommande

que des programmes d'observations astronomiques de haute précision soient organisés de sorte que

1. la maintenance des systèmes de référence astronomiques puissent être assurée au meilleur niveau de précision pour les hémisphères Nord et Sud, et
2. des observations de haute précision de la précession/nutation soient disponibles pour la confrontation avec les modèles géophysiques et pour des applications en astronomie et en géodésie.

CHAPTER V

REPORT OF THE EXECUTIVE COMMITTEE 1994-1997

The present report covers the period between the conclusion of the XXIInd General Assembly on August 24, 1994, and that of the XXIIIrd General Assembly on August 27, 1997.

EXECUTIVE COMMITTEE 1994-1997

L. Woltjer	President
R.P. Kraft	President-Elect
I. Appenzeller	General Secretary
J. Andersen	Assistant General Secretary
C.A. Anguita	Vice-President
B. Hidayat	Vice-President
D.S. Mathewson	Vice-President
F. Pacini	Vice-President
J.I. Smak	Vice-President
V.L. Trimble	Vice-President
A.A. Boyarchuk	Advisor
J. Bergeron	Advisor

1. Administrative Matters

EXECUTIVE COMMITTEE MEETINGS

The full Executive Committee met as follows during the reporting period:

66th Meeting, August 25, 1994, in The Hague, The Netherlands
67th Meeting, August 18 - 21, 1995, at Mt. Stromlo Observatory, Canberra, Australia
68th Meeting, June 23 - 25, 1996, at the Space Telescope Science Institute, Baltimore, USA

The activities of the Executive Committee are recorded in the minutes of the Executive Committee meetings. Summaries of these minutes were published in the IAU Information Bulletin (EC 65: IB 74, p. 25; EC 66: IB 74, p. 25; EC 67: IB 76, p. 5; EC 68: IB 79, p. 3).

OFFICERS MEETINGS

The IAU Officers (President, President-Elect, General Secretary, Assistant General Secretary) met at the IAU Secretariat in Paris, France, on February 25-26, 1995; February 25, 1996; and February 24, 1997.

IAU SECRETARIAT

The Secretariat continued to function at its premises at the Institut d'Astrophysique de Paris (INSU-CNRS), according to an agreement in force since November 1989. The Secretariat staff consisted, as before, of Mrs. M. Léger-Orine (Administrative Assistant) and Ms. J. Saucedo (Secretary).

IAU ARCHIVES

Among the duties of the General Secretary explicitly mentioned in our Statutes is to preserve the archives of the Union. The Executive Committee, therefore, accepted with gratitude the kind offer of Professor Adriaan Blaauw, former IAU President, to reorganise the IAU archives for the period 1919-1970.

J. Andersen (ed.), Transactions of the International Astronomical Union Volume XXIIIB, 53–71.

During the past three years, Professor Blaauw (with some help from IAP students) has done an outstanding job in sorting, labelling and describing the many documents which had accumulated in the Secretariat. It is expected that these documents will be deposited, probably in 1998, in a professional archive, where they will be properly maintained and readily accessible to future historians of science.

REVISION OF THE IAU STATUTES, BYE-LAWS, AND WORKING RULES

A significant fraction of the EC's work was devoted to a revision of the IAU Statutes, Bye-Laws and Working Rules. The successful introduction of Divisions into our organisational structure on an informal basis as decided by the XXIInd General Assembly in The Hague, and the resulting wish to make the Divisions a permanent feature of the IAU, implied a number of revisions of our basic rules.

In the course of making these amendments, the EC decided to also correct a number of deficiencies that had become apparent in the previous version of these documents. Among the proposed modifications is a provision which will explicitly allow Division Presidents to propose astronomers who are not represented by an IAU adhering body as new Individual Members of our Union. While such proposals from Commission Presidents had occasionally been accepted in the past, the actual procedure to be followed was unclear, which had led to difficulties during the XXIInd General Assembly.

Amendments of the IAU Working Rules were also proposed, adding for the first time written rules and guidelines for the drafting and submission of Resolutions at General Assemblies as well as rules concerning the formation of Working Groups of the Union.

As recorded in Chapter II of these Transactions, the proposed revisions were approved unanimously by the General Assembly. The revised Statutes, Bye-Laws, and Working Rules are printed in Chapter VII of this volume.

ADHERING COUNTRIES

In conformity with the ICSU policy regarding the States of the former Soviet Union and the former State of Yugoslavia, Azerbajan, Croatia, Georgia and Latvia became National Members of the IAU, all in Category of Adherence I, with their Academies of Science as the Adhering Bodies.

The Executive Committee decided to recommend to the General Assembly that Bolivia be admitted as an Associate Member of the Union, and that the Central American Assembly of Astronomers (CAAA) also be admitted as an Associate Member, representing the astronomers in Costa Rica, El Salvador, Guatemala, Honduras, Nicaragua, and Panama. The same rules as for normal Associate Countries will apply also to the CAAA as the Adhering Organisation.

The Executive Committee approved a change in the Category of Adherence of Australia from III to IV and of Spain from II to IV, both effective from 1996, and of Denmark from II to III, effective from 1997.

Regrettably, the dues of the Democratic People's Republic of Korea and Morocco not having been paid for over five years, the IAU membership of these countries terminated at the end of 1995 and 1996, respectively, as laid down in Article 7 of the IAU Statutes.

Number of Adhering National Organisations as of April 30, 1997:	60
Full Members	56
Associate Members	4

INDIVIDUAL MEMBERSHIP AND CONSULTANTS

Individual members admitted at the XXIInd General Assembly	674
Number of individual members as of May 12, 1997	7497
Number of consultants as of May 12, 1997	219

The Executive Committee reports, with regret, that the deaths of the following 136 Members of the Union have been reported to the Secretariat since the XXIInd General Assembly:

Alfven Hannes
Baize Paul
Bel Nicole J
Bertiau Flor C
Boigey Francoise
Bosman-Crespin D
Canavaggia Renée
Candy Michael P
Cardelli Jason A
Catarzi Marco
Chandrasekhar S
Chanmugam Ganesar
Chugajnov P F
Chuvaev K K
de Kort Jules J
de Vaucouleurs G
Demin Vladimir
Di-Sheng Zhai
Dicke Robert H
Divari N B
Doggett Leroy E
Epstein Isadore
Florentin Nielsen R
Fowler William A
Gleissberg Wolfgang
Godart Odon
Götz Woldemar
Gottlieb Kurt
Habibullin Sh T
Haug Ulrich
Herr Richard B
Hirst William P
Hooghoudt B G ir
Hubert Henri
Irwin John B
Jacchia Luigi G
Jaks Waldemar
Jaschek Mercedes
Joshi G C
Jurkevich Igor
Kaburaki Masaki
Karandikar R V
Katsis Demetrius
Kessler Karl G
King Robert Burnett
Konin V V
Kothari D S
Koyama Ko-u-ichi
Koziel Karol
Krassovsky V I
Kristian Jerome
Lahiri N C
Lipovetsky Valentin A
Lyttleton Raymond A
Loose Hans-Hermann
Lüst Rhea
Lutz Thomas E
Luyten Willem J
Macris Constantin J
Markert Thomas H
Marx Siegfried
Mayall Margaret W
McKeith Conal D
Mergentaler Jan
Middlehurst B M
Milovanovic Vladeta
Mitchell Jr Walter E
Monsignori Fossi B
Morando Bruno L
Moutsoulas Michael
Mrkos Antonin
Müller Edith A
Nadolschi V
Nagasawa Shingo
Nakano Saburo
Nemiro Andrej A
Ney Edward P
Nicolet Marcel
Ogir Maya
Oliver Bernard M
Onderlicka Bedrich
Page Thornton L
Parijskij N N
Petrov Georgij I
Pilowski K
Pohl Eckhard
Prokof'ev Vladimir K
Purcell Edward M
Rahe Jürgen
Rana Narayan C
Ribes Elizabeth
Rosenberg J
Rosino Leonida
Rubashev Boris M
Runcorn Stanley K
Rybka Przemyslaw
Rylov Valerij S
Sagan Carl
Sagnier Jean-Louis
Scaria K K
Schiller Karl
Schrutka-Rechtenstamm P
Schwarzschild Martin
Sen S N
Shah Ghanshyam A
Shaw R William
Shchegolev Dimitrij E
Shoemaker Eugene M
Shul'berg A M
Shuter William L H
Smolinski Jan
Smoluchowski Roman
Sobolev Vladislav M
Spitzer Jr Lyman
Strassl Hans L
Tarafdar Shankar P
Tayler Roger J
Thomas Richard N
Tombaugh Clyde W
Tousey Richard
Unsöld Albrecht
Urbarz H
van de Kamp Peter
Vanysek Vladimir
von Der Heide Johann
Wackernagel H Beat
Wares Gordon W
Wargau Walter F
Wattenberg D
Wehlau William H
Weigert Alfred
Xanthopoulos B C
Yakovkin N A
Yamakoshi Kazuo
Yen Jui-Lin
Zel'manov A L

DIVISIONS

The 11 Divisions introduced on a trial basis at the XXIInd General Assembly fulfilled their mission very satisfactorily, leading to their adoption on a permanent basis through the revision of the Statutes, Bye-Laws and Working Rules approved at the XXIIIrd General Assembly.

In particular, the advice of the Division Presidents was solicited and carefully considered by the Executive Committee in the selection of IAU sponsored scientific meetings. Considerations were also initiated regarding the optimum organisational structure within each Division. During the triennium, Commission 7 joined Division I, and Commission 50 joined Division III.

COMMISSIONS

Commission 6: Astronomical Telegrams

The number of IAU Circulars issued by the Central Bureau for Astronomical Telegrams was 208, 159, and 247 in 1994, 1995, and 1996, respectively. The decreasing number from 1994 to 1995 is explained by the introduction of the *Minor Planet Electronic Circulars* and other electronic services.

Commission 7: Celestial Mechanics and Dynamical Astronomy

The proposed change of name of Commission 7 from *Celestial Mechanics* to the above was approved.

Commission 20: Positions & Motions of Minor Planets, Comets & Satellites

In 1994, 1995 and 1996, *Minor Planet Circulars* were issued, under the direction of B.G. Marsden for a total of 1574, 1784 and 2432 pages, respectively. In all, 411, 550 and 615 minor planets were numbered in 1994, 1995, and 1996, respectively.

Commission 38: Exchange of Astronomers

The aim of this programme is to provide support to astronomers from developing countries who spend periods over three months at a foreign host institution to pursue their training and formation in astronomy and their scientific collaboration with other astronomers. During the triennium 1994 - 1997, 31 travel grants were awarded through this programme.

Commission 44: Space and High Energy Astrophysics

The former Commissions 44 (Astronomy from Space) and 48 (High Energy Astrophysics) merged to form this new, single Commission.

Commission 46: Teaching of Astronomy

International Schools for Young Astronomers (ISYA)

These schools are of three weeks' duration. They consist of both lectures by national and international senior astronomers, usually with emphasis on a broad astronomical theme, and an introduction to observational techniques, data reduction, and analysis, at an active observatory when possible. Attending students and young astronomers are both from the host country and nearby countries.

The following ISYA's took place during the triennium:

21st ISYA Cairo University and Kottamia Observatory, Egypt, September 18 - October 8, 1994
22nd ISYA Belo Horizonte, Brazil, 9 - 29 July, 1995
23rd ISYA Zanjan, Iran, 4 - 27 July, 1997

Teaching for Astronomy Development (TAD)

A major new educational activity introduced by the present EC is the *Teaching for Astronomy Development (TAD)* programme, which aims at developing astronomy in countries in which astronomy has been developing slowly or not at all, and where local scientists (usually physicists) wish to develop education and research in astronomy. The format of the programme is flexible, adapted to local needs, and may include the provision of teaching materials (including books, PCs, software, or instruments), visits by lecturers from abroad, or support for regional meetings.

The TAD programme was started in 1996 and 1997 with the support of educational institutions in Vietnam and Central America. TAD activities in Morocco and Sri Lanka are under consideration.

WORKING GROUPS OF THE EXECUTIVE COMMITTEE

The Executive Committee approved the creation of a Working Group on *Future Large Scale Facilities*, reporting to the Executive Committee. Its mandate is to promote the early discussion and dissemination of information on potential large scale international projects, and to set up and maintain an inventory of such projects, in order to further contacts and cooperation between different projects and to identify areas of significant duplication or lack of clearly desirable efforts. The composition of the Working Group is reported in Chapter VI of this volume.

REPRESENTATION AT OTHER INTERNATIONAL ORGANIZATIONS

The IAU was represented by the Assistant General Secretary at the following ICSU meetings: General Committee, Rabat, Morocco, October 13-15, 1994 and Chiang Mai, Thailand, October 7-9, 1995; XXVth General Assembly, Washington, DC, September, 1996. At the General Assembly, a major modernisation of ICSU's goals and organisational structure was approved. The Assistant General Secretary, elected to the ICSU Standing Committee on Membership, Structure and Statutes at the General Assembly, attended the meeting of that Committee in Paris, March 26-28, 1997, the main task of which was to prepare a corresponding revision of the ICSU Statutes and Rules of Procedure.

The IAU was represented by the General Secretary, at the XXXIst Scientific Assembly of COSPAR, the ICSU Committee for Space Research, at Birmingham, UK, July 14 - 21, 1996. Consultations between COSPAR and the IAU helped to coordinate the scientific programmes of the XXXth COSPAR Scientific Assembly and the XXIInd IAU General Assembly, both in 1994.

The IAU representatives to other international organisations are listed below.

IAU representatives to ICSU and other International Organizations 1994-1997

Acronyms	Organisations	Representatives
ICSU	International Council of Scientific Unions General Committee	*J. Andersen*
BIPM	Bureau International des Poids et Mesures	*G. Winkler*
CCDS	International Consultative Committee for the Definition of the Second	-
CCDS WG	Application of General Relativity to Metrology	*T. Fukushima*
CCIR	International Radio Consultative Committee Study Group 2	*J. Whiteoak/* *A.R. Thompson*
CIE	Compagnie Internationale de l'Eclairage	*S. Isobe*
CODATA	Committee on Data for Science & Technology	*E. Raymond*
COSPAR	Committee on Space Research COSPAR SC B COSPAR SC D COSPAR SC E COSPAR Sub. Committee E1 COSPAR Sub. Committee E2	*I. Appenzeller* *C. de Bergh* *F. Verheest* *W. Wamsteker* *R. Sunyaev* *O. Engvold*
COSTED	Committee on Science & Technology in Developing Countries	*I. Appenzeller*
CTS	Committee on the Teaching of Science	*J. Pasachoff*
FAGS	Federation of Astronomical & Geophysical Services	*P. Pâquet/J. Kovalevsky*
IAF	International Astronautical Federation	*Y. Kondo*
IERS	International Earth Rotation Service	*B. Kolaczek*
IGBP	International Geosphere-Ionosphere Programme	*J. Eddy*
IUCAF	Inter-Union Commission on Frequency Allocation for Radio Astronomy & Space Science	*B.A. Doubinsky* *M. Ishiguro* *R. Sinha* *A.R. Thompson*
IUPAP	International Union of Pure & Applied Physics	*V. Trimble*
IUWDS	International Ursigram & World Day Service	*H. Coffey*
QBSA	Quarterly Bulletin on Solar Activity	*P. Lantos*
SCOPE	Scientific Committee on Problems of Environment	*D. McNally*
SCOSTEP	Scientific Committee on Solar-Terrestrial Physics	*B. Schmieder*
URSI	Union Radio-Scientifique Internationale	*R. Ekers*
WMO	World Meteorological Union	*G. Wallerstein*

2. Scientific Matters

SCIENTIFIC MEETINGS 1994-1997

During the triennium reported on here, 20 Symposia, 18 Colloquia, 2 Regional Astronomy Meetings, and 2 Technical Workshops were held. The Union also co-sponsored 3 meetings with other international Unions and ICSU bodies. These meetings are listed below.

The Executive Committee decided to raise the amount of travel support awarded to IAU sponsored meetings, effective from 1996, to 25.000 CHF for Symposia and Regional Meetings, 16.000 CHF for Colloquia, and 3-5.000 CHF to Co-Sponsored Meetings according to the circumstances of each case.

A new edition of the Rules for IAU Sponsored Scientific Meetings was prepared by the Assistant General Secretary. They are published in Chapter VIII of this volume.

IAU Symposia:

170 ***Co: Twenty-five Years of Millimetre-Wave Spectroscopy***
Tucson, AZ, USA, May 29 - June 2, 1995

171 ***New Light on Galaxy Evolution***
Heidelberg, F.R. Germany, June 26-30, 1995

172 ***Dynamics, Ephemerides and Astrometry of the Solar System***
Paris, July 3-8, 1996

173 ***Astrophysical Applications of Gravitational Lensing***
Melbourne, Australia, July 9-14, 1995

174 ***Dynamical Evolution of Star Clusters – Confrontation of Theory and Observations-***
Tokyo, Japan, August 22-25, 1995

175 ***Extragalactic Radio Sources***
Bologna, Italy, October 10-14, 1995

176 ***Stellar Surface Structure***
Wien, Austria, October 9-13, 1995

177 ***The Carbon Phenomenon***
Antalya, Turkey, May 27-31, 1996

178 ***Molecules in Astrophysics: Probes and Processes***
Leiden, Netherlands, July 1-5, 1996,

179 ***New Horizons from Multi-Wavelength Sky Surveys***
Baltimore, USA, August 26 - 30, 1996

180 ***Planetary Nebulae***
Groningen, The Netherlands, August 26-30, 1996

181 ***Sounding Solar and Stellar Interiors***
Nice, France, September 30 - October 3, 1996,

182 ***Herbi-Haro Flows and the Birth of Low-Mass Stars***
Chamonix, France, January 20-26, 1997

183 ***Cosmological Parameters and the Evolution of the Universe***
Kyoto, Japan, August 18-22, 1997

184 ***The Central Regions of the Galaxy and Galaxies***
Kyoto, Japan, August 18-22, 1997

185 ***New Eyes to See Inside the Sun and the Stars: Pushing the Limits of Helio and Astro-Seismology with New Observations from the Ground and from Space***
Kyoto, Japan, August 18-22, 1997

186 ***Galaxy Interactions at High and Low Redshift***
Kyoto, Japan, August 18-22, 1997

187 ***Cosmic Chemical Evolution***
Kyoto, Japan, August 26-30, 1997

188 ***The Hot Universe***
Kyoto, Japan, August 26-30, 1997

189 ***Fundamental Stellar Properties: The Interaction Between Observation and Theory***
Sydney, Australia, January 13-17, 1997

IAU Colloquia:

150 ***Physics, Chemistry, and Dynamics of Interplanetary Dust***
Gainesville, FL, USA, August 14-18, 1995
151 ***Flares and Flashes – Views from the Ground and from Space***
Sonneberg, Germany, December 5-9, 1994
152 ***Astrophysics in the Extreme Ultraviolet***
Berkeley, CA, USA, March 27-30, 1995
153 ***Magnetodynamic Phenomena in the Solar Atmosphere***
Makuhari, Tokyo, Japan, May 22-27, 1995
154 ***Solar and Interplanetary Transients***
Pune, India, January 23-27, 1995
155 ***Astrophysical Applications of Stellar Pulsation***
Cape Town, South Africa, February 6-10, 1995
156 ***The Impact of Comet Shoemaker-Levy 9 on Jupiter***
Baltimore, MD, USA, May 9-12, 1995
157 ***Barred Galaxies***
Tuscaloosa, AL, USA, May 30 - June 3, 1995
158 ***Cataclysmic Variables and Related Objects***
Keele, UK, June 25-30, 1995
159 ***Emission Lines in Active Galaxies: New Methods and Techniques***
Shanghai, China, June 17-20, 1996
160 ***Pulsars: Problems and Progress***
Sydney, Australia, January 8-12, 1996
161 ***Astronomical and Biochemical Origins and the Search for Life in the Universe***
Capri, Italy, July 1-5, 1996
162 ***New Trends in Astronomy Teaching***
London and Milton Keynes, UK, July 8-12, 1996
163 ***Accretion Phenomena and Related Outflows***
Port Douglas, Australia, July 15-19, 1996
164 ***Radio Emission from Galactic and Extragalactic Compact Sources***
Socorro, New Mexico, USA, April 28 - May 2, 1997
165 ***Dynamics and Astrometry of Natural and Artificial Celestial Bodies***
Poznan, Poland, July 1-5, 1996
166 ***The Local Bubble and Beyond***
Garching b. München, Germany, April 21-25, 1997
167 ***New Perspectives of Solar Prominences***
Aussois, France, April 28 - May 4, 1997

IAU Regional Astronomy Meetings

VIIIth Latin-American Regional Astronomy Meeting
Montevideo, Uruguay, November 27 - December 1, 1996
VIIth Asian-Pacific Regional Astronomy Meeting
Pusan, Korea, August 19-23, 1996

IAU Technical Workshops

Library and Information Services in Astronomy (LISA II)
Garching b. München, May 10-12, 1995
Statistical Challenges in Modern Astronomy II
University Park, PA, USA, June 2-5, 1996

Co-Sponsored Meetings

Commsphere 95 (URSI, IAU)
Eilat, Israel, January 22-27, 1995
31st COSPAR Scientific Assembly (COSPAR, IAU)
Birmingham, UK, July 14-21, 1996

Third SOLTIP Symposium on Solar Transient and Interplanetary Regions
(SCOSTEP, COSPAR, IAU)
Beijing, China, October 14-18, 1996

IAU PUBLICATIONS

Since the expiring data of the IAU publishing contract with Kluwer Academic Publishers was December 31st, 1997, the EC issued a call for tenders for a new contract early in 1996. From four bids received by the General Secretary, the EC selected the Astronomical Society of the Pacific (ASP) as the new IAU Publisher for the period 1998 - 2003, and a six-year contract with the ASP has been signed accordingly. It has been agreed that Kluwer will publish the Proceedings of all IAU meetings held through the end of 1997, while the ASP will handle all meetings held from the beginning of 1998.

The publications that have appeared by or under the auspices of the IAU during the reporting period are listed below. Except for the Colloquium Proceedings, these were published by the IAU Publisher, Kluwer Academic Publishers, Dordrecht, The Netherlands.

IAU Information Bulletin:
Seven issues, No. 74-80

IAU Transactions XXIIB
Proceedings of the Twenty-Second General Assembly
Ed. I. Appenzeller, 1995

IAU Transactions XXIIIA
Reports on Astronomy
Ed. I. Appenzeller, 1997

Highlights of Astronomy, Vol. 10
Ed. I. Appenzeller, 1995

Membership List 1994/1997

IAU Symposium Proceedings:

160 ***Asteroids, Comets, Meteors*** 1993
Eds. A. Milani, M. di Martino & A. Cellino
161 ***Astronomy from Wide-Field Imaging***
Eds. H.T. MacGillivray, E.B. Thomson, B.M. Lasker, I.N. Reid, D.F. Malin, R.M. West & H. Lorenz
162 ***Pulsation, Rotation and Mass Loss in Early-Type Stars***
Eds. L.A. Balona, H.F. Henrichs & J.M. Le Contel
163 ***Wolf-Rayet Stars: Binaries, Colliding Winds, Evolution***
Eds. K.A. van der Hucht & P.M. Williams
164 ***Stellar Populations***
Eds. P.C. van der Kruit & G. Gilmore
165 ***Compact Stars in Binaries***
Eds. J. van Paradijs, E.P.J. van den Heuvel & E. Kuulkers
166 ***Astronomical and Astrophysical Objectives of Sub-Milliarcsecond Optical Astrometry***
Eds. E. Høg & P.K. Seidelmann
167 ***New Developments in Array Technology and Applications***
Eds. A.G. Davis Philip, K.A. Janes & A.R. Upgren
168 ***Examining the Big Bang and Diffuse Background Radiations***
Eds. M. Kafatos & Y. Kondo
169 ***Unsolved Problems of the Milky Way***
Eds. L. Blitz &P. Teuben
170 ***CO: Twenty-Five Years of Millimetre-wave Spectroscopy***
Eds. W.B. Latter, S.J.E. Radford, P.R. Jewell, J.G. Mangum & J. Bally
171 ***New Light on Galaxy Evolution***
Eds. R. Bender & R.L. Davies

172 ***Dynamics, Ephemerides & Astrometry of the Solar System***
Eds. S. Ferraz-Mello & J.-E. Arlot
173 ***Astrophysical Applications of Gravitational Lensing***
Eds. C.S. Kochanek & J.N. Hewitt
174 ***Dynamical Evolution of Star Clusters - Confrontation of Theory and Observations***
Eds. P. Hut & J. Makino
175 ***Extragalactic Radio Sources***
Eds. R. Ekers, C. Fanti & L. Padrielli
176 ***Stellar Surface Structure***
Eds. K.G. Strassmeier & J.L. Linsky
178 ***Molecules in Astrophysics: Probes and Processes***
Ed. E.F. van Dishoeck

IAU Colloquium Proceedings:

145 ***Supernovae and Supernova Remnants***
Eds. R. McCray & Z. Wang
Cambridge Univ. Press, I1996
146 ***Molecular Opacities in the Stellar Environment***
Ed. U.G. Jørgensen
Springer-Verlag, 1994
147 ***The Equation of State in Astrophysics***
Eds. G. Chabrier & E. Schatzman
Cambridge Univ. Press, 1994
148 ***Future Utilization of Schmidt Telescopes***
Eds. J. Chapman, R. Cannon, S. Harrison & B. Hidayat
ASP Conf. Ser. Vol. 84, 1995
149 ***Tridimensional Optical Spectroscopic Methods in Astrophysics***
Eds: G. Comte & M. Marcelin
ASP Conf. Ser. Vol. 71, I 1995
150 ***Physics, Chemistry, and Dynamics of Interplanetary Dust***
Eds. B.Å.S. Gustafson & M.S. Hanner
ASP Conf. Ser. Vol. 104, 1996
151 ***Flares and Flashes – Views from the Ground and from Space***
Eds. J. Greiner, H.W. Duerbeck & R.E. Gershberg
Springer-Verlag, 1994
152 ***Astrophysics in the Extreme Ultraviolet***
Eds. S. Bowyer & R.F. Malina
Kluwer Acad. Publ., 1996
153 ***Magnetodynamic Phenomena in the Solar Atmosphere***
Eds. Y. Uchida, T. Kosugi & H.S. Hudson
Kluwer Acad. Publ., 1996
154 ***Solar and Interplanetary Transients***
Eds. S. Ananthakrishnan & A. Pramesh Rao
Kluwer Acad. Publ., 1996
155 ***Astrophysical Applications of Stellar Pulsation***
Eds: R.S. Stobie & P.A. Whitelock
ASP Conf. Ser. Vol. 83, 1995
156 ***The Impact of Comet Shoemaker–Levy 9 on Jupiter***
Ed. K.S. Noll
Cambridge Univ. Press, 1996
157 ***Barred Galaxies***
Eds: R. Buta, D.A. Crocker & B.G. Elmegreen
ASP Conf. Ser. Vol. 91, 1996
158 ***Cataclysmic Variables and Related Outflows***
Eds. A. Evans & J.H. Wood
Kluwer Acad. Publ., 1996

159 ***Emission Lines in Active Galaxies: New Methods and Techniques***
Eds: B.M. Peterson, F.-Z. Cheng & A.S. Wilson
ASP Conf. Ser. Vol. 113, 1997

160 ***Pulsars: Problems and Progress***
Eds: M. Bailes, S. Johnston & M.A. Walker
ASP Conf. Ser. Vol. 105, 1996

161 ***Astronomical and Biochemical Origins and the Search for Life in the Universe***
Eds: C.S. Cosmovici, S. Bowyer & D. Werthimer
Editrice Compositori, Bologna, 1997

163 ***Accretion Phenomena and Related Outflows***
Eds: D.T. Wickramasinghe, G. Bicknell & L. Ferrario
ASP Conf. Ser. Vol. 121, 1997

165 ***Dynamics and Astrometry of Natural and Artificial Celestial Bodies***
Eds: I. Wytrzyszczak, J.H. Lieske & R.A. Feldman
Kluwer Acad. Publ., 1997

Proceedings of IAU Regional Meetings:

Proceedings of the VIIIth Latin-American Regional Astronomy Meeting
Rev. Mexicana de Astronomía y Astrofísica, Ser de Conf., Vol. 4, 1996

Proceedings of the VIIth Asian-Pacific Regional Astronomy Meeting
Journal of the Astronomical Society of Korea, Vol. 29, Supplement, 1996

3. Financial Matters

Budget 1994-1996 & 1997 approved at previous General Assemblies (CHF)

INCOME	1994	1995	1996	*1994-96*	*1997*
Unit of contribution	2580	2660	2740		2820
Number of units of contribution	248	254	254		254
ADHERING ORGANIZATIONS	**639840**	**675640**	**695960**	***2011440***	***716280***
SPECIAL CONTRIBUTIONS					
ICSU/UNESCO					
ICSU: General					
UNESCO: General	**30300**	**34000**	**37000**	***101300***	***40000***
ICSU/AEI					
UNESCO/AEI					
COSPAR/AEI					
PUBLICATIONS: ROYALTIES	**62000**	**68000**	**68000**	***198000***	***68000***
BANK INTEREST					
Current accounts					
Deposit accounts	**25000**	**25000**	**25000**	***75000***	***25000***
Saving accounts					
Money market					
TOTAL INCOME	**757140**	**802640**	**825960**	***2385740***	***849280***

Budget 1994-1996 & 1997 approved at previous General Assemblies (CHF)

EXPENDITURE	1994	1995	1996	*1994-96*	*1997*
SCIENTIFIC ACTIVITIES					
General Assembly					
Grants					*167000*
Commission Presidents					*40000*
Staff	275000			*275000*	*18000*
Operation		9000	9000	*18000*	*30000*
Executive Committee					*80000*
Meetings					
Symposia/Colloquia	143000	206000	213000	*562000*	*220000*
Regional Meetings		22000	22000	*44000*	
Research projects					
ISYA (46)		45000	45000	*90000*	
VLP (46)	28000	15000	15000	*58000*	*15000*
EC WGs	5000	18000	18000	*41000*	*18000*
Exchange of Astromers (38)	26000	30000	30000	*86000*	*30000*
Others	1500			*1500*	
Commission Activities					
Telegram Bureau (06)	3700	4000	4000	*11700*	*4000*
Minor Planet Center (20)	3700			*3700*	
Meteor Data Center (22)	1100			*1100*	
Variable Star Catalogue (27)	3700			*3700*	
Astronomy Teaching (46)	5000	5000	5000	*15000*	*5000*
Commission WGs	2000	18000	18000	*38000*	*18000*
Delegates to other Unions	9000	10000	10000	*29000*	*10000*
Total	**506700**	**382000**	**389000**	***1277700***	***655000***
DUES TO UNIONS/ORGANIZATIONS					
ICSU	22400	16900	17400	*56700*	*18000*
IERS/FAGS	7000	7000	7000	*21000*	*7500*
IUCAF	7500	7500	7500	*22500*	*7500*
CTS	600	600	600	*1800*	*600*
GLOBMET	1500			*1500*	
Total	**39000**	**32000**	**32500**	***103500***	***33600***

Budget 1994-1996 & 1997 approved at previous General Assemblies (CHF)

EXPENDITURE (ctd)	1994	1995	1996	*1994-96*	*1997*
EXECUTIVE COMMITTEE					
Executive cttee meetings		40000	41200	*81200*	
Officers meetings	8000	5000	5200	*18200*	*5400*
General Secretary expenditure	13500	22000	22700	*58200*	*23400*
President expenses	7000	7000	7000	*21000*	*7000*
Assist General Secretary expenses	4250	3000	3000	*10250*	*3000*
Local help to As General Secretary	1250			*1250*	
Total	**34000**	**77000**	**79100**	***190100***	***38800***
PUBLICATIONS					
Information Bulletin	24000	40000	42000	*106000*	*44000*
Distr. Devel. Countries	12000	12000	12000	*36000*	*13000*
Distr. Exec. Committee	5000	5000	5000	*15000*	*5500*
Other Publications	25000			*25000*	
Total	**66000**	**57000**	**59000**	***182000***	***62500***
ADMINISTRATION/SECRETARIAT					
Salaries & Charges	149000	153400	158000	*#REF!*	*162800*
Training courses	3000	4000	4000	*11000*	*4500*
General office expenses:	66000	68000	70000	*204000*	*72000*
Audit fee	3500	3000	3100	*9600*	*3200*
Bank					
Charges on internal account transfers	2500			*2500*	
Bank charges	1450	2000	2000	*5450*	*2500*
Commissions on deposits		4000	4000	*8000*	*4500*
Total	**225450**	**234400**	**241100**	***700950***	***249500***
TOTAL EXPENDITURE	**871150**	**782400**	**800700**	***2454250***	***1039400***

Excess/Loss of Income over Expenditure	**-114010**	**20240**	**25260**	***-68510***	***-190120***

1994-1996 Statement of income (CHF)

	1991-1993	1994	1995	1996	*1994-1996*
ADHERING ORGANIZATIONS	***1728100***	**619862**	**647150**	**562835**	***1829847***
SPECIAL CONTRIBUTIONS	***176300***	**184166**	**2122**	**22451**	***208739***
ICSU/UNESCO					
ICSU: General	*26090*	11880		15480	***27360***
UNESCO: General	*67926*	17424		13416	***30840***
ICSU/AEI	*20108*				
UNESCO/AEI	*36886*		5800		***5800***
COSPAR/AEI	*720*				
Total	***151730***	**29304**	**5800**	**28896**	***64000***
PUBLICATIONS: ROYALTIES	***193544***	**64114**	**62803**	**63750**	***190667***
BANK INTERESTS					
Current accounts	*2511*	109	197	43	***349***
Deposit accounts	*117283*	17279	23473	6957	***47709***
Saving accounts	*1109*				
Money market	*35305*	20706	11327	11660	***43693***
Total	***156210***	**38094**	**34997**	**18660**	***91751***
OTHER INCOME					
Refund grants	*2207*	6467			***6467***
Refunds SPM/CQM	*7496*				
Refunds bank	*509*				
Refund Union representation	*4415*				
Gain on exchange 1 Jan current year	*9039*			35404	***35404***
Gain on internal transfers	*4891*	1503	2	6310	***7815***
Total	***28557***	**7970**	**2**	**41714**	***49686***
TOTAL INCOME	***2434443***	**943510**	**752874**	**738306**	***2434690***

1994-1996 Statement of expenditure (CHF)

	1991-1993	1994	1995	1996	*1994-1996*
SCIENTIFIC ACTIVITIES					
General Assembly					
Grants (XXI)	*132685*				
Grants (XXII)	*31040*	402321			***402321***
Commission Presidents	*48687*	22997			***22997***
Staff	*19423*	2688			***2688***
Operation (XXI)	*17218*				
Operation (XXII)	*9293*	15448	163		***15611***
Operation (XXIII)			2124		***2124***
Operation (XXIV)		579	2533		***3112***
Executive Committee	*81848*	79101			***79101***
Meetings					
Symposia/Colloquia	*409398*	85000	225994	256193	***567187***
Co-sponsored Meetings	*45658*	12923	2727	8000	***23650***
Regional Meetings	*43156*		24123	25000	***49123***
AEI IAU/UNESCO meeting	*60420*				
Other meetings		15036	142	3000	***18178***
Research projects					
ISYA (46)	*65792*	31138	25183		***56321***
VLP/TAD (46)	*6094*	1839		21418	***23257***
EC WGs	*18930*	1207	10517	14242	***25966***
Exchange of Astronomers (38)	*39513*	17456	15752	17481	***50689***
Archives			2301	15355	***17656***
Others	*9636*				
Commission Activities					
Telegram Bureau (06)	*17727*	3270		4000	***7270***
Minor Planet Center (20)	*7300*	3372			***3372***
Meteor Data Center (22)	*6425*	1100	1000	1000	***3100***
Variable Star Catalogue (27)	*7300*	3728			***3728***
Astronomy Teaching (46)	*10593*	2047			***2047***
Commission WGs (20)	*11124*		5494	831	***6325***
Others	*1878*				
Representation to other Unions	*10536*	7341	8287	9119	***24747***
Total	***1111681***	**708591**	**326340**	**375639**	***1410570***
DUES TO UNIONS /ORGANIZATIONS					
ICSU	*67706*	11290	10543	12144	*33977*
IERS/FAGS	*19915*	7000	7000	7000	***21000***
IUCAF	*23389*	7500	7500	7500	***22500***
CTS	*1768*				
Total	***112778***	**25790**	**25043**	**26644**	*77477*

1994-1996 Statement of expenditure (CHF) ctd

	1991-1993	1994	1995	1996	*1994-1996*
EXECUTIVE COMMITTEE					
Executive cttee meetings	*79525*	507	35137	37319	***72963***
Officers meetings	*12698*	5222	2215	4430	***11867***
SNC expenditure	*8104*				
General Secretary expenditure	*12969*	9046	19044	18060	***46150***
Representation	*5167*	971	1851		***2822***
President expenses	*1792*				
Total	***120258***	**15746**	**58247**	**59809**	***133802***
PUBLICATIONS					
Infomation Bulletin	*167850*	23319	21118	41156	***85593***
Free Distribution	*20909*	27854	22733	34561	***85148***
IAU History	*6662*	3666			***3666***
Total	***195421***	**54839**	**43851**	**75717**	***174407***
ADMINISTRATION/SECRETARIAT					
Salaries					
Salaries & Charges	*466865*	145841	133981	161627	***441449***
Training courses	*3368*	2997	1693	1569	***6259***
General office expenses					
Leasing telex	*4021*				
Mail	*19976*	12510	4689	6892	***24091***
Telephone	*31772*	14814	12575	14236	***41625***
Telex	*9041*				
Office supplies	*37208*	10467	7557	11731	***29755***
Rental (INSU/IAP)	*5817*	5180	10328	4500	***20008***
Furniture	*23051*	4347	12513	274	***17134***
Computers	*9920*	18387	13760	18641	***50788***
Audit fee	*7156*	2332	741	1964	***5037***
Bank					
Loss on exchange 1 January of the year	*3864*	30858	42478		***73336***
Loss on internal transfers	*30612*	1003	1293		***2296***
Charges on internal transfers	*735*	529	486	87	***1102***
Bank charges	*9045*	2383	2130	2724	***7237***
Commissions on deposits	*10377*	2675	3293	4259	***10227***
Total	***672828***	***254323***	**247517**	**228504**	***730344***
ECLIPSE USSR (special account)	***33379***				
TOTAL EXPENDITURE	**2246347**	**1059289**	**700998**	**766313**	***2526600***
Excess of Income over Expenditure	***188095***	**-115780**	**51877**	**-28007**	***-91910***

Proposed budget 1998-2000 (CHF)

Notes in connection with the 1998-2000 proposed budget:

An allowance for inflation of 2% for 1998, 2.5% for 1999 and 3% for 2000 has been made
The royalties from IAU publications have been reduced in order to allow lower selling prices for the IAU books.

INCOME	*BUDGET**	PROPOSED BUDGET			*TOTAL*
	1997	1998	1999	2000	*98-2000*
Unit of contribution	*2820*	2880	2950	3040	
Number of units of contribution	*254*	254	254	254	
ADHERING ORGANIZATIONS	*716280*	731520	749300	772160	*2252980*
SPECIAL CONTRIBUTIONS					
ICSU/UNESCO	*40000*	40000	40000	40000	*120000*
PUBLICATIONS: ROYALTIES	*68000*	30000	30000	30000	*90000*
BANK INTEREST	*25000*	15000	15000	15000	*45000*
OTHER INCOME					
TOTAL INCOME	*849280*	816520	834300	857160	2507980

(*) 1997 budget as approved at the XXIInd General Assembly

1998-2000 proposed budget (CHF)

EXPENDITURE	BUDGET*	PROPOSED BUDGET			*TOTAL*
	1997	1998	1999	2000	*98-2000*
SCIENTIFIC ACTIVITIES					
General Assembly					
Grants	*167000*			180000	*180000*
Commission Presidents	*40000*			40000	*40000*
Staff	*18000*			15000	*15000*
Operation	*30000*	5000	5000	30000	*40000*
Executive Committee	*80000*			65000	*65000*
Meetings					
Symposia/Colloquia	*220000*	250000	256000	264000	*770000*
Regional Meetings		25000	25000		*50000*
Research Projects					
ISYA (46)		30000	30000		*60000*
TAD (46)	*15000*	30000	30000	30000	*90000*
EC WGs	*18000*	18000	18000	18000	*54000*
Exchange of Astromers (38)	*30000*	25000	25000	25000	*75000*
Commission Activities					
Telegram Bureau (06)	*4000*	4000	4000	4000	*12000*
Meteor Data Center (22)	*1100*	1100	1100	1100	*3300*
Astronomy Teaching (46)	*5000*	5000	5000	5000	*15000*
Commission WGs	*18000*	15000	15000	15000	*45000*
Delegates to other Unions	*10000*	10000	10000	10000	*30000*
Total SCIENTIFIC ACTIVITIES	***655000***	**418100**	**424100**	**702100**	*1544300*
DUES TO UNIONS/ORGANIZATIONS					
ICSU	*18000*	16000	16500	17000	*49500*
IERS/FAGS	*7500*	7500	7500	7500	*22500*
IUCAF	*7500*	7500	7500	7500	*22500*
CTS	*600*				
Total DUES TO UNIONS/ORGANIZATION	***33600***	**31000**	**31500**	**32000**	*94500*

(*) 1997 budget as approved at the XXIInd General Assembly

1998-2000 proposed budget (CHF)

EXPENDITURE (ctd)	BUDGET*	PROPOSED BUDGET			TOTAL
	1997	1998	1999	2000	98-2000
EXECUTIVE COMMITTEE					
Executive cttee meetings		35000	36000		71000
Officers meetings	5400	5000	5100	5200	15300
General Secretary expenditure	23400	26000	26500	27500	80000
President expenses	7000	3500	3500	3500	10500
Assist General Secretary expenses	3000	3000	3000	3000	9000
Total EXECUTIVE COMMITTEE	**38800**	**72500**	**74100**	**39200**	185800
PUBLICATIONS					
Information Bulletin	44000	32000	33000	34000	99000
Free Distribution	18500	19000	19000	19000	57000
Total PUBLICATIONS	**62500**	**51000**	**52000**	**53000**	156000
ADMINISTRATION/SECRETARIAT					
Salaries & Charges	162800	155000	159000	164000	478000
Training courses	4500	4500	4500	4500	13500
General office expenses: Mail, Telephone, Office supplies, Rental (INSU/IAP), Furniture, Computers	72000	70000	70000	72000	212000
Audit fee	3200	2000	2000	2000	6000
Bank:					
Loss on exchange					
Loss on internal transfers					
Bank charges	2500	3000	3000	4000	10000
Commissions on deposits	4500	4000	4000	4500	12500
Total ADMINISTRATION/SECRETARIAT	**249500**	**238500**	**242500**	**251000**	**732000**
TOTAL EXPENDITURE	**1039400**	**811100**	**824200**	**1077300**	**2712600**
Excess/Loss of Income over Expenditure	**-190120**	**5420**	**10100**	**-220140**	**-204620**

(*) 1997 budget as approved at the XXIInd General Assembly

CHAPTER VI

REPORT OF DIVISIONS, COMMISSIONS, AND WORKING GROUPS

DIVISION I

FUNDAMENTAL ASTRONOMY

Division I provides a focus for astronomers studying a wide range of problems related to fundamental physical phenomena such as time, the intertial reference frame, positions and proper motions of celestial objects, and precise dynamical computation of the motions of bodies in stellar or planetary systems in the Universe.

PRESIDENT: P. Kenneth Seidelmann
U.S. Naval Observatory, 3450 Massachusetts Ave NW
Washington, DC 20392-5100, US
Tel. +1 202 762 1441
Fax. +1 202 762 1516
E-mail: pks@spica.usno.navy.mil

BOARD

E.M. Standish	President Commission 4
C. Froeschle	President Commisison 7
H. Schwan	President Commisison 8
D.D. McCarthy	President Commisison 19
E. Schilbach	President Commisison 24
T. Fukushima	President Commisison 31
J. Kovalevsky	Past President Division I

PARTICIPATING COMMISSIONS:

COMMISSION 4	**EPHEMERIDES**
COMMISSION 7	**CELESTIAL MECHANICS AND DYNAMICAL ASTRONOMY**
COMMISSION 8	**POSITIONAL ASTRONOMY**
COMMISSION 19	**ROTATION OF THE EARTH**
COMMISSION 24	**PHOTOGRAPHIC ASTROMETRY**
COMMISSION 31	**TIME**

COMMISSION 4: EPHEMERIDES

President: H. Kinoshita Secretary: C.Y. Hohenkerk

Commission 4 held one business meeting.

Business Meeting, August 26:

During the triennium three members had died, Dr LeRoy Doggett, Dr Bruno Morando and Dr H. Beat Wackernagel, the meeting stood for a moment, in remembrance.

The officers elected to Commission 4 for 1997 to 2000 are, President E. Myles Standish Jr., and Vice-President Jean Chapront.

The following membership of the organizing committee was approved:

V.K. Abalakin	He Miao-Fu	H. Kinoshita	H. Schwan
J.-E. Arlot	C.Y. Hohenkerk	G.A. Krasinsky	P.K. Seidelmann
T. Fukushima	G.H. Kaplan	J.H. Lieske	

The following new members were elected:

Steven Bell	David Harper	Georgiy Eroshkin	Martin Lara

The following consultants were approved:

J.A. Bangert J. Meeus

The total membership of Commission 4, a middle sized commission, is 91. The President reported that the Commission had supported Joint Discussions 3 and 7, which resulted in resolutions B1, B2, B5. being adopted by the GA. The Chairman of the Working Group (WG) on JD and MJD reported that their work had resulted in the adoption of Resolution B1.

Report on Working Groups of Division 1

Kovalevsky reported on the decisions made by Executive Committee concerning WGs. A WG on general relativity in the framework of the space-time reference and methodology. A joint WG of the IAU, BIPM and IAG. The President would be Petit or Leschinta. The WG is to study the definitions, conventions, and notation at a sufficient level of accuracy.

A WG on general relativity in celestial mechanics and astrometry, Their President is Soffel. The WG is to consider the application of observations to celestial mechanics and astrometry, and study the concepts, algorithms, and constants used in astrometry.

A WG on the ICRS. Their President is Mignard. There are two sub-groups, chaired by Dehant and Helmut. Their job is to consider all consequences of the new reference system, the new nutation series, and the maintenance and extension of the Hipparcos Catalogue, and to liaise with the IERS.

The main tasks of the WG's were stated in resolutions B5 and B2. All those interested in participating should write or send e-mail to the appropriate WG President.

Kovalesky reported on the continuation of the following three WGs. The IAU/IUGG WG on Precession and Nutation, chaired by Dehant. The WG on Astronomical Standards, chaired by Fukushima, with sub-groups, on Best Estimates, chaired by McCarthy, and SOFA, chaired by Wallace. The WG of the IAU/IAG/COSPAR on Cartographic Coordinates and Rotational Elements of Planets and Satellites, now chaired by Seidelmann. The commission asked the chairman to thank Davies, the retiring Chairman for his work over many years.

The Future for National Nautical Almanac Offices

Yallop gave a talk about Nautical Almanac Offices. The text of his talk will be published in the Commission 4 Newletter.

Reports from various Institutions

HM Nautical Almanac Office

Hohenkerk reported that there have been several changes in staff. Yallop had retired, and Sinclair was now in charge (part-time), and Harper had been appointed.

Collaboration continued with USNO on *The Astronomical Almanac* and other publications. HMNAO continues to produce, enhance and increase its publications as required, and this now includes the financial viability as well as their usefulness and intellectual content. *The Nautical Almanac* is still the core commercial publication. The web, which by its nature freely provides information to all, is a medium that at present causes problems to HMNAO, who must cover costs through sales of almanacs and related data. There are also problems with copyright. We believe that HMNAO has a future. It will be part of

J. Andersen (ed.), Transactions of the International Astronomical Union Volume XXIIIB, 77–78.

the core of the new RGO, whose business plan is being prepared. We are continuing our work and trying to keep our products up to date as well as developing new ones. We would like to thank all those who wrote letters of support for HMNAO and the RGO.

The Bureau des Longitudes

Arlot reported on their publications, which are mostly in French, and are Government funded. An important activity was providing the public with information. This was done through the successful MINITEL system. They have now published an *Explanatory Supplement to the Connaissance des Temps*. A web service was being developed (French and English), some sections were available (e.g. physical ephemeris of IO).

Rechen-Institut

Schwan reported on "The Future of *Apparent Places of Fundamental Stars*". APFS has been published since 1941, and from 1960 this had been done by the RI. From 1988 it was computed from the final FK5 catalogue. The volume for the year 1999 will be published as before. However APFS was under increasing scrutiny. With the release of the Hipparcos catalogue the replacement of the FK5 is being considered. Certainly, apparent places for all Hipparcos stars will not be computed. The FK6 project is considering how to combine these catalogues. The problem arises due to the short period (3.5 years) of the Hipparcos mission, so many of the astrometric doubles have virtually 'instantaneous' proper motions. Thus for epochs 5 years from the mean Hipparcos epoch, the place will be computed on the basis of the FK5 proper motion (a mean motion for most stars), because it averages over about 200 years and will statistically provide better predictions that using the Hipparcos proper motions.
All remarks and suggestions and recommendations were welcome. Please send them to s25@ix.urz.uni-heidelberg.de.

Purple Mountain Observatory

The Director of the Ephemeris Division had asked Seidelmann to read their report. They have four annual publications, an Astronomical Ephemeris, a Nautical Almanac, a Surveying Almanac, and an almanac with general information. They also have almanacs on various Chinese Calendars and they produce Solar/Lunar eclipse predications. There is an ephemeris software package for PCs MICE, and graphical software for eclipses and meridian passages.

US Naval Observatory

Seidelmann reported that Paul Janiczek, head of Astronomical Applications Department, had retired, and they were in transition period. They had produced *The Astronomical Almanac*(AsA), *The Nautical Almanac*, *The Air Almanac*, *Astronomical Phenomena* each year; the software package MICA v 1.5 1990 - 2005 was available and MICA v 2.0, a major revision for Windows and Mac was being produced. There was other work for the Defense department, including a research into asteroids masses and ephemerides. They are distributing a survey about *The Astronomical Almanac*. How much should be printed, should there be a CD-ROM companion, or should it be on the web. Other considerations; should the Moon polynomial coefficients be printed; what use was the satellite section? USNO had a satellite ephemeris package. Other questions concerned the minor planets, the stars and the observatory list? All comments were welcome.

The National Astronomical Observatory of Japan

NAOJ produces an Ephemeris of about 100 pages (free), but it is also part of a larger data book on natural science, which is produced commercially. The Hydrographic Office publish the *Japanese Ephemeris* and *Nautical Almanac*.

Jet Propulsion Laboratory

Standish reported that they had a web page, from which people can get a table of ephemerides — like AsA, which was mainly used by professionals.

Message from the next president Standish said had been thinking about his role, he would be active through e-mail. We are now in the electronic age. He saw that the NAO's were facing problems. Professionals need access to ephemerides. There must be joint effort to provide them with the data they needed.

COMMISSION 7 : CELESTIAL MECHANICS (MÉCANIQUE CELESTE)

Report of Meetings, 26 August 1997

President : S. FERRAZ-MELLO
Vice-President : C. FROESCHLÉ
Secretary : A. LEMAITRE

1. Business

1.1. ELECTION OF THE ORGANIZING COMMITTEE

The commission elected the following officers and members of the Organizing Committee for the term 1997 – 2000 :

President:	C. Froeschlé
Vice-President	J. D. Hadjidemetriou
Members:	R. Dvorak
	S. Ferraz-Mello (*Past President*)
	T. Fukushima
	I. A. Gerasimov
	D. C. Heggie
	Z. Knežević
	J. H. Lieske
	A. Milani
	M. Moons
	J. C. Muzzio
	M. Soffel
	Y.-S. Sun

1.2. ELECTION OF CONSULTANT MEMBERS OF THE COMMISSION

The following non-IAU members were elected as consultant members of Commission 7 for their extended contributions to activities relevant to the Commission: A. Giorgilli (Italy), K. R. Meyer (USA), A. Neishtadt (Russia), C. Simò (Spain), and Z. Xia (USA)

1.3. ELECTION OF NEW MEMBERS OF THE COMMISSION

The following IAU members were approved for membership in Commission 7, following proposals by their National Committees or by members of the Commission's Organizing Committee:

M. Barbosu (Rumania), C. Beaugé (Argentina), E. Bois (France), L. Floria (Spain), T. Fukushima (Japan), T. Gallardo (Uruguay), I. A. Gerasimov (Russia), G. Hahn (Germany), A. Harris (USA), A. Krivov (Russia), A. Morbidelli (France), X X Newhall (USA), I. M. P. Osorio (Portugal), B. Parv (Rumania), H. Rickman (Sweden) V. Sokolov (Russia), N. Sorokin (Russia), B. A. Steves (UK), G. Tancredi (Uruguay), H. Varvoglis (Greece), N. Vassiliev (Russia), N. Watanabe (Japan), O. C. Winter (Brazil), and C. Zhao (China PR).

J. Andersen (ed.), Transactions of the International Astronomical Union Volume XXIIIB, 79–84.

1.4. DECEASED MEMBERS

At the session opening the presents revered the memory of those Commission members whose decease was brought to our knowledge: E. P. Aksenov, D. Betis, F. Boigey, V. G. Demin, L. Doggett, B. Morando, and J.L. Sagnier.

1.5. REPORT OF ACTIVITIES 1994-1997

The conclusions of the Working Group created by the Commission during the past General Assembly with the objective of redefining the goals of the Commission and to propose the necessary modifications so as to allow its insertion in one of the IAU divisions were the following:

1. The aim of the Commission may comprehend all astronomical problems where Dynamics is the main determining factor. Commission 7 should be renamed as "Celestial Mechanics and Dynamical Astronomy".
2. Commission 7 should remain outside the current IAU divisions as it shares interests with several of them.
3. The triennial report published in the IAU Transactions should continue as it is now, asking specialists to review the progresses in some topics selected in advance. Only modification is the recommendation that a broad sample of people be consulted by the Commission President before selecting the topics to be included in the report. A larger diffusion using e-mail facilities is necessary.
4. Commission 7 should have as policy to propose a Symposium every three years on a subject of actuality considered relevant and broad enough to guarantee the support of a few other IAU Commissions. Nowadays the Commission makes no proposals by itself and restricts its action to the discussion of spontaneous proposals originated from groups of members or organizations.

The Organizing Committee agreed with the proposal of renaming the Commission and communicated it to the EC. The new name "Celestial Mechanics and Dynamical Astronomy" will be in effect after this General Assembly. Taking into account the difficulties experienced for remaining outside the new IAU Divisions, the OC asked the IAU Executive Committee to be affiliated with two of them: Division 1 (Fundamental Astronomy) and Division 3 (Solar System). The answer of the EC was negative. After a long discussion, the OC decided, in 1997, to join Division 1. The OC agrees that the affiliation to one Division impairs its relationship with the others; however it also realized, during the elapsed three years, that the fact of remaining outside the divisional structure was weakening its ties with all divisions and with IAU itself; the continuation of this situation could impair efforts towards the realization of important commission goals.

After the past General Assembly the Commission has supported IAU Symposium 172 and co-supported IAU Colloquium 165 as well as 4 of the joint discussions held during this General Assembly. The Commission is also supporting IAU Colloquium 172 "The Impact of Modern Dynamics in Astronomy" and co-supporting IAU Colloquium 173 "Evolution and source regions of asteroids and comets" to be held in Namur and Stara Lesna, respectively, in 1998.

Other activities of the Commission in the period were the active support for the approval, by the NATO Scientific Affairs Division, of a new edition of the Advanced Studies Institute in Celestial Mechanics (see below). The Commission has also proposed the publication of an English version of C. V. L. Charlier's masterwork "Die Mechanik des Himmels", at the centennial of its first edition, in 2002.

1.6. CELESTIAL MECHANICS AND DYNAMICAL ASTRONOMY

J. Henrard presented a report on the current situation of the journal *Celestial Mechanics and Dynamical Astronomy.*

> The two main problems for the Journal are that it is behind schedule by about six months and that it is not in the list of "source journals" for the *Science Citation Index*. The two problems are connected. The Journal was dropped from the list of "source journals" when it changed name (and ISBN number) several years ago. Since then our efforts to get it back on the list have met with little success. The latest reason given us is that the Journal is too far behind schedule. Now the reason we are behind schedule is a dip in the submission of good papers in 1994 and 1995. We are now receiving and publishing at a correct rate but we are still behind schedule. We plan to fill the gap by scheduling fewer pages for a year or two. Let us hope that these actions will permit *Celestial Mechanics and Dynamical Astronomy* to get back on the list of source journals.
> For six months now the refereeing process of the journal has been conducted electronically. Soon the abstracts of published papers will be available on the web. Later on the full articles will be accessible from the computers of the institutions who do have a subscription to the journal.

The Commission decided to contact the Institute of Scientific Information (ISI) in order to manifest its extreme worries with the fact that the journal, still remains outside the source list of the Science Citation Index; it does not appear in the "Journal Citation Report", notwithstanding its quality standards well above those of many other journals considered and indexed by that Institute.

1.7. NATO ADVANCED STUDY INSTITUTES IN CELESTIAL MECHANICS

S.Ferraz-Mello read the report written for the Commission by Prof. A. E. Roy, current director of the NATO Advanced Study Institutes in Celestial Mechanics, on the 25 years of the Institute. An abstract of this report is given below:

> The series of NATO Institutes in Celestial Mechanics began in 1972 first under the directorship of Professor Victor Szebehely and subsequently under Professor Archie Roy. All except the latest have been held in Cortina d'Ampezzo, Italy. For the last 25 years, these Institutes have played a crucial role in the development of scientists working in Dynamical Astronomy, Orbital Dynamics, Celestial Mechanics and Space Astrodynamics. They involve the coming together for two weeks of some fifteen specialist lecturers of international reputation and some 60 to 70 participants from about 20 different countries scattered around the world. Almost all the present leading scientists in these branches of Science and Engineering Mechanics in North America and Western Europe, have attended at least one of these Institutes. Many of those now active in the field made their first international contacts at these Institutes.
> The latest ASI in the series, entitled "The Dynamics of Small Bodies in the Solar System: A Major Key to Solar System Studies", was held recently in Acquafredda di Maratea, Italy. The study of dynamics of asteroids, comets, meteor streams, natural satellites and ring systems currently provides a wealth of information concerning the history and dynamical evolution of the solar system as a whole. Further information about the Institute may be found on the web site `http://www.maths.gcal.ac.uk /natoconf`.

1.8. DIVISION I WORKING GROUPS

J. Kovalevsky, president of IAU Division I informed about the Working Groups of interest of Commission 7 (two of them chaired by members of the Commission's Organizing Committee):

- WG on Near Earth Objects (Divisions I and III). Chairman: D. Morrison(dmorrison@mail.arc.nasa.gov)
- WG on Relativity for Celestial Mechanics and Astrometry. Chairman: M. Soffel (soffel@rcs.urz.tu-dresden.de)
- WG on ICRS. Chairman: F.Mignard (mignard@ocar01.obs-azur.fr)
- WG on Astronomical Standards. Chairman: T. Fukushima (toshio@spacetime.mtk.nao.ac.jp)
- WG on Precession and Nutation (IUGG and IAU). Chairman: V. Dehant (Veronique.Dehant@ksb-orb.oma.be)

2. Scientific Reports

Three invited reports were presented and the following abstracts were provided by the lecturers:

1. "Kozai Resonance" : H.Kinoshita (Kinoshita@nao.ac.jp)
2. "Hamiltonian Systems and Tangent Space to Invariant Tori" R. Vieira Martins (rvm@on.br)
3. "The causes of chaotic motion in the Hecuba gap " J.Henrard (jhenrard@math.fundp.ac.be)

KOZAI RESONANCE

H. KINOSHITA
Tokyo Astronomical Observatory, Tokyo, Japan

Kozai (1962) found that the argument of perihelion of an asteroid with high inclination does librate and even though the initial eccentricity is very small, the eccentricity becomes very large. This phenomenon is recently called Kozai resonance, since the mean motion of the perihelion longitude is equal to that of the longitude of node. Kozai resonance is now recognized as an important factor in the chaotic dynamical evolutions of the small bodies in the solar system and the exo-planetary systems. Kozai resonance; however, it is a little bit fragile compared with the mean motion resonance and the secular resonances. Kinoshita and Nakai (1991) discussed secular perturbation of fictitious satellites of Uranus, which are disturbed by the Sun and the oblateness of Uranus. Satellites located in the region where the solar perturbation is dominant become highly eccentric due to Kozai resonance. Satellites located in the region where the oblateness perturbation is dominant keep the original small eccentricity since Kozai resonance is suppressed. Innanen *et al.* (1997) investigated the stability of the planetary orbits in binary systems. When the mutual perturbations among planets are neglected, the eccentricity becomes very large due to Kozai resonance. When, however, the mutual perturbations are included, the Kozai resonance is suppressed and the eccentricities of the planets stay small and the orbital planes of the planets move with small mutual inclinations.

References

Innanen, K.A., Zheng, J.Q., Mikkola, S. and Valtonen, M.J. (1997), *Astronomical Journal* **113**, 1915.
Kinoshita, H. and Nakai, H. (1991) *Celestial Mechanics* **52**, 293.
Kozai, Y. (1962) *Astronomical Journal* **67**, 591.

HAMILTONIAN SYSTEMS AND TANGENT SPACE TO INVARIANT TORI

R. VIEIRA MARTINS
Observatório Nacional, Rio de Janeiro, Brazil

We consider an integrable Hamiltonian system with n degrees of freedom. In very general conditions, we know that almost every trajectory of the system is dense on a torus T^n which is a manifold with dimension n immersed in the phase space which has dimension $2n$.

For every particular solution of the system, and in particular for any arbitrarily chosen solution $z(t)$ dense on a invariant torus, we can write the variational equations. It is well known that the variational equations associated to the first variation is a system of homogeneous linear equations. For the kth variation we have a non-homogeneous linear system which differs from the first variation system by an additive independent term which is function of the solutions of the $(k-1)^{th}$ variation systems. In very general conditions, we have the following result concerning the solutions of the variational equations:

"For t sufficiently large, the ratio of every arbitrary solution of the kth variation to t^k defines a vector which is sufficiently near another vector whose components can be associated to the kth derivatives of a parametrization of the invariant torus associated to $z(t)$".

Thus, with the first variation we can know the tangent space to the torus, with the second variations we cän compute the curvatures, and so on.

Some applications follow naturally from this result. For example, if we know the tangent space to the torus, we can evaluate the frequencies associated to the trajectory on the torus. We can also compute the singularities of the projections of the tangent spaces on the configuration space. So, it is possible to know the Lagrangian singularities which define the caustics of the torus in this space. With the higher derivatives of the parametrization of the invariant torus, we can know the higher order Lagrangian singularities. These singularities are interesting in the study of some properties of the Hamiltonian integrable systems orbits in the configuration space.

A proof of this result for the first variation case and its application to the computation of caustics may be found in Stuchi and Vieira Martins (1995).

References

Vieira Martins, R. and Stuchi T. (1995), *Phys. Lett. A* **201**, 179.

THE CAUSES OF CHAOTIC MOTION IN THE HECUBA GAP

J. HENRARD
Dept. Mathématique, F.U.N.D.P., Namur, Belgium

For about fifteen years now, one of the main goals of dynamicists of the Solar System has been to show that the depletion of the Kirkwood gaps in the asteroid belt is due to close encounters with the terrestrial planets. The favored scenario is that chaotic motion inside the Jovian resonances forces a large increase in the eccentricity of asteroids which then become Mars crossers or even Earth crossers or even Sun-grazers.

This search for chaotic motion as the "smoking gun" of the Kirkwood gap enigma received a dramatic uplift with the investigation of the 3/1 gap by Wisdom (1983). The mechanism responsible for the chaotic zone investigated by Wisdom is the *periodic crossing of the separatrix* of the circular averaged restricted problem (Wisdom, 1985) due to the eccentricity of Jupiter. The mechanism does not cover the full extend of the 3/1 Kirkwood gap and account only for an increase in eccentricity just large enough to force asteroids to become Mars crossers but the search was on. It was later found out that there is a path to much higher eccentricities (Ferraz-Mello and Klafke, 1991) and that a stronger mechanism, *the overlap of secular resonances*, account for a much extended chaotic zone with a much shorter time scale for eccentricity increase (Moons and Morbidelli, 1995). The overlap of secular resonance is also responsible for the depletion of several other gaps (Morbidelli and Moons, 1995).

But the first order resonances (2/1 and 3/2 being the most important one in the asteroid belt) are much stronger than higher order resonances. The separatrix crossing exists but affect only a very small volume of phase space; the overlap of secular resonances exists but affects only orbits which are high eccentric to start with (Morbidelli and Moons, 1993).

Wisdom (1987) produced strong numerical evidences that another chaotic zone exists at low eccentricity and hinted that orbits starting in this zone could reach high eccentricities by going through a high inclined episode. This chaotic zone had been discovered much earlier (Froeschlé and Scholl, 1976) but its importance overlooked because, in the planar problem, it was not responsible for any dramatic increase in the eccentricity. Later on the existence of this low eccentric chaotic zone was explained as due to *secondary resonances* between the frequency of libration and the pericentric frequency (Lemaître and Henrard, 1990) and the bridge, between this low eccentric zone and the high eccentric zone due to secular resonances, explained as being due to the ν_{16} resonance in inclination (Henrard *et al.*, 1995).

Between these two zones of chaotic motion, in both the 2/1 and the 3/2 resonances, there still remained a large size volume of phase space, at moderate eccentricities and low inclination, apparently non chaotic. It is precisely in this zone that the Hilda family is found in the 3/2 resonance, while this zone is almost devoid of asteroid in the 2/1 resonance. In the 2/1 resonance this zone is cut in two by the ν_{16} resonance in inclination (Michtchenko and Ferraz-Mello, 1997) but this does not seem to be enough to explain the depletion. In the same paper the authors make a strong point about the importance of the "great inequality", this 5/2 quasi resonance between the mean motions of Jupiter and Saturn, something which had already been hinted by Ferraz-Mello *et al.* (1996).

Very recently, a large scale numerical survey of the first order Jovian resonances (Nesvorný and Ferraz-Mello, 1997) confirms the existence and the location of all the sources of chaotic motion mentioned above, while an ad-hoc numerical exploration (Henrard, 1997) shows that high order secondary resonances with the great inequality is responsible for a weak chaoticity which leads to a slow random walk in the amplitude of libration; this is enough, on time scale of hundreds of millions of years, to dump orbits in the ν_{16} resonance in inclination; the orbit is then very quickly "lifted" to the strong chaotic zone sitting at high inclination and high eccentricity.

References

Ferraz-Mello, S. and Klafke, J.C. (1991), *Predictability, Stability and Chaos in N-body Dynamical Systems* (A.E.Roy ed.) Plenum Press (NATO Adv. Stud. Inst. Series B Phys **272**), 177.
Ferraz-Mello, S., Klafke, J.C., Michtchenko, T.A. and Nesvorný, D. (1996), *Celest. Mech. Dynam. Astron.* **64**, 93.
Froeschlé, C. and Scholl, H. (1976), *Astron. Astrophys.* **48**, 389.
Henrard, J., Watanabe, N. and Moons, M. *et al.* (1995), *Icarus* **115**, 336.
Henrard, J. (1997), *Celest. Mech. Dynam. Astron.* to appear.
Lemaître, A. and Henrard,J. (1990), *Icarus* **83** 391.
Michtchenko, T. and Ferraz-Mello, S. (1997), Planet. Spac. Sci., in press.
Moons, M. and Morbidelli, A. (1995), *Icarus* **114**, 33.
Morbidelli, A. and Moons, M. (1993), *Icarus* **102**, 316.
Morbidelli, A. and Moons, M. (1995), *Icarus* **115**, 60.
Nesvorný, D. and Ferraz-Mello, S. (1997), *Icarus* in press.
Wisdom, J. (1983), *Icarus* **56**, 51.
Wisdom, J. (1985), *Icarus* **63**, 272.
Wisdom, J. (1987), *Icarus* **72**, 241.

COMMISSION 8: POSITIONAL ASTRONOMY
COMMISSION 24: PHOTOGRAPHIC ASTROMETRY

Business Sessions: Saturday, 23 August 1997 and Tuesday, 26 August, 1997

PRESIDENT, COMMISSION 8: Thomas E. Corbin

PRESIDENT, COMMISSION 24: Catherine Turon

VICE-PRESIDENT, COMMISSION 8: Heiner Schwan

VICE-PRESIDENT, COMMISSION 24: Elena Schilbach

ORGANISING COMMITTEES

COMMISSION 8: P. Benevides-Soares, L. Helmer, Xu Jiayan, J. Kovalevsky, L. Lindegren, J. A. Lopez, L. Morrison, F. Noel, D. Polojentsev, Y. Requieme, R. Stone, Jin Wenjing, M. Yoshizawa

COMMISSION 24: C. de Vegt, P.D. Hemenway, P.A. Ianna, K.J. Johnston, I.I. Kumkova, C.E. Lopez, E. Schilbach, H.G. Walter, J.J. Wang, G.L. White

1. Commission 8: business session 23 August 1997

1.1. INTRODUCTION

The President presented the agenda for the business meetings to the members for their approval. He summarized the steps that had been taken to address the question of merger with Commission 24 over the past three years. The membership in general and the Organizing Committee in particular took an active and responsive part in this process, and this was greatly appreciated. Special thanks must go to Dr. J. Kovalevsky, President of Division 1, and Dr. C. Turon, President of Commission 24 for their collaboration, help and friendship which turned a chore into a pleasure. Great appreciation is also due to Monique Orine and Julie Saucedo, office of the IAU Secretariat, for their help with all of the support needed to run the commission these past three years. Nearly instant response with accurate information was always their hallmark.

1.2. IN MEMORIAM

A moment of silence was observed by members of commission noting the deaths of the following colleagues: V. V. Konin, A. A. Nemiro, K. Pilowski, and J. von der Heide.

1.3. MEMBERSHIP

Although it had been the intent of the previous president to reduce the size of the commission by eliminating inactive members, the list received in 1994 from the Secretariat, when corrected for some omissions, totaled 169 names. One member, Joseph M. Chamberlain, resigned. The following were proposed for membership and received unanimous approval from the members:

Anna A. Androva	Pulkovo Observatory	Russia
Christian Delmas	OCA-CERGA	France
Alexander V. Devyatkin	Pulkovo Observatory	Russia
Yu Fan	Yunan Observatory	China

J. Andersen (ed.), Transactions of the International Astronomical Union Volume XXIIIB, 85–100.

Gueorgui A. Gontcharov	Pulkovo Observatory	Russia
Irina S. Guseva	Pulkovo Observatory	Russia
Alexei E. Ilin	Pulkovo Observatory	Russia
Carme Jordi	Universitat de Barcelona	Spain
Andrei V. Kuzmin	Sternberg Astronomical Institute	Russia
Qi Li	Beijing Astronomical Observatory	China
Zheng Xing Li	Purple Mountain Observatory	China
Olexandr Molotaj	Astronomical Observatory (Kyiv)	Ukraine
Kouji Ohnishi	Nagano National College	Japan
Thomas Schildknecht	Astronomical Institute (Bern)	Switzerland
Sean Urban	U. S. Naval Observatory	USA
Miguel Vallejo	Observatorio de la Armada	Spain
Norbert Zacharias	U. S. Naval Observatory	USA
Zi Zhu	Shaanxi Observatory	China

1.4. COMMISSION 8 OFFICERS FOR THE NEXT TRIENNIUM: 1997-2000

President:	Heiner Schwan (Astronomisches Rechen Institut, Germany)
Vice-President:	Wenjing Jin (Shanghai Observatory, China)
Organising Committee:	T.E. Corbin (USA), J. Kovalevsky (France) J.A. Lopez (Argentina), L.V. Morrison (UK), F. Noel (Chile), G. Pinigin (Ukraine), D. Polojentsev (Russia), R. Stone (USA), J. Xu (China)

Their e-mail addresses are given below:

Corbin	tec@sicon.usno.navy.mil
Kovalevsky	kovalevsky@mfg.cnes.fr
Lopez	celopez@unsjfa.edu.ar
Morrison	merlp@ast.cam.ac.uk
Noel	fnoel@calan.das.uchile.cl
Pinigin	pinigin@mao.nikolaev.ua
Polojentsev	ddp@mahis.spb.su
Stone	rcs@nofs.navy.mil
Xu	TT.WW@company.BJBTA.chinamail.sprint.com

1.5. MERGER WITH COMMISSION 24

During July, 1997, a referendum on merger with Commission 24 and a vote on new Commission 8 officers were held. The President reported on the results.

Issue 1: Merger of Commissions 24 and 8. This will become effective at the conclusion of the XXIVth General Assembly in 2000. For the triennium 1997 to 2000 the two commissions will have a common vice-president who will become the first president of the new commission for the triennium 2000 to 2003. The first vice-president of the combined commission will be elected by a vote of the combined membership of the two commissions by the close of the XXIVth General Assembly. The Organizing Committee of the new commission will be formed by combining the existing Commission 24 and 8 Organizing Committees.

FOR: 55 AGAINST: 3 ABSTAIN: 0

Issue 2: Election of Commission 8 Vice-president. Based on the recommendation of the Commission 8 Organizing Committee and the comments of the Commission 8 membership, Dr. Jin Wenjing will serve as Vice-president of Commission 8 for the triennium 1997 to 2000.

FOR: 52 AGAINST: 2 ABSTAIN: 4

Issue 3: Organizing Committee. The Organizing Committee of Commission 8 is being reduced in membership for the next triennium. This is in anticipation of the merger in 2000. No new members have been added and those who have served the longest on the current committee have been dropped. The resulting committee for 1997 to 2000 would then be:

J. Kovalevsky, J. A. Lopez, L. Morrison, F. Noel, G. Pinigin, D. Polojentsev, R. Stone, Xu Jiayan and T. Corbin (as ex-president).

FOR: 55 AGAINST: 1 ABSTAIN: 2

Issue 4: Name of the combined commission. Please submit your suggestion for the name of the new commission. If there is an overwhelming preference in both commissions for a particular name, no vote will be needed at the Kyoto business sessions. Otherwise, the two or three names that receive the most support will be the candidates, and a vote will be taken.

The three names receiving the most votes: Astrometry–23, Fundamental Astrometry–3, Positional Astrometry–3.

Issue 5: Membership in the new commission. During the triennium 1997 to 2000 the members of Commissions 24 and 8 will be polled by the presidents of the two commissions. Only those requesting membership in the new commission will become members. The intent here is to eliminate those who no longer have an interest in astrometry.

FOR: 54 AGAINST 2 ABSTAIN 2

The report was approved by the members.

1.6. SPONSORSHIP OF IAU SYMPOSIA AND COLLOQUIA

- Symposium 172:
  Dynamics, Ephemerides and Astrometry in the Solar System, Paris, France, July 1995.
- Colloquium 165:
  Dynamics and Astrometry of Natural and Artificial Bodies, Poznan, Poland, July 1996.
- Symposium 179:
  New Horizons from Multi-Wavelength digital Sky Surveys, Baltimore, USA, August 1996.

2. Commission 24: business session 23 August 1997

2.1. INTRODUCTION

During the triennium 1994-1997, yearly circular letters and many circular e-mails have been circulated to Commission 24 members. The major difficulty was to update the address and electronic address lists. It is important to stress the crucial role of electronic mailings, saving a lot of effort and money to the Commission President, and allowing an immediate forward of any relevant information or request coming either from the Commission itself, from a working group or from Division 1. In addition, a WWW page was established for Commission 24, giving easy access to information, texts, reports, membership list, announcements of meetings relevant to the activity of the Commission, etc. The triennial report is also accessible from this page.

As a wish for the future, members are urged to communicate their e-mail address to the new president, if they do not have already done so, and any change in their address.

2.2. IN MEMORIAM

A minute of silence was observed in memory of the members who left us during the last triennium: Thomas E. Lutz, Willem J. Luyten, and Peter van de Kamp.

2.3. MEMBERSHIP

The following new members were approved, under the reservation that they are or become IAU members:

Frédéric Arenou	Observatoire de Paris-Meudon	France
Christine Ducourant	Observatoire de Bordeaux	France

Nina V. Kharchenko	Kiev Observatory	Ukrainia
Chunlin Lu	Purple Mountain Observatory	China
V.P. Rilkov	Pulkovo Observatory	Russia
K.V. Kuimov	Sternberg Institute	Russia

A list of the membership was circulated during 1996 to all members, asking for updating of membership, address, phone, fax and e-mail. The following members asked to be deleted from Commission 24 membership because of a change in their activities or of retirement: Jean Delhaye (France), Denis Harwood (Australia), Wulff D. Heintz (USA), Aden B. Meinel (USA), Franz V. Prochazka (Austria), Gerard Scholz (Germany).

2.4. OFFICERS FOR THE NEXT TRIENNIUM: 1997-2000

During the Business Meeting, held on Saturday 23 August, the officers proposed by the Organising Committee, and approved by a ballot organized by mail and e-mail through all members of the commission before the meeting of the General Assembly in Kyoto, were confirmed:

President:	Elena Schilbach (Germany)
Vice-President:	Jin Wenjing (China)
Organising Committee:	Michel Crézé (France), Paul D. Hemenway (USA), Irina I. Kumkova (Russia), Imants Platais (USA), Siegfried Röser (Germany), Catherine Turon as ex-President (France), Jia-Ji Wang (China)

Their e-mail addresses are given below:

M. Crézé	michel.creze@iu-vannes.fr
P.D. Hemenway	paul@astro.as.utexas.edu
W.J. Jin	jwj@center.shao.ac.cn
I.I. Kumkova	kumkova@ipa.rssi.ru
I. Platais	imants@astro.yale.edu
S. Röser	s19@ix.urz.uni-heidelberg.de
E. Schilbach	eschilbach@aip.de
C. Turon	catherine.turon@obspm.fr
J.J. Wang	wangjj@center.shao.ac.cn

2.5. MERGER WITH COMMISSION 8

A preliminary circular letter to the members of Commissions 8 and 24 was sent during the summer 1996. It was asking for the opinion of the members on the possibility of a merger between our two commissions. As a vast majority of answers were highly positive about this merger, we went on with its preparation in both Commissions: we proposed the same Vice-President for the two Commissions for the triennium 1997-2000, and reduced the number of members of the Organising Committees.

The third step was the organization of a ballot, in each Commission, last summer, before the General Assembly (as it appeared that many members will not attend the Kyoto meeting). 92 % of the answers in Commission 24 were in favour of the merger. The name 'Astrometry' is proposed for the new Commission.

The current plans are the following:

- The merger of Commissions 8 and 24 will become effective at the conclusion of the XXIVth General Assembly in 2000.
- For the triennium 1997 to 2000 the two Commissions will have a common Vice-President who will become the first President of the new Commission for the triennium 2000 to 2003.
- The first Vice-President of the combined Commission will be elected by a vote of the combined membership of the two Commissions by the close of the XXIVth General Assembly. The Organising Committee of the new Commission will be formed by combining the existing Commission 24 and 8 Organising Committees.

Results of the ballot (see above for the complete texts submitted to ballot):

Issue 1: Merger of Commissions 24 and 8, following the plans given above.

FOR: 37 AGAINST: 2 ABSTAIN: 1

Issue 2: Election of Commission 24 Vice-president. Based on the recommendation of the Commission 24 Organizing Committee and the comments of the Commission 24 membership, Dr. Jin Wenjing will serve as Vice-president of Commission 24 for the triennium 1997 to 2000.

FOR: 35 AGAINST: 1 ABSTAIN: 4

Issue 3: Organising Committee. The resulting committee for 1997 to 2000 would then be: M. Crézé, P.D. Hemenway, W.J. Jin, I.I. Kumkova, I. Platais, S. Röser, E. Schilbach, C. Turon (as ex-president), J.J. Wang.

FOR: 36 AGAINST: 1 ABSTAIN: 3

Issue 4: Name of the combined commission. The three names receiving the most votes: Astrometry–23, Fundamental Astrometry–2, Positional Astrometry–2.

Issue 5: Membership in the new commission. During the triennium 1997 to 2000 the members of Commissions 24 and 8 will be polled by the presidents of the two commissions. Only those requesting membership in the new commission will become members. The intent here is to eliminate those who no longer have an interest in astrometry.

FOR: 37 AGAINST 2 ABSTAIN 1

After one session of separate business meetings in each Commission, the three other sessions (business and science, including the Working Groups reports) were held in common, and their report is given below.

It is important to acknowledge the high spirit of cooperation with which all this process was completed, and we are looking forward to seeing the new input to astrometry given by this new Commission.

2.6. SPONSORSHIP OF IAU SYMPOSIA AND COLLOQUIA

- Symposium 179:
  New Horizons from Multi-Wavelength digital Sky Surveys, Baltimore, USA, August 1996.

ACKNOWLEDGEMENTS

The President would like to thank all members who answered the inquiries or requests for inputs for the triannual report. Such exchanges made a lively communication among the members of the Commission.

The President would also like to express her gratitude to the IAU secretariat officers, Monique Orine and Julie Saucedo, for their constant support and kindness, and their permanent patience to answer any kind of questions.

3. Joint Session of Commissions 8 and 24: 23 August 1997

The two presidents reported on the results of the votes on the merger referendum from each commission. The combined membership gave final approval to the merger and to the name of 'Astrometry' for the new commission when merger takes place in 2000. The Division 1 President, J. Kovalevsky, stated that the new commission would be referred to as Commission 8, that being the lower number, unless a different number was requested by the membership. The combined membership then approved Commission 8 (Astrometry) as the title for the new commission.

The President of Division 1, J. Kovalevsky, differentiated the objectives of two of the working groups. The IAU-BIPM Joint Committee on Relativity (IAU WG on Relativity for Reference Systems and Metrology) is to establish conventions, definitions and uniform notations in order to provide a coherent frame for space-time references for all uses (Astronomy, Geodesy and Meteorology). After consultation with the BIPM, it has been agreed that the chairman will be G. Petit. A second IAU working group, Relativity for Celestial Mechanics and Astrometry, chaired by M. Soffel, is charged with defining how these concepts are to be applied to specific problems of Astrometry and Celestial Mechanics.

4. Reports of the Working Groups: joint session 23 August 1997

4.1. ASTROGRAPHIC CATALOG PLATES - PRESENTED BY CORBIN

Jones (Chairman), Bucciarelli, Corbin, Dahn, de Vegt, Röser, Urban, Smart.

1. Great progress has been made on re-reducing the published AC Catalogue. The USNO has now re-reduced all the AC Zones and Kuzmin and colleagues in Moscow have done the same. Kuzmin presented a poster of their work at IAU Symposium 179 in Baltimore. All zones of AC data are presently being converted to the Hipparcos system within the framework of the Provisional Tycho Reference Catalogue (pTRC), a collaboration between Moscow, Copenhagen and Heidelberg. Conversion will be completed by the end of September. Publication of pTRC is expected at the end of 1997.

The pilot TRC reported at the Venice Symposium showed that AC and Tycho could yield 1.5-2.0 mas/yr proper motions for nearly all Tycho stars; but this is just an internal (random) accuracy. The same is not expected for the pTRC because Hipparcos is poorly suited for the purpose of AC calibration, being too bright and sparse. Further announcements about the pTRC are expected at Kyoto. The faint limit varies from zone to zone and is usually around B = 12. The positional errors of the different zones were established by Röser and Høg from positions on overlapping plates; they varied between 0.43 and 0.23 arcsec.

These errors arise from several causes; normal astrographs are neither normal, nor the same; the observation logs are missing or incomplete for many zones, so exposure details and meteorological data are unavailable. Different observatories used different measurement techniques; the eyepiece grid, the eyepiece scale or the eyepiece micrometer. The last seems to be the most accurate.

The name of the USNO AC catalog is 'AC 2000'. A poster paper can be found at Joint Discussion 7. The main characteristics of the catalog are the following:

- Number of stars in AC = 4.62 million
- Number of images = 8.63 million
- Zone with lowest precision/image = Sydney ($0\farcs48$ in ra, $0\farcs43$ dec)
- Zone with highest precision/image = Algiers ($0\farcs19$ in ra, $0\farcs19$ dec)

These precisions are based on overlapping plate data and are means of all stars within the zone. The USNO positions are currently being converted from the ACRS to the Hipparcos system. Further details can be found at the web site http://aries.usno.navy.mil/ad/ac.html.

2. The chart plates were not taken for all the zones. When they were taken the corresponding atlas was sometimes not published (eg Cape). The limiting magnitude of most chart plates is about B = 15. Most chart plates have three exposures in form of a small equatorial triangle. The problem of centering on these images has been attacked by several colleagues. Earlier attempts were done by the MAMA team of the Observatoire de Paris (J. Guibert and coworkers) in collaboration with M. Geffert (Bonn). The results and an astronomical application along with references have been published by Geffert et al. in A&A Supp **118**. Meanwhile other colleagues (P.Brosche, W. Dick, R. Galas, M. Hiesgen) have made new efforts mainly under a European network project 'Salvaging an Astrometric Treasure' under the leadership of P. Brosche (Bonn). The project was summarized by Hiesgen et al. in a poster displayed at IAU Symposium 179 in Baltimore. Methods and techniques will be described in detail by Ortiz-Gil et al. in the A&A Supp. series (in press). Accuracies in the range 0.1 - 0.2 arcsec are attainable.

A project has been started at Bordeaux Observatory with the aim of systematically measuring all Carte du Ciel plates of the Bordeaux zone on an automatic plate scanning machine within the next 3 years. The goals of this project are to preserve the information content of these old observations in digital form and to derive high accuracy proper motions for stars as faint as 14th or 15th magnitude. In support of the latter, a systematic re-observation of the Bordeaux zone with the Bordeaux CCD meridian circle has started in 1997 February. The project is therefore run under the title 'Meridien 2000'. As a pilot study, they are currently dealing with the derivation of proper motions in the region of the open cluster NGC 2355 and in a region intersecting the galactic disk close to the galactic anticenter.

Some zones (eg Greenwich) took single images on their chart plates. Such plates have been measured with an accuracy between 0.07 and 0.11 arcsec (see Odenkirchen et al. in A&A Supp **124**).

Fresneau is completing a study dealing with the measurements of 60 Carte du Ciel plates distributed along the Vulpecula Rift. The measurements were performed with the fast measuring machine, MAMA, located at the Paris Observatory and the plates scanned are stored in the plate vault there. The celestial coordinates are compared to the GSC survey in order to derive proper motions and the AC is used to calibrate metric and radiometric properties of these deep plates (identical to the plates used for the AC). This study rediscovers the molecular clouds associated with the Vulpecula OB association at a distance of 300 pc and locates the boundary layer between the CO maps and the IRAS 100 micron maps. The observing procedure of making three exposures is also quite useful for investigating variable stars. This study also locates the variable stars associated with nebulosities and spectroscopic follow-up observations are planned. The combination of the POSS I and II surveys can doubtless provide more accurate data. A careful analysis of a real improvement by Carte du Ciel plates has still to be made.

3. Jones has measured six plates kindly loaned by the Vatican Observatory on the RGO 10x10 PDS bought in 1974. A preliminary report on this work was presented at IAU Symposium No 179 'New Horizons from Multi-wavelength Sky Surveys'. Those reductions were made relative to the ACRS catalogue.

Now that Hipparcos is available the error budget for a single coordinate of a single image is as follows:

Reproducibility of PDS in one orientation	0.05 arcsec
Comparison between two measures at 0 and 90	0.09 arcsec
Comparison between 6 and 3 min exposures	0.11 arcsec
Comparison with Hipparcos	0.19 arcsec
Errors in Hipparcos proper motions	0.12 arcsec

The Hipparcos proper motion errors were assumed to be 1.5 mas/yr and the time-base 80 years. Subtracting the fifth quantity quadratically from the fourth yields 0.15. Thus an accuracy between 0.11 and 0.15 arcsec in one coordinate can be achieved which in conjunction with Tycho should yield proper motions in the range 1.4 - 1.9 mas/yr; comparable to the accuracy of Hipparcos. In general there are two images of every star and the majority of stars appears on two plates so that the accuracy of most star positions can be reduced to 0.08 arcsec.

The PDS used is very slow. To rasterise a row of three images on a catalogue plate takes 1.9 minutes in one orientation and 0.8 in the other. With roughly 2000 stars per plate and 1000 plates per zone it would be uneconomic to attempt to re-measure a complete zone with this machine. The above reductions were made with a simple circular Gaussian centering algorithm and a six component linear fit. Both could doubtless be improved on. Images within 8 arcsec of a reseau line were ignored. One of the more time-consuming tasks lay in deciding which images were 'good' and which were too poor to include in the fit. There is a trade-off between accuracy and limiting magnitude.

4. The availability of the original plates varies from zone to zone. Some have been destroyed and some observatory directors are hesitant to let the plates leave their observatories. The number of people who know exactly where the plates are stored is steadily decreasing. Jones visited the Cape in January 1997 and spent about a week finding the plates and their records and making a superficial assessment of their quality. A pilot programme is under way to measure a number of these plates at the USNO. There is also a whole unmeasured AC Zone at the Santiago Observatory. The Melbourne and Sydney Observatories are now closed but their plate collections are held at Macquarie University Library, in Sydney. They have about 20,000 plates which were mainly taken between 1890 and 1964. The current curator of Sydney Observatory (which has now become a museum) says these plates are available to anyone who could be interested. The Greenwich and Oxford I and II plates are held in Cambridge.

The most critical AC zone is Potsdam, which was redistributed among Oxford, Hyderabad and Uccle where plates were taken around 1930-1940. If the original Potsdam plates could be measured (they are believed to have been destroyed in World War II) it would greatly improve the proper motions in the +32 to +39 degree declination zones.

5. It is concluded that there would be real scientific benefits in re-measuring both the chart and catalogue plates. The million stars in Tycho are currently being combined with the published AC positions to derive proper motions. For the best zones the accuracy will be comparable with Hipparcos but for the worst zones the errors will be doubled. After the second Tycho processing the number of stars will rise to three million.

To add to this work both in accuracy and number of stars the first priority should be the chart plates which have never been measured before. The majority of stars will be too faint to appear in Tycho but GSC1.2 and the USNO A1 and UJ catalogues and their successors will provide good modern positions. A proper motion accuracy of 6 mas/year should be attainable. More than seven million stars are expected in this sample of which 4.6 million are in the published AC catalogue. Seven million is a 'whole sky' number and the extant chart plates cover only about half the sky. The number is critically dependent on the limiting magnitude and may be much larger.

The USNO are projecting a new survey in the Southern hemisphere to reach 16th magnitude with an accuracy of 40 mas; comparison with re-measured chart plates should yield a proper motion accuracy of 1.6 mas. The USNO intend to continue this work in the north after the completion of the one in the south.

The catalogue plates have a lower priority. The zones of lower accuracy as determined by modern re-discussion of the published positions (eg Sydney, Vatican) should be measured first. Again many stars will be too faint to appear in Tycho as presently available but most should be present in the second Tycho processing.

These priorities will be subject to the availability of the plates. However, nothing worthwhile can be achieved without a measuring machine an order of magnitude faster than a PDS, and the same astrometric accuracy, which could be allocated to this project for a period of years. Designs have been suggested for a dedicated machine but there does not seem to be any immediate hope of finding the money to build one.

6. We recommend that the Working Group continue for another three years but with a revised membership list. The following have expressed their willingness to serve on a new Working Group: Fresneau, Jones (but not as chairman), and de Vegt. We gratefully acknowledge the contributions of many colleagues to this report: Brosche, Eichhorn, Fresneau, Geffert, Hög, Kuzmin, Odenkirchen, Ortiz-Gil, Smart, and Urban.

* After the report was given there was discussion on continuing the work. It was decided by the members attending that the WG would continue with the focus on surveying AC and CdC plates available for measuring and available machines that are capable of doing the work. Jones resigned as chairman and Fresneau was elected chairman of the working group for the term 1997 - 2000.

4.2. ASTROLABES - PRESENTED BY CHOLLET

Chollet (Chairman), Benevides, Gubanov, Noel, Vondrak, Xu

Observations of the Sun: At this time astrolabes are the best instrument for measuring the solar diameter, and it is expected that they will also give the best positions of the Sun. With help of stations as CERGA, San Fernando, Malatya (Turkey), Santiago, Sao Paulo, and Rio de Janeiro, we will have a good set of observations. In the future data from additional stations will be added, particularly from Eastern Europe and China.

Reference Frames: Astrolabes will be able to make a good contribution to linking the Dynamical System to the ICRS by referring observations of Solar System objects to Hipparcos. Astrolabes can also contribute to the internal connection between different dynamical systems by observing planets, minor planets and the Sun.

Future observations: As new and better methods are developed, continue for some time the observations with astrolabes in order to connect old and new observations. Discontinue after that as we have for the star positions applied to Earth rotation.

4.3. ASTRONOMICAL STANDARDS - PRESENTED BY MCCARTHY

Fukushima (Chairman). See WGAS Notice Nos. 17 and 18 for membership.

In 1994, the IAU approved the continuation of the WG on Astronomical Standards (IAU/WGAS) for one more triennium. T. Fukushima (NAO, Japan) continued to serve as chairman. The mission of this WG were
1) to maintain the 'two-tier' mechanism on astronomical constants and the numerical values of related quantities,
2) to investigate the issue on general relativistic definitions of astronomical units and constants, and

3) to implement the proposal to collect and authorize some basic software for fundamental astronomy, to be named as SOFA (Standards Of Fundamental Astronomy).

Three sub-groups were set up in the WG:
1) Maintenance Committee of Astronomical Constants chaired by D.D. McCarthy (USNO, USA),
2) General Relativity Sub-Group chaired by V.A. Brumberg (IAA, Russia), and
3) Review Board of SOFA chaired by P.T. Wallace (STARLINK, UK).

The maintenance committee prepared the IAU (1997) Best Estimates of Astronomical Quantities following the Chapter 4 of IERS Convention (1996) (McCarthy - IERS Tech. Note No.21). For the moment, the WG sees no necessity to update the IAU (1976) System of Astronomical Constants. It is still to be used for creating long-time standards such as nautical almanacs and star catalogs.

The general relativity sub-group has exchanged opinions through e-mail extensively. In the course of discussions, the issue of the transformation of GM values was highlighted. This is one of the by-products caused by the change of astronomical time arguments adopted by the IAU in 1991.

The SOFA Board of Review has actively discussed the necessary items to prepare computational codes needed for the basic calculations and to provide them electronically. The Announcement of Opportunity for institutions to bid to become the SOFA Center was published at the beginning of 1997, in IAU Bulletin 79 and elsewhere. In June 1997 the Board endorsed Rutherford Appleton Laboratory as the SOFA Center. The Center will be operated in parallel with the Starlink service and will employ the same techniques as Starlink's 'Software Store'. It is intended that the service provided by the SOFA Center will be monitored by a management board. This board will present reports and assessments at each triennial General Assembly of the IAU. The management board will initially be the WGAS.

4.4. REFERENCE FRAMES - PRESENTED BY MORRISON

Membership: L.V. Morrison (Chairman)

Members of Working Group on Reference Frames 1994-1997

Arias, F.	FCAG La Plata, Ar	Miyamoto, M.	NAO, Jp
Carter, W.E.	NOAA, USA	Morrison, L.V.	RGO, UK
Charlot, P.	Paris Obs, Fr	Nicholson, G.D.	HRAO, Za
Corbin, T.E.	USNO, USA	Nothnagel, A.	Bonn, De
Eubanks, T.M.	USNO, USA	Preston, R.A.	JPL, USA
Feissel, M.	Paris Obs, Fr	Réquième, J.	Obs de Bordeaux, Fr
Fey, Alan	USNO, USA	Reynolds, J.	Aust Tel Nat Fac, Au
Fukushima, T.	NAO, Jp	Shuhe, W.	Shanghai Obs, Cn
Gontier, A-M.	Paris Obs, Fr	Sovers, O.	JPL, USA
Jacobs, C.	JPL, USA	Standish, E.M.	JPL, USA
Johnston, K.J.	USNO, USA	de Vegt	Hamburg Obs, De
Kovalevsky, J.	Obs Côte d'Azur, Fr	Walter, H.H.	ARI, De
Kumkova, I.	Inst Appl Astr, Ru	White G.	Univ of Sydney, Au
Lestrade, J-F.	Obs de Paris, Fr	Williams, J.G.	JPL, USA
Li, Z.	Shanghai Obs, Cn	Yatski, Ya.S.	Kiev Obs, Ue
Ma, C.	GSFC, USA	Zacharias, N.	USNO, USA
McCarthy, D.D.	USNO, USA		

Key dates and decisions:

Aug 94: Terms of reference set by IAU94 at The Hague;

- main task - produce list of VLBI positions to replace the previous frame (FK5)
- appointment of members of WG

Feb 95: Meeting of WGRF at CERGA, Grasse

- VLBI sub-group formed to produce ICRF
  Chopo Ma (GSFC), Coordinator

Felicitas Arias (FCAG), Anne-Marie Gontier (IERS)
Chris Jacobs, Ojars Sovers (JPL)
Marshall Eubanks, Alan Fey (USNO)

- recommended IERS to set up a VLBI Coordinating Centre (US National Earth Orientation Service selected)

May 95: VLBI sub-group (and Chairman) met at Paris Observatory

- laid down timetable for forming ICRF

Nov 95: Hipparcos deadline for ICRF

- ICRF fixed and preliminary subset released to Hipparcos

Apr 96: Hipparcos fixed to ICRF

- mainly by radio stars (0.6 mas; 0.25 mas/yr)

Apr 96: Issue of report on the maintenance of the Hipparcos reference frame by Jean Kovalevsky

- methods include VLBI, VLBA, MERLIN radio astrometry of Hipparcos radio stars, photographic/CCD astrometry of ICRF sources, interferometry (NPOI), and future astrometric space missions

Oct 96: Meeting of WGRF at Paris Observatory in conjunction with IERS96 Workshop

- discussion of ICRF accuracy - influence of atmospheric models, source structure
- maintenance of ICRF
- recommended that IERS and new IAU WG should promote, monitor observations associated with maintaining and extending the reference frame
- JPL ephemerides of Solar System tied to ICRF with an accuracy of $\sim$ 1 mas for planets Mercury to Mars

Oct 96: Commission 8/24 review of current and planned work on the optical-radio link

- Publication of the review

Jul 97: Issue of IERS TN23 with positions of 608 ICRF sources;

- 212 definin sources with $\sigma \sim 0.5$ mas (mainly in N. hemisphere)
- 294 candidate sources
- 102 other sources useful for the radio-optical link

Jul 97: Issue of Hipparcos & Tycho Catalogues

- Hipparcos median accuracy $\sim$ 1 mas

Aug 97: Resolution to IAU97 (Kyoto);

- adoption of ICRF as the primary reference frame
- adoption of Hipparcos Catalogue as the realization of the ICRF at optical wavelengths

4.5. STAR LISTS - PRESENTED BY HELMER

Helmer (Chairman), Abalakin, Carrasco, Corbin, Jin, Klemola, Li, Luo, Miyamoto, Morrison, Polojentsev, Requieme, van Altena, Yatskiv.

Since the last GA in den Haag two large databases have been under consideration by the WGSL.

1. The extension of the fundamental reference frame to fainter magnitudes. Stars at a density of 1 per square degree have been selected mainly from the various zones of the AC, from the north pole down to a declination of -40 degrees in the magnitude interval 11.5 to 13.0. A total of 35 600 stars were selected and put on the observing list of the Carlsberg Automatic Meridian Circle (CAMC) at La Palma for a second epoch. The observations are virtually completed, and proper motions for all these stars have been computed using the AC position, transformed to the FK5 reference frame, as first epoch. The AC positions on FK5 were kindly provided by Siegfried Röser of the Astronomisches Rechen-Institut in Heidelberg. Due to the large difference in epoch, and the accurate second epoch observation, these stars have proper motion errors of typically 0.003 arcsec/year in each coordinate. The errors of a catalogue position range

from 0.07 - 0.14 arcsec, depending on zenith distance. At zenith distances greater than 45 degrees the errors in declination become slightly worse and reach 0.20 arcsec at 70 degrees. The observations were originally made relative to the FK5 system, but when the Hipparcos catalogue became available, this was used as the reference catalogue. All the positions and proper motions for these stars are found in the Carlsberg Meridian Catalogue No. 9, which is now available on CD-ROM from the consortium running the CAMC (Royal Greenwich Observatory, Instituto y Observatorio de la Armada en San Fernando and Copenhagen University Observatory).

2. Faint reference stars surrounding extragalactic radio sources. To aid in making links between the ICRS and the optical frame, a high density of accurate, faint, reference stars is needed in the vicinity of each radio source, in order to overcome the magnitude problem. Work on this has been carried out at several observatories for some time, but without real coordination, especially in selecting the reference stars. At the XXIInd GA an informal group was formed to collect in one database reference stars suitable for photographic or CCD observations of positions of optical counterparts of such sources. The database is maintained at the Kiev University Observatory, Ukraine, and consists of stars at magnitudes 12 - 14 in 1/2 degree fields around approximately 600 radio sources from the IAU list. The fields are being observed photographically in Kiev and Bucharest, and, starting in 1996, with the automatic CCD meridian circle of Nicolaiv Observatory (Tel'nyuk-Adamchuk, V., Pinigin, G., Ukraine, Stavinschy, M., Romania and Polojentsev, D., Russia).

It was suggested that the Working Group on Star Lists should be discontinued, and reformed under the new ICRS Working Group as a subgroup with L. Helmer as its chairman. The subgroup's main task would be to coordinate efforts to extend the optical reference frame to fainter magnitudes. This was approved by the members.

5. Individual Reports: joint session 26 August 1997

5.1. CMC - MORRISON

Carlsberg Meridian Catalogue La Palma Number 9 (CMC9) contains all the data obtained with the Carlsberg Meridian Telescope in the period May 1984 to March 1995. It comprises 141 593 positions, and magnitudes of 138 603 stars north of declination −40 deg, 117 559 proper motions, and 19 585 positions and magnitudes of 97 Solar System objects. It also contains the nightly value of the atmospheric extinction (in V) and all the meteorological data. These are available by ftp and will be issued on CD-ROM in Oct 97.

About 60 000 of the stars are common to the Hipparcos Catalogue and details of the systematic differences Hipparcos–CMC9 have been derived in position and proper motion. The systematic differences are in the range ± 60 mas in RA and −40 to +120 mas in Dec, and their general behaviour is similar to Hipparcos–FK5. The systematic differences in proper motion are in the range ± 3 mas/yr in RA and Dec. These systematic differences have been removed from CMC9, thus referring the positions and proper motions to the ICRF.

Slightly under half of the stars in CMC9 are fainter than V=10.5 and are generally not in the Hipparcos or Tycho catalogues. The progress of these observations from CMC9 through CMC10 (May 84–Dec 96, to be issued in Jan 98) is reviewed here under the headings of the various observational programmes. The figures quoted are the cumulative number of observations: CMC10 includes CMC9. The lower declination limit in all cases is −40 deg, unless otherwise specified.

- Global net of reference stars; 1 star per square deg, $10.5 < mag < 13.5$; target list = 35 000 stars; positions in CMC9 = 30 000
- Reference stars in fields of ICRF sources, $10.5 < mag < 13.5$; list = 11 400 stars, CMC9 = 1 300
- Grid of reference stars in Schmidt fields, 10h to 24h RA, $-3 < \delta < +11$; two lists: mag ~ 12.5 (12 000 stars) and ~ 13.5 (6000 stars); CMC10 = 16 000
- Reference stars for Veron/Cetty galaxies, $-35 < \delta < +30$, mag ~ 12; list = 4200; CMC10 = 3500
- Luyten (NLTT), $\mu > 0.3$/yr; list = 3000, CMC10 = 2000
- Luyten (NLTT), $\mu > 0.18$/yr; list = 7500, CMC10 = 1800
- Variable stars (GCVS), $12 < mag_{min} < 14$; list = 3500, CMC10 = 2500
- Third Catalogue of Nearby Stars (CN3); $12 < mag < 14$; list = 2600, CMC9 = 2600

- Astrographic Catalogue stars not in HST GSC, V>12, $-40 < \delta < +30$; list = 3000, CMC10 = 1 300
- Reference stars CCD search for occultations by Neptune/Pluto, V $\sim$ 13; list = 1500, CMC10 = 1200
- Planets/satellites/asteroids, in support of space missions; list = 7 objects, CMC9 $\sim$ 20 000.

Since April 97 the telescope has been operated remotely via the InterNet. The nightly observations are transferred daily to the home base where the observational catalogue is compiled.

The present scanning-slit micrometer is to be replaced by a CCD camera at the beginning of 1998. The main observational programme will be associated with the transfer of the Hipparcos Catalogue reference frame to fainter magnitudes, with particular emphasis on improving the accuracy of the astrometry that can be obtained from Schmidt survey plates. The programme will also include astrometry of the major planets in the outer Solar System and asteroids, in support of space missions and the determination of masses by mutual perturbations between the latter.

5.2. PROGRESS IN ASTROMETRY AT THE USNO - CORBIN

USNO home page: http://www.usno.navy.mil/

AC 2000 - Work has been completed at the USNO on the long-term project to make new reductions of the Astrographic Catalog zones and combine them on a common system. The ACRS was used to do the individual plate reductions. The plates of each zone were then combined and the magnitudes from the plates reduced to the Tycho B system. After the zones were combined, the positions were reduced to the Hipparcos system at the mean epochs of the AC plates. The result is a catalog of positions of 4.62 million stars at an average epoch of 1907 and with errors in the range of 0.13 to 0.32 arcsec (average of two plate positions). The AC 2000 has been combined with the Tycho catalog to produce improved proper motions of 988 758 Tycho stars. The errors of these new proper motions are in the range of 1 to 4 mas/year. The resulting catalog is named ACT (Astrographic Catalog/Tycho), and the results have been referred to the epoch J2000. Both catalogs will be made available via CD-ROM in one to two months. For more information, contact S. Urban at: seu@pyxis.usno.navy.mil.

TAC (Twin Astrograph Catalog) - The TAC, with an average epoch of 1981, is the result of a four-fold astrograph program, in two colors, that extends from declinations 90 deg to -20 deg. The density is about 25 percent greater than that of Tycho. The version of TAC on the FK5 system has been available on CD-ROM since early this year. The plates are currently being given new reductions using Hipparcos that will yield positions accurate to between 50 and 60 mas, and the catalog will be available in two to three months. This version will be combined with the AC 2000 to give proper motions comparable in quality to ACT.

NPOI (Navy Prototype Optical Interferometer) - Imaging observations started in November 1996 with a three baseline configuration. During this initial phase an average of 600 observations per month has been made. Sufficient observing of several close (under 10 mas) binaries has already been made to permit orbital solutions with errors in the 0.2 mas range. In addition, limbdarkened stellar angular diameters are now being regularly achieved with accuracies of 0.1 percent. The complete imaging array is scheduled to begin operations in 1999 with a four element array that will be capable of imaging binaries separated by as little as 0.1 to 0.2 mas.

Wide angle astrometry is due to begin in the next three to four months. It is expected that a catalog of about 1000 stars will be completed in late 1998 that will give a rigid frame at the 1 - 3 mas level and will be rotated into the ICRS system via observations of radio stars. This will provide the first rigorous test of the Hipparcos proper motion system.

WDS (Washington Double Star database) - The WDS 1996.0 has been finished and distributed. This contains 451 546 measures of 78 100 systems. It is available at the USNO web site. Since then 5 461 new measures and 1 109 new systems have been added. The 3 001 new systems resulting from the Hipparcos program are also being added. A new orbit catalog is in preparation, and this will be available in January 1998.

Observations of doubles and problem stars identified in Hipparcos are being actively pursued through speckle programs and now on the optical interferometer. Success with the problem stars has not been good, and it is believed at this point that most of these are doubles with large values of delta-m. These stars will probably prove to be a challenge to double star observing for some time to come.

FASTT (Flagstaff Astrometric Scanning Transit Telescope) - The program currently consists of differential observations (calibration fields for SDSS), wide angle observations (systematic distortions in FK5 and, more recently, observations for the linkage of the radio and optical frames), and minor planets. A limiting magnitude of 17.5 is regularly reached and 50 mas achieved for differential observations using Tycho positions as reference. Preparations are now underway to begin a much more extensive minor planet program on the new 1.3 m astrometric reflector when it goes into operation in 1998.

Hamburg/USNO Radio-Optical Link Program - This program helps establish the tie between the radio and optical frames through the observations of primary and secondary reference stars and deep frames to the optical counterparts of the ICRF sources. This was one of the contributors to the Hipparcos-ICRS tie, but only about 80 fields had been done at that time. About 400 fields have now been observed and reduced, and the link for each source will be 10 to 20 mas once the reference stars of the UCAC program (see Zacharias below) are available. Globally, this will provide a link at the 1 mas level for the continued maintenance of the optical frame.

PMM (Precision measuring Machine) - A series of very high density catalogs resulting from measuring Schmidt plates at the USNO Flagstaff Station:
1.) USNO A1.0 results from measuring: the first epoch plates of the Palomar Observatory Sky Survey (POSS-I), the UK Science Research Council survey (SRC-J) plates, and the European Southern Observatory survey (ESO-R) plates. This has yielded a catalog of 488 006 860 stars with positional accuracies of 250 mas for a single image but without proper motions. It is available on a ten CD-ROM set that has been send to libraries, data centers, etc.
2.) USNO SA1.0 is a subset of the USNO A1.0. It contains 55 000 000 evenly distributed over the sky and is available on CD-ROM (see USNO web site).
3.) USNO A-2 will result from new reductions of the USNO A1.0 using the ACT catalog. It will have the same number of stars as USNO A1.0, but the single image accuracy is expected to improve to 150 mas.

Currently the PMM is being used to measure the second epoch Palomar Schmidt plates (POSS-II) and those of the SERC survey. It also has been proposed to measure the Edinburgh AAO plates to provide second epochs south of -20 degrees. When the positions resulting from all of these data are combined with the USNO A1.0 a catalog of positions and proper motions of almost half a billion stars reaching to 21st magnitude will result.

5.3. MIRA - YOSHIZAWA

MIRA is the series name of ground-based optical/infrared interferometer arrays for astrometry and astrophysics at the 1 mas – 0.1 mas level of accuracy. At present we are operating at Mitaka campus of NAOJ a two-element interferometer MIRA-I for stellar fringe detection and tracking experiments.

The main characteristics of MIRA-I are as follows:
- Baseline is south-north 4m and siderostats of 30cmϕ.
- Beam transfer is inside evacuated pipes (transfer length >30 m; beam size of 30mmϕ).
- Train-type fine/fast delay lines carrying a retro-reflector and a Cassegrain telescope with a piezo-actuating flat mirror at its Cassegrain focus.
- Fringe detection with photon-counting APD's.

MIRA-II is a multi-element (four fixed and three moveable siderostats), long baseline (longer than 500 m at the maximum) optical/infrared (up to K-band) interferometer with laser metrology units for global astrometry to be constructed at Mitaka. Our final goal is a very large array MIRA-III, which should be constructed at one of the best sites in the world for astronomical observations.

Several of the most important astronomical targets that we will try to investigate with the MIRA's are:
1. Global astrometry and construction of an optical reference frame with an accuracy better than 10 mas to 1 mas level.
2. Galactic dynamics; determination of structure, kinematic properties, and dynamical evolution of our Galaxy, through a deep survey in the near-infrared.
3. Stellar astrophysics; observations of surface structure of various kinds of stars and the determination of their radii, pulsating variations, orbits and masses of binary stars, etc.
4. Stellar formation; structure and physical processes of proto-stellar disks and proto-planetary disks by high resolution imaging and spectroscopy.

5.4. USNO POLE TO POLE PROGRAM - RAFFERTY

Between the years 1985 and 1996 the U.S. Naval Observatory carried out a program of absolute observations covering the entire sky. Two transit circles were involved, one located in Washington, DC (USA) and the other in Blenheim, New Zealand. Over 672 000 nighttime observations were made, including over 318 000 observations of International Reference Stars (IRS), and over 12 000 observations of all the major planets (except Pluto) and nine minor planets. Over 55 000 observations were obtained of day-time objects including the Sun, Mercury, Venus, Mars, and bright stars. The objective was to form a self-consistent, all-sky catalog of absolute positions which could be linked by rigid rotations to the dynamical system. However, with the success of Hipparcos, the objective has been modified in that the interest has now shifted from the positions of the stars to the positions of the planets. Instead of rotating the stellar instrumental frame into the dynamical frame as defined by the planetary observations, the planets will be rotated into the Hipparcos frame by means of the stars.

The reductions of the Pole-to-Pole observations will differ from some of the methods used in past USNO catalogs. Enough daytime observations of stars were made with the transit circle in New Zealand that daytime determinations of the clock correction will be possible. A daytime correction to the constant of refraction should also be possible for each transit circle. Using the daytime and nighttime observations of Mars should be an important step in testing the night-minus-day corrections (see Rafferty and Loader in A&A **271**). The traditional determinations of the instrumental flexure, using horizontal collimators, and of the correction to the constant of refraction, from circumpolar observations, were found to be unreliable for each instrument. New methods for determining the flexure and the correction to the constant of refraction have been developed utilizing star observations over the entire sky. Only the correction to the assumed latitude will be determined from the circumpolar observations. The traditional equator correction determined from solar system observations has been dropped. Instead the instrumental system, as defined by the star observations only, will be considered rigid and the solar system observations will be used to rotate it to the dynamical frame if desired, as discussed by Holdenried and Rafferty in A&A (in press). Or, as mentioned earlier, the planets, which are on the stellar instrumental frame, can be rotated into the Hipparcos frame.

5.5. REPORT OF THE CENTRAL BUREAU OF IERS

5.5.1. *INTRODUCTION*

The activities of the Celestial System Section of the IERS Central Bureau are performed at Paris Observatory, under the scientific responsibility of E. F. Arias (La Plata Observatory). During 1993-1997 they were mostly dedicated to the construction of the new International Celestial Reference Frame (ICRF), in close relationship with the IAU Working Group on Reference Frames (WGRF), to which four members of the IERS Central Bureau belong. The ICRF was adopted by IAU at its 23rd General Assembly (see Recommendation B2). The Hipparcos stellar reference frame was astrometrically aligned with ICRF to within $\pm 0\overset{''}{.}0006$ at the epoch 1991.25 and to within ± 0.00025 ″/year in rotation. This is discussed by Kovalevsky *et al.* in A&A **323**, and it provides the primary realization of ICRS in optical wavelengths.

5.5.2. *THE INTERNATIONAL CELESTIAL REFERENCE SYSTEM (ICRS) [Arias et al. 1995]*

Equator/Pole of ICRS: Using VLBI data, one can estimate that the pole at J2000.0 is shifted from the ICRS pole by 17.2 mas in the direction 12h and by 5.1 mas in the direction 18h, with uncertainties smaller than 1 mas. The ICRS celestial pole is consistent with that of FK5 within the uncertainty of the latter (± 50 mas). See Arias *et al.* in A&A **303**.

Origin of right ascensions of ICRS: The origin of right ascensions of ICRS was defined by fixing the right ascension of 3C273B to its FK5 value. Using VLBI and LLR data, Folkner *et al.* in A&A **287** estimated that the mean equinox at J2000.0 is offset from the origin of right ascensions of ICRS by 78 ±10 mas. The origin of right ascensions of ICRS is consistent with that of FK5 within the uncertainty of the latter (± 80 mas).

5.5.3. *THE INTERNATIONAL CELESTIAL REFERENCE FRAME (ICRF)*

A complete description of the ICRF, labelled RSC(WGRF) 95 R 01 in the IERS terminology, is given in the IERS Technical Note No 23 by Ma and Feissel, and is available on request to iers@obspm.fr. The frame is realized by the J2000.0 VLBI coordinates of 608 extragalactic radio sources evenly distributed on the sky. Objects in RSC(WGRF) 95 R 01 are divided into three categories: defining, candidate and other sources. The 212 defining sources proved to be high-quality objects over the total the period of observations; they served to set the orientation of axes in coincidence with those of ICRS. Most of the 294 candidate sources are compact sources with still too few observations; they are likely to climb up to the 'defining' category when more observations are made. Finally, 102 other objects were included in the frame either for the sake of densification or just because they contributed to the link of other frames to ICRF. The complete frame, as well as separate lists with the three categories of sources, are available via *anonymous ftp* from *hpiers.obspm.fr* (145.238.100.28), directory *iers/icrf/iau/icrf_src* (files *icrf.rsc, icrf.def, icrf.can, icrf.oth*).

IERS/CB also maintains a compilation file that gives physical information on about 2000 extragalactic radio sources, including the type of object, redshift, spectral index, visual magnitude. This information is collected from astrometric and astrophysical surveys. The file was updated with recent information (ftp file *iers/icrf/iau/icrf.car*). A complete dictionary gives, for each ICRF source, its designation as well as all the aliases currently used (ftp file *iers/icrf/iau/icrf.dico*). All these files can also be accessed through the World Wide Web site *http://hpiers.obspm.fr*

5.5.4. *Maintenance of ICRF*

The non-rotating character of the frame results from the assumption that extragalactic radio sources have no detectable proper motions; nevertheless, some objects showed apparent proper motions due to structure effects at mas-scale: radio sources may undergo unpredictable changes in their structure which can affect their positions. Regular checks of defining and candidate sources positional behaviour are necessary to assess the stability of their direction. As VLBI observations and modeling progress, more accurate radio source positions will be determined, and new sources will be incorporated to densify the frame.

In the course of the 1996 IERS annual analyses, extragalactic celestial frames were submitted by seven VLBI Analysis Centres. The source coordinates in these solutions were compared to those in ICRF in order to check possible significant changes in coordinates that could be caused by observations accumulated since the completion of ICRF or by evolution in the analysis procedures. Eight objects among the defining sources were found to have inconsistencies in some individual reference frames. These results on the defining sources should be interpreted as very preliminary, since it will be necessary a more extended period of observation and a confrontation with other analysis to give some conclusion. The same analysis will be performed on candidate sources in the next years. New sources were observed since the implementation of ICRF.

A total of 28 extragalactic radio sources not included in RSC(WGRF) 95 R 01 were provided in the new solutions. Preliminary ICRS positions were computed.

5.6. GSC 1.2 AND STARNET (4 MILLION PROPER MOTIONS) - RÖSER

GSC 1.0 has been reduced by Lasker using AGK3 in the northern, SAOC in the southern hemisphere. Röser *et al* (see IAU Symposium 179) have made a completely new reduction of GSC onto the system of PPM. The plate-dependent distortions have been reduced using a numerical filter as described by Röser *et al.* in ASP Conference Series, Vol. 84. After removal of the plate-dependent distortions, the remaining distortions were shown by Morrison in AJ **111** to be magnitude dependent. These distortions were removed using the Astrographic Catalog, which, however, has an epoch difference of about 80 years with respect to GSC. At the first glance this seems impossible or wrong, but it could be shown that there was a radial dependence of the magnitude effect from the GSC plate centres. Piling up the GSC plates on top of each other, i.e. plotting the coordinate differences GSC - AC as functions of the GSC plate coordinates, the effects of individual proper motions, solar apex motion and galactic rotation cancel out, and the systematic errors of the GSC plates remain. This new version is called GSC 1.2. The details of the catalog and the methods used are described the proceedings of IAU Symposium No. 179 (in press).

Using GSC 1.2 and the Astrographic Catalog, a new catalog of 4.3 million stars with positions and proper motions was constructed. This catalog, STARNET, has an average star density of 100 stars per square degree, a median magnitude of $B = 12.0^m$ on the southern hemisphere and $V = 11.5^m$ on the northern hemisphere. The present-day rms-accuracy of the positions is 0.3 arcsec, that of the proper motions 5 mas/year. See Röser in IAU Symposium No. 172.

5.7. THE UCAC-S PROJECT - ZACHARIAS

The U.S. Naval Observatory CCD Astrograph Catalog South (UCAC-S) project is scheduled to start in January 1998 at Cerro Tololo (CTIO), Chile, to cover the entire Southern Hemisphere in a 2-fold overlap down to 16th magnitude. The program should be completed with 2 years of observing. The goal is to produce a high density, high accuracy, astrometric catalog in the Southern Hemisphere.

A 4k by 4k CCD camera will be used in combination with the 5-element, 0.2-meter aperture lens in a 579-642nm bandpass covering a field of 61 arcminutes square at $0.''9$/pixel. The Kodak CCD chip is of high cosmetic quality with no dead pixels in the light sensitive area. Mounted in parallel to the red-corrected lens, is a yellow-corrected lens of the same aperture, which is used for guiding with an ST-4 autoguider sitting on a moveable x,y-slide. Focusing is performed with a Hartmann screen. The telescope is fully automated, and all components are under computer control.

Each field will be taken with a long ($\approx$ 120 sec) and a short ($\approx$ 30 sec) exposure to extend the dynamic range and to check for magnitude dependent systematic errors. Additional calibration observations will be made in selected fields with the telescope and camera being in various orientations with respect to the sky.

Full astrometric reductions will be run within 24 hours of the data acquisition. Tycho stars (from the ACT catalog) will be used for preliminary reductions, and a direct tie to the Hipparcos stars is feasible with block adjustment techniques. A catalog accuracy of 20 mas per coordinate is expected for stars in the 7 to 14 magnitude range. Additional longer exposures ($\approx$ 300 sec) are planned around extragalactic reference frame sources. The optical counterparts of these radio sources will be observed with larger telescopes quasi simultaneously. Thus a strong tie to the ICRF is expected, allowing for a re-determination of the radio-optical reference frame link at an epoch of about 1999. Nights with minimal sky conditions will be used to observe bright minor planets (9^m - 14^m) on the 30 mas level of accuracy.

The raw pixel data will be compressed (lossless) and saved on tapes and CD-ROMs. Over 3 Gbytes of data are expected from a single night. After 2 years, an archive of about 100 000 frames (1.7 Tbytes) will have been collected. The project will start at the South Celestial Pole and epoch differences to overlapping fields will be kept small.

This project will be a major contribution for a densification of the reference frame, with a density exceeding that of the Guide Star Catalog and an accuracy similar to the Tycho Catalogue. The UCAC-S will have numerous applications: Schmidt survey astrometric reductions, calibration of field distortions in other telescopes, accurate positioning of fiber optics, input catalog for future space missions, reference stars for minor planet observations, high accuracy proper motion determinations.

COMMISSION 19: ROTATION OF THE EARTH
ROTATION DE LA TERRE

PRESIDENT: Jan Vondrák
VICE-PRESIDENT: Dennis D. McCarthy
ORGANIZING COMMITTEE:
F. Arias, N. Capitaine, V. Dehant, J. O. Dickey, S. Dickman, M. Feissel, Y. Fumin, B. Kołaczek, S. Manabe, W. Melbourne, L. Morrison, D. Robertson, L. Rykhlova, V. Tarady, Ch. Veillet, P. Wilson

1. Introduction

Because of the large number of working groups and other cooperating bodies under Commission 19 we held two business sessions plus two Working groups sessions at the 23rd IAU General Assembly.

2. Business Meetings (Saturday, August 23, 1997)

The meetings were chaired by the President of the Commission, Jan Vondrák. The following issues were discussed:

2.1. REPORT ON THE ACTIVITY OF COMMISSION 19 SINCE THE 22ND IAU GA (J. Vondrák reported)

It was noted that the detailed report on scientific activities of Commission 19 for the years 1993.5–1996.5 had been published in I. Appenzeller (ed.) *Reports on Astronomy*, Vol. XXIIIA, IAU 1997, 81–96. The reprints were sent to all members of the Organizing Committee of the Commission plus all contributors to the Report. The following persons (in the order as they appear in the text) sent their contributions, often covering a broader region than their own institutions:

M. Feissel (France),
D. D. McCarthy (USA),
J. O. Dickey (USA),
Jin Wenjing & Yang Fumin (China),
S. Manabe (Japan),
V. Dehant & P. Pâquet (Belgium),
L. Rykhlova (Russia),
Ya. Yatskiv & A. Korsuń (Ukraine),
N. Capitaine (France),
M. Stavinschi (Rumania),
B. Kołaczek & A. Brzeziński (Poland),
M. Meinig (Germany),
H. Jochmann (Germany),
L. Morrison (UK),
S. Dickman (USA),
I. Pešek & J. Vondrák (Czech Republic).

The President thanked all of them for their input and commented briefly on the discussion of these reports that took place at the Commission Presidents' meeting with the Executive Committee. He expressed his opinion that these reports are useful, and are the important part of the work done by Commission membership and should be continued.

In the past three years the Commission co-operated closely with the following bodies:

a) IAU/IUGG International Earth Rotation Service (IERS) (chaired by Ch. Reigber, Germany),

J. Andersen (ed.), Transactions of the International Astronomical Union Volume XXIIIB, 101–105.

b) IAU WG on Astronomical Standards, supported by Commissions 4, 5, 8, 19, 24 and 31 (chaired by T. Fukushima, Japan),
c) IAU WG on Reference Frames, supported by Commissions 4, 8, 19, 24 and 31 (chaired by L. Morrison, UK),
d) IAU/IUGG WG on the Non-rigid Earth Nutation Theory, supported by Commissions 4, 7 and 19 (chaired by V. Dehant, Belgium),
e) IAU Commission 19 WG on Earth Rotation in the Hipparcos Reference Frame (chaired by J. Vondrák, Czech Republic),
f) IAG International GPS Service for Geodynamics (chaired by G. Beutler, Switzerland),
g) IAG SSG 5.173 Interaction of the Atmosphere and Oceans with the Earth's Rotational Dynamics (chaired by C. Wilson, USA).

All of these bodies were represented at the IAU GA at Kyoto, either by taking part in Joint Discussions 3 and 7, or by presenting their reports at Commission 19 Working Group Sessions.

Commission 19 had also formally supported or co-supported a number of international scientific meetings:

- IAU Colloquium No. 165 *Dynamics and Astrometry of Natural and Artificial Celestial Bodies,* July 1–5, 1996, Poznan, Poland,
- Joint Discussion 3 *Precession-nutation and Astronomical Constants for the Dawn of the 21st Century,* August 21, 1997, Kyoto, Japan,
- Joint Discussion 7 *The New International celestial Reference Frame,* August 22, 1997, Kyoto, Japan,
- Joint Discussion 12 *Electronic Publishing: Now and the Future,* August 25, 1997, Kyoto, Japan,
- Joint Discussion 14 *The First Results of HIPPARCOS and TYCHO,* August 25, 1997, Kyoto, Japan.

2.2. MEMBERSHIP MATTERS
(J. Vondrák reported)

According to the President's knowledge, two members of the Commission passed away since the last General Assembly: S. K. Runcorn (UK) in December 1995 and W. Jakś (Poland) in January 1997. The attendees observed a minute of silence in their memory.

According to the statutes of the IAU, the President accepted the following IAU members to Commission 19, during his three-year presidency:

J. Souchay, France,
P. Shelus, USA,
A. Cazenave, France,
G. Petit, France,
M. Bougeard, France.

The total number of members is 138, out of which 87 are accessible by e-mail; the number of consultants was 16, 10 of which are accessible by e-mail. There were also 'lost' members whose correspondence was returned. These included Tsao Mo, Taipei, and Han Tianqi, Wuchang, China. The President also received a list of newly proposed IAU members who expressed their wish to become members of Commission 19 and announced their names:

P. Defraigne, Belgium
A. Gozhy, Ukraine
Zhen-Nian Gu, China
Yan-ben Han, China
A. Korsuń, Ukraine
W. Kosek, Poland
Jinling Li, China
Qi Li, China
De-chun Liao, China
Chopo Ma, USA
Z. Malkin, Russia

C. Ron, Czech Republic
Jing Tian, China
Kemin Wang, China
Zhifang Zeng, China
Min Zhong, China.

From their records all appeared to be active in the field, with the exception of Qi Li who seemed to be more involved in positional astronomy. Consequently, the Vice-President consulted with the President of Commission 8 and considering that only one Commission membership is allowed for the first three-year period, Qi Li was accepted in Commission 8.

2.3. ANNOUNCEMENT OF THE NEW OFFICERS OF THE COMMISSION
(J. Vondrák reported)

Elections of new officers was organized before the General Assembly, in June 1997 in order to permit all members of the Commission to vote. The members leaving the Organizing Committee (OC), due to their second term expiring, are:

S. Dickman, USA,
W. Melbourne, USA,
D. Robertson, USA,
Ch. Veillet, France - resigned due to his new responsibilities,
P. Wilson, Germany.

The candidates for the vacancies were proposed by the existing OC members and ballots with their names (two candidates for vice-president, ten candidates for the new OC members) were sent to all Commission members for their votes. 66 ballots were returned by June 25, 1997, with the following results:

D. D. McCarthy (USA), President,
N. Capitaine (France), Vice-President,
A. Brzeziński (Poland), new OC member,
Ya. Yatskiv (Ukraine), new OC member,
G. Beutler (Switzerland), new OC member,
P. Brosche (Germany), new OC member,
R. Gross (USA), new OC member.

These are added to those who continue their OC membership:

E. F. Arias, Argentina,
V. Dehant, Belgium,
S. Manabe, Japan,
L. Morrison, UK,
L. Rykhlova, Russia,
V. Tarady, Ukraine,
Yang Fumin, China

and the three ex-officio members (the first two as representatives of the collaborating bodies):

M. Feissel (France), IERS,
C. Wilson (USA), IAG SSG 5.173,
J. Vondrák (Czech Republic), past President.

The OC members also proposed the following Consultants for the next three-year period whose names were submitted to the IAU General Secretary:

B. Chao, USA,
R. Eanes, USA,
T. Herring, USA,
H. Jochmann, Germany,
J. Kouba, Canada,
P. Mathews, India,
J. Nastula, Poland,
A. Nothnagel, Germany,

K. Nurutdinov, Ukraine,
Ch. Reigber, Germany,
B. Richter, Germany,
S. Rudenko, Ukraine,
D. Salstein, USA,
H. Schuh, Germany,
C. Wilson, USA,
V. Zharov, Russia.

2.4. SCIENTIFIC DISCUSSION ON FUTURE PLANS FOR STUDYING EARTH ROTATION

The following contributions were presented:
B. Kołaczek: *On future plans of deeper investigations of El Niño influences on Earth rotation,*
R. O. Vicente: *On terrestrial and celestial reference frames,*
Wu Hong-qiu, Yu Nanhua, Zheng Dawai: *New contribution of AAM data.*

In the discussion that followed it was decided that Commission 19 would not propose any resolution of type C since it was felt that all problems specific to this commission were covered by type B resolutions already endorsed by the first session of the GA. It was stressed that, for the sake of the future of Lunar Laser Ranging, it would be necessary to better present its importance to other members of the astronomical community. The importance of satellite methods for studying Earth orientation parameters was also discussed, as well as the necessity of studying the El Niño effects in exciting Earth orientation.

3. Working Group meetings (Tuesday, August 26, 1997)

These two extra meetings (chaired by J. Vondrák and D.D. McCarthy) were devoted mostly to the reports and information of the bodies co-operating with Commission 19. The following reports were presented:

3.1. FINAL REPORT ON THE ACTIVITY OF THE WG ON EARTH ROTATION IN THE HIPPARCOS REFERENCE FRAME
(presented by its chairman, J. Vondrák)

It was stated that the WG (that was set up by IAU Commission 19 in 1988) proposed the algorithms of the solution at the IAU GA in 1991. Its activity was then prolonged, and it was prolonged again at the 22nd IAU GA in 1994. The data from 48 instruments at 31 observatories were collected since then and re-calculated into the most recent astronomical constants and standards. In the absence of the final Hipparcos Catalogue, 11 intermediary solutions were made, first with local star catalogues, later with preliminary (H30, H37, H37C) catalogues. Very recently, when the final Hipparcos Catalogue became available, two solutions were made and the first results presented at the Venice Hipparcos symposium (May 1997) and at JD3, JD7 and JD14 of the 23rd IAU GA (as posters). The data, procedures used, and results will be described in detail in a special volume of the Publications of the Astronomical Institute of the Academy of Sciences of the Czech Republic that will be widely distributed. The chairman then proposed to close the activity of this WG since it successfully fulfilled its task.

3.2. REPORT ON THE ACTIVITY OF THE INTERNATIONAL GPS SERVICE FOR GEODYNAMICS
(presented by its chairman, G. Beutler)

G. Beutler described the recent activities of the IGS that closely cooperates with the International Earth Rotation Service. The service uses the observations of the GPS satellites to monitor the motions of observing sites on the Earth, but it also determines the orbits of the observed satellites and some of the Earth orientation parameters (polar motion, length-of-day changes on a short time scale and, most recently, also the first time derivatives of celestial pole offsets).

3.3. REPORT ON THE ACTIVITY OF THE INTERNATIONAL EARTH ROTATION SERVICE
(reported by the director of its Central Bureau, M. Feissel, in collaboration with D. Gambis and by the director of the Sub-bureau for Rapid Service, D. McCarthy)

M. Feissel gave the global overview of the service and its activities during the last three years. D. Gambis then discussed the results in more detail, describing the input of individual techniques (VLBI, LLR, SLR, GPS, DORIS). The most recent technique, DORIS, is active since 1995 and now achieves the accuracy of 2mas. The precision and accuracy of the combined solution further improved to the level of 0.2 mas, with better and quicker availability of the results. Plans for the future include improvement of the combination methods, studying geophysical effects (atmosphere, oceans, groundwater) and clarification of polar motion and nutation in subdiurnal periods. D. McCarthy spoke about rapid service that now makes use of electronic distribution (twice-weekly) and uses also IGS products. The dominant contribution in polar motion is GPS, and for UT1 it is VLBI. Predictions for one year in advance are made with a precision of 0.02" in polar motion and 0.06s in UT1.

3.4. DISCUSSION ON THE NEW WORKING GROUPS
(reported by Division 1 President, J. Kovalevsky)

J. Kovalevsky outlined the results of the recent meeting of presidents and vice-presidents of Division 1. Following the type B resolutions endorsed at the first session of the GA, several working groups are formed:

a) *Committee on General Relativity in the Framework of Space-time References and Metrology.* This body is common to IAU, BIPM and IAG and it should be concerned with definitions, conventions and notations at a sufficient level of accuracy.
b) *IAU WG on Relativity in Celestial Mechanics and Astrometry.* This WG should address relativity issues relating to celestial mechanics and astrometry in general (application, concepts, algorithms and constants).
c) *IAU WG on the International Celestial Reference System.* This WG should interact with the IERS in the maintenance and extension of the ICRS, and formulate the new definitions of the celestial reference pole, sidereal time and the transformation between the celestial and terrestrial reference frame.

In addition to these, three more WG's supported by Division 1 will continue their activity; IAU/IUGG WG on Non-rigid Earth Nutation, WG on Astronomical Standards and WG on Cartographic Coordinates and Rotational Elements of Planets and Satellites.

3.5. REPORT ON THE ACTIVITY OF THE IAG SPECIAL STUDY GROUP 5.173 AND IAU WG ON NUTATION
(reported by the chairwoman of the WG, V. Dehant)

The IAG SSG created a WWW page that contains member reports and results. They organized a number of special sessions at scientific meetings (1995 Fall AGU, 1996 EGS, 1996 Spring AGU, 1996 Fall AGU, 1997 EGS, 1997 Spring AGU, 1997 IAPSO meetings). The SSG was also active during the IERS Workshop 1996 to improve studies of the interaction of the oceans and atmosphere with the rotating Earth and made new initiatives in gravity and geocenter studies.

The IAU 1980 standard nutation theory differs from the observations by about 10 mas indicating that a new Earth's transfer function is required which takes into consideration a realistic seismic model of the rheological properties of the Earth. A hydrostatic equilibrium model does not fit with the observed Earth orientation parameters of the free-core nutation. The rigid Earth nutation which serves as the basis for any non-rigid model, also needs improvement.

COMMISSION 31 TIME (HEURE)

Minutes of the Commission 31 Business Meetings held at the XXIIIrd IAU General Assembly in Kyoto, Japan

HENRY FLIEGEL (Commission President)

PRESIDENT: Toshio Fukushima
VICE-PRESIDENT: Gerard Petit

Commission 31 held business meetings during three sessions on Tuesday, 26 August 1997. The first two sessions were devoted to reports and summaries organized by G. Petit (Chairman of the Commission 31 Working Group on Pulsars and Timing) and by B. Guinot (ex Director of the former BIH). In the third session, reports were received from agencies to which the IAU sends Representatives, and Commission members discussed our relations with those agencies, and the future role of Commission 31 Working Groups, and their relationship with Division 1 Working Groups.

Also in the third session, the following list of officers and members of the Organizing Committee, which had been nominated and elected by votes solicited by all the members by correspondence in January-April 1997, was approved without dissent.

for President: Toshio Fukushima (Japan)

for Vice-President: Gerard Petit (France)

as members of the Organizing Committee:

Donald Backer (USA)
Gerhard Beutler (Switzerland)
Victor Brumberg (Russia)
Henry Fliegel (USA)
Sigfrido Leschiutta (Italy)
John Luck (Australia)
Paul Paquet (Belgium)
Edoardo Proverbio (Italy)
Qi Guan Rong (China)
Claudine Thomas (France)
Christian Veillet (France)
Gernot Winkler (USA)
Zhai Zao-Cheng (China)

In addition, William Klepczynski (USA) was named IAU Representative to the International Telecommunication Union Radiocommunication Bureau (ITU-BR) and to its Study Group 7; and Paul Paquet (Belgium) to the Federation of Astronomical and Geophysical Data Analysis Services (FAGS). These names had been reported and accepted by the IAU Executive during the meeting of Commission Presidents, on 25 August. There was some discussion concerning nominations for the IAU Representative to the Consultative Committee for the Definition of the Second (CCDS). Jean Kovalevsky, as President of Division 1, suggested that the IAU Representative should not be a member of the Bureau International des Poids et Mesures (BIPM) or of the CIPM or CCDS. Members voted to propose Henry Fliegel (USA)

J. Andersen (ed.), Transactions of the International Astronomical Union Volume XXIIIB, 107–109.

as the IAU Representative to the CCDS/BIPM, succeeding Gernot Winkler, pending the expected invitation of such a Representative from the CCDS to the IAU, and subject to consideration by the new Commission 31 President, Toshio Fukushima.

The following are the reports given in support of the Representatives and of the Working Groups of the Commission.

1. Session 1

This session, on "Implications of Pulsars and Dynamical Time Scales" was organized by Gerard Petit.

E. Myles Standish spoke on "Relations Between the Pulsar Timing Data and the Planetary Ephemerides". Especially noted was the difficulty in separating long-term changes in pulsar periods from systematics in the Earth's ephemeris, and the importance of this issue to the problem of constructing a truly independent and dynamical time scale from pulsar measurements.

D. N. Matsakis, J. H. Taylor, and T. M. Eubanks authored a paper on "Pulsar Time Scales and Sigma - z".

G. Petit presented a summary of "Pulsars and Atomic Time Scales".

Short summaries were provided in absentia of two papers: "An Astronomical Time Scale Based on the Orbital Motion of Pulsars in a Binary System" by Yu P. Iyasov, S. N. Kopeikin, and A.E. Rodin; and "Binary Pulsars as Detectors of Ultralow Frequency Gravitational Waves" by S. M. Kopeikin.

In the discussion of these papers, questions were raised whether estimates of a possible secular drift of pulsar-based dynamical time with respect to TAI might be corrupted by uncertainties in the planetary perturbations, and whether the distribution of pulsar observations was optimal to decide such questions. It was agreed that possible timing applications have low visibility to the radio astronomical community, and that members of the Working Group might profitably examine and recompile existing data to provide a basis for further work in timing.

2. Session 2

This session, on "Time, Metrology, and Physics" was organized by Bernard Guinot.

F. Riehle of the Physikalisch - Technische Bundesanstalt [PTB], Braunschweig, Germany, opened by reviewing the definitions of time standards in the framework of the International System of Units (SI). Of the seven base units of the International System, the unit of time (the second) can be realized with the lowest uncertainty. Consequently, if the definitions of other base units can be based on the second and on known physical laws, the uncertainties of the values of those base units can be much reduced.

Michito Imae surveyed the "Research Work in Time and Frequency at the Communications Research Laboratory [CRL], Tokyo, Japan". The CRL has responsibility for time and frequency standardization in Japan. CRL is developing frequency standards (optically pumped cesium, cesium fountain, H-maser, and cesium ensemble), and time transfer techniques, and disseminates time via JJY (HF), JG2AS (LF), telephone JJY dial-up service, and NTP = Network Time Protocol. Other time-related research includes development of an Acoustic Optic Spectrometer (AOS), exploration of general relativity, GPS and space-borne techniques, and observations using the 34m radiotelescope at Kashima Space Research Center.

William Klepczynski reported the work of the CIPM Working Group on Two-Way Satellite Time Transfer (TWSTT). The major issues with which the WG is concerned were reported to be (a) exchange of data between participating stations, (b) length of the observing sessions, (c) frequency of data collection, (d) the precision of the technique, (e) the accuracy of the technique, and (f) its long term stability.

Bernard Guinot concluded the Session by summarizing the Report of the CCDS Working Group on Relativity and Metrology, recently published in Metrologia, 34, 261-290 (1997). It concludes that the base units of physics must be proper quantities provided by local standards, established directly by local experiments, and that the present definitions of the SI base units are compatible with this requirement.

3. Session 3

The following reports were presented on behalf of IAU Representatives.

The chairman presented charts summarizing the work of the Federation of Astronomical and Geophysical Data Analysis Services (FAGS), which had been prepared by Paul Paquet. Since 1997, FAGS supports 11 services, of which 6 are directly related to IAU activities: (1) International Earth Rotation Service (IERS); (2) Quarterly Bulletin on Solar Activity (QBSA); (3) International Space Environment Service (ISES); (4) Centre de Donnes Stellaires (CDS); (5) Sunspot Index Data Center (SIDC); and (6) International Geodynamic Service (IGS).

William Klepczynski reported on the work of the International Telecommunication Union,with special focus on its Study Group 7. The ITU combines the previous International Consultative Committee for Radio (CCIR) and the International Consultative Committee for Telegraph and Telephone (CCITT), to coordinate all telecommunications, radio, and wire (fiber optics). Its Study Group 7 (continuing the former CCIR Study Group 7) comprises four Working Parties: 7A, Time and Frequency; 7B, Space Applications; 7C, Earth Observations; and 7D, Radio Astronomy. Of these, Working Party 7A, chaired by Gerrit de Jong, is of primary interest to Commission 31; it provides Reports and Recommendations for Standard Frequency and Time Broadcasting Stations.

The chairman asked Jean Kovalevsky, the President of Division 1, to describe the charters of the Division Working Groups [WG's]. The Division supports five WG's: (1) The joint IAU- BIPM - IAG Committee on General Relativity in the Framework of Spacetime Relations and Metrology; (2) the IAU WG on General Relativity in Celestial Mechanics and Astrometry; (3) the WG on the Maintenance of the International Celestial Reference Frame [ICRS]; (4) the IAU WG on Non-Rigid Earth Nutation Theory; and (5) the IAU WG on Astronomical Standards.

The members were then asked to comment on proposals for Commission 31 WG's. It was voted to continue the WG on Pulsar Timing, under its present chairman, Gerard Petit. Members were asked to comment on a proposal to continue the WG on Time Transfer, with an extended charter "(a) to advise CCDS on the applicability of laser ranging techniques to the maintenance of TAI; (b) in collaboration with the CSTG SLR/LLR Subcomission, to define standards and formats for data exchange; (c) to recommend appropriate calibration and verification procedures". The members agreed with the Session chairman that we must not duplicate the work of the CCDS groups, and we cannot separate laser from radio techniques (VLBI, GPS), since often a combination of techniques is necessary to transfer time on a continuing basis - e.g., when using LASSO. Commission members recommended that the decisions whether to continue the WG on Time Transfer, and if so with what chairman, members, and charter, should be left to the discretion of the incoming Commission President, Toshio Fukushima.

DIVISION II

SUN AND HELIOSPHERE

Division II provides a forum for astronomers studying a wide range of phenomena related to the structure, radiation and activity of the Sun, and its interaction with the Earth and the rest of the solar system.

PRESIDENT:

Peter V. Foukal
CRI, Inc.,
80 Ashford St., Boston MA 02134, USA
Tel. +1 617 787 5700
Fax. +1 617 787 4488
E-mail: pfoukal@world.std.com

BOARD

Ai Guoxiang	President Commission 10
A. Benz	Vice-President Commission 10
S. Solanki	Vice-President Commission 12
F. Verheest	President Commission 49
M. Vandas	Vice-President Commission 49
O. Engvold	Past President Division II

PARTICIPATING COMMISSIONS

COMMISSION 10 **SOLAR ACTIVITY**

COMMISSION 12 **SOLAR RADIATION AND STRUCTURE**

COMMISSION 49 **THE INTERPLANETARY PLASMA AND THE HELIOSPHERE**

DIVISION II: THE SUN AND THE HELIOSPHERE

Commission 10: Solar Activity
Commission 12: Solar radiation and structure
Commission 49: The interplanetary plasma and the heliosphere

President Division II:	O. Engvold		
President (10):	O. Engvold	Vice-President:	Ai Guoxiang
President (12):	F.-L. Deubner	Vice-President:	Peter V. Foukal
President (49):	H. Ripken	Vice-President:	F. Verheest

This report was prepared by the three commissions belonging to DIVISION II and it describes their activities during the 23rd General Assembly of the IAU.

The IAU Division II meeting held on August 23, 1997 in Kyoto included presentations from the Working Groups on Eclipses, on the Quarterly Solar Activity Bulletins, on the Sunspot Index Data Center, and on a compilation of solar activity. The IAU's newly adopted divisional structure, its impact on the Commissions, and on structure of the IAU were presented. Discussion followed on ideas for increasing participation in the divisional IAU business meetings. J. Pecker urged that the new divisional structure not be allowed to split up the IAU, as he felt has happened at the IUGG.

At the second meeting of the Division II, were also discussed plans for a strong solar and heliospheric program at the next General Assembly in Manchester, U.K.

1. The reorganization of IAU Commissions

The joint business meetings reflect the consolidation of our three Commissions into IAU Division II "The Sun and the Heliosphere", which was implemented at the 22nd General Assembly of the IAU in The Hague, The Netherlands, August 1994. The increased interdisciplinary contacts and cooperations between the Commissions that have taken place under this new structure, have been most helpful. The three Commission Presidents expressed the opinion that the new structure is mutually beneficial and may serve to strengthen and further the science of our fields. It was expected that its detailed form may be improved and shaped in the next triennia ahead.

It was suggested that the name of the Division should be brought into proper English and the Executive Committee had accepted that the new name for Division II shall be *Sun and Heliosphere*. In the same vein, the President of Commission 49 had also asked that Commission 49 be called *Interplanetary Plasma and Heliosphere*, respectively, *Plasma Interplanetaire et Heliosphere*, which was subsequently granted by the EC.

2. Reports of the WG on Eclipses and the representatives of the International Programmes

2.1. WG ON SOLAR ECLIPSES

The Chairman of the Working Group Solar Eclipses of the International Astronomical Union, Prof. J. Pasachoff, presented overviews of Eclipses since 1994, and of future Eclipses towards solar cycle maximum. For information on the activities of this working group is referred to:

`http://www.williams.edu/Astronomy/IAU_eclipses/`

J. Andersen (ed.), Transactions of the International Astronomical Union Volume XXIIIB, 113–115.

2.2. ISSA

The "International Services for Solar Activity" (ISSA) is a board in charge of the distribute funds from the "Federation of Astronomical and Geophysical Data Analysis Services" (FAGS) between Quarterly Bulletin on Solar Activity (QBSA), the Sunspot Index Data Center (SIDC), and the Debrecen Photoheliographic Data (DPD) Center. The members of the ISSA Board are: Prof. Franca Chiuderi Drago, Italy (Chair), Prof. Pierre Lantos, France, Dr. Rainer Schwenn, Germany, and Ms. Helen Coffey, USA.

2.2.1. *Quartely Bulletins on Solar Activity (QBSA)*
The Quarterly Bulletin on Solar Activity (QBSA) provides definitive sunspot numbers, the daily spotted area and the central zone sunspot numbers. The definitive sunspot numbers and their North and South components are published in Solar-Geophysical Data (NOAA, Boulder).

2.2.2. *Sunspot Index Data Center (SIDC)*
Dr. Pierre Cugnon, Belgium, presented a comprehensive report on the Sunspot Index Data Center to IAU for the period 1994-97. The SIDC has collected data from 34 (July 1994) to 49 (August 1997) cooperating centers in order to calculate the provisional daily and monthly International Sunspot Number R'_i. Since 1993 the SIDC network for the provisional sunspot number R'_i appears stabilized regarding as well the quantity as the quality of the observations.

The SIDC has been accessible on the World Wide Web since 1995 at the address:

`http://www.oma.be/KSB-ORB/SIDC/index.html`

2.2.3. *Debrecen Photoheliographic Data (DPD)*
Dr. Pierre Lantos presented, on behalf of Dr. Andreas Ludmany, a summary of the recent state of our catalogue activities of the photoheliograph catalogue work in Debrecen, Hungary.

The series of the Greenwich Photoheliographic Results (GPR) has been finished with the publication of the material for the year 1976. The IAU has charged the Heliophysical Observatory of the Hungarian Academy of Sciences to continue this work.

Three scientists of the Debrecen team L. Dezso, A. Kovacs and O. Gerlei started compiling the Debrecen Photoheliographic Results with the year 1977. Their program was more ambitious than the original GPR which was restricted to the mean positions and total areas of the sunspot groups. The DPR also included single spot data, magnetic polarities of the sunspots (from external sources) and the life of the groups was also followed. This work resulted in the DPR data for the years 1977 and 1978. The data for 1977 are available both in hardcopy and by ftp, the data for 1978 are accessible by ftp.

A separate group of the Debrecen team started a parallel project under the title Debrecen Photoheliographic Data (DPD). This is essentially based on the same methodology as the DPR, so that its reliability in the position and area measurements is the same. It does not contain, however, polarity data and it does not follow the history of single spots. This work resulted in the DPD data for the year 1986 which is available both in hardcopy and by ftp, the material of the first half of the year 1987 is available by ftp as a preliminary material, the second half has yet to be completed with the area data.

A major advancement is expected from a new procedure and software developed by Dr. L. Gyori. The method uses a combination of some absolute position measurements and the automatic evaluation of the CCD-scans of the active regions; in this latter step the software finds automatically the borders of the umbrae and penumbrae and determines their positions and areas. A significant speed up may be expected from this methodological advancement. Materials of the years 1993 and 1994 are being processed presently in this way. Full coverage of the missing years will be achieved by an appropriate schedule.

The anonymous ftp address of the presently accessible materials is as follows:

`ftp fenyi.sci.klte.hu/pub/DPR, or /DPD.`

3. Election of commission officers and new Organizational Committees

At the end of the second meeting were elected new commission officers and Organizing Committees.

Members of Commission 10 elected the following Organizing Committee for the period 1997-2000: President: Ai Guoxiang, China, Vice-President: A. Benz, Switzerland, and Secretary: K.P. Dere, USA. Members of the OC are: N. Gopalswamy, USA, A. Hood, UK, B.V. Jackson, USA, I. Kim, Russia, P.C. Martens, The Netherlands, G. Poletto, Italy, J.P. Rozelot, France, A.J. Sanchez, Spain, R. Hammer, Germany, K. Shibata, Japan, L. van Driel-Geztelyi, France, and O. Engvold, Norway (Past President).

Members of Commission 12 elected the following Organizing Committee for the period 1997-2000: President: P. V. Foukal, USA, Vice-President: S. Solanki, and Secretary: J. Mariska, USA. Members of the OC are: S. Baliunas, USA, D. Deming, USA, T. Duvall, USA, D. Dravins, Sweden, C. Fang, China, V. Gaizauskas, Canada, P. Heinzel, Czechia, J. Karpen, USA, E. Kononovich, Russia, S. Koutchmy, France, D. Melrose, Australia, M. Stix, Germany, Y. Suematsu, Japan, and F.-L. Deubner, Germany (Past President).

Members of Commission 49 elected the following Organizing Committee for the period 1997-2000: President: F. Verheest, Belgium, and Vice-President: Marek Vandas, Czechia. Members of the new OC are: B. Buti, India, N. Cramer, Australia, M. Dryer, USA, S. Habbal, USA, J. Hollweg, USA, M. Huber, The Netherlands, M. Kojima, Japan, and H. Ripken, Germany (Past President).

The following new Organizing committee of the IAU Division II was elected for the period 1997-2000: President: Peter V. Foukal, USA (P., Com 12). Members of OC: Ai Guoxiang, China (P., Comm 10), F. Verheest, Belgium (P., Comm 49), A. Benz, Switzerland (V.P., Comm 10), S. Solanki, Switzerland (V.P., Comm 12), M. Vandas, Czechia (V.P. Comm 49), and O. Engvold, Norway (Past P. Div II).

4. Commission Membership

The commissions were pleased to welcome 38, 9 and 3 new members to, respectively, commissions 10, 12, and 49.

Franz Ludwig Deubner read the names of the deceased members of Commissions 10 and 12, and those attending the session rose in their honor.

DIVISION III

PLANETARY SYSTEM SCIENCES

Division III gathers astronomers engaged in the study of a comprehensive range of phenomena in the solar system and its bodies, from the major planets via comets to meteorites and interplanetary dust.

PRESIDENT: Michael F. A'Hearn
Dept. of Astronomy, University of Maryland, College Park, MD 20742, USA
Tel. +1 301 405 6076
Fax. +1 301 314 9067
E-mail: ma@astro.umd.edu

BOARD

M F. A' Hearn	President Division III Past President Commission 15
V. Zappalá	President Commission 15
C. De Bergh	President Commission 16
M. Marov	Past President Commission 16
H. Rickman	President Commission 20
D.K. Yeomans,	Past President Commission 20
S. Bowyer	President Commission 21
C. Leinert	Past President Commission 21
W.J Baggaley	President Commission 22
I.P. Williams	Past President Commission 22
F.R. Colomb	President Commission 51
J.C. Tarter	Past President Commission 51

PARTICIPATING COMMISSIONS:

COMMISSION 15	**PHYSICAL STUDY OF COMETS, MINOR PLANETS AND METEORITES**
COMMISSION 16	**PHYSICAL STUDY OF PLANETS AND SATELLITES**
COMMISSION 20	**POSITION AND MOTIONS OF MINOR PLANETS, COMETS AND SATELLITES**
COMMISSION 21	**LIGHT OF THE NIGHT SKY**
COMMISSION 22	**METEORS AND INTERPLANETARY DUST**
COMMISSION 51	**BIOASTRONOMY: SEARCH FOR EXTRATERRESTRIAL LIFE**

COMMISSION NO. 15

Physical Studies of Comets, Minor Planets, and Meteorites
L'Etude Physique des Cometes, des Petites Planetes et des Meteorites

PRESIDENT: Michael F. A'Hearn
VICE-PRESIDENT: Vincenzo Zappalá
SECRETARY: Hermann Böhnhardt

1. Commission Membership

The president opened the meeting, welcoming members and interested guests. There were 29 in attendance, including at least 25 members of the commission. The agenda as previously posted on the commission's WWW site was adopted. The members stood for a moment of silence in honor of the six esteemed members of the commission who had died: M. P. Candy 04/11/94, R.A. Lyttleton 16/05/95, R. Smoluchowski 12/01/96, A. Mrkos 29/05/96, J. Rahe 18/06/97, E. Shoemaker 18/07/97, and V. Vanysek 27/07/97. The commission then elected new members K. Muinonen (Finland), S. Price (USA), and J. Watanabe (Japan). Furthermore, the following IAU membership nominees were elected to the commission pending their confirmation as IAU members at the second session of the General Assembly: J.-P. Barriot (France), T. Bonev (Bulgaria), A. Fitzsimmons (United Kingdom), D. Foryta (Brazil), S. Ibadov (Tajikistan), Y. Ma (PR China), P. Magnusson (Sweden), F. Marzari (Italy), A. Nakamura (Japan), P. Rousselot (France), A. S. Sharma (USA), and J. Zhu (PR China).

2. Commission Officers

The commission, noting that nominees for president and vice president had already been forward to the IAU Executive Committee, elected officers for the next triennium, doing so by acclamation for president, vice-president, secretary, chairman of the working group on asteroids, and continuing members of the OC. The new officers will be: President - V. Zappalá (Italy), Vice-President - H. U. Keller (Germany), Secretary - K. Muinonen (Finland), Chair of the WG on Comets - W. Huebner, Chair of the WG on Asteroids - E. Tedesco, and Members of the OC - M. Bailey (United Kingdom), R. Binzel (USA), M. T. Capria (Italy), P. D. Feldman (USA), J. Fernandez (Uruguay), C. Lagerkvist (Sweden), A.-C. Levasseur-Regourd (France), K. Meech (USA), J. Watanabe (Japan), and R. M. West (Germany).

3. Commission Organization and Communication

The president reported on the IAU's plans for greater involvement of the new divisions in the running of the IAU. One major role expected for the divisions is in the screening and ranking of proposals for IAU-sponsored scientific meetings. He also reported very briefly on the structure of Division III - Planetary System Sciences.

No initiative was undertaken to reinstall the former WG on Meteorites. However, this subject matter will still be covered in the triennial report as far as it is related to the commission's interests.

A particular endorsement of the Working Group on Interplanetary Pollution was not deemed necessary by the commission since the report of the WG had been submitted to Division III and they had forwarded it with a divisional resolution to the IAU Resolutions Committee.

The out-going commission officers have initiated the distribution of commission communication via electronic media. The commission web page can be found at www.ss.astro.umd.edu/IAU/comm15/ and can also be reached directly from the IAU web pages. Email addresses are available for almost 80 percent of the commission members for use in internal communication. In the future, the commission newsletters will be diseminated via email in one or more, widely used formats (LATEX, DVI, POSTSCRIPT) and they will be posted on the web pages. Members for whom email addresses are not now available will receive a last mailing indicating that they should respond with an email address or indicating that they do not have email access and must continue to receive printed mailings.

J. Andersen (ed.), Transactions of the International Astronomical Union Volume XXIIIB, 119–121.

4. Resolutions

4.1. COMA PARAMETERS

The commission discussed the confusion that arises when researchers try to compare water production rates reported by different investigators due to the use of different models and different parameters for the models. It was noted that the problem is more widespread than just indices for release of water, involving a lack of a standard way of reporting release of dust and also the lack of any standard for determining the nuclear magnitude of a comet. After a proposal to create a special working group, it was decided that the Working Group on Comets, augmented as seen fit by the chairman of the working group, should be charged with addressing this problem and the commission passed, unopposed, the following resolution as a charge to the working group:

IAU Commission 15, noting the great confusion that arises when comparing molecular production rates from various papers in the refereed literature and noting that much of this confusion is due to the wide range of models and parameters used therein to interpret the fluxes from emission features, resolves that the Working Group on Comets should appoint a subgroup charged with proposing a standard set of parameters for deducing a water release index. This group is charged with considering both the physical correctness of the parameters and the ease of application by astronomers in all situations. The working group should also consider whether other parameters, such as nuclear magnitudes, indices of dust release, and indices of gas/dust ratio, should similarly be standardized. This group will report back to Commission 15 in time to consider resolutions for adoption at the 24th General Assembly.

4.2. MINOR PLANET CENTER

The funding situation of the Minor Planet Center (MPC) was discussed briefly, noting the cessation of IAU support, the increasing workload at the MPC, the declining subscription revenues, and the temporary support provided by NASA. Although the MPC reports to Commission 20, its work is crucial in enabling the members of Commission 15 to carry out their work and Commission 15 therefore passed a resolution requesting that the IAU provide financial support to the MPC and authorizing the officers of Division III to provide a final wording that would make the actions of commissions 15 and 20 self-consistent. It was noted that a formal financial proposal would be necessary before any action was taken by the IAU and that this would be done through Commission 20 and Division III.

4.3. THE LEONID METEOR STREAM

A resolution supporting airborne campaigns to study the Leonid meteor showers in 1997 was introduced. After considerable discussion, an alternative resolution was passed by a vote of 8 to 4:

Commission 15, recognizing the unique opportunity of the 1998-1999 Leonid meteor showers for meteor studies, for cometary studies, and for studying the hazard of future meteor storms to spacecraft, and considering that this is a one-of-a-kind opportunity both because of the expected intensity of the events and because precession of the orbits will inhibit such intense activity for the foreseeable future (i.e., for centuries), urges all national and international agencies and organizations with relevant resources to encourage, coordinate, and support as wide a range of studies as possible, including airborne and space-borne studies as well as amateur and professional ground-based investigations.

5. Triennial Report

The further issuing of the triennial report of the commission was discussed. Roughly one half of the commission members present find the triennial report useful for their scientific work. Several options of formats for the report were discussed, consistent with discussions at the IAU Executive committee, i.e.,

- printed report in previous format to become an IAU publication
- 2-3 pages printed summary supplemented by a reference list sorted by subject for publication by the IAU
- 2-3 pages printed summary of the report for publication by the IAU every third year supplemented by an annual detailed version to be put on the commission web page.

The exact format of the next report will be worked out between the responsible commission officers, the president and the vice-president of the commission.

6. Future Meetings

Due to the increasing number of workshops and conferences related to subjects of interest to our commission, some degree of coordination of the meetings both in subject and dates would be highly desirable for the interested scientists. Several ideas for an improvement of the situation were discussed, i.e.,

- information exchange between the organizers well in advance of the meeting
- installation of a joint webpage for advertizing the meetings of interest
- keeping a fixed 3 years meeting cycle like IAU GA in year 1, ACM in year 2, other topics in year 3.

The discussion concluded that most likely any type of coordination of meetings is at present difficult to achieve since it may put unwanted constraints on the organizers of the meetings.

The president urged the incoming president to push for at least two sessions for the business meeting at the next General Assembly since it had been impossible to even get to potential new business from the floor at the present meeting which extended fully into the break.

COMMISSION 16: PHYSICAL STUDY OF PLANETS & SATELLITES
ETUDE PHYSIQUE DES PLANETES & SATELLITES

MIKHAIL MAROV
Keldysh Institute of Applied Mathematics, Russian Academy of Sciences, Miusskaja 4, Moscow 125047, Russia, e-mail: marov@spp.keldysh.ru

Commission 16 held its Business Meeting during the 97 General Assembly on the 26th of August as it was scheduled.

The Commission accepted the report of the President on the Commission's activities for 1994-1997 and expressed its satisfaction that all main foals planned at the 1994 General Assembly were achieved, hence contributing to the revitalization of Commission 16 activities.

Five meetings were co-sponsored or organized by other bodies jointly with Commission 16, and two important events Assembly were held upon the Commissions's initiative. :

JD 6 Interactions betzeen Planets & Small Bodies

and

SPS 1 The Galileo Mission to the Jupiter System during the 1997 General

International programs of coordinated observations of Mars and Jupiter were successfully continued.

The Commission also heard and approved progress report of IAU/IAG/COSPAR Working Group on Cartographic Coordinates and Rotational Elements of the Planets & Satellites and report on the current status of activity on preventing Interplanetary Pollution, presented by Dr. K. Seidelman and Dr. C. Keay, respectively.

Organizing Committee 1997-2000

In accordance with the IAU bye-laws, Commission 16 elected new President, Vice-President, Secretary and Organizing Committee for the period 1997-200 as follows:

PRESIDENT	Catherine	de Bergh	France
VICE-PRESIDENT	Dale P	Cruikshank	USA
SECRETARY	Angioletta	Coradini	Italy
	Michael J S	Belton	USA
	Carlo	Blanco	Italy
	Guy Joseph	Consolmagno	USA
	Daniel	Gautier	France
	Mikhail Ya	Marov	Russia
	David	Morrison	USA
	Keith Stephen	Noll	USA
	Tobias C	Owen	USA
	Viktor G	Tejfel	Kazakhstan
	Andrzej	Woszczyk	Poland

Membership

For the interim time Commission 16 lost its distinguished members J. Pollack, C. Sagan and J. Rahe. Seventeen new IAU members applied for the Commission membership and were accetped.

Acknowledgement

Generous contribution of the Commission's Organising Committee members to the reviewed activities is acknowledged.

J. Andersen (ed.), Transactions of the International Astronomical Union Volume XXIIIB, 123.

COMMISSION 20: POSITION & MOTIONS OF MINOR PLANETS, COMETS & SATELLITES

POSITIONS & MOUVEMENTS DES PETITES PLANETES, DES COMETES & DES SATELLITES

Report not received

COMMISSION 21
LIGHT OF THE NIGHT SKY / LUMIÈRE DU CIEL NOCTURNE

Report of the business meeting, Kyoto, August 26, 1997

PRESIDENT: Christoph Leinert
VICE-PRESIDENT: Stuart Bowyer

The president opened the meeting by welcoming the members present. He asked for a moment of silence in remembrance of Ed Ney who had died during the past triennium.

1. REPORT OF THE PRESIDENT

Large part of the commission activity related to conferences, often organised by commission members:

"Unveiling the cosmic infrared background", College Park, April 23-25, 1995, was organised by the COBE team. The proceedings appeared in the AIP series as volume 348 with E. Dwek as editor.

"Physics, chemistry, and dynamics of interplanetary dust", Gainesville, August 14-18, 1995, IAU colloquium No. 150, was the traditional meeting of the interplanetary dust community happening since 1969 about every five years. Sponsored by commissions 21 and 22, the proceedings appeared in the ASP conference series as volume No. 104, with B. Gustafson and M. Hanner as editors.

"Diffuse infrared radiation and the IRTS", Tokyo, November 11-14, 1996, was organised by T. Matsumoto. The proceedings appeared in the ASP conference series as volume No. 124, with H. Okuda, T. Matsumoto and L.Roellig as editors.

At the general assembly in Kyoto commission 21 has been
- co-sponsoring JD5 "Preserving the astronomical windows", organiser S. Isobe (commission 50)
- sponsoring SPS 2 "Highlights of the ISO mission", organiser D. Lemke
- supporting the "Workshop on zodiacal cloud sciences", Kobe, September 1-3, 1997, organised by T. Mukai.

The next IAU colloquium on interplanetary dust and zodiacal light is planned for spring 2000 in Canterbury, England, with T. McDonnell as host.

Communication with the members was based on a yearly newsletter. A poll among the commission members confirmed the preliminary decision from the general assembly in den Haag that commission 21 should join division III (Planetary system sciences). The triennual report for the IAU transactions on the scientific activities in the field was prepared with important help from S. Bowyer and K. Mattila and sent to the members with the January 1997 newsletter.

Commission 21 was represented in the inter-commission working group on "The prevention of interplanetary pollution" (convenor C. Keay) by B. Gustafson and A.-Ch. Levasseur-Regourd. The working group report led to a resolution at the Kyoto general assembly.

2. MEMBERSHIP

By vote the following colleagues were formally added to the membership list: Peter Abraham (Hungary), Sergeij Dodonov, Alexander Kopylow, Sergeij Neizvestny (Russia), William Reach (USA), Wolfhard Schlosser (Germany). Also Jakob Staude and Jerry Weinberg returned to membership.

New consultants are the younger scientists Rebecca Bernstein (USA) and Andreas Wicenec (Germany).

At present there are 108 members and 14 consultants to the commission.

J. Andersen (ed.), Transactions of the International Astronomical Union Volume XXIIIB, 127–128.

3. ELECTION OF OFFICERS

President: Stuart Bowyer (USA)
Vice-President: Philippe Lamy (France)

Organising committee: President, Vice-President and E.Dwek (USA), Bo A. Gustafson (USA), M.S. Hanner (USA), A.-Ch. Levasseur-Regourd (France), J.S. Mikhail (Egypt),I. Mann (Germany), T. Mukai (Japan), T. Matsumoto (Japan)

It was the opinion of the commission members present that in the future membership in the organising committeee should be limited to three consecutive periods.

4. STANDARD VALUES OF NIGHT SKY BRIGHTNESS

The president reported that the summary paper "The 1997 reference of diffuse night sky brightness" had been finished and submitted to A&A supplements. He thanked all the members who had contributed and helped to close this long-standing action item for commission work, at least for the time being.

5. COMMUNICATION WITHIN THE COMMISSION

The incoming president, S. Bowyer, explained that he intended to use the possibilities offered by internet for an improved communication. In particular he plans to set up a suitable commission page with relevant informations on the World Wide Web, linked to the IAU pages.

6. TRIENNUAL REPORT

The matter of writing this report was also discussed, with an outcome similar to the one at the meeting of commission presidents a few days ago. The reports are useful and should be continued; they help to integrate the commission and are a readily available starting point for outsiders on ongoing research in the field. However, they should not try to be complete but concentrate on important developments and references. In addition much freedom should be given to those actually composing the report. The combination of a short report, giving a reliable overview for a certain date, with frequently updated information available on the internet would be a possible way to go.

7. RESOLUTIONS

After short discussion, the members present gave support to further resolutions to be promoted by division III: to encourage studies of the 1998-199 Leonid meteor showers which are expected to be exceptionally rich, and to request further IAU support for the Minor Planet Center.

8. SCIENTIFIC REPORTS

In the short time left at the end of the meeting S. Bowyer reported on the existing observational limits on the Far UV diffuse background and, in particular, on the limits from 50 nm to 110 nm. J. Overduin then discussed the implications of these background limits on the radiative decay of fundamental particles in the universe.

COMMISSION 22: METEORS AND INTERPLANETARY DUST

I.P.WILLIAMS
Astronomy Unit
Queen Mary
London E1 4NS
UK

The meeting was held on Thursday 1997 August 21 at 14.00 hours with the President of the Commission, I.P.Williams taking the chair. Also present were Vice-President W.J.Baggaley, Secretary R.L.Hawkes and about 30 other members. The President reported the death of two of the commission members, Jacchia and Yamakoshi. The meeting stood in silence for one minute in their memory.

1. Report On The Last Trienium

The report of the last Trienium has been published in Reports on Astronomy volume XXIIIA and reprints were distributed to the attendees. The report had also been circulated to members electronically. The report was formally approved by the meeting.

2. Report Of Intercommission Working Groups

2.1. THE PREVENTION OF INTERPLANETARY POLLUTION

This report has also been published in Reports on Astronomy. The commission accepted the report and recommended that Division 3 should request the IAU also to accept the report and to take further steps so that the main recommendations of the report could be implemented.

2.2. NEAR-EARTH OBJECTS

This report has also been published in Reports on Astronomy. The commission accepted the report and agreed to support the proposition that the Working Group should become a Division 3 Working Group for the next three years.

3. Report Of Commission Working Groups

3.1. PROFESSIONAL-AMATEUR COOPERATION

A verbal report of the activities of the group was given by Hawkes (present chair). It was agreed with no dissentions that the Working Group should continue, though some discussion took place regarding future activity.

4. News Of Future Meetings

4.1. MEETING IN NANJING

The president reported that a meeting entitled "Cometary Nuclei in Space and Time" was proposed to be held in Nanjing on 1998 May 18-22. He had supported the proposistion to the Executive Committee on behalf of the Commission.

4.2. MEETINGS AT TATRSANSKA LOMNIKA, SLOVAK REPUBLIC

Two adjacent meeting were proposed to be held at Tatranska Lomnika, "Meteoroids 1998" on 1998 August 16-22 and "Evolution and source regions of Comets and Asteroids" to be held on August 24-28. The membership were requested to note both of these interesting meetings.

J. Andersen (ed.), Transactions of the International Astronomical Union Volume XXIIIB, 129–130.

4.3. ACM 99

The membership were informed that this meeting would be held at Cornell University, USA.

4.4. PHYSICS AND CHEMISTRY OF INTERPLANETARY DUST

The commission agreed to support an application to the IAU for a further Colloquium in the above series to be held at the University of Kent, UK in the year 2000.

5. Elections

The following were elected unopposed:
President: W.J.Baggaley (New Zealand)
Vice President: V.Porubcan (Slovakia)
Secretary: R.L.Hawkes (Canada)
Organizing Committe: The above three plus P. Babadzhanov (Tajikstan), B. Gustafson (USA), I. Mann (Germany), G. Elford (Australia), I. Hasegawa (Japan), C. Koeberl (Austria), P. Spurney (Czeck Republic) and I. Williams (UK).

5.1. EXISTING IAU MEMBERS WISHING TO JOIN COMMISSION 22

The following were duly accepted and welcomed
Kosi, Lipschutz, Rickman, Tatum, Watanabe.

5.2. NEW MEMBERS

The following New Members of the IAU had expressed interest in becoming Members of Commission 22. They were all accepted.
Cooper, Jenniskens, Jopek, Kalenichenko and Lemaire.

5.3. CONSULTANTS

The following were elected as Consultants.
Adolfson, Dr. Lars G. (USA) Andreev , Dr. Gennadij (Russia) Asher, Dr. David (Japan) Beech, Dr. Martin (Canada) Betlem, Mr. M.J. (The Netherlands) Bone, Mr. Neil (UK) Brown, Mr. Peter (Canada) de Lignie, Mr. Marc (The Netherlands) Getman, Dr. Vladimir (USA) Jones, Dr. William (UK) Kessler, Dr. Donald (USA) Klar Renner, Mr. Gilberto (Brasil) Mathews, Dr. John D. (USA) Murad, Dr. Edmond (USA) Nakamura, Dr. Takuji (Japan) Nagasawa, Dr. Koh (Japan) Obrubov, Prof. Yuri V. (Russia) Ocenas, Mr. Daniel (Slovak Republic) Ohtsuka, Mr. K. (Japan) Pellinen-Wannberg, Dr. Asta (Sweden) Rendtel, Mr. Jurgen (Germany) Richardson, Mr. James (USA) Roggemans, Mr. Paul (Belgium) Sidorov, Prof V. (Russia) Suzuki, Mr. Satoru (Japan) Voloshchuk, Prof. Y. (Ukraine) Wood, Mr. Jeff (Australia) Wu, Zidian (Canada) Xu, Pinxin (P.R. China) Yoshida, Mr. Takatsugu (Japan) Zook, Dr. Herb (USA).

5.4. REPRESENTATION ON DIVISON WORKING GROUPS

Near Earth Objects: I. Williams, V. Porubcan.

5.5. COMMISSION WORKING GROUP

Professional-Amateur Working Group
Chair:P. Jenniskens
Members: Betlem, Bone, Brown, Cooper, Hasegawa, Hawkes, Klar-Renner, Rendtel, Richardson, Yoshida.

6. Other Business

The incoming President thanked the outgoing President, Secretary and all the other members who had worked hard for the commission over the last trienium.

REPORT OF COMMISSION 51: BIOASTRONOMY MEETING KYOTO GENERAL ASSEMBLY

J.C. TARTER
SETI Institute
2035 Landings Drive, Mountain View, California USA 94043

1. Commission Business

Attendance at the two business meetings of Commission 51 was much lighter than has been experienced in the recent past. The meetings conflicted with very interesting scientific sessions. As a result, during a subsequent meeting with other Commission Presidents and the IAU Secretariat, I suggested that in the future, specific times be set aside for Commission meetings and other business during which there would be no scientific sessions; my personal preference being one hour at the end of each day during the General Assembly. The intent of this recommendation received strong support. I look forward to the implementation strategy for the next General Assembly.

The first order of business was the election of a new slate of officers for the term 1997-2000. Although Vice President Fernando R. Colomb could not be present at Kyoto, he sent acknowledgment of his eagerness to assume the duties of Commission President. The officers and new Organizing Committee of Commission 51 are:

President - Fernando R. Colomb (Argentina)
 Vice President - Stuart Bowyer (USA)
Organizing Committee:
 Fernando R. Colomb (Argentina)
 Stuart Bowyer (USA)
 Jill Tarter (USA)
 Ivan Almar (Hungary)
 Kelvin Wellington (Australia)
 Alain Leger (France)
 Toby Owen (USA)
 Woody Sullivan (USA)
 Michel Mayor(Switzerland)
 Frank Drake (USA)
There are four new Commission 51 members. They are:
 Dr. Nirupama Raghavan (Nehru Planetarium, New Delhi, India)
 Igor Kapisinsky (Slovak Academy of Sciences, Bratislava, Slovak Republic)
 Alan Hale (South West Inst. for Space Research, Cloudcroft, New Mexico, USA)
 Yaohui Qiu (Yunnan Observatory, Kunming Yunnan, PRC)

2. Scientific Matters

Stuart Bowyer reported on IAU Colloquium 168, The Fifth International Conference on Bioastronomy held in Capri, Italy on July 1-5 , 1996. Bowyer was chair of the Scientific Organizing Committee, and Cristiano Cosmovici chaired the Local Organizing Committee. The meeting was an enormous success, attended by over 200 people from 27 countries who presented 60 papers and 44 posters. The proceedings have already been published as Astronomical and Biochemical Origins And The Search For Life In The Universe, Editrice Compositori (1997). The multi-disciplinary nature of the meeting can be seen in it's session titles:

J. Andersen (ed.), Transactions of the International Astronomical Union Volume XXIIIB, 131–132.

Astronomical Origins
Significance of Asteroid and Comet Impacts
Protostellar Structures
Discovery of Extrasolar Planets
Origins of Life
Evolution of Intelligence
Searches for Evidence of Extraterrestrial Intelligent Activity

The Sixth Bioastronomy Conference will be held August 2-6, 1999 on the Big Island of Hawaii. The chair of the LOC, Karen Meech presented a summary of various possible locations, with the Hapuna Prince Hotel being the current favorite. Every attempt will be made to keep the daily costs for participants at a minimum and less expensive alternatives to hotels will be arranged (though they will require rental cars for transportation). Tours to volcanoes and the telescopes on Mauna Kea will augment the scientific content of meeting. Representatives of the scientific community in Iceland formally announced their intention of proposing to host the Seventh Bioastronomy Conference in Reykjavik in 2002.

An outline of the Commission's Triennial Report, prepared for this General Assembly was presented with thanks to the contributing authors; Toby Owen (USA), Michel Mayor (Switzerland), Andre Brack (France), Lori Marino (USA), Jill Tarter (USA). The format of these Triennial Reports has been criticized by a number of Commission Presidents. It was subsequently decided that a very brief two-page report format could be adopted by all Commissions whose subject matter was regularly reviewed in the literature, with references provided to those reviews. For Commission Presidents who so choose, additional space will be provided so that the traditional summary of scientific and scholarly activities of their Commissions can be can continue to be captured in the Triennial Reports.

The only other business was the announcement that the Organizing Committee had decided to implement a resolution passed during the First Bioastronomy Conference in 1984 to award a Bioastronomy Medal for significant advances in our field. There were four recipients of this medal who were honored for their work in the detection of extrasolar planets; Prof. Michel Mayor and Dr. Didier Queloz of Switzerland and Prof. Geoff Marcy and Dr. Paul Butler of the USA. I presented Dr. Butler with his medal (actually an engraved glass award) during Joint Discussion 13 on the Detection of Extrasolar Planets, which Commission 51 co-sponsored. The other medals were mailed to the recipients, all of whom were on mountain tops at the time!

DIVISION IV

STARS

Division IV organises astronomers studying the characterisation, interior and atmospheric structure, and evolution of stars of all masses, ages, and chemical compositions.

PRESIDENT: Lawrence E. Cram
School of Physics, University of Sydney, Sydney, NSW 2006, Australia
Tel.: +61 2 9351 2537
Fax.: +61 2 9660 2903
E-mail: L.Cram@physics.usyd.edu.au

BOARD

H. Zinnecker	President Commission 26
B. Barbuy	President Commission 29
J.-P. Zahn	President Commission 35
R. Pallavacini	President Commission 36
M. Gerbaldi	President Commission 45
D. Lambert	Past President Division IV

PARTICIPATING COMMISSIONS:

COMMISSION 26	**DOUBLE AND MULTIPLE STARS**
COMMISSION 29	**STELLAR SPECTRA**
COMMISSION 35	**STELLAR CONSTITUTION**
COMMISSION 36	**THEORY OF STELLAR ATMOSPHERES**
COMMISSION 45	**STELLAR CLASSIFICATION**

COMMISSION 26: DOUBLE & MULTIPLE STARS/ETOILES DOUBLES ET MULTIPLES

Report by Charles E. WORLEY
President 1994-97

Organizing Committee 1997-2000

PRESIDENT:	H. Zinnecker	Gernany
VICE-PRESIDENT:	C. Scarfe	Canada
ORGANIZING COMMITTEE:	C. Allen	Mexico
*	J. Armstrong*	USA
	W. Hartkopf	USA
*	R. Mathieu*	USA
	A. Tokovinin	Russia
*	M. Valtonen*	Finland
	C. Worley (ex-officio)	USA

* Incoming member of the OC

Membership

NEW MEMBERS:

R. Hindsley (USA), C. Hummel (Germany), D. Jassur (Iran), E. Martin (Spain), R. Mathieu (USA), C. Prieto (Spain), G. Torres (USA), S. Soederhjelm (Sweden) & J. Wang (China PR).

CONSULTANT:

B. Mason (USA)

NECROLOGY:

I regret to inform you of the deaths of the follwing members: P. Baize, W. Luyten, A. Shul'berg & P. van de Kamp

Business Meeting

The attendance was small (9 members). Results of the mail ballot resulted in the election to office of those names starred above. As is customary, the former Vice-President, Dr. Zinnecker, succeeds to the Presidency.

There was a brief discussion of the future direction the Commission might take, and there was general agreement that this would undoubtedly involve the combination of the classical astrometry and spectroscopy of binary stars with rapidly growing and much more accurate modern techniques. Also, it was emphasized that a true understanding of the mechanisms of star formation cannot be attained without accounting for the large number of binary and multiple stars extant.

Two important meetings concerning binary and multiple stars occurred in 1995 (Calgary, Canada) and 1996 (Santiago de Compostela, Spain). Dr. Zinnecker tentatively proposed a future Symposium about the year 2000 on general topics of interest. Worley reported on the present status of the Washington Visual Double Star Catalog and the Orbit Catalog, as well as the progress of the speckle program and the Navy Prototype Optical Interferometer.

J. Andersen (ed.), Transactions of the International Astronomical Union Volume XXIIIB, 135.

COMMISSION 29: STELLAR SPECTRA/SPECTRES STELLAIRES

Report not received

COMMISSION 35: STELLAR CONSTITUTION/CONSTITUTION DES ETOILES

JEAN-PAUL ZAHN
Observatoire de Paris, F-92195 Meudon, France,
Tel: (33) 1 45 07 78 04, Fax: (33) 1 45 07 78 72, e-mail: zahn@obspm.fr

Kyoto, Aug. 25, 1997

The audience was extremely sparse at this business meeting, mainly because most members of the commission stayed on Kyoto only during the first week, in which were gathered most meetings of prime interest to stellar physicists, such as symposia 183, 185 and 187, and the first joint discussion.

Organizing Committee 1997-2000

The incoming President Jean-Paul Zahn reported the results of the consultation which was held by e-mail to renew the Organising Committee:

PRESIDENT	Jean-Paul	Zahn	France
VICE-PRESIDENT	Don	VandenBerg	Canada
	Romanon M.	Canal	Spain
	Cesare S.	Chiosi, ex officio	Italy
*	Wojciech	Dziembowski	Poland
*	Joyce	Guzik	USA
*	Georges	Meynet	Switzerland
	Georges J.	Michaud	Canada
	Alvio	Renzini	Italy
*	Hideyuki	Saio	Japan
	A.V.	Tutukov	Russia
	Gérard P.	Vauclair	France

* Incoming members of the OC

The president greeted the newcomers, and expressed his gratitude to the four officers who were leaving the OC: Pierre Demarque, Icko Iben, André Maeder and Ken Nomoto.

The 13 candidatures for membership of the commission were all approved, giving a total membership of 334.

Business Meeting

The president gave a brief account on the commission presidents' meeting and recalled that one of the main tasks of the commissions is to approve the symposia and colloquia which are sponsored by the IAU. Alavaro Gimenez announced that he intended to organize during the fall of 1998 a workshop on the treatment of stellar convection.

J. Andersen (ed.), Transactions of the International Astronomical Union Volume XXIIIB, 139.

COMMISSION 36: THEORY OF STELLAR ATMOSPHERES/ THEORIE DES ATMOSPHERES STELLAIRES

PRESIDENT: Lawrence Cram
VICE-PRESIDENT: Roberto Pallavicini

The business meeting of Commission 36 was held on Friday, 22 August. It was attended by the President (L.E. Cram) and Vice President R. Pallavicini, approximately 12 members of the Commission, and the President of Division IV, D. Lambert. Attendance was reduced by the fact there were several parallel sessions of direct relevance to members of the Commission.

The President reported on the main activities of the Commission over the past 3 years, and raised some issues of general interest. He described the process of generating the Commission Report, noting the existence of a view by some members of the IAU that Reports no longer have an abiding significance. During the discussion of this matter, several points were raised: (1) old Commission reports (prior to approx 1965) often provide a good way for an astronomer working in a particular field now to get a picture of the development of the field at that time, (2) the volume and diversity of work within a Commission's sphere of interest is very large and difficult to cover properly, (3) readership of the Reports may be small, (4) it would be easier to prepare contributions for the reports if authors accepted the task at the beginning of the 3-year period, and (5) it is not clear whether a contribution to the report should attempt to be "complete" or composed of "highlights." The President and Vice President noted these comments.

In response to a question about a test e-mail message sent some time ago, the President explained that this test had confirmed that the IAU Secretariat's e-mail list was so incomplete and out of date that it was not useful for contacting Commission members. Methods for repairing this situation were discussed.

The President invited R. Cayrel to speak to his proposed resolution concerning the zero point of the scale of bolometric magnitudes. After a constructive discussion which clarified certain points, the Commission formally resolved as follows:

On the Zero Point of the Scale of Bolometric Magnitudes

Noting the absence of a strict definition for the zero point of bolometric magnitudes, and the resulting proliferation of different zero points in the literature, Commission 36 resolves to define the zero point by specifying that the absolute radiative luminosity, L, of a star of absolute bolometric magnitude $M_{bol} = 0$ has the value:

$$L = 3.055 \times 10^{28} \text{ W}.$$

This choice is intended to be close to the most current practice, and its equivalent to taking the value $M_{bol} = 4.75$ (C. Allen, "Astrophysical Quantities") for the nominal bolometric luminosity adopted for the Sun by the international GONG project ($L_\odot = 3.846 \times 10^{26}$ W).

Organizing Committee 1997-2000

The President explained the rules and criteria for the nomination for the positions of President, Vice President and Organising Committee for the next triennium. The current Vice President, Roberto Pallavacini, was

J. Andersen (ed.), Transactions of the International Astronomical Union Volume XXIIIB, 141–142.

elected as President, and Dainis Dravins was elected as Vice President, unanimously. The Organising Committee was unanimously adopted to be the following list:

B. Barbuy	D. Sasselov
I. Hubeny	M. Spite
S. Owocki	K. Stepien
H. Saio	R. Wehrse

There being no further business, the President thanked Commission members for their attendance and closed the meeting.

COMMISSION 45: STELLAR CLASSIFICATION/CLASSIFICATION STELLAIRE

PRESIDENT: HUGO LEVATO
Complejo Astronomico, El Leoncito, CC 467, 5400 San Juan, Argentina, e-mail: levato@castec.edu.ar

Officers triennium 1997-2000

PRESIDENT — Michèle Gerbaldi
Université de Paris-Sud XI, Centre National de la Recherche Scientifique
Institut d'Astrophysique, 98bis Bd. Arago, F75014 PARIS, France
e-mail: gerbaldi@iap.fr

VICE-PRESIDENT — T. Thomas Lloyd Evans
S A A O, P O BOX 9, OBSERVATORY 7935, South Africa
e-mail: tle@saao.ac.za

ORGANIZING COMMITTEE

	Juan Claria	Argentine	claria@oac.uncor.edu
	S J Christopher Corbally	Vatican	corbally@as.arizona.edu
	John Drilling	USA	drilling@rouge.phys.lsu.edu
	Catherine Garmany	USA	garmany@jila.colorado.edu
	Richard O. Gray	USA	grayro@appstate.edu
	Donald Wayne Kurtz	South Africa	dkurtz@physci.uct.ac.za
ex-officio	Orlando Hugo Levato	Argentine	levato@castec.edu.ar
	Pierre North	Switzerland	north@obs.unige.ch
	Kazimeras Zdanavicius	Lithuania	astro@itpa.lp

Membership

During the triennium the Commission lost a very active member by the death of Dr. Mercedes Jaschek.

An update list of the members of the Commission has been prepared . It is accurate as far as each member with e-mail address and phone number has been contacted and replied to the request. The membership of Commission 45 is 121.

Meetings

During the last three years the commission has lent its support to the following scientific meetings:

The MK Process at 50 Years: A Powerfull Tool for Astrophysical Insight
held in Tucson, USA, in September 1993

IAU Symposium 162: Pulsation, Rotation & Mass Loss in Early Type Stars
Juan-les-Pins, October 1993

IAU Colloquium 148: Future Utilization of Schmidt Telescopes
Bandung in March 1994

J. Andersen (ed.), Transactions of the International Astronomical Union Volume XXIIIB, 143–145.

IAU Symposium 166: Astronomical & Astrophysical Objectives of Sub-milliarcsecond Optical Astrometry
held in Le Hague in August 1994

IAU Symposium 167: New Developments in Array Technology & Applications
held in Le Hague in August 1994

IAU Symposium 177: The Carbon Star Phenomenon
Antalya in May 1996

Meeting on Automatic Spectral Classification

Report by Christopher J. Corbally, Vatican Observatory

This meeting has been held during the General Assembly in Kyoto.

IMPRESSIONS

"Automated spectral classification techniques are mature," began Ted von Hippel in the opening talk of the session. This remains my dominant impression.

For details, particularly on how both stellar spectral and luminosity types can now be derived by artificial neural networks (ANNs) and weighted metric-distance methods (MDMs) with the same accuracy as that obtained by the human expert, I would refer you to the papers by von Hippel, Gulati, and Weaver. From Weaver's paper we learnt how this accuracy is achieved even for relatively low-resolution, near-infrared (NIR) spectra. In this NIR spectral region, Wing showed that TiO types correlate very well with Keenan types. These presenters joined in making the point that such heavy-duty techniques as ANNs and MDMs are needed because pattern recognition in spectra is a highly non-linear problem, but one for which humans can be readily trained to find a solution.

SHOPPING LIST

The complete training of automated systems now needs more input: i.e., a large, high-quality, homogeneous library of digital spectra that span the MK System's standard stars . This library would firmly anchor an automated classification package in the well-proven and productive MK System. Richard Gray et al. can foresee such a package lodged at telescopes in the output stream of data-acquisition computers. These computers would instantly translate spectra into rough MK classifications and so signal anything of immediate interest. Back at home would be a more sophisticated package that should include, at least via Internet, a human classifier to cream off the most peculiar spectra.

Could the automated classification techniques be trained on synthesized spectra? Yes, for an automated system has already been so trained (see von Hippel) and even yielded metallicities, besides effective temperatures and surface gravities. But these data depend on current atmospheric models, and here Kurucz cautioned that at present "observations have nothing to do with stars," (and laughter greeted this expression of his call for more atomic and molecular data and more physically precise models). For durability amid changing theory, it is much better to follow the wisdom of "M" and "K" and anchor spectral classifications in an autonomous system of standards through the proposed comprehensive library. This library should also please our galaxy-oriented colleagues who want the best possible input to synthesize stellar populations.

While the comprehensive spectral library and more line data were the main items on the session's shopping list, other needs surfaced: "red flags" on auto-classified spectra to indicate special difficulty with the classification and so special peculiarities (actually, this appeared to be a call to calibrate goodness-of-fit further); a more fully developed, metal-weak dimension to parallel the MK System; an investigation why Houk's luminosity class IV stars were not well-distinguished from class V and class III by ANNs; and an understanding of the best training strategies for ANNs and MDMs. There was general agreement that the best way to fulfil these needs would first be at the local level and only then to think of global implementation.

THE FUTURE

The session ended with a sense of optimism. The mega-spectral outputs of surveys such as the Sloan Digital Sky Survey, described in the paper by Andy Connolly, could be coped with and the potential galactic-structure and stellar-atmospheres insights retrieved. Automated spectral classification has come of age and is ready -- with sufficient funding -- to enter this new era of data acquisition. The meeting was concluded by Ch. Corbally with a certain optimistic view about the possibility of coping with the outputs of the big surveys such as Sloan Digital Sky survey. The procedures for automated spectral classification can be applied and useful results for galactic structure studies may be obtained.

Working Groups

WG ON STANDARD STARS
Chairperson: Robert Garrison

WG ON AP AND RELATED STARS
Chairperson: M. Takada-Hidai (Japan)
Ed. Newsletter: P. North (Switzerland)

WG ON PECULIAR RED GIANTS
Chairperson: Robert F. Wing

WG ON ACTIVE B STARS
Chairperson: D. Baade

DIVISION V

VARIABLE STARS

Division V provides a joint forum for the study of stellar variability in all its manifestations, whether due to pulsations, surface inhomogeneities, evolutionary changes, or to eclipses and other phenomena specifically related to double or multiple stars.

PRESIDENT

Mikolaj Jerzykiewicz — Past President Commission 27
Astronomical Institute, Wroclaw University, Ul. Kopernika 11
PL - 51 622 Wroclaw, Poland
Tel. +48 71 48 2434
Fax. +48 71 48 2434
E-mail: mjerz@astro.uni.wroc.pl

BOARD

D.W. Kurtz	President Commission 27
E. Guinan	President Commission 42
M. Rodonò	Past President Commission 42
Y. Kondo	Past President Division V
L. Balona	Commission 27 Member
J. Sahade	Commission 42 Member

PARTICIPATING COMMISSIONS

COMMISSION 27 — **VARIABLE STARS**

COMMISSION 42 — **CLOSE BINARY STARS**

COMMISSION 27: VARIABLE STARS/ETOILES VARIABLES

PRESIDENT:	Mikolaj Jerzykiewicz
VICE-PRESIDENT:	Donald W. Kurtz

Report by D.W. Kurtz

Officers Triennium 1997-2000

PRESIDENT:	D.W. Kurtz	South Africa	dkurtz@physci.uct.ac.za
VICE-PRESIDENT:	J. Christensen-Dalsgaard	Denmark	
ORGANIZING COMMITTEE MEMBERS:			
	C. Cacciari	Italy	
	P. Cottrell	New Zealand	
	P. Harmanec	Czech republic	
Ex officio	M. Jerzykiewicz	Poland	
	J. Matthews	Canada	
	P. Moskalik	Poland	
	D. Sasselov	USA	
	D. Welch	Canada	
	P. Whitelock	South-Africa	

Membership

NEW MEMBERS OF THE COMMISSION

Kem Cook, Peter Cottrell, Katy Tsvetkova, Stanislaw Zola, John Caldwell, Denis Gillet, Steven Kawaler, Howard H. Lanning, Albert Bruch, Shinichi Tamura, Eric Gosset, J. Robert Buchler, F. Kerschbaum, C. Aerts

RESIGNATION

K.K. Kwee

NEW IAU MEMBERS

The commission has supported the application of: Alisher Hojaev, Astronomical Institute, Uzbek Academy of Sciences, Tashkent, Uzbekistan,for the membership of the IAU.

Communication

The home page of Commission 27 (made by Andras Holl, Konkoly Observatory) is available from the home page of Konkoly Observatory:

http://www.konkoly.hu/IAUC27/index.html

The most important news (deadlines etc) are promised to be circulated by normal mail.

J. Andersen (ed.), Transactions of the International Astronomical Union Volume XXIIIB, 149–151.

Support of Working Groups

Commission 27 is going to continue the support of the running working groups on Active B stars, Ap and related stars and Peculiar Red Giants. More suggestion is welcome.

Professional - amateur connection

J. Percy emphasized that the role of amateurs is increasing in variable star research since not only visual but photoelectric and CCD observations are obtained by them. J. Mattei reported on the activity of AAVSO. The scientific meetings (May 1997, Switzerland and AAS meeting in June, 1997) proved the increasing connection of professional and amateur astronomers (50% - 50%). Amateur-professional common publications are submitted. More and more photoelectric networks (presently Japan+AAVSO) are established. Any other country is welcome, data are free for everybody. Suggestion: the small telescopes, getting to be closed, could be used by amateurs and professionals.

An idea of an IAU Coll organized in Canada on amateurs, teaching and connection between amateurs and professionals was mentioned by J. Percy for getting support. T. Oswald outlined the difference in the activity of AAVSO and IAPPP.

IAPPP does not have as long tradition as AAVSO but, since small observatories with more serious equipments are also involved, it could be regarded as a deeper organization.

Future meetings

According to the election in Los Alamos the next pulsation conference will be held in Hungary, 1999. Tentative proposal was presented by L. Szabados as follows:

Title: The Impact of Large-Scale Photometry on the Research of Pulsating Stars Date: August 9-13, 1999, Budapest (Total solar eclipse in Hungary on August 11, 1999).

Highlights: massive photometries, structure of the instability strip, precise light curve analyses, double-mode variables, Blazhko-variables, local reddening and stellar distribution, distance scale, etc.

Division

NEW PRESIDENT: M. Jerzykiewicz (Poland).

Mike Jerzykiewicz is the new president of division V which includes us (C27) and binary stars (C42). Division V now has a board which consists of:

The presidents of C27 and C42, D. Kurtz and E. Guinan, the past-presidents of C27 and C42, M. Jerzykiewicz and M. Rodonò, the past-president of Division V, Y. Kondo and one member each from C27 and C42, L. Balona (C27) and J. Sahada (C42).

The EC will now deal primarily with the Divisions, rather than with the commissions directly. Thus our support of Symposia and Colloquia will be channelled through our division to the EC.

Miscellaneous

IBVS

There was a report on the operation of IBVS by Katalin Olah. IBVS is a privately operated and funded journal. The editors are Katalin Olah and Laszlo Szabados; Andras Holl is the technical editor, and Zoltan Kovari is the administrative editor. The journal has an editorial/advisory board to assist the editors in scientific aspects of the editing process advise the editors in the management of the journal.

Chris Sterken has just concluded 6 years as chair of the advisory board. The new chair of the board is Petr Harmanec.

IBVS published about 150 issues a year. Rejection rates are now about 25%, and language is often a problem.

The announcement of new variables can be problematic, especially where the type of variability is not well characterised.

New ground rules for such announcements are being discussed by the advisory board now. This subject generated a lot of excitement at the business meeting - obviously many people felt threatened that they would not be able to announce their own new variables, and found this distressing. That is probably not what is going to happen to the professionals, although some types of announcements may be shortened and/or gathered into a few announcement issues per year. Note the "some" - not all by any means, so don't panic.

IBVS is now listed on ADS and the visitors to the Website increased by a factor of 5 with that.

GCVS

Nicolai Samus talked about the GCVS and their problems. They have minimal salaries and NO funds for anything else. Nevertheless, progress has been made in going electronic - ultimately full connection on the internet plus cross referencing with other data bases and services is desired.

There are scientific problems which are being addressed:

Massive surveys are discovering new variables by the tens of thousands. This is very exciting, but does generate a huge amount of work for the GCVS. Should the new survey variables be incorporated in the GCVS? Opinions vary. HIPPARCOS liased with the GCVS before publication to give traditional variable star names to the new variables. Should and will the massive photometric surveys do the same? Feelings on this subject are likely to be strong and cover a wide range.

The entire institute is under threat of closure because of the funding crisis in Russia.

They would obviously like to put the entire GCVS on-line. If you know of any funding possibilities, please help. You can check out "Astronet" at the SternbergAstronomical Institute at

/www.sai.msu.su/

although, because of the severe funding crisis you will not get into GCVS yet.

Scientific meetings

Finally, E. Guinan proposed that we (C27 and C42) consider a joint meeting of our two commissions in either 1999 or 2000. We are already well along the way to the next pulsation meeting in August 1999 in Budapest; Geza Kovac and Laszlo Szabados have produced an impressive first draft proposal

There are also binary star meetings in 1999, including one in honor of Brian Warner for his 60th birthday. So 1999 is out.

Next General Assembly

2000 is the next GA in Manchester. If we are to follow up on Ed's suggestion, then we will probably be proposing a Symposium at the time of the GA. There will need to be some stellar Symposium, so we may have a chance if we produce a good proposal.

COMMISSION No. 42 CLOSE BINARY STARS/ÉTOILES DOUBLES SERRÉES

Report of Business Meeting: August 27, 1997

PRESIDENT: Marcello Rodonò
ORGANIZING COMMITTEE: E.F. Guinan (Vice-President), Y. Kondo (Past-President), J.V. Clausen, H. rechsel, P.P. Eggleton, B.J. Geldzahler, G. Hill, S.L. Kenyon, F. Szkody, R.E. Taam, A.V. Tutukov, O. Vilhu, J.C. Wheeler, A. Yamasaki, Di-Sheng Zhai.

Organizing Committee (1997-2000)

The following composition of the OC for the triennium 1997-2000, consisting of newly elected and ex-officio members, was approved:

PRESIDENT:	Edward F.	Guinan	USA
VICE PRESIDENT:	Paula	Szkody	USA
ORGANIZING COMMITTEE:	Luciana	Bianchi	Italy
	Jens Viggo	Clausen	Denmark
	Horst	Drechsel	Germany
	Graham	Hill	UK
	Scott J	Kenyon	USA
	Virpi S	Niemela	Argentina
	Yoji	Osaki	Japan
	Marcello	Rodonò	Italy
	Christiaan L	Sterken	Belgium
	Ronald E	Taam	USA
	Osmi	Vilhu	Finland
	J Craig	Wheeler	USA
	Janet H	Wood	UK

Commission membership and new members

During the course of the past triennium the following distinguished and active members of Commission No. 42 passed away: Mario G. Fracastoro, John B. Irwin, Jürgen Rahe, Walter F. Wargau, William H. Wehlau, Igor Yurkevich & Di-Sheng Zhai.

The Commission members were standing in silence for a minute to acknowledge their valuable and outstanding contributions to the Commission activities.

The total number of Commission members in the past triennium was 348.

NEW MEMBERS named by the OC: Ho-Il Kim (Korea), Chun-Hwey Kim (Korea), Andrew R. King (U.K.), Jeffrey L. Linsky (U.S.A.), Mario Livio (Israel), Lawrence W. Ramsey (U.S.A.), P. Vivekananda Rao (India), already proposed at the previous GA, and Luciana Bianchi (Italy), Albert Bruch (Germany), Giuseppe Cutispoto (Italy), Andrea K. Dupree (U.S.A.), Alister B. Hojaev (Uzbekistan), Howard H. Lanning (U.S.A.), Antonino F. Lanza (Italy), James E. Neff (U.S.A.), Christiaan L. Sterken (Belgium).

CONSULTANTS: Luigi Stella (Italy), Corrado Trigilio (Italy), Grazia Umana (Italy)

J. Andersen (ed.), Transactions of the International Astronomical Union Volume XXIIIB, 153–155.

MEMBERS PROPOSED BY NATIONAL COMMITTEES: G. Anupama (India), Hrvoje Bozic (Croatia), Joao Canalle (Brazil), Rudolf Duemmler (Finland), Mohammad Edalati Sharbaf (Iran), Nickolas Elias (USA), Sergei Fabrika (Russia), Carole Haswell (UK), Coel Hellier (UK), Harold Kenny (Canada), Ulrich Kolb (UK), Urs Muerset (Switzerland), Grzegorz Pojmanski (Poland), Konstantjn Postnov (Russia), Frederic Rasio (USA), Frederick Ringwald (USA), Maximilian Ruffert (Germany), Roberto Turolla (Italy), Rolf Walder (Switzerland), Tae Yoon (Korea R).

The Commission members unanimously approved the above listed candidates and reiterates the traditional conditions to be met by each candidate for admission, i.e., a) Ph.D. in Astronomy or Physics , b) at least three substantive papers published in refereed internationally reckoned journals, with the candidate as first author or among the first three authors. The President is required to check that these two criteria are met before passing the list of the new Commission members to the IAU secretariat.

IAU Division structure

Commissions No. 42 and No. 27 have agreed to merge into **DIVISION V: *Variable Stars***. This merging was unanimously welcomed, as well as the proposed restructuring of IAU into Divisions. The Board of Directors set up by Yoji Kondo, Division President for the triennium just ending, is composed of Y. Kondo (ex officio), M. Jerzykiewicz and M. Rodonò (Commissions' Presidents), D. Kurtz and E. Guinan (Commissions' Vice-Presidents), L. Balona and J. Sahade, as distinguished members of the two Commissions. The Board of Directors meet on August 25 in Kyoto and elected M. Jerzykiewicz as new Division President for the 1997-2000 term. The election was unanimously approved. Final IAU approval will occur during the course of the upcoming GA.

Working Groups

The request of the Chairman of the ***Working Group on Active B Stars***, Myron Smith, that this WG be affiliated to Comm. 42, has been preliminarily accepted by the Commission President, after consultation with SOC members. The new IAU rules, however, imply that WGs approved within Commissions or Divisions should be regarded as internal organization structures. IAU WGs are promoted and are subject to the approval of the Executive Committee.

Sponsored Meetings

Following consultations of the OC members by the President, Commission sponsorship or co-sponsorship was granted to the following meetings and Joint Discussions (proposer names are given between brackets):

- Progress of Observational Accuracy and Modelling Variable Stars (Mine Takeuti), IAU Kyoto
- *Flares and Jets in Astrophysics* (Kazunari Shibata), IAU Kyoto
- *Stellar Evolution in Real Time* (Robert Koch and Edward Guinan), IAU Kyoto
- The Hot Luminous Corner of the H-R Diagram (Mariko Kato), IAU Kyoto
- *The Hot Universe* (Katsuji Koyama), IAU Kyoto
- *Hipparcos and Tycho Results* (Catherine Turon), IAU Kyoto
- Electronic Publishing: now and the future (Antony G. Hearn), IAU Kyoto
- Population Synthesis: from Planets to Black Holes and from Black Holes to the Universe (E.P.J. van den Heuvel, V.M. Lipunov), September 1997 (Moscow)
- *Accretion Phenomena and Associated Outflows* (D. Wickmasinghe), June-July 1998 (Port Douglas, Queensland, Australia)
- *Precise Radial Velocities* (J. Hearnshaw), June 1998 (Victoria BC, Canada)

Julian Date vs MJD (JD - 2 400 000.5)

IAU Commissions 4, 19, and 31 objected the *Resolution C3* passed at the IAU GA in The Hague banning the use of MJD, introduced in 1973 by *Resolution 4*.

A Working Group, chaired by Jorge Sahade and consisting of representatives of Commission 4, 19, 27, 30, 31, and 42 (H. Fiegel, T. Fukushima, M. Jerzykievicz, Y. Kondo, D. McCarthy, M. Rodonò, C. Sharfe, and P.K. Seidelmann), was formed by the IAU General Secretary to consider the issue. The WG proposed the following resolution, that was illustrated by the WG Chairman:

The International Astronomical Union,

recognizing

the need for a system of continuous dating for the purpose of analyzing time-varying astronomical data, and that both JD and MJD have been employed for this purpose in astronomy, geodesy, and geophysics,

recommends

that JD is used to record the instants of the occurrences of astronomical phenomena,

that for those cases where it is convenient to employ a day beginning at midnight, the MJD (equivalent to JD - 2 400 000.5) be used, and

that, where there is any possibility of doubt regarding the usage of MJD, care be exercited to state specifically its definition.

From the floor the question was raised whether it was an appropriate procedure to discuss and approve this resolution during the first GA, before having submitted it to the appropriate commissions for approval. Commision No. 42 expresses strong concern for the unusual and possibly unproper procedure adopted in this circumstance.

Commission Report (1993-96)

The report of the Commission President on the scientific activity in the last triennium was published in the IAU Transaction. A number of people did actively collaborated in the effort. Their substantial contrinution was warmly acknowledged by the President. Offprints of the Commission Report were distributed to all participants.

Information Bulletin on Variable Stars

(Discussion deferred to the business meeting to be held jointly with Commission No. 27)

Bibliography and Program Notes on Close Binaries

Following the resignation of Atsuma Yamasaki, to whom the warmest thanks of the Commission were presented for the wonderful job done, a new *Editor-in-chief* (Horst Drechsel) was nominated by the Commission President. The editorial board includes: J. Carrigan, O. Demircan, R. Haefner, T.J. Herczeg, V.G. Karetnikov, C. Maceroni, P. Mayer, R.G. Samec, D. Scarfe, M. Vetesnik, and A. Yamasaki.

The editorial board proposed *to dismiss the "Current Programs" section* because it appears strongly biased due to the small number of contributors (usually within the editorial board, only), and therefore rather incomplete and unrepresentative of the actual programs being done. The Commision members agree.

Accordingly, following the Editor-in-chief proposal, it was unanimously agreed to rename the *"Bibliography and Program Notes on Close Binaries"* into *"Bibliography on Close Binaries (BCB)"*.

Other business and science session

It was decided to held the second business meeting jointly with Commission No. 27 to discuss matters of common interest, such as the publication policy of the ***"Information Bulletin of Variable Stars"***, and dedicate the last thirty minutes of this meeting to the presentation of a survey paper by S. Slee, K, Jones and E. Budding on *"Binary star radio survey data"*.

DIVISION VI

INTERSTELLAR MATTER

Division VI gathers astronomers studying the diffuse matter in space between the stars, ranging from primordial intergalactic clouds via dust and neutral and ionised gas in galaxies to the densest molecular clouds and the processes by which stars are formed.

PRESIDENT:

Michael A. Dopita
Mount Stromlo & Siding Spring Observatories, Private Bag, Weston Creek PO, Canberra, ACT 2611, Australia
Tel. +61 2 6249 0212
Fax. +61 2 6249 0233
E-mail: michael.dopita@anu.edu.au

BOARD

See following page

PARTICIPATING COMMISSION:

COMMISSION 34 **INTERSTELLAR MATTER**

COMMISSION 34: INTERSTELLAR MATTER/MATIERE INTERSTELLAIRE

Division VI of the International Astronomical Union deals with Interstellar Matter, and incorporates Commission 34. It gathers astronomers studying the diffuse matter in space between the stars, ranging from primordial intergalactic clouds via dust and neutral and ionised gas in galaxies to the densest molecular clouds and the processes by which stars are formed. Following the 23rd General Assembly held at Kyoto, in August 1997, there are now ~730 Division/Commission members.

Officers 1997-2000

PRESIDENT:
Michael A. Dopita, Australia
Mount Stromlo & Siding Spring Observatories, Private Bag, Weston Creek PO. Canberra, ACT 2611, Australia
Tel. 61 6 249 0212
Fax. 61 6 249 0233
E-mail : michael.dopita@anu.edu.au
VICE-PRESIDENT:
Bo Reipurth, Chile
ESO
E-mail: breipurt@eso.org

ORGANIZING COMMITTEE:

Rafael Bachiller, Spain, 1997 <bachiller@oan.es>
Molecular clouds. Star formation. Protostars. Bipolar molecular outflows.Neutral matter in planetary, proto-planetary nebulae.Interstellar molecular lines. Interstellar chemistry.Millimeter-wave radioastronomy.

Michael Burton, Australia, 1994 <mgb@newt.phys.unsw.edu.au>
Infrared and Antarctic astronomy, star formation, excitation of molecular clouds, molecular shock waves and photodissociation regions, the Orion nebula.

Michael Dopita, (President), Australia, 1997-2000 <Michael.Dopita@anu.edu.Au>
Active Galactic Nuclei, Shock Theory, Photoionisation Modelling, Planetary Nebulae, Supernova Remnants, Star Formation & Chemical Evolution.

John Dyson, UK, 1997, <ed@ast.leeds.ac.uk>
The theory of the interstellar medium, gas dynamics, Active Galactic Nuclei, ionisation and recombination fronts.

Debra Elmegreen, USA, 1994 <elmegreen@vaxsar.vasser.edu>
Star formation and spiral structure in galaxies at optical, near-infrared, millimeter, and centimeter wavelengths.

Thomas Henning, Germany, 1997, <henning@astro.uni-jena.de>
Interstellar dust, Laboratory astrophysics, Star formation, Line spectroscopy & Radiation hydrodynamics.

Sun Kwok, Canada, 1997, <kwok@iras.ucalgary.ca>
Evolution of planetary nebulae interstellar dust, and circumstellar envelopes of AGB stars. Chair Planetary Nebula Working Group.

J. Andersen (ed.), Transactions of the International Astronomical Union Volume XXIIIB, 159–162.

Haruyudi Okuda, Japan, 1994, <kuda@astro.isas.ac.jp>
Infrared studies of interstellar matter. Infrared surveys by balloon and satellite.

Guillaume Pineau des Forets, France, 1997, <forets@obspm.fr>
Molecules and interstellar chemistry, gas dynamics and MHD shocks, photodominated regions.

John Raymond, USA, 1997, <jraymond@cfa.harvard.edu>
The theory and observation of optical, UV and X-ray emission from interstellar shock waves.

Bo Reipurth, (Vice Pres.), Chile/ESO, 1997- 2000, <breipurt@eso.org>
Pre-main sequence objects, Herbig-Haro jets and outflows.

Guillermo Tenorio-Tagle, Spain, 1997, <gtt@ast.cam.ac.uk>
Theoretical multi-dimensional hydrodynamics of HII regions and galactic outflows.

Sylvia Torres-Peimbert, Mexico, 1997, <silvia@astroscu.unam.mx>
Spectroscopy and Chemical abundances of Planetary Nebulae and HII regions.

Ewine van Dishoeck, NL, 1994, <ewine@strw.leidenuniv.nl>
Interstellar molecules; Star-formation, Basic molecular processes; Laboratory astrophysics. Secretary: Astrochemistry WG.

Sueli Viegas-Aldrovandi, Brazil, 1994, <viegas@vax.iagusp.usp.br>

Thomas Wilson, Germany, 1994, <p073twi@mpifr-bonn.mpg.de>
Sub-mm Astronomy.

Working Groups of Commission Number 34

PLANETARY NEBULA WORKING GROUP MEMBERS

Chair: Sun	Kwok	USA	kwok@iras.ucalgary.ca
Agnes	Acker	France	acker@cds6.u-strasbg.fr
Michael J.	Barlow	UK	mjb@star.ucl.ac.uk
George	Jacoby	USA	jacoby@noao.edu
Jim	Kaler	USA	kaler@astro.uiuc.edu
Walter	Maciel	Brazil	maciel@orion.iagusp.usp.br
Dipankar C.V.	Malick	India	dcvmlk@iiap.ernet.in
Mario	Perinotto	Italy	mariop@arcetri.astro.it
Stuart	Pottasch	Netherlands	s.r.pottasch@astro.rug.nl
Luis	Rodriguez	Mexico	luisfr@astrocu.unam.mx
Detlef	Schönberner	Germany	deschoenberner@aip.de
Yervant	Terzian	USA	terzian@astrosun.tn.cornell.edu
Roumald	Tylenda	Poland	tylenda@ncac.torun.pl
Peter	Wood	Australia	<wood@mso.anu.edu.au>

ASTROCHEMISTRY WORKING GROUP MEMBERS

Chair: D.A. Williams, Dept. of Physics & Astronomy, University College London, Gower Street, London WC1E 6BT, U.K., e-mail: daw@star.ucl.ac.uk, FAX: 44 171 380 7145, Tel: 44 171 391 1355

Secretary: E.F. van Dishoeck, Leiden Observatory, University of Leiden, P.O. Box 9513, 2300 RA Leiden, The Netherlands, e-mail: ewine@strw.leidenuniv.nl, FAX: 31 71 5275819, Tel: 31 71 5275814

L.W. Avery, Joint Astronomy Center, 660 N. Aohoku Place, Hilo, HI 96720U.S.A. [Regular address:Herzberg Institute of Astrophysics National Research Council of Canada CANADA], e-mail: lwa@jach.hawaii.edu FAX: 1 808 961 6516 Tel: 1 808 969 6551

J.H. Black, Onsala Rymdobservatorium, Chalmers Tekniska Hogskola, S-439 92 Onsala, Sweden, e-mail: jblack@oso.chalmers.se, FAX: 46 31 772 5590, Tel: 46 31 772 5540

V. Buch, Hebrew Univ. of Jerusalem, Fritz Haber Res. Center for Mol. Dynamics, Dept. of Physical Chemistry, Jerusalem 91904, Israel, e-mail: viki@batata.fh.huji.ac.il, FAX: 972 2 6513742, Tel: 972 2 6584223

A. Dalgarno, Center for Astrophysics, 60 Garden Street, MS 14 Cambridge, MA 02138, U.S.A., e-mail: adalgarno@cfa.harvard.edu, FAX: 1 617 495 5970, Tel: 1 617 495 4403

J.M. Greenberg, Huygens Laboratory, University of Leiden, P.O. Box 9513, 2300 RA Leiden, The Netherlands, e-mail: greenber@strw.leidenuniv.nl, FAX: 31 71 5275819, Tel: 31 71 5275804

C. Henkel, Max-Planck Institut fur Radioastronomie, Postfach 20 24 or Auf dem Huegel 69, 53010 Bonn 53121 Bonn, Germany, e-mail: p220hen@mpifr-bonn.mpg.de, FAX: 49 228 52 5229, Tel: 49 228-52 5305

W.M. Irvine, 626 Lederle Graduate Research Center, University of Massachusetts, Amherst, MA 01003, U.S.A., e-mail: irvine@fcrao1.phast.umass.edu, FAX: 1 413 545 4223, Tel: 1 413 545 0733

J.P. Maier, Univ. of Basel, Inst. fur Phys. Chemie, Klingelbergstrasse 80, CH-4056 Basel, Switzerland, e-mail: maier@ubaclu.unibas.ch, FAX: 41 61 267 3855, Tel: 41 61 267 3826

K.M. Menten, Harvard-Smithsonian Center for Astrophysics, 60 Garden Street, MS 42, Cambridge, MA 02144, U.S.A., e-mail: menten@cfa.harvard.edu, FAX: , Tel: 1 617 495 7385

Y.C. Minh, Korea Astronomy Observatory, Hwaam Yusong, Taejeon 309-348, Korea R, e-mail: minh@hanul.issa.re.kr, FAX: 82 42 861 5610, Tel: 82 42 865 3263

M. Ohishi, Nobeyama Radio Observatory, Nobeyama, Minimisaku, Nagano 384-13, Japan, e-mail: ohishi@nro.nao.ac.jpm, FAX: 81 267 98 2884, Tel: 81 267 98 4373

B. Rowe, Dept. de Phys. Atomique et Moleculaire, URA 1203 du CNRS, Univ. de Rennes, Bt. 11C, F 35042 Rennes Cedex, France, e-mail: rowe@univ-rennes1.fr, FAX: 33 99 28 6786, Tel: 33 99 28 6183

P.D. Singh, Instituto Astronomico e Geofisico, Universidade de Sao Paulo, Av. Miguel Stefano, 4200, 01051 Sao Paulo, Brazil, e-mail: pdsingh@vax.iagusp.usp.br, FAX: 55 11 577 8599, Tel: 55 11 577 8202

L.E. Snyder, Department of Astronomy, University of Illinois, 103 Astronomy Building, 1002 West Green Street, Urbana, IL 61801, U.S.A., e-mail: snyder@prairie.astro.uiuc.edu, FAX: 1 217 244 7638, Tel: 1 217 333 5530

Qin Zeng, Purple Mountain Observatory, Academia Sinica, Nanjing 210008, China P.R., e-mail: qinzeng@public1.ptt.js.cn, FAX: 86 25 33 01459, Tel: 86 25 30 3738

Divions VI Working Group Proposals

The Organising Committee of Division VI is in the process of forming two new ISM working groups to monitor progress in their fields and to help develop proposals for future IAU Symposia/Colloquia.

There already exist the Planetary Nebula Working Group and the Astrochemistry Working Group and these have served us well in organising periodic seminars in these subject areas. However, our wish to form new groups is motivated by the realisation that other areas of the ISM are not properly represented in the current organisation.

These working groups would be:

1. Working Group on the Hot and Extragalactic ISM
2. Working Group on Star Forming Regions

We propose that these each take in ex-officio members being either the Division Presidents, or else proposed by the liasing Divisions.

WORKING GROUP ON STAR FORMING REGIONS

At this time we have a list of nominations (all have been approched and are willing to serve) for, which Rafael Bachiller would chair. These are as follows:

R. Bachiller (Chair)	Spain	34	Radio (mm), low-mass SFR
M. Burton	Australia	34	Infrared, shocks
L. Cram	Australia	Pt Div. IV	
Y. Fukui	Japan	34	Radio (mm), clouds, outflows
G. Garay	Chile	40	Radio (cm), high-mass SFR
T. Henning	Germany	34	Dust, mm continuum
S. Lizano	Mexico	34	Theory
F. Palla	Italy	34	Theory, IR, intermed-mass SFR
J. Palous	Czech Rep.	33	Nominated: Div VII
B. Reipurth	USA	34 (VPt)	HH objects, optical
A. Sargent	USA	34	Radio(mm), disks, low/interm. mass
S. Strom	USA	36,27	Optical/IR

The Executive Committee is expected to formally constitute this Working Group by the end of January 1997, and to give agreement in principle to the setting up of the Working Group on the Hot and Extragalactic ISM. Anyone who wishes to seek nomination in the latter WG should contact the President of Div VI:

Meetings supported by Division VI/Commission 34

IAU Symposium on Wolf-Rayet Phenomena in Massive Stars and Starburst Galaxies
Puerto Vallarta, 3-7 November 1998

Contact: Prof. Karel A. van der Hucht <K.vanderHucht@sron.ruu.nl>

IAU Symposium on "New Views of the Magellanic Clouds"
13-19 July 1998, Victoria, Canada

Contact: Prof. You Hua Chu <chu@astro.uiuc.edu>

Condolences

It is with deep regret that we report that Dr. Tarafdar passed away on September 7, 1996.

DIVISION VII

GALACTIC SYSTEM

Division VII provides a forum for astronomers studying our home galaxy, the Milky Way, which offers a unique laboratory for exploring the detailed structure of the stellar and gaseous components of galaxies and the processes by which they form and evolve.

PRESIDENT: Kenneth C. Freeman
Mount Stromlo & Siding Spring Observatories, Private Bag,
Western Creek PO
Canberra, ACT 2611, Australia
Tel. +61 2 6249 0264
Fax. +61 2 6249 0233
E-mail: kcf@mso.anu.edu.au

BOARD

G.S. Da Costa	Vice President Commission 37
J.J. Binney	Past President Division VII
L. Blitz	
F. Matteuci	
D. Spergel	Vice President Commission 33

PARTICIPATING COMMISSIONS:

COMMISSION 33 **STRUCTURE AND DYNAMICS OF THE GALACTIC SYSTEM**

COMMISSION 37 **STAR CLUSTERS AND ASSOCIATIONS**

COMMISSION 33: STRUCTURE & DYNAMICS OF THE GALACTIC SYSTEM
STRUCTURE & DYNAMIQUE DU SYSTEME GALACTIQUE

Report not received

COMMISSION 37: STAR CLUSTERS & ASSOCIATIONS

AMAS STELLAIRES & ASSOCIATIONS

Officers 1997-2000

PRESIDENT: Dr. Gary S. Da Costa
Mount Stromlo Observatory, Private Bag Western Creek PO Canberra, ACT 2611, Australia
Phone: 61 2 6249 0236, Fax: 61 2 6249 0233, E-mail: gdc@mso.anu.edu.au

VICE-PRESIDENT: Georges Meylan
ESO, Karl Schwarzschild-Str 2, D 85748 Garching bei München, Germany, Phone: 49 89 320 06 293, Fax: 49 89 320 23 62, E-mail: gmeylan@eso.org

ORGANIZING COMMITTEE:

	Dr. Sverre Aarseth	UK
	Dr. Roberto Buonanno	Italy
	Dr. Russell Cannon	Australia
	Dr. Vittorio Castellani	Italy
	Dr. Kyle Cudworth	USA
ex officio	Alessandro Feinstein	Argentina
	Dr. Charles Lada	USA
	Dr. Ata Sarajedini	USA

Drs Cannon and Sarajedini were added to the OC in Kyoto succeeding Dr. Mermilliod and Dr. Zhao.

Membership

The following new members have been co-opted during the General Assembly: Alberto Buzzoni (Italy), Michael Chryssovergis (Greece), Elvira Covino (Italy), Kathleen DeGioia-Eastwood (USA), Mohammad Hossein Dehghani (Iran), Laurent Drissen (Canada), Miroslaw Giersz (Poland), Pavel Kroupa (Germany), Peter James T. Leonard (USA), Rosemary Mardling (Australia), Randy L. Phelps (USA), Andres Eduardo Piatti (Argentina), Charles Franklin Prosser Jr. (USA), Ata Sarajedini (USA), Evgeni Semkov (Bulgaria), Koji Takahashi (Japan), Luciano Terranegra (Italy), Polina E. Zakharova (Russia).

The total number of the members is then ~230.

Activity during the General Assembly

At Kyoto Commission 37 was a Supporting Commission for:

JD 15, The Combination of Theory, Observation, and Simulation for the Dynamics of Stars and Star Clusters in the Galaxy,

and a Co-Supporting Commission for:

Symposium 186	Galaxy Interactions at Low and High Redshift
JD 1	Abundance Ratios in the oldest Stars: Bulge and extreme Halo
JD 10	Low-Luminosity Stars
JD 14	The first Results of Hipparcos and Tycho.

J. Andersen (ed.), Transactions of the International Astronomical Union Volume XXIIIB, 167.

DIVISION VIII

GALAXIES AND THE UNIVERSE

Division VIII gathers astronomers engaged in the study of the visible and invisible matter in the Universe at large, from Local Group galaxies via distant galaxies and galaxy clusters to the large-scale structure of the Universe and the cosmic background radiation.

PRESIDENT: Peter A. Shaver
European Southern Observatory, Karl Schwarzschildstrasse 2
D - 85748 Garching b. Muenchen. Germany
Tel. +49 89 320 06233
Fax. +49 89 320 06480
E-mail: pshaver@eso.org

BOARD:

F. Bertola	present President Commission 28
J. Narlikar	past President Commission 49
S. Okamura	present Vice-President Commission 28
J. Peacock	present President Commission 49
A. Szalay	Vice-President Commission 49
V. Trimble	past President Commission 28

PARTICIPATING COMMISSIONS:

COMMISSION 28 **GALAXIES**

COMMISSION 47 **COSMOLOGY**

As elaborated in the January 1995 issue of the IAU Information Bulletin, the Division coordinates the activities of the affiliated Commissions, including proposals for new Commissions, Working Groups, IAU Symposia and Colloquia, and Joint Discussions at General Assemblies.

The possibility of creating further Commissions within Division VIII has been discussed over the last three years. Specifically, it was suggested that there could be four Commissions within the Division : Cosmology (or Formal Cosmology), Large Scale Structure, Active Galaxies, and Normal Galaxies & Clusters. It was finally decided to keep to the original two Commissions for the time being, but this issue will be re-visited in the future, and Suggestions are always welcome.

COMMISSION 28 GALAXIES

Officers 1997-2000

PRESIDENT:	Francesco	Bertola	Italy	bertola@astrpd.pd.astro.it
VICE PRESIDENT:	Sadanori	Okamura	Japan	okamura@astron.s.u-tokyo.jp
ORGANIZING COMMITTEE:	Chantal	Balkowski	France	balkowsky@obspm.fr
	David	Burstein	USA	burstein@samuri.la.asu.edu
	Sandip	Chakrabarti	India	chakraba@tifrvax.tifr.res.in
	Pieter	de Zeeuw	Netherlands	tim@strw.leidenuniv.nl
	Michael	Feast	South Africa	mwf@uctvms.uct.ac.za
	John	Huchra	USA	huchra@cfa.harvard.edu
ex-officio, past president	Virginia	Trimble	USA	vtrimble@uci.edu/ vtrimble@astro.umd.edu
	Richard	Wielebinski	Germany	p022rwi@mpifr-bonn.mpg.de
	Anatole	Zasov	Russia	zasov@sai.msu.su
	Zhen-Long	Zou	China	zouzl@sun.ihep.ac.cn

Report prepared by V. Trimble.

During the triennium

The Commission sponsored or co-sponsored several symposia and colloquia, which were prioritized by the members of the Committee annually. A single mailing went to the entire membership in January, 1996 soliciting volunteers and nominations of people to help with the triennial report, serve on the Committee, and otherwise participate in the activities of the Commission. Of the 30 or so people who responded or were nominated, several indeed helped write the report, several; are among the new Committee members, and a couple were among the suggestions from Division VIII for the special Nominating Committee.

At the General Assembly

The Commission sponsored Symposium 187 and Joint Discussions 2 and 11. It was a co-sponsor of S183, 184 and 188 and of JD 1,18,21 and 22. During the business meeting, the new slate of officers was elected and 25 current IAU members who had asked to join the commission were elected to membership. In addition, about 11 newly elected IAU memebers had Commission 28 as their first choice. All were, in principle, accepted, although in a couple of cases it was clear that they really wanted to join 29 or some other Commission. No enthusiasm was expressed for preserving any of the working groups that have existed in the past, including redshifts, Photometry and Internal Structure, Magellanic Clouds, and Supernovae; and all of these have now lapsed.

The possibility of splitting the commission into two, devoted to active and normal galaxies repectively was discussed briefly (as it had also been at the earlier meeting of Division VIII). No strong desire to do this was expressed; several members disagreed on the distinction between normal and active galaxies; and in the light of the current development of the Divisional structure, the decision was made to leave things as they are. The Commission thus currently has about 700 members.

J. Andersen (ed.), Transactions of the International Astronomical Union Volume XXIIIB, 171.

COMMISSION 47: COSMOLOGY/COSMOLOGIE

J.V. NARLIKAR
Inter-University Center for Astronomy & Astrophysics, Post Bag 4, Pune 411 077, India
e-mail: jvn@iucaa.ernet.in

Commission 47 held its Business Meeting on August 26, 1997, 09:00 hours, Room D

Attendance

J.V. Narlikar (President in the Chair), P. Shaver (Vice-President), B. Partridge (Past-President and President, Division VIII), Y. Chu (O.C. Member), J. Einasto (O.C. Member), D. Jauncey (O.C. Member), P. Crane, A. Fairall, B. Tully, M. Wanas

Condolences

The President welcomed all those present and began the proceedings by recording sorrow at the passing away of the Commission 47 members during 1994-97. He informed that as per the information supplied by the IAU Secretariat, the Commission had lost the following members:

Professor Hannes Alfvén
Dr. Nicole J. Bel
Professor Gérard de Vancouleurs
Professor Robert H. Dicke
Professor Odon Godart
Professor Roger J. Tayler
Dr. B.C. Xanthopoulos
Dr. A.L. Zel'manov

All members present stood in silence as a mark of respect to the departed.

New Members

The President presented a list of those IAU members who were admitted to the membership of Commission 47 during the triennium 1994-97. They had fulfilled the IAU criteria for commission membership.

The President further presented another list of new members admitted to the IAU at the Kyoto GA, who had expressed desire to be members of Commission 47. These were duly admitted with the proviso that if C. Kaul (India) and D. Kompaneets (Russia) wished to join C 44 (Space and High Energy Astrophysics) instead, they may be offered that option as their interest and field of expertise might lie closer with that commission.

J. Andersen (ed.), Transactions of the International Astronomical Union Volume XXIIIB, 173–175.

Division VIII Decisions

At the President's request Bruce Partridge, President, Division VIII summarized the decisions taken at the Business Meeting of Division VIIIon August 23:

A decision was taken to keep the other commission within the Division (C 28, Galaxies) as it is, so that Division VIII would continue to have only two commissions.

P. Shaver would serve as President of Division VIII for the triennium 1997-2000.

During 1994-97 the Division structure was 'on probation'. At this GA it had now been formalized and so, the Division President will have closer liason with the Executive Committee on matters relating to IAU meetings and other matters.

Officers triennium 1997-2000

PRESIDENT

As per the procedure arrived at during the last Business Meeting at the Hague IAU-GA (1994) the Vice-President takes over as President of the Commission at the end of the triennium. In 1994, Commission 47 had two Vice-Presidents, P. Shaver and A. Szalay, who decided by mutual agreement, that A. Szalay would take over as Commission President at this Business Meeting.

Accordingly A. Szalay will serve as the President of Commission 47 for the period 1997-2000.

VICE-PRESIDENT

Following the guidelines of the last Business Meeting, nominations for the Vice-President were invited from the OC members. Of the two names proposed, one withdrew and so only one name, that of J. Peacock was put up on the slate for a poll by the general membership of the Commission.

Accordingly J. Peacock was declared elected Vice-President of Commission 47 for the period 1997-2000.

ORGANIZING COMMITTEE

The 1994 Business Meeting had stipulated that the Organizing Committee should include about 12 members, including the President, Vice-President, Past President and President Division VIII as ex-officio members. Other OC-members should have two terms, unless a member is elected Vice- President or Division President.

As Y. Chu, J. Einasto, J. Peacock and A. Wolfe were elected in 1994, they would continue for another term. The guidelines of the 1994 Business Meeting further stipulated that nominations for OC should be invited from the general membership of the Commission 47 and that a ballot be taken of the membership to choose the required number from the slate, after ensuring that at least one OC member is taken from each of the following regions : (i) Asia-Pacific (ii) Europe-Africa (iii) Latin America and (iv) North America.

Accordingly a poll was conducted by the President and the following new members were declared elected to the OC of the Commission.

The entire OC of Commission 47 for 1997-2000 is as follows

PRESIDENT	A. Szalay	Hungary
VICE-PRESIDENT	J. Peacock	UK
	Y. Chu	China PR
*	L. da Costa	Brazil
	J. Einasto	Estonia
*	G.F.R. Ellis	USA
*	D. Koo	USA
*	S.J. Lilly	Canada
Past-President	J.V. Narlikar	India
President, Division VIII	P Shaver	Germany
*	R. Webster	Australia
	S. White	UK
	A. Wolfe	USA

* Incoming members

Recommended Meetings

The President informed that the following symposia had been recommended to the Executive Committee by the Organizing Committee for sponsorship of Commission 47:

Symposium entitled "The Light Elements and their Evolution" to be held in Natal, Brazil in 1998.

Symposium entitled "The Low Surface Brightness Universe" to be held in Cardiff in July 1998.

Symposium entitled "The Evolution of Galaxies on Cosmological Time Scales" to be held in Tenerife in July 1998.

Symposium entitled "Activity in Galaxies and Related Phenomena" to be held in Byurakan Astrophysical Observatory, Armenia from August 17-21, 1998.

B. Partridge, in his capacity as President, Division VIII, informed that the first symposium had been postponed by the organizers while the other three were under consideration of the IAU Executive Committee.

The President informed that at the meeting of Commission Presidents with the IAU officials on August 25, the incoming General Secretary had discussed revised guidelines that were being drafted for submitting proposals for new symposia. As per the new guidelines, a proposal would be first vetted by the concerned commission and forwarded to the division president, if found suitable for sponsorship. Based on the information supplied by the commission president(s) the division president would prioritize the symposia with appropriate gradings for final decision by the Executive Committee. The new guidelines will stipulate the time table for all these intermediate steps.

Other Matters

Amongst any other matters, the President sought the members' feedback on the triennial scientific reports of the Commission, which appear in the IAU Transactions. He mentioned that based on the feedbacks received from the various commission presidents, the incoming General Secretary was preparing new guidelines for IAU Commission Reports. The commissions would have the option of preparing a short 2-3 page report listing highlights with source material (reviews, conferences, etc.) topicwise, rather than writing more detailed accounts. Some commissions may prefer the concise option while others may opt for the detailed ones as before.

There was some discussion on this issue where members highlighted advantages of both options. It was expected that the OC will take the decision for Commission 47 in the coming year.

Concluding Remarks

The President thanked the Vice-Presidents and his other colleagues on the OC for the cooperation extended to him. He welcomed Alex Szalay as his successor and hoped that the existing and new members of the OC will continue to provide him with similar support. He specially thanked all those who wrote parts of the Trienniel Report of Commission 47, some at relatively short notice. He complimented B. Partridge for his interactive guidance to the commission as Division President, and the IAU Secretariat for help on administrative issues.

On behalf of the Commission, Bruce Partridge proposed a vote of thanks and appreciation to the outgoing President for his leadership of the Commission.

DIVISION IX

OPTICAL TECHNIQUES

Division IX provides a forum for astronomers engaged in the innovation, development, and calibration of optical instrumentation and observational procedures, including data processing.

PRESIDENT: Christiaan L. Sterken
Faculty of Sciences, Vrije Universiteit Brussels, Pleinlaan 2
B- 1050 Brussels, Belgium
Tel. +32 2 629 3469
Fax. +32 2 936 23976
E-mail: csterken@vub.ac.be

BOARD

I.S. McLean	President Commission 9
G. Lelièvre	Past President Commission 9
J.D. Landstreet	Past President Commission 25
J.B. Hearnshaw	President Commission 30
C.D. Scarfe	Past President Division IX
	Past President Commission 30

PARTICIPATING COMMISSIONS:

COMMISSION 9 **INSTRUMENTS**

COMMISSION 25 **STELLAR PHOTOMETRY AND POLARIMETRY**

COMMISSION 30 **RADIAL VELOCITIES**

COMMISSION 9: INSTRUMENTS/INSTRUMENTS & TECHNIQUES

Organizing Committee triennium 1997-2000

PRESIDENT:	Ian S.	McLean	UK
VICE-PRESIDENT:	Ding-qiang	Su	China PR
ORGANIZING COMMITTEE:	Martin	Cullum	Germany
	Michel	Dennefeld	France
	George H.	Jacoby	USA
	Gérard	Lelièvre	France
	Shiro Emer	Nishimura	Japan
	William J.	Tango	Australia
	Milcho K.	Tsvetkov	Bulgaria
	A.K.	Saxena	India

The President thanked the out-going Members for their work inside the Committee: J.C. Bhattacharyya (India), David F. Malin (Australia), Fritz Merkle (Germany) and Richard M. West (Germany).

Membership

The following members were co-opted:

Ahmad Zaharim Abdul Aziz (Malaysia), Zeljko Andreic (Croatia), Samuel Charles Barden (USA), Giovanni Bonanno (Italy), Jin Chang (China PR), Oberto Citterio (Italy), Yurii Denishchik (Ukraine), Lionid Didkovsky (Ukraine), Gerard Gillanders (Ireland), Shuichi Gunji (Japan), Donald John Hutter (USA), Abdanour Irbah (Algeria), Thomas Kentischer (Germany), Laurent Koechlin (France), Piero Madau (USA), Tadashi Nakajima (USA), Junichi Noumaru (Japan), John O'Byrne (Australia), Creidhe O'Sullivan (Ireland), Neil Parker (UK), Didier Queloz Switzerland), Walter Seifert Germany), Shengcai Shi (China PR), Dirk Soltau Germany), Alan Morton Title (USA), Munetaka Ueno (Japan), Xinghai Yin (China PR), Zhaowang Zhao (China PR).

The total number of Commission Members is thus 309.

Working Groups

Working Group on Detectors (Chairman: Martin Cullum)
Working Group on High Angular Resolution Interferometry (Chairman: J.T. Armstrong)
Working Group on Wide Field Imaging Surveys (Chairman: Noah Brosch)

Activities during the General Assembly

Commission 9 has co-supported the following events:

JD 8 Stellar Evolution in Real Time
JD 9 Future Large Scale Facilities in Astronomy
JD 12 Electronic Publishing: Now and the Future
JD 13 Detection and Study of Planets outside the Solar System

J. Andersen (ed.), Transactions of the International Astronomical Union Volume XXIIIB, 179–180.

Activities during the Triennium

The Commission has sponsored or co-sponsored the following Symposia:

166 Astronomical & Astrophysical Objectives of Sub-Milliarcsecond Optical Astrometry
The Hague, The Netherlands, August 15-19, 1994
Eds. Erik Høg & Kenneth Seidelmann
Kluwer Academic Publishers, 1995, ISBN 0-7923-3442-6

167 New Developments in Array Technology & Applications
The Hague, The Netherlands, August 23-27, 1994
Eds. A.G. Davis Philip, Kenneth A. Janes & Arthur R. Upgren
Kluwer Academic Publishers, 1995, ISBN 0-7923-3639-9

179 New Horizons from Multi-Wavelength Sky Surveys
Baltimore, USA, August 26-30, 1996
Eds. Brian J. McLean, Daniel A. Golombek, Jeffrey J.E. Hayes & Harry E. Payne
Kluwer Academic Publishers, 1998, ISBN 0-7923-4802-8/ ISBN 0-7923-4802-6

COMMISSION 25: STELLAR PHOTOMETRY AND POLARIMETRY
PHOTOMÉTRIE ET POLARIMÉTRIE STELLAIRE

PRESIDENT: John D. LANDSTREET
VICE-PRESIDENT: Chris L. STERKEN

1. Introduction

Much Commission business was transacted between General Assemblies by the Organizing Committee, which is able to meet effectively by e-mail as needed, and make recommendations to the Executive Committee, for example, about what Symposia to accept for the most recent General Assembly.

During the Kyoto meeting, two meetings of the Commission were held. About a dozen people came to each; perhaps this small attendance is a reflection of the extent to which the Symposia and Joint Discussion dominate the format of the IAU now. Below is a report on the business transacted.

2. Business meetings

2.1. MEMBERSHIP

The commission at present has a membership of slightly more than 200 members. New members who joined during the triennium include Drs S. HUBRIG, J.-L. LEROY, G. MATHYS, D. MOURARD, and J. SUDZIUS. In addition, eight new members of the IAU also joined the Commission during the 1997 General Assembly: B. CARTER, N. CRAMER, Y. EFIMOV, M. LEMKE, A. MANDAYAM, K. MASLENNIKOV, B. POKRZWKA, and M. SZYMAŃSKI.

2.2. OFFICERS

The members present (about 10 persons) unanimously elected Chris STERKEN (Vrije Universitaet Brussels, Pleinlaan 2, B-1050 Brussels, Belgium) as President, and Arlo LANDOLT (USA) as Vice-President. We wish them well during the coming triennium.

Seven members of the OC reached the ends of their terms: J. KNUDE, J. LUB, I. MCLEAN, J. MENZIES, F. VRBA, V. STRAISZYS, and A. YOUNG. Five others are continuing: S. ADELMAN, M. BREGER, D. KURTZ, E. MILONE, and T. MOFFETT. We thank the retiring OC members, and the continuing ones, for their work during the past triennium. New OC members elected are: M. BESSELL, I. GLASS, J. GRAHAM, H. HENSBERGE, K. SEKIGUCHI, J. TINBERGEN, and W. WARREN.

It was agreed by the Commission members present that before the next IAU General Assembly, we will try to have an e-mail election of Commission Officers to increase the partipation of Commission members in this important activity.

2.3. REORGANIZATION

Reorganization of the IAU following decisions made at the Hague General Assembly and confirmed at the Kyoto meeting has been proceeding. The new structure of Divisions which group together Commissions of similar interests, and which have more contact with the Executive Committee than the Commissions have had in the past, is beginning to get established. Our Commission is part of Division IX, Optical Techniques, together with Commission 9 (Detectors and techniques) and Commission 30 (Radial Velocities). During the past triennium, the Divison Board consisted of the three Presidents and the three Vice-Presidents of these Commissions, with Colin SCARFE of Commission 30 as Division President. A number of issues have come to the Division Board, among which are questions as to which proposed IAU Symposia and Colloquia should be selected. In addition, Division Presidents were invited to participate in most of the Executive Committee meetings during the General Assembly in Kyoto. I expect that Divisions will be more and more used in future to furnish advice to the Executive Committee.

J. Andersen (ed.), Transactions of the International Astronomical Union Volume XXIIIB, 181–183.

The new Division IX Board will again be made up of the Presidents and Vice-Presidents of the Commissions. The new Division President will be the President of our Commission, Chris STERKEN.

The other aspect of reorganization which has greatly changed the nature of the IAU is the decision to hold six Symposia together with the General Assembly, and to replace to a significant extent the Commission scientific meetings with Joint Discussions. In this context, our Commission supported Symposium 184, "New eyes to see inside the sun and stars", and Joint Discussions 8 "Stellar evolution in real time" and 14 "First results from Hipparchos and Tycho" during the Kyoto meeting. These scientific meetings attracted large numbers of participants, and clearly were of wide interest.

It should be noted that with the increasing importance of Joint Discussion for the scientific life of the Union, members of our Commission should start thinking about JD's that could be of general interest during the next General Assembly in Manchester. Suggestions should be sent to the new President of the Commission.

2.4. REPORTS ON ASTRONOMY

Every three years a small group (usually Commission Officers or past Officers, with help from some others) prepare a report on the state of our field for publication in Reports on Astronomy, Part A. The Executive Committee has been asking Commissions and Divisions for their opinions on the usefulness of these documents. After a discussion between most of the Commission Officers and the Officers of the Union, it seems clear that the EC will decide to keep these reports, while accepting a greater diversity of style and substance from one Commission to another, depending on the needs of each Commission. A discussion at the Commission business meeting of this question produced a similar consensus; most people present agreed that the report for Commission 25 is useful and should certainly be retained, but without being overly concerned about the exact format of the report. It was widely agreed that the report would be much more useful if it were distributed immediately upon completion to all the members of the Commission, rather than having everyone wait until it is published. The current report was in fact distributed in September 1997 by e-mail as a postscript document to the present list of members for whom e-mail addresses are available. It seems clear that this will become increasingly practical as the list of e-mail addresses becomes more complete.

2.5. E-MAIL DISTRIBUTION OF INFORMATION

Commission members who have not received any e-mail from the Officers during the summer of 1997 are urged to send their e-mail addresses to the new President, Chris STERKEN (at csterken@vub.ac.be), to be sure that our list of addresses is as complete and correct as possible.

2.6. PROPOSAL FOR DEFINITION OF BOLOMETRIC MAGNITUDE

The following resolution, proposed to the Commission by Roger CAYREL, was approved by the Commission in principle. (The present form has evolved somewhat from that which was presented at the Commission business meeting, but the substance is the same.)

Noting the absence of a strict definition for the zero point of bolometric magnitudes, and the resulting proliferation of different zero points in the literature, the Commission

Recommends to define the zero point by specifying that the absolute radiative luminosity, L, of a star of absolute bolometric magnitude $M_{bol} = 0$ has the value $L = 3.055 \times 10^{28}$ W. This choice is intended to be close to the most common practice, and is equivalent to taking the value $M_{bol} = 4.75$ (Allen, *Astrophysical Quantities*) for the nominal bolometric luminosity adopted for the Sun by the international GONG Project, ($L_{\odot} = 3.846 \times 10^{26}$ W).

2.7. CLOSURE OF SMALL TELESCOPES

A short discussion was held during the business meeting about whether photometrists could do anything useful to slow the pace of closure of small telescopes (see Reports on Astronomy, Part A, Report of Commission 25). Several suggestions were discussed; the only one which survived group criticism was that people should apply for time regularly on small telescopes in the best sites, to make it clear to time allocation committees and funding agencies that these telescopes are really still valuable to the community.

2.8. REPORTS OF WORKING GROUPS

The meeting heard short oral reports from three of its Working Groups (a fourth group, on Infrared Filters, made its report in written form in the Commission Report in Reports on Astronomy, Part A).

Business from the Working Group on Ap and Related Stars was reported by G. MATHYS. The WG organized Joint Discussion 16, on "Spectroscopy with large telescopes of chemically peculiar stars", during the Kyoto meeting; this meeting was very successful. The new Chair of the WG is Masahide TAKADA-HIDAI. The Editor of *A Peculiar Newsletter* continues to be Pierre NORTH. Members of the Commission interested in the field of this WG are reminded of the meeting of the European Working Group on CP Stars which will be held October 27 - 29, 1997 at the Vienna Observatory. Contact Pierre NORTH (Pierre.North@obs.unige.ch) for details.

The activities of the Working Group on Standard Stars were reported by Bob GARRISON. This WG is supported by Commissions 25, 29, 30, and 45. Its function is to maintain, develop, and communicate lists of suitable standard stars. This is done through physical meetings during IAU General Assemblies, and via its newletter at http://clavius.as.arizona.edu/ssn. Membership is open to all interested astronomers.

Arlo LANDOLT reported for the Working Group on Filter Standardization. He has been working on a comparison of photoelectric and CCD photometry. He finds that CCD U magnitudes are very poorly related to photoelectric values, with a scatter of order 0.2 magnitude! However, B, V, R and I magnitudes from the two types of detectors are very similar; the scatter in colours measured with the two detectors is of order 0.015 magnitude. LANDOLT is also working on a set of photometric standards at $+45^o$ in addition to those already available at 0°.

COMMISSION 30: RADIAL VELOCITIES/VITESSES RADIALES

PRESIDENT:	C.D. Scarfe	Canada
VICE-PRESIDENT:	J.B. Hearnshaw	New Zealand
ORGANIZING COMMITTEE:	W.D. Cochran	USA
	L.N. da Costa	Brazil
	A.P. Fairall	South Africa
	F.C. Fekel	USA
	K.C. Freeman	Australia
	M. Mayor	Switzerland
	B. Nordström	Denmark
	R.P. Stefanik	USA
	A.A. Tokovinin	Russia

The commission's business meeting began on August 21 at 14:00. The following items were on the agenda.

1. Report of the President

1.1. MEMBERSHIP

The commission has lost a senior member by the death of F.C. Bertiau on 1995 December 27. In addition the following have resigned their membership: M. Azzopardi, L. Balona, H. Eelsalu, N. Martin and L. Oetken. However, the commission welcomed the following new members: B. Garcia (Argentina), Y. Gnedin (Russia), S. Hubrig (Germany), D. Queloz (Switzerland), A. Rastorguev (Russia), G. Scholz (Germany) and S. Udry (Switzerland). And it has proposed that the following non-members of the IAU be named as consultants to the commission: P. Butler (U.S.A.), N. Gorynya (Russia) and G. Marcy (U.S.A.).

1.2. MEETINGS

During the last three years the commission has lent its support to a variety of scientific meetings, including I.A.U. Symposium 179 on "New Horizons from Multi-Wavelwngth Digital Sky Surveys", held in Baltimore in August 1996, and the meeting in Calgary in June 1995 on the "Origins, Evolution and Destinies of Binary Stars in Clusters", which regrettably the Executive Committee was unable to support. We have also supported the following Joint Discussions at the 1997 General Assembly: (i) JD 8, "Stellar Evolution in Real Time", (ii) JD 11, "Redshift Surveys in the Twenty-First Century", and (iii) JD 12, "Electronic Publishing, Now and in the Future". Indeed the proposal for JD 11 originated in Commission 30, despite the fact that Commission 28 is listed as its principal sponsor, and most of the work of proposing and arranging that JD was done by Drs. Fairall and Huchra, to whom the president expressed gratitude.

1.3. INTERNAL OPERATION OF THE COMMISSION

i) Members' list. In order to facilitate communication between members, it was decided early in the triennium to prepare, and distribute to all, a list of members, with postal addresses, telephone and fax numbers, and electronic mail addresses for as many as possible. Unfortunately there remain about 15 members for whom we have no electronic address. At the suggestion of the vice-president, John Hearnshaw, a code for areas of research interests, developed largely by him, was added to the information for each member, and has proved very useful.

J. Andersen (ed.), Transactions of the International Astronomical Union Volume XXIIIB, 185–188.

ii)The commission has also initiated a procedure for choosing its new officers and organizing committee members that is more democratic than that previously in effect, so as to give members a more obviously direct say in the activities of the commission. Any two members of the commission may nominate candidates, and if an election is needed all members are entitled to vote.

iii)Finally the president thanked all those who participated in the activities of the commission over the past three years, including in particular the members of the organizing committee, whose advice and help has been greatly appreciated, and above all John Hearnshaw, who he said had been a superb vice-president, and could be expected to be an outstanding president in the next triennium.

2. Election

Exactly four candidates were nominated to fill the four vacancies on the Organizing Committee, so no election was necessary. However two members were nominated for the post of vice-president, necessitating an election. The president announced the result of that election:

A.A. Tokovinin	19 votes
A.P. Fairall	18 votes

Thus the new officers and organizing committee for 1997 to 2000 will be:

PRESIDENT:	J.B. Hearnshaw	New Zealand
VICE-PRESIDENT:	A.A. Tokovinin	Russia
ORGANIZING COMMITTEE:		
A.Continuing Members:	W.D. Cochran	USA
	F.C. Fekel	USA
	B. Nordström	Denmark
	R.P. Stefanik	USA
B. New Members:	T. Mazeh	Israel
	N. Morell	Argentina
	H. Quintana	Chile
	S. Udry	Switzerland

The president thanked all candidates for their willingness to serve.

3. Galaxy radial-velocity catalogues

A.P. Fairall summarized the current situation, noting the existence of the following five major data bases.

a. NASA/IPAC Extragalactic Database (NED), maintained by H. Corwin, B. Madore et al., which currently includes about 770,000 extragalactic objects, of which over 100,000 have referenced redshifts. About 50,000 redshifts include detailed information on instrumentation, measurement and reduction. Conversion is available between geocentric, heliocentric and 3K background reference frames or between user-defined reference frames.

b. Lyon Extragalactic Database (LEDA), maintained by G. Paturel et al., for which no current details were available.

c. Strasbourg Data Centre (SIMBAD).

d. Harvard-Smithsonian Center for Astrophysics Catalog (ZCAT), maintained by J. Huchra et al., which currently includes 91022 entries, including 87904 galaxies with velocities, 74255 of which are published.

e. Southern Redshift Catalogue (SRC), maintained by A. Fairall, which includes all measures published for each galaxy, noting any large differences.

Dr. Fairall recommended that the commission include in its report a reminder to those publishing redshift data to provide information necessary for users of those data, and the commission agreed to do so, with the following statement: "Commission 30 urges all those who publish radial velocities of galaxies to publish `heliocentric' velocities; if another frame of reference is used, this should be clearly indicated. The velocities should be expressed as cz (i.e. without relativistic correction). Accurate co-ordinates should also be provided."

4. Stellar radial velocity catalogues

H. Levato is the leader of a group in Argentina who have taken over responsibility for maintaining a record of, and publishing information on, work on stellar radial velocities, as successors to the group at Marseille led for twenty years by M. Barbier, who has recently retired. Dr. Levato presented the following progress report.

The Argentine group have continued with the compilation of a bibliography for papers that contain radial velocities of stars. Up to 1970 the compilation was made by Abt and Biggs and from 1970 to 1990 by M. Barbier. The work has been continued, looking for papers with radial velocity measures since January 1st, 1991. The period 1991-1994 is already complete, with more than 13,000 new objects added, and the catalogue has been deposited at the CDS. The CDS will add other data for the stars, such as magnitudes, spectral types, positions, etc. The period 1995-1996 is 80% complete and the first semester of 1997 is 40% complete.

Papers with radial velocity data have been sought in all the major journals: ApJ, ApJS, A&A, A&AS, PASP, MNRAS, AJ, Rev. Mex. Astron. Astrofis., The Observatory, and around a dozen more. The catalogue shows the radial velocity published by the authors in the case of one measure per star or the average for several measurements if the authors published it. Also indicated are the cases of relative velocities or Coravel velocities, and other unusual details of some kinds of measurements. It is hoped that the catalogue will be available to all users at the CDS as soon as possible. The data for 1995-1996 will be ready by the end of 1997. In the future it is planned to send new material to the CDS about every six months.

5. Standard-velocity stars

R. Stefanik, who chairs the commission's Working Group on Radial Velocity Standards gave a review and progress report on the efforts of that Working Group. For some time it has been known that the official list of IAU radial velocity standards (J. Pearce, Trans. IAU, 9, 441, 1955; R. Bouigue, Trans. IAU, 15A, 409, 1973.) included velocity variables, and that the velocities were not on an absolute or even common velocity system. An observational campaign involving the Dominion Astrophysical Observatory, the Geneva Observatory and the Harvard-Smithsonian Center for Astrophysics was initiated to address this problem. The objective was to establish a new set of late-type IAU radial velocity standard stars with individual mean velocities and an absolute zero point of the system good to 100 m/s. The initial results of this campaign were presented in Trans. IAU, 21B, 269, 1991. Nine stars were eliminated from the IAU list because they showed significant velocity variations of over 1 km/s. However, there was a color dependence of the zero point comparison between the observatories. Work has continued at the three observatories to refine the sample further, and to address the color dependence problem. More than 10,000 observations covering a time span approaching two decades, and in a few cases longer, have been obtained.

a. This monitoring shows that the following stars should be eliminated as velocity standards: HD 29587 (SB1, P=1483d, K=1.00 km/s); HD 42397 (a double- line spectroscopic binary of long but unknown period); HD 114762 (SB1, P=84d, K=0.60 km/s); HD 123782 (Pulsation variable, P=494d, K=0.95 km/s); HD 140913 (SB1, P=148d, K=1.83 km/s) and HD 171232 (Long period SB1 with a 4 km/s velocity decrease during the past 14 years).

b. As a step toward establishing an absolute zero point to the velocity system the Center for Astrophysics has been monitoring the velocities of minor planets. An absolute system is defined by minor planet velocities, computed from astrometric obits by the Minor Planet Center and good to several tens of mm/s. The offset of the CfA system from this absolute system is -95 +/- 18 m/s with no trends in the residuals with time, declination, hour angle, air mass, or signal-to-noise. This comparison covers eight years with over 800 observations of 25 minor planets.

c. The color dependence in the comparison of velocities from the three observatories continues to be an unresolved problem. There does not appear to be a color dependence for solar type stars with B-V between 0.5 and 0.8. The combined mean data from CfA and DAO do not show any color dependence or a difference in zero-point between the bright and faint groups of standards. However, the differences between CfA+DAO and the mean velocities from CORAVEL are correlated with the color indices of the stars, becoming increasingly negative for redder stars. Toward resolving this problem the standard stars are being reobserved by the Geneva team using ELODIE for comparison with the CORAVEL results.

d. Little additional progress has been made in establishing a list of early spectral type standard velocity stars from that reported in Trans. IAU, 21B, 269, 1991. Fekel continues to monitor the early type standard candidates at KPNO. He reports that the following stars are variable or probably variable and should be removed from the candidate list: HD 145570 (HR 6031), HD 147394 (HR6092), HD 179761 (HR 7287) and HD 196426 (HR 7878).

In response to a question by J.B. Hearnshaw, Dr. Stefanik indicated that the absolute accuracy of the standard star system is probablv about 100 m/s.

6. Meeting proposal

J.B. Hearnshaw reported on the proposal for an IAU Symposium on Precise Stellar Radial Velocities, that he had sent to the Assistant General Secretary in November 1996. The proposal is for a meeting in Victoria, Canada, from June 21 to June 26, 1998. The major purpose of the meeting is to explore the various applications of highly precise velocity data. The scope will therefore be broad, and will include such topics as stellar pulsation and line asymmetry, as well as the search for extrasolar planets. The Scientific Organizing Committee (SOC) will be chaired by J.B. Hearnshaw, and the Local Organizing Committee (LOC) by C.D. Scarfe. Further information can be found at the World Wide Web site:

http://astrowww.phys.uvic.ca/prvs/prvs.html

that has been set up by the LOC.

(Subsequent to this meeting, the proposal was accepted by the Executive Committee not as a Symposium, but a Colloquium, numbered 170 in the IAU's series.)

7. End of meeting

There being no further business, the meeting adjourned at 15:45.

DIVISION X

RADIO ASTRONOMY

Division X provides a common theme for astronomers using radio techniques to study a vast range of phenomena in the Universe, from exploring the Earth's ionosphere or making radar measuremnts in the solar system, via mapping the distribution of gas and molecules in our own and other galaxies, to the study of previous vast explosive processes in radio galaxies and QSOs and the faint afterglow of the Big Bang itself.

PRESIDENT: James M. Moran
Center for Astrophysics, MS 42, 60 Garden St., Cambridge, MA 02138, USA
Tel. +1 617 495 7477
Fax. +1 617 495 7345
E-mail: moran@cfa.harvard.edu

BOARD

See following pages

PARTICIPATING COMMISSION

COMMISSION 40 **RADIO ASTRONOMY**

REPORT ON COMMISSION 40 BUSINESS MEETINGS

JOHN WHITEOAK
Australia Telescope National Facility CSIRO
PO Box 76
Epping NSW 2121
Australia

The 23rd IAU General Assembly was extremely busy, with many unfortunate overlaps of interesting meetings. Because Commission 40 has its own Division (Div. X), as its President I was further disadvantaged, because for the first time, Division Presidents were invited to the meetings of the IAU Executive Committee, and these extended over several days during the General Assembly. However, to our advantage, the participation of Division Presidents provided an information link between the Commissions and the Executive Committee.

Commission 40 held four 'Business Sessions' during the General Assembly, on August 20, 25 (two sessions), and 26. The first two sessions were devoted to strictly business activities, whereas the other two were scientific sessions at which a number of astronomers presented short informal scientific reports. Commission 40 Business Highlights are as follows:

1. Professor Lucia Padrielli of Italy is the new Commission 40 Vice-President. She will become President at the end of the next General Assembly, and in fact the first Italian President of the Commission.

2. I retired as President of the Commission, but as the Past President I will remain a member of the Organizing Committee until the next General Assembly.

3. My tally of Commission members listed in the latest Transactions of the IAU yielded 762 members (probably for epoch 1996 October). A total of 88 new members should have been added to this list after the latest General Assembly.

4. The following retired from the Commission 40 Organizing Committee:

Prof E. Baart (South Africa)
Dr F. Colomb (Argentina)
Dr E. Gerard (France)
Dr J. Gomez Gonzales (Spain)
Dr I. Gosachinskij (Russia)
Prof M. Morimoto (Japan)
Dr T. Velusamy (USA)
Prof S. Ye (China PR)

The new members of the Commission are:

Dr L. Bronfman (Chile)
Dr P. Dewdney (Canada)
Dr L. Litvinenko (Ukraine)
Dr J-M. Marcaide (Spain)
Dr R-D. Nan (China PR)
Dr G. Nicolson (South Africa)
Dr R. Schilizzi (Netherlands)

J. Andersen (ed.), Transactions of the International Astronomical Union Volume XXIIIB, 191–192.

To complete the membership, the continuing members are:

Dr K. Anantharamaiah (India)
Dr L. Baath (Sweden)
Dr E. Berkhuijsen (Germany)
Dr R. Davis (UK)
Dr H. Dickel (USA)
Dr K. Johnston (USA)
Dr J. Moran (USA) - President
Dr L. Padrielli (Italy) - Vice-President
Dr J. Whiteoak (Australia) - Immediate Past-President

5. Representation on other committees etc:

R. Ekers (Australia) to Chair IAU Executive Working Group for Future Large Scale Facilities

S. Ananthakrishnan (India) and K. Kawaguchi (Japan) to join B. Doubinski (Russia) and A. Thompson as IAU representatives on IUCAF

J. Moran (USA) as a new IAU representative on URSI

A. Thompson (USA) and J. Whiteoak (Australia) to represent the IAU at the International Telecommunication Union (Radiocommunication)

6. The Commission 40 Working Group on the astrophysically most important spectral lines has been inactive during the last three years. However, IUCAF now has a Working Group that has taken over the task. At the ITU's World Radiocommunication Conference in 1999 there may be an Agenda item related to a revision of radio astronomy allocations at frequencies above 70 GHz. It is important that the radio astronomy community has a well-prepared position set up by the time of this meeting.

7. After some discussion at both Commission 40 and Executive Committee meetings, the Executive Committee decided to continue with the 3-year Commission reports. However, the Committee may provide a pro-forma to facilitate the task.

8. A meeting of Observatory Directors was Chaired by R. Booth (Sweden). The main aim of the meeting was to formulate a resolution, 'The Kyoto Declaration', to be signed by the Directors. A draft document was set up during the General Assembly, and published in an issue of the daily IAU newspaper ('The Sidereal Times'). Since then it has been distributed around the observatories by email. In the document the Directors resolve to conduct educational activities related to protecting radio astronomy's spectrum requirements, and to study means to mitigate the problem of radio interference. Additional matters include an endorsement of the Radio Astronomy Working Group of the OECD's Megascience Forum, agreement to intensify participation in ITU's frequency regulation processes, and agreement to increase the level of coordination between the world's radio observatories in order to present a common position on issues of radio spectrum management.

9. Six proposals for IAU-sponsored Colloquia or Symposia in 1998 were co-sponsored by Commission 40. At an Executive Meeting towards the end of the General Assembly, the following were supported:

Symp. 190: New Views of the Magellanic Clouds - Victoria, Canada, 1998 July 13-19.
Symp. 194: Activity in Galaxies and Related Phenomena - Byurakan, Armenia, 1998 August 17-21.
Colloq.171: The Low Surface Brightness Universe - Cardiff, UK, 1998 July 6-10.

10. Two possible future meetings sponsored by Commission 40 which are already under discussion are a 1999 meeting (in India) on long-wavelength radio astronomy, and a 2000 meeting (in Manchester) involving astronomy at high angular resolution. Information on these can be obtained from V. Kapahi (vijay@gmrt.ernet.in) and R. Schilizzi (rts@nfra.nl) respectively.

11. The two sessions on short scientific contributions proved quite popular; 23 different presentations were squeezed into the limited time available.

DIVISION XI

COMMISSION 44

SPACE AND HIGH ENERGY ASTROPHYSICS

Division XI connects astronomers using space techniques or particle detectors for an extremely large range of investigations, from in-situ studies of bodies in the solar system to orbiting observatories studying the Universe in wavelengths ranging from radio waves to gamma rays, to underground detectors for cosmic neutrino radiation.

PRESIDENT: Willem Wamsteker, *ESA IUE Ground Station, Vilspa, Apdo 50727, E - 28080 Madrid, Spain*
Tel. +34 1 813 1100, Fax. +34 1 183 1139
E-mail: ww@vilspa.esa.es

BOARD:

	S. Baliunas	USA	baliunas@cfa.harvard.edu
	N. Brosch	Israel	noah@wise1.tau.ac.il
	C. Cesarsky	France	cesarsky@sapvxg.saclay.cea.fr
	Th.-J.L. Courvoisier	Switzerland	Thierry.Courvoisier@obs.unige.ch
	J.M. da Costa	Brazil	Fax 55-123 21 8743
	V. Domingo	ESA	vdoming@esa.nascom.nasa.gov
	C. Fransson	Sweden	claes@astro.su.se
	A. Fabian	UK	acf@ast.cam.ac.uk
ex officio	G. Fazio	USA	gfazio@cfa.harvard.edu
	G. Hasinger	Germany	ghasinger@aip.de
	H. Inoe	Japan	inoue@astro.isas.ac.jp
	Zhongyuan Li	PR China	lzy@ms.ess.ustc.cn
	A. Michalitsianos	USA	
	P. O'Brien	UK	pto@star.le.ac.uk
	G. Oertel	USA	goertel@stsci.edu
**	H. Okuda	Japan	okuda@astro.isas.ac.jp
	H. Quintana	Chile	hquintana@astro.puc.cl
	T.N. Rengarajan	India	renga@tifrc3.tifr.res.in
	R. Schilizzi	NL	rts@nfra.nl
	B. Shustov	Russia	bshustov@inasan.rssi.ru
*	G. Srinivasan	India	srini@rri.ernet.in
	H. Thronson	USA	hthronso@hq.nasa.gov
	O. Vilhu	Finland	osmi.vilhu@helsinki.fi
	Zhen ru Wang	China PR	zrwang@nju.edu.cn

* President, Commisison 44
** Vice-President, Commisison 44

J. Andersen (ed.), Transactions of the International Astronomical Union Volume XXIIIB, 193.

DIVISION XI/COMMISSION 44

SPACE & HIGH ENERGY ASTROPHYSICS/
ASTROPHYSIQUE SPATIALE & DES HAUTES ENERGIES

PRESIDENT: *Giovanni G. Fazio, Harvard Smithsonian Center for Astrophysics, Cambridge, MA 02138, USA (gfazio@cfa.harvard.ed)*

1. Introduction

The business meeting of Division XI/Commission 44 was held on Saturday 23 August 1997 from 11:00 AM until 12:30 PM in Room H at the XXIII General Assembly of the IAU in Kyoto, Japan. Sixteen members of the Commission were present.

Chair of the meeting was Dr. Giovanni G. Fazio, President, Division XI/Commission 44. Also present were the two Vice-Presidents, Dr. Willem Wamsteker and Dr. Ginesan Srinivasan. The meeting began with introduction of the President and Vice-Chairman, followed by a reading of the proposed agenda. The agenda was accepted.

The President reviewed the history of the merger of Commissions 44 (Space Astronomy) and 48 (High Energy Astrophyics) into the current Commission 44 and also discussed the proposed structure of the Division and the Commission for the coming year. Also presented and discussed was the need for an increased role of the Division Presidents in the deliberations of the IAU Executive Committee.

2. President's report

The President reported that at the meeting of the XIII General Assembly of the IAU the Commission co-sponsored the following Symposia:

Symposium 183. Cosmological Parameters and Evolution of the Universe
Symposium 186. Galaxy Interactions at High Redshift
Symposium 187. Hot Universe

Supported or co-sponsored the following Joint Discussions:

JD 8. Stellar Evolution in Real Time
JD 9. Future Large Scale Facilities in Astronomy
JD 12. Electronic Publishing
JD 13. Detection and Study of Planets Outside the Solar System
JD 18. High-Energy Transients
JD 19. Astronomy from the Moon

and supported one Special Session: Highlights of the ISO Mission

J. Andersen (ed.), Transactions of the International Astronomical Union Volume XXIIIB, 195–197.

For 1998 the Commission has agreed to co-sponsor four proposals for Symposia:

98-01 Population Synthesis: From Planets to Black Holes, Black Holes to the Universe
98-06 Nuclei in the Cosmos
98-07 New Views of the Magellanic Clouds
98-12 Wof-Rayet Phenomena in Massive Stars and Starburst Galaxies.

Through the efforts of Dr. Wamsteker the Commission also established a World Wide Web page at http://www.vilspa.esa.es/IAU-XI/ This page contains a summary of current astronomy space missions as well as a membership list of the Commission.

In the Commission there is one Working Group: Astronomy from the Moon This WG is chaired by Professor Yervant Terzian. The Working Group sponsored a Joint Discussion at the meeting of the General Assembly.

3. Membership Status and Issues

The membership status was reviewed and questions were answered concerning application for membership in the Commission. There are approximately 600 members in the Commission.

4. Reports on Astronomy

Prior to the General Assembly Meeting considérable discussion was held among the Division and Commission Presidents concerning the value of the IAU Reports on Astronomy and whether they should be continued. At this Business Meeting a discussion was held to view the reactions of the Commission. In general there was the reaction that the Reports, in their present form, are not very useful to the membership and are rarely referred to in the scientific literature. It was also noted that almost all of what is now in the Reports can be accessed more rapidly by use of the World Wide Web. Instead, it was proposed the Reports be used as a summary of Commission activity, identifying the direction that space research is heading and noting current trends. Such a summary does not exist anywhere else.

5. Report of the Working Group on Astronomy from the Moon

Professor Yervant Terzian, Chair, reviewed the activity of the Commission's Working Group. The Group sponsored a very successful and well attended Joint Discussion on Astronomy from the Moon at the General Assembly.

Interest in the topic and among the members is still very high. The Group will also support a Resolution to the General Assembly on Protection of the Far Side of the Moon. The Working Group has also received an invitation from COSPAR to sponsor a session at its next meeting (Nagoya, Japan).

Professor Terzian has agreed to remain Chair of the WG for the next three years.

6. Concept for a World Space Obervatory (WSO)

Vice-President Wamsteker presented a concept for a World Space Observatory as a challenge for the new millenium. The basic idea behind the WSO is that general facilities in the windows for astronomical observations which require satellite observatories, are better done through a project with a world-wide support, participation and contribution, rather than specific projects defined in a more limited national configuration. Such an observatory would also help increase the participation of developing countries in astronomy and space science by providing front line facilities and tools for their research. The Commission supported the concept and recommended further evaluation of such an observatory. The possibility was raised that the UNISPACE III Conference in Vienna in 1999, might present a good occasion to expand on the implementation of such concepts on a truly world-wide scale.

7. Election of Officers

PRESIDENT DIVISION XI:	W. Wamsteker	ww@vilspa.esa.es
PRESIDENT COMMISSION 44:	G. Srinivasan	srini@rri.ernet.in
VICE-PRESIDENT COMMISSION 44:	H. Okuda	okuda@astro.isas.ac.jp

The Division proposes, that, at the XXIV General Assembly of the IAU, Commission 44 be dissolved and that only Division XI exist, and that Vice-President H. Okuda become President of Division XI.

The proposed new name for the Division would be Space Astronomy.

ORGANIZING COMMITTEE IAU DIVISION XI / COMMISSION 44 1997-2000

	S.	Baliunas	USA	baliunas@cfa.harvard.edu
	N.	Brosch	Israel	noah@wise1.tau.ac.il
	C.	Cesarsky	France	cesarsky@sapvxg.saclay.cea.fr
	Th.J.L.	Courvoisier	Switzerland	Thierry.Courvoisier@obs.unige.ch
	J.M.	da Costa	Brazil	Fax 55-123 21 8743
	V.	Domingo	ESA	vdoming@esa.nascom.nasa.gov
	C.	Fransson	Sweden	claes@astro.su.se
	A.	Fabian	UK	acf@ast.cam.ac.uk
ex-President	G.	Fazio	USA	gfazio@cfa.harvard.edu
	G.	Hasinger	Germany	ghasinger@aip.de
	H.	Inoe	Japan	inoue@astro.isas.ac.jp
	Zhongyuan	Li	PR China	lzy@ms.ess.ustc.cn
	A.	Michalitsianos	USA	
	P.	O'Brien	UK	pto@star.le.ac.uk
	G.	Oertel	USA	goertel@stsci.edu
	H.	Quintana	Chile	hquintana@astro.puc.cl
	T.N.	Rengarajan	India	renga@tifrc3.tifr.res.in
	R.	Schilizzi	NL	rts@nfra.nl
	B.	Shustov	Russia	bshustov@inasan.rssi.ru
	H.	Thronson	USA	hthronso@hq.nasa.gov
	O.	Vilhu	Finland	osmi.vilhu@helsinki.fi
*	W.	Wamsteker	Spain	ww@vilspa.esa.es
	Zhen ru	Wang	PR China	zrwang@nju.edu.cn

8. Obituary

With great sadness it was noted that on 29 October 1997, our colleague and friend Dr. Andy Michalitsianos, recently elected as member of the Scientific Organizing Committee of Division XI, passed away, after a prolonged sickbed. Our best wishes accompany his wife and three children.

25 November 1997

COMMISSIONS OF THE EXECUTIVE COMMITTEE

COMMISSION 5	DOCUMENTATION & ASTRONOMICAL DATA
COMMISSION 6	ASTRONOMICAL TELEGRAMS
COMMISSION 14	ATOMIC AND MOLECULAR DATA
COMMISSION 38	EXCHANGE OF ASTRONOMERS
COMMISSION 46	TEACHING OF ASTRONOMY
COMMISSION 50	PROTECTION OF EXISTING AND POTENTIAL OBSERVATORY SITES

COMMISSION 5. DOCUMENTATION AND ASTRONOMICAL DATA (DOCUMENTATION ET DONNEES ASTRONOMIQUES)

PRESIDENT: B. Hauck
VICE-PRESIDENT: O. Dluzhnevskaya
ORGANIZING COMMITTEE: H.A. Abt, M.S. Bessel, M. Crézé, A.G. Hearn, H. Jenkner, Li Qi-Bin, A. Piskunov, E. Raimond, G.R. Riegler, W.H. Warren, D.C. Wells, R. Wielen, G.A. Wilkins
SECRETARY: F. Genova

1. Business meeting (August 22, 1997, Chairperson B. Hauck)

1.1. REPORT OF THE PRESIDENT FOR 1994–1997

All the Commission members received last Winter the Commission activity report which was published in Newsletter # 13 and which also appeared in IUA Transactions Vol. XXIIIA/1997.

Meetings of several WG and TG of Commission 5 have been and will be held at this GA (section 2). A Commission 5 meeting about "Ground–based observatories data handling and archiving" was organized by F. Pasian on August 21st (Section 3). Joint Discussion 12 "Electronic Publishing: Now and the Future", organized by A.G. Hearn, chairperson of the WG on Information Handling, will be held on August 25th.

The conference held in July 1996 in St Petersburg, on "International Cooperation in the Dissemination of Astronomical Data", was a great success. It allowed for example a special meeting between five representatives of astronomical data centres to take place. The proceedings have been published in Baltic Astronomy, Vol 6, Number2, 1997, and are available on the WWW at URL
`http://www.inasan.rssi.ru/~colloq/`.

Four meetings related to the scope of Commission 5 will soon be held. One, the third edition of "Library Information Services in Astronomy" (LISA III), is sponsored by our Commission. It will be held in Puerto de la Cruz in April 1998. The others are the third edition of "Astronomy with large databases" (ALD III), to be held in Postdam, the 1997 edition of "Astronomical Data Analysis Software and Systems" (ADASS'97), to be held in Sonthofen in September 1997, and the IXth Canary Island Winter School of Astrophysics on "Astrophysics with large databases in the Internet age", in November 1997 in Tenerife.

The last Business Meeting of Commission 5 in The Hague adopted a new structure for the Working Groups and Task Groups. The main argument for proposing this new structure was the very high number of Working Groups. Furthermore, it was necessary to adapt the scope of the WG to the evolution of the field covered by the Commission, and thus the Working Group on Information handling was created. Since the last General Assembly, Commission 5 has three WGs covering broad fields of interest and several TGs which cover narrower, or more technical fields. This new structure operated well, but the FITS TG insists that they would prefer by far to be a WG. For them, the difference between a WG and a TG seems to have a sociological importance, mainly due to the fact that the FITS community is much larger than the IAU TG. The President suggests that the point be discussed between the new President and Vice–President and the FITS Task Group.

Finally, from the last GA to now, 6 Newsletters have been sent to Commission members and numerous librarians.

B. Hauck thanks all the colleagues, especially the SOC members, who have helped him during his two terms as President of Commission 5.

1.2. REPORTS FROM THE CHAIRPERSONS OF THE WORKING AND TASKS GROUPS

Several WG and TG meetings were held in Kyoto. A short summary of the activities and discussions will be given in the next Section of this paper. More detailed reports will be posted on the Commission 5 WWW service if available.

J. Andersen (ed.), Transactions of the International Astronomical Union Volume XXIIIB, 201–206.

1.3. OFFICERS, ORGANIZING COMMITTEE, MEMBERS AND CONSULTANTS FOR 1997-2000

The SOC proposed to the General Assembly is composed of President O. Dluzhnevskaya, Vice-President F. Genova, and the members : M.S. Bessel, P.B. Boyce, H.R. Dickel, B. Hauck, A.G. Hearn, H. Jenkner, F. Murtagh, K. Nakajima, F. Ochsenbein, A. Piskunov, E. Raimond, G.R. Riegler, R. Wielen, Zhao Yongheng, D.C. Wells.

The following new members and consultants were proposed to the General Assembly :

- new members : P.B. Boyce (Washington DC), Guo Hongfeng (Beijing), S. Lesteven (Strasbourg), D. Lubovich (AIP), O. Malkov (Moscow), F. Murtagh (Armagh), F. Pasian (Trieste), A.H. Rots (GSFC), R. White (GSFC), Zhao Yongheng (Beijing);
- consultants for 1997-2000 : T. Banks (New Zealand), S. Borde (Paris), E. Bouton (NRAO), C. Cheung (GSFC/ADC), B. Corbin (USNO), M. Cummins (DDO), G. Eichhorn (CfA/ADS), M. Gomez (Tenerife), U. Grothkopf (ESO), M. Hamm (Strasbourg), S. Laloe (Paris), Li De-He (Lintong, Xian), M. Schmitz (IPAC/NED).

Resignations of B. Cogan, A. Heck and C. Jaschek were recorded.

Working groups and Task groups and their chairpersons for 1997-2000 :
WG Astronomical Data : E. Raimond, Vice-chair R.P. Norris
WG Information Handling: P.B. Boyce
WG Libraries: F. Murtagh/U. Grothkopf
TG Data Centres & Networks: F. Genova
TG Designations: H.R. Dickel
TG FITS: D.C. Wells, Vice-chair E. Raimond
TG UDC 52: G.A. Wilkins

1.4. RESOLUTIONS

Six resolutions were discussed and proposed to the General Assembly, which adopted three of them on August 27th, dealing with:

- proposal for registering a new acronym;
- support to data archiving and Data Centers;
- FITS evolution to take into account year 2000.

The questions raised by the recent developments about Astronomy and Astrophysics Abstracts, which had been brought to the knowledge of Commission 5 by R. Wielen, were thoroughly discussed, and a resolution about the subject was proposed to the General Assembly. The Commission of Resolutions sent this resolution to the Executive Committee for follow-up. Similar action was taken for another resolution proposed by Commission 5 about the need for data links with a sufficient bandwidth.

Quite generally, the importance of the "de facto" standards used in astronomy as been continuously stressed during the various Commission 5 meetings. In particular, a fifth resolution proposed as a result of discussions during the "Ground-based observatories data handling and archiving" meeting is taken into account at the level of Commission 5 :

Considering
the fast evolution towards fully interlinked astronomy on-line resources which rely heavily on communication standards, and the need to include observatory archives in this framework;
considering
the current development of ground- and space-based projects aimed at providing the astronomical community with large-scale observing facilities, and the need to harmonize the access to the resulting huge amount of data;

the IAU Commission 5 supports the efforts currently being carried out to define a standard protocol to access individual observatory data, and points to the need to define a generalized and unique format to describe individual observations.

Furthermore, the "bibcode" standard, first defined by the CDS and NED, and now heavily used by the ADS and the electronic journals, was put under the responsibility of the WG on Information Handling.

More work will be carried out in the coming months, on the procedures to maintain the standard, and a recommandation allowing the recognition of the bibcode as an official reference standard of the IAU should be prepared for the next GA, as proposed by G. Eichhorn.

1.5. ACTIVITIES FOR THE NEXT TRIENNIUM

O. Dluszhnevskaya presented her views on the Commission activities for the next triennium. She pointed at the growing volume of observational material, and the rapid development of electronic publications. One goal of the Commission will continue to be to encourage the observatories to open their archives. The Newsletter distribution will be actively continued, and contributions and suggestions are welcome. This can be used e.g. as a support to transmit information about standards.

A email distribution mechanism will be implemented, with support from NRAO. A WWW page describing Commission 5 activities will soon be opened at INASAN.

2. Working and Task Groups meetings and reports

As explained above, only a brief summary of the Working and Task Groups meetings and and of reports sent by the WG and TG chairpersons is given in this text. More detailed versions will be posted on the Commission 5 WWW service at INASAN when available.

2.1. TG ON DESIGNATIONS (AUGUST 21ST, 1997, CHAIRPERSON H.R. DICKEL)

A joint meeting of the TG on Designations, with members of the FITS TG, Data Centers, Journal editors, librarians, was held on the same day before the TG Business Meeting, to discuss recent developments in TG on Designation and common problems. Three main topics were discussed:

- Description of the new IAU "Registry of new Acronyms" by F. Genova (CDS). She described the "Second Dictionary of Nomenclature of Celestial Objects outside the solar system", which is maintained and updated by the CDS. She then presented examples of its use and of the new Registry of Acronyms which is part of this on-line Dictionary (there is also a Resolution re the Registry).
- Proposal to increase the minimum support length at 8 characters for the FITS keyword `OBJECT` to 32 characters. Several people spoke in support of this proposal; the main speaker was D.C. Wells of NRAO who is chair of the FITS TG.
  There is no technical reason that this cannot be changed from the original, now out-dated 8 character limit to 32 characters. D.C. Wells had put this proposal on the FITS group news but there was no response - pro or con. The Data Centers and electronic journals realize the need for this change to facilitate the exchange, search and retrieval of data; unique designations are needed for objects and it is no longer possible with 8 or even 16 characters. 24 characters would suffice for many current designations but 32 allows for the immediate future.
- Short presentations were made, including P.B. Boyce (AAS electronic journals), F. Genova (CDS), and D. Lubowich (AIP), regarding how to facilitate the proper complete designations being given in papers, data tables, before the paper is submitted to journal and/or data submitted to an archive.

2.2. WG ON LIBRARIES (AUGUST 21ST, 1997, CHAIRPERSONS B. HAUCK AND U. GROTHKOPF)

U. Grothkopf presented the report she had prepared with B.G. Corbin (to be distributed in the next Commission 5 Newsletter). An impressive list of services maintained by librarians was presented: list of astronomy Newsletters (C. Van Atta, NOAO), astronomy book reviews database (M. Cummins, U. Toronto), list of IAU colloquia and other meetings (S. Stevens-Rayburn, STScI), list of astronomy librarians and libraries (U. Grothkopf, ESO), list of international astronomy meetings (L. Bryson, CFHT), an international clearinghouse for observatory manuals and information on how to access them (L. Bryson, CFHT), the Physics-Astronomy-Mathematics (PAM) Division of the Special Libraries Association (SLA) homepage (D. Stern, Yale). Several projects were also discussed : a Distributed database of Online Astronomy Preprints and Documents (NASA Grant NAG5-3942, PI R.J. Hanisch), a catalogue of individual observatory publications (B. Corbin, USNO), a preservation effort for the series of observatory publications (D. Coletti, CfA, B. Corbin, USNO). Other important resources are the astronomy Thesaurus (R. and R. Shobbrook) and the Union List of Astronomy Serials (J.L. Bausch, Yerkes).

Two distribution lists are available, which are specifically for astronomy librarians, Astrolib, managed by E. Bouton (NRAO) and EGAL (European Group of Astronomy Librariansa, managed by I. Howard, RGO).

The third LISA conference will be hosted by the Instituto Astrofisica de Canarias (IAC), Tenerife, from April 21 through 24, 1998.

Electronic journals licensing issues, and the fate of Astronomy and Astrophysics Abstracts, were also discussed.

2.3. TG ON REVISION OF UDC 52 (AUGUST 21ST, 1997, CHAIRPERSON G.A. WILKINS)

The report presented by G.A. Wilkins (to be found on Commission 5 WWW service) was discussed. It covered the current status of revision, the procedures for review, checking and approval of drafts, the relationship between UDC 52 and the Astronomy Thesaurus, and the procedures for the maintenance of UDC 52 and Astronomy Thesaurus. The need for greater support from the astronomical community was emphasized. Conclusions can be summarized in the following way: it was recognized that UDC was no longer widely used in astronomy, but, nevertheless, it was hoped that the revision would be completed as soon as possible and that the schedule and index for astronomy would be made available on the World Wide Web. It was noted that the Astronomy Thesaurus and UDC 52 are complementary documents and so it was considered that it would be appropriate to maintain them in a cooperative manner so as to insure that they would both be up-to-date and mutually compatible.

2.4. TG DATA CENTERS AND NETWORKS (AUGUST 21ST, 1997, CHAIRPERSON F. GENOVA)

Several members of the TG took a very active part in the organization of the St Petersburg Colloquium "International Cooperation in dissemination of Astronomical Data" (July 2-9, 1996, Co-chairs O. Dluzhnevskaya and B. Hauck). The data center activities, the diffusion of data, catalogues, the new possibilities offered by electronic publication, etc were thouroughly presented. This meeting was an excellent occasion to create collaborations, in particular for the participants of Eastern Europe and FSU countries. A TG meeting was held during the Colloquium, with five of the Data Center Directors and key personnal. Collaboration, and the common standard for table description, were discussed. A resolution was proposed, assessing the need for data archiving and the role of the data centers.

A second task group meeting was held in Kyoto. C. Cheung, F. Genova, K. Nakajima, and O. Dluzhnevskaya, Directors of the American (ADC), French (CDS), Japanese (ADAC), and Russian (INASAN) Centers, presented their activities. The data centers share a common data set of catalogues and tables, and a common standard for table description, first proposed by CDS and now shared by the data centers and several major journal editors (A&A, AAS, Astronomicheskii Zhurnal). This standard is the key for data checks, exchange and transformation. The presentations illustrated the good collaboration between the Data Centers, and their impact on the dissemination of astronomical information for the world-wide astronomical community. The data centers (also the Chinese and Indian ones) also play an important role at regional level. The Data Center activities are more and more linked to the journals and the ADS, with the rapid evolution towards electronic publication, with in particular tables published in electronic form only. It is very important to have more journals joining the evolution.

2.5. TG FITS (AUGUST 21ST, 1997, CHAIRPERSON D.C. WELLS)

D.C. Wells presented a report on the activities of the FITS TG. The activities of the NSSDC FITS support office were also reviewed (report prepared by B.M. Schlesinger and presented by D.C. Wells).

It was recalled that the FITS WG was created by a General assembly resolution at Baltimore (1988); it is the "owner" of the FITS format, and controls the evolution of the FITS standard. There are three regional FITS Committees. The IAU FITS TG only considers proposed changes which have been approved by all regional committees and for which interoperability has been demonstrated, and which get approval by 3/4 majority and no "NO" vote. `BINTABLE` was approved in June 1994. The TG is currently considering year-2000 `DATE`*xxxx* proposal (a Resolution was approved in Kyoto on that topic). Probable activities for the next triennium include the approval of the year-2000 `DATE`*xxxx* agreement, in 1997; approval of the NOST V2.0 FITS standard, probably during 1998 (this will be a comprehensive compilation of the FITS standard); register of FITS with the Internet as a MIME data type, probably in 1998 after approval of the NOST standard; and to continue the progress towards a WCS (World Coordinate System) agreement.

A technical panel of the FITS support office is preparing draft for Version 2, including **IMAGE**, **BINTABLE**, blocking agreement, defining keyword value, table entry, data display formats no longer relying on FORTRAN definitions. This will be available for community review in 1–3 months. Version 4.0 of the User's Guide was released in May 1997 (348 accesses on ftp site in a couple of months). The Standard document got 246 accesses on ftp on the first half–year of 1997. WWW pages have been installed, which contains basics and information, and the list of extension types registered with the IAU TG. Expansion of the service is planned.

2.6. WG INFORMATION HANDLING (AUGUST 27TH, 1997, CHAIRPERSON A.G. HEARN)

The main WG activity in Kyoto was the organization of Joint Discussion 12, "Electronic Publishing, Now and the Future" (published in the Proceedings). P.B. Boyce will be the WG Chairperson for 1997-2000.

2.7. WG ASTRONOMICAL DATA (AUGUST 27TH, 1997, CHAIRPERSON E. RAIMOND)

The membership of the Working Group on Astronomical data comprises a relatively large fraction of the membership of Commission 5. All members have in common that they are actively working with astronomical data in a global way, i.e. more than an average astronomer would do in the course of his/her own research.

The activities of the Working Group as a whole are not so easily listed. During the past triennium a considerable amount of e-mail traffic between members of the WGAD has taken place. Many members contributed actively to compile the two-yearly report of the IAU to the ICSU organisation CODATA. A recently revised and updated version of this report reflecting the activities of the WGAD and other bodies of Commission 5 over the past three years (1994 - 1997), is available now.

The areas of work which are probably most specific to the WGAD are those concerning archiving of observed data, processing data and making the archived data available for later use. Interesting developments in this field have taken place in the last decennium or two. More and more observatories are setting up usable archives. And, although many of them have been plagued by shortage of funds and lack of active interest by the observers, the notion that a proper archive with data that can easily be used by outside users is getting more and more common.

After a learning period during which reasonably usable archives were set up, we've now reached the stage in which the design of the data archive is being made part of the design of a new telescope or a new observatory. The Italian Galileo Telescope, the ESO VLT and the Japanese Subaru telescopes are excellent examples. The special session on archiving at this General Assembly illustrates the growing interest in observatory archives.

R.P. Norris, ATNF, Australia, was elected as a vice-chair in 1997.

3. Meeting on "Ground–based observatories data handling and archiving" (August 21st, 1997, Chairperson F. Pasian)

This meeting was organized in coordination with the WG on Astronomical Data, to review the current status of data handling and archiving in active and furure ground–based observatories.

In his introduction, Fabio Pasian pointed out that most of the observing facilities of the new generation now include data archiving as one of their basic tasks, and this is a significant change with respect to the past. He reminded that the purpose of archives is twofold, i.e. technical (monitoring the performance of telescope and instruments) and scientific (re-using data for purposes different from the original ones). While the first purpose implies storing all possible data including telemetry, the scientific user should be shielded from this huge amount of data, and should access only calibrated scientific exposures. Data handling and archiving are therefore closely connected with data processing. As a matter of fact, the new projects in the optical domain (ESO, Gemini, TNG) consider the observatory data management as an "end-to-end" mechanism: starting from the observation proposal, the scheduling, observation, calibration, quality control, and archiving steps are performed; through archival research a scientist gathers information useful for proposing for observing time again. Retrofitting existing archives of observatories organized in the "traditional" way to match the new concepts has proven to be impossible in most cases.

Ernst Raimond presented the results of a survey made on the status of radio-observatories archives, made by e-mail and using the Web. Seventeen facilities were screened, ranging from mm to m wavelengths, and including VLBI networks and arrays. In general, major radio-observatories save their data, but only large institutions can afford to keep data readable on modern media. While an observations catalog is usually searchable by outside users, observations are kept off-line, and when available for external use after a proper period (18 months on average), their require staff support to be retrieved. Catalogs are usually searchable by position, and sometimes by other parameters (source name, type of observation, etc.). Results of surveys can usually be retrieved directly, or through a data center. The accessibility ranges from excellent to usable, although it sometimes happens that the archive is not easily traceable in the hierarchy of Web pages of the observatory, and hence could be "advertised" better.

The concepts behind the data handling and archiving in the Subaru telescope project, and a description of the system, were presented by Ryusuke Ogasawara. Purpose of the system, actually a supercomputer network with dedicated hardware for data storage, is to store all observational data (including the technical information on telescope operations), manipulate the data for off-line analysis, provide a single system for database handling and image processing and analysis, and support numerical simulations to be compared to actual observations. The hardware system includes a 50Gflops vector parallel processor, a 27-nodes 98-CPU scalar parallel server, 150 TeraBytes data storage machine, 2.4 TeraBytes on fast disk, 50 workstations and 100 PCs for user activity. The computers are all linked by a fast backbone network (266 Mbps), and the connection to the telescope is guaranteed by an ATM link. The system has been installed in March 1997, and the ATM link is expected to be activated in January 1998. This outstanding hardware setup has been made possible by the close cooperation among the Subaru project and several computer and peripheral manufacturers.

Françoise Genova focussed on a number of different aspects in which data centers can complement, integrate and harmonize the information produced by observatories. Task of the data centers is to hold "metadatabases" rather than observations, and to provide navigation tools among heterogeneous distributed services including observatory archives. By displaying a number of Web pages, the CDS was used as a practical example of the services which are or can be provided. The basic service of a data center is to provide full access to individual on-line catalogs (Astronomer's Bazaar at the CDS); upgrades are the possibility of using selection criteria on a library of catalogs and displaying the results (VizieR), or providing a new service by merging information from a library of different catalogs and integrating with full bibliographic information (Simbad, CDS bibliographic service). The crucial point to link the information available with the observatory archives is to be able to reference the observational data with standard keywords (i.e. an observation identifier): in this way, users can be informed that data are available, and can select and access useful data. A first step in this direction is the definition of a standard Internet-based query language (ASU - Astronomical Server URL), defined by CDS, ESO, IUE, CADC and OAT.

The discussion held at the end of the session evidenced the following items:

- there is an increasing need for the definition of standards allowing to access uniformly on-line observatory information;
- it is necessary to define mechanisms to link metadata and bibliography information to observational data;
- source names and acronyms in observatory catalogs should be standardized: this issue becomes a technical problem if data centers are used as name resolvers through network connections;
- keeping old data readable is a serious problem; staff is also required to allow access to off-line data: the availability of new storage devices and media at affordable prices may help alleviating these problems;
- there is a need for improvements in networking: a high-capacity world-wide scientific backbone will be essential in the near future;
- making calibration software available to remote users is a critical issue for observatory archives: distribution and maintenance of such software require a considerable amount of resources;
- all of the above items imply that proper staffing and funding are to be allocated by the agencies to observatory archives and data centers.

The results of the discussion were reflected in some of the resolutions Commission 5 decided to adopt in its business meeting, on the importance of data archiving, on the need of world-wide efficient data links, and on the need to define a protocol and a unique format to access observatory data.

COMMISSION 6: ASTRONOMICAL TELEGRAMS/TELEGRAMMES\ ASTRONOMIQUES

President: R. M. West
Secretary: B. G. Marsden

After the adoption of the agenda for the meeting, President West welcomed the eight other persons present and asked them to stand in memory of three members lost since the last meeting: Michael P. Candy, Antonin Mrkos and Leonida Rosino.

Given the very small number of active members of Commission 6, it was agreed that West and Marsden should serve second terms as President and Vice President, respectively. K. Aksnes, S. Nakano and E. Roemer were elected to the Organizing Committee. In the expectation that he would be elected to IAU membership, Green was transferred from consultant to membership status. C. Kouveliotou and M.K. Tsvetkov were also elected as new members of the commission, and G.R. Kastel and J. Ticha were appointed as consultants.

Remarking that Commission 6 is one of the IAU's earliest and that the rapid dissemination of urgent astronomical information by the Central Bureau for Astronomical Telegrams (CBAT) is one of the oldest and most essential services of the IAU, President West reminded those present that one of his hopes as president had been that it would be possible to disseminate the *IAU Circulars* free of charge: after all, in some countries, the cost of a subscription was a sizeable fraction of an observatory's budget. It was therefore very welcome news to read in the *Sidereal Times* that this had been accomplished that very day! Of course, the production of the *Circulars* is not without cost, and he noted that Vice President Marsden would report on how this arrangement had in fact been made. West pointed out that, since Commission 6 was outside the new IAU divisional structure, there were special demands on members of Commission 6. It was therefore rather deplorable that a certain fraction of the membership seemed to take only little interest in the commission.

Reporting on the CBAT's activity during the triennium, Marsden remarked that the downward trend evident in 1994 and 1995 had subsequently been significantly reversed, thanks to the appearance of two spectacular comets and extensive items on γ-ray bursters; furthermore, the 107 supernovae already announced in 1997 already represented a record for a full year. In discussing the financial arrangements whereby it had been possible to place the *IAU Circulars* freely in the World Wide Web just 160 minutes before the start of the meeting, Marsden noted that, during the preceding year, some 21 percent of the combined income of the CBAT and the Minor Planet Center (MPC) had come from subscribers to the Computer Service. Although there would still be a charge for e-mail delivery of *IAU Circulars* and use of the Computer Service (CS) on their own computers, the CBAT and the MPC could stand to lose this fraction of their combined income, some 65 percent of which goes for salaries and benefits, principally those of CBAT Associate Director D. W. E. Green and MPC Associate Director G. V. Williams, the latter serving also as webmaster and thus responsible for the Web CS dissemination of the *IAU Circulars*. What had made the free distribution possible was the anticipation that Williams' salary for the next four years would instead be paid by a grant from NASA. A recent editorial in *Nature* had condemned the use of the *IAU Circulars* as a forum for the publication of theoretical ideas. Marsden admitted that, although an attempt was made to weed out those sections of the items received that contained theoretical arguments, such publication occasionally happened. Certainly, it seems appropriate to use the *Circulars* to make a timely prediction that observations thereby inspired might confirm or deny (publication in *Nature*, for example, being too slow), and some background reasoning may then be necessary, but he agreed that some contributors went too far. It may not be appreciated that many of the items appearing on *the IAU Circulars* are in fact refereed, but this could always be done more thoroughly. Members of Commission 6 should be, and several of them are an important resource for refereeing activities.

J. Andersen (ed.), Transactions of the International Astronomical Union Volume XXIIIB, 207–208.

West reported on the Commission 20 vote concerning the uniqueness of comet names. There had in a fact been a tie between those who felt that names should be unique and those for whom it was sufficient that the designations be unique. The *status quo* would therefore be maintained, at least for the next three years. A second vote had caused the reinstatement of the practice that comets might have three names, rather than the two recommended by the committee set up to produce guidelines for comet naming. S. Nakano remarked on the fact that several amateur astronomers had recently made announcements of their discoveries of alleged comets, novae and supernovae over the internet. It was agreed that this was an unwise development that could deprive the rightful discoverers of credit. If would-be discoverers insist on internet announcement in this way, it was strongly recommended that they also inform the Central Bureau of their finds. It would also be useful if those who *do* make reports to the Central Bureau were to alert the Bureau to the fact that they had made an internet report. There was a strong feeling that, given the widespread availability of CCD equipment to amateurs, the latter should take some responsibility for making at least some initial confirmation of their claims. Otherwise, the whole concept of a ``discovery" could become very blurred. This was somewhat ironic, given the Commission 20 vote on reinstating the practice that comets should be named for up to three *discoverers*.

The meeting ended with a brief discussion on whether it would be appropriate, in these days of more modern communications, to change the word ``telegrams" in the names of both the commission and the Central Bureau. The prevailing view was that there is little wrong with maintaining a harmless tradition. The Bureau in fact still receives telegrams -usually by snail-mail- two or three times a year. In any case, some professional contributors frequently refer to the *Circulars* quite matter-of-factly as ``telegrams", a point that surely -and perhaps appropriately- stresses their urgency.

COMMISSION 14: ATOMIC & MOLECULAR DATA
DONNEES ATOMIQUES & MOLECULAIRES

President:	W.H. Parkinson
Vice President:	F. Rostas
Secretary	P.L. Smith

Business Meeting

The business meeting of IAU Commission 14 was called to order by F. Rostas (Vice-President) in the absence of W. H. Parkinson (President) at 14:00 on Thursday, August 21, 1997 in room H of the Kyoto International Conference Hall in Kyoto, Japan.

Present were: F. Rostas (France, Vice-President), W.C. Martin (USA, Outgoing WG 1 chair), (W.F. Huebner (USA) and R. Kurucz (USA)

W.H. Parkinson (President) and P.L. Smith (Secretary) were excused, being unable to attend due to lack of support from their national authorities. Most of the points on the agenda, especially the proposals for new officers and WG chairmen have been presented in a letter circulated by W.H. Parkinson before the meeting.

APPROVAL OF NEW OFFICERS

F. Rostas announced that P. L. Smith has agreed to serve as Vice-President and Nicole Feautrier as Secretary for the next triennium. There were no objections.

ORGANIZING COMMITTEE

According to the rules set at the 1991 GA in Buenos-Aires, members of the OC appointed in 1991 are leaving (S.J. Adelman, J. Dubau). They are replaced by K.A. Berrington, Nicole Feautrier and W.C. Martin. W.H. Parkinson remains in the OC, as ex-president and P.L. Smith as Vice-President.

The structure of Commission 14 for the triennium, 1997-2000 as it is shown on the table below is approved:

President:	François Rostas, France (1994)
Vice President:	Peter Smith, USA (1991)
Secretary:	Nicole Feautrier, France (1997)
Organizing Committee:	Keith A. Berrington, UK (1997)
	Nicolas Grevesse, Belgium (1994)
	Sveneric Johansson, Sweden (1994)
	W.C. Martin, USA (1997)
	Uffe Gråe Jørgensen, Denmark (1994)
	Helen Mason, UK (1994)
	William H. Parkinson, USA (1985), past President 1994-1997
	W.-Ü. L. Tchang-Brillet, France, (1994)

WORKING GROUPS:

W.C. Martin is retiring as chair of WG1, S. Johansson has agreed to replace him. D.R. Schultz has replaced Jean Gallagher as chair of WG3 since her death in 1995. Nicole Feautrier is becoming secretary of the Commission and has proposed Chantal Stehlé, who accepted, as new chair of WG4.

J. Andersen (ed.), Transactions of the International Astronomical Union Volume XXIIIB, 209–210.

As suggested by the executive group, vice chairpersons are being looked for by the WG Chairs. Some have already been designated as shown below.

Working Group Chair	Vice	Co- Chair
1. Atomic Spectra & Wavelengths	S. Johansson	TBD
2. Atomic Transition Probabilities	W.L. Wiese	J.R. Fuhr
3. Collision Processes	D.R. Schultz	TBD
4. Line Broadening	C. Stehlé	G. Peach
5. Molecular Structure	E. F. Van Dishoeck	J. H. Black
6. Molecular Reactions on Solid Surfaces	S. Leach	TBD

NEW MEMBERS

Nine new IAU members have chosen Commission 14 as their primary interest. These are K. Aggarwal (India), C. Barnbaum (USA), H. Boechat-Roberty (Brazil), F. Launay (France), A. Le Floch (France), H. Ozeki (Japan), C.C. Pei (China P.R.), F. Rogers (USA), J.-Y. Roncin (France).

EVOLUTION OF THE COMMISSION REPORTS

At the meeting of commission Presidents, the new General Secretary (J. Andersen) announced that the format of the Commission Reports published in Reports on Astronomy (Transactions A) will be modified: basically they would be reduced to a 2 page outline. Supplementary space would be requested by the Commissions who wish to maintain a detailed scientific report. Most presidents said they would be in that case. It has been agreed that Commission 14 would continue to publish an extensive study based on the Working Group's reports.

MEETING ON THE SPECTROSCOPY OF LARGE MOLECULES

The principle of such a meeting initiated and sponsored by Commission 14 is accepted. Collaboration with interested Commissions has to be looked for. The exact format and date will have to be worked out by the organizing committee in consultation with competent members.

COMMISSION 38: EXCHANGE OF ASTRONOMERS

Report of Business Meeting at 14:00, 22. August 1997

President: H. E. Jørgensen
Vice-president: M. Roberts
Secretary: R. M. West
Members present: President, Vice-president, J. Sahade, R. M. West, S.-H. Ye
Also present were: M. Yang, J. Zhao

Agenda:

1. Report of the President
2. Review of membership of the Commission for the next triennium including President, Vice-president and Organizing Committee
3. Review of the Guidelines for Grants
4. Other proposals from members
5. Any other business

The President opened the meeting welcoming those who attended. The agenda was adopted.

1) The President reported that during the triennium the Commission business was carried out by the Vice-president and himself mainly through E-mail. E-mail has greatly facilitated the application and granting procedure.

The President pointed out that the cooperation with the Vice-president had been excellent and he wished to express his sincere thanks to Dr. Morton Roberts.

During the triennium up to August 12, 1997, 31 grants were given and only very few replies were negative, since the applications did not meet the Guidelines for Grants. The grants that were approved went to astronomers in 14 different countries, namely:

India (6)
Argentina (4)
Russia (4)
USA (3)
Australia (2)
China (2)
Nigeria (2)
Turkey (2)
Bulgaria (1)
Egypt (1)
Israel (1)
Japan (1)
Morocco (1)
Ukrania (1)

J. Andersen (ed.), Transactions of the International Astronomical Union Volume XXIIIB, 211–213.

The number in parenthesis being the number of grantees from each particular country. The President pointed out that the program is a general program open to applicants of all nationalities. The host institute to which the grantees went were in the following countries:

USA (12)
UK (5)
India (4)
South Africa (3)
Argentina (1)
Canada (1)
Czech Republic (1)
France (1)
Germany (1)
Spain (1)
Sweden (1)

The numbers in parenthesis again being the number of grantees that went to a particular country.

These numbers deviate somewhat from the numbers in the Report of Commission 38 as published in the IAU Report on Astronomy, Vol XXIII A due to the fact that the Report was prepared in August 1996.

The President indicated the amount of money that was available in the IAU budget in the past two triennia and the amount that would be available for the next triennium as

1992 - 1994: 76.000 CHF
1995 - 1997: 88.000 CHF
1998 - 2000: 75.000 CHF

The decrease in the amount available for the next triennium compared to the past is not due to a lower priority of the granting program of Commission38. In fact the number of grants have increase considerably during the last triennium compared to previous ones. However, the price of air fares have been reduced considerably during the years.

According to the IAU Information Bulletin No. 80 the expenditures are as follow :

1991 - 1993: 39513 CHF
1994 - 1996: 50689 CHF

The budget for the next 3 years thus exceeds the expenditure in the last 3 years by approximately 50% and the President foresees no money problems.

The number of grants assigned in the last triennium is larger than in any recent triennia, namely

1976 - 1979: 18
1979 - 1982: 24
1982 - 1985: 23
1985 - 1988: 23
1988 - 1991: 27
1991 - 1994: 24
1994 - 1997: 31

Finally the President invited questions or remarks to his report. Dr. R. West asked about the negative replies and the trend. The President said that only 2-3 negative replies were given to applications. E.g. travel money was not given to go and take up a new job at another institute. Also grants could not be given to researchers in totally different fields. Furthermore, informal requests by E-mail had to be answered negatively unless the applications were in line with the Guidelines for Grants. Potential applicants were informed of the Guidelines.

Dr. F. Yang asked about the percentage of negative replies. The President mentioned that 31 grants were approved and with negative answers to only a few real applications as mentioned the percentage of negative replies was very low.

The President wished to stress that the Commission does not work as a Research Council. If an application is in agreement with the Guidelines for Grants then a grant is given.

2) The President reported that Dr. Hong-jun Su has asked to cancel his name from the member list of Commission 38 due to heavy work load. The President expressed his gratitude to Dr. Su for his support to our Commission.

The President also reported that he had approached a number of active astronomers from regions presently under-represented in Commission 38 getting advice from the Vice-president and Dr. West. In this connection Dr. Mazlan Othman (Malaysia), Dr. Russell Cannon (Australia), Prof. Vytas Straizys (Lithuania) and Dr. Massimo Capaccioli (Italy) have confirmed that they are prepared to serve as Commission 38 members. The President proposed them as new members. Furthermore Dr. S.-H. Su proposed Prof. J. Zhao (China) as a new member. Commission 38 agreed to those being proposed by the President and Dr. Ye and welcomed them as new members. No additional names were proposed and the final list as presented by the President was agreed.

The President proposed to confirm to the EC the following names for the next triennium :
- Dr. Morton Roberts as the next President
- Dr. Richard M. West as the next Vice-president

and D. M. Chitre, J. R. Ducati, G. Krisna, M. Morimoto, C. R. Tolbert, S.-H. Ye and H. E. Jorgensen as members of O. C. No objections were raised.

3) The President had received no proposals for changes of 'Guidelines for Grants'. He pointed out that the Guidelines work very well and have the necessary flexibility being guidelines and not strict rules. As mentioned once by the Vice-president: There are no standard cases - all seem to be special. The President could fully echo this statement. There were no proposals for changes of the 'Guidelines for Grants' during the meeting.

4) The President had received no proposals from members but considered this moment to be a good oppertunity to express opinions about the future work of the Commission. He said that the 'Guidelines for Grants' are now being printed in every issue of the IAU Information Bulletin. It is very important to remind the IAU members of the existence of the program, since the professional astronomers are supervisors for graduate students and can make the students aware of the program. Also the travel grant program has been mentioned in the EAS News.

To a question from Dr. West if UNESCO was informed the President said that he was not aware of that. It would be a good idea to inform UNESCO since it has offices in many countries. However, the granting program is mentioned in a number of commercially available booklets informing about grants.

5) No issues were raised.

The Vice-president thanked the President for having done an outstanding job and expressed his great pleasure to work with the President. Dr. Ye echoed the words by the Vice-president.

The President closed the meeting at 14:38.

COMMISSION 46 (THE TEACHING OF ASTRONOMY)

Report of Business Meetings: 21 and 25 August, 1997

PRESIDENT: John R.Percy

SECRETARY: Jay M. Pasachoff

1. Report of the President

Most of the work of the Commission is carried out by the Organizing Committee, and this is outlined in the Commission's Triennial Report. The president concentrated on making contacts with astronomy educators in many countries (including ones not yet adhering to the IAU), visiting as many countries as possible (including Brazil, Central America, China, Paraguay, several countries of Europe, and Canada and the US), and/or attending meetings and writing papers which publicize the work of the Commission. The president has worked to develop and maintain good links with the planetarium and science centre community, the amateur astronomy community, and with the science education community, as well as with major scientific and educational organizations.

2. Officers and Organizing Committee

The following officers and Organizing Committee were recommended for the 1997-2000 triennium: Julieta Fierro (President). Syuzo Isobe (Vice-President; Liaison, Asia-Pacific Region), Armando Arellano Ferro (Liaison, Latin America), Alan Batten (Chair, Working Group on the Worldwide Development of Astronomy), Michèle Gerbaldi (Secretary, ISYA), Edward Guinan (Assistant Secretary, ISYA), William Gutsch (Liaison, IPS), Darrel Hoff (Books and Journals Program), Barrie Jones (Newsletter Editor), Peter Martinez (Liaison, Africa), Derek McNally (Chair, TAD), Jayant Narlikar (Liaison, India), Andrew Norton (Webmaster), Jay Pasachoff (Chair, Subcommittee on Eclipses), John Percy (Past President; Coordinator, Travelling Telescope), Donat Wentzel (Secretary, TAD).

3. Membership

The Commission 46 membership records at IAU Secretariat have several problems (such as many National Representatives not being recorded as members of the Commission). The president will attend to this problem immediately after the General Assembly.

In the past, all members have received the publications of the Commission by mail, and this has been an expensive proposition. Now that the publications are available by e-mail, and on the WWW (*address : http : //physics.open.ac.uk/IAU*46/), they will be mailed only to National Representatives, and to a few other key people who do not have access to the Internet. Members of the IAU may join Commission 46 if they have a special interest in education, above and beyond the routine teaching of courses. This membership does not count as part of the IAU's "three commission limit".

4. International Schools for Young Astronomers (ISYA)

ISYA's were held in Egypt (1994), Brazil (1995) and Iran (1997). Reports on these have been published in the Newsletter, which is posted on the WWW. Possible future sites include: Romania (in conjunction with the total solar eclipse in 1999), Southeast Asia, Nigeria, Peru, Ukraine, or Vietnam, depending on what invitations are received. ISYA's are costly, since 20-30 students must be brought to the site and supported for three weeks.

J. Andersen (ed.), Transactions of the International Astronomical Union Volume XXIIIB, 215–219.

5. Teaching for Astronomical Development (TAD)

Secretary Donat Wentzel reported on behalf of Derek McNally, chair of TAD. ¿From 20 applications for the TAD program, four were chosen. Programs are underway in Central America, and Vietnam, and are being considered for Morocco and Sri Lanka, pending the choice of suitable sites and local sponsors. TAD funds are used for a variety of purposes, including visiting lecturers, providing equipment, etc. There is often a problem transporting equipment, and getting it into the country.

In September 1997, Wentzel and a colleague will run a workshop in Vietnam for 15 university instructors, and 15 physics students, to update them in astronomy, and instill a sense of inquiry. There will be practical activities in both astronomy, and astronomy teaching. The lack of audio-visual facilities is partly offset by a generous donation of textbooks by a US publisher. Two Vietnamese students are in PhD programs in Paris, and are expected to return to Vietnam to do research. The TAD program is supported by the highest levels of the government of Vietnam. Wayne Orchiston mentioned a New Zealand astronomer, now working in Vietnam, as a possible source of assistance.

A second TAD program operates in Central America. The IAU supports an annual course in observation, at the observatory in Honduras, and also the annual meeting of Central American astronomers, which is a useful form of professional development. The European Council is providing funds for a series of six MSc courses - possibly one in each of the six participating countries - though this program has some problems. There was a UN/ESA (United Nations/European Space Agency) Workshop on Basic Space Science, in Honduras, in June 1997.

6. Travelling Telescope

The travelling telescope consists of a Celestron-8 telescope with accessories and instruments, in four large shipping crates. Because of the size, weight, and delicacy of the telescope, it has been difficult to transport. For the last several years, however, the instruments have been in Paraguay, as a follow-up to the Visiting Lecturers Program. They were returned to Canada in August 1997 for refurbishment. What should be the future of the travelling telescope and its instruments. Should they be given on long-term loan? Should the telescope be sold, to raise money for other instruments which are easier to transport?

7. Newsletter (Electronic and Paper) and Web Site

Generally, the electronic version of the Newsletter is produced and sent every three months. The editor is John Percy, and the distributor of the electronic Newsletter is Armando Arellano Ferro. Every six months, two electronic issues are printed and mailed to the National Representatives and other contact people. The issues are also posted on the Commission's web site. These arrangements will continue for the next triennium. Percy will continue as Newsletter editor until April 1998, when Barrie Jones will take over. Norton will continue as webmaster. Rajesh Kochhar noted that many countries, including India, have only limited access to the WWW. The Commission Newsletter, and other such resources, should be distributed as much as possible by the National Representatives.

8. Triennial Reports

Triennial Reports on astronomy education in the countries adhering to the IAU, and several other countries, were prepared and distributed by mail to the National Representatives and other contact people. They were also posted on the web site. Several National Representatives did not provide reports, despite many reminders. Percy met with the National Representatives to the IAU General Assembly in August 1997, to urge the appointment of active National Representatives to the Commission. The Triennial Reports will continue in their present format for another triennium. Wayne Orchiston suggested including short bibliographies with the reports.

9. Solar Eclipses and Astronomy Education

There is a subcommittee of the Commission - Jay Pasachoff (chair), Julieta Fierro, and Ralph Chou - which promotes safe and successful eclipse viewing to advance astronomy education. Total solar eclipses are especially striking, but partial eclipses occur over a wider area of the Earth, and also provide a good opportunity for public education. In the next triennium, there will be a total eclipse in February 1998 visible in Panama, northern South America, and some of the islands of the Caribbean; the zone of the partial eclipse covers much of North and South America. The August 1999 eclipse will be total in a band from southern England across Europe, through Turkey, Syria, Iran and Iraq, to India and Bangladesh. It will be partial in a wide band of Europe and Asia. The annular eclipse of 1998 will bring a partial eclipse throughout Southeast Asia, and the annular eclipse of 1999 will bring a partial eclipse throughout Australia and New Zealand. Chou has been testing safe filter materials. There is now an increased availability of solar-filter materials, such as CD's, the inside of floppy discs, and even aluminized Mylar used to pack tea in South Africa and perhaps elsewhere.

10. Books and Journals

The Commission has begun a program to find and distribute surplus books and journals to suitable institutions in the developing countries. Darrel Hoff is presently co-ordinating this program, and requests help in finding suitable material. First priority goes to institutions which are taking part in other Commission programs such as ISYA, TAD, and VLP. The IAU sends complimentary copies of its publications to many needy institutions, especially in the former Soviet Union, and several other organizations send copies of their publications to a few developing countries. The problem is to find books and journals which are at a suitable technical level. These might include: *Sky and Telescope*, *Astronomy and Geophysics* (a new publication of the Royal Astronomical Society), and *The Messenger*, a publication of the European Southern Observatory.

11. Future Meetings

There was much discussion of an invitation, presented by Wayne Orchiston (New Zealand) and Maria Hunt (Australia), to hold the next proposed IAU colloquium on astronomy education in Australia, specifically in Penrith (50 km from Sydney) in the Blue Mountains, 12-16 July, 1999. This follows by one week, a joint meeting of Australian and New Zealand astronomers. There will be field trips on July 10-11, and a teachers' day on July 11. The theme of the meeting is "astronomy education in the schools, including roles of observatories, planetariums, and astronomical societies". Graeme White (University of Western Sydney) will chair the LOC, and several other members of the LOC have already been appointed. Julieta Fierro will chair the SOC. The invitation was unanimously and enthusiastically accepted by the meeting.

Other forthcoming meetings of possible interest to the Commission include: a colloquium on amateur-professional cooperation in astronomical education and research, in Toronto, July 5-7, 1999; a colloquium on light pollution and related topics, organized by IAU Commission 50 as part of UNISPACE III in Vienna, in late July 1999; a summer school of the European Association of Astronomy Educators, to be held in France in connection with the August 11, 1999 total solar eclipse; and a possible ISYA in Romania at the same time.

The traditional one-day meeting between astronomers and schoolteachers was organized, at the 1997 IAU General Assembly, by Syuzo Isobe and his colleagues. About 150 schoolteachers attended, along with several astronomers. John Percy thanked him, on behalf of the Commission, for organizing such a successful event.

12. Regional Activities

Syuzo Isobe reported that he will continue to publish the bulletin *Teaching Astronomy in the Asia-Pacific Region* every half year. Wayne Orchiston suggested that we should have some contact with the Asian-Pacific Economic Group, which is meeting soon.

There is now a Working Group on Basic Space Science in Africa. Peter Martinez is our liaison person with that group. Francois Querci (querci@astro.obs-mip.fr) reported on that group's new publication *African Skies/Cieux Africains*. See also the web site: http://www.saao.ac.za/ wgssa/

Nikolai Bochkarev (Sternberg Astronomical Institute, Russia) reported on the work of the Euro-Asian Astronomical Society, which includes the countries of the former Soviet Union. They have several branches connected with astronomical education, including planetariums. They have several popular magazines, an association of schoolchildren etc. Professor Bochkarev urged the commission to appoint a liaison person for the countries of the former Soviet Union.

13. Liaison with Other Organizations

Working Group on the Worldwide Development of Astronomy.. Alan Batten reported that there is much overlap between the work of the WGWWDA and of Commission 46, though the WGWWDA is concerned about the full range of astronomical activity - not just education. Batten, John Percy, and Donat Wentzel have travelled to several astronomically developing countries, and there has been good co-operation between the WG and the Commission - in part, because the president of one is a member of the Organizing Committee of the other. This kind of co-operation should continue. With regard to Morocco: Batten noted that the Atlas Mountains may be one of the last great observing sites in the world which is not already used. Site testing is being done, in collaboration with the Observatoire de Nice.

International Planetarium Society. William Gutsch, who is a member of the Organizing Committee of the Commission, was not able to be present, but his report was read. Tens of millions of people visit planetariums each year, so they have a major impact on formal and informal education. Gutsch's report described a project to enlist the help of astronomers as advisors and visiting lecturers in planetariums.

UN Office for Basic Space Science. Hans Haubold reported on the series of UN/ESA Workshops on Basic Space Science, held since 1991. Five were held in the five major regions of the developing world. The sixth was held in Germany, to assess the first five, and plan for the future. A seventh workshop was held in Honduras, in June 1997, on the occasion of the official opening of the observatory there. These workshops are part of the activity of the UN Committee on the Peaceful Uses of Outer Space (COPUOS). The UN thus contributes to the development of astronomy.

Donat Wentzel proposed the following resolution: "Considering the practical advances in astronomy that followed the UN/ESA workshops on basic space science, and considering the strong interest of Commission 46 in the countries affected by these workshops, Commission 46 urges that the UN Committee on the Peaceful Uses of Outer Space, and ESA continue the series of workshops." Jean-Claude Pecker, chair of the IAU Resolutions Committee, pointed out that this cannot be an IAU resolution, but should be dealt with by the IAU Executive Committee, especially as it has budgetary implications.

Alan Batten suggested that the IAU should co-sponsor the workshops. In that case, support for at least one speaker would be expected. There are no funds allocated for this in the proposed IAU budget. Wentzel suggested that an IAU-supported speaker could organize a half-day workshop, as well as giving a talk. Jean-Claude Pecker suggested that, when the resolution was transmitted to the IAU Executive Committee, it should have an appendix, saying that the Commission would like the IAU to be a co-sponsor, provided that suitable funds can be allocated to it. The resolution was passed unanimously. The president transmitted it to the IAU Executive Committee.

Haubold noted that his office receives more and more requests, from developing countries, on how to introduce astronomy in the teaching curricula of high schools, colleges, and universities. They would like to be able to provide kits to teachers, free of charge. They might even be able to translate resource material into various languages, especially if the length of any resource document could be kept to 48 pages or less. Wentzel offered to develop a booklet of this approximate length, dealing with the introduction of astronomy in university-level physics courses. The UN may be willing and able to translate it.

Haubold's address is: Program on Space Applications, Office for Outer Space Affairs, United Nations, Room F-0839, Vienna International Centre, P.O. Box 500, A-1400 Vienna, Austria. E-mail: haubold@eunet

ICSU (International Council of Scientific Unions). The IAU Assistant General Secretary Johannes Andersen has been the official representative to ICSU. There is no longer representation on an ISCU Committee for the Teaching of Science. ICSU now has a Committee on Capacity Building in Science, which is obviously of interest to the IAU and Commission 46. Furthermore, ICSU is our channel to UNESCO. Jean-Claude Pecker recommended that the Commission should make proper contacts with ICSU's General Secretary so that, at the meeting of UNESCO in November, on teaching, the importance of the teaching of astronomy will not be overlooked. Derek McNally agreed that it is very important to develop and maintain this connection with UNESCO. The head of its physics commission is sympathetic to astronomy.

Committee on Space Research: (COSPAR). Alan Batten reported on a possible IAU-COSPAR initiative to provide workshops, for astronomers from developing countries, on how to access and use databases from satellite observatories. The IAU WGWWDA has advised the IAU General Secretary that, in its opinion, the idea is a good one, but there would be practical difficulties such as the compatability of computer systems.

International Dark-Sky Association. David Crawford reported on this organization, whose goal is the preservation of the astronomical environment in both developed and developing countries. It has over

2000 members in 62 countries. Since the IAU has a commission (50) which deals with this issue, and since our vice-president Syuzo Isobe is past president of that commission, that could provide an appropriate link. We could also cross-link our web site with theirs (http://www.darksky.org).

14. Astronomy Educational Material

The Commission formerly had a project - Astronomy Education Material (AEM) - which was a listing of educational resource material in various languages. It also had a Project Contratype, which was a collection of photographs and slides, for loan. The discussion by Hans Haubold, above, led to a discussion of the continued need for resource lists, and other teaching material..

Robert Robbins, who was responsible for the English-language AEM list, reported by e-mail message. He recommended a WWW-based listing, which could include mini-reviews by users of the material. The Internet "bookstore" Amazon.com contains reviews, and people could be encouraged to submit reviews of astronomy books.

John Percy reported that the Astronomical Society of the Pacific was developing a set of 40 slides to illustrate a core unit on astronomy at the grade 5-8 level. These might be useful to developing countries.

There were several comments about possible resource materials: listings of resources, publications, audio-visual materials, kits etc. (1) Slide projectors are unavailable at many institutions; overhead projectors may also be unavailable. Perhaps a corporation such as Kodak would be willing to fund a booklet of photos of selected astronomical objects. (2) Material should be written in the language of the country involved; the UN can possibly do the translation. (3) In India, teachers want information on how to build a small telescope, and also on "exotic" topics like space probes and black holes. It may therefore not be possible to satisfy every need through a single, short document. (4) Some resource materials, including images, are available on the WWW, but most institutions in the developing world do not have access to this.

15. Other Business

Rules and Guidelines of the Commission. These were approved in 1973 and updated in 1988. The Organizing Committee has decided that, although the rules and guidelines are slightly out of date, no great effort will be made to revise them at this time. John Percy agreed to look into this matter in the near future.

International Astronomy Day/Week. The president was approached by Gary Tomlinson, the co-ordinator of International Astronomy Day (at least within North America), asking the Commission to co-sponsor or support this event. In the US, the Astronomical League, and Sky Publishing Corporation have developed a useful handbook on organizing this event locally. Although there was support in principle for the concept of International Astronomy Day (indeed, it is celebrated in many countries, though not necessarily on the same date), there was concern that the Commission had no control on the content of any local Astronomy Day event, and this could be a source of embarassment. Various members of the audience spoke about their own local experiences with Astronomy Day. Derek McNally proposed a resolution: "to encourage adhering bodies to find an appropriate local body to organize an astronomy day or week in their respective countries". There was no objection. This resolution by the Commission will be transmitted to the National Representives through the newsletter, and to other appropriate organizations; the International Union of Amateur Astronomers was one possibility.

Donat Wentzel brought forward a budget item: a request to support an annual newsletter on astronomy education, with a circulation of 4000, being distributed in the former Soviet Union. The editor is Edward Kononovich. The newsletter is in Russian. Wentzel had an English translation. Wentzel suggested that the commission provide $500 on a one-time basis. There was no objection. Jay Pasachoff suggested placing the newsletter on the commission's web site. Terry Oswalt offered to place it on the web site of the International Amateur-Professional Photoelectric Photometry web site, because that organization has many members in the FSU.

In closing, John Percy thanked the outgoing past president Lucienne Gouguenheim, who is retiring from the Organizing Committee, for her long and effective service to the Commission. The incoming president Julieta Fierro expressed her happiness at becoming president, because of the importance of the Commission's work. She expressed the thanks of the Commission to the outgoing president John Percy, for his excellent work.

COMMISSION 50: PROTECTION OF EXISTING AND POTENTIAL OBSERVATORY SITES/ PROTECTION DES SITES D'OBSERVATOIRES EXISTANTS & POTENTIELS

S. Isobe
National Astronomical Observatory
2-21-1, Osawa, Mitaka, Tokyo 181, Japan

PRESIDENT: S. Isobe
VICE-PRESIDENT: W.T. Sullivan

After the adoption of the agenda for the meeting, President Isobe welcomed the nine other persons present. It was reported by the President that the IAU one-and half day long JD5 "Preserving the Astronomical Windows" had been successfully held with a total attendance of 200 in changing from one session to the other.

For the problem of light pollution, CIE and IAU Commission 50 nicely worked out together to bring its guideline for minimizing sky glow to the CIE Division 4 resolution and the IAU resolution. For the problem of radio interference, Commission 50 did not work hard contrary to that of light pollution. A report by J. Cohen a representative of Commission 40, showed a difficult situation of radio interference and requested a strong support from Commission 50. Three resolutions were proposed at the JD5 under the directorship of Commission 50 and were discussed and approved.

Following tradition, W.T. Sullivan, who had been vice-president, was approved as President of Commission 50. As for the next vice-president, Isobe proposed J. Cohen, a radio astronomer, since the problem of radio interference should be worked on more in Commission 50. This proposal was accepted together with the proposal to set up an SOC. Commission 50 did not have an SOC, and therefore, its President and vice-president should work on the matter in most cases. This limited its activities within light pollution, and some other problems were not properly worked on.

The following SOC members were appointed considering distributions of astronomical fields and geometry in the world. They are:

D. McNally	UK
T.A. Spoelstra	Netherland
D. Crawford	USA
B. Hidayat	Indonesia
S. Isobe	Japan
S. Jiang	China
D. Malin	Australia).

Four members expressed their wish to retire from Commission 50. They are:

Dr. Victor M. Blanco (USA)
Dr. Wolfgang Mattig (Germany)
Dr. Sidney van der Bergh (Canada)
Dr. J. Dommanget (Belgium)

who are thanked for their past activities.

J. Andersen (ed.), Transactions of the International Astronomical Union Volume XXIIIB, 221–222.

There were a number of inactive members who will, we hope, be active in the future because the activities of Commission 50 relate to all the fields of astronomy. There are 4 members to whom our correspondence no longer reaches. It was decided to delete their names from the Commission 50 member list after the final contact of the new Commission president to them. One new member, Dr. Hubertus Wöhl (Germany), and one consulting member, Mr. Javier Diaz-Castro, were accepted. During the JD5, 26 people expressed their intention to be members of Commission 50. After checking who are IAU member or not, W. Sullivan will make a decision who will be members or consulting members.

Considering that a joint project between CIE and IAU on light pollution is approaching the final stage it was decided that S. Isobe would continue to be a liaison to the CIE. W. Sullivan proposed to have an IAU Colloquium in 1999, adjacent to the UNISPACE III held in Wien, July 1999. Its proposal should be sent to the EC by the new Commission 50 President. As we found at the JD5, many astronomers are interested in the activities of Commission 50. We concluded to continue to work these

COMMISSIONS NON ATTACHED TO A DIVISION

COMMISSION 41: HISTORY OF ASTRONOMY/(HISTOIRE DE L'ASTRONOMIE)

President:	S. M. R. Ansari
Vice President:	Steven J. Dick
Secretary (Business meeting):	W. Orchiston

1. **Business Sessions** (Friday, August 22 and Tuesday, August 26, 1997)

1.1 GENERAL

Commission Vice President S. Dick called the meeting to order at 1400 hours, August 22 in Room H of the Kyoto International Conference Hall. W. Orchiston was appointed Secretary, and S. Débarbat balloteer for the elections. Of the members of the 1994-1997 Organizing Committee S. Débarbat was present; the remaining members (A. Gurshtein, J. North. S. Nakayama) were unable to attend. 21 others were in attendance. A moment of silence was observed for members deceased since the last G.A., including LeRoy Doggett (U. S. Naval Observatory, Washington) and Bruno Morando (Bureau des Longitudes, Paris).

President Ansari reported on the attempts to organize a symposium on several themes, including "Astronomy in Asia and the Far East", for the current General Assembly; in the end the EC approved the present 1.5 day Joint Discussion on "The History of Oriental Astronomy". He also reported that S. Dick sent out two Commission Newsletters over the past year. In an attempt to raise the Commission membership, the President wrote a message that was posted on the World Wide Web and sent to Commission Presidents.

1.2 ELECTION OF OFFICERS, NEW MEMBERS AND CONSULTANTS

Elections duly held, taking into account absentee ballots, resulted in the following officers for the 1997-2000 triennium:

President:	S.J. Dick (USA)
Vice-President:	F.R. Stephenson UK)
Immediate Past President:	S.M.R. Ansari (India)
Organizing Committee:	Wolfgang Dick (Germany)
	Alex Gurshtein (Russia)
	Il.-S. Nha (Korea)
	Wayne Orchiston (New Zealand)
	Edoardo Proverbio (Italy)
	Woodruff T. Sullivan (USA)
	Xi Zezong (China)

New members of the Commission approved were Emmanuel Danezis (Greece), Joseph. S. Mikhail (Egypt), Dimitrios Papathanasoglou (Greece), Efstratios Theodossiou (Greece), Maria Stathopolou (Greece), Theofanis Grammenos (Greece), Ian Glass (S. Africa), John Hearnshaw (New Zealand), Alan Batten (Canada), Brian Warner (S. Africa), Kwan-Yu Chen (USA), Jang-Hae Jeong (Korea), Yong-Sam Lee (Korea), Chun-Hwey Kim (Korea), Yonggi Kim (Korea) Hi-Il Kim (Korea), Woo-Baik Lee (Korea), Kyu-Dong Oh (Korea), Tiberiu Oproiu (Romania), Y. Sobouti (Iran), Virginia Trimble (USA), Shi-zhu Cui (China PR), Roslynn Haynes (Australia), Thomas Hockey (USA), Wolfgang Kokott (Germany), Chun-yuuu Ma (China PR), Nagahoshi

J. Andersen (ed.), Transactions of the International Astronomical Union Volume XXIIIB, 223–225.

Ohashi (USA), John Perdrix (Australia), Sh. Ehgamberdiev (Uzbekistan), John Whiteoak (Australia), David Jauncey (Australia), Mike Bessell (Australia), Don Matthewson (Australia), Bruce McAdam (Australia), Ken Freeman (Australia), William Tobin (New Zealand), Kitiro Hurukawa (Japan). Of these Cui, Haynes, Hockey, Kokott, Ma, Ohashi and Perdrix were elected new members of the IAU during this General Assembly.

New consultants voted were Brenda Corbin (USA), David A. King (Germany), E. S. Kennedy (USA), K. Locher (Switzerland), Stephen McCluskey (USA), J. F. Oudet (France), Clive Ruggles (UK), and M. Yano (Japan). Consultants reelected were J. Tenn (USA), J. Bennett (UK), K. Bracher (USA), J. Evans (USA), R. Freitag (USA), A. Jones, S. R. Sarma (India), B. G. Sidharth (India), B. van Dalen (Netherlands), T. Williams (USA) and L. Wlodarczyk.

The total membership of the Commission, including the new members, stands at 147, plus 17 consultants.

1.3 RESOLUTIONS

Débarbat read the resolutions passed at the last General Assembly regarding the organizing and cataloguing of IAU archives and the preservation of relics related to F.G.W. Struve's measurement of the arc of the meridian. Regarding the first, Débarbat reported that the IAU archives have been deposited in Paris with the General Secretariat. Regarding the latter, Batten reported that the resolution has had some effect, notably by contacts with the International Union of Geodetic Surveyors, and that efforts are continuing.

New resolutions included the following:

1) Whereas historical astronomical records are important to the heritage of astronomy and may be essential to applied astronomy

   The IAU supports the recovery, inventory and preservation of astronomical archives of national and international institutions, including observatories, societies and other institutions.

2) That Commission 41 records its serious concern regarding grave losses at Pulkovo as the result of fire, and supports the assessment of these losses to the cultural heritage of astronomy.

3) That, in order to facilitate research into the history of astronomy in a country (the "host country") that was subjugated or governed by another country ("governing country"), and where the relevant source material now resides in the governing country, every attempt should be made to provide copies of such source material to the host country.

4) Noting that vital primary source material pertaining to history of astronomy in a country (the host country) that was ruled or governed by another (the governing country) resides in the governing country, it is recommended that visiting fellowships be created by IAU, European Union, and bilateral agreements between countries to enable researchers from a host country to consult source material in a governing country.

1.4 COMMISSION 41 STATUS IN IAU

The members voted to support the position that "History of astronomy is a discipline that overarches the entire field of study of the IAU, and therefore should not be confined to one Division. We wish to remain a separate Commission until such time as we can become a separate History of Astronomy Division".

Discussions with incoming General Secretary Johannes Anderson and incoming President Robert Kraft indicated that this is not a problem.

1.5 FUTURE MEETINGS

The Commission endorsed a proposal by Edoardo Proverbio that a meeting (probably an IAU colloquium) on "Polar Motion: Historical and Scientific Problems" be held in Italy on the occasion of the centenary of the International Latitude Service (ILS) in 1999. The meeting would be held immediately before or after the Italian Astronomical Society meeting in September/October, 1999, hosted by the Cagliari Astronomical Observatory. Co-sponsorship by Commissions 19 (Earth Rotation) and 31 (Time) are being sought.

Looking forward to the next General Assembly in Manchester, the chair proposed as a possible topic for a Joint Discussion "Applied History of Astronomy", especially since the world's leading scholar in the field (F. R. Stephenson) is in the UK, and he is now V.P. of Commission 41. This proposal was enthusiastically endorsed by the members.

In light of the fact that the next General Assembly will be held in the year 2000, it was suggested that Commission 41 sponsor an Invited Discourse on "The History of Astronomy in the 20th Century". President S. Dick will follow up.

1.6 OTHER NEW BUSINESS

W. Orchiston (New Zealand) discussed a proposal to form a new "Journal for the History and Heritage of Astronomy". It would replace the current Australian Journal of Astronomy, and John L. Perdrix (Australia) would remain its editor. It would not wish to be competitive with the current Journal for the History of Astronomy, but would seek to publish a wider variety of articles.

1.7 COUNTRY REPORTS AND OTHER PAPERS

At the end of the first business meeting Rajesh Kochhar spoke on the History of Astronomy in India, and Wayne Orchiston on history of astronomy activities in New Zealand. Following the Joint Discussion at the second business meeting, Suzanne Débarbat spoke on the history of astronomy activities in the Department of Fundamental Astronomy at Paris Observatory, and Il.-S. Nha reported on a new Museum of Astronomy in Korea, to open in October, 1998.

Incoming President S. Dick thanked outgoing President Ansari, Professor Yano, and the members of the Scientific Organizing Committee of JD 17 for putting together a successful meeting. He also thanked the outgoing members of the Commission 41 OC, and wished Suzanne Débarbat best wishes in her retirement.

2. SCIENTIFIC SESSIONS: JOINT DISCUSSION 17, AUGUST 25-26

The entire day Monday, August 25 and the first part of the morning August 26, were devoted to JD 17 "History of Oriental Astronomy". Approximately 100 people in attendance heard papers on the ancient and medieval periods, including: the earliest stage of Chinese astronomy (Y. Maeyama), Islamic astronomy in China (B. van Dalen), an Arabic commentary on al-Tusi's Tadhkirah and its Sanskrit translation (T. Kusuba), Ancient Indian astronomy in China (J. Xiao-Yuan), Korean star maps of the 18th century (Il-S. Nha), Knowledge of the starry sky in Indonesia (B. Hidayat), the projection method of star mapping in the Song Dynasty, astronomy in the Orient to the 12th and 13th centuries (K.-Y. Chen), Vedanga astronomy (Y. Ohashi), spherical trigonometry in the astronomy of the medieval Kerala school (K. Plofker), and astronomical dating and statistical analysis of Shang dynasty oracle bone records (K. Pang, K. Yau and H. Chou).

Among the papers in the modern astronomy session were The Drkpaksasarani: A Sanskrit version of de la Hire's Tabulae Astronomicae (D. Pingree), Modern astronomy in Indo-Persian sources (S.M.R. Ansari), Takamine and Saha's contacts with western astrophysics (D. DeVorkin), contemporary astronomy in Iran (Y. Sobouti), astronomy education in the East (S. Isobe), Kepler's law in China (K. Hashimoto), the status of astronomy in Uzbekistan (S. Ehgamberdiev), Power and politics in 19th century Australian astronomy (W. Orchiston), old Burmese sky charts (M. Nishiyama), and an overview of Oriental astronomy (S. Nakayama).

3. Other Sessions

In addition to JD 17, Commission 41 also supported JD 8 "Stellar Evolution in Real Time,", JD 20 "Enhancing Astronomical Research and Education in Developing Countries" and JD 23 "He Leonid Meteor Stream: Historical Significance and Upcoming Opportunities". S. Dick gave a paper in the latter on "Observations of the Leonids over the Last Millennium".

WORKING GROUPS OF THE EXECUTIVE COMMITTEE

WORKING GROUP	ON PLANETARY SYSTEM NOMENCLATURE
WORKING GROUP	ON ENCOURAGING THE INTERNATIONAL DEVELOPMENT OF ANTARCTIC ASTRONOMY
WORKING GROUP	FOR THE WORLDWIDE DEVELOPMENT OF ASTRONOMY
WORKING GROUP	ON FUTURE LARGE SCALE FACILITIES

ANTARCTIC ASTRONOMY

WGDAA Report for Period July 1993—June 1996

COMPILED BY M.G. BURTON
Joint Australian Centre for Astrophysical Research in Antarctica
University of New South Wales
Sydney, NSW 2052, Australia

1. Introduction

The Antarctic plateau provides the best site conditions on the Earth for a wide range of astronomical observations, both of photons and particles. This is a result of the unique combination of cold, dry and tenuous air found only there. Wintertime temperatures average below -60°C, with minimal diurnal variation, the precipitable water vapour content is below 250 μm, the katabatic wind is low on top of the plateau and there are no jet streams at high altitude. The vast quantities of pure ice can be used as an absorber for particle detectors. Secondary benefits include continuous viewing for any source visible, lack of pollution and dust in the atmosphere, and low electromagnetic interference. Considerable activity is now focussed at the South Pole on developing facilities for astronomy. Initial investigations of higher sites have begun, particularly at Dome C.

2. South Pole Observatory (2,900 m)

Two principle astronomical activities are underway at the Amundsen–Scott South Pole station, operated by the National Science Foundation of the USA, those of CARA and of AMANDA. Scientists from Australia, France, Germany, Sweden and the UK also participate in these projects.

2.1. CENTER FOR ASTROPHYSICAL RESEARCH IN ANTARCTICA

AST/RO, the 'Antarctic Sub-mm Telescope / Remote Observatory', a 1.7–m diameter sub-mm telescope, was commissioned in 1995. It has surveyed the southern Milky Way in the 492 GHz [CI] line, only observable infrequently from other good observing sites. Carbon emission has been found to be widespread and at least as extensive as regions of CO emission. The first detection of [CI] from the Magellanic Clouds has been made. Lines as weak as 0.02 K have been measured.

COBRA, the 'Cosmic Background Anisotropy Experiment', using the 0.75–m Python telescope, has reliably reproduced structure observed in the CMBR on angular scales from 0.75° to 5.5° over 4 successive austral summers. The sky coverage has been increased from 8 to 123 square degrees in that time. Significantly greater anisotropy is detected on degree scales than found by the COBE satellite at 20°.

SPIREX, the 'South Pole Infrared Explorer', a 60–cm near–IR telescope, was installed in 1994. SPIREX has achieved exceedingly dark backgrounds at 2.4 μm, as low as 23.5 mags/arcsec^2 for long integrations. It enjoyed a nearly uninterrupted view of the collisions of Comet Shoemaker–Levy with Jupiter. Only 4 of 20 events were obscured by clouds and over 3,000 images were obtained. These were also transferred back to the 'mainland' during the event over the internet, demonstrating the level of communication links that are now established to Pole.

A comprehensive series of site testing measurements have been conducted at the South Pole, demonstrating the quality of the site. The 25% quartile for ppt H_2O in winter is 0.19 mm, compared to 1.05 mm on Mauna Kea. Both SPIREX and an Australian experiment, the IRPS ('Infrared Photometer Spectrometer'), have measured the sky brightness in the 2.29-2.46 μm window, where airglow emission is minimal. It is found to be typically 100 μJy/arcsec^2, a factor $\sim$ 40 times lower than Mauna Kea. In the L–band, 3-4 μm, the sky brightness is typically 20-100 mJy/arcsec^2, a factor $\geq$ 20 times lower than good mainland

J. Andersen (ed.), Transactions of the International Astronomical Union Volume XXIIIB, 229–230.

sites. Above a surface inversion layer typically 200 m high, the mean visual seeing is 0.36". The isoplanatic angle is $\geq 5'$ at 2.4 μm.

2.2. ANTARCTIC MUON AND NEUTRINO DETECTOR ARRAY

AMANDA is designed to observe high-energy ($\sim$ 1 TeV) neutrinos. Strings of widely spaced photomultiplier tubes (PMTs) are placed into deep water-drilled holes in the ice. High energy neutrinos coming up through the earth will occasionally interact with ice or rock and create a muon. The Cerenkov photons produced are tracked by the PMTs.

During 1994 four strings were placed in the ice, with PMTs from 800–1000 m in depth ('AMANDA-A'). Four more strings were deployed in 1996 at 1600–2000 m depth ('AMANDA-B'). Optical properties of the ice have been determined. It has an extraordinarily long absorption length, $\sim$ 150 m. The scattering length is nearly two orders of magnitude higher at the lower depth, $\sim$ 25 m.

SPASE, the 'South Pole Air Shower Experiment', continues to monitor cosmic rays above 100 TeV with two arrays and is using the muon detection capability of AMANDA, together with an air-Cerenkov detection system, to measure the mass composition of cosmic rays above 1000 TeV. The second array, SPASE-II, was constructed in 1995 and placed 300 m from AMANDA, to assist with the coincidence timing in screening for neutrino detections.

3. Dome C (Circe, Concorde or Charlie; 3,200 m)

A French–Italian collaboration started construction of a station on this site in 1995. Currently an ice-core drilling operation is underway. The first winter-over is scheduled for 2,000. Daytime measurements of the micro-turbulence in the atmosphere were conducted in 1996.

4. Other Sites

On the 2,960 m elevation Hercules Névé, near the Italian Terra Nova station, measurements of the mid-IR sky emission have been carried out, as a trial for operating an instrument at Dome C, using a 0.8 m telescope and liquid ^{4}He cooled photometer.

Australia operates a cosmic ray research station at Mawson, containing a neutron monitor and muon telescope.

At the Argentinian Belgrano Base an 11" Celestron telescope has been operated, recording seeing measurements and determining atmospheric extinction coefficients.

5. Further Information

The following URL's provide links to web pages which serve as resources for further information on this subject area:

AMANDA	http://dilbert.lbl.gov/www/amanda.html
CARA	http://pen.k12.va.us/~alloyd/CARA.html
JACARA	http://www.phys.unsw.edu.au/~mgb/jacara.html

Volume 13 of the *Publications of the Astronomical Society of Australia* (1996) is devoted to articles on Antarctic astronomy. In addition to the WGDAA of the IAU, SCAR, the 'Scientific Committee for Antarctic Research', have established a sub-committee, STAR, 'Solar Terrestrial and Astrophysics Research', to coordinate astronomical activities in Antarctica.

WORKING GROUP FOR PLANETARY SYSTEM NOMENCLATURE (WGPSN) (GROUPE DE TRAVAIL POUR LA NOMENCLATURE DU SYSTEME PLANETAIRE)

PRESIDENT: K. Aksnes
MEMBERS: M. E. Davies, M. Ya. Marov, B. G. Marsden, P. Moore, T. C. Owen, V. V. Shevchenko, B. A. Smith
CONSULTANTS: G. A. Burba, L. Gaddis, P. Masson, J. Blue

1. Introduction

WGPSN held two morning sessions on 20 August 1997 in Kyoto during the 23rd IAU General Assembly. The meeting was attended by five WG members and one member of an adhering nomenclature Task Group plus four guests. The absent WG members had beforehand by e-mail commented on most of the agenda items, so that representative decisions could be made.

Classifications and names for a record number (815) of features on planetary bodies were approved at this meeting. Since the IAU-approved planetary nomenclature is scattered in many volumes of the IAU Transactions, there is a need for a comprehensive listing of this nomenclature. The U.S. Geological Survey in Flagstaff has in cooperation with WGPSN published such a listing complete up to 1994: *Gazetteer of Planetary Nomenclature 1994* (USGS Bulletin 2129).

1.1. MEMBERSHIP CHANGES

D. Morrison is replaced on the WG and as Mercury Task Group Chair by M.E. Davies. Other TG Chairs 1997-2000 are: V.V. Schechenko (Lunar TG), G.A. Burba (Venus TG), B.A. Smith (Mars TG), T.C. Owen (Outer Solar System TG), and B.G. Marsden (Small Bodies TG).

2. Nomenclature Corrections

- VENUS:
- In *IAU Trans. XIXB 1985*, change *Fedosova (crater)*, *Siddons (crater)* and *Mist Fossae* to respectively *Fedosova Patera*, *Siddons Patera* and *Mist Chasma.*
- In *IAU Trans. XXIA 1991*, change *Ba'het Patera* to *Ba'het Corona.*
- In *IAU Trans. XXIB 1991*, change crater names *Amalasthuna* and *Goppert-Mayer* to *Amalasuntha* and *Goeppert-Mayer.*
- In *IAU Trans. XXIIA 1993*, change *Ciuacoatl Corona* and *Kunapipi Corona* to *Ciuacoatl Mons* and *Kunapipi Mons.*
- In *IAU Trans. XXIIB 1994*, change *Anqet Farrum* and *Oduduva Corona* to *Anqet Farra* and *Oduduwa Corona*, and change coordinates of *Citlalpul Valles* from 57.4S and 185.0E to 53.0S and 183.0E.

- MARS:
- In *IAU Trans. XVIIB 1979*, change *Cerberus Rupes* to *Cerberus Fossae.*
- In *IAU Trans. XXIA 1991*, delete *Peneus Mons.*

J. Andersen (ed.), Transactions of the International Astronomical Union Volume XXIIIB, 231–251.

- EUROPA, GANYMEDE, CALLISTO:
- In *IAU Trans. XVIIB 1979*, change *Tyre Macula* on *Europa* to *Tyre (Large Ringed Feature)* and change coordinates and diameter of the crater *Adlinda* on *Callisto* from 56.6S, 23.1W, 274 km to 46.0S, 33.0W, 600 km.
- In *IAU Trans. XIXB 1985*, change *Punt Facula* on *Ganymede* to *Punt (crater)*.
- In *IAU Trans. XXA 1988*, delete *Sais Facula* on *Ganymede*.

3. New Feature Descriptor Terms

The following new feature descriptor terms have been introduced:

- EUROPA:
- *Lenticula* with a theme of Celtic gods and heroes (as for craters).
- *Large Ringed Feature* with a theme of Celtic stone circles.
- *Chaos* with a theme of places associated with Celtic myths.

- GANYMEDE:
- *Catena* with a theme of gods and heroes of ancient Fertile Crescent people (as for craters).

- MIRANDA:
- *Sulci* with a theme of characters and places from Shakespeare's plays.

- IO:
- Name *Montes, Plana, Regiones, Tholi, and Mensae* for nearby named features, but use names from Dante's Inferno if no nearby features have been named.
- Use themes for *Patera* and *Eruptive Centers* for *Fluctus* not located near a named feature.

4. New Nomenclature

MARS (17)

NAME	LAT	LONG	DIAM (km)	ATTRIBUTE
CRATERS (11)				
Bonestell	42.3N	30.3W	39	Chesley; Am. space artist (1888-1986).
Calahorra	26.8N	38.7W	35	Town in Spain.
Kayne	15.5S	186.4W	32	Town in Botswana.
Nier	43.1N	253.9W	46	Alfred O.C.; Am. physicist (1911-1994).
Lydda	24.6N	31.9W	32	Town in Israel.
Masursky	12.0N	32.5W	110	Harold; Am. astrogeol. (1922-1990).
Novara	25.2S	10.5W	85	Town in Italy.
Peta	21.5S	9.1W	80	Town in Greece.
Pollack	7.9S	334.7W	95	James B.; Am. physicist (1938-1994).
Swanage	26.7N	33.7W	9	Town in England.
Thira	14.5S	184.0W	20	Town on Santorini Isl., Aegean Sea.
PLANUM (1)				
Lucus Planum	0.0	160.0W	2900	Albedo feature name.
THOLI (2)				
Apollinaris Tholus	17.8S	184.2W	40	Albedo feature name.
Zephyria Tholus	19.9S	187.1W	32	Albedo feature name.
VALLES (3)				
Athabasca Vallis	9.0N	204.5W	175	River in Canada.
Durius Valles	17.8S	188.0W	210	Classical name, Douro River, Portugal.
Elaver Vallis	9.5S	49.6W	175	Classical name, Allier River, France.

M O O N (3)

NAME	LAT	LONG	DIAM (km)	ATTRIBUTE
CRATERS (3)				
Cailleux	60.8S	153.3E	50	Andre; French geologist (1907-1986).
Kozyrev	46.8S	129.3E	65	Nikolay A.; Russ. astron. (1908-1983).
Oberth	62.4N	155.4E	60	Hermann; Austr. space sc. (1894-1989).

C A L L I S T O (34)

NAME	LAT	LONG	DIAM (km)	ATTRIBUTE
CATENAE (7)				
Eikin Catena	8.5S	15.9W	191	Norse river.
Fimbulthul Catena	8.4N	65.4W	378	Norse river.
Geirvimul Catena	49.0N	347.1W	90	Norse river.
Gomul Catena	35.4N	48.0W	324	Norse river.
Gunntro Catena	19.3S	343.3W	136	Norse river.
Sid Catena	48.7N	105.4W	78	Norse river.
Svol Catena	11.0N	37.1W	140	Norse river.
CRATERS (27)				
Aegir	45.9S	104.4W	46	Norse sea god.
Agloolik	47.9S	82.9W	49	Eskimo spirit of the seal caves.
Arcas	85.0S	66.8W	41	Callisto's child by Zeus.
Audr	31.0S	81.2W	70	Ottar's ancestor.
Austri	81.3S	64.1W	9	Norse dwarf.
Barri	31.6S	71.0W	83	Ottar's ancestor.
Biflindi	53.8S	74.3W	57	Another name for Odinn.
Doh	30.4N	142.1W	55	Ketian shaman who created the earth.
Gandalfr	81.0S	63.3W	12	Norse dwarf.
Ginandi	85.7S	50.0W	30	Ottar's ancestor.
Jalkr	38.6S	83.2W	74	Another name for Odinn.
Keelut	77.3S	92.1W	47	Eskimo evil spirit resembling a dog.
Lofn	57.0S	24.0W	200	Norse goddess of marriage.
Lycaon	45.2S	5.8W	55	Callisto's father.
Nakki	56.6S	69.9W	65	Finnish water god.
Numi-Torum	50.2S	93.4W	65	Mansi creator god.
Nyctimus	62.7S	3.8W	29	Brother of Callisto.
Oluksak	47.9S	63.9W	74	Eskimo god of lakes.
Orestheus	44.5S	50.1W	30	Brother of Callisto.
Randver	72.3S	53.6W	21	Ottar's ancestor.
Reginleif	66.2S	97.2W	32	Servant of the gods.
Reifnir	50.8S	63.8W	39	Ottar's ancestor.
Uksakka	49.5S	42.3W	22	Lapp protector goddess.
Skeggold	49.6S	31.9W	39	Servant of the gods.
Thekkr	80.8S	61.6W	10	Norse dwarf.
Thorir	32.0S	67.3W	43	Ottar's ancestor.
Yuryung	54.9S	86.1W	74	Yakutian heaven god.

G A N Y M E D E (35)

NAME	LAT	LONG	DIAM (km)	ATTRIBUTE
CATENAE (3)				
Enki Catena	39.5N	13.2W	151	Principal water god of the Apsu.
Khnum Catena	32.1N	350.9W	59	Egyptian creation god.
Nanshe Catena	14.7N	355.0W	59	Water goddess, daughter of Enki.

NAME	LAT	LONG	DIAM (km)	ATTRIBUTE
CRATERS (24)				
Aleyin	16.3N	134.7W	12	Son of Ba'al, spirit of springs.
Amset	14.5S	178.2W	10	God of the dead, son of Horus.
Anhur	32.6N	193.9W	25	Egyptian warrior god.
Chrysor	16.5N	134.9W	6	Phoenician god of fishing.
Cisti	32.0S	65.0W	65	Iranian healing god.
Ea	18.7N	149.2W	20	Assyro-Bab god of water, wisdom, earth.
El	1.9N	151.2W	50	"Father of Men", preceded the gods.
En-zu	12.2N	168.6W	7	Babylonian moon god.
Epigeus	24.0N	181.0W	320	Phoenician god.
Erichthonius	15.5S	174.6W	35	Possible father of Ganymede.
Hay-tau	15.8N	133.6W	28	Nega god, forest spirit.
Khensu	1.8N	152.8W	15	Egyptian moon god.
Khepri	21.5N	148.1W	50	God for transforming the Heliopitans.
Latpon	61.0N	175.0W	45	One of the sons of El.
Lugalmeslam	23.5N	195.0W	70	Sumerian god of the underworld.
Maa	1.0N	203.8W	30	Egyptian god of the sense of sight.
Mont	43.5N	314.0W	10	Theban war god.
Mot	10.5N	166.2W	25	Harvest spirit, son of El.
Nefertum	43.0N	323.0W	30	Divine son of the Memphis triad.
Nergal	38.5N	202.3W	8	Assyro-Babyl. king of underworld.
Ningishzida	14.0N	190.5W	30	Sumerian vegetation god.
Serapis	10.0S	50.0W	155	Egyptian healing god.
Upuant	45.0N	321.5W	15	Jackal-headed warrior god of the dead.
Zakar	30.5N	335.2W	150	Assyrian supreme diety.
FACULAE (3)				
Akhmin Facula	28.0N	191.0W	225	Egyptian town of Min worship.
Heliopolis Facula	19.5N	147.6W	50	Sacred Egyptian city of the sun.
Hermopolis Facula	22.0N	196.0W	200	Place where Unut was worshipped.
SULCI (5)				
Akitu Sulcus	39.0N	197.0W	380	Place where Marduk was worshipped.
Byblus Sulcus	38.0N	202.0W	600	Ancient city of Adonis worship.
Nineveh Sulcus	26.0N	60.0W	1000	City where Ishtar was worshipped.
Philae Sulcus	68.5N	175.0W	900	Temple, chief sanctuary of Isis.
Xibalba Sulci	35.0N	80.0W	2000	Mayan "place of fright" for the dead.
		E U R O P A (12)		
CHAOS (1)				
Conamara Chaos	9.5N	273.3W	127	Place in Ireland named for Conmac.
CRATERS (3)				
Govannan	37.5S	302.6W	10	A smith and brewer, child of Don.
Manann\%an	2.0N	240.0W	30	Irish sea and fertility god.
Pwyll	26.0S	271.0W	38	Celtic god of the underworld.
LINEAE (6)				
Agave Linea	12.6N	273.0W	1250	Daughter of Harmonia and Cadmus.
Chthonius Linea	0.1N	311.3W	1850	A founder of Thebes.

E U R O P A (Cont.)

NAME	LAT	LONG	DIAM (km)	ATTRIBUTE
LINEAE				
Harmonia Linea	27.0N	168.0W	925	Wife of Cadmus.
Hyperenor Linea	3.1S	314.7W	2200	A founder of Thebes.
Ino Linea	5.0S	163.0W	1400	Daughter of Harmonia and Cadmus.
Pelagon Linea	34.0N	170.0W	800	King who sold Cadmus a Moon cow.
REGIONES (1)				
Moytura Regio	47.9S	297.1W	347	Fomorians vs Tuatha de Danan battle ground.
LARGE RINGED FEATURES (1)				
Callanish	16.0S	333.4W	100	Stone circle in the Outer Hebrides.

I O (44)

NAME	LAT	LONG	DIAM (km)	ATTRIBUTE
ERUPTIVE CENTERS (2)				
Kanehekili	18.0S	40.0W		Hawaiian thunder god.
Zamama	18.0N	174.0W		Babylonian sun, corn, and war god.
FLUCTUS (6)				
Acala Fluctus	11.0N	337.0W	300	Japanese fire god.
Fjorgynn Fluctus	11.5N	358.0W	300	Norse thunder god.
Kanehekili Fluctus	16.0S	38.0W	250	Hawaiian thunder god.
Lei-Kung Fluctus	38.0N	204.0W	400	Chinese thunder god.
Marduk Fluctus	27.0S	209.0W	150	Sumero-Akkadian fire god.
Masubi Fluctus	48.0S	60.0W	800	Japanese fire god.
MENSAE (1)				
Hermes Mensa	43.0S	247.0W	130	Freed Io from Argus.
MONTES (9)				
Caucasus Mons	33.0S	239.0W	150	Mountains where Io fled from a gadfly.
Dorian Montes	24.0S	198.0W	450	Region in ancient Greece.
Egypt Mons	41.0S	257.0W	300	Io ended her wanderings here.
Euxine Mons	27.0N	126.0W	200	Io passed by here in her wanderings.
Ionian Mons	9.0N	236.0W	150	Io crossed this sea in her wanderings.
Nile Montes	52.0N	253.0W	450	Where Zeus restored Io to a human.
Rata Mons	35.0S	201.0W	200	Maori sun hero.
Skythia Mons	26.0N	98.0W	200	Io passed by here in her wanderings.
Tohil Mons	29.0S	157.0W	300	Centr. Am. god who gave fire to man.
PATERAE (24)				
Aidne Patera	2.0S	178.0W	50	Irish creator of fire.
Altjirra Patera	34.0S	108.0W	50	Australian thunder god.
Arusha Patera	38.0S	101.0W	60	Hindu god of the rising sun.
Catha Patera	53.0S	100.0W	60	Etruscan sun god.
Dusura Patera	37.0N	119.0W	70	Nabataean sun god.
Fo Patera	40.5N	192.0W	50	Chinese fire and sun god.
Gish Bar Patera	17.0N	90.0W	150	Babylonian sun god.
Hi'iaka Patera	3.0S	80.0W	80	Sister of Pele.

NAME	LAT	LONG	DIAM (km)	ATTRIBUTE
PATERAE				
Isum Patera	29.0N	208.0W	100	Assyrian fire god.
Janus Patera	3.0S	42.5W	60	Italian sun god.
Karei Patera	2.0N	16.0W	100	Semangan (Malayan) thunder god.
Kurdalagon Patera	49.5S	218.0W	50	Ossetian celestial smith.
Laki-io Patera	37.5S	62.5W	130	Bornean hero who invented fire.
Monan Patera	19.0N	106.0W	50	Brazilian god who destroyed the world.
Mulungu Patera	17.0N	218.0W	50	African thunder god.
Pillan Patera	12.0S	244.0W	80	Araucanian thunder, fire, and volcano.
Rata Patera	35.5S	199.5W	30	Maori sun hero.
Sethlaus Patera	52.0S	194.0W	80	Etruscan celestial smith.
Shamshu Patera	8.0S	64.0W	90	Arabian sun goddess.
Sigurd Patera	5.0S	98.0W	60	Norse sun hero.
Tiermes Patera	22.5N	351.5W	50	Lapp thunder god.
Tupan Patera	18.0S	141.0W	50	Tupi-Guarani (Brazil) thunder god.
Ukko Patera	33.0N	20.0W	40	Finnish thunder god.
Zal Patera	42.0N	76.0W	130	Iranian sun god.
REGIONES (2)				
Bosphorus Regio	0.0	120.0W	1200	Where Io passed escaping the gadfly.
Illyrikon Regio	72.0S	160.0W	700	Io passed by here in her wanderings.

M I R A N D A (3)

REGIONES (1)				
Ephesus Regio	15.0S	250.0E	225	Setting for "The Comedy of Errors".
SULCI (2)				
Naples Sulcus	32.0S	260.0E	260	Destination in "The Tempest."
Syracusa Sulcus	15.0S	293.0E	40	Home of twins in "Comedy of Errors".

I D A (1)

NAME	LAT	LONG	DIAM (km)	ATTRIBUTE
CRATERS (1)				
Peacock	2.0S	52.0E	0.2	Cave in Florida.

V E N U S (666)

CRATERS (287)				
Abra	6.2N	97.4E	8.5	Ewe first name.
Adzoba	12.8N	117.0E	10.0	Ewe (Ghana) first name.
Afiba	47.1S	102.7E	11.4	Ewe first name.
Afiruwa	4.3N	3.8E	5.2	Hausa first name.
Aftenia	50.0N	324.0E	7	Moldavian first name.
Afua	15.5N	124.0E	10.0	Akan (Ghana) first name.
Agoe	13.1N	4.3E	6.3	Eve first name.
Aigul	38.2N	280.4E	6	Kalmyk first name.

V E N U S (Cont.)

NAME	LAT	LONG	DIAM (km)	ATTRIBUTE
CRATERS				
Ailar	15.8S	68.4E	8.2	Turkman first name.
Aisha	39.3N	53.3E	10.6	Kyrgyz first name.
Akosua	58.6S	18.1E	6.2	Akan first name.
Akuba	9.6N	23.0E	5.5	Eve first name.
Altana	1.4N	69.9E	6	Kalmyk first name.
Aminata	6.6N	25.2E	9.7	Mandingo first name.
Avene	40.4N	149.4E	10.0	Akan (Ghana) first name.
Ayashe	22.7N	31.4E	6.7	Hausa first name.
Ayisatu	34.6N	5.5E	7	Fulbe first name.
Bachira	26.5N	10.0E	7.3	Algerian first name.
Bahriyat	50.3N	357.5E	5	Kumyk (Daghestan) first name.
Bakisat	26.0N	356.8E	7.4	Chechen first name.
Barauka	10.6N	346.3E	12.9	Hausa (Nigeria) first name.
Bineta	57.3N	144.1E	10.7	Mandingo (Africa, Mali) first name.
Chechek	2.6S	272.3E	7.2	Tuva (Siberia) first name.
Cholpon	40.0N	290.0E	6.3	Kyrgyz first name.
Chubado	45.3N	5.6E	7	Fulbe first name.
Clio	6.3N	333.5E	11.4	Greek first name.
Dado	13.9S	87.6E	11.2	Fulbe first name.
Dafina	28.6N	244.1E	5.5	Albanian first name.
Defa	32.2N	11.3E	8.5	Fulbe first name.
Degu	27.3N	289.8E	5.5	Adygan (N. Caucasus) first name.
Domnika	18.4N	294.3E	6.7	Moldavian first name.
Dunghe	56.2S	295.3E	5.5	Kalmyk first name.
Dyasya	5.1N	297.6E	7.8	Nganasan (Samoyed) first name.
Eila	75.0S	94.6E	9.5	Finnish first name.
Eini	41.6S	96.4E	5.9	Finnish first name.
Elma	10.1S	91.1E	10.2	Finnish first name.
Emilia	26.5S	88.2E	12.5	Swedish first name.
Emma	13.7S	302.3E	11.8	German first name.
Erkeley	43.9N	103.3E	8	Altai first name.
Esmeralda	64.4N	104.5E	9.8	Gypsy first name.
Eugenia	80.6N	105.4E	6	Greek first name.
Faina	71.1N	100.7E	10.0	Turkish first name.
Faufau	18.8N	8.3E	7.8	Polynesian first name.
Fava	0.7S	87.4E	9.7	Dunghan (Kyrgyzstan) first name.
Fazu	32.4N	106.0E	6.1	Avarian (Daghestan) first name.
Feruk	64.0S	107.6E	8.3	Nivkhi (Sakhalin Isl.) first name.
Firuza	51.8N	108.0E	6	Persian first name.
Florence	15.2S	85.0E	10.5	English first name.
Flutra	68.4S	112.0E	6	Albanian first name.
Frosya	29.5N	113.4E	9.8	Russian first name.
Gahano	80.2S	77.4E	4.5	Seneca first name.
Giselle	11.8S	298.0E	10.4	French first name.
Gulchatay	20.5N	295.5E	9	Arabic first name.
Gulnara	23.7S	174.0E	5	Uzbek first name.
Guzel	57.6S	298.7E	7.3	Arabic first name.
Hadisha	39.0S	97.2E	8.9	Kazakh first name.
Halima	28.5N	14.6E	8.9	Hausa first name.

NAME	LAT	LONG	DIAM (km)	ATTRIBUTE
CRATERS				
Hanka	27.3S	114.3E	5	Czech first name.
Hapei	66.1N	178.0E	4.2	Cheyenne (Oklahoma) first name.
Helga	10.4S	116.7E	8.8	Norwegian first name.
Helvi	12.4N	82.7E	12.2	Estonian first name.
Hilkka	69.0S	72.0E	10.3	Finnish first name.
Hiriata	15.3N	23.5E	5	Polynesian first name.
Hiromi	35.2N	287.3E	6	Japanese first name.
Huarei	15.0N	32.3E	8.5	Polynesian first name.
Icheko	6.6N	97.9E	5.9	Evenk/Tungu first name.
Imagmi	48.4S	100.7E	7.6	Eskimo (Chukotka) first name.
Ines	67.1S	241.9E	11.2	Spanish first name.
Inga	38.1N	226.6E	10.0	Danish first name.
Inkeri	28.3S	223.9E	10.1	Finnish first name.
Iondra	10.5N	286.5E	7.9	Selkup (Samoyed) first name.
Iraida	27.8N	108.1E	6.5	Greek first name.
Irinuca	51.4N	121.9E	8	Romanian first name.
Irma	50.9S	122.0E	9.5	Finnish first name.
Isolde	74.5S	211.9E	11.9	English first name.
Istadoy	51.8S	132.6E	5.4	Tajik first name.
Ivne	27.0S	132.8E	9	Koryak (Kamchatka) first name.
Izakay	12.3S	210.8E	10.2	Mari first name.
Izudyr	53.9S	153.2E	6.6	Mari (Volga Finn) first name.
Jaantje	46.5N	123.0E	7.8	Dutch first name.
Jalgurik	42.3S	125.1E	7.5	Evenk/Tungu first name.
Jamila	45.8N	134.8E	7.9	Afghan first name.
Janice	87.3N	261.9E	10.0	English first name.
Janina	2.0S	135.7E	9.3	Polish first name.
Janyl	28.0S	138.8E	5.6	Kyrgyz first name.
Jitka	61.9S	70.9E	13.0	Slovakian first name.
Jodi	35.7S	68.7E	10.2	English first name.
Jumaisat	15.1S	135.6E	7.5	Kumyk (Daghestan) first name.
Jutta	0.0N	142.6E	7	Finnish first name.
Kafutchi	26.7N	16.4E	7.1	Bantu first name.
Kaisa	13.5N	293.3E	12.0	Finnish first name.
Kalombo	30.5S	34.0E	9.6	Bantu first name.
Karo	21.9N	37.2E	7	Maori first name.
Kastusha	28.6S	59.9E	13.0	Mordovian (Volga River) first name.
Katrya	29.5S	108.0E	9.2	Ukrainian first name.
Katya	57.8N	285.7E	10.5	Russian first name.
Kavtora	59.0N	23.3E	9.8	Afghan first name.
Ketzia	3.9N	300.5E	14.6	Hebrew first name.
Khadako	54.2N	139.3E	7.4	Nenets (Samoyed) first name.
Khafiza	6.0N	299.2E	7	Arabic first name.
Kimitonga	25.1S	48.3E	5	Polynesian first name.
Kodu	0.9N	338.7E	10.5	Wolof (Africa, Senegal) first name.
Koinyt	30.9S	293.2E	11.7	Nivkhi (E. Siberia) first name.
Kollado	61.0S	53.4E	5.5	Fulbe first name.
Kosi	43.9S	54.9E	7.7	Ewe first name.
Kumba	26.3N	332.7E	11.4	Fulbe (W. Africa, Guinea) first name.

V E N U S (Cont.)

NAME	LAT	LONG	DIAM (km)	ATTRIBUTE
CRATERS				
Kumudu	61.3N	154.1E	4.4	Singalese first name.
Kuro	7.8N	57.6E	8.8	Fulbe first name.
Kyen	6.2S	64.7E	5.2	Bantu first name.
Kylli	41.1N	67.0E	13.2	Finnish first name.
Lhagva	75.8S	300.1E	7.9	Mongolian first name.
Liv	21.1S	303.9E	11.2	Norwegian first name.
Loan	28.3N	60.0E	7.4	Vietnamese first name.
Lorelei	55.7N	243.9E	15.0	German first name.
Lu Zhi	42.6S	303.4E	8.3	Chinese first name.
Lyuba	1.6N	283.9E	12.4	Russian first name.
Madina	22.7N	58.0E	6.3	Kabarda first name.
Mae	40.5S	345.2E	7.5	From Margaret, Greek first name.
Mamajan	65.1S	257.3E	2	Turkman first name.
Mansa	33.9S	63.4E	8.1	Akan first name.
Marere	19.6N	65.8E	6.3	Polynesian first name.
Maret	33.3S	280.2E	11.7	Estonian first name.
Mariko	23.3S	132.9E	12.7	Japanese first name.
Marysya	53.3N	75.1E	6.3	Belorussian first name.
Masha	60.7N	88.5E	6.4	Russian first name.
Matahina	72.3S	65.9E	8.5	Polynesian first name.
Maurea	39.5S	69.1E	9.9	Polynesian first name.
Mbul'di	23.8N	74.7E	6	Fulbe/Wodabi first name.
Melanka	34.4N	19.2E	9	Ukranian first name.
Meredith	14.5S	278.9E	11.4	English first name.
Miovasu	72.1N	99.9E	4.5	Cheyenne first name.
Mosaido	17.3N	75.2E	7.4	Fulbe/Wodabi first name.
Nadeyka	54.8S	305.3E	9.3	Belorussian first name.
Nakai	61.0S	286.2E	4.5	Cheyenne first name.
Nalkuta	30.1N	307.8E	6.5	Ossetian (N. Caucasus) first name.
Namiko	43.4N	56.2E	13.0	Japanese first name.
Nastya	49.0S	275.8E	12.5	From Anastasiya, Russian first name.
Ndella	15.9S	60.7E	5.9	Wolof first name.
Neda	16.7N	313.5E	7.7	Macedonian first name.
Nedko	8.8S	317.6E	8.5	Nenets (Samoyed) first name.
Neeltje	12.4N	124.4E	10.0	Dutch first name.
Nelike	26.8S	329.2E	6.3	Nanay (E. Siberia) first name.
Ngaio	53.3S	61.8E	9.5	Maori first name.
Ngone	6.0N	331.9E	12.2	Wolof (Africa, Senegal) first name.
Nilanti	38.2S	331.4E	9.2	Singalese first name.
Ninzi	15.9N	331.7E	7.1	Burma (Myanmar) first name.
Nomeda	49.2S	55.5E	10.4	Lithuanian first name.
Nsele	6.7N	64.2E	5.1	Mandingo first name.
Nuon	78.6N	336.6E	6.5	Khmer first name.
Nutsa	27.5N	341.2E	8	Abkhazian (Georgia) first name.
Nyal'ga	17.0N	64.5E	5.5	Fulbe/Wodabi first name.
Nyele	22.7S	318.4E	11.9	Mandingo (Africa, Mali) first name.
Nyogari	46.4S	306.4E	13.0	Ewe (W. Africa, Ghana) first name.
Odarka	40.8N	138.2E	7	Ukrainian first name.
Odikha	41.6S	238.1E	10.6	Uzbek first name.

NAME	LAT	LONG	DIAM (km)	ATTRIBUTE
CRATERS				
Ogulbek	2.4N	145.0E	6.5	Turkman first name.
Oivit	73.9S	195.5E	4.8	Cheyenne first name.
Oku	64.2S	232.2E	13.3	Karelian first name.
Olena	10.9N	149.0E	7	Ukrainian first name.
Olesya	5.6N	273.3E	12.0	Ukrainian first name.
Olivia	37.2N	207.9E	10.2	Dutch first name.
Onissya	25.6S	150.2E	8.2	Komi-Permyak (Urals Finn) first name.
Opika	57.1S	151.9E	9.8	Chuvash (Volga area) first name.
Orguk	23.5S	198.2E	11.7	Nivkhi (E. Siberia) first name.
Orlette	68.1S	193.3E	12.5	French first name.
Ortensia	7.6N	155.7E	7	Italian first name.
Oshalche	29.7N	155.5E	8.3	Mari (Volga Finn) first name.
Ottavia	47.5S	187.1E	12.9	Roman first name.
Outi	61.6N	267.7E	10.5	Finnish first name.
Parishan	0.2S	146.5E	6.8	Kurdian first name.
Parvina	62.2S	153.0E	7	Tajik first name.
Pasha	42.7N	156.3E	7.2	Russian first name.
Pat	2.9N	262.6E	10.1	English first name.
Patimat	1.3S	156.5E	5.1	Avarian (Daghestan) first name.
Pavlinka	25.5S	158.7E	7.5	Belorussian first name.
Pirkko	44.8N	254.6E	12.3	Finnish first name.
Puhioia	20.6N	69.4E	5.5	Maori first name.
Purev	31.1S	46.4E	11.6	Mongolian first name.
Pychik	62.4S	33.8E	10.1	Chukcha (NE Siberia) first name.
Qarlygha	33.0S	162.9E	9.3	Kazakh first name.
Qulzhan	23.5N	165.4E	7.9	Kazakh first name.
Quslu	6.2N	166.8E	8.7	Kazakh first name.
Radhika	30.3S	166.4E	7.9	Tamil first name.
Radmila	69.1N	167.0E	5.2	Serbocroatian first name.
Rae	8.9S	58.4E	5.5	From Rachel, Hebrew first name.
Rafiga	62.9N	175.6E	5.7	Azeri first name.
Raki	49.4S	70.0E	7.5	Fulbe first name.
Rampyari	50.6N	179.3E	7.7	Hindu first name.
Raymonde	48.4N	191.5E	5.3	French first name.
Reiko	22.6N	192.1E	9.7	Japanese first name.
Retno	52.9S	192.3E	7.2	Indonesian first name.
Roptyna	62.2N	28.9E	11.5	Chukcha (NE Siberia) first name.
Royle	32.7S	193.7E	6.1	Bashkir first name.
Rufina	74.6S	195.1E	5	Greek first name.
Ruit	25.5S	72.9E	6.4	Polynesian first name.
Runak	58.5S	196.3E	7.6	Kurdian first name.
Safarmo	10.8S	161.4E	7.4	Tajik first name.
Saida	28.2N	302.0E	9.5	Arabic first name.
Sandugach	59.9N	143.5E	10.0	Tartar first name.
Sasha	38.3N	277.3E	4.6	Russian first name
Sayligul	73.6N	172.9E	4.3	Tajik first name.
Seseg	36.3S	312.6E	9.8	Buryat (Siberia) first name.
Shasenem	44.0S	258.9E	9	Turkman first name.
Sheila	19.9N	50.2E	5.6	Irish/Celtic first name.

V E N U S (Cont.)

NAME	LAT	LONG	DIAM (km)	ATTRIBUTE
CRATERS				
Shushan	43.8S	70.2E	8.5	Armenian first name.
Simbya	74.4S	130.0E	4.0	Nganasan (Samoyed) first name.
Tahia	44.3N	73.7E	6.1	Polynesian first name.
Tako	25.1N	285.3E	10.7	Fulbe first name.
Talvikki	41.9N	22.0E	12.6	Finnish first name.
Tehina	30.4S	76.4E	5.4	Polynesian first name.
Tekarohi	21.2N	76.4E	9.3	Polynesian first name.
Temou	10.0S	83.4E	9.3	Polynesian first name.
Terhi	45.7N	253.1E	10.7	Finnish first name.
Teroro	75.8S	88.1E	9.2	Polynesian first name.
Teumere	38.3S	88.1E	5.4	Polynesian first name.
Teura	12.3S	90.2E	9.3	Polynesian first name.
Tinyl	9.7N	132.1E	12.8	Chukcha (NE Siberia) first name.
Tolgonay	68.8N	271.1E	4.6	Kyrgyz first name.
Tsetsa	31.3N	317.7E	9.9	Mordovian (Volga Finn) first name.
Tsyrma	14.1S	318.5E	7.8	Buryat (Siberia) first name.
Tursunoy	80.9N	229.3E	4.7	Uzbek first name.
Tuyara	62.9S	15.5E	13.2	Yakut first name.
Ualinka	13.2N	168.6E	8.1	Ossetian (N. Caucasus) first name.
Udagan	10.7N	206.9E	11.5	Yakut first name.
Udyaka	30.9N	172.9E	7.7	Orochi (Amur River) first name.
Ugne	34.9N	205.8E	10.3	Lithuanian first name.
Ul'yana	24.3N	253.0E	12.5	Russian first name.
Uleken	33.7N	185.1E	10.9	Nanay (E. Siberia) first name.
Ulla	51.5S	184.5E	10.4	Swedish first name.
Ulpu	35.7S	179.0E	7	Finnish first name.
Uluk	62.2S	178.6E	10.3	Neghidalian (E. Siberia) first name.
Umaima	23.3S	195.4E	6.9	Arabic first name.
Umkana	53.3S	198.6E	6.2	Eskimo (Chukotka) first name.
Unay	53.5N	172.7E	11.4	Mari first name.
Unitkak	40.8N	199.5E	8	Eskimo (Chukotka) first name.
Urazbike	9.0S	202.5E	7	Tartar first name.
Ustinya	41.2S	251.6E	11.8	Russian first name.
Uyengimi	76.9S	204.9E	8.9	Khanty, Mansi (Ob R. Finn) first name.
Vaka	41.4S	8.9E	11.8	Bulgarian first name.
Vard	17.5N	314.5E	6.1	Armenian first name.
Varya	2.8N	211.8E	14.3	Russian first name.
Vasilutsa	16.5N	334.4E	5.7	Moldavian first name.
Vassi	34.4N	346.5E	8.5	Karelian first name.
Veriko	20.4N	350.1E	5.2	Georgian first name.
Veta	42.6N	349.5E	6.4	Romanian first name.
Viola	36.1S	240.5E	10.0	English first name.
Virga	26.9S	7.7E	10.3	Lithuanian first name.
Vlasta	28.4N	250.1E	10.7	Czech first name.
Volyana	60.6N	359.9E	5.3	Gypsy first name.
Wazata	33.6N	298.3E	13.9	Hausa (Nigeria) first name.
Wendla	22.5N	207.6E	5.7	Swedish first name.
Wilma	36.7N	1.7E	12.5	English first name.
Wiwi-yokpa	73.8S	228.4E	4.5	Abenaki/Algonquin (Canada) first name.

NAME	LAT	LONG	DIAM (km)	ATTRIBUTE
CRATERS				
Wynne	55.0N	53.6E	10.0	English first name.
Xenia	30.3S	249.4E	13.5	Greek first name.
Xi Wang	14.0N	208.0E	7.7	Chinese first name.
Ximena	68.2S	243.6E	12.8	Portuguese first name.
Yakyt	2.1N	170.2E	13.8	Karakalpak first name.
Yambika	32.6N	208.7E	6.5	Mari (Volga Finn) first name.
Yasuko	26.1S	169.0E	10.6	Japanese first name.
Yazruk	21.2N	160.2E	10.5	Nivkhi (E. Siberia) first name.
Yelya	47.5S	211.7E	8.6	Nenets (Samoyed) first name.
Yemysh	11.9N	214.7E	6	Mari (Volga Finn) first name.
Yenlik	16.0S	225.4E	8.6	Kazakh first name.
Yerguk	42.7N	226.8E	6.3	Neghidalian (Amur River) first name.
Yeska	27.4N	230.1E	9.1	Selkup (Samoyed) first name.
Yokhtik	50.1S	158.1E	11.4	Nivkhi (E. Siberia) first name.
Yoko	5.7S	232.0E	5	Japanese first name.
Yolanda	7.8N	152.7E	11.4	Greek first name.
Yomile	27.3S	138.7E	13.6	Bashkir first name.
Yonok	65.1S	234.1E	9.5	Korean first name.
Yonsuk	34.0S	234.8E	8.5	Korean first name.
Ytunde	49.9N	81.1E	6.1	Yoruba first name.
Yvette	7.5N	249.6E	10.6	French first name.
Zakiya	66.5S	234.1E	7.5	Arabic first name.
Zarema	16.8N	235.2E	5	Avarian (Daghestan) first name.
Zeinab	2.2S	159.6E	12.5	Persian first name.
Zemfira	46.2S	157.7E	11.4	Gypsy first name.
Zerine	29.6S	258.6E	6.5	Persian first name.
Zivile	48.8N	113.1E	13.5	Lithuanian first name.
Zosia	18.9S	109.2E	10.5	Polish first name.
Zuhrah	34.7N	357.0E	5.8	Arabic first name.
Zula	7.3N	282.0E	5	Chechen first name.
Zulfiya	18.4N	101.9E	12.9	Uzbek first name.
Zulma	7.7S	102.0E	11.0	Spanish first name.
Zumrad	32.1N	94.8E	12.9	Uzbek first name.
Zurka	12.8S	275.2E	5.5	Gypsy first name.
CHASMATA (22)				
Aikhylu Chasma	32.0N	292.0E	300	Bashkir myth's moon daughter.
Ardwinna Chasma	21.0N	197.0E	500	Continental Celtic wildwood goddess.
Artio Chasma	35.5S	39.0E	450	Celtic wildlife bear-goddess.
Chang Xi Chasmata	59.0S	17.0E	220	Chinese, gave birth to twelve moons.
Gamsilg Chasma	46.0S	64.0E	600	Chechen and Ingush evil forest deity.
Geyaguga Chasma	56.5S	70.0E	800	Cherokee moon deity.
Gui Ye Chasma	9.0S	337.1E	210	Chinese moon fairy.
Hanghepiwi Chasma	48.5S	18.0E	1100	Dakota name of the moon and night.
Kokomikeis Chasma	0.0N	85.0E	1000	Blackfoot/Algonquin moon goddess.
Kov-Ava Chasma	58.8S	21.8E	470	Mordovian (Finnish) forest mistress.
Nang-byon Chasma	4.0N	316.5E	450	Vietnamese moon goddess.
Olapa Chasma	42.0S	208.5E	650	Massai (Kenya, Tanzania) moon goddess.
Pinga Chasma	20.0S	287.0E	500	Eskimo goddess of hunt.

V E N U S (Cont.)

NAME	LAT	LONG	DIAM (km)	ATTRIBUTE
CHASMATA				
Reitia Chasma	1.0N	285.0E	1300	Venetian health and hunting goddess.
Seo-Ne Chasma	63.5S	26.0E	430	Korean moon deity, sun's wife.
Sutkatyn Chasmata	64.0S	11.0E	350	Kumyk (Daghestan) forest spirit.
Tellervo Chasma	60.0S	125.0E	600	Finnish maiden of woods.
Tkashi-mapa Chasma	13.0N	206.0E	1100	Georgian forest goddess.
Tsects Chasma	61.6S	35.0E	600	Haida spirit of forest underworld.
Xaratanga Chasma	54.0S	70.0E	1300	Tarascan (Mexico) moon goddess.
Zewana Chasma	9.0N	212.0E	900	W. Slavic/Polish hunting goddess.
Zverine Chasma	18.5N	271.0E	1300	Lithuanian hunting goddess.
COLLES (6)				
Chernava Colles	10.5S	335.5E	1000	Russian sea czar daughter.
Marake Colles	55.7N	217.8E	150	Mansi (Ob River Ugra) sea mistress.
Nahete Colles	38.0N	241.0E	400	Fon (Benin) wife of sea god Agbe.
Nuliayoq Colles	48.0N	224.0E	350	Netsilik Inuit sea mistress.
Ruad Colles	68.0S	118.0E	400	Irish female deity, sank into the sea.
Salofa Colles	63.0S	167.0E	250	Samoan tale's girl/sea turtle.
CORONAE (70)				
Ama Corona	45.7S	278.2E	300	Jukun (Nigeria) goddess of birth.
Ambar-ona Corona	70.0S	82.5E	550	Uzbek women's and fertility goddess.
Atahensik Corona	19.0S	170.0E	700	Huron/Iroquois goddess, sun and moon creator.
Bibi-Patma Corona	47.0S	302.0E	450	Turkman goddess of women.
Boala Corona	70.0S	359.0E	220	Bantu name for the first woman.
Codidon Corona	46.0S	56.0E	250	Arauakan (Colombia) mother goddess.
Demvamvit Corona	65.5S	38.0E	370	Gurage (SW Ethiopia) women's goddess.
Deohako Corona	67.5S	118.0E	300	Seneca Iroquois crops spirit.
Dhorani Corona	8.0S	243.0E	300	Thai earth and love goddess.
Didilia Corona	19.0N	38.0E	320	E. Slavic childbirth goddess.
Dunne-Musun Corona	60.0S	85.0E	630	Evenk (Tungu) earth mistress.
Dyamenyuo Corona	57.5S	42.5E	200	Enets (Samoyed) women's and childbirth deity.
Eingana Corona	5.0N	350.0E	375	Australian aboriginal snake goddess.
Ekhe-Burkhan Corona	50.0S	40.0E	600	Buryatian creator goddess.
Emegen Corona	37.5N	290.5E	180	Tuva (S. Siberia) childcare goddess.
Enekeler Corona	46.0S	264.0E	350	Altay childbirth goddess.
Erigone Corona	34.5S	284.0E	325	Greek harvest goddess.
Flidais Corona	24.5S	177.3E	150	Irish fertility goddess.
Furachoga Corona	38.0S	258.0E	550	Chibcha/Muiska earth goddess.
Hlineu Corona	38.7S	241,0E	150	Chin/Kieng (Burma) ancestor goddess.
Iang-Mdiye Corona	47.0S	86.0E	300	Ede (Vietnam) goddess of rice.
Ilyana Corona	69.5S	65.0E	300	Moldavian main female deity.
Ituana Corona	19.5N	153.5E	220	Amazon River people great goddess.
Juksakka Corona	19.5S	44.5E	320	Lapp goddess of birth.
Kaltash Corona	0.5N	75.0E	450	Mansi (Ob River Ugra) mother-goddess.
Katieleo Corona	12.5S	327.5E	210	Senufo (Burkina Faso) creator goddess.
Latmikaik Corona	64.0S	123.0E	500	Palau (Micronesia) fertility goddess.
Latta Corona	38.6S	287.0E	225	Chechen/Ingush (Caucasus) earth goddess.

V E N U S (Cont.)

NAME	LAT	LONG	DIAM (km)	ATTRIBUTE
CORONAE				
Marzyana Corona	53.0S	67.5E	550	Slavic grain and fertility goddess.
May-Enensi Corona	42.5S	68.0E	330	Teleutan (S. Altay) fertility goddess.
Moombi Corona	64.5S	235.5E	100	Gikuyu (Kenya), the first woman.
Mou-nyamy Corona	49.5S	59.0E	200	Nganasan (Samoyed) goddess of life.
Mukylchin Corona	12.5S	46.0E	525	Udmurt (Ural) fertility goddess.
Mykh-Imi Corona	73.0S	99.0E	150	Khanty (Ob River Ugra) earth goddess.
Naotsete Corona	58.3S	249.5E	200	Keresan Pueblo ancestor goddess.
Navolga Corona	48.6S	296.5E	170	Ganda (Uganda) goddess of childbirth.
Nungui Corona	42.5S	245.2E	150	Hibaro (Peru/Eq.) fertility goddess.
Nzambi Corona	45.0S	287.5E	225	Bantu goddess, mother of all beings.
Obasi-Nsi Corona	53.5S	291.0E	230	Ekoi (S. Nigeria) fertility goddess.
Obiemi Corona	31.9S	276.6E	300	Bini (Nigeria) childbirth goddess.
Obilukha Corona	81.5S	19.0E	220	E. Slavic crop protection deity.
Okhin-Tengri Corona	70.5S	40.0E	400	Kalmykan fertility goddess.
Omosi-Mama Corona	64.5N	306.0E	480	Manchoo childbirth goddess.
Pasu-Ava Corona	29.0N	319.0E	250	Mari (Volga Finn) harvest goddess.
Pugos Corona	19.0S	335.0E	180	Khanty (Ob' River) goddess of life.
Romi-Kumi Corona	81.2S	180.0E	150	Tukano (Colombia) great mother goddess.
Santa Corona	34.5S	288.0E	200	Sabine goddess of fertility/health.
Saunau Corona	1.3S	173.0E	200	Abkhazian goddess of corn milling.
Seisui Corona	62.0S	241.0E	150	Tupi/Huarani fertility goddess.
Semiramus Corona	37.0S	293.0E	375	Assyrian fertility goddess.
Shulamite Corona	38.8S	284.3E	275	Hebrew fertility goddess.
Shyv-Amashe Corona	57.0S	63.0E	410	Chuvash (Volga Region) water goddess.
Su-Anasy Corona	78.0S	39.0E	300	Tartar/Kumyk/Karachay water mother.
Tangba Corona	47.0S	258.0E	200	Lobi (Burkina Faso) earth goddess.
Tonatzin Corona	53.0S	164.0E	400	Aztec earth and childbirth goddess.
Triglava Corona	53.5S	95.0E	400	Slavic earth goddess.
Tureshmat Corona	51.5S	289.5E	150	Ainu (Japan) creator of Hokkaido.
Tutelina Corona	29.0N	348.0E	180	Roman harvest goddess.
Ugatame Corona	76.5S	255.0E	370	Kapauku (Papua) Great Mother goddess.
Umay-ene Corona	27.5S	50.5E	370	Kazakh childcare goddess.
Utset Corona	55.5S	167.0E	150	Zia (SW USA) the First Mother.
Ved-Ava Corona	33.0N	143.0E	200	Mordovian (Volga Finn) water mother.
Vesuna Corona	65.5S	275.0E	200	Italic (Umbrian) vegetation goddess.
Xcacau Corona	56.0S	131.0E	200	Quiche (Guatemala) cacao goddess.
Xcanil Corona	37.0S	43.0E	200	Aztec and Quiche maize goddess.
Xmukane Corona	28.2S	269.5E	200	Mayan mother and fertility goddess.
Yanbike Corona	1.5S	328.5E	200	Bashkir mythical first woman.
Zemire Corona	31.5N	312.5E	200	Kumyk (Daghestan) fertility goddess.
Zemlika Corona	33.5S	50.0E	150	Latvian earth goddess.
Zhivana Corona	13.0N	287.5E	180	Slavic goddess of life.
Zywie Corona	38.6S	291.2E	200	Polish goddess of life.
DORSA (53)				
Abe Mango Dorsa	47.0N	212.0E	800	Tukano (Brazil) daughter of sun god.
Achek Dorsa	37.0S	220.0E	100	Dinka (Sudan) wife of rain god Deng.
Aida-Wedo Dorsa	73.0N	214.0E	450	Haitian rainbow spirit.
Akewa Dorsa	45.5N	184.0E	900	Toba (Argentina) sun goddess.

V E N U S (Cont.)

NAME	LAT	LONG	DIAM (km)	ATTRIBUTE
DORSA				
Akuanda Dorsa	57.0N	236.0E	1000	Adygan light deity.
Alkonost Dorsa	5.0S	341.0E	730	E. Slavic wonder bird with woman's face.
Amitolane Dorsa	77.0S	335.0E	900	Zuni (SW USA) name of the Rainbow.
Anpao Dorsa	62.0N	207.0E	550	Dakota name of the Dawn.
Asiaq Dorsa	52.5S	55.0E	400	Eskimo weather goddess.
Barbale Dorsa	15.0N	143.0E	1200	Georgian sun goddess.
Biliku Dorsa	46.5S	138.0E	600	Andaman Isl. moonsoon deity.
Charykh-Keyok Dorsa	54.5N	285.0E	550	Khakasian (S. Siberia) magic bird.
Dylacha Dorsa	19.0S	76.0E	650	Evenk/Tungu (Siberia) sun goddess.
Etain Dorsa	45.0S	199.0E	1400	Irish sun and horse goddess.
Kadlu Dorsa	69.5S	188.0E	500	Eskimo thunder deity, noisy girl.
Kalm Dorsa	18.0N	309.0E	300	Mansi winged messenger from the gods.
Kastiatsi Dorsa	53.0S	245.0E	1200	Acoma (SW USA) name of the Rainbow.
Khadne Dorsa	14.0S	334.5E	220	Nenets (Samoyed) snowstorm maiden.
Kotsmanyako Dorsa	76.0S	242.0E	1900	Keresan Pueblo girl, put stars in sky.
Kuldurok Dorsa	50.4S	61.0E	1100	Uzbek thunder and lightning goddess.
Laverna Dorsa	50.0S	132.0E	1100	Roman night darkness goddess.
Lemkechen Dorsa	19.0N	65.0E	2000	Berber pole star goddess, holds camel.
Lumo Dorsa	24.5N	149.0E	500	Tibetan goddess of sky, rain, mist.
Metelitsa Dorsa	16.0N	31.0E	1300	E. Slavic snowstorm deity.
Naatse-elit Dorsa	64.0S	249.0E	950	Navajo rainbow goddess.
Naran Dorsa	56.0S	234.0E	600	Mongolian sun goddess.
Natami Dorsa	71.5S	258.0E	800	Mon (Burma/Myanmar) beauty fairy.
Norwan Dorsa	65.0N	163.0E	450	Wintun (Calif.) light goddess.
Nuvakchin Dorsa	53.0S	212.0E	2200	Hopi snow maiden ('kachina').
Ojuz Dorsa	6.0N	37.0E	800	Tajik deity of frost and cold wind.
Oshumare Dorsa	58.5S	79.0E	550	Yoruba rainbow deity.
Poludnitsa Dorsa	5.0N	179.5E	1500	E. Slavic witch, corn fields deity.
Pulugu Dorsa	65.0S	225.0E	650	Andaman Isl. moonsoon wind deity.
Ragana Dorsa	69.0S	246.0E	950	Latvian witch.
Rokapi Dorsa	55.0S	222.0E	2200	Georgian main witch.
Shishimora Dorsa	37.0N	297.0E	800	E. Slavic night and dream deity.
Siksaup Dorsa	73.0S	228.0E	650	Kachin (Burma/Myanmar) sun goddess.
Sinanevt Dorsa	67.0N	177.0E	1800	Itelmen (Kamchatka) wife of sky man.
Sogbo Dorsa	40.0S	237.0E	900	Fon (Benin) thunder goddess/god.
Spidola Dorsa	73.5S	325.0E	950	Latvian witch, flies in sky.
Sunna Dorsa	53.0S	134.0E	500	Norse sun goddess.
Tikoiwuti Dorsa	56.0N	225.0E	1000	Hopi goddess of darkness.
Tinianavyt Dorsa	51.0S	239.0E	1500	Koryak (Kamchatka) wife of sky man.
Tsovinar Dorsa	46.0S	254.0E	1100	Armenian lightning deity.
Tukwunag Dorsa	67.0S	160.0E	1000	Hopi cumulus cloud maiden ('kachina').
Unelanuhi Dorsa	12.0N	87.0E	2600	Cherokee sun goddess.
Unuk Dorsa	4.5S	351.5E	400	Eskimo (Chukotka) night maiden.
Urkuk Dorsa	12.0S	320.0E	600	Nivkhi (Sakhalin Isl.) night maiden.
Vejas-mate Dorsa	71.0S	245.0E	1600	Lithuanian "wind mother".
Wala Dorsa	17.0S	62.0E	500	Fox (US Plains) name of the Dawn.
Yalyane Dorsa	7.0N	177.0E	1200	Nenets (Samoyed) maiden of light.
Zaryanitsa Dorsa	0.0N	170.0E	1100	E. Slavic night lightning goddess.
Zimcerla Dorsa	47.5S	73.5E	850	W. Slavic dawn goddess.

V E N U S (Cont.)

NAME	LAT	LONG	DIAM (km)	ATTRIBUTE
FLUCTUS (21)				
Agrimpasa Fluctus	0.0N	175.5E	950	Scythian goddess of love.
Alpan Fluctus	7.5S	349.0E	500	Lezghin (Daghestan) fire goddess.
Arubani Fluctus	55.0S	132.0E	620	Urartu supreme goddess.
Bolotnitsa Fluctus	50.0N	160.0E	1100	E. Slavic swamp mermaid.
Cavillaca Fluctus	72.0S	340.0E	800	Huarochiri (Peru) virgin goddess.
Djabran Fluctus	43.5S	183.0E	300	Abkhazian goddess of goats.
Djata Fluctus	66.5N	307.5E	280	Ngadju (Indonesia) water goddess.
Dotetem Fluctus	6.0S	177.5E	530	Ketian (Yenisey R.) evil spirit.
Hikuleo Fluctus	52.5N	208.0E	600	Tonga (Polynesia) underworld goddess.
Juturna Fluctus	76.0S	350.0E	900	Roman nymph, wife of Janus.
Koti Fluctus	12.5N	318.0E	400	Creek (SE USA) water-frog spirit.
Mamapacha Fluctus	60.0N	185.0E	900	Inca earthquake goddess.
Medb Fluctus	56.0S	127.0E	350	Irish mother of gods, wife of Ailil.
Nambubi Fluctus	61.0S	135.0E	850	Ganda goddess, mother of god Mukasa.
Praurime Fluctus	16.0N	154.0E	750	Lithuanian fire goddess.
Rafara Fluctus	65.0S	159.0E	700	Malagasy (Madagascar) water goddess.
Sobra Fluctus	6.0N	248.0E	700	Marindanim (New Guinea) creator goddess.
Sonmunde Fluctus	60.0S	120.0E	400	Korean mountain goddess.
Tsunghi Fluctus	67.0S	130.0E	800	Hibaro (Ecuador) water goddess.
Turgmam Fluctus	56.0N	220.0E	500	Nivkhi (Sakhalin Isl.) fire mistress.
Uilata Fluctus	17.0N	314.0E	700	Cherokee stone-clad female monster.
FOSSAE (16)				
Aife Fossae	67.0N	131.0E	280	Irish warrior deity.
Ajina Fossae	45.0S	258.0E	300	Tajik evil spirit.
Albasty Fossae	9.0S	336.5E	500	Tartar evil spirit.
Brynhild Fossae	23.0S	20.0E	1800	Norse warrior maiden.
Gulaim Fossae	5.0S	329.0E	800	Karakalpak amazon leader.
Hanekasa Fossae	29.0N	148.5E	700	Sanema (Venezuela) amazon warrior.
Karra-m\-ahte Fossae	28.0N	342.0E	1800	Latvian warrior goddess.
Khosedem Fossae	13.0S	303.0E	1800	Ketian (Yenisey R.) evil goddess.
Magura Fossae	12.0S	332.5E	600	E. Slavic winged warrior maiden.
Naijok Fossae	70.2S	337.0E	450	Dinka (Sudan) evil deity.
Narundi Fossae	66.5S	329.0E	700	Elam goddess of victory.
Penthesilea Fossa	12.0S	214.0E	1700	Greek amazon queen.
Perunitsa Fossae	10.0S	307.0E	1300	E. Slavic winged warrior maiden.
Saykal Fossae	73.0N	139.0	300	Kyrgyz warrior maiden.
Yenkhoboy Fossae	48.0S	7.0E	900	Buryatian warrior sisters.
Yuzut-Arkh Fossae	48.0N	224.0E	550	Khakas (S. Siberia) evil deity.
LINEAE (4)				
Badb Linea	14.0N	15.0E	1750	Irish war goddess.
Discordia Linea	58.0S	246.5E	800	Roman war goddess, close to Bellona.
Sarykyz Linea	77.3S	200.0E	370	Uzbek evil spirit.
Sui-ur Linea	61.0S	260.0E	700	Mansi (Ob River Ugra) wife of war god.
MONTES (45)				
Abeona Mons	44.8S	273.1E	375	Roman goddess of travelers.

V E N U S (Cont.)

NAME	LAT	LONG	DIAM (km)	ATTRIBUTE
MONTES				
Aleksota Mons	9.0S	308.5E	250	Lithuanian goddess of love.
Atai Mons	22.0S	291.0E	250	Efik (Ghana) wife of sky god Abassi.
Atsyrkhus Mons	78.5S	227.0E	170	Ossetian, daughther of sun god Khur.
Awenhai Mons	60.0S	248.0E	100	Mohawk/Iroquois fertility goddess.
Bagbartu Mons	65.5N	279.0E	600	Urartu goddess, worshipped at Musasir.
Chloris Mons	45.4S	294.6E	180	Greek flower goddess.
Chuginadak Mons	38.0S	246.0E	450	Aleutian volcano goddess.
Cipactli Mons	31.5S	32.5E	200	Aztec monster, origin of Earth.
Dzalarhons Mons	0.5N	34.0E	120	Haida (NW Coast) volcano goddess.
Egle Mons	59.0S	134.0E	110	Lithuanian underwater queen.
Faravari Mons	43.5S	309.0E	500	Malagasy (Madagascar) water goddess.
Erzulie Mons	68.0S	8.0E	300	Haitian voodoo goddess of love.
Gurshi Mons	47.5S	58.5E	210	Buryatian fishing deity.
Iseghey Mons	9.0N	171.0E	500	Yakutian/Saha goddess of cows.
Katl-Imi Mons	69.0S	126.0E	120	Khanty (Ob River Ugra) sun goddess.
Kokyanwuti Mons	35.5N	212.0E	400	Hopi earth goddess - "Spider Woman".
Kshumay Mons	54.9S	58.0E	250	Nuristan (NE Afghanistan) vegetation goddess.
Laka Mons	79.9N	262.0E	220	Hawaiian uncultivated area goddess.
Lanig Mons	68.5S	91.0E	400	Semang (Malay Peninsula) creator goddess.
Loo-Wit Mons	59.5S	56.0E	150	NW Indian St. Helens volcano goddess.
Ludjatako Mons	12.0S	251.0E	500	Creek (SE USA) Giant Turtle deity.
Mem Loimis Mons	9.5N	209.0E	300	Wintun (California) goddess.
Mertseger Mons	38.1S	270.3E	450	Snake goddess of Theban necropolis.
Mielikki Mons	27.8S	280.5E	450	Finnish forest goddess.
Nahas-tsan Mons	14.0N	205.0E	500	Navajo Mother Earth.
Nayunuwi Montes	2.0N	83.0E	900	Cherokee stone-clad female monster.
Ne Ngam Mons	43.0S	257.5E	200	Lao world creator goddess.
Niola Mons	45.0N	185.0E	150	Lithuanian underworld goddess.
Ongwuti Mons	2.0S	194.5E	500	Hopi salt-woman deity; predicts seasons.
Pahto Mons	64.5S	114.5E	300	NW Indian mountain goddess.
Polik-mana Mons	24.5N	264.0E	600	Hopi butterfly maiden ('kachina').
Rakapila Mons	43.7S	321.5E	130	Malagasy (Madag.) sacred trees diety.
Rhpisunt Mons	2.5N	301.5E	250	Haida (NW Coast) Bear Mother deity.
Samodiva Mons	13.6N	291.0E	200	Bulgarian winged water deity.
Sakwap-mana Mons	4.0N	322.0E	400	Hopi maiden of blue corn ('kachina').
Siduri Mons	42.3S	297.3E	105	Babylonian goddess of wine & wisdom.
Ts'an Nu Mons	27.2S	272.9E	310	Chinese goddess of silkworms.
Tuzandi Mons	42.5S	41.5E	200	Palaun (Burma) ancestor deity.
Ua-ogrere Mons	40.5N	117.0E	200	Kivai (New Guinea) ancestor deity.
Uretsete Mons	12.0S	261.0E	500	Keresan Pueblo ancestor goddess.
Uti Hiata Mons	16.0N	69.0E	500	Pawnee Mother Corn deity.
Vostrukha Mons	6.3S	299.4E	180	Belorussian deity of home.
Yolkai-Estsan Mons	17.0N	194.0E	600	Navajo myth female deity.
Yunya-mana Mons	18.0S	285.0E	500	Hopi prickly pear cactus maiden.

V E N U S (Cont.)

NAME	LAT	LONG	DIAM (km)	ATTRIBUTE
PATERAE (18)				
Bakhtadze Patera	45.5N	219.5E	50	Kseniya; Georgian tea geneticist, (1899-1978).
Fedchenko Patera	24.0S	226.5E	75	Olga; Russian botanist (1845-1921).
Grizodubova Patera	16.7N	299.6E	50	Valentina; Soviet aviator (1910-1993).
Jaszai Patera	32.0N	305.0E	70	Mary; Hungarian actress (1850-1926).
Jotuni Patera	6.5S	214.0E	100	Maria; Finnish writer (1880-1943).
Kvasha Patera	9.5S	69.0E	50	Lidiya; Soviet mineralogist, meteorite expert (1909-1977).
Mansfield Patera	29.5N	227.5E	80	Cathrin (Cathleen Beauchamp); New Zealand writer (1888-1923).
Mehseti Patera	16.0N	311.0E	60	Ganjevi; Azeri/Persian poetess (c.1050-c.1100).
Mikhaylova Patera	26.8S	348.2E	70	Dariya (Dasha of Sevastopol); Russian (c.1830-c.1915).
Nordenflycht Patera	35.0S	266.0E	140	Hedwig; Swedish poetess (1718-1763).
Panina Patera	13.0S	309.8E	50	Varya; Gypsy/Russian singer (1872-1911).
Pchilka Patera	26.5N	234.0E	100	Olena (Olga Kosach); Ukrainian writer and ethnographer (1849-1930).
Serova Patera	20.0N	247.0E	60	Valentina (Polovikova); Soviet actress (1918-1975).
Shelikhova Patera	75.7S	162.5E	60	Natalia; Russian explorer of Alaska (c.1750-c.1800).
Shulzhenko Patera	6.5N	264.5E	60	Klavdiya; Soviet singer (1906-1984).
Villepreux-Power Patera	22.0S	210.0E	100	Jeannette; French marine biologist (1794-1871).
Vovchok Patera	38.0S	310.0E	80	Marko (Mariya Vilinskaya-Markovich); Ukrainian/Russian writer (1833-1907).
\vZemaite Patera	35.0S	263.0E	60	Julia; Lithuanian writer (1845-1921).
PLANITIAE (21)				
Aibarchin Planitia	73.0S	25.0E	1200	Uzbek "Alpamysh" epic tale heroine.
Akhtamar Planitia	27.0N	65.0E	2700	Armenian epic heroine.
Alma-Merghen Planitia	76.0S	100.0E	1500	Mongol/Tibet/Buryat "Gheser" epic tale heroine; wife of knight Gheser.
Dzerassa Planitia	15.0S	295.0E	2800	Ossetian golden-haired heroine.
Fonueha Planitia	44.0S	48.0E	3000	Samoan blind old woman, became a shark.
Gunda Planitia	16.0S	267.0E	1200	Abkhazian epic beautiful heroine.
Hinemoa Planitia	5.0N	265.0E	3700	Maori tale heroine.
Imapinua Planitia	60.0S	142.0E	2100	E. Greenland Eskimo sea mistress.
Kanykey Planitia	10.0S	350.0E	2100	Kyrgyz "Manas" epic tale heroine.
Laimdota Planitia	58.0S	117.0E	1800	Latvian myth heroine, beautiful girl.
Libu\vse Planitia	60.0N	290.0E	1200	Czech folktale heroine.
Llorona Planitia	18.0N	145.0E	2600	Mexican/Spanish folktale heroine.
Lowana Planitia	43.0N	98.0E	2700	Australian aboriginal tale heroine.
Mugazo Planitia	69.0S	60.0E	1500	Vietnamese tale heroine.
Nuptadi Planitia	73.0S	250.0E	1200	Mandan (US Plains) folk heroine.
Sologon Planitia	8.0N	107.0E	1600	Mandingo (Mali) epic heroine.

V E N U S (Cont.)

NAME	LAT	LONG	DIAM (km)	ATTRIBUTE
PLANITIA				
Tahmina Planitia	23.0S	80.0E	3000	Iranian (Farsi) epic heroine.
Tilli-Hanum Planitia	54.0N	120.0E	2300	Azeri "Ker-ogly" epic tale heroine.
Undine Planitia	13.0N	303.0E	2800	Lithuanian water nymph, mermaid.
Wawalag Planitia	30.0S	217.0E	2600	Yulengor (Arnhemland) tale heroines.
Zhibek Planitia	40.0S	157.0E	2000	Kazakh "Kyz-Zhibek" epic tale heroine.
PLANUM (1)				
Astkhik Planum	45.0S	20.0E	2000	Armenian goddess of love.
REGIONES (4)				
Dsonkwa Regio	53.0S	167.0E	1500	Kwakiutl (NW Coast) forest giantess.
Ishkus Regio	61.0S	245.0E	1000	Makah (NW Coast) forest giantess.
Neringa Regio	65.0S	288.0E	1100	Lithuanian seacoast giantess.
Vasilisa Regio	11.0S	332.0E	1200	Russian tale heroine.
TESSERAE (41)				
Adrasthea Tesserae	30.0N	55.0E	750	Greek goddess of law.
Athena Tessera	35.0N	175.0E	1800	Greek goddess of wisdom.
Chimon-mana Tessera	3.0S	270.0E	1500	Hopi (SW USA) goddess of the insane.
Clidna Tessera	42.0S	29.0E	500	Irish bird goddess of afterlife.
Cocomama Tessera	62.0S	23.0E	1600	Quechua (Peru) happiness goddess.
Dolya Tessera	8.0S	296.0E	1100	E. Slavic good fate goddess.
Dou-Mu Tesserae	60.0S	244.0E	400	Chinese life/death ruling goddess.
Gbadu Tessera	1.0S	38.0E	700	Fon (Benin) goddess of guessing.
Gegute Tessera	17.0N	121.0E	1600	Lithuanian goddess of time.
Giltine Tesserae	39.0S	250.0E	300	Lithuanian bad fate goddess.
Haasttse-baad Tessera	6.0N	127.0E	2600	Navajo good health goddess.
Hikuleo Tesserae	42.0S	54.0E	1400	Tonga underworld goddess.
Humai Tessera	53.0S	250.0E	350	Iranian happiness bird.
Husbishag Tesserae	28.0S	101.0E	1100	Semitic underworld goddess.
Kruchina Tesserae	36.0N	27.0E	1000	E. Slavic goddess of saddness.
Lahevhev Tesserae	29.0N	189.0E	1300	Melanesian dead souls goddess.
Lhamo Tessera	51.0S	15.0E	800	Tibetan time and fate goddess.
Likho Tesserae	40.0N	134.0E	1200	E. Slavic deity of bad fate.
Mago-Halmi Tesserae	70.0N	157.0E	400	Korean helping goddess.
Magu Tessera	52.0S	305.0E	300	Chinese goddess of immortality.
Mamitu Tesserae	22.0N	44.0E	900	Akkadian destiny goddess.
Minu-Anni Tessera	20.0S	30.0E	1300	Assyrian fate goddess.
Nedolya Tesserae	5.0N	294.0E	1200	E. Slavic bad fate goddess.
Norna Tesserae	50.0S	263.0E	700	Norse fate goddess.
Nortia Tesserae	50.0S	150.0E	650	Etruscan fate goddess.
Nuahine Tessera	9.0S	157.0E	1000	Rapanui (Easter Isl.) fate goddess.
Oddibjord Tessera	82.0N	85.0E	900	Scandinavian volvas (fortune deity).
Pasom-mana Tesserae	33.0S	49.0E	1200	Hopi goddess of dreams and insanity.
Senectus Tesserae	50.0N	292.0E	1400	Roman goddess of old age.
Shait Tessera	54.0S	173.5E	220	Egyptian human destiny goddess.
Snotra Tesserae	24.0N	134.0E	1000	Scandinavian goddess of wisdom.

V E N U S (Cont.)

NAME	LAT	LONG	DIAM (km)	ATTRIBUTE
TESSERAE				
Sopdet Tesserae	45.0S	243.0E	500	Egyptian goddess of Sirius star.
Sudenitsa Tesserae	33.0N	270.0E	4200	E. Slavic fate deities.
Sudice Tessera	37.0S	112.0E	500	Czech goddess of fate.
Urd Tessera	40.0S	174.5E	250	Norse fate goddess; a Norna.
Ustrecha Tesserae	40.0S	263.0E	1000	Old Russian goddess of chance.
Verpeja Tesserae	58.0S	160.0E	600	Lithuanian life thread goddess.
Vako-nana Tesserae	27.0N	40.0E	1200	Adygan wise future-teller.
Xi Wang-mu Tessera	30.0S	62.0E	1300	Chinese goddess of eternal life.
Yuki-Onne Tessera	39.0N	256.0E	1200	Japanese spirit of death.
Zirka Tessera	33.0N	300.0E	450	Belorussian happiness goddess.
THOLI (14)				
Amra Tholus	53.0N	98.0E	50	Abkhazian sun deity.
Angerona Tholus	29.8S	287.2E	200	Italian goddess of silence.
Angrboda Tholus	73.8S	116.0E	80	Norse Titaness.
Gerd Tholi	54.5S	291.5E	50	Scandinavian sky maiden.
Justitia Tholus	28.7S	296.5E	60	Roman goddess of justice.
Khotal-Ekva Tholi	9.1S	177.8E	50	Mansi (Ob River Ugra) sun goddess.
Kwannon Tholus	26.3S	296.8E	135	Japanese Buddhist goddess of mercy.
Meiboia Tholus	44.7S	281.3E	85	Greek bee goddess.
Muru Tholus	9.0S	305.5E	40	Estonian deity of meadows.
Ndara Tholus	57.5S	16.0E	70	Toraji (Indonesia) underworld and earthquake goddess.
Podaga Tholus	56.3S	2.0E	40	Slavic weather goddess.
Rohina Tholus	40.6S	295.4E	30	Hindu cow goddess.
Vilakh Tholus	6.5S	176.5E	15	Lakian/Kazikumukhan fire goddess.
Vupar Tholus	13.5S	306.0E	100	Chuvash (Volga area) evil spirit causing lunar and solar eclipses.
VALLES (42)				
Albys Vallis	29.5S	30.5E	240	Tuva/Altay river deity.
Apisuahts Vallis	66.5S	17.0E	550	Blackfoot/Algonquin name for Venus.
Austrina Vallis	49.5S	177.0E	600	Latvian name for planet Venus.
Chasca Vallis	52.8S	359.0E	400	Quechua name for planet Venus.
Dzyzlan Vallis	16.0S	182.0E	250	Abkhazian river goddess.
Fetu-ao Vallis	61.0S	254.7E	400	Samoan name for planet Venus.
Fufei Vallis	46.0N	341.0E	170	Chinese goddess of Lo River.
Gendenwitha Vallis	63.0S	259.0E	900	Iroquois name for planet Venus.
Helmud Vallis	33.9S	171.3E	280	Afghanistan river goddess.
Hoku-ao Vallis	28.0N	166.5E	450	Hawaiian name for planet Venus.
Ikhwezi Vallis	16.0N	147.8E	1700	Zulu name for planet Venus.
Jutrzenka Vallis	27.0N	155.8E	970	Polish name for planet Venus.
Khalanasy Vallis	51.0S	168.5E	320	Azeri river mermaid.
Kinsei Vallis	13.6N	141.0E	800	Japanese name for planet Venus.
Koidut\:aht Vallis	76.5S	130.0E	700	Estonian name for planet Venus.
Kumanyefie Vallis	80.5S	335.0E	600	Ewe name for planet Venus.
Kumsong Vallis	59.0S	152.5E	700	Korean name for planet Venus.
Lunang Vallis	68.2N	310.0E	250	Nuristan goddess of Parun River.
Lusaber Vallis	47.5S	164.0E	500	Armenian name for planet Venus.

V E N U S (Cont.)

NAME	LAT	LONG	DIAM (km)	ATTRIBUTE
VALLES				
Martuv Vallis	23.0N	156.0E	250	Kyrgyz river deity.
Matlalcue Vallis	33.0S	167.5E	300	Aztec fresh water goddess.
Merak Vallis	63.5S	162.0E	200	Balochi (Pakistan) river deity.
Nahid Valles	55.1S	171.0E	500	Persian name for planet Venus.
Nantosuelta Vallis	61.9S	193.0E	320	Celtic (Gaullic) river goddess.
Nepra Vallis	0.2N	23.5E	350	E. Slavic goddess of Dneper River.
Ngyandu Vallis	62.0S	12.0E	500	Swahili name for planet Venus.
Nyakaio Vallis	47.5N	339.0E	150	Shilluk (Sudan) river deity.
Olokun Vallis	81.5N	269.0E	150	Bini sea and river goddess.
Omutnitsa Vallis	33.0N	292.0E	150	E. Slavic river deeps deity.
Poranica Vallis	21.0S	178.5E	550	Slovenian name for planet Venus.
Sezibwa Vallis	44.0S	37.0E	300	Ganda river spirit.
Tai-pe Valles	11.0N	156.5E	400	Chinese name for planet Venus.
Tan-yondozo Vallis	41.5S	87.0E	800	Bashkir name for planet Venus.
Tapati Vallis	27.0N	304.0E	150	Indian Tapti (Tapi) River goddess.
Tawera Vallis	11.6S	67.5E	500	Maori name for planet Venus.
Tingoi Vallis	6.0N	318.6E	250	Mande (Sierra Leone) river spirit.
Umaga Valles	49.0S	152.0E	400	Old Tagal name for planet Venus.
Uottakh-sulus Valles	12.5N	239.0E	1100	Yakutian/Saha name for planet Venus.
Utrenitsa Vallis	55.0N	280.0E	700	Old Russian name for planet Venus.
Veden-Ema Vallis	15.0S	141.0E	300	Finnish goddess of fishing.
Xulab Vallis	57.4S	185.0E	1000	Mayan name for planet Venus.
Yuvkha Valles	10.5N	239.5E	200	Turkman river spirit.

WORKING GROUP FOR THE WORLDWIDE DEVELOPMENT OF ASTRONOMY

GROUPE DE TRAVAIL POUR LE DEVELOPPEMENT MONDIAL DE L'ASTRONOMIE

Members of the Working Group held a business session during the lunchtime break of the Joint Discussion. They considered such matters as the future organization of the Group, the continuation of the annual newsletter, and plans for further meetings along the lines of the Joint Discussion.

Composition of the Working Group

The composition of the Working Group is the following:

CHAIRMAN:		A.H.	Batten	Canada
MEMBERS:	ex officio	J.	Fierro	Mexico
		J.B.	Hearnshaw	New Zealand
		B.	Hidayat	Indonesia
		Y.	Kozai	Japan
		D.	McNally	UK
		M.C.	Pineda de Carias	Honduras
	ex officio	M.S.	Roberts	USA
		D.G.	Wentzel	USA
CONSULTANT:		S.	Raither	UNESCO

Joint Discussion 20: Enhancing Astronomy Research and Education in Developing Countries

Chairmen: A.H. Batten, J.R. Percy, H. Jørgensen and S.M.R. Ansari.

The principal meeting of the Working Group during the General Assembly was Joint Discussion 20 on Enhancing Astronomy Research and Education in Developing Countries. Some 60 to 100 persons attended the day-long session, and many of them contributed to the discussions. Twenty-two papers were pesented during the Joint Discussion. These papers will be printed in full in the forthcoming Volume 11 of Highlights in Astronomy, which will also contain a summary of the discussions. Nine poster papers were on display in connection with the Joint Discussion. The question of the relationship between the IAU and the UN/ESA series of workshops on Basic Space Science for Developing Countries was also discussed. It was planned to continue discussing these matters by correspondence, after the Assembly.

J. Andersen (ed.), Transactions of the International Astronomical Union Volume XXIIIB, 253.

WORKING GROUP ON FUTURE LARGE SCALE FACILITIES

Report not received

CHAPTER VII

STATUTES, BYE-LAWS AND WORKING RULES

INTERNATIONAL ASTRONOMICAL UNION

Kyoto, August 20, 1997

STATUTES

I DENOMINATION, OBJECTS AND DOMICILE

1 The International Astronomical Union (referred to as the Union) is a non-governmental organization, whose objects are:

(a) to develop astronomy through international co-operation,

(b) to promote the study and development of astronomy in all aspects,

(c) to further and safeguard the interests of astronomy.

2 The legal domicile of the Union is Brussels.

II ADHERENCE OF THE UNION

3 The Union adheres to the International Council of Scientific Unions

III COMPOSITION OF THE UNION

4 The Union is composed of:

(a) full members (Adhering Countries)

(b) associate members (Associate Countries)

(c) individual members (Members)

IV AFFILIATED ORGANIZATIONS

5 The Union may admit the affiliation of international non-governmental organizations which contribute to the development of astronomy.

J. Andersen (ed.), Transactions of the International Astronomical Union Volume XXIIIB, 257–283.

V ADHERING COUNTRIES

6 Countries adhere to the Union either :

(a) through the organization by which they adhere to the International Council of Scientific Unions, or through a National Committee of Astronomy approved by that organization,

or:

(b) if they do not adhere to the International Council of Scientific Unions, through a National or other appropriate Committee for Astronomy recognized by the Executive Committee of the Union,

(c) the Adhering Organizations and National or other appropriate Committees of Astronomy being referred to as adhering bodies.

7 Adherence of a country to the Union is approved, on the proposal of the Executive Committee, by the General Assembly; it terminates if the country withdraws from the Union or if the country has not paid its dues for five years.

8 Adhering Countries are classified in categories. The number of categories shall be specified in the Bye-Laws. A country requesting adherence shall specify the category in which it desires to be classed. The specification may be declined by the Executive Commitee if the category proposed is manifestly inadequate.

VI ASSOCIATE COUNTRIES

9 Countries that would like to join the Union while developing Astronomy in their territory may do so as Associate Members.

10 The Adhering Body for Associate Members may be the organization by which the country adheres to the International Council of Scientific Unions or through an institution of higher learning or a National Research Council.

11 Countries are accepted as Associate Members by the General Assembly, on the proposal of the Executive Committee, for a maximum interval of nine years, at the end of which they either become a full Member, or they resign from the Union.

12During the probationary period, the Union, if asked by the Adhering Organization, may agree to help in the development of Astronomy in that country through the Visiting Lecturers' Programme and/or any other appropriate programme.

VII MEMBERS

13 Members are admitted to the Union by the Executive Committee, on the proposal of an adhering body referred to in article 6, with regard to their achievements in some branch of astronomy. Individual astronomers who are not represented by an adhering body may be admitted by the Executive Committee on the proposal of a Division President.

VIII GENERAL ASSEMBLY

14 (a) The work of the Union is directed by the General Assembly of representatives of Adhering Countries and of Members. Each Adhering Country appoints a representative authorised to vote in its name.

(b) The General Assembly draws up Bye-Laws governing the application of the Statutes.

(c) It appoints an Executive Committee to implement the decisions of the General Assembly, and to direct the affairs of the Union in the interval between meetings of two successive ordinary General Assemblies. The Executive Committee reports to the General Assembly. The General Assembly, in accepting the report of the Executive Committee, discharges it of liability.

15 (a) On questions concerning the administration of the Union, not involving its budget, voting at the General Assembly is by Adhering Country, each country having one vote. Adhering Countries which have not paid their annual contributions up to 31 December of the year preceding the General Assembly may not participate in the voting.

(b) On questions involving the budget of the Union, voting is similarly by Adhering Country, under the same conditions and with the same reservations as in article 15(a), the number of votes for each Adhering Country being one greater than the number of its category, as defined in article 8.

(c) Adhering Countries may vote by correspondence on questions on the agenda for the General Assembly.

(d) A vote is valid only if at least two thirds of the Adhering Countries having the right to vote by virtue of article 15(a) participate in it.

(e) Associate Countries have the right to vote only on questions concerning associate membership.

16 On scientific questions not involving the budget of the Union the Members of the Union each have one vote.

17 On all questions in articles 15 and 16, decisions are taken by an absolute majority of the votes cast. However, a decision to change the Statutes is only valid if taken with the approval of at least two thirds of the votes of the Adhering Countries having the right to vote by virtue of article 15(a).

18 A motion to change the Statutes can only be discussed if it appears, in specific terms, on the agenda for the General Assembly.

IX EXECUTIVE COMMITTEE

19 The Executive Committee consists of the President of the Union, the President-elect, six Vice-Presidents, the General Secretary and the Assistant General Secretary elected by the General Assembly on the proposal of a Special Nominating Committee.

X COMMISSIONS AND DIVISIONS

20 The General Assembly forms Commissions and Divisions. The Divisions combine Commissions with related scientific or technical themes. Divisions should have approximately equal numbers of Members.

XI LEGAL REPRESENTATION OF THE UNION

21 The General Secretary is the legal representative of the Union.

XII BUDGET AND DUES

22 (a) For each ordinary General Assembly the Executive Committee prepares a budget proposal covering the period to the next ordinary General Assembly, together with the accounts of the Union for the preceding period. It submits these to the Finance Committee for consideration; this Finance Committee consists of one member nominated by each adhering body and approved by the General Assembly. At its first meeting during the General Assembly, the Finance Committee elects a President from among its members.

(b) The Finance Committee examines the accounts of the Union from the point of view of responsible expenditure within the intent of the previous General Assembly, and it considers whether the proposed budget is adequate to implement the policy of the General Assembly, as interpreted by the Executive Committee. It submits reports on these matters to the General Assembly for the approval of the accounts and decision on the budget.

(c) Each Adhering Country pays annually to the Union a number of units of contribution according to its category. The number of units of contribution for each category shall be specified in the Bye-Laws.

(d) Associate Countries pay annually one half unit of contribution.

(e) The amount of the unit of contribution is determined by the General Assembly, on the proposal of the Executive Committee and with the advice of the Finance Committee.

(f) The payment of contributions is the responsibility of the adhering bodies. The liability of each Adhering Country in respect of the Union is limited to the amount of that country's dues to the Union.

(g) An Adhering Country that ceases to adhere to the Union resigns at the same time its rights to a share in the assets of the Union.

XIII DISSOLUTION OF THE UNION

23 The decision to dissolve the Union is only valid if taken with the approval of three quarters of the votes of the Adhering Countries having the right to vote by virtue of article 15(a).

XIV EMERGENCY POWERS

24 If, through events outside the control of the Union, circumstances arise in which it is impracticable to comply with the provisions of these Statutes and of the Bye-Laws drawn up by the General Assembly, the organs and officers of the Union, in the order specified below, shall take such actions as they deem necessary for the continued operation of the Union. Such action shall be reported to a higher authority immediately this becomes practicable until such time as an extraordinary General Assembly can be convened. The following is the order of authority:

The General Assembly; an extraordinary General Assembly; the Executive Committee in meeting or by correspondence; the President of the Union; the General Secretary; or failing the practicability or availability of any of the above, one of the Vice-Presidents.

XV FINAL CLAUSES

25 These Statutes enter into force on 20 August 1997.

26 The present Statutes are being published in French and English versions. In case of doubt, the French version is the only authority.

Kyoto, August 20, 1997

BYE - LAWS

I MEMBERSHIP

1 Applications of countries for adherence to the International Astronomical Union (referred to as the Union) are examined by the Executive Committee and submitted to the General Assembly for approval.

2 Proposed changes in the list of Members are, with due regard to the suggestions of the Presidents of Divisions, submitted for advice to the Nominating Committee, consisting of one representative of each Adhering Country designated by the appropriate adhering body, before decision by the Executive Committee.

3 Commissions may, with the approval of the Executive Committee, co-opt consultants whom they consider may contribute to their work. The adherence of consultants expires on the last day of the ordinary General Assembly next following their admission, unless renewed.

4 An affiliated organization may participate in the work of the Union as mutually agreed between the organization and the Executive Committee.

II GENERAL ASSEMBLY

5 The Union meets in ordinary General Assembly, as a rule, once every three years. The place and date of the ordinary General Assembly unless determined by the General Assembly at its previous meeting, shall be fixed by the Executive Committee and communicated to the adhering bodies at least six months beforehand.

6 The President, with the consent of the Executive Committee, may summon an extraordinary General Assembly. The President must do so at the request of one third of the Adhering Countries.

7 The agenda of business for each ordinary General Assembly is determined by the Executive Committee and is communicated to the adhering bodies at least four months before the first day of the meeting. It shall include the proposal of the Executive Committee with regard to the unit of contribution as called for in article 27.

8 (a) Any motion or proposal received by the General Secretary at least five months before the first day of an ordinary General Assembly, whether from an adhering body, from a Commission of the Union, from an Inter-Union Commission on which the Union is represented, must be placed on the agenda.

(b) A motion or proposal concerning the administration of budget of the Union which does not appear on the agenda prepared by the Executive Committee, or any amendment to a motion that appears on the agenda, shall only be discussed with the prior approval of at least two thirds of the votes of Adhering Countries represented at the General Assembly and having the right to vote by virtue of Statute 15(a).

9 If there is doubt as to the administrative or scientific character of a question giving rise to a vote, the President determines the issue.

10 Where there is an equal division of votes, the President determines the issue.

11 The President may invite representatives of other organizations, scientists and young astronomers to participate in the General Assembly. Subject to the agreement of the Executive Committee the President may delegate this privilege concerning representatives of other organizations to the General Secretary, and concerning scientists and young astronomers to the adhering national organization or other appropriate bodies. Representatives of other organizations, scientists and young astronomers can participate in the General Assembly but have no voting right.

III SPECIAL NOMINATING COMMITTEE

12 (a) Proposals for elections of the President of the Union, the President-elect, six Vice-Presidents, the General Secretary and the Assistant General Secretary are submitted to the General Assembly by the Special Nominating Committee. This Committee consists of the President and past President of the Union, a member proposed by the retiring Executive Committee, and four members elected by the Nominating Committee from among twelve Members proposed by Presidents of Divisions, with due regard to an appropriate distribution over the major branches of astronomy. Except for the President and immediate past President present and former members of the Executive Committee shall not serve on the Special Nominating Committee. No two members of the Special Nominating Committee shall belong to the same country or adhering body.

(b) The General Secretary and the Assistant General Secretary participate in the work of the Special Nominating Committee in an advisory capacity.

(c) The Special Nominating Committee is appointed by the General Assembly to which it reports directly. It remains in office until the end of the ordinary General Assembly next following that of its appointment, and it may fill any vacancy occurring among its members.

IV OFFICERS AND EXECUTIVE COMMITTEE

13 (a) The President of the Union remains in office until the end of the ordinary General Assembly following election to that function. Normally, the President-elect succeeds the President at that moment. The Vice-Presidents remain in office until the end of the second ordinary General Assembly following that of their election. They may not be re-elected immediately to the same offices.

(b) The General Secretary and the Assistant General Secretary remain in office until the end of the ordinary General Assembly next following that of their election. Normally the Assistant General Secretary succeeds the General Secretary though both officers may be re-elected for another term.

(c) The election takes place at the last session of the General Assembly, the names of the candidates proposed having been announced at a previous session.

14 The retiring President and the retiring General Secretary become advisers to the Executive Committee until the end of the ordinary General Assembly next following that of their retirement. They participate in the work of the Executive Committee and attend its meetings without voting right.

15 The Executive Committee may fill any vacancy occurring among its members. Any person so appointed remains in office until the next ordinary General Assembly.

16 The Executive Committee may draw up and publish Working Rules to clarify the Statutes and Bye-Laws.

17 The Executive Committee appoints the Union's representative to the International Council of Scientific Unions; if not already an elected member of the Executive Committee, this representative will become its adviser.

18 (a) The General Secretary is responsible to the Executive Committee for not incurring expenditure in excess of the funds designated for that office.

(b) An Administrative office, under the direction of the General Secretary, conducts the correspondence, administers the funds, and preserves the archives of the Union.

V DIVISIONS AND COMMISSIONS

19 (a) The Commissions of the Union shall pursue the scientific objects of the Union by activities such as the study of special branches of astronomy, the encouragement of collective investigations, and the discussion of questions relating to international agreements or to standardization.

(b) The Commissions of the Union shall prepare reports on the work with which they are concerned.

20 Each Commission consists of:

(a) a President and at least one Vice-President elected by the General Assembly on the proposal of the Executive Committee. They remain in office until the end of the ordinary General Assembly next following that of their election. They are not normally eligible for re-election;

(b) an Organizing Committee, whose members are appointed by the Commission subject to the approval by the Executive Committee. The Organizing Committee assists the President and Vice-President(s) in their duties. A Commission may decide that it needs no Organizing Committee;

(c) Members of the Union, appointed by the President, Vice-President(s) and the Organizing Committee, in consideration of their special interests; their appointment is subject to the confirmation by the Executive Committee.

21 Between two ordinary General Assemblies, Presidents of Commissions may co-opt, from among Members of the Union, new members to the Organizing Committees and to the Commissions themselves.

22 Commissions draw up their own rules. Decisions within Commissions are taken according to the vote of their members, and they become effective once they are approved by the Executive Committee.

23 Commissions with related scientific or technical tasks form Divisions. The Divisions coordinate the activities of the affiliated Commissions, including the making of proposals for new Commissions, proposals for dissolving or combining Commissions, and proposals for Working Groups. The Divisions also endorse proposals for IAU Symposia and Colloquia and they may organize Joint Discussions during the General Assemblies.

24 Each Division is directed by a board which is composed of up to six persons including the Presidents of the constituent Commissions. This board elects the Division President.

25 The Division Presidents report to and advise the Executive Committee.

VI ADHERING BODIES

26 The functions of the Adhering Bodies are to promote and co-ordinate, in their respective territories, the study of the various branches of astronomy, more especially in relation to their international requirements. They are entitled to submit to the Executive Committee motions for discussions by the General Assembly.

VII FINANCES

27 Each Adhering Country pays annually to the Union a number of units of contribution corresponding to its category as follows; Associate Members pay annually one half unit of contribution:

Category as defined in Statute 8										
I	II	III	IV	V	VI	VII	VIII	VIII½	IX	X
Number of units of contribution										
1	2	4	6	10	14	20	27	30	35	45

If further Categories of Adherence are required in the future, the step in the number of units shall be 10 units/category.

28 The income of the Union is to be devoted to its objects, including

(a) costs of publication and expenses of administration;

(b) the promotion of astronomical enterprises requiring international co-operation;

(c) the contribution due from the Union to the International Council of Scientific Unions.

29 Funds derived from donations are used by the Union in accordance with the wishes expressed by the donors.

30 The Union has copyright to all materials printed in its publications, unless otherwise arranged.

31 Members of the Union are entitled to receive the publications of the Union free of charge or at reduced prices at the discretion of the Executive Committee taking due regard of the financial situation of the Union.

VIII FINAL CLAUSES

32 These Bye-Laws enter into force on 20 August 1997. They can be changed with the approval of an absolute majority of the votes of the Adhering Countries having the right to vote by virtue of Statute 15 (a).

33 The present Bye-Laws are being published in French and English versions. In case of doubt, the French version is the only authority.

Kyoto, August 20, 1997

WORKING RULES

I NON-DISCRIMINATION

1 The International Astronomical Union follows the International Council of Scientific Unions (ICSU) regulations and concurs with ICSU statute 5 which defines the basic tenets of non-discrimination and of the universality of science:

"In pursuing its objectives in respect of the rights and responsibilities of scientists, ICSU, as an international non-governmental body, shall observe and actively uphold the principle of the universality of science. This principle entails freedom of association, expression, information, communication and movement in connection with international scientific activities, without any discrimination on the basis of such factors as citizenship, religion, creed, political stance, ethnic origin, race, colour, language, age or sex. ICSU shall recognize and respect the independence of the internal science policies of its National Members. ICSU shall not permit any of its activities to be disturbed by statements or actions of a political nature."

Participants in IAU-sponsored activities who feel that they have been subjected to discrimination are urged in the first instance to seek immediate clarification of all aspects of the incident, which may have occurred simply because of misunderstandings due to cultural differences inherent in an international organization such as the IAU. If the attempt to seek clarification does not prove satisfactory, contact should then be made with the IAU General Secretary who will seek to resolve the issue.

In the last resort, the Chairperson or the Secretary of the ICSU Standing Committee on the Freedom in the Conduct of Science (SCFCS) should be approached. The SCFCS has been created by ICSU in 1963 in order to safeguard the principle of the universality of science and to assist in the solution of specific problems. The SCFCS has, ever since, worked vigorously to ensure that this principle is upheld by providing advice and taking appropriate measures.. The Executive Secretary of the SCFCS, Dr. P. Schindler, can be reached at the Swiss Academy of Sciences, Bärenplatz 2, 3011 Bern, Switzerland (Telephone: 41 31 312 33 75, Telefax: 41 31 312 32 91, e-mail: schindler@sanw.unibe.ch).

II MEMBERSHIP

A ADHERING COUNTRIES

2 Applications of countries for adherence to the Union are examined by the Executive Committee for:

(a) the adequacy of the category in which the country wishes to be classified;

(b) the present state and expected development of astronomy in the applying country;

(c) the degree to which the prospective adhering body is representative of its country's astronomical activity.

3 Applications proposing an adequate annual contribution to the Union shall, with the recommendation of the Executive Committee, be submitted to the General Assembly for decision.

B MEMBERS

4 Individuals proposed for Union Membership should, as a rule, be chosen from among astronomers and scientists, whose activity is closely linked with astronomy taking into account:

(a) the standard of their scientific achievement;

(b) the extent to which their scientific activity involves research in astronomy;

(c) their desire to assist in the fulfilment of the aims of the Union.

5 Young astronomers should be considered eligible for membership after they have shown their capability (as a rule Ph.D. or equivalent) of and experience (some years of successful activity) in conducting original research.

6 For full time professional astronomers the achievement in astronomy may consist either of original research or of substantial contributions to major observational programs.

7 Others are eligible for membership only if they are making original contributions closely linked with astronomical research.

8 Eight months before an ordinary General Assembly, adhering bodies will be asked to propose new Members. The proposals should reach the General Secretary not later than five months before the first session of the General Assembly. Proposals received after the closing date will only be taken into consideration if the delay is justified by exceptional circumstances.

9 Each proposal shall be prepared separately and signed by the proposer. It should include the name, first names, postal and electronic addresses of the candidates, Institute or Observatory, place and date of birth, the University and the year of Ph.D. or equivalent title, present occupation, titles and bibliographic data for two or three of the more important papers or publications, and details, if any, worthy to be considered by the Nominating Committee.

10 (a) Pursuant to article 13 of the Statutes Presidents of Union Divisions wishing to nominate candidates for Membership should address their suggestions to the General Secretary at least nine months before the first session of an ordinary General Assembly. The nominations should contain particulars as in article 9.

(b) The General Secretary notifies the adhering bodies about such suggestions.

11 The General Secretary shall prepare two lists for the Nominating Committee:

(a) one containing the candidates proposed by the adhering bodies.

(b) the other containing those suggested by Presidents of Divisions, but not included among the proposals of the adhering bodies.

12 The Nominating Committee prepares the final proposals for Union membership from the two lists as mentioned in article 11.

13 Adhering Bodies should propose the deletion of Members who have left the field of astronomy for other interests, unless they continue to contribute to astronomy. Such proposals should be announced to the Member concerned and to the General Secretary.

14 The alphabetical list of Union Members will be published by the General Secretary following each ordinary General Assembly.

III COMMISSION MEMBERSHIP

15 Members of Union Commissions are co-opted by Commissions. The rules governing the procedure of such co-option are drawn up by the Commissions themselves.

16 Commissions should choose, or approve of, Commission members taking into account their special interests, in particular their scientific activity in the appropiate fields of research and their contribution to the work of the Commission. They may:

(a) invite Union Members to become members of their Commission;

(b) remove Union members who have not contributed to the work of the Commission;

(c) accept or reject applications for membership from existing or proposed Union Members;

17 Members may not, as a rule, be members of more than three Commissions.

18 Members may apply for Commission membership by writing to the President of the Commission concerned. Such applications should only be made if the Member is actively engaged in the appropriate field of research and is prepared to contribute to the work of the Commission.

19 Members of Commissions may resign from a Commission by writing to its President.

20 Adhering Bodies, in sending in their proposals for new Members, may also suggest one Commission for each candidate.

21 The General Secretary will record and analyse the lists of members of Commissions. If necessary, the General Secretary will try to resolve any outstanding anomalies.

22 The list of Commission members will be published by the General Secretary in the Transactions of each ordinary General Assembly.

IV CONSULTANTS

23 Eligible as Consultants are non-astronomers in a position to further the interest in astronomy.

24 Proposals of Commissions for the approval of consultants should, as a rule, reach the General Secretary not later than five months before the first session of an ordinary General Assembly.

25 The General Secretary shall prepare a list of those proposed for admission as consultants and submit it to the Executive Committee for approval.

26 The Administrative Office will maintain an alphabetical list of consultants.

27 Consultants may participate in the meetings of the Union. They may have voting right in the respective Commission. They receive, free of charge, the Information Bulletin of the Union.

V SCIENTIFIC MEETINGS

28 The General Secretary shall publish rules for scientific meetings organized or sponsored by the Union.

VI PUBLICATIONS

29 The publications of the International Astronomical Union, approved in the budget by the General Assembly, are prepared by the Administrative Office of the Union.

30 Commissions of the Union may, with the approval of the Executive Committee, issue their publications independently.

31 The distribution of publications of the Union is decided, on the proposal of the General Secretary, by the Executive Committee.

32 Members may purchase the publications of the Union at reduced prices.

VII EXTERNAL CONTACTS

33 No dealings with third parties, attributable to the Union, shall be undertaken by any Member of the Union except on the authority of the General Secretary.

34 Representatives of the Union in other bodies, especially ICSU Committees and ICSU Inter-Union Committees, shall be appointed by the Executive Committee. Nominations are sought from Presidents of appropriate Commissions.

35 Expenses incurred by Representatives of the Union in other bodies will be reimbursed at the discretion of the General Secretary, within the provisions of the Budget Estimate adopted by the General Assembly. Representatives are required to obtain prior approval of the General Secretary before incurring such expenses.

VIII GENERAL ASSEMBLIES

36 The General Secretary distributes the budget prepared by the Executive Committee to National or other appropriate Committees of Astronomy and/or Adhering Organizations for comments eight months before the General Assembly.

37 The decisions and recommendations of the Union on scientific and organizational matters are expressed in its Resolutions. Resolutions are proposed, evaluated, and approved according to the following guidelines:

(a) Resolutions fall in three categories:

A: Resolutions, proposed by Adhering Bodies or by the Executive Committee,
B: Resolutions, proposed by Divisions or Commissions not attached to a Division and adopted by the General Assembly,
C: Resolutions, adopted by Divisions or Commissions, but not presented to the General Assembly.

(b) Resolutions should be submitted on standard forms appropriate to Resolutions of type A, B, and C, respectively. These forms are available from the IAU Secretariat.

(c) Resolutions of type A must be placed on the Agenda of the General Assembly and must be submitted to the General Secretary at least six months prior to the beginning of the General Assembly. Resolutions of type A or B which have implications for the budget of the Union must be submitted to the General Secretary nine months in advance in order to be considered by the General Assembly.

All other Resolutions of type B must be submitted to the General Secretary three months before the beginning of the General Assembly.

(d) In truly exceptional cases the Executive Committee may consider accepting late proposals for resolutions of type B.

(e) At its second session, each General Assembly appoints a Resolutions Committee consisting of five members of the Union, one of whom should be a member of the Executive Committee. The Resolutions Committee remains in office until the end of the following General Assembly.

(f) The Resolutions Committee will examine the content, wording, and implications of all resolutions of types A and B to be presented to the second session of the General Assembly. In particular, it will address the following points:

i Suitability of the subject for an IAU Resolution,
ii Correct and unambiguous wording,
iii Consistency with previous IAU Resolutions.

The Resolutions Committee may refer a Resolution back to the proposers for reconsideration or withdrawal, but can neither withdraw nor modify the substance of a Resolution on its own initiative. The Resolutions Committee will notify the Executive Committee of any perceived problems with the substance of a proposed Resolution.

(g) The Executive Committee will examine the substance and implications of all Resolutions proposed for adoption by the General Assembly (types A and B). The Resolutions Committee presents the proposals with the recommendations of the Executive Committee to the second session of the General Assembly for approval.

(h) Resolutions of type C have force only within the Commission or Division of origin.

IX WORKING GROUPS

38 The Executive Committee and the Divisions and Commissions may set up Working Groups for special tasks. Working Groups established by Divisions and Commissions have to be approved by the Executive Committee. All Working Groups are established initially for a period of three years. Before each General Assembly the Divisions and Commissions shall inform the Executive Committee which Working Groups are to be retained for the next 3-year period and which Working Groups are to be dissolved.

UNION ASTRONOMIQUE INTERNATIONALE

Kyoto, le 20 Août 1997

STATUTS

I DÉNOMINATION, BUTS ET DOMICILE

1 L'Union Astronomique Internationale (ci-après dénommée l'Union) est une organisation non-gouvernementale, qui a pour buts :

(a) de développer l'astronomie par la coopération internationale,

(b) d'encourager l'étude et le développement de l'astronomie sous tous ses aspects,

(c) de servir et sauvegarder les intérêts de l'astronomie.

2 L'Union a son siège légal à Bruxelles.

II AFFILIATION DE L'UNION

3 L'Union adhère au Conseil International des Unions Scientifiques.

III MEMBRES DE L'UNION

4 L'Union a pour membres :

(a) des personnes morales (Pays adhérents)

(b) des personnes morales associées (Pays associés)

(c) des membres individuels (Membres).

IV ORGANISATIONS AFFILIÉES

5 L'Union peut accepter l'affiliation d'organisations internationales non-gouvernementales qui contribuent au développement de l'astronomie.

V PAYS ADHÉRENTS

6 Les pays adhèrent à l'Union soit :

(a) par l'intermédiaire de l'organisation par laquelle ils adhèrent au Conseil International des Unions Scientifiques, ou par l'intermédiaire d'un Comité National d'Astronomie approuvé par cette organisation,

soit :

(b) s'ils n'adhèrent pas au Conseil International des Unions Scientifiques, par l'intermédiaire d'un Comité National d'Astronomie ou autre Comité approprié reconnu par le Comité Exécutif de l'Union,

(c) les organisations ou Comités mentionné(e)s à l'article 6(a) et les Comités Nationaux d'Astronomie ou autres Comités appropriés mentionnés à l'article 6(b) étant dénommés ci-après organismes adhérents.

7 L'adhésion d'un pays à l'Union est proposée par le Comité Exécutif et approuvée par l'Assemblée Générale : elle prend fin si le pays se retire de l'Union ou si le pays n'a pas payé sa contribution durant cinq ans.

8 Les Pays adhérents sont répartis en catégories. Le nombre des catégories est fixé par le Règlement. Un pays qui sollicite son adhésion indique la catégorie dans laquelle il désire être classé. La proposition peut être refusée par le Comité Exécutif si la catégorie est manifestement inadéquate.

VI PAYS ASSOCIÉS

9 Les pays souhaitant faire partie de l'Union tout en développant l'astronomie dans leur territoire peuvent le faire à titre de Membres associés.

10 L'organisme adhérent d'un pays associé peut être soit l'organisation par l'intermédiaire de laquelle le pays adhère au Conseil International des Unions Scientifiques, soit une institution d'éducation supérieure, soit un conseil scientifique national.

11 Les pays sont acceptés en qualité de Membres associés par l'Assemblée Générale, sur proposition du Comité Exécutif, pour une période maximale de neuf ans au terme de laquelle ils deviennent membres à part entière ou se retirent de l'Union.

12 Durant la période probatoire, l'Union peut accepter, à la requête de l'organisation adhérente, d'aider au développement de l'astronomie dans ce pays via le Programme de Professeurs Visiteurs et/ou de tout programme adéquat.

VII MEMBRES

13 Les Membres sont admis dans l'Union par le Comité Exécutif, sur proposition de l'un des organismes adhérents mentionnés à l'article 6, en considération de leur activité dans une branche de l'astronomie. Les astronomes qui ne sont pas représentés par un Organisme adhérent peuvent être admis par le Comité Exécutif sur proposition d'un Président de Division.

VIII ASSEMBLÉE GÉNÉRALE

14 (a) L'activité de l'Union est dirigée par l'Assemblée Générale des représentants des Pays adhérents et des Membres. Chaque Pays adhérent nomme un représentant autorisé à voter en son nom.

(b) L'Assemblée Générale rédige un Règlement qui précise les modalités d'application des Statuts.

(c) Elle nomme un Comité Exécutif chargé d'exécuter les décisions de l'Assemblée Générale, et d'administrer l'Union pendant la période séparant les réunions de deux Assemblées Générales ordinaires successives. Le Comité Exécutif rend compte de sa gestion à l'Assemblée Générale. L'Assemblée Générale, en acceptant le rapport du Comité Exécutif, le décharge de sa responsabilité.

15 (a) Sur les questions concernant l'administration de l'Union, sans implication budgétaire, le vote à l'Assemblée Générale a lieu par Pays adhérent, chaque pays disposant d'une voix. Les Pays adhérents qui ne sont pas à jour de leurs cotisations annuelles au 31 décembre de l'année précédant l'Assemblée Générale ne peuvent pas participer aux votes.

b) Sur les questions engageant le budget de l'Union, le vote a lieu de même par Pays adhérent, dans les conditions et avec les réserves prévues à l'article 15(a), le nombre de voix de chaque Pays adhérent étant égal à l'indice de sa catégorie, définie conformément à l'article 8, augmenté d'une unité.

(c) Les Pays adhérents peuvent voter par correspondance sur les questions figurant à l'ordre du jour de l'Assemblée Générale.

(d) Un scrutin n'est valable que si au moins deux tiers des Pays adhérents disposant du droit de vote en vertu de l'article 15(a) y prennent part.

(e) Les Pays associés ne peuvent voter que sur des questions concernant les Membres associés.

16 Sur les questions scientifiques n'engageant pas le budget de l'Union, les Membres de l'Union disposent chacun d'une voix.

17 Sur toutes les questions prévues aux articles 15 et 16, les décisions sont prises à la majorité absolue des suffrages. Cependant, une décision de modification des Statuts n'est valable que si elle a été prise à la majorité des deux tiers des voix des Pays adhérents qui disposent du droit de vote en vertu de l'article 15(a).

18 Une proposition de modification des Statuts ne peut être discutée que si elle figure, en tant que telle, à l'ordre du jour de l'Assemblée Générale.

IX COMITÉ EXÉCUTIF

19 Le Comité Exécutif se compose du Président de l'Union, du "President-elect", de six Vice-Présidents, du Secrétaire Général et du Secrétaire Général Adjoint, élus par l'Assemblée Générale sur la proposition du Comité Spécial des Nominations.

X COMMISSIONS ET DIVISIONS DE L'UNION

20 L'Assemblée Générale crée des Commissions et des Divisions. Les Divisions regroupent les Commissions des thèmes scientifiques ou techniques apparentés. Les Divisions comprennent un nombre de membres approximativement identique.

XI REPRÉSENTATION LÉGALE DE L'UNION

21 Le Secrétaire Général est le représentant légal de l'Union.

XII BUDGET ET COTISATIONS

22 (a) Pour chaque Assemblée Générale ordinaire, le Comité Exécutif prépare un projet de budget pour la période à courir jusqu'à l'Assemblée Générale ordinaire suivante, ainsi que les comptes de l'Union pour la période précédente. Il les soumet au Comité des Finances pour examen ; ce Comité des Finances est composé de membres nommés par les organismes adhérents, à raison d'un membre par organisme, et il est approuvé par l'Assemblée Générale. Lors de sa première séance pendant l'Assemblée Générale, le Comité des Finances élit un Président parmi ses membres.

(b) Le Comité des Finances examine les comptes de l'Union pour voir si les dépenses engagées ont été conformes aux voeux émis lors de la précédente réunion de l'Assemblée Générale et il s'assure que le budget proposé vise à la poursuite de la politique de l'Assemblée Générale, telle qu'elle est interprétée par le Comité Exécutif. Il présente des rapports sur ces questions qu'il soumet à l'Assemblée Générale pour approbation des comptes et pour décision sur le budget.

(c) Chaque Pays adhérent verse annuellement à l'Union un nombre d'unités de cotisation qui est fonction de sa catégorie. Le nombre d'unités de cotisation pour chaque catégorie est fixé par le Règlement.

(d) La cotisation annuelle des Pays associés s'élève à une demi-unité de contribution.

(e) Le montant de l'unité de cotisation est fixé par l'Assemblée Générale, sur la proposition du Comité Exécutif et avec l'avis du Comité des Finances.

(f) Le paiement des cotisations est à la charge des organismes adhérents. La responsabilité de chaque Pays adhérent envers l'Union est limitée au montant des cotisations dues par ce pays à l'Union.

(g) Un Pays adhérent qui cesse d'adhérer à l'Union renonce de ce fait à ses droits sur l'actif de l'Union.

XIII DISSOLUTION DE L'UNION

23 La décision de dissoudre l'Union n'est valable que si elle est prise à la majorité des trois quarts des voix des Pays adhérents qui disposent du droit de vote en vertu de l'article 15(a).

XIV DÉVOLUTION DE L'AUTORITÉ EN CAS DE FORCE MAJEURE

24 Si, par suite d'événements indépendants de la volonté de l'Union, des circonstances apparaissent qui rendent impossible le respect des clauses de ces Statuts et du Règlement établi par l'Assemblée Générale, les organes et membres du Comité Exécutif de l'Union, dans l'ordre fixé ci-dessous, prendront toutes dispositions qu'ils jugeront nécessaires pour la continuation du fonctionnement de l'Union. Ces dispositions devront être soumises à une autorité supérieure dès que cela deviendra possible, jusqu'à ce qu'une Assemblée Générale extraordinaire puisse être réunie. L'autorité est dévolue dans l'ordre ci-dessous :

l'Assemblée Générale ; une Assemblée Générale extraordinaire ; le Comité Exécutif, réuni ou par correspondance ; le Président de l'Union ; le Secrétaire Général ou, à défaut de la possibilité de recourir à l'une de ces autorités ou de leur disponibilité, un des Vice-Présidents.

XV CLAUSES FINALES

25 Ces Statuts entrent en vigueur le 20 août 1997.

26 Les présents Statuts sont publiés en versions française et anglaise. En cas d'incertitude, la version française fait seule autorité.

Kyoto, le 20 Août 1997

REGLEMENT

I LES MEMBRES DE L'UNION

1 Les demandes d'adhésion des pays à l'Union Astronomique Internationale (ci-après dénommée l'Union) sont examinées par le Comité Exécutif et soumises à l'approbation de l'Assemblée Générale.

2 Les propositions de modifications de la liste des Membres sont, après examen attentif des suggestions des Présidents de Divisions soumises pour avis au Comité des Nominations, composé d'un représentant de chaque Pays adhérent désigné par l'organisme adhérent habilité, avant la décision du Comité Exécutif.

3 Les Commissions peuvent, avec l'approbation du Comité Exécutif, coopter des consultants qu'elles jugent en mesure d'apporter une contribution utile à leur travail. L'adhésion des consultants a pour terme le dernier jour de la première Assemblée Générale ordinaire qui suit leur admission, à moins qu'elle ne soit renouvelée.

4 Une organisation affiliée peut participer au travail de l'Union dans les conditions fixées par accord entre l'organisation et le Comité Exécutif.

II L'ASSEMBLÉE GÉNÉRALE

5 L'Union se réunit en Assemblée Générale ordinaire régulièrement une fois tous les trois ans. Si le lieu et la date de l'Assemblée Générale ordinaire n'ont pas été décidés lors de la précédente Assemblée Générale, ils sont fixés par le Comité Exécutif et communiqués aux organismes adhérents au moins six mois à l'avance.

6 Le Président peut convoquer, avec l'accord du Comité Exécutif, une Assemblée Générale extraordinaire. Le Président est tenu de le faire à la demande du tiers des Pays adhérents.

7 L'Ordre du Jour de chaque Assemblée Générale ordinaire est arrêté par le Comité Exécutif et communiqué aux Organismes adhérents au moins quatre mois avant le premier jour de la réunion. Il devra inclure la proposition du Comité Exécutif concernant le montant de l'unité de cotisation qui permet l'application de l'article 24.

8 (a) L'Ordre du Jour doit inclure toute motion ou proposition reçue par le Secrétaire Général au moins cinq mois avant le premier jour d'une Assemblée Générale ordinaire, qu'elle émane d'un organisme adhérent, d'une Commission de l'Union ou d'une Commission mixte dans laquelle l'Union est représentée.

(b) Une motion ou proposition concernant l'administration ou le budget de l'Union qui ne figure pas à l'Ordre du Jour, préparé par le Comité Exécutif, ou tout amendement à une motion qui figure à l'Ordre du Jour, ne peut être discuté qu'avec l'accord préalable des deux tiers au moins des voix des Pays adhérents représentés à l'Assemblée Générale et disposant du droit de vote en vertu de l'article 15(a) des Statuts.

9 S'il y a doute sur le caractère administratif ou scientifique d'une question donnant lieu à un vote, l'avis du Président est prépondérant.

10 En cas de partage égal des voix, le résultat est déterminé par le Président.

11 Le Président peut inviter des représentants d'autres organisations, des scientifiques et de jeunes astronomes à participer à l'Assemblée Générale. Avec l'accord du Comité Exécutif, le Président peut déléguer ce privilège au Secrétaire Général en ce qui concerne les représentants d'autres organisations, aux Comités Nationaux d'Astronomie ou autres Comités appropriés en ce qui concerne les scientifiques et les jeunes astronomes. Des

représentants d'autres organisations, des scientifiques et des jeunes astronomes peuvent participer à l'Assemblée Générale mais ne peuvent pas participer aux votes.

III LE COMITÉ SPÉCIAL DES NOMINATIONS

12 (a) Les propositions pour les élections du Président de l'Union, du President-elect, des six Vice-Présidents, du Secrétaire Général et du Secrétaire Général Adjoint sont soumises à l'Assemblée Générale par le Comité spécial des Nominations. Ce Comité se compose du Président en fonction et du Président sortant, d'un membre proposé par le Comité Exécutif sortant et n'appartenant ni au Comité Exécutif actuel ni au Comité Exécutif précédent, et de quatre membres élus par le Comité des Nominations parmi douze membres proposés par les Présidents de Divisions en tenant compte des différentes branches de l'astronomie. A l'exception du Président en fonction et du Président sortant, les membres actuels et les anciens membres du Comité Exécutif ne doivent pas faire partie du Comité spécial des Nominations. Les membres du Comité spécial des Nominations doivent tous appartenir à des pays ou organismes adhérents différents.

(b) Le Secrétaire Général et le Secrétaire Général Adjoint participent au travail du Comité spécial des Nominations à titre consultatif.

(c) Le Comité spécial des Nominations est nommé par l'Assemblée Générale et est responsable directement devant elle. Il reste en fonction jusqu'à la fin de l'Assemblée Générale ordinaire qui suit immédiatement sa nomination, et il peut combler toute vacance survenant parmi ses membres.

IV LE COMITÉ EXÉCUTIF ET SES MEMBRES

13 (a) Le Président de l'Union reste en fonction jusqu'à la fin de l'Assemblée Générale ordinaire qui suit immédiatement celle de son élection. En principe, le President-elect succède au Président à cette date. Les Vice-Présidents restent en fonction jusqu'à la fin de la deuxième Assemblée Générale ordinaire qui suit celle de leur élection. Les Vice-Présidents ne sont pas rééligibles immédiatement pour les mêmes fonctions.

(b) Le Secrétaire Général et le Secrétaire Général Adjoint restent en fonction jusqu'à la fin de l'Assemblée Générale ordinaire qui suit immédiatement celle de leur élection. Normalement, le Secrétaire Général Adjoint succède au Secrétaire Général, mais ces deux responsables peuvent être réélus aux mêmes fonctions pour une seconde période consécutive.

(c) Les élections ont lieu au cours de la dernière réunion de l'Assemblée Générale, les noms des candidats proposés ayant été annoncés au cours d'une réunion antérieure.

14 Le Président sortant et le Secrétaire Général sortant deviennent conseillers du Comité Exécutif jusqu'à la fin de l'Assemblée Générale ordinaire qui suit immédiatement celle de la fin de leur mandat. Ces conseillers participent au travail du Comité Exécutif et assistent à ses réunions sans droit de vote.

15 Le Comité Exécutif peut combler toute vacance survenant en son sein. Toute personne ainsi nommée reste en fonction jusqu'à l'Assemblée Générale ordinaire suivante.

16 Le Comité Exécutif peut rédiger et publier des Directives pour expliciter les Statuts et le Règlement.

17 Le Comité Exécutif nomme le représentant de l'Union qui doit siéger au sein du Conseil International des Unions Scientifiques ; si ce représentant n'est pas déjà un membre élu du Comité Exécutif, il/elle devient conseiller.

18 (a) Le Secrétaire Général est responsable auprès du Comité Exécutif des dépenses engagées, qui ne doivent pas dépasser le montant des fonds mis à sa disposition.

(b) Un bureau administratif, sous la direction du Secrétaire Général, est chargé de la correspondance, de la gestion des fonds de l'Union, et de la conservation des archives.

V DIVISIONS ET COMMISSIONS

19 (a) Les Commissions de l'Union poursuivent les buts scientifiques de l'Union par des moyens tels que l'étude de domaines particuliers de l'Astronomie, l'encouragement de recherches collectives et la discussion de questions relatives aux accords internationaux et à la standardisation.

(b) Les Commissions de l'Union établissent des rapports sur les sujets qui leur ont été confiés.

20 Chaque Commission se compose de :

(a) un Président et au moins un Vice-Président élus par l'Assemblée Générale sur la proposition du Comité Exécutif. Ils/Elles demeurent en fonction jusqu'à la fin de l'Assemblée Générale ordinaire qui suit immédiatement celle de leur élection. Ils/Elles ne sont pas normalement rééligibles,

(b) un Comité d'Organisation, dont les membres sont désignés par la Commission sous réserve de l'approbation du Comité Exécutif. Le Comité d'Organisation assiste le Président et le(s) Vice-Président(s) dans leur tâche. Une Commission peut décider qu'elle n'a pas besoin de Comité d'Organisation,

(c) des membres de l'Union, nommés par les Présidents, Vice-Président(s) et Comité d'Organisation, en considération de leurs spécialités ; leur désignation est soumise à confirmation par le Comité Exécutif.

21 Entre deux Assemblées Générales ordinaires, les Présidents de Commissions peuvent coopter, parmi les Membres de l'Union, de nouveaux membres des Comités d'Organisation et des Commissions elles-mêmes.

22 Les Commissions rédigent leur propre règlement. Les décisions sont prises, à l'intérieur des Commissions, par un vote de leurs membres et elles deviennent d'application après approbation par le Comité Exécutif.

23 Les Commissions ayant pour object des tâches scientifiques ou techniques apparentées forment des Divisions. Les Divisions coordonnent les activités des Commissions qui les constituent, y compris en ce qui concerne les propositions de formation de nouvelles Commissions, de dissolution ou de regroupement de Commissions, et de création de Groupes de Travail. Les Divisions donnent leur avis sur les propositions de Symposia et Colloquia de l'IAU et peuvent organiser des Discussions Communes durant les Assemblées Générales.

24 Chaque Division est dirigée par un comité qui est composé d'un maximum de six personnes, y compris les Présidents des Commissions qui la constitute. Ce comité élit le Président de Division.

25 Les Présidents de Division rendent compte au Comité Exécutif et le conseillent.

VI ORGANISMES ADHÉRENTS

26 Le rôle des Organismes adhérents est d'encourager et de coordonner, sur leurs territoires respectifs, l'étude des diverses branches de l'astronomie, particulièrement en ce qui concerne leurs besoins sur le plan international. Ils ont le droit de soumettre au Comité Exécutif des propositions pour discussion par l'Assemblée Générale.

VII FINANCES

27 Chaque Pays adhérent verse à l'Union une cotisation annuelle, qui est multiple de l'unité de cotisation en fonction de sa catégorie, les pays associés payant une demi-unité de cotisation :

Catégories définies conformément à l'Article 8 des Statuts :

I	II	III	IV	V	VI	VII	VIII	VIII½	IX	X

Nombre respectif d'unités de contribution :

1	2	4	6	10	14	20	27	30	35	45

Si des catégories d'adhésion doivent être ajoutées ultérieurement, le pas du nombre d'unités sera de 10 unités par catégorie.

28 Les ressources de l'Union sont consacrées à la poursuite de ses buts, y compris :

(a) les frais de publication et les dépenses administratives ;

(b) l'encouragement des activités astronomiques qui nécessitent la coopération internationale ;

(c) la cotisation due par l'Union au Conseil International des Unions Scientifiques.

29 Les ressources provenant de dons sont utilisées par l'Union en tenant compte des voeux exprimés par les donateurs.

30 L'Union a la propriété littéraire de tous les textes imprimés dans ses publications, sauf accord différent.

31 Les Membres de l'Union ont le droit de recevoir les publications de l'Union gratuitement ou à prix réduit, à la discrétion du Comité Exécutif qui décide en fonction de la situation financière de l'Union.

VIII CLAUSES FINALES

32 Ce règlement entre en vigueur le 20 août 1997. Il peut être modifié avec l'approbation de la majorité absolue des voix des Pays adhérents qui disposent du droit de vote en vertu de l'article 15(a) des Statuts.

33 Le présent règlement est publié en versions française et anglaise. En cas d'incertitude, la version française fait seule autorité.

Kyoto, le 20 Août 1997

DIRECTIVES

I PRINCIPE FONDAMENTAL DE NON DISCRIMINATION

1 L'Union Astronomique Internationale observe les règlements du Conseil International des Unions Scientifiques (ICSU) et en conséquence, aplique l'article 5 des statuts de l'ICSU définissant les principes fondamentaux de non discrimination et d'universalité de la science :

"Dans la poursuite de ses buts en ce qui concerne les droits et responsabilités des scientifiques, l'ICSU, en tant qu'institution internationale non-gouvernementale, observe et défend activement le principe de l'universalité de la science. Ce principe inclut la liberté d'association, d'expression, d'information, de communication et de déplacement en relation avec des activités scientifiques internationales, sans discrimination d'aucune sorte basée sur des éléments tels que nationalité, religion, croyance, opinion politique, origine ethnique, race, couleur, langue, âge ou sexe. L'ICSU reconnaît et respecte l'idépendance des politiques scientifiques internes de ses membres nationaux. L'ICSU ne permet pas que ses activités, quelles qu'elles soient, soient perturbées par des déclarations ou actions de nature politique."

Les participants à des activités patronnées par l'UAI qui pensent avoir été l'objet d'une discrimination sont invités à rechercher une clarificaton immédiate de tous les aspects de l'incident, lequel peut n'être survenu que suite à un malentendu dû aux différences culturelles inhérentes à une organisation internationale du type de l'UAI. Si cette tentative de recherche de clarification ne s'avère pas satisfaisante, contact doit alors être pris avec le Secrétaire Général de l'UAI, lequel/laquelle cherchera à résoudre le différend.

En dernier ressort, le Président ou le Secrétaire du Comité Permanent pour la Liberté dans la Conduite de la Science (SCFCS) doit être approché(e). Le SCFCS a été créé par l'ICSU en 1963 afin de sauvegarder le principe fondamental de l'universalité de la science et d'aider au règlement de problèmes spécifiques. . Depuis lors, le SCFCS a oeuvré énergiquement pour faire en sorte que ce principe soit observé, en donnant des conseils et en prenant des mesures appropriées. Le Secrétaire Exécutif du SCFCS, Dr. P. Schindler, peut être joint à l'Académie des Sciences Suisse, Bärenplatz, 2, 3011 Bern, Switzerland (téléphone : 41 31 312 33 75, télécopie : 41 31 312 32 91 , e-mail : schindler@sanw.unibe.ch).

II APPARTENANCE À L'UNION

A PAYS ADHÉRENTS

2 Les demandes d'adhésion à l'Union formulées par les pays sont examinées par le Comité Exécutif compte tenu des points suivants :

(a) justesse du choix de la catégorie dans laquelle le pays souhaite être classé ;

(b) situation actuelle de l'Astronomie dans le pays formulant la demande, et ses possibilités de développement ;

(c) mesure dans laquelle le futur organisme adhérent est représentatif de l'activité astronomique de son pays.

3 Les demandes proposant une contribution annuelle appropriée seront soumises pour décision à l'Assemblée Générale, avec la recommandation du Comité Exécutif.

B MEMBRES

4 Les personnes proposées pour devenir Membres de l'Union doivent en principe être choisies parmi des astronomes et des chercheurs dont les activités sont liées à l'astronomie, compte tenu de :

(a) la qualité de leur oeuvre scientifique ;

(b) la mesure dans laquelle leur activité scientifique implique des recherches astronomiques ;

(c) leur désir de contribuer à la poursuite des buts de l'Union.

5 Les jeunes astronomes doivent être considérés comme pouvant devenir Membres de l'Union dès qu'ils ont fait la preuve de leur capacité (en principe par une thèse de doctorat ou son équivalent) et de leur aptitude (quelques années d'activité fructueuse) à mener une recherche personnelle.

6 Pour les astronomes professionnels, leur contribution à l'astronomie peut consister soit en des recherches personnelles, soit en une collaboration assidue à des programmes importants d'observations.

7 Les autres personnes ne peuvent devenir Membres de l'Union que si certains de leurs travaux originaux concernent étroitement la recherche astronomique.

8 Huit mois avant une Assemblée Générale ordinaire, il sera demandé aux organismes adhérents de proposer de nouveaux Membres. Les propositions devront parvenir au Secrétaire Général au moins cinq mois avant la première session de l'Assemblée Générale. Les propositions reçues après cette date limite ne seront prises en considération que si des circonstances exceptionnelles justifient le retard.

9 Chaque proposition de nouveau Membre doit être préparée séparément et indiquer le nom, les prénoms et les adresses postale et électronique du/de la candidat(e), (Institut ou Observatoire), ses date et lieu de naissance, l'Université devant laquelle sa thèse ou le diplôme équivalent a été soutenu(e), la date de soutenance, la situation actuelle du/de la candidat(e), les titres et renseignements bibliographiques de deux ou trois de ses articles ou publications les plus significatifs et, s'il y a lieu, tous les renseignements susceptibles d'être pris en considération par le Comité des Nominations.

10 (a) Conformément à l'article 13 des Statuts, les Présidents de Division qui désirent présenter des candidats doivent adresser leurs suggestions au Secrétaire Général au moins neuf mois avant la première session d'une Assemblée Générale ordinaire. Les propositions devront fournir les mêmes renseignements que ceux mentionnés à l'article 12.

(b) Le Secrétaire Général fait part de ces suggestions aux organismes adhérents intéressés.

11 Le Secrétaire Général préparera deux listes pour le Comité des Nominations

(a) l'une contenant les noms des candidats proposés par les organismes adhérents,

(b) l'autre contenant les noms des candidats proposés par les Présidents de Divisions, mais qui ne sont pas déjà inclus dans les propositions des organismes adhérents.

12 A partir des deux listes mentionnées à l'article 11, le Comité des Nominations prépare les propositions définitives de nouveaux membres de l'Union.

13 Les organismes adhérents peuvent proposer la radiation de Membres ayant abandonné le domaine de l'astronomie pour d'autres activités, à moins qu'ils ne continuent à apporter une contribution à l'astronomie. Ces propositions doivent être portées à la connaissance du Secrétaire Général et du Membre concerné.

14 Le Secrétaire Général publiera la liste alphabétique des Membres de l'Union après chaque Assemblée Générale ordinaire.

III MEMBRES DES COMMISSIONS

15 Les membres des Commissions de l'Union sont cooptés par les Commissions ; Cette procédure est régie par des règles établies par les Commissions elles-mêmes.

16 Les Commissions devraient choisir, ou approuver, la liste des membres de leurs commissions compte tenu de la spécialité de ces personnes, en particulier de leur activité scientifique dans le domaine de recherche de la Commission, et leur contribution au travail de la Commission. Elles peuvent

(a) inviter les Membres de l'Union à devenir membres de la Commission,

(b) radier les membres de la Commission qui n'ont pas contribué à son activité,

(c) accepter ou refuser les demandes présentées soit par des Membres de l'Union, soit par des personnes proposées comme Membres de l'Union,

17 Les Membres de l'Union ne peuvent pas, en règle générale, appartenir à plus de trois Commissions.

18 Les Membres de l'Union peuvent demander à être admis dans une Commission en écrivant au Président de cette Commission. Ils ne devraient faire cette demande que si leur propre activité rentre dans le cadre des recherches de la Commission et s'ils sont décidés à contribuer au travail de la Commission.

19 Les membres des Commissions peuvent se retirer d'une Commission en écrivant à son Président.

20 En envoyant leur propositions de nouveaux Membres, les organismes adhérents peuvent également suggérer le choix d'une Commission pour chaque candidat(e).

21 Le Secrétaire Général enregistrera et analysera la liste des membres des Commissions et, si cela est nécessaire, tentera de trouver une solution aux anomalies évidentes.

22 Le Secrétaire Général publiera la liste des membres des Commissions dans les Transactions de chaque Assemblée Générale ordinaire.

IV CONSULTANTS

23 Peuvent être élus Consultants des personnes qui ne sont pas astronomes, mais qui sont susceptibles de servir les intérêts de l'astronomie.

24 Les Commissions doivent en principe envoyer, pour approbation, leurs propositions de consultants au Secrétaire Général au moins cinq mois avant la première session d'une Assemblée Générale ordinaire.

25 Le Secrétaire Général préparera une liste des personnes proposées comme consultants et la soumettra pour approbation au Comité Exécutif.

26 Le Bureau administratif établira une liste alphabétique des consultants.

27 Les consultants peuvent participer aux réunions de l'Union. Ils peuvent avoir droit de vote dans leurs Commissions respectives. Ils reçoivent gratuitement le Bulletin d'Information de l'Union.

V RÉUNIONS SCIENTIFIQUES

28 Le Secrétaire Général publiera un règlement pour les réunions scientifiques organisées ou parrainées par l'Union.

VI PUBLICATIONS

29 Les publications de l'Union Astronomique International, approuvées dans le budget par l'Assemblée Générale, sont préparées par le Bureau administratif l'Union.

30 Les Commissions de l'Union peuvent, avec l'approbation du Comité Exécutif, avoir leurs propres publications.

31 Le Comité Exécutif décide, sur la proposition du Secrétaire Général, des modalités de distribution des publications de l'Union.

32 Les Membres de l'Union peuvent acquérir les publications de l'Union à un prix réduit.

VII CONTACTS EXTÉRIEURS

33 Nul ne peut se prévaloir de son appartenance à l'Union pour traiter avec des tiers, sans autorisation du Secrétaire Général.

34 Les représentants de l'Union dans d'autres organisations, en particulier les Comités de l'ICSU et les Commissions Inter-Unions, seront désignés par le Comité Exécutif. Les noms sont proposés par les Présidents des Commissions concernées.

35 Les dépenses encourues par les représentants de l'Union dans d'autres organisations seront remboursées à la discrétion du Secrétaire Général, dans les limites du Budget adopté par l'Assemblée Générale. Les représentants sont priés d'obtenir l'accord préalable du Secrétaire Général avant d'engager ces dépenses.

VIII ASSEMBLÉES GÉNÉRALES

36 Huit mois avant l'Assemblée Générale, le Secrétaire Général envoie aux Comités Nationaux d'Astronomie ou autres Comités appropriés et aux Organisations adhérentes le budget préparé par le Comité Exécutif, pour commentaires.

37 Les décisions et recommandations de l'Union sur des sujets d'ordre scientifique ou organisationnel sont formulées dans ses Résolutions. Les Résolutions sont proposées, évaluées et approuvées selon les directives suivantes :

(a) Les Résolutions sont de 3 catégories différentes:

A: Résolutions proposées par des Membres Adhérents ou par le Comité Exécutif,

B: Résolutions proposées par des Divisions ou des Commissions non rattachées à une Division et adoptées par l'Assemblée Générale,

C: Résolutions adoptées par des Divisions ou Commissions, mais non soumises à l'Assemblée Générale.

(b) Les Résolutions doivent être soumises en utilisant les fomulaires appropriés aux Résolutions de type A, B et C, respectivement. Ces fomulaires sont disponibles auprès du Secrétariat.

(c) Les Résolutions de type A doivent être inscrites à l'ordre du jour de l'Assemblée Générale et soumises au Secrétaire Général au moins 6 mois avant la date de début de l'Assemblée Générale. Les Résolutions de type A ou B qui ont des répercussions budgétaires pour l'Union doivent être soumises au Secrétaire Général neuf mois à l'avance afin d'être prises en compte par l'Assemblée Générale.

Toutes les autres Résolutions de type B doivent être soumises au Secrétaire Général trois mois avant le début de l'Assemblée Générale.

(d) Face à des circonstances véritablement exceptionnelles, le Comité Exécutif peut envisager d'accepter des propositions de dernière minute de type B.

(e) Lors de sa seconde session, chaque Assemblée Générale nomme un Comité des Résolutions composé de 5 membres de l'Union dont l'un choisi parmi les membres du Comité Exécutif. Le Comité des Résolutions demeure en activité jusqu'à la fin de l'Assemblée Générale suivante.

(f) Le Comité des Résolutions examinera le contenu, la formulation et les conséquences de toutes les Résolutions de type A et B qui seront présentées lors de la seconde session de l'Assemblée Générale. En particulier, il considérera les points suivants:

i L'adéquation du sujet à une Résolution de l'UAI,
ii Sa formulation correcte et sans ambiguïté,
iii Sa cohérence avec les Résolutions antérieures de l'UAI.

Le Comité des Résolutions peut renvoyer une Résolution à ses proposants pour reconsidération ou retrait mais ne peut ni annuler ni modifier son fond de sa propre initiative. Le Comité des Résolutions avisera le Comité Exécutif de tout problème potentiel induit par le fond d'une Résolution.

(g) Le Comité Exécutif examinera le contenu et les conséquences de toutes les Résolutions (de type A et B) proposées pour adoption par l'Assemblée Générale. Le Comité des Résolutions présente les propositions accompagnées des recommandations du Comité Exécutif à la seconde session de l'Assemblée Générale pour approbation.

(h) Les Résolutions de type C ne sont applicables qu'au sein de leur Commission ou Divison d'origine.

IX GROUPES DE TRAVAIL

38 Le Comité Exécutif, les Divisions et les Commissions peuvent créer des Groupes de Travail pour des tâches spécifiques. Les Groupes de Travail créés par les Divisions et les Commissions doivent être approuvés par le Comité Exécutif. Tous les Groupes de Travail sont créés pour une duré initiale de 3 ans. Avant chaque Assemblée Générale, les Divisions et Commissions doivent informer le Comité Exécutif des Groupes de Travail qui doivent être maintenus pour le triennium suivant et de ceux qui doivent être dissous.

CHAPTER VIII

RULES FOR IAU SPONSORED SCIENTIFIC MEETINGS

General

The programme of scientific meetings is one of the most important means by which the IAU pursues its goal of promoting the science of astronomy through international collaboration. A large fraction of the Union's budget is devoted to the support of the IAU scientific meetings. The Executive Committee (EC) places great emphasis on maintaining high scientific standards, coverage of a balanced spectrum of subjects, and an appropriately international flavour for the programme of meetings sponsored by the IAU.

Types of meetings sponsored by the IAU

The IAU currently supports the following types of meeting:

a. Symposia
b. Colloquia
c. Regional Astronomy Meetings
d. Joint Discussions during General Assemblies (GAs)
e. Co-sponsored meetings

These types of meeting are characterised as follows:

a. The IAU Symposium Series is the scientific flagship of the IAU meeting programme. Symposia are organised on suitably broad, yet well-defined scientific themes of considerable general interest. They are intended to significantly advance the field by seeking answers to current key questions and/or clarify emerging concepts. Symposium proposals are forwarded to the EC by an IAU Division, or by an IAU Commission acting jointly with one or more supporting Commissions. Symposia receive a fixed allocation from the IAU to support the attendance of qualified scientists from all parts of the world (see below). The amount of support appears on the current Application Form. The Proceedings of IAU Symposia are published in the standard IAU Symposium series by the IAU Publisher. In years when a GA is held, most or all Symposia are attached to the GA and held at the same venue. In these cases, the local organisation is covered by the Local Organising Committee (LOC) for the GA itself, and travel support for the Symposia is coordinated with that for the GA.

b. For IAU Colloquia, equally high scientific standards are expected as for Symposia, but their subjects may be somewhat narrower or more specialised in scope. Accordingly, Colloquium proposals may be forwarded to the EC with the support of a single Commission, and IAU travel support is lower (see Application Form). Proceedings of IAU Colloquia are normally published, but their format and the choice of publisher are left to the discretion of the Scientific Organising Committee (SOC); there is no regular series of IAU Colloquium volumes. In the years of a GA, Colloquia are scheduled no closer than three months to the GA itself.

J. Andersen (ed.), Transactions of the International Astronomical Union Volume XXIIIB, 285–298.

c. IAU Regional Astronomy Meetings, currently held in the Latin American and Asian-Pacific regions in years between General Assemblies, are usually held at the invitation of a national astronomical society in the region. Their purpose, in addition to the discussions of specific scientific subjects, is to promote contacts between scientists in the region concerned, especially amongst young astronomers. Therefore, both a much wider range of scientific topics, a larger SOC, and a larger total attendance are accepted than normal for Symposia and Colloquia. IAU travel support is provided at a similar level as for Symposia. Presentations by young astronomers, including Ph.D. students, are particularly encouraged. Proceedings are usually published in a regional astronomical publication series.

d. Joint Discussions, taking place during General Assemblies, address scientific themes of interest to more than one IAU Commission. They require the approval of the EC and normally last between 0.5 and 1.5 days at the GA. No separate financial support is allocated to Joint Discussions, but participants may apply for General Assembly Travel Grants, and the proceedings of Joint Discussions are published in the Highlights of Astronomy without cost to the organisers.

e. The IAU may decide to co-sponsor meetings which are organised by other Scientific Unions. Main organisational and financial responsibility for such meetings rests with the main sponsoring Union. The IAU expects to be represented in the relevant SOC(s) and to be consulted about publication of the proceedings and other major issues, and the IAU may make a financial contribution to the travel expenses of participants at the meeting.

The number of meetings which available funds enable the IAU to sponsor each year is very limited, about six Symposia and six Colloquia, a Regional Meeting in years without a General Assembly, and 1-2 Co-Sponsored Meetings. Accordingly, not all scientifically valid proposals can be awarded IAU sponsorship.

Meeting proposals for a given year are ranked by the EC according to scientific criteria, taking into account the advice of Division Presidents on the priority of proposals within their respective fields. Proposals of adequate scientific merit are then approved in order of ranking, up to the limit imposed by available funds.

Application procedures

Normally, the initiative to propose a scientific meeting for IAU sponsorship originates from a group of scientists in the field who, in collaboration with colleagues worldwide, prepare a draft scientific programme and nominations for a Scientific Organising Committee (SOC), who will be responsible for the scientific aspects of the meeting from its inception to its conclusion.

Responsibility for the preparation and timely submission of the final proposal to the EC rests with the Chairperson of the proposed Scientific Organising Committee (SOC). The practical preparation of the submission may be delegated to another person, normally another SOC member or the Chairperson of the proposed Local Organising Committee (LOC), but ultimate responsibility for the scientific and other factual contents of a proposal remains with the SOC Chairperson.

Before submitting an application for an IAU meeting the proposers should send, at the earliest possible date, a letter of intent to the Assistant General Secretary (AGS) stating the topic and proposed dates of the meeting. The AGS will then be able to inform the proposers of any pending plans for other meetings in the same or a similar field in order to avoid unintended competition between similar proposals.

Application procedures have been designed so as to ensure that the information necessary for the evaluation of the proposals by the EC is complete and in a uniform format that allows direct comparison between proposals as far as possible. Therefore, proposals are to be submitted on the official IAU Symposium and Colloquium application forms, which are available from the IAU WEb page, but can also be requested from the Secretariat or from AGS.

In all cases, an IAU Commission should accept to act as the main scientific sponsor of the proposed meeting (the "Proposing Commission"). If the Proposing Commission belongs to an IAU Division, the proposed SOC Chairperson of the meeting should submit the proposal to the President of that Division before the deadline, with a copy to the Presidents of the Proposing and all supporting Commissions.

Each Division President will then ascertain the degree of Commission support within the Division, and forward all proposals to the EC via the AGS with a suggested priority ranking and overall scientific quality assessment, as well as an appropriately detailed motivation for this evaluation. Commissions not belonging to a Division, or Working Groups of the EC, may forward meeting proposals, or support of proposals, directly to the AGS. The AGS forwards all proposals, with the recommendations of the Divisions Presidents as outlined above, to the EC for final decision.

In order to allow sufficient time for evaluation of the proposals by Division Presidents before final review and decision by the EC, and for timely announcement of the approved meetings, proposals must be submitted before the deadline published annually in the IAU Information Bulletin. Normally, this deadline will be in February or March of the year preceding the year of the meeting. If desired, proposals may also be submitted one year earlier.

The application forms, which should be filled in completely, are mostly self-explanatory. However, the following points deserve clarifying comments:

(10) Scientific Organising Committee (SOC)
The Scientific Organising Committee has overall responsibility for the scientific standards of the meeting. It exercises this responsibility in three main respects: *(i)* Definition of the scientific programme of the meeting, including the choice and distribution of topics for individual sessions, and the selection of invited reviews and shorter papers, and contributed and poster papers; *(ii)* Choice of key speakers for invited reviews, and *(iii)* Proposals for the allocation of individual travel grants from the IAU (and possibly other) funds provided for the meeting. The Chairperson and other members of the SOC are appointed by the EC as part of the approval process.

The composition of the proposed SOC is therefore a key element in assessing the potential scientific value of an application. Thus, the SOC should cover the principal subjects to be treated in the discussions. As experience shows that large committees tend be inefficient, the SOC should be **no larger than ten persons in total** (Regional Meetings excepted), but at the same time appropriately composed with regard to geography, gender, etc. The composition of the SOC should thus conform, in a constructive manner, to the intent of the ICSU *Statement on Freedom in the Conduct of Science* (see below). Normally, any one institution should not be represented on the SOC by more than one person. It is customary, but not required, that SOC members are also members of the IAU.

(13) Editor(s) of Proceedings
As a rule, it is considered important that the Proceedings of IAU meetings be published as a valuable record of the event, for future reference. Arrangements for the publication of Proceedings from the various types of meetings are somewhat different and are summarised in a separate section below.

In all cases, the proposal should make clear whether it is intended to publish the Proceedings. If so, the name(s) of the proposed Editor(s) must be given. In the case of two or more Editors, it has been found rational to designate one of these as Chief Editor, with primary responsibility for the contacts to the IAU and the Publisher. The proposal should identify the person who will serve in this capacity. Confirmation of the name(s) is part of the approval process; any change of Editor(s) after a meeting has been accepted for IAU sponsorship requires the prior approval of the EC, through the AGS.

(15) Registration fees
Every effort should be made to keep registration fees and administrative expenses low so as to make the meeting accessible to all. Such efforts include the use of low-cost meeting facilities and other sources of local support or sponsorship. The acceptable level of a registration fee will depend on circumstances, and proposers should carefully specify which services are covered by the fee. Currently, about 150 USD should be considered an absolute upper

limit for IAU sponsored Symposia and Colloquia, and the EC may reject or withhold approval of otherwise valid proposals if the proposed registration fee is exorbitant.

Substantially lower sums should be aimed at, using the following guidelines: Social events and meals during the conference should be optional rather than paid from the mandatory registration fee. If found desirable by the SOC, a copy of the Proceedings may be included in the fee. In such cases, the SOC should negotiate a substantial discount from the publishers.

(16) Price of hotels and/or other accommodation
Again, in the interest of enabling interested and qualified colleagues from all parts of the world to attend the meeting, affordable accommodation should be available. It is recognised that in some cases, conference centres offering meeting rooms, meals, and accommodation in one location may offer an environment which is favourable to the overall scientific outcome of the meeting. However, such centres tend to be expensive. In these cases, efforts should be made to secure additional affordable accommodation in shared rooms, neighbouring, cheaper establishments, or by any other suitable arrangements.

(19) Free access to meeting by all qualified scientists
The statement that the ICSU rules on non-discrimination in the access to the meeting (see below) will be strictly observed **MUST** be explicitly confirmed before any proposal will receive final approval by the EC. A summary of the measures taken to ensure this should be given, and the signatures of both the SOC and LOC Chairpersons are required.

Unrestricted participation in IAU sponsored meetings

Participation in Symposia or Colloquia is by invitation of the SOC Chairperson. Invitations may be sought by suitably qualified scientists working in the field.

It is the policy of the IAU to promote the full participation of astronomers worldwide in its meeting programme. Symposia and Colloquia must be open to all who are qualified to participate. It is recognised that there may be occasions where logistical constraints force the SOC to limit the total number of participants in a meeting, but it is essential that no restriction based on sex, race, colour, nationality, or religious or political affiliation be imposed on the full participation of all *bona fide* scientists in any aspect of the organisation and conduct of IAU sponsored meetings, either by the organisers of Symposia, Colloquia, or Regional Meetings, or by the authorities of the host country.

Approval of a proposal for an IAU sponsored meeting requires explicit guarantees that this principle will be respected. Failure to honour such a commitment may result in cancellation of IAU sponsorship, and indeed of the meeting.

The commitment of the IAU to this principle is consistent with and backed by its membership of the International Council of Scientific Unions (ICSU). ICSU is the oldest existing non-governmental body committed to international scientific cooperation for the benefit of humanity, and is recognised as such by the relevant United Nations organisations. In this capacity, ICSU pursues a consistent and vigorous policy of non-discrimination in matters of international scientific cooperation, expressed in the *ICSU Statement on Freedom in the Conduct of Science*. The central paragraphs of this document are reproduced below (October 1995 version):

"...One of the basic principles in [ICSU's] Statutes is that of the universality of science (see Statute 5), which affirms the right and freedom of scientists to associate in international scientific activity without regard to such factors as citizenship, religion, creed, political stance, ethnic origin, race, colour, language, or sex. Such rights are embodied in a variety of articles in the International Bill of Human Rights.

ICSU seeks to protect and promote awareness of the rights and fundamental freedoms of scientists in their scientific pursuits. ICSU has a well-established non-political tradition which is central to its character and operations, and it does not permit any of its activities to be disturbed by statements or actions of a political nature.

As the intrinsic nature of science is universal, its success depends on cooperation, interaction and exchange, often beyond national boundaries. Therefore, ICSU strongly supports the principle that scientists must have free access to each other and to scientific data and information. It is only through such access that international scientific cooperation flourishes and science thus progresses.

On these grounds, ICSU works to resolve such cases as do, nevertheless, arise from time to time when such open access is denied or restricted and in cases primarily involving members of the ICSU family. In most cases, private consultations involving members of the ICSU family have been successful. Where private consultations have failed, ICSU has publicised acts of discrimination against scientists and taken steps to prevent their repetition, including, if necessary, such measures as encouraging members of the ICSU family to decline invitations to hold or attend meetings in the country concerned.

On the basis of its firm and unwavering commitment to the principle of the universality of science, ICSU reaffirms its opposition to any actions which weaken or undermine this principle."

Contact persons

The organisation of the scientific programme of the General Assemblies, including the programme of Joint Discussions, is the responsibility of the General Secretary (GS). Contacts concerning proposals for Regional Meetings should also be made with the GS.

Inquiries and proposals concerning Symposia, Colloquia and Co-sponsored Meetings, and correspondence on approved Regional Meetings, should be directed to the Assistant General Secretary (AGS).

In order to avoid last-minute complications, prospective meeting organisers are strongly encouraged to contact these persons well in advance of the announced proposal deadlines.

Selection criteria for IAU sponsorship

The following general guidelines for the selection of meetings for IAU sponsorship will be useful to prospective proposers:

First, the scientific merit of a proposal is evaluated by the EC, taking the comments and advice of Commission and Division Presidents into consideration. An IAU sponsored meeting should have a well-defined scientific theme, be scheduled at a propitious time for significant progress in the field, and be of interest to young researchers as well as senior experts. After reviewing the scientific content of a proposal, the EC may decide to move the meeting to a different category from that proposed, e.g. from Symposium to Colloquium or vice versa.

Second, since the IAU embraces all major fields in astronomy, the IAU meeting programme should maintain a broad and balanced scope and cover the main active fields at appropriate intervals. Accordingly, even scientifically strong proposals in the same or largely overlapping fields cannot be approved at very short intervals.

Third, given the international nature of the Union, IAU sponsored meetings must be internationally oriented. This implies a well-balanced geographical distribution of both organisers and key speakers; normally, substantially less than half of the proposed SOC and/or key speakers should come from any single country.

For Regional and Co-sponsored Meetings, less restrictive criteria apply. As outlined above, the stimulation of contacts between (especially young) astronomers in the region concerned is an important function of the Regional Meetings. Accordingly, much broader subject areas are acceptable than for Symposia. In Co-sponsored Meetings, main responsibility for the scientific content and organisation of the meeting lies with the main sponsoring Scientific Union, and the IAU assumes only an advisory role on the relevant aspects of the overall programme.

Educational aspects of scientific meetings

At some IAU meetings, "Teachers' Workshops" or similar educational activities have been organised adjacent to the scientific meeting itself. By taking advantage of the presence of many distinguished national and foreign scientists, one- or two-day events have been organised for the benefit of university and high-school astronomy educators in or near the country hosting the meeting. These initiatives have generally been very successful and well received by their audiences.

Stimulating and improving the teaching of science, and of astronomy in particular, is becoming increasingly urgent, and parallel educational activities of the above kind in connection with IAU sponsored meetings are encouraged. While the quality of the proposed scientific programme will remain the primary selection criterion for IAU sponsorship, a good parallel educational programme will certainly add to the overall merit of a proposal.

Approval of meeting proposals

After review of the slate of proposals, the EC decides the programme of approved meetings for the following year, including the details of the arrangements for the individual meetings. The decisions of the EC on each meeting, including any conditions to be fulfilled before final approval, are communicated to the proposers by the contact person for the corresponding proposal.

After final approval, a letter of award is issued, accompanied by an official form listing the essential facts of the meeting as approved by the EC. Any change of the circumstances recorded on this form require approval by the IAU EC, obtained through the appropriate contact person.

Publication of Proceedings

Normally, the Proceedings of IAU sponsored meetings should be published as a record of the scientific results achieved. It is desirable that these proceedings appear as soon as possible in order to reflect the current status of the field. While practical arrangements will be negotiated in each case between the Editor(s) and the IAU or other Publisher, the completed manuscript should be delivered to the publishers within about three months after the end of the meeting. At the same time, one copy of the complete manuscript should be sent to the AGS for information.

Procedures for publication differ somewhat for the different types of meeting. For Symposia, proceedings are published by the IAU Publisher in the IAU Symposium Series at no cost to the organisers. For Colloquia there is no regular IAU series, and the SOC is free to propose, based on scientific and financial considerations, whether to publish the proceedings, and if so where and in which format. Proceedings of Regional Meetings are normally also published, under the responsibility of the SOC.

For all three types of IAU sponsored meeting, the format and editorship of the proceedings are fixed by the EC as part of the approval process, and any changes require prior approval by the EC, through the Assistant General Secretary. For Co-sponsored meetings, the format and editorship of the proceedings are decided by the main sponsoring Scientific Union in consultation with the IAU.

The IAU Publisher will automatically supply a copy of the Proceedings of all IAU Symposia to the IAU archive. For IAU Colloquia and all other IAU sponsored meetings, the Editors should send one copy of the published Proceedings to the IAU Secretariat in Paris.

Travel grants to IAU-sponsored meetings

IAU funds allocated to IAU-sponsored meetings are intended to cover travel costs for participants and, if necessary, board and lodging at the conference, but not for administrative expenses. Since the IAU funds are very limited, it is expected that the organisers will attract substantial financial support from other sources also.

Proposals for the distribution of IAU Travel Grants to individual participants are made by the Scientific Organising Committee (SOC) and sent to the Assistant General Secretary (AGS) for approval. IAU priority is to support qualified scientists to whom few or no other means of support are available, e.g. colleagues from countries in economic difficulties and young scientists. Also, a reasonable geographical distribution is expected: Normally, no more than 50% of the funds should be allocated to a single country or region. Moreover, the IAU support should preferably carry significant weight in ensuring the participation of the selected grantees, rather than adding comfort for colleagues whose attendance is already assured.

Within these general guidelines, it is left to the scientific judgment of the SOC whether to allocate funds in a few major grants to ensure the participation of key speakers, or broaden attendance at the meeting by supporting several participants, maintaining the overall scientific standard of the meeting as the primary criterion.

The recommendations of the SOC should be sent to the AGS (e-mail is accepted and preferred), specifying for each person: Name, nationality, full mailing address, e-mail address when useful, amount of proposed grant (in Swiss Francs), place from which journey will be made, and title and nature of contribution (review talk, thesis presentation, ...). A sample application form is found below, summarising the required information; use of the form itself is optional. It will help to avoid difficult last-minute reallocations if the SOC verifies that grantees expect to be able to attend with the proposed amounts of support. The SOC recommendation should reach the AGS:

> *No later than **THREE MONTHS** before the meeting in any case, but **FIVE MONTHS** in the case of participants from the Former Soviet Union, China, or other countries where completion of visa formalities may require extra time.*

The above deadlines have been found necessary in order to ensure timely notification to participants, completion of visa formalities, etc. Upon approval of the proposal, individual grant letters are mailed to the recipients by the AGS, with a copy to the SOC or LOC Chair. On request, an early informal message (by fax or e-mail) may be sent to grantees who may need extra time to initiate visa or other formalities.

The normal administrative procedure is to transfer the entire IAU grant to the LOC bank account. Individual grants are then paid to each recipient upon arrival and registration at the meeting. This procedure has been found safe and convenient, especially in accommodating the occasional, but inevitable cases of late cancellations. Currency conversions are made at the official ICSU exchange rates in force at that time.

There are cases, however, when a cash advance is needed for a participant to be able to travel to the meeting at all. Such cases should be argued in reasonable detail to the SOC and AGS, and individual cheques may then be sent directly from the Paris Secretariat. These procedures are explained in the individual grant letters.

Reporting after meetings

Within one month after the meeting, a copy of the complete scientific programme, list of participants, signed receipts for the individual IAU grants, and a brief report to the Executive Committee on the scientific results of the meeting, should be sent to the AGS by the SOC and/or LOC Chairpersons. For convenience, an optional report form is attached.

For all other meetings than IAU Symposia, the Editors should send one copy of the published Proceedings to the IAU Secretariat in Paris.

INTERNATIONAL ASTRONOMICAL UNION
UNION ASTRONOMIQUE INTERNATIONALE

PROPOSAL FOR AN IAU SYMPOSIUM

(1) Title:

(2) Date and duration:

(3) Location:

(4) Coordinating IAU Division:

(5) Proposing Commission:

(6) Supporting Commission(s):

N.B. *Letters from the Presidents of the relevant IAU Commissions must accompany the proposal.*

(7) Other ICSU body co-sponsoring the meeting, if any:

(8) Other supporting organisations, if any:

(9) Contact address:

Telephone:
Telefax:
E-mail:

(10) Proposed Scientific Organising Committee

	Name:	Country:	Name:	Country:
Chairperson:				
Other Members: (up to ten)				

(11) Proposed Chairperson of the Local Organising Committee:

(12) Proposed names of other LOC Members:

(13) Proposed Editor(s) of the Proceedings:
(Identify the Chief Editor)

(14) Expected or maximum number of participants:

(15) Registration fee: .... CHF (or approximately .... USD).
Expenses covered: Transportation from airport to hotel (....); from conference to hotel (....); coffee breaks (....), closing dinner (....), proceedings (....), any other items: ....

(16) Expected price of hotels and/or other accommodations:

(17) Amount requested for travel support from the IAU:
(Maximum currently 25,000 CHF)

(18) Topics in the Preliminary Scientific Programme (max. 10 lines):
(For announcement in the IAU Information Bulletin)

Please append a detailed scientific rationale and draft programme.

(19) We confirm that attendance from ALL countries is guaranteed, in accordance with the ICSU Rules on Freedom in the Conduct of Science (see note, and describe steps taken to this end):

(20) Signature of Chairperson of SOC:

(21) Signature of Chairperson of LOC:

(22) Signature of proposer:

Date and Place:

Proposals for and all other correspondence concerning IAU Symposia and Colloquia should be sent to the IAU Assistant General Secretary:

H. Rickman, Astronomical Observatory, Box 515, S-751 20 Uppsala, Sweden.
Fax. +46 18 52 7583, E-mail: hans@astro.uu.se

INTERNATIONAL ASTRONOMICAL UNION
UNION ASTRONOMIQUE INTERNATIONALE

PROPOSAL FOR AN IAU COLLOQUIUM

(1) Title:

(2) Date and duration:

(3) Location:

(4) Coordinating IAU Division:

(5) Proposing Commission:

(6) Supporting Commission(s):

N.B. *Letters from the Presidents of the relevant IAU Commissions must accompany the proposal.*

(7) Other ICSU body co-sponsoring the meeting, if any:

(8) Other supporting organisations, if any:

(9) Contact address:

Telephone:
Telefax:
E-mail:

(10) Proposed Scientific Organising Committee

	Name:	Country:	Name:	Country:
Chairperson:				
Other Members: (up to ten)				

(11) Proposed Chairperson of the Local Organising Committee:

(12) Proposed names of other LOC Members:

INTERNATIONAL ASTRONOMICAL UNION
UNION ASTRONOMIQUE INTERNATIONALE

APPLICATION FOR AN IAU TRAVEL GRANT

(1) **IAU Symposium/Colloquium No.:**

(2) **Title of meeting:**

(3) **Location (city, country):**

(4) **Dates of meeting:**

(5) **Name of applicant:**

(6) **Nationality:**

(7) **Position:**

(8) **Address:**
Telephone:
Telefax:
E-mail:

(9) **Starting point of journey** (if different from home address):

(10) **Type of contribution** (e.g. review talk, thesis presentation, ...):

(11) **Title of contribution:**

(12) **Total amount of IAU support applied for (CHF):**
Subtotal for travel (max.: economy air fare or equivalent):
Subtotal for subsistence (if no other funds available):

(13) **Other sources of support applied to:**
Amount requested/granted:

(14) **Signature of applicant:**
Date and place:

(15) **For Ph.D. students:**
Signature of Thesis Director/Supervisor:
Institution:

This form, or one including the same essential information, should be submitted to the Chairperson of the Scientific Organising Committee by the specified deadline (cf. Guidelines for IAU Travel Grants).

INTERNATIONAL ASTRONOMICAL UNION
UNION ASTRONOMIQUE INTERNATIONALE

SCIENTIFIC REPORT ON IAU SPONSORED MEETING

To be submitted to the IAU Assistant General Secretary within one month after the meeting.

The following documents should be attached:

(i) Final programme of the meeting
(ii) Final list of participants
(iii) Receipts signed by the recipients of IAU Travel Grants

For IAU Colloquia, one copy of the published Proceedings should be sent to the IAU Secretariat in Paris.

(1) **IAU Symposium/Colloquium No.:**

(2) **Title of meeting:**

(3) **Dedication of meeting** (if any):

(4) **Location (city, country):**

(5) **Dates of meeting:**

(6) **Number of participants:**

(7) **Number of countries represented:**

(8) **Report submitted by:**

Date and place:

Signature of SOC Chairperson

(9) **Summary of scientific highlights of the meeting (1-2 pages)**
(May be submitted on separate sheets; electronic mail acceptable)

CHAPTER IX

MEMBERSHIP

LIST OF ADHERING ORGANIZATIONS

(Countries in italics are *Associate Members*, as opposed to Full Members)

☎: Telephone - ✆: Fax

Year of Adherence		Members
1988	**Algeria** CRAAG BP 63 Bouzareah Alger ☎: 213 2 79 14 43/79 16 06 ✆: 213 2 79 14 43	3
1927	**Argentina** CONICET Rivadavia 1917 1033 Buenos Aires ☎: 54 1 953 72 30/72 34 ✆: 54 1 953 43 45	90
1935/1994	**Armenia** President, Armenian National Academy of Sciences Marshal Baghramian av. 24 375019 Yerevan ☎: 3742 52 70 31 ✆: 3742 50 06 867	31
1939	**Australia** Australian Academy of Sciences Ian Potter House Gordon St G.P.O. Box 783 Canberra ACT 2601 ☎: 61 6 247 57 77 ✆: 61 6 257 46 20	191

1955 **Austria** 31
Die österreichische Akademie
der Wissenschaften
Dr-Ignaz-Seipel-Platz 2
A 1010 Wien
☎: 43 1 52 15 86
✆: 43 1 52 15 85 35

1935/1993 **Azerbajan** 8
Azerbajan Academy of Sciences
10 Kommunisticheskaja St
Baku 37061
☎: 994 12 92 35 29
✆: 994 12 92 56 99

1920 **Belgium** 88
Chef du service "Activités Internationales"
Affaires scientifiques, techniques et culturelles
8, rue de la Science
B 1000 Brussels
☎: 32 2 238 34 11
✆: 32 2 230 59 12

1998 ***Bolivia*** -
Academia Nacional de Ciencias de Bolivia
Av. 16 de Julio No. 1732
La Paz
☎: 591 2 86 89 90
✆: 591 2 37 96 81

1961 **Brazil** 109
CNPq
Av W3 Norte Quadra 507 B
Caixa Postal 11-1142
70740 Brasilia DF
☎: 55 61 274 11 55
✆: 55 61 274 19 50

1957 **Bulgaria** 50
Bulgarian Science Academy
15 Noemvri Street 1
BG 1040 Sofia
☎: 359 2 87 7087
✆: 359 2 80 30 23

1957 **Canada** 201
NRC, International Affairs
Montreal Road
Ottawa ON K1A OR6
☎: 1 613 990 60 91
✆: 1 613 952 99 07

1947 **Chile** 45
Chilean Academy of Science
CONICYT
Casilla 297 V
Santiago de Chile
☎: 56 2 74 45 37
✆: 56 2 49 67 29

1935 **China Beijing** **369**
Chinese Astronomical Society
Purple Mountian
Academia Cinica
Nanjing 210008
☎: 86 25 63 75 51
✆: 86 25 30 27 28

1959 **China Taipei** **23**
Academica Sinica
Taiwan
Taipei 11529
☎: 886 2 782 42 04
✆: 886 2 785 38 52

1935/1994 **Croatia** 13
CASA
Hrvatsko Astronomsko Drustvo
Kaciceva 26
41000 Zagreb Hrvatska
☎: 38 41 442 600
✆: 38 41 445 410

1922/1993 **Czech Rep** 71
Academy of Sciences of the Czech Republic
Narodni tr 3
111 42 Praha 1
☎: 42 2 242 40 531
✆: 42 2 243 40 513

1922 **Denmark** 50
Kongelige Danske Videnskabernes Selskab
H.C. Andersen Bvd 35
DK 1553 Copenhagen V
☎: 45 33 11 32 40
✆: 45 33 91 07 36

1925 **Egypt** 39
ASRT
101 Kasr El-Einy Street
Cairo
☎: 20 2 354 83 63
✆: 20 2 356 28 20

1935/1992 **Estonia** 22
Tartu Observatory
Toeravere
EE 2444 Tartumaa
☎: 372 7 410 265
✆: 372 7 41 02 05

1948 **Finland** 37
Secretary General
Deleg Finnish Acad Sci/Letters
Mariankatu 5
SF 00170 Helsinki
☎: 358 9 63 30 05
✆: 358 9 66 10 65
🖂:

1920 **France** 608
Academie des Sciences
COFUSI
23, quai Conti
F 75006 Paris
☎: 33 1 44 41 43 99
✆: 33 1 44 41 44 04
🖂:

1935/1994 **Georgia** 19
Georgian Academy of Sciences
52 Rustaveli Avenue
Tbilisi 380060
☎: 995 32 37 52 26
✆: 995 32 98 70 00
🖂:

1951 **Germany** 490
Deutsche Forschungsgemeinschaft
Kennedyallee 40
D 53170 Bonn
☎: 49 228 665 23 46
✆: 49 228 885 2550

1920 **Greece** 89
Dpt Ministry of Development
International Organisations
General Secretariat for Research and Technology
14-18 Athens P.O. Box 14631
GR 10679 Athens
☎: 30 1 360 0207/362 67 17
✆: 30 1 363 48 06

1947 **Hungary** 41
Hungarian Academy of Sciences
Nador Street 7
H 1051 Budapest
☎: 36 1 117 25 75
✆: 36 1 117 25 75/62 15

1988 **Iceland** 4
Icelandic Ministry of Education
c/o The Icelandic Astronomical Society
Raunvisindastofnunin, Dunhaga 3
IS 107 Reykjavik
☎: 354 525 48 09
✆: 354 552 88 01/89 11

1964 **India** 226
Indian National Science Academy
Bahadur Shah Zafar Marg
New Delhi 110 002
☎: 91 11 323 20 66/323 20 75/323 10 38
✆: 91 11 323 56 48

1979 **Indonesia** 12
Indonesian Institute of Sciences (LIPI)
Gedung Widya Graha
Jl. Jenderal Gatot Subroto 10
Jakarta 12710
☎: 62 21 525 57 11/525 15 42
✆: 62 21 520 72 26

1969 **Iran** 15
University of Tehran
Office of International Relations
Enghlab Ave
Tehran
☎: 98 21 646 98 07
✆: 98 21 640 93 48

1947 **Ireland** 33
The Royal Irish Academy
19 Dawson Street
Dublin 2
☎: 353 1 676 25 70/42 22
✆: 353 1 676 23 46

1954 **Israel** 44
Israel Academy of Sciences and Humanities
43 Jabotinsky Street
P.O. Box 4040
91040 Jerusalem
☎: 972 2 63 62 11/62 14
✆: 972 2 66 60 59

1920 **Italy** 408
Consiglio Nazionale delle Ricerche
Servizio Relazioni Internazionali
Piazzale Aldo Moro 7
I 00100 Roma
☎: 39 6 4993 33 49
✆: 39 6 446 98 33

1920 **Japan** 446
Official Science Information and International Affairs Division
Science Council of Japan
22-34, Roppongi 7 chome
Tokyo 106
☎: 81 3 3403 62 91
✆: 81 3 3403 19 82

1973 Korea Rep 51
Korean Astronomical Society
Dpt of Astron Coll Ntl Science
Seoul National University
Seoul 151 742
☎: 82 2 880 6621
✆: 82 2 887 1435

1996 **Latvia** 8
Latvian Academy of Sciences
Turgeneva iela 19
Riga LV 1524
☎: 371 2 22 15 24
✆: 371 2 2 22 87 84/371 782 11 53

1935/1993 **Lithuania** 11
Lithuanian Academy of Sciences
MTP-1
Gedimino Prospekt 3
232600 Vilnius
☎: 370 61 40 41
✆: 370 61 84 64

1988 **Malaysia** 6
Space Science Studies Division
Ministry of Science, Technology and Environment, Planetarium Negara
Lot 53 Jalan Perdana
50480 Kuala Lumpur
☎: 60 03 273 54 84/5/6
✆: 60 03 273 54 88

1921 **Mexico** 82
Instituto de Astronomia
Universidad Nacional Autonoma de Mexico
Apdo Postal 70 264 Cd. Universitaria
Mexico 04510 DF
☎: 525 616 2601/2726/1412
✆: 525 616 0653

1922 **Netherlands** 165
Koninklijke Nederlandse Akademie
van Wetenschappen
Kloveniersburgwal 29 Box 19121
NL 1000 GC Amsterdam
☎: 31 20 551 07 00
✆: 31 20 620 49 41

1964 **New Zealand** 27
Royal Society of New Zealand
P.O. Box 598
Wellington
☎: 64 4 472 74 21
✆: 64 4 473 18 41

1922 **Norway** 21
Norske Videnskaps-Akademi
Drammensveien 78
N 0271 Oslo 2
☎: 47 22 44 42 96
✆: 47 22 56 26 56

1988 **Peru** 1
CONCYTEC
Parque Universitario s/n
Apdo Postal 1984
Lima 100
☎:
✆: 51 1 22 42947

1922 **Poland** 117
Polish Academy of Sciences
Palac Kultury i Nauki
Room 22-03
PL 00 901 Warsaw
☎: 48 22 65 62 89
✆: 48 22 65 22 567

1924 **Portugal** 17
SPUIAGG
Inst Portugues de Cartografia e Cadastro
Rua da Artilharia 1 No. 107
P 1070 Lisboa
☎: 351 1 381 96 00
✆: 351 1 381 96 99

1992 **Romania** 36
Romanian Academy
Astronomical Institute
Ul. Cutitul de Argint 5
RO 75212 Bucuresti
☎: 40 1 312 33 91
✆: 40 1 312 33 91/52 24

1935/1992 **Russian Federation** 341
Russian Academy of Sciences
Foreign Relations Dpt
Leninskij Prospekt 14
117071 Moscow
☎: 7 095 237 28 22
✆: 7 095 237 91 07/230 27 41

1988 **Saudi Arabia** 10
International Cooperation Department
KACS
Box 6086
Riyadh 11442
☎: 966 1 488 35 55/34 44
✆: 966 1 488 37 44

1922/1993	**Slovak Rep** Scientific Secretary Slovak Academy of Sciencies Stefanikova 49 81438 Bratislava ☎: 42 7 492 75 19 ✆: 42 7 496 849	27
1938	**South Africa** South African ICSU Secretariat Foundation for Research Development Box 2600 Pretoria 0001 ☎: 27 12 481 40 93/40 28/41 10 ✆: 27 12 481 40 07	46
1922	**Spain** Comision Nacional de Astronomia Instituto Geografico/Cadastral General Ibanez de Ibero 3 E 28003 Madrid ☎: 34 91 533 31 21 ✆: 34 91 554 67 43	205
1925	**Sweden** The Foreign Secretary Royal Swedish Academy of Sciences Box 50005 S 10405 Stockholm ☎: 46 8 673 95 00 ✆: 46 8 15 56 70	95
1923	**Switzerland** Schweizerische Akademie der Naturwissenschaft Baerenplatz 2 CH 3011 Bern ☎: 41 31 312 33 75 ✆: 41 31 312 32 91	69
1993	**Tadjikistan** Tajik Academy of Sciences Prospect Lenine 33 Dushanbe 734025 ☎: 37 72 22 68 51 ✆: 37 72 22 50 17	7
1961	**Turkey** Türkish Astronomical Society Science Faculty Department of Astronomy and Space Sciences 06100 Tandogan Ankara ☎: 90 312 212 67 20/13 12 ✆: 90 312 223 23 95	53

1961 **UK** 535
The Royal Society
6 Carlton House Terrace
London SW1Y 5AG
☎: 44 171 839 55 61
✆: 44 171 980 21 70

1935/1993 **Ukraine** 119
Science Academy of Ukraine
Ulitza Volodimirskaya 54
252601 Kiev
☎: 7 44 225 63 66/221 64 44
✆: 7 44 224 32 43

1970 **Uruguay** 3
CONICYT
Avenida 18 de Julio 1082
Casilla de Correo 1869 CP 11 100
Montevideo
☎: 598 2 91 25 25/26, 90 59 39
✆: 598 2 92 48 70

1920 **USA** 2232
National Research Council
Committee on International Organizations and Programs
2101 Constitution Avenue NW
Washington DC 20418
☎: 1 202 334 28 07
✆: 1 202 334 22 31

1932 **Vatican City State** 5
Governatorato Citta del Vaticano
V 00120 Citta del Vaticano
☎: 39 6 69 88 31 95/34 51/54 16
✆: 39 6 69 88 52 18

1953 **Venezuela** 11
CIDA
Apartado Postal 264
Merida 5101 A
☎: 58 74 79 18 993
✆: 58 74 71 24 59

COMMISSION MEMBERSHIP

Composition of Commission 4

Ephemerides/Ephémérides

President: Standish E Myles

Vice-President: Chapront Jean

Secretary: Hohenkerk Catherine

Organizing Committee:

Schwan Heiner
Abalakin Victor K
Arlot Jean-Eudes
Fukushima Toshio
He Miaofu
Kaplan George H
Kinoshita Hiroshi
Krasinsky George A
Lieske Jay H
Seidelmann P Kenneth

Members:

Aoki Shinko
Bandyopadhyay A
Bec-Borsenberger Annick
Bell Steven
Bhatnagar Ashok Kumar
Bretagnon Pierre
Brumberg Victor A
Capitaine Nicole
Catalan Manuel
Chapront-Touze Michelle
Chollet Fernand
Coma Juan Carlos
Davies Merton E
de Castro Angel
de Greiff J Arias
Deprit Andre
Di Xiaohua
Dickey Jean O'Brien
Duncombe Raynor L
Dunham David W
Eroshkin Georgiy I
Fiala Alan D
Fominov Alexandr M
Fursenko Margarita A
Glebova Nina I
Gondolatsch Friedrich
Harper David
Henrard Jacques
Hilton James Lindsay
Ilyas Mohammad
Janiczek Paul M
Johnston Kenneth J
King Robert Wilson
Klepczynski William J
Kolaczek Barbara
Kubo Yoshio
Lara Martin
Laskar Jacques
Lederle Trudpert
Lehmann Marek
Li Gi Man
Li Hyok Ho
Li Neng-Yao
Liu Bao-Lin
Majid Abdul Bin A H
Morrison Leslie V
Mueller Ivan I
Newhall X X
Nguyen Mau Tung
O'Handley Douglas A
Oesterwinter Claus
Reasenberg Robert D
Romero Perez M Pilar
Rossello Gaspar
Salazar Antonio
Shapiro Irwin I
Shiryaev Alexander A
Simon Jean-Louis
Sinzi Akira M
Sochilina Alla S
Soma Mitsuru
Thuillot William
Ting Yeou-Tswen
Tong Fu
Van Flandern Tom
Wielen Roland
Wilkins George A
Williams Carol A
Williams James G
Winkler Gernot M R
Wytrzyszczak Iwona
Xian Ding-Zhang
Yallop Bernard D
Yamazaki Akira
Zambrano Alejandro

Composition of Commission 5

Documentation and Astronomical Data/Documentation et données astronomiques

President: Dluzhnevskaya Olga B

Vice-President: Genova Francoise

Organizing Committee:

Bessell Michael S
Boyce Peter B
Dickel Helene R
Hauck Bernard
Hearn Anthony G
Jenkner Helmut
Murtagh Fionn
Nakajima Koichi
Ochsenbein Francois
Piskunov Anatoly E
Raimond Ernst
Riegler Guenter R
Wells Donald C
Wielen Roland
Zhao Yongheng

Members:

A'Hearn Michael F
Abalakin Victor K
Abt Helmut A
Aizenman Morris L
Albrecht Miguel A
Alvarez Pedro
Andernach Heinz
Baker Norman H
Benacchio Leopoldo
Benn Chris R
Bouska Jiri
Chu Yaoquan
Cogan Bruce C
Coluzzi Regina
Creze Michel
Davis Morris S
Davis Robert J
de Boer Klaas Sjoerds
Dewhirst David W
Dixon Robert S
Dubois Pascal
Ducati Jorge Ricardo
Duncombe Raynor L
Egret Daniel
Garstang Roy H
Green David
Griffin Roger F
Grosbol Preben Johnson
Guibert Jean
Guo Hongfeng
Hanisch Robert J
Harvel Christopher Alvin
Heck Andre
Hefele Herbert
Heinrich Inge
Heintz Wulff D
Helou George
Huang Bi-kun
Jaschek Carlos O R
Kadla Zdenka I
Kalberla Peter
Kaplan George H
Kharin Arkadiy S
Kleczek Josip
Kuin Nicolaas Paulus M
Lantos Pierre
Lederle Trudpert
Lequeux James
Lesteven Soizick
Li Qibin
Linde Peter
Liu Jinming
Lonsdale Carol J
Lortet Marie-Claire
Malkov Oleg Yu
Matz Steven Micheal
McLean Brian J
McNally Derek
Mcnamara Delbert H
Mead Jaylee Montague
Meadows A Jack
Mein Pierre
Mermilliod Jean-Claude
Mitton Simon
Nishimura Shiro Emer
Pamyatnikh Alexsey A
Pasian Fabio
Pasinetti Laura E
Paturel Georges
Pecker Jean-Claude
Philip A G Davis
Pizzichini Graziella
Polechova Pavla
Pucillo Mauro
Quintana Hernan
Ratnatunga Kavan U
Remy Battiau Liliane G A
Renson P F M
Roessiger Siegfried
Roman Nancy Grace
Rots Arnold H
Russo Guido
Sarasso Maria
Schilbach Elena
Schlesinger Barry M
Schlueter A
Schmadel Lutz D
Schmidt K H
Schneider Jean
Serrano Alfonso
Shakeshaft John R
Shcherbina-Samojlova I
Sokolsky Andrej G

Spite Francois M
Terashita Yoichi
Teuben Peter J
Tritton Susan Barbara
Tsvetkov Milcho K
Turner Kenneth C
Uesugi Akira
Wallace Patrick T
Warren Wayne H
Wayman Patrick A
Weidemann Volker
Wenger Marc
Westerhout Gart
White Richard Allan
Wilkins George A
Worley Charles E
Wright Alan E
Wu Zhiren

Composition of Commission 6

Astronomical Telegrams/Télégrammes astronomiques

President: West Richard M

Vice-President: Marsden Brian G

Organizing Committee:

Aksnes Kaare
Nakano Makoto
Roemer Elizabeth

Members:

Biraud Francois
Filippenko Alexei V
Gilmore Alan C
Green Daniel William E
Grindlay Jonathan E
Hers Jan
Isobe Syuzo
Liu Jinming
Mathieu Robert D
Nakano Syuichi
Phillips Mark M
Pounds Kenneth A
Rosino Leonida
Sharov A S
Tholen David J
West Richard M

Composition of Commission 7

Celestial Mechanics and Dynamical Astronomy/Mécanique céleste & astronomie dynamique

President: Froeschle Claude

Vice-President: Hadjidemetriou John D

Secretary: Lemaitre Anne

Organizing Committee:

Dvorak Rudolf
Ferraz-Mello Sylvio
Fukushima Toshio
Gerasimov Igor A
Heggie Douglas C
Knezevic Zoran
Lieske Jay H
Milani Andrea
Moons Michele B M M
Muzzio Juan C
Soffel Michael
Sun Yisui

Members:

Abad Alberto J
Abalakin Victor K
Ahmed Mostafa Kamal
Akim Efraim L
Aksnes Kaare
Alexander Murray E
Altavista Carlos A
Anosova Joanna
Antonacopoulos Greg
Aoki Shinko
Balmino Georges G
Barberis Bruno
Barbosu Mihail
Batrakov Yuri V
Beauge Christian
Bec-Borsenberger Annick
Benest Daniel
Bettis Dale G
Beutler Gerhard
Bhatnagar K B
Bois Eric
Borderies Nicole
Boss Alan P
Bozis George
Branham Richard L
Bretagnon Pierre
Brieva Eduardo
Brookes Clive J
Broucke Roger
Brumberg Victor A
Brunini Adrian
Calame Odile
Caranicolas Nicholas
Carpino Mario
Cefola Paul J
Chapront Jean
Chapront-Touze Michelle
Chen Zhen
Choi Kyu-Hong
Cid Palacios Rafael
Contopoulos George
Cook Alan H
Counselman Charles C
Cui Douxing
Danby J M Anthony
Davis Morris S
Deprit Andre
Dikova Smiliana D
Dormand John Richard
Dourneau Gerard
Drozyner Andrzej
Duncombe Raynor L
Duriez Luc
Edelman Colette
Eichhorn Heinrich K
El Bakkali Larbi
Elipe Sanchez Antonio
Emelianov Nikolaj V
Erdi B
Farinella Paolo
Fernandez Silvia M
Ferrer Martinez Sebastian
Fiala Alan D
Floria Luis
Fong Chugang
Galibina Irini V
Galletto Dionigi
Gaposchkin Edward M
Garfinkel Boris
Gaska Stanislaw
Giacaglia Giorgio E
Goldreich Peter
Gomes Rodney D S
Gonzalez Camacho Antonio
Goudas Constantine L
Grebenikov Evgenij A
Greenberg Richard
Groushinsky Nikolai P
Hamid S El Din
Hanslmeier Arnold
He Miaofu
Helali Yhya E
Henon Michel C
Henrard Jacques
Hori Genichiro
Huang Cheng
Huang Tianyi
Hut Piet
Ivanova Violeta
Izvekov Vladimir A
Janiczek Paul M
Jefferys William H
Journet Alain
Jovanovic Bozidar
Jupp Alan H
Kammeyer Peter C
Kaula William M
Kholshevnikov Konstatin V
King-Hele Desmond G
Kinoshita Hiroshi
Klokocnik Jaroslav
Kovalevsky Jean
Kozai Yoshihide
Krasinsky George A
Krivov Alexander

Kustaanheimo Paul E
Lala Petr
Laskar Jacques
Lazovic Jovan P
Liao Xinhao
Lin Douglas N C
Lissauer Jack J
Lu Benkui
Lundquist Charles A
Maciejewski Andrzej J
Magnaradze Nina G
Makhlouf Amar
Marchal Christian
Markellos Vassilis V
Marsden Brian G
Martinet Louis
Matas Vladimir R
Mavraganis A G
Meire Raphael
Melbourne William G
Merman G A
Message Philip J
Mignard Francois
Mikkola Seppo
Mioc Vasile
Morbidelli Alessandro
Mulholland J Derral
Musen Peter
Myachin Vladimir F
Nacozy Paul E
Nahon Fernand
Nobili Anna M
Noskov Boris N
Novoselov Victor S
O'Handley Douglas A
Oesterwinter Claus
Ollongren A
Omarov Tuken B
Orellana Rosa Beatriz
Orus Juan J
Osorio Isabel Maria T V P
Osorio Jose J S P
Pal Arpad
Parv Bazil
Pauwels Thierry
Peale Stanton J
Petit Jean-Marc
Petrovskaya Margarita S
Pierce A Keith
Robinson William J
Rodriguez-Eillamil R
Roy Archie E
Ryabov Yu A
Sansaturio Maria E
Sato Massae
Scholl Hans
Schubart Joachim
Sconzo Pasquale
Segan Stevo
Sehnal Ladislav
Seidelmann P Kenneth
Sein-Echaluce M Luisa
Sessin Wagner
Shapiro Irwin I
Sharaf Sh G
Sidlichovsky Milos
Sima Zdislav
Simon Jean-Louis
Sinclair Andrew T
Siry Joseph W
Skripnichenko Vladimir
Sokolov Viktor G
Sokolsky Andrej G
Sorokin Nikolai A
Souchay Jean
Standish E Myles
Stellmacher Irene
Steves Bonita Alice
Sultanov G F
Szebehely Victor G
Taborda Jose Rosa
Tatevyan S K
Tawadros Maher Jacoub
Taylor Donald Boggia
Thiry Yves R
Thuillot William
Tong Fu
Tremaine Scott Duncan
Tsuchida Masayoshi
Valsecchi Giovanni B
Valtonen Mauri J
Varvoglis H
Vashkov'Yak Sof-Ya N
Vassiliev Nikolay N
Veillet Christian
Vieira Martins Roberto
Vienne Alain
Vilhena De Moraes R
Vondrak Jan
Walch Jean-Jacques
Walker Ian Walter
Watanabe Noriaki
Whipple Arthur L
Williams Carol A
Winter Othon Cabo
Wnuk Edwin
Wu Lianda
Wytrzyszczak Iwona
Xu Jihong
Xu Pinxin
Yarov-Yarovoj M S
Yi Zhaohua
Yokoyama Tadashi
Yoshida Haruo
Yoshida Junzo
Yuasa Manabu
Zafiropoulos Basil
Zare Khalil
Zhang Zheng-Pan
Zhao Changyin
Zhdanov Valery
Zheng Jia-Qing
Zheng Xuetang
Zhou Hongnan
Zhu Wenyao

Composition of Commission 8

Positional astronomy/Astronomie de position

President: Schwan Heiner

Vice-President: Jin Wenjing

Organizing Committee:

Corbin Thomas Elbert
Kovalevsky Jean
Lopez Jose A
Morrison Leslie V
Noel Fernando
Pinigin Gennadij I
Polozhentsev Dimitrij D
Stone Ronald Cecil
Xu Jiayan

Members:

Andronova Anna A
Anguita Claudio
Argyle Robert William
Arias Elisa Felicitas
Bacchus Pierre
Backer Donald Ch
Bagildinskij Bronislav K
Bakry Abdel-aziz
Bem Jerzy
Benevides Soares P
Bernstein Hans Heinrich
Bien Reinhold
Bougeard Mireille L
Brouw W N
Bucciarelli Beatrice
Bykov Mikle F
Carrasco Guillermo
Catalan Manuel
Cha Du Jin
Chamberlain Joseph M
Chiumiento Giuseppe
Chlistovsky Franca
Chollet Fernand
Costa Edgardo
Counselman Charles C
Crifo Francoise
de Vegt Chr
Debarbat Suzanne V
Dejaiffe Rene J
Delmas Christian
Devyatkin Alexander V
Dick Steven J
Djurovic Dragutin M
Dravskikh Alexander F
Duma Dmitrij P
Duncombe Raynor L
Eichhorn Heinrich K
Einicke Ole H
Fabricius Claus V
Fan Yu
Feissel Martine
Fomin Valery A
Froeschle Michel
Fujishita Mitsumi
Gao Buxi
Gauss F Stephen
Gontcharov Gueorgui A
Grudler Pierre
Gubanov Vadim S
Gulyaev Albert P
Guseva Irina S
Heintz Wulff D
Helmer Leif
Hemenway Paul D
Hering Roland
Hoeg Erik
Hu Ning-Sheng
Hua Yingmin
Ilin Alexei E
Jackson Paul
Jiang Chongguo
Johnston Kenneth J
Jordi Carme
Journet Alain
Kaplan George H
Kharin Arkadiy S
Klock B L
Kokurin Yurij L
Kosin Gennadij S
Kurzynska Krystyna
Kuzmin Andrei V
Laclare Francis
Lattanzi Mario G
Lederle Trudpert
Lehmann Marek
Lenhardt Helmut
Li Dongming
Li Neng-Yao
Li Qi
Li Zhengxing
Li Zhifang
Li Zhigang
Lindegren Lennart
Loyola Patricio
Lu Chunlin
Luo Dingjiang
Manrique Walter T
Mao Wei
Mavridis L N
Melchior Paul J
Mitic Ljubisa A
Miyamoto Masanori
Molotaj Olexandr
Muinos Jose L
Murray C Andrew
Nakajima Koichi

Nefedeva Antonina I
Nikoloff Ivan
Ohnishi Kouji
Olsen Fogh H J
Osorio Jose J S P
Pakvor Ivan
Perryman Michael A C
Petrov G M
Pham-Van Jacqueline
Polnitzky Gerhard
Poma Angelo
Proverbio Edoardo
Pugliano Antonio
Qi Guanrong
Qian Zhihan
Rafferty Theodore J
Raimond Ernst
Reiz Anders
Requieme Yves
Reynolds John
Roeser Siegfried
Rousseau Jean-Michel
Russell Jane L
Rusu I
Sadzakov Sofija
Saletic Dusan
Sanchez Manuel
Sarasso Maria
Sato Koichi
Schildknecht Thomas
Schmeidler F
Schwekendiek Peter
Sevarlic Branislav M
Shen Kaixian
Shi Guang-Chen
Sims Kenneth P
Soderhjelm Staffan
Solaric Nikola
Soma Mitsuru
Spoelstra T A Th
Stange Lothar
Stavinschi Magdalena
Taff Laurence G
Tan Detong
Teixeira Ramachrisna
Telnyuk-Adamchuk V
Thoburn Christine
Thomas David V
Turon Catherine
Urban Sean Eugene
Vallejo Miguel
van Altena William F
van Leeuwen Floor
Volyanskaya Margarita Yu
Wallace Patrick T
Walter Hans G
Westerhout Gart
Wielen Roland
Xia Yifei
Xie Liangyun
Xu Bang-Xin
Xu Tong-Qi
Yamazaki Akira
Yan Haojian
Yasuda Haruo
Yatskiv Ya S
Ye Shuhua
Yoshizawa Masanori
Yu Kyung-Loh
Zacharias Norbert
Zhang Hui
Zhu Zi

Composition of Commission 9

Insturments/Instruments et techniques

President: McLean Ian S

Vice-President: Su Dingqiang

Organizing Committee:

Cullum Martin
Dennefeld Michel
Jacoby GeorgeH
Lelievre Gerard
Nishimura Shiro Emer
Saxena AK
Tango WilliamJ

Members:

Abdul Aziz Ahmad Zaharim
Ables Harold D
Ai GuoxiangAime Claude
Albrecht Rudolf
Alvarez Pedro
Andreic Zeljko
Aparici Juan
Arnaud Jean Paul
Arp Halton
Ashley Michael
Ashok N M
Assus Pierre
Atherton Paul David
Baba Naoshi
Babcock Horace W
Baffa Carlo
Bao Keren
Baranne Andre
Barbieri Cesare
Barcia Alberto
Barden Samuel Charles
Barroso JR Jair
Baruch John
Barwig Heinz
Baum William A
Becklin Eric E
Beer Reinhard
Bensammar Slimane
Bhattacharyya J C
Bingham Richard G
Blitzstein William
Bonanno Giovanni
Bonanno Giovanni
Bonneau Daniel
Borgnino Julien
Boyce Peter B
Brault James W
Breckinridge James B
Brejdo Izabella I
Bridgeland Michael
Burton W Butler
Cao Changxin
Carter Bradley Darren
Chang Jin
Chelli Alain
Choudary Debi Prasad
Christy James Walter
Citterio Oberto
Clarke David
Cohen Richard S
Cooke John Alan
Cornejo Alejandro A
Crawford David L
Currie Douglas G
Dall-Oglio Giorgio
Dan Xhixiang
Davis John
Denishchik Yurii
Desai Jyotindra N
Didkovsky Lionid
Diego Francisco
Dobronravin Peter
Dokuchaeva Olga D
Douglas Nigel
Dravins Dainis
Dreher John W
Duchesne Maurice
Dunkelman Lawrence
Edwin Roger P
Engels Dieter
Engvold Oddbjoern
Fabricant Daniel G
Fehrenbach Charles
Felenbok Paul
Fletcher J Murray
Fomenko Alexandr F
Ford jr W Kent
Fort Bernard P
Foy Renaud
Fu Delian
Galan Maximino J
Gao Bilie
Gauss F Stephen
Gay Jean
Gibson David Michael
Gillanders Gerard
Gillingham Peter
Glass Ian Stewart
Gong Shou-shen
Gray Peter Murray
Griffiths Richard E
Grigorjev Victor M
Grosbol Preben Johnson
Guibert Jean
Gunji Shuichi
Gutcke Dietrich

Hadley Brian W
Hallam Kenneth L
Hammerschlag Robert H
Hanisch Robert J
Hao Yunxiang
Harmer Charles F W
Harmer Dianne L
Heckathorn Harry M
Henden Arne Anthon
Heudier Jean-Louis
Hewitt Anthony V
Hilliard Ron
Hodapp Klaus-Werner
Honeycutt R Kent
Hough James
Hu Jingyao
Hu Ning-Sheng
Huang Tieqin
Humphries Colin M
Hutter Donald John
Hysom Edmund J
Ilyas Mohammad
Irbah Abdanour
Ivchenko Vasily
Iyengar Srinivasan Rama
Jayarajan A P
Jeffers Stanley
Jenkner Helmut
Jiang Shengtao
Jiang Shi-Yang
Jones Barbara
Jordan Carole
Joseph Charles Lynn
Karachentsev Igor D
Karpinskij Vadim N
Kentischer Thomas
Kipper Tõnu
Kissell Kenneth E
Klock B L
Klocok Lubomir
Koechlin Laurent
Koehler H
Koehler Peter
Kopylov Ivan M
Korovyakovskij Yurij P
Kovachev B J
Kreidl Tobias J N
Kuehne Christoph F
Kulkarni Prabhakar V
Labeyrie Antoine
Laques Pierre
Lasker Barry M
Lemaitre Gerard R
Li Depei
Li Ting
Li Zhigang
Livingston William C
Lochman Jan
Lu Ruwei
Lynch David K
Mack Peter
Madau Piero
Mahra H S
Maillard Jean-Pierre
Major John
Malin David F
Malkamaeki Lauri J
Mariotti Jean Marie
Martins Donald Henry
Matz Steven Micheal
McGregor Peter John
McMullan Dennis
Megevand Denis
Meng Xinmin
Merkle Fritz
Mertz Lawrence N
Mikhelson Nikolaj N
Millikan Allan G
Minarovjech Milan
Mitchell Peter
Morgan Brian Lealan
Morris Michael C
Morton Donald C
Murray Stephen S
Nakai Yoshihiro
Nakajima Tadashi
Nelson Jerry E
Newton Gavin
Niemi Aimo
Nishikawa Jun
Noumaru Junichi
Nunes Rogerio S de sousa
O'Byrne John
O'Dell Charles R
O'Sullivan Creidhe
Odgers Graham J
Ohtsubo Junji
Parker Neil
Pasian Fabio
Pati A K
Penny Alan John
Perryman Michael A C
Petford A David
Petrov Peter P
Picat Jean-Pierre
Povel Hanspeter
Pritchet Christopher J
Prokof'eva Valentina V
Pucillo Mauro
Qiu Puzhang
Queloz Didier
Racine Rene
Rakos Karl D
Ramsey Lawrence W
Reay Newrick K
Redfern Michael R
Richardson E Harvey
Ring James
Robertson Norna
Robinson Lloyd B
Roddier Claude
Roddier Francois
Rosch Jean
Rountree Janet
Ruder Hanns
Rupprecht Gero
Rusconi Luigia
Saha Swapan Kumar
Sault Robert
Schneider Nicholas M
Schroeder Daniel J
Schultz Alfred Bernard
Schultz G V
Schumann Joerg Dieter
Sedmak Giorgio
Seifert Walter
Servan Bernard
Shakhbazyan Yurij L
Shcheglov P V
Shen Changjun
Shen Parnan
Shi Shengcai
Shivanandan Kandiah
Shortridge Keith
Sim Mary E
Slovak Mark Haines
Smyth Michael J
Snezhko Leonid I
Soltau Dirk
Steshenko N V
Storey John W V
Swings Jean-Pierre
Tapde Suresh Chandra
Title Alan Morton
Traub Wesley Arthur
Tsvetkov Milcho K
Tueg Helmut
Tull Robert G
Ueno Munetaka
Ulich Bobby Lee
Vakili Farrokh
Valnicek Boris
Valtonen Mauri J
van Citters Gordon W
Velkov Kiril

Vladimirov Simeon
Vrba Frederick J
Walker Alistair Robin
Walker David Douglas
Walker Gordon A H
Walker Merle F
Wallace Patrick T
Wampler E Joseph
Wang Chuanjin
Wang Lan-Juan
Wang Ya'nan
Wang Yiming
Wang Yong
Wang Zhengming
Ward Henry
Watson Frederick Garnett
Weber Joseph
Weiss Werner W
West Richard M
Westphal James A
Windhorst Rogier A
Wlerick Gerard
Woehl Hubertus
Worden Simon P
Worswick Susan
Wu Linxiang
Wyller Arne A
Yang Shijie
Yao Zhengqiu
Ye Binxun
Yin Xinhui
Zacharov Igor
Zealey William J
Zhang Xiuzhong
Zhang Youyi
Zhao Zhaowang
Zhou Bifang
Zhu Nenghong

Composition of Commission 10

Solar Activity/Activité solaire

President: Ai Guoxiang

Vice-President: Benz Arnold O

Secretary: Dere Kenneth Paul

Organizing Committee:

Engvold Oddbjoern
Gopalswamy N
Hammer Reiner
Hood Alan
Jackson Bernard V
Kim Iraida S
Martens Petrus C
Poletto Giannina
Rozelot Jean-Pierre
Sanchez Almeida Jorge
Shibata Kazunari
van Driel Gesztzlyi L

Members:

Abbasov Alik R
Abdelatif Toufik
Aboudarham Jean
Abramenko Valentina
Abrami Alberto
Ahluwalia Harjit Singh
Alissandrakis C
Almleaky Yasseen
Altrock Richard C
Altschuler Martin D
Aly Jean Jacques
Amari Tahar
Ambastha A K
Ambroz Pavel
Anderson Kinsey A
Antalova Anna
Antiochos Spiro Kosta
Antonucci Ester
Anzer Ulrich
Aschwanden Markus
Atac Tamer
Athay R Grant
Aurass Henry
Avignon Yvette
Babin Arthur
Babin Valerij G
Bagare S P
Ballester Jose Luis
Balli Edibe
Baranovsky Edward A
Barrow Colin H
Batchelor David Allen
Beckers Jacques M
Bedding Timothy
Beebe Herbert A
Bell Barbara
Belvedere Gaetano
Berger Mitchell
Bergeron Jacqueline A
Berrilli Francesco
Bhatnagar Arvind
Bocchia Romeo
Bohn Horst-Ulrich
Bommier Veronique
Bondal Krishna Raj
Bornmann Patricia L
Bougeret Jean Louis
Boyer Rene
Brajsa Roman
Brandenburg Axel
Brandt Peter N
Braun Douglas Clifford
Bray Robert J
Brekke Pal Ording lie
Brosius Jeffrey William
Brown John C
Browning Philippa
Brueckner Guenter E
Bruner Marilyn E
Bruns Andrey V
Bruzek Anton
Buecher Alain
Buechner Joerg
Bumba Vaclav
Buyukliev Georgi
Cadez Vladimir
Cally Paul S
Cane Hilary Vivien
Cargill Peter J
Carlqvist Per A
Cauzzi Gianna
Chambe Gilbert
Chandra Suresh
Chaoudhuri Arnab
Chapman Gary A
Charbonneau Paul
Chen Zhencheng
Cheng Chung-chieh
Chernov Gennadij P
Chertok Ilia M
Chertoprud Vladim E
Chiuderi-Drago Franca Pr
Chiueh Tzihong
Choudary Debi Prasad
Choudhuri Arnab Rai
Chupp Edward L
Cliver Edward W
Coffey Helen E
Collados Manuel
Cook John W

Correia Emilia
Costa Joaquim E R
Coutrez Raymond A J
Covington Arthur E
Craig Ian Jonathan D
Cramer Neil
Crannell Carol Jo
Csada Imre K
Culhane Leonard
Datlowe Dayton
Davila Joseph
de Groot T
de Jager Cornelis
Del Toro Iniesta Jose
Demoulin Pascal
Dennis Brian Roy
Dermendjiev Vladimir
Deubner Franz-Ludwig
Dezso Lorant
Dialetis Dimitris
Ding Mingde
Ding Youji
Dinulescu Simona
Dizer Muammer
Dollfus Audouin
Dryer Murray
Dubau Jacques
Dubois Marc A
Duldig Marcus Leslie
Dulk George A
Dumitrache Christana
Duncan Robert A
Dunn Richard B
Dwivedi Bhola Nath
Dzubenko Nikolai
Eddy John A
Elste Gunther H
Elwert Gerhard
Emslie A Gordon
Enome Shinzo
Falchi Ambretta
Falciani Roberto
Fang Cheng
Farnik Frantisek
Feibelman Walter A
Feng Kejia
Ferriz-Mas Antonio
Fisher George Hewitt
Foing Bernard H
Fontenla Juan Manuel
Forbes Terry G
Fortini Teresa
Fossat Eric
Friedman Herbert
Fu Qijun
Gabriel Alan H
Gaizauskas Victor
Galal A A
Galloway David
Galsgaard Klaus
Gan Weiqun
Garcia Howard A
Garcia Jose I de la rosa
Gary Gilmer Allen
Gelfreikh Georgij B
Gergely Tomas Esteban
Ghizaru Mihai
Gibson David Michael
Gilliland Ronald Lynn
Gilman Peter A
Glatzmaier Gary A
Godoli Giovanni
Goedbloed Johan P
Gokhale Moreshwar Hari
Goossens Marcel
Gopasyuk S I
Grandpierre Attila
Gray Norman
Gu Xiaoma
Guhathakurta Madhulika
Gurman Joseph B
Gyori Lajos
Hagen John P
Hagyard Mona June
Hanaoka Yoichiro
Hanasz Jan
Hansen Richard T
Hanslmeier Arnold
Harrison Richard A
Harvey John W
Hasan Saiyid Strajul
Hathaway David H
Hayward John
Heinzel Petr
Henoux Jean-Claude
Hermans Dirk
Hiei Eijiro
Hildebrandt Joachim
Hildner Ernest
Hirayama Tadashi
Hoeksema Jon Todd
Hohenkerk Catherine
Hollweg Joseph V
Holman Gordon D
Holzer Thomas Edward
Hong Hyon Ik
Hood Alan
Houdebine Eric
Howard Robert F
Hoyng Peter
Huang Youran
Hudson Hugh S
Hurford Gordon James
Hyder C L
Ioshpa Boris A
Ireland Jack
Isliker Heinz
Ivanchuk Victor I
Ivanov Evgeny I
Ivchenko Vasily
Jain Rajmal
Jakimiec Jerzy
Jardine Moira Mary
Jiang Yaotiao
Jimenez Mancebo A j
Jockers Klaus
Jones Harrison Price
Jordan Stuart D
Joselyn Jo Ann c
Jovanovic Bozidar
Kaburaki Osamu
Kahler Stephen W
Kalman Bela
Kane Sharad R
Kang Jin Sok
Karlicky Marian
Karpen Judith T
Kaufmann Pierre
Keppens Rony
Kim Kap-sung
Kiplinger Alan L
Kitai Reizaburo
Kjeldseth-Moe Olav
Kleczek Josip
Klein Karl Ludwig
Kliem Bernhard
Klimchuk James A
Klvana Miroslav
Knoska Stefan
Kopecky Miloslav
Kostik Roman I
Kotrc Pavel
Koutchmy Serge
Kovacs Agnes
Krivsky Ladislav
Krueger Albrecht
Kryvodubskyj Valery
Kubota Jun
Kucera Ales
Kuenzel Horst
Kuklin Georgly V
Kundu Mukul R
Kuperus Max
Kurochka L N
Kurokawa Hiroki

Landman Donald Alan
Lang Kenneth R
Lantos Pierre
Lawrence John Keeler
Leibacher John
Leroy Bernard
Leroy Jean-Louis
Li Chun-Sheng
Li Hui
Li Kejun
Li Son Jae
Li Wei
Li Weibao
Lin Yuanzhang
Liritzis Ioannis
Liu Xinping
Livshits Mikhail A
Longbottom Aaron
Low Boon Chye
Lozitskij Vsevolod
Lundstedt Henrik
Luo Baorong
Luo Xianhan
Lustig Guenter
Machado Marcos
MacKinnon Alexander L
MacQueen Robert M
Makarov Valentine I
Makita Mitsugu
Malherbe Jean Marie
Malitson Harriet H
Maltby Per
Malville J Mckim
Manabe Seiji
Mandrini Cristina Hemilse
Mann Gottfried
Maris Georgeta
Mariska John Thomas
Martres Marie-Josephe
Mason Glenn M
Masuda Satoshi
Matsuura Oscar T
Mattig W
Maxwell Alan
McCabe Marie K
McIntosh Patrick S
McKenna Lawlor Susan
McLean Donald J
Meerson Baruch
Mein Pierre
Melrose Donald B
Mendes Da Costa Aracy
Messerotti Mauro
Michalitsianos Andrew
Mogilevskij Eh I
Moiseev I G
Moreno-Insertis Fernando
Moreton G E
Moriyama Fumio
Morozhenko N N
Motta Santo
Muller Richard
Musatenko Sergij
Musielak Zdzislaw E
Nakagawa Yoshinari
Nakajima Hiroshi
Nakariakov Valery
Namba Osamu
Narain Udit
Neidig Donald F
Nelson Graham John
Neukirch Thomas
Neupert Werner M
Nishi Keizo
Nocera Luigi
Noens Jacques-Clair
Noyes Robert W
Nussbaumer Harry
Obridko Vladimir N
Oekten Adnan
Ofman Leon
Ohki Kenichiro
Ozguc Atila
Padmanabhan Janardhan
Paletou Frederic
Pallavicini Roberto
Palle Pere-Lluis
Palus Pavel
Pan Liande
Pap Judit
Parfinenko Leonid D
Park Young-Deuk
Parkinson John H
Parkinson William H
Pasachoff Jay M
Paterno Lucio
Pedersen Bent M
Peres Giovanni
Petrosian Vahe
Petrovay Kristof
Pevtsov Alexei A
Pflug Klaus
Phillips Kenneth J H
Pick Monique
Piddington Jack H Res Fel
Pneuman Gerald W
Poedts Stefaan
Poland Arthur I
Poquerusse Michel
Porter Jason G
Preka-Papadema P
Priest Eric R
Prokakis Theodore J
Raadu Michael A
Rabin Douglas Mark
Rao A Pramesh
Raoult Antoinette
Rayrole Jean R
Rees David Elwyn
Reeves Edmond M
Reeves Hubert
Regulo Clara
Rieger Erich
Rijnbeek Richard
Robinson Richard D
Roca Cortes Teodoro
Roemer Max
Romanchuk Pavel R
Rompolt Bogdan
Rosa Dragan
Rosch Jean
Roudier Thierry
Rovira Marta Graciela
Roxburgh Ian W
Ruediger Guenther
Rusin Vojtech
Rust David M
Ruzdjak Vladimir
Ruzickova-Topolova B
Rybansky Milan
Ryutova Margarita P
Saemundson Thorsteinn
Sahal-Brechot Sylvie
Saito Kuniji
Sakao Taro
Sakurai Kunitomo
Sakurai Takashi
Saniga Metod
Sawyer Constance B
Schatten Kenneth H
Schindler Karl
Schlueter A
Schmahl Edward J
Schmelz Joan T
Schmidt H U
Schmieder Brigitte
Schober Hans J
Schrijver C J
Schroeter Egon H
Schuessler Manfred
Semel Meir
Shapley Alan H
Shea Margaret A
Sheeley Neil R
Shi Zhongxian

Shibasaki Kiyoto
Shine Richard A
Silberberg Rein
Simnett George M
Simon Guy
Sinha K
Sitnik G F
Slonim E M
Smaldone Luigi Antonio
Smith Dean F
Smol-Kov Gennadij Ya
Sobotka Michal
Solanki Sami K
Soliman Mohamed Ahmed
Somov Boris V
Sotirovski Pascal
Spadaro Daniele
Spicer Daniel Shields
Spruit Henk C
Steiner Oskar
Stellmacher Goetz
Stenflo Jan O
Stepanian N N
Stepanov Alexander V
Steshenko N V
Stewart Ronald T
Stix Michael
Stoker Pieter H
Strong Keith T
Sturrock Peter A
Subramanian K R
Sukartadiredja Darsa
Sun Kai
Svestka Zdenek
Sykora Julius
Sylwester Barbara
Sylwester Janusz
Szalay Alex
Takakura Tatsuo Emer
Takano Toshiaki
Talon Raoul
Tamenaga Tatsuo
Tandberg-Hanssen Einar A
Tandon Jagdish Narain
Tang Yuhua
Tapping Kenneth F
Ternullo Maurizio
Teske Richard G
Thomas John H
Thomas Roger J
Tifrea Emilia
Tlamicha Antonin
Tomczak Michal
Trellis Michel
Treumann Rudolf A
Tritakis Basil P
Trottet Gerard
Tsinganos Kanaris
Tsubaki Tokio
Tsubota Yukimasa
Tsuneta Saku
Tuominen Ilkka V
Underwood James H
Urpo Seppo I
Valnicek Boris
van Allen James A
van den Oord Bert H J
van der Linden Ronald
van Hoven Gerard
van't Veer Frans
Vaughan Arthur H
Veck Nicholas
Vekstein Gregory
Velkov Kiril
Velli Marco
Velli Marco
Venkatesan Doraswamy
Ventura Rita
Vergez Madeleine
Verheest Frank
Verma V K
Vial Jean-Claude
Vilmer Nicole
Vinluan Renato
Vinod S Krishan
Vitinskij Yurij I
Vrsnak Bojan
Vyalshin Gennadij F
Waldmeier Max
Walsh Robert
Wang Haimin
Wang Jia-Long
Wang Jingxiu
Wang Yi-ming
Webb David F
Wentzel Donat G
White Stephen Mark
Wiehr Eberhard
Wild John Paul
Wilson Peter R
Wittmann Axel D
Woehl Hubertus
Wolfson Richard
Woltjer Lodewijk
Wu Hong'ao
Wu Linxiang
Wu Mingchan
Wu Shi Tsan
Xu Aoao
Xu Zhentao
Yao Jinxing
Ye Shi-hui
Yeh Tyan
Yoshimura Hirokazu
You Jianqi
Yun Hong-Sik
Zachariadis Theodosios
Zappala Rosario Aldo
Zelenka Antoine
Zhang Bairong
Zhang Heqi
Zhang Zhenda
Zhao Renyang
Zharkova Valentina
Zhelyazkov Ivan
Zhong Hongqi
Zhou Aihua
Zhou Daoqi
Zhugzhda Yuzef D
Zirin Harold
Zlobec Paolo
Zou Yixin
Zwaan Cornelis

Composition of Commission 12

Solar Radiation and Structure/Radiation et structure solaires

President: Foukal Peter V

Vice-President: Solanki Sami K

Secretary: Mariska John Thomas

Organizing Committee:

Baliunas Sallie L
Deubner Franz-Ludwig
Dravins Dainis
Duvall Thomas L
Fang Cheng
Gaizauskas Victor
Heinzel Petr
Kononovich Edward V
Koutchmy Serge
Melrose Donald B
Stix Michael
Suematsu Yoshinori

Members:

Aboudarham Jean
Acton Loren W
Adam Madge G
Ai Guoxiang
Aime Claude
Alissandrakis C
Altrock Richard C
Altschuler Martin D
Ansari S M Razaullah
Antia H M
Arnaud Jean Paul
Artzner Guy
Athay R Grant
Ayres Thomas R
Balthasar Horst
Beard David B
Beckers Jacques M
Beckman John E
Beebe Herbert A
Bendlin Cornelia
Benford Gregory
Bhatnagar Arvind
Bhattacharyya J C
Billings Donald E
Blackwell Donald E
Blamont Jacques Emile
Bocchia Romeo
Boehm Karl-Heinz
Boehm-Vitense Erika
Bohn Horst-Ulrich
Bommier Veronique
Bonnet Roger M
Book David L
Bornmann Patricia L
Bougeret Jean Louis
Brandt Peter N
Brault James W
Bray Robert J
Breckinridge James B
Brosius Jeffrey William
Brueckner Guenter E
Bruner Marilyn E
Bruning David H
Bruzek Anton
Bumba Vaclav
Cadez Vladimir
Cavallini Fabio
Ceppatelli Guido
Chambe Gilbert
Chan Kwing Lam
Chapman Gary A
Chen Biao
Cheng Chung-chieh
Chertok Ilia M
Chistyakov Vladimir E
Christensen-Dalsgaard J
Chvojkova Woyk E
Clark Thomas Alan
Clette Frederic
Collados Manuel
Cook John W
Cox Arthur N
Craig Ian Jonathan D
Cram Lawrence Edward
Cramer Neil
Dara Helen
de Jager Cornelis
Degenhardt Detlev
Del Toro Iniesta Jose
Delbouille Luc
Deliyannis Jean
Demarque Pierre
Deming Leo Drake
Dezso Lorant
Ding Mingde
Diver Declan Andrew
Dogan Nadir
Dumont Simone
Dunkelman Lawrence
Dunn Richard B
Dzubenko Nikolai
Ehgamberdiev Shurat
Einaudi Giorgio
Elliott Ian
Elste Gunther H
Epstein Gabriel Leo
Esser Ruth

Evans J V
Falciani Roberto
Feldman Uri
Fiala Alan D
Fisher George Hewitt
Fofi Massimo
Fomichev Valeri V
Fontenla Juan Manuel
Fossat Eric
Frazier Edward N
Friedman Herbert
Froehlich Claus
Gabriel Alan H
Garcia Howard A
Garcia-Berro Enrique
Gaur V P
Glatzmaier Gary A
Godoli Giovanni
Gokdogan Nuzhet
Goldman Martin V
Gomez Maria Theresa
Gopalswamy N
Gopasyuk S I
Gordon Charlotte
Grevesse Nicolas
Grigorieva Virginia P
Gu Xiaoma
Guhathakurta Madhulika
Gulyaev Albert P
Hagyard Mona June
Hammer Reiner
Harvey John W
Hein Righini Giovanna
Hejna Ladislav
Hiei Eijiro
Hildner Ernest
Hill Frank
Hirayama Tadashi
Hiroyasu Ando
Hoang Binh Dy
Holweger Hartmut
Horton Brian H
Hotinli Metin
House Lewis L
Howard Robert F
Hoyng Peter
Huang Guangli
Illing Rainer M E
Ivanov Evgeny I
Jabbar Sabeh Rhaman
Jackson Bernard V
Jefferies Stuart
Jones Harrison Price
Jordan Carole
Jordan Stuart D
Kalkofen Wolfgang
Kalman Bela
Karlicky Marian
Karpen Judith T
Karpinskij Vadim N
Kato Shoji
Kaufmann Pierre
Kawaguchi Ichiro
Keil Stephen L
Khetsuriani Tsiala S
Kim Iraida S
Klein Karl Ludwig
Kneer Franz
Knoelker Michael
Kopecky Miloslav
Kostik Roman I
Kotov Valery
Kotrc Pavel
Koyama Shin
Kraemer Gerhard
Krueger Albrecht
Kryvodubskyj Valery
Kubicela Aleksandar
Kucera Ales
Kuklin Georgly V
Kulcar Ladislav
Kundu Mukul R
Kuperus Max
Kurochka L N
La Bonte Barry James
Labs Dietrich
Landi Degl-Innocenti E
Landman Donald Alan
Landolfi Marco
Lantos Pierre
Leibacher John
Leroy Jean-Louis
Li Linghuai
Lin Yuanzhang
Linsky Jeffrey L
Livingston William C
Locke Jack L
Lopez-Arroyo M
Luest Reimar
Lustig Guenter
Makarov Valentine I
Makarova Elena A
Makita Mitsugu
Mandrini Cristina Hemilse
Marik Miklos
Marilli Ettore
Marmolino Ciro
Martinez Pillet Valentin
Mattig W
McKenna Lawlor Susan
Mein Pierre
Mewe R
Meyer Friedrich
Michard Raymond
Mihalas Dimitri
Milkey Robert W
Monteiro Mario J P F G
Moore Ronald L
Moreno-Insertis Fernando
Moriyama Fumio
Mouradian Zadig M
Muller Richard
Munro Richard H
Namba Osamu
Neckel Heinz
Nesis Anastasios
New Roger
Nicolas Kenneth Robert
Nishi Keizo
Nordlund Ake
Noyes Robert W
Oster Ludwig F
Owocki Stanley Peter
Padmanabhan Janardhan
Palle Pere-Lluis
Palus Pavel
Pande Mahesh Chandra
Papathanasoglou D
Parkinson William H
Pasachoff Jay M
Pecker Jean-Claude
Petrovay Kristof
Peyturaux Roger H
Pflug Klaus
Phillips Kenneth J H
Pierce A Keith
Poquerusse Michel
Povel Hanspeter
Priest Eric R
Prokakis Theodore J
Rabin Douglas Mark
Radick Richard R
Raoult Antoinette
Rees David Elwyn
Reeves Edmond M
Regulo Clara
Roberti Giuseppe
Roca Cortes Teodoro
Roddier Francois
Rodriguez Hildago Ines l
Roland Ginette
Roudier Thierry
Rovira Marta Graciela
Rusin Vojtech
Rutten Robert J

Rybansky Milan
Sakai Junichi
Sakurai Takashi
Samain Denys
Sanchez Almeida Jorge
Saniga Metod
Sauval A Jacques
Schleicher Helmold
Schmahl Edward J
Schmidt Wolfgang
Schmieder Brigitte
Schmitt Dieter
Schober Hans J
Schuessler Manfred
Schwartz Steven Jay
Seaton Michael J
Semel Meir
Severino Giuseppe
Shallis Michael J
Shchukina Nataliya
Sheeley Neil R
Shen Longxiang
Shine Richard A
Simon George W
Simon Guy
Singh Jagdev
Sinha K
Sitnik G F
Sivaraman K R
Skumanich Andre
Smith Peter L
Song Mutao
Sotirovski Pascal
Souffrin Pierre B
Spicer Daniel Shields
Stathopoulou Maria
Staude Juergen
Stebbins Robin
Steffen Matthias
Steiner Oskar
Stenflo Jan O
Svestka Zdenek
Swensson John W
Tandberg-Hanssen Einar A
Teplitskaya R B
Thomas John H
Torelli M
Tripathi B M
Trujillo Bueno Javier
Tsap T T
Tsiropoula Georgia
Tsubaki Tokio
Uchida Yutaka
Unno Wasaburo
Uus Undo
van Hoven Gerard
Vasileva Galina J
Vaughan Arthur H
Venkatakrishnan P
Vial Jean-Claude
Vilmer Nicole
Vitinskij Yurij I
Volonte Sergio
von der Luehe Oskar
Vukicevic K M
Waldmeier Max
Wang Jingxiu
Wang Zhenyi
Warwick James W
Weiss Nigel O
Wentzel Donat G
Wilson Peter R
Wittmann Axel D
Woehl Hubertus
Worden Simon P
Wu Hsin-Heng
Wu Linxiang
Wyller Arne A
Yoshimura Hirokazu
You Jianqi
Youssef Nahed H
Yun Hong-Sik
Zampieri Luca
Zarro Dominic M
Zelenka Antoine
Zhou Daoqi
Zhugzhda Yuzef D
Zirin Harold
Zirker Jack B
Zwaan Cornelis

Composition of Commission 14

Atomic and Molecular Data/Données atomiques et moléculaires

President: Rostas Francois

Vice-President: Smith Peter L

Secretary: Feautrier Nicole

Organizing Committee:

Berrington Keith Adrian
Grevesse Nicolas
Johansson Sveneric
Jorgensen Uffe Grae
Martin William C
Mason Helen E
Parkinson William H
Tchang-Brillet Lydia

Members:

Adelman Saul J
Aggarwal Kanti Mal
Allard Nicole
Arduini-Malinovsky Monique
Artru Marie-Christine
Barnbaum Cecilia
Barrow Richard F
Bartaya R A
Bely-Dubau Francoise
Biemont Emile
Black John Harry
Boechat-Roberty Heloisa M
Bommier Veronique
Branscomb L M
Brault James W
Bromage Gordon E
Burgess Alan
Carbon Duane F
Carroll P Kevin
Carver John H
Chance Kelly V
Cook Alan H
Corliss C H
Cornille Marguerite
Czyzak Stanley J
d'Hendecourt Louis
Dalgarno Alexander
Davis Sumner P
de Frees Douglas J
Delsemme Armand H
Desesquelles Jean
Diercksen Geerd H F
Dimitrijevic Milan
Dressler Kurt
Dubau Jacques
Dufay Maurice
Edlen Bengt
Eidelsberg Michele
Epstein Gabriel Leo
Faucher Paul
Federici Luciana
Federman Steven Robert
Fink Uwe
Flower David R
Fuhr Jeffrey Robert
Gabriel Alan H
Gallagher Jean W
Gargaud Muriel
Garstang Roy H
Garton W R S
Glagolevskij Juri V
Goldbach Claudine Mme
Grant Ian P
Green Louis C
Heddle Douglas W O
Herold Heinz
Herzberg Gerhard
Hesser James E
Hoang Binh Dy
House Lewis L
Huber Martin C E
Huebner Walter F
Iliev Ilian
Irwin Alan W
Jamar Claude A J
Johnson Donald R
Johnson Fred M
Joly Francois
Jordan Carole
Jordan H L direktor
Jorgensen Henning E
Kato Takako
Kennedy Eugene T
Kielkopf John F
Kim Zong Dok
Kingston Arthur E
Kipper Tonu
Kirby Kate P
Kohl John L
Kroto Harold
Kurucz Robert L
Lagerqvist Albin
Lambert David L
Landman Donald Alan
Lang James
Langhoff Stephen Robert
Launay Francoise
Launay Jean-Michel
Lawrence G M
Layzer David
Le Bourlot Jacques
Le Dourneuf Maryvonne
Le Floch Andre

Leach Sydney
Leger Alain
Lemaire Jean-louis
Lesage Alain
Loulergue Michelle
Lovas Francis John
Lutz Barry L
Maillard Jean-Pierre
McWhirter R W Peter
Mewe R
Mickelson Michael E
Mumma Michael Jon
Newsom Gerald H
Nicholls Ralph W
Nollez Gerard
Nussbaumer Harry
Obi Shinya
Oetken L
Oka Takeshi
Omont Alain
Orton Glenn S
Ozeki Hiroyuki
Peach Gillian
Pei Chunchuan
Petrini Daniel
Petropoulos Basil Ch
Pettini Marco
Pfennig Hans H
Phillips John G
Piacentini Ruben
Pradhan Anil
Querci Francois R
Rao K Narahari
Richter Johannes
Rogers Forrest J
Roncin Jean-Yves
Ross John E R
Roueff Evelyne M A
Ruder Hanns
Rudzikas Zenonas B
Sahal-Brechot Sylvie
Savanov Igor S
Schrijver Johannes
Seaton Michael J
Sharp Christopher
Shore Bruce W
Sinha K
Smith Geoffrey
Smith Wm Hayden
Somerville William B
Sorensen Gunnar
Spielfiedel Annie
Stark Glen
Steenman-Clark Lois
Stehle Chantal
Strachan Leonard
Strelnitski Vladimir
Summers Hugh P
Swings Jean-Pierre
Takayanagi Kazuo
Tatum Jeremy B
Tozzi Gian Paolo
Tran Minh Nguyet
Trefftz Eleonore E
van Dishoeck Ewine F
van Regemorter Henri
van Rensbergen Walter
Varshalovich Dimitrij
Voelk Heinrich J
Volonte Sergio
Vujnovic Vladis
Weniger Schame
Wiese Wolfgang L
Wilson Robert
Winnewisser Gisbert
Wunner Guenter
Yoshino Kouichi
Young Louise Gray
Yu Yan
Zeippen Claude
Zeng Qin
Zirin Harold

Composition of Commission 15

Physical Study of Comets, Minor Planets and meteorites/
Etude physique des comètes des petites planètes & des météorites

President: Zappala Vincenzo

Vice-President: Keller Horst Uwe

Secretary: Muinonen Karri

Organizing Committee:

A'Hearn Michael F
Bailey Mark Edward
Binzel Richard P
Capria Maria Teresa
Feldman Paul Donald
Fernandez Julio A
Huebner Walter F
Lagerkvist Claes-Ingvar
Levasseur-Regourd A.-C.
Meech Karen
Tedesco Edward F
Watanabe Jun-ichi
West Richard M

Members:

Allegre Claude
Andrienko Dmitry A
Arnold James R
Arpigny Claude
Axford W Ian
Babadzhanov Pulat B
Barker Edwin S
Barriot Jean-Pierre
Barucci Maria A
Beard David B
Bell Jeffrey F
Belton Michael J S
Birch Peter
Blamont Jacques Emile
Bockelee-Morvan Dominique
Boehnhardt Hermann
Boice Daniel Craig
Bonev Tanyu
Bouska Jiri
Bowell Edward L G
Brandt John C
Brecher Aviva
Brown Robert Hamilton
Brownlee Donald E
Brunk William E
Buie Marc W
Buratti Bonnie J
Burlaga Leonard F
Burns Joseph A
Campins Humberto
Capaccioni Fabrizio
Carruthers George R
Carsenty Uri
Carusi Andrea
Cellino Alberto
Ceplecha Zdenek
Cerroni Priscilla
Chandrasekhar T
Chapman Clark R
Chapman Robert D
Chen Daohan
Cherednichenko V I
Chernykh N S
Clairemidi Jacques
Clayton Geoffrey C
Clayton Robert N
Clube S V M
Cochran Anita L
Cochran William David
Combi Michael R
Consolmagno Guy Joseph
Cosmovici Batalli C
Cristescu Cornelia G
Crovisier Jacques
Cruikshank Dale P
Cuypers Jan
Danks Anthony C
de Almeida Amaury A
de Pater Imke
de Sanctis Giovanni
Debehogne Henri Sc
Delsemme Armand H
Dermott Stanley F
Deutschman William A
Di Martino Mario
Donn Bertram D
Dossin F
Dryer Murray
Dzhapiashvili Victor P
Encrenaz Therese
Ershkovich Alexander
Eviatar Aharon
Farinella Paolo
Ferrin Ignacio
Festou Michel C
Fitzsimmons Alan
Forti Giuseppe
Foryta Dietmar William
Froeschle Christiane D
Fujiwara Akira
Fulchignoni Marcello
Gammelgaard Peter Mog
Gehrels Tom
Geiss Johannes
Gerard Eric
Gibson James
Giovane Frank
Gradie Jonathan Carey
Green Simon F
Greenberg J Mayo
Greenberg Richard
Grossman Lawrence
Grudzinska Stefania
Gruen Eberhard
Gustafson Bo A S
Hajduk Anton
Halliday Ian
Hanner Martha S
Hapke Bruce W
Harris Alan William
Hartmann William K

Harwit Martin
Hasegawa Ichiro
Haser Leo N K
Haupt Hermann F
Helin Eleanor Francis
Herzberg Gerhard
Hu Zhongwei
Hughes David W
Huntress Wesley T
Ibadinov Khursandkul
Ibadov Subhon
Ip Wing-huen
Irvine William M
Isobe Syuzo
Ivanova Violeta
Jackson William M
Jockers Klaus
Johnson Torrence V
Karttunen Hannu
Keay Colin S l
Keil Klaus
Kiselev Nikolai N
Kliem Bernhard
Knacke Roger F
Knezevic Zoran
Koeberl Christian
Kohoutek Lubos
Konopleva Varvara P
Kowal Charles Thomas
Kozasa Takashi
Kresakova Margita
Krishna Swamy K S
Kristensen Leif Kahl
Lamy Philippe
Lancaster Brown Peter
Lane Arthur Lonne
Larson Harold P
Larson Stephen M
Lebofsky Larry Allen
Lee Thyphoon
Liller William
Lillie Charles F
Lindsey Charles Allan
Lipschutz Michael E
Lissauer Jack J
Liu Linzhong
Liu Zongli
Lopes-Gautier Rosaly
Lumme Kari A
Lupishko Dmitrij F
Lutz Barry L
Ma Yuehua
Magnusson Per
Malaise Daniel J
Maran Stephen P
Marcialis Robert
Marsden Brian G
Marzari Francesco
Matson Dennis L
Matsuura Oscar T
McCord Thomas B
McCrosky Richard E
McDonnell J A M
McFadden Lucy Ann
McKenna Lawlor Susan
Meisel David D
Mendis Devamitta Asoka
Michalowski Tadeusz
Milani Andrea
Milet Bernard L
Miller Freeman D
Millis Robert L
Moehlmann Diedrich
Moore Elliott P
Moroz Vasilis I
Morrison David
Mukai Tadashi
Mumma Michael Jon
Nakamura Akiko M
Nakamura Tsuko
Napier William M
Neff John S
Neukum G
Newburn Ray L
Niedner Malcolm B
Ninkov Zoran
Noll Keith Stephen
O'Dell Charles R
O'Keefe John A
Paolicchi Paolo
Parisot Jean-Paul
Pellas Paul
Pendleton Yvonne Jean
Perez de Tejada H A
Pilcher Carl Bernard
Pillinger Colin
Pittich Eduard M
Prialnik-Kovetz Dina
Proisy Paul E
Reitsema Harold J
Remy Battiau Liliane G A
Revelle Douglas Orson
Rickman Hans
Roemer Elizabeth
Rousselot Philippe
Sagdeev Roald Z
Saito Takao
Saito Takao
Scaltriti Franco
Schleicher David G
Schloerb F Peter
Schmidt H U
Schmidt Maarten
Schober Hans J
Scholl Hans
Schubart Joachim
Sekanina Zdenek
Sharma A Surjalal
Sharp Christopher
Shimizu Mikio
Shkodrov V G
Shor Viktor A
Shul-man L M
Sivaraman K R
Smith Bradford A
Snyder Lewis E
Solc Martin
Spinrad Hyron
Steel Duncan I
Stern S Alan
Surdej Jean M G
Svoren Jan
Swade Daryl Allen
Sykes Mark Vincent
Szego Karoly
Tacconi-Garman Lowell E
Takeda Hidenori
Tanabe Hiroyoshi
Tancredi Gonzalo
Tatum Jeremy B
Terentjeva Alexandra K
Tholen David J
Tomita Koichiro
Toth Imre
Valsecchi Giovanni B
Van Flandern Tom
Vanysek Vladimir
Veeder Glenn J
Veverka Joseph
Vilas Faith
Walker Alistair Robin
Wallis Max K
Wang Sichao
Wasson John T
Wdowiak Thomas J
Weaver Harold F
Wehinger Peter A
Weidenschilling S J
Weissman Paul Robert
Wells Eddie Neil
Wetherill George W
Whipple Fred L
Wilkening Laurel L
Williams Iwan P
Wood John A

Woolfson Michael M
Woszczyk Andrzej
Wyckoff Susan
Yabushita Shin A
Yavnel Alexander A
Yeomans Donald K
Zarnecki Jan Charles
Zellner Benjamin H
Zhu Jin

Composition of Commission 16

Physical Study of Planets and Satellites/Etude physique des planètes et des satellites

President: de Bergh Catherine

Vice-President: Cruikshank Dale P

Secretary: Coradini Angioletta

Organizing Committee:

Belton Michael J S
Blanco Carlo
Consolmagno Guy Joseph
Gautier Daniel
Marov Mikhail Ya
Morrison David
Noll Keith Stephen
Owen Tobias C
Tejfel Viktor G
Woszczyk Andrzej

Members:

Akabane Tokuhide
Akimov Leonid
Alexandrov Yuri V
Appleby John F
Arthur David W G
Atkinson David H
Atreya Sushil K
Barrow Colin H
Batson Raymond Milner
Battaner Eduardo
Baum William A
Bazilevsky Alexandr T
Beebe Reta Faye
Beer Reinhard
Bell III James F
Ben-Jaffel Lofti
Bender Peter L
Berge Glenn L
Bergstralh Jay T
Bertaux Jean Loup
Bhatia R K
Binzel Richard P
Blamont Jacques Emile
Bondarenko Ludmila N
Bosma Pieter B
Boss Alan P
Boyce Peter B
Brahic Andre
Brecher Aviva
Broadfoot A Lyle
Brown Robert Hamilton
Brunk William E
Buie Marc W
Buratti Bonnie J
Burns Joseph A
Calame Odile
Caldwell John James
Cameron Winifred S
Camichel Henri
Campbell Donald B
Capria Maria Teresa
Carsmaru Maria M
Catalano Santo
Chamberlain Joseph W
Chapman Clark R
Chen Daohan
Chevrel Serge
Clairemidi Jacques
Cochran Anita L
Colic Petar-Kasimir
Colombo G
Combi Michael R
Connes Janine
Counselman Charles C
Davies Merton E
de Pater Imke
Dermott Stanley F
Dickel John R
Dickey Jean O'Brien
Dollfus Audouin
Drake Frank D
Drossart Pierre
Durrance Samuel T
Dzhapiashvili Victor P
El Baz Farouk
Elliot James L
Elston Wolfgang E
Encrenaz Therese
Epishev Vitali P
Eshleman Von R
Esposito Larry W
Farinella Paolo
Fielder Gilbert
Fink Uwe
Fox Kenneth
Fox W E
Fujiwara Akira
Geake John E
Gehrels Tom
Geiss Johannes
Gerard Jean-Claude M C
Giclas Henry L
Gierasch Peter J
Gold Thomas
Goldreich Peter
Goldstein Richard M
Goody R M
Gor'kavyi Nikolai
Gorenstein Paul
Goudas Constantine L

Green Jack
Grossman Lawrence
Guerin Pierre
Guest John E
Gulkis Samuel
Gurshtein Alexander A
Haenninen Jyrki
Hagfors Tor
Halliday Ian
Hammel Heidi B
Harris Alan William
Harris Alan William
Herzberg Gerhard
Hide Raymond
Holberg Jay B
Horedt Georg Paul
Hovenier J W
Hu Zhongwei
Hubbard William B
Hunt G E
Hunten Donald M
Irvine William M
Iwasaki Kyosuke
Johnson Torrence V
Jurgens Raymond F
Kaula William M
Kiladze R I
Kim Yongha
Kislyuk Vitalij S
Kowal Charles Thomas
Ksanfomaliti Leonid V
Kumar Shiv S
Kurt Vitaliy G
Kuzmin Arkadii D
Lane Arthur Lonne
Larson Harold P
Larson Stephen M
Leikin Grigerij A
Lewis J S
Lissauer Jack J
Lockwood G Wesley
Lopes-Gautier Rosaly
Lopez-Moreno Jose Juan
Lopez-Puertas Manuel
Lopez-Valverde M A
Lumme Kari A
Lutz Barry L
Mahra H S
Marcialis Robert
Matson Dennis L
Matsui Takafumi
Mayer Cornell H
McCord Thomas B
McElroy M B
McGrath Melissa Ann
McKinnon William Beall
Meadows A Jack
Mickelson Michael E
Mikhail Joseph Sidky
Millis Robert L
Miyamoto Sigenori
Moehlmann Diedrich
Molina Antonio
Moore Patrick
Moreno Fernando
Moroz Vasilis I
Morozhenko A V
Mosser Benoit
Mulholland J Derral
Mumma Michael Jon
Murphy Robert E
Nakagawa Yoshitsugu
Ness Norman F
Neukum G
O'Keefe John A
Ottelet I J
Pang Kevin
Paolicchi Paolo
Petit Jean-Marc
Petropoulos Basil Ch
Pettengill Gordon H
Pillinger Colin
Pokorny Zdenek
Potter Andrew E
Predeanu Irina
Rao M N
Rodionova Janna F
Rodrigo Rafael
Roques Francoise
Rosch Jean
Ruskol Eugenia L
Safronov Victor S
Saissac Joseph
Sanchez-Lavega Agustin
Schleicher David G
Schloerb F Peter
Schneider Nicholas M
Shapiro Irwin I
Shevchenko Vladislav V
Shimizu Mikio
Shimizu Tsutomu Emer
Shkuratov Yurii
Sicardy Bruno
Sinton William M
Sjogren William L
Smith Bradford A
Soderblom Larry
Sonett Charles P
Sprague Ann Louise
Stern S Alan
Stoev Alexei
Stone Edward C
Strobel Darrell F
Strom Robert G
Strong John D
Synnott Stephen P
Tchouikova Nadejda A
Terrile Richard John
Tholen David J
Trafton Laurence M
Tran-Minh Francoise
Troitsky V S
Tyler G Leonard
van Allen James A
Van Flandern Tom
Veiga Carlos Henrique
Veverka Joseph
Walker Alta Sharon
Walker Robert M A
Wallace Lloyd V
Wamsteker Willem
Wasserman Lawrence H
Wasson John T
Weidenschilling S J
Weimer Theophile P F
Wells Eddie Neil
Wetherill George W
Whitaker Ewen A
Wildey Robert L
Williams Iwan P
Williams James G
Wood John A
Woolfson Michael M
Yoder Charles F
Young Andrew T
Young Louise Gray
Zharkov Vladimir N

Composition of Commission 19

Rotation of the Earth/Rotation de la Terre

President: McCarthy Dennis D

Vice-President: Capitaine Nicole

Organizing Committee:

Arias Elisa Felicitas
Beutler Gerhard
Brosche Peter
Brzezinski Aleksander
Dehant Veronique
Feissel Martine
Gross Richard Sewart
Manabe Seiji
Morrison Leslie V
Rykhlova Lidija V
Tarady Vladimir K
Vondrak Jan
Wilson P
Yang Fumin
Yatskiv Ya S

Members:

Arabelos Dimitrios
Bang Yong Gol
Banni Aldo
Barlier Francois E
Bender Peter L
Blinov Nikolai S
Boucher Claude
Bougeard Mireille L
Boytel Jorge Del Pino
Cazenave Anny
Chiumiento Giuseppe
Debarbat Suzanne V
Defraigne Pascale
Dejaiffe Rene J
Dickey Jean O'Brien
Dickman Steven R
Djurovic Dragutin M
El Shahawy Mohamad
Fliegel Henry F
Fong Chugang
Fujishita Mitsumi
Gambis Daniel
Gao Buxi
Gaposchkin Edward M
Gayazov Iskander S
Gozhy Adam
Groten Erwin
Gu Zhennian
Guinot Bernard R
Hall R Glenn
Han Tianqi
Han Yanben
Hefty Jan
Hemmleb Gerhard
Hua Yingmin
Iijima Shigetaka
Jin Wenjing
Kakuta Chuichi
Kameya Osamu
Klepczynski William J
Knowles Stephen H
Kolaczek Barbara
Korsun Alla
Kosek Wieslaw
Lehmann Marek
Li Jinling
Li Zhengxin
Li Zhian
Liao De-chun
Lieske Jay H
Liu Ciyuan
Luo Dingjiang
Luo Shi-Fang
Ma Chopo
Malkin Zinovy M
Meinig Manfred
Melbourne William G
Melchior Paul J
Merriam James B
Monet Alice K B
Morgan Peter
Mueller Ivan I
Naumov Vitalij A
Newhall X X
Niemi Aimo
Ooe Masatsugu
Paquet Paul Eg
Pesek Ivan
Petit Gerard
Picca Domenico
Pilkington John D H
Poma Angelo
Popelar Josef
Proverbio Edoardo
Randic Leo
Ray James R
Robertson Douglas S
Rochester Michael G
Ron Cyril
Ruder Hanns
Rusu I
Sadzakov Sofija
Sanchez Manuel
Sasao Tetsuo
Sato Koichi
Schutz Bob Ewald
Sekiguchi Naosuke
Sevarlic Branislav M
Sevilla Miguel J
Shapiro Irwin I
Shelus Peter J
Sidorenkov Nikolay S
Smith Humphry M

Soffel Michael
Souchay Jean
Stanila George
Stavinschi Magdalena
Stephenson F Richard
Sugawa Chikara
Tapley Byron D
Tian Jing
Tsao Mo
Veillet Christian
Vicente Raimundo O
Wang Kemin
Wang Zhengming
Wilkins George A
Williams James G
Wu Bin
Wu Shouxian
Wuensch Johann Jakob
Xiao Naiyuan
Xu Jiayan
Xu Tong-Qi
Ye Shuhua
Yokoyama Koichi
Yumi Shigeru
Zeng Zhifang
Zhang Guo-Dong
Zhao Ming
Zheng Dawei
Zhong Min
Zhu Yaozhong
Zhu Yonghe

Composition of Commission 20

Position and Motions of Minor Planets, Comets and Satellites/
Positions et mouvements des petites planètes des comètes et des satellites

President: Rickman Hans

Vice-President: Bowell Edward L G

Secretary: Valsecchi Giovanni B

Organizing Committee:

Aksnes Kaare
Arlot Jean-Eudes
Carusi Andrea
Lemaitre Anne
Marsden Brian G
Shor Viktor A
Wasserman Lawrence H
Yeomans Donald K
Zhang Jiaxiang

Members:

A'Hearn Michael F
Abalakin Victor K
Aikman G Chris l
Babadzhanov Pulat B
Bailey Mark Edward
Batrakov Yuri V
Bec-Borsenberger Annick
Benest Daniel
Bien Reinhold
Blanco Carlo
Blow Graham L
Boerngen Freimut
Branham Richard L
Burns Joseph A
Calame Odile
Carpino Mario
Chapront-Touze Michelle
Chernykh N S
Chio Chol Zong
Chodas Paul Winchester
Cristescu Cornelia G
de Pascual Martinez M
de Sanctis Giovanni
Debehogne Henri Sc
Delsemme Armand H
Dollfus Audouin
Donnison John Richard
Dourneau Gerard
Doval Jorge Perez
Dunham David W
Dvorak Rudolf
Edelman Colette
Edmondson Frank K
Elliot James L
Elst Eric Walter
Emelianov Nikolaj V
Epishev Vitali P
Fernandez Julio A
Ferraz-Mello Sylvio
Ferreri Walter
Forti Giuseppe
Franklin Fred A
Freitas Mourao R r
Froeschle Claude
Garfinkel Boris
Gehrels Tom
Gibson James
Giclas Henry L
Gilmore Alan C
Gong Xiangdong
Greenberg Richard
Hahn Gerhard J
Hainaut Olivier R
Harper David
Harris Alan William
Hasegawa Ichiro
Haupt Hermann F
He Miaofu
Helin Eleanor Francis
Hemenway Paul D
Henrard Jacques
Hers Jan
Heudier Jean-Louis
Hurnik Hieronim
Hurukawa Kiitiro
Ianna Philip A
Isobe Syuzo
Ivanova Violeta
Izvekov Vladimir A
Jacobson Robert A
Khatisashvili Alfez Sh
Kiang Tao
Kilmartin Pamela
Kinoshita Hiroshi
Kisseleva Tamara P
Klemola Arnold R
Knezevic Zoran
Kohoutek Lubos
Kosai Hiroki
Kowal Charles Thomas
Kozai Yoshihide
Krasinsky George A
Kristensen Leif Kahl
Krolikowska-Soltan Malgorzata
Kulikova Nelli V
Lagerkvist Claes-Ingvar
Lazzaro Daniela

Lieske Jay H
Lin Qinchang
Lindblad Bertil A
Lomb Nicholas Ralph
Lovas Miklos
Machado Luiz E da silva
Mahra H S
Manara Alessandro A
Matese John J
McCrosky Richard E
McNaught Robert H
Medvedev Yuriy D
Message Philip J
Milani Andrea
Milet Bernard L
Millis Robert L
Mintz Blanco Betty
Monet Alice K B
Moons Michele B M M
Morris Charles S
Muinonen Karri
Mulholland J Derral
Murray Carl D
Nacozy Paul E
Nakamura Tsuko
Nakano Syuichi
Nobili Anna M
Oterma Liisi
Overbeek Michiel Daniel
Owen William Mann
Pandey A K
Pascu Dan
Pauwels Thierry
Pierce David Allen
Pittich Eduard M
Porubcan Vladimir
Protich Milorad B
Rajamohan R
Raju Vasundhara
Rapaport Michel
Reitsema Harold J
Roemer Elizabeth
Roeser Siegfried
Russell Kenneth S
Sato Massae
Schmadel Lutz D
Schober Hans J
Scholl Hans
Schubart Joachim
Schuster William John
Seidelmann P Kenneth
Sekanina Zdenek
Shelus Peter J
Shen Kaixian
Shkodrov V G
Sinclair Andrew T
Sitarski Grzegorz
Sokolsky Andrej G
Solovaya Nina A
Soma Mitsuru
Standish E Myles
Steel Duncan I
Stellmacher Irene
Sultanov G F
Svoren Jan
Synnott Stephen P
Tancredi Gonzalo
Tatum Jeremy B
Taylor Donald Boggia
Tholen David J
Tomita Koichiro
Torres Carlos
Tsuchida Masayoshi
Vaghi Sergio
Van Flandern Tom
van Houten C J
van Houten-Groeneveld I
Vavrova Zdenka
Veillet Christian
Vieira Martins Roberto
Vienne Alain
Vu Duong Tuyen
Weissman Paul Robert
West Richard M
Whipple Arthur L
Whipple Fred L
Wild Paul
Williams Iwan P
Williams James G
Wroblewski Herbert
Yabushita Shin A
Yoshikawa Makoto
Yuasa Manabu
Zadunaisky Pedro E
Zagretdinov Renat
Zappala Vincenzo
Zhu Jin
Ziolkowski Krzysztof

Composition of Commission 21

Light of the Night Sky/Lumière du ciel nocture

President: Bowyer C Stuart

Vice-President: Lamy Philippe

Organizing Committee:

Dwek Eli
Gustafson Bo A S
Hanner Martha S
Leinert Christoph
Levasseur-Regourd A-C
Mann Ingrid
Matsumoto Toshio
Mikhail Joseph Sidky
Mukai Tadashi

Members:

Abraham Peter
Angione Ronald J
Baggaley William J
Banos Cosmas J
Belkovich Oleg I
Blamont Jacques Emile
Broadfoot A Lyle
Chamberlain Joseph W
Clairemidi Jacques
d'Hendecourt Louis
Dermott Stanley F
Dodonov Serguej
Dubin Maurice
Dufay Maurice
Dumont Rene
Dunkelman Lawrence
Elsaesser Hans
Feldman Paul Donald
Fujiwara Akira
Gadsden Michael
Galperin Yuri I
Giovane Frank
Greenberg J Mayo
Gruen Eberhard
Harwit Martin
Hauser Michael G
Hecht James H
Henry Richard C
Hofmann Wilfried
Hong Seung Soo
Hurwitz Mark V
Ivanov Kholodny Goz S
Jackson Bernard V
James John F
Joubert Martine
Karygina Zoya V
Kopylov Alexander
Koutchmy Serge
Kulkarni Prabhakar V
Leger Alain
Lemke Dietrich
Lillie Charles F
Lopez-Gonzalez Maria J
Lopez-Moreno Jose Juan
Lopez-Puertas Manuel
Lumme Kari A
Maihara Toshinori
Mather John Cromwell
Mattila Kalevi
Maucherat J
McDonnell J A M
Mikhail Fahmy I
Misconi Nebil Yousif
Morgan David H
Muinonen Karri
Mukai Sonoyo
Nawar Samir
Neizvestny Sergei
Nishimura Tetsuo
Paresce Francesco
Perrin Jean Marie
Pfleiderer Jorg
Radoski Henry R
Reach William
Robley Robert
Rodrigo Rafael
Rozhkovskij Dimitrij A
Sanchez Francisco
Sanchez-Saavedra M Luisa
Saxena P P
Schlosser Wolfhard
Schwehm Gerhard
Shefov Nicolai N
Soberman Robert K
Sparrow James G
Staude Hans Jakob
Sykes Mark Vincent
Tanabe Hiroyoshi
Toller Gary N
Toroshlidze Teimuraz I
Tyson John Anthony
van de Hulst H C
Vrtilek Jan M
Wallis Max K
Weinberg J L
Wesson Paul S
Wilson P
Witt Adolf N
Wolstencroft Ramon D
Woolfson Michael M
Yamamoto Tetsuo
Yamashita Kojun
Zerull Reiner H

Composition of Commission 22

Meteors & Interplanetary Dust/Météores & poussière interplanétaire

President: Baggaley William J

Vice-President: Porubcan Vladimir

Secretary: Hawkes Robert Lewis

Organizing Committee:

Babadzhanov Pulat B
Elford William Graham
Gustafson Bo A S
Hasegawa Ichiro
Koeberl Christian
Mann Ingrid
Spurny Pavel
Williams Iwan P

Members:

Abbott William N
Beard David B
Belkovich Oleg I
Bhandari N
Bibarsov Ravil-sh
Borovicka Jiri
Brownlee Donald E
Carusi Andrea
Ceplecha Zdenek
Cevolani Giordano
Clifton Kenneth St
Clube S V M
Cooper Timothy
Djorgovski Stanislav
Dubin Maurice
Forti Giuseppe
Glass Billy Price
Goswami J N
Gruen Eberhard
Hajduk Anton
Hajdukova Maria
Halliday Ian
Hanner Martha S
Harvey Gale A
Hawkins Gerald S
Helin Eleanor Francis
Hey James Stanley
Hodge Paul W
Hong Seung Soo
Hughes David W
Jenniskens Petrus M M
Jennison Roger C
Jones James
Jopek Tadeusz Jan
Kalenichenko Valentin
Kapisinsky Igor
Kashscheev B L
Keay Colin S L
Kramer Kh N
Kresakova Margita
Kruchinenko Vitaliy G
Lamy Philippe
Lemaire Joseph F
Levasseur-Regourd Annie Chantal
Lindblad Bertil A
Lovell Sir Bernard
Marvin Ursula B
Mason John William
McCrosky Richard E
McDonnell J A M
McIntosh Bruce A
Meisel David D
Miles Howard G
Misconi Nebil Yousif
Murray C Andrew
Murray Carl D
Nakazawa Kiyoshi
Napier William M
Newburn Ray L
Nuth Joseph A III
O'Keefe John A
Pecina Petr
Pillinger Colin
Plavec Zdenka
Polnitzky Gerhard
Poole Graham
Quesada Vinicio
Rajchl Jaroslav
Revelle Douglas Orson
Rickman Hans
Ripken Hartmut W
Russell John A
Sekanina Zdenek
Shao Cheng-yuan
Simek Milos
Soberman Robert K
Steel Duncan I
Svestka Jiri
Tatum Jeremy B
Taylor Andrew
Tedesco Edward F
Terentjeva Alexandra K
Tomita Koichiro
Verniani Franco
Wang dechang
Watanabe Jun-ichi
Weinberg J L
Wetherill George W
Whipple Fred L
Wood John A
Woolfson Michael M
Xu Pinxin
Yavnel Alexander A
Yeomans Donald K
Zvolankova Judita

Composition of Commission 24

Photographic Astrometry/Astrométrie photographique

President: Schilbach Elena

Vice-President: Jin Wenjing

Organizing Committee:

Creze Michel
Hemenway Paul D
Kumkova Irina I
Platais Imants K
Roeser Siegfried
Turon Catherine
Wang Jiaji

Members:

Abad Carlos
Abhyankar Krishna D
Arenou Frederic
Argue A Noel
Ballabh G M
Bastian Ulrich
Benedict George F
Blaauw Adriaan
Bouigue Roger
Branham Richard L
Breakiron Lee Allen
Bronnikova Nina M
Brosche Peter
Bunclark Peter Stephen
Chiu Liang-Tai George
Christy James Walter
Clube S V M
Connes Pierre
Corbin Thomas Elbert
Crifo Francoise
Cudworth Kyle Mccabe
Dahn Conard Curtis
de Vegt Chr
Dick Wolfgang
Dommanget J
Douglass Geoffrey G
Ducourant Christine
Eichhorn Heinrich K
Elsmore Bruce
Fallon Frederick W
Fanselow John Lyman
Firneis Friedrich J
Firneis Maria G
Franz Otto G
Fredrick Laurence W
Fresneau Alain
Gallouet Louis
Gatewood George
Gauss F Stephen
Geffert Michael
Giclas Henry L
Goyal A N
Guibert Jean
Hanson Robert B
Hartkopf William I
Hershey John L
Heudier Jean-Louis
Hill Graham
Hoffleit E Dorrit
Ianna Philip A
Irwin Michael John
Jahreiss Hartmut
Jefferys William H
Johnston Kenneth J
Jones Burton
Jones Derek H P
Kanayev Ivan I
Kharchenko Nina
Kislyuk Vitalij S
Klemola Arnold R
Klock B L
Kolchinskij I G
Kovalevsky Jean
Lapushka K K
Latypov A A
Le Poole Rudolf S
Lippincott Sarah Lee
Lopez Carlos
Lozinskij Alexander M
Lu Chunlin
Lu Phillip K
MacConnell Darrell J
Machado Luiz E da silva
Marschall Laurence A
McAlister Harold A
McLean Brian J
Mennessier Marie-Odile
Miyamoto Masanori
Monet David G
Morrison Leslie V
Murray C Andrew
Nunez Jorge
Oja Tarmo
Onegina A B
Osborn Wayne
Pascu Dan
Perryman Michael A C
Pizzichini Graziella
Polozhentsev Dimitrij D
Potter Heino I
Qin Dao
Rafferty Theodore J
Requieme Yves
Rizvanov Naufal G
Roemer Elizabeth
Ruder Hanns

Russell Jane L
Sanders Walt L
Scholz Ralf Dieter
Shelus Peter J
Shi Guang-Chen
Sims Kenneth P
Stein John William
Steinert Klaus Guenter
Stock Jurgen D
Stone Ronald Cecil
Strand Kaj Aa
Thomas David V
Tucholke Hans-Joachim
Upgren Arthur R
Valbousquet Armand
van Altena William F
Vilkki Erkki U
Walter Hans G
Wan Lai
Wasserman Lawrence H
Westerhout Gart
White Graeme Lindsay
Williams Carol A
Worley Charles E
Wroblewski Herbert
Yang Tinggao
Younis Saad M
Zacharias Norbert

Composition of Commission 25

Stellar Photometry and Polarimetry/Photométrie stellaire et polarimétrie

President: Sterken Christiaan Leo

Vice-President: Landolt Arlo U

Organizing Committee:

Adelman Saul J
Bessell Michael S
Breger Michel
Glass Ian Stewart
Graham John A
Hensberge Herman
Kurtz Donald Wayne
Landstreet John D
Milone Eugene F
Moffett Thomas J
Sekiguchi Kazuhiro
Tinbergen Jaap
Warren Wayne H

Members:

Ables Harold D
Albrecht Rudolf
Anandaram Mandayam N
Angel J Roger P
Angione Ronald J
Anthony-Twarog Barbara J
Argue A Noel
Arnaud Jean Paul
Arsenijevic Jelisaveta
Ashok N M
Aspin Colin
Axon David
Baldinelli Luigi
Balona Luis Antero
Barnes III Thomas G
Barrett Paul Everett
Beck Rainer
Becker Wilhelm
Behr Alfred
Blanco Victor M
Blecha Andre Boris G
Bookmyer Beverly B
Borgman Jan
Borra Ermanno F
Brown Douglas Nason
Bruck Hermann A
Buser Roland
Carney Bruce William
Carter Brian
Castelaz Micheal W
Celis Leopoldo
Connolly Leo Paul
Cousins A W J
Coyne George V
Cramer Noel
Crawford David L
Cuypers Jan
Dachs Joachim
Dahn Conard Curtis
Danford Stephen C
Denoyelle Jozef Kic
Deshpande M R
Dolan Joseph F
Dubout Renee
Ducati Jorge Ricardo
Ducourant Christine
Edwards Paul J
Eelsalu Heino
Efimov Yuri
Fabregat Juan
Feinstein Alejandro
Fernie J Donald
Forte Juan Carlos
Gallouet Louis
Gehrz Robert Douglas
Genet Russel M
Gerbaldi Michele
Ghosh S K
Golay Marcel
Goy Gerald
Grauer Albert D
Grenon Michel
Grewing Michael
Guetter Harry Hendrik
Gutierrez-Moreno A
Hall Douglas S
Hauck Bernard
Hayes Donald S
Heck Andre
Hilditch Ronald W
Hill Philip W
Holmberg Erik B
Hu Jingyao
Huang Lin
Hubrig Swetlana
Huovelin Juhani
Hyland A R Harry
Irwin Alan W
Iyengar K V K
Jerzykiewicz Mikolaj
Jordi Carme
Joshi Suresh Chandra
Joshi U C
Kawara Kimiaki
Kepler S O
Kilkenny David
King Ivan R
Knude Jens Kirkeskov
Koch Robert H
Kulkarni Prabhakar V
Kunkel William E
Labhardt Lukas
Laskarides Paul G

Lasker Barry M
Lemke Michael
Lenzen Rainer
Leroy Jean-Louis
Li Sin Hyong
Linde Peter
Lockwood G Wesley
Lub Jan
Luna Homero G
Maitzen Hans M
Manfroid Jean
Markkanen Tapio
Marraco Hugo G
Masani A
Maslennikov Kirill L
Mathys Gautier
Mayer Pavel
McCarthy Martin F
McLean Ian S
Mendoza V Eugenio E
Menzies John W
Mianes Pierre
Miller Joseph S
Mintz Blanco Betty
Moreno Hugo
Morris Stephen C
Mourard Denis
Muller Andre B
Mumford George S
Nicolet Bernard
Noguchi Kunio
Notni P
Oblak Edouard
Oestreicher Roland
Orsatti Ana M
Page Arthur
Pedreros Mario
Pel Jan Willem
Penny Alan John
Perry Charles L
Pfau Werner
Pfeiffer Raymond J
Philip A G Davis
Piirola Vilppu E
Platais Imants K
Pokrzywka Bartlomiej
Rao P Vivekananda
Raveendran A V
Reglero-Velasco Victor
Richardson Lorna Logan
Robb Russell M
Robinson Edward Lewis
Roslund Curt
Rufener Fredy G
Rydgren Alfred Eric
Sarma M B K
Schmidt Edward G
Schoeneich W
Schuster William John
Shakhovskoj Nikolay M
Shawl Stephen J
Smyth Michael J
Steinlin Uli
Stock Jurgen D
Stockman Hervey S
Stone Remington P S
Straizys V
Strohmeier Wolfgang
Sudzius Jokubas
Sullivan Denis John
Szkody Paula
Szymanski Michal
Tandon S N
Tapia-Perez Santiago
Todoran Ioan
Tokunaga Alan Takashi
Tolbert Charles R
Ulrich Bruce T
Ureche Vasile
Vardanian R A
Vaughan Arthur H
Verma R P
Visvanathan Natarajan
Vrba Frederick J
Walker Alistair Robin
Wallenquist Aake A E
Walraven Th
Wang Chuanjin
Weiss Werner W
Weistrop Donna
Wesselius Paul R
White Nathaniel M
Wielebinski Richard
Willstrop Roderick V
Winiarski Maciej
Woo Jong Ok
Wramdemark Stig S o
Yamashita Yasumasa
Yin Jisheng
Young Andrew T
Ziznovsky Jozef

Composition of Commission 26

Double and Multiple Stars/Etoiles doubles et multiples

President: Zinnecker Hans

Vice-President: Scarfe Colin D

Organizing Committee:

Allen Christine
Armstrong John Thomas
Hartkopf William I
Mathieu Robert D
Tokovinin Andrej A
Trimble Virginia L
Valtonen Mauri J
Worley Charles E

Members:

Abt Helmut A
Anosova Joanna
Argue A Noel
Argyle P E
Bacchus Pierre
Bagnuolo William G
Balega Yuri Yu
Batten Alan H
Beavers Willet I
Bernacca Pietio L
Bonneau Daniel
Brosche Peter
Cabrita Ezequiel
Campbell Alison
Cester Bruno
Chen Zhen
Couteau Paul
Culver Roger Bruce
Dadaev Aleksandr N
Docobo Jose A Durantez
Dommanget J
Douglass Geoffrey G
Dunham David W
Eichhorn Heinrich K
Fekel Francis C
Ferrer Osvaldo Eduardo
Fletcher J Murray
Franz Otto G
Fredrick Laurence W
Freitas Mourao R r
Gatewood George
Gaudenzi Silvia
Geyer Edward H
Hakkila Jon Eric
Halbwachs Jean Louis
Herrera Miguel Angel
Hershey John L
Hidayat Bambang
Hill Graham
Hindsley Robert Bruce
Hummel Christian Aurel
Hummel Wolfgang
Ianna Philip A
Jaschek Carlos O R
Jassur Davoud MZ
Kiselyov Alexej A
Kumsishvili J I
Lampens Patricia
Latham David W
Lattanzi Mario G
Leinert Christoph
Ling J
Lippincott Sarah Lee
Loden Kerstin R
Loden Lars Olof
Martin Eduardo
McAlister Harold A
Meyer Claude
Mikkola Seppo
Mohan Chander
Morbey Christopher L
Morbidelli Roberto
Morel Pierre Jacques
Muller Paul
Oblak Edouard
Oswalt Terry D
Pannunzio Renato
Peterson Deane M
Popovic Georgije
Poveda Arcadio
Prieto Cristina
Rakos Karl D
Russell Jane L
Salukvadze G N
Scardia Marco
Schmidtke Paul C
Sinachopoulos D
Smak Joseph I
Soderhjelm Staffan
Sowell James Robert
Stein John William
Strand Kaj Aa
Szabados Laszlo
Torres Guillermo
Tsay Wean-Shun
Upgren Arthur R
Valbousquet Armand
van Altena William F
van der Hucht Karel A
van Dessel Edwin Ludo
Walker Richard L
Wang Jiaji
Weis Edward W
Yan Lin-shan

Composition of Commission 27

Variable Stars/Etoiles variables

President: Kurtz Donald Wayne

Vice-President: Christensen-Dalsgaard J

Organizing Committee:

Cacciari Carla
Cottrell Peter Ledsam
Harmanec Petr
Jerzykiewicz Mikolaj
Matthews Jaymie
Moskalik Pawel
Sasselov Dimitar D
Welch Douglas L
Whitelock Patricia Ann

Members:

Aerts Conny
Aizenman Morris L
Albinson James
Albrow Michael
Alfaro Emilio Javier
Allan David W
Alpar Ali
Andrievsky Sergei
Antipova Lyudmila I
Antonello Elio
Antov Alexandar
Arellano Ferro Armando
Arkhipova Vesa P
Arsenijevic Jelisaveta
Asteriadis Georgios
Avgoloupis Stavros
Baade Dietrich
Baglin Annie
Baker Norman H
Balona Luis Antero
Barnes III Thomas G
Bartolini Corrado
Barwig Heinz
Bastien Pierre
Bateson Frank M OBE
Bath Geoffrey T
Bauer Wendy Hagen
Bedogni Roberto
Belmonte Aviles J A
Belserene Emilia P
Belvedere Gaetano
Benson Priscilla J
Berdnikov Leonid N
Bersier David
Berthomieu Gabrielle
Bessell Michael S
Bianchini Antonio
Bochonko D Richard
Bolton C Thomas
Bond Howard E
Bopp Bernard W
Boulon Jacques J
Bowen George H
Boyarchuk Alexander A
Boyarchuk Margarita E
Breger Michel
Brown Douglas Nason
Buchler J Robert
Burki Gilbert
Busko Ivo C
Butler C John
Butler Dennis
Caldwell John A R
Cameron Andrew Collier
Casares Valazquez Jorge
Catchpole Robin M
Chavira Enrique Sr
Cherepashchuk Analily M
Christy Robert F
Cohen Martin
Connolly Leo Paul
Contadakis Michael E
Cook Kem Holland
Coulson Iain M
Coutts-Clement Christine
Cox Arthur N
Cutispoto Giuseppe
Cuypers Jan
Danford Stephen C
de Groot Mart
Delgado Antonio Jesus
Demers Serge
Deupree Robert G
Dickens Robert J
Donahue Robert Andrew
Downes Ronald A
Dunlop Storm
Dupuy David L
Dziembowski Wojciech
Edwards Paul J
Efremov Yuri N
El Basuny Ahmed Alawy
Eskioglu A Nihat
Evans Aneurin
Evans Nancy Remage
Evren Serdar
Fadeyev Yuri A
Feast Michael W
Feibelman Walter A
Ferland Gary Joseph
Fernie J Donald
Fitch Walter S
Friedjung Michael
Frolov Mikhail S
Gahm Goesta F
Gallagher John S
Garrido Rafael
Gascoigne S C B
Gasparian Lazar

Genet Russel M
Gershberg R E
Geyer Edward H
Gieren Wolfgang P
Gies Douglas R
Gillet Denis
Glagolevskij Juri V
Godoli Giovanni
Gorbatsky Vitalij G
Gosset Eric
Gough Douglas O
Goupil Marie Jose
Graham John A
Grasdalen Gary L
Grinin Vladimir P
Groenewegen Martin
Grygar Jiri
Guerrero Gianantonio
Guinan Edward Francis
Gurm Hardev S
Gursky Herbert
Guzik Joyce Ann
Hackwell John A
Haefner Reinhold
Haisch Bernhard Michael
Halbwachs Jean Louis
Hall Douglas S
Hamdy M A M
Hansen Carl J
Hawley Suzanne Louise
Heiser Arnold M
Henden Arne Anthon
Herbig George H
Hers Jan
Hesser James E
Hill Philip W
Hill Henry Allen
Hiroyasu Ando
Hoffleit E Dorrit
Hojaev Alisher S
Houk Nancy
Howell Steve Bruce
Huenemoerder David P
Hutchings John B
Iben Icko
Iijima Takashi
Ishida Toshihito
Jablonski Francisco
Jarzebowski Tadeusz
Jeffery Christopher S
Jewell Philip R
Jiang Shi-Yang
Jones Albert F
Kadouri Talib Hadi
Kambe Eiji
Kanamitsu Osamu
Kanbur Shashi
Kanyo Sandor
Karitskaya Eugenia A
Karovska Margarita
Karp Alan Hersh
Kawaler Steven D
Kepler S O
Khaliullin Khabibrachman F
Kim Tu Hwan
Kim Chulhee
Kiplinger Alan L
Kippenhahn Rudolf
Kjeldsen Hans
Kjurkchieva Diana
Koen Marthinus
Kollath Zoltan
Kopylov Ivan M
Kraft Robert P
Krautter Joachim
Kreiner Jerzy Marek
Krisciunas Kevin
Krzeminski Wojciech
Kubiak Marcin A
Kuhi Leonard V
Kumsishvili J I
Kunkel William E
Lago Maria Teresa V T
Landolt Arlo U
Laney Clifton D
Lanning Howard Hugh
Lanza Antonino Francesco
Laskarides Paul G
Lawson Warrick
Lazaro Carlos
Leite Scheid Paulo
Leung Kam Ching
Li Yan
Li Zhiping
Little-Marenin Irene R
Liu Zongli
Lockwood G Wesley
Longmore Andrew J
Lopez De Coca M D P
Lub Jan
Madore Barry Francis
Maeder Andre
Maffei Paolo
Mahmoud Farouk M A B
Mahra H S
Makarenko Ekaterina N
Mannino Giuseppe
Mantegazza Luciano
Margrave Thomas Ewing
Martin William L
Martinez Peter
Masani A
Mattei Janet Akyuz
Mauche Christopher W
Mavridis L N
McGraw John T
Mcnamara Delbert H
Melikian Norair D
Mennessier Marie-Odile
Metz Klaus
Michel Eric
Milone Luis A
Milone Eugene F
Minikulov Nasridin K
Mirzoyan Ludwik V
Moffett Thomas J
Mohan Chander
Morrison Nancy Dunlap
Mukai Koji
Mumford George S
Murdin Paul G
Nather R Edward
Neff John S
Niarchos Panayiotis
Nicolov Nikolai S
Nikolov Andrej
Nugis Tiit
O'Donoghue Darragh
Odgers Graham J
Olah Katalin
Opolski Antoni
Oswalt Terry D
Papaloizou John C B
Paparo Margit
Papousek Jiri
Parsamyan Elma S
Parthasarathy M
Paterno Lucio
Pavlovski Kresimir
Percy John R
Petersen J O
Petrov Peter P
Pettersen Bjoern Ragnvald
Piirola Vilppu E
Pijpers Frank Peter
Pollard Karen
Pop Vasile
Pringle James E
Provost Janine
Pskovskij Juri P
Pugach Alexander F
Rakos Karl D
Rao N Kameswara
Reinsch Klaus
Renson P F M

Richter G A
Robinson Edward Lewis
Rodgers Alex W
Rodono Marcello
Rodriguez Eloy
Romano Giuliano
Romanov Yuri S
Rountree Janet
Russev Ruscho
Sadik Aziz R
Saha Abhijit
Samus Nikolai N
Sandmann William Henry
Sanyal Ashit
Sareyan Jean-Pierre
Sarma M B K
Sato Naonobu
Schlegel Eric Matthew
Schmidt Edward G
Schwartz Philip R
Schwarzenberg-Czerny A
Schwope Axel
Scuflaire Richard
Seeds Michael August
Shahul Hameed Mohin
Shakhovskaya Nadejda I
Shara Michael
Sharma Dharma Pal
Sherwood William A
Shobbrook Robert R
Sinvhal Shambhu Dayal
Smak Joseph I
Smeyers Paul
Smith Myron A
Soliman Mohamed Ahmed
Srivastava Ram Kumar
Starrfield Sumner
Stellingwerf Robert F
Stepien Kazimierz
Sterken Christiaan Leo
Stobie Robert S
Strassmeier Klaus G
Strohmeier Wolfgang
Strom Karen M
Strom Stephen E
Szabados Laszlo
Szatmary Karoly
Szecsenyi-Nagy Gabor
Szeidl Bela
Szkody Paula
Takeuti Mine
Tammann Gustav Andreas
Tamura Shin'ichi
Tempesti Piero
Terzan Agop
Tjin-a-Djie Herman R E
Torres Carlos Alberto
Tremko Jozef
Tsioumis Alexandros
Tsvetkov Milcho K
Tsvetkova Katya
Turner David G
Tutukov A V
Tylenda Romuald
Usher Peter D
Valtier Jean-Claude
van Genderen Arnoud M
van Hoolst Tim
Ventura Rita
Verheest Frank
Viotti Roberto
Vogt Nikolaus
Waelkens Christoffel
Walker William S G
Walker Edward N
Walker Merle F
Wallerstein George
Walraven Th
Warner Brian
Watson Robert
Webbink Ronald F
Wehlau Amelia
Weiss Werner W
Wenzel W
Williamon Richard M
Willson Lee Anne
Wilson Lionel
Wing Robert F
Wood Peter R
Xiong Darun
Yao Baoan
Yudin Boris F
Zijlstra Albert
Zola Stanislaw
Zsoldos Endre
Zuckerman Ben M

Composition of Commission 28

Galaxies

President: Bertola Francesco

Vice-President: Okamura Sadanori

Organizing Committee:

Balkowski-Mauger Chantal
Burstein David
Chakrabarti Sandip K
de Zeeuw Pieter T
Feast Michael W
Huchra John Peter
Trimble Virginia L
Wielebinski Richard
Zasov Anatole V
Zou Zhenlong

Members:

Ables Harold D
Abrahamian Hamlet V
Adler David Scott
Afanas'ev Viktor L
Aguero Estela L
Aguilar Luis A Chiu
Ahmad Farooq
Alcaino Gonzalo
Alighieri Sparello Di Serego
Alladin Saleh Mohamed
Allen Ronald J
Alloin Danielle
Alonso Maria Victoria
Amram Philippe
Andernach Heinz
Andrillat Yvette
Ann Hong Bae
Anosova Joanna
Aparicio Antonio
Ardeberg Arne L
Arkhipova Vesa P
Athanassoula Evangelia
Ayani Kazuya
Azzopardi Marc
Bahcall John N
Bailey Mark Edward
Bajaja Esteban
Baldwin Jack A
Ballabh G M
Banhatti Dilip Gopal
Barbon Roberto
Barcons Xavier
Barthel Peter
Bassino Lilia P
Basu Baidyanath
Battaner Eduardo
Battinelli Paolo
Baum Stef Alison
Baum William A
Beck Rainer
Becker Peter Adam
Begeman Kor G
Bender Ralf
Bendinelli Orazio
Benedict George F
Benevides Soares P
Berczik Peter
Bergeron Jacqueline A
Bergvall Nils Ake Sigvard
Berkhuijsen Elly M
Berman Vladimir
Bettoni Daniela
Bian Yulin
Bica Eduardo L D
Biermann Peter L
Bijaoui Albert
Binette Luc
Binggeli Bruno
Binney James J
Biretta John Anthony
Birkinshaw Mark
Bland-Hawthorn Jonathan
Blitz Leo
Block David Lazar
Blumenthal George R
Boerngen Freimut
Boeshaar Gregory Orth
Boisson Catherine
Boksenberg Alec
Bomans Dominik J
Bontekoe Romke
Borchkhadze Tengiz M
Bosma Albert
Bottinelli Lucette
Bouchet Patrice
Bower Gary Allen
Braccesi Alessandro
Braun Robert
Brecher Kenneth
Briggs Franklin
Brinkmann Wolfgang
Brinks Elias
Brodie Jean P
Brosch Noah
Brosche Peter
Brouillet Nathalie
Bruzual Gustavo
Burbidge Eleanor Margaret
Burbidge Geoffrey R
Burns Jack O'Neal
Buta Ronald J
Butcher Harvey R
Byrd Gene G
Calzetti Daniela
Cameron Luzius Martin
Campos Ana

Campusano Luis E
Cannon Russell D
Cao Shenglin
Cao Xinwu
Capaccioli Massimo
Carigi Leticia
Carignan Claude
Carranza Gustavo J
Carrillo Moreno Rene
Carswell Robert F
Carter David
Casertano Stefano
Casoli Fabienne
Cellone Sergio Aldo
Cepa Jordi
Chalabaev Almas
Chamaraux Pierre
Chatterjee Tapan K
Chen Jiansheng
Chen Zhencheng
Chiba Masashi
Chincarini Guido L
Chokshi Arati
Chou Chih-Kang
Chu Yaoquan
Chugai Nikolai N
Clavel Jean
Cohen Ross D
Colin Jacques
Combes Francoise
Comte Georges
Considere Suzanne
Contopoulos George
Cook Kem Holland
Corwin Harold G
Couch Warrick
Courtes Georges
Courvoisier Thierry J-L
Cowsik Ramanath
Crane Philippe
d'Odorico Sandro
da Costa Luiz A N
Danks Anthony C
Davidge Timothy J
Davidsen Arthur Falnes
Davies Rodney D
Davis Jonathan Ivor
Davis Marc
de Boer Klaas Sjoerds
de Bruyn A Ger
de Carvalho Reinaldo
de la Noe Jerome
de Robertis M M
de Silva L N K
Dejonghe Herwig Bert
Dekel Avishai
Del Olmo Orozco A
Demers Serge
Deng Zugan
Dennefeld Michel
Dettmar Ralf-Juergen
Di Fazio Alberto
Diaz Angeles Isabel
Dickens Robert J
Dickey John M
Dietrich Matthias
Doi Mamoru
Dokuchaev Vyacheslav I
Donas Jose
Donner Karl Johan
Dopita Michael A
Dottori Horacio A
Doyon Rene
Dressel Linda L
Dressler Alan
Drinkwater Michael J
Dubois Pascal
Dufour Reginald James
Dultzin-Hacyan D
Durret Florence
Duval Marie-France
Ebisuzaki Toshikazu
Edelson Rick
Edmunds Michael Geoffrey
Efstathiou George
Einasto Jaan
Ekers Ronald D
Ellis Richard S
Elmegreen Debra Meloy
Elvis Martin S
Elvius Aina M
Emerson David
English Jayanne
Espey Brian Russell
Evans Robert
Evans Roger G
Fabbiano Giuseppina
Faber Sandra M
Fabricant Daniel G
Fairall Anthony P
Falco-Acosta Emilio E
Fall S Michael
Feinstein Carlos
Feitzinger Johannes
Ferland Gary Joseph
Ferrini Federico
Field George B
Filippenko Alexei V
Firmani Claudio A
Flin Piotr
Florsch Alphonse
Foltz Craig B
Forbes Duncan Alan
Ford Holland C Res
Ford W Kent
Forte Juan Carlos
Fouque Pascal
Fraix-Burnet Didier
Francis Paul
Freedman Wendy L
Freeman Kenneth C
Fricke Klaus
Fried Josef Wilhelm
Fritze Klaus
Fritze-von Alvensleben Ute
Frogel Jay Albert
Ftaclas Christ
Fuchs Burkhard
Fujimoto Masayuki
Fukugita Masataka
Gallagher John S
Galletta Giuseppe
Gamaleldin Abdulla I
Garcia Lambas Diego
Garcia-Burillo Santiago
Garilli Bianca
Gascoigne S C B
Gavazzi Giuseppe
Geller Margaret Joan
Georgiev Tsvetan
Gerhard Ortwin
Ghigo Francis D
Ghosh P
Giacani Elsa Beatriz
Gigoyan Kamo
Giovanardi Carlo
Giovanelli Riccardo
Glass Ian Stewart
Godlowski Wlodzimierz
Gonzalez Serrano J I
Goodrich Robert W
Gorgas Garcia Javier
Goss W Miller
Gottesman Stephen T
Gouguenheim Lucienne
Graham John A
Grandi Steven Aldridge
Grasdalen Gary L
Gregg Michael David
Griffiths Richard E
Gunn James E
Gurzadian Grigor A
Gyulbudaghian Armen L
Hagen-Thorn Vladimir A
Hagio Fumihiko

Hamabe Masaru
Hambrayan Valeri V
Hammer Francois
Hanami Hitoshi
Handa Toshihiro
Hara Tetsuya
Hardy Eduardo
Harms Richard James
Harnett Julieine
Hasan Hashima
He Xiangtao
Heckman Timothy M
Heeschen David S
Heidmann Jean
Heidt Jochen
Held Enrico V
Heller Clayton
Helou George
Henry Richard B C
Hensler Gerhard
Hernquist Lars Eric
Hewitt Adelaide
Hewitt Anthony V
Hickson Paul
Hintzen Paul Michael N
Hiotelis Nicolaos
Hjalmarson Ake G
Hjorth Jens
Hjorth Jens
Hodge Paul W
Holmberg Erik B
Hopp Ulrich
Hough James
Hoyle Fred
Hu Fuxing
Hua Chon Trung
Huang Jiehao
Huang Keliang
Huang Song-nian
Huang Yongwei
Huchtmeier Walter K
Hummel Edsko
Humphreys Roberta M
Hunstead Richard W
Hunter Christopher
Hunter James H
Ichikawa Shin-ichi
Ichikawa Takashi
Illingworth Garth D
Impey Christopher D
Infante Leopoldo
Irwin Judith
Israel Frank P
Issa Issa Aly
Iwanowska Wilhelmina
Iye Masanori
Izotov Yuri
Jaffe Walter Joseph
Jerjen Helmut
Jog Chanda J
Joly Monique
Jones Paul
Jones Thomas Walter
Jorgensen Inger
Joshi Mohan N
Joshi U C
Joy Marshall J
Jugaku Jun
Junkes Norbert
Junkkarinen Vesa T
Kalafi Manoucher
Kalinkov Marin P
Kalloglian Arsen T
Kandalian Rafik A
Kandrup Henry Emil
Kaneko Noboru
Kapahi Vijay K
Karachentsev Igor D
Karachentseva Valentina
Karoji Hiroshi
Katgert Peter
Kaufman Michele
Keel William C
Kellermann Kenneth I
Kennicutt Robert C
Kerr Frank J
Khachikian E Ye
Khanna Ramon
Khare Pushpa
King Ivan R
Kinman Thomas D
Kirshner Robert Paul
Klein Ulrich
Knapen Johan Hendrik
Knapp Gillian R
Kochhar R K
Kodaira Keiichi
Kogoshvili Natela G
Kojoian Gabriel
Kollatschny Wolfram
Kontizas Evangelos
Kontorovich Victor
Koo David C-Y
Koratkar Anuradha P
Koribalski Baerbel Silvia
Kormendy John
Kotilainen Jari
Kraan-Korteweg Renee C
Krause Marita
Krishna Gopal
Kron Richard G
Krumm Nathan Allyn
Kumai Yasuki
Kunchev Peter
Kunth Daniel
Kustaanheimo Paul E
Lafon Jean-Pierre J
Lancon Ariane
Larson Richard B
Lasker Barry M
Lauberts Andris
Laurikainen Eija
Lausberg Andre
Lawrence Andrew
Layzer David
Le Fevre Olivier
Leacock Robert Jay
Ledlow Michael James
Lehto Harry J
Leibundgut Bruno
Lelievre Gerard
Lequeux James
Li Jing
Li Qibin
Li Xiaoqing
Liller William
Lilly Simon J
Lin Chia C
Lindblad Per Olof
Liu Ruliang
Liu Yongzhen
Lo Kwok-Yung
Lonsdale Carol J
Lopez Rosario
Lopez-Cruz Omar
Lord Steven Donald
Lorenz Hilmar
Lortet Marie Claire
Loup Cecile
Low Frank J
Lu Limin
Lucey John
Lugger Phyllis M
Luminet Jean-Pierre
Lynden-Bell Donald
Lynds Beverly T
Lynds Roger C
Ma Er
Macalpine Gordon M
Maccagni Dario
Macchetto Ferdinando
Madden Suzanne
Madore Barry Francis
Magris C Gladis
Mahtessian Abraham P

Malagnini Maria Lucia
Mannucci Filippo
Marcelin Michel
Marques Dos Santos P
Marston Anthony Philip
Martin Maria Cristina
Martinet Louis
Masegosa Gallego J
Mathewson Donald S
Mauersberger Rainer
Maurice Eric N
Mavrides Stamatia
Mayya Divakara
McBreen Brian Philip
McGaugh Stacy Sutton
Mediavilla Evencio
Meier David L
Meikle William P S
Meisenheimer Klaus
Mendes de Oliveira Claudia
Menon T K
Meusinger Helmut
Milet Bernard L
Miley George K
Miller Hugh R
Miller Joseph S
Miller Richard H
Mills Bernard Y
Mineva Veneta
Mirabel Igor Felix
Mizuno Takao
Moles Mariano J
Molinari Emilio
Moody Joseph Ward
Moorwood Alan F M
Moreno Edmundo
Moss Christopher
Mould Jeremy R
Mueller Volker
Munoz-Tunon Casiana
Muratorio Gerard
Murray Stephen S
Mushotzky Richard
Muzzio Juan C
Nair Sunita
Nakai Naomasa
Namboodiri P M S
Narlikar Jayant V
Navarro Julio Fernando
Ninkovic Slobodan
Nityananda R
Noguchi Masafumi
Noonan Thomas W
Norman Colin A
Nulsen Paul
O'Connell Robert West
O'Dea Christopher P
Oemler Augustus
Ohta Kouji
Oke J Beverley
Oleak H
Olofsson Kjell
Oosterloo Thomas
Osman Anas Mohamed
Osterbrock Donald E
Owen Frazer Nelson
Pacholczyk Andrzej G
Palmer Philip
Palumbo Giorgio G C
Pan Rong-Shi
Papayannopoulos Th
Parker Quentin
Pastoriza Miriani G
Paturel Georges
Peimbert Manuel
Pello Roser Descayre
Peng Qiuhe
Perea-Duarte Jaime D
Perez Enrique
Perez Fournon Ismael
Perry Judith J
Peters William L III
Peterson Bradley Michael
Peterson Charles John
Petit Jean-Marc
Petrosian Artashes R
Petrov Georgy Trendafilov
Petuchowski Samuel J
Pfenniger Daniel
Phillips Mark M
Pickles Andrew John
Pikichian Hovhannes
Pismis de Recillas Paris
Pogge Richard William
Popescu Cristina Carmen
Popov Vasil Nikolov
Poveda Arcadio
Prabhu Tushar P
Prendergast Kevin H
Press William H
Prevot-Burnichon Marie Louise
Prieur Jean-louis
Pritchet Christopher J
Pronik I I
Pronik V I
Proust Dominique
Puerari Ivanio
Quinn Peter
Quintana Hernan
Rafanelli Piero
Rampazzo Roberto
Raychaudhury Somak
Reaves Gibson
Rebeirot Edith
Reichert Gail Anne
Rephaeli Yoel
Richer Harvey B
Richstone Douglas O
Richter Gotthard
Roberts Morton S
Roberts William W
Robertson James Gordon
Rocca-Volmerange Brigitte
Rodriguez-Espinosa Jose M
Roeser Hermann-Josef
Rood Herbert J
Roos Nicolaas
Rosa Michael Richard
Rosado Margarita
Rose James Anthony
Rots Arnold H
Rozas Maite
Rubin Vera C
Rudnicki Konrad
Rydbeck Gustaf H B
Ryder Stuart
Sackett Penny
Sadat Rachida
Sadler Elaine Margaret
Sadun Alberto Carlo
Sahibov Firuz H
Sakai Shoko
Sakhibov Firouz
Sala Ferran
Salvador-Sole Eduardo
Sanahuja Blas
Sancisi Renzo
Sanders David B
Sanders Robert
Sanroma Manuel
Sapre A K
Sarazin Craig L
Sargent Wallace L W
Sasaki Minoru
Sasaki Toshiyuki
Saslaw William C
Sastry Shankara K
Savage Ann
Sawa Takeyasu
Schechter Paul L
Schmidt K H
Schmidt Maarten
Schuecking E L
Schultz G V
Schulz Hartmut

Schwarz Ulrich J
Schweizer Francois
Sciama Dennis W
Scorza de Appl Cecilia
Scoville Nicholas Z
Searle Leonard
Sedrakian David
Seiden Philip E
Sekiguchi Maki
Sellwood Jerry A
Seshadri Sridhar
Setti Giancarlo
Shakeshaft John R
Shakhbazian Romelia K
Sharples Ray
Shaver Peter A
Shaya Edward J
Sherwood William A
Shields Gregory A
Shields Joseph C
Shimasaku Kazuhiro
Shore Steven N
Shostak G Seth
Sigurdsson Steinn
Sil'chenko Olga K
Sillanpaa Aimo Kalevi
Simien Francois
Simkin Susan M
Singh Kulinder Pal
Sitko Michael L
Skillman Evan D
Sleath John
Slezak Eric
Smecker-Hane Tammy A
Smith Bruce F
Smith Eric Philip
Smith Haywood C
Smith Malcolm G
Smith Harding E
Soares Domingos S L
Sobouti Yousef
Soltan Andrzej Maria
Song Guoxuan
Sparks William Brian
Spinrad Hyron
Srinivasan G
Staveley-Smith Lister
Steiman-Cameron Thomas
Stepanian A A
Stepanian Jivan A
Stiavelli Massimo
Stirpe Giovanna M
Stone Remington P S
Storchi-Bergman Thaisa
Strom Richard G
Strom Robert G
Subrahmanyam P V
Sugai Hajime
Sulentic Jack W
Sullivan Woodruff T
Sun Wei-Hsin
Sundelius Bjoern
Syer David
Tacconi Linda J
Tacconi-Garman Lowell E
Tagger Michel
Takalo Leo O
Takami Hideki
Takase Bunshiro
Takato Naruhisa
Talbot Raymond J
Tammann Gustav Andreas
Tanaka Yutaka D
Taniguchi Yoshiaki
Telesco Charles M
Telles Eduardo
Terlevich Roberto Juan
Terzian Yervant
Thakur Ratna Kumar
Theis Christian
Thomas Peter A
Thomasson Magnus
Thonnard Norbert
Thuan Trinh Xuan
Tiersch Heinz
Tifft William G
Tong Yi
Tonry John
Toomre Alar
Tovmassian Hrant M
Toyama Kiyotaka
Traat Peeter
Treder H J
Tremaine Scott Duncan
Trinchieri Ginevra
Tsuchiya Toshio
Tully Richard Brent
Turner Edwin L
Tyson John Anthony
Udry Stephane
Ulrich Marie-Helene D
Unger Stephen
Urbanik Marek
Valentijn Edwin A
Valtonen Mauri J
van Albada tjeerd S
van den Bergh Sidney
van der Hulst Jan M
van der Kruit Pieter C
van der Laan Harry
van der Marel Roeland P
van Driel Willem
van Genderen Arnoud M
van Gorkom Jacqueline H
van Moorsel Gustaaf
van Woerden Hugo
Varma Ram Kumar
Vauglin Isabelle
Veilleux Sylvain
Verdes-Montenegro Lourdes
Vermeulen Rene Cornelis
Veron Marie-Paule
Veron Philippe
Vigroux Laurent
Visvanathan Natarajan
Voglis Nikos
Vrtilek Jan M
Wada Keiichi
Wagner Stefan
Wakamatsu Ken-ichi
Wallin John Frederick
Walterbos Rene A M
Wang Tinggui
Ward Martin John
Warner John W
Weedman Daniel W
Wehinger Peter A
Weiler Kurt W
Welch Gary A
Westerlund Bengt E
White Richard Allan
White Simon David Manion
Whitford Albert E
Whitmore Bradley C
Wielen Roland
Wiita Paul Joseph
Wild Paul
Wilkinson Althea
Williams Barbara A
Williams Robert E
Williams Theodore B
Wills Beverley J
Wills Derek
Wilson Albert G
Wilson Andrew S
Windhorst Rogier A
Wisotzki Lutz
Wlerick Gerard
Woosley Stanley E
Worrall Diana Mary
Wozniak Herve
Wrobel Joan Marie
Wynn-Williams C G
Xia Xiaoyang
Yamada Yoshiyaki

Yamagata Tomohiko
Yoshida Michitoshi
Young Judith Sharn
Zamorano Jaime
Zavatti Franco
Zeilinger Werner W
Zepf Stephen Edward
Zhang Xiaolei
Zhang Yang
Zheng Wei
Zhou Xu
Zhou Youyuan
Ziegler Harald
Zinn Robert J

Composition of Commission 29

Stellar Spectra/Spectres stellaires

President: Barbuy Beatriz

Vice-President: Mathys Gautier

Secretary:

Organizing Committee:

Cottrell Peter Ledsam
Gray David F
Heber Ulrich
Khokhlova Vers L
Lambert David L
Nissen Poul E
Peterson Ruth Carol
Sneden Christopher A

Members:

Abhyankar Krishna D
Abt Helmut A
Adelman Saul J
Aikman G Chris l
Ake Thomas Bellis
Alecian Georges
Aller Lawrence Hugh
Ambruster Carol
Andrillat Yvette
Appenzeller Immo
Arkharov Arkadi A
Artru Marie-Christine
Atac Tamer
Baade Dietrich
Baliunas Sallie L
Ballereau Dominique
Baratta Giovanni Battista
Basri Gibor B
Batalha Celso Correa
Bauer Wendy Hagen
Beckman John E
Bellas-Velidis Ioannis
Berger Jacques G
Bessell Michael S
Bikmaev Ilfan
Boesgaard Ann M
Boggess Albert
Bohlender David
Bond Howard E
Bonsack Walter K
Bopp Bernard W
Bouvier Jerome
Bouvier Pierre
Boyarchuk Alexander A
Brandi Elisande Estela
Breysacher Jacques
Brown Douglas Nason
Brown Paul James Frank
Bruhweiler Fred C
Bruning David H
Bues Irmela D
Burkhart Claude
Butler Keith
Carney Bruce William
Carpenter Kenneth G
Carter Bradley Darren
Cassatella Angelo
Castelli Fiorella
Catala Claude
Catalano Santo
Catchpole Robin M
Cayrel Roger
Cayrel de Strobel Giusa
Chavez-Dagostino Miguel
Cidale Lydia Sonia
Climenhaga John L
Coluzzi Regina
Conti Peter S
Corbally Christopher
Cornide Manuel
Cowley Anne P
Cowley Charles R
Crowther Paul
Cunha Katia
Damineli Neto Augusto
Dawanas Djoni N
de Araujo Francisco X
de Castro Elisa
de Groot Mart
Divan Lucienne
Doazan Vera
Dobronravin Peter
Dolidze Madona V
Dravins Dainis
Duncan Douglas Kevin
Dworetsky Michael M
Eenens Philippe
Faraggiana Rosanna
Feast Michael W
Felenbok Paul
Fernandez-Figueroa M J
Fitzpatrick Edward L
Floquet Michele
Foing Bernard H
Foy Renaud
Franchini Mariagrazia
Francois Patrick
Frandsen Soeren
Freire Ferrero Rubens G
Friedjung Michael
Friel Eileen D
Fujita Yoshio
Garcia Lopez Ramon J

Garmany Catherine D
Garrison Robert F
Gautier Daniel
Gehren Thomas
Gerbaldi Michele
Gershberg R E
Gesicki Krzysztof
Ghosh Kajal Kumar
Giampapa Mark S
Gilra Daya P
Giovannelli Franco
Glagolevskij Juri V
Glushneva Irina N
Goebel John H
Grady Carol Anne
Gratton Raffaela G
Greenstein Jesse L
Griffin Rita E M
Griffin Roger F
Gustafsson Bengt
Guthrie Bruce N G
Hack Margherita
Hanuschik Reinhard
Harmer Charles F W
Harmer Dianne L
Hartmann Lee William
Hashimoto Osama
Hearnshaw John B
Heintze J R W
Henrichs Hubertus F
Herbig George H
Heske Astrid
Hessman Frederic Victor
Hill Grant
Hinkle Kenneth H
Hirai Masanori
Hirata Ryuko
Hoeflich Peter
Houk Nancy
Houziaux Leo
Hron Josef
Huang Changchung
Hubeny Ivan
Hubert-Delplace A -M
Hubrig Swetlana
Huenemoerder David P
Hunger Kurt
Hyland A R Harry
Izumiura Hideyuki
Jiang Shi-Yang
Johnson Hollis R
Jordan Carole
Joshi Suresh Chandra
Jugaku Jun
Keenan Philip C
Kipper Tonu
Kitchin Christopher R
Klochkova Valentina
Kodaira Keiichi
Kogure Tomokazu
Kolka Indrek
Kotnik-Karuza Dubravka
Koubsky Pavel
Kovachev B J
Kraft Robert P
Krempec-Krygier Janina
Kwok Sun
Lago Maria Teresa V T
Lagrange Anne-Marie
Laird John B
Lamers Henny J G L M
Lamontagne Robert
Landstreet John D
Langer George Edward
Lanz Thierry
Le Contel Jean-Michel
Lebre Agnes
Leckrone David S
Leedjaerv Laurits
Lester John B
Levato Orlando Hugo
Liebert James W
Little-Marenin Irene R
Locanthi Dorothy Davis
Lopes Dalton De faria
Luck R Earle
Lundstrom Ingemar
Magain Pierre
Magazzu Antonio
Maillard Jean-Pierre
Maitzen Hans M
Malaroda Stella M
Marilli Ettore
Massey Philip L
McGregor Peter John
Mcnamara Delbert H
Megessier Claude
Mickaelian Areg Martin
Mikulasek Zdenek
Mishenina Tamara
Moffat Anthony F J
Molaro Paolo
Moos Henry Warren
Morossi Carlo
Morrison Nancy Dunlap
Napiwotzki Ralf
Neckel Heinz
Nicholls Ralph W
Niedzielski Andrzej
Niemela Virpi S
Nishimura Shiro Emer
Norris John
Nugis Tiit
Okazaki Atsuo T
Orlov Mikhail
Owocki Stanley Peter
Pagel Bernard E J
Parsons Sidney B
Parthasarathy M
Pasinetti Laura E
Pavlenko Yakov V
Pedoussaut Andre
Peery Benjamin F
Perrin Marie-Noel
Peters Geraldine Joan
Pilachowski Catherine
Pintado Olga Ines
Plavec Mirek J
Plez Bertrand
Polcaro V F
Polidan Ronald S
Praderie Francoise
Prinja Raman
Querci Francois R
Querci Monique
Rao N Kameswara
Rautela B S
Rebolo Rafael
Rego Fernandez M
Reimers Dieter
Ringuelet Adela E
Rodgers Alex W
Rose James Anthony
Rossi Corinne
Rossi Lucio
Rutten Robert J
Ryan Sean Gerard
Sadakane Kozo
Sahade Jorge
Sanchez Almeida Jorge
Sanwal Basant Ballabh
Sareyan Jean-Pierre
Schild Rudolph E
Scholz Gerhard
Schroeder Klaus Peter
Seggewiss Wilhelm
Shore Steven N
Simon Theodore
Sinnerstad Ulf E
Slettebak Arne
Smalley Barry
Smith Graeme H
Smith Keith Colin
Smith Myron A
Smith Verne V

Snow Theodore P
Soderblom David R
Sonneborn George
Sonti Sreedhar Rao
Spite Francois M
Spite Monique
St-Louis Nicole
Stalio Roberto
Stawikowski Antoni
Stecher Theodore P
Steffen Matthias
Stefl Stanislav
Stencel Robert Edward
Suntzeff Nicholas B
Svolopoulos Sotirios
Swensson John W
Swings Jean-Pierre
Takada-Hidai Masahide
Talavera A
Thevenin Frederic
Tomov Toma V
Torrejon Jose Miguel
Tuominen Ilkka V
Underhill Anne B
Utsumi Kazuhiko
Valtier Jean-Claude
van der Hucht Karel A
van Winckel Hans
van't Veer-Menneret Claude
Vasu-Mallik Sushma
Vilhu Osmi
Viotti Roberto
Vladilo Giovanni
Vogt Nikolaus
Vogt Steven Scott
Vreux Jean Marie
Wahlgren Glenn Michael
Wallerstein George
Waterworth Michael
Wegner Gary Alan
Wehinger Peter A
Weiss Werner W
Weniger Schame
Williams Peredur M
Wing Robert F
Wolf Bernhard
Wolff Sidney C
Wood H J
Wyckoff Susan
Yamashita Yasumasa
Yoshioka Kazuo
Zhao Gang
Zorec Juan
Zverko Juraj

Composition of Commission 30

Radial Velocities/Vitesses radiales

President: Hearnshaw John B

Vice-President: Tokovinin Andrej A

Organizing Committee:

Cochran William David
Fekel Francis C
Mazeh Tsevi
Morrell Nidia
Nordstroem Birgitta
Quintana Hernan
Scarfe Colin D
Stefanik Robert
Udry Stephane

Members:

Abt Helmut A
Andersen Johannes
Balona Luis Antero
Barbier-Brossat Madeleine
Batten Alan H
Beavers Willet I
Beers Timothy C
Boulon Jacques J
Breger Michel
Burki Gilbert
Burnage Robert
Carney Bruce William
Carquillat Jean-Michel
Crampton David
da Costa Luiz A N
Davis Marc
Davis Robert J
de Jonge J K
Dubath Pierre
Duflot Marcelle
Edmondson Frank K
Eelsalu Heino
Fairall Anthony P
Fehrenbach Charles
Fletcher J Murray
Florsch Alphonse
Foltz Craig B
Freeman Kenneth C
Garcia Beatriz Elena
Georgelin Yvon P
Gilmore Gerard Francis
Giovanelli Riccardo
Gnedin Yurij N
Gouguenheim Lucienne
Grenon Michel
Griffin Roger F
Halbwachs Jean Louis
Heintze J R W
Hewett Paul
Hilditch Ronald W
Hill Graham
Hrivnak Bruce J
Huang Changchung
Hube Douglas P
Hubrig Swetlana
Huchra John Peter
Imbert Maurice
Irwin Alan W
Kadouri Talib Hadi
Karachentsev Igor D
Kraft Robert P
Latham David W
Levato Orlando Hugo
Lewis Brian Murray
Lindgren Harri
Marschall Laurence A
Maurice Eric N
Mayor Michel
McClure Robert D
McMillan Robert S
Melnick Gary J
Mermilliod Jean-Claude
Meylan Georges
Missana Marco
Morbey Christopher L
Oetken L
Pedoussaut Andre
Pellegrini Paulo S S
Perry Charles L
Peterson Ruth Carol
Philip A G Davis
Popov Victor S
Preston George W
Prevot Louis
Rastorguev Alexey S
Ratnatunga Kavan U
Rebeirot Edith
Romanov Yuri S
Rubin Vera C
Samus Nikolai N
Sanwal N B
Scholz Gerhard
Smith Myron A
Solivella Gladys Rebecca
Stock Jurgen D
Suntzeff Nicholas B
Tonry John
van Dessel Edwin Ludo
Verschueren Werner
Walker Gordon A H
Wegner Gary Alan
Willstrop Roderick V
Yang Stephenson L S
Yoss Kenneth M

Composition of Commission 31

Time/L'Heure

President: Fukushima Toshio

Vice-President: Petit Gerard

Organizing Committee:

Backer Donald Ch
Beutler Gerhard
Brumberg Victor A
Fliegel Henry F
Leschiutta S
Luck John M
Paquet Paul Eg
Proverbio Edoardo
Qi Guanrong
Thomas Claudine
Veillet Christian
Winkler Gernot M R
Zhai Zaocheng

Members:

Abele Maris K
Afanasjeva Praskovya M
Allan David W
Alley Carrol O
Aoki Shinko
Benavente Jose
Bender Peter L
Blinov Nikolai S
Breakiron Lee Allen
Carter William Eugene
Catalan Manuel
Chamberlain Joseph M
Dehant Veronique
Dickey Jean O'Brien
Douglas R J
Enslin Heinz
Fallon Frederick W
Feissel Martine
Fujimoto Masa-Katsu
Gaignebet Jean
Gambis Daniel
Granveaud Michel
Grudler Pierre
Guinot Bernard R
Hall R Glenn
Han Tianqi
Hellwig Helmut Wilhelm
Hemmleb Gerhard
Hers Jan
Iijima Shigetaka
Ilyasov Yuri P
Jin Wenjing
Kakuta Chuichi
Klepczynski William J
Kolaczek Barbara
Kovalevsky Jean
Liang Zhonghuan
Lieske Jay H
Liu Jinming
Lu Benkui
Luo Dingchang
Luo Shi-Fang
Mathur B S
Matsakis Demetrios N
McCarthy Dennis D
Meinig Manfred
Melbourne William G
Melchior Paul J
Miao Yongrui
Morgan Peter
Mueller Ivan I
Naumov Vitalij A
Newhall X X
Niimi Yukio
Noel Fernando
Pilkington John D H
Popelar Josef
Pushkin Sergey B
Ray James R
Robertson Douglas S
Schuler Walter
Smith Humphry M
Smylie Douglas E
Song Jin'an
Stanila George
Tsuchiya Atsushi
Vicente Raimundo O
Webrova Ludmila
Wilkins George A
Wu Guichen
Wu Shouxian
Xu Bang-Xin
Yang Ke-jun
Yatskiv Ya S
Ye Shuhua
Zhang Jintong
Zhao Gang
Zheng Ying
Zhuang Qixiang

Composition of Commission 33

Structure and Dynamics of the Galactic System/
Structure and dynamique du système galactique

President: Freeman Kenneth C

Vice-President: Spergel David N

Organizing Committee:

Binney James J
Fridman Aleksey M
Gerhard Ortwin
Grindlay Jonathan E
Kalnajs Agris J
Matteucci Francesca
Palous Jan
Perryman Michael A C
Whitelock Patricia Ann

Members:

Aarseth Sverre J
Adamson Andrew
Afanas'ev Viktor L
Agekjan Tateos A
Aguilar Luis A Chiu
Aizu Ko
Altenhoff Wilhelm J
Ambastha A K
Andersen Johannes
Antonov Vadim A
Aoki Shinko
Aparicio Antonio
Ardeberg Arne L
Asteriadis Georgios
Athanassoula Evangelia
Bahcall John N
Baier Frank
Balazs Lajos G
Balbus Steven A
Baldwin John E
Banhatti Dilip Gopal
Barbanis Basil
Barberis Bruno
Bartasiute Stanislava
Bash Frank N
Basu Baidyanath
Baud Boudewijn
Becker Wilhelm
Berkhuijsen Elly M
Bienayme Olivier
Blaauw Adriaan
Blanco Victor M
Blitz Leo
Bloemen Johannes B G M
Boulon Jacques J
Brand Jan
Bronfman Leonardo
Burke Bernard F
Burton W Butler
Caldwell John A R
Cane Hilary Vivien
Carpintero Daniel Diego
Carrasco Luis
Caswell James L
Cesarsky Catherine J
Cesarsky Diego A
Chapman Jessica
Chen Li
Chen Zhen
Christodoulou Dmitris
Churchwell Edward B
Cincotta Pablo Miguel
Ciurla Tadeusz
Clemens Dan P
Clube S V M
Cohen Richard S
Colin Jacques
Comins Neil Francis
Contopoulos George
Costa Edgardo
Courtes Georges
Crampton David
Crawford David L
Creze Michel
Cubarsi Rafael
Cudworth Kyle Mccabe
Cuperman Sami
Davies Rodney D
Dawson Peter
de Jong Teije
Dejonghe Herwig Bert
Dekel Avishai
Denoyelle Jozef Kic
Dickel Helene R
Dickel John R
Dickman Robert L
Dieter Conklin Nannielou H
Djorgovski Stanislav
Downes Dennis
Drilling John S
Ducati Jorge Ricardo
Ducourant Christine
Dzigvashvili R M
Edmondson Frank K
Efremov Yuri N
Egret Daniel
Einasto Jaan
Elmegreen Debra Meloy
Elsaesser Hans
Evangelidis E

Evans Wyn
Faber Sandra M
Fall S Michael
Feast Michael W
Fehrenbach Charles
Feitzinger Johannes
Fenkart Rolf P
Figueras Francesca
Fitzgerald M Pim
Flynn Chris
Fuchs Burkhard
Fujimoto Masa-Katsu
Fujiwara Takao
Fukunaga Masataka
Galletto Dionigi
Garzon Francisco
Gemmo Alessandra
Genkin Igor L
Genzel Reinhard
Georgelin Yvon P
Georgelin Yvonne M
Gilmore Gerard Francis
Goldreich Peter
Gomez Ana E
Gordon Mark A
Gottesman Stephen T
Grayzeck Edwin J
Grenon Michel
Gupta Sunil K
Gyldenkerne Kjeld
Habe Asao
Habing Harm J
Hakkila Jon Eric
Hamajima Kiyotoshi
Hanami Hitoshi
Hartkopf William I
Hawkins Michael R S
Hayli Avram
Heiles Carl
Henon Michel C
Herbst William
Herman Jacobus
Hernandez-Pajares Manuel
Hobbs Robert W
Hori Genichiro
Hozumi Shunsuke
Hron Josef
Huang Song-nian
Hughes Victor A
Hulsbosch A N M
Humphreys Roberta M
Hunter Christopher
Ikeuchi Satoru
Inagaki Shogo
Innanen Kimmo A
Isobe Syuzo
Israel Frank P
Iwaniszewska Cecylia
Iwanowska Wilhelmina
Iye Masanori
Jablonka Pascale
Jackson Peter Douglas
Jahreiss Hartmut
Jaschek Carlos O R
Jasniewicz Gerard
Jiang Dongrong
Joench-Soerensen Helge
Jog Chanda J
Johnson Hugh M
Jonas Justin Leonard
Jones Derek H P
Kalandadze N B
Kang Yong Hee
Kasumov Fikret K O
Kato Shoji
Kerr Frank J
Kharadze E K
King Ivan R
Kinman Thomas D
Klare Gerhard
Knapp Gillian R
Kolesnik Igor G
Kolesnik L N
Kormendy John
Kulsrud Russell M
Kutuzov Sergei A
Lafon Jean-Pierre J
Larson Richard B
Latham David W
Lecar Myron
Lee Hyung Mok
Lee Sang Gak
Leisawitz David
Li Jing
Liebert James W
Lin Chia C
Lindblad Per Olof
Lockman Felix J
Loden Kerstin R
Loden Lars Olof
Lu Phillip K
Lunel Madeleine
Lynden-Bell Donald
MacConnell Darrell J
Macrae Donald A
Manchester Richard N
Mark James Wai-Kee
Marochnik L S
Martinet Louis
Mathewson Donald S
Mavridis L N
Mayor Michel
McCarthy Martin F
McGregor Peter John
Meatheringham Stephen
Mennessier Marie-Odile
Mezger Peter G
Mikkola Seppo
Miller Richard H
Mirabel Igor Felix
Mirzoyan Ludwik V
Miyamoto Masanori
Moffat Anthony F J
Monet David G
Monnet Guy J
Morris Mark Root
Muench Guido
Murray C Andrew
Muzzio Juan C
Nahon Fernand
Namboodiri P M S
Neckel Th
Nelson Alistair H
Ninkovic Slobodan
Nishida Minoru
Nishida Mitsugu
Nordstroem Birgitta
Norman Colin A
Nuritdinov Salakhutdin
Oblak Edouard
Oh Kap-Soo
Oja Tarmo
Okuda Haruyuki
Olano Carlos Alberto
Ollongren A
Ostriker Jeremiah P
Palmer Patrick E
Pandey A K
Papayannopoulos Th
Pauls Thomas Albert
Peimbert Manuel
Perek Lubos
Perry Charles L
Pesch Peter
Petrovskaya Irina
Philip A G Davis
Pier Jeffrey R
Pismis de Recillas Paris
Polyachenko Valerij L
Polymilis Chronis
Price R Marcus
Priester Wolfgang
Rabolli Monica
Raharto Moedji
Ratnatunga Kavan U

Rebeirot Edith
Reid Neill
Reif Klaus
Rich Robert Michael
Riegel Kurt W
Roberts Morton S
Roberts William W
Robin Annie C
Robinson Brian J
Rohlfs Kristen
Rong Jianxiang
Rubin Vera C
Ruelas-Mayorga R A
Ruiz Maria Teresa
Rybicki George B
Saar Enn
Sala Ferran
Sanchez-Saavedra M Luisa
Sandqvist Aage
Sanz I Subirana Jaume
Sargent Anneila I
Schechter Paul L
Schmidt Hans
Schmidt K H
Schmidt Maarten
Schmidt-Kaler Theodor
Seggewiss Wilhelm
Seimenis John
Sellwood Jerry A
Serabyn Eugene
Shane William W
Sharov A S
Sher David
Shimizu Tsutomu Emer
Shu Frank H
Simonson S Christian
Slettebak Arne
Sobouti Yousef
Solomon Philip M
Song Guoxuan
Soubiran Caroline
Sparke Linda
Spiegel E
Stecker Floyd W
Steinlin Uli
Stephenson C Bruce
Stibbs Douglas W N
Strobel Andrzej
Sturch Conrad R
Surdin Vladimir G
Svolopoulos Sotirios
Sygnet Jean Francois
Szebehely Victor G
Tammann Gustav Andreas
Terzides Charalambos
The Pik-Sin
Thielheim Klaus O
Thomas Claudine
Tinney Christopher
Tobin William
Tomisaka Kohji
Tong Yi
Toomre Alar
Toomre Juri
Torra Jordi
Tosa Makoto
Trefzger Charles F
Tsioumis Alexandros
Tsujimoto Takuji
Turon Catherine
Upgren Arthur R
Valtonen Mauri J
van der Kruit Pieter C
van Woerden Hugo
Vandervoort Peter O
Vega E Irene
Venugopal V R
Vergne Maria Marcela
Verschuur Gerrit L
Vetesnik Miroslav
Voroshilov V I
Wachlin Felipe Carlos
Wayman Patrick A
Weaver Harold F
Weistrop Donna
Westerhout Gart
Westerlund Bengt E
White Raymond E
Whiteoak John B
Whittet Douglas C B
Wielebinski Richard
Wielen Roland
Woltjer Lodewijk
Woodward Paul R
Wouterloot Jan Gerard A
Wramdemark Stig S o
Wyse Rosemary F
Xiang Delin
Yamagata Tomohiko
Yoshii Yuzuru
Younis Saad M
Yuan Chi
Zachilas Loukas
Zhang Bin
Zhao Junliang

Composition of Commission 34

Interstellar Matter/Matière interstellaire

President: Dopita Michael A

Vice-President: Reipurth Bo

Organizing Committee:

Bachiller Rafael
Burton Michael G
Dyson John E
Elmegreen Bruce Gordon
Flower David R
Henning Thomas
Kwok Sun
Okuda Haruyuki
Pineau des Forets Guillaume
Raymond John Charles
Tenorio-Tagle Guillermo
Torres-Peimbert Silvia
van Dishoeck Ewine F
Viegas Aldrovandi S M
Wilson Thomas L

Members:

Aannestad Per Arne
Acker Agnes
Adams Fred
Aiad A Zaki
Aitken David K
Akabane Kenji A
Alcolea Javier
Aller Lawrence Hugh
Altenhoff Wilhelm J
Anantharamaiah K R
Andersson Bengt Goeran
Andrillat Yvette
Andronov Ivan
Anglada Guillem
Arkhipova Vesa P
Arny Thomas T
Arthur Jane
Avery Lorne W
Axford W Ian
Azcarate Ismael N
Baars Jacob W M
Baart Edward E
Baldwin John E
Baluteau Jean-Paul
Bania Thomas Michael
Barlow Michael J
Barnes Aaron
Bash Frank N
Baudry Alain
Becklin Eric E
Beckman John E
Beckwith Steven V W
Bedogni Roberto
Benaydoun Jean-Jacques
Bergeron Jacqueline A
Bergman Per-Goeran
Berkhuijsen Elly M
Bernat Andrew Plous
Bertout Claude
Bhatt H C
Bianchi Luciana
Bieging John Harold
Bignell R Carl
Binette Luc
Birkle Kurt
Black John Harry
Blades John Chris
Blair Guy Norman
Blair William P
Bless Robert C
Blitz Leo
Bloemen Johannes B G M
Bobrowsky Matthew
Bochkarev Nikolai G
Bode Michael F
Bodenheimer Peter
Boeshaar Gregory Orth
Boggess Albert
Bohlin Ralph C
Boisse Patrick
Boland Wilfried
Borgman Jan
Boulanger Francois
Bouvier Jerome
Brand Jan
Brand Peter W J l
Braunsfurth Edward
Breitschwerdt Dieter
Brinkmann Wolfgang
Bromage Gordon E
Brouillet Nathalie
Brown Ronald D
Bruhweiler Fred C
Bryce Myfanwy
Bujarrabal Valentin
Burgess Alan
Burke Bernard F
Burton W Butler
Bystrova Natalija V
Bzowski Maciej
Canto Jorge
Caplan James
Cappa de Nicolau Cristina
Capriotti Eugene R
Capuzzo Dolcetta Roberto
Carruthers George R
Castaneda Hector
Caswell James L
Cecchi-Pellini Cesare
Centurion Martin Miriam
Cernicharo Jose
Cerruti'sola Monica
Cersosimo Juan Carlos

Cesarsky Catherine J
Cesarsky Diego A
Chandra Suresh
Chen Yang
Cheng Kwang Ping
Chevalier Roger A
Chini Rolf
Chopinet Marguerite
Chu You-Hua
Churchwell Edward B
Ciardullo Robin
Clark Frank Oliver
Clegg Robin E S
Code Arthur D
Cohen Marshall H
Colangeli Luigi
Collin-Souffrin Suzy
Colomb Fernando R
Combes Francoise
Corbelli Edvige
Costero Rafael
Courtes Georges
Cowie Lennox Lauchlan
Cox Donald P
Cox Pierre
Coyne George V
Crane Philippe
Crawford Ian Andrew
Crovisier Jacques
Cruvellier Paul E
Cudaback David D
Cugnon Pierre
Czyzak Stanley J
Dahn Conard Curtis
Dalgarno Alexander
Danks Anthony C
Danly Laura
Davies Rodney D
Davis Jonathan Ivor
de Almeida Amaury A
de Boer Klaas Sjoerds
de Gouveia Dal Pino E M
de Jong Teije
de la Noe Jerome
Deguchi Shuji
Deharveng Lise
Deiss Bruno M
Dennefeld Michel
Dewdney Peter E F
d'Hendecourt Louis
Dickel Helene R
Dickel John R
Dickey John M
Di Fazio Alberto
Dinerstein Harriet L
Disney Michael J
d'Odorico Sandro
Dokuchaev Vyacheslav I
Dokuchaeva Olga D
Donn Bertram D
Dorschner Johann
Dottori Horacio A
Downes Dennis
Draine Bruce T
Drapatz Siegfried W
Dreher John W
Dubner Gloria
Dubout Renee
Dufour Reginald James
Duley Walter W
Dupree Andrea K
Duvert Gilles
Dwek Eli
Elitzur Moshe
Elliott Kenneth H
Elmegreen Debra Meloy
Elvius Aina M
Emerson James P
Encrenaz Pierre J
Escalante Vladimir
Esipov Valentin F
Esteban Lopez Cesara
Evans Aneurin
Evans Neal J
Falgarone Edith
Falk Sydney W
Falle Samuel A
Fan Ying
Faulkner Donald J
Federman Steven Robert
Feibelman Walter A
Feitzinger Johannes
Felli Marcello
Felten James E
Ferlet Roger
Ferrini Federico
Fesen Robert A
Fiebig Dirk
Field David
Field George B
Fierro Julieta
Fischer Jacqueline
Flannery Brian Paul
Fleck Robert Charles
Ford Holland C Res
Forster James Richard
Fridlund Malcolm
Friedemann Christian
Frisch Priscilla
Fuente Asuncion
Fukui Yasuo
Furniss Ian
Gardner Francis F
Garnett Donald Roy
Gaustad John E
Gay Jean
Geballe Thomas R
Gehrels Tom
Genzel Reinhard
Georgelin Yvon P
Gerard Eric
Gerin Maryvonne
Gerola Humberto
Gezari Daniel Ysa
Ghanbari Jamshid
Giacani Elsa Beatriz
Gilra Daya P
Giovanelli Riccardo
Godfrey Peter Douglas
Goebel John H
Goldes Guillermo Victor
Goldreich Peter
Goldsmith Donald W
Goldstein Samuel J
Goldsworthy Frederick A
Golovatyj Volodymyr
Gomez Gonzalez Jesus
Gonzales-Alfonso Eduardo
Goodman Alyssa Ann
Gordon Courtney P
Gordon Mark A
Gosachinskij Igor V
Goss W Miller
Graham David A
Grasdalen Gary L
Gredel Roland
Greenberg J Mayo
Gregorio-Hetem Jane
Grewing Michael
Guelin Michel
Guertler Joachin
Guesten Rolf
Gull Theodore R
Gurzadian Grigor A
Guseinov O H
Habing Harm J
Hackwell John A
Hanami Hitoshi
Hardebeck Ellen G
Harrington J Patrick
Harris Alan William
Harris Stella
Harten Ronald H
Hartl Herbert
Hartquist Thomas Wilbur

Harvey Paul Michael
Hayashi Saeko S
Haynes Raymond F
Hecht James H
Heiles Carl
Hein Righini Giovanna
Helfer H Lawrence
Helou George
Henkel Christian
Henney William John
Herbstmeier Uwe
Herzberg Gerhard
Heydari-Malayeri Mohammad
Hidayat Bambang
Higgs Lloyd A
Hildebrand Roger H
Hippelein Hans H
Hirano Naomi
Hiromoto Norihisa
Hjalmarson Ake G
Hjellming Robert M
Hobbs Lewis M
Hoeglund Bertil
Hollenbach David John
Hollis Jan Michael
Hong Seung Soo
Houziaux Leo
Hovhannessian Rafik
Hua Chon Trung
Huggins Patrick J
Hughes Victor A
Hulsbosch A N M
Hutchings John B
Hutsemekers Damien
Hyung Siek
Irvine William M
Isobe Syuzo
Israel Frank P
Issa Issa Aly
Itoh Hiroshi
Iyengar K V K
Jabir Niama Lafta
Jacoby George H
Jacq Thierry
Jaffe Daniel T
Jenkins Edward B
Jenkins Louise F
Jennings R E
Johnson Fred M
Johnson Hugh M
Johnston Kenneth J
Jones Frank Culver
Jourdain de Muizon M
Jura Michael
Just Andreas
Kafatos Minas
Kaftan May A
Kahn Franz D
Kaifu Norio
Kaler James B
Kamijo Fumio
Kazes Ilya
Keene Jocelyn Betty
Kegel Wilhelm H
Kennicutt Robert C
Kerr Frank J
Kharadze E K
Khromov Gavriil S
Kimura Hiroshi
Kimura Toshiya
King David Leonard
Kirkpatrick Ronald C
Kirshner Robert Paul
Knacke Roger F
Knapp Gillian R
Knude Jens Kirkeskov
Koempe Carsten
Koeppen Joachim
Kohoutek Lubos
Koike Chiyoe
Kolesnik Igor G
Kondo Yoji
Koornneef Jan
Kostyakova Elena B
Kozasa Takashi
Krautter Joachim
Kravchuk Sergei
Kreysa Ernst
Krishna Swamy K S
Kuiper Thomas B H
Kumar C Krishna
Kundu Mukul R
Kunth Daniel
Kunze Ruediger
Kutner Marc Leslie
Kwitter Karen Beth
Kylafis Nikolaos D
Lada Charles Joseph
Lafon Jean-Pierre J
Langer William David
Lasker Barry M
Latter William B
Laureijs Rene J
Laurent Claudine
Le Squeren Anne-Marie
Lee Terence J
Leger Alain
Leisawitz David
Lepine Jacques R D
Lequeux James
Leto Giuseppe
Leung Chun Ming
Likkel Lauren Jones
Liller William
Lin Chia C
Linke Richard Alan
Lis Dariusz C
Liseau Rene
Liszt Harvey Steven
Lizano-Soberon Susana
Lo Kwok-Yung
Lockman Felix J
Lopez Jose Alberto
Loren Robert Bruce
Lortet Marie Claire
Louise Raymond
Lovas Francis John
Low Frank J
Lozinskaya Tat-yana A
Lucas Robert
Luo Shaoguang
Lynds Beverly T
Mac Low Mordecai-Mark
Maciel Walter J
MacLeod John M
Maihara Toshinori
Mallik D C V
Mampaso Antonio
Manchado Arturo
Manchester Richard N
Manfroid Jean
Marston Anthony Philip
Martin Peter G
Martin Robert N
Martin-Pintado Jesus
Masson Colin R
Mather John Cromwell
Mathews William G
Mathewson Donald S
Mathis John S
Matsuhara Hideo
Matsumura Masafumi
Mattila Kalevi
Mauersberger Rainer
McCall Marshall Lester
McCray Richard
McCrea J Dermott
McGee Richard X
McGregor Peter John
McKee Christopher F
McNally Derek
Meaburn John
Mebold Ulrich
Mehringer David Michael
Meier Robert R

Mellema Garrelt
Melnick Gary J
Mendez Roberto H
Mennella Vito
Menon T K
Menzies John W
Meszaros Peter
Mezger Peter G
Millar Thomas J
Miller Joseph S
Milne Douglas K
Minin Igor N
Minn Young Key
Mitchell George F
Miyama Syoken
Mizuno Shun
Mo Jing-er
Monin Jean-louis
Montmerle Thierry
Moreno Corral Marco A
Morgan David H
Morimoto Masaki
Morris Mark Root
Morton Donald C
Mouschovias Telemachos Ch
Muench Guido
Mufson Stuart Lee
Murthy Jayant
Myers Philip C
Nagata Tetsuya
Nakada Yoshikazu
Nakagawa Takao
Nakamoto Taishi
Nakano Makoto
Nakano Takenori
Nandy Kashinath
Neugebauer Gerry
Nguyen-Quang Rieu
Nishi Ryoichi
Nordh Lennart H
Norman Colin A
Nulsen Paul
Nussbaumer Harry
Nuth Joseph A III
O'Dell Charles R
O'Dell Stephen L
Ohtani Hiroshi
Olofsson Hans
Omont Alain
Onaka Takashi
Onello Joseph S
Opendak Michael
Osaki Toru
Osborne John L
Osterbrock Donald E
Ozernoy Leonid M
Pagano Isabella
Pagel Bernard E J
Palla Francesco
Palmer Patrick E
Panagia Nino
Pankonin Vernon Lee
Parker Eugene N
Parthasarathy M
Pauls Thomas Albert
Pecker Jean-Claude
Peimbert Manuel
Pena Miriam
Pendleton Yvonne Jean
Penzias Arno A
Pequignot Daniel
Perault Michel
Perinotto Mario
Persi Paolo
Peters William L III
Petrosian Vahe
Petuchowski Samuel J
Phillips John Peter
Phillips Thomas Gould
Pismis de Recillas Paris
Poeppel Wolfgang G l
Pollacco Don
Pongracic Helen
Porceddu Ignazio E P
Pottasch Stuart R
Pouquet Annick
Prasad Sheo S
Preite-Martinez Andrea
Price R Marcus
Pronik I I
Prusti Timo
Pskovskij Juri P
Puget Jean-Loup
Qin Zhihai
Radhakrishnan V
Raimond Ernst
Ratag Mezak Arnold
Rawlings Jonathan
Rengarajan T N
Reynolds Ronald J
Rickard Lee J
Robbins R Robert
Roberge Wayne G
Roberts William W
Robinson Brian J
Robinson Garry
Roche Patrick F
Rodriguez Luis F
Roelfsema Peter
Roeser Hans-peter
Roger Robert S
Rogers Alan E E
Rohlfs Kristen
Rosa Michael Richard
Rosado Margarita
Rose William K
Rouan Daniel
Roxburgh Ian W
Rozhkovskij Dimitrij A
Rozyczka Michal
Rubin Robert Howard
Russell Stephen
Sabano Yutaka
Sabbadin Franco
Sahu Kailash C
Salinari Piero
Salpeter Edwin E
Salter Christopher John
Sanchez-Saavedra M Luisa
Sancisi Renzo
Sandell Goran Hans l
Sandqvist Aage
Sarazin Craig L
Sargent Anneila I
Sarma N V G
Sato Fumio
Sato Shuji
Savage Blair D
Savedoff Malcolm P
Scalo John Michael
Scappini Flavio
Scarrott Stanley M
Schatzman Evry
Scherb Frank
Scheuer Peter A G
Schmid-Burgk J
Schmidt Thomas
Schmidt-Kaler Theodor
Schultz G V
Schulz Rolf Andreas
Schwartz Philip R
Schwartz Richard D
Schwarz Ulrich J
Scott Eugene Howard
Scoville Nicholas Z
Seaton Michael J
Seki Munezo
Sellgren Kristen
Shane William W
Shao Cheng-yuan
Shapiro Stuart L
Sharpless Stewart
Shaver Peter A
Shawl Stephen J
Shcheglov P V

Shematovich Valery I
Sherwood William A
Shields Gregory A
Shmeld Ivar
Shu Frank H
Shull John Michael
Shull Peter Otto
Shustov Boris M
Silberberg Rein
Silich Sergey
Silk Joseph I
Silvestro Giovanni
Simons Stuart
Singh Patan Deen
Sitko Michael L
Sivan Jean-Pierre
Skilling John
Skulskyj Mychajlo Y
Sloan Gregory Clayton
Smith Barham W
Smith Craig H
Smith Michael
Smith Peter L
Smith Robert G
Snell Ronald L
Snow Theodore P
Sobolev Andrej M
Sobolev V V
Sofia Sabatino
Sofue Yoshiaki
Solc Martin
Solomon Philip M
Somerville William B
Stahler Steven W
Stanga Ruggero
Stanghellini Letizia
Stasinska Grazyna
Stecher Theodore P
Stecklum Bringfried
Stenholm Bjoern
Stenholm Lars
Stone James McLellan
Strom Richard G
Suh Kyung-Won
Sun Jin
Suzuki Yoshimasa
Swade Daryl Allen
Sylvester Roger
Szczerba Ryszard
Takakubo Keiya
Takano Toshiaki
Tamura Motohide
Tamura Shin'ichi
Tanaka Masuo
Taylor Kenneth N R
Terzian Yervant
Thaddeus Patrick
The Pik-Sin
Thompson A Richard
Thonnard Norbert
Thronson Harley Andrew
Tokarev Yurij V
Torrelles Jose M
Tosi Monica
Townes Charles Hard
Treffers Richard R
Turner Barry E
Turner Kenneth C
Ulrich Marie-Helene D
Urasin Lirik A
van de Hulst H C
van der Hulst Jan M
van der Laan Harry
van Gorkom Jacqueline H
van Woerden Hugo
Vanden Bout Paul A
Varshalovich Dimitrij
Verschuur Gerrit L
Viala Yves
Viallefond Francois
Vidal Jean-Louis
Vidal-Madjar Alfred
Visvanathan Natarajan
Vogel Manfred
Volk Kevin
Voshchinnikov Nicolai
Vrba Frederick J
Wakker Bastiaan Pieter
Waldhausen Silvia
Walker Gordon A H
Walmsley C Malcolm
Walton Nicholas A
Wang Junjie
Wannier Peter Gregory
Watt Graeme David
Weaver Harold F
Weiler Kurt W
Weinberger Ronald
Weisheit Jon C
Wendker Heinrich J
Wesselius Paul R
Weymann Ray J
White Glenn J
White Richard L
Whitelock Patricia Ann
Whiteoak John B
Whittet Douglas C B
Whitworth Anthony Peter
Wickramasinghe N C
Williams David A
Williams Robert E
Williams Robin
Willis Allan J
Willner Steven Paul
Wilson Christine
Wilson Robert W
Winnberg Anders
Winnewisser Gisbert
Witt Adolf N
Wolfire Mark Guy
Wolstencroft Ramon D
Wolszczan Alexander
Woltjer Lodewijk
Wood Douglas O S
Woodward Paul R
Woolf Neville J
Wootten Henry Alwyn
Wouterloot Jan Gerard A
Wright Edward L
Wu Chi Chao
Wynn-Williams C G
Xiang Delin
Yabushita Shin A
Yamamoto Satoshi
Yamashita Takuya
Yan Jun
Yang Ji
York Donald G
Yorke Harold W
Yoshida Shigeomi
Younis Saad M
Zavagno Annie
Zealey William J
Zeilik Michael Ii
Zeng Qin
Zhang Cheng-Yue
Zhou ZhenpPu
Zimmermann Helmut
Zuckerman Ben M

Composition of Commission 35

Stellar Constitution/Constitution des étoiles

President: Zahn Jean-Paul

Vice-President: Vandenberg Don

Organizing Committee:

Canal Ramon M
Chiosi Cesare S
Dziembowski Wojciech
Guzik Joyce Ann
Meynet Georges
Michaud Georges J
Renzini Alvio
Saio Hideyuki
Tutukov A V
Vauclair Gerard P

Members:

Adams Mark T
Aiad A Zaki
Aizenman Morris L
Anand S P S
Angelov Trajko
Antia H M
Appenzeller Immo
Arai Kenzo
Arimoto Nobuo
Arnett W David
Arnould Marcel L
Audouze Jean
Baglin Annie
Baker Norman H
Baym Gordon Alan
Beaudet Gilles
Becker Stephen A
Belmonte Aviles J A
Benz Willy
Berthomieu Gabrielle
Bisnovatyi-Kogan Grennadij S
Bludman Sidney A
Bocchia Romeo
Bodenheimer Peter
Boehm Karl-Heinz
Bondi Hermann
Bono Giuseppe
Boss Alan P
Bravo Eduardo
Brownlee Robert R
Bruenn Stephen W
Buchler J Robert
Burbidge Geoffrey R
Callebaut Dirk K
Caloi Vittoria
Cameron Alastair G W
Caputo Filippina
Carson T R
Castellani Vittorio
Castor John I
Caughlan Georgeanne R
Chan Kwing Lam
Chan Roberto
Chechetkin Valerij M
Chevalier Claude
Chitre Shashikumar M
Chkhikvadze Iakob N
Christensen-Dalsgaard J
Christy Robert F
Cohen Jeffrey M
Connolly Leo Paul
Cowan John J
D'Antona Francesca
Das Mrinal Kanti
Davis Cecil G
de Greve Jean-Pierre
de Jager Cornelis
de Loore Camiel
de Medeiros Jose Renan
Dearborn David Paul K
Deinzer W
Deliyannis Constantine P
Demarque Pierre
Denissenkov Pavel
Despain Keith Howard
Deupree Robert G
Dluzhnevskaya Olga B
Durisen Richard H
Edwards Alan Ch
Edwards Terry W
Eggleton Peter P
Eminzade T A
Endal Andrew S
Ergma E V
Eriguchi Yoshiharu
Ezer-Eryurt Dilhan
Fadeyev Yuri A
Faulkner Donald J
Faulkner John
Flannery Brian Paul
Fontaine Gilles
Forbes J E
Forestini Manuel
Fossat Eric
Foukal Peter V
Frantsman Yu L
Fujimoto Masayuki
Gabriel Maurice R
Gallino Roberto
Garcia Domingo
Gautschy Alfred
Geroyannis Vassilis S
Giannone Pietro
Gimenez Alvaro
Gingold Robert Arthur
Giridhar Sunetra

Glatzmaier Gary A
Gong Shumo
Gough Douglas O
Goupil Marie Jose
Graham Eric
Greggio Laura
Guenther David
Gurm Hardev S
Hachisu Izumi
Hammond Gordon L
Han Zhanwen
Hashimoto Masa-aki
Hayashi Chushiro
Henry Richard B C
Hernanz Margarita
Hilf Eberhard R H
Hitotsuyanagi Juichi
Hollowell David Earl
Hoshi Reiun
Hoyle Fred
Huang Runqian
Huggins Patrick J
Humphreys Roberta M
Iben Icko
Iliev Ilian
Imbroane Alexandru
Imshennik Vladimir S
Isaak George R
Isern Jorge
Ishizuka Toshihisa
Itoh Naoki
James Richard A
Kaehler Helmuth
Kaminishi Keisuke
Kato Mariko
Khozov Gennadij V
Kiguchi Masayoshi
King David S
Kippenhahn Rudolf
Kiziloglu Nilguen
Knoelker Michael
Kochhar R K
Koester Detlev
Kosovichev Alexander
Kovetz Attay
Kozlowski Maciej
Kumar Shiv S
Kushwaha R S
Kwok Sun
Labay Javier
Lamb Susan Ann
Lamb Donald Quincy
Langer Norbert
Lapuente Pilar Ruiz
Larson Richard B
Laskarides Paul G
Lasota Jean-Pierre
Latour Jean J
Lattanzio John
Lebovitz Norman R
Lebreton Yveline
Lee Thyphoon
Lepine Jacques R D
Li Zongwei
Liebert James W
Linnell Albert P
Littleton John E
Livio Mario
Luo Guoquan
Maeder Andre
Maheswaran Murugesapillai
Mallik D C V
Marx Gyorgy
Masani A
Massevich Alla G
Matteucci Francesca
Mazurek Thaddeus John
Mazzitelli Italo
McCrea J Dermott
Melik-Alaverdian Yu
Mestel Leon
Meyer-Hofmeister Eva
Mitalas Romas Assoc
Miyaji Shigeki
Moellenhoff Claus
Mohan Chander
Monaghan Joseph J
Monier Richard
Moore Daniel R
Morgan John Adrian
Morris Stephen C
Moskalik Pawel
Moss David L
Mueller Ewald
Nadyozhin Dmittris K
Nakamura Takashi
Nakano Takenori
Nakazawa Kiyoshi
Narasimha Delampady
Narita Shinji
Newman Michael John
Nishida Minoru
Noels Arlette
Nomoto Ken'ichi
Odell Andrew P
Ohyama Noboru
Okamoto Isao
Osaki Yoji
Ostriker Jeremiah P
Oswalt Terry D
Paczynski Bohdan
Pamyatnikh Alexsey A
Pande Girish Chandra
Papaloizou John C B
Pearce Gillian
Phillips Mark M
Pines David
Pinotsis Antonis D
Plavec Mirek J
Pongracic Helen
Porfir'ev Vladimir V
Poveda Arcadio
Prentice Andrew J R
Prialnik-Kovetz Dina
Proffitt Charles R
Provost Janine
Qu Qinyue
Raedler K H
Ramadurai Souriraja
Ray Alak
Rayet Marc
Reeves Hubert
Reiz Anders
Ritter Hans
Rood Robert T
Rouse Carl A
Roxburgh Ian W
Ruben G
Sackmann Ingrid Juliana
Sadollah Nasiri Gheidari
Sakashita Shiro
Salpeter Edwin E
Santos Filipe D
Sato Katsuhiko
Savedoff Malcolm P
Savonije Gerrit Jan
Scalo John Michael
Schatten Kenneth H
Schatzman Evry
Schild Hansruedi
Schoenberner Detlef
Schramm David N
Schutz Bernard F
Scuflaire Richard
Sears Richard Langley
Seidov Zakir F
Sengbusch Kurt V
Shaviv Giora
Shibahashi Hiromoto
Shibata Yukio
Shustov Boris M
Sienkiewicz Ryszard
Signore Monique
Silvestro Giovanni
Sion Edward Michael

Smeyers Paul
Smith Robert Connon
Sobouti Yousef
Sofia Sabatino
Souffrin Pierre B
Sparks Warren M
Spiegel E
Sreenivasan S Ranga
Starrfield Sumner
Stellingwerf Robert F
Stibbs Douglas W N
Strittmatter Peter A
Suda Kazuo
Sugimoto Daiichiro
Sweet Peter A
Sweigart Allen V
Taam Ronald Everett
Takahara Mariko
Tassoul Monique
Thielemann Friedrich-Karl
Thomas Hans -Christoph
Tjin-a-Djie Herman R E
Tohline Joel Edward
Toomre Juri
Trimble Virginia L
Truran James W
Tscharnuter Werner M
Tuominen Ilkka V
Turck-Chieze Sylvaine
Uchida Juichi
Ulrich Roger K
Unno Wasaburo
Uus Undo
van den Heuvel Edward P J
van Der Borght Rene
van der Raay Herman B
van Horn Hugh M
van Riper Kenneth A
Vardya M S
Vila Samuel C
Vilhu Osmi
Vilkoviskij Emmanuil Y
Ward Richard A
Weaver Thomas A
Webbink Ronald F
Weiss Achim
Weiss Nigel O
Wheeler J Craig
Willson Lee Anne
Wilson Robert E
Winkler Karl-Heinz A
Wood Matthew Alan
Wood Peter R
Woosley Stanley E
Xiong Darun
Yamaoka Hitoshi
Yorke Harold W
Yungelson Lev R
Zhevakin S A
Ziolkowski Janusz

Composition of Commission 36

Theory of Stellar Atmospheres/Théorie des atmosphères stellaires

President: Pallavicini Roberto

Vice-President: Dravins Dainis

Organizing Committee:

Barbuy Beatriz
Cram Lawrence Edward
Hubeny Ivan
Owocki Stanley Peter
Saio Hideyuki
Sasselov Dimitar D
Spite Monique
Stepien Kazimierz

Members:

Abbott David C
Abhyankar Krishna D
Aller Lawrence Hugh
Altrock Richard C
Arpigny Claude
Athay R Grant
Auer Lawrence H
Auman Jason R
Avrett Eugene H
Baade Dietrich
Baird Scott R
Baliunas Sallie L
Balona Luis Antero
Baschek Bodo
Basri Gibor B
Becker Sylvia
Bell Roger A
Bernat Andrew Plous
Bertout Claude
Blanco Carlo
Bless Robert C
Blomme Ronny
Bodo Gianluigi
Boehm Karl-Heinz
Boehm-Vitense Erika
Boesgaard Ann M
Bopp Bernard W
Bowen George H
Brown Alexander
Brown Douglas Nason
Bues Irmela D
Carbon Duane F
Carlsson Mats
Carson T R
Cassinelli Joseph P
Castor John I
Catala Claude
Catalano Francesco A
Catalano Santo
Cayrel Roger
Cayrel de Strobel Giusa
Chan Kwing Lam
Chen Peisheng
Chugai Nikolai N
Cidale Lydia Sonia
Conti Peter S
Cowley Charles R
Crivellari Lucio
Cugier Henryk
Cuntz Manfred
Cuny Yvette J
Davis Cecil G
Dimitrijevic Milan
Doazan Vera
Domke Helmut
Donati Jean-Francois
Doyle John Gerard
Drake Stephen A
Dufton Philip L
Dupree Andrea K
Edvardsson Bengt
Elste Gunther H
Eriksson Kjell
Evangelidis E
Faraggiana Rosanna
Faurobert-Scholl Marianne
Fitzpatrick Edward L
Fontaine Gilles
Fontenla Juan Manuel
Foy Renaud
Freire Ferrero Rubens G
Friend David B
Frisch Helene
Frisch Uriel
Froeschle Christiane D
Gail Hans-Peter
Gallino Roberto
Gebbie Katharine B
Gesicki Krzysztof
Giampapa Mark S
Gigas Detlef
Glebocki Robert
Gokdogan Nuzhet
Gonzales Jean-Francois
Gordon Charlotte
Gough Douglas O
Grant Ian P
Gray David F
Greenstein Jesse L
Grevesse Nicolas
Grinin Vladimir P
Gussmann E A
Gustafsson Bengt
Hack Margherita
Haisch Bernhard Michael
Hall Douglas S
Hamann Wolf-Rainer
Hartmann Lee William
Harutyunian Haik A

Hauschildt Peter H
Hearn Anthony G
Heasley James Norton
Heber Ulrich
Hekela Jan
Herold Heinz
Hitotsuyanagi Juichi
Hoare Melvin
Hoeflich Peter
Holweger Hartmut
Holzer Thomas Edward
Hotinli Metin
House Lewis L
Hunger Kurt
Husfeld Dirk
Hutchings John B
Ito Yutaka
Ivanov Vsevolod V
Jahn Krzysztof
Jatenco-Pereira Vera
Johnson Hollis R
Jordan Stefan
Judge Philip
Kadouri Talib Hadi
Kalkofen Wolfgang
Kamp Lucas Willem
Kandel Robert S
Karp Alan Hersh
Khokhlova Vers L
Kiselman Dan
Klein Richard I
Kodaira Keiichi
Koester Detlev
Kolesov Alexander K
Kondo Yoji
Kontizas Evangelos
Krikorian Ralph
Krishna Swamy K S
Kubat Jiri
Kudritzki Rolf-Peter
Kuhi Leonard V
Kumar Shiv S
Kurucz Robert L
Kushwaha R S
Lambert David L
Lamers Henny J G L M
Landstreet John D
Leibacher John
Liebert James W
Linnell Albert P
Linsky Jeffrey L
Liu Caipin
Luck R Earle
Luttermoser Donald
Lyubimkov Leonid S
Madej Jerzy
Magnan Christian
Marlborough J M
Mashonkina Lyudmila
Massaglia Silvano
Mathys Gautier
Matsumoto Masamichi
Michaud Georges J
Mihalas Dimitri
Miyamoto Sigenori
Mnatsakanian Mamikon A
Muench Guido
Mukai Sonoyo
Musielak Zdzislaw E
Mutschlecner J Paul
Nagendra K N
Nagirner Dmitrij I
Narasimha Delampady
Nariai Kyoji
Neff John S
Nikoghossian Arthur G
Nordlund Ake
O'Mara Bernard J
Oxenius Joachim
Pacharin-Tanakun P
Pagel Bernard E J
Panek Robert J
Pasinetti Laura E
Pecker Jean-Claude
Peraiah Annamaneni
Peters Geraldine Joan
Phillips John G
Pinto Philip Alfred
Piskunov Nikolai E
Pogodin Mikhail A
Pottasch Stuart R
Praderie Francoise
Puls Joahim
Querci Francois R
Querci Monique
Rachkovsky D N
Ramsey Lawrence W
Rangarajan K E
Rao D Mohan
Rauch Thomas
Reimers Dieter
Rodono Marcello
Ross John E R
Rostas Francois
Rovira Marta Graciela
Rucinski Slavek M
Rutten Robert J
Ryabchikova Tanya
Rybicki George B
Saito Kuniji
Sakhibullin Nail A
Sapar Arved
Savanov Igor S
Schaerer Daniel
Scharmer Goeran Bjarne
Schmalberger Donald C
Schmid-Burgk J
Schmutz Werner
Schoenberner Detlef
Scholz M
Schrijver C J
Seaton Michael J
Sedlmayer Erwin
Shcherbakov Alexander
Shine Richard A
Shipman Harry L
Simon Klaus Peter
Simon Theodore
Simonneau Eduardo
Sitnik G F
Skumanich Andre
Snezhko Leonid I
Snijders Mattheus A J
Sobolev V V
Souffrin Pierre B
Spiegel E
Spite Francois M
Spruit Henk C
Stalio Roberto
Steffen Matthias
Stein Robert F
Stibbs Douglas W N
Strom Stephen E
Swihart Thomas L
Szecsenyi-Nagy Gabor
Takeda Yoichi
Thejll Peter Andreas
Toomre Juri
Traving Gerhard
Tsuji Takashi
Tuominen Ilkka V
Ueno Sueo
Uesugi Akira
Ulmschneider Peter
Underhill Anne B
Unno Wasaburo
Vakili Farrokh
van Regemorter Henri
van't Veer Frans
van't Veer-Menneret Claude
Vardavas Ilias Mihail
Vardya M S
Vasu-Mallik Sushma
Vaughan Arthur H
Velusamy T

Viik Tonu
Vilhu Osmi
Walter Frederick M
Watanabe Tetsuya
Waters Laurens B F M
Weber Stephen Vance
Wehrse Rainer
Weidemann Volker
Wellmann Peter
Werner Klaus
White Richard L
Wickramasinghe N C
Willson Lee Anne
Wilson Peter R
Wilson S J
Woehl Hubertus
Wolff Sidney C
Wyller Arne A
Yanovitskij Edgard G
Yengibarian Norair
Yorke Harold W
Zahn Jean-Paul
Zwaan Cornelis

Composition of Commission 37

Star Clusters and Associations/ Amas stellaires et associations

President: Da Costa Gary Stewart

Vice-President: Meylan Georges

Organizing Committee:

Aarseth Sverre J
Buonanno Roberto
Cannon Russell D
Castellani Vittorio
Cudworth Kyle Mccabe
Feinstein Alejandro
Lada Charles Joseph
Sarajedini Ata

Members:

Abou el-ella Mohamed S
Agekjan Tateos A
Aiad A Zaki
Alcaino Gonzalo
Alfaro Emilio Javier
Alksnis Andrejs
Allen Christine
Aparicio Antonio
Armandroff Taft E
Auriere Michel
Balazs Bela A
Bell Roger A
Bijaoui Albert
Blaauw Adriaan
Bouvier Pierre
Burkhead Martin S
Butler Dennis
Buzzoni Alberto
Byrd Gene G
Callebaut Dirk K
Caloi Vittoria
Caputo Filippina
Capuzzo Dolcetta Roberto
Carney Bruce William
Chavarria'k Carlos
Cheng Kwang Ping
Chiosi Cesare S
Christian Carol Ann
Chryssovergis Michael
Chun Mun-suk
Claria Juan
Colin Jacques
Covino Elvira
Cuffey J
D'Antona Francesca
Danford Stephen C
Dapergolas A
Daube-Kurzemniece I A
DeGioia-Eastwood Kathleen
Dehghani Mohammad Hossein
Dejonghe Herwig Bert
Demarque Pierre
Demers Serge
Di Fazio Alberto
Dickens Robert J
Dluzhnevskaya Olga B
Drissen Laurent
Efremov Yuri N
Einasto Jaan
Einasto Jaan
Einasto Jaan
El Basuny Ahmed Alawy
Elmegreen Bruce Gordon
Fall S Michael
Feast Michael W
Fitzgerald M Pim
Forbes Douglas
Forte Juan Carlos
Freeman Kenneth C
Friel Eileen D
Fusi Pecci Flavio
Gascoigne S C B
Geffert Michael
Geisler Douglas P
Giersz Miroslaw
Golay Marcel
Gratton Raffaela G
Green Elizabeth M
Griffiths William K
Grindlay Jonathan E
Grubissich C
Guetter Harry Hendrik
Hanes David A
Harris Gretchen L H
Harris Hugh C
Harris William E
Harvel Christopher Alvin
Hassan S M
Hatzidimitriou Despina
Hawarden Timothy G
Hazen Martha L
Heggie Douglas C
Henon Michel C
Herbst William
Hesser James E
Heudier Jean-Louis
Hills Jack G
Hodapp Klaus-Werner
Hut Piet
Iben Icko
Illingworth Garth D
Inagaki Shogo
Ishida Keiichi
Janes Kenneth A
Jones Derek H P
Joshi U C
Kadla Zdenka I
Kamp Lucas Willem
Kandrup Henry Emil

Kilambi G C
King Ivan R
Kontizas Evangelos
Kontizas Mary
Kraft Robert P
Kroupa Pavel
Kun Maria
Landolt Arlo U
Lapasset Emilio
Larsson-Leander Gunnar
Latham David W
Laval Annie
Leisawitz David
Leonard Peter James T
Lloyd Evans Thomas Harry
Loden Lars Olof
Lu Phillip K
Lynden-Bell Donald
Maeder Andre
Makino Junichiro
Mardling Rosemary
Markkanen Tapio
Marraco Hugo G
Marshall Kevin P
Martins Donald Henry
Matteucci Francesca
Mayor Michel
Mendez Mariano
Menon T K
Menzies John W
Mermilliod Jean-Claude
Mikkola Seppo
Milone Eugene F
Moffat Anthony F J
Mohan Vijay
Mould Jeremy R
Murray C Andrew
Muzzio Juan C
Nemec James
Nesci Roberto
Newell Edward B
Ninkov Zoran
Nissen Poul E
Ogura Katsuo
Ortolani Sergio
Osman Anas Mohamed
Pandey A K
Parsamyan Elma S
Pedreros Mario
Penny Alan John
Peterson Charles John
Petrovskaya Margarita S
Phelps Randy L
Philip A G Davis
Piatti Andres Eduardo
Pilachowski Catherine
Piskunov Anatoly E
Platais Imants K
Poveda Arcadio
Pritchet Christopher J
Prosser Charles Franklin
Qian Bochen
Renzini Alvio
Richer Harvey B
Richtler Tom
Roth-Hoppner Maria Luise
Rountree Janet
Ruprecht Jaroslav
Russeva Tatjana
Salukvadze G N
Samus Nikolai N
Sanders Walt L
Schild Hansruedi
Semkov Evgeni
Sharov A S
Shawl Stephen J
Sher David
Shobbrook Robert R
Shu Chenggang
Simoda Mahiro
Smith Graeme H
Spurzem Rainer
Stauffer John Richard
Stetson Peter B
Sugimoto Daiichiro
Suntzeff Nicholas B
Szecsenyi-Nagy Gabor
Takahashi Koji
Terranegra Luciano
Terzan Agop
The Pik-Sin
Tornambe Amedeo
Tripicco Michael J
Trullols I Farreny Enric
Tsvetkov Milcho K
Tsvetkova Katya
Tucholke Hans-Joachim
Turner David G
Twarog Bruce A
Upgren Arthur R
van Altena William F
van den Bergh Sidney
Vandenberg Don
Vazquez Ruben Angel
Verschueren Werner
von Hippel Theodore A
Walker Gordon A H
Walker Merle F
Wan Lai
Warren Wayne H
Weaver Harold F
Wehlau Amelia
White Raymond E
Wielen Roland
Wramdemark Stig S o
Wu Hsin-Heng
Xiradaki Evangelia
Zakharova Polina E
Zhao Junliang
Zinn Robert J

Composition of Commission 38

Exchange of Astronomers/Echange d'astronomes

President: Roberts Morton S

Vice-President: West Richard M

Organizing Committee:

Chitre Dattakumar M
Ducati Jorge Ricardo
Jorgensen Henning E
Krishna Gopal
Morimoto Masaki
Tolbert Charles R
Ye Shuhua

Members:

Al-Naimiy Hamid M K
Al-Sabti Abdul Adim
Boyarchuk Alexander A
Caccin Bruno
Florsch Alphonse
Haupt Hermann F
Kozai Yoshihide
Leung Kam Ching
Macrae Donald A
Marik Miklos
Nha Il-Seong
Ninkovic Slobodan
Okoye Samuel E
Routly Paul M
Sahade Jorge
Smith Francis Graham
Swarup Govind
van den Heuvel E P J
Wang Shouguan
Wood Frank Bradshaw

Composition of Commission 40

Radio Astronomy/Radioastronomie

President: Moran James M

Vice-President: Padrielli Lucia

Organizing Committee:

Anantharamaiah K R
Baath Lars B
Berkhuijsen Elly M
Bronfman Leonardo
Davis Richard J
Dewdney Peter E F
Dickel Helene R
Johnston Kenneth J
Litvinenko Leonid N
Marcaide Juan-Maria
Nan Rendong
Nicolson George D
Schilizzi Richard T
Whiteoak John B

Members:

Abdulla Shaker Abdul Aziz
Ables John G
Abrami Alberto
Ade Peter A R
Aizu Ko
Akabane Kenji A
Akujor Chidi E
Alexander Joseph K
Alexander Paul
Allen Ronald J
Aller Hugh D
Aller Margo F
Altenhoff Wilhelm J
Ambrosini Roberto
Andernach Heinz
Aparici Juan
Arnal Marcelo Edmundo
Aschwanden Markus
Assousa George Elias
Aubier Monique G
Aurass Henry
Avery Lorne W
Avignon Yvette
Axon David
Baan Willem A
Baars Jacob W M
Baart Edward E
Bachiller Rafael
Backer Donald Ch
Bagri Durgadas S
Bajaja Esteban
Baker Joanne
Balasubramanian V
Baldwin John E
Balklavs A E
Ball Lewis
Bally John
Balonek Thomas J
Banhatti Dilip Gopal
Barrow Colin H
Bartel Norbert Harald
Barthel Peter
Barvainis Richard
Bash Frank N
Basu Dipak
Batty Michael
Baudry Alain
Baum Stef Alison
Beck Rainer
Benaglia Paula
Benn Chris R
Bennett Charles L
Benson Priscilla J
Benz Arnold O
Berge Glenn L
Bhandari Rajendra
Bhonsle Rajaram V
Bieging John Harold
Biermann Peter L
Biggs James
Bignell R Carl
Biraud Francois
Biretta John Anthony
Birkinshaw Mark
Blair David Gerald
Blandford Roger David
Bloemhof Eric E
Bockelee-Morvan Dominique
Booth Roy S
Boriakoff Valentin
Bos Albert
Bottinelli Lucette
Bowers Phillip F
Bracewell Ronald N
Braude Semion Ya Ag
Breahna Iulian
Bregman Jacob D Ir
Bridle Alan H
Brinks Elias
Broderick John
Broten Norman W
Brouw W N
Browne Ian W A
Bujarrabal Valentin
Burbidge Geoffrey R
Burke Bernard F
Carilli Christopher L
Caroubalos C A
Carr Thomas D
Casoli Fabienne
Castets Alain
Caswell James L
Cernicharo Jose

Chan Kwing Lam
Chandler Claire
Charlot Patrick
Chen Hongsheng
Chikada Yoshihiro
Chini Rolf
Cho Se Hyung
Christiansen Wayne A
Christiansen Wilbur
Chu Hanshu
Chyzy Krzysztof Tadeusz
Clark Barry G
Clark David H
Clark Frank Oliver
Clemens Dan P
Cohen Marshall H
Cohen Raymond J
Cohen Richard S
Cole Trevor William
Coleman Paul Henry
Colomb Fernando R
Colomer Francisco
Combes Francoise
Combi Jorge Ariel
Condon James J
Conklin Edward K
Conway John
Conway Robin G
Cordes James M
Cotton William D
Coutrez Raymond A J
Covington Arthur E
Crane Patrick C
Croom David L
Crovisier Jacques
Crutcher Richard M
Cudaback David D
Dagkesamansky Rustam D
Daintree Edward J
Daishido Tsuneaki
Davies Rodney D
Davis Michael M
Davis Robert J
de Bergh Catherine
de Groot T
de Jager Cornelis
de la Noe Jerome
de Ruiter Hans Rudolf
de Vicente Pablo
de Young David S
Degaonkar S S
Delannoy Jean
Denisse Jean-Francois
Dent William A
Deshpande Avinash
Despois Didier
Dhawan Vivek
Dickel John R
Dickey John M
Dickman Robert L
Dieter Conklin Nannielou H
Dixon Robert S
Doubinskij Boris A
Douglas James N
Downes Dennis
Downs George S
Drake Frank D
Drake Stephen A
Dravskikh Alexander F
Dreher John W
Duffett-Smith Peter James
Dulk George A
Dwarakanath K S
Dyson Freeman J
Edelson Rick
Ekers Ronald D
Elgaroy Oystein
Ellis G R A
Elsmore Bruce
Elwert Gerhard
Emerson Darrel Trevor
Emerson David
Enome Shinzo
Epstein Eugene E
Erickson William C
Eriksen Gunnar
Eshleman Von R
Evans Kenton Dower
Ewing Martin S
Facondi Silvia Rosa
Fanti Roberto
Feigelson Eric D
Feix Gerhard
Feldman Paul A
Felli Marcello
Felten James E
Feretti Luigina
Ferrari Attilio
Fey Alan Lee
Field George B
Finkelstein Andrej M
Fleischer Robert
Florkowski David R
Foley Anthony
Fomalont Edward B
Fort David Norman
Forveille Thierry
Fouque Pascal
Frail Dale Andrew
Franco Mantovani
Frater Robert H
Friberg Per
Friedman Herbert
Frisk Urban
Fuerst Ernst
Fukui Yasuo
Gallego Juan Daniel
Galt John A
Garay Guido
Gardner Francis F
Garrett Michael
Garrington Simon
Gaylard Michael John
Gebler Karl-heinz
Geldzahler Bernard J
Gelfreikh Georgij B
Gent Hubert
Genzel Reinhard
Gerard Eric
Gergely Tomas Esteban
Ghigo Francis D
Gil Janusz A
Ginzburg Vitaly L
Gioia Isabella M
Giovannini Gabriele
Gold Thomas
Goldstein Samuel J
Goldwire Henry C
Gomez Gonzalez Jesus
Gonze Roger F J
Gopalswamy N
Gordon Mark A
Gorgolewski Stanislaw Pr
Gorschkov Alexander G
Gosachinskij Igor V
Goss W Miller
Gottesman Stephen T
Gower J F R
Graham David A
Green Anne
Green David
Gregorini Loretta
Gregorio-Hetem Jane
Gregory Philip C
Grewing Michael
Gubchenko Vladimir M
Guelin Michel
Guesten Rolf
Guidice Donald A
Gulkis Samuel
Gull Stephen F
Gupta Yashwant
Gurvits Leonid
Gwinn Carl R
Haddock Fred T

Hagen John P
Hall Peter J
Hamilton P A
Han Fu
Han Jinlin
Han Wenjun
Hanasz Jan
Hanbury Brown Robert
Handa Toshihiro
Hanisch Robert J
Hankins Timothy Hamilton
Hardee Philip
Harnett Julieine
Harris Daniel E
Harten Ronald H
Haschick Aubrey
Hasegawa Tetsuo
Haslam C Glyn T
Hayashi Masahiko
Haynes Martha P
Haynes Raymond F
Hazard Cyril
Heeschen David S
Heidmann Jean
Heiles Carl
Helou George
Henkel Christian
Heske Astrid
Hewish Antony
Hey James Stanley
Higgs Lloyd A
Hilf Eberhard R H
Hills Richard E
Hirabayashi Hisashi
Hjalmarson Ake G
Hjellming Robert M
Ho Paul T P
Hoang Binh Dy
Hobbs Robert W
Hoegbom Jan A
Hoeglund Bertil
Hogg David E
Hollis Jan Michael
Hong Xiaoyu
Howard William E III
Huang Fuquan
Huchtmeier Walter K
Hughes Philip
Hughes Victor A
Hulsbosch A N M
Hunstead Richard W
Ikhsanov Robert N
Ikhsanova Vera N
Inatani Junji
Inoue Makoto
Ipatov Alexander V
Irvine William M
Ishiguro Masato
Iwata Takahiro
Jackson Neal
Jacq Thierry
Jaffe Walter Joseph
Janssen Michael Allen
Jauncey David L
Jenkins Charles R
Jennison Roger C
Jewell Philip R
Ji Shuchen
Jin Shenzeng
Johansson Lars Erik B
Johnson Donald R
Joly Francois
Jonas Justin Leonard
Jones Dayton L
Jones Paul
Joshi Mohan N
Kaftan May A
Kahlmann Hans Cornelis
Kahn Franz D
Kaifu Norio
Kakinuma Takakiyo T
Kalberla Peter
Kameya Osamu
Kandalian Rafik A
Kang Gon Ik
Kapahi Vijay K
Kardashev Nicolay S
Kasuga Takashi
Kaufmann Pierre
Kawabata Kinaki
Kawabe Ryohei
Kawaguchi Kentarou
Kazes Ilya
Kellermann Kenneth I
Kenderdine Sidney
Kerr Frank J
Kesteven Michael J l
Kijak Jaroslaw
Killeen Neil
Kim Kwang-tae
Kim Tu Hwan
Kislyakov Albert G
Klein Karl Ludwig
Klein Ulrich
Ko Hsien C
Kobayashi Hideyuki
Kojima Masayoshi
Kojoian Gabriel
Konovalenko Olexandr
Korzhavin Anatoly
Kotelnikov Vladimir A
Kraus John D
Kreysa Ernst
Krishna Gopal
Krishnamohan S
Krishnan Thiruvenkata
Kronberg Philipp
Kruegel Endrik
Krueger Albrecht
Krygier Bernard
Kuijpers H Jan M E
Kuiper Thomas B H
Kulkarni Prabhakar V
Kulkarni Shrinivas R
Kulkarni Vasant K
Kumkova Irina I
Kundt Wolfgang
Kundu Mukul R
Kuril'chik Vladimir N
Kus Andrzej Jan
Kutner Marc Leslie
Kuzmin Arkadii D
Kwok Sun
Lada Charles Joseph
Laing Robert
Landecker Thomas L
Lang Kenneth R
Langer William David
Lantos Pierre
Large Michael I
Lasenby Anthony
Lawrence Charles R
Le Squeren Anne-Marie
Leahy J Patrick
Leblanc Yolande
Legg Thomas H
Lepine Jacques R D
Lequeux James
Lesch Harold
Lestrade Jean-Francois
Leung Chun Ming
Levreault Russell M
Li Chun-Sheng
Li Gyong Won
Li Hong-Wei
Liang Shiguang
Likkel Lauren Jones
Lilley Edward A
Linke Richard Alan
Lis Dariusz C
Liseau Rene
Little Leslie T
Liu Yuying
Lo Kwok-Yung
Locke Jack L

Lockman Felix J
Loiseau Nora
Longair Malcolm S
Loren Robert Bruce
Lorenz Hilmar
Lovell Sir Bernard
Lozinskaya Tat-yana A
Lu Yang
Lubowich Donald A
Luks Thomas
Luo Xianhan
Lyne Andrew G
Macchetto Ferdinando
MacDonald Geoffrey H
MacDonald James
Machalski Jerzy
MacLeod John M
Macrae Donald A
Maehara Hideo
Malofeev Valerij M
Malumian Vigen
Manchester Richard N
Mandolesi Nazzareno
Maran Stephen P
Marques Dos Santos P
Marscher Alan Patrick
Martin Robert N
Martin-Pintado Jesus
Masheder Michael
Maslowski Jozef
Masson Colin R
Matheson David Nicholas
Matsakis Demetrios N
Matsuo Hiroshi
Matthews Henry E
Mattila Kalevi
Matveyenko Leonid I
Mauersberger Rainer
Maxwell Alan
May J
Mayer Cornell H
McAdam W Bruce
McConnell David
McCulloch Peter M
McKenna Lawlor Susan
McLean Donald J
Mebold Ulrich
Meeks M Littleton
Meier David L
Menon T K
Mezger Peter G
Michalec Adam
Miley George K
Mills Bernard Y
Milne Douglas K
Milogradov-Turin Jelena
Mirabel Igor Felix
Mitchell Kenneth J
Miyoshi Makoto
Mizuno Akira
Moiseev I G
Molchanov Andrea P
Montmerle Thierry
Morganti Raffaella
Morimoto Masaki
Morison Ian
Morita Kazuhiko
Morita Koh-ichiro
Moriyama Fumio
Morras Ricardo
Morris David
Morris Mark Root
Mundy Lee G
Murdoch Hugh S
Mutel Robert Lucien
Muxlow Thomas
Myers Philip C
Nadeau Daniel
Nagnibeda Valery G
Nakano Takenori
Neff John S
Nguyen-Quang Rieu
Nicastro Luciano
Nicholls Jennifer
Nishio Masanori
Norris Raymond Paul
O'Dea Christopher P
O'Sullivan John David
Oezel Mehmet Emin
Ogawa Hideo
Ohishi Masatoshi
Okoye Samuel E
Okumura Sachiko
Olberg Michael
Onuora Lesley Irene
Osterbrock Donald E
Otmianowska-Mazur Katarzyna
Owen Frazer Nelson
Pacholczyk Andrzej G
Padman Rachael
Palmer Patrick E
Pankonin Vernon Lee
Papagiannis Michael D
Paredes Jose Maria
Parijskij Yuri N
Parker Edward A
Parma Paola
Parrish Allan
Pasachoff Jay M
Pauliny Toth Ivan K K
Pauls Thomas Albert
Payne David G
Pearson Timothy J
Pedlar Alan
Peng Bo
Peng Yunlou
Penzias Arno A
Perez Fournon Ismael
Perley Richard Alan
Peters William L III
Pettengill Gordon H
Phillips Thomas Gould
Pick Monique
Planesas Pere
Ponsonby John E B
Pooley Guy
Porcas Richard
Preston Robert Arthur
Preuss Eugen
Price R Marcus
Priester Wolfgang
Puschell Jeffery John
Qian Shanjie
Qiu Yuhai
Radford Simon John E
Radhakrishnan V
Raimond Ernst
Ramaty Reuven
Rao A Pramesh
Raoult Antoinette
Ray Thomas P
Razin Vladimir A
Readhead Anthony C S
Reber Grote
Reich Wolfgang
Reid Mark Jonathan
Reif Klaus
Reyes Francisco
Reynolds John
Ribes Jean-Claude
Richer John
Rickard Lee J
Rickett Barnaby James
Riihimaa Jorma J
Riley Julia M
Rizzo Jose Ricardo
Roberts David Hall
Roberts Morton S
Robertson Douglas S
Robertson James Gordon
Robinson Brian J
Robinson Richard D
Rodriguez Luis F
Roelfsema Peter
Roennaeng Bernt O

Roeser Hans-peter
Roger Robert S
Rogers Alan E E
Rogstad David H
Rohlfs Kristen
Romero Gustavo Esteban
Romney Jonathan D
Rowson Barrie
Rubin Robert Howard
Rubio Monica
Rudnick Lawrence
Rudnitskij Georgij M
Russell Jane L
Rydbeck Gustaf H B
Rydbeck Olof E H
Rys Stanislaw
Sadollah Nasiri Gheidari
Saikia Dhruba Jyoti
Sakamoto Seiichi
Salpeter Edwin E
Salter Christopher John
Sanamian V A
Sandell Goran Hans l
Sanders David B
Sargent Anneila I
Sarma N V G
Sastry Ch V
Sato Fumio
Saunders Richard D E
Savage Ann
Sawant Hanumant S
Scalise Eugenio
Schaal Ricardo E
Scheuer Peter A G
Schlickeiser Reinhard
Schmidt Maarten
Schuch Nelson Jorge
Schultz G V
Schulz Rolf Andreas
Schwartz Philip R
Schwarz Ulrich J
Scott John S
Scott Paul F
Seaquist Ernest R
Seielstad George A
Seiradakis John Hugh
Sekimoto Yutaro
Setti Giancarlo
Shaffer David B
Shakeshaft John R
Shaposhnikov Vladimir E
Shaver Peter A
Sheridan K V
Shevgaonkar R K
Shibata Katsunori M
Shimmins Albert John
Shitov Yuri P
Shmeld Ivar
Sholomitsky G B
Shone David
Shulga Valery
Sieber Wolfgang
Sinha Rameshwar P
Skillman Evan D
Slade Martin A III
Slee O B
Slysh Viacheslav I
Smith Alex G
Smith Dean F
Smith Francis Graham
Smith Niall
Smol-Kov Gennadij Ya
Soboleva N S
Sodin Leonid
Sofue Yoshiaki
Sokolov Konstantin
Sorochenko R L
Spencer John Howard
Spencer Ralph E
Spoelstra T A Th
Sramek Richard A
Stahr-Carpenter M
Stanley G J
Stannard David
Steffen Matthias
Stewart Paul
Stewart Ronald T
Stone R G
Storey Michelle
Stotskii Alexander A
Strom Richard G
Strukov Igor A
Subrahmanya C R
Sugitani Koji
Sukumar Sundarajan
Sullivan Woodruff T
Sunada Kazuyoshi
Swarup Govind
Swenson George W
Tabara Hiroto
Takaba Hiroshi
Takagi Kojiro
Takakubo Keiya
Takakura Tatsuo Emer
Takano Toshiaki
Tanaka Riichiro
Tarter Jill C
Tatematsu Ken-ichi
Taylor A R
te Lintel Hekkert Peter
Terzian Yervant
Thomasson Peter
Thompson A Richard
Thum Clemens
Tlamicha Antonin
Tofani Gianni
Tolbert Charles R
Tomasi Paolo
Torrelles Jose M
Tovmassian Hrant M
Townes Charles Hard
Tritton Keith P
Troitsky V S
Troland Thomas Hugh
Truong Bach
Trushkin Sergei Anatol'evich
Tsuboi Masato
Tsutsumi Takahiro
Turlo Zygmunt
Turner Barry E
Turner Jean L
Turner Kenneth C
Turtle A J
Tzioumis Anastasios
Udal'tsov V A
Udaya Shankar N
Ukita Nobuharu
Ulrich Bruce T
Ulrich Marie-Helene D
Ulvestad James Scott
Unger Stephen
Unwin Stephen C
Urpo Seppo I
Uson Juan M
Val'tts Irina E
Vallee Jacques P
Valtaoja Esko
Valtonen Mauri J
van de Hulst H C
van der Hulst Jan M
van der Kruit Pieter C
van der Laan Harry
van Driel Willem
van Gorkom Jacqueline H
van Langevelde Huib Jan
van Nieuwkoop JIR
van Woerden Hugo
Vanden Bout Paul A
Vats Hari OM
Vaughan Alan
Velusamy T
Venturi Tiziana
Venugopal V R
Verkhodanov Oleg
Vermeulen Rene Cornelis

Veron Philippe
Verschuur Gerrit L
Verter Frances
Vilas Faith
Vilas-Boas Jose W
Vivekanand M
Vogel Stuart Newcombe
Walker Robert C
Wall Jasper V
Walmsley C Malcolm
Walsh Dennis
Wan Tongshan
Wang Jingsheng
Wang Shouguan
Wannier Peter Gregory
Wardle John F C
Warmels Rein Herm
Warner Peter J
Warwick James W
Watson Robert
Wehrle Ann Elizabeth
Wei Mingzhi
Weigelt Gerd
Weiler Edward J
Weiler Kurt W
Welch William J
Wellington Kelvin
Wendker Heinrich J
Westerhout Gart
Westfold Kevin C
Wickramasinghe N C
Wielebinski Richard
Wiklind Tommy
Wild John Paul
Wilkinson Peter N
Willis Anthony Gordon
Wills Beverley J
Wills Derek
Willson Robert Frederick
Wilner David J
Wilson Andrew S
Wilson Robert W
Wilson Thomas L
Wilson William J
Windhorst Rogier A
Wink Joern Erhard
Winnberg Anders
Winnewisser Gisbert
Witzel Arno
Wolszczan Alexander
Woltjer Lodewijk
Wood Douglas O S
Woodsworth Andrew W
Wootten Henry Alwyn
Wright Alan E
Wrobel Joan Marie
Wu Huai-Wei
Wu Nailong
Wu Shengyin
Wu Xinji
Wu Yuefang
Xia Zhiguo
Xu Peiyuan
Xu Zhicai
Yang Jian
Yang Zhigen
Ye Shuhua
Yin Qi-Feng
Younis Saad M
Yu Zhiyao
Zabolotny Vladimir F
Zaitsev Valerii V
Zensus J-Anton
Zhang Fujun
Zhang Jian
Zhang Jin
Zhang Xizhen
Zhao Jun Hui
Zhelezniakov Vladimir V
Zheng Xinwu
Zheng Yijia
Zhou Ti-jian
Zieba Stanislaw
Zlobec Paolo
Zlotnik Elena Ya
Zuckerman Ben M
Zylka Robert

Composition of Commission 41

History of Astronomy/Histoire de l'Astronomie

President: Dick Steven J

Vice-President: Stephenson F Richard

Organizing Committee:

Ansari S M Razaullah
Dick Wolfgang
Gurshtein Alexander A
Nha Il-Seong
Orchiston Wayne
Proverbio Edoardo
Sullivan Woodruff T
Xi Zezong

Members:

Aoki Shinko
Badolati Ennio
Bandyopadhyay A
Batten Alan H
Benson Priscilla J
Berendzen Richard
Bessell Michael S
Bishop Roy L
Bo Shuren
Bonoli Fabrizio
Botez Elvira
Brooks Randall C
Brunet Jean-Pierre
Carlson John B
Chen Kwan-yu
Cornejo Alejandro A
Cui Shizhu
Cui Zhenhua
Dadic Zarko
Danezis Emmanuel
Debarbat Suzanne V
Deeming Terence J
Dekker E
Devorkin David H
Dewhirst David W
Dobrzycki Jerzy
Dumont Simone
Eddy John A
Edmondson Frank K
Eelsalu Heino
Ehgamberdiev Shurat
Erpylev Nikolaj P
Esteban Lopez Cesara
Fernie J Donald
Firneis Maria G
Florides Petros S
Fodera Seriio Giorgia
Freeman Kenneth C
Freitas Mourao R r
Gingerich Owen
Glass Ian Stewart
Hawkins Gerald S
Hayli Avram
Haynes Raymond F
Haynes Roslynn
Hearnshaw John B
Hemenway Mary Kay M
Herrmann Dieter
Hockey Thomas Arnold
Hoskin Michael A
Hurukawa Kiitiro
Hysom Edmund J
Idliš Grigorij M
Jackisch Gerhard
Jauncey David L
Jeong Jang Hae
Jiang Xiaoyuan
Kennedy John E
Khromov Gavriil S
Kiang Tao
Kim Chun Hwey
Kim Yonggi
King David S
King Henry C
Kochhar R K
Kokott Wolfgang
Krisciunas Kevin
Krupp Edwin C
Kunitzsch Paul
Lang Kenneth R
Lee Woo-baik
Lee Yong-Sam
Levy Eugene H
Li Zhisen
Liu Ciyuan
Lopes-Gautier Rosaly
Lopez Carlos
Ma Chun-yu
Mathewson Donald S
McAdam W Bruce
McKenna Lawlor Susan
Mickelson Michael E
Mikhail Joseph Sidky
Moesgaard Kristian P
Nadal Robert
Nakayama Shigeru
Nicolaidis Efthymios
North John David
Oh Kyu Dong
Ohashi Nagayoshi
Ohashi Yukio
Oproiu Tiberiu
Osterbrock Donald E
Papathanasoglou D
Pedersen Olaf
Perdrix John
Peterson Charles John
Petri Winfried
Pigatto Luisa

Pingree David
Polozhentsev Dimitrij D
Porter Neil A
Poulle Emmanuel
Prokakis Theodore J
Pustylnik Izold B
Quan Hejun
Reaves Gibson
Ronan Colin A
Sbirkova-Natcheva T
Shukla K
Signore Monique
Sima Zdislav
Sobouti Yousef
Solc Martin
Stathopoulou Maria
Stavinschi Magdalena
Stoev Alexei
Sundman Anita
Svolopoulos Sotirios
Swerdlow Noel
Taton Rene
Theodossiou Efstratios
Tobin William
Trimble Virginia L
Vass Gheorghe
Verdet Jean-Pierre
Volyanskaya Margarita Yu
Wang dechang
Warner Brian
Whitaker Ewen A
White Graeme Lindsay
Whiteoak John B
Whitrow Gerald James
Wilson Curtis A
Wright Helen Greuter
Xu Zhentao
Yabuuti Kiyoshi
Yau Kevin K C
Yeomans Donald K
Zhanf Shouzhong
Zhang Peiyu
Zhuang Weifeng
Zosimovich Irina D

Composition of Commission 42

Close Binary Stars/Etoiles binaires serrées

President: Guinan Edward Francis

Vice-President: Szkody Paula

Organizing Committee:

Bianchi Luciana
Clausen Jens Viggo
Drechsel Horst
Hill Graham
Kenyon Scott J
Niemela Virpi S
Osaki Yoji
Rodono Marcello
Sterken Christiaan Leo
Taam Ronald Everett
Vilhu Osmi
Wood Janet H

Members:

Abhyankar Krishna D
Al-Naimiy Hamid M K
Andersen Johannes
Antipova Lyudmila I
Antonopoulou E
Anupama G C
Aquilano Roberto Oscar
Awadalla Nabil Shoukry
Barone Fabrizio
Bartolini Corrado
Bateson Frank M OBE
Bath Geoffrey T
Batten Alan H
Bell Steven
Blair William P
Blitzstein William
Bolton C Thomas
Bonazzola Silvano
Bookmyer Beverly B
Bopp Bernard W
Boyle Stephen
Bozic Hrvoje
Bradstreet David H
Brandi Elisande Estela
Breinhorst Robert A
Broglia Pietro
Brownlee Robert R
Bruch Albert
Bruhweiler Fred C
Budding Edwin
Bunner Alan N
Busso Maurizio
Callanan Paul
Canalle Joao B G
Catalano Santo
Cester Bruno
Chambliss Carlson R
Chapman Robert D
Chaubey Uma Shankar
Chen Kwan-yu
Cherepashchuk Analily M
Chochol Drahomir
Choi Kyu-Hong
Cillie G G
Claria Juan
Collins George W ii
Cowley Anne P
Cristaldi Salvatore
Cutispoto Giuseppe
D'Antona Francesca
Dadaev Aleksandr N
De Garcia Maria J M
de Greve Jean-Pierre
de Groot Mart
de Loore Camiel
Delgado Antonio Jesus
Demircan Osman
Dorfi Ernst Anton
Dougherty Sean
Doughty Noel A
Duemmler Rudolf
Duerbeck Hilmar W
Dupree Andrea K
Durisen Richard H
Duschl Wolfgang J
Eaton Joel A
Edalati Sharbaf Mohammad Taghi
Eggleton Peter P
Elias Nicholas
Etzel Paul B
Fabrika Sergei
Faulkner John
Fekel Francis C
Ferluga Steno
Ferrer Osvaldo Eduardo
Firmani Claudio A
Flannery Brian Paul
Frank Juhan
Frantsman Yu L
Fredrick Laurence W
Friedjung Michael
Garmany Catherine D
Geldzahler Bernard J
Geyer Edward H
Giannone Pietro
Gimenez Alvaro
Giovannelli Franco
Giuricin Giuliano
Goldman Itzhak
Gosset Eric
Grygar Jiri
Gulliver Austin Fraser

Gursky Herbert
Guseinov O H
Gyldenkerne Kjeld
Hadrava Petr
Hall Douglas S
Hammerschlag-Hensberge G
Hanawa Tomoyuki
Hantzios Panayiotis
Harmanec Petr
Hassall Barbara J M
Haswell Carole A
Hazlehurst John
Heintz Wulff D
Hellier Coel
Helt Bodil E
Hensler Gerhard
Herczeg Tibor J
Hilditch Ronald W
Hills Jack G
Hjellming Robert M
Hoffmann Martin
Holt Stephen S
Honeycutt R Kent
Horiuchi Ritoku
Hric Ladislav
Hrivnak Bruce J
Huang Runqian
Hube Douglas P
Hutchings John B
Ibanoglu Cafir
Imamura James
Imbert Maurice
Jabbar Sabeh Rhaman
Jaschek Carlos O R
Jasniewicz Gerard
Jeong Jang Hae
Joss Paul Christopher
Kadouri Talib Hadi
Kaitchuck Ronald H
Kaluzny Janusz
Kandpal Chandra D
Kang Young Woon
Karetnikov Valentin G R
Kawabata Shusaku
Kenny Harold
Khalesseh Bahram
Kim Chun Hwey
Kim Ho Il
King Andrew R
Kitamura M
Kjurkchieva Diana
Koch Robert H
Kolb Ulrich
Kondo Yoji
Koubsky Pavel
Kraft Robert P
Kraicheva Zdravska
Krautter Joachim
Kreiner Jerzy Marek
Kriz Svatopluk
Kruchinenko Vitaliy G
Kruszewski Andrzej
Krzeminski Wojciech
Kumsiashvily Mzia I
Kurpinska-Winiarska M
Kwee K K
la Dous Constanze A
Lacy Claud H
Lamb Donald Quincy
Landolt Arlo U
Lanning Howard Hugh
Lapasset Emilio
Larsson Stefan
Larsson-Leander Gunnar
Lavrov Mikhail I
Lee Woo-baik
Lee Yong-Sam
Leedjaerv Laurits
Leung Kam Ching
Li Zhongyuan
Linnell Albert P
Linsky Jeffrey L
Liu Qingyao
Liu Xuefu
Livio Mario
Lloyd Huw
Lucy Leon B
Lyuty Victor M
MacDonald James
Maceroni Carla
Malasan Hakim Luthfi
Mammano Augusto
Mardirossian Fabio
Marilli Ettore
Marino Brian F
Markworth Norman Lee
Marsh Thomas
Mathieu Robert D
Mattei Janet Akyuz
Mauder Horst
Mayer Pavel
Mazeh Tsevi
McCluskey George E
Melia Fulvio
Meyer-Hofmeister Eva
Mezzetti Marino
Mikolajewska Joanna
Mikulasek Zdenek
Milano Leopoldo
Milone Eugene F
Mineshige Shin
Miyaji Shigeki
Mochnacki Stephan W
Morgan Thomas H
Mouchet Martine
Muerset Urs
Mumford George S
Mutel Robert Lucien
Nakamura Yasuhisa
Nariai Kyoji
Nather R Edward
Neff James Edward
Nelson Burt
Newsom Gerald H
Nha Il-Seong
Niarchos Panayiotis
Nordstroem Birgitta
Oezkan Mustafa Tuerker
Oh Kyu Dong
Okazaki Akira
Olah Katalin
Oliver John Parker
Olson Edward C
Paczynski Bohdan
Padalia T D
Pandey Uma Shankar
Park Hong Suh
Parthasarathy M
Patkos Laszlo
Pavlenko Elena
Pavlovski Kresimir
Peters Geraldine Joan
Piccioni Adalberto
Piirola Vilppu E
Plavec Mirek J
Pojmanski Grzegorz
Polidan Ronald S
Popper Daniel M
Postnov Konstantin A
Pringle James E
Pustyl'Nik Izold B
Qiao Guojun
Rafert James Bruce
Rahunen Timo
Rakos Karl D
Ramsey Lawrence W
Rao P Vivekananda
Rasio Frederic A
Refsdal Sjur
Reglero-Velasco Victor
Richards Mercedes T
Ringwald Frederick Arthur
Ritter Hans
Robb Russell M
Robertson John Alistair

Robinson Edward Lewis
Rovithis Peter
Rovithis-Livaniou Helen
Roxburgh Ian W
Rucinski Slavek M
Ruffert Maximilian
Russo Guido
Sadik Aziz R
Sahade Jorge
Saijo Keiichi
Samec Ronald G
Sanwal N B
Sanyal Ashit
Savonije Gerrit Jan
Scaltriti Franco
Scarfe Colin D
Schiller Stephen
Schmid Hans Martin
Schmidt Hans
Schmidtke Paul C
Schober Hans J
Schoeffel Eberhard F
Seggewiss Wilhelm
Semeniuk Irena
Shafter Allen W
Shakura Nicholaj I
Shaviv Giora
Shen Liangzhao
Shu Frank H
Sima Zdislav
Simmons John Francis l
Sinvhal Shambhu Dayal
Sion Edward Michael
Sistero Roberto F
Skopal Augustin
Slovak Mark Haines
Smak Joseph I
Smith Robert Connon
Sobieski Stanley
Soderhjelm Staffan
Solheim Jan Erik
Sonti Sreedhar Rao
Sparks Warren M
Srivastava J B
Srivastava Ram Kumar
Stagg Christopher
Starrfield Sumner
Steiman-Cameron Thomas
Steiner Joao E
Stencel Robert Edward
Strohmeier Wolfgang
Sugimoto Daiichiro
Sundman Anita
Svechnikova Maria A
Szabados Laszlo
Szafraniec Rozalia
Tan Huisong
Teays Terry J
Todoran Ioan
Tout Christopher
Tremko Jozef
Trimble Virginia L
Turolla Roberto
Tutukov A V
Ureche Vasile
van den Heuvel Edward P J
van Hamme Walter
van Paradijs Johannes
van't Veer Frans
Vaz Luiz Paulo Ribeiro
Vetesnik Miroslav
Wade Richard Alan
Walder Rolf
Walker Richard L
Walker William S G
Ward Martin John
Warner Brian
Webbink Ronald F
Weiler Edward J
Wellmann Peter
Wheeler J Craig
Williamon Richard M
Williams Glen A
Williams Robert E
Wilson Robert E
Wood Frank Bradshaw
Yamasaki Atsuma
Yoon Tae
Zeilik Michael Ii
Zhai Disheng
Zhang Er-Ho
Zhang Jintong
Zhou Daoqi
Zhou Hongnan
Zhu Cisheng
Ziolkowski Janusz
Zola Stanislaw
Zuiderwijk Edwardus J
Zwitter Tomaz

Composition of Commission 44

Space and High Energy Astrophysics/Astrophysique spatiale et des hautes énergies

President: Srinivasan G

Vice-President: Okuda Haruyuki

Organizing Committee:

Baliunas Sallie L
Brosch Noah
Cesarsky Catherine J
Courvoisier Thierry J-L
da Costa Jose Marques
Domingo Vicente
Fabian Andrew C
Fazio Giovanni G
Fransson Claes
Hasinger Guenter
Inoue Hajime
Li Zhongyuan
Michalitsianos Andrew
O'Brien Paul Thomas
Oertel Goetz K
Quintana Hernan
Rengarajan T N
Schilizzi Richard T
Shustov Boris M
Thronson Harley Andrew
Vilhu Osmi
Wamsteker Willem
Wang Zhenru

Members:

Abramowicz Marek
Acharya Bannanje S
Acton Loren W
Adams David J
Agrawal P C
Aguiar Odylio Denys
Ahluwalia Harjit Singh
Ahmad Imad Aldean
Aizu Ko
Alexander Joseph K
Almleaky Yasseen
Apparao K M V
Arafune Jiro
Arnaud Monique
Arnould Marcel L
Arons Jonathan
Aschenbach Bernd
Asseo Estelle
Audouze Jean
Awaki Hisamitsu
Axford W Ian
Ayres Thomas R
Baan Willem A
Bade Norbert
Badiali Massimo
Barnothy Jeno
Barstow Martin Adrian
Basu Dipak
Baym Gordon Alan
Becker Robert Howard
Begelman Mitchell Craig
Belloni Tomaso
Benedict George F
Benford Gregory
Bennett Charles L
Bennett Kevin
Benvenuto Omar
Bergeron Jacqueline A
Bernacca Pietio L
Bhattacharjee Pijushpani
Bianchi Luciana
Bicknell Geoffrey V
Biermann Peter L
Bignami Giovanni F
Biswas Sukumar
Blamont Jacques Emile
Blandford Roger David
Bleeker Johan A M Ir
Bless Robert C
Bloemen Johannes B G M
Blondin John M
Bludman Sidney A
Boggess Albert
Boggess Nancy W
Bohlin Ralph C
Boksenberg Alec
Bonazzola Silvano
Bonnet Roger M
Bonnet-Bidaud Jean Marc
Bonometto Silvio A
Borozdin Konstantin
Bougeret Jean Louis
Bowyer C Stuart
Boyarchuk Alexander A
Boyd Robert L F
Bradley Arthur J
Braga Joao
Brandt John C
Brecher Kenneth
Breslin Ann
Brinkman Bert C
Brown Alexander
Brueckner Guenter E
Bruhweiler Fred C
Bruner Marilyn E
Bumba Vaclav
Bunner Alan N
Burbidge Geoffrey R
Burger Marijke
Burke Bernard F
Burrows Adam Seth
Burrows David Nelson
Burton William M
Butler C John

Butterworth Paul
Camenzind Max
Cameron Alastair G W
Campbell Murray F
Cardini Daniela
Carpenter Kenneth G
Carroll P Kevin
Carver John H
Cash Webster C
Casse Michel
Catura Richard C
Caughlan Georgeanne R
Cavaliere Alfonso G
Chakrabarti Sandip K
Chakraborty Deo K
Chapman Robert D
Charles Philip Allan
Chechetkin Valerij M
Cheng Kwong-sang
Chengmo Zhang
Chian Abraham Chian-Long
Chikawa Michiyuki
Chitre Shashikumar M
Chochol Drahomir
Chubb Talbot A
Chupp Edward L
Churazov Eugene M
Clark George W
Clark Thomas Alan
Clay Roger
Code Arthur D
Cohen Jeffrey M
Collin-Souffrin Suzy
Comastri Andrea
Condon James J
Corbet Robin Henry D
Corcoran Michael Francis
Cordova France A D
Courtes Georges
Cowie Lennox Lauchlan
Cowsik Ramanath
Crannell Carol Jo
Cruise Adrian Michael
Culhane Leonard
Curir Anna
D'Amico Flavio
da Costa Antonio A
Dadhich Naresh
Dai Zigao
Damle S V
Dautcourt G
Davidsen Arthur Falnes
Davidson William
Davis Michael M
Davis Robert J
Davis Leverett
Dawson Bruce
de Felice Fernando
de Jager Cornelis
de Martino Domitilla
de Young David S
Debrunner Hermann
Dempsey Robert C
Dennis Brian Roy
Dermer Charles Dennison
Dewitt Bryce S
Di Cocco Guido
Digel Seth William
Disney Michael J
Dokuchaev Vyacheslav I
Dolan Joseph F
Dotani Tadayasu
Drake Frank D
Drury Luke O'Connor
Dunkelman Lawrence
Duorah Hira Lal
Dupree Andrea K
Durouchoux Philippe
Duthie Joseph G
Edelson Rick
Edwards Paul J
Eichler David
Eilek Jean
El Raey Mohamed E
Elvis Martin S
Emanuele Alessandro
Evans W Doyle
Fabricant Daniel G
Fang Li-zhi
Faraggiana Rosanna
Feldman Paul Donald
Felten James E
Fenton K B
Ferrari Attilio
Fichtel Carl E
Field George B
Fisher Philip C
Fishman Gerald J
Fitton Brian
Foing Bernard H
Fomin Valery
Forman William Richard
Franceschini Alberto
Frandsen Soeren
Frank Juhan
Fredga Kerstin
Friedman Herbert
Frisk Urban
Fu Chengqi
Furniss Ian
Gabriel Alan H
Gaisser Thomas K
Galeotti Piero
Garcia Howard A
Garmire Gordon P
Gehrels Neil
Gezari Daniel Ysa
Ghisellini Gabriele
Giacconi Riccardo
Gilra Daya P
Ginzburg Vitaly L
Gioia Isabella M
Glaser Harold
Gold Thomas
Goldsmith Donald W
Gondhalekar Prabhakar
Gonzales'a Walter D
Grebenev Sergei A
Greenhill John
Greisen Kenneth I
Grenier Isabelle
Grewing Michael
Greyber Howard D
Griffiths Richard E
Grindlay Jonathan E
Gull Theodore R
Gunn James E
Gursky Herbert
Guseinov O H
Hack Margherita
Haddock Fred T
Hakkila Jon Eric
Hall Andrew Norman
Hallam Kenneth L
Han Zhengzhong
Hang Hengrong
Harms Richard James
Harris Daniel E
Harvey Christopher C
Harvey Paul Michael
Harwit Martin
Hatsukade Isamu
Haubold Hans Joachim
Hauser Michael G
Hawkes Robert Lewis
Hawking Stephen W
Hawkins Isabel
Haymes Robert C
Hearn Anthony G
Heckathorn Harry M
Hein Righini Giovanna
Heise John
Helfand David John
Helmken Henry F
Helou George

Henoux Jean-Claude
Henriksen Richard N
Henry Richard C
Hensberge Herman
Heske Astrid
Hoffman Jeffrey Alan
Holberg Jay B
Holloway Nigel J
Holt Stephen S
Houziaux Leo
Howarth Ian Donald
Hoyle Fred
Hoyng Peter
Hu Wenrui
Huang Keliang
Huber Martin C E
Hunt Leslie
Hurley Kevin C
Hutchings John B
Ichimaru Setsuo
Imamura James
Imhoff Catherine L
Inoue Makoto
Ipser James R
Ishida Manabu
Israel Werner
Ito Kensai A
Itoh Masayuki
Iyengar K V K
Jackson John Charles
Jaffe Walter Joseph
Jamar Claude A J
Janka Hans Thomas
Jenkins Edward B
Jokipii J R
Jones Frank Culver
Jones Thomas Walter
Jordan Carole
Jordan Stuart D
Joss Paul Christopher
Juliusson Einar
Kafatos Minas
Kafka Peter
Kahabka Peter
Kahn Franz D
Kapoor Ramesh Chander
Karpinskij Vadim N
Kasturirangan K
Katz Jonathan I
Kawai Nobuyuki
Kellermann Kenneth I
Kellogg Edwin M
Kembhavi Ajit K
Kessler Martin F
Kii Tsuneo
Killeen Neil
Kim Yonggi
Kimble Randy A
Kirk John
Klinkhamer Frans
Klose Sylvio
Koch-Miramond Lydie
Kocharov Grant E
Koide Shinji
Kojima Yasufumi
Kolb Edward W
Kondo Masaaki
Kondo Yoji
Koshiba Masa-Toshi
Koupelis Theodoros
Koyama Katsuji
Kozlowski Maciej
Kraemer Gerhard
Kraushaar William L
Kreisel E
Kristiansson Krister
Kulsrud Russell M
Kumagai Shiomi
Kundt Wolfgang
Kurt Vitaliy G
Kusunose Masaaki
Lamb Frederick K
Lamb Susan Ann
Lamb Donald Quincy
Lamers Henny J G L M
Lampton Michael
Lasher Gordon Jewett
Lattimer James M
Lea Susan Maureen
Leckrone David S
Leighly Karen Marie
Lemaire Philippe
Lewin Walter H G
Li Qibin
Li Tipei
Li Yuanjie
Li Zhifang
Li Zongwei
Liang Edison P
Lindblad Bertil A
Linsky Jeffrey L
Linsley John
Liu Ruliang
Lochner James Charles
Long Knox S
Longair Malcolm S
Lovelace Richard V E
Lovell Sir Bernard
Lu Jufu
Lu Tan
Luest Reimar
Luminet Jean-Pierre
Lynden-Bell Donald
Ma YuQian
Maccacaro Tommaso
Macchetto Ferdinando
Maggio Antonio
Makarov Valeri
Malaise Daniel J
Malitson Harriet H
Malkan Matthew Arnold
Manara Alessandro A
Mandolesi Nazzareno
Maran Stephen P
Marar T M k
Marov Mikhail Ya
Martin Inacio Malmonge
Masai Kuniaki
Mason Glenn M
Mason Keith Owen
Mather John Cromwell
Matsumoto Ryoji
Matsuoka Masaru
Matz Steven Micheal
Mazurek Thaddeus John
McBreen Brian Philip
McCluskey George E
McCray Richard
McWhirter R W Peter
Mead Jaylee Montague
Medina Jose
Meier David L
Meiksin Avery Abraham
Melia Fulvio
Melnick Gary J
Melrose Donald B
Mestel Leon
Meszaros Peter
Mewe R
Meyer Friedrich
Meyer Jean-Paul
Micela Giuseppina
Michel F Curtis
Miller Guy Scott
Miller John C
Miyaji Shigeki
Miyamoto Sigenori
Mizumoto Yoshihiko
Mizutani Kohei
Modisette Jerry L
Monet David G
Monfils Andre G
Montmerle Thierry
Moon Shin Haeng
Moos Henry Warren

Morgan Thomas H
Morrison Philip
Morton Donald C
Murakami Hiroshi
Murakami Toshio
Murdock Thomas Lee
Murtagh Fionn
Murthy Jayant
Naidenov Victor O
Nakayama Kunji
Neeman Yuval
Neff Susan Gale
Ness Norman F
Neupert Werner M
Nichols-Bohlin Joy
Nicollier Claude
Nityananda R
Nomoto Ken'ichi
Norci Laura
Nordh Lennart H
Norman Colin A
Novick Robert
Noyes Robert W
Nulsen Paul
O'Connell Robert F
O'Mongain Eon
O'Sullivan Denis F
Oda Minoru
Oda Naoki
Oezel Mehmet Emin
Ogawara Yoshiaki
Ogelman Hakki B
Okeke Pius N
Okoye Samuel E
Okuda Toru
Olthof Hindericus
Oohara Ken-ichi
Osborne Julian P
Ostriker Jeremiah P
Ostrowski Michal
Owen Tobias C
Ozernoy Leonid M
Pacholczyk Andrzej G
Paciesas William S
Pacini Franco
Page Clive G
Paltani Stephane
Palumbo Giorgio G C
Pandey Uma Shankar
Papagiannis Michael D
Park Myeong-gu
Parker Eugene N
Parkinson John H
Parkinson William H
Pauliny Toth Ivan K K
Peacock Anthony
Peng Qiuhe
Perola Giuseppe C
Perry Peter M
Peters Geraldine Joan
Peterson Bruce A
Peterson Laurence E
Pethick Christopher J
Petro Larry David
Petrosian Vahe
Phillips Kenneth J H
Piddington Jack H Res Fel
Pinkau K
Pinto Philip Alfred
Pipher Judith L
Piro Luigi
Polidan Ronald S
Porter Neil A
Pounds Kenneth A
Poutanen Juri
Prasanna A R
Preuss Eugen
Price Stephan Donald
Proszynski Mieczyslaw
Protheroe Raymond J
Qu Qinyue
Radhakrishnan V
Ramadurai Souriraja
Rao Arikkala Raghurama
Rao Ramachandra V
Raubenheimer Barend C
Razdan Hiralal
Rees Martin J
Rees Martin J
Reeves Edmond M
Reeves Hubert
Reichert Gail Anne
Reig Pablo
Rense William A
Riegler Guenter R
Roman Nancy Grace
Rosendhal Jeffrey D
Rosner Robert
Rovero Adrian Carlos
Ruben G
Ruder Hanns
Ruffini Remo
Sabau-Graziati Lola
Sagdeev Roald Z
Sahade Jorge
Salpeter Edwin E
Salvati Marco
Sanders Wilton Turner III
Santos Nilton Oscar
Sartori Leo
Saslaw William C
Sato Katsuhiko
Savage Blair D
Savedoff Malcolm P
Scargle Jeffrey D
Schaefer Gerhard
Schatten Kenneth H
Schatzman Evry
Scheuer Peter A G
Schmidt K H
Schnopper Herbert W
Schoeneich W
Schramm David N
Schreier Ethan J
Schultz G V
Schwartz Daniel A
Schwartz Steven Jay
Schwehm Gerhard
Sciama Dennis W
Sciortino Salvatore
Scott John S
Seielstad George A
Selvelli Pierluigi
Sequeiros Juan
Setti Giancarlo
Seward Frederick D
Shaham Jacob
Shakura Nicholaj I
Shapiro Maurice M
Shaver Peter A
Shaviv Giora
Sheffield Charles
Shibai Hiroshi
Shibazaki Noriaki
Shields Gregory A
Shigeyama Toshikazu
Shivanandan Kandiah
Sholomitsky G B
Shukre C S
Signore Monique
Sikora Marek
Silberberg Rein
Silvestro Giovanni
Simon Paul C
Singh Kulinder Pal
Skilling John
Smale Alan Peter
Smith Barham W
Smith Bradford A
Smith Linda J
Smith Peter L
Snow Theodore P
Sofia Sabatino
Sonneborn George
Sood Ravi

Spada Gianfranco
Speer R J
Srivastava Dhruwa
Stachnik Robert V
Staubert Ruediger Prof
Stecher Theodore P
Stecker Floyd W
Steigman Gary
Steiner Joao E
Stencel Robert Edward
Stepanian A A
Stephens S A
Stern Robert Allan
Stevens Ian
Stier Mark T
Stockman Hervey S
Stone R G
Straumann Norbert
Stringfellow Guy Scott
Strong Ian B
Sturrock Peter A
Sun Wei-Hsin
Sunyaev Rashid A
Suzuki Hideyuki
Svensson Roland
Swank Jean Hebb
Tagliaferri Gianpiero
Takahara Fumio
Takahashi Masaaki
Takahashi Tadayuki
Takakura Tatsuo Emer
Tanaka Yasuo
Tanaka Yasuo
Tashiro Makoto
Terrell Nelson James
Thomas Roger J
Thorne Kip S
Tomimatsu Akira
Totsuka Yoji
Tovmassian Hrant M
Traub Wesley Arthur
Trimble Virginia L
Truemper Joachim
Truran James W
Trussoni Edoardo
Tsunemi Hiroshi
Tsuru Takeshi
Tsuruta Sachiko
Tsygan Anatolii I
Tylka Allan J
Underhill Anne B
Underwood James H
Upson Walter L II
Vahia Mayank N
Valnicek Boris
Valtonen Mauri J
van Beek Frank
van de Hulst H C
van den Heuvel Edward P J
van der Hucht Karel A
van der Walt D J
van Duinen R J
van Riper Kenneth A
van Speybroeck Leon P
Vial Jean-Claude
Vidal Nissim V
Vidal-Madjar Alfred
Viotti Roberto
Voelk Heinrich J
von Montigny Corinna
Vrtilek Saeqa Dil
Walsh Dennis
Wanas Mamdouh Ishaac
Wang Deyu
Wang Shouguan
Wang Shui
Wang Yi-ming
Warner John W
Weaver Thomas A
Webster Adrian S
Wehrle Ann Elizabeth
Weiler Edward J
Weiler Kurt W
Weinberg J L
Weisheit Jon C
Weisskopf Martin Ch
Wells Donald C
Wentzel Donat G
Wesselius Paul R
Westfold Kevin C
Westphal James A
Wheeler J Craig
Wheeler John A
White Nicholas Ernest
Will Clifford M
Willis Allan J
Willner Steven Paul
Wilson Andrew S
Wilson James R
Wilson Robert
Winkler Christoph
Wolfendale Arnold W
Wolstencroft Ramon D
Wolter Anna
Woltjer Lodewijk
Worrall Diana Mary
Wray James D
Wu Chi Chao
Wu Xuejun
Wunner Guenter
Yamada Shoichi
Yamamoto Yoshiaki
Yamashita Kojun
Yamauchi Makoto
Yamauchi Shigeo
Yang Lantian
Yang Pibo
Yock Philip
Yoshida Atsumasa
You Junhan
Zamorani Giovanni
Zarnecki Jan Charles
Zdziarski Andrzej
Zhang Heqi
Zhang Jialu
Zhang Zhen-Jiu
Zheng Wei
Zombeck Martin V
Zou Huicheng

Composition of Commission 45

Stellar classification/Classification stellaire

President: Gerbaldi Michele

Vice-President: Lloyd Evans Thomas Harry

Organizing Committee:

Claria Juan
Corbally Christopher
Drilling John S
Garmany Catherine D
Gray Richard O
Kurtz Donald Wayne
Levato Orlando Hugo
North Pierre
Zdanavicius Kazimeras

Members:

Albers Henry
Ardeberg Arne L
Arellano Ferro Armando
Babu G S D
Baglin Annie
Barbier-Brossat Madeleine
Bartaya R A
Bartkevicius Antanas
Bell Roger A
Bidelman William P
Blanco Victor M
Buscombe William
Buser Roland
Celis Leopoldo
Cester Bruno
Cherepashchuk Analily M
Christy James Walter
Coluzzi Regina
Cowley Anne P
Crawford David L
Divan Lucienne
Duflot Marcelle
Egret Daniel
Faraggiana Rosanna
Feast Michael W
Fehrenbach Charles
Fitzpatrick Edward L
Fukuda Ichiro
Garrison Robert F
Geyer Edward H
Glagolevskij Juri V
Golay Marcel
Grenon Michel
Guetter Harry Hendrik
Gurzadian Grigor A
Hack Margherita
Hallam Kenneth L
Hauck Bernard
Hayes Donald S
Houk Nancy
Huang Lin
Humphreys Roberta M
Jaschek Carlos O R
Kato Ken-ichi
Keenan Philip C
Kharadze E K
Kurtanidze Omar
Kurtz Michael Julian
Labhardt Lukas
Lasala Gerald J
Lee Sang Gak
Loden Kerstin R
Low Frank J
Lu Phillip K
Luri Xavier
Lutz Julie H
MacConnell Darrell J
Maehara Hideo
Malagnini Maria Lucia
Malaroda Stella M
McCarthy Martin F
McClure Robert D
Mcnamara Delbert H
Mead Jaylee Montague
Mendoza V Eugenio E
Morossi Carlo
Morrell Nidia
Nandy Kashinath
Nicolet Bernard
Notni P
Oja Tarmo
Olsen Erik H
Osborn Wayne
Oswalt Terry D
Parsons Sidney B
Pasinetti Laura E
Perry Charles L
Philip A G Davis
Pizzichini Graziella
Preston George W
Rautela B S
Roman Nancy Grace
Rountree Janet
Rudkjobing Mogens
Sanwal N B
Schild Rudolph E
Schmidt-Kaler Theodor
Seitter Waltraut C
Sharpless Stewart
Shore Steven N
Sinnerstad Ulf E
Sion Edward Michael
Slettebak Arne
Sonti Sreedhar Rao
Steinlin Uli
Stephenson C Bruce
Stock Jurgen D
Straizys V
Strobel Andrzej

Upgren Arthur R
von Hippel Theodore A
Walborn Nolan R
Walker Gordon A H
Warren Wayne H
Weaver William Bruce
Wesselius Paul R
Westerlund Bengt E
Williams John A
Wing Robert F
Wu Hsin-Heng
Wyckoff Susan
Yamashita Yasumasa
Yoss Kenneth M

Composition of Commission 46

Teaching of astronomy/Enseignement de l'astronomie

President: Fierro Julieta

Vice-President: Isobe Syuzo

Organizing Committee:

Acker Agnes
Batten Alan H
Gerbaldi Michele
Guinan Edward Francis
Hoff Darrel Barton
Jones Barrie W
Martinez Peter
McNally Derek
Narlikar Jayant V
Pasachoff Jay M
Percy John R
Wentzel Donat G

Members:

Aguilar Maria Luisa
Aiad A Zaki
Alexandrov Yuri V
Alvarez Pomares A O
Anandaram Mandayam N
Andrews Frank
Ansari S M Razaullah
Arellano Ferro Armando
Aubier Monique G
Bajaja Esteban
Benson Priscilla J
Bernabeu Guillermo
Black Adam Robert S
Bobrowsky Matthew
Bochonko D Richard
Borchkhadze Tengiz M
Botez Elvira
Bottinelli Lucette
Braes L L E
Brieva Eduardo
Brosch Noah
Bruck Mary T
Budding Edwin
Buscombe William
Calvet Nuria
Catala Poch M A
Chamberlain Joseph M
Codina Landaberry Sayd J
Cottrell Peter Ledsam
Couper Heather Miss
Cui Zhenhua
Daniel Jean Yves
DeGioia-Eastwood Kathleen
Dupuy David L
Duval Marie-France
Dworetsky Michael M
Emerson David
Fairall Anthony P
Feng Kejia
Fernandez Julio A
Fernandez-Figueroa M J
Fienberg Richard T
Fleck Robert Charles
Gallino Roberto
Ghobros Roshdy Azer
Gingerich Owen
Gouguenheim Lucienne
Gurm Hardev S
Hafizi Mimoza
Haupt Hermann F
Havlen Robert J
Hawkins Isabel
Haywood J
Hearnshaw John B
Hemenway Mary Kay M
Heudier Jean-Louis
Hidayat Bambang
Hill Philip W
Hockey Thomas Arnold
Houziaux Leo
Huang Tianyi
Ilyas Mohammad
Impey Christopher D
Inglis Michael
Iwaniszewska Cecylia
Jarrett Alan H
Keller Hans Ulrich
Kennedy John E
Kitchin Christopher R
Kolka Indrek
Kononovich Edward V
Kourganoff Vladimir
Kreiner Jerzy Marek
Krupp Edwin C
Lago Maria Teresa V T
Lai Sebastiana
Lanciano Nicoletta
Leung Chun Ming
Leung Kam Ching
Li Qibin
Little-Marenin Irene R
Lomb Nicholas Ralph
Luck John M
Ma Er
Ma Xingyuan
Maciel Walter J
Maddison Ronald Ch
Marsh Julian C D
Martinet Louis
Mavridis L N
Maza Jose
McCarthy Martin F
Meidav Meir
Mizuno Takao
Momchev Gospodin
Moreels Guy

Muzzio Juan C
Nha Il-Seong
Nicolov Nikolai S
Nicolson Iain
Noels Arlette
Oja Heikki
Okoye Samuel E
Olsen Fogh H J
Onuora Lesley Irene
Osborn Wayne
Osorio Jose J S P
Oswalt Terry D
Othman Mazlan
Owaki Naoaki
Pandey Uma Shankar
Parisot Jean-Paul
Penston Margaret
Pokorny Zdenek
Ponce G A
Prabhakaran Nayar S R
Proverbio Edoardo
Quamar Jawaid
Querci Francois R
Raghavan Nirupama
Ramadurai Souriraja
Rijsdijk Case
Robbins R Robert
Robinson Leif J
Rodgers Alex W
Ros Rosa M
Roslund Curt
Roy Archie E
Safko John L
Sanahuja Blas
Sandqvist Aage
Saxena P P
Sbirkova-Natcheva T
Schleicher David G
Schlosser Wolfhard
Schmidt Thomas
Schmitter Edward F
Schroeder Daniel J
Seeds Michael August
Shen Chun-Shan
Shipman Harry L
Siroky Jaromir
Solheim Jan Erik
Stefl Vladimir
Stenholm Bjoern
Stoev Alexei
Sukartadiredja Darsa
Svestka Jiri
Szecsenyi-Nagy Gabor
Szostak Roland
Taborda Jose Rosa
Tolbert Charles R
Torres-Peimbert Silvia
Troche-Boggino A E
Tsubota Yukimasa
van Santvoort Jacques
Vauclair Sylvie D
Vladimirov Simeon
Vujnovic Vladis
Wang Shunde
Ward Richard A
Whitelock Patricia Ann
Williamon Richard M
Woo Jong Ok
Zealey William J
Zeilik Michael Ii
Zimmermann Helmut

Composition of Commission 47

Cosmology/Cosmologie

President: Szalay Alex

Vice-President: Peacock John Andrew

Organizing Committee:

Chu Yaoquan
da Costa Luiz A N
Einasto Jaan
Ellis George F R
Koo David C-Y
Lilly Simon J
Narlikar Jayant V
Shaver Peter A
Webster Rachel
White Simon David Manion
Wolfe Arthur M

Members:

Aizu Ko
Allan Peter M
Amendola Luca
Andreani Paola Michela
Audouze Jean
Auluck Faqir Chand
Azuma Takahiro
Bahcall Neta A
Bajtlik Stanislaw
Baldwin John E
Banerji Sriranjan
Banhatti Dilip Gopal
Barberis Bruno
Barbuy Beatriz
Bardeen James M
Barnothy Jeno
Barrow John David
Bartelmann Matthias
Barthel Peter
Basu Dipak
Bechtold Jill
Beckman John E
Beesham Aroonkumar
Belinsky Vladimir A
Bennett Charles L
Bergeron Jacqueline A
Berman Marcelo S
Bertola Francesco
Bertschinger Edmund
Betancort-Rijo Juan
Bhavsar Suketu P
Bicknell Geoffrey V
Bignami Giovanni F
Birkinshaw Mark
Blanchard Alain
Bleyer Ulrich
Bludman Sidney A
Boksenberg Alec
Bond John Richard
Bondi Hermann
Bonnor W B
Borgeest Ulf
Bouchet Francois R
Boyle Brian
Brecher Kenneth
Burbidge Geoffrey R
Burns JR Jack O'Neal
Calvani Massimo
Cappi Alberto
Carr Bernard John
Castagnino Mario
Cavaliere Alfonso G
Chang Kyongae
Charlot Stephane
Chen Jiansheng
Cheng Fuhua
Cheng Fuzhen
Chincarini Guido L
Chitre Dattakumar M
Chodorowski Michal
Claria Juan
Cocke William John
Cohen Jeffrey M
Cohen Ross D
Coles Peter
Colless Matthew
Colombi Stephane
Condon James J
Crane Patrick C
Crane Philippe
Cristiani Stefano
Da Costa Gary Stewart
Dadhich Naresh
Danese Luigi
Das P K
Datta Bhaskar
Davidson William
Davies Paul Charles W
Davies Roger L
Davis Marc
Davis Michael M
de Lapparent-Gurriet Valerie
de Ruiter Hans Rudolf
de Zotti Gianfranco
Dekel Avishai
Demaret Jacques
Demianski Marek
Dhurandhar Sanjeev
Dionysiou Demetrios
Djorgovski Stanislav
Doroshkevich Andrei G
Dressler Alan
Drinkwater Michael J

Dultzin-Hacyan D
Dunlop James
Dunsby Peter
Dyer Charles Chester
Efstathiou George
Ehlers Jurgen
Ellis Richard S
Elvis Martin S
Enginol Turan B
Faber Sandra M
Fairall Anthony P
Falk jr Sydney W
Fall S Michael
Fang Li-zhi
Felten James E
Field George B
Filippenko Alexei V
Firmani Claudio A
Florides Petros S
Focardi Paola
Fomin Piotr Ivanovich
Fong Richard
Ford Holland C Res
Forman William Richard
Fouque Pascal
Franceschini Alberto
Frenk Carlos S
Fujimoto Mitsuaki
Fukugita Masataka
Fukui Takao
Galletto Dionigi
Garilli Bianca
Garrison Robert F
Geller Margaret Joan
Giallongo Emanuele
Gioia Isabella M
Giuricin Giuliano
Gold Thomas
Goldsmith Donald W
Gong Shumo
Gonzales Alejandro
Goret Philippe
Gosset Eric
Gottloeber Stefan
Gouda Naoteru
Goyal Ashok Kumar
Gray Richard O
Gregory Stephen Albert
Greyber Howard D
Grishchuk Leonid P
Gudmundsson Einar H
Gunn James E
Guzzo Luigi
Hagen Hans-Juergen
Hamilton Andrew J S
Hardy Eduardo
Harms Richard James
Harrison Edward R
Hawking Stephen W
Hayashi Chushiro
He Xiangtao
Heavens Alan
Heidmann Jean
Hellaby Charles William
Heller Michael
Henriksen Mark Jeffrey
Hewett Paul
Hewitt Adelaide
Hnatyk Bohdan
Hoyle Fred
Hu Esther M
Huchra John Peter
Hwang Jai-chan
Icke Vincent
Ikeuchi Satoru
Impey Christopher D
Iovino Angela
Ishihara Hideki
Iyer B R
Jannuzi Buell Tomasson
Jaroszynski Michal
Jauncey David L
Jedamzik Karsten
Jiang Shuding
Jones Bernard J T
Joshi Mohan N
Junkkarinen Vesa T
Juszkiewicz Roman
Kajino Toshitaka
Kandrup Henry Emil
Kapoor Ramesh Chander
Karachentsev Igor D
Kasper U
Kato Shoji
Kaul Chaman
Kawabata Kinaki
Kawabata Kiyoshi
Kawasaki Masahiro
Kayser Rainer
Kellermann Kenneth I
Kembhavi Ajit K
Khare Pushpa
Kim Jik Su
Kirilova Daniela
Kneib Jean-Paul
Kodama Hideo
Kokkotas Konstantinos
Kolb Edward W
Kompaneets Dmitriy A
Kormendy John
Kovetz Attay
Kozai Yoshihide
Kozlovsky B Z
Krasinski Andrzej
Kriss Gerard A
Kunth Daniel
Kustaanheimo Paul E
Lacey Cedric
Lachieze-Rey Marc
Lahav Ofer
Lake Kayll William
Larionov Mikhael G
Lasota Jean-Pierre
Lausberg Andre
Layzer David
Lequeux James
Li Xiaoqing
Liddle Andrew
Liebscher Dierck-E
Lilje Per Vidar Barth
Liu Liao
Liu Yongzhen
Longair Malcolm S
Lonsdale Carol J
Lu Tan
Lucchin Francesco
Lukash Vladimir N
Luminet Jean-Pierre
Lynden-Bell Donald
Maccagni Dario
MacCallum Malcolm A H
Maddox Stephen
Maeda Kei-ichi
Mandolesi Nazzareno
Mandzhos Andrej V
Mansouri Reza
Marano Bruno
Mardirossian Fabio
Marek John
Martinez-Gonzalez E
Materne Juergen
Mather John Cromwell
Mathez Guy
Matsumoto Toshio
Matzner Richard A
Mavrides Stamatia
McCrea J Dermott
Mellier Yannick
Melott Adrian L
Merighi Roberto
Meszaros Attila
Meszaros Peter
Meyer David M
Meylan Georges
Mezzetti Marino

Misner Charles W
Miyoshi Shigeru
Mo Houjun
Moreau Olivier
Morrison Philip
Moscardini Lauro
Muecket Jan P
Muller Richard A
Murakami Izumi
Nambu Yasusada
Narasimha Delampady
Neeman Yuval
Neves de Araujo Jose Carlos
Nicoll Jeffrey Fancher
Nishida Minoru
Noerdlinger Peter D
Noh Hyerim
Noonan Thomas W
Norman Colin A
Nottale Laurent
Novello Mario
Novikov Igor D
Novosyadlyj Bohdan
Novotny Jan
O'Connell Robert West
Oemler jr Augustus
Okoye Samuel E
Olowin Ronald Paul
Omnes Roland
Onuora Lesley Irene
Ott Heinz-Albert
Oukbir Jamila
Ozernoy Leonid M
Ozsvath I
Pachner Jaroslav
Padmanabhan T
Padrielli Lucia
Page Don Nelson
Pan Rong-Shi
Parnovsky Sergei
Partridge Robert B
Pecker Jean-Claude
Peebles P James E
Pello Roser Descayre
Penzias Arno A
Perryman Michael A C
Persides Sotirios C
Peterson Bruce A
Petitjean Patrick
Petrosian Vahe
Plionis Manolis
Press William H
Puget Jean-Loup
Qu Qinyue
Ramella Massimo
Rawlings Steven
Raychaudhuri Amalkumar
Rebolo Rafael
Rees Martin J
Reeves Hubert
Refsdal Sjur
Riazi Nematollah
Rindler Wolfgang
Rivolo Arthur Rex
Roberts David Hall
Robinson I
Rocca-Volmerange Brigitte
Rosquist Kjell
Rowan-Robinson Michael
Roxburgh Ian W
Rubin Vera C
Rudnick Lawrence
Rudnicki Konrad
Ruffini Remo
Saar Enn
Sadat Rachida
Sahni Varun
Salvador-Sole Eduardo
Salzer John Joseph
Sanz Jose L
Sapar Arved
Sargent Wallace L W
Sasaki Misao
Sasaki Shin
Sato Humitaka
Sato Katsuhiko
Sato Shinji
Savage Ann
Sazhin Michail
Schatzman Evry
Schechter Paul L
Scheuer Peter A G
Schindler Sabine
Schmidt Maarten
Schneider Donald P
Schneider Jean
Schneider Peter
Schramm David N
Schramm K Jochen
Schramm Thomas
Schuch Nelson Jorge
Schuecker Peter
Schuecking E L
Schultz G V
Schumacher Gerard
Sciama Dennis W
Scott Douglas
Scott Elizabeth L
Segal Irving E
Seiden Philip E
Seielstad George A
Setti Giancarlo
Shandarin Sergei F
Shanks Thomas
Shao Zhengyi
Shaviv Giora
Shaya Edward J
Shibata Masaru
Shivanandan Kandiah
Signore Monique
Silk Joseph I
Simon Rene L E
Sironi Giorgio
Sistero Roberto F
Smette Alain
Smith Rodney M
Smith jr Harding E
Smoot III George F
Sokolowski Lech
Song Doo Jong
Souriau Jean-Marie
Spyrou Nicolaos
Stecker Floyd W
Steigman Gary
Stewart John Malcolm
Stoeger William R
Straumann Norbert
Struble Mitchell F
Strukov Igor A
Subrahmanya C R
Suginohara Tatsushi
Sugiyama Naoshi
Sunyaev Rashid A
Surdej Jean M G
Sutherland William
Suto Yasushi
Takahara Fumio
Tammann Gustav Andreas
Tanabe Kenji
Tarter Jill C
Thuan Trinh Xuan
Tifft William G
Tipler Frank Jennings
Tomimatsu Akira
Tomita Kenji
Tonry John
Treder H J
Tremaine Scott Duncan
Trevese Dario
Trimble Virginia L
Tsamparlis Michael
Tully Richard Brent
Turner Edwin L
Turner Michael S
Turnshek David A

Tyson John Anthony
Tytler David
Umemura Masayuki
Uson Juan M
Vagnetti Fausto
Vaidya P C
Valls-Gabaud David
van der Laan Harry
van Haarlem Michiel
Vanysek Vladimir
Vedel Henrik
Vettolani Giampaolo
Vishniac Ethan T
Vishveshwara C V
Voglis Nikos
von Borzeszkowski H H
Wagoner Robert V
Wainwright John
Wambsganss Joachim
Wanas Mamdouh Ishaac
Wang Renchuan
Webster Adrian S
Weinberg Steven
Wesson Paul S
West Michael
Wheeler John A
Whitrow Gerald James
Wilkinson David T
Will Clifford M
Wilson Albert G
Wilson Andrew S
Windhorst Rogier A
Woltjer Lodewijk
Woszczyna Andrzej
Wright Edward L
Wu Xiangping
Xiang Shouping
Xu Chongming
Yang Lantian
Yokoyama Jun-ichi
Yoshii Yuzuru
Yoshimura Motohiko
Zamorani Giovanni
Zhang Jialu
Zhang Zhen-Jiu
Zhou Youyuan
Zhu Shichang
Zhu Xingfeng
Zieba Stanislaw
Zou Zhenlong
Zuiderwijk Edwardus J

Composition of Commission 49

The Interplanetary Plansma and the Heliosphere/Plasma interplanétaire et héliosphère

President: Verheest Frank

Vice-President: Vandas Marek

Organizing Committee:

Buti Bimla
Cramer Neil
Dryer Murray
Habbal Shadia Rifai
Hollweg Joseph V
Huber Martin C E
Kojima Masayoshi
Ripken Hartmut W

Members:

Ahluwalia Harjit Singh
Ananthakrishnan S
Anderson Kinsey A
Barnes Aaron
Barrow Colin H
Barth Charles A
Benz Arnold O
Bertaux Jean-Loup
Blackwell Donald E
Blandford Roger David
Blum Peter
Bochsler Peter
Bonnet Roger M
Brandt John C
Buechner Joerg
Burlaga Leonard F
Chamberlain Joseph W
Chashei Igor V
Chassefiere Eric
Chen Biao
Chitre Dattakumar M
Chitre Shashikumar M
Chou Chih-Kang
Couturier Pierre
Cuperman Sami
de Jager Cornelis
Dinulescu Simona
Dolginov Arkady Z
Duldig Marcus Leslie
Durney Bernard
Dyson John E
Ergma E V
Eshleman Von R
Eviatar Aharon
Fahr Hans Joerg
Feynman Joan
Field George B
Goldman Martin V
Gosling John T
Grzedzielski Stanislaw Pr
Harvey Christopher C
Heras Ana M
Heynderickx Daniel
Heyvaerts Jean
Holzer Thomas Edward
Humble John Edmund
Inagaki Shogo
Ivanov Evgeny I
Jokipii J R
Joselyn Jo Ann c
Kakinuma Takakiyo T
Keller Horst Uwe
Lafon Jean-Pierre J
Lai Sebastiana
Levy Eugene H
Lotova Natalia A
Luest Reimar
Lundstedt Henrik
MacQueen Robert M
Mangeney Andre
Manoharan P K
Mason Glenn M
Matsuura Oscar T
Mavromichalaki Helen
Meister Claudia Veronika
Melrose Donald B
Mendis Devamitta Asoka
Mestel Leon
Michel F Curtis
Moussas Xenophon
Nakagawa Yoshinari
Paresce Francesco
Parker Eugene N
Perkins Francis W
Pflug Klaus
Pneuman Gerald W
Quemerais Eric
Raadu Michael A
Readhead Anthony C S
Reay Newrick K
Rickett Barnaby James
Riddle Anthony C
Rosner Robert
Roxburgh Ian W
Rucinski Daniel
Russell Christopher T
Sagdeev Roald Z
Sarris Emmanuel T
Sastri Hanumath J
Sawyer Constance B
Schatzman Evry
Scherb Frank
Schindler Karl
Schmidt H U
Schreiber Roman
Schwartz Steven Jay

Setti Giancarlo
Shea Margaret A
Smith Dean F
Sonett Charles P
Stone R G
Sturrock Peter A
Suess Steven T
Tritakis Basil P
Vainstein L A
van Allen James A
Vinod S Krishan
von Steiger Rudolf
Vucetich Hector
Wang Shunde
Wang Yi-ming
Watanabe Takashi
Weller Charles S
Wild John Paul
Wu Shi Tsan
Yeh Tyan

Composition of Commission 50

Protection of Existing and Potential Observatory Sites/ Protection des sites d'observatoires existants et potentiels

Teaching of astronomy/Enseignement de l'astronomie

President: Sullivan Woodruff T

Vice-President: Cohen Raymond J

Organizing Committee:

Crawford David L
Hidayat Bambang
Isobe Syuzo
Jiang Shi-Yang
Malin David F
McNally Derek
Spoelstra T A Th

Members:

Ardeberg Arne L
Arsenijevic Jelisaveta
Barreto Luiz Muniz
Bensammar Slimane
Bhattacharyya J C
Blanco Carlo
Blanco Victor M
Brown Robert Hamilton
Burstein David
Cayrel Roger
Costero Rafael
Coyne George V
Davis John
Dawe John Alan
de Greiff J Arias
Dommanget J
Dunkelman Lawrence
Edwards Paul J
Galan Maximino J
Gergely Tomas Esteban
Gibson David Michael
Goebel Ernst
Helmer Leif
Hoag Arthur A
Huang Yinliang
Kahlmann Hans Cornelis
Kontizas Evangelos
Kontizas Mary
Kovalevsky Jean
Kozai Yoshihide
Leibowitz Elia M
Mahra H S
Markkanen Tapio
Mattig W
McCarthy Martin F
Menzies John W
Murdin Paul G
Nelson Burt
Oezel Mehmet Emin
Osorio Jose J S P
Owen Frazer Nelson
Pankonin Vernon Lee
Sanchez Francisco
Schilizzi Richard T
Shcheglov P V
Smith Francis Graham
Suntzeff Nicholas B
Torres Carlos
Torres Carlos Alberto
Tremko Jozef
Upgren Arthur R
van den Bergh Sidney
Vetesnik Miroslav
Walker Merle F
Wayman Patrick A
Whiteoak John B
Woolf Neville J
Woszczyk Andrzej
Wu Mingchan
Yano Hajime
Zhang Bairong

Composition of Commission 51

Bioastronomy: Search for Extraterrestrial life/Bioastronomie: Recherche de la vie extraterrestre

President: Colomb Fernando R

Vice-President: Bowyer C Stuart

Organizing Committee:

Almar Ivan
Drake Frank D
Leger Alain
Mayor Michel
Owen Tobias C
Sullivan Woodruff T
Tarter Jill C
Wellington Kelvin
Wilson Thomas L

Members:

Al-Naimiy Hamid M K
Alsabti Abdul Athem
Balazs Bela A
Ball John A
Bania Thomas Michael
Barbieri Cesare
Basu Dipak
Basu Baidyanath
Baum William A
Beaudet Gilles
Beckman John E
Beckwith Steven V W
Beebe Reta Faye
Benest Daniel
Berendzen Richard
Bernacca Pietio L
Billingham John
Biraud Francois
Bless Robert C
Boyce Peter B
Bracewell Ronald N
Broderick John
Brown Ronald D
Burke Bernard F
Calvin William H
Campusano Luis E
Carlson John B
Carr Thomas D
Chaisson Eric J
Chou Kyong Chol
Clark Thomas A
Connes Pierre
Couper Heather Miss
Currie Douglas G
Daigne Gerard
Davis Michael M
Dawe John Alan
de Graaff W
de Jager Cornelis
de Jonge J K
de Loore Camiel
de Vincenzi Donald
Delsemme Armand H
Dick Steven J
Dixon Robert S
Dorschner Johann
Doubinskij Boris A
Downs George S
Dyson Freeman J
Eccles Michael J
Ellis George F R
Epstein Eugene E
Evans Neal J
Fazio Giovanni G
Fejes Istvan
Feldman Paul A
Field George B
Firneis Friedrich J
Firneis Maria G
Fisher Philip C
Fredrick Laurence W
Freire Ferrero Rubens G
Fujimoto Masa-Katsu
Fujimoto Mitsuaki
Gatewood George
Gehrels Tom
Ghigo Francis D
Ginzburg Vitaly L
Giovannelli Franco
Godoli Giovanni
Goldsmith Donald W
Gott J Richard
Goudis Christos D
Greenberg J Mayo
Greenstein Jesse L
Gregory Philip C
Gulkis Samuel
Gunn James E
Gurm Hardev S
Haddock Fred T
Haisch Bernhard Michael
Hajduk Anton
Hale Alan
Harrison Edward R
Hart Michael H
Heck Andre
Heeschen David S
Heidmann Jean
Herczeg Tibor J
Hershey John L
Heudier Jean-Louis
Hinners Noel W
Hirabayashi Hisashi

Hiroyasu Ando
Hoang Binh Dy
Hoegbom Jan A
Hollis Jan Michael
Horowitz Paul
Hunten Donald M
Hunter James H
Hysom Edmund J
Idlis Grigorij M
Irvine William M
Israel Frank P
Jastrow Robert
Jeffers Stanley
Jennison Roger C
Jones Eric M
Jugaku Jun
Kafatos Minas
Kafka Peter
Kapisinsky Igor
Kardashev Nicolay S
Kaufmann Pierre
Keay Colin S l
Keller Hans Ulrich
Kellermann Kenneth I
Klein Michael J
Knowles Stephen H
Kocer Durcun
Koch Robert H
Koeberl Christian
Kraus John D
Ksanfomaliti Leonid V
Kuiper Thomas B H
Kuzmin Arkadii D
Lafon Jean-Pierre J
Laques Pierre
Lee Sang Gak
Levasseur-Regourd A-C
Lilley Edward A
Lippincott Sarah Lee
Loden Lars Olof
Lovell Sir Bernard
Maffei Paolo
Margrave Jr Thomas Ewing
Marov Mikhail Ya
Martin Anthony R
Martin Maria Cristina
Marx Gyorgy
Matsakis Demetrios N
Matsuda Takuya
Matthews Clifford
Mavridis L N
McAlister Harold A
McDonough Thomas R
Mendoza V Eugenio E
Milet Bernard L
Minn Young Key
Mirabel Igor Felix
Morimoto Masaki
Moroz Vasilis I
Morris Mark Root
Morrison David
Morrison Philip
Muller Richard A
Nakagawa Yoshinari
Nelson Robert M
Niarchos Panayiotis
Norris Raymond Paul
Oda Minoru
Ollongren A
Ostriker Jeremiah P
Pacini Franco
Papagiannis Michael D
Parijskij Yuri N
Pasinetti Laura E
Perek Lubos
Ponsonby John E B
Prochazka Franz V
Qiu Yaohui
Qiu Puzhang
Quintana Hernan
Quintana Jose M
Raghavan Nirupama
Rajamohan R
Reay Newrick K
Rees Martin J
Riihimaa Jorma J
Robinson Leif J
Rood Robert T
Rowan-Robinson Michael
Rubin Robert Howard
Russell Jane L
Sakurai Kunitomo
Sancisi Renzo
Scargle Jeffrey D
Schatzman Evry
Schild Rudolph E
Schneider Jean
Schober Hans J
Schuch Nelson Jorge
Seeger Charles Louis III
Seielstad George A
Seiradakis John Hugh
Shapiro Maurice M
Shen Chun-Shan
Shimizu Mikio
Shostak G Seth
Singh H P
Sivaram C
Slysh Viacheslav I
Snyder Lewis E
Sofue Yoshiaki
Stalio Roberto
Stein John William
Straizys V
Sturrock Peter A
Takaba Hiroshi
Takada-Hidai Masahide
Tavakol Reza
Tedesco Edward F
Tejfel Viktor G
Terzian Yervant
Thaddeus Patrick
Tolbert Charles R
Toro Tibor
Tovmassian Hrant M
Townes Charles Hard
Trimble Virginia L
Turner Edwin L
Turner Kenneth C
Valbousquet Armand
Vallee Jacques P
Van Flandern Tom
Varshalovich Dimitrij
Vauclair Gerard P
Vazquez Manuel
Venugopal V R
Verschuur Gerrit L
Vogt Nikolaus
von Hoerner Sebastian
Wallis Max K
Watson Frederick Garnett
Welch William J
Wesson Paul S
Wetherill George W
Wielebinski Richard
Williams Iwan P
Willson Robert Frederick
Wolstencroft Ramon D
Wright Alan E
Ye Shuhua
Zuckerman Ben M

DIVISION AND COMMISSION WORKING GROUPS

Division Working Groups

Division(s)

BIPM/IAU Joint Committee on General Relativity for space-time reference systems and metrology — I

G. Petit, Chair — gpetit@bipm.fr

WG on General Relativity in Celestial Mechanics and Astrometry RCMA — I
http://rcswww.urz.tu-dresden.de/~lohrmobs/iauwg.html

M.H. Soffel, Chair — soffel@rcs.urz.tu-dresden.de

WG on the International Celestial Reference System (ICSR) — I

F. Mignard, Chair — mignard@obs-azur.fr

Sub-groups		Chair
Maintenance and extension of the ICRS	C. Ma	cma@gemini.gsfc.nasa.gov
Relation with IERS	F. Arias	felicitas@fcaglp.edu.ar
Densification in optics	S. Urban	seu@pyxis.usno.navy.mil
Link to the dynamical system	M. Standish	ems@smyles.jpl.nasa.gov
Computational consequences	N. Capitaine	capitaine@obspm.fr
Ties with previous and new catalogues	F. Mignard	mignard@obs-azur.fr

WG on Solar Eclipses — II
http://www.williams.edu/Astronomy/IAU_eclipses/

J.M. Pasachoff, Chair — jay.m.pasachoff@williams.edu

Small Bodies Names Committee — III

M.F. A'Hearn, Chair — ma@astro.umd.edu

WG on Hot Massive Stars — IV

P. Eenens, Chair — eenens@carina.astro.ugto.mx

WG on Star Forming Regions — VI
http://www.oan.es/sfwg/

R. Bachiller, Chair — bachiller@oan.es

The Planetary Nebula WG — VI

S. Kwok, Chair — kwok@iras.ucalgary.ca

Astrochemistry WG — VI
http://www.strw.leidenuniv.nl/~iau34/

D.A. Williams, Chair — daw@star.ucl.ac.uk
E.F. van Dishoeck, Secretary — ewine@strw.leidenuniv.nl

WG for astronomy from the moon — XI

Y. Terzian, Chair — terzian@astrosun.tn.cornell.edu

Inter Division Working Groups

WG on Near Earth Objects **I & III**

D. Morrison, Chair david.morrison@arc.nasa.gov

Commission Working Groups

TG on Data Centers & Networks **5**
http://set.inasan.rssi.ru/IAU/WGS/TGDCN/

F. Genova, Chair genova@ u-strasbg.fr

WG on Astronomical Data **5**
http://set.inasan.rssi.ru/IAU/WGS/WGAD/

E. Raimond, Chair exr@nfra.nl
R.P. Norris, Vice-Chair rnorris@rp.csiro.au

TG on Designations **5**
http://set.inasan.rssi.ru/IAU/WGS/TGD/

H.R. Dickel, Chair lanie@astro.uiuc.edu

SG 0: Clearing House	H.R. Dickel, Leader	lanie@astro.uiuc.edu
SG 1: Journals/Newsletters		
SG 2: Libraries		
SG 3: Data Centers	F. Ochsenbein, Leader	francois@simbad.u-strasbg.fr
SG 4: FITS/other	F. Genova, Liasons with CDS	genova@ u-strasbg.fr
SG 5: Observatories	J.M. Dickey, Co-Leader	john@ast1.spa.umn.edu
	H.J. Andernach, Co-Leader	heinz@polaris.astro.ugto.mx
SG 6: IAU Commissions/Officials		

WG on Revision of UDC52 **5**
http://set.inasan.rssi.ru/IAU/WGS/TGU/

G. Wilkins, Chair G.A.Wilkins@exeter.ac.uk

WG on Libraries **5**
http://set.inasan.rssi.ru/IAU/WGS/WGL/

U. Grothkopf, Co-Chair esolib@eso.org
F. Murtagh. Co-Chair fmurtagh@cdsxb6.u-strasbg.fr

TG on FITS IAU FWG **5**
http://set.inasan.rssi.ru/IAU/WGS/TGF/

D.C. Wells, Chair dwells@nrao.edu
E. Raimond, Vice-Chair exr@nfra.nl

WG on Information Handling **5**
http://set.inasan.rssi.ru/IAU/WGS/WGIH/

P. Boyce, Chair pboyce@aas.org

WG on Astrolabes 8

F. Chollet Chollet@mesiob.obspm.fr

WG on Atomic Spectra & Wavelengths 14

S. Johansson, Chair atomsej@seldc52

WG on Atomic Transition Probabilities 14

W.L. Wiese, Chair wolfgang.wiese@nist.gov
J.R. Fuhr, Co-Chair fuhr@atm.nist.gov

WG on Collision Processes 14

D.R. Schultz, Chair schultz@mail.phy.ornl.gov
P.C. Stancil, vice-Chair stancil@mail.phy.ornl.gov

WG on Line Broadening 14

C. Stehlé Chantal.Stehle@obspm.fr
G. Peach ucap22g@ucl.ac.uk

WG on Molecular Structure 14

E.F. Van Dishoeck ewine@strw.leidenuniv.nl
J.H. Black jblack@oso.chalmers.se

WG on Molecular Reactions on Solid Surfaces 14

S. Leach leach@obspm.fr

WG for Water Release Rates from Comets 15

W. Huebner, Chair whuebner@swri.edu

WG on Asteroids 15

E F. Tedesco, Chair etedesco@TerraSys.com

WG of the IAU/IAG/IUGG/COSPAR on Cartographic Coordinates and Rotational Elements of Planets and Satellites 16

K. Seidelmann, Chair pks@spica.usno.navy.mil

WG on Comets 20

G.B. Valsecchi, Chair giovanni@ias.rm.cnr.it

WG on Satellites 20
http://www.bdl.fr/iauwg.html

J.E. Arlot, Chair arlot@bdl.fr

Minor Planet Center and Institute for Theoretical Astronomy (MPC & ITA) Consultative Group 22

E.L.G. Bowell, chair ebowell@lowell.edu

WG Amateur-Professional co-operation in Meteors 22

P. Jenniskins, Chair peter@max.arc.nasa.gov

WG on Archiving Spectroscopic Data 29

E. Griffin, Chair e.griffin1@physics.oxford.ac.uk

WG for radial-velocity standard stars 30

R. Stefanik stefanik@cfa.harvard.edu

WG for the cataloguing and bibliography of stellar radial velocities 30

H. Levato, Chair levato@castec.edu.ar

WG on Pulsar Timing 31

G. Petit, Chair gpetit@bipm.fr

WG on Time Transfer 31

T. Fukushinma, Chair toshio@spacetime.mtk.nao.ac.jp

Sub-Committee on Solar Eclipes & Astronomy Education 46

J. Pasachoff, Chair jmp@wiliams.edu

INTER COMMISSION WORKING GROUPS

IAU/IUGG WG on the Non-rigid Earth Nutation Theory 4, 7 & 19
http://www.oma.be/KSB-ORB/D1/EARTH_ROT/wgnut.html

V. Dehant, Chair veronique.dehant@ksb-orb.oma.be

WG on Astronomical Standards (WGAS) 4, 5, 8, 19, 24 & 31

T. Fukushima, Chair toshio@spacetime.mtk.nao.ac.jp

D. McCarthy, Chair, Sub-group on Best Estimates
P.T. Wallace, Chair, Sub-Group SOFA

WG of the IAU/IAG/IUGG/COSPAR on Cartographic Coordinates and Rotational Elements of Planets and Satellites 4 & 16
http://www.ari.uni-heidelberg.de/interessantes/iaucommission8/wg/seidelmann.html

K. Seidelmann, Chair pks@spica.usno.navy.mil

WG dealing with the one-century old plates of the Astrographic Program 8 & 24
http://astro.u-strasbg.fr/~fresneau/workgroup.html

A. Fresneau, Chair fresneau@astro.u-strasbg.fr

WG on Standard Stars 25, 29, 30 & 45
http://clavius.as.arizona.edu/ssn/

R.F. Garrison, Chair garrison@astro.utoronto.ca
C. Corbally corbally@as.arizona.edu
Ed. of the Standard Star Newsletter

WG on Ap and related stars **27 & 45**

M. Takada-Hidai, Chair hidai@keyaki.cc.u-tokai.ac.jp
P. North Pierre.north@obs.unige.ch
Ed. of the Ap and Related Star Newsletters

WG on Active B Stars **27 & 45**

D. Baade, Chair dbaade@dgaeso51.bitnet
G.J. Peters gjpeters@mucen.usc.edu
Ed. of the Be Star Newsletter

WG on Peculiar Red Giants **27, 29 & 45**

R.F. Wing, Chair wing@astronomy,ohio-state.edu
S. Yorka yorka@cc.denison.edu
Ed. of the Newsletter of Chemically Peculiar Red Giant Stars
Electronic version distributed by S. Yorka

Paper version distributed by R.F. Wing

INDIVIDUAL MEMBERSHIP FROM ADHERING AND *ASSOCIATE* COUNTRIES

Algeria

Abdelatif Toufik
Irbah Abdanour
Makhlouf Amar

Argentina

Abadi Mario
Aguero Estela L
Alonso Maria Victoria
Altavista Carlos A
Aquilano Roberto Oscar
Arias Elisa Felicitas
Arnal Marcelo Edmundo
Azcarate Ismael N
Bajaja Esteban
Barba Rodolfo Hector
Bassino Lilia P
Beauge Christian
Benaglia Paula
Benvenuto Omar
Brandi Elisande Estela
Branham Richard L
Cappa de Nicolau Cristina
Carpintero Daniel Diego
Carranza Gustavo J
Castagnino Mario
Cellone Sergio Aldo
Cidale Lydia Sonia
Cincotta Pablo Miguel
Claria Juan
Colomb Fernando R
Combi Jorge Ariel
Costa Andrea
Dubner Gloria
Feinstein Alejandro
Feinstein Carlos
Fernandez Silvia M
Ferrer Osvaldo Eduardo
Filloy Emilio Manuel E E
Forte Juan Carlos
Garcia Beatriz Elena
Garcia Lambas Diego
Giacani Elsa Beatriz
Goldes Guillermo Victor
Gomez Mercedes
Hernandez Carlos Alberto
Iannini Gualberto
Lapasset Emilio
Levato Orlando Hugo
Lopez Carlos
Lopez Garcia Zulema
Lopez Jose A
Lopez-Garcia Francisco
Luna Homero G
Machado Marcos
Malaroda Stella M
Mandrini Cristina Hemilse
Manrique Walter T
Marabini Rodolfo Jose
Marraco Hugo G
Martin Maria Cristina
Mauas Pablo
Mendez Mariano
Milone Luis A
Morras Ricardo
Morrell Nidia
Muriel Hernan
Muzzio Juan C
Niemela Virpi S
Nunez Josue Arturo
Olano Carlos Alberto
Orellana Rosa Beatriz
Orsatti Ana M
Perdomo Raul
Piacentini Ruben
Piatti Andres Eduardo
Pintado Olga Ines
Plastino Angel Ricardo
Poeppel Wolfgang G
Rabolli Monica
Ringuelet Adela E
Rizzo Jose Ricardo
Romero Gustavo Esteban
Rovero Adrian Carlos
Rovira Marta Graciela
Sahade Jorge
Sistero Roberto F
Solivella Gladys Rebecca
Vazquez Ruben Angel
Vega E Irene
Vergne Maria Marcela
Villada Monica Maria
Vucetich Hector
Wachlin Felipe Carlos
Waldhausen Silvia
Zadunaisky Pedro E

Armenia

Abrahamian Hamlet V
Arshakian Tigran G
Gasparian Lazar
Gigoyan Kamo
Gurzadian Grigor A
Gyulbudaghian Armen L
Hambrayan Valeri V
Harutyunian Haik A
Hovhannessian Rafik

Kalloglian Arsen T
Kandalian Rafik A
Khachikian E Ye
Magakian Tigran Y
Mahtessian Abraham P
Malumian Vigen
Melik-Alaverdian Yu
Melikian Norair D
Mickaelian Areg Martin
Mirzoyan Ludwik V
Mnatsakanian Mamikon A
Nikoghossian Arthur G
Parsamyan Elma S
Petrosian Artashes R
Pikichian Hovhannes
Sanamian V A
Sedrakian David
Shakhbazian Romelia K
Shakhbazyan Yurij L
Stepanian Jivan A
Vardanian R A
Yengibarian Norair

Australia

Ables John G
Ashley Michael
Bailey Jeremy A
Ball Lewis
Batty Michael
Bedding Timothy
Bersier David
Bessell Michael S
Bicknell Geoffrey V
Biggs James
Birch Peter
Blair David Gerald
Bland-Hawthorn Jonathan
Boyle Brian
Bray Robert J
Brouw W N
Brown Ronald D
Burton Michael G
Cally Paul S
Cane Hilary Vivien
Cannon Russell D
Carter Bradley Darren
Carter David
Carver John H
Caswell James L
Chapman Jessica
Christiansen Wilbur
Clay Roger
Cogan Bruce C
Cole Trevor William
Colless Matthew
Couch Warrick
Cram Lawrence Edward
Cramer Neil
Da Costa Gary Stewart
Davis John
Dawe John Alan
Dawson Bruce
Dopita Michael A
Drinkwater Michael J
Duldig Marcus Leslie
Duncan Robert A
Durrant Christopher J
Edwards Paul J
Ekers Ronald D
Elford William Graham
Ellis G R A
Erickson William C
Evans Robert
Faulkner Donald J
Fenton K B
Francis Paul
Frater Robert H
Freeman Kenneth C
Galloway David
Gardner Francis F
Gascoigne S C B
Gingold Robert Arthur
Godfrey Peter Douglas
Gollnow H
Gray Peter Murray
Green Anne
Green Elizabeth M
Greenhill John
Hall Peter J
Hamilton P A
Harnett Julieine
Hatzidimitriou Despina
Haynes Raymond F
Haynes Roslynn
Horton Brian H
Hosking Roger J
Humble John Edmund
Hunstead Richard W
Hyland A R Harry
Jauncey David L
Jerjen Helmut
Jones Paul
Kalnajs Agris J
Keay Colin S l
Kennedy Hans Daniel
Kesteven Michael J l
Killeen Neil
Koribalski Baerbel Silvia
Lambeck Kurt
Large Michael I
Lattanzio John
Lawson Warrick
Liffman Kurt
Lomb Nicholas Ralph
Luck John M
Malin David F
Manchester Richard N
Mardling Rosemary
Mathewson Donald S
McAdam W Bruce
McConnell David
McCulloch Peter M
McGee Richard X
McGregor Peter John
McLean Donald J
McNaught Robert H
Meatheringham Stephen
Melrose Donald B
Mills Bernard Y
Milne Douglas K
Minnet Harry C
Mitchell Peter
Monaghan Joseph J
Moreton G E
Morgan Peter
Mould Jeremy R
Mullaly Richard F
Murdoch Hugh S
Nelson Graham John
Newell Edward B
Nicholls Jennifer
Nikoloff Ivan
Norris John
Norris Raymond Paul
Nulsen Paul
O'Byrne John
O'Mara Bernard J
O'Sullivan John David
Oosterloo Thomas
Page Arthur
Perdrix John
Peterson Bruce A
Piddington Jack H Res Fel

Pongracic Helen
Prentice Andrew J R
Protheroe Raymond J
Quinn Peter
Reber Grote
Rees David Elwyn
Reynolds John
Robertson James Gordon
Robinson Brian J
Robinson Garry
Rodgers Alex W
Ross John E R
Russell Kenneth S
Ryan Sean Gerard
Sadler Elaine Margaret
Sault Robert
Savage Ann
Sharma Dharma Pal
Sheridan K V
Shimmins Albert John
Shobbrook Robert R
Shortridge Keith
Sims Kenneth P
Slee O B
Smith Craig H
Smith Robert G
Sood Ravi
Sparrow James G
Staveley-Smith Lister
Steel Duncan I
Stewart Ronald T
Stibbs Douglas W N
Storey John W V
Storey Michelle
Tango William J
Taylor Andrew
Taylor Keith
Taylor Kenneth N R
te Lintel Hekkert Peter
Tinney Christopher
Tuohy Ian R
Turtle A J
Tzioumis Anastasios
van Der Borght Rene
Vaughan Alan
Visvanathan Natarajan
Waterworth Michael
Watson Frederick Garnett
Watson Robert
Webb John
Webster Rachel
Wellington Kelvin
Westfold Kevin C
White Graeme Lindsay
Whiteoak John B
Wickramasinghe D T
Wild John Paul
Wilson Brian G
Wilson Peter R
Wood Peter R
Wright Alan E
Zealey William J

Austria

Auner Gerhard
Breger Michel
Dorfi Ernst Anton
Dvorak Rudolf
Firneis Friedrich J
Firneis Maria G
Goebel Ernst
Hanslmeier Arnold
Hartl Herbert
Haubold Hans Joachim
Haupt Hermann F
Hron Josef
Jackson Paul
Kerschbaum Franz
Koeberl Christian
Lala Petr
Lustig Guenter
Maitzen Hans M
Nittmann Johann
Pfleiderer Jorg
Polnitzky Gerhard
Prochazka Franz V
Rakos Karl D
Schnell Anneliese
Schober Hans J
Schroll Alfred
Stift Martin Johannes
Strassmeier Klaus G
Weinberger Ronald
Weiss Werner W
Zeilinger Werner W

Azerbajan

Abbasov Alik R
Aslanov I A
Eminzade T A
Guseinov O H
Gusejnov Ragim Eh
Kasumov Fikret K O
Seidov Zakir F
Sultanov G F

Belgium

Aerts Conny
Arnould Marcel L
Arpigny Claude
Baeck Nicole A l
Biemont Emile
Blomme Ronny
Burger Marijke
Callebaut Dirk K
Clette Frederic
Coutrez Raymond A J
Cugnon Pierre
Cuypers Jan
David Marc
De Cuyper Jean-Pierre M
de Greve Jean-Pierre
de Loore Camiel
de Rop Yves
Debehogne Henri Sc
Defraigne Pascale
Dehant Veronique
Dejaiffe Rene J
Dejonghe Herwig Bert
Delbouille Luc
Demaret Jacques

Denis Carlo
Denoyelle Jozef Kic
Dommanget J
Dossin F
Elst Eric Walter
Gabriel Maurice R
Gerard Jean-Claude M C
Gonze Roger F J
Goossens Marcel
Gosset Eric
Grevesse Nicolas
Henrard Jacques
Hensberge Herman
Heynderickx Daniel
Houziaux Leo
Hutsemekers Damien
Jamar Claude A J
Jorissen Alain
Lampens Patricia
Lausberg Andre
Lemaire Joseph F
Lemaitre Anne
Magain Pierre
Malaise Daniel J
Manfroid Jean
Meire Raphael
Melchior Paul J
Moerdijk Willy G
Monfils Andre G
Moons Michele B M M
Noels Arlette
Ottelet I J
Oxenius Joachim
Paquet Paul Eg
Pauwels Thierry
Perdang Jean M
Poedts Stefaan
Rayet Marc
Remy Battiau Liliane G A
Renson P F M
Robe H A G
Roland Ginette
Sauval A Jacques
Scuflaire Richard
Simon Paul C
Simon Rene L E
Sinachopoulos D
Smeyers Paul
Sterken Christiaan Leo
Steyaert Herman
Svalgaard Leif
Swings Jean-Pierre
van der Linden Ronald
van Dessel Edwin Ludo
van Hoolst Tim
van Rensbergen Walter
van Santvoort Jacques
van Winckel Hans
Verbeek Paul
Verheest Frank
Verschueren Werner
Vreux Jean Marie
Waelkens Christoffel
Zander Rodolphe

Bolivia: No member

Brazil

Abraham Zulema
Aguiar Odylio Denys
Aldrovandi Ruben
Barbuy Beatriz
Barreto Luiz Muniz
Barroso Jair
Batalha Celso Correa
Benevides Soares P
Berman Marcelo S
Bica Eduardo L D
Boechat-Roberty Heloisa M
Braga Joao
Brunini Adrian
Canalle Joao B G
Capelato Hugo Vicente
Chan Roberto
Chian Abraham Chian-Long
Codina Landaberry Sayd J
Correia Emilia
Costa Joaquim E R
Cunha Katia
D'Amico Flavio
da Costa Jose Marques
da Costa Luiz A.N.
da Rocha Vieira E
da Silva Licio
Damineli Neto Augusto
de Almeida Amaury A
de Araujo Francisco X
de Carvalho Reinaldo
de Freitas Pacheco J A
de Gouveia Dal Pino E M
de la Reza Ramiro
de Medeiros Jose Renan
de Souza Ronaldo
Dottori Horacio A
Ducati Jorge Ricardo
Faundez-Abans M
Ferraz-Mello Sylvio
Foryta Dietmar William
Freitas Mourao R r
Giacaglia Giorgio E
Gomes Alercio M
Gomes Rodney D S
Gomide Fernando de Mello
Gonzales'a Walter D
Gregorio-Hetem Jane
Gruenwald Ruth
Jablonski Francisco
Janot-Pacheco Eduardo
Jatenco-Pereira Vera
Jayanthi Udaya B
Kaufmann Pierre
Kepler S O
Kotanyi Christophe
Lazzaro Daniela
Leister Nelson Vani
Leite Scheid Paulo
Lepine Jacques R D
Lima Botti Luiz Claudio
Lopes Dalton De faria
Machado Luiz E da silva
Maciel Walter J
Magalhaes Antonio Mario
Maia Marcio A G
Marques Dos Santos P
Martin Inacio Malmonge
Matsuura Oscar T
Mendes Da Costa Aracy
Mendes de Oliveira Claudia
Neves de Araujo Jose Carlos
Novello Mario
Oliveira Grijo A K
Opher Reuven
Palmeira Ricardo A R
Pastoriza Miriani G
Pellegrini Paulo S S
Piazza Liliana Rizzo

Quarta Maria Lucia
Quast Germano Rodrigo
Rao K Ramanuja
Santos Nilton Oscar
Sato Massae
Sawant Hanumant S
Scalise Eugenio
Schaal Ricardo E
Schuch Nelson Jorge
Sessin Wagner
Singh Patan Deen
Soares Domingos S L
Sodre Laerte
Steiner Joao E
Storchi-Bergman Thaisa
Takagi Shigetsugu
Tateyama Claudio Eiichi
Teixeira Ramachrisna
Telles Eduardo
Torres Carlos Alberto
Tsuchida Masayoshi
Vaz Luiz Paulo Ribeiro
Veiga Carlos Henrique
Viegas Aldrovandi S M
Vieira Martins Roberto
Vilas-Boas Jose W
Vilhena De Moraes R
Villela Thyrso Neto
Willmer Christopher N A
Winter Othon Cabo
Yokoyama Tadashi

Bulgaria

Antov Alexandar
Bonev Tanyu
Borisova Jordanka
Buyukliev Georgi
Dermendjiev Vladimir
Dikova Smiliana D
Filipov Latchezar
Georgiev Leonid
Georgiev Tsvetan
Golev Valery K
Iliev Ilian
Ivanov Georgi R
Ivanova Violeta
Kalinkov Marin P
Kaltcheva Nadia
Kirilova Daniela
Kjurkchieva Diana
Kolev Dimitar Zdravkov
Komitov Boris
Kovachev B J
Kraicheva Zdravska
Kunchev Peter
Markova Nevjana
Mineva Veneta
Momchev Gospodin
Nicolov Nikolai S
Nikolov Andrej
Panov Kiril
Petrov Georgy Trendafilov
Popov Vasil Nikolov
Raikova Donka
Russev Ruscho
Russeva Tatjana
Sbirkova-Natcheva T
Semkov Evgeni
Shkodrov V G
Spasova Nedka Marinova
Stoev Alexei
Strigatchev Anton
Tomov Nikolai
Tomov Toma V
Tsvetkov Milcho K
Tsvetkov Tsvetan
Tsvetkova Katya
Umlenski Vasil
Velkov Kiril
Vladimirov Simeon
Yankulova Ivanka
Zhekov Svetozar A
Zhelyazkov Ivan

Canada

Aikman G Chris l
Alexander Murray E
Argyle P E
Auman Jason R
Barker Paul K
Bartel Norbert Harald
Bastien Pierre
Basu Dipak
Batten Alan H
Beaudet Gilles
Bell Morley B
Bishop Roy L
Bochonko D Richard
Bohlender David
Bolton C Thomas
Bond John Richard
Borra Ermanno F
Brandie George W
Brooks Randall C
Broten Norman W
Burke J Anthony
Caldwell John James
Cannon Wayne H
Carignan Claude
Carlberg Raymond Gary
Chau Wai Y
Clark Thomas Alan
Clarke Thomas R
Clement Maurice J
Climenhaga John L
Clutton-Brock Martin
Colombi Stephane
Couchman Hugh M P
Coutts-Clement Christine
Covington Arthur E
Crabtree Dennis
Crampton David
Davidge Timothy J
Dawson Peter
de Robertis M M
Demers Serge
Dewdney Peter E F
Doherty Lorne H
Douglas R J
Doyon Rene
Drissen Laurent
Duley Walter W
Dyer Charles Chester
English Jayanne
Fahlman Gregory G
Feldman Paul A
Fernie J Donald
Fich Michel
Fitzgerald M Pim
Fletcher J Murray
Fontaine Gilles
Forbes Douglas

Friedli Daniel
Gaizauskas Victor
Galt John A
Garrison Robert F
Gower Ann C
Gower J F R
Gray David F
Gregory Philip C
Griffith John S
Guenther David
Gulliver Austin Fraser
Halliday Ian
Hanes David A
Hardy Eduardo
Harris Gretchen L H
Harris William E
Harrower George A
Hartwick F David A
Hasegawa Tetsuo
Hawkes Robert Lewis
Henriksen Richard N
Herzberg Gerhard
Hesser James E
Hickson Paul
Higgs Lloyd A
Hube Douglas P
Hughes Victor A
Hutchings John B
Innanen Kimmo A
Irwin Alan W
Irwin Judith
Israel Werner
Jeffers Stanley
Joncas Gilles
Jones James
Kamper Karl W
Kennedy John E
Kenny Harold
Knee Lewis
Koehler James A
Kronberg Philipp
Kwok Sun
Lake Kayll William
Lamontagne Robert
Landecker Thomas L
Landstreet John D
Leahy Denis A
Legg Thomas H
Lester John B
Lilly Simon J
Locke Jack L
Lowe Robert P
Ma Chun-yu
MacLeod John M
Macrae Donald A
Mann Patrick J
Marlborough J M
Martin Peter G
Matthews Jaymie
McCall Marshall Lester
McClure Robert D
McCutcheon William H
McDonald J K Petrie
McIntosh Bruce A
Menon T K
Merriam James B
Michaud Georges J
Milone Eugene F
Mitalas Romas Assoc
Mitchell George F
Mochnacki Stephan W
Moffat Anthony F J
Moffat John W
Moorhead James M
Morbey Christopher L
Morris Simon
Morris Stephen C
Morton Donald C
Nadeau Daniel
Naqvi S I H
Nemec James
Nicholls Ralph W
Odgers Graham J
Oke J Beverley
Pachner Jaroslav
Page Don Nelson
Pathria Raj K
Percy John R
Pineault Serge
Poeckert Roland H
Popelar Josef
Pritchet Christopher J
Pryce Maurice H l
Purton Christopher R
Racine Rene
Rice John B
Richardson E Harvey
Richer Harvey B
Robb Russell M
Rochester Michael G
Roger Robert S
Rogers Christopher
Routledge David
Roy Jean-Rene
Scarfe Colin D
Scott Douglas
Scrimger J Norman
Seaquist Ernest R
Smylie Douglas E
Sreenivasan S Ranga
St-Louis Nicole
Stagg Christopher
Stetson Peter B
Sukumar Sundarajan
Sutherland Peter G
Tapping Kenneth F
Tassoul Jean-louis
Tassoul Monique
Tatum Jeremy B
Taylor A R
Turner David G
Underhill Anne B
Vallee Jacques P
van den Bergh Sidney
Vandenberg Don
Venkatesan Doraswamy
Volk Kevin
Wainwright John
Walker Gordon A H
Wehlau Amelia
Welch Douglas L
Welch Gary A
Wesemael Francois
Wesson Paul S
West Michael
Willis Anthony Gordon
Wilson Christine
Woodsworth Andrew W
Woolsey E G
Yang Stephenson L S
Yee Howard K C
Zhang Zheng-Pan
Zhuang Qixiang

Chile

Alcaino Gonzalo
Alvarez Hector
Anguita Claudio
Aparici Juan
Baldwin Jack A
Bouchet Patrice
Bronfman Leonardo
Campusano Luis E
Carrasco Guillermo
Celis Leopoldo
Costa Edgardo
Fouque Pascal
Garay Guido
Geisler Douglas P
Gieren Wolfgang P
Gredel Roland
Gutierrez-Moreno A
Haikala Lauri K
Hainaut Olivier R
Hamuy Mario
Infante Leopoldo
Ingerson Thomas
Krzeminski Wojciech
Kunkel William E
Liller William
Lindgren Harri
Loyola Patricio
Mathys Gautier
May J
Maza Jose
Melnick Jorge
Mintz Blanco Betty
Moreno Hugo
Noel Fernando
Pedreros Mario
Phillips Mark M
Quintana Hernan
Roth Miguel R
Rubio Monica
Ruiz Maria Teresa
Smith Malcolm G
Storm Jesper
Suntzeff Nicholas B
Torres Carlos
Walker Alistair Robin
Wroblewski Herbert

China Nanjing

Ai Guoxiang
Bao Keren
Bian Yulin
Bo Shuren
Cao Changxin
Cao Lihong
Cao Shenglin
Cao Xinwu
Chang Jin
Chen Biao
Chen Daohan
Chen Hongsheng
Chen Jiansheng
Chen Li
Chen Li
Chen Peisheng
Chen Xiaozhong
Chen Yang
Chen Zhen
Chen Zhencheng
Cheng Fuhua
Cheng Fuzhen
Chengmo Zhang
Chu Hanshu
Chu Yaoquan
Cui Douxing
Cui Shizhu
Cui Xiangqun
Cui Zhenhua
Dai Zigao
Dan Xhixiang
Deng Zugan
Di Xiaohua
Ding Mingde
Ding Youji
Fan Ying
Fan Yu
Fang Cheng
Feng Hesheng
Feng Kejia
Fong Chugang
Fu Chengqi
Fu Delian
Fu Qijun
Gan Weiqun
Gao Bilie
Gao Buxi
Gong Shou-shen
Gong Shumo
Gong Xiangdong
Gu Xiaoma
Gu Zhennian
Guo Hongfeng
Guo Nei-shu
Guo Quanshi
Han Fu
Han Jinlin
Han Tianqi
Han Wenjun
Han Yanben
Han Zhanwen
Han Zhengzhong
Hang Hengrong
Hao Yunxiang
He Miaofu
He Xiangtao
Hong Xiaoyu
Hsiang Yan-Yu
Hu Fuxing
Hu Jingyao
Hu Ning-Sheng
Hu Wenrui
Hu Zhongwei
Hua Yingmin
Huang Bi-kun
Huang Changchung
Huang Cheng
Huang Fuquan
Huang Guangli
Huang Jiehao
Huang Keliang
Huang Kunyi
Huang Lin
Huang Runqian
Huang Tianyi
Huang Tieqin
Huang Yinliang
Huang Yongwei
Huang Youran
Ji Hongqing
Ji Shuchen
Jiang Chongguo
Jiang Dongrong
Jiang Shi-Yang
Jiang Shuding
Jiang Xiaoyuan
Jiang Yaotiao
Jiang Zhaoji
Jin Biaoren

Jin Shenzeng
Jin Wenjing
Kimura Hiroshi
Li Chun-Sheng
Li Depei
Li Dongming
Li Guoping
Li Hui
Li Jing
Li Jinling
Li Kejun
Li Linghuai
Li Neng-Yao
Li Qi
Li Qibin
Li Ting
Li Tipei
Li Wei
Li Weibao
Li Xiaoqing
Li Yan
Li Yuanjie
Li Zhengxin
Li Zhengxing
Li Zhian
Li Zhifang
Li Zhigang
Li Zhiping
Li Zhisen
Li Zhongyuan
Li Zongwei
Liang Shiguang
Liang Zhonghuan
Liao De-chun
Liao Xinhao
Lin Qinchang
Lin Yuanzhang
Liu Bao-Lin
Liu Caipin
Liu Ciyuan
Liu Jinming
Liu Liao
Liu Lin
Liu Linzhong
Liu Qingyao
Liu Ruliang
Liu Xinping
Liu Xuefu
Liu Yongzhen
Liu Yuying
Liu Zongli
Lu Benkui
Lu Chunlin
Lu Jufu
Lu Ruwei
Lu Tan
Lu Yang
Luo Baorong
Luo Dingchang
Luo Dingjiang
Luo Guoquan
Luo Shaoguang
Luo Shi-Fang
Luo Xianhan
Ma Er
Ma Xingyuan
Ma Yuehua
Ma YuQian
Mao Wei
Meng Xinmin
Miao Yongkuan
Miao Yongrui
Mo Jing-er
Nan Rendong
Pan Junhua
Pan Liande
Pan Ning-Bao
Pan Rong-Shi
Pei Chunchuan
Peng Bo
Peng Qiuhe
Peng Yunlou
Qi Guanrong
Qian Bochen
Qian Shanjie
Qian Zhihan
Qiao Guojun
Qin Dao
Qin Songnian
Qin Zhihai
Qiu Puzhang
Qiu Yaohui
Qiu Yuhai
Qu Qinyue
Quan Hejun
Rong Jianxiang
Shao Zhengyi
Shen Changjun
Shen Kaixian
Shen Liangzhao
Shen Longxiang
Shen Parnan
Shi Guang-Chen
Shi Shengcai
Shi Zhongxian
Shu Chenggang
Song Guoxuan
Song Jin'an
Song Mutao
Su Dingqiang
Su Hongjun
Sun Jin
Sun Kai
Sun Yisui
Sun Yongxiang
Tan Detong
Tan Huisong
Tang Yuhua
Tian Jing
Tong Fu
Tong Yi
Wan Lai
Wan Tongshan
Wang Chuanjin
Wang dechang
Wang Deyu
Wang Gang
Wang Jia-Long
Wang Jiaji
Wang Jingsheng
Wang Jingxiu
Wang Junjie
Wang Kemin
Wang Lan-Juan
Wang Renchuan
Wang Shouguan
Wang Shui
Wang Shunde
Wang Sichao
Wang Tinggui
Wang Ya'nan
Wang Yiming
Wang Yong
Wang Zhengming
Wang Zhenru
Wang Zhenyi
Wu Bin
Wu Guichen
Wu Hong'ao
Wu Huai-Wei
Wu Lianda
Wu Linxiang
Wu Mingchan
Wu Shengyin
Wu Shouxian
Wu Xiangping
Wu Xinji
Wu Xuejun
Wu Yuefang
Wu Zhiren
Xi Zezong
Xia Xiaoyang
Xia Yifei
Xia Zhiguo
Xian Ding-Zhang

Xiang Delin
Xiang Shouping
Xiao Naiyuan
Xie Guangzhong
Xie Liangyun
Xiong Darun
Xu Aoao
Xu Bang-Xin
Xu Chongming
Xu Jiayan
Xu Jihong
Xu Peiyuan
Xu Pinxin
Xu Tong-Qi
Xu Wenli
Xu Zhentao
Xu Zhicai
Yan Haojian
Yan Jun
Yan Lin-shan
Yang Fumin
Yang Ji
Yang Jian
Yang Lantian
Yang Pibo
Yang Shijie
Yang Tinggao
Yang Zhigen
Yao Baoan
Yao Jinxing
Yao Zhengqiu
Ye Binxun
Ye Shi-hui
Ye Shuhua
Ye Wenwei
Yi Meiliang
Yi Zhaohua
Yin Jisheng
Yin Xinhui
You Jianqi
You Junhan
Yu Xin Alfred
Yu Zhiyao
Yue Zengyuan
Zeng Qin
Zeng Zhifang
Zhai Disheng
Zhai Zaocheng
Zhang Bairong
Zhang Bin
Zhang Fujun
Zhang Guo-Dong
Zhang Heqi
Zhang Hui
Zhang Jialu
Zhang Jian
Zhang Jiaxiang
Zhang Jin
Zhang Jintong
Zhang Peiyu
Zhang Xiuzhong
Zhang Xizhen
Zhang Yang
Zhang Youyi
Zhang Zhen-Jiu
Zhang Zhenda
Zhao Changyin
Zhao Gang
Zhao Gang
Zhao Junliang
Zhao Ming
Zhao Renyang
Zhao Yongheng
Zhao Zhaowang
Zheng Dawei
Zheng Xinwu
Zheng Xuetang
Zheng Yijia
Zheng Ying
Zhong Hongqi
Zhong Min
Zhou Aihua
Zhou Bifang
Zhou Daoqi
Zhou Hongnan
Zhou Ti-jian
Zhou Xu
Zhou Youyuan
Zhou ZhenpPu
Zhu Cisheng
Zhu Jin
Zhu Nenghong
Zhu Shichang
Zhu Wenyao
Zhu Xingfeng
Zhu Yaozhong
Zhu Yonghe
Zhu Yongtian
Zhu Zi
Zhuang Weifeng
Zou Huicheng
Zou Yixin
Zou Zhenlong

China Taipei

Chen Wen Ping
Chiueh Tzihong
Chou Chih-Kang
Chou Dean-Yi
Fu-Shong Kuo
Gir Be Young
Hsiang-Kuang Tseng
Huang Yi-Long
Huang Yinn-Nien
Hwang Woei-yann P
Lee Jong Truenliang
Lee Thyphoon
Ling Chih-Bing
Nee Tsu-Wei
Ng Kin-Wang
Shen Chun-Shan
Sun Wei-Hsin
Ting Yeou-Tswen
Tsai Chang-Hsien
Tsao Mo
Tsay Wean-Shun
Wu Hsin-Heng
Yuan Kuo-Chuan

Croatia

Andreic Zeljko
Bozic Hrvoje
Brajsa Roman
Colic Petar-Kasimir
Dadic Zarko
Kotnik-Karuza Dubravka
Pavlovski Kresimir
Randic Leo
Rosa Dragan
Ruzdjak Vladimir
Solaric Nikola
Vrsnak Bojan
Vujnovic Vladis

Czech Rep

Ambroz Pavel
Bicak Jiri
Borovicka Jiri
Bouska Jiri
Bumba Vaclav
Bursa Milan
Ceplecha Zdenek
Chvojkova Woyk E
Farnik Frantisek
Fischer Stanislav
Grygar Jiri
Hadrava Petr
Harmanec Petr
Heinzel Petr
Hejna Ladislav
Hekela Jan
Horsky Jan
Hudec Rene
Karas Vladimir
Karlicky Marian
Kleczek Josip
Klokocnik Jaroslav
Klvana Miroslav
Kopecky Miloslav
Kotrc Pavel
Koubsky Pavel
Krivsky Ladislav
Kriz Svatopluk
Kubat Jiri
Letfus Vojtech
Lochman Jan
Mayer Pavel
Meszaros Attila
Mikulasek Zdenek
Neuzil Ludek
Novotny Jan
Odstrcil Dusan
Ouhrabka Miroslav
Palous Jan
Papousek Jiri
Pecina Petr
Perek Lubos
Pesek Ivan
Pokorny Zdenek
Polechova Pavla
Rajchl Jaroslav
Ron Cyril
Ruprecht Jaroslav
Ruzickova-Topolova B
Sehnal Ladislav
Sidlichovsky Milos
Sima Zdislav
Simek Milos
Siroky Jaromir
Sobotka Michal
Solc Martin
Spurny Pavel
Stefl Stanislav
Stefl Vladimir
Svestka Jiri
Tlamicha Antonin
Valnicek Boris
Vandas Marek
Vanysek Vladimir
Vavrova Zdenka
Vetesnik Miroslav
Vokrouhlicky David
Vondrak Jan
Vykutilova Marie
Webrova Ludmila
Wolf Marek
Zacharov Igor

Denmark

Andersen Johannes
Baerentzen Jorn
Christensen Per R
Christensen-Dalsgaard J
Clausen Jens Viggo
Einicke Ole H
Fabricius Claus V
Frandsen Soeren
Gammelgaard Peter Mog
Gyldenkerne Kjeld
Hansen Leif
Helmer Leif
Helt Bodil E
Hjorth Jens
Hoeg Erik
Joench-Soerensen Helge
Johansen Karen T
Jones Bernard J T
Jones Janet E
Jorgensen Henning E
Jorgensen Uffe Grae
Kjaergaard Per
Kjeldsen Hans
Knude Jens Kirkeskov
Kristensen Leif Kahl
Kustaanheimo Paul E
Lacey Cedric
Lund Niels
Madsen Jes
Makarov Valeri
Moesgaard Kristian P
Nissen Poul E
Nordlund Ake
Nordstroem Birgitta
Norgaard-Nielsen Hans U
Olsen Erik H
Olsen Fogh H J
Pedersen Holger
Pedersen Olaf
Petersen J O
Pethick Christopher J
Pijpers Frank Peter
Rudkjobing Mogens
Sodemann M
Sommer-Larsen Jesper
Sorensen Gunnar
Thejll Peter Andreas
Thomsen Bjarne B Lect

Ulfbeck Ole
Vedel Henrik
Westergaard Niels J

Eygpt

Abd El Hady Ahmed A
Abd El Hamid Rabab
Abdelkawi M Abubakr
Abou'el-ella Mohamed S
Abulazm Mohamed Samir
Ahmed Imam Ibrahim
Ahmed Mostafa Kamal
Aiad A Zaki
Awad Mervat El-Said
Bakry Abdel-aziz
El Basuny Ahmed Alawy
El Nawaway Mohamed Saleh
El Raey Mohamed E
El Shahawy Mohamad
El-Sharawy Mohamed Bahgat
Galal A A
Gamaleldin Abdulla I
Ghobros Roshdy Azer
Hamdy M A M
Hamid S El Din
Hassan S M
Helali Yhya E
Issa Issa Aly
Kamel Osman M
Mahmoud Farouk M A B
Marie M A
Mikhail Fahmy I
Mikhail Joseph Sidky
Morcos Abd El Fady B
Nawar Samir
Osman Anas Mohamed
Shalabiea Osama M A
Shaltout Mosalam A M
Sharaf Mohamed Adel
Soliman Mohamed Ahmed
Tawadros Maher Jacoub
Wanas Mamdouh Ishaac
Yousef Shahinaz M
Youssef Nahed H

Estonia

Eelsalu Heino
Einasto Jaan
Einasto Maret
Ergma E V
Gramann Mirt
Haud Urmas
Joeveer Mihkel
Kipper Tonu
Kolka Indrek
Leedjaerv Laurits
Malyuto Valeri
Nugis Tiit
Pelt Jaan
Pustylnik Izold B
Saar Enn
Sapar Arved
Tago Erik
Traat Peeter
Uus Undo
Veismann Uno
Vennik Jaan
Viik Tonu

Finland

Donner Karl Johan
Duemmler Rudolf
Flynn Chris
Haemeen Anttila Kaarle A
Haenninen Jyrki
Harju Jorma Sakari
Huovelin Juhani
Jaakkola Toivo S
Karttunen Hannu
Kultima Johannes
Laurikainen Eija
Lehto Harry J
Lumme Kari A
Markkanen Tapio
Mattila Kalevi
Mikkola Seppo
Muinonen Karri
Niemi Aimo
Oja Heikki
Oterma Liisi
Piirola Vilppu E
Rahunen Timo
Raitala Jouko T
Riihimaa Jorma J
Roos Matts
Salo Heikki
Sillanpaa Aimo Kalevi
Takalo Leo O
Teerikorpi Veli Pekka
Tiuri Martti
Tuominen Ilkka V
Urpo Seppo I
Valtaoja Esko
Valtaoja Leena
Valtonen Mauri J
Vilhu Osmi
Zheng Jia-Qing

France

Aboudarham Jean
Acker Agnes
Aime Claude
Alecian Georges
Allard Nicole
Allegre Claude
Alloin Danielle
Aly Jean Jacques
Amari Tahar
Amram Philippe
Andrillat Yvette
Arduini-Malinovsky Monique

Arenou Frederic
Arlot Jean-Eudes
Arnaud Jean Paul
Arnaud Monique
Artru Marie-Christine
Artzner Guy
Asseo Estelle
Assus Pierre
Athanassoula Evangelia
Aubier Monique G
Audouze Jean
Augarde Renee
Auriere Michel
Auvergne Michel
Avignon Yvette
Azzopardi Marc
Bacchus Pierre
Baglin Annie
Balkowski-Mauger Chantal
Ballereau Dominique
Balmino Georges G
Baluteau Jean-Paul
Baranne Andre
Barbier-Brossat Madeleine
Barge Pierre
Barlier Francois E
Barriot Jean-Pierre
Barucci Maria A
Baudry Alain
Bec-Borsenberger Annick
Bely-Dubau Francoise
Ben-Jaffel Lofti
Benaydoun Jean-Jacques
Benest Daniel
Bensammar Slimane
Bergeat Jacques G
Berger Christiane
Berger Jacques G
Berruyer-Desirotte Nicole
Bertaux Jean Loup
Berthomieu Gabrielle
Bertout Claude
Bienayme Olivier
Bijaoui Albert
Binette Luc
Biraud Francois
Blamont Jacques Emile
Blanchard Alain
Blazit Alain
Bocchia Romeo
Bockelee-Morvan Dominique
Bois Eric
Boisse Patrick
Boisson Catherine
Bommier Veronique
Bonazzola Silvano
Bonneau Daniel
Bonnet Roger M
Bonnet-Bidaud Jean Marc
Borgnino Julien
Bosma Albert
Bottinelli Lucette
Boucher Claude
Bouchet Francois R
Bougeard Mireille L
Bougeret Jean Louis
Bouigue Roger
Boulanger Francois
Boulesteix Jacques
Boulon Jacques J
Bouvier Jerome
Boyer Rene
Brahic Andre
Bretagnon Pierre
Briot Danielle
Brouillet Nathalie
Brunet Jean-Pierre
Bruston Paul
Bryant John
Buecher Alain
Burkhart Claude
Burnage Robert
Calame Odile
Camichel Henri
Capitaine Nicole
Caplan James
Carquillat Jean-Michel
Casoli Fabienne
Casse Michel
Castets Alain
Catala Claude
Cayrel Roger
Cayrel de Strobel Giusa
Cazenave Anny
Celnikier Ludwik
Cesarsky Catherine J
Cesarsky Diego A
Chalabaev Almas
Chamaraux Pierre
Chambe Gilbert
Chapront Jean
Chapront-Touze Michelle
Charlot Patrick
Charlot Stephane
Chassefiere Eric
Chauvineau Bertrand
Chevalier Claude
Chevrel Serge
Chollet Fernand
Chopinet Marguerite
Clairemidi Jacques
Colin Jacques
Collin-Souffrin Suzy
Combes Francoise
Combes Michel
Comte Georges
Connes Janine
Connes Pierre
Considere Suzanne
Cornille Marguerite
Coupinot Gerard
Courtes Georges
Courtin Regis
Couteau Paul
Couturier Pierre
Cox Pierre
Creze Michel
Crifo Francoise
Crovisier Jacques
Cruvellier Paul E
Cruzalebes Pierre
Cuny Yvette J
d'Hendecourt Louis
Daigne Gerard
Daniel Jean Yves
Davoust Emmanuel
de Bergh Catherine
de la Noe Jerome
de Lapparent-Gurriet Valerie
Debarbat Suzanne V
Deharveng Jean-Michel
Deharveng Lise
Delaboudiniere Jean-Pierre
Delannoy Jean
Delmas Christian
Demoulin Pascal
Denisse Jean-Francois
Dennefeld Michel
Desesquelles Jean
Despois Didier
Divan Lucienne
Doazan Vera
Dolez Noel
Dollfus Audouin
Donas Jose
Donati Jean-Francois
Dourneau Gerard
Downes Dennis
Drossart Pierre
Dubau Jacques
Dubois Marc A
Dubois Pascal
Dubout Renee
Duchesne Maurice
Ducourant Christine

Dufay Maurice
Duflot Marcelle
Dulk George A
Dumont Rene
Dumont Simone
Duriez Luc
Durouchoux Philippe
Durret Florence
Duval Marie-France
Duvert Gilles
Edelman Colette
Egret Daniel
Eidelsberg Michele
Encrenaz Pierre J
Encrenaz Therese
Falgarone Edith
Faucher Paul
Faurobert-Scholl Marianne
Feautrier Nicole
Fehrenbach Charles
Feissel Martine
Felenbok Paul
Ferlet Roger
Fernandez Jean-Claude
Festou Michel C
Floquet Michele
Florsch Alphonse
Forestini Manuel
Fort Bernard P
Forveille Thierry
Fossat Eric
Foy Renaud
Fraix-Burnet Didier
Francois Patrick
Freire Ferrero Rubens G
Fresneau Alain
Friedjung Michael
Fringant Anne-Marie
Frisch Helene
Frisch Uriel
Froeschle Christiane D
Froeschle Claude
Froeschle Michel
Fulchignoni Marcello
Gabriel Alan H
Gaignebet Jean
Gallouet Louis
Gambis Daniel
Gargaud Muriel
Garnier Robert
Gautier Daniel
Gay Jean
Genova Francoise
Georgelin Yvon P
Georgelin Yvonne M
Gerard Eric
Gerbal Daniel
Gerbaldi Michele
Gerin Maryvonne
Gillet Denis
Giraud Edmond
Goldbach Claudine Mme
Gomez Ana E
Gonczi Georges
Gonzales Jean-Francois
Gordon Charlotte
Goret Philippe
Gouguenheim Lucienne
Goupil Marie Jose
Gouttebroze Pierre
Granveaud Michel
Grec Gerard
Grenier Isabelle
Grenier Suzanne
Greve Albert
Grudler Pierre
Gry Cecile
Guelin Michel
Guerin Pierre
Guibert Jean
Guiderdoni Bruno
Guinot Bernard R
Halbwachs Jean Louis
Hammer Francois
Harvey Christopher C
Hayli Avram
Heck Andre
Hecquet Josette
Heidmann Jean
Henon Michel C
Henoux Jean-Claude
Heudier Jean-Louis
Heydari-Malayeri Mohammad
Heyvaerts Jean
Hoang Binh Dy
Hoffman Jeffrey Alan
Hua Chon Trung
Hubert-Delplace A -M
Imbert Maurice
Irigoyen Maylis
Israel Guy Marcel
Jablonka Pascale
Jacq Thierry
Jacquinot Pierre
Jasniewicz Gerard
Joly Francois
Joly Monique
Joubert Martine
Journet Alain
Jung Jean
Kahane Claudine
Kandel Robert S
Kazes Ilya
Klein Karl Ludwig
Kneib Jean-Paul
Koch-Miramond Lydie
Koechlin Laurent
Kourganoff Vladimir
Koutchmy Serge
Kovalevsky Jean
Krikorian Ralph
Kunth Daniel
Labeyrie Antoine
Labeyrie Jacques
Lachieze-Rey Marc
Laclare Francis
Laffineur Marius
Lafon Jean-Pierre J
Lagrange Anne-Marie
Lallement Rosine
Lamy Philippe
Lancon Ariane
Lannes Andre
Lantos Pierre
Laques Pierre
Laskar Jacques
Lasota Jean-Pierre
Latour Jean J
Launay Francoise
Launay Jean-Michel
Laurent Claudine
Laval Annie
Lazareff Bernard
Le Borgne Jean Francois
Le Bourlot Jacques
Le Contel Jean-Michel
Le Dourneuf Maryvonne
Le Fevre Olivier
Le Floch Andre
Le Squeren Anne-Marie
Leach Sydney
Leblanc Yolande
Lebre Agnes
Lebreton Yveline
Lefebvre Michel
Lefevre Jean
Leger Alain
Lelievre Gerard
Lemaire Jean-Louis
Lemaire Philippe
Lemaitre Gerard R
Lena Pierre J
Leorat Jacques
Lequeux James
Leroy Bernard

Leroy Jean-Louis
Lesage Alain
Lesteven Soizick
Lestrade Jean-Francois
Levasseur-Regourd Annie Chantal
Levy Jacques R
Lortet Marie-Claire
Losco Lucette
Loucif Mohammed Lakhdar
Louise Raymond
Loulergue Michelle
Loup Cecile
Lucas Robert
Luminet Jean-Pierre
Lunel Madeleine
Madden Suzanne
Magnan Christian
Maillard Jean-Pierre
Malherbe Jean Marie
Mangeney Andre
Marcelin Michel
Marchal Christian
Mariotti Jean Marie
Martin Francois
Martin Jean-Michel P
Martres Marie-Josephe
Masnou Francoise
Masnou Jean Louis
Mathez Guy
Maucherat J
Maurice Eric N
Maurogordato Sophie
Mauron Nicolas
Mavrides Stamatia
Mazure Alain
McCarroll Ronald
Megessier Claude
Mein Nicole
Mein Pierre
Mekarnia Djamel
Mellier Yannick
Meneguzzi Maurice M
Mennessier Marie-Odile
Merat Parviz
Mercier Claude
Meyer Claude
Meyer Jean-Paul
Mianes Pierre
Michard Raymond
Michel Eric
Mignard Francois
Milet Bernard L
Millet Jean
Milliard Bruno
Mirabel Igor Felix
Mochkovitch Robert
Monier Richard
Monin Jean-louis
Monnet Guy J
Montes Carlos
Montmerle Thierry
Morbidelli Alessandro
Moreau Olivier
Moreels Guy
Morel Pierre Jacques
Morris David
Mosser Benoit
Mouchet Martine
Mouradian Zadig M
Mourard Denis
Mulholland J Derral
Muller Paul
Muller Richard
Muratorio Gerard
Murtagh Fionn
Nadal Robert
Nahon Fernand
Nguyen-Quang Rieu
Noens Jacques-Clair
Nollez Gerard
Nottale Laurent
Oblak Edouard
Ochsenbein Francois
Omnes Roland
Omont Alain
Oukbir Jamila
Paletou Frederic
Parcelier Pierre
Parisot Jean-Paul
Paturel Georges
Paul Jacques
Pecker Jean-Claude
Pedersen Bent M
Pedoussaut Andre
Pellas Paul
Pellet Andre
Pelletier Guy
Pello Roser Descayre
Pequignot Daniel
Perault Michel
Perrier Christian
Perrin Jean Marie
Perrin Marie-Noel
Petit Gerard
Petit Jean-Marc
Petitjean Patrick
Peton Alain
Petrini Daniel
Peyturaux Roger H
Pham-Van Jacqueline
Picat Jean-Pierre
Pick Monique
Pierre Marguerite
Pineau des Forets Guillaume
Pollas Christian
Poquerusse Michel
Poulle Emmanuel
Pouquet Annick
Poyet Jean-Pierre
Praderie Francoise
Prantzos Nikos
Prevot Louis
Prevot-Burnichon Marie Louise
Prieur Jean-louis
Proisy Paul E
Proust Dominique
Provost Janine
Puget Jean-Loup
Quemerais Eric
Querci Francois R
Querci Monique
Rabbia Yves
Raoult Antoinette
Rapaport Michel
Rayrole Jean R
Rebeirot Edith
Reeves Hubert
Reinisch Gilbert
Requieme Yves
Ribes Jean-Claude
Ricort Gilbert
Rieutord Michel
Robillot Jean-Maurice
Robin Annie C
Robley Robert
Rocca-Volmerange Brigitte
Roncin Jean-Yves
Roques Francoise
Roques Sylvie
Rosch Jean
Rostas Francois
Rothenflug Robert
Rouan Daniel
Roudier Thierry
Roueff Evelyne M A
Rousseau Jean-Michel
Rousseau Jeanine
Rousselot Philippe
Rozelot Jean Pierre
Sadat Rachida
Sahal-Brechot Sylvie
Saissac Joseph
Samain Denys
Sareyan Jean-Pierre

Schaerer Daniel
Schatzman Evry
Scheidecker Jean-Paul
Schmieder Brigitte
Schneider Jean
Scholl Hans
Schumacher Gerard
Semel Meir
Servan Bernard
Sharp Christopher
Sibille Francois
Sicardy Bruno
Signore Monique
Simien Francois
Simon Guy
Simon Jean-Louis
Simonneau Eduardo
Sirousse Zia Haydeh
Sivan Jean-Pierre
Slezak Eric
Snijders Mattheus A J
Sol Helene
Soru-Escaut Irina
Sotirovski Pascal
Soubiran Caroline
Soucail Genevieve
Souchay Jean
Souffrin Pierre B
Soulie Guy
Souriau Jean-Marie
Spielfiedel Annie
Spite Francois M
Spite Monique
Stasinska Grazyna
Steenman-Clark Lois
Stehle Chantal
Stellmacher Goetz
Stellmacher Irene
Sygnet Jean Francois
Tagger Michel
Talon Raoul
Tarrab Irene
Taton Rene
Tchang-Brillet Lydia
Terzan Agop
Thevenin Frederic
Thiry Yves R
Thomas Claudine
Thuillot William
Tran Minh Nguyet
Tran-Minh Francoise
Trellis Michel
Trottet Gerard
Truong Bach
Tully John A
Turck-Chieze Sylvaine
Turon Catherine
Vakili Farrokh
Valbousquet Armand
Valiron Pierre
Valls-Gabaud David
Valtier Jean-Claude
van Driel Willem
van Driel Gesztzlyi L
van Regemorter Henri
van't Veer Frans
van't Veer-Menneret Claude
Vapillon Loic J
Vauclair Gerard P
Vauclair Sylvie D
Vauglin Isabelle
Velli Marco
Verdet Jean-Pierre
Vergez Madeleine
Veron Marie-Paule
Veron Philippe
Vial Jean-Claude
Viala Yves
Viallefond Francois
Vidal Jean-Louis
Vidal-Madjar Alfred
Vienne Alain
Vigier Jean-Pierre
Vigroux Laurent
Vilkki Erkki U
Vilmer Nicole
Viton Maurice
Volonte Sergio
Vu Duong Tuyen
Vuillemin Andre
Walch Jean-Jacques
Weimer Theophile P F
Wenger Marc
Weniger Schame
Widemann Thomas
Wink Joern Erhard
Wlerick Gerard
Woltjer Lodewijk
Wozniak Herve
Zahn Jean-Paul
Zambon Giulio
Zavagno Annie
Zeippen Claude
Zorec Juan

Georgia

Bartaya R A
Borchkhadze Tengiz M
Chkhikvadze Iakob N
Dolidze Madona V
Dzhapiashvili Victor P
Dzigvashvili R M
Kalandadze N B
Kharadze E K
Khatisashvili Alfez Sh
Khetsuriani Tsiala S
Kiladze R I
Kogoshvili Natela G
Kumsiashvily Mzia I
Kumsishvili J I
Kurtanidze Omar
Lominadze Jumber
Magnaradze Nina G
Salukvadze G N
Toroshlidze Teimuraz I

Germany

Abraham Peter
Albrecht Miguel A
Albrecht Rudolf
Altenhoff Wilhelm J
Anzer Ulrich
Appenzeller Immo
Arp Halton
Aschenbach Bernd
Aurass Henry
Axford W Ian
Baade Dietrich
Bade Norbert
Baessgen Martin
Bahner Klaus
Baier Frank

Balthasar Horst
Barrow Colin H
Bartelmann Matthias
Barwig Heinz
Baschek Bodo
Bastian Ulrich
Beck Rainer
Becker Sylvia
Beckwith Steven V W
Behr Alfred
Bender Ralf
Bendlin Cornelia
Benvenuti Piero
Bergeron Jacqueline A
Berkhuijsen Elly M
Bernstein Hans Heinrich
Beuermann Klaus P
Bien Reinhold
Biermann Peter L
Birkle Kurt
Bleyer Ulrich
Blum Peter
Boehnhardt Hermann
Boerner Gerhard
Boerngen Freimut
Bohn Horst-Ulrich
Bohrmann Alfred
Bomans Dominik J
Borgeest Ulf
Boschan Peter
Brandt Peter N
Brauninger Heinrich
Braunsfurth Edward
Breinhorst Robert A
Breitschwerdt Dieter
Breysacher Jacques
Brinkmann Wolfgang
Brosche Peter
Bruch Albert
Bruzek Anton
Buechner Joerg
Bues Irmela D
Butler Keith
Camenzind Max
Carsenty Uri
Che-Bohnenstengel Anne
Chini Rolf
Crane Philippe
Cullum Martin
d'Odorico Sandro
Dachs Joachim
Danziger I John
Dautcourt G
de Boer Klaas Sjoerds
de Kool Marthijn
de Vegt Chr
Degenhardt Detlev
Deinzer W
Deiss Bruno M
Dettmar Ralf-Juergen
Deubner Franz-Ludwig
Dick Wolfgang
Diercksen Geerd H F
Dietrich Matthias
Domke Helmut
Dorenwendt Klaus
Dorschner Johann
Drapatz Siegfried W
Drechsel Horst
Duerbeck Hilmar W
Duschl Wolfgang J
Ehlers Jurgen
El Eid Mounib
Elsaesser Hans
Elwert Gerhard
Enard Daniel
Engels Dieter
Enslin Heinz
Fahr Hans Joerg
Feitzinger Johannes
Feix Gerhard
Fiebig Dirk
Fosbury Robert A E
Freudling Wolfram
Fricke Klaus
Fried Josef Wilhelm
Friedemann Christian
Fritze Klaus
Fritze-von Alvensleben Ute
Fuchs Burkhard
Fuerst Ernst
Gail Hans-Peter
Gebler Karl-heinz
Geffert Michael
Gehren Thomas
Genzel Reinhard
Geyer Edward H
Giacconi Riccardo
Gilmozzi Roberto
Glatzel Wolfgang
Gondolatsch Friedrich
Gottloeber Stefan
Graham David A
Grahl Bernd H
Grewing Michael
Groenewegen Martin
Groote Detlef
Grosbol Preben Johnson
Grossmann-Doerth U
Groten Erwin
Gruen Eberhard
Guenther Eike
Guertler Joachin
Guesten Rolf
Gussmann E A
Gutcke Dietrich
Hachenberg Otto
Haefner Reinhold
Haerendel Gerhard
Hagen Hans-Juergen
Hagfors Tor
Hahn Gerhard J
Hamann Wolf-Rainer
Hammer Reiner
Hanuschik Reinhard
Harris Alan William
Hartquist Thomas Wilbur
Haser Leo N K
Hasinger Guenter
Haslam C Glyn T
Haupt Wolfgang
Hazlehurst John
Heber Ulrich
Hefele Herbert
Heidt Jochen
Heinrich Inge
Heller Clayton
Hemmleb Gerhard
Henkel Christian
Henning Thomas
Hensler Gerhard
Herbstmeier Uwe
Hering Roland
Herman Jacobus
Herold Heinz
Herrmann Dieter
Hessman Frederic Victor
Hildebrandt Joachim
Hilf Eberhard R H
Hillebrandt Wolfgang
Hippelein Hans H
Hirth Wolfgang Ernst
Hoeflich Peter
Hoffmann Martin
Hofmann Wilfried
Holweger Hartmut
Hopp Ulrich
Horedt Georg Paul
House Franklin C
Hubrig Swetlana
Huchtmeier Walter K
Hummel Wolfgang
Hunger Kurt
Husfeld Dirk
Ip Wing-huen

Isserstedt Joerg
Jackisch Gerhard
Jaeger Friedrich W
Jahreiss Hartmut
Jedamzik Karsten
Jensch A
Jockers Klaus
Jordan H L direktor
Jordan Stefan
Junkes Norbert
Just Andreas
Kaehler Helmuth
Kafka Peter
Kalberla Peter
Kanbach Gottfried
Kasper U
Kaufmann Jens Peter
Kayser Rainer
Kegel Wilhelm H
Keller Hans Ulrich
Keller Horst Uwe
Kendziorra Eckhard
Kentischer Thomas
Khanna Ramon
Kilian-Montenbruck Judith
Kippenhahn Rudolf
Kirk John
Klare Gerhard
Klein Ulrich
Kliem Bernhard
Klinkhamer Frans
Klose Sylvio
Kneer Franz
Koehler H
Koehler Peter
Koehler-Vogel Suzanne
Koempe Carsten
Koeppen Joachim
Koester Detlev
Kohoutek Lubos
Kokott Wolfgang
Kollatschny Wolfram
Kraemer Gerhard
Krause Marita
Krautter Joachim
Kreisel E
Kreysa Ernst
Kroupa Pavel
Kruegel Endrik
Krueger Albrecht
Kudritzki Rolf-Peter
Kuehne Christoph F
Kuehr Helmut
Kuenzel Horst
Kundt Wolfgang
Kunitzsch Paul
Kunze Ruediger
la Dous Constanze A
Labs Dietrich
Lamla Erich E
Langer Norbert
Lederle Trudpert
Lehman Holger
Leibundgut Bruno
Leinert Christoph
Lemke Dietrich
Lemke Michael
Lenhardt Helmut
Lenzen Rainer
Lesch Harold
Liebscher Dierck E
Lorenz Hilmar
Lucy Leon B
Luest Reimar
Luks Thomas
Mac Low Mordecai-Mark
Mandel Holger
Mann Gottfried
Mann Ingrid
Mannheim Karl
Matas Vladimir R
Materne Juergen
Mattig W
Mauder Horst
Mebold Ulrich
Meinig Manfred
Meisenheimer Klaus
Meister Claudia Veronika
Mendez Roberto H
Merkle Fritz
Metz Klaus
Meusinger Helmut
Meyer Friedrich
Meyer-Hofmeister Eva
Meylan Georges
Mezger Peter G
Minniti Dante
Mo Houjun
Moehler Sabine
Moehlmann Diedrich
Moellenhoff Claus
Moeller Palle
Moorwood Alan F M
Muecket Jan P
Mueller Ewald
Mueller Volker
Mundt Reinhard
Napiwotzki Ralf
Neckel Heinz
Neckel Th
Nesis Anastasios
Neukum G
Notni P
Oestreicher Roland
Oetken L
Oleak H
Ott Heinz-Albert
Paresce Francesco
Patnaik Alok Ranjan
Pauldrach Adalbert W A
Pauliny Toth Ivan K K
Petri Winfried
Pfau Werner
Pfennig Hans H
Pflug Klaus
Pinkau K
Pitz Eckhart
Porcas Richard
Preuss Eugen
Priester Wolfgang
Puls Joahim
Raedler K H
Rauch Thomas
Refsdal Sjur
Reich Wolfgang
Reif Klaus
Reimers Dieter
Reinsch Klaus
Richter G A
Richter Gotthard
Richter Johannes
Richtler Tom
Rieger Erich
Ripken Hartmut W
Ritter Hans
Roemer Max
Roeser Hans-peter
Roeser Hermann-Josef
Roeser Siegfried
Roessiger Siegfried
Rohlfs Kristen
Rosa Dorothea
Rosa Michael Richard
Roth-Hoppner Maria Luise
Ruben G
Ruder Hanns
Ruediger Guenther
Rupprecht Gero
Saglia Roberto Philip
Sakhibov Firouz
Sams Bruce Jones III
Schaefer Gerhard
Schaifers Karl
Scheffler Helmut
Schilbach Elena

Schindler Karl
Schindler Sabine
Schleicher Helmold
Schlickeiser Reinhard
Schlosser Wolfhard
Schlueter A
Schlueter Dieter
Schmadel Lutz D
Schmeidler F
Schmid Hans Martin
Schmid-Burgk J
Schmidt H U
Schmidt Hans
Schmidt K H
Schmidt Thomas
Schmidt Wolfgang
Schmidt-Kaler Theodor
Schmitt Dieter
Schneider Peter
Schoeffel Eberhard F
Schoenberner Detlef
Schoeneich W
Schoenfelder Volker
Scholz Gerhard
Scholz M
Scholz Ralf Dieter
Schramm K Jochen
Schramm Thomas
Schroeder Klaus Peter
Schroeder Rolf
Schroeter Egon H
Schruefer Eberhard
Schubart Joachim
Schuecker Peter
Schuessler Manfred
Schultz G V
Schulz Hartmut
Schulz Rolf Andreas
Schumann Joerg Dieter
Schutz Bernard F
Schwan Heiner
Schwartz Rolf
Schwekendiek Peter
Schwope Axel
Scorza de Appl Cecilia
Sedlmayer Erwin
Seggewiss Wilhelm
Seifert Walter
Seitter Waltraut C
Sengbusch Kurt V
Serafin Richard A
Shaver Peter A
Sherwood William A
Sieber Wolfgang
Simon Klaus Peter
Smith Michael
Soffel Michael
Solf Josef
Sollazzo Claudio
Soltau Dirk
Spruit Henk C
Spurzem Rainer
Stahl Otmar Richard
Stange Lothar
Staubert Ruediger Prof
Staude Hans Jakob
Staude Juergen
Stecklum Bringfried
Steffen Matthias
Steiner Oskar
Steinert Klaus Guenter
Steinle Helmut
Stix Michael
Strohmeier Wolfgang
Stumpff Peter
Stutzi Juergen
Syer David
Szostak Roland
Tacconi Linda J
Tacconi-Garman Lowell E
Tarenghi Massimo
Theis Christian
Thielheim Klaus O
Thomas Hans -Christoph
Tiersch Heinz
Traving Gerhard
Treder H J
Trefftz Eleonore E
Treumann Rudolf A
Truemper Joachim
Tscharnuter Werner M
Tucholke Hans-Joachim
Tueg Helmut
Ulmschneider Peter
Ulrich Bruce T
Ulrich Marie-Helene D
Voelk Heinrich J
Vogt Nikolaus
Voigt Hans H
Volland H
von Appen-Schnur Gerhard F
von Borzeszkowski H H
von der Luehe Oskar
von Hoerner Sebastian
von Montigny Corinna
von Weizsaecker C F
Wagner Stefan
Walmsley C Malcolm
Walter Hans G
Wambsganss Joachim
Warmels Rein Herm
Wehrse Rainer
Weidemann Volker
Weigelt Gerd
Weiss Achim
Wellmann Peter
Wendker Heinrich J
Wenzel W
Werner Klaus
West Richard M
Wiehr Eberhard
Wielebinski Richard
Wielen Roland
Wilson P
Wilson Thomas L
Winnewisser Gisbert
Wisotzki Lutz
Wittmann Axel D
Witzel Arno
Woehl Hubertus
Wolf Bernhard
Wolf Rainer E A
Wouterloot Jan Gerard A
Wuensch Johann Jakob
Wunner Guenter
Yamada Shoichi
Yorke Harold W
Zekl Hans Wilhelm
Zerull Reiner H
Zickgraf Franz Josef
Ziegler Harald
Zimmermann Helmut
Zinnecker Hans
Zylka Robert

Greece

Abbott William N
Alissandrakis C
Antonacopoulos Greg
Antonopoulou E
Arabelos Dimitrios
Asteriadis Georgios
Avgoloupis Stavros
Banos Cosmas J
Banos George J

Barbanis Basil
Bellas-Velidis Ioannis
Bozis George
Caranicolas Nicholas
Caroubalos C A
Chryssovergis Michael
Contadakis Michael E
Contopoulos George
Danezis Emmanuel
Dapergolas A
Dara Helen
Deliyannis Jean
Dialetis Dimitris
Dionysiou Demetrios
Geroyannis Vassilis S
Goudas Constantine L
Goudis Christos D
Hadjidemetriou John D
Hantzios Panayiotis
Hiotelis Nicolaos
Isliker Heinz
Kazantzis Panayotis
Kokkotas Konstantinos
Kontizas Evangelos
Kontizas Mary
Korakitis Romylos
Kylafis Nikolaos D
Laskarides Paul G
Liritzis Ioannis
Markellos Vassilis V
Mathioudakis Mihalis
Mavraganis A G
Mavridis L N
Mavromichalaki Helen
Merzanides Constantinos
Metaxa Margarita
Moussas Xenophon
Niarchos Panayiotis
Nicolaidis Efthymios
Papaelias Philip
Papathanasoglou D
Papayannopoulos Th
Persides Sotirios C
Petropoulos Basil Ch
Pinotsis Antonis D
Plionis Manolis
Polymilis Chronis
Poulakos Constantine
Preka-Papadema P
Prokakis Theodore J
Rovithis Peter
Rovithis-Livaniou Helen
Sarris Eleftherios
Sarris Emmanuel T
Seimenis John
Seiradakis John Hugh
Spithas Elefterios N
Spyrou Nicolaos
Stathopoulou Maria
Svolopoulos Sotirios
Terzides Charalambos
Theodossiou Efstratios
Tritakis Basil P
Tsamparlis Michael
Tsikoudi Vassiliki
Tsinganos Kanaris
Tsioumis Alexandros
Tsiropoula Georgia
Vardavas Ilias Mihail
Varvoglis H
Veis George
Ventura Joseph
Vlachos Demetrius G
Vlahos Loukas
Voglis Nikos
Xiradaki Evangelia
Zachariadis Theodosios
Zachilas Loukas
Zafiropoulos Basil
Zikides Michael C

Hungary

Almar Ivan
Bagoly Zsolt
Balazs Bela A
Balazs Lajos G
Barcza Szabolcs
Barlai Katalin
Csada Imre K
Dezso Lorant
Erdi B
Fejes Istvan
Gerlei Otto
Grandpierre Attila
Gyori Lajos
Horvath Andras
Ill Marton J
Illes Almar Erzsebet
Jankovics Istvan
Kalman Bela
Kanyo Sandor
Kelemen Janos
Kollath Zoltan
Kovacs Agnes
Kovacs Geza
Kun Maria
Lovas Miklos
Ludmany Andras
Marik Miklos
Marx Gyorgy
Olah Katalin
Paparo Margit
Patkos Laszlo
Petrovay Kristof
Szabados Laszlo
Szatmary Karoly
Szecsenyi-Nagy Gabor
Szego Karoly
Szeidl Bela
Toth Imre
Veres Ferenc
Vinko Jozsef
Zsoldos Endre

Iceland

Bjoernsson Gunnlaugur
Gudmundsson Einar H
Juliusson Einar
Saemundson Thorsteinn

India

Abhyankar Krishna D
Acharya Bannanje S
Aggarwal Kanti Mal
Agrawal P C
Ahmad Farooq
Alladin Saleh Mohamed
Alurkar S K
Ambastha A K
Anandaram Mandayam N
Ananthakrishnan S
Anantharamaiah K R
Ansari S M Razaullah
Antia H M
Anupama G C
Apparao K M V
Ashok N M
Auluck Faqir Chand
Babu G S D
Bagare S P
Balasubramanian V
Ballabh G M
Bandyopadhyay A
Banerji Sriranjan
Banhatti Dilip Gopal
Basu Baidyanath
Bhandari N
Bhandari Rajendra
Bhat Chaman Lal
Bhat Narayana P
Bhatia Prem K
Bhatia R K
Bhatia V B
Bhatnagar Arvind
Bhatnagar Ashok Kumar
Bhatnagar K B
Bhatt H C
Bhattacharjee Pijushpani
Bhattacharya Dipankar
Bhattacharyya J C
Bhonsle Rajaram V
Biswas Sukumar
Boddapati G Anandarao
Bondal Krishna Raj
Chakrabarti Sandip K
Chakraborty Deo K
Chandra Suresh
Chandrasekhar T
Chaoudhuri Arnab
Chaubey Uma Shankar
Chitre Shashikumar M
Chokshi Arati
Choudary Debi Prasad
Choudhuri Arnab Rai
Cowsik Ramanath
Dadhich Naresh
Damle S V
Das Mrinal Kanti
Das P K
Datta Bhaskar
Degaonkar S S
Desai Jyotindra N
Deshpande Avinash
Deshpande M R
Dhurandhar Sanjeev
Duorah Hira Lal
Durgaprasad N
Dwarakanath K S
Dwivedi Bhola Nath
Gaur V P
Ghosh Kajal Kumar
Ghosh P
Ghosh S K
Giridhar Sunetra
Gokhale Moreshwar Hari
Gopala Rao U V
Goswami J N
Goyal A N
Goyal Ashok Kumar
Gupta Sunil K
Gupta Yashwant
Gurm Hardev S
Hasan Saiyid Strajul
Iyengar K V K
Iyengar Srinivasan Rama
Iyer B R
Jain Rajmal
Jayarajan A P
Jog Chanda J
Joshi Mohan N
Joshi Suresh Chandra
Joshi U C
Kandpal Chandra D
Kapahi Vijay K
Kapoor Ramesh Chander
Kasturirangan K
Kaul Chaman
Kembhavi Ajit K
Khare Pushpa
Kilambi G C
Kochhar R K
Krishna Gopal
Krishna Swamy K S
Krishnamohan S
Krishnan Thiruvenkata
Kulkarni Prabhakar V
Kulkarni Vasant K
Kushwaha R S
Lal Devendra
Mahra H S
Mallik D C V
Manchanda R K
Manoharan P K
Marar T M k
Mathur B S
Mitra A P
Mohan Chander
Mohan Vijay
Nagendra K N
Namboodiri P M S
Narain Udit
Naranan S
Narasimha Delampady
Narayana J V
Narlikar Jayant V
Nath Mishra Kameshwar
Nityananda R
Padalia T D
Padmanabhan Janardhan
Padmanabhan T
Pande Girish Chandra
Pande Mahesh Chandra
Pandey A K
Pandey S K
Pandey Uma Shankar
Parthasarathy M
Pati A K
Peraiah Annamaneni
Prabhakaran Nayar S R
Prabhu Tushar P
Prasanna A R
Pratap R
Punetha Lalit Mohan
Raghavan Nirupama
Rajamohan R
Raju P K
Raju Vasundhara
Rakshit H
Ram Sagar
Ramadurai Souriraja
Ramamurthy Swaminathan
Ramana Murthy P V
Rangarajan K E
Rao A Pramesh
Rao Arikkala Raghurama
Rao D Mohan
Rao M N
Rao N Kameswara
Rao P Vivekananda
Rao Ramachandra V

Rautela B S
Raveendran A V
Ray Alak
Raychaudhuri Amalkumar
Raychaudhury Somak
Razdan Hiralal
Rengarajan T N
Saha Swapan Kumar
Sahni Varun
Saikia Dhruba Jyoti
Sanwal Basant Ballabh
Sanwal N B
Sapre A K
Sarma M B K
Sarma N V G
Sastri Hanumath J
Sastry Ch V
Sastry Shankara K
Saxena A K
Saxena P P
Seshadri Sridhar
Shahul Hameed Mohin
Shastri Prajval
Shevgaonkar R K
Shukla K
Shukre C S
Singh H P
Singh Jagdev
Singh Kulinder Pal
Sinha K
Sinvhal Shambhu Dayal
Sivaram C
Sivaraman K R
Sonti Sreedhar Rao
Sreekantan B V
Srinivasan G
Srivastava Dhruwa
Srivastava J B
Srivastava Ram Kumar
Stephens S A
Subrahmanya C R
Subrahmanyam P V
Subramanian K R
Subramanian Kandaswamy
Swarup Govind
Talwar Satya P
Tandon Jagdish Narain
Tandon S N
Tapde Suresh Chandra
Thakur Ratna Kumar
Tonwar Suresh C
Trehan Surindar K
Tripathi B M
Udaya Shankar N
Vahia Mayank N
Vaidya P C
Vardya M S
Varma Ram Kumar
Vasu-Mallik Sushma
Vats Hari OM
Venkatakrishnan P
Venugopal V R
Verma R P
Verma Satya Dev
Verma V K
Vinod S Krishan
Vishveshwara C V
Vivekanand M
Vivekananda Rao

Indonesia

Dawanas Djoni N
Hidayat Bambang
Ibrahim Jorga
Malasan Hakim Luthfi
Radiman Iratius
Raharto Moedji
Ratag Mezak Arnold
Siregar Suryadi
Sukartadiredja Darsa
Sutantyo Winardi
Wiramihardja Suhardja
Wiyanto Paulus

Iran

Adjabshirizadeh Ali
Ardebili M Reza
Dehghani Mohammad Hossein
Edalati Sharbaf Mohammad T
Ghanbari Jamshid
Jassur Davoud M Z
Kalafi Manoucher
Khalesseh Bahram
Kiasatpoor Ahmad
Malakpur Iradj
Mansouri Reza
Riazi Nematollah
Sadollah Nasiri Gheidari
Sobouti Yousef
Teherany D

Ireland

Breslin Ann
Callanan Paul
Carroll P Kevin
Cawley Michael
Drury Luke O'Connor
Elliott Ian
Fegan David J
Florides Petros S
Gillanders Gerard
Haywood J
Hoey Michael J
Kennedy Eugene T
Kiang Tao
Lang Mark
McBreen Brian Philip
McCrea J Dermott
McKeith Niall Enda
McKenna Lawlor Susan
Meurs Evert
Norci Laura
O'Connor Seamus L
O'Mongain Eon
O'Sullivan Creidhe
O'Sullivan Denis F
Porter Neil A
Ray Thomas P
Redfern Michael R
Russell Stephen
Shearer Andrew
Smith Niall
van Breda Ian G
Wayman Patrick A
Wrixon Gerard T

Israel

Barkat Zalman
Bekenstein Jacob D
Braun Arie
Brosch Noah
Cuperman Sami
Dekel Avishai
Ershkovich Alexander
Eviatar Aharon
Finzi Arrigo
Glasner Shimon Ami
Goldman Itzhak
Goldsmith S
Gradsztajn Eli
Harpaz Amos
Horwitz Gerald
Ibbetson Peter Aaron
Joseph J H
Katz Joseph
Kovetz Attay
Kozlovsky B Z
Leibowitz Elia M
Livio Mario
Mazeh Tsevi
Meerson Baruch
Meidav Meir
Mekler Yuri
Neeman Yuval
Netzer Hagai
Ohring George
Pekeris Chaim Leib
Prialnik-Kovetz Dina
Rakavy Gideon
Regev Oded
Rephaeli Yoel
Sack Noam
Sadeh D
Segaluvitz Alexander
Shaviv Giora
Shiryaev Alexander A
Steinitz Raphael
Tuchman Ytzhak
Vager Zeev
Vidal Nissim V
Yeivin Y

Italy

Abrami Alberto
Aiello Santi
Alighieri Sparello Di Serego
Altamore Aldo
Ambrosini Roberto
Amendola Luca
Andreani Paola Michela
Angeletti Lucio
Anile Angelo M
Antonello Elio
Antonucci Ester
Antonuccio-Delogu Vincenzo
Auriemma Giulio
Badiali Massimo
Badolati Ennio
Baffa Carlo
Baldinelli Luigi
Ballario M C
Bandiera Rino
Banni Aldo
Baratta Giovanni Battista
Baratta Giuseppe Antonio
Barbaro G
Barberis Bruno
Barbieri Cesare
Barbon Roberto
Barletti Raffaele
Barone Fabrizio
Bartolini Corrado
Battinelli Paolo
Battistini Pierluigi
Bedogni Roberto
Belvedere Gaetano
Benacchio Leopoldo
Bendinelli Orazio
Bernacca Pietio L
Berrilli Francesco
Bertelli Gianpaolo
Bertin Giuseppe
Bertola Francesco
Bettoni Daniela
Bianchini Antonio
Bignami Giovanni F
Blanco Carlo
Bodo Gianluigi
Bonaccini Domenico
Bonanno Giovanni
Bonanno Giovanni
Bonifazi Angelo
Bono Giuseppe
Bonoli Fabrizio
Bonometto Silvio A
Braccesi Alessandro
Brand Jan
Brini Domenico
Brocato Enzo
Broglia Pietro
Bucciarelli Beatrice
Buonanno Roberto
Buson Lucio M
Busso Maurizio
Buzzoni Alberto
Cacciani Alessandro
Caccin Bruno
Calamai Giovanni
Caloi Vittoria
Calvani Massimo
Cantu Alberto M
Capaccioli Massimo
Capaccioni Fabrizio
Cappellaro Enrico
Cappi Alberto
Capria Maria Teresa
Caprioli Giuseppe
Caputo Filippina
Capuzzo Dolcetta Roberto
Cardini Daniela
Carpino Mario
Carusi Andrea
Cassatella Angelo
Castellani Vittorio
Castelli Fiorella
Catalano Francesco A
Catalano Santo
Cauzzi Gianna
Cavaliere Alfonso G
Cavallini Fabio
Cazzola Paolo
Cecchi-Pellini Cesare
Cellino Alberto
Ceppatelli Guido
Cerroni Priscilla
Cerruti'sola Monica
Cester Bruno
Cevolani Giordano
Chincarini Guido L
Chiosi Cesare S
Chiuderi Claudio
Chiuderi-Drago Franca Pr
Chiumiento Giuseppe
Chlistovsky Franca
Ciatti Franco

Citterio Oberto
Colangeli Luigi
Colombo G
Coluzzi Regina
Comastri Andrea
Comoretto Giovanni
Conconi Paolo
Coradini Angioletta
Corbelli Edvige
Cosmovici Batalli C
Costa Enrico
Covino Elvira
Cristaldi Salvatore
Cristiani Stefano
Crivellari Lucio
Cugusi Leonino
Curir Anna
Cutispoto Giuseppe
D'Antona Francesca
Dall-Oglio Giorgio
Dallaporta N
Danese Luigi
de Biase Giuseppe A
de Felice Fernando
de Martino Domitilla
de Ruiter Hans Rudolf
de Sabbata Vittorio
de Sanctis Giovanni
de Zotti Gianfranco
Della Valle Massimo
Di Cocco Guido
Di Fazio Alberto
Di Martino Mario
Di Tullio Graziella
Einaudi Giorgio
Emanuele Alessandro
Facondi Silvia Rosa
Falchi Ambretta
Falciani Roberto
Falomo Renato
Fanti Carla Giovannini
Fanti Roberto
Faraggiana Rosanna
Farinella Paolo
Federici Luciana
Felli Marcello
Feretti Luigina
Ferluga Steno
Ferrari Attilio
Ferreri Walter
Ferrini Federico
Ficarra Antonino
Focardi Paola
Fodera Seriio Giorgia
Fofi Massimo
Forti Giuseppe
Fortini Teresa
Franceschini Alberto
Franchini Mariagrazia
Franco Mantovani
Fulle Marco
Fusco-Femiano Roberto
Fusi Pecci Flavio
Galeotti Piero
Galletta Giuseppe
Galletto Dionigi
Galliano Pier Giorgio
Gallino Roberto
Garilli Bianca
Gaudenzi Silvia
Gavazzi Giuseppe
Gemmo Alessandra
Ghisellini Gabriele
Giachetti Riccardo
Giallongo Emanuele
Giannone Pietro
Giannuzzi Maria A
Giovanardi Carlo
Giovannelli Franco
Giovannini Gabriele
Giuricin Giuliano
Godoli Giovanni
Gomez Maria Theresa
Gratton Raffaela G
Greggio Laura
Gregorini Loretta
Grubissich C
Grueff Gavril
Guarnieri Adriano
Guerrero Gianantonio
Guzzo Luigi
Hack Margherita
Held Enrico V
Huang Song-nian
Hunt Leslie
Iijima Takashi
Iovino Angela
Kotilainen Jari
La Padula Cesare
Lai Sebastiana
Lanciano Nicoletta
Landi Degl-Innocenti E
Landini Massimo
Landolfi Marco
Lanza Antonino Francesco
Lari Carlo
Lattanzi Mario G
Leone Francesco
Leschiutta S
Leto Giuseppe
Lisi Franco
Lucchin Francesco
Maccacaro Tommaso
Maccagni Dario
Maceroni Carla
Maffei Paolo
Magazzu Antonio
Maggio Antonio
Magni Gianfranco
Malagnini Maria Lucia
Mammano Augusto
Manara Alessandro A
Mancuso Santi
Mandolesi Nazzareno
Mannino Giuseppe
Mannucci Filippo
Mantegazza Luciano
Marano Bruno
Maraschi Laura
Mardirossian Fabio
Margoni Rino
Marilli Ettore
Marmolino Ciro
Martini Aldo
Marzari Francesco
Masani A
Massaglia Silvano
Matteucci Francesca
Mazzitelli Italo
Mazzoni Massimo
Mazzucconi Fabrizio
Mennella Vito
Merighi Roberto
Messerotti Mauro
Messina Antonio
Mezzetti Marino
Micela Giuseppina
Milani Andrea
Milano Leopoldo
Missana Marco
Missana Natale
Molaro Paolo
Molinari Emilio
Morbidelli Lorenzo
Morbidelli Roberto
Morganti Raffaella
Morossi Carlo
Moscardini Lauro
Motta Santo
Mureddu Leonardo
Natali Giuliano
Natta Antonella
Nesci Roberto
Nicastro Luciano
Nobili Anna M

Nobili L
Nocera Luigi
Noci Giancarlo
Occhionero Franco
Oliva Ernesto
Ortolani Sergio
Pacini Franco
Padrielli Lucia
Pagano Isabella
Palagi Francesco
Palla Francesco
Pallavicini Roberto
Palumbo Giorgio G C
Pannunzio Renato
Paolicchi Paolo
Parma Paola
Pasian Fabio
Pasinetti Laura E
Pastori Livio
Paterno Lucio
Patriarchi Patrizio
Peres Giovanni
Perinotto Mario
Perola Giuseppe C
Persi Paolo
Pettini Marco
Picca Domenico
Piccioni Adalberto
Pigatto Luisa
Pinto Girolamo
Piotto Giampaollo
Piro Luigi
Pirronello Valerio
Pizzella G
Pizzichini Graziella
Polcaro V F
Poletto Giannina
Poma Angelo
Porceddu Ignazio E P
Poretti Ennio
Preite-Martinez Andrea
Proverbio Edoardo
Pucillo Mauro
Pugliano Antonio
Quesada Vinicio
Rafanelli Piero
Ramella Massimo
Rampazzo Roberto
Ranieri Marcello
Renzini Alvio
Richichi Andrea
Righini Alberto
Roberti Giuseppe
Rodono Marcello
Romano Giuliano
Rosino Leonida
Rossi Corinne
Rossi Lucio
Ruffini Remo
Rusconi Luigia
Russo Guido
Sabbadin Franco
Saggion Antonio
Salinari Piero
Salvati Marco
Santamaria Raffaele
Santin Paolo
Sarasso Maria
Saverio Delli Santi
Scaltriti Franco
Scappini Flavio
Scardia Marco
Sciama Dennis W
Sciortino Salvatore
Secco Luigi
Sedmak Giorgio
Selvelli Pierluigi
Semenzato Roberto
Serio Salvatore
Setti Giancarlo
Severino Giuseppe
Silvestro Giovanni
Sironi Giorgio
Smaldone Luigi Antonio
Smriglio Filippo
Spada Gianfranco
Spadaro Daniele
Spagna Alessandro
Stagni Ruggero
Stalio Roberto
Stanga Ruggero
Stanghellini Letizia
Stefano Andreon
Stiavelli Massimo
Stirpe Giovanna M
Strafella Francesco
Strazzulla Giovanni
Taffara Salvatore
Tagliaferri Gianpiero
Tagliaferri Giuseppe
Tanzella-Nitti Giuseppe
Tanzi Enrico G
Tempesti Piero
Ternullo Maurizio
Terranegra Luciano
Tofani Gianni
Tomasi Paolo
Torelli M
Tornambe Amedeo
Torricelli Guidetta
Tosi Monica
Tozzi Gian Paolo
Trevese Dario
Trinchieri Ginevra
Trussoni Edoardo
Turatto Massimo
Turolla Roberto
Ubertini Pietro
Uras Silvano
Vagnetti Fausto
Valsecchi Giovanni B
Velli Marco
Ventura Rita
Venturi Tiziana
Vergnano A
Verniani Franco
Vettolani Giampaolo
Vietri Mario
Vigotti Mario
Viotti Roberto
Virgopia Nicola
Vittone Alberto Angelo
Vittorio Nicola
Vladilo Giovanni
Wolter Anna
Zamorani Giovanni
Zaninetti Lorenzo
Zappala Rosario Aldo
Zappala Vincenzo
Zavatti Franco
Zitelli Valentina
Zlobec Paolo
Zuccarello Francesca

Japan

Aikawa Toshiki
Aizu Ko
Akabane Kenji A
Akabane Tokuhide
Aoki Shinko
Arafune Jiro

Arai Kenzo
Arimoto Nobuo
Awaki Hisamitsu
Ayani Kazuya
Azuma Takahiro
Baba Naoshi
Chiba Masashi
Chikada Yoshihiro
Chikawa Michiyuki
Daishido Tsuneaki
Deguchi Shuji
Doi Mamoru
Dotani Tadayasu
Ebisuzaki Toshikazu
Enome Shinzo
Eriguchi Yoshiharu
Fujimoto Masa-Katsu
Fujimoto Masayuki
Fujimoto Mitsuaki
Fujishita Mitsumi
Fujita Yoshio
Fujiwara Akira
Fujiwara Takao
Fukuda Ichiro
Fukue Jun
Fukugita Masataka
Fukui Takao
Fukui Yasuo
Fukunaga Masataka
Fukushima Toshio
Gouda Naoteru
Gunji Shuichi
Habe Asao
Hachisu Izumi
Hagio Fumihiko
Hamabe Masaru
Hamajima Kiyotoshi
Hanami Hitoshi
Hanaoka Yoichiro
Hanawa Tomoyuki
Handa Toshihiro
Hara Tadayoshi
Hara Tetsuya
Hasegawa Ichiro
Hasegawa Tatsuhiko
Hashimoto Masa-aki
Hashimoto Osama
Hatsukade Isamu
Hayashi Chushiro
Hayashi Masahiko
Hiei Eijiro
Hirabayashi Hisashi
Hirai Masanori
Hirano Naomi
Hirata Ryuko
Hirayama Tadashi
Hiromoto Norihisa
Hiroyasu Ando
Hitotsuyanagi Juichi
Hori Genichiro
Horiuchi Ritoku
Hoshi Reiun
Hosokawa Yoshimasa H
Hozumi Shunsuke
Hudson Hugh S
Hurukawa Kiitiro
Ichikawa Shin-ichi
Ichikawa Takashi
Ichimaru Setsuo
Iijima Shigetaka
Ikeuchi Satoru
Inagaki Shogo
Inatani Junji
Inoue Hajime
Inoue Makoto
Inoue Takeshi
Iriyama Jun
Ishida Keiichi
Ishida Manabu
Ishida Toshihito
Ishiguro Masato
Ishihara Hideki
Ishizawa Toshiaki A
Ishizuka Toshihisa
Isobe Syuzo
Ito Kensai A
Ito Yutaka
Itoh Hiroshi
Itoh Masayuki
Itoh Naoki
Iwasaki Kyosuke
Iwata Takahiro
Iye Masanori
Izumiura Hideyuki
Jugaku Jun
Kaburaki Osamu
Kaburaki Osamu
Kaifu Norio
Kajino Toshitaka
Kakinuma Takakiyo T
Kakuta Chuichi
Kambe Eiji
Kameya Osamu
Kamijo Fumio
Kaminishi Keisuke
Kanamitsu Osamu
Kaneko Noboru
Karoji Hiroshi
Kasuga Takashi
Kato Ken-ichi
Kato Mariko
Kato Shoji
Kato Takako
Kawabata Kinaki
Kawabata Kiyoshi
Kawabata Shusaku
Kawabe Ryohei
Kawaguchi Ichiro
Kawaguchi Kentarou
Kawai Nobuyuki
Kawara Kimiaki
Kawasaki Masahiro
Kawata Yoshiyuki
Kiguchi Masayoshi
Kii Tsuneo
Kikuchi Sadaemon
Kimura Toshiya
Kinoshita Hiroshi
Kitai Reizaburo
Kitamoto Shunji
Kitamura M
Kobayashi Eisuke
Kobayashi Hideyuki
Kobayashi Yukisayu
Kodaira Keiichi
Kodama Hideo
Kogure Tomokazu
Koide Shinji
Koike Chiyoe
Kojima Masayoshi
Kojima Yasufumi
Kondo Masaaki
Kondo Masayuki
Kosai Hiroki
Koshiba Masa-Toshi
Kosugi Takeo
Koyama Katsuji
Koyama Shin
Kozai Yoshihide
Kozasa Takashi
Kubo Yoshio
Kubota Jun
Kumagai Shiomi
Kumai Yasuki
Kunieda Hideyo
Kurokawa Hiroki
Kusunose Masaaki
Maeda Kei-ichi
Maeda Koitiro
Maehara Hideo
Maihara Toshinori
Makino Fumiyoshi
Makino Junichiro
Makishima Kazuo
Makita Mitsugu

Manabe Seiji
Masai Kuniaki
Masuda Satoshi
Matsuda Takuya
Matsuhara Hideo
Matsui Takafumi
Matsumoto Masamichi
Matsumoto Ryoji
Matsumoto Toshio
Matsumura Masafumi
Matsuo Hiroshi
Matsuoka Masaru
Mikami Takao
Mineshige Shin
Mitsuda Kazuhisa
Miyaji Shigeki
Miyama Syoken
Miyamoto Masanori
Miyamoto Sigenori
Miyoshi Makoto
Miyoshi Shigeru
Mizumoto Yoshihiko
Mizuno Akira
Mizuno Shun
Mizuno Takao
Mizutani Kohei
Morimoto Masaki
Morita Kazuhiko
Morita Koh-ichiro
Moriyama Fumio
Mukai Sonoyo
Mukai Tadashi
Murakami Hiroshi
Murakami Izumi
Murakami Toshio
Nagase Fumiaki
Nagata Tetsuya
Nakada Yoshikazu
Nakagawa Naoya
Nakagawa Takao
Nakagawa Yoshinari
Nakagawa Yoshitsugu
Nakai Naomasa
Nakai Yoshihiro
Nakajima Hiroshi
Nakajima Koichi
Nakajima Tadashi
Nakamoto Taishi
Nakamura Akiko M
Nakamura Takashi
Nakamura Tsuko
Nakamura Yasuhisa
Nakano Makoto
Nakano Syuichi
Nakano Takenori
Nakayama Kunji
Nakayama Shigeru
Nakazawa Kiyoshi
Nambu Yasusada
Narita Shinji
Niimi Yukio
Nishi Keizo
Nishi Ryoichi
Nishida Minoru
Nishida Mitsugu
Nishikawa Jun
Nishimura Jun
Nishimura Masaki
Nishimura Shiro Emer
Nishio Masanori
Noguchi Kunio
Noguchi Masafumi
Nomoto Ken'ichi
Obi Shinya
Oda Minoru
Oda Naoki
Ogawa Hideo
Ogawara Yoshiaki
Ogura Katsuo
Ohashi Takaya
Ohashi Yukio
Ohishi Masatoshi
Ohki Kenichiro
Ohnishi Kouji
Ohta Kouji
Ohtani Hiroshi
Ohtsubo Junji
Ohyama Noboru
Okamoto Isao
Okamura Sadanori
Okazaki Akira
Okazaki Atsuo T
Okazaki Seichi
Okuda Haruyuki
Okuda Toru
Okumura Sachiko
Onaka Takashi
Ono Yoro
Ooe Masatsugu
Oohara Ken-ichi
Osaki Toru
Osaki Yoji
Owaki Naoaki
Ozeki Hiroyuki
Sabano Yutaka
Sadakane Kozo
Saijo Keiichi
Saio Hideyuki
Saito Kuniji
Saito Mamoru
Saito Sumisaburo
Saito Takao
Saito Takao
Sakai Junichi
Sakamoto Seiichi
Sakao Taro
Sakashita Shiro
Sakurai Kunitomo
Sakurai Takashi
Sakurai Takeo T
Sasaki Minoru
Sasaki Misao
Sasaki Shin
Sasaki Toshiyuki
Sasao Tetsuo
Sato Fumio
Sato Humitaka
Sato Katsuhiko
Sato Koichi
Sato Naonobu
Sato Shinji
Sato Shuji
Sato Yuzo
Sawa Takeyasu
Seki Munezo
Sekiguchi Maki
Sekiguchi Naosuke
Sekimoto Yutaro
Shibahashi Hiromoto
Shibai Hiroshi
Shibasaki Kiyoto
Shibata Katsunori M
Shibata Kazunari
Shibata Masaru
Shibata Shinpei
Shibata Yukio
Shibazaki Noriaki
Shigeyama Toshikazu
Shimasaku Kazuhiro
Shimizu Mikio
Shimizu Tsutomu Emer
Simoda Mahiro
Sinzi Akira M
Sofue Yoshiaki
Soma Mitsuru
Suda Kazuo
Suematsu Yoshinori
Sugai Hajime
Sugawa Chikara
Sugimoto Daiichiro
Suginohara Tatsushi
Sugitani Koji
Sugiyama Naoshi
Sunada Kazuyoshi
Suto Yasushi

Suzuki Hideyuki
Suzuki Yoshimasa
Tabara Hiroto
Takaba Hiroshi
Takada-Hidai Masahide
Takagi Kojiro
Takahara Fumio
Takahara Mariko
Takahashi Koji
Takahashi Masaaki
Takahashi Tadayuki
Takakubo Keiya
Takakura Tatsuo Emer
Takami Hideki
Takano Toshiaki
Takarada Katsuo
Takase Bunshiro
Takato Naruhisa
Takayanagi Kazuo
Takeda Hidenori
Takeda Yoichi
Takenouchi Tadao
Takeuti Mine
Tamenaga Tatsuo
Tamura Motohide
Tamura Shin'ichi
Tanabe Hiroyoshi
Tanabe Kenji
Tanabe Toshihiko
Tanaka Masuo
Tanaka Riichiro
Tanaka Wataru
Tanaka Yasuo
Tanaka Yasuo
Tanaka Yutaka D
Taniguchi Yoshiaki
Tanikawa Kiyotaka
Tashiro Makoto
Tatematsu Ken-ichi
Tawara Yuzuru
Terashita Yoichi
Tomimatsu Akira
Tomisaka Kohji
Tomita Kenji
Tomita Koichiro
Torao Masahisa
Tosa Makoto
Totsuka Yoji
Toyama Kiyotaka
Tsubaki Tokio
Tsuboi Masato
Tsubokawa Ietsune
Tsubota Yukimasa
Tsuchiya Atsushi
Tsuchiya Toshio
Tsuji Takashi
Tsujimoto Takuji
Tsunemi Hiroshi
Tsuneta Saku
Tsuru Takeshi
Tsutsumi Takahiro
Uchida Juichi
Uchida Yutaka
Ueno Munetaka
Ueno Sueo
Uesugi Akira
Ukita Nobuharu
Umemura Masayuki
Unno Wasaburo
Utsumi Kazuhiko
Wada Keiichi
Wakamatsu Ken-ichi
Wako Kojiro
Washimi Haruichi
Watanabe Jun-ichi
Watanabe Noriaki
Watanabe Takashi
Watanabe Tetsuya
Yabushita Shin A
Yabuuti Kiyoshi
Yamada Yoshiyaki
Yamagata Tomohiko
Yamaguchi Shichiro
Yamamoto Satoshi
Yamamoto Tetsuo
Yamamoto Yoshiaki
Yamaoka Hitoshi
Yamasaki Atsuma
Yamashita Kojun
Yamashita Takuya
Yamashita Yasumasa
Yamauchi Makoto
Yamauchi Shigeo
Yamazaki Akira
Yasuda Haruo
Yokosawa Masayoshi
Yokoyama Jun-ichi
Yokoyama Koichi
Yoneyama Tadaoki
Yoshida Atsumasa
Yoshida Haruo
Yoshida Junzo
Yoshida Michitoshi
Yoshida Shigeomi
Yoshii Yuzuru
Yoshikawa Makoto
Yoshimura Hirokazu
Yoshimura Motohiko
Yoshioka Kazuo
Yoshizawa Masanori
Yuasa Manabu
Yumi Shigeru

Korea Rep

Ann Hong Bae
Chang Kyongae
Cho Se Hyung
Choe Seung Urn
Choi Kyu-Hong
Chou Kyong Chol
Chun Mun-suk
Chung Hyun Soo
Hong Seung Soo
Hwang Jai-chan
Hyun Jong-June
Hyung Siek
Jeong Jang Hae
Kang Yong Hee
Kang Young Woon
Kim Chulhee
Kim Chun Hwey
Kim Ho Il
Kim Kap-sung
Kim Kwang-tae
Kim Tu Hwan
Kim Yonggi
Kim Yongha
Koo Bon Chul
Lee Dong Hun
Lee Hyung Mok
Lee Myung Gyoon
Lee Sang Gak
Lee See-woo
Lee Woo-baik
Lee Yong-Sam
Lee Young Wook
Minh Young Chol
Minn Young Key
Moon Shin Haeng
Nha Il-Seong
Noh Hyerim
Oh Kap-Soo
Oh Kyu Dong

Park Changbom
Park Hong Suh
Park Myeong-gu
Park Seok Jae
Park Young-Deuk
Ryu Dongsu
Song Doo Jong
Suh Kyung-Won
Woo Jong Ok
Yoon Tae
Yu Kyung-Loh
Yun Hong-Sik

Latvia

Abele Maris K
Alksnis Andrejs
Balklavs A E
Daube-Kurzemniece I A
Frantsman Yu L
Lapushka K K
Shmeld Ivar
Zhagar Youri H

Lithuania

Bartasiute Stanislava
Bartkevicius Antanas
Bogdanovicius Pavelas
Cernis Kazimieras
Kupliauskiene Alicija
Rudzikas Zenonas B
Sperauskas Julius
Straizys V
Sudzius Jokubas
Tautvaisiene Grazina
Zdanavicius Kazimeras

Malaysia

Abdul Aziz Ahmad Zaharim
Ilyas Mohammad
Mahat Rosli H
Majid Abdul Bin A H
Mohd Zambri Zainuddin
Othman Mazlan

Mexico

Aguilar Luis A Chiu
Allen Christine
Andernach Heinz
Arellano Ferro Armando
Arthur Jane
Bisiacchi Gianfranco
Brinks Elias
Canto Jorge
Cardona Octavio
Carigi Leticia
Carraminana Alberto
Carrasco Berth Esperanza
Carrasco Luis
Carrillo Moreno Rene
Chatterjee Tapan K
Chavarria'k Carlos
Chavez-Dagostino Miguel
Chavira Enrique Sr
Chelli Alain
Colin Pedro
Cornejo Alejandro A
Costero Rafael
Cruz-Gonzalez Irene
Curiel Salvador
Daltabuit Enrique
de La Herran Jose V
Dultzin-Hacyan D
Echeverria Roman Juan M
Eenens Philippe
Escalante Vladimir
Fierro Julieta
Firmani Claudio A
Franco Jose
Galindo Trejo Jesus
Garcia-Barreto Jose A
Gomez Yolanda
Gonzales Alejandro
Gonzalez G
Gonzalez J Jesus
Guichard Jose
Henney William John
Herrera Miguel Angel
Klapp Jaime
Koenigsberger Gloria
Kraan-Korteweg Renee C
Lekht Evueni
Lizano-Soberon Susana
Lopez Jose Alberto
Lopez-Cruz Omar
Malacara Daniel
Martinez Mario
Mayya Divakara
Mendez Manuel
Mendoza V Eugenio E
Mendoza-Torres Jose-Eduar
Migenes Victor
Moreno Corral Marco A
Moreno Edmundo
Obregon Diaz Octavio J
Page Dany
Peimbert Manuel
Pena Jose
Pena Miriam
Peniche Rosario
Perez de Tejada H A
Perez-Peraza Jorge A
Pismis de Recillas Paris
Poveda Arcadio
Puerari Ivanio
Recillas-Cruz Elsa
Rodriguez Luis F
Rosado Margarita
Ruelas-Mayorga R A
Sarmiento-Galan A F
Schuster William John

Serrano Alfonso
Tapia Mauricio
Torres-Peimbert Silvia
Tovmassian Gaghik
Tovmassian Hrant M
Vazquez-Semadeni Enrique
Warman Josef

Netherlands

Achterberg Abraham
Atanasijevic Ivan
Barthel Peter
Baud Boudewijn
Begeman Kor G
Beintema Douwe A
Belloni Tomaso
Bennett Kevin
Bijleveld Willem
Blaauw Adriaan
Bleeker Johan A M Ir
Bloemen Johannes B G M
Boland Wilfried
Bontekoe Romke
Borgman Jan
Bos Albert
Bosma Pieter B
Braes L L E
Braun Robert
Bregman Jacob D Ir
Briggs Franklin
Brinkman Bert C
Burger J J
Burton W Butler
Butcher Harvey R
de Bruyn A Ger
de Geus Eugene
de Graaff W
de Graauw Th
de Groot T
de Jager Cornelis
de Jong Teije
de Korte Pieter A J
de Vries Cornnelis
de Zeeuw Pieter T
Dekker E
Deul Erik
Douglas Nigel
Fitton Brian
Foing Bernard H
Foing-Ehrenfreund Pascale
Foley Anthony
Franx Marijn
Fridlund Malcolm
Fritzova-Svestka L
Garrett Michael
Goedbloed Johan P
Greenberg J Mayo
Gurvits Leonid
Habing Harm J
Hammerschlag Robert H
Hammerschlag-Hensberge G
Hearn Anthony G
Heintze J R W
Heise John
Henrichs Hubertus F
Heras Ana M
Hermsen Willem
Heske Astrid
Hoekstra Roel
Houdebine Eric
Hovenier J W
Hoyng Peter
Hubenet Henri
Huber Martin C E
Hulsbosch A N M
Hummel Edsko
Icke Vincent
Israel Frank P
Ives John Christopher
Jaffe Walter Joseph
Jakobsen Peter
Kaastra Jelle S
Kahabka Peter
Kahlmann Hans Cornelis
Katgert Peter
Katgert-Merkelijn J K
Keppens Rony
Koornneef Jan
Kuijpers H Jan M E
Kuperus Max
Kuyken Koenraad H
Kwee K K
Lamers Henny J G L M
Lanz Thierry
Le Poole Rudolf S
Lub Jan
Martens Petrus C
Metcalfe Leo
Mewe R
Miley George K
Muller Andre B
Muller C A
Namba Osamu
Nieuwenhuijzen Hans
North John David
Ollongren A
Olnon Friso
Olthof Hindericus
Parmar Arvind Nicholas
Peacock Anthony
Pel Jan Willem
Peletier Reynier Frans
Perryman Michael A C
Pottasch Stuart R
Radhakrishnan V
Raimond Ernst
Roelfsema Peter
Roos Nicolaas
Rutten Robert J
Sackett Penny
Sancisi Renzo
Sanders Robert
Savonije Gerrit Jan
Schadee Aert
Scheepmaker Anton
Schilizzi Richard T
Schrijver C J
Schrijver Johannes
Schutte Willem Albert
Schwarz Ulrich J
Schwehm Gerhard
Smit J A
Spoelstra T A Th
Stevens Gerard A
Strom Richard G
Svestka Zdenek
Swanenburg B N
Takens Roelf Jan
Tauber Jan
The Pik-Sin
Tinbergen Jaap
Tjin-a-Djie Herman R E
Vaghi Sergio
Valentijn Edwin A
van Albada tjeerd S
van Beek Frank
van Bueren Hendrik G
van de Hulst H C
van de Stadt Herman
van den Heuvel Edward P J
van den Oord Bert H J
van der Hucht Karel A
van der Hulst Jan M

van der Klis Michiel
van der Kruit Pieter C
van der Laan Harry
van der Werf Paul P
van Diggelen J
van Dishoeck Ewine F
van Duinen R J
van Genderen Arnoud M
van Haarlem Michiel
van Herk Gijsbert
van Houten C J
van Houten-Groeneveld I
van Langevelde Huib Jan
van Nieuwkoop JIR
van Paradijs Johannes
van Woerden Hugo
Verbunt Franciscus
Vermeulen Rene Cornelis
Waters Laurens B F M
Wesselius Paul R
Wijnbergen Jan
Winkler Christoph
Zwaan Cornelis

New Zealand

Albrow Michael
Allen William
Andrews Frank
Baggaley William J
Bateson Frank M
Blow Graham L
Budding Edwin
Carter Brian
Cottrell Peter Ledsam
Craig Ian Jonathan D
Dodd Richard J
Doughty Noel A
Gilmore Alan C
Gulyaev Sergei A
Hearnshaw John B
Hill Graham
Jones Albert F
Kerr Roy P
Kilmartin Pamela
Marino Brian F
Orchiston Wayne
Pollard Karen
Rumsey Norman J
Sullivan Denis John
Tobin William
Walker William S G
Yock Philip

Norway

Aksnes Kaare
Brahde Rolf
Brekke Pal Ording lie
Carlsson Mats
Elgaroy Oystein
Engvold Oddbjoern
Eriksen Gunnar
Hansteen Viggo
Hauge Oivind
Havnes Ove
Jensen Eberhart
Kjeldseth-Moe Olav
Leer Egil
Lilje Per Vidar Barth
Maltby Per
Oestgaard Erlend
Pettersen Bjoern Ragnvald
Ringnes Truls S
Solheim Jan Erik
Stabell Rolf
Trulsen Jan K

Peru

Aguilar Maria Luisa

Poland

Bajtlik Stanislaw
Bem Jerzy
Borkowski Kazimierz M
Brzezinski Aleksander
Bzowski Maciej
Chodorowski Michal
Choloniewsski Jacek
Chyzy Krzysztof Tadeusz
Ciurla Tadeusz
Cugier Henryk
Czerny Bozena
Demianski Marek
Dobrzycki Jerzy
Drozyner Andrzej
Dziembowski Wojciech
Flin Piotr
Gaska Stanislaw
Gesicki Krzysztof
Giersz Miroslaw
Gil Janusz A
Glebocki Robert
Godlowski Wlodzimierz
Gorgolewski Stanislaw Pr
Grabowski Boleslaw
Grudzinska Stefania
Grzedzielski Stanislaw Pr
Haensel Pawel
Hanasz Jan
Heller Michael
Hurnik Hieronim
Iwaniszewska Cecylia
Iwanowska Wilhelmina
Jahn Krzysztof
Jakimiec Jerzy
Jaroszynski Michal
Jarzebowski Tadeusz
Jerzykiewicz Mikolaj
Jopek Tadeusz Jan
Juszkiewicz Roman
Kaluzny Janusz
Kijak Jaroslaw
Kluzniak Wlodzimiere

Kolaczek Barbara
Kosek Wieslaw
Kozlowski Maciej
Krasinski Andrzej
Kreiner Jerzy Marek
Krelowski Jacek
Krempec-Krygier Janina
Krolikowska-Soltan Malgorzata
Kruszewski Andrzej
Krygier Bernard
Kubiak Marcin A
Kurpinska-Winiarska M
Kurzynska Krystyna
Kus Andrzej Jan
Lehmann Marek
Machalski Jerzy
Maciejewski Andrzej J
Madej Jerzy
Maslowski Jozef
Michalec Adam
Michalowski Tadeusz
Mietelski Jan S
Mikolajewska Joanna
Moskalik Pawel
Niedzielski Andrzej
Opolski Antoni
Ostrowski Michal
Otmianowska-Mazur Katarzyna
Paczynski Bohdan
Pigulski Andrzej
Pojmanski Grzegorz
Pokrzywka Bartlomiej
Proszynski Mieczyslaw
Rompolt Bogdan
Rozyczka Michal
Rucinski Daniel
Rudak Bronislaw
Rudnicki Konrad
Rys Stanislaw
Sarna Marek Jacek
Schreiber Roman
Schwarzenberg-Czerny A
Semeniuk Irena
Sienkiewicz Ryszard
Sikora Marek
Sikorski Jerzy
Sitarski Grzegorz
Smak Joseph I
Sokolowski Lech
Soltan Andrzej Maria
Stawikowski Antoni
Stepien Kazimierz
Strobel Andrzej
Sylwester Barbara
Sylwester Janusz
Szafraniec Rozalia
Szczerba Ryszard
Szymanski Michal
Tomczak Michal
Turlo Zygmunt
Tylenda Romuald
Udalski Andrzej
Urbanik Marek
Usowics Jerzy Bogdan
Winiarski Maciej
Wnuk Edwin
Woszczyk Andrzej
Woszczyna Andrzej
Wytrzyszczak Iwona
Zdunik Julian
Zdziarski Andrzej
Zieba Stanislaw
Ziolkowski Janusz
Ziolkowski Krzysztof
Zola Stanislaw

Portugal

Braga da Costa Campos L M
Cabrita Ezequiel
Coelho Balsa Mario C
da Costa Antonio A
da Silva A V C S
Lago Maria Teresa V T
Magalhaes Antonio A S
Marques Manuel N
Monteiro Mario J P F G
Nunes Rogerio S de sousa
Osorio Isabel Maria T V P
Osorio Jose J S P
Pascoal Antonio J B
Santos Filipe D
Taborda Jose Rosa
Tavares J T l
Vicente Raimundo O

Romania

Barbosu Mihail
Botez Elvira
Breahna Iulian
Burs Lucian
Carsmaru Maria M
Chis Gheorghe Dorin
Cristescu Cornelia G
Dinescu A
Dinulescu Simona
Dramba C
Dumitrache Christana
Dumitrescu Alexandru
Ghizaru Mihai
Imbroane Alexandru
Lungu Nicolaie
Maris Georgeta
Mihaila Ieronim
Mioc Vasile
Oprescu Gabriela
Oproiu Tiberiu
Pal Arpad
Parv Bazil
Pop Vasile
Popescu Cristina Carmen
Popescu Petre
Predeanu Irina
Radu Eugenia
Rusu I
Rusu L
Stanila George
Stavinschi Magdalena
Suran Marian Doru
Tifrea Emilia
Todoran Ioan
Toro Tibor
Ureche Vasile
Vass Gheorghe

Russia

Abalakin Victor K
Afanas'ev Viktor L
Afanasjeva Praskovya M
Agekjan Tateos A
Akim Efraim L
Andreev Vladimir V
Andronova Anna A
Antipova Lyudmila I
Antonov Vadim A
Arkharov Arkadi A
Arkhipova Vesa P
Artyukh Vadim S
Atkinson David H
Babadzhanianc Michail K
Babin Valerij G
Bagildinskij Bronislav K
Balega Yuri Yu
Baryshev Yuri V
Batrakov Yuri V
Bazilevsky Alexandr T
Belinsky Vladimir A
Belkovich Oleg I
Belyaev Nikolaj A
Berdnikov Leonid N
Bikmaev Ilfan
Bisikalo Dmitrij V
Bisnovatyi-Kogan Grennadij S
Blinov Nikolai S
Bochkarev Nikolai G
Bogod Vladimir M
Bondarenko Ludmila N
Borozdin Konstantin
Boyarchuk Alexander A
Boyarchuk Margarita E
Brejdo Izabella I
Bronnikova Nina M
Brumberg Victor A
Burdyuzha Vladimir V
Bystrova Natalija V
Chashei Igor V
Chechetkin Valerij M
Cherepashchuk Analily M
Chernin Arthur D
Chernov Gennadij P
Chertok Ilia M
Chertoprud Vladim E
Chistyakov Vladimir E
Chugai Nikolai N
Churazov Eugene M
Dadaev Aleksandr N
Dagkesamansky Rustam D
Danilov Vladimir M
Denissenkov Pavel
Devyatkin Alexander V
Dluzhnevskaya Olga B
Dodonov Serguej
Dokuchaev Vyacheslav I
Dokuchaeva Olga D
Dolginov Arkady Z
Doroshkevich Andrei G
Doubinskij Boris A
Dravskikh Alexander F
Efremov Yuri N
Efremov Yuri I
Emelianov Nikolaj V
Eroshkin Georgiy I
Erpylev Nikolaj P
Esipov Valentin F
Fabrika Sergei
Fadeyev Yuri A
Finkelstein Andrej M
Fomenko Alexandr F
Fomichev Valeri V
Fomin Valery A
Fominov Alexandr M
Fridman Aleksey M
Frolov Mikhail S
Fursenko Margarita A
Galeev Albert A
Galibina Irini V
Galperin Yuri I
Gayazov Iskander S
Gelfreikh Georgij B
Gerasimov Igor A
Gilfanov Marat R
Ginzburg Vitaly L
Glagolevskij Juri V
Glebova Nina I
Glushneva Irina N
Gnedin Yurij N
Gontcharov Gueorgui A
Gorbatsky Vitalij G
Gorschkov Alexander G
Gosachinskij Igor V
Grachev Stanislav I
Grebenev Sergei A
Grebenikov Evgenij A
Grigorieva Virginia P
Grigorjev Victor M
Grishchuk Leonid P
Groushinsky Nikolai P
Gubanov Vadim S
Gubchenko Vladimir M
Gulyaev Albert P
Gulyaev Rudolf A
Gurshtein Alexander A
Guseva Irina S
Hagen-Thorn Vladimir A
Idlis Grigorij M
Ikhsanov Robert N
Ikhsanova Vera N
Ilin Alexei E
Ilyasov Yuri P
Imshennik Vladimir S
Ioshpa Boris A
Ipatov Alexander V
Ivanov Evgeny I
Ivanov Vsevolod V
Ivanov Kholodny Goz S
Izvekov Vladimir A
Kadla Zdenka I
Kanayev Ivan I
Karachentsev Igor D
Kardashev Nicolay S
Karitskaya Eugenia A
Karpinskij Vadim N
Khaliullin Khabibrachman F
Khokhlova Vers L
Kholshevnikov Konstatin V
Kholtygin Alexander F
Khozov Gennadij V
Khromov Gavriil S
Kim Iraida S
Kirian Tatiana R
Kiselyov Alexej A
Kislyakov Albert G
Kisseleva Tamara P
Klochkova Valentina
Kocharov Grant E
Kokurin Yurij L
Kolesov Alexander K
Komberg Boris V
Kompaneets Dmitriy A
Kononovich Edward V
Kopylov Alexander
Kopylov Ivan M
Korchak Alexander A
Korovyakovskij Yurij P
Korzhavin Anatoly
Kosin Gennadij S
Kostina Lidija D
Kostyakova Elena B
Kotelnikov Vladimir A
Krasinsky George A
Krivov Alexander
Ksanfomaliti Leonid V

Kuklin Georgly V
Kulikova Nelli V
Kumajgorodskaya Raisa N
Kumkova Irina I
Kuril'chik Vladimir N
Kurt Vitaliy G
Kutuzov Sergei A
Kuzmin Andrei V
Kuzmin Arkadii D
Larionov Mikhael G
Lavrov Mikhail I
Lavrukhina Augusta K
Leikin Grigerij A
Lipunov Vladimir M
Livshits Mikhail A
Lotova Natalia A
Lozinskaya Tat-yana A
Lozinskij Alexander M
Lukash Vladimir N
Makarov Valentine I
Makarova Elena A
Malkin Zinovy M
Malkov Oleg Yu
Malofeev Valerij M
Mandzhos Andrej V
Marov Mikhail Ya
Mashonkina Lyudmila
Maslennikov Kirill L
Massevich Alla G
Matveyenko Leonid I
Medvedev Yuriy D
Merman G A
Merman Natalia V
Mikhelson Nikolaj N
Mingaliev Marat G
Minin Igor N
Mitrofanova Lyudmila A
Mogilevskij Eh I
Molchanov Andrea P
Moroz Vasilis I
Myachin Vladimir F
Nadyozhin Dmittris K
Nagirner Dmitrij I
Nagnibeda Valery G
Nagovitsyn Yuri A
Naidenov Victor O
Naumov Vitalij A
Nefedeva Antonina I
Neizvestny Sergei
Nezlin Mikhail
Nikitin Alexsey A
Noskov Boris N
Novikov Igor D
Novikov Sergej B
Novoselov Victor S
Obridko Vladimir N
Pamyatnikh Alexsey A
Panchuk Vladimir E
Parfinenko Leonid D
Parijskij Yuri N
Pavlinsky Mikhail A
Petrov Gennadij M
Petrovskaya Irina
Petrovskaya Margarita S
Piskunov Anatoly E
Pogodin Mikhail A
Poliakov Eugene V
Polozhentsev Dimitrij D
Polyachenko Valerij L
Popov Michkail V
Popov Victor S
Porfir'ev Vladimir V
Postnov Konstantin A
Potter Heino I
Prodan Yuri I
Prokof-Eva Irina A
Pskovskij Juri P
Pushkin Sergey B
Pyatunina Tamara B
Rastorguev Alexey S
Razin Vladimir A
Rizvanov Naufal G
Rodionova Janna F
Rudnitskij Georgij M
Ruskol Eugenia L
Ryabchikova Tanya
Ryabov Yu A
Rykhlova Lidija V
Ryutova Margarita P
Rzhiga Oleg N
Safronov Victor S
Sagdeev Roald Z
Sakhibullin Nail A
Samus Nikolai N
Sazhin Michail
Shakura Nicholaj I
Shandarin Sergei F
Shaposhnikov Vladimir E
Sharaf Sh G
Sharov A S
Shcheglov P V
Shcherbina-Samojlova I
Sheffer Eugene K
Shefov Nicolai N
Shematovich Valery I
Shevchenko Vladislav V
Shishov Vladimir I
Shitov Yuri P
Sholomitsky G B
Shor Viktor A
Shulov Oleg S
Shustov Boris M
Sidorenkov Nikolay S
Sil'chenko Olga K
Silant-ev Nikolai
Sitnik G F
Skripnichenko Vladimir
Skulachov Dmitry
Slysh Viacheslav I
Smirnov Michael
Smol-Kov Gennadij Ya
Snezhko Leonid I
Sobolev Andrej M
Sobolev V V
Soboleva N S
Sochilina Alla S
Sokolov Viktor G
Sokolsky Andrej G
Solovaya Nina A
Somov Boris V
Sorochenko R L
Sorokin Nikolai A
Stankevich Kazimir S
Stepanov Alexander V
Stotskii Alexander A
Strukov Igor A
Sunyaev Rashid A
Surdin Vladimir G
Svechnikova Maria A
Sveshnikov Mikhail
Tatevyan S K
Tchouikova Nadejda A
Teplitskaya R B
Terekhov Oleg V
Terentjeva Alexandra K
Tokarev Yurij V
Tokovinin Andrej A
Troitsky V S
Trushkin Sergei Anatol'evich
Trutse Yu L
Tsarevsky Gregory
Tseytlin Naum M
Tsygan Anatolii I
Tutukov A V
Udal'tsov V A
Urasin Lirik A
Vainstein L A
Val'tts Irina E
Valyaev Valery
Varshalovich Dimitrij
Vashkov'Yak Sof-Ya N
Vasileva Galina J
Vassiliev Nikolay N
Vereshchagin Sergei V
Verkhodanov Oleg

Vitinskij Yurij I
Vityazev Andrei
Vityazev Veneamin V
Voshchinnikov Nicolai
Vyalshin Gennadij F
Yakovlev Dmitry
Yarov-Yarovoj M S
Yavnel Alexander A
Yershov Vladimir N
Yudin Boris F
Yungelson Lev R
Zabolotny Vladimir F
Zagretdinov Renat
Zaitsev Valerii V
Zakharova Polina E
Zasov Anatole V
Zharkov Vladimir N
Zhelezniakov Vladimir V
Zhevakin S A
Zhugzhda Yuzef D
Zlotnik Elena Ya

Saudi Arabia

Goharji Adan
Al-Malki M B
Almleaky Yasseen
Basurah Hassan
Boydag-Yildizdogdu F S
Brosterhus Elmer B F
Eker Zeki
Hamzaoglu Esat E H
Malawi Abdulrahman
Niazy Adnan Mohammad
Topaktas Latif A

Slovak R

Antalova Anna
Chochol Drahomir
Dzifcakova Elena
Hajduk Anton
Hajdukova Maria
Hefty Jan
Hric Ladislav
Kapisinsky Igor
Klocok Lubomir
Knoska Stefan
Kresakova Margita
Kucera Ales
Kulcar Ladislav
Minarovjech Milan
Palus Pavel
Pittich Eduard M
Porubcan Vladimir
Rusin Vojtech
Rybansky Milan
Saniga Metod
Skopal Augustin
Svoren Jan
Sykora Julius
Tremko Jozef
Ziznovsky Jozef
Zverko Juraj
Zvolankova Judita

South Africa

Baart Edward E.
Balona Luis Antero
Beesham Aroonkumar
Block David Lazar
Buckley David
Caldwell John A R
Cillie G G
Cooper Timothy
Cousins A W J
de Jager Gerhard
de Jager Ocker C
Dunsby Peter
Ellis George F R
Engelbrecht Christian
Fairall Anthony P
Feast Michael
Flanagan Claire Susan
Gaylard Michael John
Glass Ian Stewart
Hellaby Charles William
Hers Jan
Hughes Arthur R W
Hutcheon Richard J
Jarrett Alan H
Jonas Justin Leonard
Kilkenny David
Koen Marthinus
Kurtz Donald Wayne
Laney Clifton D
Lloyd Evans Thomas Harry
Martinez Peter
Menzies John W
Nicolson George D
O'Donoghue Darragh
Overbeek Michiel Daniel
Poole Graham
Raubenheimer Barend C
Rijsdijk Case
Smits Derck P
Soltynski Maciej
Stobie Robert S
Stoker Pieter H
van der Walt D J
Walraven Th
Warner Brian
Whitelock Patricia Ann

Spain

Abad Alberto J
Alcolea Javier
Alfaro Emilio Javier
Alvarez Pedro
Anglada Guillem
Arribas Santiago
Bachiller Rafael
Ballester Jose Luis
Barcia Alberto

Barcons Xavier
Battaner Eduardo
Beckman John E
Belmonte Aviles J A
Benavente Jose
Benn Chris R
Bernabeu Guillermo
Betancort-Rijo Juan
Boloix Rafael
Bonet Jose A
Bravo Eduardo
Buitrago Jesus
Bujarrabal Valentin
Calvo Manuel
Camarena Badia Vicente
Campos Ana
Canal Ramon M
Cardus Almeda J O
Casares Valazquez Jorge
Castaneda Hector
Catala Poch M A
Catalan Manuel
Centurion Martin Miriam
Cepa Jordi
Cernicharo Jose
Cid Palacios Rafael
Clavel Jean
Clement Rosa Maria
Codina Vidal J M
Collados Manuel
Colomer Francisco
Coma Juan Carlos
Cornide Manuel
Costa Victor
Cubarsi Rafael
de Castro Angel
de Castro Elisa
De Garcia Maria J M
de Pascual Martinez M
de San Francisco Eiroa C
de Vicente Pablo
Del Olmo Orozco A
Del Rio Gerardo
Del Toro Iniesta Jose
Delgado Antonio Jesus
Diaz Angeles Isabel
Docobo Jose A Durantez
Elipe Sanchez Antonio
Estalella Robert
Esteban Lopez Cesara
Fabregat Juan
Fernandez-Figueroa M J
Ferrer Martinez Sebastian
Ferriz-Mas Antonio
Figueras Francesca
Floria Luis
Fuensalida Jimenez J
Fuente Asuncion
Galan Maximino J
Gallego Juan Daniel
Garcia Domingo
Garcia Jose I de la rosa
Garcia Lopez Ramon J
Garcia-Berro Enrique
Garcia-Burillo Santiago
Garcia-Pelayo Jose
Garrido Rafael
Garzon Francisco
Gimenez Alvaro
Gomez Gonzalez Jesus
Gonzales-Alfonso Eduardo
Gonzalez Camacho Antonio
Gonzalez Serrano J I
Gonzalez-Riestra R
Gorgas Garcia Javier
Harlaftis Emilios
Hernandez-Pajares Manuel
Hernanz Margarita
Herrero Davo Artemio
Hidalgo Miguel A
Isern Jorge
Jaschek Carlos O R
Jimenez Mancebo A j
Jordi Carme
Jourdain de Muizon M
Kessler Martin F
Labay Javier
Lahulla J Fornies
Lapuente Pilar Ruiz
Lara Martin
Laureijs Rene J
Lazaro Carlos
Ling J
Loiseau Nora
Lopez De Coca M D P
Lopez Rosario
Lopez-Arroyo M
Lopez-Gonzalez Maria J
Lopez-Moreno Jose Juan
Lopez-Puertas Manuel
Lopez-Valverde M A
Luri Xavier
Mampaso Antonio
Manchado Arturo
Marcaide Juan-Maria
Martin Eduardo
Martin-Diaz Carlos
Martin-Loron M
Martin-Pintado Jesus
Martinez Pillet Valentin
Martinez Roger Carlos
Martinez-Gonzalez E
Masegosa Gallego J
Massaguer Josep
Mediavilla Evencio
Medina Jose
Moles Mariano J
Molina Antonio
Morales-Duran Carmen
Moreno Fernando
Moreno-Insertis Fernando
Muinos Jose L
Munoz-Tunon Casiana
Nunez Jorge
Orte Alberto
Orus Juan J
Palle Pere-Lluis
Paredes Jose Maria
Pensado Jose
Perea-Duarte Jaime D
Perez Enrique
Perez Fournon Ismael
Planesas Pere
Prieto Cristina
Prieto Mercedes
Prusti Timo
Quintana Jose M
Rebolo Rafael
Reglero-Velasco Victor
Rego Fernandez M
Regulo Clara
Roca Cortes Teodoro
Rodrigo Rafael
Rodriguez Eloy
Rodriguez Hildago Ines l
Rodriguez-Eillamil R
Rodriguez-Espinosa Jose M
Rodriguez-Franco Arturo
Rolland Angel
Romero Perez M Pilar
Ros Rosa M
Rossello Gaspar
Rozas Maite
Rutten Renee G M
Sabau-Graziati Lola
Sala Ferran
Salazar Antonio
Salvador-Sole Eduardo
Sanahuja Blas
Sanchez Almeida Jorge
Sanchez Filomeno
Sanchez Francisco
Sanchez Manuel
Sanchez-Lavega Agustin
Sanchez-Saavedra M Luisa

Sanroma Manuel
Sansaturio Maria E
Sanz I Subirana Jaume
Sanz Jose L
Schwarz Hugo E
Sein-Echaluce M Luisa
Sequeiros Juan
Sevilla Miguel J
Simo Charles
Steppe Hans
Talavera A
Tenorio-Tagle Guillermo
Thum Clemens
Torra Jordi
Torrejon Jose Miguel
Torrelles Jose M
Torroja J
Trujillo Bueno Javier
Trullols I Farreny Enric
Vallejo Miguel
Vazquez Manuel
Verdes-Montenegro Lourdes
Vila Samuel C
Vilchez Medina Jose M
Vives Teodoro Jose
Walton Nicholas A
Wamsteker Willem
Watson Robert
Zambrano Alejandro
Zamorano Jaime

Sweden

Abramowicz Marek
Adolfsson Tord
Ardeberg Arne L
Artymowicz Pawel
Baath Lars B
Bergman Per-Goeran
Bergvall Nils Ake Sigvard
Bjornsson Claes-Ingvar
Black John Harry
Booth Roy S
Carlqvist Per A
Cato B Torgny
Conway John
Dravins Dainis
Edlen Bengt
Edvardsson Bengt
Ellder Joel
Elvius Aina M
Eriksson Kjell
Faelthammar Carl Gunne
Fransson Claes
Fredga Kerstin
Frisk Urban
Gahm Goesta F
Gustafsson Bengt
Hansson Nils
Hjalmarson Ake G
Hoegbom Jan A
Hoeglund Bertil
Holmberg Erik B
Joersaeter Steven
Johansson Lars Erik B
Johansson Lennart
Johansson Sveneric
Kiselman Dan
Kollberg Erik L
Kristenson Henrik
Kristiansson Krister
Lagerkvist Claes-Ingvar
Lagerqvist Albin
Larsson Stefan
Larsson-Leander Gunnar
Lauberts Andris
Laurent Bertel E
Lehnert B P
Lindblad Bertil A
Lindblad Per Olof
Linde Peter
Lindegren Lennart
Liseau Rene
Loden Kerstin R
Loden Lars Olof
Lundqvist Peter
Lundstedt Henrik
Lundstrom Ingemar
Magnusson Per
Mellema Garrelt
Nilson Peter
Nordh Lennart H
Nyman Lars-Aake
Oehman Yngve
Oja Tarmo
Olberg Michael
Olofsson Goeran S
Olofsson Hans
Olofsson Kjell
Piskunov Nikolai E
Plez Bertrand
Poutanen Juri
Raadu Michael A
Reiz Anders
Rickman Hans
Roennaeng Bernt O
Roslund Curt
Rosquist Kjell
Rydbeck Gustaf H B
Rydbeck Olof E H
Sandqvist Aage
Scharmer Goeran Bjarne
Sinnerstad Ulf E
Soderhjelm Staffan
Stenholm Bjoern
Stenholm Lars
Sundelius Bjoern
Sundman Anita
Svensson Roland
Swensson John W
Thomasson Magnus
van Groningen Ernst
Wahlgren Glenn Michael
Wallenquist Aake A E
Wallinder Frederick
Westerlund Bengt E
Wiedling Tor
Wiklind Tommy
Winnberg Anders
Wramdemark Stig S

Switzerland

Babel Jacques
Bartholdi Paul
Becker Wilhelm
Benz Arnold O
Berthet Stephane
Beutler Gerhard
Binggeli Bruno
Blecha Andre Boris G
Bochsler Peter
Bouvier Pierre
Burki Gilbert
Buser Roland

Cameron Luzius Martin
Chmielewski Yves
Courvoisier Thierry J-L
Cramer Noel
Debrunner Hermann
Dressler Kurt
Dubath Pierre
Duerst Johannes
Fenkart Rolf P
Froehlich Claus
Gautschy Alfred
Geiss Johannes
Gerhard Ortwin
Golay Marcel
Goy Gerald
Grenon Michel
Guedel Manuel
Hauck Bernard
Labhardt Lukas
Maeder Andre
Maetzler Christian
Magun Andreas
Martinet Louis
Mayor Michel
Megevand Denis
Mermilliod Jean-Claude
Meynet Georges
Muerset Urs
Nicolet Bernard
North Pierre
Nussbaumer Harry
Paltani Stephane
Pfenniger Daniel
Povel Hanspeter
Queloz Didier
Rufener Fredy G
Schaller Gerard
Schanda Erwin
Schild Hansruedi
Schildknecht Thomas
Schmutz Werner
Schuler Walter
Solanki Sami K
Spaenhauer Andreas Martin
Steinlin Uli
Stenflo Jan O
Straumann Norbert
Tammann Gustav Andreas
Thielemann Friedrich-Karl
Trefzger Charles F
Udry Stephane
Vogel Manfred
von Steiger Rudolf
Walder Rolf
Waldmeier Max
Walter Roland
Wild Paul
Zelenka Antoine

Tajikistan

Babadzhanov Pulat B
Bibarsov Ravil-sh
Ibadinov Khursandkul
Ibadov Subhon
Kiselev Nikolai N
Minikulov Nasridin K
Sahibov Firuz H

Turkey

Akan Mustafa Can
Akcayli Melek M A
Akyol Mustafa Unal
Alpar Ali
Aslan Zeki
Atac Tamer
Avcioglu Kamuran
Aydin Cemal
Balli Edibe
Bolcal Cetin
Bozkurt Sukru
Demircan Osman
Denizman Levent
Derman I Ethem
Dizer Muammer
Dogan Nadir
Engin Semanur
Enginol Turan B
Ercan E Nihal
Ertan A Yener
Eskioglu A Nihat
Evren Serdar
Ezer-Eryurt Dilhan
Goelbasi Orhan
Gokdogan Nuzhet
Gudur N
Gulmen Omur
Gulsecen Hulusi
Hazer S
Hotinli Metin
Ibanoglu Cafir
Kandemir Guelcin
Karaali Salih
Keskin Varol
Kiral Adnan
Kirbiyik Halil
Kiziloglu Nilguen
Kiziloglu Uemit
Kocer Durcun
Marsoglu A
Mentese Huseyin
Oekten Adnan
Oezel Mehmet Emin
Oezkan Mustafa Tuerker
Ozguc Atila
Pekuenlue E Rennan
Saygac A Talat
Sezer Cengiz
Tektunali H Gokmen
Tufekcioglu Zeki
Tunca Zeynel
Yilmaz Fatma
Yilmaz Nihal

UK

Aarseth Sverre J
Adam Madge G
Adams David J
Adamson Andrew
Ade Peter A R
Aitken David K

Albinson James
Alexander Paul
Allan Peter M
Allen Anthony John
Anderson Bryan
Andrews David A
Andrews Peter J
Ardavan Houshang
Argue A Noel
Argyle Robert William
Atherton Paul David
Axon David
Bailey Mark Edward
Baker Joanne
Baldwin John E
Barlow Michael J
Barocas Vinicio
Barrow John David
Barrow Richard F
Barstow Martin Adrian
Baruch John
Bastin John A
Bates Brian
Bath Geoffrey T
Beale John S
Beggs Denis W
Bell Kenneth Lloyd
Bell Steven
Bell Burnell Jocelyn S
Berger Mitchell
Berrington Keith Adrian
Bingham Richard G
Binney James J
Birkinshaw Mark
Black Adam Robert S
Blackman Clinton Paul
Blackwell Donald E
Bode Michael F
Boksenberg Alec
Bondi Hermann
Bonnor W B
Boyd Robert L F
Boyle Stephen
Brand Peter W J l
Brandenburg Axel
Branduardi-Raymont G
Branson Nicholas J B A
Bridgeland Michael
Bromage Gordon E
Brookes Clive J
Brown John C
Brown Paul James Frank
Browne Ian W A
Browning Philippa
Bruck Hermann A
Bruck Mary T
Bryce Myfanwy
Bunclark Peter Stephen
Burgess Alan
Burgess David D
Burton William M
Butchins Sydney Adair
Butler C John
Byrne Patrick B
Cameron Andrew Collier
Campbell James W
Cargill Peter J
Carr Bernard John
Carson T R
Carswell Robert F
Catchpole Robin M
Chan Siu Kuen Josphine
Chandler Claire
Charles Philip Allan
Clark David H
Clarke Catherine
Clarke David
Clegg Peter E
Clegg Robin E S
Clube S V M
Coe Malcolm
Cohen Raymond J
Coles Peter
Conway Robin G
Cook Alan H
Cooke B A
Cooke John Alan
Couper Heather Miss
Crawford Ian Andrew
Croom David L
Crowther Paul
Cruise Adrian Michael
Culhane Leonard
Czerny Michal
Daintree Edward J
Davidson William
Davies Paul Charles W
Davies Rodney D
Davies Roger L
Davis Jonathan Ivor
Davis Richard J
de Groot Mart
Dennison P A
Dewhirst David W
Dhillon Vikram Singh
Dickens Robert J
Diego Francisco
Disney Michael J
Diver Declan Andrew
Donnison John Richard
Dormand John Richard
Dougherty Sean
Downes Ann Juliet B
Doyle John Gerard
Drew Janet
Duffett-Smith Peter James
Dufton Philip L
Dunlop James
Dunlop Storm
Dworetsky Michael M
Dyson John E
Eccles Michael J
Edge Alastair
Edmunds Michael Geoffrey
Edwin Patricia
Edwin Roger P
Efstathiou George
Eggleton Peter P
Elliott Kenneth H
Ellis Richard S
Elsmore Bruce
Elson Rebecca Anne Wood
Elsworth Yvonne
Emerson David
Emerson James P
Evangelidis E
Evans Aneurin
Evans Dafydd Wyn
Evans Kenton Dower
Evans Roger G
Evans Wyn
Fabian Andrew C
Falle Samuel A
Fawell Derek R
Fiedler Russell
Field David
Fielder Gilbert
Fitzsimmons Alan
Flower David R
Fludra Andrzej
Fong Richard
Forbes Duncan Alan
Fox W E
Frenk Carlos S
Furniss Ian
Gadsden Michael
Galsgaard Klaus
Garrington Simon
Garton W R S
Geake John E
Gent Hubert
Gietzen Joseph W
Gilmore Gerard Francis
Glencross William M
Godwin Jon Gunnar

Goldsworthy Frederick A
Gondhalekar Prabhakar
Gough Douglas O
Grainger John F
Grant Ian P
Gray Norman
Green David
Green Robin M
Green Simon F
Griffin Ian Paul
Griffin Matthew J
Griffin Rita E M
Griffin Roger F
Griffiths William K
Guest John E
Gull Stephen F
Guthrie Bruce N G
Hadley Brian W
Hall Andrew Norman
Hanbury Brown Robert
Haniff Christopher
Harper David
Harris Stella
Harrison Richard A
Hassall Barbara J M
Haswell Carole A
Hawarden Timothy G
Hawking Stephen W
Hawkins Michael R S
Hayward John
Hazard Cyril
Heavens Alan
Heddle Douglas W O
Heggie Douglas C
Hellier Coel
Hermans Dirk
Hewett Paul
Hewish Antony
Hey James Stanley
Hide Raymond
Hilditch Ronald W
Hill Philip W
Hills Richard E
Hoare Melvin
Hohenkerk Catherine
Holloway Nigel J
Hood Alan
Hood Alan
Horne Keith
Hoskin Michael A
Hough James
Hough James
Howarth Ian Donald
Hoyle Fred
Hughes David W
Hughes Shaun
Humphries Colin M
Hunt G E
Hysom Edmund J
Inglis Michael
Ireland Jack
Ireland John G
Irwin Michael John
Isaak George R
Jackson John Charles
Jackson Neal
James John F
James Richard A
Jameson Richard F
Jardine Moira Mary
Jeffery Christopher S
Jenkins Charles R
Jennings R E
Jennison Roger C
Johnstone Roderick
Jones Barrie W
Jones Derek H P
Jones Michael
Jordan Carole
Jorden Paul Richard
Joseph Robert D
Jupp Alan H
Kahn Franz D
Kanbur Shashi
Kenderdine Sidney
Kibblewhite Edward J
King Andrew R
King David Leonard
King Henry C
King-Hele Desmond G
Kingston Arthur E
Kitchin Christopher R
Knapen Johan Hendrik
Kolb Ulrich
Kroto Harold
Lahav Ofer
Laing Robert
Lancaster Brown Peter
Lang James
Lasenby Anthony
Lawrence Andrew
Leahy J Patrick
Lee Terence J
Liddle Andrew
Little Leslie T
Liu Xiaowei
Lloyd Huw
Longair Malcolm S
Longbottom Aaron
Lovell Sir Bernard
Lucey John
Lynas-Gray Anthony E
Lynden-Bell Donald
Lyne Andrew G
MacCallum Malcolm A H
MacDonald Geoffrey H
MacGillivray Harvey T
MacKay Craig D
MacKinnon Alexander L
Maddison Ronald Ch
Maddox Stephen
Major John
Mallia Edward A
Marek John
Marsh Julian C D
Marsh Thomas
Martin Anthony R
Martin Derek H
Martin William L
Masheder Michael
Mason Helen E
Mason John William
Mason Keith Owen
Matheson David Nicholas
McDonnell J A M
McHardy Ian Michael
McMahon Richard
McMullan Dennis
McNally Derek
McWhirter R W Peter
Meaburn John
Meadows A Jack
Meikle William P S
Meiksin Avery Abraham
Message Philip J
Mestel Leon
Miles Howard G
Millar Thomas J
Miller John C
Mills Allan A
Mitton Jacqueline
Mitton Simon
Moffatt Henry Keith
Monteiro Tania S
Moore Daniel R
Moore Patrick
Morgan Brian Lealan
Morgan David H
Morison Ian
Morris Michael C
Morrison Leslie V
Moss Christopher
Moss David L
Murdin Paul G
Murray C Andrew

Murray Carl D
Murray John B
Muxlow Thomas
Nair Sunita
Nakariakov Valery
Nandy Kashinath
Napier William M
Nelson Alistair H
Neukirch Thomas
New Roger
Newton Gavin
Nicolson Iain
O'Brien Paul Thomas
O'Brien Tim
Onuora Lesley Irene
Osborne John L
Osborne Julian P
Padman Rachael
Page Clive G
Pagel Bernard E J
Palmer Philip
Papaloizou John C B
Parker Edward A
Parker Neil
Parker Quentin
Parkinson John H
Paxton Harold J B R
Peach Gillian
Peach John V
Peacock John Andrew
Pearce Gillian
Pedlar Alan
Penny Alan John
Penston Margaret
Perry Judith J
Petford A David
Pettini Max
Phillips John Peter
Phillips Kenneth J H
Pike Christopher David
Pilkington John D H
Pillinger Colin
Podsiadlowski Philipp
Pollacco Don
Ponman Trevor
Ponsonby John E B
Pooley Guy
Pounds Kenneth A
Priest Eric R
Pringle James E
Prinja Raman
Pye John P
Quenby John J
Raine Derek J
Rapley Christopher G
Rawlings Jonathan
Rawlings Steven
Reay Newrick K
Rees Martin J
Regoes Enikoe
Reig Pablo
Richardson Kevin J
Richardson Lorna Logan
Richer John
Rijnbeek Richard
Riley Julia M
Ring James
Robertson John Alistair
Robertson Norna
Robinson Andrew
Robinson William J
Roche Patrick F
Ronan Colin A
Rowan-Robinson Michael
Rowson Barrie
Roxburgh Ian W
Roy Archie E
Ruffert Maximilian
Sanford Peter William
Saunders Richard D E
Scarrott Stanley M
Scheuer Peter A G
Schwartz Steven Jay
Scott Paul F
Seaton Michael J
Seymour P A H
Shakeshaft John R
Shallis Michael J
Shanks Thomas
Sharples Ray
Shone David
Sim Mary E
Simmons John Francis l
Simnett George M
Simons Stuart
Sinclair Andrew T
Sisson George M
Skillen Ian
Skilling John
Skinner Gerald
Sleath John
Smalley Barry
Smith Francis Graham
Smith Geoffrey
Smith Humphry M
Smith Keith Colin
Smith Linda J
Smith Robert Connon
Smith Rodney M
Smyth Michael J
Somerville William B
Sorensen Soren-Aksel
Speer R J
Spencer Ralph E
Stannard David
Steele Colin D C
Stephenson F Richard
Stevens Ian
Steves Bonita Alice
Stewart John Malcolm
Stewart Paul
Stickland David J
Summers Hugh P
Sutherland William
Sweet Peter A
Sykes-Hart Avril B
Sylvester Roger
Tavakol Reza
Taylor Donald Boggia
Ter Haar Dirk
Terlevich Elena
Terlevich Roberto Juan
Thoburn Christine
Thomas David V
Thomas Peter A
Thomasson Peter
Thomson Robert
Tout Christopher
Tozer David C
Tritton Keith P
Tritton Susan Barbara
Turner Martin J l
Twiss R Q
Tworkowski Andrzej S
Unger Stephen
van der Raay Herman B
van Leeuwen Floor
Veck Nicholas
Vekstein Gregory
Walker David Douglas
Walker Edward N
Walker Helen J
Walker Ian Walter
Wall J W
Wall Jasper V
Wallace Patrick T
Wallis Max K
Walsh Dennis
Walsh Robert
Ward Henry
Ward Martin John
Warner Peter J
Warwick Robert S
Watson Micheal G
Weiss Nigel O

Wellgate G Bernard
White Glenn J
White Simon David Manion
Whitrow Gerald James
Whitworth Anthony Peter
Wickramasinghe N C
Wilkins George A
Wilkinson Althea
Wilkinson Peter N
Williams David A
Williams Iwan P
Williams Peredur M
Williams Robin
Willis Allan J
Willmore A Peter
Willstrop Roderick V
Wilson Lionel
Wilson Michael John
Wilson Robert
Woan Graham
Wolfendale Arnold W
Wolstencroft Ramon D
Wood Janet H
Wood Roger
Woolfson Michael M
Worrall Diana Mary
Worrall Gordon
Worswick Susan
Wright Andrew
Yallop Bernard D
Zarnecki Jan Charles
Zharkova Valentina
Zijlstra Albert
Zuiderwijk Edwardus J

Ukraine

Abramenko Valentina
Akimov Leonid
Alexandrov Yuri V
Andrienko Dmitry A
Andrievsky Sergei
Andronov Ivan
Babin Arthur
Baranovsky Edward A
Berczik Peter
Bratijchuk Matrona V
Braude Semion Ya Ag
Bruns Andrey V
Cherednichenko V I
Chernykh N S
Denishchik Yurii
Didkovsky Lionid
Dobronravin Peter
Duma Dmitrij P
Dzubenko Nikolai
Efimov Yuri
Epishev Vitali P
Fedorov Petro
Fomin Piotr Ivanovich
Fomin Valery
Gershberg R E
Golovatyj Volodymyr
Gopasyuk S I
Gor'kavyi Nikolai
Gordon Isaac M
Gozhy Adam
Grinin Vladimir P
Hnatyk Bohdan
Ivanchuk Victor I
Ivchenko Vasily
Izotov Yuri
Kalenichenko Valentin
Karachentseva Valentina
Karetnikov Valentin G R
Kashscheev B L
Kharchenko Nina
Kharin Arkadiy S
Kislyuk Vitalij S
Klymyshyn I A
Kolchinskij I G
Kolesnik Igor G
Kolesnik L N
Komarov N S
Konopleva Varvara P
Konovalenko Olexandr
Kontorovich Victor
Korsun Alla
Kostik Roman I
Kotov Valery
Koval I K
Kravchuk Sergei
Kruchinenko Vitaliy G
Kryvodubskyj Valery
Kurochka L N
Litvinenko Leonid N
Lozitskij Vsevolod
Lupishko Dmitrij F
Lyubimkov Leonid S
Lyuty Victor M
Makarenko Ekaterina N
Medvedev Yuri A
Men A V
Mironov Nikolay T
Mishenina Tamara
Moiseev I G
Molotaj Olexandr
Morozhenko A V
Morozhenko N N
Musatenko Sergij
Nesterov Nikolai S
Novosyadlyj Bohdan
Onegina A B
Orlov Mikhail
Parnovsky Sergei
Pavlenko Elena
Pavlenko Yakov V
Petrov G M
Petrov Peter P
Pilyugin Leonid
Pinigin Gennadij I
Polosukhina-Chuvaeva Nina
Prokof'eva Valentina V
Pronik I I
Pronik V I
Pugach Alexander F
Rachkovsky D N
Romanchuk Pavel R
Romanov Yuri S
Savanov Igor S
Shakhovskaya Nadejda I
Shakhovskoj Nikolay M
Shcherbakov Alexander
Shchukina Nataliya
Shkuratov Yurii
Shul-man L M
Shulga Valery
Silich Sergey
Skulskyj Mychajlo Y
Sodin Leonid
Sokolov Konstantin
Stepanian A A
Stepanian N N
Steshenko N V
Tarady Vladimir K
Telnyuk-Adamchuk V
Terebizh Valery Yu
Tsap T T
Volyanskaya Margarita Yu
Voroshilov V I
Yanovitskij Edgard G
Yatskiv Ya S
Zakhozhaj Volodimir
Zhdanov Valery

Zhilyaev Boris
Zosimovich Irina D
Romero Perez M Pilar
Ros Rosa M
Rossello Gaspar
Rozas Maite
Rutten Renee G M
Sabau-Graziati Lola
Sala Ferran
Salazar Antonio
Salvador-Sole Eduardo
Sanahuja Blas
Sanchez Almeida Jorge
Sanchez Filomeno
Sanchez Francisco
Sanchez Manuel
Sanchez-Lavega Agustin
Sanchez-Saavedra M Luisa
Sanroma Manuel
Sansaturio Maria E
Sanz I Subirana Jaume
Sanz Jose L
Schwarz Hugo E
Sein-Echaluce M'Luisa
Sequeiros Juan
Sevilla Miguel J
Simo Charles
Steppe Hans
Talavera A
Tenorio-Tagle Guillermo
Thum Clemens
Torra Jordi
Torrejon Jose Miguel
Torrelles Jose M
Torroja J
Trujillo Bueno Javier
Trullols I Farreny Enric
Vallejo Miguel
Vazquez Manuel
Verdes-Montenegro Lourdes
Vila Samuel C
Vilchez Medina Jose M
Vives Teodoro Jose
Walton Nicholas A
Wamsteker Willem
Watson Robert
Zambrano Alejandro
Zamorano Jaime

Uruguay

Abad Carlos
Fernandez Julio A
Gallardo Castro Carlos Tabare
Magris C Gladis
Tancredi Gonzalo

USA

A'Hearn Michael F
Aannestad Per Arne
Abbott David C
Ables Harold D
Abt Helmut A
Acton Loren W
Adams Fred
Adams Mark T
Adams Thomas F
Adams James H
Adelman Saul J
Adler David Scott
Ahluwalia Harjit Singh
Ahmad Imad Aldean
Aizenman Morris L
Ake Thomas Bellis
Albers Henry
Alexander David R
Alexander Joseph K
Allan David W
Allen Ronald J
Aller Hugh D
Aller Lawrence Hugh
Aller Margo F
Alley Carrol O
Alpher Ralph Asher
Altrock Richard C
Altschuler Daniel R
Altschuler Martin D
Alvarez Manuel
Ambruster Carol
Anand S P S
Anderson Christopher M
Anderson Kinsey A
Anderson Kurt S
Andersson Bengt Goeran
Angel J Roger P
Angione Ronald J
Anosova Joanna
Anthony-Twarog Barbara J
Antiochos Spiro Kosta
Aparicio Antonio
Appleby John F
Appleton Philip Noel
Armandroff Taft E
Armstrong John Thomas
Arnett W David
Arnold James R
Arny Thomas T
Arons Jonathan
Arthur David W G
Aschwanden Markus
Aspin Colin
Assousa George Elias
Athay R Grant
Atreya Sushil K
Auer Lawrence H
Augason Gordon C
Avery Lorne W
Avrett Eugene H
Ayres Thomas R
Baan Willem A
Baars Jacob W M
Babcock Horace W
Backer Donald Ch
Bagnuolo William G
Bagri Durgadas S
Bahcall John N
Bahcall Neta A
Baird Scott R
Baker James Gilbert
Baker Norman H
Balachandran Suchitra C
Balbus Steven A
Baldwin Ralph B
Balick Bruce
Baliunas Sallie L
Ball John A
Bally John
Balonek Thomas J
Bandermann L W
Bania Thomas Michael
Bardeen James M
Barden Samuel Charles
Barker Edwin S
Barker Timothy
Barnbaum Cecilia

Barnes Aaron
Barnes III Thomas G
Barnothy Jeno
Barrett Paul Everett
Barsony Mary
Barth Charles A
Barvainis Richard
Basart John P
Bash Frank N
Basri Gibor B
Bastian Timothy Stephen
Batchelor David Allen
Batson Raymond Milner
Bauer Carl A
Bauer Wendy Hagen
Baum Stef Alison
Baum William A
Baustian W W
Bautz Laura P
Baym Gordon Alan
Beard David B
Beavers Willet I
Bechtold Jill
Becker Peter Adam
Becker Robert A
Becker Robert Howard
Becker Stephen A
Beckers Jacques M
Becklin Eric E
Beebe Herbert A
Beebe Reta Faye
Beer Reinhard
Beers Timothy C
Begelman Mitchell Craig
Bell Barbara
Bell Jeffrey F
Bell Roger A
Bell III James F
Belserene Emilia P
Belton Michael J S
Bender Peter L
Benedict George F
Benford Gregory
Bennett Charles L
Benson Priscilla J
Benz Willy
Berendzen Richard
Berg Richard A
Berge Glenn L
Bergstralh Jay T
Berman Robert Hiram
Berman Vladimir
Bernat Andrew Plous
Bertschinger Edmund
Bettis Dale G
Bhavsar Suketu P
Bianchi Luciana
Bidelman William P
Bieging John Harold
Bignell R Carl
Billingham John
Billings Donald E
Binzel Richard P
Biretta John Anthony
Blades John Chris
Blaha Milan
Blair Guy Norman
Blair William P
Blanco Victor M
Blandford Roger David
Blasius Karl Richard
Bless Robert C
Blitz Leo
Blitzstein William
Bloemhof Eric E
Blondin John M
Bludman Sidney A
Blumenthal George R
Bobrowsky Matthew
Bodenheimer Peter
Boehm Karl-Heinz
Boehm-Vitense Erika
Boesgaard Ann M
Boeshaar Gregory Orth
Boggess Albert
Boggess Nancy W
Bohannan Bruce Edward
Bohlin J David
Bohlin Ralph C
Boice Daniel Craig
Boldt Elihu
Boley Forrest I
Bond Howard E
Bonsack Walter K
Book David L
Bookbinder Jay A
Bookmyer Beverly B
Booth Andrew J
Bopp Bernard W
Bord Donald John
Borderies Nicole
Boriakoff Valentin
Bornmann Patricia L
Boss Alan P
Bowell Edward L G
Bowen George H
Bower Gary Allen
Bowers Phillip F
Bowyer C Stuart
Boyce Peter B
Boynton Paul Edward
Bracewell Ronald N
Bradley Arthur J
Bradstreet David H
Branch David R
Brandt John C
Branscomb L M
Brault James W
Braun Douglas Clifford
Breakiron Lee Allen
Brecher Aviva
Brecher Kenneth
Breckinridge James B
Bregman Joel N
Bridle Alan H
Broadfoot A Lyle
Broderick John
Brodie Jean P
Brosius Jeffrey William
Broucke Roger
Brown Alexander
Brown Douglas Nason
Brown Robert Hamilton
Brown Robert L
Brownlee Donald E
Brownlee Robert R
Brucato Robert J
Brueckner Guenter E
Bruenn Stephen W
Brugel Edward W
Bruhweiler Fred C
Bruner Marilyn E
Bruning David H
Brunk William E
Buchler J Robert
Buff James S
Buhl David
Buie Marc W
Bunner Alan N
Buratti Bonnie J
Burbidge Eleanor Margaret
Burbidge Geoffrey R
Burke Bernard F
Burkhead Martin S
Burlaga Leonard F
Burns Joseph A
Burns Jack O'Neal
Burrows Adam Seth
Burrows David Nelson
Burstein David
Buscombe William
Busko Ivo C
Buta Ronald J
Buti Bimla
Butler Dennis

Butterworth Paul
Byard Paul L
Byrd Gene G
Cacciari Carla
Cahn Julius H
Caillault Jean Pierre
Calvin William H
Calzetti Daniela
Cameron Alastair G W
Cameron Winifred S
Campbell Alison
Campbell Belva G S
Campbell Donald B
Campbell Murray F
Campins Humberto
Canfield Richard C
Canizares Claude R
Capen Charles F
Capriotti Eugene R
Carbon Duane F
Carilli Christopher L
Carleton Nathaniel P
Carlson John B
Carney Bruce William
Caroff Lawrence J
Carpenter Kenneth G
Carpenter Lloyd
Carr John Sherman
Carr Thomas D
Carruthers George R
Carter William Eugene
Casertano Stefano
Cash Webster C
Cassinelli Joseph P
Castelaz Micheal W
Castelli John P
Castor John I
Caton Daniel B
Catura Richard C
Caughlan Georgeanne R
Cefola Paul J
Centrella Joan M
Cersosimo Juan Carlos
Chaffee Frederic H
Chaisson Eric J
Chamberlain Joseph M
Chamberlain Joseph W
Chambliss Carlson R
Chan Kwing Lam
Chance Kelly V
Chandra Subhash
Chapman Clark R
Chapman Gary A
Chapman Robert D
Charbonneau Paul
Chen Kwan-yu
Cheng Chung-chieh
Cheng Kwang Ping
Chevalier Roger A
Chitre Dattakumar M
Chiu Hong Yee
Chiu Liang-Tai George
Chodas Paul Winchester
Christian Carol Ann
Christiansen Wayne A
Christodoulou Dmitris
Christy James Walter
Christy Robert F
Chu You-Hua
Chubb Talbot A
Chupp Edward L
Churchwell Edward B
Ciardullo Robin
Clark Barry G
Clark Frank Oliver
Clark George W
Clark Thomas A
Clark Alfred
Clarke John T
Claussen Mark J
Clayton Donald D
Clayton Geoffrey C
Clayton Robert N
Clemens Dan P
Clifton Kenneth St
Cline Thomas L
Cliver Edward W
Cochran Anita L
Cochran William David
Cocke William John
Code Arthur D
Coffeen David L
Coffey Helen E
Cohen Jeffrey M
Cohen Judith
Cohen Leon
Cohen Marshall H
Cohen Martin
Cohen Richard S
Cohen Ross D
Cohn Haldan N
Colburn David S
Coleman Paul Henry
Colgate Stirling A
Collins George W ii
Combi Michael R
Comins Neil Francis
Condon James J
Conklin Edward K
Connolly Leo Paul
Consolmagno Guy Joseph
Conti Peter S
Cook John W
Cook Kem Holland
Corbally Christopher
Corbet Robin Henry D
Corbin Thomas Elbert
Corcoran Michael Francis
Cordes James M
Cordova France A D
Corliss C H
Corwin Harold G
Cotton William D
Coulson Iain M
Counselman Charles C
Cowan John J
Cowie Lennox Lauchlan
Cowley Anne P
Cowley Charles R
Cox Arthur N
Cox Donald P
Craine Eric Richard
Crane Patrick C
Crannell Carol Jo
Crawford David L
Crenshaw Daniel Michael
Crocker Deborah Ann
Crotts Arlin Pink
Cruikshank Dale P
Crutcher Richard M
Cudaback David D
Cudworth Kyle Mccabe
Cuffey J
Culver Roger Bruce
Cuntz Manfred
Currie Douglas G
Czyzak Stanley J
Dahn Conard Curtis
Dalgarno Alexander
Daly Ruth Agnes
Danby J M Anthony
Danford Stephen C
Danks Anthony C
Danly Laura
Dappen Werner
Datlowe Dayton
David Laurence P
Davidsen Arthur Falnes
Davidson Kris
Davies Merton E
Davila Joseph
Davis Marc
Davis Michael M
Davis Morris S
Davis Robert J

Davis Sumner P
Davis Cecil G
Davis Leverett
de Frees Douglas J
de Jonge J K
de Pater Imke
de Vincenzi Donald
de Young David S
Dearborn David Paul K
Deeming Terence J
DeGioia-Eastwood Kathleen
Deliyannis Constantine P
Delsemme Armand H
Demarque Pierre
Deming Leo Drake
Dempsey Robert C
Dennis Brian Roy
Dennison Edwin W
Dent William A
Deprit Andre
Dere Kenneth Paul
Dermer Charles Dennison
Dermott Stanley F
Despain Keith Howard
Deupree Robert G
Deutschman William A
Devinney Edward J
Devorkin David H
Dewey Rachel J
Dewitt Bryce S
Dewitt-John H
Dewitt-Morette Cecile Pr
Dhawan Vivek
Diamond Philip John
Dick Steven J
Dickel Helene R
Dickel John R
Dickey Jean O'Brien
Dickey John M
Dickinson Dale F
Dickman Robert L
Dickman Steven R
Dieter Conklin Nannielou H
Digel Seth William
Dinerstein Harriet L
Dixon Robert S
Djorgovski Stanislav
Doherty Lowell R
Dolan Joseph F
Domingo Vicente
Donahue Robert Andrew
Donn Bertram D
Doschek George A
Douglas James N
Douglass Geoffrey G
Downes Ronald A
Downs George S
Doyle Laurance R
Draine Bruce T
Drake Frank D
Drake Jeremy
Drake Stephen A
Dreher John W
Dressel Linda L
Dressler Alan
Drever Ronald W P
Drilling John S
Dryer Murray
Dubin Maurice
Dufour Reginald James
Duncan Douglas Kevin
Duncombe Raynor L
Dunham David W
Dunkelman Lawrence
Dunn Richard B
Dupree Andrea K
Dupuy David L
Durisen Richard H
Durney Bernard
Durrance Samuel T
Duthie Joseph G
Duvall Thomas L
Dwek Eli
Dyck Melvin
Dyer Edward R
Dyson Freeman J
Eaton Joel A
Eddy John A
Edelson Rick
Edmondson Frank K
Edwards Alan Ch
Edwards Terry W
Eichhorn Heinrich K
Eichler David
Eilek Jean
El Baz Farouk
Elias Nicholas
Elitzur Moshe
Elliot James L
Elmegreen Bruce Gordon
Elmegreen Debra Meloy
Elste Gunther H
Elston Wolfgang E
Elvis Martin S
Emerson Darrel Trevor
Emslie A Gordon
Endal Andrew S
Epps Harland Warren
Epstein Eugene E
Epstein Gabriel Leo
Epstein Richard I
Eshleman Von R
Eskridge Paul B
Espey Brian Russell
Esposito F Paul
Esposito Larry W
Esser Ruth
Etzel Paul B
Eubanks Thomas Marshall
Evans Ian Nigel
Evans J V
Evans John W
Evans Nancy Remage
Evans Neal J
Evans W Doyle
Ewen Harold I
Ewing Martin S
Fabbiano Giuseppina
Faber Sandra M
Fabricant Daniel G
Fahey Richard P
Falco-Acosta Emilio E
Falk Sydney W
Fall S Michael
Faller James E
Fallon Frederick W
Fang Li-zhi
Fanselow John Lyman
Faulkner John
Fay Theodore D
Fazio Giovanni G
Federman Steven Robert
Feibelman Walter A
Feigelson Eric D
Fekel Francis C
Feldman Paul Donald
Feldman Uri
Felten James E
Ferland Gary Joseph
Fesen Robert A
Fey Alan Lee
Feynman Joan
Fiala Alan D
Fichtel Carl E
Fiedler Ralph L
Field George B
Fienberg Richard T
Filippenko Alexei V
Fink Uwe
Finn Lee Samuel
Firor John W
Fischel David
Fischer Jacqueline
Fisher George Hewitt
Fisher J Richard

Fisher Philip C
Fisher Richard R
Fishman Gerald J
Fitch Walter S
Fitzpatrick Edward L
Fix John D
Flannery Brian Paul
Fleck Robert Charles
Fleischer Robert
Fleming Thomas Anthony
Fliegel Henry F
Florkowski David R
Fogarty William G
Foltz Craig B
Fomalont Edward B
Fontenla Juan Manuel
Forbes J E
Forbes Terry G
Ford Holland C Res
Ford W Kent
Forman William Richard
Forrest William John
Forster James Richard
Fort David Norman
Foster Roger S
Foukal Peter V
Fox Kenneth
Frail Dale Andrew
Frank Juhan
Franklin Fred A
Franz Otto G
Frazier Edward N
Fredrick Laurence W
Freedman Wendy L
French Richard G
Friberg Per
Friedlander Michael
Friedman Herbert
Friedman Scott David
Friel Eileen D
Friend David B
Frisch Priscilla
Frogel Jay Albert
Frost Kenneth J
Fruchter Andrew S
Fruscione Antonella
Frye Glenn M
Ftaclas Christ
Fuhr Jeffrey Robert
Gaetz Terrance J
Gaisser Thomas K
Gallagher Jean W
Gallagher John S
Gallet Roger M
Gaposchkin Edward M
Garcia Howard A
Garcia Michael R
Garfinkel Boris
Garlick George F
Garmany Catherine D
Garmire Gordon P
Garnett Donald Roy
Garstang Roy H
Gary Dale E
Gary Gilmer Allen
Gatewood George
Gatley Ian
Gaume Ralph A
Gauss F Stephen
Gaustad John E
Geballe Thomas R
Gebbie Katharine B
Gehrels Neil
Gehrels Tom
Gehrz Robert Douglas
Geldzahler Bernard J
Geller Margaret Joan
Genet Russel M
Gergely Tomas Esteban
Gerola Humberto
Gezari Daniel Ysa
Ghigo Francis D
Giampapa Mark S
Gibson David Michael
Gibson James
Giclas Henry L
Gierasch Peter J
Gies Douglas R
Gigas Detlef
Gilliland Ronald Lynn
Gillingham Peter
Gilman Peter A
Gilra Daya P
Gingerich Owen
Gioia Isabella M
Giovane Frank
Giovanelli Riccardo
Glaser Harold
Glaspey John W
Glass Billy Price
Glassgold Alfred E
Glatzmaier Gary A
Goebel John H
Gold Thomas
Goldman Martin V
Goldreich Peter
Goldsmith Donald W
Goldsmith Paul F
Goldstein Richard M
Goldstein Samuel J
Goldwire Henry C
Golub Leon
Goode Philip R
Goodman Alyssa Ann
Goodrich Robert W
Goody R M
Gopalswamy N
Gordon Courtney P
Gordon Kurtiss J
Gordon Mark A
Gorenstein Marc V
Gorenstein Paul
Gosling John T
Goss W Miller
Gott J Richard
Gottesman Stephen T
Gottlieb Carl A
Gould Robert J
Graboske Harold C
Gradie Jonathan Carey
Grady Carol Anne
Graham Eric
Graham John A
Grandi Steven Aldridge
Grasdalen Gary L
Grauer Albert D
Gray Richard O
Grayzeck Edwin J
Green Daniel William E
Green Jack
Green Louis C
Green Richard F
Greenberg Richard
Greenhill Lincoln J
Greenhouse Matthew A
Greenstein George
Greenstein Jesse L
Gregg Michael David
Gregory Stephen Albert
Greisen Kenneth I
Greyber Howard D
Griffin I P
Griffiths Richard E
Grindlay Jonathan E
Grinspoon David Harry
Gross Peter G
Gross Richard Sewart
Grossman Allen S
Grossman Lawrence
Gudehus Donald Henry
Guetter Harry Hendrik
Guhathakurta Madhulika
Guidice Donald A
Guinan Edward Francis
Gulkis Samuel

Gull Theodore R
Gunn James E
Gurman Joseph B
Gursky Herbert
Gustafson Bo A S
Guzik Joyce Ann
Gwinn Carl R
Habbal Shadia Rifai
Hackwell John A
Haddock Fred T
Hagen John P
Hagyard Mona June
Haisch Bernhard Michael
Hakkila Jon Eric
Hale Alan
Hall Donald N
Hall Douglas S
Hall R Glenn
Hallam Kenneth L
Hamilton Andrew J S
Hammel Heidi B
Hammond Gordon L
Hanisch Robert J
Hankins Timothy Hamilton
Hanner Martha S
Hansen Carl J
Hansen Richard T
Hanson Robert B
Hapke Bruce W
Hardebeck Ellen G
Hardee Philip
Harmer Charles F W
Harmer Dianne L
Harms Richard James
Harnden Frank R
Harrington J Patrick
Harris Alan William
Harris Daniel E
Harris Hugh C
Harrison Edward R
Hart Michael H
Harten Ronald H
Hartkopf William I
Hartmann Dieter H
Hartmann Lee William
Hartmann William K
Hartoog Mark Richard
Harvel Christopher Alvin
Harvey Gale A
Harvey John W
Harvey Paul Michael
Harwit Martin
Hasan Hashima
Haschick Aubrey
Hathaway David H
Hatzes Artie P
Hauschildt Peter H
Hauser Michael G
Havlen Robert J
Hawkins Gerald S
Hawkins Isabel
Hawley Suzanne Louise
Hayashi Saeko S
Hayes Donald S
Haymes Robert C
Haynes Martha P
Hazen Martha L
Heap Sara R
Heasley James Norton
Hecht James H
Heckathorn Harry M
Heckman Timothy M
Hedeman E Ruth
Heeschen David S
Hegyi Dennis J
Heiles Carl
Hein Righini Giovanna
Heintz Wulff D
Heiser Arnold M
Helfand David John
Helfer H Lawrence
Helin Eleanor Francis
Hellwig Helmut Wilhelm
Helmken Henry F
Helou George
Hemenway Mary Kay M
Hemenway Paul D
Henden Arne Anthon
Henriksen Mark Jeffrey
Henry Richard B C
Henry Richard C
Herbig George H
Herbst Eric
Herbst William
Herczeg Tibor J
Hernquist Lars Eric
Hershey John L
Hertz Paul L
Hewitt Adelaide
Hewitt Anthony V
Hibbs Albert R
Hildebrand Roger H
Hildner Ernest
Hill Frank
Hill Grant
Hill Henry Allen
Hilliard Ron
Hills Jack G
Hilton James Lindsay
Hindsley Robert Bruce
Hinkle Kenneth H
Hinners Noel W
Hintzen Paul Michael N
Hjellming Robert M
Ho Paul T P
Hoag Arthur A
Hobbs Lewis M
Hobbs Robert W
Hockey Thomas Arnold
Hodapp Klaus-Werner
Hodge Paul W
Hoeksema Jon Todd
Hoessel John Greg
Hoff Darrel Barton
Hoffleit E Dorrit
Hogan Craig J
Hogg David E
Holberg Jay B
Hollenbach David John
Hollis Jan Michael
Hollowell David Earl
Hollweg Joseph V
Holman Gordon D
Holt Stephen S
Holzer Thomas Edward
Honeycutt R Kent
Horowitz Paul
Houck James R
Houk Nancy
House Lewis L
Howard Robert F
Howard W Michael
Howard William E III
Howell Steve Bruce
Hrivnak Bruce J
Hu Esther M
Hubbard William B
Hubeny Ivan
Huchra John Peter
Huebner Walter F
Huenemoerder David P
Huggins Patrick J
Hughes John P
Hughes Philip
Huguenin G Richard
Hummel Christian Aurel
Humphreys Roberta M
Hundhausen Arthur
Hunten Donald M
Hunter Christopher
Hunter Deidre Ann
Hunter James H
Huntress Wesley T
Hurford Gordon James
Hurley Kevin C

Hurwitz Mark V
Hut Piet
Hutter Donald John
Hyder C L
Ianna Philip A
Iben Icko
Illing Rainer M E
Illingworth Garth D
Imamura James
Imhoff Catherine L
Impey Christopher D
Ipser James R
Irvine William M
Jackson Bernard V
Jackson Peter Douglas
Jackson William M
Jacobs Kenneth C
Jacobsen Theodor S
Jacobson Robert A
Jacoby George H
Jaffe Daniel T
Janes Kenneth A
Janiczek Paul M
Janka Hans Thomas
Jannuzi Buell Tomasson
Janssen Michael Allen
Jastrow Robert
Jefferies Stuart
Jefferys William H
Jenkins Edward B
Jenkins Louise F
Jenkner Helmut
Jenner David C
Jenniskens Petrus Matheus Marie
Jewell Philip R
Jiang Shengtao
Johnson Donald R
Johnson Fred M
Johnson Hollis R
Johnson Hugh M
Johnson Torrence V
Johnston Kenneth J
Jokipii J R
Jones Barbara
Jones Burton
Jones Dayton L
Jones Eric M
Jones Frank Culver
Jones Harrison Price
Jones Thomas Walter
Jordan Stuart D
Jorgensen Inger
Joselyn Jo Ann c
Joseph Charles Lynn
Joss Paul Christopher
Joy Marshall J
Judge Philip
Junkkarinen Vesa T
Jura Michael
Jurgens Raymond F
Kafatos Minas
Kaftan May A
Kahler Stephen W
Kaitchuck Ronald H
Kaler James B
Kalkofen Wolfgang
Kammeyer Peter C
Kamp Lucas Willem
Kandrup Henry Emil
Kane Sharad R
Kaplan George H
Kaplan J
Kaplan Lewis D
Karovska Margarita
Karp Alan Hersh
Karpen Judith T
Katz Jonathan I
Kaufman Michele
Kaula William M
Kawaler Steven D
Keel William C
Keenan Philip C
Keene Jocelyn Betty
Keil Klaus
Keil Stephen L
Keller Charles F
Keller Christoph U
Keller Geoffrey
Kellermann Kenneth I
Kellogg Edwin M
Kemball Athol
Kennicutt Robert C
Kent Stephen M
Kenyon Scott J
Kerr Frank J
Khare Bishun N
Kielkopf John F
Kimble Randy A
King David S
King Ivan R
King Robert Wilson
Kinman Thomas D
Kinney Anne L
Kiplinger Alan L
Kirby Kate P
Kirkpatrick Ronald C
Kirshner Robert Paul
Kissell Kenneth E
Klarmann Joseph
Klein Michael J
Klein Richard I
Kleinmann Douglas E
Klemola Arnold R
Klemperer W K
Klepczynski William J
Klimchuk James A
Klinglesmith Daniel A
Kliore Arvydas Joseph
Klock B L
Knacke Roger F
Knapp Gillian R
Kniffen Donald A
Knoelker Michael
Knowles Stephen H
Ko Hsien C
Koch David G
Koch Robert H
Kofman Lev
Kohl John L
Kojoian Gabriel
Kolb Edward W
Kondo Yoji
Konigl Arieh
Koo David C-Y
Kopp Greg
Kopp Roger A
Koratkar Anuradha P
Kormendy John
Kosovichev Alexander
Koupelis Theodoros
Kouveliotou Chryssa
Kovar N S
Kovar Robert P
Kowal Charles Thomas
Kraft Robert P
Kramer Kh N
Kraus John D
Kraushaar William L
Kreidl Tobias J N
Krieger Allen S
Krisciunas Kevin
Kriss Gerard A
Krogdahl W S
Krolik Julian H
Kron Richard G
Krumm Nathan Allyn
Krupp Edwin C
Kuhi Leonard V
Kuhn Jeffery Richard
Kuin Nicolaas Paulus M
Kuiper Thomas B H
Kulkarni Shrinivas R
Kulsrud Russell M
Kumar C Krishna

Kumar Shiv S
Kundu Mukul R
Kurfess James D
Kurtz Michael Julian
Kurucz Robert L
Kutner Marc Leslie
Kutter G Siegfried
Kwitter Karen Beth
La Bonte Barry James
Lacy Claud H
Lacy John H
Lada Charles Joseph
Laird John B
Lamb Frederick K
Lamb Richard C
Lamb Susan Ann
Lamb Donald Quincy
Lambert David L
Lampton Michael
Lande Kenneth
Landecker Peter Bruce
Landman Donald Alan
Landolt Arlo U
Lane Adair P
Lane Arthur Lonne
Lang Kenneth R
Langer George Edward
Langer William David
Langhoff Stephen Robert
Lanning Howard Hugh
Larson Harold P
Larson Richard B
Larson Stephen M
Lasala Gerald J
Lasher Gordon Jewett
Lasker Barry M
Latham David W
Latter William B
Lattimer James M
Lautman D A
Lawrence Charles R
Lawrence G M
Lawrence John Keeler
Lawrie David G
Layden Andrew Choisy
Layzer David
Lea Susan Maureen
Leacock Robert Jay
Lebofsky Larry Allen
Lebovitz Norman R
Lecar Myron
Leckrone David S
Ledlow Michael James
Lee Paul D
Leibacher John
Leighly Karen Marie
Leisawitz David
Leitherer Claus
Leonard Peter James T
Lepp Stephen H
Lester Daniel F
Leung Chun Ming
Leung Kam Ching
Levine Randolph H
Levison Harold F
Levreault Russell M
Levy Eugene H
Lewin Walter H G
Lewis Brian Murray
Lewis J S
Li Hong-Wei
Liang Edison P
Libbrecht Kenneth G
Liddell U
Liebert James W
Lieske Jay H
Likkel Lauren Jones
Lilley Edward A
Lillie Charles F
Lin Chia C
Lin Douglas N C
Lincoln J Virginia
Lindsey Charles Allan
Lingenfelter Richard E
Linke Richard Alan
Linnell Albert P
Linsky Jeffrey L
Linsley John
Lippincott Sarah Lee
Lipschutz Michael E
Lis Dariusz C
Lissauer Jack J
Liszt Harvey Steven
Little-Marenin Irene R
Littleton John E
Litvak Marvin M
Liu Sou-Yang
Livingston William C
Lo Kwok-Yung
Locanthi Dorothy Davis
Lochner James Charles
Lockman Felix J
Lockwood G Wesley
Long Knox S
Longmore Andrew J
Lonsdale Carol J
Lopes-Gautier Rosaly
Lord Steven Donald
Loren Robert Bruce
Lovas Francis John
Lovelace Richard V E
Low Boon Chye
Low Frank J
Lu Limin
Lu Phillip K
Lubowich Donald A
Luck R Earle
Lucke Peter B
Lugger Phyllis M
Lundquist Charles A
Luttermoser Donald
Lutz Barry L
Lutz Julie H
Lynch David K
Lynds Beverly T
Lynds Roger C
Ma Chopo
Macalpine Gordon M
Macchetto Ferdinando
MacConnell Darrell J
MacDonald James
Mack Peter
MacQueen Robert M
Macy William Wray
Madau Piero
Madore Barry Francis
Magnani Loris Alberto
Malina Roger Frank
Malitson Harriet H
Malkamaeki Lauri J
Malkan Matthew Arnold
Malville J Mckim
Mangum Jeffrey Gary
Mansfield Victor N
Maran Stephen P
Marcialis Robert
Margon Bruce H
Margrave Thomas Ewing
Mariska John Thomas
Mark James Wai-Kee
Markowitz William
Markworth Norman Lee
Marochnik L S
Marschall Laurence A
Marscher Alan Patrick
Marsden Brian G
Marshall Herman Lee
Marston Anthony Philip
Martin Robert N
Martin William C
Martins Donald Henry
Marvin Ursula B
Mason Glenn M
Massa Derck Louis
Massey Philip L

Masson Colin R
Matese John J
Mather John Cromwell
Mathews William G
Mathieu Robert D
Mathis John S
Matsakis Demetrios N
Matson Dennis L
Mattei Janet Akyuz
Matthews Clifford
Matthews Henry E
Matthews Thomas A
Mattox John
Matz Steven Micheal
Matzner Richard A
Mauche Christopher W
Mauersberger Rainer
Max Claire E
Maxwell Alan
Mayer Cornell H
Mayfield Earle B
Mazurek Thaddeus John
McAlister Harold A
McCabe Marie K
McCammon Dan
McCarthy Dennis D
McClain Edward F
McClintock Jeffrey E
McCluskey George E
McCord Thomas B
McCray Richard
McCrosky Richard E
McDonald Frank B
McDonough Thomas R
McElroy M B
McFadden Lucy Ann
McGaugh Stacy Sutton
McGimsey Ben Q
McGrath Melissa Ann
McGraw John T
McIntosh Patrick S
McKee Christopher F
McKinnon William Beall
McLaren Robert A
McLean Brian J
McLean Ian S
McMahan Robert Kenneth
McMillan Robert S
Mcnamara Delbert H
Mead Jaylee Montague
Meech Karen
Meeks M Littleton
Mehringer David Michael
Meier David L
Meier Robert R
Meisel David D
Melbourne William G
Melia Fulvio
Melnick Gary J
Melott Adrian L
Mendis Devamitta Asoka
Mertz Lawrence N
Meszaros Peter
Meyer David M
Meyers Karie Ann
Michalitsianos Andrew
Michel F Curtis
Mickelson Michael E
Mighell Kenneth John
Mihalas Barbara R Weibel
Mihalas Dimitri
Mikesell Alfred H
Milkey Robert W
Miller Freeman D
Miller Guy Scott
Miller Hugh R
Miller Joseph S
Miller Richard H
Milligan J E
Millikan Allan G
Millis Robert L
Misconi Nebil Yousif
Misner Charles W
Mitchell Kenneth J
Modali Sarma B
Modisette Jerry L
Moffett Thomas J
Molnar Michael R
Monet Alice K B
Monet David G
Moody Joseph Ward
Mook Delo E
Moore Elliott P
Moore Ronald L
Moos Henry Warren
Moran James M
Morgan John Adrian
Morgan Thomas H
Moriarty-Schieven G H
Morris Charles S
Morris Mark Root
Morris Steven
Morrison David
Morrison Nancy Dunlap
Morrison Philip
Morton G A
Motz Lloyd
Mouschovias Telemachos Ch
Mozurkewich David
Mueller Ivan I
Muench Guido
Mufson Stuart Lee
Mukai Koji
Mukherjee Krishna
Mullan Dermott J
Muller Richard A
Mumford George S
Mumma Michael Jon
Mundy Lee G
Munro Richard H
Murdock Thomas Lee
Murphy Brian William
Murphy Robert E
Murray Stephen David
Murray Stephen S
Murthy Jayant
Musen Peter
Mushotzky Richard
Musielak Zdzislaw E
Musman Steven
Mutel Robert Lucien
Mutschlecner J Paul
Myers Philip C
Nacozy Paul E
Narayan Ramesh
Nariai Kyoji
Nather R Edward
Navarro Julio Fernando
Neff James Edward
Neff John S
Neff Susan Gale
Neidig Donald F
Nelson Burt
Nelson George Driver
Nelson Jerry E
Nelson Robert M
Nemiroff Robert
Ness Norman F
Neugebauer Gerry
Neupert Werner M
Newburn Ray L
Newhall X X
Newman Michael John
Newsom Gerald H
Newton Robert R
Nichols-Bohlin Joy
Nicolas Kenneth Robert
Nicoll Jeffrey Fancher
Nicollier Claude
Niedner Malcolm B
Niell Arthur E
Nilsson Carl
Ninkov Zoran
Nishimura Tetsuo
Noerdlinger Peter D

Noll Keith Stephen
Noonan Thomas W
Noriega-Crespo Alberto
Norman Colin A
Noumaru Junichi
Novick Robert
Noyes Robert W
Nuth Joseph A III
O'Connell Robert West
O'Connell Robert F
O'Dea Christopher P
O'Dell Charles R
O'Dell Stephen L
O'Handley Douglas A
O'Keefe John A
O'Leary Brian T
Odell Andrew P
Odenwald Sten F
Oegelman Hakki B
Oegerle William R
Oemler Augustus
Oertel Goetz K
Oesterwinter Claus
Ofman Leon
Ogelman Hakki B
Ohashi Nagayoshi
Oka Takeshi
Oliver Bernard M
Oliver John Parker
Olivier Scot Stewart
Olmi Luca
Olowin Ronald Paul
Olsen Kenneth H
Olson Edward C
Onello Joseph S
Opendak Michael
Orlin Hyman
Ormes Jonathan F
Orton Glenn S
Osborn Wayne
Osmer Patrick S
Oster Ludwig F
Osterbrock Donald E
Ostriker Jeremiah P
Ostro Steven J
Oswalt Terry D
Owen Frazer Nelson
Owen Tobias C
Owen William Mann
Owocki Stanley Peter
Ozernoy Leonid M
Ozsvath I
Pacholczyk Andrzej G
Paciesas William S
Padovani Paolo
Paerels Frederik B S
Palmer Patrick E
Pan Xiao-Pei
Panagia Nino
Panek Robert J
Pang Kevin
Pankonin Vernon Lee
Pap Judit
Papagiannis Michael D
Papaliolios Costas
Parise Ronald A
Parker Eugene N
Parker Robert A R
Parkinson Truman
Parkinson William H
Parrish Allan
Parsons Sidney B
Partridge Robert B
Pasachoff Jay M
Pascu Dan
Pauls Thomas Albert
Payne David G
Peale Stanton J
Pearson Timothy J
Peebles P James E
Peery Benjamin F
Pellerin Charles J
Pendleton Yvonne Jean
Penzias Arno A
Perkins Francis W
Perley Richard Alan
Perry Charles L
Perry Peter M
Pesch Peter
Peters Geraldine Joan
Peters William L III
Peterson Bradley Michael
Peterson Charles John
Peterson Deane M
Peterson Laurence E
Peterson Ruth Carol
Petre Robert
Petro Larry David
Petrosian Vahe
Pettengill Gordon H
Petuchowski Samuel J
Pevtsov Alexei A
Pfeiffer Raymond J
Phelps Randy L
Philip A G Davis
Phillips John G
Phillips Thomas Gould
Pickles Andrew John
Pier Jeffrey R
Pierce A Keith
Pierce David Allen
Pilachowski Catherine
Pilcher Carl Bernard
Pines David
Pingree David
Pinsonneault Marc Howard
Pinto Philip Alfred
Pipher Judith L
Platais Imants K
Plavec Mirek J
Plavec Zdenka
Pneuman Gerald W
Pogge Richard William
Poland Arthur I
Polidan Ronald S
Ponnamperuma Cyril
Popper Daniel M
Porter Jason G
Potter Andrew E
Pradhan Anil
Prasad Sheo S
Pravdo Steven H
Prendergast Kevin H
Press William H
Preston George W
Preston Robert Arthur
Price Michael J
Price R Marcus
Price Stephan Donald
Prince Helen Dodson
Probstein R F
Proffitt Charles R
Prosser Charles Franklin
Protheroe William M
Pryor Carlton Philip
Puche Daniel
Puetter Richard C
Puschell Jeffery John
Pyper Smith Diane M
Quirk William J
Rabin Douglas Mark
Radford Simon John E
Radick Richard R
Radoski Henry R
Rafert James Bruce
Rafferty Theodore J
Ramaty Reuven
Ramsey Lawrence W
Rank David M
Rankin Joanna M
Rao K Narahari
Rasio Frederic A
Ratnatunga Kavan U
Ray James R
Raymond John Charles

Reach William
Readhead Anthony C S
Reames Donald V
Reasenberg Robert D
Reaves Gibson
Reed B Cameron
Reeves Edmond M
Reichert Gail Anne
Reid Mark Jonathan
Reid Neill
Reipurth Bo
Reitsema Harold J
Rense William A
Revelle Douglas Orson
Reyes Francisco
Reynolds John H
Reynolds Ronald J
Reynolds Stephen P
Rhodes Edward J
Rich Robert Michael
Richards Mercedes T
Richardson R S
Richstone Douglas O
Rickard James Joseph
Rickard Lee J
Ricker George R
Rickett Barnaby James
Riddle Anthony C
Riegel Kurt W
Riegler Guenter R
Rindler Wolfgang
Ringwald Frederick Arthur
Rivolo Arthur Rex
Roark Terry P
Robbins R Robert
Roberge Wayne G
Roberts David Hall
Roberts Morton S
Roberts William W
Robertson Douglas S
Robinson Edward Lewis
Robinson I
Robinson Leif J
Robinson Lloyd B
Robinson Richard D
Robson Ian E
Roddier Claude
Roddier Francois
Rodman Richard B
Roemer Elizabeth
Rogers Alan E E
Rogers Forrest J
Rogerson John B
Rogstad David H
Roman Nancy Grace
Romani Roger William
Romanishin William
Romney Jonathan D
Rood Herbert J
Rood Robert T
Roosen Robert G
Rose James Anthony
Rose William K
Rosen Edward
Rosendhal Jeffrey D
Rosner Robert
Ross Dennis K
Rots Arnold H
Rountree Janet
Rouse Carl A
Routly Paul M
Rubin Robert Howard
Rubin Vera C
Rucinski Slavek M
Ruderman Malvin A
Rudnick Lawrence
Rugge Hugo R
Rule Bruce H
Russell Christopher T
Russell Jane L
Russell John A
Rust David M
Rybicki George B
Ryder Stuart
Rydgren Alfred Eric
Sackmann Ingrid Juliana
Sadun Alberto Carlo
Safko John L
Sage Leslie John
Saha Abhijit
Sahai Raghvendra
Sahu Kailash C
Sakai Shoko
Salas Luis
Salisbury J W
Salpeter Edwin E
Salter Christopher John
Salzer John Joseph
Samec Ronald G
Sampson Douglas H
Sandage Allan
Sandell Goran Hans I
Sanders David B
Sanders Walt L
Sanders Wilton Turner III
Sandford Maxwell T II
Sandford Scott Alan
Sandmann William Henry
Sanyal Ashit
Sarajedini Ata
Sarazin Craig L
Sargent Anneila I
Sargent Wallace L W
Sartori Leo
Saslaw William C
Sasselov Dimitar D
Savage Blair D
Savedoff Malcolm P
Sawyer Constance B
Scalo John Michael
Scargle Jeffrey D
Schatten Kenneth H
Schechter Paul L
Scherb Frank
Scherrer Philip H
Schild Rudolph E
Schiller Stephen
Schlegel Eric Matthew
Schleicher David G
Schlesinger Barry M
Schloerb F Peter
Schmahl Edward J
Schmalberger Donald C
Schmelz Joan T
Schmidt Edward G
Schmidt Maarten
Schmidtke Paul C
Schneider Donald P
Schneider Glenn H
Schneider Nicholas M
Schneps Matthew H
Schnopper Herbert W
Schoolman Stephen A
Schramm David N
Schreier Ethan J
Schroeder Daniel J
Schuecking E L
Schulte D H
Schulte-Ladbeck Regina E
Schultz Alfred Bernard
Schutz Bob Ewald
Schwartz Daniel A
Schwartz Philip R
Schwartz Richard D
Schweizer Francois
Sconzo Pasquale
Scott Elizabeth L
Scott Eugene Howard
Scott John S
Scoville Nicholas Z
Searle Leonard
Sears Richard Langley
Seeds Michael August
Seeger Charles Louis III
Seeger Philip A

Segal Irving E
Seidelmann P Kenneth
Seiden Philip E
Seielstad George A
Sekanina Zdenek
Sekiguchi Kazuhiro
Sellgren Kristen
Sellwood Jerry A
Serabyn Eugene
Seward Frederick D
Shaffer David B
Shafter Allen W
Shaham Jacob
Shane William W
Shao Cheng-yuan
Shapero Donald C
Shapiro Irwin I
Shapiro Maurice M
Shapiro Stuart L
Shapley Alan H
Shara Michael
Sharma A Surjalal
Sharpless Stewart
Shaw James Scott
Shaw John H
Shawl Stephen J
Shaya Edward J
Shea Margaret A
Sheeley Neil R
Sheffield Charles
Shelus Peter J
Shen Benjamin S P
Sher David
Shields Gregory A
Shields Joseph C
Shine Richard A
Shipman Harry L
Shivanandan Kandiah
Shlosman Isaac
Shore Bruce W
Shore Steven N
Shostak G Seth
Shu Frank H
Shull John Michael
Shull Peter Otto
Sigurdsson Steinn
Silberberg Rein
Silk Joseph I
Silverberg Eric C
Simkin Susan M
Simon George W
Simon Michal
Simon Norman R
Simon Theodore
Simonson S Christian
Sinha Rameshwar P
Sinton William M
Sion Edward Michael
Siry Joseph W
Sitko Michael L
Sjogren William L
Skalafuris Angelo J
Skillman Evan D
Skumanich Andre
Slade Martin A III
Slettebak Arne
Sloan Gregory Clayton
Slovak Mark Haines
Smale Alan Peter
Smecker-Hane Tammy A
Smette Alain
Smith Alex G
Smith Barham W
Smith Bradford A
Smith Bruce F
Smith Dean F
Smith Eric Philip
Smith Graeme H
Smith Haywood C
Smith Myron A
Smith Peter L
Smith Verne V
Smith Wm Hayden
Smith Harding E
Smoot III George F
Sneden Christopher A
Snell Ronald L
Snow Theodore P
Snyder Lewis E
Soberman Robert K
Sobieski Stanley
Soderblom David R
Soderblom Larry
Sofia Sabatino
Soifer Baruch T
Solanes Majua Jose M
Solomon Philip M
Sonett Charles P
Sonneborn George
Soon Willie H
Sowell James Robert
Sparke Linda
Sparks Warren M
Sparks William Brian
Spencer John Howard
Spergel David N
Spicer Daniel Shields
Spiegel E
Spinrad Hyron
Sprague Ann Louise
Sramek Richard A
Stacey Gordon J
Stachnik Robert V
Stahler Steven W
Stahr-Carpenter M
Standish E Myles
Stanford Spencer A
Stanley G J
Stark Antony A
Stark Glen
Starrfield Sumner
Stauffer John Richard
Stebbins Robin
Stecher Theodore P
Stecker Floyd W
Stefanik Robert
Steiger W R
Steigman Gary
Steiman-Cameron Thomas
Stein John William
Stein Robert F
Stein Wayne A
Steinolfson Richard S
Stellingwerf Robert F
Stencel Robert Edward
Stephenson C Bruce
Stepinski Tomasz
Stern Robert Allan
Stern S Alan
Stier Mark T
Stinebring Daniel R
Stocke John T
Stockman Hervey S
Stockton Alan N
Stone Edward C
Stone James McLellan
Stone R G
Stone Remington P S
Stone Ronald Cecil
Strachan Leonard Jr
Strand Kaj Aa
Strelnitski Vladimir
Stringfellow Guy Scott
Strittmatter Peter A
Strobel Darrell F
Strom Karen M
Strom Robert G
Strom Stephen E
Strong Ian B
Strong John D
Strong Keith T
Struble Mitchell F
Struck-Marcell Curtis J
Stryker Linda L
Sturch Conrad R

Sturrock Peter A
Suess Steven T
Sulentic Jack W
Sullivan Woodruff T
Surdej Jean M G
Sutton Edmund Charles
Swade Daryl Allen
Swank Jean Hebb
Sweigart Allen V
Sweitzer James Stuart
Swenson George W
Swerdlow Noel
Swihart Thomas L
Sykes Mark Vincent
Synnott Stephen P
Szalay Alex
Szebehely Victor G
Szkody Paula
Taam Ronald Everett
Tademaru Eugene
Taff Laurence G
Talbot Raymond J
Tandberg-Hanssen Einar A
Tapia-Perez Santiago
Tapley Byron D
Tarnstrom Guy
Tarter C Bruce
Tarter Jill C
Taylor Donald J
Taylor Gregory Benjamin
Taylor Joseph H
Teays Terry J
Tedesco Edward F
Telesco Charles M
Terrell Nelson James
Terrile Richard John
Terzian Yervant
Teske Richard G
Teuben Peter J
Thaddeus Patrick
Tholen David J
Thomas John H
Thomas Roger J
Thompson A Richard
Thonnard Norbert
Thorne Kip S
Thorsett Stephen Erik
Thorstensen John R
Thronson Harley Andrew
Thuan Trinh Xuan
Tifft William G
Timothy J Gethyn
Tipler Frank Jennings
Title Alan Morton
Tohline Joel Edward
Tokunaga Alan Takashi
Tolbert Charles R
Toller Gary N
Tomasko Martin G
Toner Clifford George
Tonry John
Toomre Alar
Toomre Juri
Torres Guillermo
Torres Dodgen Ana V
Townes Charles Hard
Trafton Laurence M
Traub Wesley Arthur
Treffers Richard R
Tremaine Scott Duncan
Trexler James H
Trimble Virginia L
Tripicco Michael J
Troland Thomas Hugh
Truran James W
Tsuruta Sachiko
Tsvetanov Zlatan I
Tucker Wallace H
Tull Robert G
Tully Richard Brent
Turner Barry E
Turner Edwin L
Turner Jean L
Turner Kenneth C
Turner Michael S
Turnshek David A
Twarog Bruce A
Tyler G Leonard
Tylka Allan J
Tyson John Anthony
Tytler David
Ulich Bobby Lee
Ulmer Melville P
Ulrich Roger K
Ulvestad James Scott
Underwood James H
Unwin Stephen C
Uomoto Alan K
Upgren Arthur R
Upson Walter L II
Upton E K l
Urban Sean Eugene
Urry Claudia Megan
Usher Peter D
Uson Juan M
van Allen James A
van Altena William F
van Blerkom David J
van Breugel Wil
van Citters Gordon W
van der Marel Roeland P
van der Veen Wilhelmus EC
van Dorn Bradt Hale
van Dyk Schuyler
Van Flandern Tom
van Gorkom Jacqueline H
van Hamme Walter
van Horn Hugh M
van Hoven Gerard
van Moorsel Gustaaf
van Riper Kenneth A
van Speybroeck Leon P
Vanden Bout Paul A
Vandervoort Peter O
Vaughan Arthur H
Veeder Glenn J
Veillet Christian
Veilleux Sylvain
Velusamy T
Verschuur Gerrit L
Verter Frances
Vesecky J F
Veverka Joseph
Vilas Faith
Vishniac Ethan T
Vogel Stuart Newcombe
Vogt Steven Scott
von Hippel Theodore A
Vorpahl Joan A
Vrba Frederick J
Vrtilek Jan M
Vrtilek Saeqa Dil
Waddington C Jake
Wade Richard Alan
Wagner Raymond L
Wagner Robert M
Wagner William J
Wagoner Robert V
Wakker Bastiaan Pieter
Walborn Nolan R
Walker Alta Sharon
Walker Merle F
Walker Richard L
Walker Robert M A
Walker Robert C
Walker Arthur B C
Wallace Lloyd V
Wallace Richard K
Waller William H
Wallerstein George
Wallin John Frederick
Walter Frederick M
Walterbos Rene A M
Wampler E Joseph
Wang Haimin

Wang Qingde Daniel
Wang Yi-ming
Wang Zhong
Wannier Peter Gregory
Ward Richard A
Ward William R
Wardle John F C
Warner John W
Warren Wayne H
Warwick James W
Wasserman Lawrence H
Wasson John T
Watson William D
Watt Graeme David
Wdowiak Thomas J
Weaver Harold F
Weaver Thomas A
Weaver William Bruce
Webb David F
Webber John C
Webbink Ronald F
Weber Joseph
Weber Stephen Vance
Webster Adrian S
Weedman Daniel W
Weekes Trevor C
Wegner Gary Alan
Wehinger Peter A
Wehrle Ann Elizabeth
Wei Mingzhi
Weidenschilling S J
Weiler Edward J
Weiler Kurt W
Weill Gilbert M
Weinberg J L
Weinberg Steven
Weis Edward W
Weisberg Joel Mark
Weisheit Jon C
Weisskopf Martin Ch
Weissman Paul Robert
Weistrop Donna
Welch William J
Weller Charles S
Wells Donald C
Wells Eddie Neil
Wentzel Donat G
Westerhout Gart
Westphal James A
Wetherill George W
Weymann Ray J
Wheeler J Craig
Wheeler John A
Whipple Arthur L
Whipple Fred L
Whitaker Ewen A
White Nathaniel M
White Nicholas Ernest
White Oran R
White R Stephen
White Raymond E
White Raymond Edwin III
White Richard L
White Richard Allan
White Richard E
White Stephen Mark
Whitford Albert E
Whitmore Bradley C
Whitney Charles A
Whittet Douglas C B
Whittle D Mark
Widing Kenneth G
Wiese Wolfgang L
Wiita Paul Joseph
Wildey Robert L
Wilkening Laurel L
Wilkes Belinda J
Wilkinson David T
Will Clifford M
Williamon Richard M
Williams Barbara A
Williams Carol A
Williams Glen A
Williams James G
Williams John A
Williams Robert E
Williams Theodore B
Willner Steven Paul
Wills Beverley J
Wills Derek
Willson Lee Anne
Willson Robert Frederick
Wilner David J
Wilson Albert G
Wilson Andrew S
Wilson Curtis A
Wilson James R
Wilson Richard
Wilson Robert E
Wilson Robert W
Wilson William J
Winckler John R
Windhorst Rogier A
Wing Robert F
Winget Donald E
Winkler Gernot M R
Winkler Karl-Heinz A
Winkler Paul Frank
Withbroe George L
Witt Adolf N
Witten Louis
Wolfe Arthur M
Wolff Sidney C
Wolfire Mark Guy
Wolfson C Jacob
Wolfson Richard
Wolszczan Alexander
Wood Douglas O S
Wood Frank Bradshaw
Wood H J
Wood John A
Wood Matthew Alan
Woodward Paul R
Woolf Neville J
Woosley Stanley E
Wootten Henry Alwyn
Worden Simon P
Worley Charles E
Wray James D
Wright Edward L
Wright Frances W
Wright Helen Greuter
Wright James P
Wright Melvyn C H
Wrobel Joan Marie
Wu Chi Chao
Wu Nailong
Wu Shi Tsan
Wuelser Jean-Pierre
Wyckoff Susan
Wyller Arne A
Wynn-Williams C G
Wyse Rosemary F
Yahil Amos
Yang Ke-jun
Yano Hajime
Yaplee B S
Yau Kevin K C
Yeh Tyan
Yeomans Donald K
Yin Qi-Feng
Yoder Charles F
York Donald G
Yoshino Kouichi
Yoss Kenneth M
Young Andrew T
Young Arthur
Young Judith Sharn
Young Louise Gray
Yu Yan
Yuan Chi
Zabriskie F R
Zacharias Norbert
Zampieri Luca
Zare Khalil

Zarro Dominic M
Zayer Igor
Zeilik Michael Ii
Zellner Benjamin H
Zensus J-Anton
Zepf Stephen Edward
Zhanf Shouzhong
Zhang Cheng-Yue
Zhang Er-Ho
Zhang Xiaolei
Zhao Jun Hui
Zheng Wei
Zinn Robert J
Zirin Harold
Zirker Jack B
Ziurys Lucy Marie
Zombeck Martin V
Zuckerman Ben M

Vatican City State

Boyle Richard P
Casanovas Juan
Coyne George V
McCarthy Martin F
Stoeger William R

Venezuela

Bruzual Gustavo
Calvet Nuria
Ferrin Ignacio
Ferro Ramos Isabel
Fuenmayor Francisco J
Ibanez S Miguel H
Mendoza Claudio
Parravano Antonio
Stock Jurgen D

INDIVIDUAL MEMBERSHIP FROM NON-ADHERING COUNTRIES

Albania

Hafizi Mimoza

Bahrain

Awadalla Nabil

Columbia

Brieva Eduardo
de Greiff J Arias
Marshall Kevin P

Cuba

Alvarez Pomares A O
Boytel Jorge Del Pino
Doval Jorge Perez
Garcia Eduardo Del pozo
Taboada Ramon Rodriguez

Honduras

Pineda de Carias Maria Cristina
Ponce G A

Iraq

Abdulla Shaker Abdul Aziz
Al-Sabti Abdul Adim
Jabbar Sabeh Rhaman
Jabir Niama Lafta
Kadouri Talib Hadi
Sadik Aziz R
Younis Saad M

Jordan

Al-Naimiy Hamid M K

Kazakhstan

Denisyuk Edvard K
Genkin Igor L
Karygina Zoya V
Kharitonov Andrej V
Obashev Saken O
Omarov Tuken B
Rozhkovskij Dimitrij A
Tejfel Viktor G
Vilkoviskij Emmanuil Y

Korea Dem Rep

Baek Chang Ryong
Bang Yong Gol
Cha Du Jin
Cha Gi Ung
Chio Chol Zong
Choi Won Chol
Dong Il Zun
Hong Hyon Ik
Kang Gon Ik
Kang Jin Sok
Kim Jik Su
Kim Yong Hyok
Kim Yong Uk
Kim Yul
Kim Zong Dok
Li Gi Man
Li Gyong Won
Li Hyok Ho
Li Sin Hyong
Li Son Jae

Lebanon

Plassard J

Mauritius

Golpa Kumar

Morocco

El Bakkali Larbi
Touma Hamid

Nigeria

Akujor Chidi E
Okeke Pius N
Okoye Samuel E
Schmitter Edward F

Pakistan

Quamar Jawaid

Paraguay

Troche-Boggino A E

Philippines

Vinluan Renato

Singapore

Wan Fook Sun
Wilson S J

Slovenia

Cadez Andrej
Dintinjana Bojan
Dominko Fran
Kilar Bogdan
Zwitter Tomaz

Sri Lanka

de Silva L N K
Maheswaran Murugesapillai

Thailand

Pacharin-Tanakun P
Songsathaporn Ruangsak
Soonthornthum Boonrucksar

Uzbekistan

Bykov Mikle F
Ehgamberdiev Shurat
Hojaev Alisher S
Kalmykov A M
Latypov A A
Nuritdinov Salakhutdin
Slonim E M
Yuldashbaev Taimas S

Vietnam

Huan Ngeyen Dush
Nguyen Mau Tung

Yugoslavia Fed Rep

Angelov Trajko
Arsenijevic Jelisaveta
Cadez Vladimir
Dimitrijevic Milan
Djurasevic Gojko
Djurovic Dragutin M
Jovanovic Bozidar
Knezevic Zoran
Kubicela Aleksandar
Kuzmanoski Mike
Lazovic Jovan P
Lukacevic Ilija S
Milogradov-Turin Jelena
Mitic Ljubisa A
Ninkovic Slobodan
Pakvor Ivan
Popovic Georgije
Protich Milorad B
Sadzakov Sofija
Saletic Dusan
Segan Stevo
Sevarlic Branislav M
Simovljevitch Jovan L
Vince Istvan
Vukicevic K M

INTERNATIONAL ASTRONOMICAL UNION

98bis Bd Arago 75014 PARIS France

Tel. 33 1 43 25 83 58 - Fax 33 1 43 25 26 16

e-mail: iau@iap.fr - www: http://www.iau.org

ALPHABETICAL LIST OF IAU MEMBERS

(As per August 31, 1997)

☎ Telephone

✆ Fax

✉ E-mail

ALPHABETICAL LIST OF MEMBERS

A

A'Hearn Michael F
Astronomy Dpt
Univ of Maryland
College Park MD 20742 2421
USA
☎ 1 301 405 6076
📠 1 301 314 9067
✉ ma@astro.umd.edu

Aannestad Per Arne
Physics Dpt
Arizona State Univ
Astronomy Program
Tempe AZ 85287 1504
USA
☎ 1 602 965 3644
📠 1 602 965 7954

Aarseth Sverre J
Inst of Astronomy
The Observatories
Madingley Rd
Cambridge CB3 0HA
UK
☎ 44 12 23 337 548
📠 44 12 23 337 523

Abad Alberto J
Dpt Fisica Teorica
Univ Zaragoza
E 50009 Zaragoza
Spain
☎ 34 7 676 1000
📠 34 7 676 1140

Abad Carlos
CIDA
Box 264
Merida 5101 A
Venezuela
☎ 598 7 471 2780/3883
📠 598 7 471 2459
✉ abad@cida.ve

Abadi Mario
Observ Astronomico
de Cordoba
Laprida 854
5000 Cordoba
Argentina
☎ 54 51 214 059
📠 54 51 21 0613
✉ abadi@uncbob.edu.ar

Abalakin Victor K
Pulkovo Observ
Acad Sciences
10 Kutuzov Quay
196140 St Petersburg
Russia
☎ 7 812 123 3392
📠 7 812 315 1701
✉ vicahal@gao.spb.su

Abbasov Alik R
Scientif/Industrial Ass
of Cosmic Res
Lenin Pspt 159
370106 Baku
Azerbajan

Abbott David C
APAS
Univ of Colorado
Campus Box 391
Boulder CO 80309 0391
USA

Abbott William N
Univ of Athens
Michalacopoulou 42
GR 115 28 Athens
Greece
☎ 30 1 721 3352

Abd El Hady Ahmed A
Astronomy Dpt
Fac of Sciences
Cairo Univ
Geza
Egypt
☎ 20 2 572 7022
📠 20 2 572 7556

Abd El Hamid Rabab
Ntl Res Institute
Astron/Geophysics
Cairo 02
Egypt
☎ 20 2 78 0046/0645
📠 20 2 78 2683

Abdelatif Toufik
CRAAG
Observ Alger
BP 63
Bouzareah 16340
Algeria
☎ 213 2 901 572/424
📠 213 2 94 11 57

Abdelkawi M Abubakr
ASRT
101 Kasr El-eini Str
Cairo 11516
Egypt
☎ 20 2 355 7972

Abdul Aziz Ahmad Zaharim
Physics Dpt
Univ Ke Bangsaan
Bangi 43600
Malaysia
☎ 60 6 829 3849
📠 60 6 829 2880
✉ ohzaha@pkris.oc.ukm.my

Abdulla Shaker Abdul Aziz
SARC
Scientific Res Council
Box 2441
Jadiriyah Baghdad
Iraq
☎ 964 1 776 5127

Abele Maris K
Astronomical Observ
Latvian State Univ
Rainis Bul 19
LV 226098 Riga
Latvia
☎ 371 722 3149
📠 371 782 0180

Abhyankar Krishna D
Astronomy Dpt
Univ of Osmania
Hyderabad 500 007
India
☎ 91 85 1672

Ables Harold D
1512 W Univ Hights Dr N
Flagstaff AZ 86001 8966
USA
☎ 1 520 774 7857
✉ hables@nofs.navy.mil

Ables John G
CSIRO
Div Radiophysics
Box 76
Epping NSW 2121
Australia
☎ 61 2 868 0222
📠 61 2 868 0310

Abou'el-ella Mohamed S
NAIGR
Helwan Observ
Cairo 11421
Egypt
☎ 20 78 0645/2683
📠 20 62 21 405 297

Aboudarham Jean
Observ Paris Meudon
DASOP
Pl J Janssen
F 92195 Meudon PPL Cdx
France
☎ 33 1 45 07 7784
📠 33 1 45 07 7469

Abraham Peter
MPI Astronomie
Koenigstuhl
D 69117 Heidelberg
Germany
☎ 49 622 152 8355
📠 49 622 152 8356
✉ abraham@mpia-hd.
mpg.de

Abraham Zulema
IAG
Univ Sao Paulo
CP 9638
01065 970 Sao Paulo SP
Brazil
☎ 55 11 577 8599
📠 55 11 276 3848

Abrahamian Hamlet V
Byurakan Astrophys Observ
Armenian Acad Sci
378433 Byurakan
Armenia
☎ 374 88 52 28 3453/4142
📠 374 88 52 52 3640
✉ byurakan@adonis.ias.msk.su

Abramenko Valentina
Crimean Astrophys Obs
Ukrainian Acad Science
Nauchny
334413 Crimea
Ukraine
☎ 380 6554 71161
📠 380 6554 40704
✉ avi@crao.crimea.ua

Abrami Alberto
Dpt Astronomia
Univ d Trieste
Via Tiepolo 11
I 34131 Trieste
Italy
☎ 39 40 79 3921
✆ 39 40 30 9418

Abramowicz Marek
Dpt Astron/Astrophys
Chalmers Technical Univ
S 412 96 Goteborg
Sweden
☎ 46 31 772 3135
✆ 46 31 772 3204
✉ marek@tfa.fy.chalmers.se

Abt Helmut A
NOAO/KPNO
Box 26732
950 N Cherry Av
Tucson AZ 85726 6732
USA
☎ 1 520 325 9215
✆ 1 520 323 4183
✉ apj@noao.edu

Abulazm Mohamed Samir
NAIGR
Helwan Observ
Cairo 11421
Egypt
☎ 20 78 0645/2683
✆ 20 62 21 405 297

Acharya Bannanje S
TIFR/Cosmic Ray Gr
Homi Bhabha Rd
Colaba
Bombay 400 005
India
☎ 91 22 215 2971
✆ 91 22 495 2110

Achterberg Abraham
Sterrekundig Inst Utrecht
Box 80000
NL 3508 TA Utrecht
Netherlands
☎ 31 30 253 5212
✆ 31 30 253 1601
✉ a.achterberg@astro.uu.nl

Acker Agnes
Observ Strasbourg
11 r Universite
F 67000 Strasbourg
France
☎ 33 3 88 15 0724
✆ 33 3 88 25 0160

Acton Loren W
Physics Dpt
Montana State Univ
Bozeman MT 59717 0350
USA
☎ 1 406 994 6072
✆ 1 406 994 4452

Adam Madge G
Astrophysics Dpt
Univ of Oxford
Keble Rd
Oxford OX1 3RQ
UK
☎ 44 186 527 3999
✆ 44 186 527 3947

Adams David J
Astronomy Dpt
Univ Leicester
University Rd
Leicester LE1 7RH
UK
☎ 44 116 252 2073
✆ 44 113 252 2200

Adams Fred
Physics Dpt
Univ of Michigan
Dennison Bldg
Ann Arbor MI 48109 1090
USA
☎ 1 313 747 4320
✆ 1 313 764 2211
✉ fadams@umiphys

Adams Mark T
McDonald Observ
Univ of Texas
Box 1337
Fort Davis TX 79734 1337
USA
☎ 1 915 426 3263
✆ 1 915 426 3641
✉ mta@astro.as.utexas.edu

Adams Thomas F
LLNL
L 170
Box 808
Livermore CA 94551 9900
USA
☎ 1 925 422 1248
✆ 1 925 422 3389
✉ adams35@llnl.gov

Adams James H
NRL
Code 4154 2
4555 Overlook Av SW
Washington DC 20375 5000
USA
☎ 1 202 767 2747

Adamson Andrew
School of Physics/Astro
Lancashire Polytechnic
Corporation St
Preston PR1 2TQ
UK

Ade Peter A R
Astrophysics Group
QMWC
Mile End Rd
London E1 4NS
UK
☎ 44 171 975 5555
✆ 44 181 975 5500

Adelman Saul J
Physics Dpt
The Citadel
Charleston SC 29409
USA
☎ 1 803 792 6943
✉ adelmans@citadel.edu

Adjabshirizadeh Ali
Ctr for Astron Res
Univ of Tabriz
Tabriz 51664
Iran
☎ 98 41 32564

Adler David Scott
STScI/CSC
Homewood Campus
3700 San Martin Dr
Baltimore MD 21218
USA
☎ 1 301 338 4458
✆ 1 301 338 4767
✉ adler@stsci.edu

Adolfsson Tord
Krageholmsgatan 12
S 216 19 Malmoe
Sweden
☎ 46 40 15 7586

Aerts Conny
Inst v Sterrenkunde
Katholicke Univ Leuven
Celestijnenlaan 200B
B 3001 Leuven
Belgium
☎ 32 16 32 70 28
✆ 32 16 32 7999
✉ conny@ster.kuleuven.ac.be

Afanas'ev Viktor L
SAO
Acad Sciences
Nizhnij Arkhyz
357147 Karachaevo
Russia
☎ 7 878 789 2501

Afanasjeva Praskovya M
Pulkovo Observ
Acad Sciences
10 Kutuzov Quay
196140 St Petersburg
Russia
☎ 7 812 298 2242
✆ 7 812 315 1701

Agekjan Tateos A
Astronomical Observ
St Petersburg Univ
Bibliotechnaja Pl 2
199178 St Petersburg
Russia
☎ 7 812 428 7129
✆ 7 812 428 4259

Aggarwal Kanti Mal
Dpt Pure/Appl Physics
Queen's Univ
Belfast BT7 1NN
UK
☎ 91 12 32 273 239
✆ 91
✉ K.Aggarwal@qub.ac.uk

Agrawal P C
TIFR
Homi Bhabha Rd
Colaba
Bombay 400 005
India
☎ 91 22 495 2311*393
✆ 91 22 495 2110

Aguero Estela L
Observ Astronomico
de Cordoba
Laprida 854
5000 Cordoba
Argentina
☎ 54 51 23 0491/236876
✆ 54 51 21 0613
✉ aguero@astro.edu.ar

Aguiar Odylio Denys
INPE
Dpt Astronomia
CP 515
12227 010 S Jose dos Campos
Brazil
☎ 55 123 41 8977*689
✆ 55 123 21 8743
✉ inpedas@brfapesp

Aguilar Luis A Chiu
Observ Astronomico Nacional
UNAM
Apt 877
Ensenada BC 22800
Mexico
☎ 52 6 174 4580
✆ 52 6 174 4607
✉ aguilar@bufadora.astroscu.unam.mx

Aguilar Maria Luisa
UNSM
Fac Ciencias Fisicas
Av Arica 830
Lima 5
Peru
☎ 51 2 43 961/52 1343
✉ mlah@unmsm.pe

Ahluwalia Harjit Singh
Dpt Phys/Astronomy
Univ New Mexico
800 Yale Blvd Ne
Albuquerque NM 87131
USA
☎ 1 505 277 2941
✆ 1 505 660 461

Ahmad Farooq
Physics Dpt
Univ of Kashmir
Srinagar 190 006
India
☎ 91 71559

Ahmad Imad Aldean
Imad-ad Dean Inc
4323 Rosedale Av
Bethesda MD 20814
USA
☎ 1 301 565 4714

Ahmed Imam Ibrahim
Astronomy Dpt
Fac of Sciences
Cairo Univ
Geza
Egypt
☎ 20 2 572 7022
✆ 20 2 572 7556

Ahmed Mostafa Kamal
Astronomy Dpt
Fac of Sciences
Cairo Univ
Geza
Egypt
☎ 20 2 572 7022
✆ 20 2 572 7556

Ai Guoxiang
Beijing Astronomical Obs
CAS
Beijing 100080
China PR
☎ 86 1 255 1968*1261071
✆ 86 1 256 10855
✉ aigx@sun10.bao.ac.cn

Aiad A Zaki
Astronomy Dpt
Fac of Sciences
Cairo Univ
Geza
Egypt
☎ 20 2 572 7022
✆ 20 2 572 7556

Aiello Santi
Dpt Astronomia
Univ d Firenze
Largo E Fermi 5
I 50125 Firenze
Italy
☎ 39 55 43 78540
✆ 39 55 43 5939

Aikawa Toshiki
Astron Institute
Tohoku Gakuin Univ
Ichinazaka Izumi-ku
Sendai 981 31
Japan
☎ 81 223 75 1111*318
✆ 81 223 75 4040
✉ aikawa@izcc.tohoku-gakuin.ac.jp

Aikman G Chris l
NRCC/HIA
DAO
5071 W Saanich Rd
Victoria BC V8X 4M6
Canada
☎ 1 250 363 0008
✆ 1 250 363 0045
✉ aikman@dao.nrc.ca

Aime Claude
Dpt Astrophysique
Universite Nice
Parc Valrose
F 06034 Nice Cdx
France
☎ 33 4 93 51 9100
✆ 33 4 93 52 9806

Aitken David K
5 Dewsbury Cottages
Bishophill
York YO1 1HB
UK
✉ dave@phadfa.ph.adfa.oz.au

Aizenman Morris L
NSF
Div Astron Sci
4201 Wilson Blvd
Arlington VA 22230
USA
☎ 1 703 306 1820
✆ 1 703 306 0525
✉ maizenman@nsf.gov

Aizu Ko
3-24-3 Katahira
Asao Ku
Kawasaki
Japan

Akabane Kenji A
Tokyo Astronomical Obs
NAOJ
Osawa Mitaka
Tokyo 181
Japan
☎ 81 267 98 2831
✆ 81 422 34 3793

Akabane Tokuhide
Hida Observ
Kyoto Univ
Kamitakara
Gifu 506 13
Japan
☎ 81 578 62 311

Akan Mustafa Can
Dpt Astron/Space Sci
Ege Univ Observ
Box 21
35100 Bornova Izmir
Turkey
☎ 90 232 388 0110*2322
✉ efeast04@vm3090.ege.edu.tr

Akcayli Melek M A
Dpt Astron/Space Sci
Ege Univ
Box 21
35100 Bornova Izmir
Turkey
☎ 90 232 388 0110*2322

Ake Thomas Bellis
NASA GSFC
Code 681
Greenbelt MD 20771
USA
☎ 1 301 286 3924
✆ 1 301 286 1752
✉ hrsake@hrs.dsfc.nasa.gov

Akim Efraim L
Keldysh Inst Applied Maths
Acad Sciences
Miusskaja Sq 4
125047 Moscow
Russia
☎ 7 095 251 3739

Akimov Leonid
Astronomical Observ
Kharkiv State Univ
Sumskaja Ul. 35
310002 Kharkiv
Ukraine
☎ 380 5724 32428
✆ 380 5724 32428
✉ akimov@astron.kharkov.ua

Aksnes Kaare
Inst Theor Astrophys
Univ of Oslo
Box 1029
N 0315 Blindern Oslo 3
Norway
☎ 47 22 856 515
✆ 47 22 85 6505
✉ kaare.aksnes@astro.uio.no

Akujor Chidi E
School Physical Sci
Imo State Univ
PMB 2000
Owerri
Nigeria

Akyol Mustafa Unal
Fac of Education
Selcuk Univ
42090 Konya
Turkey

Al-Malki M B
Astronomy Dpt
King Saud Univ
Box 2455
Riyadh 11451
Saudi Arabia
☎ 966 1 4676 314/23/12
✆ 966 1 467 4253
✉ f40a004@saksu00

Al-Naimiy Hamid M K
Al al-Bayt Univ
Astron/Space Sci
PO Box 130302
Mafraq
Jordan
☎ 962 6 457 1101
✆ 962 6 569 9270
✉ alnaimiy@yahoo.com

Al-Sabti Abdul Adim
Physics Dpt
Univ of Baghdad
Science College
Jadiriyah Baghdad
Iraq
☎ 964 1 555 2340

Albers Henry
62 Prospect St
Falmouth MA 02540
USA
☎ 1 508 540 0978

Albinson James
Physics Dpt
Univ of Keele
Keele ST5 5BG
UK
☎ 44 178 262 1111
✆ 44 178 271 1093
✉ jsa@uk.ac.kl.ph.star

Albrecht Miguel A
ESO
Karl Schwarzschildstr 2
D 85748 Garching
Germany
☎ 49 893 200 6346
✆ 49 893 200 6480
✉ malbrecht@eso.org

Albrecht Rudolf
Space Telescope ECF
ESO Room 419
Karl Schwarzschild Str 2
D 85748 Garching
Germany
☎ 49 893 200 6287
✆ 49 893 200 6480
✉ ralbrech@eso.org

Albrow Michael
Dpt Phys/Astronomy
Univ of Canterbury
Private Bag 4800
Christchurch 1
New Zealand
☎ 64 3 364 2987*7579
✆ 64 3 364 2469
✉ m.albrow@phys.canterbury.ac.nz

Alcaino Gonzalo
Instituto Isaac Newton
Casilla 8 9
Santiago 9
Chile
☎ 56 2 217 2013
✆ 56 2 217 2352

Alcolea Javier
Ctr Astron d Yebes
OAN
Apt 148
E 19080 Guadalajara
Spain
☎ 34 1 129 0311
✆ 34 1 129 0063
✉ alcolea@cay.es

Aldrovandi Ruben
Inst Fisica Teorica
Rua Pamplona 15
01065 970 Sao Paulo SP
Brazil
☎ 55 11 288 5643

Alecian Georges
Observ Paris Meudon
DAF
Pl J Janssen
F 92195 Meudon PPL Cdx
France
☎ 33 1 45 07 74 20
✆ 33 1 45 07 7414
✉ alecian@obspm.fr

Alexander David R
Physics Dpt
Wichita State Univ
Wichita KS 67260
USA
☎ 1 316 689 3190
✆ 1 316 689 3770
✉ dra@twsuvm.uc.twsu.edu

Alexander Joseph K
NASA Headquarters
Code P
Office of Chief Scientist
Washington DC 20546
USA

Alexander Murray E
NRC
Inst for Biodiagnostics
435 Ellice Av
Winnipeg MB R3B 1Y6
Canada

Alexander Paul
MRAO
Cavendish Laboratory
Madingley Rd
Cambridge CB3 0HE
UK
☎ 44 12 23 337 294
✆ 44 12 23 354 599
✉ pa25@uk.ac.cam.phx

Alexandrov Yuri V
Str Groznenskaya 32
Apt 117
310124 Kharkiv
Ukraine
☎ 380 5724 57537/43 2428
✉ alex@astron.kharkov.ua

Alfaro Emilio Javier
Inst Astrofisica
Andalucia Apt 3004
Prof Albareda 1
E 18080 Granada
Spain
☎ 34 5 812 1311
✆ 34 5 881 4530
✉ emilio@iaa.es

Alighieri Sperello Di Serego
Osserv Astrofis Arcetri
Univ d Firenze
Largo E Fermi 5
I 50125 Firenze
Italy
☎ 39 55 43 78540
✆ 39 55 43 5939

Alissandrakis C
Astrophysics Dpt
Ntl Univ Athens
Panepistimiopolis
GR 157 84 Zografos
Greece
☎ 30 1 724 3414
✆ 30 1 723 5122
✉ calissan@cc.uoi.gr

Alksnis Andrejs
Radioastrophys Observ
Latvian Acad Sci
Turgeneva 19
LV 226098 Riga
Latvia
☎ 371 13 2 932088
✆ 371 132 22 8784
✉ astra@lza.riga.lv

Alladin Saleh Mohamed
Astronomy Dpt
Univ of Osmania
Hyderabad 500 007
India
☎ 91 71 116

Allan David W
Box 66
Fountain Green UT 84632
USA
✆ 1 801 445 3215
✉ allanstime@yvax.byu.edu

Allan Peter M
Astronomy Dpt
Univ Manchester
Oxford Rd
Manchester M13 9PL
UK
☎ 44 161 275 4224
✆ 44 161 275 4223

Allard Nicole
IAP
98bis bd Arago
F 75014 Paris
France
☎ 33 1 44 32 8082
✆ 33 1 44 32 8001

Allegre Claude
IPG
4 Pl Jussieu
F 75005 Paris
France

Allen Anthony John
Astronomy Unit
QMWC
Mile End Rd
London E1 4NS
UK
☎ 44 171 975 5454
✆ 44 181 981 9587
✉ allen@uk.ac.qmc.maths

Allen Christine
Instituto Astronomia
UNAM
Apt 70 264
Mexico DF 04510
Mexico
☎ 52 5 622 3901
✆ 52 5 616 0653
✉ chris@astroscu.unam.mx

Allen Ronald J
STScI
Homewood Campus
3700 San Martin Dr
Baltimore MD 21218
USA
☎ 1 301 338 4574
✆ 1 301 338 4596
✉ rjallen@stsci.edu

Allen William
Adams Lane Observ
46 Adams Lane
Blenheim
New Zealand
☎ 64 057 87258

Aller Hugh D
Astronomy Dpt
Univ of Michigan
Dennison Bldg
Ann Arbor MI 48109 1090
USA
☎ 1 313 764 3466
✆ 1 313 764 2211

Aller Lawrence Hugh
Dpt Astronomy/Phys
UCLA
Box 951562
Los Angeles CA 90025 1562
USA
☎ 1 310 825 3515
✆ 1 310 206 2096

Aller Margo F
Astronomy Dpt
Univ of Michigan
Dennison Bldg
Ann Arbor MI 48109 1090
USA
☎ 1 313 764 3465
✆ 1 313 764 2211

Alley Carrol O
Physics Dpt
Univ of Maryland
College Park MD 20742 2421
USA
☎ 1 301 454 3405
✆ 1 301 314 9547

Alloin Danielle
DAPNIA/SAP
CEA Saclay
Orme d Merisiers Bt 709
F 91191 Gif s Yvette Cdx
France
☎ 33 1 69 08 9263
✆ 33 1 69 08 9266

Almar Ivan
Konkoly Observ
Thege U 13/17
Box 67
H 1525 Budapest
Hungary
☎ 36 1 375 4122
✆ 36 1 275 4668

Almleaky Yasseen
Astronomy Dpt
Box 9028
KAAU
Jeddah 21413
Saudi Arabia
☎ 966 2 695 2285
✆ 966 2 640 0736
✉ scf3017@sakaau03

Alonso Maria Victoria
Observ Astronomico
de Cordoba
Laprida 854
5000 Cordoba
Argentina
☎ 54 51 331 006
✆ 54 51 331 006
✉ vicky@oac.uncor.edu

Alpar Ali
Physics Dpt
Middle East Tech Univ
06531 Ankara
Turkey
☎ 90 41 22 337100/ 3259
✆ 90 41 22 36945
✉ alpar@trmetu

Alpher Ralph Asher
Physics Dpt
Union College
Schenectady NY 12308
USA
☎ 1 518 388 6345
✆ 1 518 388 6947
✉ alpherr@gar.union.edu

Altamore Aldo
Istt Astronomico
Univ d Roma La Sapienza
Via G M Lancisi 29
I 00161 Roma
Italy
☎ 39 6 44 03734
✆ 39 6 44 03673

Altavista Carlos A
Observ Astronomico
Paseo d Bosque S/n
1900 La Plata (Bs As)
Argentina
☎ 54 21 21 7308
✆ 54 21 211 761

Altenhoff Wilhelm J
RAIUB
Univ Bonn
auf d Huegel 69
D 53121 Bonn
Germany
☎ 49 228 525 293
✆ 49 228 525 229

Altrock Richard C
AFGL
NSO
Box 62
Sunspot NM 88349
USA
☎ 1 505 434 7016
✆ 1 505 434 7029
✉ raltrock@sunspot.noao.edu

Altschuler Daniel R
Arecibo Observ
Box 995
Arecibo PR 00613
USA
☎ 1 809 878 2612
✆ 1 809 878 1861
✉ daniel@naic.edu

Altschuler Martin D
Dpt Rad Therapy Box 522
Hosp Univ Pennsylvania
3400 Spruce St
Philadelphia PA 19104
USA
☎ 1 215 662 6472

Alurkar S K
PRL
Navrangpura
Ahmedabad 380 009
India
☎ 91 272 46 2129
✆ 91 272 44 5292

Alvarez Hector
Dpt Astronomia
Univ Chile
Casilla 36 D
Santiago
Chile
☎ 56 2 229 4101
✆ 56 2 229 4002

Alvarez Manuel
Observ Astronomico
Nacional
Box 439027
San Ysidro CA 92027
USA
☎ 1 706 674 4580
✆ 1 706 667 4607
✉ alvarez@alfa.astroscu.unam.mx

Alvarez Pedro
IAC
Observ d Teide
via Lactea s/n
E 38200 La Laguna
Spain
☎ 34 2 232 9100
✆ 34 2 232 9117
✉ pam@iac.es

Alvarez Pomares A O
Inst Geophys/Astron
C 212 N 2906/29 Y 31
Lisa
La Habana
Cuba
☎ 53 21 8435
✉ biotec@ceniai.cu

Aly Jean-Jacques
DAPNIA/SAP
CEA Saclay
BP 2
F 91191 Gif s Yvette Cdx
France
☎ 33 1 69 08 4030
✆ 33 1 69 08 9266
✉ jjaly@solar

Amari Tahar
Observ Paris Meudon
DASOP
Pl J Janssen
F 92195 Meudon PPL Cdx
France
☎ 33 1 45 07 7760
✆ 33 1 45 07 7469
✉ amari@obspm.fr

Ambastha A K
Udaipur Solar Observ
11 Vidya Marg
Udaipur 313 001
India
☎ 91 25 626

Ambrosini Roberto
Istt Radioastronomia
CNR
Via P Gobetti101
I 40129 Bologna
Italy
☎ 39 51 639 9385
✆ 39 51 639 9431
✉ ambrosini@astbo1.bo.cnr.it

Ambroz Pavel
Astronomical Institute
Czech Acad Sci
Fricova 1
CZ 251 65 Ondrejov
Czech R
☎ 420 204 85 314
✆ 420 2 88 1611
✉ pambroz@asu.cas.cz

Ambruster Carol
Astronomy Dpt
Villanova Univ
Mendel Hall
Villanova PA 19085
USA
☎ 1 215 645 4822
✆ 1 215 645 7465
✉ ambruster@vuvaxcom

Amendola Luca
OAR
Via d Parco Mellini 84
I 00136 Roma
Italy
☎ 39 6 34 7056
✆ 39 6 349 8236
✉ amendola@oarhp1.rm.astro.it

Amram Philippe
Observ Marseille
2 Pl Le Verrier
F 13248 Marseille Cdx 04
France
☎ 33 4 91 95 9088
✆ 33 4 91 62 1190
✉ amram@observatoire.cnrs.mrs.fr

Anand S P S
Applied Res Corp
8201 Corporate Dr
Suite 920
Landover MD 20785
USA
☎ 1 301 459 8442
✆ 1 301 459 0761

Anandaram Mandayam N
Physics Dpt
Bangalore Univ
Jnanabharathi Campus
Bangalore 560 056
India
☎ 91 300 4001*275/080 671 3955
✆ 91 080 338 9295
✉ libn@bnguni.kar.nic.in

Ananthakrishnan S
NCRA/TIFR
Pune Univ Campus Pb 3
Ganeshkhind
Pune 411 007
India
☎ 91 212 33 6105
✆ 91 212 35 7257
✉ ananth@ncra.tift.res.in

Anantharamaiah K R
RRI
Sadashivanagar
CV Raman Av
Bangalore 560 080
India
☎ 91 80 334 0122
✆ 91 80 334 0492
✉ anantha@rri.ernet.in

Andernach Heinz
Dpt Astronomia
Univ Guanajuato
Apdo 144
Guanajuato GTO 36000
Mexico
☎ 52 4 732 9548
✆ 52 4 732 0253
✉ heinz@astro.ugto.mx

Andersen Johannes
Astronomical Observ, NBIfAFG
Copenhagen Univ
Juliane Maries Vej 30
DK 2100 Copenhagen
Denmark
☎ 45 35 32 5934
✆ 45 35 32 5989
✉ ja@astro.ku.dk

Anderson Bryan
NRAL
Univ Manchester
Jodrell Bank
Macclesfield SK11 9DL
UK
☎ 44 14 777 1321
✆ 44 147 757 1618

Anderson Christopher M
Washburn Observ
Univ Wisconsin
475 N Charter St
Madison WI 53706
USA
☎ 1 608 262 0492
✆ 1 608 263 0361

Anderson Kinsey A
Space Sci Laboratory
Univ of California
Grizzly Peak Blvd
Berkeley CA 94720 7950
USA
☎ 1 415 642 1313

Anderson Kurt S
Astronomy Dpt
NMSU
Box 30001 Dpt 4500
Las Cruces NM 88003
USA
☎ 1 505 646 1032
✆ 1 505 646 1602

Andersson Bengt Goeran
Dpt Phys/Astronomy
JHU
Charles/34th St
Baltimore MD 21218
USA
☎ 1 410 516 8378
✆ 1 410 516 5494
✉ bg@pha.jhu.edu

Andreani Paola Michela
Dpt Astronomia
Univ d Padova
Vic d Osservatorio 5
I 35122 Padova
Italy
☎ 39 49 829 3442
✆ 39 49 875 9840

Andreev Vladimir V
Engelhardt Astronom
Observ
Observatoria Station
422526 Kazan
Russia
☎ 7 32 4827
✆ 7 38 0924
✉ eao@astro.kazan.su

Andreic Zeljko
Ruder Boskovic Inst
Physics Dpt
Bijenicka 54
HR 10000 Zagreb
Croatia
☎ 385 1 456 1222
✆ 385 1 425 497
✉ andreic@rudjer.irb.hr

Andrews David A
Armagh Observ
College Hill
Armagh BT61 9DG
UK
☎ 44 18 61 522 928
✆ 44 18 61 527 174

Andrews Frank
Carter Observ
Ntl Observ New Zealand
Box 2909
Wellington 1
New Zealand
☎ 64 4 472 8167
✆ 64 4 472 8320

Andrews Peter J
Royal Greenwich Obs
Madingley Rd
Cambridge CB3 0EZ
UK
☎ 44 1223 374 000
✆ 44 12 23 374 700

Andrienko Dmitry A
Astronomical Observ
Kyiv State Univ
Observatornaya Ul 3
252053 Kyiv
Ukraine
☎ 380 4421 62691

Andrievsky Sergei
Astronomy Dpt
Odessa State Univ
Shevchenko Park
270014 Odessa
Ukraine
☎ 380 4822 28442
✆ 380 4822 28442
✉ root@astro.odessa.ua

Andrillat Yvette
OMP
14 Av E Belin
F 31400 Toulouse Cdx
France
☎ 33 5 61 27 3131
✆ 33 5 61 53 2840

Andronov Ivan
Astronomy Dpt
Odessa State Univ
Shevchenko Park
270014 Odessa
Ukraine
☎ 380 4822 28442
✆ 380 4822 28442
✉ root@astro.odessa.ua

Andronova Anna A
Pulkovo Observ
Acad Sciences
10 Kutuzov Quay
196140 St Petersburg
Russia
☎ 7 812 123 4252
℻ 7 812 123 4922
✉ gaoran@mail.wplus.net

Angel J Roger P
Steward Observ
Univ of Arizona
Tucson AZ 85721
USA
☎ 1 520 621 6541
✉ rangel@as.arizona.edu

Angeletti Lucio
OAR
Via d Parco Mellini 84
I 00136 Roma
Italy
☎ 39 6 34 7056
℻ 39 6 349 8236

Angelov Trajko
Astronomy Dpt
Univ of Belgrade
Studentski Trg 16
11000 Beograd
Yugoslavia FR
☎ 381 11 638 715
℻ 381 11 630 151

Angione Ronald J
Astronomy Dpt
San Diego State Univ
San Diego CA 92182
USA
☎ 1 619 265 6183
℻ 1 619 594 1413
✉ angione@mintaka.
sdsu.edu

Anglada Guillem
Dpt Astronom Meteo
Univ Barcelona
Avd Diagonal 647
E 08028 Barcelona
Spain
☎ 34 3 402 1121
℻ 34 3 411 0873

Anguita Claudio
Dpt Astronomia
Univ Chile
Casilla 36 D
Santiago
Chile
☎ 56 2 229 4002
℻ 56 2 229 4002
✉ claudio@das.uchile.cl
anguita@das.uchile.cl

Anile Angelo M
Dpt Matematica
Citta Universitaria
Via A Doria 6
I 95125 Catania
Italy
☎ 39 95 733 0533
℻ 39 95 33 0592

Ann Hong Bae
Dpt Earth Sciences
Pusan Ntl Univ
Kum Jong Ku
Pusan 609 735
Korea RP
☎ 82 515 10 2705
℻ 82 515 13 7495

Anosova Joanna
Aerospace Eng Dpt
Univ of Texas
WRW 414
Austin TX 78712
USA
☎ 1 512 471 5863
℻ 1 512 471 3788
✉ anosova@clyde.as.
utexas.edu

Ansari S M Razaullah
Physics Dpt
Aligarh Muslim Univ
Aligarh Up 202 002
India
☎ 91 57 140 1001
℻ 91 571 400 466

Antalova Anna
Astronomical Institute
Slovak Acad Sciences
SK 059 60 Tatranska Lomni
Slovak Republic
☎ 421 969 96 7866
℻ 421 969 96 7656

Anthony-Twarog Barbara J
Dpt Phys/Astronomy
Univ of Kansas
Lawrence KS 66045
USA
☎ 1 913 864 4933
℻ 1 913 864 5262
✉ anthony@kuphsx.phsx.
ukans.edu

Antia H M
TIFR
Homi Bhabha Rd
Colaba
Bombay 400 005
India
☎ 91 22 495 2311
℻ 91 22 495 2110

Antiochos Spiro Kosta
NRL
Code 7675
4555 Overlook Av SW
Washington DC 20375 5352
USA
☎ 1 202 767 6199
℻ 1 202 404 7997
✉ santiochos@solar.
stanford.edu

Antipova Lyudmila I
Inst of Astronomy
Acad Sciences
Pyatnitskaya Ul 48
109017 Moscow
Russia
☎ 7 095 231 0680
℻ 7 095 230 2081

Antonacopoulos Greg
Astronomy Dpt
Univ of Patras
GR 261 10 Rion
Greece
☎ 30 61 99 7572
℻ 30 61 99 7636

Antonello Elio
Osserv Astronomico
d Milano
Via E Bianchi 46
I 22055 Merate
Italy
☎ 39 990 6412
℻ 39 990 8492

Antonopoulou E
Astronomy Dpt
Ntl Univ Athens
Panepistimiopolis
GR 157 84 Zografos
Greece
☎ 30 1 724 3211
℻ 30 1 723 5122

Antonov Vadim A
Inst Theoret Astron
Acad Sciences
N Kutuzova 10
191187 St Petersburg
Russia
☎ 7 812 278 88 35
℻ 7 812 272 79 68
✉ ita@iipah.spb.su

Antonucci Ester
Istt Fisica
Univ d Torino
Corso d Azeglio 46
I 10125 Torino
Italy
☎ 39 11 657 694

Antonuccio-Delogu Vincenzo
Osserv Astronomico
d Catania
Via A Doria 6
I 95125 Catania
Italy
☎ 39 95 733 2244
℻ 39 95 33 0592
✉ vantonuccio@astrct.ct.
astro.it

Antov Alexandar
Astronomy Dpt
Bulgarian Acad Sci
72 Lenin Blvd
BG 1784 Sofia
Bulgaria
☎ 359 2 75 8927
℻ 359 2 75 8927

Anupama G C
IIA
Koramangala
Sarjapur Rd
Bangalore 560 034
India
☎ 91 80 553 0672
℻ 91 80 553 4043
✉ gca@iiap.ernet.in

Anzer Ulrich
MPA
Karl Schwarzschildstr 1
D 85748 Garching
Germany
☎ 49 893 299 00
℻ 49 893 299 3235

Aoki Shinko
Tokyo Astronomical Obs
NAOJ
Osawa Mitaka
Tokyo 181
Japan
☎ 81 422 32 5111
℻ 81 422 34 3793
✉ baoki2x@c1.mtk.nao.ac.jp

Aparici Juan
Dpt Astronomia
Univ Chile
Casilla 36 D
Santiago
Chile
☎ 56 2 229 4101
℻ 56 2 229 4002

Aparicio Antonio
Carnegie Observatories
813 Santa Barbara St
Pasadena CA 91101 1292
USA
☎ 1 818 577 1122
℻ 1 818 795 8136
✉ aaj@cyclone.ociw.edu

Apparao K M V
TIFR
Homi Bhabha Rd
Colaba
Bombay 400 005
India
☎ 91 22 219 111*341
℻ 91 22 215 2110

Appenzeller Immo
Landessternwarte
Koenigstuhl
D 69117 Heidelberg
Germany
☎ 49 622 150 9292
℻ 49 622 150 9202
✉ iappenze@lsw.uni-
heidelberg.de

Appleby John F
APL
JHU
Johns Hopkins Rd
Laurel MD 20723 6099
USA
☎ 1 301 953 5243
℻ 1 301 953 5969
✉ John.Appleby@jhuapl.edu

Appleton Philip Noel
Physics Dpt
Iowa State Univ
Ames IA 50011
USA
☎ 1 515 294 3667
℻ 1 515 294 3262

Aquilano Roberto Oscar
Instituto Fisica
Rosario Conicet Unr
Bv 27 de Febrero 210 bis
2000 Rosario
Argentina
☎ 54 41 82 1769/72
✆ 54 41 25 7164

Arabelos Dimitrios
Dpt Geodesy/Survey
Univ Thessaloniki
GR 540 06 Thessaloniki
Greece
☎ 30 31 99 2693
✆ 30 31 99 6408
✉ arab@eng.auth.gr

Arafune Jiro
Inst Cosmic Ray Res
Univ of Tokyo
Midoricho Tanashi
Tokyo 188
Japan
☎ 81 424 61 4131
✆ 81 424 68 1438

Arai Kenzo
Physics Dpt
Kumamoto Univ
2-39-1 Kurokami
Kumamoto 860
Japan
☎ 81 963 442111

Ardavan Houshang
Inst of Astronomy
The Observatories
Madingley Rd
Cambridge CB3 0HA
UK
☎ 44 12 23 337 548
✆ 44 12 23 337 523

Ardeberg Arne L
Lund Observ
Box 43
S 221 00 Lund
Sweden
☎ 46 46 222 7290
✆ 46 46 222 4614
✉ arne.ardeberg@astro.lu.se

Ardebili M Reza
Box 47415 341
Babolsar
Iran

Arduini-Malinovsky Monique
CNES
2 Pl Maurice Quentin
F 75039 Paris Cdx 01
France
☎ 33 1 44 76 7500
✆ 33 1 44 76 7676

Arellano Ferro Armando
Instituto Astronomia
UNAM
Apt 70 264
Mexico DF 04510
Mexico
☎ 52 5 622 3906
✆ 52 4 732 0253
✉ armando@astroscu.unam.mx

Arenou Frederic
Observ Paris Meudon
DASGAL
Pl J Janssen
F 92195 Meudon PPL Cdx
France
☎ 33 1 45 07 7849
✆ 33 1 45 07 7878
✉ frederic.arenou@obspm.fr

Argue A Noel
Inst of Astronomy
The Observatories
Madingley Rd
Cambridge CB3 0HA
UK
☎ 44 12 23 337 548
✆ 44 12 23 337 523
✉ cis@ast.cam.ac.uk

Argyle P E
Suite 401
5880 Hampton Place
Vancouver BC V6T 2E9
Canada

Argyle Robert William
Royal Greenwich Obs
Madingley Rd
Cambridge CB3 0EZ
UK
☎ 44 12 23 37 4783
✆ 44 12 23 374 700
✉ rwa@ast.cam.ac.uk

Arias Elisa Felicitas
Observ Astronomico
Paseo d Bosque S/n
1900 La Plata (Bs As)
Argentina
☎ 54 21 217 308
✆ 54 21 211 761
✉ felicitas@fcaglp.fcaglp.unlp.edu.ar

Arimoto Nobuo
Inst Astronomy
Univ of Tokyo
Mitaka
Tokyo 181
Japan
☎ 81 422 34 3732
✆ 81 422 34 3749
✉ arimoto@mtk.ioa.s.u-tokyo.ac.jp

Arkharov Arkadi A
Pulkovo Observ
Acad Sciences
10 Kutuzov Quay
196140 St Petersburg
Russia
☎ 7 812 123 4431
✆ 7 812 123 1922
✉ arkharov@gao.pnpi.spb.su

Arkhipova Vesa P
SAI
Acad Sciences
Universitetskij Pr 13
119899 Moscow
Russia
☎ 7 095 139 2657
✆ 7 095 939 0126

Arlot Jean-Eudes
BDL
77 Av Denfert Rochereau
F 75014 Paris
France
☎ 33 1 40 51 2267
✆ 33 1 46 33 2834
✉ arlot@bdl.fr

Armandroff Taft E
NOAO/KPNO
Box 26732
950 N Cherry Av
Tucson AZ 85726 6732
USA
☎ 1 520 325 9382
✆ 1 520 325 9360
✉ armand@noao.edu

Armstrong John Thomas
USNO
Astrometry Dpt
3450 Massachusetts Av NW
Washington DC 20392 5100
USA
☎ 1 202 653 1769
✉ tarmstr@atlas.usno.navy.mil

Arnal Marcelo Edmundo
IAR
CC 5
1894 Villa Elisa (Bs As)
Argentina
☎ 54 21 4 3793
✆ 54 21 211 761
✉ arnal@irma.edu.ar

Arnaud Jean-Paul
OMP
14 Av E Belin
F 31400 Toulouse Cdx
France
☎ 33 5 61 33 2929
✆ 33 5 61 53 6722
✉ arnaud@obs-mip.fr

Arnaud Monique
DAPNIA/SAP
CEA Saclay
BP 2
F 91191 Gif s Yvette Cdx
France
☎ 33 1 69 08 7017
✆ 33 1 69 08 6577

Arnett W David
Steward Observ
Univ of Arizona
Tucson AZ 85721
USA
☎ 1 520 621 9587
✆ 1 520 621 1532
✉ dave@bohr.physics.arizona.edu

Arnold James R
Dpt Chemstry
UCSD
Box 017
La Jolla CA 92093 0216
USA
☎ 1 619 534 2908
✆ 1 619 534 2294

Arnould Marcel L
Inst Astrophysique
Univ Bruxelles
Campus Plaine CP 226
B 1050 Brussels
Belgium
☎ 32 2 650 2864
✆ 32 2 650 4226
✉ marnould@astro.ulb.ac.be

Arny Thomas T
Dpt Phys/Astronomy
Univ Massachusetts
GRC B
Amherst MA 01003
USA
☎ 1 413 545 2194

Arons Jonathan
Astronomy Dpt
Univ of California
601 Campbell Hall
Berkeley CA 94720 3411
USA
☎ 1 415 642 4730
✆ 1 510 642 3411

Arp Halton
MPA
Karl Schwarzschildstr 1
D 85748 Garching
Germany
☎ 49 893 299 880
✆ 49 893 299 3235

Arpigny Claude
Inst Astrophysique
Universite Liege
Av Cointe 5
B 4000 Liege
Belgium
☎ 32 4 254 7510
✆ 32 4 254 7511
✉ arpigny@astro.ulg.ac.be

Arribas Santiago
IAC
Observ d Teide
via Lactea s/n
E 38200 La Laguna
Spain
☎ 34 2 232 9100
✆ 34 2 232 9117

Arsenijevic Jelisaveta
Astronomical Observ
Volgina 7
11050 Beograd
Yugoslavia FR
☎ 381 1 419 357/421 875
✆ 381 1 419 553
✉ jarsenijevic@aob.aob.bg.ac.yu

Arshakian Tigran G
Byurakan Astrophys Observ
Armenian Acad Sci
378433 Byurakan
Armenia
☎ 374 88 52 28 3453/4142
✆ 374 88 52 52 3640
✉ tigar@bao.sci.am

Arthur David W G
USGS
Br of Astrogeology
2255 N Gemini Dr
Flagstaff AZ 86001
USA

Arthur Jane
Instituto Astronomia
UNAM
Apt 70 264
Mexico DF 04510
Mexico
☎ 52 5 622 3910
℻ 52 5 616 0653
✉ jane@astrosmo.unam.mx

Artru Marie-Christine
CRAL
ENS
46 All Italie
F 69364 Lyon Cdx 07
France
☎ 33 4 72 72 8000
℻ 33 4 72 72 8080

Artymowicz Pawel
Stockholm Observ
Royal Swedish Acad Sciences
S 133 36 Saltsjoebaden
Sweden
☎ 46 8 716 4461
℻ 46 8 717 4719
✉ pawel@astro.su.se

Artyukh Vadim S
Lebedev Physical Inst
Acad Sciences
Leninsky Pspt 53
117924 Moscow
Russia
☎ 7 095 923 3558
℻ 7 095 135 7880
✉ artyukh@rasfian.
serpukhov.su

Artzner Guy
IAS
Bt 121
Universite Paris XI
F 91405 Orsay Cdx
France
☎ 33 1 69 85 8584/8525
℻ 33 1 69 85 8675

Aschenbach Bernd
MPE
Postfach 1603
D 85740 Garching
Germany
☎ 49 893 299 00
℻ 49 893 299 3569

Aschwanden Markus
LMMS
Dpt H112 Bg 252
3251 Hanover St.
Palo Alto CA 94304 1191
USA
☎ 1 650 424 4001
℻ 1 650 424 3994
✉ aschwanden@sag.
1msal.com

Ashley Michael
School of Physics
UNSW
Sydney NSW 2052
Australia
☎ 61 2 9385 5465
℻ 61 2 663 3420
✉ mcba@newt.phys.unsw.
edu.au

Ashok N M
PRL
Navrangpura
Ahmedabad 380 009
India
☎ 91 272 46 2129
℻ 91 272 44 5292
✉ ashok@prl.ernet.in

Aslan Zeki
Akdeniz Universitesi
Fen-Edebiyat Fakultesi
Fizik Bolumu PK 510
Antalya
Turkey

Aslanov I A
Shemakha Astrophysical
Observ
Azerbajan Acad Sci
373243 Shemakha
Azerbajan
☎ 994 89 22 39 8248

Aspin Colin
Joint Astronomy Ctr
660 N A'ohoku Pl
University Park
Hilo HI 96720
USA
☎ 1 808 961 3756
℻ 1 808 961 6516
✉ caa@jach.hawaii.edu

Asseo Estelle
CTP
Ecole Polytechnique
F 91128 Palaiseau Cdx
France
☎ 33 1 69 41 8200

Assousa George Elias
545 Boylston St
Suite 901
Boston MA 02116
USA

Assus Pierre
OCA
Observ Nice
BP 139
F 06304 Nice Cdx 4
France
☎ 33 4 92 00 3086
℻ 33 4 92 00 3033

Asteriadis Georgios
Dpt Geodesy/Astron
Univ Thessaloniki
GR 540 06 Thessaloniki
Greece
☎ 30 31 99 2693

Atac Tamer
Kandilli Observ
Bogazici Univ
Cengelkoy
81220 Istanbul
Turkey
☎ 90 216 308 0514
℻ 90 216 332 1711
✉ atac@boun.edu.tr

Atanasijevic Ivan
Fac of Sciences
NL 6500 GL Nijmegen
Netherlands

Athanassoula Evangelia
Observ Marseille
2 Pl Le Verrier
F 13248 Marseille Cdx 04
France
☎ 33 4 91 95 9088
℻ 33 4 91 62 1190

Athay R Grant
1818 N Sundai
Nesa AZ 85205
USA
✉ athay@ncar.ucar.edu

Atherton Paul David
Astrophysics Gr
Imperial Coll
Prince Consort Rd
London SW7 2BZ
UK
☎ 44 171 594 7771
℻ 44 171 594 7772

Atkinson David H
Dpt Electric Engin
Univ of Idaho
Moscow ID 83844 1023
Russia
☎ 7 208 885 6870
℻ 7 208 885 7579
✉ atkinson@maxwell.ee.
uidaho.edu

Atreya Sushil K
Space Physics Res Lb
Univ of Michigan
2455 Hayward St
Ann Arbor MI 48109 2143
USA
☎ 1 734 763 6234
℻ 1 734 764 5137
✉ atreya@umich.edu

Aubier Monique G
Observ Paris Meudon
ARPEGES
Pl J Janssen
F 92195 Meudon PPL Cdx
France
☎ 33 1 45 34 7755
℻ 33 1 45 07 7971

Audouze Jean
IAP
98bis bd Arago
F 75014 Paris
France
☎ 33 1 44 32 8133
℻ 33 1 44 32 8001

Auer Lawrence H
LANL
MS F665 ESS 5
Box 1663
Los Alamos NM 87545 2345
USA
☎ 1 505 667 5824
℻ 1 505 665 4055

Augarde Renee
Observ Marseille
2 Pl Le Verrier
F 13248 Marseille Cdx 04
France
☎ 33 4 91 95 9088
℻ 33 4 91 62 1190

Augason Gordon C
NASA/ARC
MS 245 6
Moffett Field CA 94035 1000
USA
☎ 1 415 694 4156
℻ 1 415 604 6779

Auluck Faqir Chand
Dpt Phys/Astrophys
Univ of Delhi
New Delhi 110 007
India
☎ 91 11 291 8993

Auman Jason R
Dpt Geophys/Astronomy
UBC
2075 Wesbrook Pl
Vancouver BC V6T 1W5
Canada
☎ 1 604 228 2892
℻ 1 604 228 6047

Auner Gerhard
Inst Astronomie
Univ Wien
Tuerkenschanzstr 17
A 1180 Wien
Austria
☎ 43 1 345 3600
℻ 43 1 470 6015

Aurass Henry
Zntrlinst Astrophysik
Sternwarte Babelsberg
Rosa Luxemburg Str 17a
D 14473 Potsdam
Germany
☎ 49 331 762225

Auriemma Giulio
Dpt Fisica
Univ d Roma
Pl A Moro 2
I 00185 Roma
Italy
☎ 39 6 49 76336
℻ 39 6 202 3507

Auriere Michel
OMP
14 Av E Belin
F 31400 Toulouse Cdx
France
☎ 33 5 62 95 1969
℻ 33 5 61 27 3179

Auvergne Michel
Observ Paris Meudon
DASGA
Pl J Janssen
F 92195 Meudon PPL Cdx
France
☎ 33 1 45 07 7847
℻ 33 1 45 07 7878

Avcioglu Kamuran
Univ Observ
Univ of Istanbul
University 34452
34452 Istanbul
Turkey
☎ 90 212 522 3597
℻ 90 212 519 0834

Avery Lorne W
Joint Astronomy Ctr
660 N A'ohoku Pl
University Park
Hilo HI 96720
USA
☎ 1 808 969 6551
✆ 1 808 961 6516
✉ lwa@jach.hawaii.edu

Avgoloupis Stavros
Astronomy Lab
Univ Thessaloniki
GR 540 06 Thessaloniki
Greece
☎ 30 31 99 1357

Avignon Yvette
Observ Paris Meudon
F 92195 Meudon PPL Cdx
France

Avrett Eugene H
CfA
HCO/SAO
60 Garden St
Cambridge MA 02138 1516
USA
☎ 1 617 495 7423
✆ 1 617 495 7356
✉ avrett@cfa.harvard.edu

Awad Mervat El-Said
Astronomy Dpt
Fac of Sciences
Cairo Univ
Geza
Egypt
☎ 20 2 572 7022
✆ 20 2 572 7556

Awadalla Nabil Shoukry
Physics Dpt
Univ of Bahrain
Box 32038
Bahrain
State of Bahrain
☎ 9739 681 234
✆ 9739 682 582

Awaki Hisamitsu
Physics Dpt
Kyoto Univ
Sakyo-ku
Kyoto 606 01
Japan
☎ 81 757 53 3851
✆ 81 75 701 5377
✉ awaki@cr.scphys.kyoto.u.ac.jp

Axford W Ian
MPI Aeronomie
Max Planck Str 2
Postfach 20
D 37189 Katlenburg Lindau
Germany
☎ 49 555 641 414
✆ 49 555 965 527

Axon David
NRAL
Univ Manchester
Jodrell Bank
Macclesfield SK11 9DL
UK
☎ 44 14 777 1321
✆ 44 147 757 1618
✉ dja@ast.man.ac.uk

Ayani Kazuya
Bisei Astronomcal Obs
Ohkura
BISEI
Okayama 714 14
Japan
☎ 81 866 87 4222
✆ 81 866 87 4224
✉ ayani@bao.go.jp

Aydin Cemal
Astronomy Dpt
Univ of Ankara
Fen Fakultesi
06100 Besevler
Turkey
☎ 90 41 23 2105*94
✆ 90 312 223 2395

Ayres Thomas R
CASA
Univ of Colorado
Campus Box 389
Boulder CO 80309 0391
USA
☎ 1 303 492 5320
✆ 1 303 492 5941

Azcarate Ismael N
IAR
CC 5
1894 Villa Elisa (Bs As)
Argentina
☎ 54 21 740 230
✆ 54 21 254 909
✉ azcarate@irma.iar.unlp.edu.ar

Azuma Takahiro
Dokkyo Univ
SOKA
Saitama 340
Japan
☎ 81 489 42 1111

Azzopardi Marc
Observ Marseille
2 Pl Le Verrier
F 13248 Marseille Cdx 04
France
☎ 33 4 91 95 9088
✆ 33 4 91 62 11 90
✉ azzopardi@obmara.cnrs-mrs.fr

B

Baade Dietrich
ESO
ST/ECF
Karl Schwarzschildstr 2
D 85748 Garching
Germany
☎ 49 893 200 6388
✆ 49 893 202 362

Baan Willem A
Arecibo Observ
Box 995
Arecibo PR 00613
USA
☎ 1 787 878 2612
✆ 1 787 878 1861
✉ willem@naic.edu

Baars Jacob W M
LMT Project Office
Univ Massachusetts
815 Lederle Grad Res Ctr
Amherst MA 01003
USA
☎ 1 413 545 2404
✆ 1 413 545 3192
✉ jbaars@phast.umass.de

Baart Edward E
Dpt Phys/Electronics
Rhodes Univ
Box 94
6140 Grahamstown
South Africa
☎ 27 461 318 454/318450
✆ 27 461 250 49

Baath Lars B
Onsala Space Observ
Chalmers Technical Univ
S 439 92 Onsala
Sweden
☎ 46 31 772 5500
✆ 46 31 772 5550

Baba Naoshi
Applied Phys Dpt
Hokkaido Univ
N 13 W 8
Sapporo 060
Japan
☎ 81 117 06 6627
✆ 81 117 16 6175
✉ babal@optica.huap.
hokudai.ac.jp

Babadzhanianc Michail K
Astronomical Observ
St Petersburg Univ
Bibliotechnaja Pl 2
198904 St Petersburg
Russia
☎ 7 812 257 9491
✆ 7 812 428 4259

Babadzhanov Pulat B
Astrophys Inst
Uzbek Acad Sci
Sviridenko Ul 22
734670 Dushanbe
Tajikistan
☎ 7 3770 23 1432
✆ 7 3770 27 5483

Babcock Horace W
Carnegie Observatories
813 Santa Barbara St
Pasadena CA 91101 1292
USA
☎ 1 818 577 1122
✆ 1 818 795 8136

Babel Jacques
36 r d Battieux
CH 2000 Neuchatel
Switzerland
☎ 41 31 323 2381
✉ babelj@bluewin.ch

Babin Arthur
Crimean Astrophys Obs
Ukrainian Acad Science
Nauchny
334413 Crimea
Ukraine
☎ 380 6554 71161
✆ 380 6554 40704

Babin Valerij G
Sibizmir
Acad Sciences
664697 Irkutsk 33
Russia
☎ 7 395 262 9383

Babu G S D
IIA
Koramangala
Sarjapur Rd
Bangalore 560 034
India
☎ 91 80 356 9179
✆ 91 80 553 4043

Bacchus Pierre
40 r Haute
F 77130 La Gde Paroisse
France
☎ 33 1 64 32 1315

Bachiller Rafael
Observ Astronomico
Nacional
Apt 1143
E 28800 Alcala d Henares
Spain
☎ 34 1 885 5060
✆ 34 1 885 5062
✉ bachiller@oan.es

Backer Donald Ch
Radio Astronomy Lab
Univ of California
601 Campbell Hall
Berkeley CA 94720
USA
☎ 1 415 642 5128
✆ 1 415 642 3411
✉ dbacker@bkypsr.
berkerely.edu

Bade Norbert
Hamburger Sternwarte
Univ Hamburg
Gojensbergsweg 112
D 21029 Hamburg
Germany
☎ 49 407 252 4137
✆ 49 407 252 4198
✉ nbade@hs.uni.hamburg.de

Badiali Massimo
IAS
Area d Ricerca CNR
Via Fosso Cavaliere 100
I 00133 Roma
Italy
☎ 39 6 4993 4473
✆ 39 6 2066 0188
✉ massimo@saturn.ias.
rm.cnr.it

Badolati Ennio
Via G Cotronei 11
I 80129 Napoli
Italy
☎ 39 81 243 245

Baeck Nicole A I
Math Physics/Astronomy
Universiteit Gent
Krijgslaan 281 S9
B 9000 Gent
Belgium
☎ 32 92 64 4764
✆ 32 92 64 4989
✉ nicole.baeck@izar.rug.ac.be

Baek Chang Ryong
Physics Dpt
Kim Il Sung Univ
Taesong district
Pyongyang
Korea DPR

Baerentzen Jorn
Elmehojvej 66
DK 8270 Hojbjerg
Denmark
☎ 45 86 27 2428

Baessgen Martin
Astronomisches Institut
Univ Tuebingen
Waldhaeuserstr 64
D 72076 Tuebingen
Germany
☎ 49 707 129 5470
✆ 49 707 129 3458

Baffa Carlo
Osserv Astrofis Arcetri
Univ d Firenze
Largo E Fermi 5
I 50125 Firenze
Italy
☎ 39 55 27 52 298
✆ 39 55 22 0039
✉ baffa@arcetri.astro.it

Bagare S P
IIA
Koramangala
Sarjapur Rd
Bangalore 560 034
India
☎ 91 80 356 6585/497
✆ 91 80 553 4043

Baggaley William J
Dpt Phys/Astronomy
Univ of Canterbury
Private Bag 4800
Christchurch 1
New Zealand
☎ 64 3 366 7001*7559
✆ 64 3 364 2469
✉ phys051@csc.canterbury.
ac.nz

Bagildinskij Bronislav K
Pulkovo Observ
Acad Sciences
10 Kutuzov Quay
196140 St Petersburg
Russia
☎ 7 812 298 2242
✆ 7 812 315 1701

Baglin Annie
Observ Paris Meudon
DASGAL
Pl J Janssen
F 92195 Meudon PPL Cdx
France
☎ 33 1 47 05 7855
✆ 33 1 45 07 7878

Bagnuolo William G
CHARA
Georgia State Univ
Atlanta GA 30303 3083
USA
☎ 1 404 651 2932
✆ 1 404 542 2492

Bagoly Zsolt
Dpt Atomic Physics
Eotvos Univ
Pushkin U 5 7
H 1088 Budapest
Hungary
☎ 36 1 187 902
✆ 36 1 118 0206

Bagri Durgadas S
NRAO
Box 0
Socorro NM 87801 0387
USA
☎ 1 505 772 4011
✆ 1 505 835 7027

Bahcall John N
Inst Advanced Study
School Natural Science
Olden LN Bg E
Princeton NJ 08540
USA
☎ 1 609 734 8054
✆ 1 609 924 7592

Bahcall Neta A
Peyton Hall
Princeton Univ
Princeton NJ 08544 1001
USA
☎ 1 609 258 6065
✆ 1 609 258 1020
✉ Neta@astro.princeton.edu

Bahner Klaus
MPI Astronomie
Adolf Kolping Str 5
D 6903 Neckargemuend
Germany
☎ 49 62 23 3735

Baier Frank
Zntrlinst Astrophysik
Sternwarte Babelsberg
Rosa Luxemburg Str 17a
D 14482 Potsdam
Germany
☎ 49 331 749 9208
✆ 49 331 749 9309
✉ fbaier@aip.de

Bailey Jeremy A
AAO
Box 296
Epping NSW 2121
Australia
☎ 61 2 868 1666
✆ 61 2 9876 8536

Bailey Mark Edward
Armagh Observ
College Hill
Armagh BT61 9DG
UK
☎ 44 18 61 522 928
✆ 44 18 61 527 174
✉ meb@star.arm.ac.uk

Baird Scott R
Dpt Phys/Astronomy
Benedictine College
N 14
Atchison KS 66002 1499
USA
☎ 1 913 367 5340

Bajaja Esteban
IAR
CC 5
1894 Villa Elisa (Bs As)
Argentina
☎ 54 21 4 3793
✆ 54 21 211 761

Bajtlik Stanislaw
Copernicus Astron Ctr
Polish Acad Sci
Ul Bartycka 18
PL 00 716 Warsaw
Poland
☎ 48 22 41 1086
✆ 48 22 41 0828
✉ bajtlik@camk.edu.pl

Baker James Gilbert
14 French Dr
Bedford NH 03102
USA
☎ 1 603 472 5860

Baker Joanne
MRAO
Cavendish Laboratory
Madingley Rd
Cambridge CB3 0HE
UK
☎ 44 12 23 337 298
✆ 44 12 23 354 599
✉ jcb@mrao.cam.ac.uk

Baker Norman H
Astronomy Dpt
Columbia Univ
538 W 120th St
New York NY 10027
USA
☎ 1 212 280 3280
✆ 1 212 316 9504

Bakry Abdel-aziz
El Azhar Univ
Fac Sciences
Astronomy Dpt
Cairo
Egypt
☎ 20 2 262 9357/58
✆ 20 2 261 1404
✉ zcc3159@kaau.edu.sa

Balachandran Suchitra C
Astronomy Dpt
Univ of Maryland
College Park MD 20742 2421
USA
☎ 1 301 405 3101
✆ 1 301 314 9067
✉ suchitra@astro.umd.edu

Balasubramanian V
TIFR/Radio Astronomy Ctr
Box 8
Udhagamandalam 643 001
India
☎ 91 423 2032

Balazs Bela A
Astronomy Dpt
Eotvos Univ
Ludovika ter 2
H 1083 Budapest
Hungary
☎ 36 1 114 1019
✆ 36 1 210 1089
✉ bb@innin.elte.hu

Balazs Lajos G
Konkoly Observ
Thege U 13/17
Box 67
H 1525 Budapest
Hungary
☎ 36 1 375 4122
✆ 36 1 275 4668
✉ balazs@ogyalla.
konkoly.hu

Balbus Steven A
University Station
Univ of Virginia
Box 3818
Charlottesville VA 22903 0818
USA
☎ 1 804 924 4897
✉ sb@virginia.edu

Baldinelli Luigi
CP 1630
I 40100 Bologna
Italy
☎ 39 51 22 7002

Baldwin Jack A
NOAO
CTIO
Casilla 603
La Serena
Chile
☎ 56 51 22 5415
✆ 56 51 20 5342

Baldwin John E
MRAO
Cavendish Laboratory
Madingley Rd
Cambridge CB3 0HE
UK
☎ 44 12 23 337 294
✆ 44 12 23 354 599

Baldwin Ralph B
6190 Gatehouse Dr SE
Grand Rapids MI 49506
USA
☎ 1 619 949 6190

Balega Yuri Yu
SAO
Acad Sciences
Nizhnij Arkhyz
357147 Karachaevo
Russia
☎ 7 878 789 2501
✆ 7 901 498 2931
✉ balega@sao.ru

Balick Bruce
Astronomy Dpt
Univ of Washington
Box 351580
Seattle WA 98195 1580
USA
☎ 1 425 543 7683
✆ 1 425 685 0403

Baliunas Sallie L
CfA
HCO/SAO MS 15
60 Garden St
Cambridge MA 02138 1516
USA
☎ 1 617 495 7415
✆ 1 617 495 7049
✉ baliunas@cfa.harvard.edu

Balklavs A E
Radioastrophys Observ
Latvian Acad Sci
Turgeneva 19
LV 226524 Riga
Latvia

Balkowski-Mauger Chantal
Observ Paris Meudon
DAF
Pl J Janssen
F 92195 Meudon PPL Cdx
France
☎ 33 1 45 34 7556
✆ 33 1 45 07 7469

Ball John A
Haystack Observ
Westford MA 01886
USA
☎ 1 617 692 4764
✆ 1 617 981 0590

Ball Lewis
Theoretical Astrophysics
Univ of Sydney
Sydney NSW 2006
Australia
☎ 61 2 9351 2621
✆ 61 2 660 2903
✉ ball@physics.usyd.
edu.au

Ballabh G M
Astronomy Dpt
Univ of Osmania
Hyderabad 500 007
India
☎ 91 71 951*247

Ballario M C
Osserv Astrofis Arcetri
Univ d Firenze
Largo E Fermi 5
I 50125 Firenze
Italy
☎ 39 55 27 521
✆ 39 55 22 0039

Ballereau Dominique
Observ Paris Meudon
DASGAL
Pl J Janssen
F 92195 Meudon PPL Cdx
France
☎ 33 1 45 07 7854
✆ 33 1 45 07 7878

Ballester Jose Luis
Dpt Fisica
Univ Las Islas
Baleares
E 07071 Palma de Mallorca
Spain
☎ 34 7 117 3228
✆ 34 7 117 3426
✉ dfsjlb0@ps.uib.es

Balli Edibe
Univ Rasathanesi
Istanbul
Turkey

Bally John
CASA
Univ of Colorado
Campus Box 389
Boulder CO 80309 0389
USA
☎ 1 303 492 5786
℻ 1 303 492 7178
✉ bally@janos.colorado.edu

Balmino Georges G
CNES/GRGS/BGI
18 Av E Belin
F 31055 Toulouse Cdx
France
☎ 33 5 61 27 4427
℻ 33 5 61 27 3179

Balona Luis Antero
SAAO
PO Box 9
7935 Observatory
South Africa
☎ 27 21 47 0025
℻ 27 21 47 3639
✉ lab@saao.ac.za

Balonek Thomas J
Dpt Phys/Astronomy
Colgate Univ
Hamilton NY 13346
USA
☎ 1 315 228 7767
℻ 1 315 228 7187
✉ tbalonek@colgateu.edu

Balthasar Horst
AIP
SOE
Telegrafenberg
D 14473 Potsdam
Germany
☎ 49 331 288 2341
℻ 49 331 288 2310
✉ hbalthasar@aip.de

Baluteau Jean-Paul
Observ Marseille
2 Pl Le Verrier
F 13248 Marseille Cdx 04
France
☎ 33 4 91 95 9088
℻ 33 4 91 62 1190
✉ obsmrs@fromrs51

Bandermann L W
21131 Grenola Dr
Cupertino CA 95014
USA

Bandiera Rino
Osserv Astrofis Arcetri
Univ d Firenze
Largo E Fermi 5
I 50125 Firenze
Italy
☎ 39 55 43 78540
℻ 39 55 43 5939
✉ bandiera@astrfi.infn.it

Bandyopadhyay A
Res Division
Birla Planetarium
96 Jawaharlal Nehru Rd
Calcutta 700 071
India
☎ 91 33 281 515

Banerji Sriranjan
Physics Dpt
Univ of Burdwan
Golopbag
Burdwan 713 104
India

Bang Yong Gol
Pyongyang Astron Obs
Acad Sciences DPRK
Taesong district
Pyongyang
Korea DPR
☎ 850 5 3134/5 & 5 3239

Banhatti Dilip Gopal
School of Physics
Madurai Kamaraj Univers
Palkalainagar
Madurai 625021
India
☎ 91 85 252

Bania Thomas Michael
Astronomy Dpt
Boston Univ
725 Commonwealth Av
Boston MA 02215
USA
☎ 1 617 353 3652
℻ 1 617 353 3200

Banni Aldo
Osserv Astronomico
d Cagliari
Poggio d Pini 54
I 09012 Capoterra
Italy
☎ 39 70 725 246
℻ 39 70 72 5425
✉ banni@ca.astro.it

Banos Cosmas J
Astronomical Institute
Ntl Observ Athens
Box 20048
GR 118 10 Athens
Greece
☎ 30 1 346 1191
℻ 30 1 346 4566

Banos George J
Physics Dpt
Univ of Ioannina
GR 451 10 Ioannina
Greece
☎ 30 651 98 471
℻ 30 65145 697

Bao Keren
Nanjing Astronomical
Instrument Factory
Box 846
Nanjing
China PR
☎ 86 25 46191
℻ 86 25 71 1256

Baranne Andre
Observ Marseille
2 Pl Le Verrier
F 13248 Marseille Cdx 04
France
☎ 33 4 91 95 9088
℻ 33 4 91 62 1190

Baranovsky Edward A
Crimean Astrophys Obs
Ukrainian Acad Science
Nauchny
334413 Crimea
Ukraine
☎ 380 6554 71161
℻ 380 6554 40704

Baratta Giovanni Battista
OAR
Via d Parco Mellini 84
I 00136 Roma
Italy
☎ 39 6 34 7056
℻ 39 6 349 8236

Baratta Giuseppe Antonio
Osserv Astronomico
d Catania
Via A Doria 6
I 95125 Catania
Italy
☎ 39 95 733 21111
℻ 39 95 33 0592

Barba Rodolfo Hector
Observ Astronomico
Paseo d Bosque S/n
1900 La Plata (Bs As)
Argentina
☎ 54 21 217 308
℻ 54 21 211 761
✉ rbarba@fcaglp.edu.ar

Barbanis Basil
Astronomy Lab
Univ Thessaloniki
GR 540 06 Thessaloniki
Greece
☎ 30 31 99 1357
℻ 30 31 99 8211

Barbaro G
Osserv Astronom Padova
Univ d Padova
Vic d Osservatorio 5
I 35122 Padova
Italy
☎ 39 49 829 3411
℻ 39 49 875 9840

Barberis Bruno
Ist Fisica/Matematica
Univ d Torino
Via C Alberto 10
I 10123 Torino
Italy
☎ 39 11 539 214

Barbier-Brossat Madeleine
Observ Marseille
2 Pl Le Verrier
F 13248 Marseille Cdx 04
France
☎ 33 4 91 95 9088
℻ 33 4 91 62 1190
✉ barbier@obmara.cnrs-mrs.fr

Barbieri Cesare
Dpt Astronomia
Univ d Padova
Vic d Osservatorio 5
I 35122 Padova
Italy
☎ 39 49 829 3411
℻ 39 49 875 9840
✉ barbieri@astropd.pd.astro.it

Barbon Roberto
Osserv Astrofisico
Univ d Padova
Via d Osservatorio 8
I 36012 Asiago
Italy
☎ 39 424 462 665
℻ 39 424 462 884

Barbosu Mihail
Fac of Mathematics
Univ of Cluj Napoca
Ul Kogalniceanu 1
RO 3400 Cluj Napoca
Rumania
☎ 40 64 194 315
℻ 40 64 111 905
✉ mbarbosu@math.ubbcluj.ro

Barbuy Beatriz
IAG
Univ Sao Paulo
CP 9638
01065 970 Sao Paulo SP
Brazil
☎ 55 11 577 8599*230
℻ 55 11 276 3848
✉ barbuy@orion.iagusp.usp.br

Barcia Alberto
Ctr Astron d Yebes
OAN
Apt 148
E 19080 Guadalajara
Spain
☎ 34 1 129 0311
℻ 34 1 129 0063
✉ barcia@cay.oan.es

Barcons Xavier
Istt Fisica d Cantabria
C/o Faculdad Ciencias
Avda Los Castros s/n
E 39005 Santander
Spain
☎ 34 4 220 1461
℻ 34 4 220 1459
✉ barcons@ifca.unican.es

Barcza Szabolcs
Konkoly Observ
Thege U 13/17
Box 67
H 1525 Budapest
Hungary
☎ 36 1 375 4122
℻ 36 1 275 4668

Bardeen James M
Physics Dpt
Univ of Washington
FM 15
Seattle WA 98195 1580
USA
☎ 1 425 545 2394
✆ 1 425 685 0403

Barden Samuel Charles
NOAO
950 N Cherry Av
Tucson AZ 85719
USA
☎ 1 520 318 8263
✆ 1 520 318 8360
✉ barden@noao.edu

Barge Pierre
LAS
Traverse du Siphon
Les Trois Lucs
F 13376 Marseille Cdx 12
France
☎ 33 4 91 05 5900
✆ 33 4 91 66 1855
✉ barge@astrsp-mrs.fr

Barkat Zalman
Racah Inst of Phys
Hebrew Univ Jerusalem
Jerusalem 91904
Israel
☎ 972 2 584 490
✆ 972 2 658 4374

Barker Edwin S
McDonald Observ
Univ of Texas
Box 1337
Fort Davis TX 79734 1337
USA
☎ 1 915 426 3263
✆ 1 915 426 3641
✉ esb@astro.as.utexas.edu

Barker Paul K
Dpt Electrophysics
Box 1143 Station B
London ON N6A 5K2
Canada
☎ 1 519 668 2871
✆ 1 519 668 2871

Barker Timothy
Dpt Phys/Astronomy
Wheaton College
Norton MA 02766
USA
☎ 1 508 285 7722

Barlai Katalin
Konkoly Observ
Thege U 13/17
Box 67
H 1525 Budapest
Hungary
☎ 36 1 375 4122
✆ 36 1 275 4668

Barletti Raffaele
Osserv Astrofis Arcetri
Univ d Firenze
Largo E Fermi 5
I 50125 Firenze
Italy
☎ 39 55 27 52278
✆ 39 55 22 0039
✉ barletti@arcetri.astro.it

Barlier Francois E
OCA
CERGA
F 06130 Grasse
France
☎ 33 4 93 36 5849
✆ 33 4 93 40 5333
✉ barlier@obs-azur.fr

Barlow Michael J
Dpt Phys/Astronomy
UCLO
Gower St
London WC1E 6BT
UK
☎ 44 171 387 7050
✆ 44 171 380 7145

Barnbaum Cecilia
STScI
Homewood Campus
3700 San Martin Dr
Baltimore MD 21218
USA
☎ 1 301 338 1059
✆ 1 301 338 4767
✉ barnbaum@stci.edu

Barnes Aaron
NASA/ARC
MS 245 3
Moffett Field CA 94035 1000
USA
☎ 1 415 694 5506
✆ 1 415 604 6779

Barnes III Thomas G
McDonald Observ
Univ of Texas
Rlm 15 308
Austin TX 78712 1083
USA
☎ 1 512 471 1301
✆ 1 512 426 3641
✉ tgb@astro.as.utexas.edu

Barocas Vinicio
11 Yewlands Av
Fulword
Preston PR2 4QR
UK
☎ 44 177 271 9249

Barone Fabrizio
Dpt Scienze Fisiche
Univ d Napoli
Mostra D Oltremare Pad 19
I 80125 Napoli
Italy
☎ 39 81 725 3447
✆ 39 81 61 4508
✉ fbarone@napoli.infn.it

Barreto Luiz Muniz
Observ Nacional
Rua Gl Bruce 586
Sao Cristovao
20921 030 Rio de Janeiro RJ
Brazil
☎ 55 21 580 7313
✆ 55 21 580 0332

Barrett Paul Everett
NASA GSFC
Code 668 1
Greenbelt MD 20771
USA
✉ paul.e.barrett.1@gsfc.nasa.gov

Barriot Jean-Pierre
OMP
14 Av E Belin
F 31400 Toulouse Cdx
France
☎ 33 5 61 33 2894
✆ 33 5 61 25 3098
✉ barriot@sc2000.cst.cnes.fr

Barroso Jair
Observ Nacional
Rua Gl Bruce 586
Sao Cristovao
20921 030 Rio de Janeiro RJ
Brazil
☎ 55 21 580 7313*273
✆ 55 21 580 0332

Barrow Colin H
MPI Aeronomie
Max Planck Str 2
Postfach 20
D 37189 Katlenburg Lindau
Germany
☎ 49 555 640 11
✆ 49 555 640 1240

Barrow John David
Astronomy Centre
Univ of Sussex
Falmer
Brighton BB1 9QH
UK
☎ 44 12 73 60 6755
✆ 44 12 73 678 097

Barrow Richard F
Physical Chemistry Lab
Univ of Oxford
Keble Rd
Oxford OX1 3QH
UK
☎ 44 186 553 322
✆ 44 186 527 3947

Barsony Mary
Physics Dpt
Univ of California
Riverside
Riverside CA 92521
USA
☎ 1 909 787 3984
✆ 1 909 787 4529
✉ fun@nusun.ucr.edu

Barstow Martin Adrian
Dpt Phys/X-Ray Astron
Univ Leicester
University Rd
Leicester LE1 7RH
UK
☎ 44 116 252 2073
✆ 44 116 250 182
✉ mab@uk.ac.le.star

Bartasiute Stanislava
Astronomical Observ
Ciurlionio 29
Vilnius 2009
Lithuania
☎ 370 2 633 343
✆ 370 2 223 563
✉ bartasiute@ff.vu.lt

Bartaya R A
Abastumani Astrophysical Observ
Georgian Acad Sci
380060 Tbilisi
Georgia
☎ 995 88 32 37 5226
✆ 995 88 32 98 5017

Bartel Norbert Harald
Physics Dpt
York Univ
4700 Keele St
North York ON M3J 1P3
Canada
☎ 1 416 736 5424
✆ 1 416 736 5516

Bartelmann Matthias
MPA
Karl Schwarzschildstr 1
D 85748 Garching
Germany
☎ 49 893 299 3236
✆ 49 893 299 3235
✉ mbartelmann@mpa.garching.mpg.de

Barth Charles A
LASP
Univ of Colorado
Campus Box 392
Boulder CO 80309 0392
USA
☎ 1 303 492 7502
✆ 1 303 492 6946

Barthel Peter
Kapteyn Sterrekundig Inst
Univ Groningen
Postbus 800
NL 9700 AV Groningen
Netherlands
☎ 31 50 363 4073
✆ 31 50 363 6100

Bartholdi Paul
Observ Geneve
Chemin d Maillettes 51
CH 1290 Sauverny
Switzerland
☎ 41 22 755 2611
✆ 41 22 755 3983
✉ bartho@obs.unige.ch

Bartkevicius Antanas
Inst Theor Physics/
Astronomy
Gostauto 12
Vilnius 2600
Lithuania
☎ 370 2 613 440
℻ 370 2 224 694
✉ bart@itpa.fi.lt

Bartolini Corrado
Dpt Astronomia
Univ d Bologna
Via Zamboni 33
I 40126 Bologna
Italy
☎ 39 51 259 301
℻ 39 51 259 407

Barucci Maria A
Observ Paris Meudon
DESPA
Pl J Janssen
F 92195 Meudon PPL Cdx
France
☎ 33 1 45 07 7539
℻ 33 1 45 07 7469

Baruch John
Dpt Industrial Technology
Univ of Bradford
Bradford BD7 1DP
UK
☎ 44 12 74 38 4024
℻ 44 12 74 38 4270
✉ j.e.f.baruch@bradford.
ac.uk

Barvainis Richard
Haystack Observ
Westford MA 01886
USA
☎ 1 617 692 4764
℻ 1 617 981 0590

Barwig Heinz
Inst Astron/Astrophysik
Univ Sternwarte
Scheinerstr 1
D 81679 Muenchen
Germany
☎ 49 899 890 21
℻ 49 899 220 9427

Baryshev Yuri V
Astron Inst St Petersburg
State Univ
Bibliotechnaya Pl 2
St Petersburg 198904
Russia
☎ 7 812 428 7129
℻ 7 812 428 4259
✉ yuba@aispbu.spb.su

Basart John P
Physics Dpt
Iowa State Univ
Ames IA 50011
USA
☎ 1 515 294 2667
℻ 1 515 294 3262

Baschek Bodo
Inst Theor Astrophysik
d Univer
Im Neuenheimer Feld 561
D 69120 Heidelberg
Germany
☎ 49 622 156 2837
℻ 49 622 154 4221

Bash Frank N
Astronomy Dpt
Univ of Texas
Rlm 15 308
Austin TX 78712 1083
USA
☎ 1 512 471 4461
℻ 1 512 471 6016
✉ fnb@astro.as.utexas.edu

Basri Gibor B
Astronomy Dpt
Univ of California
601 Campbell Hall
Berkeley CA 94720 3411
USA
☎ 1 415 642 8198
℻ 1 510 642 3411

Bassino Lilia P
Observ Astronomico
Paseo d Bosque S/n
1900 La Plata (Bs As)
Argentina
☎ 54 21 217 308
℻ 54 21 258 985
✉ lbassino@fcaglp.edu.ar

Bastian Timothy Stephen
NRAO
Box 0
Socorro NM 87801 0387
USA
☎ 1 505 835 7259
℻ 1 505 835 7027
✉ tbastian@nrao.edu

Bastian Ulrich
ARI
Moenchhofstr 12-14
D 69120 Heidelberg
Germany
☎ 49 622 140 5152
℻ 49 622 140 5297
✉ s01@ix.urz.uni-
heidelberg.de

Bastien Pierre
Dpt d Physique
Universite Montreal
CP 6128 Succ A
Montreal QC H3C 3J7
Canada
☎ 1 514 343 7355
℻ 1 514 343 2071

Bastin John A
Astrophysics Group
QMWC
Mile End Rd
London E1 4NS
UK
☎ 44 171 975 5555
℻ 44 181 975 5500

Basu Baidyanath
75/3A Hazra Rd
Calcutta 700 029
India
☎ 91 474 5489

Basu Dipak
Physics Dpt
Carleton Univ
Ottawa ON K1S 5B6
Canada
☎ 1 613 788 4377
℻ 1 613 788 4061
✉ basu@garm.physics.
carleton.ca

Basurah Hassan
Astronomy Dpt
Box 9028
KAAU
Jeddah 21413
Saudi Arabia

Batalha Celso Correa
Observ Nacional
Rua Gl Bruce 586
Sao Cristovao
20921 030 Rio de Janeiro RJ
Brazil
☎ 55 21 580 7181
℻ 55 21 580 0332
✉ ccb@lnccum

Batchelor David Allen
NASA GSFC
Code 632
Greenbelt MD 20771
USA
☎ 1 301 286 2988

Bates Brian
Dpt Pure/Appl Physics
Queen's Univ
Belfast BT7 1NN
UK
☎ 44 12 32 24 5133
℻ 44 12 32 43 8918
✉ b.bates@queens-belfast.
ac.uk

Bateson Frank M OBE
Astronomical Res Ltd
Box 3093
Greerton Tauranga
New Zealand
☎ 64 75 410 216

Bath Geoffrey T
Astrophysics Dpt
Univ of Oxford
Keble Rd
Oxford OX1 3RQ
UK
☎ 44 186 551 1336
℻ 44 186 527 3947

Batrakov Yuri V
Inst Theoret Astron
Acad Sciences
N Kutuzova 10
191187 St Petersburg
Russia
☎ 7 812 272 40 23
℻ 7 812 272 7968

Batson Raymond Milner
USGS
Br of Astrogeology
2255 N Gemini Dr
Flagstaff AZ 86001
USA
☎ 1 602 556 7260
℻ 1 602 556 7090

Battaner Eduardo
Dpt Fisica Teorica/Cosmos
c/o Fac Ciencias
Avda Fuentenueva
E 18002 Granada
Spain
☎ 34 5 824 3305
℻ 34 5 824 4012

Batten Alan H
NRCC/HIA
DAO
5071 W Saanich Rd
Victoria BC V8X 4M6
Canada
☎ 1 250 363 0009
℻ 1 250 363 0045
✉ alan.batten@hia.nrc.ca

Battinelli Paolo
OAR
Via d Parco Mellini 84
I 00136 Roma
Italy
☎ 39 6 34 7056
℻ 39 6 349 8236

Battistini Pierluigi
Osserv Astronom Bologna
Univ d Bologna
Via Zamboni 33
I 40126 Bologna
Italy
☎ 39 51 51 9593
℻ 39 51 25 9407

Batty Michael
School Maths/Physics
Computing/ Electron
Macquarie Univ
Macquarie 2109
Australia

Baud Boudewijn
Fokker Bv
Space Division
Box 7600
NL 1117 ZJ Shiphol
Netherlands
☎ 31 20 544 9111

Baudry Alain
Observ Bordeaux
BP 89
F 33270 Floirac
France
☎ 33 5 56 86 4330
℻ 33 5 56 40 4251

Bauer Carl A
Astronomy Dpt
Pennsylvania State Univ
525 Davey Lab
University Park PA 16802
USA
☎ 1 814 865 3631
℻ 1 814 863 7114

Bauer Wendy Hagen
Whitin Observ
Wellesley Coll
Wellesley MA 02181
USA
☎ 1 617 235 0320
℻ 1 617 283 3667
✉ w bauer@lucy.wellesley.edu

Baum Stef Alison
STScI
Homewood Campus
3700 San Martin Dr
Baltimore MD 21218
USA
☎ 1 301 338 4797
℻ 1 301 338 4767
✉ sbaum@stsci.edu

Baum William A
2124 Ne Park Rd
Seattle WA 98105
USA

Baustian W W
NOAO/KPNO
Box 26732
950 N Cherry Av
Tucson AZ 85726 6732
USA
☎ 1 520 327 5511
℻ 1 520 325 9360

Bautz Laura P
NSF
Div Astron Sci
4201 Wilson Blvd
Arlington VA 22230
USA
☎ 1 202 357 9488
℻ 1 703 306 0525

Baym Gordon Alan
Physics Dpt
Univ of Illinois
1110 W Green St
Urbana IL 61801 3080
USA
☎ 1 217 333 4363
℻ 1 217 333 9819
✉ baym@uinpla.npl.
uiuc.edu

Bazilevsky Alexandr T
Vernadsky Inst Geochem/
Analytical Chemistry
Kosygin Str 19
117334 Moscow
Russia
☎ 7 095 137 7538

Beale John S
231 Marlborough Rd
Swindon SN3 1NN
UK
☎ 44 17 933 4725

Beard David B
Gorham House
50 New Portland Rd
Gorham ME 04038
USA

Beaudet Gilles
Dpt d Physique
Universite Montreal
CP 6128 Succ A
Montreal QC H3C 3J7
Canada
☎ 1 514 343 6669
℻ 1 514 343 2071

Beauge Christian
Observ Astronomico
de Cordoba
Laprida 854
5000 Cordoba
Argentina
☎ 54 51 23 0491
℻ 54 51 331 063
✉ beauge@oac.uncor.edu

Beavers Willet I
11 Captain Forbush Lane
Acton MA 01720
USA
☎ 1 515 294 3776

Bec-Borsenberger Annick
BDL
77 Av Denfert Rochereau
F 75014 Paris
France
☎ 33 1 40 51 2273
℻ 33 1 46 33 2834
✉ borsen@bdl.fr

Bechtold Jill
Steward Observ
Univ of Arizona
Tucson AZ 85721
USA
☎ 1 520 621 6533
℻ 1 520 428 2854

Beck Rainer
RAIUB
Univ Bonn
auf d Huegel 71
D 53121 Bonn
Germany
☎ 49 228 525 320
℻ 49 228 525 229
✉ r.beck@mpifr-bonn.
mpg.de

Becker Peter Adam
Physics Dpt
George Mason Univ
Pairpax VA 22030
USA
☎ 1 703 993 3619
℻ 1 703 993 1980
✉ pbecker@hubble.gmu.edu

Becker Robert A
Box 4609
Carmel CA 93921
USA

Becker Robert Howard
Physics Dpt
Univ of California
Davis CA 95616
USA
☎ 1 916 752 6921
✉ bob@igpp.llnl.gov

Becker Stephen A
LANL
MS B220
Box 1663
Los Alamos NM 87545 2345
USA
☎ 1 505 667 8931
℻ 1 505 665 4055

Becker Sylvia
Inst Astron/Astrophysik
Univ Sternwarte
Scheinerstr 1
D 81679 Muenchen
Germany
☎ 49 899 220 9439
℻ 49 899 220 9427
✉ becker@usm.uni-
muenchen.de

Beckers Jacques M
NOAO/NSO
Box 26732
950 N Cherry Av
Tucson AZ 85726 6732
USA
☎ 1 520 318 8328
℻ 1 520 318 8278
✉ jbeckers@noao.edu

Becklin Eric E
Dpt Astronomy/Phys
UCLA
Box 951562
Los Angeles CA 90025 1562
USA
☎ 1 310 206 0208
℻ 1 310 206 2096
✉ becklin@uclastro.edu

Beckman John E
IAC
Observ d Teide
via Lactea s/n
E 38200 La Laguna
Spain
☎ 34 2 232 9100
℻ 34 2 232 9117
✉ jeb@iac.es

Beckwith Steven V W
MPI Astronomie
Koenigstuhl
D 69117 Heidelberg
Germany
☎ 49 622 152 8211
℻ 49 622 152 8246
✉ svwb@mpia-hd.mpg.de

Bedding Timothy
School of Physics
UNSW
Sydney NSW 2006
Australia
☎ 61 2 9351 2680
℻ 61 2 9351 7726
✉ bedding@physics.usyd.
edu.au

Bedogni Roberto
Dpt Astronomia
Univ d Bologna
Via Zamboni 33
I 40126 Bologna
Italy
☎ 39 51 259 301
℻ 39 51 259 407

Beebe Herbert A
Astronomy Dpt
NMSU
Box 30001 Dpt 4500
Las Cruces NM 88003
USA
☎ 1 505 646 4438
℻ 1 505 646 1602

Beebe Reta Faye
Astronomy Dpt
NMSU
Box 30001 Dpt 4500
Las Cruces NM 88003
USA
☎ 1 505 646 1938
℻ 1 505 646 1602

Beer Reinhard
CALTECH/JPL
MS 183 301
4800 Oak Grove Dr
Pasadena CA 91109 8099
USA
☎ 1 818 354 4748
℻ 1 818 393 6030

Beers Timothy C
Physics/Astronomy Dpt
Michigan State Univ
East Lansing MI 48824
USA
☎ 1 517 353 4541
℻ 1 517 535 4500
✉ beers@msupa.pa.msu.edu

Beesham Aroonkumar
Dpt Maths
Univ of Zululand
Private Bag X1001
3886 Kwa Dlangezwa
South Africa
☎ 27 35 193 911
℻ 27 35 193 735
✉ abeesham@pan.uzulu.
ac.za

Begelman Mitchell Craig
JILA
Univ of Colorado
Campus Box 440
Boulder CO 80309 0440
USA
☎ 1 303 492 7856
℻ 1 303 492 5235
✉ mitch@jila.colorado.edu

Begeman Kor G
Kapteyn Sterrekundig Inst
Univ Groningen
Postbus 800
NL 9700 AV Groningen
Netherlands
☎ 31 50 363 4073
℻ 31 50 363 6100
✉ kgb@rugfx4.rug.nl

Beggs Denis W
Inst of Astronomy
The Observatories
Madingley Rd
Cambridge CB3 0HA
UK
☎ 44 12 23 337 548
℻ 44 12 23 337 523

Behr Alfred
Eschenweg 3
D 3406 Bovenden
Germany
☎ 49 551 889 7

Beintema Douwe A
SRON
Univ Groningen
Postbus 800
NL 9700 AV Groningen
Netherlands
☎ 31 50 363 4073
℻ 31 50 363 6100

Bekenstein Jacob D
Racah Inst of Phys
Hebrew Univ Jerusalem
Jerusalem 91904
Israel
☎ 972 2 658 4605
℻ 972 2 61 1519
✉ bekenste@vms.huji.ac.il

Belinsky Vladimir A
Landau Inst Theor Physics
Acad Sciences
Leninsky Pspt 53
117940 Moscow
Russia
☎ 7 095 137 3244
℻ 7 095 135 7880

Belkovich Oleg I
Engelhardt Astronom
Observ
Observatoria Station
422526 Kazan
Russia
☎ 7 32 4827
✉ eao@astro.kazan.su

Bell Barbara
CfA
HCO/SAO MS 43
60 Garden St
Cambridge MA 02138 1516
USA
☎ 1 617 495 2688

Bell Jeffrey F
Inst of Geophysics
Univ of Hawaii
2525 Correa Rd
Honolulu HI 96822
USA
☎ 1 808 956 3136
℻ 1 808 956 6322
✉ bell@kahana.pgd.
hawaii.edu

Bell Kenneth Lloyd
Dpt Appl Maths/Theor Phys
Queen's Univ
Belfast BT7 1NN
UK
☎ 44 12 32 24 5133
℻ 44 12 32 23 9182

Bell Morley B
NRCC/HIA
100 Sussex Dr
Ottawa ON K1A 0R6
Canada
☎ 1 613 993 6060
℻ 1 613 952 6602

Bell Roger A
Astronomy Program
Univ of Maryland
College Park MD 20742 2421
USA
☎ 1 301 454 6282
℻ 1 301 314 9067

Bell Steven
Royal Greenwich Obs
Madingley Rd
Cambridge CB3 0EZ
UK
☎ 44 12 23 374 774
℻ 44 12 23 374 700
✉ nao@ast.cam.ac.uk

Bell Burnell Jocelyn S
Physics Dpt
The Open University
Walton Hall
Milton Keynes MK7 6AA
UK
☎ 44 190 827 4066
℻ 44 190 865 4192
✉ s.j.b.burnell@uk.ac.
open

Bell III James F
Astronomy Dpt
Cornell Univ
512 Space Sc Bldg
Ithaca NY 14853 6801
USA
☎ 1 607 255 5911
℻ 1 607 255 9002
✉ jimbo@cuspif.tn.
cornell.edu

Bellas-Velidis Ioannis
Astronomical Institute
Ntl Observ Athens
Box 20048
GR 118 10 Athens
Greece
☎ 30 1 613 2066
℻ 30 1 804 0453
✉ ibellas@astro.noa.gr

Belloni Tomaso
Astronomical Institute
Univ of Amsterdam
Kruislaan 403
NL 1098 SJ Amsterdam
Netherlands
☎ 31 20 525 7478
℻ 31 20 525 7484
✉ tmb@astro.uva.nl

Belmonte Aviles J A
IAC
Observ d Teide
via Lactea s/n
E 38200 La Laguna
Spain
☎ 34 2 232 9100
℻ 34 2 232 9117

Belserene Emilia P
421 E Ahlvers Rd
Port Angeles WA 98362
USA
☎ 1 206 457 3806
✉ emiliab@olympus.net

Belton Michael J S
NOAO/KPNO
Box 26732
950 N Cherry Av
Tucson AZ 85726 6732
USA
☎ 1 520 318 8000
℻ 1 520 318 8360
✉ mbelton@noao.edu

Belvedere Gaetano
Istt Astronomia
Citta Universitaria
Via A Doria 6
I 95125 Catania
Italy
☎ 39 95 733 2236
℻ 39 95 33 0592
✉ gbelvedere@astrct.ct.
astro.it

Bely-Dubau Francoise
OCA
Observ Nice
BP 139
F 06304 Nice Cdx 4
France
☎ 33 4 93 89 0420
℻ 33 4 92 00 3033

Belyaev Nikolaj A
Inst Theoret Astron
Acad Sciences
N Kutuzova 10
191187 St Petersburg
Russia
☎ 7 812 279 0667
℻ 7 812 272 7968

Bem Jerzy
Astronomical Institute
Wroclaw Univ
Ul Kopernika 11
PL 51 622 Wroclaw
Poland
☎ 48 71 372 9373/74
℻ 48 71 372 9378
✉ pres@astro.uni.wroc.pl

Ben-Jaffel Lofti
IAP
98bis bd Arago
F 75014 Paris
France
☎ 33 1 44 32 8076
℻ 33 1 44 32 8001
✉ bjaffel@iap.fr

Benacchio Leopoldo
Osserv Astronom Padova
Univ d Padova
Vic d Osservatorio 5
I 35122 Padova
Italy
☎ 39 49 829 3411
℻ 39 49 875 9840

Benaglia Paula
IAR
CC 5
1894 Villa Elisa (Bs As)
Argentina
☎ 54 21 254 909
℻ 54 21 254 909
✉ pbenagli@irma.edu.ar

Benavente Jose
Urbanizacion Las Redes
Oceano Alantico 11
E 11500 Puerto Santa Mari
Spain

Benaydoun Jean-Jacques
Observ Grenoble
Lab Astrophysique
BP 53x
F 38041 S Martin Heres Cdx
France
☎ 33 4 76 51 4914
℻ 33 4 76 44 8821

Bender Peter L
JILA
Univ of Colorado
Campus Box 440
Boulder CO 80309 0440
USA
☎ 1 303 492 6793
℻ 1 303 492 5235
✉ pbender@jila.colorado.edu

Bender Ralf
Inst Astron/Astrophysik
Univ Sternwarte
Scheinerstr 1
D 81679 Muenchen
Germany
☎ 49 899 220 9426
℻ 49 899 220 9427
✉ bender@usm.uni-
muenchen.de

Bendinelli Orazio
Dpt Astronomia
Univ d Bologna
Via Zamboni 33
I 40126 Bologna
Italy
☎ 39 51 259 301
℻ 39 51 259 407

Bendlin Cornelia
Astronomisches Institut
Univ Wuerzburg
am Hubland
D 97074 Wuerzburg
Germany
☎ 49 931 888 5035
℻ 49 931 888 4603
✉ bendlin@astro.uni-
wuerzburg.de

Benedict George F
McDonald Observ
Univ of Texas
Rlm 15 308
Austin TX 78712 1083
USA
☎ 1 512 471 3448
℻ 1 512 471 6016
✉ fritz@dorrit.as.utexas.edu

Benest Daniel
OCA
Observ Nice
BP 139
F 06304 Nice Cdx 4
France
☎ 33 4 92 00 3108
℻ 33 4 92 00 3033
✉ benest@obs-nice.fr

Benevides Soares P
IAG
Univ Sao Paulo
CP 9638
01065 970 Sao Paulo SP
Brazil
☎ 55 11 275 3720
✆ 55 11 276 3848
✉ pbs@vax.iagusp.usp.br

Benford Gregory
Dpt Phys/Astronomy
UCI
Irvine CA 92697 4575
USA
☎ 1 714 824 5147
✆ 1 714 824 2174

Benn Chris R
Royal Greenwich Obs
Apt 321
Santa Cruz de La Palma
E 38780 Santa Cruz
Spain
☎ 34 2 240 5500
✆ 34 2 240 5646/405 501
✉ crb@ing.iac.es

Bennett Charles L
NASA GSFC
Code 685
Greenbelt MD 20771
USA
☎ 1 301 286 3902

Bennett Kevin
ESA/ESTEC
SSD
Box 299
NL 2200 AG Noordwijk
Netherlands
☎ 31 71 565 6555
✆ 31 71 565 4690
✉ kbennett@astro.estec.esa.nl

Bensammar Slimane
Observ Paris Meudon
DASGAL
Pl J Janssen
F 92195 Meudon PPL Cdx
France
☎ 33 1 45 07 7835
✆ 33 1 45 07 7971

Benson Priscilla J
Whitin Observ
Wellesley Coll
Wellesley MA 02181
USA
☎ 1 617 235 0320
✆ 1 617 283 3667
✉ pbenson@lucy.wellesley.edu

Benvenuti Piero
ESO
ST/ECF
Karl Schwarzschildstr 2
D 85748 Garching
Germany
☎ 49 893 200 6291
✆ 49 893 202 362

Benvenuto Omar
Observ Astronomico
Paseo d Bosque S/n
1900 La Plata (Bs As)
Argentina
☎ 54 21 217 308
✆ 54 21 255 004

Benz Arnold O
Inst Astronomie
ETH Zentrum
CH 8092 Zuerich
Switzerland
☎ 41 1 632 4223
✆ 41 1 632 12 05
✉ benz@astro.phys.ethz.ch

Benz Willy
CfA
HCO/SAO
60 Garden St
Cambridge MA 02138 1516
USA
☎ 1 617 495 9889
✆ 1 617 495 7356

Berczik Peter
Main Astronomical Obs
Ukrainian Acad Science
Golosiiv
252650 Kyiv 22
Ukraine
☎ 380 4426 63110
✆ 380 4426 62147
✉ berczik@mao.kiev.ua

Berdnikov Leonid N
SAI
Acad Sciences
Universitetskij Pr 13
119899 Moscow
Russia
☎ 7 095 939 1622
✆ 7 095 932 8841
✉ berdnik@sai.msu.su

Berendzen Richard
Physics Dpt
The American Univ
Washington DC 20016 8058
USA
☎ 1 202 885 2121

Berg Richard A
Ntl Imagery/Map Agency
Stop D 82
4600 Sangamore
Bethesda MD 20816
USA
☎ 1 301 227 3334
✆ 1 301 277 3332
✉ bergr@nima.mil

Berge Glenn L
421 Mariposa Av
Sierra Madre CA 91024
USA

Bergeat Jacques G
Observ Lyon
Av Ch Andre
F 69561 S Genis Laval cdx
France
☎ 33 4 78 56 0705
✆ 33 4 72 39 9791

Berger Christiane
OCA
CERGA
F 06130 Grasse
France
☎ 33 4 93 40 5389
✆ 33 4 93 40 5353
✉ berger@ocar01.obs-azur.fr

Berger Jacques G
Observ Paris
61 Av Observatoire
F 75014 Paris
France
☎ 33 1 40 51 2247
✆ 33 1 44 54 1804

Berger Mitchell
Dpt Applied Maths
Univ of St Andrews
North Haugh
St Andrews Fife KY16 9SS
UK
☎ 44 1334 76161
✆ 44 1334 74487

Bergeron Jacqueline A
ESO
Karl Schwarzschildstr 2
D 85748 Garching
Germany
☎ 49 893 200 60
✆ 49 893 202 362
✉ jbergero@eso.org

Bergman Per-Goeran
Onsala Space Observ
Chalmers Technical Univ
S 439 92 Onsala
Sweden
☎ 46 3 1772 5552
✆ 46 3 1772 5590
✉ bergman@oso.chalmers.se

Bergstralh Jay T
NASA Headquarters
MC SLC
600 Independence Av SW
Washington DC 20546
USA
☎ 1 202 358 0313
✆ 1 202 358 3097
✉ jbergstr@nhqvax.hq.nasa.gov

Bergvall Nils Ake Sigvard
Astronomical Observ
Box 515
S 751 20 Uppsala
Sweden
☎ 46 18 53 0265
✆ 46 18 527 583

Berkhuijsen Elly M
RAIUB
Univ Bonn
auf d Huegel 69
D 53121 Bonn
Germany
☎ 49 228 525 0
✆ 49 228 525 229

Berman Marcelo S
Rua Candido Hartman 575
Apt 17 Ed Renoir
80430 Curitiba PR
Brazil
☎ 55 412 24 6426
✆ 55 41 226 1679

Berman Robert Hiram
Physics Dpt
MIT
Box 165
Cambridge MA 02139 4307
USA
☎ 1 617 253 1000
✆ 1 617 253 9798

Berman Vladimir
STScI
Homewood Campus
3700 San Martin Dr
Baltimore MD 21218
USA
☎ 1 301 486 3518
✆ 1 301 338 4767
✉ berman@stsci.edu

Bernabeu Guillermo
Dto Fisica e Ing de
Esc Politecnica Superior
Apt 99 Univ Alicante
E 03080 Alicante
Spain
☎ 34 6 590 3682
✉ bernabeu@disc.ua.es

Bernacca Pietio L
Osserv Astrofisico
Univ d Padova
Via d Osservatorio 8
I 36012 Asiago
Italy
☎ 39 424 462 665
✆ 39 424 462 884

Bernat Andrew Plous
Dpt Computer Science
Univ of Texas
El Paso TX 79968
USA
☎ 1 915 747 5480
✆ 1 915 747 5030
✉ abernat@cs.utep.edu

Bernstein Hans Heinrich
ARI
Moenchhofstr 12-14
D 69120 Heidelberg
Germany
☎ 49 622 140 5252
✆ 49 622 140 5297
✉ s03@ix.urz.uni-heidelberg.de

Berrilli Francesco
Dpt Fisica/Astrofisica
Univ d Roma Tor Vergata
Via Ric Scientifica
I 00133 Roma
Italy
☎ 39 6 7259 4552
✆ 39 6 202 3507
✉ berrilli@roma2.infn.it

Berrington Keith Adrian
Dpt Appl Maths/Theor Phys
Queen's Univ
Belfast BT7 1NN
UK
☎ 44 12 32 24 5133
✆ 44 12 32 23 9182

Berruyer-Desirotte Nicole
OCA
Observ Nice
BP 139
F 06304 Nice Cdx 4
France
☎ 33 4 92 00 3011
✆ 33 4 92 00 3033

Bersier David
MSSSO
Weston Creek
Private Bag
Canberra ACT 2611
Australia
☎ 61 262 490 290
✆ 61 262 490 233
✉ bersier@mso.anu.
edu.au

Bertaux Jean-Loup
Service Aeronomie
BP 3
F 91371 Verrieres Buisson
France
☎ 33 1 69 20 3116

Bertelli Gianpaolo
Dpt Astronomia
Univ d Padova
Vic d Osservatorio 5
I 35122 Padova
Italy
☎ 39 49 829 3411
✆ 39 49 875 9840

Berthet Stephane
Office Federal Education
& Science
Box 5675
CH 3001 Bern
Switzerland
☎ 41 31 322 9967
✆ 41 31 322 7854
✉ stephane.berthet@f2.bbw0.
adwin-ch.admin-

Berthomieu Gabrielle
OCA
Observ Nice
BP 139
F 06304 Nice Cdx 4
France
☎ 33 4 93 89 0420
✆ 33 4 92 00 3033

Bertin Giuseppe
Scuola Normale Superiore
Classe d Scienze
Piazza dei Cavalieri 7
I 56126 Pisa
Italy
☎ 39 50 509 265
✆ 39 50 563 513
✉ bertin@sns.it

Bertola Francesco
Dpt Astronomia
Univ d Padova
Vic d Osservatorio 5
I 35122 Padova
Italy
☎ 39 49 829 3436
✆ 39 49 875 9840
✉ bertola@astrpd.pd.
astro.it

Bertout Claude
IAP
98bis bd Arago
F 75014 Paris
France
☎ 33 1 44 32 8000
✆ 33 1 44 32 8001

Bertschinger Edmund
Physics Dpt
MIT Rm 6 207
Box 165
Cambridge MA 02139 4307
USA
☎ 1 617 253 5083
✆ 1 617 253 9798
✉ edbert@arcturus.mit.edu

Bessell Michael S
MSSSO
Weston Creek
Private Bag
Canberra ACT 2611
Australia
☎ 61 262 490 268
✆ 61 262 490 233
✉ bessell@merlin.anu.
edu.au

Betancort-Rijo Juan
IAC
Observ d Teide
via Lactea s/n
E 38200 La Laguna
Spain
☎ 34 2 232 9100
✆ 34 2 232 9117

Bettis Dale G
TICOM
Univ of Texas
Rlm 15 308
Austin TX 78712 1083
USA

Bettoni Daniela
Osserv Astronom Padova
Univ d Padova
Vic d Osservatorio 5
I 35122 Padova
Italy
☎ 39 49 829 3411
✆ 39 49 875 9840

Beuermann Klaus P
Univ Sternwarte
Goettingen
Geismarlandstr 11
D 37083 Goettingen
Germany
☎ 49 551 395 041
✆ 49 551 395 043
✉ beuermann@uswo50.
dnet.gwdg.de

Beutler Gerhard
Astronomisches Institut
Univ Bern
Sidlerstr 5
CH 3012 Bern
Switzerland
☎ 41 31 631 8591
✆ 41 31 631 3869
✉ beutler@aiub.unibe.ch

Bhandari N
PRL
Navrangpura
Ahmedabad 380 009
India
☎ 91 272 46 2129
✆ 91 272 44 5292

Bhandari Rajendra
RRI
Sadashivanagar
CV Raman Av
Bangalore 560 080
India
☎ 91 80 336 0122
✆ 91 80 334 0492

Bhat Chaman Lal
Nuclear Res Lab
Bhabha Atomic Res Ctr
Trombay
Bombay 400 085
India
☎ 91 22 55 4225
✆ 91 22 556 0750
✉ clbhat@magnum.barct1.
ernet.in

Bhat Narayana P
TIFR
Homi Bhabha Rd
Colaba
Bombay 400 005
India
☎ 91 22 495 2311
✆ 91 22 495 2110

Bhatia Prem K
Mathematics Dpt
Univ of Jodhpur
Jodhpur 342 001
India

Bhatia R K
Astronomy Dpt
Univ of Osmania
Hyderabad 500 007
India
☎ 91 71 951

Bhatia V B
Dpt Phys/Astrophys
Univ of Delhi
New Delhi 110 007
India
☎ 91 11 291 8993

Bhatnagar Arvind
Udaipur Solar Observ
11 Vidya Marg
Udaipur 313 001
India
☎ 91 25 626/23 861

Bhatnagar Ashok Kumar
Positional Astr Ctr
P 546 Block N 1st fl
New Alipore
Calcutta 700 053
India
☎ 91 33 450 321/493 541

Bhatnagar K B
IA/47 C
Ashok Vihar
Delhi 110 052
India
☎ 91 743 0366
✆ 91 724 2427

Bhatt H C
IIA
Koramangala
Sarjapur Rd
Bangalore 560 034
India
☎ 91 80 356 6585
✆ 91 80 553 4043

Bhattacharjee Pijushpani
IIA
Koramangala
Sarjapur Rd
Bangalore 560 034
India
☎ 91 80 553 0672/0676
✆ 91 80 553 4043
✉ pijush@iiap.ernet.in

Bhattacharya Dipankar
RRI
Sadashivanagar
CV Raman Av
Bangalore 560 080
India
☎ 91 80 334 0122
✆ 91 80 334 0492
✉ dipankar@rri.ernet.in

Bhattacharyya J C
IIA
Koramangala
Sarjapur Rd
Bangalore 560 034
India
☎ 91 80 356 6583/6585
✆ 91 80 553 4043

Bhavsar Suketu P
Dpt Phys/Astronomy
Univ Kentucky
Lexington KY 40506 0055
USA
☎ 1 606 257 6722
✆ 1 606 323 2846

Bhonsle Rajaram V
PRL
Navrangpura
Ahmedabad 380 009
India
☎ 91 272 46 2129
✆ 91 272 44 5292

Bian Yulin
Beijing Astronomical Obs
CAS
Beijing 100080
China PR
☎ 86 1 28 1698
✆ 86 1 256 10855
✉ bmabao@ica.beijing.canet.cn

Bianchi Luciana
STScI
Homewood Campus
3700 San Martin Dr
Baltimore MD 21218
USA
☎ 1 301 516 5009
📠 1 410 338 7967
✉ bianchi@stsci.edu

Bianchini Antonio
Osserv Astrofisico
Univ d Padova
Via d Osservatorio 8
I 36012 Asiago
Italy
☎ 39 424 462 665
📠 39 424 462 884

Bibarsov Ravil-sh
Astrophys Inst
Uzbek Acad Sci
Sviridenko Ul 22
734670 Dushanbe
Tajikistan
☎ 7 3770 23 1432
📠 7 3770 27 5483

Bica Eduardo L D
Instituto Fisica
UFRGS
CP 15051
91501 900 Porto Alegre RS
Brazil
☎ 55 512 36 4677

Bicak Jiri
Dpt Theor Physics
Charles Univ
V Holesovickach 2
CZ 180 00 Praha 8
Czech R
☎ 420 2 8576 2493
📠 420 2 688 5095
✉ bicak@hp03.troja.mff.cuni.cz

Bicknell Geoffrey V
MSSSO
Weston Creek
Private Bag
Canberra ACT 2611
Australia
☎ 61 262 490 245
📠 61 262 490 233
✉ geoff@mso.anu.edu.au

Bidelman William P
Physics Dpt
CWRU
Rock Bdg
Cleveland OH 44106
USA
☎ 1 216 368 6699

Bieging John Harold
Steward Observ
Univ of Arizona
Tucson AZ 85721
USA
☎ 1 520 621 4878
📠 1 520 621 1532
✉ jbieging@as.arizona.edu

Biemont Emile
Inst Astrophysique
Universite Liege
Av Cointe 5
B 4000 Liege
Belgium
☎ 32 4 254 7510
📠 32 4 254 7511
✉ biemont@astro.ulg.ac.be

Bien Reinhold
ARI
Moenchhofstr 12-14
D 69120 Heidelberg
Germany
☎ 49 622 140 50
📠 49 622 140 5297
✉ s04@mvs.urz.uni-heidelberg.de

Bienayme Olivier
Observ Strasbourg
11 r Universite
F 67000 Strasbourg
France
☎ 33 3 88 15 0710
📠 33 3 88 25 0160
✉ bienayme@cdsxb6.u-strasbg.fr

Biermann Peter L
RAIUB
Univ Bonn
auf d Huegel 69
D 53121 Bonn
Germany
☎ 49 228 525 279
📠 49 228 525 229
✉ plbiermann@mpifr-bonn.mpg

Bifang Zhou
Ctr for Astronomy
Instrumen Res
CAS
Nanjing 210042
China PR
☎ 86 25 64 6191
📠 86 25 71 1256
✉ zbfnairc@pub.jlonline.com

Biggs James
Perth Observ
Walnut Rd
Bickley WA 6076
Australia
☎ 61 8 92 93 8255
📠 61 8 92 93 8138
✉ rbiggsjd@cc.curtin.edu.au

Bignami Giovanni F
IFCTR CNR
Univ d Milano
Via E Bassini 15
I 20133 Milano
Italy
☎ 39 2 236 3542
📠 39 2 266 5753
✉ gfb@ifctr.mi.cnr.it

Bignell R Carl
NRAO
VLA
Box 0
Socorro NM 87801 0387
USA
☎ 1 505 772 4242
📠 1 505 835 7027

Bijaoui Albert
OCA
Observ Nice
BP 139
F 06304 Nice Cdx 4
France
☎ 33 4 93 89 0420
📠 33 4 92 00 3033

Bijleveld Willem
Omniversum Space Theatre
Pres Kennedylaan 5
NL 2517 JK The Hague
Netherlands
☎ 31 70 54 7479
📠 31 70 52 4280

Bikmaev Ilfan
Astronomy Dpt
Kazan State Univ
Kremlevskaya Str 18
Kazan 420008
Russia
☎ 7 843 264 3092
✉ bikmaev@astro.ksu.ras.ru

Billingham John
SETI Institute
2035 Landings Dr
Mountain View CA 94043
USA
☎ 1 415 961 6633
📠 1 415 961 7099

Billings Donald E
Crabtree
14557 Old Rivers Rd S
Statesboro CA 30458
USA
☎ 1 912 764 7625

Binette Luc
Observ Lyon
Av Ch Andre
F 69561 St Genis Laval Cdx
France
☎ 33 4 78 56 0705
📠 33 4 72 39 9791
✉ binette@obs.Univ-lyon1.fr

Binggeli Bruno
Astronomisches Institut
Univ Basel
Venusstr 7
CH 4102 Binningen
Switzerland
☎ 41 61 271 7711/12
📠 41 61 205 5455
✉ binggeli@astro.unibas.ch

Bingham Richard G
Dpt Phys/Astronomy
UCLO
Gower St
London WC1E 6BT
UK
☎ 44 171 419 3513
📠 44 171 380 7153
✉ rgb@star.ucl.ac.uk

Binney James J
Dpt Th Physics
Univ of Oxford
Keble Rd
Oxford OX1 3NP
UK
☎ 44 186 527 3979
📠 44 186 527 3947
✉ binney@thphys.ox.ac.uk

Binzel Richard P
Dpt Earth Science
MIT Rm 54 426
Box 165
Cambridge MA 02139
USA
☎ 1 617 253 6486
✉ rpb@astron.mit.edu

Biraud Francois
Observ Paris Meudon
ARPEGES
Pl J Janssen
F 92195 Meudon PPL Cdx
France
☎ 33 1 45 07 7602
📠 33 1 45 07 7971

Birch Peter
Perth Observ
Walnut Rd
Bickley WA 6076
Australia
☎ 61 8 92 93 8255
📠 61 8 92 93 8138
✉ pvbirch@iinet.net.au

Biretta John Anthony
STScI
Homewood Campus
3700 San Martin Dr
Baltimore MD 21218
USA
☎ 1 301 338 4917
📠 1 301 338 4767
✉ biretta@stsci.edu

Birkinshaw Mark
Physics Dpt
Univ of Bristol
Tyndall Av
Bristol BS8 1TL
UK
☎ 44 11 79 288 775
📠 44 11 79 255 624
✉ Mark.Birkinshaw@bristol.ac.uk

Birkle Kurt
MPI Astronomie
Koenigstuhl
D 69117 Heidelberg
Germany
☎ 49 622 152 80
📠 49 622 152 8246

Bishop Roy L
Avonport NS B0P 1B0
Canada

Bisiacchi Gianfranco
Instituto Astronomia
UNAM
Apt 70 264
Mexico DF 04510
Mexico
☎ 52 5 622 3906
℻ 52 5 616 0653

Bisikalo Dmitrij V
Inst of Astronomy
Acad Sciences
Pyatnitskaya Ul 48
109017 Moscow
Russia
☎ 7 095 231 7375
℻ 7 095 230 2081
✉ biskalo@inasan.rssi.ru

Bisnovatyi-Kogan Grennadij S
Space Res Inst
Acad Sciences
Profsojuznaya Ul 84/32
117810 Moscow
Russia
☎ 7 095 333 3122
℻ 7 095 310 7023

Biswas Sukumar
TIFR/Cosmic Ray Gr
Homi Bhabha Rd
Colaba
Bombay 400 005
India
☎ 91 22 219 111
℻ 91 22 215 2110

Bjoernsson Gunnlaugur
Science Institute
Univ of Iceland
Dunhaga 3
IS 107 Reykjavik
Iceland
☎ 354 525 4800
℻ 354 552 8911
✉ gulli@raunvis.hi.is

Bjornsson Claes-Ingvar
Stockholm Observ
Royal Swedish Acad Sciences
S 133 36 Saltsjoebaden
Sweden
☎ 46 8 717 0195
℻ 46 8 717 4719
✉ bjornsson@astro.su.se

Blaauw Adriaan
Kapteyn Sterrekundig Inst
Univ Groningen
Postbus 800
NL 9700 AV Groningen
Netherlands
☎ 31 50 363 4073
℻ 31 50 363 6100
✉ blaauw@astro.rug.nl

Black Adam Robert S
CUP
West Coast Office, Press Building
Stanford Univ
Stanford CA 94305 2235
USA
☎ 44 650 723 0663
℻ 44 650 723 0625
✉ ablack@cup.stanford.edu

Black John Harry
Onsala Space Observ
Chalmers Technical Univ
S 439 92 Onsala
Sweden
☎ 46 31 772 5500
℻ 46 31 772 5590
✉ jblack@oso.chalmers.se

Blackman Clinton Paul
DERA
Chobham Lane
Chertsey
Surray KT16 0EE
UK
☎ 44 134 463 3149
✉ cpblackman@dra.hmg.gb

Blackwell Donald E
Astrophysics Dpt
Univ of Oxford
Keble Rd
Oxford OX1 3RQ
UK
☎ 44 186 551 1336
℻ 44 186 527 3947

Blades John Chris
STScI
Homewood Campus
3700 San Martin Dr
Baltimore MD 21218
USA
☎ 1 301 338 4805
℻ 1 301 338 4767
✉ blades@stsci.edu

Blaha Milan
NRL
Code 4720
4555 Overlook Av SW
Washington DC 20375 5000
USA
☎ 1 202 878 6700

Blair David Gerald
Physics Dpt
Univ W Australia
Nedlands WA 6009
Australia
☎ 61 9 380 2738
℻ 61 9 380 1014

Blair Guy Norman
MS 63 196
Box 1000
Wilsonville OR 97070
USA
✉ guy_blair@ccm.hf.intel.com

Blair William P
Dpt Phys/Astronomy
JHU
Charles/34th St
Baltimore MD 21218
USA
☎ 1 301 516 8447
℻ 1 301 516 8260
✉ wpb@hut4.pha.jhu.edu

Blamont Jacques-Emile
CNES
2 Pl Maurice Quentin
F 75039 Paris Cdx 01
France
☎ 33 1 45 08 7612
℻ 33 1 44 76 7676

Blanchard Alain
Observ Strasbourg
11 r Universite
F 67000 Strasbourg
France
☎ 33 3 88 15 0731
℻ 33 3 88 25 0160
✉ blanchard@cdsxb6.u-strasbu.fr

Blanco Carlo
Istt Astronomia
Citta Universitaria
Via A Doria 6
I 95125 Catania
Italy
☎ 39 95 733 2245
℻ 39 95 33 0592
✉ cblanco@alpha4.ct.astro.it

Blanco Victor M
636 Flamevine Lane
Vero Beach FL 32963
USA
✉ vblanco@noao.edu

Bland-Hawthorn Jonathan
AAO
Box 296
Epping NSW 2121
Australia
☎ 61 2 9372 4851
℻ 61 2 9372 4880

Blandford Roger David
CALTECH
MS 130 33
Theroetical Astrophysics
Pasadena CA 91125
USA
☎ 1 818 395 4200
℻ 1 818 796 5675
✉ rdb@tapir.caltech.edu

Blasius Karl Richard
Santa Barbara
Research CtR B 31/40
75 Coromar Dr
Goleta CA 93117
USA
☎ 1 805 968 3511

Blazit Alain
OCA
Observ Calern/Caussols
2130 r Observatoire
F 06460 S Vallier Thiey
France
☎ 33 4 93 42 6270

Blecha Andre Boris G
Observ Geneve
Chemin d Maillettes 51
CH 1290 Sauverny
Switzerland
☎ 41 22 755 2611
℻ 41 22 755 3983
✉ andre.blecha@obs.unige.ch

Bleeker Johan A M Ir
SRON
Postbus 800
Sorbonnelaan 2
NL 3584 CA Utrecht
Netherlands
☎ 31 30 253 5732
℻ 31 30 254 0860
✉ jbleeker@sron.ruu.nl

Bless Robert C
Astronomy Dpt
Univ Wisconsin
475 N Charter St
Madison WI 53706 1582
USA
☎ 1 608 262 1715
℻ 1 608 263 0361
✉ bless@sal.wisc.edu

Bleyer Ulrich
URANIA Berlin eV
An d Urania 17
D 10787 Berlin
Germany
☎ 49 302 189 091
℻ 49 302 110 998

Blinov Nikolai S
SAI
Acad Sciences
Universitetskij Pr 13
119899 Moscow
Russia
☎ 7 095 939 1049
℻ 7 095 932 8841
✉ japet@sai.msu.su

Blitz Leo
Astronomy Program
Univ of Maryland
College Park MD 20742 2421
USA
☎ 1 301 405 6650
℻ 1 301 314 9067
✉ blitz@astro.umd.edu

Blitzstein William
Dpt Astron/Astrophys
Univ of Pennsylvania
David Rittenhouse Lab E1
Philadelphia PA 19104
USA
☎ 1 215 898 7899
℻ 1 215 898 9336

Block David Lazar
Dpt Comput/Appl Maths
Witwatersrand Univ
Private Bag 3
2050 Wits
South Africa
☎ 27 11 716 3761
℻ 27 11 339 7965
✉ block@gauss.cam.wits.ac.za

Bloemen Johannes B G M
SRON
Postbus 800
Sorbonnelaan 2
NL 3584 CA Utrecht
Netherlands
☎ 31 30 253 8572
✆ 31 30 254 0860
✉ H.Bloemen@sron.ruu.nl

Bloemhof Eric E
CALTECH
Palomar Obs
MS 105 24
Pasadena CA 91125
USA
☎ 1 818 356 4000
✆ 1 818 568 9352
✉ eeb@astro.caltech.edu

Blomme Ronny
Koninklijke Sterrenwacht
van Belgie
Ringlaan 3
B 1180 Brussels
Belgium
☎ 32 2 373 0284
✆ 32 2 374 9822
✉ Ronny.Blomme@oma.be

Blondin John M
Physics Dpt
N Carolina State Univ
Raleigh NC 27695 8202
USA
☎ 1 919 515 7096
✆ 1 919 515 6538
✉ john_blondin@ncsu.edu

Blow Graham L
Sky Data Services
Box 2241
Wellington
New Zealand
☎ 64 4 479 2504
✉ blow_g@kosmos.wcc.govt.nz

Bludman Sidney A
Physics Dpt
Univ of Pennsylvania
Philadelphia PA 19104
USA
☎ 1 215 898 8151
✆ 1 215 898 2010
✉ bludman@bludman.hep.upenn.edu

Blum Peter
IAEF
Univ Bonn
auf d Huegel 71
D 53121 Bonn
Germany
☎ 49 228 73 3665
✆ 49 228 73 3672

Blumenthal George R
Lick Observ
Univ of California
Santa Cruz CA 95064
USA
☎ 1 831 429 2005
✆ 1 831 426 3115

Bo Shuren
Inst History Nat Science
137 Cao Neistreel
Beijing 100080
China PR
☎ 86 1 55 7180

Bobrowsky Matthew
CTA Inc
4601 Forbes Blvd
Suite 210
Lanham MD 20706
USA
☎ 1 301 982 5414
✆ 1 301 459 3304
✉ mattb@cta.com

Bocchia Romeo
Observ Bordeaux
BP 89
F 33270 Floirac
France
☎ 33 5 56 86 4330
✆ 33 5 56 40 4251

Bochkarev Nikolai G
SAI
Acad Sciences
Universitetskij Pr 13
119899 Moscow
Russia
☎ 7 095 939 1672/932 8844
✆ 7 095 939 0126
✉ boch@sai.msk.su

Bochonko D Richard
Dpt Maths/Astronomy
Univ of Manitoba
Winnipeg MB R3T 2M8
Canada
☎ 1 204 474 9501

Bochsler Peter
Physikalisches Institut
Univ Bern
Sidlerstr 5
CH 3012 Bern
Switzerland
☎ 41 31 65 4429

Bockelee-Morvan Dominique
Observ Paris Meudon
ARPEGES
Pl J Janssen
F 92195 Meudon PPL Cdx
France
☎ 33 1 45 07 7605
✆ 33 1 45 07 7971

Boddapati G Anandarao
PRL
Navrangpura
Room 760
Ahmedabad 380 009
India
☎ 91 272 46 2129
✆ 91 272 44 5292

Bode Michael F
Astrophysics Group
Liverpool J M Univ
Byrom St
Liverpool L3 3AF
UK
☎ 44 151 231 2337
✆ 44 151 231 2337

Bodenheimer Peter
Lick Observ
Univ of California
Santa Cruz CA 95064
USA
☎ 1 831 429 2064
✆ 1 831 426 3115

Bodo Gianluigi
Osserv Astronomico
d Torino
St Osservatorio 20
I 10025 Pino Torinese
Italy
☎ 39 11 461 9000
✆ 39 11 461 9030

Boechat-Roberty Heloisa M
Univ Fed Rio d Janeiro
Observ d Valongo
Ladeira Pedro Antonio 43
20080 090 Rio de Janeiro
Brazil
☎ 55 21 263 0685
✆ 55 21 263 0685
✉ heloisa@ov.ufrj.br

Boehm Karl-Heinz
Astronomy Dpt
Univ of Washington
Box 351580
Seattle WA 98195 1580
USA
☎ 1 425 543 2888
✆ 1 425 685 0403

Boehm-Vitense Erika
Astronomy Dpt
Univ of Washington
Box 351580
Seattle WA 98195 1580
USA
☎ 1 425 543 4858
✆ 1 425 685 0403

Boehnhardt Hermann
Inst Astron/Astrophysik
Univ Sternwarte
Scheinerstr 1
D 81679 Muenchen
Germany
☎ 49 899 220 9446
✆ 49 899 220 9427
✉ hermann@vlt.usm.uni-muenchen.d

Boerner Gerhard
MPA
Karl Schwarzschildstr 1
D 85748 Garching
Germany
☎ 49 893 299 00
✆ 49 893 299 3235

Boerngen Freimut
Pfarrgartenstr 1
D 07751 Isserstedt
Germany

Boesgaard Ann M
Inst for Astronomy
Univ of Hawaii
2680 Woodlawn Dr
Honolulu HI 96822
USA
☎ 1 808 956 8756
✆ 1 808 988 2790

Boeshaar Gregory Orth
STScI
Homewood Campus
3700 San Martin Dr
Baltimore MD 21218
USA
☎ 1 301 338 4700
✆ 1 301 338 4767

Bogdanovicius Pavelas
Inst Theor Physics/
Astronomy
Gostauto 12
Vilnius 2600
Lithuania
☎ 370 2 620 949
✆ 370 2 225 361
✉ bogd@itpa.fi.lt

Boggess Albert
2420 Balsam Dr
Boulder CO 80304
USA
✉ hrsboggess@stars.gsfc.nasa.gov

Boggess Nancy W
319 Stonington Rd
Silver Spring MD 20902
USA

Bogod Vladimir M
Pulkovo Observ
Acad Sciences
10 Kutuzov Quay
196140 St Petersburg
Russia
☎ 7 812 123 4200
✆ 7 812 314 3360
✉ vbog@radiosun.spb.su

Bohannan Bruce Edward
NOAO/KPNO
Box 26732
950 N Cherry Av
Tucson AZ 85726 6732
USA
☎ 1 520 327 5511
✆ 1 520 325 9360

Bohlender David
DAO
NRCC/HIA
5071 West Saanich Rd
Victoria BC V8X 4M6
Canada
☎ 1 250 363 0025
✆ 1 250 363 0045
✉ david.bohlender@hia.nrc.ca

Bohlin J David
NASA Headquarters
Code EZ
600 Independence Av SW
Washington DC 20546
USA
☎ 1 202 453 1466

Bohlin Ralph C
STScI
Homewood Campus
3700 San Martin Dr
Baltimore MD 21218
USA
☎ 1 301 338 4804
✆ 1 301 338 4767

Bohn Horst-Ulrich
Elzerberg 21
D 82541 Muensing Ammerlan
Germany
☎ 49 817 792 030
✆ 49 817 792 032
✉ ubohn@alpha.connectnet.de

Bohrmann Alfred
Hamburger Sternwarte
Univ Hamburg
Gojensbergsweg 112
D 21029 Hamburg
Germany
☎ 49 407 252 4112
✆ 49 407 252 4198

Boice Daniel Craig
Southwest Res Inst
Div 15
6220 Culebra Rd
San Antonio TX 78228 0551
USA
☎ 1 210 522 3782
✆ 1 210 647 4325
✉ boice@swri.space.
swri.edu

Bois Eric
Observ Bordeaux
BP 89
F 33270 Floirac
France
☎ 33 5 57 77 6125
✆ 33 5 57 77 6110
✉ bois@observ.u-
bordeaux.fr

Boisse Patrick
DEMIRM
ENS
24 r Lhomond
F 75231 Paris Cdx 05
France
☎ 33 1 43 29 1225
✆ 33 1 45 87 3489

Boisson Catherine
Observ Paris Meudon
DAEC
Pl J Janssen
F 92195 Meudon PPL Cdx
France
☎ 33 1 45 07 7436
✆ 33 1 45 07 7469
✉ boisson@obspm.fr

Boksenberg Alec
Inst of Astronomy
The Observatories
Madingley Rd
Cambridge CB3 0HA
UK
☎ 44 12 23 339 909
✆ 44 12 23 33 9910
✉ boksy@ast.cam.ac.uk

Boland Wilfried
NFRA
Postbus 2
NL 7990 AA Dwingeloo
Netherlands
☎ 31 521 59 5100
✆ 31 52 159 7332

Bolcal Cetin
Physics Dpt
Univ of Istanbul
University 34452
34459 Vezneciler
Turkey
☎ 90 212 332 0240
✆ 90 212 519 0834

Boldt Elihu
NASA GSFC
Code 661
Greenbelt MD 20771
USA
☎ 1 301 286 5853

Boley Forrest I
7 River Woods Dr
C 114
Exeter NH 03833
USA

Boloix Rafael
ROA
Cecilio Pujazon 22-3 A
E 11110 San Fernando
Spain
☎ 34 5 688 3548
✆ 34 5 659 9366

Bolton C Thomas
David Dunlap Observ
Univ of Toronto
Box 360
Richmond Hill ON L4C 4Y6
Canada
☎ 1 905 884 9562
✆ 1 905 884 2672
✉ bolton@astro.utoronto.ca

Bomans Dominik J
Astronomisches Institut
Ruhr Univ Bochum
Postfach 102148
D 44780 Bochum
Germany
☎ 49 234 700 2335
✆ 49 234 709 4169
✉ bomans@astro.ruhr-uni-
bochum.de

Bommier Veronique
Observ Paris Meudon
DAMAP
Pl J Janssen
F 92195 Meudon PPL Cdx
France
☎ 33 1 45 07 7454
✆ 33 1 45 07 7472

Bonaccini Domenico
Osserv Astrofis Arcetri
Univ d Firenze
Largo E Fermi 5
I 50125 Firenze
Italy
☎ 39 55 43 78540
✆ 39 55 43 5939

Bonanno Giovanni
Osserv Astronomico
d Catania
Via A Doria 6
I 95125 Catania
Italy
☎ 39 95 733 2204
✆ 39 95 33 0592
✉ gbo@sunct.ct.astro.it

Bonanno Giovanni
Osserv Astronomico
d Catania
Via A Doria 6
I 95125 Catania
Italy
☎ 39 95 733 21111
✆ 39 95 33 0592

Bonazzola Silvano
Observ Paris Meudon
DARC
Pl J Janssen
F 92195 Meudon PPL Cdx
France
☎ 33 1 45 07 7429
✆ 33 1 45 07 7971

Bond Howard E
STScI
Homewood Campus
3700 San Martin Dr
Baltimore MD 21218
USA
☎ 1 301 338 4718
✆ 1 301 338 4767
✉ bond@stsci.edu

Bond John Richard
Cita Mclennan Labs
Univ of Toronto
60 St George St
Toronto ON M5S 1A1
Canada
☎ 1 416 978 6874
✆ 1 416 978 3921
✉ bond@utorphys.edu

Bondal Krishna Raj
Uttar Pradesh State
Observ
Po Manora Peak 263 129
Nainital 263 129
India
☎ 91 59 42 2136/2583

Bondarenko Ludmila N
SAI
Acad Sciences
Universitetskij Pr 13
119899 Moscow
Russia
☎ 7 095 139 3721
✆ 7 095 939 0126

Bondi Hermann
69 Mill Lane
Impington Cambs CB4 4XN
UK
☎ 44 122 323 5075
✆ 44 122 333 6180

Bonet Jose A
IAC
Observ d Teide
via Lactea s/n
E 38200 La Laguna
Spain
☎ 34 2 232 9100
✆ 34 2 232 9117

Bonev Tanyu
Astronomy Dpt
Bulgarian Acad Sci
72 Lenin Blvd
BG 1784 Sofia
Bulgaria
☎ 359 2 75 8927
✆ 359 2 75 8927
✉ tbonev@phys.acad.bg

Bonifazi Angelo
Osserv Astronom Bologna
Univ d Bologna
Via Zamboni 33
I 40126 Bologna
Italy
☎ 39 51 259 401
✆ 39 51 259 407
✉ lino@astbo3.bo.astro.it

Bonneau Daniel
OCA
Observ Calern/Caussols
2130 r Observatoire
F 06460 S Vallier Thiey
France
☎ 33 4 93 40 5454
✆ 33 4 93 40 5433

Bonnet Roger M
ESA
8 10 r Mario Nikis
F 75738 Paris Cdx 15
France
☎ 33 1 53 69 7654
✆ 33 1 53 69 7236

Bonnet-Bidaud Jean-Marc
DAPNIA/SAP
CEA Saclay
BP 2
F 91191 Gif s Yvette Cdx
France
☎ 33 1 69 08 9259
✆ 33 1 69 08 6577

Bonnor W B
1 South Bank Terrace
Surbiton Surrey KT6 6DG
UK
☎ 44 139 91 103

Bono Giuseppe
OAT
Box Succ Trieste 5
Via Tiepolo Ii
I 34131 Trieste
Italy
☎ 39 40 31 99233
✆ 39 40 30 9418

Bonoli Fabrizio
Osserv Astronom Bologna
Univ d Bologna
Via Zamboni 33
I 40126 Bologna
Italy
☎ 39 51 259 401
✆ 39 51 259 407

Bonometto Silvio A
Dpt Fisica G Galilei
Univ d Padova
Via Marzolo 8
I 35131 Padova
Italy
☎ 39 49 84 4111
✆ 39 49 84 4245

Bonsack Walter K
Suite 298
5100 1b Clayton Rd
Concord CA 94521
USA

Bontekoe Romke
Bontekoe Data Consultancy
Herengracht 47
NL 2312 LC Leiden
Netherlands
☎ 31 71 512 0491
📠 31 71 512 0491
✉ romke

Book David L
NRL
Code 4040
4555 Overlook Av SW
Washington DC 20375 5000
USA
☎ 1 202 767 6700

Bookbinder Jay A
CfA
HCO/SAO MS 58
60 Garden St
Cambridge MA 02138 1516
USA
☎ 1 617 495 7058
📠 1 617 495 7356
✉ bookbind@cfa227.
harvard.edu

Bookmyer Beverly B
Dpt Phys/Astronomy
Clemson Univ
Clemson SC 29634 1911
USA
☎ 1 803 656 3417

Booth Andrew J
CALTECH/JPL
MS 171 113
4800 Oak Grove Dr
Pasadena CA 91109 8099
USA
☎ 1 213 351 1321
📠 1 818 393 6030
✉ booth@physics.usyd.
edu.au

Booth Roy S
Onsala Space Observ
Chalmers Technical Univ
S 439 92 Onsala
Sweden
☎ 46 31 772 5500
📠 46 31 772 5550

Bopp Bernard W
Dpt Phys/Astronomy
Univ of Toledo
2801 W Bancroft St
Toledo OH 43606
USA
☎ 1 419 537 2274
📠 1 419 530 2723

Borchkhadze Tengiz M
Abastumani Astrophysical
Observ
Georgian Acad Sci
380060 Tbilisi
Georgia
☎ 995 88 32 37 5226
📠 995 88 32 98 5017
✉ tenat@dtapha.kheta.ge

Bord Donald John
Dpt Natural Sci
Univ of Michigan
Dearborn
Dearborn MI 48128
USA
☎ 1 313 593 5483

Borderies Nicole
CALTECH/JPL
MS 301 150
4800 Oak Grove Dr
Pasadena CA 91109 8099
USA
☎ 1 818 354 8211
📠 1 818 393 6030

Borgeest Ulf
Hamburger Sternwarte
Univ Hamburg
Gojensbergsweg 112
D 21029 Hamburg
Germany
☎ 49 407 252 4121
📠 49 407 252 4198
✉ st40010@dhhuni4

Borgman Jan
Nieuwe Parklaan 4
NL 2597 LC The Hague
Netherlands

Borgnino Julien
Dpt Astrophysique
Universite Nice
Parc Valrose
F 06034 Nice Cdx
France
☎ 33 4 93 51 9100
📠 33 4 93 52 9806

Boriakoff Valentin
PL/GPS
Philips Lab. Geophysics
Hanscom AFB
Bedford MA 01731 3010
USA
☎ 1 617 377 9666
📠 1 617 377 3160
✉ boriakoff@amenra.
plhaf.mil

Borisova Jordanka
Astronomy Dpt
Bulgarian Acad Sci
72 Lenin Blvd
BG 1784 Sofia
Bulgaria
☎ 359 2 75 8927
📠 359 2 75 8927

Borkowski Kazimierz M
Inst Radio Astronomy
N Copernicus Univ
Ul Gagarina 11
PL 87 100 Torun
Poland
☎ 48 56 78 3327
📠 48 56 11 651
✉ kb@astro.uni.torun.pl

Bornmann Patricia L
NOAA ERL R/E/SE
Space Environment Lab
325 Broadway
Boulder CO 80303
USA
☎ 1 303 497 3532

Borovicka Jiri
Astronomical Institute
Czech Acad Sci
Fricova 1
CZ 251 65 Ondrejov
Czech R
☎ 420 204 85 7153
📠 420 2 88 1611
✉ borovic@asu.cas.cz

Borozdin Konstantin
Space Res Inst
Profsoyuznaya 84/32
Moscow 117810
Russia
☎ 7 095 333 4523
📠 7 095 333 5377
✉ kbor@hea.iki.rssi.ru

Borra Ermanno F
Dpt d Physique
Universite Laval
Ste Foy QC G1K 7P4
Canada
☎ 1 418 656 7405
📠 1 418 656 2040
✉ borra@astro.phys.
ulaval.ca

Bos Albert
NFRA
Postbus 2
NL 7990 AA Dwingeloo
Netherlands
☎ 31 521 59 5100
📠 31 52 159 7332

Boschan Peter
Astronomisches Institut
Univ Muenster
Wilhelm Klemm Str 10
D 48149 Muenster
Germany
☎ 49 251 833 561
📠 49 251 833 669
✉ boschan@cygnus.uni-
muenster.de

Bosma Albert
Observ Marseille
2 Pl Le Verrier
F 13248 Marseille Cdx 04
France
☎ 33 4 91 95 9088
📠 33 4 91 62 1190

Bosma Pieter B
Dpt Phys/Astronomy
Free University
De Boelelaan 1081
NL 1081 HV Amsterdam
Netherlands
☎ 31 20 444 7957
📠 31 20 444 7899
✉ bosma@nat.vu.nl

Boss Alan P
Dpt Terrestr Magnetism
Carnegie Inst Washington
5241 Bd Branch Rd NW
Washington DC 20015 1305
USA
☎ 1 202 684 4370
📠 1 202 364 8726
✉ boss@ciw.ciw.edu

Botez Elvira
Astronomical Observ
Romanian Acad Sciences
Ul Ciresilor 19
RO 3400 Cluj Napoca
Rumania
☎ 40 64 194 592
📠 40 64 19 2820
✉ obsastr@mail.soroscj.ro

Bottinelli Lucette
Observ Paris Meudon
ARPEGES
Pl J Janssen
F 92195 Meudon PPL Cdx
France
☎ 33 1 45 07 7604
📠 33 1 45 07 7971

Boucher Claude
IGN ENSG LAREG
6/8 Av Blaise Pacal
F 77455 Marnes la Vallee Cdx 2
France
☎ 33 1 64 15 3250
📠 33 1 6415 3253
✉ boucher@ensg.ign.fr

Bouchet Francois R
IAP
98bis bd Arago
F 75014 Paris
France
☎ 33 1 44 32 8095
📠 33 1 44 32 8001
✉ bouchet@iap.fr

Bouchet Patrice
ESO
La Silla Observ
Casilla 19001
Santiago 19
Chile
☎ 56 2 698 8757
📠 56 2 228 5132

Bougeard Mireille L
Observ Paris
DANOF
61 Av Observatoire
F 75014 Paris
France
☎ 33 1 40 51 2226
📠 33 1 40 51 2291
✉ bougeard@hpvlbi.
obspm.fr

Bougeret Jean-Louis
Observ Paris Meudon
DESPA
Pl J Janssen
F 92195 Meudon PPL Cdx
France
☎ 33 1 45 07 7704
📠 33 1 45 07 2806

Bouigue Roger
14 Av V Hugo
F 09500 Mirepoix
France

Boulanger Francois
DEMIRM
ENS
24 r Lhomond
F 75231 Paris Cdx 05
France
☎ 33 1 43 29 1225
📠 33 1 40 51 2002

Boulesteix Jacques
Observ Marseille
2 Pl Le Verrier
F 13248 Marseille Cdx 04
France
☎ 33 4 91 95 9088
📠 33 4 91 62 1190
✉ boulesteix@obmara.
cnrs-mrs.fr

Boulon Jacques J
Observ Paris
61 Av Observatoire
F 75014 Paris
France
☎ 33 1 40 51 2253
📠 33 1 40 51 2232
✉ boulon@obspm.fr

Bouska Jiri
Astronomical Institute
Charles Univ
V Holesovickack 2
CZ 180 00 Praha 8
Czech R
☎ 420 2 2191 2572
📠 420 2 688 5095
✉ bouska@mbox.cesnet.cz

Bouvier Jerome
Observ Grenoble
Lab Astrophysique
BP 53x
F 38041 S Martin Heres Cdx
France
☎ 33 4 76 51 4790
📠 33 4 76 44 8821

Bouvier Pierre
Observ Geneve
Chemin d Maillettes 51
CH 1290 Sauverny
Switzerland
☎ 41 22 755 2611
📠 41 22 755 3983

Bowell Edward L G
Lowell Observ
1400 W Mars Hill Rd
Box 1149
Flagstaff AZ 86001
USA
☎ 1 520 774 3358
📠 1 520 774 6296
✉ ebowell@lowell.edu

Bowen George H
Physics Dpt
Iowa State Univ
Ames IA 50011
USA
☎ 1 515 294 7659
📠 1 515 294 3262

Bower Gary Allen
NOAO/KPNO
Box 26732
950 N Cherry Av
Tucson AZ 85726 6732
USA
☎ 1 520 318 8285
📠 1 520 318 8360
✉ gbower@noao.edu

Bowers Phillip F
Lowell Observ
1400 W Mars Hill Rd
Box 1149
Flagstaff AZ 86001
USA
✉ bowers@sextans.
lowell.edu

Bowyer C Stuart
Astronomy Dpt
Univ of California
601 Campbell Hall
Berkeley CA 94720 3411
USA
☎ 1 510 642 1648
📠 1 510 643 8303
✉ bowyer@ssl.berkeley.edu

Boyarchuk Alexander A
Inst of Astronomy
Acad Sciences
Pyatnitskaya Ul 48
109017 Moscow
Russia
☎ 7 095 951 0924
📠 7 095 230 2081
✉ aboyar@inasan.rssi.ru

Boyarchuk Margarita E
Inst of Astronomy
Acad Sciences
Pyatnitskaya Ul 48
109017 Moscow
Russia
☎ 7 095 231 5461
📠 7 095 230 2081

Boyce Peter B
AAS
2000 Florida Av Nw
Suite 400
Washington DC 20009 1231
USA
☎ 1 202 328 2010
📠 1 202 234 2560
✉ pboyce@aas.org

Boyd Robert L F
41 Church St
Littlehampton BN17 5PU
UK

Boydag-Yildizdogdu F S
KACST
Box 2455
Riyadh 11495
Saudi Arabia
☎ 966 1 467 6324
📠 966 1 467 4253

Boyer Rene
Observ Paris Meudon
DASOP
Pl J Janssen
F 92195 Meudon PPL Cdx
France
☎ 33 1 45 07 7741
📠 33 1 45 07 7959

Boyle Brian
AAO
Box 296
Epping NSW 2121
Australia
☎ 61 2 868 1666
📠 61 2 9876 8536

Boyle Richard P
Vatican Observ
I 00120 Citta del Vaticano
Vatican City State
☎ 39 6 698 3411/5266
📠 39 6 698 84671

Boyle Stephen
UCLO
ULO
Mill Hill Park
London NW7 2QS
UK
☎ 44 181 959 0421
📠 44 181 388 1450
✉ sjb@uk.ac.ucl.star

Boynton Paul Edward
Astronomy Dpt
Univ of Washington
Box 351580
Seattle WA 98195 1580
USA
☎ 1 425 543 2888
📠 1 425 685 0403

Boytel Jorge Del Pino
CENAILS
Calle 17 No 61 E4 y 6
Santiago de Cuba
Vista Alegre 90400
Cuba
☎ 53 226 41 623
📠 53 226 41 579
✉ lasersat@ceniai.inf.cu

Bozic Hrvoje
Hvar Observ
Fac Geodesy
Kaciceva 26
HR 10000 Zagreb
Croatia
☎ 385 1 456 1222
📠 385 1 445 410
✉ hrvoje.bozic@x400.
srce.hr

Bozis George
Dpt Th Mechanics
Univ Thessaloniki
GR 540 06 Thessaloniki
Greece
☎ 30 31 99 2845
📠 30 31 99 8211

Bozkurt Sukru
Dpt Astron/Space Sci
Ege Univ
Box 21
35100 Bornova Izmir
Turkey
☎ 90 232 388 0110*2322

Braccesi Alessandro
Dpt Astronomia
Univ d Bologna
Via Zamboni 33
I 40126 Bologna
Italy
☎ 39 51 259 301
📠 39 51 259 407

Bracewell Ronald N
Stanford Univ
Durand 329 A
Stanford CA 94305
USA
☎ 1 415 497 3545
📠 1 415 723 4840

Bradley Arthur J
Box 91
Annapolis Junct MD 20701
USA
☎ 1 301 901 6052
📠 1 301 901 6086
✉ Bradley.Art@lmmail.hst.
nasa.gov

Bradstreet David H
Dpt Physical Sci
Eastern College
St Davids PA 19087
USA
☎ 1 215 341 5945

Braes L L E
Leiden Observ
Box 9513
NL 2300 RA Leiden
Netherlands
☎ 31 71 527 2727
📠 31 71 527 5819

Braga Joao
INPE
Dpt Astronomia
CP 515
12227 010 S Jose dos Campos
Brazil
☎ 55 123 41 8977*679
📠 55 123 21 8743
✉ inpedas@brfapesr

Braga da Costa Campos L M
Inst Superior Tecnico
Compl Interdisciplinar
Av Rovisco Pais
P 1096 Lisboa Codex
Portugal
☎ 351 1 3524 303
📠 351 1 3524 372

Brahde Rolf
Inst Theor Astrophys
Univ of Oslo
Box 1029
N 0315 Blindern Oslo 3
Norway
☎ 47 22 856 508
📠 47 22 85 6505

Brahic Andre
Observ Paris Meudon
DAEC
Pl J Janssen
F 92195 Meudon PPL Cdx
France
☎ 33 1 45 07 7402
✆ 33 1 45 07 7971

Brajsa Roman
Hvar Observ
Fac Geodesy
Kaciceva 26
HR 10000 Zagreb
Croatia
☎ 385 1 456 1222
✆ 385 1 445 410
✉ romanb@geodet.geof.hr

Branch David R
Dpt Phys/Astronomy
Univ of Oklahoma
Norman OK 73019
USA
☎ 1 405 325 3961
✆ 1 405 325 7557

Brand Jan
Istt Radioastronomia
CNR
Via P Gobetti101
I 40129 Bologna
Italy
☎ 39 51 639 9372
✆ 39 51 639 9431
✉ brand@astbo1.bo.cnr.it

Brand Peter W J l
Royal Observ
Blackford Hill
Edinburgh EH9 3HJ
UK
☎ 44 131 667 3321
✆ 44 131 668 8356

Brandenburg Axel
School Maths/Statistics
Univ of Newcastle
Newcastle/Tyne NE1 7RU
UK
☎ 44 191 222 7411
✆ 44 191 261 1182
✉ Axel.Brandenburg@Newcastle.ac.uk

Brandi Elisande Estela
Observ Astronomico
Paseo d Bosque S/n
1900 La Plata (Bs As)
Argentina
☎ 54 21 217 308
✆ 54 21 211 761

Brandie George W
Environmental Eng
Queen's Univ
Kingston ON K7L 3N6
Canada

Brandt John C
LASP
Univ of Colorado
Campus Box 392
Boulder CO 80309 0392
USA
☎ 1 303 492 3215
✆ 1 303 492 6946
✉ brabdt@lyrae.colorado.edu

Brandt Peter N
Kiepenheuer Inst
Sonnenphysik
Schoeneckstr 6
D 79104 Freiburg Breisgau
Germany
☎ 49 761 319 8250
✆ 49 761 319 8111
✉ pnb@kis.uni-freiburg.de

Branduardi-Raymont G
Mullard Space Science Lab
Univ College London
Holmbury St Mary
Dorking Surrey RH5 6NT
UK
☎ 44 13 06 702 92
✆ 44 14 83 278 312

Branham Richard L
Centro Regional Invest
Cientifica/Technologicas
CC 131
5500 Mendoza
Argentina
☎ 54 61 288 314
✆ 54 61 520 005
✉ rlb@lanet.com.ar

Branscomb L M
NBS
Washington DC 20025
USA

Branson Nicholas J B A
University Registry
The Old Schools
Cambridge CB2 1TN
UK
☎ 44 12 23 33 2250
✆ 44 12 23 33 2332

Bratijchuk Matrona V
Satelite Observ
Uzhgorod State Univ
Horkiy 46
294000 Uzhgorod
Ukraine
☎ 380 3122 36065
✆ 380 3122 36136

Braude Semion Ya Ag
Inst Radio Astron
Ukrainian Acad Science
4 Chervonopraporna st
310002 Kharkiv
Ukraine
☎ 380 5724 41092

Brault James W
1006 Hineysuckle Lane
Louisville CO 80027
USA
✉ brault@noao.edu

Braun Arie
Racah Inst of Phys
Hebrew Univ Jerusalem
Jerusalem 91904
Israel
☎ 972 2 58 4521
✆ 972 2 61 1519

Braun Douglas Clifford
HAO
NCAR
Box 3000
Boulder CO 80307 3000
USA
☎ 1 303 497 2558
✆ 1 303 497 1589
✉ dbraun@solar.stanford.edu

Braun Robert
NFRA
Postbus 2
NL 7990 AA Dwingeloo
Netherlands
☎ 31 521 59 5100
✆ 31 52 159 7332
✉ rbraun@nfra.nl

Brauninger Heinrich
MPE
Postfach 1603
D 85740 Garching
Germany
☎ 49 893 299 566
✆ 49 893 299 3569

Braunsfurth Edward
Im Haarmannsboch 99 A
D 46300 Bochum 1
Germany

Bravo Eduardo
Dpt Fisica y Eng Nuclear
Uni Politecnica Catalunya
Av Diagonal 647
E 08028 Barcelona
Spain
☎ 34 3 401 6565
✆ 34 3 401 6600
✉ bravo@fen.upc.es

Bray Robert J
31/126 Crimea Rd
MarSField NSW 2122
Australia

Breahna Iulian
Astronomical Observ
Allea Sadoveanu 5
RO 6600 Iasi
Rumania
☎ 40 32 140 305

Breakiron Lee Allen
USNO
Time Service Directte
3450 Massachusetts Av NW
Washington DC 20392 5420
USA
☎ 1 202 762 1092
✆ 1 202 762 1511
✉ lab@tycho.usno.navy.mil

Brecher Aviva
35 Madison St
Belmont MA 02178
USA
☎ 1 617 489 1386

Brecher Kenneth
Astronomy Dpt
Boston Univ
725 Commonwealth Av
Boston MA 02215
USA
☎ 1 617 353 3423
✆ 1 617 353 5704
✉ brecher@bu.edu

Breckinridge James B
CALTECH/JPL
MS 183 301
4800 Oak Grove Dr
Pasadena CA 91109 8099
USA
☎ 1 213 354 6785
✆ 1 818 393 6030

Breger Michel
Inst Astronomie
Univ Wien
Tuerkenschanzstr 17
A 1180 Wien
Austria
☎ 43 1 345 3605
✆ 43 1 470 6015
✉ breger@procyon.ast.univie.ac.at

Bregman Jacob D Ir
NFRA
Postbus 2
NL 7990 AA Dwingeloo
Netherlands
☎ 31 521 59 5100
✆ 31 52 159 7332

Bregman Joel N
Astronomy Dpt
Univ of Michigan
Dennison Bldg
Ann Arbor MI 48109 1090
USA
☎ 1 313 764 3440
✆ 1 313 764 2211
✉ jbregman@astro.lsa.umich.edu

Breinhorst Robert A
Astronomisches Institut
Univ Bonn
auf d Huegel 71
D 53121 Bonn
Germany
☎ 49 228 73 3660
✆ 49 228 73 3672

Breitschwerdt Dieter
MPE
Postfach 1603
D 85740 Garching
Germany
☎ 49 893 299 3221
✆ 49 893 299 3569
✉ breitsch@rosat.mpe-garching.mpg.de

Brejdo Izabella I
Pulkovo Observ
Acad Sciences
10 Kutuzov Quay
196140 St Petersburg
Russia
☎ 7 812 297 9459
✆ 7 812 315 1701

Brekke Pal Ording lie
Inst Theor Astrophys
Univ of Oslo
Box 1029
N 0315 Blindern Oslo 3
Norway
☎ 47 22 856 508
℻ 47 22 856 505
✉ paalb@astro.uio.no

Breslin Ann
Experim Physics Dpt
Univ College
Belfield
Dublin 4
Ireland
☎ 353 1 706 2230/706 2213
℻ 353 1 2837 275
✉ acb@ferdia.ucd.ie

Bretagnon Pierre
BDL
77 Av Denfert Rochereau
F 75014 Paris
France
☎ 33 1 40 51 2269
℻ 33 1 46 33 2834
✉ pierre@bdl.fr

Breysacher Jacques
ESO
Karl Schwarzschildstr 2
D 85748 Garching
Germany
☎ 49 893 200 6224
℻ 49 893 202 362

Bridgeland Michael
Inst of Astronomy
The Observatories
Madingley Rd
Cambridge CB3 0HA
UK
☎ 44 12 23 337 548
℻ 44 12 23 337 523

Bridle Alan H
NRAO
520 Edgemont Rd
Charlottesville VA 22903
USA
☎ 1 804 296 0375

Brieva Eduardo
Observ Astronomico Ncl
Apdo 2584
Bogota 1 DE
Columbia
☎ 57 1 342 3786
℻ 57 1 342 3786

Briggs Franklin
Kapteyn Sterrekundig Inst
Univ Groningen
Postbus 800
NL 9700 AV Groningen
Netherlands
☎ 31 50 363 4073
℻ 31 50 363 6100
✉ fbriggs@astro.rug.nl

Brini Domenico
Istt d Fisica
Univ d Bologna
Via Irnerio 46
I 40126 Bologna
Italy
☎ 39 51 232 856

Brinkman Bert C
SRON
Postbus 800
Sorbonnelaan 2
NL 3584 CA Utrecht
Netherlands
☎ 31 30 253 5600
℻ 31 30 254 0860

Brinkmann Wolfgang
MPE
Postfach 1603
D 85740 Garching
Germany
☎ 49 893 299 877
℻ 49 893 299 3569

Brinks Elias
Dpt Astronomia
Univ Guanajuato
Apdo 144
Guanajuato GTO 36000
Mexico
☎ 52 4 732 9548
℻ 52 4 732 0253
✉ ebrinks@astro.ugto.mx

Briot Danielle
Observ Paris
DASGAL
61 Av Observatoire
F 75014 Paris
France
☎ 33 1 40 51 2239
℻ 33 1 44 54 1804

Broadfoot A Lyle
Lunar/Planetary Lab
Room 325 Bg 92
Univ of Arizona
Tucson AZ 85721 0092
USA
☎ 1 520 621 4301
℻ 1 520 621 4933

Brocato Enzo
Osserv Astron Collurani
Teramo
Viale Maggini
I 64100 Teramo
Italy
☎ 39 861 21 0490
℻ 39 86 121 0492

Broderick John
Physics Dpt
Virginia Tech
Blacksburg VA 24061
USA
☎ 1 703 231 5321

Brodie Jean P
Lick Observ
Univ of California
Santa Cruz CA 95064
USA
☎ 1 831 459 2987
℻ 1 831 426 3115
✉ brodie@lick.ucsc.edu

Broglia Pietro
Osserv Astronomico
d Milano
Via E Bianchi 46
I 22055 Merate
Italy
☎ 39 990 6412
℻ 39 990 8492

Bromage Gordon E
Rutherford Appleton Lab
Space/Astrophysics Div
Bg R25/R68
Chilton Didcot OX11 0QX
UK
☎ 44 12 35 821 900
℻ 44 12 35 44 5808

Bronfman Leonardo
Dpt Astronomia
Univ Chile
Casilla 36 D
Santiago
Chile
☎ 56 2 228 1941
℻ 56 2 229 4002

Bronnikova Nina M
Pulkovo Observ
Acad Sciences
10 Kutuzov Quay
196140 St Petersburg
Russia
☎ 7 812 1234 285
℻ 7 812 123 1922

Brookes Clive J
Mathematics Dpt
Earth/satellite Res Unit
Aston Univ
Birmingham B4 7ET
UK
☎ 44 12 13 59 3611

Brooks Randall C
Physics Sciences
NRCC
Box 9724
Ottawa ON K1G 5A3
Canada
☎ 1 613 990 2804
℻ 1 613 991 3636
✉ RBrooks@nmstc.ca

Brosch Noah
Wise Observ
Tel Aviv Univ
Ramat Aviv
Tel Aviv 69978
Israel
☎ 972 3 640 7413
℻ 972 3 640 8179
✉ noah@wise1.tau.ac.il

Brosche Peter
Observatorium Hoher List
Univ Sternwarte Bonn
D 54550 Daun
Germany
☎ 49 65 92 2150
℻ 49 65 92 2937
✉ pbrosche@astro.uni-bonn.de

Brosius Jeffrey William
NASA GSFC
Code 682 1
Greenbelt MD 20771
USA
☎ 1 301 286 6200
℻ 1 301 286 1617

Brosterhus Elmer B F
Lockheed City
Box 6308
Jeddah
Saudi Arabia
☎ 966 2 656 2501*355
℻ 966 2 611519

Broten Norman W
48 Pineglen Crescent
Nepean ON K2E 6X9
Canada
✉ broten@hiasras.hia.nrc.ca

Broucke Roger
7203 Running Rope Circle
Austin TX 78731
USA
☎ 1 512 345 6435

Brouillet Nathalie
Observ Bordeaux
BP 89
F 33270 Floirac
France
☎ 33 5 56 86 4330
℻ 33 5 56 40 4251
✉ nathalie@observ.u-bordeaux.fr

Brouw W N
CSIRO
ATNF
Box 76
Epping NSW 2121
Australia
☎ 61 2 9372 4316
℻ 61 2 9372 4310
✉ wbrouw@atnf.csiro.au

Brown Alexander
CASA
Univ of Colorado
Campus Box 389
Boulder CO 80309 0389
USA
☎ 1 303 492 7810
℻ 1 303 492 7178
✉ ab@echidna.colorado.edu

Brown Douglas Nason
Astronomy Dpt
Univ of Washington
Box 351580
Seattle WA 98195 1580
USA
☎ 1 425 543 4313*2888
℻ 1 425 685 0403

Brown John C
Dpt Phys/Astronomy
Univ of Glasgow
Glasgow G12 8QQ
UK
☎ 44 141 339 8855
℻ 44 141 334 9029

Brown Paul James Frank
Physics Dpt
Queen's Univ
Belfast BT7 1NN
UK
☎ 44 12 32 273 648
℻ 44 12 32 438 918
✉ p.brown@qub.ac.uk

Brown Robert Hamilton
CALTECH/JPL
MS 183 501
4800 Oak Grove Dr
Pasadena CA 91109 8099
USA
☎ 1 818 354 2517
✆ 1 818 393 6030

Brown Robert L
NRAO
520 Edgemont Rd
Charlottesville VA 22903
USA
☎ 1 804 296 0222
✆ 1 804 296 0385
✉ rbrown@nrao.edu

Brown Ronald D
Chemistry Dpt
Monash Univ
Wellington Rd
Clayton VIC 3168
Australia
☎ 61 3 565 4550
✆ 61 3 565 4597
✉ rdbrown@vaxc.cc.monash.edu.au

Browne Ian W A
NRAL
Univ Manchester
Jodrell Bank
Macclesfield SK11 9DL
UK
☎ 44 14 777 1321
✆ 44 147 757 1618

Browning Philippa
Dpt Pure/Appl Physics
UMIST
Box 88
Manchester M60 1QD
UK
☎ 44 161 200 3677
✆ 44 161 200 3669
✉ mccppb@uk.ac.umrcc.cms

Brownlee Donald E
Astronomy Dpt
Univ of Washington
Box 351580
Seattle WA 98195 1580
USA
☎ 1 425 543 2888
✆ 1 425 685 0403
✉ brownlee@astro.washington.edu

Brownlee Robert R
4879 Franklin Av
Loveland CO 80538
USA
☎ 1 970 663 0646
✆ 1 970 663 0647
✉ rrb2@IX.netcom.com

Brucato Robert J
CALTECH
Palomar Obs
MS 105 24
Pasadena CA 91125
USA
☎ 1 818 356 4035
✆ 1 818 568 1517

Bruch Albert
Astronomisches Institut
Univ Muenster
Wilhelm Klemm Str 10
D 48149 Muenster
Germany
☎ 49 251 833 561
✆ 49 251 833 669
✉ albert@cygnus@uni-muenster.de

Bruck Hermann A
Craigower
Penicuik EH26 9LA
UK
☎ 44 196 87 5918

Bruck Mary T
Craigower
Penicuik
Midlotgian EH26 9LA
UK
☎ 44 196 867 5918

Brueckner Guenter E
NRL
Code 7660
4555 Overlook Av SW
Washington DC 20375 5352
USA
☎ 1 202 767 3287
✆ 1 202 767 5636

Bruenn Stephen W
Physics Dpt
Florida Atlantic Univ
Boca Raton FL 33431 0991
USA
☎ 1 561 367 3385
✆ 1 561 367 2662
✉ bruenn@acc.fau.edu

Brugel Edward W
CASA
Univ of Colorado
Campus Box 389
Boulder CO 80309 0391
USA
☎ 1 303 492 4054
✆ 1 303 492 5941
✉ brugel@cygnus.colorado.edu

Bruhweiler Fred C
Catholic Univ of America
Astrophysics Program
Physics Dpt
Washington DC 20064
USA
☎ 1 202 319 5315
✆ 1 202 319 4448
✉ bruhweiler@cua.edu

Brumberg Victor A
Inst Appl Astronomy
Acad Sciences
Zhdanovskaya Ul 8
197042 St Petersburg
Russia
☎ 7 812 230 7414
✆ 7 812 230 7413
✉ brumberg@ipa.rssi.ru

Bruner Marilyn E
Lockheed Palo Alto Res Lb
Dpt 91 30 Bg 256
3251 Hanover St
Palo Alto CA 94304 1191
USA
☎ 1 415 424 3273
✆ 1 415 424 3994

Brunet Jean-Pierre
OMP
14 Av E Belin
F 31400 Toulouse Cdx
France
☎ 33 5 61 25 2101
✆ 33 5 61 27 3179

Bruning David H
Stellar Res/Edu
Box 1223
Waukesha WI 53187
USA
✉ dhbrun01@ulkyvx

Brunini Adrian
IAG
Univ Sao Paulo
CP 9638
01065 970 Sao Paulo SP
Brazil
☎ 55 11 577 8599
✆ 55 11 276 3848

Brunk William E
NASA Headquarters
Code Sl off Space Science
400 Maryland Av SW
Washington DC 20546
USA
☎ 1 202 453 1596

Bruns Andrey V
Crimean Astrophys Obs
Ukrainian Acad Science
Nauchny
334413 Crimea
Ukraine
☎ 380 6554 71119
✆ 380 6554 40704
✉ bruns@crao.crimea.ua

Bruston Paul
EPCOS/LPCE
Univ Paris Val de Marne
Av Gl de Gaulle
F 94010 Creteil Cedex
France
☎ 33 1 42 07 1285
✆ 33 1 42 07 7012

Bruzek Anton
Schwaighofstr 7
D 79104 Freiburg Breisgau
Germany
☎ 49 761 785 22
✆ 49 761 3198 111

Bruzual Gustavo
CIDA
Box 264
Merida 5101 A
Venezuela
☎ 58 7 471 2780/3883
✆ 58 7 471 2459
✉ gbruzual@cide.ve

Bryant John
47 Av Felix Faure
F 75015 Paris
France
☎ 33 1 45 57 7647

Bryce Myfanwy
Astronomy Dpt
Univ Manchester
Oxford Rd
Manchester M13 9PL
UK
☎ 44 161 275 4226
✆ 44 161 275 4223
✉ mbryce@uk.ac.man.ast

Brzezinski Aleksander
Space Res Ctr
Polish Acad Sci
Ul Bartycka 18 A
PL 00 716 Warsaw
Poland
☎ 48 22 40 3766
✆ 48 22 36 8961
✉ alek@cbk.waw.pl

Bucciarelli Beatrice
Osserv Astronomico
d Torino
St Osservatorio 20
I 10025 Pino Torinese TO
Italy
☎ 39 11 461 9024
✆ 39 11 461 9030
✉ bucc@to.astro.it

Buchler J Robert
Physics Dpt
Univ of Florida
Williamson 220A
Gainesville FL 32611 8440
USA
☎ 1 904 392 0507
✆ 1 352 392 5339
✉ buchler@phys.ufl.edu

Buckley David
SAAO
PO Box 9
7935 Observatory
South Africa
☎ 27 21 47 0025
✆ 27 21 47 3639
✉ dibnob@saao.ac.za

Budding Edwin
Carter Observ
Ntl Observ New Zealand
Box 2909
Wellington 1
New Zealand
☎ 64 4 472 8167
✆ 64 4 472 8320
✉ budding@matai.vuw.ac.nz

Buecher Alain
Residence Renoir
14 passage d Allies
F 64000 Pau
France
☎ 33 5 59 02 0940

Buechner Joerg
MPI Extraterrestisch
Auss Berlin Geb 1616/1625
Rudower Chaussee 5
D 12489 Berlin Adlershof
Germany
☎ 49 306 392 3937
✆ 49 306 392 3937
✉ jb@mpe.fta-berlin.de

Bues Irmela D
Dr Remeis Sternwarte
Univ Erlangen-Nuernberg
Sternwartstr 7
D 96049 Bamberg
Germany
☎ 49 951 577 08
✆ 49 951 952 2222
✉ bues@sternwarte.uni-erlangen.d

Buff James S
Dpt Phys/Astronomy
Dartmouth College
6127 Wilder Lab
Hanover NH 03755
USA
☎ 1 603 646 2359
✆ 1 603 646 1446

Buhl David
NASA GSFC
Code 693
Greenbelt MD 20771
USA
☎ 1 301 286 8810

Buie Marc W
Lowell Observ
1400 W Mars Hill Rd
Box 1149
Flagstaff AZ 86001
USA
☎ 1 602 774 3358
✆ 1 602 774 6296
✉ buie@lowell.edu

Buitrago Jesus
IAC
Observ d Teide
via Lactea s/n
E 38200 La Laguna
Spain
☎ 34 2 232 9100
✆ 34 2 232 9117

Bujarrabal Valentin
Ctr Astron d Yebes
OAN
Apt 148
E 19080 Guadalajara
Spain
☎ 34 1 122 3358
✆ 34 1 129 0063

Bumba Vaclav
Astronomical Institute
Czech Acad Sci
Fricova 1
CZ 251 65 Ondrejov
Czech R
☎ 420 204 85 7157
✆ 420 2 88 1611
✉ bumba@asu.cas.cz

Bunclark Peter Stephen
Royal Greenwich Obs
Madingley Rd
Cambridge CB3 0EZ
UK
☎ 44 12 23 33 7548
✆ 44 12 23 337 523
✉ psb@ast.cam.ac.uk

Bunner Alan N
Perkin Elmer Corp
MS 897
100 Wooster Heights Rd
Danbury CT 06810 7859
USA
☎ 1 203 797 6339

Buonanno Roberto
OAR
Via d Parco Mellini 84
I 00136 Roma
Italy
☎ 39 6 34 7056
✆ 39 6 349 8236

Buratti Bonnie J
CALTECH/JPL
MS 183 501
4800 Oak Grove Dr
Pasadena CA 91109 8099
USA
☎ 1 818 354 7427
✆ 1 818 393 6030

Burbidge Eleanor Margaret
CASS
UCSD
C 011
La Jolla CA 92093 0216
USA
☎ 1 619 452 4477
✆ 1 619 534 2294

Burbidge Geoffrey R
CASS
UCSD
C 011
La Jolla CA 92093 0216
USA
☎ 1 619 452 6626
✆ 1 619 534 2294

Burdyuzha Vladimir V
Space Res Inst
Acad Sciences
Profsojuznaya Ul 84/32
117810 Moscow
Russia
☎ 7 095 333 3366
✆ 7 095 310 7023
✉ burdyuzh@dpc.asc.rssi.ru

Burger Marijke
Koninklijke Sterrenwacht van Belgie
Ringlaan 3
B 1180 Brussels
Belgium
☎ 32 2 373 0240
✆ 32 2 374 9822
✉ marijke.burger@oma.be

Burgess Alan
Dpt Appl Maths/Theor Phys
Silver Street
Cambridge CB3 9EW
UK
☎ 44 12 23 337 00
✆ 44 12 23 337 918

Burgess David D
Astrophysics Gr
Imperial Coll
Prince Consort Rd
London SW7 2BZ
UK
☎ 44 171 594 7771
✆ 44 171 594 7772

Burke Bernard F
Physics Dpt
MIT Rm 26 335
Box 165
Cambridge MA 02139 4307
USA
☎ 1 617 253 2572
✆ 1 617 253 9798

Burke J Anthony
Physics Dpt
Univ of Victoria
Box 1700
Victoria BC V8W 2Y2
Canada
☎ 1 250 721 7743
✆ 1 250 721 7715

Burkhart Claude
Observ Lyon
Av Ch Andre
F 69561 S Genis Laval cdx
France
☎ 33 4 78 86 8383
✆ 33 4 78 86 8386
✉ burkhart@obs.univ-lyon1.fr

Burkhead Martin S
Astronomy Dpt
Indiana Univ
Swain W 319
Bloomington IN 47405
USA
☎ 1 812 335 6917
✆ 1 812 855 8725

Burki Gilbert
Observ Geneve
Chemin d Maillettes 51
CH 1290 Sauverny
Switzerland
☎ 41 22 755 2611
✆ 41 22 755 3983
✉ gilbert.burki@obs.unige.ch

Burlaga Leonard F
NASA GSFC
Code 692
Greenbelt MD 20771
USA
✆ 1 301 286 3271

Burnage Robert
OHP
F 04870 S Michel Obs
France
☎ 33 4 92 76 6368
✆ 33 4 92 76 6295
✉ burnage@obshpx.obs-hp.fr

Burns Joseph A
Astronomy Dpt
Cornell Univ
512 Space Sc Bldg
Ithaca NY 14853 6801
USA
☎ 1 607 255 7186
✆ 1 607 255 6354
✉ jab16@cornell.edu

Burns Jack O'Neal
Univ of Missouri Columbia
Office of Research
205 Jesse Hall
Columbia MO 65211
USA
☎ 1 573 882 9500
✆ 1 573 884 8371
✉ burns@research.missouri.edu

Burrows Adam Seth
Dpt Phys/Astronomy
Steward Observ
Univ of Arizona
Tucson AZ 85721
USA
☎ 1 520 621 4359
✆ 1 520 621 4933
✉ burrow@ccir.arizona.edu

Burrows David Nelson
Astronomy Dpt
Pennsylvania State Univ
525 Davey Lab
University Park PA 16802
USA
☎ 1 814 863 2466
✆ 1 814 863 3399
✉ burrows@astro.psu.edu

Burs Lucian
Astronomical Observ
P Ta Axente Sever 1
RO 1900 Timisoara
Rumania
☎ 40 56 162 838

Bursa Milan
Astronomical Institute
Czech Acad Sci
Bocni II 1401
CZ 141 31 Praha 4
Czech R
☎ 420 2 67 10 3060
✆ 420 2 76 90 23
✉ astdss@csearn

Burstein David
Physics Dpt
Arizona State Univ
Astronomy Program
Tempe AZ 85287 1504
USA
☎ 1 602 965 3561
✆ 1 605 965 7954

Burton Michael G
School of Physics
UNSW
Sydney NSW 2025
Australia
☎ 61 2 9385 5618
✆ 61 2 660 2903
✉ mgb@newt.phys.unsw.edu.au

Burton W Butler
Leiden Observ
Box 9513
NL 2300 RA Leiden
Netherlands
☎ 31 71 527 5832
✆ 31 71 527 5819
✉ burton@strw.leidenuniv.nl

Burton William M
Rutherford Appleton Lab
Space/Astrophysics Div
Bg R25/R68
Chilton Didcot OX11 0QX
UK
☎ 44 12 35 821 900
✆ 44 12 35 445 808

Buscombe William
1231 Asbury Av
Evanston IL 60202 1101
USA
✆ 1 708 491 9982

Buser Roland
Astronomisches Institut
Univ Bern
Venusstr 7
CH 4102 Binningen
Switzerland
☎ 41 61 271 7711/12
✆ 41 61 205 5455
✉ buser1@ubaclu.unibas.ch

Busko Ivo C
STScI/SSC
Homewood Campus
3700 San Martin Dr
Baltimore MD 21218
USA
☎ 1 301 338 4700
✆ 1 301 338 4767
✉ busko@stsci.edu

Buson Lucio M
Osserv Astronom Padova
Univ d Padova
Vic d Osservatorio 5
I 35122 Padova
Italy
☎ 39 49 829 3411
✆ 39 49 875 9840
✉ buson@astrpd.infnet

Busso Maurizio
Osserv Astronomico
d Torino
St Osservatorio 20
I 10025 Pino Torinese
Italy
☎ 39 11 461 9000
✆ 39 11 461 9030

Buta Ronald J
Dpt Phys/Astronomy
Univ of Alabama
Box 870324
Tuscaloosa AL 35487 0324
USA
☎ 1 205 348 3792
✆ 1 205 348 5051
✉ buta@sarah.astr.ua.edu

Butcher Harvey R
NFRA
Postbus 2
NL 7990 AA Dwingeloo
Netherlands
☎ 31 521 59 5100
✆ 31 52 159 7332
✉ Butcher@nfra.nl

Butchins Sydney Adair
Dpt Civil Aviation Stu
Fac of Maths
The Minories
Tower Hill EC3N 1JY
UK
☎ 44 171 722 7344

Buti Bimla
CALTECH/JPL
MS 169 506
4800 Oak Grove Dr
Pasadena CA 91109 8099
USA
☎ 1 818 354 8366
✆ 1 818 354 8895
✉ bbuti@jplsp.jpl.nasa.gov

Butler C John
Armagh Observ
College Hill
Armagh BT61 9DG
UK
☎ 44 18 61 522 928
✆ 44 18 61 527 174

Butler Dennis
Astronomy Dpt
Yale Univ
Box 208101
New Haven CT 06520 8101
USA
☎ 1 203 432 3000
✆ 1 203 432 5048

Butler Keith
Inst Astron/Astrophysik
Univ Sternwarte
Scheinerstr 1
D 81679 Muenchen
Germany
☎ 49 899 890 21
✆ 49 899 220 9427

Butterworth Paul
NASA/GSFC
Code 664
Greenbelt MD 20771
USA
☎ 1 301 286 3995
✆ 1 301 286 1629

Buyukliev Georgi
Ntl Astronomical Obs
Bulgarian Acad Sci
Box 136
BG 4700 Smoljan
Bulgaria
☎ 359 7 341 599

Buzzoni Alberto
Osserv Astronomico
d Milano
Via E Bianchi 46
I 22055 Merate
Italy
☎ 39 990 6412
✆ 39 990 8492
✉ buzzoni@merate.mi.astro.it

Byard Paul L
Astronomy Dpt
Ohio State Univ
174 W 18th Av
Columbus OH 43210 1106
USA
☎ 1 614 292 9532
✆ 1 614 292 2928
✉ byard@payne.mps.ohio-state.edu

Bykov Mikle F
Astronomical Institute
Uzbek Acad Sci
Astronomicheskaya Ul 33
700000 Tashkent
Uzbekistan
☎ 7 3712 35 8102

Byrd Gene G
Dpt Phys/Astronomy
Univ of Alabama
Box 1921
University AL 35487 0324
USA
☎ 1 205 348 5050
✆ 1 205 348 5051

Bystrova Natalija V
SAO/St Petersburg Br
Acad Sciences
10 Kutuzov Quay
196140 St Petersburg
Russia
☎ 7 812 123 4038/4372
✆ 7 812 315 1701
✉ bnv@sai.sob.su

Bzowski Maciej
Space Res Ctr
Polish Acad Sci
Ul Bartycka 18A
PL 00 716 Warsaw
Poland
☎ 48 22 40 3766
✆ 48 22 36 8961
✉ bzowski@cbk.waw.pl

C

Cabrita Ezequiel
OAL
Tapada d Ajuda
P 1300 Lisboa
Portugal
☎ 351 1 363 7351
✆ 351 1 362 1722

Cacciani Alessandro
Dpt Fisica
Univ d Roma
Pl A Moro 2
I 00185 Roma
Italy
☎ 39 6 49 76265
✆ 39 6 202 3507

Cacciari Carla
STScI
Homewood Campus
3700 San Martin Dr
Baltimore MD 21218
USA
☎ 1 301 338 5096
✆ 1 301 338 4767
✉ cacciari@stsci.edu

Caccin Bruno
Dpt Fisica
Via Raimondo Snc
Univ Tor Vergata
I 00173 Roma
Italy
☎ 39 6 79 79 2323
✆ 39 6 202 3507

Cadez Andrej
Physics Dpt
Univ E Kardelj
Jadranska 19
Ljubljana
Slovenia
☎ 386 61 265 061
✆ 386 61 217 281
✉ andrej.cadez@uni-lj.ac.mail.yu

Cadez Vladimir
Inst of Physics
Box 57
11001 Beograd
Yugoslavia FR
☎ 381 11 107 107
✆ 381 11 108198

Cahn Julius H
Astronomy Dpt
Tesuque NM 87574
USA
☎ 1 505 983 0691

Caillault Jean Pierre
Dpt Phys/Astronomy
Univ of Georgia
Athens GA 30602 2451
USA
☎ 1 706 542 2883
✆ 1 706 542 2492
✉ jpc@jove.physast.uga.edu

Calamai Giovanni
Osserv Astrofis Arcetri
Univ d Firenze
Largo E Fermi 5
I 50125 Firenze
Italy
☎ 39 55 27 52310
✆ 39 55 220 039
✉ vanni@arcetri.astro.it

Calame Odile
OCA
CERGA
F 06130 Grasse
France
☎ 33 4 93 36 5849
✆ 33 4 93 40 5353

Caldwell John A R
SAAO
PO Box 9
7935 Observatory
South Africa
☎ 27 21 47 0025
✆ 27 21 47 3639
✉ jac@saao.ac.za

Caldwell John James
Physics Dpt
York Univ
4700 Keele St
North York ON M3J 1P3
Canada
☎ 1 416 736 2100
✆ 1 416 736 5386

Callanan Paul
Physics Dpt
Univ College Cork
Cork
Ireland

Callebaut Dirk K
Physics Dpt
Univ Instelling Antwerpen
Universiteitsplein 1
B 2610 Wilrijk
Belgium
☎ 32 3 820 2457
✆ 32 3 820 2245
✉ calldirk@uia.ua.ac.be

Cally Paul S
Mathematics Dpt
Monash Univ
Wellington Rd
Clayton VIC 3168
Australia
☎ 61 3 565 4471
✉ apm150f@vaxc.cc.monash.edu.au

Caloi Vittoria
IAS
Area d Ricerca CNR
Via Fosso Cavaliere 100
I 00133 Roma
Italy
☎ 39 6 4993 4473
✆ 39 6 2066 0188

Calvani Massimo
Dpt Astronomia
Univ d Padova
Vic d Osservatorio 5
I 35122 Padova
Italy
☎ 39 49 829 3431
✆ 39 49 875 9840
✉ calvani@pd.astro.it

Calvet Nuria
CIDA
Box 264
Merida 5101 A
Venezuela
☎ 58 7 471 2780/3883
✆ 58 7 471 2459

Calvin William H
Physics Dpt
Univ of Washington
Box 351800
Seattle WA 98195 1580
USA
☎ 1 425 328 1192
✆ 1 425 720 1989
✉ wcalvin@u.washington.edu

Calvo Manuel
Dpt Astronomia
Univ Zaragoza
E 50009 Zaragoza
Spain
☎ 34 7 676 1000
✆ 34 7 676 1140

Calzetti Daniela
STScI
Homewood Campus
3700 San Martin Dr
Baltimore MD 21218
USA
☎ 1 301 338 4518
✆ 1 301 338 4767
✉ calzetti@stsci.edu

Camarena Badia Vicente
Dpt Mat Aplicada
Univ Zaragoza
E 50009 Zaragoza
Spain
☎ 34 7 676 1000
✆ 34 7 676 1140

Camenzind Max
Landessternwarte
Koenigstuhl
D 69117 Heidelberg
Germany
☎ 49 622 150 9262
✆ 49 622 150 9202
✉ M.Camenzind@lsw.uni-heidelberg.de

Cameron Alastair G W
CfA
HCO/SAO
60 Garden St
Cambridge MA 02138 1516
USA
☎ 1 617 495 5374
✆ 1 617 495 7356
✉ cameron@cfa.harvard.edu

Cameron Andrew Collier
School of Physics/Astro
Univ of St Andrews
North Haugh
St Andrews Fife KY16 9SS
UK
☎ 44 1334 76161
✆ 44 1334 74487
✉ Andrew.Cameron@st-and.ac.uk

Cameron Luzius Martin
Astronomisches Institut
Univ Basel
Venusstr 7
CH 4102 Binningen
Switzerland
☎ 41 61 271 7711/12
✆ 41 61 205 5455

Cameron Winifred S
La Ranehita de La Luna Bg 26
200 Rojo Dr
Sedona AZ 86336
USA

Camichel Henri
24 Av C Flammarion
F 31500 Toulouse
France
☎ 33 5 61 48 9691
✆ 33 5 56 40 4251

Campbell Alison
Dpt Phys/Astronomy
JHU
Charles/34th St
Baltimore MD 21218
USA
☎ 1 301 338 5186
℻ 1 301 338 8260

Campbell Belva G S
Dpt Phys/Astronomy
Univ New Mexico
800 Yale Blvd Ne
Albuquerque NM 87131
USA
☎ 1 505 277 5148

Campbell Donald B
Astronomy Dpt
Cornell Univ
512 Space Sc Bldg
Ithaca NY 14853 6801
USA
☎ 1 607 255 5274
℻ 1 607 255 1767

Campbell James W
Royal Observ
Blackford Hill
Edinburgh EH9 3HJ
UK
☎ 44 131 667 3321
℻ 44 131 668 8356

Campbell Murray F
Physics/Astronomy Dpt
Colby College
Waterville ME 04901
USA
☎ 1 207 872 3251
℻ 1 207 872 3255
✉ mfcampbe@colby.edu

Campins Humberto
Astronomy Dpt
Univ of Florida
211 SSRB
Gainesville FL 32611
USA
☎ 1 904 392 3066
℻ 1 904 392 5089
✉ campins@astro.ufl.edu.

Campos Ana
Observ Astronomico
Nacional
Apt 1143
E 28800 Alcala d Henares
Spain
☎ 34 1 885 5060
℻ 34 1 885 5062
✉ ana@oan.es

Campusano Luis E
Dpt Astronomia
Univ Chile
Casilla 36 D
Santiago
Chile
☎ 56 2 229 4101
℻ 56 2 229 4002

Canal Ramon M
Dpt Fisica Atmosfera
Univ Barcelona
Avd Diagonal 645
E 08028 Barcelona
Spain
☎ 34 3 402 1125
℻ 34 3 402 1133

Canalle Joao B G
Instituto Fisica
Univ Estado R de Janeiro
RS Francisco Xavier 534/3002D
20550 013 Rio de Janeiro
Brazil
☎ 55 21 587 7447
℻ 55 21 587 7447
✉ canalle@vmesa.uerj.br

Cane Hilary Vivien
Physics Dpt
Univ Tasmania
GPO Box 252c
Hobart TAS 7001
Australia
☎ 61 2 202 401
℻ 61 2 202 186
✉ hilary.cane@phys.
utas.edu.au

Canfield Richard C
Physics Dpt
Montana State Univ
Bozeman MT 59717 0350
USA
☎ 1 406 994 6874
℻ 1 406 994 4452

Canizares Claude R
Ctr for Space Res
MIT Rm 37 241
Box 165
Cambridge MA 02139 4307
USA
☎ 1 617 253 7480
℻ 1 617 253 0861
✉ brendap@space.mit.edu

Cannon Russell D
AAO
Box 296
Epping NSW 2121
Australia
☎ 61 2 868 1666
℻ 61 2 9876 8536

Cannon Wayne H
Dpt Phys/Earth/Atm Sci
York Univ
4700 Keele St
Downsview ON M3J 1P3
Canada
☎ 1 416 667 6410

Canto Jorge
Instituto Astronomia
UNAM
Apt 70 264
Mexico DF 04510
Mexico
☎ 52 5 622 3906
℻ 52 5 616 0653

Cantu Alberto M
Istt Cibern/Biofisica
CNR
I 16032 Camogli
Italy
☎ 39 18 577 0646

Cao Changxin
Nanjing Astronomical
Instrument Factory
Box 846
Nanjing
China PR
☎ 86 25 46191
℻ 86 25 71 1256

Cao Lihong
Purple Mountain Obs
CAS
Nanjing 210008
China PR
☎ 86 25 46700
℻ 86 25 301 459

Cao Shenglin
Astronomy Dpt
Beijing Normal Univ
Beijing 100875
China PR
☎ 86 1 201 2255
℻ 86 1 201 3929

Cao Xinwu
Shanghai Observ
CAS
80 Nandan Rd
Shanghai 200030
China PR
☎ 86 21 6438 6191*42
℻ 86 21 6438 4618
✉ cxw@center.shao.ac.cn

Capaccioli Massimo
Dpt Astronomia
Univ d Padova
Vic d Osservatorio 5
I 35122 Padova
Italy
☎ 39 49 829 3411
℻ 39 49 875 9840

Capaccioni Fabrizio
IAS
Rep Planetologia
Via d universita 11
I 00133 Roma
Italy
☎ 39 6 4993 4473
℻ 39 6 2066 0188
✉ fabrizio@saturn.ias.
fra.cnr.it

Capelato Hugo Vicente
INPE
Dpt Astronomia
CP 515
12227 010 S Jose dos Campos
Brazil
☎ 55 123 22 9977
℻ 55 123 21 8743

Capen Charles F
Solis Lacus Observ
RT 2
Box 262 E
Cuba MO 65453
USA

Capitaine Nicole
Observ Paris
DANOF
61 Av Observatoire
F 75014 Paris
France
☎ 33 1 40 51 2231
℻ 33 1 40 51 2291
✉ capitaine@obspm.fr

Caplan James
Observ Marseille
2 Pl Le Verrier
F 13248 Marseille Cdx 04
France
☎ 33 4 91 95 9088
℻ 33 4 91 62 1190
✉ caplan@obmara.cnrs-
mrs.fr

Cappa de Nicolau Cristina
IAR
CC 5
1894 Villa Elisa (Bs As)
Argentina
☎ 54 21 254 909
℻ 54 21 254 909
✉ ccappa@irma.edu.ar

Cappellaro Enrico
Osserv Astronom Padova
Univ d Padova
Vic d Osservatorio 5
I 35122 Padova
Italy
☎ 39 49 829 3411
℻ 39 49 875 9840

Cappi Alberto
Osserv Astronom Bologna
Univ d Bologna
Via Zamboni 33
I 40126 Bologna
Italy
☎ 39 51 259 406
℻ 39 51 259 407
✉ cappi@astbo3.bo.
astro.it

Capria Maria Teresa
IAS
Rep Planetologia
Via d universita 11
I 00133 Roma
Italy
☎ 39 6 4993 4473
℻ 39 6 2066 0188
✉ capria@saturn.ias.fra.
cnr.it

Caprioli Giuseppe
OAR
Via d Parco Mellini 84
I 00136 Roma
Italy
☎ 39 6 34 7056
℻ 39 6 349 8236

Capriotti Eugene R
Astronomy Dpt
Ohio State Univ
174 W 18th Av
Columbus OH 43210 1106
USA
☎ 1 614 422 1773
℻ 1 614 292 2928

Caputo Filippina
Osserv d Capodimonte
Via Moiariello 16
I 80131 Napoli
Italy
☎ 39 81 44 0101
✆ 39 81 45 6710
🖃 caputo@astrna.na.
astro.it

Capuzzo Dolcetta Roberto
Istt Astronomico
Univ d Roma La Sapienza
Via G M Lancisi 29
I 00161 Roma
Italy
☎ 39 6 44 03734
✆ 39 6 44 03673
🖃 dolcetta@astrmb.rm.
astro.it

Caranicolas Nicholas
Astronomy Lab
Univ Thessaloniki
GR 540 06 Thessaloniki
Greece
☎ 30 31 99 1357/59

Carbon Duane F
NASA/ARC
MS 258 5
Moffett Field CA 94035 1000
USA
☎ 1 415 604 4413
✆ 1 415 604 6779
🖃 dcarbon@nas.nasa.gov

Cardini Daniela
IAS
Area d Ricerca CNR
Via Fosso Cavaliere 100
I 00133 Roma
Italy
☎ 39 6 4993 4473
✆ 39 6 2066 0188
🖃 cardini@saturn.ias.
rm.cnr.it

Cardona Octavio
INAOE
Tonantzintlaz
Apdo 51 y 216
Puebla PUE 72000
Mexico
☎ 52 2 247 2011
✆ 52 2 247 2231

Cardus Almeda J O
Osserv d Ebro
URL
E 43520 Roquetes
Spain
☎ 34 7 750 0511
✆ 34 7 750 4660

Cargill Peter J
Space/Atmosph Phys
Imperial Coll
Prince Consort Rd
London SW7 2BZ
UK
☎ 44 171 594 7773
✆ 44 171 594 7772
🖃 p.cargill@ic.ac.uk

Carigi Leticia
Instituto Astronomia
UNAM
Apt 70 264
Mexico DF 04510
Mexico
☎ 52 5 622 3906
✆ 52 5 616 0653
🖃 carigi@astroscu.
unam.mx

Carignan Claude
Dpt d Physique
Universite Montreal
CP 6128 Succ A
Montreal QC H3C 3J7
Canada
☎ 1 514 343 7355
✆ 1 514 343 2071
🖃 carignan@cc.
umontreal.ca

Carilli Christopher L
NRAO
Box 0
Socorro NM 87801 0387
USA
☎ 1 505 835 7000
✆ 1 505 835 7027
🖃 ccarilli@nrao.edu

Carlberg Raymond Gary
Physics Dpt
York Univ
4700 Keele St
North York ON M3J 1P3
Canada
☎ 1 416 667 3851
✆ 1 416 736 5386

Carleton Nathaniel P
CfA
HCO/SAO
60 Garden St
Cambridge MA 02138 1516
USA
☎ 1 617 495 7405
✆ 1 617 495 7356

Carlqvist Per A
Dpt Plasma Physics
Royal Inst of Technology
S 100 44 Stockholm 70
Sweden
☎ 46 87 87 7697

Carlson John B
Archeoastron Ctr
Box X
College Park MD 20741 3022
USA
☎ 1 301 864 6637

Carlsson Mats
Inst Theor Astrophys
Univ of Oslo
Box 1029
N 0315 Blindern Oslo 3
Norway
☎ 47 22 856 536
✆ 47 22 856 505
🖃 mats.carlsson@astro.
uio.no

Carney Bruce William
Dpt Phys/Astronomy
Univ North Carolina
204 Phillips Hall 039a
Chapel Hill NC 27514 3255
USA
☎ 1 919 962 3023
✆ 1 919 962 0480
🖃 bruce@sloth.astro.
unc.edu

Caroff Lawrence J
NASA/ARC
Space Sci Div
MS 245 6
Moffett Field CA 94035 1000
USA
☎ 1 415 694 5523
✆ 1 415 604 6779

Caroubalos C A
Lab Electronic Physics
Ntl Univ Athens
Kthria Typa-ilissia
GR 157 01 Athens
Greece
☎ 30 1 724 4096/11119

Carpenter Kenneth G
NASA GSFC
Code 681
Greenbelt MD 20771
USA
☎ 1 301 286 3453
✆ 1 301 286 1753
🖃 hrscarpenter@tma1.gsfc.
nasa.gov

Carpenter Lloyd
13902 Resin Ct
Bowie MD 20720
USA

Carpino Mario
Osserv Astronomico
d Brera
Via Brera 28
I 20121 Milano
Italy
☎ 39 2 723 20306
✆ 39 2 720 01600
🖃 carpino@brera.mi.
astro.it

Carpintero Daniel Diego
FCAGS
Paseo d Bosque S/n
1900 La Plata (Bs As)
Argentina
☎ 54 21 217 308
✆ 54 21 211 761
🖃 ddc@fcaglp.fcaglp.
unlp.edu.ar

Carquillat Jean-Michel
OMP
14 Av E Belin
F 31400 Toulouse Cdx
France
☎ 33 5 61 25 2101
✆ 33 5 61 27 3179

Carr Bernard John
School Mathemat Sc
QMWC
Mile End Rd
London E1 4NS
UK
☎ 44 171 980 4811
✆ 44 181 975 5500

Carr John Sherman
NRL
Code 7217
4555 Overlook Av SW
Washington DC 20375 5351
USA
☎ 1 202 767 0670
✆ 1 202 404 8894
🖃 carr@mriga.nrl.navy.mil

Carr Thomas D
Astronomy Dpt
Univ of Florida
211 SSRB
Gainesville FL 32611
USA
☎ 1 904 392 2066
✆ 1 904 392 5089
🖃 tacarr@uffsc

Carraminana Alberto
INAOE
Tonantzintlaz
Apdo 51 y 216
Puebla PUE 72000
Mexico
☎ 52 2 247 2011
✆ 52 2 247 2231
🖃 elsare@inaoep.mx

Carranza Gustavo J
Observ Astronomico
de Cordoba
Laprida 854
5000 Cordoba
Argentina
☎ 54 51 23 0491
✆ 54 51 21 0613
🖃 carranza@astro.edu.ar

Carrasco Berth Esperanza
INAOE
Tonantzintlaz
Apdo 51 y 216
Puebla PUE 72000
Mexico
☎ 52 2 247 2011
✆ 52 2 247 2231

Carrasco Guillermo
Dpt Astronomia
Univ Chile
Casilla 36 D
Santiago
Chile
☎ 56 2 229 4101
✆ 56 2 229 4002

Carrasco Luis
Instituto Astronomia
UNAM
Apt 70 264
Mexico DF 04510
Mexico
☎ 52 5 622 3906
✆ 52 5 616 0653

Carrillo Moreno Rene
Instituto Astronomia
UNAM
Apt 70 264
Mexico DF 04510
Mexico
☎ 52 5 622 3906
✆ 52 5 616 0653
✉ rene@astroscu.unam.mx

Carroll P Kevin
Physics Dpt
Univ College
Belfield
Dublin 4
Ireland
☎ 353 1 693 244
✆ 353 1 694 409

Carruthers George R
NRL
Code 7123
4555 Overlook Av SW
Washington DC 20375 5000
USA
☎ 1 202 767 2764
✆ 1 202 404 7296

Carsenty Uri
DFVLR
Ne Oe Pe
Oberpfaffenhofen
D 8031 Wessling
Germany
☎ 49 815 328 1328
✆ 49 815 32476

Carsmaru Maria M
Astronomical Institute
Romanian Acad Sciences
Cutitul de Argint 5
RO 75212 Bucharest
Rumania
☎ 40 1 641 3686
✆ 40 1 312 3391
✉ carsmaru@roastro.astro.ro

Carson T R
Dpt Phys/Astronomy
Univ of St Andrews
North Haugh
St Andrews Fife KY16 9SS
UK
☎ 44 1334 76161
✆ 44 1334 74487

Carswell Robert F
Inst of Astronomy
The Observatories
Madingley Rd
Cambridge CB3 0HA
UK
☎ 44 12 23 337 548
✆ 44 12 23 337 523

Carter Bradley Darren
Fac of Sciences
Univ of SQLD
Toowoomba QLD 4320
Australia
☎ 61 7 46 312801
✆ 61 7 46 31312721
✉ carterb@usq.edu.au

Carter Brian
Carter Observ
Ntl Observ New Zealand
Box 2909
Wellington 1
New Zealand
☎ 64 4 472 8167
✆ 64 4 472 8320
✉ brian.carter@vuw.ac.nz

Carter David
MSSSO
Weston Creek
Private Bag
Canberra ACT 2611
Australia
☎ 61 262 881 111
✆ 61 262 490 233

Carter William Eugene
Dpt Civil Eng
Univ of Florida
345 Weil Hall
Gainesville FL 32611
USA
☎ 1 904 392 5003
✆ 1 904 392 4957
✉ bcarter@ce.ufl.edu

Carusi Andrea
IAS
Rep Planetologia
Via d universita 11
I 00133 Roma
Italy
☎ 39 6 4993 4473
✆ 39 6 2066 0188
✉ carusi@saturn.ias.fra.cnr.it

Carver John H
Australian Ntl Univ
Res School Phys Science
Box 4
Canberra ACT 2600
Australia
☎ 61 2 62 49 2476

Casanovas Juan
Vatican Observ
I 00120 Citta del Vaticano
Vatican City State
☎ 39 6 698 3411/5266
✆ 39 6 698 84671

Casares Valazquez Jorge
IAC
Observ d Teide
via Lactea s/n
E 38200 La Laguna
Spain
☎ 34 2 232 9100
✆ 34 2 232 9117

Casertano Stefano
Dpt Astron/Physics
Univ of Illinois
1011 W Springfield Av
Urbana IL 61801
USA
☎ 1 217 333 9380
✆ 1 217 244 7638
✉ stefano@rigel.astro.uluc.edu

Cash Webster C
CASA
Univ of Colorado
Campus Box 389
Boulder CO 80309 0389
USA
☎ 1 303 492 4056
✆ 1 303 492 4052

Casoli Fabienne
DEMIRM
ENS
24 r Lhomond
F 75231 Paris Cdx 05
France
☎ 33 1 43 29 1225
✆ 33 1 40 51 2002
✉ casoli@frulm11

Cassatella Angelo
IAS
Area d Ricerca CNR
Via Fosso Cavaliere 100
I 00133 Roma
Italy
☎ 39 6 4993 4473
✆ 39 6 2066 0188
✉ angelo@saturn.ias.rm.cnr.it

Casse Michel
DAPNIA/SAP
CEA Saclay
BP 2
F 91191 Gif s Yvette Cdx
France
☎ 33 1 69 08 3917
✆ 33 1 69 08 6577

Cassinelli Joseph P
Astronomy Dpt
Univ Wisconsin
475 N Charter St
Madison WI 53706 1582
USA
☎ 1 608 262 1752
✆ 1 608 263 0361

Castagnino Mario
IAFE
CC 67 Suc 28
1428 Buenos Aires
Argentina
☎ 54 1 781 6755
✆ 54 1 814 4299
✉ castagnino@iafe.edu.ar

Castaneda Hector
IAC
Observ d Teide
via Lactea s/n
E 38200 La Laguna
Spain
☎ 34 2 232 9100
✆ 34 2 232 9117

Castelaz Micheal W
Physics Dpt
E Tennessee State Univ
Box 70652
Johnson City TN 37614
USA
☎ 1 615 461 7064
✉ castelam@etsuarts.etsu-tn.edu

Castellani Vittorio
Dpt Fisica
Univ d Pisa
Piazza Torricelli 2
I 56100 Pisa
Italy
☎ 39 50 443 343
✆ 39 50 482 77

Castelli Fiorella
OAT
Box Succ Trieste 5
Via Tiepolo 11
I 34131 Trieste
Italy
☎ 39 40 31 99255
✆ 39 40 30 9418

Castelli John P
125 Hillside Av
Arlington MA 02174
USA

Castets Alain
Observ Grenoble
Lab Astrophysique
BP 53x
F 38041 S Martin Heres Cdx
France
☎ 33 4 76 51 4786
✆ 33 4 76 44 8821

Castor John I
LLNL
L 58
Box 808
Livermore CA 94551 9900
USA
☎ 1 415 422 4664
✆ 1 415 422 5102
✉ castor1@llnl.gov

Caswell James L
CSIRO
Div Radiophysics
Box 76
Epping NSW 2121
Australia
☎ 61 2 868 0222
✆ 61 2 868 0310
✉ jcaswell@atnf.csiro.au

Catala Claude
Observ Paris Meudon
Pl J Janssen
F 92195 Meudon PPL Cdx
France
☎ 33 1 45 07 7663
✆ 33 1 45 07 7971

Catala Poch M A
M Asuncion
Muntaner 83 B 3/3
E 08011 Barcelona
Spain
☎ 34 4 5 32 569

Catalan Manuel
ROA
Cecilio Pujazon 22-3 A
E 11110 San Fernando
Spain
☎ 34 5 688 3548
✆ 34 5 659 9366
✉ catalan@czv1.uca.es

Catalano Francesco A
Istt Astronomia
Citta Universitaria
Via A Doria 6
I 95125 Catania
Italy
☎ 39 95 733 21111
✆ 39 95 33 0592

Catalano Santo
Istt Astronomia
Citta Universitaria
Via A Doria 6
I 95125 Catania
Italy
☎ 39 95 733 21111
✆ 39 95 33 0592

Catchpole Robin M
Royal Greenwich Obs
Madingley Rd
Cambridge CB3 0EZ
UK
☎ 44 12 23 374 000
✆ 44 12 23 374 700
✉ cathpole@uk.ac.cam.
ast-star

Cato B Torgny
Nordisk Telesatellitstat
Box 107
S 457 00 Tanumshede
Sweden
☎ 46 52 52 9155

Caton Daniel B
Dpt Phys/Astronomy
Appalachian State Univ
Boone NC 28608
USA
☎ 1 704 262 2446
✉ catondb@appstate.edu

Catura Richard C
1008 Columbus St
Half Moon Bay CA 94019
USA
✉ catura@sag.space.
lockeed.com

Cauzzi Gianna
Osserv d Capodimonte
Via Moiariello 16
I 80131 Napoli
Italy
☎ 39 81 557 5111
✆ 39 81 45 6710
✉ gcauzzi@cerere.
na.astro.it

Cavaliere Alfonso G
Dpt Fisica/Astrofisica
Univ d Roma Tor Vergata
Via Ric Scientifica
I 00133 Roma
Italy
☎ 39 6 725 94301
✆ 39 6 202 3507

Cavallini Fabio
Osserv Astrofis Arcetri
Univ d Firenze
Largo E Fermi 5
I 50125 Firenze
Italy
☎ 39 55 43 78540
✆ 39 55 43 5939

Cawley Michael
Physics Dpt
St Patrick's College
Maynooth
Maynooth Co Kildare
Ireland
☎ 353 1 285 222*499

Cayrel Roger
Observ Paris
DASGAL
61 Av Observatoire
F 75014 Paris
France
☎ 33 1 40 51 2251
✆ 33 1 40 51 2232
✉ cayrel@obspm.fr

Cayrel de Strobel Giusa
Observ Paris Meudon
DASGAL
Pl J Janssen
F 92195 Meudon PPL Cdx
France
☎ 33 1 45 07 7863
✆ 33 1 45 07 7971

Cazenave Anny
CNES/GRGS/BGI
18 Av E Belin
F 31055 Toulouse Cdx
France
☎ 33 5 61 27 4011
✆ 33 5 61 25 3205
✉ anny.cazenave@to.
astro.it

Cazzola Paolo
Osserv Astronom Padova
Univ d Padova
Vic d Osservatorio 5
I 35122 Padova
Italy
☎ 39 49 829 3411
✆ 39 49 875 9840

Cecchi-Pellini Cesare
Osserv d Capodimonte
Via Moiariello 16
I 80131 Napoli
Italy
☎ 39 81 298 384
✆ 39 81 45 6710
✉ cesarecp@cosmic.
na.astro.it

Cefola Paul J
Mail Station 64
C S Draper Lab
555 Technology Sq
Cambridge MA 02139
USA
☎ 1 617 258 1787

Celis Leopoldo
Dpt Astronomia
Univ Catolica
Casilla 36 D
Santiago
Chile
☎ 56 2 229 4101
✆ 56 2 229 4002

Cellino Alberto
Osserv Astronomico
d Torino
St Osservatorio 20
I 10025 Pino Torinese
Italy
☎ 39 11 461 9000
✆ 39 11 461 9030
✉ cellino@astto2.infn.it

Cellone Sergio Aldo
Complejo Astronomico
El Leoncito
CC 467
5400 San Juan
Argentina
☎ 54 64 213 653
✆ 54 64 213 693
✉ scellone@castec.
edu.ar

Celnikier Ludwik
Observ Paris Meudon
DAEC
Pl J Janssen
F 92195 Meudon PPL Cdx
France
☎ 33 1 45 07 7410
✆ 33 1 45 07 7971

Centrella Joan M
Physics Dpt
Drexel Univ
Philadelphia PA 19104
USA
☎ 1 215 895 2715

Centurion Martin Miriam
IAC
Observ d Teide
via Lactea s/n
E 38200 La Laguna
Spain
☎ 34 2 232 9100
✆ 34 2 232 9117
✉ mcm@iac.es

Cepa Jordi
IAC
Observ d Teide
via Lactea s/n
E 38200 La Laguna
Spain
☎ 34 2 232 9100
✆ 34 2 232 9117

Ceplecha Zdenek
Astronomical Institute
Czech Acad Sci
Fricova 1
CZ 251 65 Ondrejov
Czech R
☎ 420 204 85 7155
✆ 420 2 88 1611
✉ ceplecha@asu.cas.cz

Ceppatelli Guido
Osserv Astrofis Arcetri
Univ d Firenze
Largo E Fermi 5
I 50125 Firenze
Italy
☎ 39 55 43 78540
✆ 39 55 43 5939

Cernicharo Jose
Ctr Astron d Yebes
OAN
Apt 148
E 19080 Guadalajara
Spain
☎ 34 1 129 0311
✆ 34 1 129 0063

Cernis Kazimieras
Inst Theor Physics/
Astronomy
Gostauto 12
Vilnius 2600
Lithuania
☎ 370 2 613 440
✆ 370 2 224 694
✉ astro@itpa.fi.lt

Cerroni Priscilla
IAS
Rep Planetologia
Via d universita 11
I 00133 Roma
Italy
☎ 39 6 4993 4473
✆ 39 6 2066 0188
✉ priscio@saturn.ias.
fra.cnr.it

Cerruti'sola Monica
Osserv Astrofis Arcetri
Univ d Firenze
Largo E Fermi 5
I 50125 Firenze
Italy
☎ 39 55 43 78540
✆ 39 55 43 5939

Cersosimo Juan Carlos
Dpt Phys/Electronic
Univ of Puerto Rico
Cuh Station
Humacao PR 00791
USA
☎ 1 809 850 9381
✆ 1 809 852 4638

Cesarsky Catherine J
CEA/DSM
F 91191 Gif s Yvette Cdx
France
☎ 33 1 69 08 7515
✆ 33 1 69 08 4004
✉ catherine.cesarsky
@cea.fr

Cesarsky Diego A
IAS
Bt 121
Universite Paris XI
F 91405 Orsay Cdx
France

Cester Bruno
OAT
Box Succ Trieste 5
Via Tiepolo 11
I 34131 Trieste
Italy
☎ 39 40 31 99255
✆ 39 40 30 9418

Cevolani Giordano
FISBAT
CNR
Via Castagnoli 1
I 40126 Bologna
Italy
☎ 39 51 239 593/94

Cha Du Jin
Pyongyang Astron Obs
Acad Sciences DPRK
Taesong district
Pyongyang
Korea DPR
☎ 850 5 3134/5 & 5 3239

Cha Gi Ung
Pyongyang Astron Obs
Acad Sciences DPRK
Taesong district
Pyongyang
Korea DPR
☎ 850 5 3134/5 & 5 3239

Chaffee Frederic H
W M Keck Observ
Box 220
65 1120 Mamalahoa Hwy
Kamuela HI 96743
USA
☎ 1 808 885 7887
℻ 1 808 885 7887
✉ fchaffee@keck.
hawaii.edu

Chaisson Eric J
STScI
Homewood Campus
3700 San Martin Dr
Baltimore MD 21218
USA
☎ 1 301 338 4757
℻ 1 301 338 4767

Chakrabarti Sandip K
TIFR
Homi Bhabha Rd
Colaba
Bombay 400 005
India
☎ 91 22 215 2971*305
℻ 91 22 495 2110
✉ chakraba@tifrvax.
tifr.res.in

Chakraborty Deo K
Physics Dpt
Univ of Ravishankar
Raipur 492 010
India
☎ 91 27 064

Chalabaev Almas
OHP
F 04870 S Michel Obs
France
☎ 33 4 92 76 6368
℻ 33 4 92 76 6295
✉ chalabaev@froni51

Chamaraux Pierre
Observ Paris Meudon
ARPEGES
Pl J Janssen
F 92195 Meudon PPL Cdx
France
☎ 33 1 45 07 7594
℻ 33 1 45 07 7971

Chambe Gilbert
Observ Paris Meudon
DASOP
Pl J Janssen
F 92195 Meudon PPL Cdx
France
☎ 33 1 45 34 7793
℻ 33 1 45 07 7959

Chamberlain Joseph M
510 W Thousand Oaks Dr
Peoria IL 61615 1395
USA
☎ 1 309 691 1442

Chamberlain Joseph W
510 W Thousand Oaks Dr
Peoria IL 61615 1394
USA
☎ 1 309 691 1442
✉ jchambe584@aol.com

Chambliss Carlson R
Dpt Physical Sci
Kutztown Univ
Kutztown PA 19530
USA
☎ 1 215 683 4439

Chan Kwing Lam
Applied Res Corp
8201 Corporate Dr
Suite 920
Landover MD 20785
USA
☎ 1 301 459 8442
℻ 1 301 459 0761

Chan Roberto
Observ Nacional
Rua Gl Bruce 586
Sao Cristovao
20921 030 Rio de Janeiro RJ
Brazil
☎ 55 21 580 7313
℻ 55 21 580 0332
✉ userchan@lnccvm

Chan Siu Kuen Josphine
Inst of Astronomy
The Observatories
Madingley Rd
Cambridge CB3 0HA
UK
☎ 44 12 23 337 548
℻ 44 12 23 337 548
✉ sjchan@ast.cam.ac.uk

Chance Kelly V
CfA
HCO/SAO
60 Garden St
Cambridge MA 02138 1516
USA
☎ 1 617 495 7389
℻ 1 617 495 7389
✉ kelly@cfa.harvard.edu

Chandler Claire
MRAO
Cavendish Laboratory
Madingley Rd
Cambridge CB3 0HE
UK
☎ 44 12 23 339 992
℻ 44 12 23 354 599
✉ cjc@mrao.cam.ac.uk

Chandra Subhash
Mis Philips Labs
345 Scarborough Rd
Briar Cliff NY 10510
USA

Chandra Suresh
School of Physical Sci
SRTM Univ
Nanded 431 606
India
☎ 91 2462 26203
℻ 91 2462 26119

Chandrasekhar T
PRL
Navrangpura
Ahmedabad 380 009
India
☎ 91 272 46 2129
℻ 91 272 44 5292

Chang Jin
Purple Mountain Obs
CAS
Nanjing 210008
China PR
☎ 86 25 330 3583
℻ 86 25 330 1459

Chang Kyongae
Dpt Physics/Optics
Chungju Univ
Chungju
Korea RP
☎ 82 431 51 8466
℻ 82 431 212 5842
✉ kcchang@soback.
kornet.nm.kr

Chaoudhuri Arnab
Physics Dpt
IISc
Karnataka
Bangalore 560 012
India
☎ 91 81 234 4411
℻ 91 81 234 1683
✉ arnab@physics.iisc.
ernet.in

Chapman Clark R
Southwest Res Inst
1050 Walnut St
Suite 426
Boulder CO 80302
USA

Chapman Gary A
Dpt Phys/Astronomy
San Fernando Observ
California State Univ.
Northridge CA 91330 8268
USA
☎ 1 818 885 2775

Chapman Jessica
CSIRO
ATNF
Box 296
Epping NSW 2121
Australia
☎ 61 2 868 0222
℻ 61 2 868 0310
✉ jchapman@atnf.csiro.au

Chapman Robert D
10976 Swanfields Rd
Columbia MD 21044
USA
☎ 1 301 596 4617

Chapront Jean
Observ Paris
DANOF
61 Av Observatoire
F 75014 Paris
France
☎ 33 1 40 51 2271
℻ 33 1 44 54 1804
✉ jean.chapront@danof.
obspm.fr

Chapront-Touze Michelle
BDL
77 Av Denfert Rochereau
F 75014 Paris
France
☎ 33 1 40 51 2266
℻ 33 1 46 33 2834

Charbonneau Paul
HAO
NCAR
Box 3000
Boulder CO 80307
USA
☎ 1 303 497 1594
℻ 1 303 497 1589
✉ paulchar@hao.
ucar.edu

Charles Philip Allan
Astrophysics Dpt
Univ of Oxford
Keble Rd
Oxford OX1 3RQ
UK
☎ 44 186 527 3374
℻ 44 186 527 3418
✉ pac@astro.ox.ac.uk

Charlot Patrick
Observ Bordeaux
BP 89
F 33270 Floirac
France
☎ 33 5 57 77 6141
℻ 33 5 577 7 6110
✉ charlot@observ.u-
bordeaux.fr

Charlot Stephane
IAP
98bis bd Arago
F 75014 Paris
France
☎ 33 1 44 32 8146
℻ 33 1 44 32 8001
✉ charlot@iap.fr

Chashei Igor V
Lebedev Physical Inst
Leninsky Pspt 53
Moscow 117924
Russia
☎ 7 096 7732 757
℻ 7 096 7732 482
✉ chashey@prao.psn.ru

Chassefiere Eric
Service Aeronomie
BP 3
F 91371 Verrieres Buisson
France
☎ 33 1 64 47 4211

Chatterjee Tapan K
Fac Ciencias
Univ A Puebla
Apt 1316
Puebla PUE 72000
Mexico
☎ 52 2 233 0455
℻ 52 2 244 8947

Chau Wai Y
Physics Dpt
Queen's Univ
Kingston ON K7L 3N6
Canada
☎ 1 613 547 3526
℻ 1 613 545 6463

Chaubey Uma Shankar
Uttar Pradesh State
Observ
Po Manora Peak 263 129
Nainital 263 129
India
☎ 91 59 42 2136/2583

Chauvineau Bertrand
OCA
CERGA
F 06130 Grasse
France
☎ 33 4 93 36 5849
℻ 33 4 93 40 5353

Chavarria'k Carlos
Instituto Astronomia
UNAM
Apt 70 264
Mexico DF 04510
Mexico
☎ 52 5 622 3906
℻ 52 5 616 0653

Chavez-Dagostino Miguel
INAOE
Tonantzintlaz
Apdo 51 y 216
Puebla PUE 72000
Mexico
☎ 52 2 247 2011
℻ 52 2 247 2231
✉ mchavez@inaoep.mx

Chavira Enrique Sr
INAOE
Tonantzintlaz
Apdo 51 y 216
Puebla PUE 72000
Mexico
☎ 52 2 247 2011
℻ 52 2 247 2231
✉ chavira@tonali.
inaoep.mx

Che-Bohnenstengel Anne
Hamburger Sternwarte
Univ Hamburg
Gojensbergsweg 112
D 21029 Hamburg
Germany
☎ 49 407 252 4112
℻ 49 407 252 4198

Chechetkin Valerij M
Keldysh Inst Applied Maths
Acad Sciences
Miusskaja Sq 4
125047 Moscow
Russia
☎ 7 095 250 7847
✉ chech@int.keldysh.ru

Chelli Alain
Instituto Astronomia
UNAM
Apt 70 264
Mexico DF 04510
Mexico
☎ 52 5 622 3906
℻ 52 5 616 0653

Chen Biao
Purple Mountain Obs
CAS
Nanjing 210008
China PR
☎ 86 25 46700
℻ 86 25 301 459

Chen Daohan
Purple Mountain Obs
CAS
Nanjing 210008
China PR
☎ 86 25 31096
℻ 86 25 301 459

Chen Hongsheng
Beijing Astronomical Obs
CAS
Beijing 100080
China PR
☎ 86 1 28 1698
℻ 86 1 256 10855

Chen Jiansheng
Beijing Astronomical Obs
CAS
Beijing 100080
China PR
☎ 86 1 28 1698
℻ 86 1 256 10855

Chen Kwan-yu
Astronomy Dpt
Univ of Florida
211 SSRB
Gainesville FL 32611
USA
☎ 1 904 392 2055
℻ 1 904 392 5089

Chen Li
Shanghai Observ
CAS
80 Nandan Rd
Shanghai 200030
China PR
☎ 86 21 6438 6191
℻ 86 21 6438 4618
✉ chenli@center.shao.ac.cn

Chen Li
Astronomy Dpt
Beijing Normal Univ
Beijing 100875
China PR
☎ 86 1 201 2288/2918
℻ 86 1 201 3929
✉ zhouzf@bnu.ihep.ac.cn

Chen Peisheng
Yunnan Observ
CAS
Kunming 650011
China PR
☎ 86 871 2035
℻ 86 871 717 1845

Chen Wen Ping
Inst Phys/Astronomy
Ntl Central Univ
Chung Li Taiwan 32054
China R
☎ 886 3 422 3424
℻ 886 3 426 2304
✉ when@phyast.phy.
ncu.edu.tw

Chen Xiaozhong
Beijing Planetarium
138 Xi Wai St
Beijing 100044
China PR
☎ 86 1 683 61691
℻ 86 10 683 53003

Chen Yang
Astronomy Dpt
Nanjing Univ
Nanjing 210093
China PR
☎ 86 25 663 7551*2882
℻ 86 25 330 2728
✉ ygchen@nju.edu.cn

Chen Zhen
Purple Mountain Obs
CAS
Nanjing 210008
China PR
☎ 86 25 46700
℻ 86 25 301 459

Chen Zhencheng
Beijing Astronomical Obs
CAS
Beijing 100080
China PR
☎ 86 1 28 1698
℻ 86 1 256 10855

Cheng Fuhua
Astrophysics Division
Univ Science/Technology
Hefei 230026
China PR
☎ 86 551 33 1134*526
℻ 86 551 33 1760

Cheng Fuzhen
Astrophysics Division
Univ Science/Technology
Hefei 230026
China PR
☎ 86 551 33 1134*987
℻ 86 551 33 1760

Cheng Kwang Ping
14204 Castle Blvd
Silver Spting MD 20904
USA
☎ 1 301 286 3019
℻ 1 301 286 1753
✉ cheng@stars.gsfc.
nasa.gov

Cheng Kwong-sang
Physics Dpt
Univ Hong Kong
Hong Kong
China PR
☎ 852 285 92 368
℻ 852 255 99 152
✉ hrspksc@hkucc.hku.hk

Chengmo Zhang
Inst High Energy Phys
CAS
Beijing 100039
China PR
☎ 86 1 6821 3344*2149
℻ 86 10 6821 3374
✉ zhangcm@heal2.ihep.
ac.cn

Cherednichenko V I
Kyiv Polytechnical Inst
252056 Kyiv
Ukraine

Cherepashchuk Analily M
SAI
Acad Sciences
Universitetskij Pr 13
119899 Moscow
Russia
☎ 7 095 939 2858
℻ 7 095 932 8841
✉ boch@astronomy.
msk.su

Chernin Arthur D
SAI
Acad Sciences
Universitetskij Pr 13
119899 Moscow
Russia
☎ 7 095 939 2858
℻ 7 095 939 0126

Chernov Gennadij P
ITMIRWP
Acad Sciences
142092 Troitsk
Russia
☎ 7 095 334 0902
℻ 7 095 334 0124

Chernykh N S
Crimean Astrophys Obs
Ukrainian Acad Science
Nauchny
334413 Crimea
Ukraine
☎ 380 6554 71161
℻ 380 6554 40704

Chertok Ilia M
IZMIRAN
Acad Sciences
Moscow Region
142092 Troitsk
Russia
☎ 7 095 334 0902
℻ 7 095 334 0124
✉ ichertok@izmiran.troitsk.ru

Chertoprud Vladim E
Hydrometeorologic Res Ctr
Bolshoi Predtechensky Per. 9-11
123342 Moscow
Russia
☎ 7 095 255 5026
℻ 7 095 255 1582

Chevalier Claude
OHP
F 04870 S Michel Obs
France
☎ 33 4 92 76 6368
℻ 33 4 92 76 6295

Chevalier Roger A
University Station
Univ of Virginia
Box 3818
Charlottesville VA 22903 0818
USA
☎ 1 804 924 4889
✉ rac5x@virginia.edu

Chevrel Serge
OMP
14 Av E Belin
F 31400 Toulouse Cdx
France
☎ 33 5 61 33 2963
℻ 33 5 61 25 3205
✉ chevrel@selenix.cnes.fr

Chian Abraham Chian-Long
INPE
Dpt Astronomia
CP 515
12227 010 S Jose dos Campos
Brazil
☎ 55 123 22 9977
℻ 55 123 21 8743

Chiba Masashi
Ntl Astronomical Obs
Osawa 2-21-1
Mitaka
Tokyo 181
Japan
☎ 81 422 34 3619
℻ 81 422 34 3793
✉ chibams@cc.nao.ac.jp

Chikada Yoshihiro
Nobeyama Radio Obs
NAOJ
Minamimaki Mura
Nagano 384 13
Japan
☎ 81 267 98 2831
℻ 81 267 98 2884

Chincarini Guido L
Osserv Astronomico
d Milano
Via E Bianchi 46
I 22055 Merate
Italy
☎ 39 990 6412
℻ 39 990 8492

Chini Rolf
RAIUB
Univ Bonn
auf d Huegel 69
D 53121 Bonn
Germany
☎ 49 228 525 0
℻ 49 228 525 229

Chio Chol Zong
Pyongyang Astron Obs
Acad Sciences DPRK
Taesong district
Pyongyang
Korea DPR
☎ 850 5 3134/5 & 5 3239

Chiosi Cesare S
Dpt Astronomia
Univ d Padova
Vic d Osservatorio 5
I 35122 Padova
Italy
☎ 39 49 829 3422/829 3411
℻ 39 49 875 9840
✉ chiosi@astrpd.pd.astro.it

Chis Gheorghe Dorin
Astronomical Observ
Romanian Acad Sciences
Ul Ciresilor 19
RO 3400 Cluj Napoca
Rumania
☎ 40 64 194 592
℻ 40 64 194 592

Chistyakov Vladimir E
Ussuriisk Solar Station
Primorsky Kray
692533 Gornotaezhnoe
Russia
☎ 7

Chitre Dattakumar M
Computer Sciences Corp
System Sciences div
8728 Colesville Rd
Silver Spring MD 20910
USA

Chitre Shashikumar M
TIFR
Homi Bhabha Rd
Colaba
Bombay 400 005
India
☎ 91 22 219 111
℻ 91 22 495 2110

Chiu Hong Yee
NASA GSFC
Code 914
Greenbelt MD 20771
USA
☎ 1 301 286 8256
℻ 1 301 286 1663
✉ chiu@oz.gsfc.nasa.gov

Chiu Liang-Tai George
IBM
T J Watson Res Ctr
Box 218
Yorktown Heights NY 10598
USA
☎ 1 914 945 2436
✉ glt@watson.ibm.com

Chiuderi Claudio
Dpt Astronomia
Univ d Firenze
Largo E Fermi 5
I 50125 Firenze
Italy
☎ 39 55 27 521
℻ 39 55 22 0039

Chiuderi-Drago Franca Pr
Osserv Astrofis Arcetri
Univ d Firenze
Largo E Fermi 5
I 50125 Firenze
Italy
☎ 39 55 43 78540
℻ 39 55 43 5939

Chiueh Tzihong
Inst Phys/Astronomy
Ntl Central Univ
Chung Li Taiwan 32054
China R
☎ 886 3 422 7151*5341
℻ 886 3 425 1175
✉ chiueh@phyast.dnet.ncu.edu.tw

Chiumiento Giuseppe
Osserv Astronomico
d Torino
St Osservatorio 20
I 10025 Pino Torinese
Italy
☎ 39 11 461 9000
℻ 39 11 810 1930
✉ chiumiento@to.astro.it

Chkhikvadze Iakob N
Abastumani Astrophysical
Observ
Georgian Acad Sci
383762 Abastumani
Georgia
☎ 995 88 32 95 5367
℻ 995 88 32 98 5017

Chlistovsky Franca
Osserv Astronomico
d Brera
Via Brera 28
I 20121 Milano
Italy
☎ 39 2 874 444
℻ 39 2 720 01600

Chmielewski Yves
Observ Geneve
Chemin d Maillettes 51
CH 1290 Sauverny
Switzerland
☎ 41 22 755 2611
℻ 41 22 755 3983
✉ yves.chimielewski@obs.unige.ch

Cho Se Hyung
ISSA
Yoosung Koon
Daejeon 305 348
Korea RP
☎ 82 428 51 1281
℻ 82 428 61 5610

Chochol Drahomir
Astronomical Institute
Slovak Acad Sci
SK 059 60 Tatranska Lomnica
Slovak Republic
☎ 421 969 967 866
℻ 421 969 96 7656
✉ chochol@auriga.ta3.sk

Chodas Paul Winchester
CALTECH/JPL
MS 301 150
4800 Oak Grove Dr
Pasadena CA 91109 8099
USA
☎ 1 818 354 7795
℻ 1 818 393 1159
✉ paul.chodas@jpl.nasa.gov

Chodorowski Michal
Copernicus Astron Ctr
Polish Acad Sci
Ul Bartycka 18
PL 00 716 Warsaw
Poland
☎ 48 22 41 1086
℻ 48 22 41 0828
✉ michal@camk.edu.pl

Choe Seung Urn
Dpt Earth Science/Ed
Seoul Ntl Univ
Sinlim Dong Gwang Gu
Seoul 151 742
Korea RP

Choi Kyu-Hong
Dpt Astron/Meteorology
Yonsei Univ
Seodaemum ku
Seoul 120 749
Korea RP
☎ 82 239 20 131
℻ 82 231 35 033

Choi Won Chol
Pyongyang Astron Obs
Acad Sciences DPRK
Taesong district
Pyongyang
Korea DPR
☎ 850 5 3134/5 & 5 3239

Chokshi Arati
IIA
Koramangala
Sarjapur Rd
Bangalore 560 034
India
☎ 91 80 553 0672/676
℻ 91 80 553 4043
✉ chokshi@iiap.ernet.in

Chollet Fernand
Observ Paris
DANOF
61 Av Observatoire
F 75014 Paris
France
☎ 33 1 40 51 2205
℻ 33 1 44 54 1804
✉ chollet@mesiob.osbpm.fr

Choloniewsski Jacek
Astronomical Observ
Warsaw Univ
Al Ujazdowskie 4
PL 00 478 Warsaw
Poland
☎ 48 22 29 4011
℻ 48 22 29 4697
✉ jch@plwauw61

Chopinet Marguerite
57 r Thiers
F 92100 Boulogne
France
☎ 33 1 47 61 1144

Chou Chih-Kang
Inst Phys/Astronomy
Ntl Central Univ
Chung Li Taiwan 32054
China R
☎ 886 3 425 1175
℻ 886 3 426 2304

Chou Dean-Yi
Physics Dpt
Ntl Tsing Hua Univ
Hsin Chu 300043
China R
✉ dychou@twnctu01

Chou Kyong Chol
Korean Space/Envirnt
1402 Life officetel Bldg
Toido Dong Yongdongpo Ku
Seoul 150 010
Korea RP
☎ 82 276 10 031
℻ 82 276 10 032

Choudhary Debi Prasad
Udaipur Solar Observ
11 Vidya Marg
Udaipur 313 001
India
☎ 91 294 560 626
℻ 91 294 525 959
✉ debi@uso.ernet.in

Choudhuri Arnab Rai
Physics Dpt
IISc
Bangalore 560 012
India
☎ 91 80 334 4411
℻ 91 80 334 1683

Christensen Per R
Theoretical Astrophysics Centre
Juliane Maries Vej 30
DK 2100 Copenhagen
Denmark
☎ 45 35 32 5913
℻ 45 35 32 5913
✉ perrex@nbivax.nbi.dk

Christensen-Dalsgaard J
Inst Phys/Astronomy
Univ of Aarhus
Ny Munkegade
DK 8000 Aarhus C
Denmark
☎ 45 89 42 3614
℻ 45 86 12 0740
✉ jcd@obs.aau.dk

Christian Carol Ann
STScI
Homewood Campus
3700 San Martin Dr
Baltimore MD 21218
USA
☎ 1 301 338 4764
℻ 1 301 338 4579
✉ carolc@stsci.edu

Christiansen Wayne A
Dpt Phys/Astronomy
Univ North Carolina
204 Phillips Hall 039a
Chapel Hill NC 27514
USA
☎ 1 919 962 3011
℻ 1 919 962 0480

Christiansen Wilbur
42 The Grange
67 Mac Gregor St
Deakin ACT 2600
Australia
☎ 61 06 281 5576

Christodoulou Dmitris
1617 S Beretania St 801
Woburn MA 01801
USA

Christy James Walter
Hughes Aircraft Co
1720 W Niona Pl
Tucson AZ 85704
USA
☎ 1 520 297 1377

Christy Robert F
CALTECH
Palomar Obs
MS 105 24
Pasadena CA 91125
USA
☎ 1 213 795 6811
℻ 1 818 568 1517

Chryssovergis Michael
22 Spartis St
GR Callithea
Greece

Chu Hanshu
Purple Mountain Obs
CAS
Nanjing 210008
China PR
☎ 86 25 301096
℻ 86 25 301 459

Chu Yaoquan
Astrophysics Division
Univ Science/Technology
Hefei 230026
China PR
☎ 86 551 33 1134
℻ 86 551 33 1760
✉ yqchu@hpe25.nic.ustc.edu.cn

Chu You-Hua
Dpt Astron/Physics
Univ of Illinois
1002 W Green St
Urbana IL 61801
USA
☎ 1 217 333 5535
℻ 1 217 244 7638
✉ chu@astro.uiuc.edu

Chubb Talbot A
5023 N 38th St
Arlington VA 22207
USA
☎ 1 703 536 4427

Chugai Nikolai N
Inst of Astronomy
Acad Sciences
Pyatnitskaya Ul 48
109017 Moscow
Russia
☎ 7 095 231 2129
℻ 7 095 230 2081

Chun Mun-suk
Dpt Astron/Meteorology
Yonsei Univ
Seodaemum ku
Seoul 120 749
Korea RP
☎ 82 236 12 685
℻ 82 231 35 033
✉ mschun@galaxy.yonsei.ac.kr

Chung Hyun Soo
Korea Astronomy Obs/ISSA
36 1 Whaam Dong
Yuseong Gu
Taejon 305 348
Korea RP
☎ 82 428 65 3282
℻ 82 428 65 3282
✉ hschung@hanul.issa.re.kr

Chupp Edward L
Physics Dpt
Univ of New Hampshire
Demeritt Hall
Durham NH 03824
USA
☎ 1 603 862 2750
℻ 1 603 862 2998
✉ elc@christa.unh.edu

Churazov Eugene M
Space Res Inst
Profsoyuznaya 84/32
Moscow 117810
Russia
☎ 7 095 333 3377
℻ 7 095 333 5377
✉ churazov@hea.iki.rssi.ru

Churchwell Edward B
Washburn Observ
Univ Wisconsin
475 N Charter St
Madison WI 53706
USA
☎ 1 608 262 7857
℻ 1 608 263 0361

Chvojkova Woyk E
Rektorska 13
Malesice
CZ 108 00 Praha 10
Czech R

Chyzy Krzysztof Tadeusz
Astronomical Observ
Krakow Jagiellonian Univ
Ul Orla 171
PL 30 244 Krakow
Poland
☎ 48 12 25 1294
℻ 48 12 251 318
✉ chris@oa.uj.edu.pl

Ciardullo Robin
Dpt Astron/Astrophys
Pennsylvania State Univ
Penn State Univ
University Park PA 16802
USA
☎ 1 814 865 6601
℻ 1 814 863 3399
✉ rbc@astro.psu.edu

Ciatti Franco
Osserv Astrofisico
Univ d Padova
Via d Osservatorio 8
I 36012 Asiago
Italy
☎ 39 424 462 665
℻ 39 424 462 884

Cid Palacios Rafael
Dpt Astronomia
Univ Zaragoza
E 50009 Zaragoza
Spain
☎ 34 7 676 1000
℻ 34 7 676 1140

Cidale Lydia Sonia
Observ Astronomico
Paseo d Bosque S/n
1900 La Plata (Bs As)
Argentina
☎ 54 21 217 308
℻ 54 21 211 761
✉ lydia@fcaglp.edu.ar

Cillie G G
4 Minserie St
7600 Stellenbosch
South Africa
☎ 27 02 231 3515

Cincotta Pablo Miguel
FCAGS
Paseo d Bosque S/n
1900 La Plata (Bs As)
Argentina
☎ 54 21 217 308/3 8810
℻ 54 21 211 761/25 8985
✉ pmc@fcaglp.fcaglp.unlp.edu.ar

Citterio Oberto
Osserv Astronomico
d Milano
Via E Bianchi 46
I 22055 Merate
Italy
☎ 39 990 6412
℻ 39 990 8492
✉ citterio@merate.mi.astro.it

Ciurla Tadeusz
Astronomical Institute
Wroclaw Univ
Ul Kopernika 11
PL 51 622 Wroclaw
Poland
☎ 48 71 372 9373/74
℻ 48 71 372 9378

Clairemidi Jacques
Observ Besancon
41bis Av Observatoire
BP 1615
F 25000 Besancon Cdx
France
☎ 33 3 81 66 6900
℻ 33 3 81 66 6944
✉ jclairemidi@obs-besancon.fr

Claria Juan
Observ Astronomico
de Cordoba
Laprida 854
5000 Cordoba
Argentina
☎ 54 51 23 0491
℻ 54 51 21 0613
✉ claria@astro.edu.ar

Clark Barry G
NRAO
VLA
Box 0
Socorro NM 87801 0387
USA
☎ 1 505 772 4011
✆ 1 505 835 7027

Clark David H
Science Division
Sci/Engineer Res Counci
North Star Av
Swindon SN22 1SZ
UK
☎ 44 17 932 6222
✆ 44 17 934 42002

Clark Frank Oliver
Dpt Phys/Astronomy
Univ Kentucky
Lexington KY 40506 0055
USA
☎ 1 606 257 3376
✆ 1 606 323 2846

Clark George W
Physics Dpt
MIT
Box 165
Cambridge MA 02139 4307
USA
☎ 1 617 253 5842
✉ gwc@space.mit.edu

Clark Thomas A
NASA GSFC
Code 974
Greenbelt MD 20771
USA
☎ 1 301 286 5957

Clark Thomas Alan
Physics Dpt
Univ of Calgary
2500 University Dr NW
Calgary AB T2N 1N4
Canada
☎ 1 403 284 5392
✆ 1 403 289 3331

Clark Alfred
Dpt Mech/Aerospace Sci
Univ of Rochester
Rochester NY 14627 0171
USA

Clarke Catherine
Inst of Astronomy
The Observatories
Madingley Rd
Cambridge CB3 0HA
UK
☎ 44 12 23 337 535
✆ 44 12 23 337 523
✉ cclarke@castr.ast.cam.ac.uk

Clarke David
Astronomy Dpt
Univ of Glasgow
Glasgow G12 8QQ
UK
☎ 44 141 339 8855
✆ 44 141 334 9029

Clarke John T
NASA GSFC
Code 681
Greenbelt MD 20771
USA
☎ 1 301 286 5781

Clarke Thomas R
McLaughlin Planetarium
Royal Ontario Museum
100 Queens Park Crescent
Toronto ON M5S 2C6
Canada
☎ 1 416 998 8551

Clausen Jens Viggo
Astronomical Observ, NBIfAFG
Copenhagen Univ
Juliane Maries Vej 30
DK 2100 Copenhagen
Denmark
☎ 45 35 32 5926
✆ 45 35 32 5989
✉ jvc@astro.ku.dk

Claussen Mark J
NRL
Code 4210
4555 Overlook Av SW
Washington DC 20375 5000
USA
☎ 1 202 767 0670

Clavel Jean
ESA
Apt 50727 Villafranca
E 28080 Madrid
Spain
☎ 34 1 813 1251
✆ 34 1 813 1172
✉ jclavel@iso.vilspa.esa.es

Clay Roger
Dpt Physics/
Math Physics
Univ Adelaide
Adelaide SA 5005
Australia
☎ 61 8 8303 5996
✆ 61 8 8303 4380
✉ rclay@physics.adelaide.edu.au

Clayton Donald D
Dpt Phys/Astronomy
Clemson Univ
Clemson SC 29634 1911
USA
☎ 1 803 656 5299
✉ clayton@clemson

Clayton Geoffrey C
Dpt Phys/Astronomy
Louisiana State Univ
Baton Rouge LA 70803 4001
USA
☎ 1 504 388 8275
✆ 1 504 388 5855
✉ gclayton@fenway.phys.lsu.edu

Clayton Robert N
Enrico Fermi Inst
Univ of Chicago
5640 S Ellis Av
Chicago IL 60637
USA
☎ 1 312 702 7777
✆ 1 312 702 8212

Clegg Peter E
Astrophysics Group
QMWC
Mile End Rd
London E1 4NS
UK
☎ 44 171 975 5038
✆ 44 181 975 5500
✉ pec@star.qmw.ac.uk

Clegg Robin E S
Particle Physics/Astronomy
Polaris House
North Star Av
Swindon SN22 1SZ
UK
☎ 44 17 934 42000
✆ 44 17 934 42002

Clemens Dan P
Astronomy Dpt
Boston Univ
725 Commonwealth Av
Boston MA 02215
USA
☎ 1 617 353 6140
✆ 1 617 353 5704

Clement Maurice J
Astronomy Dpt
Univ of Toronto
60 St George St
Toronto ON M5S 1A1
Canada
☎ 1 416 978 4833
✆ 1 416 978 3921

Clement Rosa Maria
Dpt Astron/Astrofisica
Univ Valencia
Dr Moliner 50
E 46100 Burjassot
Spain
☎ 34 6 398 3073
✆ 34 6 398 3084
✉ clement@deneb.matapl.uv.es

Clette Frederic
Koninklijke Sterrenwacht
van Belgie
Ringlaan 3
B 1180 Brussels
Belgium
☎ 32 2 373 0233
✆ 32 2 374 9822
✉ frederic.clette@oma.be

Clifton Kenneth St
NASA/MSFC
Space Science Lab
Code ES 63
Huntsville AL 35812
USA
☎ 1 205 453 2305
✆ 1 205 547 7754

Climenhaga John L
Dpt Phys/Astronomy
Univ of Victoria
Box 3055
Victoria BC V8W 3P6
Canada
☎ 1 250 721 7741
✆ 1 250 721 7715

Cline Thomas L
NASA GSFC
Code 661
Greenbelt MD 20771
USA
☎ 1 301 286 8375

Cliver Edward W
AFPL
Space Physics Div
Hanscom AFB
Bedford MA 01731 3010
USA
☎ 1 617 861 3975

Clube S V M
Astrophysics Dpt
Univ of Oxford
Keble Rd
Oxford OX1 3RH
UK
☎ 44 186 527 3999
✆ 44 186 527 3947

Clutton-Brock Martin
Dpt Maths/Astronomy
Univ of Manitoba
Winnipeg MB R3t 2N2
Canada
☎ 1 204 474 9501

Cochran Anita L
Astronomy Dpt
Univ of Texas
Rlm 15 308
Austin TX 78712 1083
USA
☎ 1 512 471 1471
✆ 1 512 471 6016
✉ anita@zinfandel.as.utexas.edu

Cochran William David
Astronomy Dpt
Univ of Texas
Rlm 15 308
Austin TX 78712 1083
USA
☎ 1 512 471 4461
✆ 1 512 471 6061
✉ wdc@shiraz.as.utexas.edu

Cocke William John
Steward Observ
Univ of Arizona
Tucson AZ 85721
USA
☎ 1 520 621 6540
✆ 1 520 621 1532
✉ jcocke@as.arizona.edu

Code Arthur D
Washburn Observ
Univ Wisconsin
475 N Charter St
Madison WI 53706
USA
☎ 1 608 262 9594
✆ 1 608 263 0361
✉ code@madraf.astro.
wisc.edu

Codina Landaberry Sayd J
Observ Nacional
Rua Gl Bruce 586
Sao Cristovao
20921 030 Rio de Janeiro RJ
Brazil
☎ 55 21 580 7313*267
✆ 55 21 580 0332

Codina Vidal J M
Fabra Observ
Gran Via de Los Cortes
Catalanes 679
E 08013 Barcelona
Spain
☎ 34 3 245 4766

Coe Malcolm
Physics Dpt
Southampton Univ
The Univ
Southampton SO9 5NH
UK
☎ 44 170 359 2108
✆ 44 170 358 5813
✉ mjc@phastr.soton.
ac.uk

Coelho Balsa Mario C
Rua Trindade Coelho 21
2o Dto
P 3000 Coimbra
Portugal

Coffeen David L
Box 151
Hastings Hudson NY 10706
USA
☎ 1 914 478 2594

Coffey Helen E
NOAA ERL R/E/SE
Space Environment Lab
325 Broadway
Boulder CO 80303
USA
☎ 1 303 497 6223
✉ hcoffey@ngdc.
noaa.gov

Cogan Bruce C
MSSSO
Weston Creek
Private Bag
Canberra ACT 2611
Australia
☎ 61 262 881 111
✆ 61 262 490 233

Cohen Jeffrey M
Physics Dpt
Univ of Pennsylvania
Philadelphia PA 19104
USA
☎ 1 215 898 8176
✆ 1 215 898 9336

Cohen Judith
CALTECH
Palomar Obs
MS 105 24
Pasadena CA 91125
USA
☎ 1 818 356 4005
✆ 1 818 568 1517
✉ jlc@deimos.caltech.edu

Cohen Leon
Physics Dpt
Hunters Coll
695 Park Av
New York NY 10021
USA
☎ 1 212 570 5696
✆ 1 212 772 5393

Cohen Marshall H
CALTECH
Palomar Obs
MS 105 24
Pasadena CA 91125
USA
☎ 1 213 356 4000
✆ 1 818 568 1517

Cohen Martin
Radio Astronomy Lab
Univ of California
601 Campbell Hall
Berkeley CA 94720
USA
☎ 1 415 642 2833
✆ 1 415 642 6424

Cohen Raymond J
NRAL
Univ Manchester
Jodrell Bank
Macclesfield SK11 9DL
UK
☎ 44 147 757 1321
✆ 44 147 757 1618
✉ rjc@jb.man.ac.uk

Cohen Richard S
Inst Space Studies
2880 Broadway
New York NY 10025
USA
☎ 1 212 678 5611

Cohen Ross D
CASS
UCSD
C 011
La Jolla CA 92093 0216
USA
☎ 1 619 534 2664
✆ 1 619 534 2294

Cohn Haldan N
Astronomy Dpt
Indiana Univ
Swain W 319
Bloomington IN 47405
USA
☎ 1 812 335 4174
✆ 1 812 855 8725

Colangeli Luigi
Osserv d Capodimonte
Via Moiariello 16
I 80131 Napoli
Italy
☎ 39 81 298 384
✆ 39 81 45 6710
✉ colangeli@astrna.na.
astro.it

Colburn David S
1944 Waverley St
Palo Alto CA 94301
USA
✆ 400 414 327 7183
✉ colburn@galileo.arc.
nasa.gov

Cole Trevor William
School of Electrical Eng
UNSW
Sydney NSW 2006
Australia
☎ 61 2 692 2682
✆ 61 2 660 2903

Coleman Paul Henry
Physics Dpt
Univ of Puerto Rico
Box 23343
Rio Piedras PR 00931 3343
USA
☎ 1 787 764 0000*7346
✆ 1 787 764 4063
✉ gruff@astro1.cnnet.
clu.edu

Coles Peter
Astronomy Unit
QMWC
Mile End Rd
London E1 4NS
UK
☎ 44 171 975 5481
✆ 44 181 981 9587
✉ pcoles@uk.ac.qmw.
starlink

Colgate Stirling A
LANL
MS B275
Box 1663
Los Alamos NM 87545 2345
USA
☎ 1 505 667 2897
✆ 1 505 665 4055

Colic Petar-Kasimir
Hvar Observ
Fac Geodesy
Kaciceva 26
HR 10000 Zagreb
Croatia
☎ 385 1 456 1222
✆ 385 1 445 410
✉ kresimir.colic@public.
srce.hr

Colin Jacques
Observ Bordeaux
BP 89
F 33270 Floirac
France
☎ 33 5 56 86 4330
✆ 33 5 56 40 4251

Colin Pedro
Instituto Astronomia
UNAM
Apt 70 264
Mexico DF 04510
Mexico
☎ 52 5 622 3906
✆ 52 5 616 0653
✉ colin@astroscu.
unam.mx

Collados Manuel
IAC
Observ d Teide
via Lactea s/n
E 38200 La Laguna
Spain
☎ 34 2 232 9100
✆ 34 2 232 9117

Colless Matthew
MSSSO
Weston Creek
Private Bag
Canberra ACT 2611
Australia
☎ 61 262 798 030
✆ 61 262 490 233/0260
✉ colless@mso.anu.
edu.au

Collin-Souffrin Suzy
Observ Paris Meudon
DAEC
Pl J Janssen
F 92195 Meudon PPL Cdx
France
☎ 33 1 45 07 7967
✆ 33 1 45 07 7469

Collins George W ii
Astronomy Dpt
Ohio State Univ
174 W 18th Av
Columbus OH 43210 1106
USA
☎ 1 614 422 5467
✆ 1 614 292 2928

Colomb Fernando R
IAR
CC 5
1894 Villa Elisa (Bs As)
Argentina
☎ 54 21 254 909
✆ 54 21 254 909
✉ rcolomb@irma.edu.ar

Colombi Stephane
Cita Mclennan Labs
Univ of Toronto
60 St Georges St.
Toronto ON M5S 1A1
Canada
☎ 1 416 978 1776
✆ 1 416 978 3921
✉ colombi@cita.utoronto.ca

Colombo G
Istt Meccanica Appl
Univ d Padova
Via F Marzolo 9
I 35122 Padova
Italy
☎ 39 49 66 1499
✆ 39 49 875 9840

Colomer Francisco
Observ Astronomico
Nacional
Apt 1143
E 28800 Alcala d Henares
Spain
☎ 34 1 885 5060
✆ 34 1 885 5062
✉ colomer@oan.es

Coluzzi Regina
OAR
Via d Parco Mellini 84
I 00136 Roma
Italy
☎ 39 6 34 7056
✆ 39 6 349 8236
✉ coluzzi@astrmp.mporzio.astro.it

Coma Juan Carlos
ROA
Cecilio Pujazon 22-3 A
E 11110 San Fernando
Spain
☎ 34 5 688 3548
✆ 34 5 659 9366
✉ ccgeneral@czv1.uca.es

Comastri Andrea
Osserv Astronom Bologna
Univ d Bologna
Via Zamboni 33
I 40126 Bologna
Italy
☎ 39 51 259 419
✆ 39 51 259 407
✉ comastri@astbo3.bo.astro.it

Combes Francoise
Observ Paris
DEMIRM
61 Av Observatoire
F 75014 Paris
France
☎ 33 1 40 51 2077
✆ 33 1 40 51 2002
✉ bottaro@obspm.fr

Combes Michel
Observ Paris
PDT
61 Av Observatoire
F 75014 Paris
France
☎ 33 1 40 51 2157
✆ 33 1 44 54 1804

Combi Jorge Ariel
IAR
CC 5
1894 Villa Elisa (Bs As)
Argentina
☎ 54 21 254 909
✆ 54 21 254 909
✉ combi@irma.edu.ar

Combi Michael R
Space Physics Res Lb
Univ of Michigan
2455 Hayward St
Ann Arbor MI 48109 2143
USA
☎ 1 313 764 7226
✆ 1 313 747 3083
✉ combi@sprlc.spl.umich.edu

Comins Neil Francis
Dpt Phys/Astronomy
Univ of Maine
Bennett Hall
Orono ME 04469
USA
☎ 1 207 581 1037

Comoretto Giovanni
Osserv Astrofis Arcetri
Univ d Firenze
Largo E Fermi 5
I 50125 Firenze
Italy
☎ 39 55 43 78540
✆ 39 55 43 5939

Comte Georges
Observ Marseille
2 Pl Le Verrier
F 13248 Marseille Cdx 04
France
☎ 33 4 91 95 9088
✆ 33 4 91 62 1190

Conconi Paolo
Osserv Astronomico
d Milano
Via E Bianchi 46
I 22055 Merate
Italy
☎ 39 990 6412
✆ 39 990 8492

Condon James J
NRAO
520 Edgemont Rd
Charlottesville VA 22903
USA
☎ 1 804 296 0322
✆ 1 804 296 0278
✉ jcondon@nrao.edu

Conklin Edward K
2959 Kalaaua Av
N 1004
Honolulu HI 96815
USA

Connes Janine
CIRCE
BP 63
F 91400 Orsay
France
☎ 33 1 69 28 7675

Connes Pierre
Service Aeronomie
BP 3
F 91371 Verrieres Buisson
France
☎ 33 1 64 47 4277

Connolly Leo Paul
Physics Dpt
California State Univ
5500 University Parkway
San Bernardino CA 92407
USA
☎ 1 714 880 5400

Considere Suzanne
Observ Besancon
41bis Av Observatoire
BP 1615
F 25010 Besancon Cdx
France
☎ 33 3 81 66 69 16
✆ 33 3 81 66 6944
✉ sc@obs-besancon.fr

Consolmagno Guy Joseph
Vatican Observ Res Gp
Steward Observ
Univ of Arizona
Tucson AZ 85721
USA
☎ 1 520 621 7855
✆ 1 520 621 1532
✉ gjc@as.arizona.edu

Contadakis Michael E
Dpt Geodesy/Survey
Univ Thessaloniki
GR 540 06 Thessaloniki
Greece
☎ 30 31 99 6134
✆ 30 31 99 6408
✉ kodadaki@eng.auth.gr

Conti Peter S
JILA
Univ of Colorado
Campus Box 440
Boulder CO 80309 0440
USA
☎ 1 303 492 7789
✆ 1 303 492 5235
✉ pconti@jila

Contopoulos George
Astronomy Dpt
Ntl Univ Athens
Panepistimiopolis
GR 157 84 Zografos
Greece
☎ 30 1 724 3211
✆ 30 1 723 5122

Conway John
Onsala Space Observ
Chalmers Technical Univ
S 439 92 Onsala
Sweden
☎ 46 31 772 5503
✆ 46 31 772 5590
✉ jconway@oso.chalmers.se

Conway Robin G
NRAL
Univ Manchester
Jodrell Bank
Macclesfield SK11 9DL
UK
☎ 44 14 777 1321
✆ 44 147 757 1618

Cook Alan H
Dpt Phys/Univ Cambridge
The Master's Lodge
Selwyn College
Cambridge CB3 9DQ
UK
☎ 44 12 23 335 889
✆ 44 12 23 335 837

Cook John W
8032 Sleepy View Ln
Springfield VA 22153
USA
☎ 1 202 767 2161

Cook Kem Holland
LLNL
L 413
Box 808
Livermore CA 94551 9900
USA
☎ 1 415 423 4634
✆ 1 415 423 0238
✉ kcook@llnl.gov

Cooke B A
Dpt Phys/X-Ray Astron
Univ Leicester
University Rd
Leicester LE1 7RH
UK
☎ 44 116 252 2073
✆ 44 116 250 182

Cooke John Alan
Royal Observ
Blackford Hill
Edinburgh EH9 3HJ
UK
☎ 44 131 667 3221
✆ 44 131 668 8356

Cooper Timothy
ASSA
Box 14740
1623 Bredell
South Africa
☎ 27 11 967 2250
✆ 27 11 929 4349
✉ tpcooper@ilink.nis.za

Coradini Angioletta
IAS
Rep Planetologia
Via d universita 11
I 00133 Roma
Italy
☎ 39 6 4993 4473
✆ 39 6 2066 0188
✉ coradini@saturn.ias.fra.cnr.it

Corbally Christopher
Steward Observ
Univ of Arizona
Vatican Obs Res Gp
Tucson AZ 85721
USA
☎ 1 520 621 3225
✆ 1 520 621 1532
✉ corbally@as.arizona.edu

Corbelli Edvige
Osserv Astrofis Arcetri
Univ d Firenze
Largo E Fermi 5
I 50125 Firenze
Italy
☎ 39 55 43 78540
✆ 39 55 43 5939
✉ edvige@sisifo.arcetri.astro.it

Corbet Robin Henry D
NASA/GSFC
Code 662
Greenbelt MD 20771
USA
☎ 1 301 286 2851
✆ 1 301 286 1684
✉ corbet@lheamail.
gsfc.naa.gov

Corbin Thomas Elbert
USNO
Astrometry Dpt
3450 Massachusetts Av NW
Washington DC 20392 5420
USA
☎ 1 202 653 1557
✆ 1 202 762 1516
✉ tec@sicon.usno.
navy.mil

Corcoran Michael Francis
NASA/GSFC
Code 660 2
Greenbelt MD 20771
USA
☎ 1 301 286 5576
✆ 1 301 286 1629
✉ corcoran@barnegat.
gsfc.nasa.gov

Cordes James M
Astronomy Dpt
Cornell Univ
512 Space Sc Bldg
Ithaca NY 14853 6801
USA
☎ 1 607 256 3734
✆ 1 607 255 1767

Cordova France A D
NASA Headquarters
MC AH
600 Independence Av SW
Washington DC 20546 0001
USA
☎ 1 202 358 1809
✆ 1 202 358 2810
✉ cordova@admingw.
hq.nasa.gov

Corliss C H
Forest Hills Lab
2955 Albemarle St NW
Washington DC 20008
USA
☎ 1 202 362 6085

Cornejo Alejandro A
INAOE
Tonantzintlaz
Apdo 51 y 216
Puebla PUE 72000
Mexico
☎ 52 2 247 2011
✆ 52 2 247 2231

Cornide Manuel
Dpt Astrofisica
Fac Fisica
Univ Complutense
E 28040 Madrid
Spain
☎ 34 1 449 5316
✆ 34 1 394 5195

Cornille Marguerite
Observ Paris Meudon
DARC
Pl J Janssen
F 92195 Meudon PPL Cdx
France
☎ 33 1 45 07 7455
✆ 33 1 45 07 7971

Correia Emilia
CRAAE INPE
EPUSP/PTR
CP 61548
01065 970 Sao Paulo SP
Brazil
☎ 55 11 815 6289
✆ 55 11 815 4272
✉ ecorreia@brusp.ansp.br

Corwin Harold G
CALTECH
MS 100 22
IPAC
Pasadena CA 91125
USA
☎ 1 818 397 9537
✆ 1 818 397 9600
✉ hgcjr@ipac.caltech.edu

Cosmovici Batalli C
IAS
Area d Ricerca CNR
Via Fosso Cavaliere 100
I 00133 Roma
Italy
☎ 39 6 4993 4473
✆ 39 6 2066 0188

Costa Andrea
IAFE
CC 67 Suc 28
1428 Buenos Aires
Argentina
☎ 54 1 781 6755
✆ 54 1 786 8114
✉ costa@iafe.edu.ar

Costa Edgardo
Dpt Astronomia
Univ Chile
Casilla 36 D
Santiago
Chile
☎ 56 2 229 4101
✆ 56 2 229 4002

Costa Enrico
IAS
Area d Ricerca CNR
Via Fosso Cavaliere 100
I 00133 Roma
Italy
☎ 39 6 4993 4004
✆ 39 6 2066 0188
✉ costa@saturn.ias.
rm.cnr.it

Costa Joaquim E R
CRAAE INPE
EPUSP/PTR
CP 61548
01065 970 Sao Paulo SP
Brazil
☎ 55 11 815 6289
✆ 55 11 815 4272
✉ jercosta@brusp.ansp.br

Costa Victor
Inst Astrofisica
Andalucia Apt 3004
Prof Albareda 1
E 18080 Granada
Spain
☎ 34 5 812 1311
✆ 34 5 881 4530

Costero Rafael
Instituto Astronomia
UNAM
Apt 70 264
Mexico DF 04510
Mexico
☎ 52 5 622 3922
✆ 52 5 616 0653
✉ costero@astroscu.
unam.mx

Cotton William D
NRAO
520 Edgemont Rd
Charlottesville VA 22903
USA
☎ 1 804 296 0319

Cottrell Peter Ledsam
Dpt Phys/Astronomy
Univ of Canterbury
Private Bag 4800
Christchurch 1
New Zealand
☎ 64 3 482 009
✆ 64 3 364 2469
✉ p.cottrell@csc.
canterbury.ac.nz

Couch Warrick
School of Physics
UNSW
BOX 1
Kensingron NSW 2033
Australia
☎ 61 2 385 4578
✆ 61 6 268 8786
✉ wjc@newt.phys.unsw.
edu.au

Couchman Hugh M P
Dpt Phys/Astronomy
Univ W Ontario
London ON N6A 3K7
Canada
☎ 1 519 661 3183
✆ 1 519 661 2009

Coulson Iain M
Joint Astronomy Ctr
660 N A'ohoku Pl
University Park
Hilo HI 96720
USA
☎ 1 808 961 3756
✆ 1 808 961 6516

Counselman Charles C
Dpt Earth/Planet Sci
MIT Rm 37 552
Box 165
Cambridge MA 02139
USA
☎ 1 617 253 7902
✆ 1 617 253 7939
✉ ccc@space.mit.edu

Couper Heather Miss
Collins Cottage Lower Rd
Loosley Row
Bucks HP17 0PF
UK

Coupinot Gerard
OMP
9 r Pont de La mouette
F 65200 Bagneres Bigorre
France
☎ 33 5 62 95 1969
✆ 33 5 62 95 1969

Courtes Georges
LAS
Traverse du Siphon
Les Trois Lucs
F 13376 Marseille Cdx 12
France
☎ 33 4 91 05 5900
✆ 33 4 91 66 1855

Courtin Regis
Observ Paris Meudon
DESPA
Pl J Janssen
F 92195 Meudon PPL Cdx
France
☎ 33 1 45 07 7729
✆ 33 1 45 07 7469
✉ courtin@obspm.fr

Courvoisier Thierry J.-L.
INTEGRAL
Science Data Ctr
16 Chemin Ecogia
CH 1290 Versoix
Switzerland
☎ 41 22 950 9101
✆ 41 22 950 9133
✉ thierry.courvoisier@obs.
unige.ch

Cousins A W J
SAAO
PO Box 9
7935 Observatory
South Africa
☎ 27 21 47 0025
✆ 27 21 47 3639

Couteau Paul
OCA
Observ Nice
BP 139
F 06304 Nice Cdx 4
France
☎ 33 4 93 89 0420
✆ 33 4 92 00 3033

Coutrez Raymond A J
Univ Brussels
6 r Egide Bouvier
B 1160 Brussels
Belgium
☎ 32 2 672 4802

Coutts-Clement Christine
Astronomy Dpt
Univ of Toronto
60 St George St
Toronto ON M5S 1A1
Canada
☎ 1 416 978 5186
✆ 1 416 978 3921

Couturier Pierre
CFHT Corp
Box 1597
Kamuela HI 96743
USA
☎ 1 808 885 7944
✆ 1 808 885 7288

Covington Arthur E
15 Forest Dr
Kingston ON K7L 4L1
Canada

Covino Elvira
Osserv d Capodimonte
Via Moiariello 16
I 80131 Napoli
Italy
☎ 39 81 557 5537
✆ 39 81 45 6710
✉ covino@astrna.na.
astro.it

Cowan John J
Dpt Phys/Astronomy
Univ of Oklahoma
Norman OK 73019
USA
☎ 1 405 325 3961
✆ 1 405 325 7557

Cowie Lennox Lauchlan
Inst for Astronomy
Univ of Hawaii
2680 Woodlawn Dr
Honolulu HI 96822
USA
☎ 1 808 956 8566
✆ 1 808 956 8312
✉ cowie@uhifa.ifa.
hawaii.edu

Cowley Anne P
Physics Dpt
Arizona State Univ
Astronomy Program
Tempe AZ 85287 1504
USA
☎ 1 602 965 2919
✆ 1 605 965 7954

Cowley Charles R
Astronomy Dpt
Univ of Michigan
Dennison Bldg
Ann Arbor MI 48109 1090
USA
☎ 1 313 764 3437
✆ 1 313 764 2211

Cowsik Ramanath
TIFR
Homi Bhabha Rd
Colaba
Bombay 400 005
India
☎ 91 22 219 111
✆ 91 22 215 2110

Cox Arthur N
LANL
Box 1663
Los Alamos NM 87545 2345
USA
☎ 1 505 667 7648
✆ 1 505 665 4055

Cox Donald P
Astronomy Dpt
Univ Wisconsin
475 N Charter St
Madison WI 53706 1582
USA
☎ 1 608 262 5916
✆ 1 608 263 0361

Cox Pierre
IAS
Bt 121
Universite Paris XI
F 91405 Orsay Cdx
France
☎ 33 1 69 85 8737/8513
✆ 33 1 69 85 8675
✉ cox@iaslab.ias.fr

Coyne George V
Direttore della Specola
Vaticana
I 00120 Citta del Vaticano
Vatican City State
☎ 39 6 698 3411/5266
✆ 39 6 698 84671
✉ gcoyne@as.arizona.edu

Crabtree Dennis
NRCC/HIA
DAO
5071 W Saanich Rd
Victoria BC V8X 4M6
Canada
☎ 1 250 388 0025
✆ 1 250 363 0045
✉ crabtree@dao.nrc.ca

Craig Ian Jonathan D
Applied Maths Dpt
Univ of Waikato
Private Bag
Hamilton
New Zealand
☎ 64 62889

Craine Eric Richard
Western Res Co
2127 E Speedway
Suite 209
Tucson AZ 85719
USA
☎ 1 520 325 4505
✆ 1 520 327 1388
✉ craine@noao.edu

Cram Lawrence Edward
School of Physics
UNSW
Sydney NSW 2006
Australia
☎ 61 2 9351 2537
✆ 61 2 9660 2903
✉ L.Cram@physics.usyd.
edu.au

Cramer Neil
School of Physics
UNSW
Sydney NSW 2006
Australia
☎ 61 2 692 3162
✆ 61 2 9660 2903

Cramer Noel
Observ Geneve
Chemin d Maillettes 51
CH 1290 Sauverny
Switzerland
☎ 41 22 755 2611
✆ 41 22 755 3983
✉ noel.cramer@obs.
unige.ch

Crampton David
NRCC/HIA
DAO
5071 W Saanich Rd
Victoria BC V8X 4M6
Canada
☎ 1 250 363 0010
✆ 1 250 363 0045
✉ david@crampton.hia.
nrc.ca

Crane Patrick C
Interferometrics Inc
8928 Cottongrass St
Waldorf MA 20603 4952
USA
☎ 1 202 404 1506
✆ 1 202 767 5636
✉ crane@susim.nrl.navy.mil

Crane Philippe
ESO
Karl Schwarzschildstr 2
D 85748 Garching
Germany
☎ 49 897 920 98
✆ 49 893 202 362

Crannell Carol Jo
NASA GSFC
Code 682
Greenbelt MD 20771
USA
☎ 1 301 286 5007
✆ 1 301 286 1617

Crawford David L
NOAO/KPNO
Box 26732
950 N Cherry Av
Tucson AZ 85726 6732
USA
☎ 1 520 318 8346
✆ 1 520 318 8360
✉ dcrawford@noao.edu

Crawford Ian Andrew
Dpt Phys/Astronomy
UCLO
Gower St
London WC1E 6BT
UK
☎ 44 171 387 7050*3498
✆ 44 171 380 7145
✉ iac@star.ucl.ac.uk

Crenshaw Daniel Michael
NASA GSFC
Code 681
Greenbelt MD 20771
USA
☎ 1 301 286 0871
✆ 1 301 286 1752

Creze Michel
IUP Vannes
8 r Montaigne
BP 561
F 56017 Vannes Cdx
France
☎ 33 2 97 46 3145
✆ 33 2 97 46 3176
✉ creze@iu-vannes.fr

Crifo Francoise
Observ Paris Meudon
DASGAL
Pl J Janssen
F 92195 Meudon PPL Cdx
France
☎ 33 1 45 07 7834
✆ 33 1 45 07 7878
✉ francoise.crifo@obspm.fr

Cristaldi Salvatore
Osserv Astronomico
d Catania
Via A Doria 6
I 95125 Catania
Italy
☎ 39 95 733 21111
✆ 39 95 33 0592

Cristescu Cornelia G
Astronomical Institute
Romanian Acad Sciences
Cutitul de Argint 5
RO 75212 Bucharest
Rumania
☎ 40 1 641 3686
✆ 40 1 312 3391

Cristiani Stefano
Dpt Astronomia
Univ d Padova
Vic d Osservatorio 5
I 35122 Padova
Italy
☎ 39 49 829 3411
✆ 39 49 875 9840

Crivellari Lucio
OAT
Box Succ Trieste 5
Via Tiepolo 11
I 34131 Trieste
Italy
☎ 39 40 31 99255
✆ 39 40 30 9418

Crocker Deborah Ann
Dpt Phys/Astronomy
Univ of Alabama
Box 870324
Tuscaloosa AL 35387 0324
USA
☎ 1 205 348 3758
✆ 1 205 348 5051
✉ crock@kudzu.astr.ua.edu

Croom David L
Rutherford Appleton Lab
Space/Astrophysics Div
Bg R25/R68
Chilton Didcot OX11 0QX
UK
☎ 44 12 35 821 900
✆ 44 12 35 44 5808

Crotts Arlin Pink
Astronomy Dpt
Columbia Univ
538 W 120th St
New York NY 10027
USA
☎ 1 212 854 7899
Fax 1 212 316 9504
Email arlin@eureka.columbia.edu

Crovisier Jacques
Observ Paris Meudon
ARPEGES
Pl J Janssen
F 92195 Meudon PPL Cdx
France
☎ 33 1 45 07 7599
Fax 33 1 45 07 7939
Email crovisie@obspm.fr

Crowther Paul
Dpt Phys/Astronomy
UCLO
Gower St
London WC1E 6BT
UK
☎ 44 171 387 7050 *3474
Fax 44 171 380 7145
Email pac@star.ucl.ac.uk

Cruikshank Dale P
NASA/ARC
MS 245 6
Moffett Field CA 94035 1000
USA
☎ 1 415 604 4244
Fax 1 415 604 6779
Email cruikshank@ssa1.arc.nasa.gov

Cruise Adrian Michael
Rutherford Appleton Lab
Chilbolton Observ
Ditton Park
Slough SL3 9JX
UK

Crutcher Richard M
Dpt Astron/Physics
Univ of Illinois
1011 W Springfield Av
Urbana IL 61801
USA
☎ 1 217 333 9581
Fax 1 217 244 7638

Cruvellier Paul E
Observ Marseille
2 Pl Le Verrier
F 13248 Marseille Cdx 04
France
☎ 33 4 91 10 7482
Fax 33 4 91 10 7484
Email igrap@obmara.cnrs-mrs.fr

Cruz-Gonzalez Irene
Instituto Astronomia
UNAM
Apt 70 264
Mexico DF 04510
Mexico
☎ 52 5 622 3906
Fax 52 5 616 0653

Cruzalebes Pierre
OCA
Dpt Fresnel
Av Copernic
F 06130 Grasse
France
☎ 33 4 93 40 5329
Fax 33 4 93 40 5333
Email cruzalesbes@obs-azur.fr

Csada Imre K
Konkoly Observ
Thege U 13/17
Box 67
H 1525 Budapest
Hungary
☎ 36 1 375 4122
Fax 36 1 275 4668

Cubarsi Rafael
Dpt Mat Aplicada
U P Cataluna
Box 30002
E 08080 Barcelona
Spain
☎ 34 3 401 6799
Fax 34 3 401 6801
Email matrcm@mat.upc.es

Cudaback David D
Radio Astronomy Lab
Univ of California
601 Campbell Hall
Berkeley CA 94720
USA
☎ 1 415 642 5724
Fax 1 415 642 6424
Email cudaback@bkyast.berkerley.edu

Cudworth Kyle Mccabe
Yerkes Observ
Univ of Chicago
Box 258
Williams Bay WI 53191
USA
☎ 1 414 245 5555
Fax 1 414 245 9805
Email kmc@yerkes.uchicago.edu

Cuffey J
Dpt Earth Sci/Astron
New Mexico State Univ
University Park NM 88001
USA

Cugier Henryk
Astronomical Institute
Wroclaw Univ
Ul Kopernika 11
PL 51 622 Wroclaw
Poland
☎ 48 71 372 9373/74
Fax 48 71 372 9378

Cugnon Pierre
Koninklijke Sterrenwacht
van Belgie
Ringlaan 3
B 1180 Brussels
Belgium
☎ 32 2 373 0276
Fax 32 2 374 9822
Email pierre.cugnon@oma.be

Cugusi Leonino
Dpt Scienze Fisiche
Univ d Cagliari
Via Ospedale 72
I 09100 Cagliari
Italy
☎ 39 70 66 4770
Fax 39 70 72 5425

Cui Douxing
Changchun Artificial
Satellite Observ
Box 1067
Changchun 130117
China PR
☎ 86 42859

Cui Shizhu
Beijing Planetarium
138 Xi Wai St
Beijing 100044
China PR
☎ 86 1 683 10804
Fax 86 10 683 53003/652 42246

Cui Xiangqun
Ctr for Astronomy
Instrumen Res
CAS
Nanjing 210042
China PR
☎ 86 25 540 5562
Fax 86 25 540 5562
Email xcui@public1.ptt.js.cn

Cui Zhenhua
Beijing Planetarium
138 Xi Wai St
Beijing 100044
China PR
☎ 86 1 683 61691
Fax 86 10 683 53003

Culhane Leonard
Mullard Space Science Lab
Univ College London
Holmbury St Mary
Dorking Surrey RH5 6NT
UK
☎ 44 14 83 274 111
Fax 44 14 83 278 312
Email jlc@mssl.ulc.uk

Cullum Martin
ESO
Karl Schwarzschildstr 2
D 85740 Garching
Germany
☎ 49 893 200 60
Fax 49 893 202 362
Email mcullum@eso.org

Culver Roger Bruce
Dpt Physics
Colorado State Univ
Fort Collins CO 80523
USA
☎ 1 303 491 6206

Cunha Katia
Observ Nacional
Rua Gl Bruce 586
Sao Cristovao
20921 030 Rio de Janeiro RJ
Brazil
☎ 55 21 580 7181
Fax 55 21 580 7181
Email katia@on.br

Cuntz Manfred
CSPAR
Univ of Alabama
Huntsville AL 35899
USA
☎ 1 205 895 6660

Cuny Yvette J
Observ Paris Meudon
DASGAL
Pl J Janssen
F 92195 Meudon PPL Cdx
France
☎ 33 1 45 07 7838
Fax 33 1 45 07 7971

Cuperman Sami
Dpt Phys/Astronomy
Tel Aviv Univ
Ramat Aviv
Tel Aviv 69978
Israel
☎ 972 3 42021/425 697
Fax 972 3 640 8179

Curiel Salvador
Instituto Astronomia
UNAM
Apt 70 264
Mexico DF 04510
Mexico
☎ 52 5 622 3906
Fax 52 5 616 0653
Email scuriel@astroscu.unam.mx

Curir Anna
Osserv Astronomico
d Torino
St Osservatorio 20
I 10025 Pino Torinese
Italy
☎ 39 11 461 9002
Fax 39 11 461 9030
Email curir@astto2.infn.it

Currie Douglas G
Physics Dpt
Univ of Maryland
College Park MD 20742 2421
USA
☎ 1 301 405 6046
Fax 1 301 314 9531
Email currie@hubble.physics.umd.edu

Cutispoto Giuseppe
Osserv Astronomico
d Catania
Via A Doria 6
I 95125 Catania
Italy
☎ 39 95 733 21111
Fax 39 95 33 0592

Cuypers Jan
Koninklijke Sterrenwacht
van Belgie
Ringlaan 3
B 1180 Brussels
Belgium
☎ 32 2 373 0234
Fax 32 2 374 9822
Email jan.cuypers@oma.be

Czerny Bozena
Copernicus Astron Ctr
Polish Acad Sci
Ul Bartycka 18
PL 00 716 Warsaw
Poland
☎ 48 22 41 1086
✆ 48 22 410 046
✉ bcz@camk.edu.pl

Czerny Michal
Astronomy Dpt
Univ Leicester
University Rd
Leicester LE1 7RH
UK
☎ 44 116 252 2073
✆ 44 113 252 2200

Czyzak Stanley J
800 North Maple Av
Fairborn OH 45324
USA

D

D'Amico Flavio
INPE Astrophys Div
Av dos Astronautas 1758
Jardim d Granja
12227 010 S Jose dos Campos
Brazil
☎ 55 12 325 6745
℻ 55 12 325 6750
✉ damico@das.inpe.br

D'Antona Francesca
Osserv Astronomico
d Roma
Via d osservatorio 5
I 00040 Monteporzio
Italy
☎ 39 6 944 9019
℻ 39 6 944 7243

d'Hendecourt Louis
IAS
Bt 121
Universite Paris XI
F 91405 Orsay Cdx
France
☎ 33 1 69 85 8640
℻ 33 1 69 85 8675

d'Odorico Sandro
ESO
Karl Schwarzschildstr 2
D 85748 Garching
Germany
☎ 49 893 200 60
℻ 49 893 202 362

da Costa Antonio A
Inst Superior Tecnico
Compl Interdisciplinar
Av Rovisco Pais
P 1096 Lisboa Codex
Portugal
☎ 351 1 3524 303
℻ 351 1 3524 372

Da Costa Gary Stewart
MSSSO
Weston Creek
Private Bag
Canberra ACT 2611
Australia
☎ 61 2 62 49 0236
℻ 61 26 249 0233
✉ gdc@mso.anu.edu.au

da Costa Jose Marques
INPE
Dpt Astronomia
CP 515
12227 010 S Jose dos Campos
Brazil
☎ 55 123 22 9977
℻ 55 123 21 8743

da Costa Luiz A N
Observ Nacional
Rua Gl Bruce 586
Sao Cristovao
20921 030 Rio de Janeiro RJ
Brazil
☎ 55 21 580 7181
℻ 55 21 580 0332
✉ lndc@lnccvm

da Rocha Vieira E
R Sao Manoel 315/501
90620 110 Porto Alegre RS
Brazil
✉ secret@lna.br

da Silva A V C S
Observ Astronomico
Univ Coimbra
Caixa Postal 147
P 3002 Coimbra
Portugal
☎ 351 3 981 4947

da Silva Licio
Observ Nacional
Rua Gl Bruce 586
Sao Cristovao
20921 030 Rio de Janeiro RJ
Brazil
☎ 55 21 580 7313
℻ 55 21 580 0332

Dachs Joachim
im Schoenblick 57
D 72076 Tuebingen
Germany

Dadaev Aleksandr N
Pulkovo Observ
Acad Sciences
10 Kutuzov Quay
196140 St Petersburg
Russia
☎ 7 812 298 2242
℻ 7 812 315 1701

Dadhich Naresh
IUCAA
PO Box 4
Ganeshkhind
Pune 411 007
India
☎ 91 212 351 414/5
℻ 91 212 350 760
✉ root@iucaa.ernet.in

Dadic Zarko
Zavod Za Povijest
Znanosti Jazu
Ante Kovacica 5
HR 10000 Zagreb
Croatia
☎ 385 41 440 124

Dagkesamansky Rustam D
Lebedev Physical Inst
Acad Sciences
Leninsky Pspt 53
117924 Moscow
Russia
☎ 7 095 135 1429
℻ 7 095 135 7880

Dahn Conard Curtis
USNO
Flagstaff Station
Box 1149
Flagstaff AZ 86002 1149
USA
☎ 1 520 779 5132
℻ 1 520 774 3626
✉ cdahn@nofs.navy.mil

Dai Zigao
Astronomy Dpt
Nanjing Univ
Nanjing 210093
China PR
☎ 86 25 321 5880
℻ 86 25 332 6467
✉ postcstd@nju.edu.cn

Daigne Gerard
Observ Bordeaux
BP 89
F 33270 Floirac
France
☎ 33 5 56 86 4330
℻ 33 5 56 40 4251
✉ daigne@observ.
u-bordeaux.fr

Daintree Edward J
NRAL
Univ Manchester
Jodrell Bank
Macclesfield SK11 9DL
UK
☎ 44 14 777 1321
℻ 44 147 757 1618

Daishido Tsuneaki
Dpt Science
Waseda Univ
Shinjuku Ku
Tokyo 160
Japan
☎ 81 220 34 141

Dalgarno Alexander
CfA
HCO/SAO
60 Garden St
Cambridge MA 02138 1516
USA
☎ 1 617 495 4403
℻ 1 617 495 7356
✉ dalgarno@cfa7

Dall-Oglio Giorgio
Dpt Fisica
Univ d Roma
Pl A Moro 2
I 00185 Roma
Italy
☎ 39 6 49 91 4271
℻ 39 6 495 7697

Dallaporta N
Dpt Astronomia
Univ d Padova
Vic d Osservatorio 5
I 35122 Padova
Italy
☎ 39 49 829 3411
℻ 39 49 875 9840

Daltabuit Enrique
Instituto Astronomia
UNAM
Apt 70 264
Mexico DF 04510
Mexico
☎ 52 5 622 3906
℻ 52 5 616 0653

Daly Ruth Agnes
Physics Dpt
Princeton Univ
Princeton NJ 08544 1001
USA
☎ 1 609 258 4413
℻ 1 609 258 6853
✉ daly@pupgg.
princeton.edu

Damineli Neto Augusto
IAG
Univ Sao Paulo
CP 9638
01065 970 Sao Paulo SP
Brazil
☎ 55 11 577 8599
℻ 55 11 276 3848

Damle S V
TIFR
Homi Bhabha Rd
Colaba
Bombay 400 005
India
☎ 91 22 495 2311
℻ 91 22 495 2110

Dan Xhixiang
Shanghai Observ
CAS
80 Nandan Rd
Shanghai 200030
China PR
☎ 86 21 6438 6191
℻ 86 21 6438 4618

Danby J M Anthony
Mathematics Dpt
N Carolina State Univ
Raleigh NC 27695 8205
USA
☎ 1 919 737 3210

Danese Luigi
Osserv Astronom Padova
Univ d Padova
Vic d Osservatorio 5
I 35122 Padova
Italy
☎ 39 49 829 3411
✆ 39 49 875 9840

Danezis Emmanuel
Astrophysics Dpt
Ntl Univ Athens
Panepistimiopolis
GR 157 84 Zografos
Greece
☎ 30 1 724 3414
✆ 30 1 723 5122

Danford Stephen C
Dpt Phys/Astronomy
Univ N Carolina
At Greensboro
Greenboro NC 27412
USA
☎ 1 919 334 5669
✉ danford@uncg.edu

Daniel Jean-Yves
IAP
98bis bd Arago
F 75014 Paris
France
☎ 33 1 44 32 8147
✆ 33 1 44 32 8001

Danilov Vladimir M
Astronomy Dpt
Uralskij State Univ
Lenin Pr 51
629983 Sverdlovsk
Russia
☎ 7 343 222 3386
✆ 7 343 222 3386

Danks Anthony C
1315 Peachtree Ct
Bowie MD 20721
USA
☎ 1 301 249 8206

Danly Laura
STScI
Homewood Campus
3700 San Martin Dr
Baltimore MD 21218
USA
☎ 1 301 338 4422
✆ 1 301 338 5090
✉ danly@stsci.edu

Danziger I John
ESO
Karl Schwarzschildstr 2
D 85748 Garching
Germany
☎ 49 893 200 60
✆ 49 893 202 362

Dapergolas A
Astronomical Institute
Ntl Observ Athens
Box 20048
GR 118 10 Athens
Greece
☎ 30 1 346 1191
✆ 30 1 804 0453
✉ adaperg@astro.noa.gr

Dappen Werner
Dpt Phys/Astronomy
USC
Los Angeles CA 90089 1342
USA
☎ 1 213 740 1316
✆ 1 213 740 6653
✉ wdappen@solar.
stanford.edu

Dara Helen
Res Ctr Astronomy
Acad Athens
14 Anagnostopoulou St
GR 106 73 Athens
Greece
☎ 30 1 361 3589
✆ 30 1 363 1606
✉ exakazo@grathun1

Das Mrinal Kanti
Physics Dpt/Delhi Univ
Sri Venkateswara College
Dhaula Kuan
New Delhi 110 021
India

Das P K
IIA
Koramangala
Sarjapur Rd
Bangalore 560 034
India
☎ 91 80 356 6585
✆ 91 80 553 4043

Datlowe Dayton
Lockheed Palo Alto Res Lb
Dpt 91 20 Bg 255
3251 Hanover St
Palo Alto CA 94304 1191
USA
☎ 1 415 858 4074
✆ 1 415 424 3994

Datta Bhaskar
IIA
Koramangala
Sarjapur Rd
Bangalore 560 034
India
☎ 91 80 356 6585/6497
✆ 91 80 553 4043

Daube-Kurzemniece I A
Radioastrophys Observ
Latvian Acad Sci
Turgeneva 19
LV 226524 Riga
Latvia
☎ 371 722 6796

Dautcourt G
MPI Gravitational Physik
Albert Einstein Inst
Schlaatzweg 1
D 14473 Potsdam
Germany
☎ 49 331 275 3731
✆ 49 331 275 3798
✉ daut@aei-potdam.
mpg.de

David Laurence P
CfA
HCO/SAO MS 4
60 Garden St
Cambridge MA 02138 1516
USA
☎ 1 617 495 7245
✆ 1 617 495 7356
✉ david@cfa

David Marc
Astrophysics Res Gr
Univ Ctr Antwerpen
Groenenborgerlaan 171
B 2020 Antwerpen
Belgium
☎ 32 3 218 0355
✆ 32 3 218 0204
✉ david@ruca.ua.ac.be

Davidge Timothy J
NRCC/HIA
DAO
5071 W Saanich Rd
Victoria BC V8X 4M6
Canada
☎ 1 250 363 0047
✆ 1 250 363 0045
✉ davidge@dao.nrc.ca

Davidsen Arthur Falnes
Dpt Phys/Astronomy
JHU
Charles/34th St
Baltimore MD 21218
USA
☎ 1 301 516 7370
✆ 1 301 516 8260
✉ afd@pha.jhu.edu

Davidson Kris
School Phys/Astronomy
Univ Minnesota
116 Church St SE
Minneapolis MN 55455
USA
☎ 1 612 373 7795
✆ 1 612 624 2029
✉ kd@aps2.spa.
umn.edu

Davidson William
80 West Close
Fernhurst Haslemere
Surrey GU27 3JT
UK

Davies Merton E
The Rand Corp
1700 Main St
Santa Monica CA 90406
USA
☎ 1 310 393 0411*7428
✆ 1 310 451 6960
✉ davies@hyrax.rand.org

Davies Paul Charles W
School of Physics
Univ of Newcastle
Newcastle/Tyne NE1 7RU
UK
☎ 44 191 222 7411
✆ 44 191 261 1182

Davies Rodney D
NRAL
Univ Manchester
Jodrell Bank
Macclesfield SK11 9DL
UK
☎ 44 14 777 1321
✆ 44 147 757 1618
✉ rdd@jb.man.ac.uk

Davies Roger L
Physics Dpt
Univ of Durham
South Rd
Durham DH1 3LE
UK
☎ 44 191 374 2000
✆ 44 191 374 3749

Davila Joseph
4927 Pale Orchis Ct
Columbia MD 21044
USA
☎ 1 301 286 8366

Davis John
School of Physics
UNSW
A28
Sydney NSW 2006
Australia
☎ 61 2 692 2544
✆ 61 2 660 2903
✉ davis@physics.
su.oz.au

Davis Jonathan Ivor
Dpt Phys/Astronomy
Univ Wales College
Box 913
Cardiff CF2 3YB
UK
☎ 44 12 22 874 458
✆ 44 12 22 874 056
✉ jid@astro.cf.ac.uk

Davis Marc
Astronomy Dpt
Univ of California
601 Campbell Hall
Berkeley CA 94720 3411
USA
☎ 1 415 642 5156
✆ 1 510 642 3411
✉ marc@coma.
berkeley.edu

Davis Michael M
Arecibo Observ
Box 995
Arecibo PR 00613
USA
☎ 1 809 878 2612
✆ 1 809 878 1861
✉ mdavis@naic.edu

Davis Morris S
Dpt Phys/Astronomy
Univ North Carolina
204 Phillips Hall 039a
Chapel Hill NC 27514
USA
☎ 1 919 962 7214
℻ 1 919 962 0480

Davis Richard J
NRAL
Univ Manchester
Jodrell Bank
Macclesfield SK11 9DL
UK
☎ 44 14 777 1321
℻ 44 147 757 1618

Davis Robert J
CfA
HCO/SAO MS 20
60 Garden St
Cambridge MA 02138 1516
USA
☎ 1 617 496 7906
℻ 1 617 495 7356
✉ rdavis@cfa.harvard.edu

Davis Sumner P
Physics Dpt
Univ of California
366 LeConte Hall
Berkeley CA 94720
USA
☎ 1 415 642 4857
℻ 1 510 642 3411

Davis Cecil G
LANL
Group P 23 H 803
Box 1663
Los Alamos NM 87545 2345
USA
☎ 1 505 667 5908
✉ cgi@lanl.gov

Davis Leverett
1772 N Grand Oaks Av
Altadena CA 91001
USA

Davoust Emmanuel
OMP
14 Av E Belin
F 31400 Toulouse Cdx
France
☎ 33 5 61 33 2868
℻ 33 5 61 33 2840
✉ davoust@obs-mip.fr

Dawanas Djoni N
Astronomy Dpt
Bandung Instte Techn
Jl Ganesha 10
Bandung 40132
Indonesia
☎ 62 22 244 0252
℻ 62 22 243 8338

Dawe John Alan
ANU
SSO
Private Bag
Coonabarabran NSW 2857
Australia
☎ 61 68 426 221

Dawson Bruce
Dpt Physics/
Math Physics
Univ Adelaide
Adelaide SA 5005
Australia
☎ 61 8 8303 5996
℻ 61 8 8303 4380
✉ bdawson@physics.
adelaide.edu.au

Dawson Peter
Physics Dpt
Trent Univ
Peterborough K9J 1B8
Canada
☎ 1 705 748 1225
✉ pdawson@trentu.ca

de Almeida Amaury A
IAG
Univ Sao Paulo
CP 9638
01065 970 Sao Paulo SP
Brazil
☎ 55 11 577 8599*38
℻ 55 11 276 3848

de Araujo Francisco X
Observ Nacional
Rua Gl Bruce 586
Sao Cristovao
20921 030 Rio de Janeiro RJ
Brazil
☎ 55 21 580 0235
℻ 55 21 580 0332
✉ userfxa@lncc

de Bergh Catherine
Observ Paris Meudon
DESPA
Pl J Janssen
F 92195 Meudon PPL Cdx
France
☎ 33 1 45 07 7666
℻ 33 1 45 07 2806
✉ debergh@obspm.fr

de Biase Giuseppe A
OAR
Via d Parco Mellini 84
I 00136 Roma
Italy
☎ 39 6 34 7056
℻ 39 6 349 8236

de Boer Klaas Sjoerds
Astronomisches Institut
Univ Bonn
auf d Huegel 71
D 53121 Bonn
Germany
☎ 49 228 73 3656
℻ 49 228 73 3672
✉ kdeboer@astro.uni-
bonn.de

de Bruyn A Ger
NFRA
Postbus 2
NL 7990 AA Dwingeloo
Netherlands
☎ 31 521 59 5100
℻ 31 52 159 7332

de Carvalho Reinaldo
Observ Nacional
Rua Gl Bruce 586
Sao Cristovao
20921 030 Rio de Janeiro RJ
Brazil
☎ 55 21 580 7181
℻ 55 21 580 0332
✉ reinaldo@maxwell. on.br

de Castro Angel
Observ Astronomico
Nacional
Alfonso XII-3
E 28014 Madrid
Spain
☎ 34 1 227 1935

de Castro Elisa
Dpt Astrofisica
Fac Fisica
Univ Complutense
E 28040 Madrid
Spain
☎ 34 1 449 5316
℻ 34 1 394 5195

De Cuyper Jean-Pierre M
Koninklijke Sterrenwacht
van Belgie
Ringlaan 3
B 1180 Brussels
Belgium
☎ 32 2 373 0222
℻ 32 2 374 9822
✉ Jean-Pierre.DeCuyper@
oma.be

de Felice Fernando
Dpt Fisica G Galilei
Univ d Padova
Via Marzolo 8
I 35131 Padova
Italy
☎ 39 49 84 4278
℻ 39 49 84 4245

de Frees Douglas J
IBM Almaden Res Cent
Dpt K 84/801
650 Harry Rd
San Jose CA 95120 6099
USA
☎ 1 408 927 2854
℻ 1 408 927 2100
✉ defrees@almvma

de Freitas Pacheco J A
IAG
Univ Sao Paulo
CP 9638
01065 970 Sao Paulo SP
Brazil
☎ 55 21 717 3518
℻ 55 11 276 3848

De Garcia Maria J M
Dpt Fisica Aplicada
Ronda de Valencia 3
E 28012 Madrid
Spain
☎ 34 1 336 7686
℻ 34 1 530 9244

de Geus Eugene
Kluwer Academic Publishers
Science/Technology Div
Box 17
NL 3300 AA Dordrecht
Netherlands
☎ 31 78 639 2315
℻ 31 78 639 2254
✉ eugene.degeus@wkap.nl

de Gouveia Dal Pino E M
IAG
Univ Sao Paulo
CP 9638
01065 970 Sao Paulo SP
Brazil
☎ 55 11 577 8599
℻ 55 11 276 3848
✉ dalpino@astro2.iagusp.
usp.br

de Graaff W
Appelgaarde 117
NL 3992 JD Houten
Netherlands

de Graauw Th
SRON
Univ Groningen
Postbus 800
NL 9700 AV Groningen
Netherlands
☎ 31 50 363 4074
℻ 31 50 363 4033
✉ thijsdg@sron.rug.nl

de Greiff J Arias
Observ Astronomico Ncl
Apdo 2584
Bogota 1 DE
Columbia
☎ 57 1 342 3786
℻ 57 1 342 3786
✉ felicitas@fcaglp.
edu.ar

de Greve Jean-Pierre
Astrofysisch Inst
Vrije Univ Brussel
Pleinlaan 2
B 1050 Brussels
Belgium
☎ 32 2 629 3498
℻ 32 2 629 3424
✉ jpdgreve@vnet3.
vub.ac.be

de Groot Mart
Armagh Observ
College Hill
Armagh BT61 9DG
UK
☎ 44 18 61 522 928
℻ 44 18 61 527 174

de Groot T
Sterrekundig Inst Utrecht
Box 80000
NL 3508 TA Utrecht
Netherlands
☎ 31 30 253 5200
℻ 31 30 253 1601

de Jager Cornelis
SRON
Postbus 800
Sorbonnelaan 2
NL 3584 CA Utrecht
Netherlands
☎ 31 30 253 5723
℻ 31 30 254 0860
✉ xmkeesj@hutruuo

de Jager Gerhard
Electrical Eng Dpt
Univ of Cape Town
Private Bag
7700 Rondebosch
South Africa
☎ 27 21 650 2791
℻ 27 21 650 3465
✉ gerhard@eleceng.
uct.ac.za

de Jager Ocker C
Physics Dpt
Potchefstroom Univ
for CHE
2520 Potchefstroom
South Africa
☎ 27 148 299 2418
℻ 27 148 299 2421

de Jong Teije
Astronomical Institute
Univ of Amsterdam
Kruislaan 403
NL 1098 SJ Amsterdam
Netherlands
☎ 31 20 525 7491
℻ 31 20 525 7484
✉ teije@sron.rug.nl

de Jonge J K
Astronomy Dpt
Univ of Pittsburgh
Riverview Park
Pittsburgh PA 15214
USA
☎ 1 215 898 8176
℻ 1 215 898 9336

de Kool Marthijn
MPA
Karl Schwarzschildstr 1
D 85748 Garching
Germany
☎ 49 893 299 3221
℻ 49 893 299 3235
✉ dekool@mpa-garching.
mpg.de

de Korte Pieter A J
SRON
Postbus 800
Sorbonnelaan 2
NL 3584 CA Utrecht
Netherlands
☎ 31 30 253 5600
℻ 31 30 254 0860

de La Herran Jose V
Instituto Astronomia
UNAM
Apt 70 264
Mexico DF 04510
Mexico
☎ 52 5 622 3906
℻ 52 5 616 0653

de la Noe Jerome
Observ Bordeaux
BP 89
F 33270 Floirac
France
☎ 33 5 56 86 4330
℻ 33 5 56 40 4251

de la Reza Ramiro
Observ Nacional
Rua Gl Bruce 586
Sao Cristovao
20921 030 Rio de Janeiro RJ
Brazil
☎ 55 21 580 7313
℻ 55 21 580 0332

de Lapparent-Gurriet Valerie
IAP
98bis bd Arago
F 75014 Paris
France
☎ 33 1 44 32 8188
℻ 33 1 44 32 8001
✉ lapparen@iap.fr

de Loore Camiel
Astrofysisch Inst
Vrije Univ Brussel
Pleinlaan 2
B 1050 Brussels
Belgium
☎ 32 2 629 3497
℻ 32 2 629 3424

de Martino Domitilla
Osserv d Capodimonte
Via Moiariello 16
I 80131 Naples
Italy
☎ 39 81 557 5111
℻ 39 81 45 6710
✉ demartino@astrna.
na.astro.it

de Medeiros Jose Renan
Dpt Fisica/CCE/UFRN
Univ Federal
Do Rio Grande d Norte
59072 970 Natal RN
Brazil
☎ 55 84 231 9586
℻ 55 84 231 9749
✉ renan@dfte.ufrn.br

de Pascual Martinez M
Observ Astronomico
Nacional
Alfonso XII-3/5
E 28014 Madrid
Spain
☎ 34 1 227 0107

de Pater Imke
Astronomy Dpt
Univ of California
601 Campbell Hall
Berkeley CA 94720 3411
USA
☎ 1 415 642 1947
℻ 1 510 642 3411

de Robertis M M
Physics Dpt
York Univ
4700 Keele St
North York ON M3J 1P3
Canada
☎ 1 416 736 2100*7761
℻ 1 416 736 5386
✉ fs300141@yusol

de Rop Yves
Inst Astrophysique
Universite Liege
Av Cointe 5
B 4000 Liege
Belgium
☎ 32 4 254 7510
℻ 32 4 254 7511
✉ derop@astro.ulg.
ac.be

de Ruiter Hans Rudolf
Osserv Astronom Bologna
Univ d Bologna
Via Zamboni 33
I 40126 Bologna
Italy
☎ 39 51 259 422
℻ 39 51 259 407
✉ deruiter@astbo3.bo.
astro.it

de Sabbata Vittorio
Istt d Fisica
Univ d Bologna
Via Irnerio 46
I 40126 Bologna
Italy
☎ 39 51 26 0991*051

de San Francisco Eiroa C
Dpt Fisica Teorica
Fac Ciencias
UAM Cantoblanco
E 28049 Madrid
Spain
☎ 34 91 397 5567
℻ 34 91 397 3936
✉ carlos@xiada.ft.
uam.es

de Sanctis Giovanni
Osserv Astronomico
d Torino
St Osservatorio 20
I 10025 Pino Torinese
Italy
☎ 39 11 461 9000
℻ 39 11 461 9030

de Silva L N K
Mathematics Dpt
Univ of Colombo
Colombo 3
Sri Lanka

de Souza Ronaldo
IAG
Univ Sao Paulo
CP 9638
01065 970 Sao Paulo SP
Brazil
☎ 55 11 577 8599
℻ 55 11 276 3848

de Vegt Chr
Hamburger Sternwarte
Univ Hamburg
Gojensbergsweg 112
D 21029 Hamburg
Germany
☎ 49 407252 4128/4112
℻ 49 407 252 4198
✉ cdevegt@hs.uni-hamburg.de

de Vicente Pablo
Observ Astronomico
Nacional
Apt 1143
E 28800 Alcala d Henares
Spain
☎ 34 1 885 5060
℻ 34 1 885 5062
✉ vicente@cay.oan.es

de Vincenzi Donald
NASA/ARC
MS 245 1
Moffett Field CA 94035 1000
USA
☎ 1 415 604 5028
℻ 1 415 604 6779

de Vries Cornnelis
SRON
Postbus 800
Sorbonnelaan 2
NL 3584 CA Ultrecht
Netherlands
☎ 31 30 253 5600
℻ 31 30 254 0860
✉ C.deVries@sron.ruu.nl

de Young David S
NOAO/KPNO
Box 26732
950 N Cherry Av
Tucson AZ 85726 6732
USA
☎ 1 520 327 5511
℻ 1 520 318 8360
✉ deyoung@noao.edu

de Zeeuw Pieter T
Rijkuniversiteit te
Huygens Lab
Box 9504
NL 2300 RA Leiden
Netherlands
☎ 31 71 527 5879/5832
℻ 31 71 527 5819
✉ tim@strw.leidenuniv.nl

de Zotti Gianfranco
Osserv Astronom Padova
Univ d Padova
Vic d Osservatorio 5
I 35122 Padova
Italy
☎ 39 49 829 3444
℻ 39 49 875 9840
✉ dezotti@astrpd.pd.astro.it

Dearborn David Paul K
LLNL
L 22
Box 808
Livermore CA 94551 9900
USA
☎ 1 415 423 0666
℻ 1 415 423 0238
✉ ddearborn@llnl.gov

Debarbat Suzanne V
Observ Paris
DANOF
61 Av Observatoire
F 75014 Paris
France
☎ 33 1 40 51 2209
✆ 33 1 43 54 1804
✉ debarbat@danof.obspm.fr

Debehogne Henri Sc
Koninklijke Sterrenwacht van Belgie
Ringlaan 3
B 1180 Brussels
Belgium
☎ 32 2 373 0212
✆ 32 2 374 9822
✉ henri.debehogne@oma.be

Debrunner Hermann
Physikalisches Institut
Univ Bern
Sidlerstr 5
CH 3012 Bern
Switzerland
☎ 41 31 65 4051

Deeming Terence J
Icarus Res
Box 540205
Houston TX 77254
USA
☎ 1 713 772 8414

Defraigne Pascale
Koninklijke Sterrenwacht van Belgie
Ringlaan 3
B 1180 Brussels
Belgium
☎ 32 2 373 0260
✆ 32 2 374 9822
✉ pascale.defraigne@oma.be

Degaonkar S S
PRL
Navrangpura
Ahmedabad 380 009
India
☎ 91 272 46 2129
✆ 91 272 44 5292

Degenhardt Detlev
Rechenzentrum Univ
Hermann herder Str 10
D 79104 Freiburg Breisgau
Germany
☎ 49 761 203 4648
✆ 49 761 203 4643
✉ degenhar@ruf.de

DeGioia-Eastwood Kathleen
Dpt Phys/Astronomy
N Arizona Univ
Box 6010
Flagstaff AZ 86011 6010
USA
☎ 1 520 523 7159
✆ 1 520 523 1371
✉ Kathy.Eastwood@nau.edu

Deguchi Shuji
Nobeyama Radio Obs
NAOJ
Minamimaki Mura
Nagano 384 13
Japan
☎ 81 267 98 2831
✆ 81 267 98 2884

Dehant Veronique
Koninklijke Sterrenwacht van Belgie
Ringlaan 3
B 1180 Brussels
Belgium
☎ 32 2 373 0266
✆ 32 2 374 9822
✉ veronique.dehant@oma.be

Deharveng Jean-Michel
LAS
Traverse du Siphon
Les Trois Lucs
F 13376 Marseille Cdx 12
France
☎ 33 4 91 05 5900
✆ 33 4 91 66 1855
✉ jmd@frlasm51

Deharveng Lise
Observ Marseille
2 Pl Le Verrier
F 13248 Marseille Cdx 04
France
☎ 33 4 91 95 9088
✆ 33 4 91 62 1190

Dehghani Mohammad Hossein
Physics Dpt
Shiraz 71454
Iran
☎ 98 71 24609
✆ 98 71 20 027
✉ dehghani@phyics.sci.shirazu.ac.ir

Deinzer W
Univ Sternwarte Goettingen
Geismarlandstr 11
D 37083 Goettingen
Germany
☎ 49 551 395 044
✆ 49 551 395 043

Deiss Bruno M
Inst Theor Physics
Univ Frankfurt
Robert Mayer Str 8-10
D 60054 Frankfurt A M
Germany
☎ 49 69 798 2636
✆ 49 69 798 8350
✉ deiss@astro.uni-frankfurt.d400.de

Dejaiffe Rene J
Koninklijke Sterrenwacht van Belgie
Ringlaan 3
B 1180 Brussels
Belgium
☎ 32 2 373 0232
✆ 32 2 374 9822
✉ rene.dejaiffe@oma.be

Dejonghe Herwig Bert
Sterrekundig Observ
Universiteit Gent
Krijgslaan 281 S9
B 9000 Gent
Belgium
☎ 32 92 64 4761
✆ 32 92 64 4989
✉ herwig.dejonghe@rug.ac.be

Dekel Avishai
Physics Dpt
Hebrew Univ Jerusalem
Jerusalem 91904
Israel
☎ 972 2 584 605
✆ 972 2 61 1519

Dekker E
Meidoornlaan 13
NL 3461 ES Linschoten
Netherlands
☎ 31 34 801 5406

Del Olmo Orozco A
Inst Astrofisica
Andalucia Apt 3004
Prof Albareda 1
E 18080 Granada
Spain
☎ 34 5 812 1311
✆ 34 5 881 4530
✉ chony@iaa.es

Del Rio Gerardo
Observ Astronomico Nacional
Alfonso XII-3
E 28014 Madrid
Spain
☎ 34 1 227 0107/1935

Del Toro Iniesta Jose
IAC
Observ d Teide
via Lactea s/n
E 38200 La Laguna
Spain
☎ 34 2 232 9100
✆ 34 2 232 9117

Delaboudiniere Jean-Pierre
IAS
Bt 121
Universite Paris XI
F 91405 Orsay Cdx
France
☎ 33 1 69 85 8625/8425
✆ 33 1 69 85 8675

Delannoy Jean
IRAM
300 r La Piscine
F 38406 S Martin Heres Cdx
France
☎ 33 4 76 42 3383
✆ 33 4 76 51 5938

Delbouille Luc
Inst Astrophysique
Universite Liege
Av Cointe 5
B 4000 Liege
Belgium
☎ 32 4 254 7510
✆ 32 4 254 7511
✉ delbouil@astro.ulg.ac.be

Delgado Antonio Jesus
Inst Astrofisica
Andalucia Apt 3004
Prof Albareda 1
E 18080 Granada
Spain
☎ 34 5 812 1300
✆ 34 5 881 4530

Deliyannis Constantine P
Astronomy Dpt
Indiana Univ
Swain W 319
Bloomington IN 47405 4201
USA
☎ 1 812 856 5197
✆ 1 812 855 8725
✉ con@athena.astro.indiana.edu

Deliyannis Jean
Astrophysics Dpt
Ntl Univ Athens
Panepistimiopolis
GR 157 84 Zografos
Greece
☎ 30 1 723 7924
✆ 30 1 723 8413
✉ jdeligia@atlas.uoa.ariadne-t.gr

Della Valle Massimo
Dpt Astronomia
Univ d Padova
Vic d Osservatorio 5
I 35122 Padova
Italy
☎ 39 49 829 3411
✆ 39 49 875 9840

Delmas Christian
OCA
CERGA
F 06130 Grasse
France
☎ 33 4 93 40 5353
✆ 33 4 93 40 5333
✉ delmas@obs-azur.fr

Delsemme Armand H
Dpt Phys/Astronomy
Univ of Toledo
2801 W Bancroft St
Toledo OH 43606
USA
☎ 1 419 537 2654
✆ 1 419 530 2723

Demaret Jacques
Inst Astrophysique
Universite Liege
Av Cointe 5
B 4000 Liege
Belgium
☎ 32 4 254 7510
✆ 32 4 254 7511
✉ demaret@astro.ulg.ac.be

Demarque Pierre
Astronomy Dpt
Yale Univ
Box 208101
New Haven CT 06520 8101
USA
☎ 1 203 432 3024
✆ 1 203 432 5048
✉ demarque@yalastro

Demers Serge
Dpt d Physique
Universite Montreal
CP 6128 Succ A
Montreal QC H3C 3J7
Canada
☎ 1 514 343 6718
✆ 1 514 343 2071

Demianski Marek
Inst Theor Physics
Warsaw Univ
Hoza 69
PL 00 681 Warsaw
Poland
☎ 48 22 628 3031* 245
✉ mde@fuw.edu.pl

Deming Leo Drake
NASA GSFC
Code 693
Greenbelt MD 20771
USA
☎ 1 301 286 6519
✆ 1 301 286 3271

Demircan Osman
Astronomy Dpt
Univ of Ankara
Fen Fakultesi
06100 Besevler
Turkey
☎ 90 41 212 6720
✆ 90 312 223 2395

Demoulin Pascal
Observ Paris Meudon
DASOP
Pl J Janssen
F 92195 Meudon PPL Cdx
France
☎ 33 1 45 07 7816
✆ 33 1 45 07 7959
✉ demoulin@obspm.fr

Dempsey Robert C
STScI
Homewood Campus
3700 San Martin Dr
Baltimore MD 21218
USA
☎ 1 301 338 1334
✆ 1 301 338 4767
✉ Dempsey@stsci.edu

Deng Zugan
Graduate School
Univ Sci/Techn
Box 3908
Beijing 100039
China PR
☎ 86 1 28 9461*89

Denis Carlo
Inst Astrophysique
Universite Liege
Av Cointe 5
B 4000 Liege
Belgium
☎ 32 4 254 7510
✆ 32 4 254 7511
✉ denis@astro.ulg.ac.be

Denishchik Yurii
State Institute
12 Naberezhnaya St
B 114043
349114 Alchevsk Lugansk r
Ukraine
☎ 380 6442 23159
✆ 380 6442 22057
✉ orion@dgmi.lugansk.ua

Denisse Jean-Francois
48 r Mr Le Prince
F 75006 Paris
France
☎ 33 1 43 29 4874

Denissenkov Pavel
Astron Inst St Petersburg
State Univ
Bibliotechnaya Pl 2
St Petersburg 198904
Russia
☎ 7 812 428 4162
✆ 7 812 428 6649
✉ dpa@aispbu.spb.su

Denisyuk Edvard K
Astrophys Inst
Kazakh Acad Sci
480068 Alma Ata
Kazakhstan
☎ 7 62 4040

Denizman Levent
Physics Dpt
Univ of Istanbul
University 34452
34454 Istanbul
Turkey
☎ 90 212 522 3597
✆ 90 212 519 0884
✉ ik131@triuuml1

Dennefeld Michel
IAP
98bis bd Arago
F 75014 Paris
France
☎ 33 1 44 32 8116
✆ 33 1 44 32 8001
✉ dennefeld@aip.fr

Dennis Brian Roy
NASA GSFC
Code 682 2
Greenbelt MD 20771
USA
☎ 1 301 286 7983
✆ 1 301 286 1617
✉ Brian.r.dennis@
gsfc.nasa.gov

Dennison Edwin W
985 Cynthia Av
Pasadena CA 91107
USA
☎ 1 818 351 8751

Dennison P A
Trinity Hall
Univ of Cambridge
Cambridge CB3
UK

Denoyelle Jozef Kic
Koninklijke Sterrenwacht
van Belgie
Ringlaan 3
B 1180 Brussels
Belgium
☎ 32 2 376 6887
✆ 32 2 374 9822
✉ jozef.denoyelle@
oma.be

Dent William A
Dpt Phys/Astronomy
Univ Massachusetts
GRC B
Amherst MA 01003
USA
☎ 1 413 545 3665

Deprit Andre
NBS
Applied Math Ctr
Gaithersburg MD 20899
USA
☎ 1 301 921 2631
✉ deprit@enh.nist.gov

Dere Kenneth Paul
NRL
Code 4163
4555 Overlook Av SW
Washington DC 20375 5000
USA
☎ 1 202 767 2517

Derman I Ethem
Astronomy Dpt
Univ of Ankara
Fen Fakultesi
06100 Besevler
Turkey
☎ 90 41 23 6550/0109
✆ 90 312 223 2395

Dermendjiev Vladimir
Astronomy Dpt
Bulgarian Acad Sci
72 Lenin Blvd
BG 1784 Sofia
Bulgaria
☎ 359 2 75 8927
✆ 359 2 75 5019

Dermer Charles Dennison
NRL
EO Holburt Ctr Space Sci
Code 7653
Washington DC 20375 5352
USA
☎ 1 202 767 2965
✆ 1 202 767 6473
✉ dermer@osse.nrl.navy.mil

Dermott Stanley F
Astronomy Dpt
Univ of Florida
224 SSRB
Gainesville FL 32611
USA
☎ 1 904 392 2361
✆ 1 904 392 9605
✉ dermott@jupiter.astro.ufl.edu

Desai Jyotindra N
PRL
Navrangpura
Room 763
Ahmedabad 380 009
India
☎ 91 272 46 2129
✆ 91 272 44 5292

Desesquelles Jean
Universite Lyon 1
Campus La Doua
43 bd 11 novembre
F 69621 Villeurbanne
France
☎ 33 4 78 89 8124

Deshpande Avinash
RRI
Sadashivanagar
CV Raman Av
Bangalore 560 080
India
☎ 91 80 336 0122
✆ 91 80 334 0492
✉ desh@rri.ernet.in

Deshpande M R
PRL
Navrangpura
Ahmedabad 380 009
India
☎ 91 272 46 2129
✆ 91 272 44 5292
✉ mrd@prl.ernet.in

Despain Keith Howard
LANL
MS B220 X 2
Box 1663
Los Alamos NM 87545 2345
USA
☎ 1 505 667 2388
✆ 1 505 665 4055

Despois Didier
Observ Bordeaux
BP 89
F 33270 Floirac
France
☎ 33 5 56 86 4330
✆ 33 5 56 40 4251

Dettmar Ralf-Juergen
Astronomisches Institut
Ruhr Univ Bochum
Postfach 102148
D 44780 Bochum
Germany
☎ 49 234 700 3454
✆ 49 234 709 4169
✉ dettmar@astro.ruhr-uni-
bochum.de

Deubner Franz-Ludwig
Astronomisches Institut
Univ Wuerzburg
am Hubland
D 97074 Wuerzburg
Germany
☎ 49 931 888 5030
✆ 49 931 888 4603
✉ deubner@astro.uni-
wuerzburg.de

Deul Erik
Leiden Observ
Box 9513
NL 2300 RA Leiden
Netherlands
☎ 31 71 527 5880
✆ 31 71 527 5819

Deupree Robert G
LANL
MS F665 Ess 5
Box 1663
Los Alamos NM 87545 2345
USA
☎ 1 505 667 8215
✆ 1 505 665 4055

Deutschman William A
Oregon Laser Consultants
455 Hillside Av
Klamath Falls OR 97601
USA
☎ 1 503 882 3295

Devinney Edward J
100 Union Av
Delanco NJ 08075
USA
☎ 1 609 764 1250

Devorkin David H
Ntl Air/Space Museum
Smithsonian Institution
Room 3530
Washington DC 20560
USA
☎ 1 202 357 2828
✉ nasdsh03@sivm.si.edu

Devyatkin Alexander V
Pulkovo Observ
Acad Sciences
10 Kutuzov Quay
196140 St Petersburg
Russia
☎ 7 812 123 4371
✆ 7 812 123 1922
✉ adev@mahis.spb.su

Dewdney Peter E F
DRAO
NRCC/HIA
Box 248
Penticton BC V2A 6K3
Canada
☎ 1 250 493 2277
✆ 1 250 493 7767

Dewey Rachel J
Physics Dpt
Princeton Univ
Jadwin Hall
Princeton NJ 08544 1001
USA
☎ 1 609 258 4365
✆ 1 609 258 6853
✉ rachel@pulsar.
princeton.edu

Dewhirst David W
Inst of Astronomy
The Observatories
Madingley Rd
Cambridge CB3 0HA
UK
☎ 44 12 23 337 548
✆ 44 12 23 337 523

Dewitt Bryce S
Astronomy Dpt
Univ of Texas
Rlm 15 308
Austin TX 78712 1083
USA
☎ 1 512 471 5055
✆ 1 512 471 6016

Dewitt- John H
3602 Hoods Hill Rd
Nashville TN 37215
USA
☎ 1 615 383 8272

Dewitt-Morette Cecile Pr
Physics Dpt
Univ of Texas
Rlm 15 308
Austin TX 78712 1083
USA
☎ 1 512 471 1052
✆ 1 512 471 6016

Dezso Lorant
Debrecen Heliophys Observ
Acad Sciences
Box 30
H 4010 Debrecen
Hungary
☎ 36 5 231 1015

Dhawan Vivek
NRAO
Box 0
Socorro NM 87801 0387
USA
☎ 1 505 835 7378
✆ 1 505 835 7027
✉ vdhawan@nrao.edu

Dhillon Vikram Singh
Physics Dpt
The University
Sheffield S3 7Rh
UK
☎ 44 114 222 4528
✆ 44 114 272 8079
✉ vik.Dhillon@sheffield.
ac.uk

Dhurandhar Sanjeev
IUCAA
PO Box 4
Ganeshkhind
Pune 411 007
India
☎ 91 212 351 414/5
✆ 91 212 350 760
✉ sdh@iucaa.ernet.in

Di Xiaohua
Purple Mountain Obs
CAS
Nanjing 210008
China PR
☎ 86 25 37609
✆ 86 25 301 459

Di Cocco Guido
TeSRE
CNR
Via P Gobetti101
I 40129 Bologna
Italy
☎ 39 51 639 8665
✆ 39 51 639 8724

Di Fazio Alberto
OAR
Via d Parco Mellini 84
I 00136 Roma
Italy
☎ 39 6 34 7056
✆ 39 6 349 8236

Di Martino Mario
Osserv Astronomico
d Torino
St Osservatorio 20
I 10025 Pino Torinese
Italy
☎ 39 11 461 9000
✆ 39 11 461 9030

Di Tullio Graziella
Osserv Astronom Padova
Univ d Padova
Vic d Osservatorio 5
I 35122 Padova
Italy
☎ 39 49 829 3411
✆ 39 49 875 9840

Dialetis Dimitris
Astronomical Institute
Ntl Observ Athens
Box 20048
GR 118 10 Athens
Greece
☎ 30 1 346 1191
✆ 30 1 346 4566

Diamond Philip John
NRAO
Box 0
Socorro NM 87801 0387
USA
☎ 1 505 835 7900
✆ 1 505 835 7027
✉ pdiamond@nrao.edu

Diaz Angeles Isabel
Dpt Fisica Teorica
Fac Ciencias
UAM Cantoblanco
E 28049 Madrid
Spain
☎ 34 91 397 4223
✆ 34 91 397 3936
✉ adiaz@emduam11

Dick Steven J
USNO
3450 Massachusetts Av NW
Washington DC 20392 5100
USA
☎ 1 202 762 0379
✆ 1 202 762 1516
✉ dick@ariel.usno.
navy.mil

Dick Wolfgang
Bund Kartograp/Geodesy
Aussenstelle Potsdam
Postfach 60 08 08
D 14408 Potsdam
Germany
☎ 49 331 316 618
✆ 49 331 316 602
✉ wdi@potsdam.ifag.de

Dickel Helene R
Astronomy Dpt
Univ of Illinois
103 UI Astronomical Bldg
Urbana IL 61801
USA
☎ 1 217 244 7044
✆ 1 217 244 7638
✉ lanie@astro.uiuc.edu

Dickel John R
Astronomy Dpt
Univ of Illinois
103 UI Astronomical Bldg
Urbana IL 61801
USA
☎ 1 217 333 5532
✆ 1 217 244 7638
✉ johnd@astro.uiuc.edu

Dickens Robert J
Rutherford Appleton Lab
Space/Astrophysics Div
Bg R25/R68
Chilton Didcot OX11 0QX
UK
☎ 44 12 35 821 900
✆ 44 12 35 44 5808

Dickey Jean O'Brien
CALTECH/JPL
MS 138 208
4800 Oak Grove Dr
Pasadena CA 91109 8099
USA
☎ 1 818 354 3235
✆ 1 818 393 6980
✉ jod@logos.jpl.nasa.gov

Dickey John M
School Phys/Astronomy
Univ Minnesota
116 Church St SE
Minneapolis MN 55455
USA
☎ 1 612 373 3308
✆ 1 612 624 2029
✉ john@ast1.spa.umn.edu

Dickinson Dale F
Lockheed Palo Alto Res Lb
Dpt 92 20 Bg 205
3251 Hanover St
Palo Alto CA 94304 1191
USA
☎ 1 415 424 2701
✆ 1 415 424 3994

Dickman Robert L
NSF
Div Astron Sci
4201 Wilson Blvd
Arlington VA 22230
USA
☎ 1 703 306 1822
✆ 1 703 306 0525
✉ rdickman@nsf.gov

Dickman Steven R
Dpt Geological Sci
State Univ of New York
Binghamton NY 13901
USA
☎ 1 607 777 4378
✆ 1 607 777 2288
✉ dickman@bingvmb.cc.
binghamton.edu

Didkovsky Leonid
Crimean Astrophys Obs
Ukrainian Acad Science
Nauchny
334413 Crimea
Ukraine
☎ 380 6554 71115
✆ 380 6554 40704
✉ didkovsky@crao.crimea.ua

Diego Francisco
Dpt Phys/Astronomy
UCLO
Gower St
London WC1E 6BT
UK
☎ 44 171 387 7050*3512
✆ 44 171 380 7145
✉ fd@uk.ac.ucl.starlink

Diercksen Geerd H F
MPA
Karl Schwarzschildstr 1
D 85748 Garching
Germany
☎ 49 893 299 00
✆ 49 893 299 3235

Dieter Conklin Nannielou H
Clay Rd
N Theford VT 05054
USA
☎ 1 802 333 4079
✉ nan.conklin@valley.net

Dietrich Matthias
Landessternwarte
Koenigstuhl
D 69117 Heidelberg
Germany
☎ 49 622 150 9256
✆ 49 622 150 9202
✉ mdietric@lsw.uni.heidelberg.de

Digel Seth William
NASA/GSFC
Code 660 1
Greenbelt MD 20771
USA
☎ 1 301 286 8709
✆ 1 301 286 1681
✉ digel@gsfc.nasa.gov

Dikova Smiliana D
Astronomy Dpt
Bulgarian Acad Sci
72 Lenin Blvd
BG 1784 Sofia
Bulgaria
☎ 359 2 75 8927
✆ 359 2 75 5019

Dimitrijevic Milan
Astronomical Observ
Volgina 7
11050 Beograd
Yugoslavia FR
☎ 381 1 419 357/421 875
✆ 381 1 419 553

Dinerstein Harriet L
Astronomy Dpt
Univ of Texas
Rlm 15 308
Austin TX 78712 1083
USA
☎ 1 512 471 3449
✆ 1 512 471 6016

Ding Mingde
Astronomy Dpt
Nanjing Univ
Nanjing 210008
China PR
☎ 86 25 663 7551 2882
✆ 86 25 330 2728

Ding Youji
Yunnan Observ
CAS
Kunming 650011
China PR
☎ 86 871 2035
✆ 86 871 717 1845

Dintinjana Bojan
Astron Observ
Univ E Kardelj
Jadranska 19
Ljubljana
Slovenia
☎ 386 61 265 061
✆ 386 61 217 281
✉ bojan.dintinjana@uni-lj.ac.mai

Dinulescu Simona
Calea 13 Sept 126
Bl P34 Sc L Et 3 Apt 11
RO 76125 Bucharest S5
Rumania

Dionysiou Demetrios
Hellenic AF Academy
Dekelia Attica
18 Amassias Str
GR 116 34 Athens
Greece
☎ 30 1 723 8436

Disney Michael J
Physics Dpt
Univ Wales College
Box 913
Cardiff CF1 3TH
UK
☎ 44 12 22 874 785
✆ 44 12 22 371 921

Divan Lucienne
IAP
98bis bd Arago
F 75014 Paris
France
☎ 33 1 44 32 8000
✆ 33 1 44 32 8001

Diver Declan Andrew
Dpt Phys/Astronomy
Univ of Glasgow
Glasgow G12 8QQ
UK
☎ 44 141 339 8855
✆ 44 141 330 5183
✉ diver@astro.gla.ac.uk

Dixon Robert S
Radio Observ
Ohio State Univ
2015 Neil Av
Columbus OH 43210
USA
☎ 1 614 422 6789
✉ dixon-r@osu-20.ircc.ohio-state.edu

Dizer Muammer
Kandilli Observ
Bogazici Univ
Cengelkov
81220 Istanbul
Turkey
☎ 90 1 332 0277

Djorgovski Stanislav
CALTECH
Palomar Obs
MS 105 24
Pasadena CA 91125
USA
☎ 1 818 356 4415
✆ 1 818 568 9352
✉ george@deimos.caltech.edu

Djurasevic Gojko
Astronomical Observ
Volgina 7
11050 Beograd
Yugoslavia FR
☎ 381 1 419 553
✆ 381 1 419 553

Djurovic Dragutin M
Astronomy Dpt
Univ of Belgrade
Studentski Trg 16
11000 Beograd
Yugoslavia FR
☎ 381 11 638 715
✆ 381 11 630 151
✉ dragutin@matf.bg.ac.yu

Dluzhnevskaya Olga B
Inst of Astronomy
Acad Sciences
Pyatnitskaya Ul 48
109017 Moscow
Russia
☎ 7 095 231 5461
✆ 7 095 230 2081
✉ olgad@inasan.rssi.ru
olga@astro.free.net

Doazan Vera
Observ Paris
DASGAL
61 Av Observatoire
F 75014 Paris
France
☎ 33 1 40 51 2235
✆ 33 1 44 54 1804

Dobronravin Peter
Crimean Astrophys Obs
Ukrainian Acad Science
Nauchny
334413 Crimea
Ukraine
☎ 380 6554 71161
✆ 380 6554 40704

Dobrzycki Jerzy
History of Science
Polish Acad Sci
Gwiazdzista 27/169
PL 01 814 Warsaw
Poland
☎ 48 22 33 2203

Docobo Jose A Durantez
Observ Astronomico
Ramon Maria Aller
Avd de Las Ciencias S/n
E Santiago de Compostela
Spain
☎ 34 8 159 2747
✆ 34 8 156 2569

Dodd Richard J
Carter Observ
Ntl Observ New Zealand
Box 2909
Wellington 1
New Zealand
☎ 64 4 4728 167
✆ 64 4 47 28 320

Dodonov Serguej
SAO
Acad Sciences
Nizhnij Arkhyz
357147 Karachaevo
Russia
☎ 7 878 784 6223
✆ 7 96 908 2861
✉ dodo@sao.ru

Dogan Nadir
Astronomy Dpt
Univ of Ankara
Fen Fakultesi
06100 Besevler
Turkey
☎ 90 41 216 6720
✆ 90 312 223 2395

Doherty Lorne H
108 1900 Marquia Av
Gloucester ONT K1J 8J2
Canada

Doherty Lowell R
1883 Forest Circle
Santa Fe NM 87505 4506
USA

Doi Mamoru
Astronomy Dpt
Univ of Tokyo
7-3-1 Hongo Bunkyo-ku
Tokyo 113
Japan
☎ 81 338 13 9224
✆ 81 338 13 9439
✉ doi@astron.s.u-tokyo.ac.jp

Dokuchaev Vyacheslav I
Inst Nuclear Res
Acad Sciences
60th Aniv Oct Pspt 7a
117312 Moscow
Russia
☎ 7 095 382 7678
✆ 7 095 292 6511
✉ dokuchaev@inr.npd.ac.ru

Dokuchaeva Olga D
SAI
Acad Sciences
Universitetskij Pr 13
119899 Moscow
Russia
☎ 7 095 939 2858
℻ 7 095 939 0126

Dolan Joseph F
NASA GSFC
Code 681
Greenbelt MD 20771
USA
☎ 1 301 286 5920
℻ 1 301 286 1753
✉ tejfd@splvin.gsfc.nasa.gov

Dolez Noel
OMP
14 Av E Belin
F 31400 Toulouse Cdx
France
☎ 33 5 61 25 2101
℻ 33 5 61 27 3179

Dolginov Arkady Z
Ioffe Physical Tech Inst
Acad Sciences
Polytechnicheskaya Ul 26
194021 St Petersburg
Russia
☎ 7 812 247 9167
℻ 7 812 247 1017

Dolidze Madona V
Abastumani Astrophysical Observ
Georgian Acad Sci
383762 Abastumani
Georgia
☎ 995 88 32 95 5367
℻ 995 88 32 98 5017

Dollfus Audouin
Observ Paris Meudon
DASOP
Pl J Janssen
F 92195 Meudon PPL Cdx
France
☎ 33 1 45 34 7530
℻ 33 1 45 07 7971

Domingo Vicente
ESA
c/o NASA/GSFC
MS 682
Greenbelt MD 20771
USA
☎ 1 301 286 4144
℻ 1 301 286 1617
✉ vdomingo@esa.nascom.nasa.gov

Dominko Fran
Saranoviceva 11
Ljubljana
Slovenia
☎ 386 61 322 210

Domke Helmut
Zntrlinst Astrophysik
Sternwarte Babelsberg
Rosa Luxemburg Str 17a
D 14473 Potsdam
Germany
☎ 49 331 275 3731

Dommanget J
Koninklijke Sterrenwacht van Belgie
Ringlaan 3
B 1180 Brussels
Belgium
☎ 32 2 373 0241
℻ 32 2 374 9822
✉ omer.nys@oma.be

Donahue Robert Andrew
CfA
HCO/SAO MS 15
60 Garden St
Cambridge MA 02138 1596
USA
☎ 1 617 495 7184
℻ 1 617 495 7095
✉ donahue@cfa.harvard.edu

Donas Jose
LAS
Traverse du Siphon
Les Trois Lucs
F 13376 Marseille Cdx 12
France
☎ 33 4 91 05 5900
℻ 33 4 91 66 1855
✉ donas@frlasm51

Donati Jean-Francois
OMP
14 Av E Belin
F 31400 Toulouse Cdx
France
☎ 33 5 61 33 2917
℻ 33 5 61 33 2840
✉ donati@obs-mip.fr

Dong Il Zun
Pyongyang Astron Obs
Acad Sciences DPRK
Taesong district
Pyongyang
Korea DPR
☎ 850 5 3134/5 & 5 3239

Donn Bertram D
NASA GSFC
Code 691
Greenbelt MD 20771
USA
☎ 1 301 286 6859

Donner Karl Johan
Helsinky Univ Observ
Box 14
FIN 00014 Helsinki
Finland
☎ 358 19 12 2940
℻ 358 19 12 2952

Donnison John Richard
Dpt Math Sci
Goldsmiths Coll
New Cross
London SE14 6NW
UK
☎ 44 181 716 927 171

Dopita Michael A
MSSSO
Weston Creek
Private Bag
Canberra ACT 2611
Australia
☎ 61 262 49 0212
℻ 61 26 249 0233
✉ michael.dopita@anu.edu.au

Dorenwendt Klaus
Physikalisch-technisches Bundesanstalt
Bundesallee 100
D 3300 Braunschweig
Germany
☎ 49 531 592 1210

Dorfi Ernst Anton
Inst Astronomie
Univ Wien
Tuerkenschanzstr 17
A 1180 Wien
Austria
☎ 43 1 345 3600
℻ 43 1 470 6015

Dormand John Richard
Mathematics Dpt
Teesside Polytechnic
Middlesbrough
Cleveland TS1 3BA
UK
☎ 44 16 42 218 121*4365

Doroshkevich Andrei G
Keldysh Inst Applied Maths
Acad Sciences
Miusskaja Sq 4
125047 Moscow
Russia
☎ 7 095 258 1314

Dorschner Johann
Astrophysik Inst
Univ Sternwarte
Schillergaesschen 2
D 07745 Jena
Germany
☎ 49 364 163 0323
℻ 49 364 163 0417

Doschek George A
NRL
Code 4170
4555 Overlook Av SW
Washington DC 20375 5000
USA
☎ 1 202 767 6473

Dossin F
Inst Astrophysique
Universite Liege
r du Travail
B 4102 Seraing
Belgium
☎ 32 4 337 3584

Dotani Tadayasu
ISAS
3 1 1 Yoshinodai
Sagamihara
Kanagawa 229 8510
Japan
☎ 81 427 51 3911
℻ 81 427 59 4253
✉ dotani@astro.isas.ac.jp

Dottori Horacio A
Instituto Fisica
UFRGS
CP 15051
91501 900 Porto Alegre RS
Brazil
☎ 55 512 36 4677

Doubinskij Boris A
SC Radio Astronomy
Acad Sciences
Marx Av 18
Moscow Centre GSP 3
Russia
☎ 7 095 202 8286
℻ 7 095 203 8414
✉ astronom@ire.rc.ac.ru

Dougherty Sean
Astrophysics Group
Liverpool J M Univ
Byrom St
Liverpool L3 3AF
UK
☎ 44 151 231 4103
℻ 44 151 231 2337
✉ smd@staru1.livjm.ac.uk

Doughty Noel A
Dpt Phys/Astronomy
Univ of Canterbury
Private Bag 4800
Christchurch 1
New Zealand
☎ 64 3 366 7001
℻ 64 3 364 2469

Douglas James N
Astronomy Dpt
Univ of Texas
Rlm 15 308
Austin TX 78712 1083
USA
☎ 1 512 471 4461
℻ 1 512 471 6016

Douglas Nigel
Kapteyn Sterrekundig Inst
Univ Groningen
Postbus 800
NL 9700 AV Groningen
Netherlands
☎ 31 50 363 4088
℻ 31 50 363 6100
✉ ndouglas@astro.rug.nl

Douglas R J
Physics Division M 36
NRCC
100 Sussex Dr
Ottawa ON K1A 0S1
Canada
☎ 1 613 993 6060
℻ 1 613 952 6602
✉ rob.douglas@nrc.ca

Douglass Geoffrey G
USNO
Astrometry Dpt
3450 Massachusetts Av NW
Washington DC 20392 5420
USA
☎ 1 202 653 1438
℻ 1 202 653 0944
✉ ggd@cruz.usno.navy.mil

Dourneau Gerard
Observ Bordeaux
BP 89
F 33270 Floirac
France
☎ 33 5 56 86 4330
℗ 33 5 56 40 4251

Doval Jorge Perez
Inst Geophys/Astron
C 212 N 2906/29 Y 31
Rpto Coromela/la Lisa
La Habana
Cuba
☎ 53 21 8435/0644

Downes Ann Juliet B
MRAO
Cavendish Laboratory
Madingley Rd
Cambridge CB3 0HE
UK
☎ 44 12 23 337 274
℗ 44 12 23 354 599

Downes Dennis
IRAM
300 r La Piscine
F 38406 S Martin Heres Cdx
France
☎ 33 4 76 82 4900
℗ 33 4 76 51 5938
✉ downes@iram.grenet.fr

Downes Ronald A
STScI/USB
Homewood Campus
3700 San Martin Dr
Baltimore MD 21218
USA
☎ 1 301 338 4700
℗ 1 301 338 4767
✉ downes@stsci.edu

Downs George S
MIT Lincoln Lab
MS S4 241
Box 73
Lexington MA 02173
USA
☎ 1 617 863 5500
✉ downs@ll.mit.edu

Doyle John Gerard
Armagh Observ
College Hill
Armagh BT61 9DG
UK
☎ 44 18 61 522 928
℗ 44 18 61 527174

Doyle Laurance R
NASA/ARC
SETI
MS 245 7
Moffett Field CA 94035 1000
USA
☎ 1 415 604 3372
℗ 1 650 604 1088

Doyon Rene
Dpt d Physique
Universite Montreal
CP 6128 Succ A
Montreal QC H3C 3J7
Canada
☎ 1 514 343 6111*3204
℗ 1 514 343 2071
✉ doyon@astro.umontreal.ca

Draine Bruce T
Princeton Univ Obs
Peyton Hall
Princeton NJ 08544 1001
USA
☎ 1 609 452 3574
℗ 1 609 258 1020

Drake Frank D
Lick Observ
Univ of California
Santa Cruz CA 95064
USA
☎ 1 831 459 4885
℗ 1 831 426 3115
✉ fdrake@ucolick.org

Drake Jeremy
CfA
HCO/SAO
60 Garden St
Cambridge MA 02138 1516
USA
☎ 1 617 496 7850
℗ 1 617 495 7356
✉ jdrake@cfa.harvard.edu

Drake Stephen A
NASA GSFC
Code 668
Greenbelt MD 20771
USA
☎ 1 301 286 6962

Dramba C
Astronomical Institute
Romanian Acad Sciences
Cutitul de Argint 5
RO 75212 Bucharest
Rumania
☎ 40 1 641 3686
℗ 40 1 312 3391

Drapatz Siegfried W
MPE
Postfach 1603
D 85740 Garching
Germany
☎ 49 893 299 880
℗ 49 893 299 3569

Dravins Dainis
Lund Observ
Box 43
S 221 00 Lund
Sweden
☎ 46 46 222 7297
℗ 46 46 222 4614
✉ dainis@astro.lu.se

Dravskikh Alexander F
Pulkovo Observ
Acad Sciences
10 Kutuzov Quay
196140 St Petersburg
Russia
☎ 7 812 297 9452
℗ 7 812 315 1701

Drechsel Horst
Dr Remeis Sternwarte
Univ Erlangen-Nuernberg
Sternwartstr 7
D 96049 Bamberg
Germany
☎ 49 951 952 2215
℗ 49 951 952 2222
✉ drechsel@sternwarte.uni-erlangen.de

Dreher John W
Physics Dpt
MIT Rm 26 315
Box 165
Cambridge MA 02139 4307
USA
☎ 1 617 253 8519
℗ 1 617 253 9798

Dressel Linda L
Applied Res Corp
8201 Corporate Dr
Suite 920
Landover MD 20785
USA
☎ 1 301 459 8442
℗ 1 301 459 0761

Dressler Alan
Carnegie Observatories
813 Santa Barbara St
Pasadena CA 91101 1292
USA
☎ 1 818 577 1122
℗ 1 818 795 8136

Dressler Kurt
Lab Physik Chemie
ETH Zentrum
CH 8092 Zuerich
Switzerland
☎ 41 1 632 4441
℗ 41 1 632 1205

Drever Ronald W P
CALTECH
MS 130 33
Pasadena CA 91125
USA
☎ 1 818 395 4597
℗ 1 818 795 1547

Drew Janet
Astrophysics Gr
Imperial Coll
Prince Consort Rd
London SW7 2BZ
UK
☎ 44 171 594 7771
℗ 44 171 594 7772

Drilling John S
Dpt Phys/Astronomy
Louisiana State Univ
Baton Rouge LA 70803 4001
USA
☎ 1 504 388 6795
℗ 1 504 388 5855
✉ drilling@rouge.phys.lsu.edu

Drinkwater Michael J
School of Physics
UNSW
Sydney NSW 2052
Australia
☎ 61 2 9385 5572
℗ 61 2 9385 6060
✉ mjd@roen.phys.unsw.edu.au

Drissen Laurent
Dpt d Physique
Universite Laval
Quebec G1K 7P4
Canada
☎ 1 418 656 2131*5641
℗ 1 418 656 2040
✉ ldrissen@phy.ulaval.ca

Drossart Pierre
Observ Paris Meudon
DESPA
Pl J Janssen
F 92195 Meudon PPL Cdx
France
☎ 33 1 45 07 7664
℗ 33 1 45 07 2806
✉ drossart@obspm.fr

Drozyner Andrzej
Inst of Astronomy
N Copernicus Univ
Ul Chopina 12/18
PL 87 100 Torun
Poland
☎ 48 56 26 018
℗ 48 56 24 602

Drury Luke O'Connor
DIAS
School Cosmic Phys
5 Merrion Sq
Dublin 2
Ireland
☎ 353 1 662 1333
℗ 353 1 662 1477
✉ ld@cp.dias.ie

Dryer Murray
NOAA ERL R/E/SE
Space Environment Lab
325 Broadway
Boulder CO 80303
USA
☎ 1 303 497 3978
℗ 1 303 798 2124
✉ mdryer@sec.noaa.gov
murraydryer@msn.com

Dubath Pierre
INTEGRAL
Science Data Ctr
16 Chemin Ecogia
CH 1290 Versoix
Switzerland
☎ 41 22 950 9124
℗ 41 22 950 9133
✉ pierre.dubath@obs.unige.ch

Dubau Jacques
Observ Paris Meudon
DARC
Pl J Janssen
F 92195 Meudon PPL Cdx
France
☎ 33 1 45 07 7456
℗ 33 1 45 07 7971

Dubin Maurice
NASA GSFC
Code 616
Greenbelt MD 20771
USA
☎ 1 301 286 5475
Fax 1 301 286 2630

Dubner Gloria
IAR
CC 5
1894 Villa Elisa (Bs As)
Argentina
☎ 54 21 4 3793
Fax 54 21 211 761

Dubois Marc A
DAPNIA/SAP
CEA Saclay
BP 2
F 91191 Gif s Yvette Cdx
France
☎ 33 1 69 08 7418
Fax 33 1 69 08 8786
Email mad@spec.saclay.cea.fr

Dubois Pascal
Observ Strasbourg
11 r Universite
F 67000 Strasbourg
France
☎ 33 3 88 15 0734
Fax 33 3 88 25 0160
Email dubois@simbad.u-strasbg.fr

Dubout Renee
Observ Marseille
2 Pl Le Verrier
F 13248 Marseille Cdx 04
France
☎ 33 4 91 95 9088
Fax 33 4 91 62 1190
Email dubout@observatoire.cnrs-mrs.fr

Ducati Jorge Ricardo
Instituto Fisica
UFRGS
CP 15051
91501 900 Porto Alegre RS
Brazil
☎ 55 512 36 4677
Fax 55 513 36 1762
Email ducati@if.1.ufrgs-anrs.br

Duchesne Maurice
Observ Paris
61 Av Observatoire
F 75014 Paris
France
☎ 33 1 43 20 1210
Fax 33 1 44 54 1804

Ducourant Christine
Observ Bordeaux
BP 89
F 33270 Floirac
France
☎ 33 5 56 86 4330
Fax 33 5 56 40 4251
Email ducourant@observ.u-bordeaux.fr

Duemmler Rudolf
Astronomy Dpt
Univ of Oulu
Box 333
FIN 90570 Oulu
Finland
☎ 358 81 553 1932
Fax 358 81 553 1984
Email rolf@hiisi.oulu.fi

Duerbeck Hilmar W
Hembrich 16
D 54552 Schalkenmehren
Germany

Duerst Johannes
Langwies
CH 8824 Schoenenberg
Switzerland
☎ 41 1 788 1785

Dufay Maurice
Universite Lyon 1
Campus La Doua
43 bd 11 novembre
F 69621 Villeurbanne
France
☎ 33 4 78 89 8124

Duffett-Smith Peter James
MRAO
Cavendish Laboratory
Madingley Rd
Cambridge CB3 0HE
UK
☎ 44 12 23 337 274
Fax 44 12 23 354 599
Email pjds@mrao.cam.ac.uk

Duflot Marcelle
Observ Marseille
2 Pl Le Verrier
F 13248 Marseille Cdx 04
France
☎ 33 4 91 95 9088
Fax 33 4 91 62 1190
Email duflot@obsmara.cnrs-mrs.fr

Dufour Reginald James
Dpt Phys/Astronomy
Rice Univ
Box 1892
Houston TX 77251 1982
USA
☎ 1 713 527 8101
Fax 1 713 285 5143

Dufton Philip L
Dpt Pure/Appl Physics
Queen's Univ
Belfast BT7 1NN
UK
☎ 44 12 32 24 5133
Fax 44 12 32 43 8918

Duldig Marcus Leslie
Austral Antarctic Div
Univ Tasmania
GPO Box 252 21
Hobart TAS 7001
Australia
☎ 61 3 6226 2022
Fax 61 3 6223 3057
Email marc.duldig@phys.utas.edu.au

Duley Walter W
Physics Dpt
Univ Waterloo
Waterloo ON N2L 3G1
Canada
☎ 1 519 885 1211*3108
Fax 1 519 746 8115
Email duley@physics.watstar.uwaterloo.ca

Dulk George A
Observ Paris Meudon
DESPA
Pl J Janssen
F 92195 Meudon PPL Cdx
France
☎ 33 1 45 34 7691
Fax 33 1 45 34 7691

Dultzin-Hacyan D
Instituto Astronomia
UNAM
Apt 70 264
Mexico DF 04510
Mexico
☎ 52 5 622 3906
Fax 52 5 616 0653
Email deborah@astroscu.unam.mx

Duma Dmitrij P
Main Astronomical Obs
Ukrainian Acad Science
Golosiiv
252650 Kyiv 22
Ukraine
☎ 380 4426 63110
Fax 380 4426 62147
Email duma@mao.kiev.ua

Dumitrache Christana
Astronomical Institute
Romanian Acad Sciences
Cutitul de Argint 5
RO 75212 Bucharest
Rumania
☎ 40 1 641 3686
Fax 40 1 312 3391
Email crisd@imar.ro

Dumitrescu Alexandru
Astronomical Institute
Romanian Acad Sciences
Cutitul de Argint 5
RO 75212 Bucharest
Rumania
☎ 40 1 641 3686
Fax 40 1 312 3391
Email adumitrescu@imar.ro

Dumont Rene
Observ Bordeaux
BP 89
F 33270 Floirac
France
☎ 33 5 56 86 4330
Fax 33 5 56 40 4251

Dumont Simone
IAP
98bis bd Arago
F 75014 Paris
France
☎ 33 1 44 32 8000
Fax 33 1 44 32 8001

Duncan Douglas Kevin
Astronomy/Astrophys Ctr
Univ of Chicago
5640 S Ellis Av
Chicago IL 60637
USA
☎ 1 312 702 8207
Fax 1 312 702 8212
Email duncan@oddjob.uchicago.edu

Duncan Robert A
CSIRO
Div Radiophysics
Box 76
Epping NSW 2121
Australia
☎ 61 2 868 0222
Fax 61 2 868 0310
Email rduncan@atnf.csiro.au

Duncombe Raynor L
Aerospace Eng Dpt
Univ of Texas
WRW 414
Austin TX 78712
USA
☎ 1 512 471 4250
Email duncombe@utcsr.ae.utexas.edu

Dunham David W
APL
JHU
Johns Hopkins Rd
Laurel MD 20723 6099
USA
☎ 1 301 953 5000
Fax 1 301 953 1093
Email Daivd.Dunham@jhuapl.edu

Dunkelman Lawrence
618 E Camino Ateza
Tucson AZ 85704 5804
USA
Fax 1 520 293 4588
Email dunkL@aol.com

Dunlop James
Astrophysics Group
Liverpool J M Univ
Byrom St
Liverpool L3 3AF
UK
☎ 44 151 231 2039
Fax 44 151 231 2337
Email jsd@uk.ac.livjm.starul

Dunlop Storm
140 Stocks Lane
East Wittering Chichester
West Sussex PO20 8NT
UK
☎ 44 124 36 70354
Fax 44 124 367 0400
Email sdunlop@star.maps.susx.ac.uk

Dunn Richard B
AFGL
NSO
Sunspot NM 88349
USA
☎ 1 505 434 1390
Fax 1 504 434 7029

Dunsby Peter
Applied Maths Dpt
Univ of Cape Town
Private Bag
7700 Rondebosch
South Africa
☎ 27 21 650 3209
℻ 27 21 650 2334
✉ peter@shiva.mth.uct.ac.za

Duorah Hira Lal
Physics Dpt
Univ of Gauhati
Guwahati 781014
India
☎ 91 361 88 531
℻ 91 361 570 133

Dupree Andrea K
CfA
HCO/SAO
60 Garden St
Cambridge MA 02138 1516
USA
☎ 1 617 495 7489
℻ 1 617 495 7049
✉ dupree@cfa.harvard.edu

Dupuy David L
Physics Dpt
Virginia Military Inst
Lexington VA 24450
USA
☎ 1 703 464 7504

Duriez Luc
Laboratoire Astronomie
Universite Lille
1 Impasse Observatoire
F 59000 Lille
France
☎ 33 3 20 52 4424
℻ 33 3 20 58 0328

Durisen Richard H
Astronomy Dpt
Indiana Univ
Swain W 319
Bloomington IN 47405
USA
☎ 1 812 335 6921
℻ 1 812 855 8725

Durney Bernard
NOAO/NSO
Box 26732
950 N Cherry Av
Tucson AZ 85726 6732
USA
☎ 1 520 327 5511
℻ 1 520 325 9360

Durouchoux Philippe
DAPNIA/SAP
CEA Saclay
BP 2
F 91191 Gif s Yvette Cdx
France
☎ 33 1 69 08 3376
℻ 33 1 69 08 6577
✉ durouchoux@sapvxg.saclay.cea.fr

Durrance Samuel T
Dpt Phys/Astronomy
JHU
Charles/34th St
Baltimore MD 21218
USA
☎ 1 301 338 8707
℻ 1 301 546 7279

Durrant Christopher J
Applied Maths Dpt
Univ of Sydney
Sydney NSW 2006
Australia
☎ 61 2 692 3373
℻ 61 2 660 2903

Durret Florence
IAP
98bis bd Arago
F 75014 Paris
France
☎ 33 1 44 32 8093
℻ 33 1 44 32 8001

Duschl Wolfgang J
Inst Theor Astrophysik
d Univer
Tiergartenstr 15
D 69121 Heidelberg
Germany
☎ 49 622 154 8967
℻ 49 622 154 4221
✉ wjd@ita.uni-heidelberg.de

Duthie Joseph G
Dpt Phys/Astronomy
Univ of Rochester
Rochester NY 14627 0171
USA
☎ 1 716 275 8527
℻ 1 716 275 4351

Duval Marie-France
Observ Marseille
2 Pl Le Verrier
F 13248 Marseille Cdx 04
France
☎ 33 4 91 95 9088
℻ 33 4 91 62 1190

Duvall Thomas L
Stanford Univ
HEPL A209
Stanford CA 94305 4085
USA
✉ duvall@quake.stanford.edu

Duvert Gilles
Observ Grenoble
Lab Astrophysique
BP 53x
F 38041 S Martin Heres Cdx
France
☎ 33 4 76 51 4885
℻ 33 4 76 44 8821
✉ Gilles.Duvert@laog.obs.ujf-grenoble.fr

Dvorak Rudolf
Inst Astronomie
Univ Wien
Tuerkenschanzstr 17
A 1180 Wien
Austria
☎ 43 1 345 3600
℻ 43 1 470 6015

Dwarakanath K S
RRI
Sadashivanagar
CV Raman Av
Bangalore 560 080
India
☎ 91 80 334 0122
℻ 91 80 334 0492

Dwek Eli
NASA GSFC
Code 697
Greenbelt MD 20771
USA
☎ 1 301 286 6209

Dwivedi Bhola Nath
Applied Phys Dpt
Univ Banaras Hindu
Varanasi 221 005
India
℻ 91 542 312 059

Dworetsky Michael M
Dpt Phys/Astronomy
UCLO
Gower St
London WC1E 6BT
UK
☎ 44 171 387 7050
℻ 44 171 380 7145

Dyck Melvin
Dpt Phys/Astronomy
Univ of Wyoming
Box 3905
Laramie WY 82071
USA
☎ 1 307 766 6150
℻ 1 307 766 2652
✉ meldyck@corral.uwyo.edu

Dyer Charles Chester
Scarborough Coll
Univ of Toronto
Phys Scs Gr Rm S 650
Toronto ON M1C 1A4
Canada
☎ 1 416 284 3318

Dyer Edward R
3626 Davis St Nw
Washington DC 20007
USA

Dyson Freeman J
Inst Advanced Study
School Natural Science
Princeton NJ 08540
USA
☎ 1 609 734 8055

Dyson John E
Astronomy Dpt
Univ Manchester
Oxford Rd
Manchester M13 9PL
UK
☎ 44 161 275 4235
℻ 44 161 275 4223
✉ ed@ast.leeds.ac.uk

Dzhapiashvili Victor P
Abastumani Astrophysical Observ
Georgian Acad Sci
383762 Abastumani
Georgia
☎ 995 88 32 95 5367
℻ 995 88 32 98 5017

Dziembowski Wojciech
Copernicus Astron Ctr
Polish Acad Sci
Ul Bartycka 18
PL 00 716 Warsaw
Poland
☎ 48 22 41 1086
℻ 48 22 41 0828

Dzifcakova Elena
Dpt Astron/Astrophys
Comenius Univ
Mlynska Dolina 1
SK 842 15 Bratislava
Slovak Republic
☎ 421 7 72 400
℻ 421 7 725 882
✉ dzifcakova@fmph.uniba.sk

Dzigvashvili R M
Abastumani Astrophysical Observ
Georgian Acad Sci
383762 Abastumani
Georgia
☎ 995 88 32 95 5367
℻ 995 88 32 98 5017

Dzubenko Nikolai
Astronomy Dpt
Univ of Kyiv
Acad Glushkov Str 6
252022 Kyiv
Ukraine
☎ 380 4426 64457
℻ 380 4426 64507
✉ kotsarenko@univ.kiev.ua

E

Eaton Joel A
Ctr Excellence Inform Sys
Tennessee State Univ
330 10th Av North
Nashville TN 37203 3401
USA
☎ 1 615 963 7302
✆ 1 615 963 7027

Ebisuzaki Toshikazu
Komaba Meguro-ku
Tokyo 153
Japan
☎ 81 334 67 1171*665
✆ 81 334 65 3925

Eccles Michael J
Ballencrieff Toll
Sunnyside
Bathgate EH48 4LD
UK
☎ 44 15 06 53 989

Echeverria Roman Juan M
Observ Astronomico Nacional
UNAM
Apt 877
Ensenada BC 22800
Mexico
☎ 52 6 174 4580
✆ 52 6 174 4607

Edalati Sharbaf Mohammad Taghi
Physics Dpt
School of Sciences
Univ of Ferdowsi
Mashhad
Iran
☎ 98 51 832 021/4
✆ 98 51 838 032

Eddy John A
3460 Ash Av
Boulder CO 80303
USA
☎ 1 303 497 1680

Edelman Colette
BDL
77 Av Denfert Rochereau
F 75014 Paris
France
☎ 33 1 40 51 2272
✆ 33 1 46 33 2834

Edelson Rick
CASA
Univ of Colorado
Campus Box 389
Boulder CO 80309 0391
USA
☎ 1 303 492 6784
✆ 1 303 492 5941

Edge Alastair
Inst of Astronomy
The Observatories
Madingley Rd
Cambridge CB3 0HA
UK
☎ 44 12 23 330 803
✆ 44 12 23 337 523
✉ ace@mail.ast.cam.ac.uk

Edmondson Frank K
Astronomy Dpt
Indiana Univ
Swain W 319
Bloomington IN 47405
USA
☎ 1 812 335 6918
✆ 1 812 855 8725

Edmunds Michael Geoffrey
Physics Dpt
Univ Wales College
Box 913
Cardiff CF1 3TH
UK
☎ 44 12 22 874 785
✆ 44 12 22 371 921

Edvardsson Bengt
Astronomical Observ
Box 515
S 751 20 Uppsala
Sweden
☎ 46 18 51 2488
✆ 46 18 52 7583
✉ bengt.edvardsson@astro.uu.se

Edwards Alan Ch
CALTECH/JPL
MS 238 600
4800 Oak Grove Dr
Pasadena CA 91109 8099
USA
☎ 1 818 354 4321
✆ 1 818 393 4965
✉ chad@logos.jpl.nasa.gov

Edwards Paul J
MSSSO
Weston Creek
Private Bag
Canberra ACT 2611
Australia
☎ 61 262 881 111
✆ 61 262 490 233

Edwards Terry W
Dpt Phys/Astronomy
Univ of Missouri
Columbia MO 65211
USA
☎ 1 314 882 3036

Edwin Patricia
The Open Univ in Scottland
10 Drumsheugh Gardens
Edinburgh EH3 70J
UK
☎ 44 133 476 161
✆ 44 13 346 3748
✉ p.m.edwin@st-andrews.ac.uk

Edwin Roger P
Dpt Phys/Astronomy
Univ of St Andrews
North Haugh
St Andrews Fife KY16 9SS
UK
☎ 44 1334 76161
✆ 44 1334 74487

Eelsalu Heino
Tartu Observ
Estonian Acad Sci
Toravere
EE 2444 Tartumaa
Estonia
☎ 372 1 434 281 63
✆ 372 7 410 205
✉ eelsalu@aai.ee

Eenens Philippe
Dpt Astronomia
Univ Guanajuato
Apdo 144
Guanajuato GTO 36000
Mexico
☎ 52 4 732 9548
✆ 52 4 732 0253
✉ eenens@carina.astro.ugto.mx

Efimov Yuri
Crimean Astrophys Obs
Ukrainian Acad Science
Nauchny
334413 Crimea
Ukraine
☎ 380 6554 40704
✆ 380 6554 40704
✉ etimov@crao.crimea.ua

Efremov Yuri I
SAI
Acad Sciences
Universitetskij Pr 13
119899 Moscow
Russia
☎ 7 095 939 1622
✆ 7 095 932 8841
✉ efremov@sai.msu.su

Efremov Yuri N
Keldysh Inst Appl Maths
Acad Sciences
Miusskaja Sq 4
125047Moscow
Russia
☎ 7 095 932 8841
✆ 7 095 939 1622
✉ efremov@sai.msu.su

Efstathiou George
Inst of Astronomy
The Observatories
Madingley Rd
Cambridge CB3 0HA
UK
☎ 44 12 23 337 548
✆ 44 12 23 337 523

Eggleton Peter P
Inst of Astronomy
The Observatories
Madingley Rd
Cambridge CB3 0HA
UK
☎ 44 12 23 337 548
✆ 44 12 23 337 523

Egret Daniel
Observ Strasbourg
11 r Universite
F 67000 Strasbourg
France
☎ 33 3 88 15 0711
✆ 33 3 88 25 0160
✉ egret@cdxba.u-strasbg.fr

Ehgamberdiev Shurat
Astronomical Institute
Uzbek Acad Sci
Astronomicheskaya Ul 33
700052 Tashkent
Uzbekistan
☎ 7 3712 35 8102
✆ 7 37 12 32 7789
✉ shuhrat@kumbel.silk.glas.apc.org

Ehlers Jurgen
MPA
Karl Schwarzschildstr 1
D 85748 Garching
Germany
☎ 49 893 299 9444
✆ 49 893 299 3235

Eichhorn Heinrich K
Box 112055
Gainesville FL 32611 2055
USA
☎ 1 904 392 7745
✉ eichhorn@astro.ufl.edu

Eichler David
Astronomy Magazine
21027 Crossroads Circle
Waukesha WI 53187
USA

Eidelsberg Michele
Observ Paris Meudon
DAMAP
Pl J Janssen
F 92195 Meudon PPL Cdx
France
☎ 33 1 45 07 7562
📠 33 1 45 07 7469

Eilek Jean
Physics Dpt
New Mexico Tech
Campus Station
Socorro NM 87801
USA
☎ 1 505 835 5238
📠 1 505 835 5707
✉ jeilek@nraso.edu

Einasto Jaan
Tartu Observ
Estonian Acad Sci
Toravere
EE 2444 Tartumaa
Estonia
☎ 372 7 410 151
📠 372 7 410 205
✉ maret@jupiter.aai.ee

Einasto Maret
Tartu Observ
Estonian Acad Sci
Toravere
EE 2444 Tartumaa
Estonia
☎ 372 3 410 450
📠 372 7 410 205

Einaudi Giorgio
Dpt Astronomia
Univ d Firenze
Largo E Fermi 5
I 50125 Firenze
Italy
☎ 39 55 27 521
📠 39 55 22 0039

Einicke Ole H
Astronomical Observ, NBIfAFG
Copenhagen Univ
Juliane Maries Vej 30
DK 2100 Copenhagen
Denmark
☎ 45 35 32 5977
📠 45 35 32 5989
✉ ohe@astro.ku.dk

Eker Zeki
KACST
Box 2455
Riyadh 11451
Saudi Arabia
☎ 966 1 4676 315
✉ f40a010@saksu00

Ekers Ronald D
CSIRO
Div Radiophysics
Box 76
Epping NSW 2121
Australia
☎ 61 2 9372 4300
📠 61 2 9372 4310
✉ rekers@atnf.csiro.au

El Bakkali Larbi
Fac de Sciences
Box 2121
Tetouan
Morocco
☎ 212 974509
📠 212 971763

El Basuny Ahmed Alawy
NAIGR
Helwan Observ
Cairo 11421
Egypt
☎ 20 78 0645/2683
📠 20 62 21 405 297

El Baz Farouk
ITEK Corp
10 Maguire Rd
Lexington MA 02173
USA
☎ 1 617 276 2532

El Eid Mounib
Schillerstr 67
D 37083 Goettingen
Germany
☎ 49 551 395 042
📠 49 551 395 043

El Nawaway Mohamed Saleh
ASRT
101 Kasr El-eini Str
Cairo 11516
Egypt
☎ 20 2 355 7972

El Raey Mohamed E
Dpt Environmt Studies
Inst Graduate Stud/Res
Univ of Alexandria
Alexandria
Egypt

El Shahawy Mohamad
Astronomy Dpt
Fac of Sciences
Cairo Univ
Geza
Egypt
☎ 20 2 572 7022
📠 20 2 572 7556

El-Sharawy Mohamed Bahgat
NAIGR
Helwan Observ
Cairo 11421
Egypt
☎ 20 78 0645/2683
📠 20 62 21 405 297

Elford William Graham
Physics Dpt
Univ of Adelaide
Box 498
Adelaide SA 5001
Australia
☎ 61 8 228 5321

Elgaroy Oystein
Inst Theor Astrophys
Univ of Oslo
Box 1029
N 0315 Blindern Oslo 3
Norway
☎ 47 22 856 504
📠 47 22 856 505

Elias Nicholas
USNO
Flagstaff Station
Box 1149
Flagstaff AZ 86002 1149
USA
✉ nme@sextans.
lowell.edu

Elipe Sanchez Antonio
Gr Mecan Espacial
Univ Zaragoza
E 50009 Zaragoza
Spain
☎ 34 7 676 1000
📠 34 7 676 1140
✉ elipe@msf.unizar.es

Elitzur Moshe
Dpt Phys/Astronomy
Univ Kentucky
Lexington KY 40506 0055
USA
☎ 1 606 257 4720
📠 1 606 323 2846

Ellder Joel
Onsala Space Observ
Chalmers Technical Univ
S 439 92 Onsala
Sweden
☎ 46 31 772 5500
📠 46 31 772 5550

Elliot James L
Dpt Earth/Planet Sci
MIT Rm 54 422a
Box 165
Cambridge MA 02139
USA
☎ 1 617 253 6308

Elliott Ian
Dunsink Observ
DIAS
Castleknock
Dublin 15
Ireland
☎ 353 1 38 7959
📠 353 1 38 7090

Elliott Kenneth H
Dpt Space Res
Univ Birmingham
Box 363
Birmingham B15 2TT
UK
☎ 44 12 14 72 1301

Ellis G R A
Physics Dpt
Univ Tasmania
GPO Box 252c
Hobart TAS 7001
Australia
☎ 61 2 202 411
📠 61 2 202 410

Ellis George F R
Applied Maths Dpt
Univ of Cape Town
Private Bag
7700 Rondebosch
South Africa
☎ 27 21 650 2332/2340
📠 27 21 650 2334

Ellis Richard S
Physics Dpt
Univ of Durham
South Rd
Durham DH1 3LE
UK
☎ 44 191 374 2000
📠 44 191 374 3749

Elmegreen Bruce Gordon
IBM
T J Watson Res Ctr
Box 218
Yorktown Heights NY 10598
USA
☎ 1 914 945 2448
📠 1 914 945 2141
✉ bge@yktvmt

Elmegreen Debra Meloy
Dpt Phys/Astronomy
Vassar College
Poughkeepsie
Poughkeepsie NY 12601
USA
☎ 1 914 437 7356
✉ elmegreen@vaxsar.
vassar.edu

Elsaesser Hans
MPI Astronomie
Koenigstuhl
D 69117 Heidelberg
Germany
☎ 49 622 152 8200
📠 49 622 152 8246

Elsmore Bruce
MRAO
Cavendish Laboratory
Madingley Rd
Cambridge CB3 0HE
UK
☎ 44 12 23 337 294
📠 44 12 23 354 599

Elson Rebecca Anne Wood
Inst of Astronomy
The Observatories
Madingley Rd
Cambridge CB3 0HA
UK
☎ 44 12 23 337 541
📠 44 12 23 337 523
✉ elson@mail.ast.cam.ac.uk

Elst Eric Walter
Koninklijke Sterrenwacht
van Belgie
Ringlaan 3
B 1180 Brussels
Belgium
☎ 32 2 373 0309
📠 32 2 374 9822
✉ eric.elst@oma.be

Elste Gunther H
Astronomy Dpt
Univ of Michigan
Dennison Bldg
Ann Arbor MI 48109 1090
USA
☎ 1 313 764 3444
✆ 1 313 764 2211

Elston Wolfgang E
Dpt Geology
Univ New Mexico
800 Yale Blvd Ne
Albuquerque NM 87131
USA
☎ 1 505 277 5339

Elsworth Yvonne
School Phys/Astron
Univ Birmingham
Box 363
Birmingham B15 2TT
UK
☎ 44 12 14 14 6453
✆ 44 12 14 143 722
✉ ype@star.sr.bham.ac.uk

Elvis Martin S
CfA
HCO/SAO
60 Garden St
Cambridge MA 02138 1516
USA
☎ 1 617 495 7442
✆ 1 617 495 7356
✉ elvis@cfa.harvard.edu

Elvius Aina M
Astronomical Observ
Box 515
S 752 29 Uppsala
Sweden
☎ 46 18 50 0857
✆ 46 18 52 7583

Elwert Gerhard
Lehrstuhl Theoret Astro
Physik Univ Tuebingen
Auf d Morgenstelle 12 C
D 72076 Tuebingen
Germany
☎ 49 707 129 6483

Emanuele Alessandro
IAS
Area d Ricerca CNR
Via Fosso Cavaliere 100
I 00133 Roma
Italy
☎ 39 6 4993 4473
✆ 39 6 2066 0188
✉ capaneo@saturn.rm.fra.cnr.it

Emelianov Nikolaj V
SAI
Acad Sciences
Universitetskij Pr 13
119899 Moscow
Russia
☎ 7 095 939 3764
✆ 7 095 939 0126
✉ emelia@sai.msu.su

Emerson Darrel Trevor
NRAO
Campus Bg 65
949 N Cherry Av
Tucson AZ 85721 0655
USA
☎ 1 520 882 8250*117
✆ 1 520 882 7955
✉ demerson@nrao.edu

Emerson James P
Astrophysics Group
QMWC
Mile End Rd
London E1 4NS
UK
☎ 44 171 975 5040
✆ 44 181 975 5500
✉ j.p.emerson@qmw.ac.uk

Eminzade T A
Shemakha Astrophysical
Observ
Azerbajan Acad Sci
373243 Shemakha
Azerbajan
☎ 994 89 22 39 8248

Emslie A Gordon
Physics Dpt
Univ of Alabama
Huntsville AL 35899
USA
☎ 1 205 895 6167
✆ 1 205 895 6790

Enard Daniel
ESO
Karl Schwarzschildstr 2
D 85748 Garching
Germany
☎ 49 893 200 6251
✆ 49 893 202 362

Encrenaz Pierre J
DEMIRM
ENS
24 r Lhomond
F 75231 Paris Cdx 05
France
☎ 33 1 43 29 1235
✆ 33 1 40 51 2002

Encrenaz Therese
Observ Paris Meudon
DESPA
Pl J Janssen
F 92195 Meudon PPL Cdx
France
☎ 33 1 45 07 7691
✆ 33 1 45 07 7971

Endal Andrew S
Applied Res Corp
8201 Corporate Dr
Suite 920
Landover MD 20785
USA
☎ 1 301 459 8442
✆ 1 301 459 3761

Engelbrecht Christian
Univ Stellenbosc
0001 Pretoria
South Africa

Engels Dieter
Hamburger Sternwarte
Univ Hamburg
Gojensbergsweg 112
D 21029 Hamburg
Germany
☎ 49 407 252 4136
✆ 49 407 252 4198
✉ dengels@hs.uni-hamburg.de

Engin Semanur
Astronomy Dpt
Univ of Ankara
Fen Fakultesi
06100 Besevler
Turkey
☎ 90 41 216 6720
✆ 90 312 223 2395

Enginol Turan B
Inst Graduate Studies
In Science/Eng
Bogazici Univ
34452 Istanbul
Turkey
☎ 90 1 163 1500
✆ 90 1 165 8480
✉ enginol@f.tr.boun

English Jayanne
Physics Dpt
Queen's Univ
Stirling Hall
Kingston ONT K7L 3N6
Canada
☎ 1 613 545 6000*4781
✆ 1 613 545 6463
✉ english@astro.queensu.ca

Engvold Oddbjoern
Inst Theor Astrophys
Univ of Oslo
Box 1029
N 0315 Blindern Oslo 3
Norway
☎ 47 22 856 521
✆ 47 22 856 505
✉ oddbjorn.engvold@astro.uio.no

Enome Shinzo
NAOJ
VSOP
Osawa 2- 21-2
Mitaka 181
Japan
☎ 81 422 34 3914
✆ 81 442 34 3869
✉ enome@hotaka.mtk.nao.ac.jp

Enslin Heinz
Alstedderstr 180
D 44534 Luenen
Germany

Epishev Vitali P
Lab Cosmic Investigat
Univ
Dalyokaya Str 2a
294000 Uzhgorod
Ukraine
☎ 380 3122 36065
✆ 380 3122 36136

Epps Harland Warren
Dpt Astronomy/Phys
UCLA
Box 951562
Los Angeles CA 90025 1562
USA
☎ 1 310 825 3025
✆ 1 310 206 2096

Epstein Eugene E
LANL
MS D 436
Box 1663
Los Alamos NM 87545 2345
USA
☎ 1 505 667 5127
✆ 1 505 665 4414
✉ epstein@lanl.gov

Epstein Gabriel Leo
NASA GSFC
Code 682
Greenbelt MD 20771
USA
☎ 1 301 286 0722
✆ 1 301 286 1617

Epstein Richard I
LANL
MS 436
Box 1663
Los Alamos NM 87545 2345
USA
☎ 1 505 667 9595
✆ 1 505 665 4055

Ercan E Nihal
Kandilli Observ
Bogazici Univ
Cengelkoy
81220 Istanbul
Turkey
☎ 90 1 332 0240/41
✆ 90 1 265 7131

Erdi B
Astronomy Dpt
Eotvos Univ
Ludovika ter 2
H 1083 Budapest
Hungary
☎ 36 1 114 1019
✆ 36 1 210 1089

Ergma E V
Dpt Th Phys/Astrophys
Tartu Univ
Uelikooli 18
EE 2400 Tartumaa
Estonia
☎ 372 73 775

Erickson William C
Physics Dpt
Univ Tasmania
GPO Box 252c
Hobart TAS 7001
Australia
☎ 61 2 202 401
✆ 61 2 202 186

Eriguchi Yoshiharu
Dpt Earth Sci/Astron
Univ of Tokyo
Meguro Ku
Tokyo 153
Japan
☎ 81 346 71 171*439
✆ 81 334 85 2904

Eriksen Gunnar
Inst Theor Astrophys
Univ of Oslo
Box 1029
N 0315 Blindern Oslo 3
Norway
☎ 47 22 856 515
✆ 47 22 85 6505

Eriksson Kjell
Astronomical Observ
Box 515
S 751 20 Uppsala
Sweden
☎ 46 18 51 2488
✆ 46 18 52 7583
✉ Kjell.Eriksson@astro.uu.se

Eroshkin Georgiy I
Inst Theoret Astron
Acad Sciences
10 Nab Kutuzova
191187 St Petersburg
Russia
☎ 7 812 275 0360
✆ 7 812 272 7968
✉ 1036@ita.spb.su

Erpylev Nikolaj P
Inst of Astronomy
Acad Sciences
Pyatnitskaya Ul 48
109017 Moscow
Russia
☎ 7 095 231 5461
✆ 7 095 230 2081

Ershkovich Alexander
Dpt Geophys/Planet Sci
Tel Aviv Univ
Ramat Aviv
Tel Aviv 69978
Israel
☎ 972 3 641 3505
✆ 972 3 640 9282

Ertan A Yener
Dpt Astron/Space Sci
Ege Univ
Box 21
35100 Bornova Izmir
Turkey
☎ 90 232 388 0110*2322

Escalante Vladimir
Instituto Astronomia
UNAM
Apt 70 264
Mexico DF 04510
Mexico
☎ 52 5 622 3908
✆ 52 5 616 0653
✉ vladimir@astroscu.unam.mx

Eshleman Von R
Stanford Univ
Durand 221
Stanford CA 94305
USA
☎ 1 415 497 3531
✆ 1 415 723 4840

Esipov Valentin F
SAI
Acad Sciences
Universitetskij Pr 13
119899 Moscow
Russia
☎ 7 095 939 2858
✆ 7 095 939 0126

Eskioglu A Nihat
Ozanlar Astrophys Lab
Sakarya Fakultesi
Ars Gor Metin Saltik
Adapazari
Turkey

Eskridge Paul B
CfA
HCO/SAO MS 81
60 Garden St
Cambridge MA 02138 1516
USA
☎ 1 617 496 7585
✆ 1 617 495 7356
✉ eskridge@cfa.harvard.edu

Espey Brian Russell
CAS
JHU
Charles/34th St
Baltimore MD 21218 2695
USA
☎ 1 301 516 5514
✆ 1 410 516 8260
✉ espey@pha.jhu.edu

Esposito F Paul
Physics Dpt
Univ of Cincinnati
210 Braunstein Ml 11
Cincinnati OH 45221 0111
USA
☎ 1 513 475 2233
✆ 1 513 556 3425

Esposito Larry W
LASP
Univ of Colorado
Campus Box 392
Boulder CO 80309 0392
USA
☎ 1 303 492 7325
✆ 1 303 492 6946

Esser Ruth
CfA
HCO/SAO
60 Garden St
Cambridge MA 02138 1516
USA
☎ 1 617 496 7566
✆ 1 617 495 7049
✉ esser@cfa.harvard.edu

Estalella Robert
Dpt Fisica Atmosfera
Univ Barcelona
Avd Diagonal 645
E 08028 Barcelona
Spain
☎ 34 3 330 7311*298
✆ 34 3 402 1133

Esteban Lopez Cesara
IAC
Observ d Teide
via Lactea s/n
E 38200 La Laguna
Spain
☎ 34 2 232 9100
✆ 34 2 232 9117

Etzel Paul B
Astronomy Dpt
San Diego State Univ
San Diego CA 92182
USA
☎ 1 619 495 6169
✆ 1 619 594 5485
✉ etzel@mintaka.sdsu.edu

Eubanks Thomas Marshall
7243 Archlaw Dr
Clifton VA 22024 2126
USA
✉ tme@cygx3.usno.navy.mil

Evangelidis E
1 Monks Horton Way
St Albans AL1 4HA
UK

Evans Aneurin
Physics Dpt
Univ of Keele
Keele ST5 5BG
UK
☎ 44 178 262 1111
✆ 44 178 271 1093

Evans Dafydd Wyn
Royal Greenwich Obs
Madingley Rd
Cambridge CB3 0EZ
UK
☎ 44 1223 374 000
✆ 44 12 23 374 700
✉ dwe@mail.ast.cam.ac.uk

Evans Ian Nigel
CfA
HCO/SAO MS 27
60 Garden St
Cambridge MA 02138 1516
USA
☎ 1 617 496 7846
✆ 1 617 495 7040
✉ evans_i@head-cfa.harvard.edu

Evans J V
Comsat Corp
6560 Rock Spring Dr
Bethesda MD 20817
USA
☎ 1 301 214 3211
✆ 1 301 214 7130

Evans John W
1 Baya Rd
Eldorado
Santa Fe NM 87503
USA

Evans Kenton Dower
Dpt Phys/X-Ray Astron
Univ Leicester
University Rd
Leicester LE1 7RN
UK
☎ 44 116 252 2073
✆ 44 116 250 182

Evans Nancy Remage
CfA
HCO/SAO MS 4
60 Garden St
Cambridge MA 02138 1516
USA
☎ 1 617 495 3627

Evans Neal J
Astronomy Dpt
Univ of Texas
Rlm 15 308
Austin TX 78712 1083
USA
☎ 1 512 471 4461
✆ 1 512 471 6016

Evans Robert
Villa 7
1 Gendarrah St
Hazelbrook NSW 2779
Australia
☎ 61 47 589 053

Evans Roger G
Rutherford Appleton Lab
Space/Astrophysics Div
Bg R25/R68
Chilton Didcot OX11 0QX
UK
☎ 44 12 35 821 900
✆ 44 12 35 44 5808

Evans W Doyle
10115 S 69th East Av
Tulsa OK 74133 6719
USA

Evans Wyn
49 Herford Av.
London SW14 8EH
UK
☎ 44 1633 867 551

Eviatar Aharon
Dpt Geophys/Planet Sci
Tel Aviv Univ
Ramat Aviv
Tel Aviv 69978
Israel
☎ 972 3 642 0620
✆ 972 3 640 9282

Evren Serdar
Dpt Astron/Space Sci
Ege Univ
Box 21
35100 Bornova Izmir
Turkey
☎ 90 232 388 0110*2322
✉ sevren@bornova.ege.edu.tr

M62

Ewen Harold I
Hillcrest Dr 60
Beaver
South Deerfield MA 01373
USA

Ewing Martin S
Astronomy Dpt
Yale Univ
Box 208267
New Haven CT 06520 8267
USA
☎ 1 203 432 4243
✆ 1 203 432 2797
✉ martin.ewing@yale.edu

Ezer-Eryurt Dilhan
Physics Dpt
Middle East Tech Univ
06531 Ankara
Turkey
☎ 90 41 23 7100*3255

F

Fabbiano Giuseppina
CfA
HCO/SAO
60 Garden St
Cambridge MA 02138 1516
USA
☎ 1 617 495 7204
📠 1 617 495 7356

Faber Sandra M
Lick Observ
Univ of California
Santa Cruz CA 95064
USA
☎ 1 831 429 2944
📠 1 831 426 3115

Fabian Andrew C
Inst of Astronomy
The Observatories
Madingley Rd
Cambridge CB3 0HA
UK
☎ 44 12 23 337 548
📠 44 12 23 337 523
✉ acf@ast.cam.ac.uk

Fabregat Juan
Dpt Astron/Astrofisica
Univ Valencia
Dr Moliner 50
E 46100 Burjassot
Spain
☎ 34 6 398 3071
📠 34 6 398 3084
✉ juan@pleione.uv.es

Fabricant Daniel G
CfA
HCO/SAO
60 Garden St
Cambridge MA 02138 1516
USA
☎ 1 617 495 7398
📠 1 617 495 7356
✉ dgf@cfa.harvard.edu

Fabricius Claus V
Astronomical Observ, NBIfAFG
Copenhagen Univ
Juliane Maries Vej 30
DK 2100 Copenhagen
Denmark
☎ 45 35 32 5966
📠 45 35 32 5989
✉ cf@astro.ku.dk

Fabrika Sergei
SAO
Acad Sciences
Nizhnij Arkhyz
357147 Karachaevo
Russia
☎ 7 878 784 6155
📠 7 96 908 2861
✉ fabrika@sao.ru

Facondi Silvia Rosa
Istt Radioastronomia
CNR
Via P Gobetti101
I 40129 Bologna
Italy
☎ 39 51 639 9383
📠 39 51 630 5700
✉ sfacondi@astbo1.bo.cnr.it

Fadeyev Yuri A
Inst of Astronomy
Acad Sciences
Pyatnitskaya Ul 48
109017 Moscow
Russia
☎ 7 095 231 5461
📠 7 095 230 2081

Faelthammar Carl Gunne
Dpt Plasma Physics
Royal Inst of Technology
S 100 44 Stockholm 70
Sweden
☎ 46 86 87 7685

Fahey Richard P
NASA/GSFC
Code 684
Greenbelt MD 20771
USA
☎ 1 301 286 9877
📠 1 301 286 1753
✉ fahey@stars.gsfc.nasa.gov

Fahlman Gregory G
Dpt Geophys/Astronomy
UBC
2075 Wesbrook Pl
Vancouver BC V6T 1W5
Canada
☎ 1 604 228 4891
📠 1 604 228 6047

Fahr Hans Joerg
IAEF
Univ Bonn
auf d Huegel 71
D 53121 Bonn
Germany
☎ 49 228 73 3677
📠 49 228 73 3672

Fairall Anthony P
Astronomy Dpt
Univ of Cape Town
Private Bag
7700 Rondebosch
South Africa
☎ 27 21 650 2392
📠 27 21 650 3352
✉ fairall@uctvms.uct.ac.za

Falchi Ambretta
Osserv Astrofis Arcetri
Univ d Firenze
Largo E Fermi 5
I 50125 Firenze
Italy
☎ 39 55 27 52236
📠 39 55 220 039
✉ falchi@arcetri.astro.it

Falciani Roberto
Dpt Astronomia
Univ d Firenze
Largo E Fermi 5
I 50125 Firenze
Italy
☎ 39 55 27 521
📠 39 55 22 0039

Falco-Acosta Emilio E
CfA
HCO/SAO MS 19
60 Garden St
Cambridge MA 02138 1516
USA
☎ 1 617 495 7131
📠 1 617 495 7467
✉ falco@cfa.harvard.edu

Falgarone Edith
DEMIRM
ENS
24 r Lhomond
F 75231 Paris Cdx 05
France
☎ 33 1 43 29 1225
📠 33 1 40 51 2002
✉ edith@ensapa.ens.fr

Falk Sydney W
1011 Shelley Av
Austin TX 78703
USA

Fall S Michael
STScI
Homewood Campus
3700 San Martin Dr
Baltimore MD 21218
USA
☎ 1 301 338 4700
📠 1 301 338 4767

Falle Samuel A
Dpt Appl Maths
Univ of Leeds
Leeds LS2 9JT
UK
☎ 44 113 233 5110
📠 44 113 242 9525

Faller James E
JILA/NBS
Univ of Colorado
Campus Box 400
Boulder CO 80309 0440
USA
☎ 1 303 492 8509

Fallon Frederick W
NGS
N/cg 114 Noaa
6010 Executive Blvd
Rockville MD 20852
USA
☎ 1 301 443 8424
✉ ffallon@dc.infi.net

Falomo Renato
Osserv Astronom Padova
Univ d Padova
Vic d Osservatorio 5
I 35122 Padova
Italy
☎ 39 49 829 3464
📠 39 49 875 9840
✉ falomo@astrpd.pd.astro.it

Fan Ying
Astronomy Dpt
Beijing Normal Univ
Beijing 100875
China PR
☎ 86 1 65 3531*6285
📠 86 1 201 3929

Fan Yu
Yunnan Observ
CAS
Kunming 650011
China PR
☎ 86 871 718 7153
📠 86 871 717 1845
✉ ynao@public.km.yn.cn

Fang Cheng
Astronomy Dpt
Nanjing Univ
Nanjing 210093
China PR
☎ 86 25 34651*2882
📠 86 25 330 2728

Fang Li-zhi
Dpt Phys/Astronomy
Steward Observ
Univ of Arizona
Tucson AZ 85721
USA
☎ 1 520 621 4721
📠 1 520 621 4721
✉ fanglz@time.physics.arizona.edu

Fanselow John Lyman
CALTECH/JPL
MS 169 315
4800 Oak Grove Dr
Pasadena CA 91109 8099
USA
☎ 1 213 354 6323
✆ 1 818 393 6030

Fanti Carla Giovannini
Istt Radioastronomia
CNR
Via P Gobetti101
I 40129 Bologna
Italy
☎ 39 51 639 9385
✆ 39 51 639 9385

Fanti Roberto
Istt d Fisica
Univ d Bologna
Via Irnerio 46
I 40126 Bologna
Italy
☎ 39 51 232 856/57

Faraggiana Rosanna
OAT
Box Succ Trieste 5
Via Tiepolo 11
I 34131 Trieste
Italy
☎ 39 40 31 99255
✆ 39 40 30 9418

Farinella Paolo
Dpt Matematica
Univ d Pisa
Via Buonarroti 2
I 56127 Pisa
Italy
☎ 39 50 844 254
✆ 39 50 844 224
✉ paolof@dm.unipi.it

Farnik Frantisek
Astronomical Institute
Czech Acad Sci
Fricova 1
CZ 251 65 Ondrejov
Czech R
☎ 420 204 85 7329
✆ 420 2 88 1611
✉ ffarnik@asu.cas.cz

Faucher Paul
OCA
Observ Nice
BP 139
F 06304 Nice Cdx 4
France
☎ 33 4 93 89 0420
✆ 33 4 92 00 3033

Faulkner Donald J
MSSSO
Weston Creek
Private Bag
Canberra ACT 2611
Australia
☎ 61 262 881 111
✆ 61 262 490 233

Faulkner John
Lick Observ
Univ of California
Santa Cruz CA 95064
USA
☎ 1 831 429 2815
✆ 1 831 426 3115

Faundez-Abans M
Observ Nacional
Rua Coronel Renno 07
CP 21
37500 Itajuba Mg
Brazil
☎ 55 356 22 0788
✉ lna@eu.ansp.br

Faurobert-Scholl Marianne
OCA
Observ Nice
BP 139
F 06304 Nice Cdx 4
France
☎ 33 4 92 00 3011
✆ 33 4 92 00 3033

Fawell Derek R
UCLO
ULO
Mill Hill Park
London NW7 2QS
UK
☎ 44 181 959 0421
✆ 44 171 380 7145

Fay Theodore D
Teledyne Brown Eng
Cummings Res Park
MS 19
Huntsville AL 35807
USA

Fazio Giovanni G
CfA
HCO/SAO
60 Garden St
Cambridge MA 02138 1516
USA
☎ 1 617 495 7458
✆ 1 617 495 7490
✉ gfazio@cfa.harvard.edu

Feast Michael W
SAAO
PO Box 9
7935 Observatory
South Africa
☎ 27 21 47 0025
✆ 27 21 47 3639

Feautrier Nicole
Observ Paris Meudon
DAMAP
Pl J Janssen
F 92195 Meudon PPL Cdx
France
☎ 33 1 45 07 7552
✆ 33 1 45 07 7971

Federici Luciana
Dpt Astronomia
Univ d Bologna
Via Zamboni 33
I 40126 Bologna
Italy
☎ 39 51 259 301
✆ 39 51 259 407

Federman Steven Robert
Dpt Phys/Astronomy
Univ of Toledo
2801 W Bancroft St
Toledo OH 43606
USA
☎ 1 419 537 2652
✆ 1 419 530 2723

Fedorov Petro
Astronomical Observ
Kharkiv State Univ
Sumska Str 35
310002 Kharkiv
Ukraine
☎ 380 5724 32428
✉ pnf@astron.kharkov.ua

Fegan David J
Physics Dpt
Univ College
Belfield
Dublin 4
Ireland
☎ 353 1 693 244
✆ 353 1 671 1759

Fehrenbach Charles
Les Magnanarelles
Lourmarin
F 84160 Cadenet
France
☎ 33 4 90 68 0028

Feibelman Walter A
NASA GSFC
Code 685
Greenbelt MD 20771
USA
☎ 1 301 286 5272
✆ 1 301 286 1617

Feigelson Eric D
Astronomy Dpt
Pennsylvania State Univ
525 Davey Lab
University Park PA 16802
USA
☎ 1 814 865 0162
✆ 1 814 863 7114
✉ edf@astro.psu.edu

Feinstein Alejandro
Observ Astronomico
Paseo d Bosque S/n
1900 La Plata (Bs As)
Argentina
☎ 54 21 25 8985
✆ 54 21 211 761
✉ afeinstein@fcaglp.
edu.ar

Feinstein Carlos
Observ Astronomico
Paseo d Bosque S/n
1900 La Plata (Bs As)
Argentina
☎ 54 21 217 308
✆ 54 21 258 985
✉ cfeinstein@fcaglp.
edu.ar

Feissel Martine
Observ Paris
DANOF
61 Av Observatoire
F 75014 Paris
France
☎ 33 1 40 51 2015
✆ 33 1 40 51 2291
✉ feissel@obspm.fr

Feitzinger Johannes
Astronomisches Institut
Ruhr Univ Bochum
Postfach 102148
D 44780 Bochum
Germany
☎ 49 234 700 3450
✆ 49 234 709 4169

Feix Gerhard
Inst Theor Physik
Ruhr Univ Bochum
Postfach 102148
D 46047 Bochum
Germany
☎ 49 234 700 2051
✆ 49 234 709 4169

Fejes Istvan
Fomi Satellite
Geodetic Observ
Box 546
H 1373 Budapest
Hungary

Fekel Francis C
Ctr Excellence Inform Sys
Tennessee State Univ
330 10th Av North
Nashville TN 37203 3401
USA
☎ 1 615 963 7302
✆ 1 615 963 7027
✉ fekel@coe.tnstate.edu

Feldman Paul A
NRCC/HIA
100 Sussex Dr
Ottawa ON K1A 0R6
Canada
☎ 1 613 993 6060
✆ 1 613 952 6004
✉ paf@hiaras.hia.nrc.ca

Feldman Paul Donald
Dpt Phys/Astronomy
JHU
Charles/34th St
Baltimore MD 21218
USA
☎ 1 301 516 7339
✆ 1 301 338 5494
✉ pdf@pha.jhu.edu

Feldman Uri
NRL
Holburt Ctr for Spaces Re
4555 Overlook Av SW
Washington DC 20375 5000
USA
☎ 1 202 767 3286

Felenbok Paul
Observ Paris Meudon
DAEC
Pl J Janssen
F 92195 Meudon PPL Cdx
France
☎ 33 1 45 07 7523
✆ 33 1 45 07 7511
🖂 felenbok@obspm.fr

Felli Marcello
Osserv Astrofis Arcetri
Univ d Firenze
Largo E Fermi 5
I 50125 Firenze
Italy
☎ 39 55 27 52240
✆ 39 55 22 0039
🖂 felli@arcetri.astro.it

Felten James E
NASA GSFC
Code 685
Greenbelt MD 20771
USA
☎ 1 301 286 6364
✆ 1 301 286 1617
🖂 felten@stars.gsfc.
nasa.gov

Feng Hesheng
Yunnan Observ
CAS
Kunming 650011
China PR
☎ 86 871 2035
✆ 86 871 717 1845

Feng Kejia
Astronomy Dpt
Beijing Normal Univ
Beijing 100875
China PR
☎ 86 1 65 3531*6967
✆ 86 1 201 3929

Fenkart Rolf P
Astronomisches Institut
Univ Basel
Venusstr 7
CH 4102 Binningen
Switzerland
☎ 41 61 271 7711/12
✆ 41 61 205 5455

Fenton K B
Physics Dpt
Univ Tasmania
GPO Box 252c
Hobart TAS 7001
Australia
☎ 61 2 202 411
✆ 61 2 202 410

Feretti Luigina
Istt Radioastronomia
CNR
Via P Gobetti101
I 40129 Bologna
Italy
☎ 39 51 639 9412
✆ 39 51 639 9431
🖂 lferetti@astbo1.bo.cnr.it

Ferland Gary Joseph
Dpt Phys/Astronomy
Univ Kentucky
Lexington KY 40506 0055
USA
☎ 1 606 257 6722
✆ 1 606 323 2846
🖂 gary@asta.pa.uky.edu

Ferlet Roger
IAP
98bis bd Arago
F 75014 Paris
France
☎ 33 1 44 32 8074
✆ 33 1 44 32 8001
🖂 ferlet@iap.fr

Ferluga Steno
Dpt Astronomia
Univ d Trieste
Via Tiepolo 11
I 34131 Trieste
Italy
☎ 39 40 76 3912
✆ 39 40 30 9418

Fernandez Jean-Claude
OCA
Observ Nice
BP 139
F 06304 Nice Cdx 4
France
☎ 33 4 93 89 0420
✆ 33 4 92 00 3033

Fernandez Julio A
Dpt Astronomia
Fac Humani/Ciencia
Tristan Naraja 1674
Montevideo 11200
Uruguay
☎ 598 2 419 087/089
✆ 598 2 409 973

Fernandez Silvia M
Observ Astronomico
de Cordoba
Laprida 854
5000 Cordoba
Argentina
☎ 54 51 23 0491
✆ 54 51 21 0613
🖂 silfer@astro.edu.ar

Fernandez-Figueroa M J
Dpt Astrofisica
Fac Fisica
Univ Complutense
E 28040 Madrid
Spain
☎ 34 1 449 5316
✆ 34 1 394 5195

Fernie J Donald
David Dunlap Observ
Univ of Toronto
Box 360
Richmond Hill ON L4C 4Y6
Canada
☎ 1 905 884 9562
✆ 1 905 884 2672
🖂 fernie@astro.utoronto.ca

Ferrari Attilio
Dpt Fisica Generale
Univ d Torino
Via P Giuria 1
I 10125 Torino
Italy
☎ 39 11 670 7457/810 1902
✆ 39 11 670 7457
🖂 ferrari@to.astro.it

Ferraz-Mello Sylvio
IAG
Univ Sao Paulo
CP 9638
01065 970 Sao Paulo SP
Brazil
☎ 55 11 577 8599*218
✆ 55 11 577 0270/276 3848
🖂 sylvio@usp.br
sylvio@iagusp.usp.br

Ferrer Martinez Sebastian
Dpt Fis Tierra/Cosmos
Univ Zaragoza
E 50009 Zaragoza
Spain
☎ 34 7 676 1000
✆ 34 7 676 1140

Ferrer Osvaldo Eduardo
Observ Astronomico
Paseo d Bosque S/n
1900 La Plata (Bs As)
Argentina
☎ 54 21 217 308
✆ 54 21 211 761

Ferreri Walter
Osserv Astronomico
d Torino
St Osservatorio 20
I 10025 Pino Torinese
Italy
☎ 39 11 461 9000
✆ 39 11 461 9030

Ferrin Ignacio
Dpt Fisica
Univ Los Andes
Merida 5101 A
Venezuela
☎ 58 7 463 9930/7477
✆ 58 7 444 1723

Ferrini Federico
Dpt Fisica
Univ d Pisa
Piazza Torricelli 2
I 56100 Pisa
Italy
☎ 39 50 911 291
✆ 39 50 48 277
🖂 federico@astr2pi.unipi.it

Ferriz-Mas Antonio
IAC
Observ d Teide
via Lactea s/n
E 38200 La Laguna
Spain
☎ 34 2 232 9100
✆ 34 2 232 9117
🖂 afm@ll.iac.es

Ferro Ramos Isabel
Apt P 1517
Carmelitas 1010
Caracas
Venezuela
🖂 iferro@skynet.usb.ve

Fesen Robert A
Dpt Phys/Astronomy
Dartmouth College
6127 Wilder Lab
Hanover NH 03755
USA
☎ 1 603 646 2949
✆ 1 603 646 1446
🖂 fesen@parsec.
dartmouth.edu

Festou Michel C
OMP
14 Av E Belin
F 31400 Toulouse Cdx
France
☎ 33 5 61 25 2101
✆ 33 5 61 27 3179
🖂 festou@astro.obs-mip.fr

Fey Alan Lee
USNO
Code EO
3450 Massachusetts Av NW
Washington DC 20392 5420
USA
☎ 1 202 762 1517
✆ 1 202 762 1563
🖂 afey@alf.usno.navy.mil

Feynman Joan
CALTECH/JPL
MS 144 218
4800 Oak Grove Dr
Pasadena CA 91109 8099
USA
☎ 1 818 354 3454
✆ 1 818 393 6030

Fiala Alan D
USNO
Astronomical Applicat Dpt
3450 Massachusetts Av NW
Washington DC 20392 5420
USA
☎ 1 202 653 1274
✆ 1 202 653 0179
🖂 adf@newcomb.usno.
navy.mil

Ficarra Antonino
Istt Radioastronomia
CNR
Via P Gobetti101
I 40129 Bologna
Italy
☎ 39 51 639 9385
✆ 39 51 639 9385

Fich Michel
Physics Dpt
Univ Waterloo
Waterloo ON N2L 3G1
Canada
☎ 1 519 885 1572
✆ 1 519 746 8115

Fichtel Carl E
NASA GSFC
Code 660
Greenbelt MD 20771
USA
☎ 1 301 286 6281
✉ fichtel@ileavx.gsfc.
nasa.gov

Fiebig Dirk
Inst Theor Astrophysik
d Univer
Tiergartenstr 15
D 69121 Heidelberg
Germany
☎ 49 622 154 8974
📠 49 622 154 4221
✉ fiebig@tethys.ita.uni-
heidelberg.de

Fiedler Ralph L
NRL
Code 4210
4555 Overlook Av SW
Washington DC 20375 5000
USA
☎ 1 202 767 0644
✉ fiedler@rira.nrl.
navy.mil

Fiedler Russell
Mathematics Dpt
Univ of St Andrews
St Andrews North Haugh
Fife KY16 9SS
UK
☎ 44 13 34 7661
📠 44 13 347 4487

Field David
School of Chemistry
Cantocks Close
Bristol BS8 1TS
UK
☎ 44 11 79 24161*505

Field George B
CfA
HCO/SAO
60 Garden St
Cambridge MA 02138 1516
USA
☎ 1 617 495 4721
📠 1 617 495 7356
✉ field@cfa.harvard.edu

Fielder Gilbert
Willow Tree
Eldroth Austwick
Lancaster LA2 8AH
UK

Fienberg Richard T
Sky/Telescope
Box 9111
Belmont MA 02178 9111
USA
☎ 1 617 864 7360
📠 1 617 864 6117
✉ rfienberg@skypub.com

Fierro Julieta
Instituto Astronomia
UNAM
Apt 70 264
Mexico DF 04510
Mexico
☎ 52 5 622 3906
📠 52 5 616 0653
✉ julieta@astroscu.unam.mx

Figueras Francesca
Dpt Astronom Meteo
Univ Barcelona
Avd Diagonal 647
E 08028 Barcelona
Spain
☎ 34 3 402 1125
📠 34 3 402 1133
✉ d3faffs0@eb0ub011

Filipov Latchezar
Space Res Inst
Bulgarian Acad Sci
Moskova St 6
BG 1000 Sofia
Bulgaria
☎ 359 2 87 0978

Filippenko Alexei V
Astronomy Dpt
Univ of California
601 Campbell Hall
Berkeley CA 94720 3411
USA
☎ 1 415 642 1813
📠 1 510 642 3411
✉ alex@bkyast

Filloy Emilio Manuel E E
IAR
CC 5
1894 Villa Elisa (Bs As)
Argentina
☎ 54 21 4 3793
📠 54 21 211 761

Fink Uwe
Lunar/Planetary Lab
Room 325 Bg 92
Univ of Arizona
Tucson AZ 85721 0092
USA
☎ 1 520 621 2736
📠 1 520 621 4933

Finkelstein Andrej M
Inst Appl Astronomy
Acad Sciences
Zhdanovskaya Ul 8
197042 St Petersburg
Russia
☎ 7 812 230 7414
📠 7 812 230 7413

Finn Lee Samuel
Dpt Phys/Astronomy
NW Univ
2131 Sheridan Rd
Evanston IL 60208 3112
USA
☎ 1 708 491 4568
📠 1 708 491 9982
✉ lsf@holmes.astro.nwu.edu

Finzi Arrigo
Mathematics Dpt
IIT
Technion City
Haifa 32000
Israel
☎ 972 4 293 020

Firmani Claudio A
Instituto Astronomia
UNAM
Apt 70 264
Mexico DF 04510
Mexico
☎ 52 5 622 3906
📠 52 5 616 0653
✉ firmani@aleph.cinstrum.
unam.mx

Firneis Friedrich J
Comp Ctr Acad Sci
Sonnenfelgasse 19/1
A 1010 Wien
Austria

Firneis Maria G
Inst Astronomie
Univ Wien
Tuerkenschanzstr 17
A 1180 Wien
Austria
☎ 43 1 345 3600
📠 43 1 470 6015

Firor John W
HAO
NCAR
Box 3000
Boulder CO 80307 3000
USA
☎ 1 303 497 1600
📠 1 303 497 1568

Fischel David
6072 Warm Stone Ct
Columbia MD 21045
USA
☎ 1 410 730 8660
📠 1 410 730 8660
✉ dfischel@eosat.com

Fischer Jacqueline
NRL
Remote Sensing Div 7217
4555 Overlook Av SW
Washington DC 20375 5000
USA
☎ 1 202 767 3058
📠 1 202 404 8894
✉ jfischer@irfp8.nrl.
navy.mil

Fischer Stanislav
Astronomical Institute
Czech Acad Sci
Bocni Ii 1401
CZ 141 31 Praha 4
Czech R
☎ 420 2 67 10 3062
📠 420 2 76 90 23
✉ fischer@ig.cas.cz

Fisher George Hewitt
Space Sci Laboratory
Univ of California
Grizzly Peak Blvd
Berkeley CA 94720 7950
USA
☎ 1 510 642 7297
📠 1 510 643 8302
✉ finsher@sunspot.ssl.
berkeley.edu

Fisher J Richard
NRAO
Box 2
Green Bank WV 24944
USA
☎ 1 304 456 2011
📠 1 304 456 2271

Fisher Philip C
Ruffner Associates
Box 1867
Santa Fe NM 87504 1867
USA

Fisher Richard R
NASA GSFC
Code 682
Greenbelt MD 20771
USA
☎ 1 301 286 8811

Fishman Gerald J
NASA/MSFC
Space Science Lab
Code ES 81
Huntsville AL 35812
USA
☎ 1 205 544 7691
📠 1 205 544 5800
✉ fishman@ssl.msfc.
nasa.gov

Fitch Walter S
Box 100
Oracle AZ 85623
USA
☎ 1 602 896 2911

Fitton Brian
ESA/ESTEC
Astrophysics Division
Box 299
NL 2200 AG Noordwijk
Netherlands
☎ 31 71 565 6555
📠 31 71 565 4532

Fitzgerald M Pim
Physics Dpt
Univ Waterloo
Waterloo ON N2L 3G1
Canada
☎ 1 519 885 1572
📠 1 519 746 8115

Fitzpatrick Edward L
Princeton Univ Obs
Peyton Hall
Princeton NJ 08544 1001
USA
☎ 1 609 258 3702
📠 1 609 258 1020
✉ fitz@astro.princeton.edu

Fitzsimmons Alan
Physics Dpt
Queen's Univ
Belfast BT7 1NN
UK
☎ 44 12 32 273 124
✆ 44 12 32 438 918
✉ a.fitzsimmons@qub.ac.uk

Fix John D
Dpt Phys/Astronomy
Univ of Iowa
605 Brookland Park Dr
Iowa City IA 52242 1479
USA
☎ 1 319 353 7064
✆ 1 319 335 1753

Flanagan Claire Susan
HartRAO
FRD
PO Box 443
1740 Krugersdorp
South Africa
☎ 27 11 642 4692
✆ 27 11 642 2424
✉ claire@bootes.hartrao.ac.za

Flannery Brian Paul
Exxon Res/Eng
Route 22 East
Annandale NJ 08801
USA
☎ 1 201 730 2540

Fleck Robert Charles
Dpt Math/Phys Sci
Embry Riddle Aeron Univ
Daytona FL 32114
USA
☎ 1 904 226 6612
✆ 1 904 226 6713

Fleischer Robert
108 Overlook St
Route 2 Box 515
Moorefield WV 26836
USA

Fleming Thomas Anthony
Steward Observ
Univ of Arizona
Tucson AZ 85721
USA
☎ 1 520 621 5049
✆ 1 520 621 1532
✉ taf@as.arizona.edu

Fletcher J Murray
NRCC/HIA
DAO
5071 W Saanich Rd
Victoria BC V8X 4M6
Canada
☎ 1 250 363 0017
✆ 1 250 363 0045
✉ murray.fletcher@hia.nrc.ca

Fliegel Henry F
Box 8682
La Crescenta CA 91224
USA
☎ 1 310 336 1710
✆ 1 310 336 5076
✉ fliegel@courier1.aero.org

Flin Piotr
Pedagogical Univ
Inst of Physics
Ul Podchorazych 2
PL 30 084 Krakow
Poland
☎ 48 12 37 4777*280
✆ 48 12 372 2243
✉ sfflin@plkrcy11

Floquet Michele
Observ Paris Meudon
DASGAL
Pl J Janssen
F 92195 Meudon PPL Cdx
France
☎ 33 1 45 07 7851
✆ 33 1 45 07 7872

Floria Luis
Gr Mecan Espacial
Univ Zaragoza
E 50009 Zaragoza
Spain
☎ 34 7 676 1000
✆ 34 7 676 1140

Florides Petros S
TCD
Physics Dpt
College Greem
Dublin 2
Ireland
☎ 353 1 608 1675
✆ 353 1 671 1759

Florkowski David R
USNO
3450 Massachusetts Av NW
Washington DC 20392 5100
USA
☎ 1 202 762 1569
✆ 1 202 762 1563
✉ drf@maia.usno.navy.mil

Florsch Alphonse
Observ Strasbourg
11 r Universite
F 67000 Strasbourg
France
☎ 33 3 88 15 0735
✆ 33 3 88 25 0160

Flower David R
Physics Dpt
Univ of Durham
South Rd
Durham DH1 3LE
UK
☎ 44 191 374 2145
✆ 44 191 374 3749
✉ david.flower@durham.ac.uk

Fludra Andrzej
Rutherford Appleton Lab
Space/Astrophysics Div
Bg R25/R68
Chilton Didcot OX11 0QX
UK
☎ 44 12 35 44 6497
✆ 44 12 35 44 6509
✉ af@ast.star.rl.ac.uk

Flynn Chris
Turku Univ
Tuorla Observ
Vaeisaelaentie 20
FIN 21500 Piikkio
Finland
☎ 358 2 274 4244
✆ 358 2 243 3767
✉ cflynn@astro.utu.fi

Focardi Paola
Osserv Astronom Bologna
Univ d Bologna
Via Zamboni 33
I 40126 Bologna
Italy
☎ 39 51 259 401
✆ 39 51 259 407

Fodera Serio Giorgia
Osserv Astronomico
Univ d Palerno
Piazza Parlemento 1
I 90134 Palermo
Italy
☎ 39 91 657 0451
✆ 39 91 48 8900

Fofi Massimo
OAR
Via d Parco Mellini 84
I 00136 Roma
Italy
☎ 39 6 34 7056
✆ 39 6 349 8236

Fogarty William G
IBM
NCMD
411 East Wisconsin Av
Milwaukee WI 53202
USA

Foing Bernard H
ESA/ESTEC
SSD
Box 299
NL 2200 AG Noordwijk
Netherlands
☎ 31 71 565 5647
✆ 31 71 565 4698
✉ bfoing@estec.esa.nl

Foing-Ehrenfreund Pascale
Leiden Observ
Box 9513
NL 2300 RA Leiden
Netherlands
☎ 31 71 527 5872
✆ 31 71 5275 819
✉ pascale@rulhl1.leidenuniv.nl

Foley Anthony
NFRA
Postbus 2
NL 7990 AA Dwingeloo
Netherlands
☎ 31 521 59 5100
✆ 31 52 159 7332
✉ folred@nfra.nl

Foltz Craig B
Mult Mirror
Telescope Obs
Univ of Arizona
Tucson AZ 85721
USA
☎ 1 520 621 1269
✆ 1 520 621 4933
✉ cfoltz@as.arizona.edu

Fomalont Edward B
NRAO
Box 0
Socorro NM 87801 0387
USA
☎ 1 505 772 4011
✆ 1 505 835 7027

Fomenko Alexandr F
SAO
Acad Sciences
Nizhnij Arkhyz
357147 Karachaevo
Russia
☎ 7 878 789 2501

Fomichev Valeri V
ITMIRWP
Acad Sciences
142092 Troitsk
Russia
☎ 7 095 334 0120
✆ 7 095 334 0124

Fomin Piotr Ivanovich
Metrologicheskaya St 14b
252143 Kyiv
Ukraine

Fomin Valery
Crimean Astrophys Obs
Ukrainian Acad Science
Nauchny
334413 Crimea
Ukraine
☎ 380 6554 71625
✆ 380 6554 40704
✉ fomin@crao.crimea.ua

Fomin Valery A
Pulkovo Observ
Acad Sciences
10 Kutuzov Quay
196140 St Petersburg
Russia
☎ 7 812 298 2242
✆ 7 812 315 1701

Fominov Alexandr M
Inst Theoret Astron
Acad Sciences
N Kutuzova 10
191187 St Petersburg
Russia
☎ 7 812 278 8898
✆ 7 812 272 7968

Fong Chugang
Shanghai Observ
CAS
80 Nandan Rd
Shanghai 200030
China PR
☎ 86 21 6438 6191
✆ 86 21 6438 4618
✉ fcg@center.shao.ac.cn

Fong Richard
Physics Dpt
Univ of Durham
South Rd
Durham DH1 3LE
UK
☎ 44 191 374 2157
✆ 44 191 374 3749

Fontaine Gilles
Dpt d Physique
Universite Montreal
CP 6128 Succ A
Montreal QC H3C 3J7
Canada
☎ 1 514 343 6680
✆ 1 514 343 2071

Fontenla Juan Manuel
NASA/MSFC
Space Science Lab
Code ES 52
Huntsville AL 35812
USA
☎ 1 205 544 7690
✆ 1 205 547 7754

Forbes Douglas
Sir Wilfred Grenfell Coll
Memorial Univ
Newfoundland
Corner Brook NF A2H 6P9
Canada
☎ 1 709 637 6295
✆ 1 709 637 6390
✉ dforbes@kean.ucs.
mun.ca

Forbes Duncan Alan
School Phys/Astron
Univ Birmingham
Edgbaston
Birmingham B15 2TT
UK
☎ 44 12 1414 6474
✆ 44 12 14 14 3722
✉ forbes@star.sr.bham.
ac.uk

Forbes J E
Box 88120
Indianapolis IN 46208
USA

Forbes Terry G
Space Science Ctr/EOS
Univ of New Hampshire
Demeritt Hall
Durham NH 03857
USA
☎ 1 603 862 3872
✉ t_forbes@unhh

Ford Holland C Res
STScI
Homewood Campus
3700 San Martin Dr
Baltimore MD 21218
USA
☎ 1 301 338 4803
✆ 1 301 338 4767

Ford W Kent
Dpt Terrestr Magnetism
Carnegie Inst Washington
5241 Bd Branch Rd NW
Washington DC 20015 1305
USA
☎ 1 202 966 0863
✆ 1 202 364 8726

Forestini Manuel
Observ Grenoble
Lab Astrophysique
BP 53x
F 38041 S Martin Heres Cdx
France
☎ 33 4 76 63 5620
✆ 33 4 76 44 8821
✉ manuel.forestini@obs.
ujf-grenoble.fr

Forman William Richard
CfA
HCO/SAO
60 Garden St
Cambridge MA 02138 1516
USA
☎ 1 617 495 7210
✉ wrf@cfa200.harvard.edu

Forrest William John
Dpt Phys/Astronomy
Univ of Rochester
Rochester NY 14627 0171
USA
☎ 1 716 275 4343
✆ 1 716 275 4351
✉ forrest@pas.rochester.edu

Forster James Richard
Hat Creek Radio Observ
RT 2
Box 500
Cassel CA 96016
USA
☎ 1 916 335 2364
✆ 1 916 335 3968

Fort Bernard P
IAP
98bis bd Arago
F 75014 Paris
France
☎ 33 1 44 32 8016
✆ 33 1 44 32 8001
✉ fort@iap.fr

Fort David Norman
CALTECH/JPL
4800 Oak Grove Dr
Pasadena CA 91109 8099
USA
☎ 1 818 354 9132
✆ 1 818 393 6030

Forte Juan Carlos
Observ Astronomico
Paseo d Bosque S/n
1900 La Plata (Bs As)
Argentina
☎ 54 21 217 308
✆ 54 21 211 761
✉ forte@iafe.uba.ar

Forti Giuseppe
Osserv Astrofis Arcetri
Univ d Firenze
Largo E Fermi 5
I 50125 Firenze
Italy
☎ 39 55 27 52236
✆ 39 55 22 0039
✉ forti@arcetri.astro.it

Fortini Teresa
Via f D Guerrazzi 19
I 00152 Roma
Italy

Forveille Thierry
Observ Grenoble
Lab Astrophysique
BP 53x
F 38041 S Martin Heres Cdx
France
☎ 33 4 76 51 4567
✆ 33 4 76 44 8821

Foryta Dietmar William
Dpt Fisica
Univ Fed d Parana
CP 19081
81531 990 Curitiba
Brazil
☎ 55 41 366 2323
✆ 55 41 267 4236
✉ foryta@fisica.ufpr.br

Fosbury Robert A E
ESO
ST/ECF
Karl Schwarzschildstr 2
D 85748 Garching
Germany
☎ 49 893 200 6235
✆ 49 893 202 362

Fossat Eric
OCA
Observ Nice
BP 139
F 06304 Nice Cdx 4
France
☎ 33 4 93 52 9805
✆ 33 4 92 00 3033

Foster Roger S
NRL
Code 7210
4555 Overlook Av SW
Washington DC 20375 5000
USA
☎ 1 202 767 0669
✆ 1 202 404 8894
✉ foster@rira.nrl.navy.mil

Foukal Peter V
CRI Inc
80 Ashford St
Boston MA 02134
USA
☎ 1 617 787 5700
✆ 1 617 787 4488
✉ pfoukal@world.std.com

Fouque Pascal
Gr Astrofisica
Univ Catolica
Casilla 104
Santiago 22
Chile
☎ 56 2 552 2375
✆ 56 2 552 5692
✉ pfouque@astrouc.puc.cl

Fox Kenneth
Dpt Phys/Astronomy
Univ Tennessee
503 Physics
Knoxville TN 37996 1200
USA
☎ 1 615 974 2288

Fox W E
British Astronomical Ass
40 Windsor Rd
Newark Nottinghamshire
UK
☎ 44 163 670 4932

Foy Renaud
Observ Lyon
Av Ch Andre
F 69561 S Genis Laval cdx
France
☎ 33 4 78 86 8546
✆ 33 4 78 86 8386
✉ foy@image.univ-lyon1.fr

Frail Dale Andrew
NRAO
VLA
Box 0
Socorro NM 87801 0387
USA
☎ 1 505 835 7338
✆ 1 505 835 7027
✉ d.frail@nrao.edu

Fraix-Burnet Didier
Observ Grenoble
Lab Astrophysique
BP 53x
F 38041 S Martin Heres Cdx
France
☎ 33 4 76 63 5727
✆ 33 4 76 44 8821
✉ fraix@gag.observ-gr.fr

Franceschini Alberto
Dpt Fisica G Galilei
Univ d Padova
Via Marzolo 8
I 35131 Padova
Italy
☎ 39 49 84 4111
✆ 39 49 84 4245

Franchini Mariagrazia
OAT
Box Succ Trieste 5
Via Tiepolo 11
I 34131 Trieste
Italy
☎ 39 40 31 99255
✆ 39 40 30 9418

Francis Paul
MSSSO
Weston Creek
Private Bag
Canberra ACT 2611
Australia
☎ 61 262 798 039
✆ 61 2 62490 233
✉ pfrancis@mso.anu.edu.au

Franco Jose
Instituto Astronomia
UNAM
Apt 70 264
Mexico DF 04510
Mexico
☎ 52 5 622 3906
✆ 52 5 548 0653
✉ jjfranco@alfa.astroscu.unam.mx

Franco Mantovani
Istt Radioastronomia
CNR
Via P Gobetti101
I 40129 Bologna
Italy
☎ 39 51 639 9385
✆ 39 51 639 9385

Francois Patrick
Observ Paris Meudon
DASGAL
Pl J Janssen
F 92195 Meudon PPL Cdx
France
☎ 33 1 45 97 7867
✆ 33 1 45 07 7472

Frandsen Soeren
Inst Phys/Astronomy
Univ of Aarhus
Ny Munkegade
DK 8000 Aarhus C
Denmark
☎ 45 86 12 8899
✆ 45 86 20 2711

Frank Juhan
Dpt Phys/Astronomy
Louisiana State Univ
Baton Rouge LA 70803 4001
USA
☎ 1 504 388 6845
✆ 1 504 388 5855
✉ frank@rouge.phys.lsu.edi

Franklin Fred A
CfA
HCO/SAO
60 Garden St
Cambridge MA 02138 1516
USA
☎ 1 617 495 7230
✆ 1 617 495 7356

Fransson Claes
Stockholm Observ
Royal Swedish Acad Sciences
S 133 36 Saltsjoebaden
Sweden
☎ 46 8 716 4469
✆ 46 8 717 4719
✉ claes@astro.su.se

Frantsman Yu L
Radioastrophys Observ
Latvian Acad Sci
Turgeneva 19
LV 226524 Riga
Latvia
☎ 371 722 6006

Franx Marijn
Kapteyn Sterrekundig Inst
Univ Groningen
Postbus 800
NL 9700 AV Groningen
Netherlands
☎ 31 50 363 4073
✆ 31 50 363 6100

Franz Otto G
Lowell Observ
1400 W Mars Hill Rd
Box 1149
Flagstaff AZ 86001
USA
☎ 1 520 774 3358
✆ 1 520 774 3358
✉ ogf@lowell.edu

Frater Robert H
CSIRO
Div Radiophysics
Box 76
Epping NSW 2121
Australia
☎ 61 2 868 0222
✆ 61 2 868 0310

Frazier Edward N
TRW Space/Technology
1 Space Park
Redondo Beach CA 90278
USA
☎ 1 213 535 4723

Fredga Kerstin
Swedish Ntl Space Board
Box 4006
S 171 04 Solna
Sweden
☎ 46 8 627 6486
✆ 46 8 627 6480

Fredrick Laurence W
University Station
Univ of Virginia
Box 3818
Charlottesville VA 22903 0818
USA
☎ 1 804 924 7494
✆ 1 804 924 3104
✉ lwf@virginia.edu

Freedman Wendy L
Carnegie Observatories
813 Santa Barbara St
Pasadena CA 91101 1292
USA
☎ 1 818 304 0204
✆ 1 818 795 8136
✉ wendy@ociw.edu

Freeman Kenneth C
MSSSO
Weston Creek
Private Bag
Canberra ACT 2611
Australia
☎ 61 2 62 490 264
✆ 61 26 249 0233
✉ kcf@mso.anu.edu.au

Freire Ferrero Rubens G
Observ Strasbourg
11 r Universite
F 67000 Strasbourg
France
☎ 33 3 88 15 0736
✆ 33 3 88 25 0160
✉ freire@cdsxb6.u-strasbg.fr

Freitas Mourao R r
Av de Execito 105
San Cristovao
20910 Rio de Janeiro RJ
Brazil
☎ 55 21 580 7154*7204
✆ 55 21 580 0332

French Richard G
Dpt Earth/Planet Sci
MIT Rm 54 422
Box 165
Cambridge MA 02139
USA
☎ 1 617 253 3392

Frenk Carlos S
Physics Dpt
Univ of Durham
South Rd
Durham DH1 3LE
UK
☎ 44 191 374 2141
✆ 44 191 374 3749

Fresneau Alain
Observ Strasbourg
11 r Universite
F 67000 Strasbourg
France
☎ 33 3 88 15 0737
✆ 33 3 88 25 0160
✉ fresneau@astro.u-strasbg.fr

Freudling Wolfram
ESO
Karl Schwarzschildstr 2
D 85748 Garching
Germany
☎ 49 893 200 6425
✆ 49 893 200 6480
✉ wfreudli@eso.org

Friberg Per
Joint Astronomy Ctr
660 N A'ohoku Pl
University Park
Hilo HI 96720
USA
☎ 1 808 969 6522
✆ 1 808 961 6516
✉ friberg@jach.hawaii.edu

Fricke Klaus
Univ Sternwarte
Goettingen
Geismarlandstr 11
D 37083 Goettingen
Germany
☎ 49 551 39 5051
✆ 49 551 395 043

Fridlund Malcolm
ESA/ESTEC
Astrophysics Division
Box 299
NL 2200 AG Noordwijk
Netherlands
☎ 31 71 565 4768
✆ 31 71 565 4690
✉ mfridlun@astro.estec.esa.nl

Fridman Aleksey M
Inst of Astronomy
Acad Sciences
Pyatnitskaya Ul 48
109017 Moscow
Russia
☎ 7 095 231 5461
✆ 7 095 230 2081
✉ iaas@node.ias.msk.su

Fried Josef Wilhelm
MPI Astronomie
Koenigstuhl
D 69117 Heidelberg
Germany
☎ 49 622 152 81
✆ 49 622 152 8246

Friedemann Christian
Astrophysik Inst
Univ Sternwarte
Schillergaesschen 2
D 07745 Jena
Germany
☎ 49 364 163 0323
✆ 49 364 163 0417

Friedjung Michael
IAP
98bis bd Arago
F 75014 Paris
France
☎ 33 1 44 32 8111
✆ 33 1 44 32 8001
✉ friedjung@friap51

Friedlander Michael
Physics Dpt
Washington Univ
Campus Box 1105
St Louis MO 63130
USA
☎ 1 314 889 6279
✆ 1 314 935 4083

Friedli Daniel
Dpt d Physique
Universite Laval
Ste Foy QC G1K 7P4
Canada
☎ 1 418 656 2131
✆ 1 418 656 2040
✉ dfriedli@phy.ulaval.ca

Friedman Herbert
2643 N Upshur St
Arlington VA 22207
USA

Friedman Scott David
Dpt Phys/Astronomy
JHU
Charles/34th St
Baltimore MD 21218
USA
☎ 1 301 338 5317
✆ 1 301 338 5494

Friel Eileen D
NSF
Div Astron Sci
4201 Wilson Blvd
Arlington VA 22230
USA
☎ 1 703 306 1825
✆ 1 703 306 0525
✉ efriel@nsf.gov

Friend David B
Dpt Phys/Astronomy
Univ of Montana
Missoula MT 59812
USA
☎ 1 406 243 2073
✆ 1 406 243 40766

Fringant Anne-Marie
Observ Paris
61 Av Observatoire
F 75014 Paris
France
☎ 33 1 40 51 2248
✆ 33 1 44 54 1804

Frisch Helene
OCA
Observ Nice
BP 139
F 06304 Nice Cdx 4
France
☎ 33 4 93 89 0420
✆ 33 4 92 00 3033

Frisch Priscilla
Astronomy/Astrophys Ctr
Univ of Chicago
5640 S Ellis Av
Chicago IL 60637
USA
☎ 1 312 702 0181
✆ 1 312 702 8211
✉ frisch@oddjob.uchicago.edu

Frisch Uriel
OCA
Observ Nice
BP 139
F 06304 Nice Cdx 4
France
☎ 33 4 93 89 0420
✆ 33 4 92 00 3033

Frisk Urban
Swedish Ntl Space Board
Box 4006
S 171 04 Solna
Sweden
☎ 46 8 627 6200
✆ 46 8 627 6480

Fritze Klaus
Astrophysik Inst
Potsdam Univ
an d Sternwarte 16
D 14482 Potsdam
Germany
☎ 49 331 749 9202
✆ 49 331 749 9200
✉ kfritze@aip.de

Fritze-von Alvensleben Ute
Univ Sternwarte
Goettingen
Geismarlandstr 11
D 37083 Goettingen
Germany
☎ 49 551 395 055
✆ 49 551 395 043
✉ ufritze@uni-sw.gwdg.de

Fritzova-Svestka L
Dopperstraat 147
NL 3752 JC Bunschoten
Netherlands
☎ 31 34 998 4403

Froehlich Claus
World Radiation Ctr
Physikalisch meteorol Obs
Posfach 173
CH 7260 Davos Dorf
Switzerland
☎ 41 81 46 2131

Froeschle Christiane D
OCA
Observ Nice
BP 139
F 06003 Nice Cdx 4
France
☎ 33 4 93 89 0420
✆ 33 4 92 00 3033

Froeschle Claude
OCA
Observ Nice
BP 139
F 06003 Nice Cdx 4
France
☎ 33 4 92 00 3024
✆ 33 4 92 00 3033
✉ claude@obs-nice.fr

Froeschle Michel
OCA
CERGA
F 06130 Grasse
France
☎ 33 4 93 35 5849
✆ 33 4 93 40 5353

Frogel Jay Albert
Astronomy Dpt
Ohio State Univ
174 W 18th Av
Columbus OH 43210 1106
USA
☎ 1 614 292 5651
✆ 1 614 292 2928
✉ frogel@ohstpy

Frolov Mikhail S
Inst of Astronomy
Acad Sciences
Pyatnitskaya Ul 48
109017 Moscow
Russia
☎ 7 095 231 5461
✆ 7 095 230 2081

Frost Kenneth J
NASA GSFC
Code 600 2
Greenbelt MD 20771
USA
☎ 1 301 286 8824

Fruchter Andrew S
Astronomy Dpt
Univ of California
601 Campbell Hall
Berkeley CA 94720 3411
USA
☎ 1 510 643 8142
✆ 1 510 642 3411
✉ asf@orestes.berkeley.edu

Fruscione Antonella
Ctr for EUV Astrophys
Univ of California
2150 Kittredge St
Berkeley CA 94720 5030
USA
☎ 1 510 643 6484
✆ 1 415 643 5660
✉ antonella@cea.berkeley.edu

Frye Glenn M
Physics Dpt
CWRU
Rock Bdg
Cleveland OH 44106
USA
☎ 1 216 368 2997

Ftaclas Christ
Perkin Elmer Corp
MS 897
100 Wooster Heights Rd
Danbury CT 06810 7859
USA
☎ 1 203 797 6448

Fu Chengqi
Shanghai Observ
CAS
80 Nandan Rd
Shanghai 200030
China PR
☎ 86 21 6438 6191
✆ 86 21 6438 4618

Fu Delian
Beijing Astronomical Obs
CAS

W Suburb
Beijing 100080
China PR
☎ 86 1 28 2070
✆ 86 1 256 10855

Fu Qijun
Beijing Astronomical Obs
CAS
Beijing 100080
China PR
☎ 86 1 256 9840
✆ 86 1 256 10855

Fu-Shong Kuo
Physics Dpt
Ntl Central Univ
Chung Li Taiwan 32054
China R
☎ 886 3 462 2302
✆ 886 3 426 2304

Fuchs Burkhard
ARI
Moenchhofstr 12-14
D 69120 Heidelberg
Germany
☎ 49 622 140 5126
✆ 49 622 140 5297
✉ fuchs@ari.uni-heidelberg.de

Fuenmayor Francisco J
Dpt Fisica
Univ Los Andes
Merida 5101 A
Venezuela
☎ 58 7 463 9930/7477
✆ 58 7 444 1723

Fuensalida Jimenez J
IAC
Observ d Teide
via Lactea s/n
E 38200 La Laguna
Spain
☎ 34 2 232 9100
✆ 34 2 232 9117

Fuente Asuncion
Observ Astronomico
Nacional
Apt 1143
E 28800 Alcala d Henares
Spain
☎ 34 1 885 5060
✆ 34 1 885 5062
✉ fuente@oan.es

Fuerst Ernst
RAIUB
Univ Bonn
auf d Huegel 69
D 53121 Bonn
Germany
☎ 49 228 525 0
✆ 49 228 525 229

Fuhr Jeffrey Robert
NIST
DCATP
Gaithersburg MD 20899
USA
☎ 1 301 975 3204
✆ 1 301 990 1350
✉ fuhr@atm.nist.gov

Fujimoto Masa-Katsu
Tokyo Astronomical Obs
NAOJ
Osawa Mitaka
Tokyo 181
Japan
☎ 81 422 32 5111
✆ 81 422 34 3793
✉ fujimoto@gravity.mtk.nao.ac.jp

Fujimoto Masayuki
Fac of Education
Niigata Univ
8050 Ikarashi 2
Niigata 950 21
Japan
☎ 81 252 62 7269

Fujimoto Mitsuaki
Physics Dpt
Nagoya Univ
Furocho Chikusa Ku
Nagoya 464 01
Japan
☎ 81 527 89 2840
✆ 81 527 89 2845

Fujishita Mitsumi
Kyushu Tokai Univ
9-1-1 Toroku
Kumamoto 862
Japan
☎ 81 963 86 2659
✆ 81 96 381 7956
✉ mfuji@ktmail.ktokai-u.ac.uk

Fujita Yoshio
Astronomy Dpt
Univ of Tokyo
Bunkyo Ku
Tokyo 113
Japan
☎ 81 423 74 4186
✆ 81 338 13 9439

Fujiwara Akira
ISAS
3 1 1 Yoshinodai
Sagamihara
Kanagawa 229 8510
Japan
☎ 81 427 51 3911
✆ 81 427 59 4253
✉ fujiwara@planeta.sci.isas.ac.jp

Fujiwara Takao
Kyoto City Univ of Arts
Nishikyo Ku
Kyoto 610 11
Japan
☎ 81 753 32 0701
✆ 81 75 332 0709
✉ Fujiwara@kcua.ac.jp

Fukuda Ichiro
Kanazawa Techn Inst
7 1 Ogigaoka
Nonoichimachi
Ishikawa 921
Japan
☎ 81 762 48 1100

Fukue Jun
Astronomical Institute
Osaka Kyoiku Univ
4 88 Minamikawahoricho
Osaka 543
Japan
☎ 81 729 76 3211 *3124
✆ 81 729 76 3269

Fukugita Masataka
Fundamental Phys
Kyoto Univ
Sakyo Ku
Kyoto 606 01
Japan
☎ 81 757 11 1381
✆ 81 75 753 7010

Fukui Takao
Dpt Liberal Arts
Dokkyo Univ
1-1 Gakuencho
Soka Saitama 340
Japan
☎ 81 489 42 1111
✆ 81 489 41 6621

Fukui Yasuo
Physics Dpt
Nagoya Univ
Furocho Chikusa Ku
Nagoya 464 01
Japan
☎ 81 527 815 111
✆ 81 527 89 2845
✉ fukui@a.phys.nagoya-u.ac.jp

Fukunaga Masataka
Astronomical Institute
Tohoku Univ
Sendai Aoba
Miyagi 980
Japan
☎ 81 222 22 1800
✆ 81 22 262 6609

Fukushima Toshio
Astrometry Cel Mech Div
NAO
Osawa Mitaka
Tokyo 181
Japan
☎ 81 422 34 3613
✆ 81 422 34 3793
✉ toshio@spacetime.mtk.nao.ac.jp

Fulchignoni Marcello
Observ Paris Meudon
DESPA
Pl J Janssen
F 92195 Meudon PPL Cdx
France
☎ 33 1 45 07 7539
✆ 33 1 45 07 7469

Fulle Marco
OAT
Box Succ Trieste 5
Via Tiepolo 11
I 34131 Trieste
Italy
☎ 39 40 31 99255
✆ 39 40 30 9418
✉ fulle@ts.astro.it

Furniss Ian
Dpt Phys/Astronomy
UCLO
Gower St
London WC1E 6BT
UK
☎ 44 171 387 7050
✆ 44 171 380 7145

Fursenko Margarita A
Inst Theoret Astron
Acad Sciences
N Kutuzova 10
191187 St Petersburg
Russia
☎ 7 812 278 8898
✆ 7 812 272 7968
✉ 1118@iipah.spb.su

Fusco-Femiano Roberto
IAS
Area d Ricerca CNR
Via Fosso Cavaliere 100
I 00133 Roma
Italy
☎ 39 6 4993 4473
✆ 39 6 2066 0188
✉ dario@saturn.rm.fra.cnr.it

Fusi Pecci Flavio
Dpt Astronomia
Univ d Bologna
Via Zamboni 33
I 40126 Bologna
Italy
☎ 39 51 259 301
✆ 39 51 259 407

G

Gabriel Alan H
IAS
Bt 121
Universite Paris XI
F 91405 Orsay Cdx
France
☎ 33 1 69 85 8500
℻ 33 1 69 85 8675

Gabriel Maurice R
Inst Astrophysique
Universite Liege
Av Cointe 5
B 4000 Liege
Belgium
☎ 32 4 254 7510
℻ 32 4 254 7511
✉ gabriel@astro.ulg.ac.be

Gadsden Michael
12 Keir St
Perth PH2 7HJ
UK

Gaetz Terrance J
CfA
HCO/SAO MS 81
60 Garden St
Cambridge MA 02138 1516
USA
☎ 1 617 496 7584
℻ 1 617 495 7356
✉ gaetz@cfa.harvard.edu

Gahm Goesta F
Stockholm Observ
Royal Swedish Acad Sciences
S 133 36 Saltsjoebaden
Sweden
☎ 46 8 717 0637
℻ 46 8 717 4719
✉ gahm@astro.su.se

Gaignebet Jean
OCA
CERGA
F 06130 Grasse
France
☎ 33 4 93 36 5849
℻ 33 4 93 40 5353

Gail Hans-Peter
Inst Theor Astrophysik
d Univer
Tiergartenstr 15
D 69121 Heidelberg
Germany
☎ 49 622 154 8982
℻ 49 622 154 4221
✉ gail@ita.uni-heidelberg.de

Gaisser Thomas K
Bartol Res Inst
Univ of Delaware
Newark DE 19716
USA
☎ 1 302 451 8111
℻ 1 302 831 1843

Gaizauskas Victor
NRCC/HIA
100 Sussex Dr
Ottawa ON K1A 0R6
Canada
☎ 1 613 993 7395
℻ 1 613 952 6602
✉ vgaizauskas@solar.stanford.edu

Galal A A
NAIGR
Helwan Observ
Cairo 11421
Egypt
☎ 20 78 0645/2683
℻ 20 62 21 405 297

Galan Maximino J
M/G Engs
S Martin de Porres 45
E 28035 Madrid
Spain
☎ 34 1 216 0995

Galeev Albert A
Space Res Inst
Acad Sciences
Profsojuznaya Ul 84/32
117810 Moscow
Russia
☎ 7 095 333 2588
℻ 7 095 310 7023

Galeotti Piero
Istt Cosmo Geofisica
CNR
Corso Fiume 4
I 10133 Torino
Italy
☎ 39 11 658 979

Galibina Irini V
Inst Theoret Astron
Acad Sciences
N Kutuzova 10
191187 St Petersburg
Russia
☎ 7 812 278 8898
℻ 7 812 272 7968

Galindo Trejo Jesus
Instituto Astronomia
UNAM
Apt 70 264
Mexico DF 04510
Mexico
☎ 52 5 622 3906
℻ 52 5 616 0653
✉ jgal@astroscu.unam.mx

Gallagher Jean W
Astronomy Dpt
Univ Wisconsin
475 N Charter St
Madison WI 53706 1582
USA
☎ 1 608 263 2456
℻ 1 608 263 0361
✉ jsg@jayg.astro.wisc.edu

Gallagher John S
Lowell Observ
1400 W Mars Hill Rd
Box 1149
Flagstaff AZ 86001
USA
☎ 1 602 734 3358
℻ 1 520 774 3358

Gallardo Castro Carlos Tabare
Dpt Astronomia
Fac Humani/Ciencia
Tristan Naraja 1674
Montevideo 11200
Uruguay
☎ 598 2 418 004
℻ 598 2 409 973
✉ gallardo@fisica.edu.uy

Gallego Juan Daniel
Ctr Astron d Yebes
OAN
Apt 148
E 19080 Guadalajara
Spain
☎ 34 1 129 0311
℻ 34 11 29 0063
✉ gallego@cay.oan.es

Gallet Roger M
964 7th St
Boulder CO 80302
USA

Galletta Giuseppe
Dpt Astronomia
Univ d Padova
Vic d Osservatorio 5
I 35122 Padova
Italy
☎ 39 49 829 3411
℻ 39 49 875 9840

Galletto Dionigi
Ist Fisica/Matematica
Univ d Torino
Via C Alberto 10
I 10123 Torino
Italy
☎ 39 11 539 214

Galliano Pier Giorgio
Istt Elettronico Nle
St della Cacce 91
I 10135 Torino
Italy
☎ 39 11 348 8933

Gallino Roberto
Istt Fisica
Univ d Torino
Corso d Azeglio 46
I 10125 Torino
Italy
☎ 39 11 655 103

Gallouet Louis
Observ Paris
61 Av Observatoire
F 75014 Paris
France
☎ 33 1 40 51 2207
℻ 33 1 44 54 1804

Galloway David
Applied Maths Dpt
Univ of Sydney
Sydney NSW 2006
Australia
☎ 61 2 692 2222
℻ 61 2 660 2903

Galperin Yuri I
Space Res Inst
Acad Sciences
Profsojuznaya Ul 84/32
117810 Moscow
Russia
☎ 7 095 333 3122
℻ 7 095 310 7023

Galsgaard Klaus
Mathematical Inst
Univ of St Andrews
North Haugh
St Andrews Fife KY16 9SS
UK
☎ 44 1334 76161
℻ 44 1334 74487
✉ klaus@dcs.st-and.ac.uk

Galt John A
DRAO
NRCC/HIA
Box 248
Penticton BC V2A 6K3
Canada
☎ 1 250 493 2277
℻ 1 250 493 7767

Gamaleldin Abdulla I
NAIGR
Helwan Observ
Cairo 11421
Egypt
☎ 20 78 0645/2683
✆ 20 62 21 405 297

Gambis Daniel
Observ Paris
DANOF
61 Av Observatoire
F 75014 Paris
France
☎ 33 1 40 51 2226
✆ 33 1 40 51 2232
✉ gambis@obspm.fr

Gammelgaard Peter Mog
Inst Phys/Astronomy
Univ of Aarhus
Ny Munkegade
DK 8000 Aarhus C
Denmark
☎ 45 89 42 2899
✆ 45 86 12 0740
✉ pg@obs.aau.dk

Gan Weiqun
Purple Mountain Obs
CAS
Nanjing 210008
China PR
☎ 86 25 331 7874
✆ 86 25 330 1459

Gao Bilie
Nanjing Astronomical
Instrument Factory
Box 846
Nanjing
China PR
☎ 86 25 46191
✆ 86 25 71 1256

Gao Buxi
Inst Geodesy/
Geophysics
Xu Dong Lu
Wu Han Hubei 430077
China PR
☎ 86 27 678 3841
✆ 86 27 678 3841
✉ gbx@asch.whigg.ac.cn

Gaposchkin Edward M
55 Farmcrest Av
Lexington MA 02173
USA
☎ 1 617 862 2538
✉ gaposche@ma.
ultranet.com

Garay Guido
Dpt Astronomia
Univ Chile
Casilla 36 D
Santiago
Chile
☎ 56 2 229 4101
✆ 56 2 229 4002

Garcia Beatriz Elena
Centro Regional Invest
Cientifica/Technologicas
CC 131
5500 Mendoza
Argentina
☎ 54 61 24 11794
✉ cricyt@planet.losandes.
com.ar

Garcia Domingo
Dpt Fisica/Enginyeria
Univ Barcelona
Avd Diagonal 647
E 08028 Barcelona
Spain
☎ 34 3 401 7421
✆ 34 3 402 1133
✉ domingo@f-720.upc.es

Garcia Eduardo Del pozo
Inst Geophys/Astron
C 212 N 2906/29 Y 31
Lisa
La Habana
Cuba
☎ 53 21 8435

Garcia Howard A
NOAA ERL R/E/SE3
Space Environment Lab
325 Broadway
Boulder CO 80303
USA
☎ 1 303 497 3916
✆ 1 303 497 3645
✉ hgarcia@solar.
stanford.edu

Garcia Jose I de la rosa
IAC
Observ d Teide
via Lactea s/n
E 38200 La Laguna
Spain
☎ 34 2 232 9100
✆ 34 2 232 9117

Garcia Lambas Diego
Observ Astronomico
de Cordoba
Laprida 854
5000 Cordoba
Argentina
☎ 54 51 23 0491
✆ 54 51 21 0613

Garcia Lopez Ramon J
IAC
Observ d Teide
via Lactea s/n
E 38200 La Laguna
Spain
☎ 34 2 232 9100
✆ 34 2 232 9117
✉ rgl@iac.es

Garcia Michael R
CfA
HCO/SAO
60 Garden St
Cambridge MA 02138 1516
USA
☎ 1 617 495 7169
✆ 1 617 495 7356
✉ garcia@cfa200.harvard.edu

Garcia-Barreto Jose A
Observ Astronomico Nacional
UNAM
Apt 877
Ensenada BC 22800
Mexico
☎ 52 6 174 4580
✆ 52 6 174 4607

Garcia-Berro Enrique
Dpt Fisica Aplicada
U P Cataluna
Campus Nord Ed C3
E 08034 Barcelona
Spain
☎ 34 3 401 6898
✆ 34 3 401 6801
✉ garcia@etseccpb.upc.es

Garcia-Burillo Santiago
Observ Astronomico
Nacional
Apt 1143
E 28800 Alcala d Henares
Spain
☎ 34 1 885 5060
✆ 34 1 885 5062
✉ burillo@oan.es

Garcia-Pelayo Jose
Inst Astrofisica
Andalucia Apt 3004
Prof Albareda 1
E 18080 Granada
Spain
☎ 34 5 825 6103
✆ 34 5 881 4530

Gardner Francis F
201b Beecroft Rd
Cheltenham NSW 2119
Australia

Garfinkel Boris
Astronomy Dpt
Yale Univ
Box 208101
New Haven CT 06520 8101
USA
☎ 1 203 436 3460
✆ 1 203 432 5048

Gargaud Muriel
Observ Bordeaux
BP 89
F 33270 Floirac
France
☎ 33 5 56 86 4330
✆ 33 5 56 40 4251
✉ muriel@frobor51

Garilli Bianca
IFCTR CNR
Univ d Milano
Via E Bassini 15
I 20133 Milano
Italy
☎ 39 2 236 3542
✆ 39 2 266 5753
✉ bianca@ifctr.mi.cnr.it

Garlick George F
267 South Beloit Av
Los Angeles CA 90049
USA
☎ 1 213 472 3512

Garmany Catherine D
JILA
Univ of Colorado
Campus Box 440
Boulder CO 80309 0440
USA
☎ 1 303 492 7836
✆ 1 303 492 5235

Garmire Gordon P
Astronomy Dpt
Pennsylvania State Univ
525 Davey Lab
University Park PA 16802
USA
☎ 1 814 865 0418
✆ 1 814 863 7114

Garnett Donald Roy
School Phys/Astronomy
Univ Minnesota
116 Church St SE
Minneapolis MN 55455
USA
☎ 1 612 624 1084
✆ 1 612 626 2029
✉ garnett@oldstyle.spa.
umn.edu

Garnier Robert
Observ Lyon
Av Ch Andre
F 69561 S Genis Laval cdx
France
☎ 33 4 78 56 0705
✆ 33 4 72 39 9791

Garrett Michael
NFRA
Postbus 2
NL 7990 AA Dwingeloo
Netherlands
☎ 31 521 59 5100
✆ 31 52 159 7332
✉ mag@jive.nfra.nl

Garrido Rafael
Inst Astrofisica
Andalucia Apt 3004
Prof Albareda 1
E 18080 Granada
Spain
☎ 34 5 812 1311
✆ 34 5 881 4530

Garrington Simon
NRAL
Univ Manchester
Jodrell Bank
Macclesfield SK11 9DL
UK
☎ 44 14 777 1321
✆ 44 147 757 1618
✉ janet stg@uk.ac.man.
jb.star

Garrison Robert F
David Dunlap Observ
Univ of Toronto
Box 360
Richmond Hill ONT L4C 4Y6
Canada
☎ 1 905 884 9562
✆ 1 905 884 2672
✉ garrison@astro.
utoronto.ca

Garstang Roy H
JILA
Univ of Colorado
Campus Box 440
Boulder CO 80309 0440
USA
☎ 1 303 492 7795
✆ 1 303 492 5235
✉ garstang@earthlink.net

Garton W R S
Astrophysics Gr
Imperial Coll
Prince Consort Rd
London SW7 2BZ
UK
☎ 44 171 594 7771
✆ 44 171 594 7772

Gary Dale E
CALTECH
MS 264 33
SA
Pasadena CA 91125
USA
☎ 1 818 356 3863

Gary Gilmer Allen
NASA/MSFC
Space Science Lab
Code ES 82
Huntsville AL 35812
USA
☎ 1 205 544 7609
✆ 1 205 544 5862
✉ allen.gary@msfc.
nasa.gov

Garzon Francisco
IAC
Observ d Teide
via Lactea s/n
E 38200 La Laguna
Spain
☎ 34 2 232 9100
✆ 34 2 232 9117

Gascoigne S C B
MSSSO
Weston Creek
Private Bag
Canberra ACT 2611
Australia
☎ 61 262 881 111
✆ 61 262 490 233

Gaska Stanislaw
Inst of Astronomy
N Copernicus Univ
Ul Chopina 12/18
PL 87 100 Torun
Poland
☎ 48 56 26 018
✆ 48 56 24 602

Gasparian Lazar
Byurakan Astrophys Observ
Armenian Acad Sci
378433 Byurakan
Armenia
☎ 374 88 52 28 3453/4142
✆ 374 88 52 52 3640
✉ byurakan@pnas.sci.am

Gatewood George
Allegheny Observ
Observ Station
Pittsburgh PA 15214
USA
☎ 1 412 321 2400
✆ 1 412 321 0606
✉ gatewood@vms.cis.
pitt.edu

Gatley Ian
NOAO
Box 26732
950 N Cherry Av
Tucson AZ 85726 6732
USA
☎ 1 520 327 5511
✆ 1 520 325 9360

Gaudenzi Silvia
Istt Astronomico
Univ d Roma La Sapienza
Via G M Lancisi 29
I 00161 Roma
Italy
☎ 39 6 44 03734
✆ 39 6 44 03673

Gaume Ralph A
USNO
3450 Massachusetts Av NW
Washington DC 20392 5420
USA
☎ 1 202 762 1519
✆ 1 202 762 1563
✉ gaume@nh3.usno.
navy.mil

Gaur V P
Uttar Pradesh State
Observ
Po Manora Peak 263 129
Nainital 263 129
India
☎ 91 59 42 2136

Gauss F Stephen
USNO
3450 Massachusetts Av NW
Washington DC 20392 5420
USA
☎ 1 202 762 1470
✆ 1 202 653 0944
✉ fsg@sicon.usno.navy.mil

Gaustad John E
Astronomy Dpt
Swarthmore Coll
Swarthmore PA 19081
USA
☎ 1 215 328 8271
✆ 1 215 328 8673
✉ jgausta1@cc.
swarthmore.edu

Gautier Daniel
Observ Paris Meudon
DESPA
Pl J Janssen
F 92195 Meudon PPL Cdx
France
☎ 33 1 45 07 7707
✆ 33 1 45 07 7971

Gautschy Alfred
Astronomisches Institut
Univ Basel
Venusstr 7
CH 4102 Binningen
Switzerland
☎ 41 61 271 7711/12
✆ 41 61 271 7810
✉ gautschy@astro.unibas.ch

Gavazzi Giuseppe
Osserv Astronomico
d Brera
Via Brera 28
I 20121 Milano
Italy
☎ 39 2 874 444
✆ 39 2 720 01600

Gay Jean
OCA
CERGA
F 06130 Grasse
France
☎ 33 4 93 36 5849
✆ 33 4 93 40 5353

Gayazov Iskander S
Inst Theoret Astron
Acad Sciences
N Kutuzova 10
191187 St Petersburg
Russia
☎ 7 812 275 4414
✆ 7 812 272 7968
✉ gayazov@iipah.spb.su

Gaylard Michael John
HartRAO
FRD
PO Box 443
1740 Krugersdorp
South Africa
☎ 27 11 642 4692
✆ 27 11 642 2446

Geake John E
Physics Dpt
UMIST
Box 88
Manchester M60 1QD
UK
☎ 44 161 200 3677
✆ 44 161 200 3669

Geballe Thomas R
Joint Astronomy Ctr
660 N A'ohoku Pl
University Park
Hilo HI 96720
USA
☎ 1 808 961 3756
✆ 1 808 961 6516

Gebbie Katharine B
JILA
Univ of Colorado
Campus Box 440
Boulder CO 80309 0440
USA
☎ 1 303 492 7825
✆ 1 303 492 5235

Gebler Karl-heinz
RAIUB
Univ Bonn
auf d Huegel 71
D 53121 Bonn
Germany
☎ 49 228 73 3662
✆ 49 228 73 3672

Geffert Michael
Observatorium Hoher List
Univ Sternwarte Bonn
D 54550 Daun
Germany
☎ 49 65 92 2150
✆ 49 65 92 9851 40
✉ geffert@astro.uni-
bonn.de

Gehrels Neil
NASA/GSFC
Code 661
Greenbelt MD 20771
USA
☎ 1 301 286 6546
✆ 1 301 286 1684
✉ gehrels@gsfc.nasa.gov

Gehrels Tom
Lunar/Planetary Lab
Room 325 Bg 92
Univ of Arizona
Tucson AZ 85721 0092
USA
☎ 1 520 621 6970
✆ 1 520 621 4933
✉ gehrels@lpl.arizona.edu

Gehren Thomas
Inst Astron/Astrophysik
Univ Sternwarte
Scheinerstr 1
D 81679 Muenchen
Germany
☎ 49 899 890 21
✆ 49 899 220 9427

Gehrz Robert Douglas
Dpt Phys/Astronomy
Univ of Wyoming
Box 3905
Laramie WY 82071
USA
☎ 1 307 766 6176
✆ 1 307 766 2652
✉ gehrz@astro.spa.
umn.edu

Geisler Douglas P
NOAO
CTIO
Casilla 603
La Serena
Chile
☎ 56 51 22 5415*208
✆ 56 51 20 5342
✉ dgeisler@noao.edu

Geiss Johannes
Physikalisches Institut
Univ Bern
Sidlerstr 5
CH 3012 Bern
Switzerland
☎ 41 31 65 8645

Geldzahler Bernard J
NRL
Code 4121 6
4555 Overlook Av SW
Washington DC 20375 5000
USA
☎ 1 202 767 6700

Gelfreikh Georgij B
Pulkovo Observ
Acad Sciences
10 Kutuzov Quay
196140 St Petersburg
Russia
☎ 7 812 298 2242
✆ 7 812 315 1701

Geller Margaret Joan
CfA
HCO/SAO
60 Garden St
Cambridge MA 02138 1516
USA
☎ 1 617 495 7409
✆ 1 617 495 7356

Gemmo Alessandra
Dpt Astronomia
Univ d Padova
Vic d Osservatorio 5
I 35122 Padova
Italy
☎ 39 49 829 3411
✆ 39 49 875 9840

Genet Russel M
Fairborn Observ
3435 E Edgewood Av
Mesa AZ 85204
USA
☎ 1 602 988 6561*223
✉ genet@regasus.la.asu.edu

Genkin Igor L
Fac of Physics
Kazakh State Univ
Tole Li 96
480012 Alma Ata
Kazakhstan
☎ 7 67 7018

Genova Francoise
Observ Strasbourg
11 r Universite
F 67000 Strasbourg
France
☎ 33 3 88 15 0721
✆ 33 3 88 15 0760
✉ genova@astro.u-strasbg.fr

Gent Hubert
Prospect House
Sherwood Lane
Worcester WR2 4NX
UK
☎ 44 142 2186

Genzel Reinhard
MPA
Karl Schwarzschildstr 1
D 85748 Garching
Germany
☎ 49 893 299 00
✆ 49 893 299 3235
✉ genzel@mpe.mpe-garching.mpg.de

Georgelin Yvon P
Observ Marseille
2 Pl Le Verrier
F 13248 Marseille Cdx 04
France
☎ 33 4 91 95 9088
✆ 33 4 91 62 1190
✉ ypgeorgelin@observatoire.cnrs-mrs.fr

Georgelin Yvonne M
Observ Marseille
2 Pl Le Verrier
F 13248 Marseille Cdx 04
France
☎ 33 4 91 95 9088
✆ 33 4 91 62 1190

Georgiev Leonid
Astronomy Dpt
Univ of Sofia
Anton Ivanov St 5
BG 1126 Sofia
Bulgaria
☎ 359 2 68 8176
✆ 359 2 68 9085

Georgiev Tsvetan
Astronomy Dpt
Bulgarian Acad Sci
72 Lenin Blvd
BG 1784 Sofia
Bulgaria
☎ 359 2 88 3503
✆ 359 2 75 5019

Gerard Eric
Observ Paris Meudon
ARPEGES
Pl J Janssen
F 92195 Meudon PPL Cdx
France
☎ 33 1 45 07 7607
✆ 33 1 45 07 7971
✉ daigne@obspm.fr

Gerard Jean-Claude M C
Inst Astrophysique
Universite Liege
Av Cointe 5
B 4000 Liege
Belgium
☎ 32 4 254 7510
✆ 32 4 254 7511
✉ gerard@astro.ulg.ac.be

Gerasimov Igor A
SAI
Acad Sciences
Universitetskij Pr 13
119899 Moscow
Russia
☎ 7 095 939 1625
✆ 7 095 939 0126
✉ snn@sai.msk.su

Gerbal Daniel
IAP
98bis bd Arago
F 75014 Paris
France
☎ 33 1 44 32 8000
✆ 33 1 44 32 8001

Gerbaldi Michele
IAP
98bis bd Arago
F 75014 Paris
France
☎ 33 1 44 32 8041
✆ 33 1 44 32 8001
✉ gerbaldi@iap.fr

Gergely Tomas Esteban
NSF
Div Astron Sci
4201 Wilson Blvd
Arlington VA 22230
USA
☎ 1 703 306 1823
✆ 1 703 306 0525
✉ tgergely@nsf.gov

Gerhard Ortwin
Astronomisches Institut
Univ Basel
Venusstr 7
CH 4102 Binningen
Switzerland
☎ 41 61 205 5419/454
✆ 41 61 205 5455
✉ gehrard@astro.unibas.ch

Gerin Maryvonne
DEMIRM
ENS
24 r Lhomond
F 75231 Paris Cdx 05
France
☎ 33 1 43 29 1225
✆ 33 1 40 51 2002
✉ gerin@frulm11

Gerlei Otto
Debrecen Heliophys Observ
Acad Sciences
Box 30
H 4010 Debrecen
Hungary
☎ 36 5 231 1015

Gerola Humberto
20390 Knollwood Dr
Saratoga CA 95070
USA

Geroyannis Vassilis S
Astronomy Dpt
Univ of Patras
GR 261 10 Rion
Greece
☎ 30 61 99 7572
✆ 30 61 99 7636

Gershberg R E
Crimean Astrophys Obs
Ukrainian Acad Science
Nauchny
334413 Crimea
Ukraine
☎ 380 6554 71161
✆ 380 6554 40704

Gesicki Krzysztof
Inst of Astronomy
N Copernicus Univ
Ul Chopina 12/18
PL 87 100 Torun
Poland
☎ 48 56 26 08*50
✆ 48 56 24 602
✉ gesicki@astri.uni.torun.pl

Geyer Edward H
Observatorium Hoher List
Univ Sternwarte Bonn
D 54550 Daun
Germany
☎ 49 65 92 2150
✆ 49 65 92 9851 40

Gezari Daniel Ysa
NASA GSFC
Code 685
Greenbelt MD 20771
USA
☎ 1 301 286 3432

Ghanbari Jamshid
Physics Dpt
School of Sciences
Univ of Ferdowsi
Mashhad
Iran
☎ 98 51 832 021/4
✆ 98 51 838 032
✉ ghanbari@science1.um.ac

Ghigo Francis D
NRAO
Box 2
Green Bank WV 24944
USA
☎ 1 304 456 2011
✆ 1 304 456 2271
✉ fghigo@nrao.edu

Ghisellini Gabriele
Osserv Astronomico
d Milano
Via E Bianchi 46
I 22055 Merate
Italy
☎ 39 990 6412
✆ 39 990 8492
✉ gabriele@merate.mi.astro.it

Ghizaru Mihai
Astronomical Institute
Romanian Acad Sciences
Cutitul de Argint 5
RO 75212 Bucharest
Rumania
☎ 40 1 335 6892
✆ 40 1 312 3391
✉ mghizaru@roimar.imar.ro

Ghobros Roshdy Azer
NAIGR
Helwan Observ
Cairo 11421
Egypt
☎ 20 78 0645/2683
✆ 20 62 21 405 297

Ghosh Kajal Kumar
IIA
Koramangala
Sarjapur Rd
Bangalore 560 034
India
☎ 91 80 356 6585
✆ 91 80 553 4043

Ghosh P
TIFR
Homi Bhabha Rd
Colaba
Bombay 400 005
India
☎ 91 22 219 111*260
✆ 91 22 495 2110

Ghosh S K
TIFR
Homi Bhabha Rd
Colaba
Bombay 400 005
India
☎ 91 22 215 2311
✆ 91 22 495 2110
✉ swarna@tifrvax.tifr.res.in

Giacaglia Giorgio E
CRAAE INPE
EPUSP/PTR
CP 61548
01065 970 Sao Paulo SP
Brazil
☎ 55 11 815 9322
✆ 55 11 815 4272

Giacani Elsa Beatriz
IAR
CC 5
1894 Villa Elisa (Bs As)
Argentina
☎ 54 21 4 3793
✆ 54 21 211 761

Giacconi Riccardo
ESO
Karl Schwarzschildstr 2
D 85748 Garching
Germany
☎ 49 893 200 60
✆ 49 893 202 362

Giachetti Riccardo
Osserv Astrofis Arcetri
Univ d Firenze
Largo E Fermi 2
I 50125 Firenze
Italy
☎ 39 55 22 98141
✆ 39 55 435939

Giallongo Emanuele
Osserv Astronomico
d Roma
Via d osservatorio 5
I 00040 Monteporzio
Italy
☎ 39 6 944 9019
✆ 39 6 944 7243

Giampapa Mark S
NOAO/NSO
Box 26732
950 N Cherry Av
Tucson AZ 85726 6732
USA
☎ 1 520 327 5511
✆ 1 520 325 9278

Giannone Pietro
OAR
Via d Parco Mellini 84
I 00136 Roma
Italy
☎ 39 6 34 7056
✆ 39 6 349 8236

Giannuzzi Maria A
Dpt Matematica
Univ d Roma La Sapienza
Piazza Gramsci 5
I 00041 Albano/laziale
Italy
☎ 39 6 932 1101

Gibson David Michael
MIT/Lincoln Laboratory
KMR Field Site
Box 58
Apo AP 96555
USA
☎ 1 805 355 1257
✆ 1 805 355 3833
✉ gibson@kmrmail.kmr.ll.mit.edu

Gibson James
6838 Greeley St
Tujunga CA 91042 2809
USA

Giclas Henry L
120 E Elm Av
Flagstaff AZ 86001
USA
☎ 1 520 774 4769
✆ 1 520 774 6296

Gierasch Peter J
Astronomy Dpt
Cornell Univ
512 Space Sc Bldg
Ithaca NY 14853 6801
USA
☎ 1 607 256 3507
✆ 1 607 255 1767

Gieren Wolfgang P
Dpt Fisica
Univ Concepcion
Casilla 4009
Concepcion
Chile
☎ 56 41 204 500
✆ 56 41 245 622
✉ wgieren@coma.cfm.udec.cl

Giersz Miroslaw
Copernicus Astron Ctr
Polish Acad Sci
Ul Bartycka 18
PL 00 716 Warsaw
Poland
☎ 48 22 41 1086
✆ 48 22 41 0828
✉ mig@camk.edu.pl

Gies Douglas R
Dpt Phys/Astronomy
Georgia State Univ
Atlanta GA 30303 3083
USA
☎ 1 404 651 1366
✆ 1 404 651 1389
✉ gies@chara.gsu.edu

Gietzen Joseph W
Royal Greenwich Obs
Madingley Rd
Cambridge CB3 0EZ
UK
☎ 44 12 23 374 000
✆ 44 12 23 374 700

Gigas Detlef
NASA GSFC
Code 910 2
Greenbelt MD 20771
USA

Gigoyan Kamo
Byurakan Astrophys Observ
Armenian Acad Sci
378433 Byurakan
Armenia
☎ 374 88 52 28 3453/4142
✆ 374 88 52 52 3640
✉ kgigoyan@helios.sci.am

Gil Janusz A
The Astronomical Ctr
Lubuska 2
PL 65 001 Zielona Gora
Poland
☎ 48 68 20 2863
✆ 48 68 202 863

Gilfanov Marat R
Space Res Inst
Acad Sciences
Profsojuznaya Ul 84/32
117810 Moscow
Russia
☎ 7 095 333 3377
✆ 7 095 333 5377
✉ mgilfano@hea.iki.rssi.ru

Gillanders Gerard
Physics Dpt
Univ College
Galway
Ireland
☎ 353 91 524 411*2529
✆ 353 91 525 700
✉ gary.gillanders@ucg.ie

Gillet Denis
OHP
F 04870 S Michel Obs
France
☎ 33 4 92 76 6368
✆ 33 4 92 76 6295

Gilliland Ronald Lynn
STScI
Homewood Campus
3700 San Martin Dr
Baltimore MD 21218
USA
☎ 1 301 338 4700
✆ 1 301 338 4767

Gillingham Peter
W M Keck Observ
Box 220
65 1120 Mamalahoa Hwy
Kamuela HI 96743
USA
☎ 1 808 885 7887
✆ 1 808 885 4464
✉ peter@keck.hawaii.edu

Gilman Peter A
HAO
NCAR
Box 3000
Boulder CO 80307 3000
USA
☎ 1 303 497 1560
✆ 1 303 497 1568

Gilmore Alan C
Mt John Observ
Box 56
Lake Tekapo 8770
New Zealand
☎ 64 64 5 056 813

Gilmore Gerard Francis
Inst of Astronomy
The Observatories
Madingley Rd
Cambridge CB3 0HA
UK
☎ 44 12 23 337 548
✆ 44 12 23 337 523
✉ gil@ast.cam.ac.uk

Gilmozzi Roberto
ESO
Karl Schwarzschildstr 2
D 85748 Garching
Germany
☎ 49 893 200 60
✆ 49 893 202 362

Gilra Daya P
SM Systems/Res Co
8401 Corporate Dr
Suite 450
Landover MD 20785
USA
☎ 1 301 763 4483

Gimenez Alvaro
ESA
Apt 50727 Villafranca
E 28080 Madrid
Spain
☎ 34 1 675 0700
✆ 34 1 813 1139

Gingerich Owen
CfA
HCO/SAO
60 Garden St
Cambridge MA 02138 1516
USA
☎ 1 617 495 7216
✆ 1 617 496 7564
✉ ginger@cfa.harvard.edu

Gingold Robert Arthur
Australian Ntl Univ
ANU Super cptr Facility
Box 4
Canberra ACT 2600
Australia
☎ 61 2 62 493 437

Ginzburg Vitaly L
Lebedev Physical Inst
Acad Sciences
Leninsky Pspt 53
117924 Moscow
Russia
☎ 7 095 135 2250
✆ 7 095 135 7880

Gioia Isabella M
Inst for Astronomy
Univ of Hawaii
2680 Woodlawn Dr
Honolulu HI 96822
USA
☎ 1 808 956 9845
✆ 1 808 956 9590
🖂 gioia@galileo.ifa.
hawaii.edu

Giovanardi Carlo
Osserv Astrofis Arcetri
Univ d Firenze
Largo E Fermi 5
I 50125 Firenze
Italy
☎ 39 55 27 52239
✆ 39 55 22 0039

Giovane Frank
NRL
Code 7601
4555 Overlock Av SW
Washington DC 20375 5320
USA
☎ 1 202 767 3188
✆ 1 202 404 7296
🖂 giovane@nrl.navy.mil

Giovanelli Riccardo
Astronomy Dpt
Cornell Univ
512 Space Sc Bldg
Ithaca NY 14853 6801
USA
☎ 1 607 255 6505
✆ 1 607 255 8803
🖂 riccardo@astrosun.tn.
cornell.edu

Giovannelli Franco
IAS
Area d Ricerca CNR
Via Fosso Cavaliere 100
I 00133 Roma
Italy
☎ 39 6 4993 4473
✆ 39 6 2066 0188
🖂 frano@saturn.rm.
fra.cnr.it

Giovannini Gabriele
Istt Radioastronomia
CNR
Via P Gobetti101
I 40129 Bologna
Italy
☎ 39 51 639 9385
✆ 39 51 639 9431
🖂 ggiovannini@astbo1.
bo.cnr.it

Gir Be Young
Dpt Earth Sciences
Ntl Taiwan Normal Univ
Taiwan
China R

Giraud Edmond
CPT
CNRS
Luminy Case 07
F 13288 Marseille Cdx
France
☎ 33 4 91 26 9519

Giridhar Sunetra
IIA
Koramangala
Sarjapur Rd
Bangalore 560 034
India
☎ 91 80 356 6585
✆ 91 80 553 4043

Giuricin Giuliano
Dpt Astronomia
Univ d Trieste
Via Tiepolo 11
I 34131 Trieste
Italy
☎ 39 40 76 8005
✆ 39 40 30 9418

Glagolevskij Juri V
SAO
Acad Sciences
Nizhnij Arkhyz
357147 Karachaevo
Russia
☎ 7 878 789 3577

Glaser Harold
1346 Bonita St
Berkeley CA 94709
USA
☎ 1 415 527 1860

Glasner Shimon Ami
Racah Inst of Phys
Hebrew Univ Jerusalem
Jerusalem 91904
Israel
☎ 972 2 584 521
✆ 972 2 658 4374

Glaspey John W
NOAO/KPNO
Box 26732
950 N Cherry Av
Tucson AZ 85726 6732
USA
☎ 1 520 318 8701
✆ 1 520 318 8724
🖂 glaspey@noao.edu

Glass Billy Price
Dpt Geology
Univ of Delaware
Newark DE 19716
USA
☎ 1 302 451 8458
✆ 1 302 831 1843

Glass Ian Stewart
SAAO
PO Box 9
7935 Observatory
South Africa
☎ 27 21 47 0025
✆ 27 21 47 3639
🖂 isg@saao.ac.za

Glassgold Alfred E
Physics Dpt
New York Univ
4 Washington Place
New York NY 10003
USA
☎ 1 212 598 2020
✆ 1 212 995 4016

Glatzel Wolfgang
Univ Sternwarte
Goettingen
Geismarlandstr 11
D 37083 Goettingen
Germany
☎ 49 551 395 042
✆ 49 551 395 043

Glatzmaier Gary A
LANL
MS F665 ESS 5
Box 1663
Los Alamos NM 87545 2345
USA
☎ 1 505 667 7647
✆ 1 505 665 4055

Glebocki Robert
Inst Theor Physics
Univ of Gdansk
Ul Wita Stwosza 57
PL 80 952 Gdansk
Poland
☎ 48 41 5241*188

Glebova Nina I
Inst Theoret Astron
Acad Sciences
N Kutuzova 10
191187 St Petersburg
Russia
☎ 7 812 278 8898
✆ 7 812 272 7968
🖂 100073.3636@
compuserve.com

Glencross William M
Dpt Phys/Astronomy
UCLO
Gower St
London WC1E 6BT
UK
☎ 44 171 387 7050
✆ 44 171 380 7145

Glushneva Irina N
SAI
Acad Sciences
Universitetskij Pr 13
119899 Moscow
Russia
☎ 7 095 139 2046
✆ 7 095 939 0126

Gnedin Yurij N
Ioffe Physical Tech Inst
Acad Sciences
Polytechnicheskaya Ul 26
194021 St Petersburg
Russia
☎ 7 812 123 4493
✆ 7 812 314 3360
🖂 gnedin@pulkovo.
spb.su

Godfrey Peter Douglas
Chemistry Dpt
Monash Univ
Wellington Rd
Clayton VIC 3168
Australia
☎ 61 3 541 0811

Godlowski Wlodzimierz
Astronomical Observ
Krakow Jagiellonian Univ
Ul Orla 171
PL 30 244 Krakow
Poland
☎ 48 12 25 1294
✆ 48 12 251 318
🖂 godlows@oa.uj.edu.pl

Godoli Giovanni
Dpt Astronomia
Univ d Firenze
Largo E Fermi 5
I 50125 Firenze
Italy
☎ 39 55 27 521
✆ 39 55 22 0039

Godwin Jon Gunnar
Astrophysics Dpt
85 Cherwell Dr
Oxford OX3 0ND
UK
☎ 44 186 572 2313
✆ 44 1865 72 1000

Goebel Ernst
Inst Astronomie
Univ Wien
Tuerkenschanzstr 17
A 1180 Wien
Austria
☎ 43 1 345 360186
✆ 43 1 470 6015

Goebel John H
NASA/ARC
Space Sci Div
MS 244 7
Moffett Field CA 94035 1000
USA
☎ 1 415 694 6525
✆ 1 415 604 4003

Goedbloed Johan P
FOM
Box 1207
NL 3430 BE Nieuwegein
Netherlands
☎ 31 30 603 1224
✆ 31 30 6031204
🖂 goedbloed@sara.nl

Goelbasi Orhan
Fac of Sciences
Inonu Univ
44069 Malatya
Turkey
☎ 90 821 21871
✆ 90 821 18133

Gokdogan Nuzhet
Univ Observ
Univ of Istanbul
University 34452
34452 Istanbul
Turkey
☎ 90 212 522 3597
✆ 90 212 519 0834

Gokhale Moreshwar Hari
IIA
Koramangala
Sarjapur Rd
Bangalore 560 034
India
☎ 91 80 356 6585
℻ 91 80 553 4043

Golap Kumar
Fac of Sciences
Univ of Mauritius
Reduit
Mauritius
☎ 230 454 1041
℻ 230 454 9642
✉ kumar@f1.n726.
z5.fidonet.org

Golay Marcel
Observ Geneve
Chemin d Maillettes 51
CH 1290 Sauverny
Switzerland
☎ 41 22 755 2611
℻ 41 22 755 3983
✉ marcel.golay@obs.
unige.ch

Gold Thomas
CRSR
Cornell Univ
306 Space Sc Bldg
Ithaca NY 14853 6801
USA
☎ 1 607 256 5284
℻ 1 607 255 9888

Goldbach Claudine
IAP
98bis bd Arago
F 75014 Paris
France
☎ 33 1 44 32 8141
℻ 33 1 44 32 8001

Goldes Guillermo Victor
Observ Astronomico
de Cordoba
Laprida 854
5000 Cordoba
Argentina
☎ 54 51 331 006
℻ 54 51 331 063
✉ goldes@oac.uncor.edu

Goldman Itzhak
Dpt Phys/Astronomy
Tel Aviv Univ
Ramat Aviv
Tel Aviv 69978
Israel
☎ 972 3 545 0303
℻ 972 3 640 8179
✉ goldman@vm.tau.ac.il

Goldman Martin V
CASA
Univ of Colorado
Campus Box 389
Boulder CO 80309 0391
USA
☎ 1 303 492 8896
℻ 1 303 492 6946

Goldreich Peter
CALTECH
Pasadena CA 91125
USA
☎ 1 213 356 6193
℻ 1 818 397 9600
✉ peter@ciideimo

Goldsmith Donald W
Interstellar Media
2153 Rusell St
Berkeley CA 94705
USA
☎ 1 415 848 1989

Goldsmith Paul F
Leadmine Hill Rd 5
Amherst MA 01002
USA
☎ 1 413 545 0925

Goldsmith S
Dpt Phys/Astronomy
Tel Aviv Univ
Ramat Aviv
Tel Aviv 69978
Israel
☎ 972 3 420 303
℻ 972 3 640 8179

Goldstein Richard M
CALTECH/JPL
MS 300 227
4800 Oak Grove Dr
Pasadena CA 91109 8099
USA
☎ 1 818 354 6999
℻ 1 818 393 6030

Goldstein Samuel J
University Station
Univ of Virginia
Box 3818
Charlottesville VA 22903 0818
USA
☎ 1 804 924 7494
✉ sjg@astsun.astro.
virginia.edu

Goldsworthy Frederick A
Dpt Appl Maths
Univ of Leeds
Leeds LS2 9JT
UK
☎ 44 113 233 5110
℻ 44 113 242 9525

Goldwire Henry C
LLNL
L 451
Box 808
Livermore CA 94551 9900
USA
☎ 1 415 423 0160
℻ 1 415 423 0238

Golev Valery K
Dpt Astron/Fac of Phys
St Kliment of Okhrid Uni
Univ Astron Obs Box 36
BG 1504 Sofia
Bulgaria
☎ 359 2 66 2324
✉ valgol@phys.uni-sofia.bg

Gollnow H
MSSSO
Weston Creek
Private Bag
Canberra ACT 2611
Australia
☎ 61 2 62 881 111

Golovatyj Volodymyr
Astronomical Observ
Lviv State Univ
Kiril/Mefodiy Str 8
290005 Lviv
Ukraine
☎ 380 3227 29088
℻ 380 3227 29088
✉ gol@astro.lviv.ua

Golub Leon
CfA
HCO/SAO
60 Garden St
Cambridge MA 02138 1516
USA
☎ 1 617 495 7177
℻ 1 617 495 7356

Gomes Alercio M
R Gaviao Peixoto 13
Apt 1401
Icarai 24000
Niteroj Erj
Brazil

Gomes Rodney D S
Observ Nacional
Rua Gl Bruce 586
Sao Cristovao
20921 030 Rio de Janeiro RJ
Brazil
☎ 55 21 589 2955
℻ 55 21 580 0332
✉ rodney@on.br

Gomez Ana E
Observ Paris Meudon
DASGAL
Pl J Janssen
F 92195 Meudon PPL Cdx
France
☎ 33 1 45 07 7843
℻ 33 1 45 07 7971

Gomez Gonzalez Jesus
Paseo Imperial 29 6h
E Madrid 5
Spain

Gomez Maria Theresa
Osserv d Capodimonte
Via Moiariello 16
I 80131 Napoli
Italy
☎ 39 81 44 0101
℻ 39 81 45 6710

Gomez Mercedes
Observ Astronomico
de Cordoba
Laprida 854
5000 Cordoba
Argentina
☎ 54 51 236 876
℻ 54 51 21 0613
✉ mercedes@uncbob.edu.ar

Gomez Yolanda
Instituto Astronomia
J J Tablada 1006
C L de Santa Maria
Morelia Michoacan 58090
Mexico
☎ 52 4 323 6162
℻ 52 4 323 6165
✉ gocy@astrosmo.unam.mx

Gomide Fernando de Mello
R Angelica Lopes de
Castro 71
25655 430 Petropolis RJ
Brazil

Gonczi Georges
OCA
Observ Nice
BP 139
F 06304 Nice Cdx 4
France
☎ 33 4 93 89 0420
℻ 33 4 92 00 3033

Gondhalekar Prabhakar
Rutherford Appleton Lab
Space/Astrophysics Div
Bg R25/R68
Chilton Didcot OX11 0QX
UK
☎ 44 12 35 821 900
℻ 44 12 35 44 5808

Gondolatsch Friedrich
ARI
Moenchhofstr 12-14
D 69120 Heidelberg
Germany
☎ 49 622 140 50
℻ 49 622 140 5297

Gong Shou-shen
Shanghai Observ
CAS
80 Nandan Rd
Shanghai 200030
China PR
☎ 86 21 6438 6191
℻ 86 21 6438 4618

Gong Shumo
Purple Mountain Obs
CAS
Nanjing 210008
China PR
☎ 86 25 46700
℻ 86 25 301 459

Gong Xiangdong
Shanghai Observ
CAS
80 Nandan Rd
Shanghai 200030
China PR
☎ 86 21 6438 6191
℻ 86 21 6438 4618
✉ yangfm@center.shao.ac.cn

Gontcharov Gueorgui A
Pulkovo Observ
Acad Sciences
10 Kutuzov Quay
196140 St Petersburg
Russia
☎ 7 812 153 0925
℻ 7 812 123 4922
✉ gaoran@mail.wplus.net

Gonzales Alejandro
INAOE
Tonantzintlaz
Apdo 51 y 216
Puebla PUE 72000
Mexico
☎ 52 2 247 2011*2307
✆ 52 2 247 2231
✉ alegs@inaoep.mx

Gonzales Jean-Francois
CRAL
ENS
46 All Italie
F 69364 Lyon Cdx 07
France
☎ 33 4 72 72 8466
✆ 33 4 72 72 8080
✉ jfgonzal@physique.ens-lyon.fr

Gonzales'a Walter D
INPE
Dpt Astronomia
CP 515
12227 010 S Jose dos Campos
Brazil
☎ 55 123 22 9977
✆ 55 123 21 8743

Gonzales-Alfonso Eduardo
Observ Astronomico Nacional
Apt 1143
E 28800 Alcala d Henares
Spain
☎ 34 1 885 5060
✆ 34 1 885 5062
✉ eduardo@oan.es

Gonzalez Camacho Antonio
Inst Astron/Geodesia
Fac Ciencias Matemat
Univ Complutense
E 28040 Madrid
Spain
☎ 34 1 244 2501
✆ 34 1 394 5195

Gonzalez G
INAOE
Tonantzintlaz
Apdo 51 y 216
Puebla PUE 72000
Mexico
☎ 52 2 247 2011
✆ 52 2 247 2231

Gonzalez J Jesus
Instituto Astronomia
UNAM
Apt 70 264
Mexico DF 04510
Mexico
☎ 52 5 622 3906
✆ 52 5 616 0653
✉ jesus@astroscu.unam.mx

Gonzalez Serrano J I
Istt Fisica d Cantabria
C/o Facultad Ciencias
Avda Los Castros s/n
E 39005 Santander
Spain
☎ 34 4 220 1450
✆ 34 4 220 1402
✉ gserrano@ccucvx.unican.es

Gonzalez-Riestra R
ESA
Apt 50727 Villafranca
E 28080 Madrid
Spain
☎ 34 1 813 1100
✆ 34 1 813 1139

Gonze Roger F J
Koninklijke Sterrenwacht van Belgie
Ringlaan 3
B 1180 Brussels
Belgium
☎ 32 2 373 0294
✆ 32 2 373 9822
✉ roger.gonze@oma.be

Goode Philip R
Physics Dpt
NJ Inst of Technology
323 High St
Newark NJ 07102
USA
☎ 1 201 596 3562

Goodman Alyssa Ann
Astronomy Dpt
Harvard Univ MS 46
60 Garden St
Cambridge MA 02138 1516
USA
☎ 1 617 495 9278
✆ 1 617 495 7345
✉ agoodman@cfa.harvard.edu

Goodrich Robert W
W M Keck Observ
Box 220
65 1120 Mamalahoa Hwy
Kamuela HI 96743
USA
☎ 1 808 885 7887
✆ 1 808 885 4464
✉ rwg@deimos.caltech.edu

Goody R M
CEPP
Pierce Hall
29 Oxford St
Cambridge MA 02138
USA
☎ 1 617 495 4517

Goossens Marcel
Ctr Plasma Astrophysics
Katholicke Univ Leuven
Celestijnenlaan 200B
B 3001 Heverlee
Belgium
☎ 32 16 32 7012
✆ 32 16 32 7998
✉ marcel.goossens@wis.kuleuven.ac.be

Gopala Rao U V
Satellite Meteorology
Indian Meteorologic Dpt
Lodi Rd/mausam Bhavan
New Delhi 110 003
India

Gopalswamy N
Astronomy Program
Univ of Maryland
College Park MD 20742 2421
USA
☎ 1 301 454 6649
✆ 1 301 314 9067

Gopasyuk S I
Crimean Astrophys Obs
Ukrainian Acad Science
Nauchny
334413 Crimea
Ukraine
☎ 380 6554 71161
✆ 380 6554 40704

Gor'kavyi Nikolai
Crimean Astrophys Obs
Simeiz
334242 Yalta
Ukraine
☎ 380 6542 40077
✆ 380 6554 40704
✉ gorka@crao.crimea.ua

Gorbatsky Vitalij G
Astronomical Observ
St Petersburg Univ
Bibliotechnaja Pl 2
198904 St Petersburg
Russia
☎ 7 812 428 7129
✆ 7 812 428 4259
✉ vayak@astro.lgu.spb.su

Gordon Charlotte
11 r Tournefort
F 75005 Paris
France

Gordon Courtney P
Astronomy Dpt
Hampshire Coll
Amherst MA 01002
USA
☎ 1 413 545 2290

Gordon Isaac M
Inst Radio Astron
Ukrainian Acad Science
4 Chervonopraporna st
310085 Kharkiv
Ukraine

Gordon Kurtiss J
Astronomy Dpt
Hampshire Coll
Amherst MA 01002
USA
☎ 1 413 549 4600

Gordon Mark A
NRAO
Campus Bg 65
949 N Cherry Av
Tucson AZ 85721 0655
USA
☎ 1 520 882 8250
✆ 1 520 882 7955

Gorenstein Marc V
CfA
HCO/SAO MS 42
60 Garden St
Cambridge MA 02138 1516
USA
☎ 1 617 495 9296
✆ 1 617 495 7356

Gorenstein Paul
CfA
HCO/SAO
60 Garden St
Cambridge MA 02138 1516
USA
☎ 1 617 495 7250
✆ 1 617 495 7356

Goret Philippe
DAPNIA/SAP
CEA Saclay
BP 2
F 91191 Gif s Yvette Cdx
France
☎ 33 1 69 08 4463
✆ 33 1 69 08 6577

Gorgas Garcia Javier
Dpt Astrofisica
Fac Fisica
Univ Complutense
E 28040 Madrid
Spain
☎ 34 1 549 5316
✆ 34 1 394 5195

Gorgolewski Stanislaw Pr
Inst Radio Astron
N Copernicus Univ
Ul Gagarina 11
PL 87 100 Torun
Poland
☎ 48 56 78 3327
✆ 48 56 11 651

Gorschkov Alexander G
SAI
Acad Sciences
Universitetskij Pr 13
119899 Moscow
Russia
☎ 7 095 939 3838
✆ 7 095 932 8841
✉ algor@sai.msu.su

Gosachinskij Igor V
SAO/St Petersburg Br
Acad Sciences
10 Kutuzov Quay
196140 St Petersburg
Russia
☎ 7 812 123 4372
✆ 7 812 123 2042
✉ gos@fsao.spb.su

Gosling John T
LANL
MS D438 ESS8
Box 1663
Los Alamos NM 87545 2345
USA
☎ 1 505 667 5389
✆ 1 505 665 4055

Goss W Miller
NRAO
VLA
Box 0
Socorro NM 87801 0387
USA
☎ 1 505 772 4011
✆ 1 505 835 7027

Gosset Eric
Inst Astrophysique
Universite Liege
Av Cointe 5
B 4000 Liege
Belgium
☎ 32 4 254 7518
✆ 32 4 254 7511
✉ gosset@astro.ulg.ac.bc

Goswami J N
PRL
Navrangpura
Ahmedabad 380 009
India
☎ 91 272 46 2129
✆ 91 272 44 5292

Gott J Richard
Dpt Astrophysical Sci
Princeton Univ
Princeton NJ 08544 1001
USA
☎ 1 609 452 3813
✆ 1 609 258 1020

Gottesman Stephen T
Astronomy Dpt
Univ of Florida
211 SSRB
Gainesville FL 32611
USA
☎ 1 904 392 2050/2052
✆ 1 904 392 5089

Gottlieb Carl A
Inst Space Studies
2880 Broadway
New York NY 10025
USA
☎ 1 212 678 5566

Gottloeber Stefan
Astrophysik Inst
Potsdam Univ
an d Sternwarte 16
D 14482 Potsdam
Germany
☎ 49 331 749 90
✆ 49 331 749 9200
✉ sgottloeber@aip.de

Gouda Naoteru
Dpt Earth Sp Sciences
Fac Sciences
Osaka Univ
Toyonaka Osaka 560
Japan
☎ 81 684 41 151 *4117
✆ 81 685 31 787
✉ gouda@oskastro.kek.ac.jp

Goudas Constantine L
Mathematics Dpt
Univ of Patras
GR 261 10 Rion
Greece
☎ 30 61 99 7572
✆ 30 61 99 7636

Goudis Christos D
Physics Dpt
Univ of Patras
GR 261 10 Rion
Greece
☎ 30 61 99 7572
✆ 30 61 99 7636

Gough Douglas O
Inst of Astronomy
The Observatories
Madingley Rd
Cambridge CB3 0HA
UK
☎ 44 12 23 337 548
✆ 44 12 23 337 523

Gouguenheim Lucienne
Observ Paris Meudon
ARPEGES
Pl J Janssen
F 92195 Meudon PPL Cdx
France
☎ 33 1 45 07 7604
✆ 33 1 45 07 7939
✉ gouguenheim@
obspm.fr

Gould Robert J
Physics Dpt
UCSD
B 01
La Jolla CA 92093 0216
USA
☎ 1 619 452 3649
✆ 1 619 534 2294

Goupil Marie-Jose
Observ Paris Meudon
DASGAL
Pl J Janssen
F 92195 Meudon PPL Cdx
France
☎ 33 1 45 07 7880
✆ 33 1 45 07 7472

Gouttebroze Pierre
IAS
Bt 121
Universite Paris XI
F 91405 Orsay Cdx
France
☎ 33 1 69 85 8621
✆ 33 1 69 85 8675

Gower Ann C
Physics Dpt
Univ of Victoria
Box 1700
Victoria BC V8W 2Y2
Canada
☎ 1 250 721 7700
✆ 1 250 721 7715

Gower J F R
1615 Mctavish Rd Rr 2
Sidney BC V8l 3S1
Canada
☎ 1 250 656 5457

Goy Gerald
Observ Geneve
Chemin d Maillettes 51
CH 1290 Sauverny
Switzerland
☎ 41 22 755 2611
✆ 41 22 755 3983

Goyal A N
Mathematics Dpt
Univ of Rajasthan
Jaipur 302 004
India
☎ 91 74 060

Goyal Ashok Kumar
Dpt Phys/Astrophys
Univ of Delhi
Delhi 110 007
India
☎ 91 11 558 0982
✉ agoyal@ducos.ernet.in

Gozhy Adam
Poltava Gravimetrical Obs
Mjasojedova 27/29
314029 Poltava
Ukraine
☎ 380 5322 72039
✆ 380 5322 74903
✉ gozhy@geo.kot.
poltava.ua

Graboske Harold C
LLNL
L 23
Box 808
Livermore CA 94551 9900
USA
☎ 1 415 422 7262
✆ 1 415 423 0238

Grabowski Boleslaw
Inst of Physics
Pedagogical Univ
Ul Oleska 48
PL 45 951 Opole
Poland
☎ 48 12 37 4777*280
✆ 48 12 3 2243

Grachev Stanislav I
Astronomical Institute
St Petersburg Univ
Bibliotechnaja Pl 2
198904 St Petersburg
Russia
☎ 7 812 428 4163
✆ 7 812 428 6649
✉ vayak@astro.lgu.spb.su

Gradie Jonathan Carey
Inst of Geophysics
Univ of Hawaii
2525 Correa Rd
Honolulu HI 96822
USA
☎ 1 808 956 6488
✆ 1 808 988 2790

Gradsztajn Eli
Dpt Phys/Astronomy
Tel Aviv Univ
Ramat Aviv
Tel Aviv 69978
Israel
☎ 972 3 640 8208
✆ 972 3 640 8179

Grady Carol Anne
Physics Dpt
The Catholic Univ
of America
Washington DC 20064
USA
☎ 1 202 319 5315
✆ 1 202 319 5579

Graham David A
RAIUB
Univ Bonn
auf d Huegel 69
D 53121 Bonn
Germany
☎ 49 228 525 282
✆ 49 228 525 229

Graham Eric
Direct Algorithms
Box 2301
Tubac AZ 85646
USA

Graham John A
Dpt Terrestr Magnetism
Carnegie Inst Washington
5241 Bd Branch Rd NW
Washington DC 20015 1305
USA
☎ 1 202 966 0863
✉ graham@jag.ciw.edu

Grahl Bernd H
RAIUB
Univ Bonn
auf d Huegel 69
D 53121 Bonn
Germany
☎ 49 225 731 12
✆ 49 228 525 229

Grainger John F
Physics Dpt
UMIST
Box 88
Manchester M60 1QD
UK
☎ 44 161 200 3677
✆ 44 161 200 3669

Gramann Mirt
Tartu Observ
Estonian Acad Sci
Toravere
EE 2444 Tartumaa
Estonia
☎ 372 7 410 265
✆ 372 7 410 205

Grandi Steven Aldridge
NOAO
Box 26732
950 N Cherry Av
Tucson AZ 85726 6732
USA
☎ 1 520 327 5511
✆ 1 520 325 9360

Grandpierre Attila
Konkoly Observ
Thege U 13/17
Box 67
H 1525 Budapest
Hungary
☎ 36 1 375 4122
✆ 36 1 275 4668
✉ grandp@ogyalla.konkoly.hu

Grant Ian P
Pembroke College
Oxford OX1 1DW
UK
☎ 44 186 524 2271

Granveaud Michel
Observ Paris
LPTF
61 Av Observatoire
F 75014 Paris
France
☎ 33 1 43 20 1210
📠 33 1 44 32 8001

Grasdalen Gary L
Dpt Phys/Astronomy
Univ of Wyoming
Box 3905
Laramie WY 82071
USA
☎ 1 307 766 4385
📠 1 307 766 2652

Gratton Raffaela G
Osserv Astronomico
d Roma
Via d osservatorio 5
I 00040 Monteporzio
Italy
☎ 39 6 944 9019
📠 39 6 944 7243

Grauer Albert D
Dpt Phys/Astronomy
UALR
33rd/Univ
Little Rock AR 72204
USA
☎ 1 501 569 3275

Gray David F
Dpt Phys/Astronomy
Univ W Ontario
London ON N6A 3K7
Canada
☎ 1 519 679 2111*6715
📠 1 519 661 2033
✉ dfgray@uwo.ca

Gray Norman
Dpt Phys/Astronomy
Univ of Glasgow
Glasgow G12 8QQ
UK
☎ 44 141 339 8855*4153
📠 44 141 334 9029
✉ norman@astro.gla.ac.uk

Gray Peter Murray
AAO
Box 296
Epping NSW 2121
Australia
☎ 61 2 868 1666
📠 61 2 9876 8536

Gray Richard O
Dpt Phys/Astronomy
Appalachian State Univ
Boone NC 28608
USA
☎ 1 704 262 2430
📠 1 704 262 2049
✉ grayro@conrad.
appstate.edu

Grayzeck Edwin J
Physics Dpt
Univ of Nevada
4505 S Maryland Parkway
Las Vegas NV 89154
USA
☎ 1 702 739 3507
📠 1 702 739 0804

Grebenev Sergei A
Space Res Inst
Profsoyuznaya 84/32
Moscow 117810
Russia
☎ 7 095 333 3377
📠 7 095 333 5377
✉ grebenev@hea.iki.
rssi.ru

Grebenikov Evgenij A
Lomonosov State Univer
117234 Moscow
Russia

Grec Gerard
Dpt Astrophysique
Universite Nice
Parc Valrose
F 06034 Nice Cdx
France
☎ 33 4 93 51 9100
📠 33 4 93 52 9806

Gredel Roland
Ctr Astron Hispano Aleman
Reina 66 9b
Apt C 511
E 040480 Almeria
Spain
☎ 34 9 502 30988
📠 34 9 502 30373
✉ gredel@caha.es

Green Anne
School of Physics
UNSW
Dpt Astrophysics
Sydney NSW 2006
Australia
☎ 61 2 9351 2727
📠 61 2 9351 7726
✉ agreen@physics.usyd.
edu.au

Green Daniel William E
CfA
HCO/SAO
60 Garden St
Cambridge MA 02138 1516
USA
☎ 1 617 495 7440
📠 1 617 495 7231
✉ green@cfa.harvard.edu

Green David
MRAO
Cavendish Laboratory
Madingley Rd
Cambridge CB3 0HE
UK
☎ 44 12 23 337 305
📠 44 12 23 354 599
✉ D.A.Green@mrao.
cam.ac.uk

Green Elizabeth M
MSSSO
Weston Creek
Private Bag
Canberra ACT 2611
Australia
☎ 61 262 881 111
📠 61 262 490 233

Green Jack
Dpt Geology
Calif State Univ
Long Beach CA 90840
USA
☎ 1 213 498 4809

Green Louis C
3300 Darby Rd
Apt 1208
Haverford PA 19041
USA
☎ 1 215 649 0265

Green Richard F
NOAO/KPNO
Box 26732
950 N Cherry Av
Tucson AZ 85726 6732
USA
☎ 1 520 325 9299
📠 1 520 325 9360

Green Robin M
Astronomy Dpt
Univ of Glasgow
Glasgow G12 8QQ
UK
☎ 44 141 339 8855
📠 44 141 334 9029

Green Simon F
Physics Lab
Univ of Kent
Canterbury CT2 7NR
UK
☎ 44 12 27 764 000*3780
📠 44 12 27 762 616

Greenberg J Mayo
Rijkuniversiteit te
Huygens Lab
Box 9504
NL 2300 RA Leiden
Netherlands
☎ 31 71 527 5700
📠 31 71 527 5819

Greenberg Richard
Lunar/Planetary Lab
Room 325 Bg 92
Univ of Arizona
Tucson AZ 85721 0092
USA
☎ 1 520 621 6940
📠 1 520 621 4933

Greenhill John
Physics Dpt
Univ Tasmania
GPO Box 252c
Hobart TAS 7001
Australia
☎ 61 2 202 429
📠 61 2 202 410
✉ grnhil@physvax.phys.
utas.edu.a

Greenhill Lincoln J
CfA
HCO/SAO MS 42
60 Garden St
Cambridge MA 02138 1516
USA
☎ 1 617 495 7194
📠 1 617 495 7345
✉ greenhill@cfa.harvard.edu

Greenhouse Matthew A
Ntl Air/Space Museum
Smithsonian Institution
MRC 321
Washington DC 20560
USA
☎ 1 202 357 1319
📠 1 202 633 8174
✉ matt@nasm.edu

Greenstein George
Astronomy Dpt
Amherst College
Amherst MA 01002
USA
☎ 1 413 542 2075
✉ gsgreenstein@amherst.edu

Greenstein Jesse L
CALTECH
Palomar Obs
MS 105 24
Pasadena CA 91125
USA
☎ 1 818 356 4006
📠 1 818 568 1517

Gregg Michael David
LLNL
L 413
Box 808
Livermore CA 94551 9900
USA
☎ 1 415 423 0666
📠 1 415 423 0238
✉ gregg@fuego.llnl.gov

Greggio Laura
Dpt Astronomia
Univ d Bologna
Via Zamboni 33
I 40126 Bologna
Italy
☎ 39 51 259 413
📠 39 51 259 407

Gregorini Loretta
Istt Radioastronomia
CNR
Via P Gobetti101
I 40129 Bologna
Italy
☎ 39 51 639 9385
📠 39 51 639 9385

Gregorio-Hetem Jane
IAG
Univ Sao Paulo
CP 9638
01065 970 Sao Paulo SP
Brazil
☎ 55 11 577 8599*235
📠 55 11 276 3848
✉ jane@vax.iagusp.usp.br

Gregory Philip C
Physics Dpt
UBC
224 Agricultural Rd
Vancouver BC V6T 1W5
Canada
☎ 1 604 228 6417
✆ 1 604 228 5324

Gregory Stephen Albert
Dpt Phys/Astronomy
Bowling Green State Univ
Bowling Green OH 43403
USA
☎ 1 419 372 2421
✆ 1 419 372 9938

Greisen Kenneth I
379 Savage Farm Rd
Ithaca NY 14850 6505
USA

Grenier Isabelle
DAPNIA/SAP
CEA Saclay
BP 2
F 91191 Gif s Yvette Cdx
France
☎ 33 1 69 08 4400
✆ 33 1 69 08 6577

Grenier Suzanne
Observ Paris Meudon
DASGA
Pl J Janssen
F 92195 Meudon PPL Cdx
France
☎ 33 1 45 07 7841
✆ 33 1 45 07 7971

Grenon Michel
Observ Geneve
Chemin d Maillettes 51
CH 1290 Sauverny
Switzerland
☎ 41 22 755 2611
✆ 41 22 755 3983
✉ michel.grenon@obs.
unige.ch

Greve Albert
IRAM
300 r La Piscine
F 38406 S Martin Heres Cdx
France
☎ 33 4 76 82 4931
✆ 33 4 76 51 5938

Grevesse Nicolas
Inst Astrophysique
Universite Liege
Av Cointe 5
B 4000 Liege
Belgium
☎ 32 4 254 7510
✆ 32 4 254 7511
✉ grevesse@astro.
ulg.ac.be

Grewing Michael
Astronomisches Institut
Univ Tuebingen
Waldhaeuserstr 64
D 72076 Tuebingen
Germany
☎ 49 707 129 2486
✆ 49 707 129 3458
✉ grewing@iram.fr

Greyber Howard D
10123 Falls Rd
Potomac MD 20854
USA

Griffin I P
Astronaut Mem Planet/Obs
Brevard Comm College
1519 Clearlake Rd
Cocoa FL 32922
USA
☎ 1 407 632 1111*63503
✆ 1 407 633 4565
✉ griffin.i@a1.brevard.
cc.fl.us

Griffin Ian Paul
Armagh Planetarium
College Hill
Armagh BT61 9DB
UK
☎ 44 18 61 524 725
✆ 44 18 61 526 187
✉ ipg@star.arm.ac.uk

Griffin Matthew J
Astrophysics Group
QMWC
Mile End Rd
London E1 4NS
UK
☎ 44 171 975 5555
✆ 44 181 975 5500
✉ mjg@uk.ac.qmc.star

Griffin Rita E M
Inst of Astronomy
The Observatories
Madingley Rd
Cambridge CB3 0HA
UK
☎ 44 12 23 337 548
✆ 44 12 23 337 523

Griffin Roger F
Inst of Astronomy
The Observatories
Madingley Rd
Cambridge CB3 0HA
UK
☎ 44 12 23 337 548
✆ 44 12 23 337 523
✉ rfg@ast.cam.ac.uk

Griffith John S
Dpt Math Sci
Lakehead Univ
Thunder Bay ON P7B 5E1
Canada
☎ 1 807 343 8227
✉ griffith@lakheadu.ca

Griffiths Richard E
Physics Dpt
Carnegie Mellon Univ
Wean Hall 8311
Pittsburgh PA 15213 3890
USA
☎ 1 412 268 1886
✆ 1 412 681 0648
✉ griffith@astro.phys.
cmu.edu

Griffiths William K
Physics Dpt
Univ of Leeds
Leeds LS2 9JT
UK
☎ 44 113 233 3860
✆ 44 113 233 3846

Grigorieva Virginia P
SAI
Acad Sciences
Universitetskij Pr 13
119899 Moscow
Russia
☎ 7 095 939 1030
✆ 7 095 939 1661
✉ vega@sai.msu.su

Grigorjev Victor M
Sibizmir
Acad Sciences
664697 Irkutsk 33
Russia
☎ 7 395 262 9388

Grindlay Jonathan E
CfA
HCO/SAO
60 Garden St
Cambridge MA 02138 1516
USA
☎ 1 617 495 7204
✆ 1 617 495 7356

Grinin Vladimir P
Crimean Astrophys Obs
Ukrainian Acad Science
Nauchny
334413 Crimea
Ukraine
☎ 380 6554 71161
✆ 380 6554 40704

Grinspoon David Harry
LASP
Univ of Colorado
Campus Box 392
Boulder CO 80309 0392
USA
☎ 1 303 492 6230
✆ 1 303 492 6946
✉ david@sunra.
colorado.edu

Grishchuk Leonid P
SAI
Acad Sciences
Universitetskij Pr 13
119899 Moscow
Russia
☎ 7 095 139 5006
✆ 7 095 939 0126

Groenewegen Martin
MPA
Karl Schwarzschildstr 1
D 85748 Garching
Germany
☎ 49 893 299 3252
✆ 49 893 299 3235
✉ groen@mpa-garching.
mpg.de

Groote Detlef
Hamburger Sternwarte
Univ Hamburg
Gojensbergsweg 112
D 21029 Hamburg
Germany
☎ 49 407 252 4112
✆ 49 407 252 4198

Grosbol Preben Johnson
ESO
Karl Schwarzschildstr 2
D 85748 Garching
Germany
☎ 49 893 200 6237
✆ 49 893 202 362
✉ pgrosbol@eso.org

Gross Peter G
714 Oxford Rd
Bala Cynwyd PA 19004
USA

Gross Richard Sewart
CALTECH/JPL
MS 238 332
4800 Oak Grove Dr
Pasadena CA 91109 8099
USA
☎ 1 818 354 4010
✆ 1 818 393 6890
✉ rsg@logos.jpl.nasa.gov

Grossman Allen S
Erwin W Fick Observ
Iowa State Univ
Ames IA 50011
USA
☎ 1 515 294 3666

Grossman Lawrence
Dpt Geophysical Sci
Univ of Chicago
5734 S Ellis Av
Chicago IL 60637
USA
☎ 1 312 962 8153

Grossmann-Doerth U
Kiepenheuer Inst
Sonnenphysik
Schoeneckstr 6
D 79104 Freiburg Breisgau
Germany
☎ 49 761 328 64
✆ 49 761 319 8111

Groten Erwin
Inst Physik Geod
Univ of Technology
Petersenstr 13
D 6100 Darmstadt
Germany
☎ 49 615 116 3109
✆ 49 615 116 4512
✉ groten@ipgs.ipg.verm.
tu-darmstadt.de

Groushinsky Nikolai P
SAI
Acad Sciences
Universitetskij Pr 13
119899 Moscow
Russia
☎ 7 095 939 2858
✆ 7 095 939 0126

Grubissich C
Vi Aosta 34/5
I 35142 Padova
Italy
☎ 39 49 38 301

Grudler Pierre
OCA
CERGA
F 06130 Grasse
France
☎ 33 4 93 40 5424
✆ 33 4 93 40 5353

Grudzinska Stefania
Inst of Astronomy
N Copernicus Univ
Ul Chopina 12/18
PL 87 100 Torun
Poland
☎ 48 56 26 018
✆ 48 56 24 602

Grueff Gavril
Istt Radioastronomia
CNR
Via P Gobetti101
I 40129 Bologna
Italy
☎ 39 51 639 9385
✆ 39 51 639 9385

Gruen Eberhard
MPI Kernphysik
Postfach 103 980
D 69029 Heidelberg
Germany
☎ 49 622 151 6478
✆ 49 622 151 5640
✉ gruen@dusty.mpi-hp.mpg.de

Gruenwald Ruth
IAG
Univ Sao Paulo
CP 9638
01065 970 Sao Paulo SP
Brazil
☎ 55 11 577 8599
✆ 55 11 276 3848

Gry Cecile
LAS
Traverse du Siphon
Les Trois Lucs
F 13376 Marseille Cdx 12
France
☎ 33 4 91 05 5900
✆ 33 4 91 66 1855
✉ cecile@frlasm51

Grygar Jiri
Inst of Physics
Czech Acad Sci
Na Slovance 2
CZ 180 00 Praha 8
Czech R
☎ 420 2 6605 2660
✆ 420 2 858 5443
✉ grygar@fzu.cz

Grzedzielski Stanislaw Pr
Space Res Ctr
Polish Acad Sci
Ul Ordona 21
PL 01 237 Warsaw
Poland
☎ 48 22 40 3766
✆ 48 22 36 8961

Gu Xiaoma
Yunnan Observ
CAS
Kunming 650011
China PR
☎ 86 871 2035
✆ 86 871 717 1845

Gu Zhennian
Shanghai Observ
CAS
80 Nandan Rd
Shanghai 200030
China PR
☎ 86 21 6438 6191
✆ 86 21 6438 4618
✉ gzn@center.shao.ac.cn

Guarnieri Adriano
Osserv Astronom Bologna
Univ d Bologna
Via Zamboni 33
I 40126 Bologna
Italy
☎ 39 51 51 9593
✆ 39 51 25 9407

Gubanov Vadim S
Astrometry/Geodin Lab
Inst Appl Astron
Zhdanovskaya St 8
197110 St Petersburg
Russia
☎ 7 812 230 7414
✆ 7 812 230 7413
✉ gubanov@isida.ipa.rssi.ru

Gubchenko Vladimir M
Instte Appl Physics RAS
46 Ulyanov St
Nizhny Novgorod 603600
Russia
☎ 7 831 238 4381
✆ 7 831 236 2061
✉ ua3thw@appl.sci-nnov.ru

Gudehus Donald Henry
CHARA
Georgia State Univ
Atlanta GA 30303 3083
USA
☎ 1 404 658 2932
✆ 1 404 651 1389
✉ gudehus@chara.gsu.edu

Gudmundsson Einar H
Raunvisindastofnun
Haskolans
Dunhaga 3
IS 107 Reykjavik
Iceland
☎ 354 525 4800
✆ 354 552 8911
✉ einar@raunvis.hi.is

Gudur N
Dpt Astron/Space Sci
Ege Univ
Box 21
35100 Bornova Izmir
Turkey
☎ 90 232 388 0110*2322

Guedel Manuel
Paul Scherrer Inst
Wuerenlingen/Villigen
CH 5232 Villigen Psi
Switzerland
☎ 41 16 32 7129
✆ 41 16 32 1205
✉ guedel@astro.phys.ethz.ch

Guelin Michel
IRAM
300 r La Piscine
F 38406 S Martin Heres Cdx
France
☎ 33 4 76 42 3383
✆ 33 4 76 51 5938

Guenther David
Astronomy Dpt
St Mary's Univ
Halifax NS B3H 3C3
Canada
☎ 1 902 420 5832
✉ dguenther@ap.stmarys.ca

Guenther Eike
Thueringer Landessternwarte
Sternwarte 5
D 07778 Tautenburg
Germany
☎ 49 364 278 6355
✆ 49 364 278 6329
✉ guenther@tls-tautenburg.de

Guerin Pierre
IAP
98bis bd Arago
F 75014 Paris
France
☎ 33 1 44 32 8007
✆ 33 1 44 32 8001

Guerrero Gianantonio
Osserv Astronomico
d Milano
Via E Bianchi 46
I 22055 Merate
Italy
☎ 39 990 6412
✆ 39 990 8492

Guertler Joachin
Astrophysik Inst
Univ Sternwarte
Schillergaesschen 2
D 07745 Jena
Germany
☎ 49 364 163 0323
✆ 49 364 163 0417

Guest John E
UCLO
ULO
Mill Hill Park
London NW7 2QS
UK
☎ 44 181 959 7367
✆ 44 171 380 7145

Guesten Rolf
RAIUB
Univ Bonn
auf d Huegel 69
D 53121 Bonn
Germany
☎ 49 228 525 379
✆ 49 228 525 229

Guetter Harry Hendrik
USNO
Flagstaff Station
Box 1149
Flagstaff AZ 86002 1149
USA
☎ 1 602 779 5132
✆ 1 520 774 3626
✉ hguetter@nofs.navy.mil

Guhathakurta Madhulika
NASA GSFC
Code 682
Greenbelt MD 20771
USA
☎ 1 301 286 0722
✆ 1 301 286 1617
✉ likae@madhu.gsfc.nasa.gov

Guibert Jean
Observ Paris
DASGAL
61 Av Observatoire
F 75014 Paris
France
☎ 33 1 40 51 2098
✆ 33 1 40 51 2090
✉ jean.guibert@obspm.fr

Guichard Jose
INAOE
Tonantzintlaz
Apdo 51 y 216
Puebla PUE 72000
Mexico
☎ 52 2 247 2011
✆ 52 2 247 2231

Guiderdoni Bruno
IAP
98bis bd Arago
F 75014 Paris
France
☎ 33 1 44 32 8098
✆ 33 1 44 32 8001
✉ guider@iap.fr

Guidice Donald A
AFPL
Space Physics Div
Hanscom AFB
Bedford MA 01731 3010
USA
☎ 1 617 861 3989

Guinan Edward Francis
Astronomy Dpt
Villanova Univ
800 Lancaster Av
Villanova PA 19085
USA
☎ 1 610 519 4823
✆ 1 610 519 6132
✉ guinan@ucis.vill.edu

Guinot Bernard R
2 r d Soupirs
F 77590 Chartrettes
France
☎ 33 1 64 81 1133
✉ bguinot@
compuserve.com

Gulkis Samuel
CALTECH/JPL
MS 169 506
4800 Oak Grove Dr
Pasadena CA 91109 8099
USA
☎ 1 818 354 5708
✆ 1 818 354 2946

Gull Stephen F
MRAO
Cavendish Laboratory
Madingley Rd
Cambridge CB3 0HE
UK
☎ 44 12 23 337 274
✆ 44 12 23 354 599

Gull Theodore R
NASA GSFC
Code 681
Greenbelt MD 20771
USA
☎ 1 301 286 6184
✆ 1 301 286 1753
✉ gull@sea.gsfc.nasa.gov

Gulliver Austin Fraser
Dpt Phys/Astronomy
Univ of Brandon
270 18th Street
Brandon MB R7A 6A9
Canada
☎ 1 204 727 7441
✆ 1 204 728 7346
✉ gulliver@brandonu.ca

Gulmen Omur
Dpt Astron/Space Sci
Ege Univ
Box 21
35100 Bornova Izmir
Turkey
☎ 90 232 388 0110*2322

Gulsecen Hulusi
Univ Observ
Univ of Istanbul
University 34452
34452 Istanbul
Turkey
☎ 90 212 522 3597
✆ 90 212 519 0834
✉ ik001@triuvm11

Gulyaev Albert P
SAI
Acad Sciences
Universitetskij Pr 13
119899 Moscow
Russia
☎ 7 095 139 1970
✆ 7 095 939 0126

Gulyaev Rudolf A
ITMIRWP
Acad Sciences
142092 Troitsk
Russia
☎ 7 095 334 0284
✆ 7 095 334 0124
✉ rgulyaev@izmiran.
rssi.ru

Gulyaev Sergei A
9/729 Remuera Rd
Auckland 1005
New Zealand
☎ 7 9 522 4531
✆ 7 9 522 4531
✉ sergei_gulyaev@
hotmail.com

Gunji Shuichi
Physics Dpt/Faculty of Sci
Yamagata Univ
1-4-12 Kojirakawa
Yamagata 990
Japan
☎ 81 236 28 4555
✆ 81 236 28 4567
✉ gunji@ksvax1.kj.
yamagata-u.ac.jp

Gunn James E
Dpt Astrophysical Sci
Princeton Univ
Peyton Hall
Princeton NJ 08544 1001
USA
☎ 1 609 452 3802
✆ 1 609 258 1020

Guo Hongfeng
Beijing Astronomical Obs
CAS
Beijing 100080
China PR
☎ 86 10 6254 5179
✆ 86 1 256 10855
✉ ghf@bao01.bao.ac.cn

Guo Nei-shu
Nanjing Astronomical
Instrument Factory
Box 846
Nanjing
China PR
☎ 86 25 64 6191
✆ 86 25 71 1256

Guo Quanshi
Purple Mountain Obs
CAS
Nanjing 210008
China PR
☎ 86 25 46700
✆ 86 25 301 459

Gupta Sunil K
TIFR
Homi Bhabha Rd
Colaba
Bombay 400 005
India
☎ 91 22 215 2971*545
✆ 91 22 495 2110
✉ guptask@tifrvax

Gupta Yashwant
NCRA/TIFR
Pune Univ Campus Pb 3
Ganeshkhind
Pune 411 007
India
☎ 91 212 357 107
✆ 91 212 355 149
✉ ygupta@ncra.tifr.
res.in

Gurm Hardev S
Dpt Astronomy/Space Sci
Univ of Panjabi
Patiala 147 002
India
☎ 91 73262*96

Gurman Joseph B
NASA GSFC
Code 602 6
Greenbelt MD 20771
USA
☎ 1 301 286 7599

Gurshtein Alexander A
Mesa State College
Box 2674
Grand Junction CO 81502
Russia
✉ alex@mesa7.mesa.colorado.ed
u

Gursky Herbert
NRL
Code 4100
4555 Overlook Av SW
Washington DC 20375 5000
USA
☎ 1 202 767 6343

Gurvits Leonid
JIVE
Postbus 2
NL 7991 PD Dwingeloo
Netherlands
☎ 31 521 59 6514
✆ 31 52 159 7332
✉ lgurvits@jive.nfra.nl

Gurzadian Grigor A
Byurakan Astrophys Observ
Armenian Acad Sci
378433 Byurakan
Armenia
☎ 374 88 52 28 3453/4142
✆ 374 88 52 52 3640
✉ gurzadyan@vx1.yerphi.am

Guseinov O H
Inst of Physics
Akademgorodoc
Narimanov Ul 33
370143 Baku
Azerbajan
☎ 994 39 3951

Gusejnov Ragim Eh
Shemakha Astrophysical
Observ
Azerbajan Acad Sci
373243 Shemakha
Azerbajan
☎ 994 89 22 39 8248

Guseva Irina S
Pulkovo Observ
Acad Sciences
10 Kutuzov Quay
196140 St Petersburg
Russia
☎ 7 812 123 4386
✆ 7 812 123 1922
✉ isg@gaoran.spb.su

Gussmann E A
Zntrlinst Astrophysik
Sternwarte Babalsberg
Rosa Luxemburg Str 17a
D 14473 Potsdam
Germany
☎ 49 331 275 3731

Gustafson Bo A S
Astronomy Dpt
Univ of Florida
211 SSRB
Gainesville FL 32611
USA
☎ 1 904 392 7677
✆ 1 904 392 5089
✉ gustaf@astro.ufl.edu

Gustafsson Bengt
Astronomical Observ
Box 515
S 751 20 Uppsala
Sweden
☎ 46 18 53 0265
✆ 46 18 52 7583

Gutcke Dietrich
Carl Zeiss Jena Gmbh
Tatzens Promenade 1a
Postfach 125
D 07740 Jena
Germany
☎ 49 364 1640
✆ 49 364 164 2542

Guthrie Bruce N G
5 Arden St
Edinburgh EH9 1BR
UK
☎ 44 131 229 4957

Gutierrez-Moreno A
Dpt Astronomia
Univ Chile
Casilla 36 D
Santiago
Chile
☎ 56 2 229 4101/4002
✆ 56 2 229 4002

Guzik Joyce Ann
LANL
MS B220 X 2
Box 1663
Los Alamos NM 87545 2345
USA
☎ 1 505 667 8927
✆ 1 505 665 4080
✉ joy@beta.lanl.gov

Guzzo Luigi
Osserv Astronomico
d Milano
Via E Bianchi 46
I 22055 Merate
Italy
☎ 39 990 6412
✆ 39 990 8492
✉ guzzo@merate.mi.
astro.it

Gwinn Carl R
Physics Dpt
Univ of California
Santa Barbara CA 93106
USA
☎ 1 805 961 2814
✆ 1 805 961 4170
✉ cgwinn@voodoo.
ucsb.edu

Gyldenkerne Kjeld
Observator Gyldenkernes
Vej 13
DK 4340 Tollose
Denmark

Gyori Lajos
Heliophysical Observ
Box 93
H 5701 Gyula
Hungary
☎ 36 3 615 53

Gyulbudaghian Armen L
Byurakan Astrophys Observ
Armenian Acad Sci
378433 Byurakan
Armenia
☎ 374 88 52 28 3453/4142
✆ 374 88 52 52 3640
✉ byurakan@adonis.ias.
msk.su

Goharji Adan
Astronomy Dpt
KAAU
Box 9028
Jeddah 21413
Saudi Arabia
☎ 966 2 695 2285
✆ 966 2 640 0736

H

Habbal Shadia Rifai
CfA
HCO/SAO
60 Garden St
Cambridge MA 02138 1516
USA
☎ 1 617 495 7348
✆ 1 617 495 7356

Habe Asao
Physics Dpt
Hokkaido Univ
Kita 10 Nishi 8
Sapporo 060
Japan
☎ 81 117 11 2111
✆ 81 222 85 132

Habing Harm J
Leiden Observ
Box 9513
NL 2300 RA Leiden
Netherlands
☎ 31 71 527 2727
✆ 31 71 527 5819

Hachenberg Otto
RAIUB
Univ Bonn
auf d Huegel 69
D 53121 Bonn
Germany
☎ 49 228 73 3658
✆ 49 228 73 3672

Hachisu Izumi
Dpt Earth Sciences
Univ of Tokyo
Meguro Ku
Tokyo 153
Japan
☎ 81 354 546 615
✆ 81 334 85 2904
✉ hachisu@chianti.c.u-tokyo.ac.jp

Hack Margherita
OAT
Box Succ Trieste 5
Via Tiepolo 11
I 34131 Trieste
Italy
☎ 39 40 31 99255
✆ 39 40 30 9418

Hackwell John A
Aerospace Corp
Box 92957
Los Angeles CA 90009 2957
USA
☎ 1 310 766 6296
✆ 1 310336 7055

Haddock Fred T
Astronomy Dpt
Univ of Michigan
Dennison Bldg
Ann Arbor MI 48109 1090
USA
☎ 1 313 764 3430
✆ 1 313 763 6317

Hadjidemetriou John D
Physics Dpt
Univ Thessaloniki
GR 540 06 Thessaloniki
Greece
☎ 30 31 99 8211
✆ 30 31 99 8211
✉ hadjidem@physics.auth.gr

Hadley Brian W
Royal Observ
Blackford Hill
Edinburgh EH9 3HJ
UK
☎ 44 131 668 8296
✆ 44 131 668 8356
✉ bwh@roe.ac.uk

Hadrava Petr
Astronomical Institute
Czech Acad Sci
Fricova 1
CZ 251 65 Ondrejov
Czech R
☎ 420 204 85 7141
✆ 420 2 88 1611
✉ had@sunstel.asu.cas.cz

Haefner Reinhold
Inst Astron/Astrophysik
Univ Sternwarte
Scheinerstr 1
D 81679 Muenchen
Germany
☎ 49 899 890 21
✆ 49 899 220 9427

Haemeen Anttila Kaarle A
Astronomy Dpt
Univ of Oulu
Box 333
FIN 90570 Oulu
Finland
☎ 358 81 553 2022
✆ 358 81 553 1934

Haenninen Jyrki
Dpt Geosc/Astronomy
Univ of Oulu
Box 333
FIN 90570 Oulu
Finland
☎ 358 81 553 1937
✆ 358 81 553 1934
✉ jyrki@seita.oulu.fi

Haensel Pawel
Copernicus Astron Ctr
Polish Acad Sci
Ul Bartycka 18
PL 00 716 Warsaw
Poland
☎ 48 22 41 1086
✆ 48 22 41 0828

Haerendel Gerhard
MPE
Postfach 1603
D 85740 Garching
Germany
☎ 49 893 299 3516
✆ 49 893 299 3569

Hafizi Mimoza
Fak I Shkencave Natyre
Katedra Fizikes Teorike
Tirana
Albania
☎ 355 42 40 805
✆ 355 42 30 747
✉ mhafizi@fshn.tirana.al

Hagen Hans-Juergen
Hamburger Sternwarte
Univ Hamburg
Gojensbergsweg 112
D 21029 Hamburg
Germany
☎ 49 407 252 4136
✆ 49 407 252 4198

Hagen John P
2920 Ashby Av
Las Vegas NV 89102 1945
USA

Hagen-Thorn Vladimir A
Astronomical Observ
St Petersburg Univ
Bibliotechnaja Pl 2
198904 St Petersburg
Russia
☎ 7 812 257 9491
✆ 7 812 428 4259

Hagfors Tor
MPI Aeronomie
Max Planck Str 2
Postfach 20
D 37189 Kaltenburg
Germany
☎ 49 555 697 90
✆ 49 555 697 9240

Hagio Fumihiko
Kumamoto Inst Techn
Ikeda 4
Kumamoto 860
Japan
☎ 81 963 26 3111
✆ 81 963 26 3000

Hagyard Mona June
NASA/MSFC
Space Science Lab
Code ES 52
Huntsville AL 35812
USA
☎ 1 205 453 5687
✆ 1 205 547 7754

Hahn Gerhard J
DLR
Inst f Planetenerkundg
Rudower Chaussee 5
D 12489 Berlin
Germany
☎ 49 306 705 5417
✆ 49 306 705 5386
✉ gerhard.hahn@dlr.de

Haikala Lauri K
ESO
Santiago Office
Casilla 19001
Santiago 19
Chile
☎ 56 2 228 5006
✆ 56 1 228 5132
✉ lhaikala@eso.org

Hainaut Olivier R
ESO
Santiago Office
Casilla 19001
Santiago 19
Chile
☎ 56 2 228 5006
✆ 56 2 228 5132
✉ ohainaut@sc.eso.org

Haisch Bernhard Michael
Lockheed Palo Alto Res Lb
Dpt 91 30 Bg 252
3251 Hanover St
Palo Alto CA 94304 1191
USA
☎ 1 415 858 4073
✆ 1 415 424 3994
✉ haisch@sag.dnet.nasa.gov

Hajduk Anton
Astronomical Institute
Slovak Acad Sci
Dubravska 9
SK 842 28 Bratislava
Slovak Republic
☎ 421 7 37 5157
✆ 421 7 37 5157
✉ astropor@savba.savba.sk

Hajdukova Maria
Dpt Astron/Astrophys
Comenius Univ
Mlynska Dolina
SK 842 15 Bratislava
Slovak Republic
☎ 421 7 72 400
✆ 421 7 32 5882

Hakkila Jon Eric
Dpt Phys/Astronomy
Mankato State Univ
Box 41
Mankato MN 56002 8400
USA
☎ 1 507 389 1840
℻ 1 161 200 3669
✉ jhakk@msus1.msus.edu

Halbwachs Jean Louis
Observ Strasbourg
11 r Universite
F 67000 Strasbourg
France
☎ 33 3 88 15 0738
℻ 33 3 88 15 0760
✉ halbwachs@astro.
u-strasbg.fr

Hale Alan
Southwest Inst for
Space Research
15 E Spur Rd
Cloudcroft NM 88317
USA
☎ 1 505 687 2075
℻ 1 505 687 2075
✉ ahale@nmsu.edu

Hall Andrew Norman
Astrophysics Dpt
Univ of Oxford
Keble Rd
Oxford OX1 3RQ
UK
☎ 44 186 527 3999
℻ 44 186 527 3947

Hall Donald N
Inst for Astronomy
Univ of Hawaii
2680 Woodlawn Dr
Honolulu HI 96822
USA
☎ 1 808 956 8312
℻ 1 808 988 2790

Hall Douglas S
Dyer Observ
Vanderbilt Univ
Box 1803
Nashville TN 37235
USA
☎ 1 615 373 4897
℻ 1 615 343 7263
✉ hallxxds@ctrvax.
vanderbilt.edu

Hall Peter J
CSIRO
ATNF
Locked Bag 194
Narrabri NSW 2390
Australia
☎ 61 67 959 205
℻ 61 67 959 255
✉ phall@atnf.csiro.au

Hall R Glenn
3612 Spring St
Chevy Chase MD 20815
USA
☎ 1 301 652 7221

Hallam Kenneth L
NASA GSFC
Code 680
Greenbelt MD 20771
USA
☎ 1 301 286 8811
℻ 1 301 286 8709

Halliday Ian
825 Killeen Av
Ottawa ON K2A 2X8
Canada
☎ 1 613 728 1497

Hamabe Masaru
Inst of Astronomy
Univ of Tokyo
Osawa 2 Chome
Mitaka 181 0015
Japan
☎ 81 422 34 3802
℻ 81 422 34 3749
✉ hamabe@mtk.ioa.s.u-
tokyo.ac.jp

Hamajima Kiyotoshi
171 Kami No Kura 1 chome
Midori Ku
Nagoya 458
Japan

Hamann Wolf-Rainer
Inst Theor Physics
Univ Potsdam
Postfach 60 15 53
D 24415 Potsdam
Germany
☎ 49 331 977 1054
℻ 49 331 977 1107
✉ wrh@carina.astro.
physik.uni-potsdam.de

Hambrayan Valeri V
Byurakan Astrophys Observ
Armenian Acad Sci
378433 Byurakan
Armenia
☎ 374 88 52 28 3453/4142
℻ 374 88 52 52 3640
✉ byurakan@adonis.ias.
msk.su

Hamdy M A M
NAIGR
Helwan Observ
Cairo 11421
Egypt
☎ 20 78 0645/2683
℻ 20 62 21 405 297

Hamid S El Din
Astronomy Dpt
Fac of Sciences
Cairo Univ
Geza
Egypt
☎ 20 2 572 7022
℻ 20 2 572 7556

Hamilton Andrew J S
JILA
Univ of Colorado
Campus Box 440
Boulder CO 80309
USA
☎ 1 303 492 7833
℻ 1 303 492 5235
✉ ajsh@wild.colorado.edu

Hamilton P A
Geelong Campus
Pro Vice Chancellor
Deaklin Univ
Geelong VIC 3217
Australia
☎ 61 3 5227 1147
℻ 61 2 5227 2175
✉ pip.hamilton@deakin.
edu.au

Hammel Heidi B
Dpt Earth Science
MIT Rm 54 416
Box 165
Cambridge MA 02139
USA
☎ 1 617 253 7568
℻ 1 617 253 2886
✉ hbh@astron.mit.edu

Hammer Francois
Observ Paris Meudon
DAEC
Pl J Janssen
F 92195 Meudon PPL Cdx
France
☎ 33 1 45 07 7408
℻ 33 1 45 07 2806
✉ hammer@obspm.fr

Hammer Reiner
Kiepenheuer Inst
Sonnenphysik
Schoeneckstr 6
D 79104 Freiburg Breisgau
Germany
☎ 49 761 319 8216
℻ 49 761 3198 111
✉ hammer@kis.uni-
freiburg.de

Hammerschlag Robert H
Sterrekundig Inst Utrecht
Box 80000
NL 3508 TA Utrecht
Netherlands
☎ 31 30 253 5218
℻ 31 30 253 1601
✉ R.H. Hammerschlag@
astro.uu.nl

Hammerschlag-Hensberge G
Astronomical Institute
Univ of Amsterdam
Kruislaan 403
NL 1098 SJ Amsterdam
Netherlands
☎ 31 20 525 7491/7492
℻ 31 20 525 7484
✉ r.h.hammerschlag@
fys.ruu.nl

Hammond Gordon L
Mathematics Dpt
Univ of S Florida
Astronomy Program
Tampa FL 33620
USA
☎ 1 813 783 1226

Hamuy Mario
NOAO
CTIO
Casilla 603
La Serena
Chile
☎ 56 51 225 415
℻ 56 51 20 5342
✉ mhumuy@noao.edu

Hamzaoglu Esat E H
KACST
Box 2455
Riyadh 11453
Saudi Arabia
☎ 966 1 467 6324
℻ 966 1 467 4253

Han Fu
Purple Mountain Obs
CAS
Nanjing 210008
China PR
☎ 86 25 33738
℻ 86 25 301 459

Han Jinlin
Beijing Astronomical Obs
CAS
Beijing 100080
China PR
☎ 86 10 6 262 5498
℻ 86 1 256 10855
✉ hjl@bao01.bao.ac.cn

Han Tianqi
Inst Geodesy/
Geophysics
Xu Dong Lu
Wuchang 430077
China PR
☎ 86 81 3712*570
✉ bwu@asch.whigg.ac.cn

Han Wenjun
Beijing Astronomical Obs
CAS
Beijing 100080
China PR
☎ 86 1 28 1698
℻ 86 1 256 10855

Han Yanben
Beijing Astronomical Obs
CAS
Beijing 100080
China PR
☎ 86 10 6254 5179
℻ 86 1 256 10855
✉ hyb@bao01.bao.ac.cn

Han Zhanwen
Yunnan Observ
CAS
Kunming 650011
China PR
☎ 86 871 2035
℻ 86 871 717 1845

Han Zhengzhong
Purple Mountain Obs
CAS
Nanjing 210008
China PR
☎ 86 25 330 3921
℻ 86 25 330 1459

Hanami Hitoshi
College Human Social Sc
Iwate Univ
Ueda 3
Morioka 020
Japan
☎ 81 196 23 5171*2284
℻ 81 196 54 2289
✉ d12697@jpnkudpc

Hanaoka Yoichiro
Nobeyama Radio Obs
NAOJ
Minamimaki Mura
Nagano 384 13
Japan
☎ 81 267 63 4381
℻ 81 267 98 2506

Hanasz Jan
Inst of Astronomy
N Copernicus Univ
Ul Chopina 12/18
PL 87 100 Torun
Poland
☎ 48 56 26 018
℻ 48 56 24 602

Hanawa Tomoyuki
Astrophysics Dpt
Nagoya Univ
Furocho Chikusa Ku
Nagoya 464 01
Japan
☎ 81 527 81 6769
℻ 81 527 89 2845

Hanbury Brown Robert
White Cottage
Penton Mewsey
Hants SP11 ORQ
UK
☎ 44 126 477 2334

Handa Toshihiro
Inst Astro/Fac Science
Univ of Tokyo
Osawa Mitaka
Tokyo 181
Japan
☎ 81 422 34 3735
℻ 81 422 34 3749
✉ handa@ghz.mtk.iao.s.u-tokyo.ac.jp

Hanes David A
Physics Dpt
Queen's Univ
Kingston ON K7L 3N6
Canada
☎ 1 613 547 5750
℻ 1 613 545 6463

Hang Hengrong
Purple Mountain Obs
CAS
Nanjing 210008
China PR
☎ 86 25 330 3921
℻ 86 25 330 1459

Haniff Christopher
MRAO
Cavendish Laboratory
Madingley Rd
Cambridge CB3 0HE
UK
☎ 44 12 23 337 363
℻ 44 12 23 354 599
✉ cah@mrao.cam.ac.uk

Hanisch Robert J
STScI
Homewood Campus
3700 San Martin Dr
Baltimore MD 21218
USA
☎ 1 301 338 4910
℻ 1 301 338 4767
✉ hanisch@stsci.edu

Hankins Timothy Hamilton
Dpt Phys/Astronomy
New Mexico Tech
Campus Station
Socorro NM 87801
USA
☎ 1 505 476 8011
℻ 1 505 835 7027
✉ thankins@nrao.edu

Hanner Martha S
CALTECH/JPL
MS 183 601
4800 Oak Grove Dr
Pasadena CA 91109 8099
USA
☎ 1 818 354 4321
℻ 1 818 393 4605
✉ msh@scn2.jpl.nasa.gov

Hansen Carl J
JILA
Univ of Colorado
Campus Box 440
Boulder CO 80309 0440
USA
☎ 1 303 492 7811
℻ 1 303 492 5235

Hansen Leif
Astronomical Observ, NBIfAFG
Copenhagen Univ
Juliane Maries Vej 30
DK 2100 Copenhagen
Denmark
☎ 45 35 32 5983
℻ 45 35 32 5989
✉ leif@astro.ku.dk

Hansen Richard T
Eng 138
VAMC
150 S Huntington Av
Boston MA 02130
USA
☎ 1 617 734 2534

Hanslmeier Arnold
Inst Astronomie
Karl-Franzens-Univ
Universitaetsplatz 5
A 8010 Graz
Austria
☎ 43 316 380 5275
℻ 43 316 384 091

Hanson Robert B
Lick Observ
Univ of California
Santa Cruz CA 95064
USA
☎ 1 831 459 2755
℻ 1 831 426 3115
✉ hanson@ucolock.org

Hansson Nils
Lund Observ
Box 43
S 221 00 Lund
Sweden
☎ 46 46 222 7000
℻ 46 46 222 4614

Hansteen Viggo
Inst Theor Astrophys
Univ of Oslo
Box 1029
N 0315 Blindern Oslo 3
Norway
☎ 47 22 856 501
℻ 47 22 85 6505

Hantzios Panayiotis
Astronomical Institute
Ntl Observ Athens
Box 20048
GR 118 10 Athens
Greece
☎ 30 1 346 1191
℻ 30 1 346 4566

Hanuschik Reinhard
Astronomisches Institut
Ruhr Univ Bochum
Postfach 102148
D 44780 Bochum
Germany
☎ 49 234 700 3450
℻ 49 234 709 4169
✉ rwh@astro.ruhr-uni-bochum.de

Hao Yunxiang
Astronomy Dpt
Beijing Normal Univ
Beijing 100875
China PR
☎ 86 1 65 6531*6285
℻ 86 1 201 3929

Hapke Bruce W
Dpt Geol/Planetary Sci
Univ of Pittsburgh
321 Old Eng Hall
Pittsburgh PA 15235
USA
☎ 1 412 624 4719
℻ 1 412 624 3914

Hara Tadayoshi
National Astronomical Obs
2 10 Hoshigaoka
Mizusawa Shi
Iwate 023
Japan
☎ 81 197 24 7111
℻ 81 197 22 7120

Hara Tetsuya
Physics Dpt
Kyoto Sangyo Univ
Kamigamo
Kyoto 603
Japan
☎ 81 757 01 2151
℻ 81 75 705 1640
✉ hara@jpnksuvx

Hardebeck Ellen G
3106 Tumbleweed Rd
Bishop CA 93514
USA

Hardee Philip
Dpt Phys/Astronomy
Univ of Alabama
Box 1921
University AL 35487 0324
USA
☎ 1 205 348 5050
℻ 1 205 348 5050

Hardy Eduardo
Dpt d Physique
Universite Laval
Fac Sciences/Genie
Laval QC G1K 7P4
Canada
☎ 1 418 656 2960
℻ 1 418 656 2040
✉ hardy@phy.ulaval.ca

Harju Jorma Sakari
Helsinky Univ
Observ
Box 14
FIN 00014 Helsinki
Finland
☎ 358 19 12 2940
℻ 358 19 12 2952

Harlaftis Emilios
Royal Greenwich Obs
Apt 321
Santa Cruz de La Palma
E 38780 Santa Cruz
Spain
☎ 34 2 240 5500
℻ 34 2 240 5646/405 501
✉ ehh@lpve.ing.iac.es

Harmanec Petr
Astronomical Institute
Czech Acad Sci
Fricova 1
CZ 251 65 Ondrejov
Czech R
☎ 420 204 85 7143
℻ 420 2 88 1611
✉ hec@sunstel.asu.cas.cz

Harmer Charles F W
NOAO/KPNO
Box 26732
950 N Cherry Av
Tucson AZ 85726 6732
USA
☎ 1 520 318 8342
℻ 1 520 318 8360
✉ charmer@noao.edu

Harmer Dianne L
NOAO
Box 26732
950 N Cherry Av
Tucson AZ 85726 6732
USA
☎ 1 520 325 9218
℻ 1 520 325 9360
✉ diharmer@noao.edu

Harms Richard James
RJH Scientifc Inc
One Beltway Ctr St 401
5904 Richmond Highway
Alexandria VA 22303
USA
☎ 1 703 329 4151
✉ harms@rjhsci.com

Harnden Frank R
CfA
HCO/SAO
60 Garden St
Cambridge MA 02138 1516
USA
☎ 1 617 495 7143
℻ 1 617 495 7356

Harnett Julieine
School of Physics
UNSW
Sydney NSW 2006
Australia
☎ 61 2 692 2727
℻ 61 2 660 2903
✉ jhatnett@robin.rp.
csiro.au

Harpaz Amos
Dpt Physics/Space Res
IIT
Technion City
Haifa 32000
Israel
☎ 972 4 293 521
✉ phr89ah@technion

Harper David
36 Hollytrees
Bar Hill
Cambridge CB3 8SF
UK
☎ 44 19 54 78 2352
✉ david@obliquity.u-
net.com

Harrington J Patrick
Astronomy Program
Univ of Maryland
College Park MD 20742 2421
USA
☎ 1 301 454 5944
℻ 1 301 314 9067

Harris Alan William
CALTECH/JPL
MS 183 501
4800 Oak Grove Dr
Pasadena CA 91109 8099
USA
☎ 1 818 354 6741
℻ 1 818 354 0966
✉ awharris@lithos.jpl.
nasa.gov

Harris Alan William
DLR
Inst f Planetenerkundg
Rudower Chaussee 5
D 12489 Berlin
Germany
☎ 49 306 705 5324
℻ 49 306 705 5386
✉ harris@terra.pe.ba.
dlr.de

Harris Daniel E
CfA
HCO/SAO MS 3
60 Garden St
Cambridge MA 02138 1516
USA
☎ 1 617 495 7148
℻ 1 617 495 7356
✉ harris@cfa.harvard.edu

Harris Gretchen L H
Physics Dpt
Univ Waterloo
Waterloo ON N2L 3G1
Canada
☎ 1 519 885 1211
℻ 1 519 746 8115
✉ glharris@astro.
waterloo.edu

Harris Hugh C
USNO
Flagstaff Station
Box 1149
Flagstaff AZ 86002 1149
USA
☎ 1 602 779 5132
℻ 1 520 774 3626

Harris Stella
Astrophysics Group
QMWC
Mile End Rd
London E1 4NS
UK
☎ 44 171 975 5555
℻ 44 181 975 5500

Harris William E
Physics Dpt
Mcmaster Univ
Hamilton ON L8S 4M1
Canada
☎ 1 416 525 9140
℻ 1 416 546 1252
✉ harris@physun.physics.
mcmaster.ca

Harrison Edward R
Dpt Phys/Astronomy
Univ Massachusetts
GRC
Amherst MA 01003
USA
☎ 1 413 545 2194

Harrison Richard A
Rutherford Appleton Lab
Space/Astrophysics Div
Bg R25/R68
Chilton Didcot OX11 0QX
UK
☎ 44 12 35 446 884
℻ 44 12 35 44 6509
✉ rah@uk.ac.rl.astro

Harrower George A
204 1033 Belmont Av
Victoria BC V8S 3T4
Canada

Hart Michael H
7301 Masonville Dr
Annandale VA 22003
USA

Harten Ronald H
RCA Astro Electronic
TB 1
Box 8oo
Princeton NJ 08540
USA
☎ 1 609 426 3551

Hartkopf William I
CHARA
Georgia State Univ
Atlanta GA 30303 3083
USA
☎ 1 404 651 2932
℻ 1 404 651 1389
✉ hartkopf@chara.
gsu.edu

Hartl Herbert
Inst Astronomie
Technikerstr 15
A 6020 Innsbruck
Austria
☎ 43 512 507 6039
℻ 43 512 507 2923
✉ Herbert.Hartl@
uibk.ac.at

Hartmann Dieter H
202 Tamassee Dr
Clemson SC 29633
USA
☎ 1 803 656 5298
℻ 1 803 656 0805
✉ hartmann@biophy.
phys.clemson.edu

Hartmann Lee William
CfA
HCO/SAO
60 Garden St
Cambridge MA 02138 1516
USA
☎ 1 617 495 7487
℻ 1 617 495 7356

Hartmann William K
Planetary Science Inst
620 N Sixth Av
Tucson AZ 85705 8331
USA
☎ 1 520 662 6300
℻ 1 520 622 8060
✉ psikey@psi.edu

Hartoog Mark Richard
Lick Observ
Univ of California
Santa Cruz CA 95064
USA
☎ 1 831 459 2513
℻ 1 831 426 3115

Hartquist Thomas Wilbur
MPA
Karl Schwarzschildstr 1
D 85748 Garching
Germany
☎ 49 893 299 838
℻ 49 893 299 3235

Hartwick F David A
Physics Dpt
Univ of Victoria
Box 1700
Victoria BC V8W 2Y2
Canada
☎ 1 250 721 7742
℻ 1 250 721 7715

Harutyunian Haik A
Byurakan Astrophys Observ
Armenian Acad Sci
378433 Byurakan
Armenia
☎ 374 88 52 28 3453/4142
℻ 374 88 52 52 3640

Harvel Christopher Alvin
6161 Steven's Forest Rd
Columbia MD 21045
USA
☎ 1 301 964 0211

Harvey Christopher C
Observ Paris Meudon
DESPA
Pl J Janssen
F 92195 Meudon PPL Cdx
France
☎ 33 1 45 07 7669
℻ 33 1 45 07 7971

Harvey Gale A
NASA Langley Res Ctr
Atmosph Sci Div
MS 401a
Hampton VA 23665 5225
USA
☎ 1 804 965 2000

Harvey John W
NOAO/NSO
Box 26732
950 N Cherry Av
Tucson AZ 85726 6732
USA
☎ 1 520 327 5511
℻ 1 520 325 9278
✉ jharvey@noao.edu

Harvey Paul Michael
Astronomy Dpt
Univ of Texas
Rlm 15 308
Austin TX 78712 1083
USA
☎ 1 512 471 4461
℻ 1 512 471 6016
✉ pmh@astro.as.utexas.edu

Harwit Martin
511 H. St SW
Washington DC 20024 2725
USA
☎ 1 202 479 6877
℻ 1 202 484 2654
✉ martin.harwit@ibm.net

Hasan Hashima
STScI
Homewood Campus
3700 San Martin Dr
Baltimore MD 21218
USA
☎ 1 301 338 4519
℻ 1 301 338 4767
✉ hasan@stsci

Hasan Saiyid Strajul
IIA
Koramangala
Sarjapur Rd
Bangalore 560 034
India
☎ 91 80 356 6585
℻ 91 80 553 4043

Haschick Aubrey
Haystack Observ
Westford MA 01886
USA
☎ 1 617 692 4764
℻ 1 617 981 0590

Hasegawa Ichiro
4-18-5 Fujiwaradai Kita
Kiya Ku
Kobe 651 13
Japan
☎ 81 789 82 5255

Hasegawa Tatsuhiko
33 Odakamachi
Yonago-shi
Tottori ken 683
Japan
☎ 81 859 22 3331
✉ hasegawa@ap.stmarys.ca

Hasegawa Tetsuo
Dpt Phys/Astronomy
Univ of Calgary
Calgary AB T2N 1N4
Canada
☎ 1 403 220 5385
✉ hasegawa@iras.
ucalgary.ca

Haser Leo N K
MPE
Postfach 1603
D 85740 Garching
Germany
☎ 49 893 299 803
℻ 49 893 299 3569

Hashimoto Masa-aki
Physics Dpt
College of Gnl Edu
Kyushu Univ Ropponmatsu
Fukuoka 810
Japan
☎ 81 927 71 4161*360
℻ 81 092 731 8745
✉ e76051ajpnccku

Hashimoto Osama
Gunma Astron Obs
1 18 7 Ohdomo
Maebashi
Gunma 371 0847
Japan
☎ 81 27 254 2882
℻ 81 27 254 2883
✉ osamu@astron.pref.
gunma.jp

Hasinger Guenter
Astrophysik Inst
Potsdam Univ
an d Sternwarte 16
D 14482 Potsdam
Germany
☎ 49 331 749 90
℻ 49 331 749 9200

Haslam C Glyn T
RAIUB
Univ Bonn
auf d Huegel 69
D 53121 Bonn
Germany
☎ 49 228 525 0
℻ 49 228 525 229

Hassall Barbara J M
Ctr for Astrophysics
Univ of Central
Lancashire
Preston PR1 2HE
UK
☎ 44 177 289 3569
℻ 44 177 289 2903
✉ b.j.m.hassall@uclan.
ac.uk

Hassan S M
NAIGR
Helwan Observ
Cairo 11421
Egypt
☎ 20 78 0645/2683
℻ 20 62 21 405 297

Haswell Carole A
Astronomy Centre
Univ of Sussex
Falmer
Brighton BN1 9QJ
UK
☎ 44 12 73 678 455
℻ 44 12 73 678 097
✉ chaswell@star.maps.
susx.ac.uk

Hathaway David H
NASA/MSFC
Space Science Lab
Code ES 52
Huntsville AL 35812
USA
☎ 1 205 544 7610
℻ 1 205 547 7754

Hatsukade Isamu
Fac of Eng
Miyazaki Univ
Gakuen Kibanadai Nishi
Miyazaki 889 21
Japan
☎ 81 985 58 2867
℻ 81 985 55 3058
✉ hatukade@astro.
miyazaki-u.ac.jp

Hatzes Artie P
Astronomy Dpt
Univ of Texas
Rlm 15 308
Austin TX 78712 1083
USA
☎ 1 512 471 1473
℻ 1 512 471 6016
✉ artie@astro.as.utexas.edu

Hatzidimitriou Despina
AAO
Epping Laboratory
Box 296
Epping NSW 2121
Australia
☎ 61 2 868 1666
℻ 61 2 9876 8536/542
✉ dh@aaoepp.oz.au

Haubold Hans Joachim
Office for Outer Space
UN Office in Vienna
Bg F 0839 Box 500
A 1400 Wien
Austria
☎ 43 1 21131 4949
℻ 43 1 213 45 5830
✉ hjh2@aip.org

Hauck Bernard
Institut Astronomie
Univ Lausanne
CH 1290 Chavannes d Bois
Switzerland
☎ 41 22 755 2611
℻ 41 22 779 1580
✉ Bernard.Hauck@obs.
unige.ch

Haud Urmas
Tartu Observ
Estonian Acad Sci
Toravere
EE 2444 Tartumaa
Estonia
☎ 372 7 410 305
℻ 372 7 410 205
✉ urmas@aai.ee

Hauge Oivind
Inst Theor Astrophys
Univ of Oslo
Box 1029
N 0315 Blindern Oslo 3
Norway
☎ 47 22 856 502
℻ 47 22 85 6505

Haupt Hermann F
Inst Astronomie
Karl-Franzens-Univ
Universitaetsplatz 5
A 8010 Graz
Austria
☎ 43 316 380 5271
℻ 43 316 384 091

Haupt Wolfgang
Girondelle 105
D 46300 Bochum 1
Germany

Hauschildt Peter H
Dpt Phys/Astronomy
Univ of Georgia
Athens GA 30602 2451
USA
☎ 1 706 542 2485
℻ 1 706 542 2492
✉ yeti@hal.physast.
uga.edu

Hauser Michael G
NASA GSFC
Code 680
Greenbelt MD 20771
USA
☎ 1 301 286 8701
℻ 1 301 286 1753
✉ hauser@stars.gsfc.
nasa.gov

Havlen Robert J
ASP
390 Ashton Av
San Francisco CA 94112
USA
☎ 1 415 337 1100
℻ 1 415 337 5205
✉ rhavlen@stars.sfsu.edu

Havnes Ove
Auroral Observ
Univ of Tromso
Box 953
N 9001 Tromso
Norway
☎ 47 83 86 060

Hawarden Timothy G
Royal Observ
Blackford Hill
Edinburgh EH9 3HJ
UK
☎ 44 131 667 3321
℻ 44 131 668 8356

Hawkes Robert Lewis
Physics Dpt
Mount Allison Univ
Sackville Nb E0A 3C0
Canada
☎ 1 506 364 2582
℻ 1 506 364 2580
✉ rhawkes@mta.ca

Hawking Stephen W
Dpt Appl Maths/Theor Phys
Silver Street
Cambridge CB3 9EW
UK
☎ 44 12 23 337 00
℻ 44 12 23 337 918

Hawkins Gerald S
Consul 906
2400 Virginia Av NW
Washington DC 20037
USA
☎ 1 202 485 2050

Hawkins Isabel
Ctr for EUV Astrophys
Univ of California
2150 Kittredge St
Berkeley CA 94720 5030
USA
☎ 1 510 643 5662
℻ 1 510 643 5660
✉ isabelh@cea.berkeley.edu

Hawkins Michael R S
Royal Observ
Blackford Hill
Edinburgh EH9 3HJ
UK
☎ 44 131 667 3321
℻ 44 131 668 8356

Hawley Suzanne Louise
Physics/Astronomy Dpt
Michigan State Univ
East Lansing MI 48824
USA
☎ 1 517 353 8667
℻ 1 517 535 4500
✉ hawley@pa.msu.edu

Hayashi Chushiro
Momoyama Yogoro-cho 1
Fushimi Ku
Kyoto 612
Japan
☎ 81 756 11 1062

Hayashi Masahiko
Astronomy Dpt
Univ of Tokyo
Bunkyo Ku
Tokyo 113
Japan
☎ 81 381 22 111
℻ 81 338 13 9439

Hayashi Saeko S
Subaru Telescope
NAOJ
650 N A'Ohoku Place
Hilo HI 96720
USA
☎ 1 808 934 5947
℻ 1 808 934 5984
✉ saeko@subaru.naoj.org

Hayes Donald S
Box 1907
Scotsdale AZ 85252
USA

Hayli Avram
Observ Lyon
Av Ch Andre
F 69561 S Genis Laval Cdx
France
☎ 33 4 78 56 0705
℻ 33 4 72 39 9791

Haymes Robert C
Dpt Space Phys/Astron
Rice Univ
Box 1892
Houston TX 77251 1892
USA
☎ 1 713 227 4962
℻ 1 713 285 5954
✉ rhaymes@alfven.rice.edu

Haynes Martha P
Astronomy Dpt
Cornell Univ
512 Space Sc Bldg
Ithaca NY 14853 6801
USA
☎ 1 607 256 3734
℻ 1 607 255 1767

Haynes Raymond F
CSIRO
Div Radiophysics
Box 76
Epping NSW 2121
Australia
☎ 61 2 868 0 276
℻ 61 2 868 0457
✉ rhaynes@rpepping.oz.au

Haynes Roslynn
School of Physics
UNSW
Sydney NSW 2052
Australia
☎ 61 2 9487 2138
℻ 61 2 9489 6058
✉ R.Haynes@unsw.edu.au

Hayward John
Dpt Maths/Computing
Polytechnic of Wales
Pontypridd
Mid Glamorgan CF38 2PJ
UK

Haywood J
TCD
Physics Dpt
College Greem
Dublin 2
Ireland
☎ 353 1 608 1675
℻ 353 1 671 1759

Hazard Cyril
Inst of Astronomy
The Observatories
Madingley Rd
Cambridge CB3 0HA
UK
☎ 44 12 23 337 548
℻ 44 12 23 337 523

Hazen Martha L
CfA
HCO/SAO
60 Garden St
Cambridge MA 02138 1516
USA
☎ 1 617 495 3362
℻ 1 617 495 7356

Hazer S
Dpt Astron/Space Sci
Ege Univ
Box 21
35100 Bornova Izmir
Turkey
☎ 90 232 388 0110*2322

Hazlehurst John
Hamburger Sternwarte
Univ Hamburg
Gojensbergsweg 112
D 21029 Hamburg
Germany
☎ 49 407 252 4137
℻ 49 407 252 4198

He Miaofu
Shanghai Observ
CAS
80 Nandan Rd
Shanghai 200030
China PR
☎ 86 21 6438 6191
℻ 86 21 6438 4618

He Xiangtao
Astronomy Dpt
Beijing Normal Univ
Beijing 100875
China PR
☎ 86 1 65 6531*6285
℻ 86 1 201 3929

Heap Sara R
NASA GSFC
Code 681
Greenbelt MD 20771
USA
☎ 1 301 286 8701
✉ hrsheap@stars.gsfc.nasa.gov

Hearn Anthony G
Sterrekundig Inst Utrecht
Box 80000
NL 3508 TA Utrecht
Netherlands
☎ 31 30 253 5202
℻ 31 30 253 1601
✉ ahearn@astro.uu.nl

Hearnshaw John B
Dpt Phys/Astronomy
Univ of Canterbury
Private Bag 4800
Christchurch 1
New Zealand
☎ 64 3 364 253371
℻ 64 3 364 2469
✉ j.hearnshaw@phys.canterbury.ac.nz

Heasley James Norton
Inst for Astronomy
Univ of Hawaii
2680 Woodlawn Dr
Honolulu HI 96822
USA
☎ 1 808 956 6826
℻ 1 808 988 2790

Heavens Alan
Royal Observ
Blackford Hill
Edinburgh EH9 3HJ
UK
☎ 44 131 668 8352
℻ 44 131 668 8356
✉ afh@roe.ac.uk

Heber Ulrich
Dr Remeis Sternwarte
Univ Erlangen-Nuernberg
Sternwartstr 7
D 96049 Bamberg
Germany
☎ 49 951 952 2214
℻ 49 951 952 2222

Hecht James H
Aerospace Corp
MS M2 255
Box 92957
Los Angeles CA 90009 2957
USA
☎ 1 310 336 7017
℻ 1 310 336 1636

Heck Andre
Observ Strasbourg
11 r Universite
F 67000 Strasbourg
France
☎ 33 3 88 15 0743
℻ 33 3 88 49 1255
✉ heck@astro.u-strasbg.fr

Heckathorn Harry M
NRL
Code 7604
4555 Overlook Av SW
Washington DC 20375 5000
USA
☎ 1 202 767 4198
℻ 1 202 404 8445
✉ harry@vader.nrl.navy.mil

Heckman Timothy M
Astronomy Program
Univ of Maryland
College Park MD 20742 2421
USA
☎ 1 301 454 3001
℻ 1 301 314 9067

Hecquet Josette
OMP
14 Av E Belin
F 31400 Toulouse Cdx
France
☎ 33 5 61 25 2101
℻ 33 5 61 27 3179

Heddle Douglas W O
Physics Dpt
Royal Holloway College
Univ of London
Egham Surrey TW20 0EX
UK
☎ 44 17 843 5351

Hedeman E Ruth
3440 St Jefferson St
Falls Church VA 22041
USA

Heeschen David S
NRAO
520 Edgemont Rd
Charlottesville VA 22903
USA

Hefele Herbert
ARI
Moenchhofstr 12-14
D 69120 Heidelberg
Germany
☎ 49 622 140 50
℻ 49 622 140 5297
✉ hefele@urz.uni-heidelberg.de

Hefty Jan
Observ of the Slovak
Technical Univ
Radlinskeho 11
SK 813 68 Bratislava
Slovak Republic
☎ 421 7 398 047
℻ 421 7 325 476
✉ hefty@cvt.stuba.cs

Heggie Douglas C
Dpt Maths/Statistics
Univ of Edinburgh
King-s Buildings
Edinburgh EH9 3JZ
UK
☎ 44 131 650 5035
℻ 44 131 650 6553
✉ d.c.heggie@ed.ac.uk

Hegyi Dennis J
Randall Laboratory
Univ of Michigan
Dennison Bldg
Ann Arbor MI 48109 1090
USA
☎ 1 313 764 5448
ⓓ 1 313 764 2211

Heidmann Jean
Observ Paris Meudon
ARPEGES
Pl J Janssen
F 92195 Meudon PPL Cdx
France
☎ 33 1 45 07 7598
ⓓ 33 1 45 07 7971

Heidt Jochen
Landessternwarte
Koenigstuhl
D 69117 Heidelberg
Germany
☎ 49 622 150 9204
ⓓ 49 622 150 9202
✉ jheidt@lsw.uni-heidelberg.de

Heiles Carl
Astronomy Dpt
Univ of California
601 Campbell Hall
Berkeley CA 94720 3411
USA
☎ 1 415 642 4510
ⓓ 1 510 642 3411

Hein Righini Giovanna
531 Main St
Roosevelt Island NY 10044
USA

Heinrich Inge
ARI
Moenchhofstr 12-14
D 69120 Heidelberg
Germany
☎ 49 622 140 50
ⓓ 49 622 140 5297
✉ s31@mvs.urz.uni-heidelberg.de

Heintz Wulff D
Astronomy Dpt
Swarthmore Coll
Swarthmore PA 19081
USA
☎ 1 215 447 7265
ⓓ 1 610 328 8660
✉ wheintz1@swarthmore.edu

Heintze J R W
Sterrekundig Inst Utrecht
Box 80000
NL 3508 TA Utrecht
Netherlands
☎ 31 30 253 5235
ⓓ 31 30 253 1601
✉ heintze@astro.uu.nl

Heinzel Petr
Astronomical Institute
Czech Acad Sci
Fricova 1
CZ 251 65 Ondrejov
Czech R
☎ 420 204 85 7233
ⓓ 420 2 88 1611
✉ pheinzel@asu.cas.cz

Heise John
SRON
Postbus 800
Sorbonnelaan 2
NL 3584 CA Utrecht
Netherlands
☎ 31 30 253 5727
ⓓ 31 30 254 0860
✉ j.heise@sron.ruu.nl

Heiser Arnold M
Dyer Observ
Vanderbilt Univ
Box 1803
Nashville TN 37235
USA
☎ 1 615 373 4897
ⓓ 1 615 343 7263

Hejna Ladislav
Astronomical Institute
Czech Acad Sci
Fricova 1
CZ 251 65 Ondrejov
Czech R
☎ 420 204 85 201
ⓓ 420 2 88 1611
✉ lhejna@cspguk11

Hekela Jan
Astronomical Institute
Czech Acad Sci
Fricova 1
CZ 251 65 Ondrejov
Czech R
☎ 420 204 85 7147
ⓓ 420 2 88 1611

Helali Yhya E
NAIGR
Helwan Observ
Cairo 11421
Egypt
☎ 20 78 0645/2683
ⓓ 20 62 21 405 297

Held Enrico V
Osserv Astronom Bologna
Univ d Bologna
Via Zamboni 33
I 40126 Bologna
Italy
☎ 39 51 259 301
ⓓ 39 51 259 407
✉ held@astbo3.bo.astro.it

Helfand David John
Astronomy Dpt
Columbia Univ
538 W 120th St
New York NY 10027
USA
☎ 1 212 854 2150
ⓓ 1 212 854 8121
✉ djh@astro.columbia.edu

Helfer H Lawrence
Dpt Phys/Astronomy
Univ of Rochester
Rochester NY 14627 0171
USA
☎ 1 716 275 4377
ⓓ 1 716 275 4351

Helin Eleanor Francis
CALTECH/JPL
MS 183 501
4800 Oak Grove Dr
Pasadena CA 91109 8099
USA
☎ 1 818 354 4606
ⓓ 1 818 393 6030
✉ efh051@iph.jpl.nasa.gov

Hellaby Charles William
Applied Maths Dpt
Univ of Cape Town
Pricate Bag
7700 Rondebosch
South Africa
☎ 27 21 650 2347
ⓓ 27 21 650 2334
✉ cwh@math.uct.ac.za

Heller Clayton
Univ Sternwarte
Goettingen
Geismarlandstr 11
D 37083 Goettingen
Germany
☎ 49 551 395 055
ⓓ 49 551 395 043
✉ cheller@uni-sw.gwdg.de

Heller Michael
Powstancow Warsawy 13/94
PL 33 110 Tarnow
Poland

Hellier Coel
Physics Dpt
Keele Univ
Staffordshire ST5 5BG
UK
☎ 44 178 25 84243
ⓓ 44 178 271 1093
✉ ch@astro.keele.ac.uk

Hellwig Helmut Wilhelm
Dpt Air Force
SAF/AQR Suite 100
1919 S Eads St
Arlington VA 22202 3053
USA
☎ 1 703 602 9301
ⓓ 1 703 602 4845

Helmer Leif
Astronomical Observ, NBIfAFG
Copenhagen Univ
Juliane Maries Vej 30
DK 2100 Copenhagen
Denmark
☎ 45 35 32 5931
ⓓ 45 35 32 5989
✉ camc@astro.ku.dk

Helmken Henry F
CfA
HCO/SAO
60 Garden St
Cambridge MA 02138 1516
USA
☎ 1 617 495 4984
ⓓ 1 617 495 7356

Helou George
CALTECH
MS 100 22
IPAC
Pasadena CA 91125
USA
☎ 1 818 397 9555
ⓓ 1 818 397 9600
✉ helou@ipac.caltech.edu

Helt Bodil E
Astronomical Observ, NBIfAFG
Copenhagen Univ
Juliane Maries Vej 30
DK 2100 Copenhagen
Denmark
☎ 45 35 32 5964
ⓓ 45 35 32 5989
✉ bodil@astro.ku.dk

Hemenway Mary Kay M
Astronomy Dpt
Univ of Texas
Rlm 15 308
Austin TX 78712 1083
USA
☎ 1 512 471 1309
ⓓ 1 512 471 6016
✉ marykay@astro.as.utexas.edu

Hemenway Paul D
Astronomy Dpt
Univ of Texas
Rlm 15 308
Austin TX 78712 1083
USA
☎ 1 512 471 4461
ⓓ 1 512 471 6016
✉ paul@skye.as.utexas.edu

Hemmleb Gerhard
Zeppelinstr 157
D 14473 Potsdam
Germany
☎ 49 331 974 148
✉ matti@fpk.ti-berlin.de

Henden Arne Anthon
USNO
Flagstaff Station
Box 1149
Flagstaff AZ 86002 1149
USA
☎ 1 602 779 5132
ⓓ 1 602 774 3626
✉ aah@nofs.navy.mil

Henkel Christian
RAIUB
Univ Bonn
auf d Huegel 69
D 53121 Bonn
Germany
☎ 49 228 525 0
ⓓ 49 228 525 229

Henney William John
Instituto Astronomia
J J Tablada 1006
C L de Santa Maria
Morelia Michoacan 58090
Mexico
☎ 52 4 323 6162
℻ 52 4 323 6165
✉ will@astrosmo.
unam.mx

Henning Thomas
Astrophysik Inst
Univ Sternwarte
Schillergaesschen 2
D 07745 Jena
Germany
☎ 49 364 163 0323
℻ 49 364 163 0417
✉ henning@astro.uni-
jena.de

Henon Michel C
OCA
Observ Nice
BP 139
F 06304 Nice Cdx 4
France
☎ 33 4 93 89 0420
℻ 33 4 92 00 3033

Henoux Jean-Claude
Observ Paris Meudon
DASOP
Pl J Janssen
F 92195 Meudon PPL Cdx
France
☎ 33 1 45 07 7803
℻ 33 1 45 07 7971

Henrard Jacques
Dpt Mathematique
FUNDP
Rempart de la Vierge 8
B 5000 Namur
Belgium
☎ 32 81 72 4903
℻ 32 81 72 4914
✉ jhenrard@math.fundp.
ac.be

Henrichs Hubertus F
Astronomical Institute
Univ of Amsterdam
Kruislann 403
NL 1098 SJ Amsterdam
Netherlands
☎ 31 20 525 7491
℻ 31 20 525 7484

Henriksen Mark Jeffrey
Physics Dpt
Univ N Dakota
Grand Forks ND 58208 7129
USA
☎ 1 701 777 4709
℻ 1 701 777 3650
✉ mahernrik@plains.
nodak.edu

Henriksen Richard N
Physics Dpt
Queen's Univ
Kingston ON K7L 3N6
Canada
☎ 1 613 545 2719
℻ 1 613 545 6463
✉ henriksen@astro.queensu.ca

Henry Richard B C
Dpt Phys/Astronomy
Univ of Oklahoma
Norman OK 73019
USA
☎ 1 405 325 3961
℻ 1 405 325 7557

Henry Richard C
Dpt Phys/Astronomy
JHU
Charles/34th St
Baltimore MD 21218
USA
☎ 1 301 516 7350
℻ 1 301 516 4109
✉ rch@pha.jhu.edu

Hensberge Herman
Koninklijke Sterrenwacht
van Belgie
Ringlaan 3
B 1180 Brussels
Belgium
☎ 32 2 373 0284
℻ 32 2 374 9822
✉ herman.hensberge@
omo.be

Hensler Gerhard
Inst Theor Phys/
Sternwarte Univ Kiel
Olshausenstr 40
D 24098 Kiel
Germany
☎ 49 431 880 4125
℻ 49 431 880 4100
✉ hensler@astrophysik.uni-
kiel.de

Heras Ana M
ESA/ESTEC
SSD
Box 299
NL 2200 AG Noordwijk
Netherlands
☎ 31 71 565 6555
℻ 31 71 565 4690

Herbig George H
Inst for Astronomy
Univ of Hawaii
2680 Woodlawn Dr
Honolulu HI 96822
USA
☎ 1 808 956 8312
℻ 1 808 988 2790

Herbst Eric
Physics Dpt
Ohio State Univ
174 W 18th Av
Columbus OH 43210 1106
USA
☎ 1 614 292 2653

Herbst William
Astronomy Dpt
Van Vleck Observ
Wesleyan Univ
Middletown CT 06457
USA
☎ 1 203 347 9411
℻ 1 860 865 2131

Herbstmeier Uwe
MPI Astronomie
Koenigstuhl
D 69117 Heidelberg
Germany
☎ 49 622 152 8355
℻ 49 622 152 8246
✉ herbst@astro

Herczeg Tibor J
Dpt Phys/Astronomy
Univ of Oklahoma
Norman OK 73019
USA
☎ 1 405 325 3961
℻ 1 405 325 7557

Hering Roland
ARI
Moenchhofstr 12-14
D 69120 Heidelberg
Germany
☎ 49 622 140 5157
℻ 49 622 140 5297
✉ s11@ix.urz.uni-
heidelberg.de

Herman Jacobus
SOS
Gutenbergstr 9
Postfach 1449
D 82199 Gilching
Germany
☎ 49 815 328 1167
✉ jaap.herman@dlr.de

Hermans Dirk
School Maths/Statistics
Univ Birmingham
Box 363
Birmingham B15 2TT
UK
☎ 44 12 14 14 3961
℻ 44 12 14 14 3907
✉ d.f.h.hermans@uk.
ac.bham

Hermsen Willem
SRON
Postbus 800
Sorbonnelaan 2
NL 3584 CA Utrecht
Netherlands
☎ 31 30 253 5600
℻ 31 30 254 0860

Hernandez Carlos Alberto
Observ Astronomico
Paseo d Bosque S/n
1900 La Plata (Bs As)
Argentina
☎ 54 21 217 308
℻ 54 21 211 761

Hernandez-Pajares Manuel
Dpt Fisica Aplicada
U P Cataluna
Campus Nord Ed C3
E 08034 Barcelona
Spain
☎ 34 3 401 6029
✉ matmhp@mat.upc.es

Hernanz Margarita
IEEC/CSIC
Edifici Nexus 104
C/Gran Capita 2-4
E 08034 Barcelona
Spain
☎ 34 3 280 2088
℻ 34 3 280 6395
✉ hernanz@ieec.fcr.es

Hernquist Lars Eric
Lick Observ
Univ of California
Santa Cruz CA 95064
USA
☎ 1 831 425 4733
℻ 1 831 426 3115
✉ lars@helios.ucsu.edu

Herold Heinz
Lehrstuhl Theoret Astro
Physik Univ Tuebingen
Auf d Morgenstelle 12 C
D 72076 Tuebingen
Germany
☎ 49 707 129 2043

Herrera Miguel Angel
Instituto Astronomia
UNAM
Apt 70 264
Mexico DF 04510
Mexico
☎ 52 5 622 3906
℻ 52 5 616 0653
✉ mike@astroscu.unam.mx

Herrero Davo Artemio
IAC
Observ d Teide
via Lactea s/n
E 38200 La Laguna
Spain
☎ 34 2 232 9100
℻ 34 2 232 9117

Herrmann Dieter
Archenhold Sternwarte
Alt Treptow 1
D 1193 Berlin
Germany
☎ 49 305 348 080
℻ 49 305 348 083

Hers Jan
Box 48
6573 Sedgefield
South Africa
☎ 27 44 55 736

Hershey John L
STScI/CSC
Homewood Campus
3700 San Martin Dr
Baltimore MD 21218
USA
☎ 1 301 338 4902
℻ 1 301 338 1592
✉ hershey@stsci.edu

Hertz Paul L
NRL
Code 4121 5
4555 Overlook Av SW
Washington DC 20375 5000
USA
☎ 1 202 767 2438

Herzberg Gerhard
NRCC/HIA
100 Sussex Dr
Ottawa ON K1A 0R6
Canada
☎ 1 613 990 0917
✆ 1 613 952 6602

Heske Astrid
ESA/ESTEC
Astrophysics Division
Box 299
NL 2200 AG Noordwijk
Netherlands
☎ 31 71 565 6555
✆ 31 17 19 84690
✉ aheske@estec.esa.nl

Hesser James E
NRCC/HIA
DAO
5071 W Saanich Rd
Victoria BC V8X 4M6
Canada
☎ 1 250 363 0007
✆ 1 250 363 6970
✉ hesser@hia.nrc.ca

Hessman Frederic Victor
Univ Sternwarte
Goettingen
Geismarlandstr 11
D 37083 Goettingen
Germany
☎ 49 551 395 052
✆ 49 551 395 043
✉ hessman@uni-sw.gwdg.de

Heudier Jean-Louis
OCA
Observ Nice
BP 139
F 06304 Nice Cdx 4
France
☎ 33 4 92 00 3012
✆ 33 4 93 13 0134
✉ heudier@obs-nice.fr

Hewett Paul
Inst of Astronomy
The Observatories
Madingley Rd
Cambridge CB3 0HA
UK
☎ 44 12 23 337 548
✆ 44 12 23 337 523
✉ phewett@ast.cam.ac.uk

Hewish Antony
MRAO
Cavendish Laboratory
Madingley Rd
Cambridge CB3 0HE
UK
☎ 44 12 23 337 294
✆ 44 12 23 354 599

Hewitt Adelaide
CASS
UCSD
C 011
La Jolla CA 92093 0216
USA
☎ 1 619 534 6627
✆ 1 619 534 2294
✉ hewitt@ucsd.edu

Hewitt Anthony V
4021 N Cerro de Falcon
Tucson AZ 85718 6724
USA
✉ anthony.hewitt@ibm.net

Hey James Stanley
4 Shortlands Close
Eastbourne BN22 0JE
UK

Heydari-Malayeri Mohammad
Observ Paris
DEMIRM
61 Av Observatoire
F 75014 Paris
France
☎ 33 1 40 51 2116
✆ 33 1 40 51 2002

Heynderickx Daniel
BIRA
Ringlaan 3
B 1180 Brussels
Belgium
☎ 32 2 373 0417
✆ 32 2 374 8423
✉ d.heynderickx@oma.be

Heyvaerts Jean
Observ Strasbourg
11 r Universite
F 67000 Strasbourg
France
☎ 33 3 88 15 0744
✆ 33 3 88 15 0760
✉ heyvaerts@astro.u-strasbg.fr

Hibbs Albert R
781 Prospect Bl
Pasadena CA 91103
USA

Hickson Paul
Dpt Geophys/Astronomy
UBC
2219 Main Mall
Vancouver BC V6T 1W5
Canada
☎ 1 604 228 2267
✆ 1 604 228 6047

Hidalgo Miguel A
Fac Ciencias Fisicas
Univ Zaragoza
E 50009 Zaragoza
Spain
☎ 34 7 676 1000
✆ 34 7 676 1140

Hidayat Bambang
Bosscha Observ
ITB Lembang
Lembang 40391
Indonesia
☎ 62 22 278 6001/6027
✆ 62 22 278 7289
✉ bhidaya@ibm.net

Hide Raymond
Geophysical Fluid
Dynamics Lab
Meteorological Office
Bracknell Berks RG12 2SZ
UK
☎ 44 13 44 42242

Hiei Eijiro
Tokyo Astronomical Obs
NAOJ
Osawa Mitaka
Tokyo 181
Japan
☎ 81 422 32 5111
✆ 81 422 34 3793

Higashi Michiyuki
Kinki Univ
Kowakae
Chikawa
Osaka 577
Japan
☎ 81 6 721 2332
✆ 81 6 727 4301
✉ chikawa@phys.kindai.ac.jp

Higgs Lloyd A
DRAO
NRCC/HIA
Box 248
Penticton BC V2A 6K3
Canada
☎ 1 250 493 2277
✆ 1 250 493 7767
✉ lah@drao.nrc.ca

Hildebrand Roger H
Enrico Fermi Inst
Univ of Chicago
5640 S Ellis Av
Chicago IL 60637
USA
☎ 1 312 702 8203
✆ 1 312 702 8212

Hildebrandt Joachim
Astrophysik Inst
Potsdam Univ
an d Sternwarte 16
D 14482 Potsdam
Germany
☎ 49 331 771 38
✆ 49 331 751 05
✉ jhildebrandt@aip.de

Hilditch Ronald W
Dpt Phys/Astronomy
Univ of St Andrews
North Haugh
St Andrews Fife KY16 9SS
UK
☎ 44 1334 76161
✆ 44 1334 74487
✉ rwh@st-and.ac.uk

Hildner Ernest
NOAA ERL R/E/SE3
Space Environment Lab
325 Broadway
Boulder CO 80303
USA
☎ 1 303 497 3311
✆ 1 303 497 3645
✉ ehildner@sel.noaa.gov

Hilf Eberhard R H
Pestrupsweg 30
D 2900 Oldenburg
Germany

Hill Frank
NOAO/NSO
Box 26732
950 N Cherry Av
Tucson AZ 85726 6732
USA
☎ 1 520 323 4138
✆ 1 520 325 9278
✉ fhill@noao.edu

Hill Graham
18A Stratford St
Parnell
Aucklund 1
New Zealand
☎ 64 9 307 9052
✆ 64 9 307 7917
✉ richjoan@iprolink.co.nz

Hill Grant
McDonald Observ
Univ of Texas
Box 1337
Fort Davis TX 79734
USA
☎ 1 915 426 4184
✆ 1 915 426 3641
✉ grant@astro.as.utexas.edu

Hill Henry Allen
Dpt Phys/Astronomy
Steward Observ
Univ of Arizona
Tucson AZ 85721
USA
☎ 1 520 621 6784
✆ 1 520 621 4933

Hill Philip W
Dpt Phys/Astronomy
Univ of St Andrews
North Haugh
St Andrews Fife KY16 9SS
UK
☎ 44 1334 76161
✆ 44 1334 74487
✉ pwhill@st-andrews.ac.uk

Hillebrandt Wolfgang
MPA
Karl Schwarzschildstr 1
D 85748 Garching
Germany
☎ 49 893 299 9409
✆ 49 893 299 3235

Hilliard Ron
Optomechanics Res Inc
PO Box 87
Vail AZ 85641
USA
☎ 1 602 647 3332
✆ 1 602 647 3312
✉ ronomr@primenet.com

Hills Jack G
LANL
MS B228
Box 1663
Los Alamos NM 87545 2345
USA
☎ 1 505 667 9152
✆ 1 505 665 4055

Hills Richard E
MRAO
Cavendish Laboratory
Madingley Rd
Cambridge CB3 0HE
UK
☎ 44 12 23 337 294
✆ 44 12 23 354 599

Hilton James Lindsay
USNO
3450 Massachusetts Av NW
Washington DC 20392 5100
USA
☎ 1 202 653 1568
✆ 1 202 653 1744
✉ hil@ham.usno.navy.mil

Hindsley Robert Bruce
4b Southnay Rd
Greenbelt MD 20770
USA
☎ 1 202 762 1523
✆ 1 202 653 1497
✉ rbh@fornax.usno.navy.mil

Hinkle Kenneth H
NOAO/KPNO
Box 26732
950 N Cherry Av
Tucson AZ 85726 6732
USA
☎ 1 520 327 5511
✆ 1 520 325 9360

Hinners Noel W
Lockheed Martin Astronautics
MS S8000
Box 179
Denver CO 80201
USA
☎ 1 303 971 1581
✆ 1 303 971 2390
✉ noel.w.hinners@den.mmn.com

Hintzen Paul Michael N
NASA GSFC
Code 681
Greenbelt MD 20771
USA
☎ 1 301 286 5101

Hiotelis Nicolaos
Astronomy Dpt
Ntl Univ Athens
Panepistimiopolis
GR 157 84 Zografos
Greece
☎ 30 1 724 3414
✆ 30 1 723 5122
✉ nhiot@grathun1

Hippelein Hans H
MPI Astronomie
Koenigstuhl
D 69117 Heidelberg
Germany
☎ 49 622 152 80
✆ 49 622 152 8246

Hirabayashi Hisashi
Nobeyama Radio Obs
NAOJ
Minamimaki Mura
Nagano 384 13
Japan
☎ 81 267 98 2831
✆ 81 267 98 2884

Hirai Masanori
Dpt Earth Sci/Astron
Fukuoka Univ of Edu
729 Munakata
Fukuoka 811 41
Japan
☎ 81 940 32 2381

Hirano Naomi
Laboratory of
Astronomy/Geophysics
Hitotsubashi Univ
Kunitachi
Tokyo 186
Japan
☎ 81 425 72 1101*5389
✆ 81 425 71 1893
✉ hirano@higashi.hit-u.ac.jp

Hirata Ryuko
Astronomy Dpt
Kyoto Univ
Sakyo Ku
Kyoto 606 01
Japan
☎ 81 75 753 7008
✆ 81 75 753 7010
✉ hirata@kuastro.kyoto-u.ac.jp

Hirayama Tadashi
Tokyo Astronomical Obs
NAOJ
Osawa Mitaka
Tokyo 181
Japan
☎ 81 422 32 5111
✆ 81 422 34 3793

Hiromoto Norihisa
Communications Res Lab
4-2-1 Nukuikitamachi
Koganei
Tokyo 184
Japan
☎ 81 423 27 7548
✆ 81 423 27 6667
✉ hiromoto@bc.crl.go.jp

Hiroyasu Ando
Ntl Astronomical Obs
of Japan
Osawa Mitaka
Tokyo 181
Japan
☎ 81 422 34 3601
✆ 81 422 34 3608
✉ oandoxx@c1.mtk.nao.ac.jp

Hirth Wolfgang Ernst
Theodor-heus Str 18
D 53540 Weilerswist
Germany

Hitotsuyanagi Juichi
Katahira 1-4-6-401
Sendai 980
Japan
☎ 81 222 27 9351

Hjalmarson Ake G
Onsala Space Observ
Chalmers Technical Univ
S 439 92 Onsala
Sweden
☎ 46 31 772 5500
✆ 46 31 772 5550

Hjellming Robert M
NRAO
Box 0
Socorro NM 87801 0387
USA
☎ 1 505 835 7273
✆ 1 505 853 7027
✉ rhjellmi@nrao.edu

Hjorth Jens
Astronomical Observ, NBIfAFG
Copenhagen Univ
Juliane Maries Vej 30
DK 2100 Copenhagen
Denmark
☎ 45 35 32 5928
✆ 45 35 32 5989
✉ jens@astro.ku.dk

Hnatyk Bohdan
Inst Appl Problems In
Mech/Mathematics
Naukov Str 3-6
290053 Lviv
Ukraine
☎ 380 3227 41114

Ho Paul T P
CfA
HCO/SAO
60 Garden St
Cambridge MA 02138 1516
USA
☎ 1 617 495 3627

Hoag Arthur A
4410 E 14th St
Tucson AZ 85711
USA
☎ 1 520 795 8644

Hoang Binh Dy
Observ Paris Meudon
DAMAP
Pl J Janssen
F 92195 Meudon PPL Cdx
France
☎ 33 1 45 07 7445
✆ 33 1 45 07 7100

Hoare Melvin
Physics Dpt
Univ of Leeds
Leeds LS2 9JT
UK
☎ 44 113 233 3860
✆ 44 113 233 3846
✉ mgh@mpia-hd.mpg.de

Hobbs Lewis M
Yerkes Observ
Univ of Chicago
Box 258
Williams Bay WI 53191
USA
☎ 1 414 245 5555
✆ 1 414 245 9805

Hobbs Robert W
CTA Inc
6116 Executive Blvd
Suite 800
Rockville MD 20852
USA

Hockey Thomas Arnold
Dpt Earth Science
Univ Northern Iowa
Cedar Falls IA 50614 7124
USA
☎ 1 319 273 2065
✆ 1 319 273 7124
✉ hockey@uni.edu

Hodapp Klaus-Werner
Inst for Astronomy
Univ of Hawaii
2680 Woodlawn Dr
Honolulu HI 96822
USA
☎ 1 808 956 8968
✆ 1 808 988 2790
✉ hodapp@ifa.hawaii.edu

Hodge Paul W
Astronomy Dpt
Univ of Washington
Box 351580
Seattle WA 98195 1580
USA
☎ 1 425 543 2888
✆ 1 425 685 0403

Hoeflich Peter
MPA
Karl Schwarzschildstr 1
D 85748 Garching
Germany
☎ 49 893 299 3249
✆ 49 893 299 3235
✉ pah@dgaipp1s

Hoeg Erik
Astronomical Observ, NBIfAFG
Copenhagen Univ
Juliane Maries Vej 30
DK 2100 Copenhagen
Denmark
☎ 45 35 32 5975
✆ 45 35 32 5989
✉ erik@astro.ku.dk

Hoegbom Jan A
Stockholm Observ
Royal Swedish Acad Sciences
S 133 36 Saltsjoebaden
Sweden
☎ 46 8 716 4445
✆ 46 8 717 4719
✉ janh@astro.su.se

Hoeglund Bertil
Onsala Space Observ
Chalmers Technical Univ
S 439 92 Onsala
Sweden
☎ 46 31 772 5500
℻ 46 31 772 5550

Hoeksema Jon Todd
HEPL
Stanford Univ
Annex B213
Stanford CA 94305 4085
USA
☎ 1 415 723 1506
℻ 1 415 725 2333
✉ todd@solar.stanford.edu

Hoekstra Roel
TPD/TNO/TH
Box 155
NL 2600 AD Delft
Netherlands

Hoessel John Greg
Washburn Observ
Univ Wisconsin
475 N Charter St
Madison WI 53706
USA
☎ 1 608 262 1752
℻ 1 608 263 0361

Hoey Michael J
Physics Dpt
Univ College
Belfield
Dublin 4
Ireland
☎ 353 1 693 244
℻ 353 1 671 1759

Hoff Darrel Barton
Box 313
Calmar IA 52132
USA

Hoffleit E Dorrit
Astronomy Dpt
Yale Univ
Box 208101
New Haven CT 06520 8101
USA
☎ 1 203 432 3032
℻ 1 203 432 5048

Hoffman Jeffrey Alan
NASA European Repr
A 230
2 Av Gabriel
F 75008 Paris
France
☎ 33 1 4312 2100
℻ 33 1 4265 8768
✉ jeffrey.hoffman@
hq.nasa.gov

Hoffmann Martin
Observatorium Hoher List
Univ Sternwarte Bonn
D 54550 Daun
Germany
☎ 49 65 92 2150
℻ 49 65 92 9851 40

Hofmann Wilfried
ARI
Moenchhofstr 12-14
D 69120 Heidelberg
Germany
☎ 49 622 140 50
℻ 49 622 140 5297

Hogan Craig J
Astronomy Dpt
Univ of Washington
Box 351580
Seattle WA 98195 1580
USA
☎ 1 425 543 2888
℻ 1 425 685 0403

Hogg David E
NRAO
520 Edgemont Rd
Charlottesville VA 22903
USA
☎ 1 804 296 0220

Hohenkerk Catherine
Royal Greenwich Obs
Nautical Almanac office
Madingley Rd
Cambridge CB3 0EZ
UK
☎ 44 1223 374 738
℻ 44 12 23 374 700
✉ cyh@mail.ast.cam.
ac.uk

Hojaev Alisher S
Astronomical Institute
Uzbek Acad Sci
Astronomicheskaya Ul 33
700052 Tashkent
Uzbekistan
☎ 7 3712 35 8102
℻ 7 37 12 33 3200
✉ hojaev@yahoo.com

Holberg Jay B
Lunar/Planetary Lab
Room 325 Bg 92
Univ of Arizona
Tucson AZ 85721 0092
USA
☎ 1 520 621 4301
℻ 1 520 621 4933

Hollenbach David John
NASA/ARC
MS 245 6
Moffett Field CA 94035 1000
USA
☎ 1 415 997 6426
℻ 1 415 604 6779

Hollis Jan Michael
NASA GSFC
Code 930
Greenbelt MD 20771
USA

Holloway Nigel J
Safety/Reliability dir
Wigshaw Lane
Culcheth
Warrington WA3 4NE
UK
☎ 44 19 53 1244

Hollowell David Earl
LANL
MS B220 X 2
Box 1663
Los Alamos NM 87545 2345
USA
☎ 1 505 667 2388
℻ 1 505 665 4055
✉ u095874@lanl.gov

Hollweg Joseph V
Physics Dpt
Univ of New Hampshire
Demeritt Hall
Durham NH 03824
USA
☎ 1 603 862 3869

Holman Gordon D
NASA GSFC
Code 682
Greenbelt MD 20771
USA
☎ 1 301 286 4636
✉ holman@stars.gsfc.
nasa.gov

Holmberg Erik B
Eneliden 2
S 433 00 Partille
Sweden
☎ 46 31 26 5842

Holt Stephen S
NASA GSFC
Code 600
Greenbelt MD 20771
USA
☎ 1 301 286 8801
℻ 1 301 286 1772
✉ holt@lheavx.gsfc.
nasa.gov

Holweger Hartmut
Inst Theor Phys/
Sternwarte Univ Kiel
Olshausenstr 40
D 24098 Kiel
Germany
☎ 49 431 88 04 107
℻ 49 431 880 4432

Holzer Thomas Edward
HAO
NCAR
Box 3000
Boulder CO 80307 3000
USA
☎ 1 303 497 1536
℻ 1 303 497 1568

Honeycutt R Kent
Astronomy Dpt
Indiana Univ
Swain W 319
Bloomington IN 47405
USA
☎ 1 812 335 6916
℻ 1 812 855 8725

Hong Hyon Ik
Physics Dpt
Kim Il Sung Univ
Taesong district
Pyongyang
Korea DPR

Hong Seung Soo
Astronomy Dpt
Seoul Ntl Univ
Kwanak Ku
Seoul 151
Korea RP
☎ 82 288 06 626
℻ 82 288 71435
✉ sshong@stroism.snu.ac.kr

Hong Xiaoyu
Shanghai Observ
CAS
80 Nandan Rd
Shanghai 200030
China PR
☎ 86 21 6438 6191
℻ 86 21 6438 4618

Hood Alan
Mathematics Dpt
Univ of St Andrews
St Andrews North Haugh
Fife KY16 9SS
UK
☎ 44 13 34 7661
℻ 44 13 347 4487
✉ amsah@savb.st-and.ac.uk.

Hopp Ulrich
Inst Astron/Astrophysik
Univ Sternwarte
Scheinerstr 1
D 81679 Muenchen
Germany
☎ 49 899 220 9425
℻ 49 899 220 9427
✉ hopp@usm.uni-muenchen.de

Horedt Georg Paul
DFVLR
Ne Oe Pe
Oberpfaffenhofen
D 8031 Wessling
Germany
☎ 49 815 328 1328

Hori Genichiro
Astronomy Dpt
Univ of Tokyo
Bunkyo Ku
Tokyo 113
Japan
☎ 81 381 22 111*4251
℻ 81 338 13 9439

Horiuchi Ritoku
Ntl Institute
Fusion Science
Nagoya 464 01
Japan
☎ 81 527 81 511
℻ 81 527 82 7106

Horne Keith
Dpt Phys/Astronomy
Univ of St Andrews
North Haugh
St Andrews Fife KY16 9SS
UK
☎ 44 1334 76161
℻ 44 1334 74487
✉ keith.horne@st-
and.ac.uk

Horowitz Paul
Physics Dpt
HCO/SAO
60 Garden St
Cambridge MA 02138 1516
USA
☎ 1 617 495 3265
℻ 1 617 495 7356

Horsky Jan
Dpt Th Phys/Astrophys
Masaryk Univ
Kotlarska 2
CZ 611 37 Brno
Czech R
☎ 420 5 4112 9381
℻ 420 5 4121 1214
✉ horsky@elanor.sci.
muni.cs

Horton Brian H
Jacabri Ent
Box 309
Goolwa SA 5214
Australia
☎ 61 85 553 376

Horvath Andras
Tit Planetarium &
Urania Observ
Box 46
H 1476 Budapest
Hungary
☎ 36 1 334 25

Horwitz Gerald
Racah Inst of Phys
Hebrew Univ Jerusalem
Jerusalem 91904
Israel
☎ 972 2 58 4592
℻ 972 2 658 4374

Hoshi Reiun
Physics Dpt
Rikkyo Univ
Nishi-ikebukuro
Toshima Ku Tokyo 171
Japan
☎ 81 398 52 414

Hoskin Michael A
Churchill College
Cambridge CB3 0DS
UK
☎ 44 12 23 336 049
℻ 44 12 23 336 180
✉ mah15@cus.cam.ac.uk

Hosking Roger J
James Cook Univ
of NQLD
Townsville Qld 4811
Australia
☎ 61 77 814 1113

Hosokawa Yoshimasa H
Sakigaoka 3 4 9
Funabashi City
Chiba Prefecture 274
Japan
☎ 81 474 48 6679

Hotinli Metin
Univ Rasathanesi
Istanbul
Turkey

Houck James R
Astronomy Dpt
Cornell Univ
512 Space Sc Bldg
Ithaca NY 14853 6801
USA
☎ 1 607 256 4806
℻ 1 607 255 2365

Houdebine Eric
ESA/ESTEC
SSD
Box 299
NL 2200 AG Noordwijk
Netherlands
☎ 31 71 565 6555
℻ 31 71 565 4697
✉ erh@so.estec.esa.nl

Hough James
Physical Sci Div
Univ Hertfordshire
College Lane
Hatfield Herts AL10 9AB
UK
☎ 44 170 728 4607
℻ 44 170 728 4644
✉ jhh@star.herts.ac.uk

Hough James
Dpt Phys/Astronomy
Univ of Glasgow
Glasgow G12 8QQ
UK
☎ 44 141 339 8855
℻ 44 1413 349 029
✉ gla.ph.i1@uk.ac.rl

Houk Nancy
Astronomy Dpt
Univ of Michigan
Dennison Bldg
Ann Arbor MI 48109 1090
USA
☎ 1 313 764 3436
℻ 1 313 763 6317
✉ houk@astro.lsa.
umich.edu

House Franklin C
Heidenreichstr 42
D 6100 Darmstadt
Germany
☎ 49 615 142 2412

House Lewis L
2034 Eisenhower Dr
Louisville CO 80027
USA

Houziaux Leo
Dpt Astrophysique
Univ Mons Hainaut
r de la Halle 15
B 7000 Mons
Belgium
☎ 32 86 38 9818
℻ 32 86 38 9818
✉ leo.houziaux@ulg.ac.be

Hovenier J W
Dpt Phys/Astronomy
Free University
De Boelelaan 1081
NL 1081 HV Amsterdam
Netherlands
☎ 31 20 540 2414
℻ 31 20 444 7899

Hovhannessian Rafik
Byurakan Astrophys Observ
Armenian Acad Sci
378433 Byurakan
Armenia
☎ 374 88 52 28 3453/4142
℻ 374 88 52 52 3640
✉ rhovhan@bao.sci.am

Howard Robert F
NOAO/NSO
Box 26732
950 N Cherry Av
Tucson AZ 85726 6732
USA
☎ 1 520 327 5511
℻ 1 520 325 9278

Howard W Michael
LLNL
L 297
Box 808
Livermore CA 94551 9900
USA
☎ 1 415 422 4138
℻ 1 415 423 0238
✉ howvin@physics.
llnl.gov

Howard William E III
Universities Space Res
Association
300 D Street SW Suite 801
Washington DC 20024
USA
☎ 1 202 488 5128
℻ 1 202 479 2613
✉ whoward@usra.edu

Howarth Ian Donald
Dpt Phys/Astronomy
UCLO
Gower St
London WC1E 6BT
UK
☎ 44 171 387 7050
℻ 44 171 380 7145

Howell Steve Bruce
Dpt Phys/Astronomy
Univ of Wyoming
Box 3905
Laramie WY 82071
USA
☎ 1 307 766 6150
℻ 1 307 766 2652
✉ showell@kaya.
uwyo.edu

Hoyle Fred
102 Admirals Walk
West Cliff Rd Bouremouth
Dorset BH2 5HF
UK

Hoyng Peter
SRON
Postbus 800
Sorbonnelaan 2
NL 3584 CA Utrecht
Netherlands
☎ 31 30 253 5600
℻ 31 30 254 0860
✉ phoyng@sron.ruu.nl

Hozumi Shunsuke
Shiga Univ
2-5-1 Hiratsu
Otsu
Shiga 520
Japan
☎ 81 775 37 7835
℻ 81 775 37 7840
✉ hozumi@sue.shiga-u.ac.jp

Hric Ladislav
Astronomical Institute
Slovak Acad Sci
SK 059 60 Tatranska Lomni
Slovak Republic
☎ 421 969 96 7866
℻ 421 969 96 7656

Hrivnak Bruce J
Physics Dpt
Valparaiso Univ
Valparaiso IN 46383
USA
☎ 1 219 464 5379
℻ 1 219 464 5489
✉ bhrivnak@exodus.valpo.edu

Hron Josef
Inst Astronomie
Univ Wien
Tuerkenschanzstr 17
A 1180 Wien
Austria
☎ 43 1 345 3600
℻ 43 1 470 6015

Hsiang Yan-Yu
Beijing Astronomical Obs
CAS
Beijing 100080
China PR
☎ 86 1 28 1698
℻ 86 1 256 10855

Hsiang-Kuang Tseng
Physics Dpt
Ntl Central Univ
Chung Li Taiwan 32054
China R
☎ 886 3 462 2302
℻ 886 3 426 2304

Hu Esther M
Inst for Astronomy
Univ of Hawaii
2680 Woodlawn Dr
Honolulu HI 96822
USA
☎ 1 808 956 7190
℻ 1 808 988 2790
✉ hu@ifa.hawaii.edu

Hu Fuxing
Purple Mountain Obs
CAS
Nanjing 210008
China PR
☎ 86 25 46700
℻ 86 25 301 459

Hu Jingyao
Beijing Astronomical Obs
CAS
Beijing 100080
China PR
☎ 86 1 28 1698
℻ 86 1 256 10855

Hu Ning-Sheng
Nanjing Astronomical
Instrument Factory
Box 846
Nanjing
China PR
☎ 86 25 46191
✆ 86 25 71 1256

Hu Wenrui
Inst of Mechanics
CAS
W Suburb
Beijing 100080
China PR
☎ 86 1 28 4185
✆ 86 1 256 10855

Hu Zhongwei
Astronomy Dpt
Nanjing Univ
Nanjing 210093
China PR
☎ 86 25 37651
✆ 86 25 330 2728

Hua Chon Trung
LAS
Traverse du Siphon
Les Trois Lucs
F 13376 Marseille Cdx 12
France
☎ 33 4 91 05 5932
✆ 33 4 91 66 1855
✉ hua@frlasm51

Hua Yingmin
Shanghai Observ
CAS
80 Nandan Rd
Shanghai 200030
China PR
☎ 86 21 6438 6191
✆ 86 21 6438 4618
✉ hym@center.shao.ac.cn

Huan Ngeyen Dush
Univ of Vinh
Province Ngh^e An
Vietnam

Huang Bi-kun
Purple Mountain Obs
CAS
Nanjing 210008
China PR
☎ 86 25 307521
✆ 86 25 301 459

Huang Changchung
Purple Mountain Obs
CAS
Nanjing 210008
China PR
☎ 86 25 46700/42817
✆ 86 25 301 459

Huang Cheng
Shanghai Observ
CAS
80 Nandan Rd
Shanghai 200030
China PR
☎ 86 21 6438 6191
✆ 86 21 6438 4618

Huang Fuquan
Purple Mountain Obs
CAS
Nanjing 210008
China PR
☎ 86 25 330 8624
✆ 86 25 330 6982

Huang Guangli
Purple Mountain Obs
CAS
Nanjing 210008
China PR
☎ 86 25 443 2817
✆ 86 25 330 1459
✉ guangli.huang@cetp.ipsl.fr

Huang Jiehao
Astronomy Dpt
Nanjing Univ
Nanjing 210093
China PR
☎ 86 25 663 7551*2882
✆ 86 25 330 2728

Huang Keliang
Astronomy Dpt
Beijing Normal Univ
Beijing 100875
China PR
☎ 86 1 65 6531*6285
✆ 86 1 201 3929

Huang Kunyi
Purple Mountain Obs
CAS
Nanjing 210008
China PR
☎ 86 25 32893
✆ 86 25 301 459

Huang Lin
Beijing Astronomical Obs
CAS
Beijing 100080
China PR
☎ 86 1 28 1698
✆ 86 1 256 10855
✉ hlin@bao01.bao.ac.cn

Huang Runqian
Yunnan Observ
CAS
Kunming 650011
China PR
☎ 86 871 2035
✆ 86 871 717 1845

Huang Song-nian
ICTP
St Costiera 11
Miramare
I 34014 Trieste
Italy
☎ 39 40 22 40111
✆ 39 40 22 4163

Huang Tianyi
Astronomy Dpt
Nanjing Univ
Nanjing 210093
China PR
☎ 86 25 663 7551*2882
✆ 86 25 330 2728

Huang Tieqin
Nanjing Astronomical
Instrument Factory
Box 846
Nanjing
China PR
☎ 86 25 46191
✆ 86 25 71 1256

Huang Yi-Long
Inst of History
Ntl Tsing Hua Univ
Hsin Chu 300043
China R
☎ 886 35 71 6780

Huang Yinliang
Tianjin Inst of Tech
Hong Qi Nan Rd
Tianjin 300191
China PR
☎ 86 36 878 7256

Huang Yinn-Nien
Dir Gen Telecommunicat
Mnstr Transport/Comm
31 Al Kuo East Rd
Taipei 106
China R
☎ 886 2 344 3604
✆ 886 2 356 0259

Huang Yongwei
Beijing Astronomical Obs
CAS
Beijing 100080
China PR
☎ 86 1 28 1698
✆ 86 1 256 10855

Huang Youran
Astronomy Dpt
Nanjing Univ
Nanjing 210093
China PR
☎ 86 25 663 7551*2882
✆ 86 25 330 2728

Hubbard William B
Lunar/Planetary Lab
Room 325 Bg 92
Univ of Arizona
Tucson AZ 85721 0092
USA
☎ 1 520 621 6942
✆ 1 520 621 4933
✉ hubbard@lpl.arizona.edu

Hube Douglas P
Physics Dpt
Univ of Alberta
Edmonton AB T6G 2J1
Canada
☎ 1 403 492 5410
✆ 1 403 492 0714
✉ hube@phys.ualberta.ca

Hubenet Henri
Sterrekundig Inst Utrecht
Box 80000
NL 3508 TA Utrecht
Netherlands
☎ 31 30 253 5200
✆ 31 30 253 1601

Hubeny Ivan
NASA GSFC
Code 681
Greenbelt MD 20771
USA
☎ 1 301 286 6072
✆ 1 301 286 1753
✉ hubeny@deblur.gsfc.nasa.gov

Huber Martin C E
ESA/ESTEC
SSD
Box 299
NL 2200 AG Noordwijk
Netherlands
☎ 31 71 565 3552
✆ 31 71 565 4699
✉ mhuber@astro.estec.esa.nl

Hubert-Delplace A.-M.
Observ Paris Meudon
DASGAL
Pl J Janssen
F 92195 Meudon PPL Cdx
France
☎ 33 1 45 34 7856
✆ 33 1 45 07 7511

Hubrig Swetlana
Astrophysik Inst
Potsdam Univ
an d Sternwarte 16
D 14482 Potsdam
Germany
☎ 49 331 749 9224
✆ 49 331 749 9309
✉ shubrig@astro.physik.uni-potsdam.de

Huchra John Peter
CfA
HCO/SAO MS 20
60 Garden St
Cambridge MA 02138 1516
USA
☎ 1 617 495 7375
✆ 1 617 495 7467
✉ huchra@cfa.harvard.edu

Huchtmeier Walter K
RAIUB
Univ Bonn
auf d Huegel 69
D 53121 Bonn
Germany
☎ 49 228 525 215
✆ 49 228 525 229
✉ hunchtmeier@mpifr-bonn.mpg.de

Hudec Rene
Astronomical Institute
Czech Acad Sci
Fricova 1
CZ 251 65 Ondrejov
Czech R
☎ 420 204 85 7128
✆ 420 2 88 1611
✉ rhudec@asu.cas.cz

Hudson Hugh S
ISAS
3 1 1 Yoshinodai
Sagamihara
Kanagawa 229 8510
Japan
☎ 81 427 69 4532
✆ 81 427 69 4531
✉ hudson@isass6.solar.
isas.ac.jp

Huebner Walter F
Southwest Res Inst
PO Drawer 28510
San Antonio TX 78228 0510
USA
☎ 1 210 522 2730
✆ 1 210 543 0052
✉ whuebner@swri.edu

Huenemoerder David P
Astronomy Dpt
Pennsylvania State Univ
525 Davey Lab
University Park PA 16802
USA
☎ 1 814 865 6601
✆ 1 814 863 7114
✉ dph@astro.psu.edu

Huggins Patrick J
Physics Dpt
New York Univ
4 Washington Place
New York NY 10003
USA
☎ 1 212 998 7717
✆ 1 212 995 4016
✉ patrick.huggins@
nyu.edu

Hughes Arthur R W
Univ of Natal
Physics Dpt
King George V Av
4001 Durban
South Africa
☎ 27 31 260 2775
✆ 27 31 261 6550
✉ hughes@ph.und.ac.za

Hughes David W
Physics Dpt
The University
Sheffield S3 7RH
UK
☎ 44 174 278 555

Hughes John P
CfA
HCO/SAO
60 Garden St
Cambridge MA 02138 1516
USA
☎ 1 617 495 7142
✆ 1 617 495 7356

Hughes Philip
Astronomy Dpt
Univ of Michigan
Dennison Bldg
Ann Arbor MI 48109 1090
USA
☎ 1 313 764 3430
✆ 1 313 764 2211

Hughes Shaun
Royal Greenwich Obs
Madingley Rd
Cambridge CB3 0EZ
UK
☎ 44 12 23 374 000
✆ 44 12 23 374 700
✉ hugnes@mail.ast.
cam.ac.uk

Hughes Victor A
Physics Dpt
Queen's Univ
Kingston ON K7L 3N6
Canada
☎ 1 613 533 6000
✆ 1 613 533 6463
✉ hughes@physics.
queensu.ca

Huguenin G Richard
Millimetrix LLC
100 Venture Way
Hadley MA 01035
USA
☎ 1 413 582 9600
✆ 1 413 582 9610

Hulsbosch A N M
Sterrenkunde
Universiteit Nijmegen
Toernooiveld
NL 6525 ED Nijmegen
Netherlands
☎ 31 24 365 2080
✆ 31 24 365 2191

Humble John Edmund
School of Physics
Univ Tasmania
GPO Box 252 21
Hobart TAS 7001
Australia
☎ 61 3 6226 2396
✆ 61 3 6226 2410
✉ john.humble@
utas.edu.au

Hummel Christian Aurel
USNO
AD5 Bg 22 Rm 220
3450 Massachusetts Av NW
Washington DC 20392 5100
USA
☎ 1 202 762 0314
✆ 1 202 762 1514
✉ cah@fornax.usno.navy.mil

Hummel Edsko
Kapteyn Sterrekundig Inst
Univ Groningen
Postbus 800
NL 9700 AV Groningen
Netherlands
☎ 31 50 363 4073
✆ 31 50 363 6100
✉ secr@astro.rug.nl

Hummel Wolfgang
Inst Astron/Astrophysik
Univ Sternwarte
Scheinerstr 1
D 81679 Muenchen
Germany
☎ 49 899 220 9440
✆ 49 899 220 9427
✉ hummel@usm.uni-
muenchen.de

Humphreys Roberta M
School Phys/Astronomy
Univ Minnesota
116 Church St SE
Minneapolis MN 55455
USA
☎ 1 612 373 9747
✆ 1 612 624 2029

Humphries Colin M
11 Craigmount Grove
North
Edinburgh EH12
UK
☎ 44 131 667 3321

Hundhausen Arthur
HAO
NCAR
Box 3000
Boulder CO 80307 3000
USA
☎ 1 303 497 1000
✆ 1 303 497 1568

Hunger Kurt
Inst Theor Phys/
Sternwarte Univ Kiel
Olshausenstr 40
D 24098 Kiel
Germany
☎ 49 431 880 4110
✆ 49 431 880 4432

Hunstead Richard W
School of Physics
UNSW
Sydney NSW 2006
Australia
☎ 61 2 692 3871
✆ 61 2 660 2903

Hunt G E
Elbury 37 Blenheim Rd
Raynes Park
London SW20 9BA
UK
☎ 44 181 542 2374

Hunt Leslie
Osserv Astrofis Arcetri
Univ d Firenze
Largo E Fermi 5
I 50125 Firenze
Italy
☎ 39 55 27 5 2296
✆ 39 55 22 0039
✉ hunt@sisifo.arcetri.
astro.it

Hunten Donald M
Lunar/Planetary Lab
Room 325 Bg 92
Univ of Arizona
Tucson AZ 85721 0092
USA
☎ 1 520 621 4002
✆ 1 520 621 4933

Hunter Christopher
Mathematics Dpt
Florida State Univ
Tallahassee FL 32306
USA
☎ 1 904 644 2488

Hunter Deidre Ann
Dpt Terrestr Magnetism
Carnegie Inst Washington
5241 Bd Branch Rd NW
Washington DC 20015 1305
USA
☎ 1 202 966 0863
✆ 1 202 364 8726

Hunter James H
Astronomy Dpt
Univ of Florida
211 SSRB
Gainesville FL 32611
USA
☎ 1 904 392 1078
✆ 1 904 392 5089

Huntress Wesley T
NASA Headquarters
Code SL
600 Independence Av SW
Washington DC 20546
USA
☎ 1 202 453 1588
✆ 1 202 426 1023

Huovelin Juhani
Helsinky Univ
Observ
Box 14
FIN 00014 Helsinki
Finland
☎ 358 19 12 2948
✆ 358 19 12 2952
✉ johani.huovelin@
helsinki.fi

Hurford Gordon James
CALTECH
MS 264 33
Pasadena CA 91125
USA
☎ 1 818 356 3866
✆ 1 909 866 4240

Hurley Kevin C
Space Sci Laboratory
Univ of California
Grizzly Peak Blvd
Berkeley CA 94720 7950
USA
☎ 1 510 643 9173
✆ 1 510 643 8602
✉ khurley@sunspot.ssl.
berkerly.edu

Hurnik Hieronim
Astronomical Observ
A Mickiewicz Univ
Ul Sloneczna 36
PL 60 286 Poznan
Poland
☎ 48 61 67 9670
✆ 48 61 53 6536

Hurukawa Kiitiro
1 3 29 Osawa
Mitaka
Tokyo 181
Japan
☎ 81 422 32 3834
✆ 81 422 33 8514

Hurwitz Mark V
Ctr for EUV Astrophys
Univ of California
2150 Kittredge St
Berkeley CA 94720 5030
USA
☎ 1 510 643 5665
✆ 1 415 643 5660
✉ markh@ssl.berkeley.edu

Husfeld Dirk
Inst Astron/Astrophysik
Univ Sternwarte
Scheinerstr 1
D 81679 Muenchen
Germany
☎ 49 899 220 9440
✆ 49 899 220 9427

Hut Piet
Inst Advanced Study
School Natural Science
Princeton NJ 08540
USA
☎ 1 609 734 8075
✆ 1 609 924 8399
✉ piet@guinness.ias.edu

Hutcheon Richard J
Plasma Physics Div
Atomic Energy Lab
Pelindara PB X256
0001 Pretoria
South Africa

Hutchings John B
NRCC/HIA
DAO
5071 W Saanich Rd
Victoria BC V8X 4M6
Canada
☎ 1 250 363 0018
✆ 1 250 363 0045
✉ hutchings@dao.nrc.ca

Hutsemekers Damien
Inst Astrophysique
Universite Liege
Av Cointe 5
B 4000 Liege
Belgium
☎ 32 4 254 7510
✆ 32 4 254 7511
✉ hutsemek@astro.
ulg.ac.be

Hutter Donald John
USNO
AD5 Bg 22 Rm 220
3450 Massachusetts Av NW
Washington DC 20392 5100
USA
☎ 1 202 762 1549
✆ 1 202 762 1514
✉ djh@fornax.usno.
navy.mil

Hwang Jai-chan
Dpt Astron/Space Sci
Kyunghee Univ
Yong In Kun
Kyunggee 449 701
Korea RP
☎ 82 539 50 6360
✆ 82 539 50 6359
✉ jchan@hanul.issa.
re.kr

Hwang Woei-yann P
Physics Dpt
Ntl Taiwan Univ
Taipei 10764
China R
☎ 886 2 363 0231*3159
✆ 886 2 363 7204/9984

Hyder C L
C/o Rozamme Smith
2837 Alvarado Ne
Albuquerque NM 87110
USA

Hyland A R Harry
Southern Cross Univty
Box 157
Lismore NSW 2480
Australia
☎ 61 66 20 3000
✆ 61 66 22 1300
✉ h.hyland@unsw.edu.au

Hysom Edmund J
8 East Dr
Caldecote
Cambridge CB3 7NZ
UK
☎ 44 19 54 211 137

Hyun Jong-June
Astronomy Dpt
Seoul Ntl Univ
Kwanak Ku
Seoul 151
Korea RP
☎ 82 877 30 10*2542
✆ 82 288 71435

Hyung Siek
Korea Astronomy Observ
Hwaam-dong Yusung-gu
San 36 1
Taejeon
Korea RP
☎ 82 563 301 014
✆ 82 563 369 450
✉ hyung@bohyun.boao.re.kr

I

Ianna Philip A
University Station
Univ of Virginia
Box 3818
Charlottesville VA 22903 0818
USA
☎ 1 804 924 7494
✆ 1 804 924 3104
✉ pai@fermi.clas.virginia.edu

Iannini Gualberto
Observ Astronomico
de Cordoba
Laprida 854
5000 Cordoba
Argentina
☎ 54 51 23 0491
✆ 54 51 21 0613

Ibadinov Khursandkul
Astrophys Inst
Uzbek Acad Sci
Sviridenko Ul 22
734670 Dushanbe
Tajikistan
☎ 7 3770 23 1432
✆ 7 3770 27 5483

Ibadov Subhon
Astrophys Inst
Uzbek Acad Sci
Sviridenko Ul 22
734670 Dushanbe
Tajikistan
☎ 7 3770 23 1432
✆ 7 3770 27 5483
✉ ibadov@td.silk.org

Ibanez S Miguel H
Dpt Fisica
Univ Los Andes
Merida 5101 A
Venezuela
☎ 58 7 463 9930/7477
✆ 58 7 444 1723

Ibanoglu Cafir
Dpt Astron/Space Sci
Ege Univ
Box 21
35100 Bornova Izmir
Turkey
☎ 90 232 388 0110*2322

Ibbetson Peter Aaron
Wise Observ
Tel Aviv Univ
Ramat Aviv
Tel Aviv 69978
Israel
☎ 972 3 641 3788
✆ 972 3 640 8179

Iben Icko
Dpt Astron/Physics
Univ of Illinois
1011 W Springfield Av
Urbana IL 61801
USA
☎ 1 217 333 3090
✆ 1 217 244 7638

Ibrahim Jorga
Astronomy Dpt
Bandung Instte Techn
Jl Tamansari 64
Bandung 40132
Indonesia

Ichikawa Shin-ichi
Tokyo Astronomical Obs
NAOJ
Osawa Mitaka
Tokyo 181
Japan
☎ 81 422 41 3604
✆ 81 422 41 3608
✉ ichikawa@c1.mtk.
nao.ac.jp

Ichikawa Takashi
Astron Institute
Tohoku Univ
Aoba
Sendai 980 77
Japan
☎ 81 22 217 6500
✆ 81 222 17 6513
✉ ichikawa@astr.tohoku.
ac.jp

Ichimaru Setsuo
Physics Dpt
Univ of Tokyo
Bunkyo Ku
Tokyo 113
Japan
☎ 81 381 22 111
✆ 81 338 13 9439

Icke Vincent
Leiden Observ
Box 9513
NL 2300 RA Leiden
Netherlands
☎ 31 71 527 5843
✆ 31 71 527 5819
✉ icke@strw.LeidenUniv.nl

Idlis Grigorij M
Inst Hist of Sci/Techn
Acad Sciences
Staropansky 1/5
103012 Moscow
Russia
☎ 7 095 228 1969

Iijima Shigetaka
4 23 6 Osawa
Mitaka
Tokyo 181
Japan
☎ 81 422 31 6031
✆ 81 422 31 6031

Iijima Takashi
Osserv Astrofisico
Univ d Padova
Via d Osservatorio 8
I 36012 Asiago
Italy
☎ 39 424 460 0030
✆ 39 424 60 0023
✉ iijima@astras.pd.
astro.it

Ikeuchi Satoru
Nagoya Univ
Physics Dpt
Nagoya 464 01
Japan
☎ 81 527 89 2427
✆ 81 527 89 2838
✉ ikeuchi@a.phys.
nagoya-u.ac.jp

Ikhsanov Robert N
Pulkovo Observ
Acad Sciences
10 Kutuzov Quay
196140 St Petersburg
Russia
☎ 7 812 298 2242
✆ 7 812 315 1701

Ikhsanova Vera N
Pulkovo Observ
Acad Sciences
10 Kutuzov Quay
196140 St Petersburg
Russia
☎ 7 812 298 2242
✆ 7 812 315 1701

Iliev Ilian
Ntl Astronomical Obs
Bulgarian Acad Sci
Box 136
BG 4700 Smoljan
Bulgaria
☎ 359 7 341 559

Ilin Alexei E
Pulkovo Observ
Acad Sciences
10 Kutuzov Quay
196140 St Petersburg
Russia
☎ 7 812 123 4387
✆ 7 812 123 1922
✉ alexei@ilin.spb.su

Ill Marton J
Baja Astronom Observ
Toth Kalman U 19
H 6501 Baja
Hungary
☎ 36 7 912 110

Illes Almar Erzsebet
Konkoly Observ
Thege U 13/17
Box 67
H 1525 Budapest
Hungary
☎ 36 1 375 4122
✆ 36 1 275 4668

Illing Rainer M E
Ball Aerospace Systems
Division
Box 1062
Boulder CO 80306
USA
☎ 1 303 939 5888

Illingworth Garth D
Lick Observ
Univ of California
Santa Cruz CA 95064
USA
☎ 1 831 459 2513
✆ 1 831 426 3115

Ilyas Mohammad
Sheikh Tahir Astron Ctr
Falak Ctr Bldg
Univ Science of Malaysia
11800 Penang
Malaysia
☎ 60 4 657 7888*2115
✆ 60 4 657 6155
✉ milyas@usm.my

Ilyasov Yuri P
Astro Space Ctr
Lebedev Physical Inst
Profsoyuznaya 84/32
Moscow 117924
Russia
☎ 7 096 773 2746
✆ 7 096 773 2482
✉ ilyasov@prao.psn.ru

Imamura James
Physics Dpt
Univ of Oregon
Eugene OR 97403
USA
☎ 1 541 346 5212
✆ 1 541 346 5217
✉ imamura@herb.
uoregon.edu

Imbert Maurice
Observ Marseille
2 Pl Le Verrier
F 13248 Marseille Cdx 04
France
☎ 33 4 91 95 9088
✆ 33 4 91 62 1190
🖂 imbert@obmara.cnrs-mrs.fr

Imbroane Alexandru
Astronomical Observ
Univ of Cluj Napoca
Ul Ciresilor 19
RO 3400 Cluj Napoca
Rumania
☎ 40 64 194 592
✆ 40 64 19 2820
🖂 alex@geogr.ubbcluj.ro

Imhoff Catherine L
Computer Sciences Corp
Science Programes
10000A Aerospace Rd
Lanham Seabrook MD 20706
USA
☎ 1 301 794 1470
✆ 1 301 459 4482
🖂 imhoff@iuegtc.dnet.nasa.gov

Impey Christopher D
Steward Observ
Univ of Arizona
Tucson AZ 85721
USA
☎ 1 520 621 6522
✆ 1 520 621 1532
🖂 impey@solpl.as.arizona.edu

Imshennik Vladimir S
Inst Theor/Exper Phys
Cheremushkinskaya Ul 25
117259 Moscow
Russia
☎ 7 095 123 0292

Inagaki Shogo
Astronomy Dpt
Kyoto Univ
Sakyo Ku
Kyoto 606 01
Japan
☎ 81 757 51 2111
✆ 81 75 753 7010
🖂 inagaki@kuastro.kyoto-u.ac.jp

Inatani Junji
Nobeyama Radio Obs
NAOJ
Minamimaki Mura
Nagano 384 13
Japan
☎ 81 267 98 2831
✆ 81 267 98 2884

Infante Leopoldo
Dpt Astron y Astrofis
Pontificia Univ
Vic Mac 4860 Cas 104
Santiago 22
Chile
☎ 56 2 686 4939
✆ 56 2 686 4948
🖂 linfante@astro.puc.cl

Ingerson Thomas
NOAO
CTIO
Casilla 603
La Serena
Chile
☎ 56 51 22 5415
✆ 56 51 20 5342
🖂 tingerson@noao.edu

Inglis Michael
Physical Sci Div
Univ Hertfordshire
College Lane
Hatfield Herts AL10 9AB
UK
☎ 44 170 728 6143
✆ 44 170 728 4644
🖂 mdi@star.herts.ac.uk

Innanen Kimmo A
Physics Dpt
York Univ
4700 Keele St
North York ON M3J 1P3
Canada
☎ 1 416 667 3837
✆ 1 416 736 5386

Inoue Hajime
ISAS
3 1 1 Yoshinodai
Sagamihara
Kanagawa 229 8510
Japan
☎ 81 427 513 911
✆ 81 427 594 253
🖂 inoue@astro.isas.ac.jp

Inoue Makoto
VSOP Project
Ntl Astronomical Obs
Osawa Mitaka
Tokyo 181
Japan
☎ 81 422 34 3807
✆ 81 422 34 3869
🖂 inoue@nao.ac.jp

Inoue Takeshi
Physics Dpt
Kyoto Sangyo Univ
Kamigamo
Kyoto 603
Japan
☎ 81 757 01 2151
✆ 81 75 705 1640

Ioshpa Boris A
ITMIRWP
Acad Sciences
142092 Troitsk
Russia
☎ 7 095 232 1921
✆ 7 095 334 0124

Iovino Angela
Osserv Astronomico
d Brera
Via Brera 28
I 20121 Milano
Italy
☎ 39 2 723 201
✆ 39 2 720 01600
🖂 iovino@brera.mi.astro.it

Ip Wing-huen
MPI Astronomie
Max Planck Str 2
Postfach 20
D 37189 Katlenburg Lindau
Germany
☎ 49 555 641 6
✆ 49 555 697 9240

Ipatov Alexander V
Inst Appl Astronomy
Acad Sciences
Zhdanovskaya St 8
197042 St Petersburg
Russia
☎ 7 812 235 3497/230 7414
✆ 7 812 230 7413
🖂 iparan@sovam.com

Ipser James R
Physics Dpt
Univ of Florida
Williamson 220A
Gainesville FL 32611 8440
USA
☎ 1 904 392 0521
🖂 ipser@possum.phys.ufl.edu

Irbah Abdanour
CRAAG
Observ Alger
BP 63
Bouzareah 16340
Algeria
☎ 213 2 901 572/424
✆ 213 2 901 424
🖂 irbah@boulega.unice.fr

Ireland Jack
School of Maths/Comp Sci
Univ of St Andrews
North Haugh
St Andrews Fife KY16 9SS
UK
☎ 44 1334 76161
✆ 44 1334 74487
🖂 jack@dcs.st-and.ac.uk

Ireland John G
C/o 13 Gorden Rd
Belvedere Kent DA17 6EA
UK

Irigoyen Maylis
Universite Paris II
12 Pl du Pantheon
F 75005 Paris
France

Iriyama Jun
Fac of Eng
Chubu Univ
1200 Matsumoto
Kasugai shi aichi 487
Japan
☎ 81 568 51 1111

Irvine William M
Five College
RAO
B619 Lederle Grad Res Twr
Amherst MA 01003
USA
☎ 1 413 545 0733
✆ 1 413 545 4223
🖂 irvine@fcrao1.phast.umass.edu

Irwin Alan W
Physics Dpt
Univ of Victoria
Box 1700
Victoria BC V8W 2Y2
Canada
☎ 1 250 721 7700
✆ 1 250 721 7715
🖂 irwin@uvastro.phys.uvic.ca

Irwin Judith
Physics Dpt
Queen's Univ
Stirling Hall Rm 308e
Kingston ON K7L 3N6
Canada
☎ 1 613 547 2712
✆ 1 613545 6463

Irwin Michael John
Inst of Astronomy
The Observatories
Madingley Rd
Cambridge CB3 0HA
UK
☎ 44 12 23 337 548
✆ 44 12 23 337 523
🖂 mike@ast.cam.ac.uk

Isaak George R
Physics Dpt
Univ Birmingham
Box 363
Birmingham B15 2TT
UK
☎ 44 12 14 72 1301
✆ 44 12 1414 45 77

Isern Jorge
C/Sepulveda 83 6 3a
E 08015 Barcelona
Spain

Ishida Keiichi
Tokyo Astronomical Obs
NAOJ
Osawa Mitaka
Tokyo 181
Japan
☎ 81 422 32 5211
✆ 81 422 34 3793

Ishida Manabu
ISAS
3 1 1 Yoshinodai
Sagamihara
Kanagawa 229 8510
Japan
☎ 81 427 51 3911
✆ 81 427 59 4253
🖂 ishida@astro.isas.ac.jp

Ishida Toshihito
Nishi Harima Astronomical
Observ
Ohnadesan Sayo Cho
Hyogo 679 53
Japan
☎ 81 790 82 3886*142
✆ 81 790 82 3514
🖂 ishida@nhao.go.jp

Ishiguro Masato
Nobeyama Radio Obs
NAOJ
Minamimaki Mura
Nagano 384 13
Japan
☎ 81 267 98 2831
℻ 81 267 98 2884
✉ ishiguro@nro.nao.ac.jp

Ishihara Hideki
Physics Dpt
Kyoto Univ
Sakyo Ku
Kyoto 606 01
Japan
☎ 81 757 53 3850
℻ 81 75 753 7010
✉ ishihara@nws841.
scphys.kyotoua

Ishizawa Toshiaki A
Astronomy Dpt
Kyoto Univ
Sakyo Ku
Kyoto 606 01
Japan
☎ 81 757 51 2111
℻ 81 75 753 7010
✉ ishizawa@kusastro.
kyoto-u.ac.j

Ishizuka Toshihisa
Physics Dpt
Ibaraki Univ
Bunkyo
Mito 310
Japan
☎ 81 292 26 1621

Isliker Heinz
Astronomy Lab
Univ Thessaloniki
GR 540 06 Thessaloniki
Greece
☎ 30 31 99 80 62
℻ 30 31 99 53 84
✉ isliker@helios.astro.
auth.fr

Isobe Syuzo
National Astronomical Obs
2-21-1 Osawa
Mitaka
Tokyo 181
Japan
☎ 81 422 34 3645
℻ 81 422 34 3641
✉ isobesz@cc.nao.ac.jp

Israel Frank P
Leiden Observ
Box 9513
NL 2300 RA Leiden
Netherlands
☎ 31 71 527 5891
℻ 31 71 527 5819
✉ israel@strw.leidenuniv.nl

Israel Guy Marcel
Service Aeronomie
BP 3
F 91371 Verrieres Buisson
France
☎ 33 1 64 47 4289

Israel Werner
Physics Dpt
Univ of Alberta
Edmonton AB T6G 2J1
Canada
☎ 1 403 432 3552
℻ 1 403 432 4256

Issa Issa Aly
NAIGR
Helwan Observ
Cairo 11421
Egypt
☎ 20 78 0645/2683
℻ 20 2 78 26 83

Isserstedt Joerg
Astronomisches Institut
Univ Wuerzburg
am Hubland
D 97074 Wuerzburg
Germany
☎ 49 931 888 5301
℻ 49 931 888 4603

Ito Kensai A
Physics Dpt
Rikkyo Univ
Nishi-ikebukuro
Tokyo 171
Japan
☎ 81 398 52 384

Ito Yutaka
Coll of Edu Akita Univ
Cosmic/Earth Science
Tegata Gakuencho
Akita 010
Japan
☎ 81 188 892 654
℻ 81 188 892 655

Itoh Hiroshi
Minamida-cho 27-160
Jodoji Sakyo-ku
Kyoto 606
Japan
☎ 81 757 51 2111
℻ 81 75 753 7010

Itoh Masayuki
Fac of Human Dvpt
Kobe Univ
3 11 Tsurukabuto Nada
Kanagawa 229
Japan
☎ 81 78 803 0913
℻ 81 78 803 0261
✉ mitah@kobe-u.ac.jp

Itoh Naoki
Physics Dpt
Sophia Univ
7-1 Kioi-cho Chiyoda Ku
Tokyo 102
Japan
☎ 81 323 83 431

Ivanchuk Victor I
Astronomical Observ
Kyiv State Univ
Observatornaya Ul 3
252053 Kyiv
Ukraine
☎ 380 4421 62691

Ivanov Evgeny I
ITMIRWP
Acad Sciences
142092 Troitsk
Russia
☎ 7 095 334 0282
℻ 7 095 334 0124
✉ ivanov@lars.izmiran.
troitsk.su

Ivanov Georgi R
Astronomy Dpt
Univ of Sofia
Anton Ivanov St 5
BG 1126 Sofia
Bulgaria
☎ 359 2 68 8176
℻ 359 2 68 9085

Ivanov Vsevolod V
Astronomical Observ
St Petersburg Univ
Bibliotechnaja Pl 2
198904 St Petersburg
Russia
☎ 7 812 257 9491
℻ 7 812 428 4259

Ivanov Kholodny Goz S
ITMIRWP
Acad Sciences
142092 Troitsk
Russia
☎ 7 095 334 0120
℻ 7 095 334 0124

Ivanova Violeta
Astronomy Dpt
Bulgarian Acad Sci
72 Lenin Blvd
BG 1784 Sofia
Bulgaria
☎ 359 2 75 8927
℻ 359 2 75 5019

Ivchenko Vasily
Astronomy Dpt
Univ of Kyiv
Acad Glushkov Str 6
252022 Kyiv
Ukraine
☎ 380 4426 64457
℻ 380 4426 64507
✉ ivchenko@astron.univ.
kiev.ua

Ives John Christopher
ESA/ESTEC
SSD
Box 299
NL 2200 AG Noordwijk
Netherlands
☎ 31 71 565 6555
℻ 31 71 565 4532

Iwaniszewska Cecylia
Inst of Astronomy
N Copernicus Univ
Ul Chopina 12/18
PL 87 100 Torun
Poland
☎ 48 56 26 018
℻ 48 56 24 602
✉ cecylia@cc.uni.torun.pl

Iwanowska Wilhelmina
Inst of Astronomy
N Copernicus Univ
Ul Chopina 12/18
PL 87 100 Torun
Poland
☎ 48 56 26 018
℻ 48 56 24 602

Iwasaki Kyosuke
Kyoto Gakuen Univ
NANJO
Sogabecho Kameoka
Kyoto 621
Japan
☎ 81 771 22 2001
℻ 81 771 24 8150

Iwata Takahiro
Kashima Space Res C
Commun Res Lab
893 1 Hirai
Kashima Ibaraki 314
Japan
☎ 81 299 84 7140
℻ 81 299 84 7159
✉ iwat@crl.go.jp

Iye Masanori
Ntl Astronomical Obs
NAOJ
Osawa Mitaka
Tokyo 181
Japan
☎ 81 422 34 3703
℻ 81 422 34 3608
✉ iye@optik.mtk.nao.ac.jp

Iyengar K V K
IIA
Koramangala
Sarjapur Rd
Bangalore 560 034
India
☎ 91 80 353 0672/0676
℻ 91 80 553 4043
✉ kvki@iiap.ernet.in

Iyengar Srinivasan Rama
IIA
Koramangala
Sarjapur Rd
Bangalore 560 034
India
☎ 91 80 553 0672
℻ 91 80 553 4043

Iyer B R
RRI
Sadashivanagar
CV Raman Av
Bangalore 560 080
India
☎ 91 80 336 0122
℻ 91 80 334 0492

Izotov Yuri
Main Astronomical Obs
Ukrainian Acad Science
Golosiiv
252650 Kyiv 22
Ukraine
☎ 380 4426 63110
℻ 380 4426 62147

Izumiura Hideyuki
Astronomy Dpt
Tokyo Gakugei Univ
Nukuikita 4 1 1
Koganei Tokyo 184
Japan
☎ 81 423 25 2111 *2680
✆ 81 423 24 9832
✉ izumiura@yamabuki.u-gakugei.ac.jp

Izvekov Vladimir A
Inst Theoret Astron
Acad Sciences
N Kutuzova 10
191187 St Petersburg
Russia
☎ 7 812 272 40 23
✆ 7 812 272 7968

J

Jaakkola Toivo S
Helsinky Univ
Observ
Box 14
FIN 00014 Helsinki
Finland
☎ 358 19 12 2907
℻ 358 19 12 2952

Jabbar Sabeh Rhaman
SARC
Scientific Res Council
Box 2441
Jadiriyah Baghdad
Iraq
☎ 964 1 776 5127

Jabir Niama Lafta
SARC
Scientific Res Council
Box 2441
Jadiriyah Baghdad
Iraq
☎ 964 1 776 5127

Jablonka Pascale
Observ Paris Meudon
DAEC
Pl J Janssen
F 92195 Meudon PPL Cdx
France
☎ 33 4 45 07 7419
℻ 33 4 45 07 7469
✉ jablonka@gin.obspm.fr

Jablonski Francisco
INPE
Dpt Astronomia
CP 515
12227 010 S Jose dos Campos
Brazil
☎ 55 123 41 8977
℻ 55 123 21 8743

Jackisch Gerhard
Zntrlinst Astrophysik
Sternwarte Sonneberg
an d Sternwarte 16
D 96515 Sonneberg
Germany
☎ 49 96 74 2287

Jackson Bernard V
CASS
UCSD
C 011
La Jolla CA 92093 0216
USA
☎ 1 619 534 3358
℻ 1 619 534 2294

Jackson John Charles
16 The Park
Newark NG24 1S0
UK

Jackson Neal
NRAL
Univ Manchester
Jodrell Bank
Macclesfield SK11 9DL
UK
☎ 44 14 777 1321
℻ 44 147 757 1618
✉ njj@jb.man.ac.uk

Jackson Paul
Inst Astronomie
Univ Wien
Tuerkenschanzstr 17
A 1180 Wien
Austria
☎ 43 1 345 3600
℻ 43 1 470 6015

Jackson Peter Douglas
Hughes STX Corp
7601 Ora Glen Dr
Suite 100
Greenbelt MD 20770
USA
☎ 1 301 513 7735
℻ 1 301 513 7726
✉ jackson@tonga.gsfc.nasa.gov

Jackson William M
Chemistry Dpt
Univ of California
Room 214
Davis CA 95616
USA
☎ 1 916 752 0503

Jacobs Kenneth C
Physics Dpt
Hollins College
Box 9661
Roanoke VA 24020
USA
☎ 1 703 362 6478

Jacobsen Theodor S
Astronomy Dpt
Univ of Washington
12501 Greenwood Av NC117
Seattle WA 98195 1580
USA

Jacobson Robert A
CALTECH/JPL
4800 Oak Grove Dr
Pasadena CA 91109 8099
USA
☎ 1 818 354 7201
℻ 1 818 393 6030
✉ raj@murphy.jpl.nasa.gov

Jacoby George H
NOAO/KPNO
Box 26732
950 N Cherry Av
Tucson AZ 85726 6732
USA
☎ 1 520 325 9292
℻ 1 520 325 9360
✉ gjacoby@noao.edu

Jacq Thierry
Observ Bordeaux
BP 89
F 33270 Floirac
France
☎ 33 5 57 77 6163
℻ 33 5 57 77 6155
✉ jacq@observ.u-bordeaux.fr

Jacquinot Pierre
Laboratoire Aime Cotton
Bt 505
Universite Paris XI
F 91405 Orsay Cdx
France

Jaeger Friedrich W
Zntrlinst Astrophysik
Sternwarte Babelsberg
Rosa Luxemburg Str 17a
D 14473 Potsdam
Germany
☎ 49 331 762225

Jaffe Daniel T
Space Sci Laboratory
Univ of California
Grizzly Peak Blvd
Berkeley CA 94720 7950
USA
☎ 1 415 642 1930
✉ dtj@astro.as.utexas.edu

Jaffe Walter Joseph
Leiden Observ
Box 9513
NL 2300 RA Leiden
Netherlands
☎ 31 71 527 5862
℻ 31 71 527 5819
✉ jaffe@strw.leidenuniv.nl

Jahn Krzysztof
Astrononomical Observ
Warsaw Univ
Al Ujazdowskie 4
PL 00 478 Warsaw
Poland
☎ 48 22 29 4011
℻ 48 22 629 4967
✉ crj@astrouw.edu.pl

Jahreiss Hartmut
ARI
Moenchhofstr 12-14
D 69120 Heidelberg
Germany
☎ 49 622 140 5119
℻ 49 622 140 5297
✉ s12@ix.uni-heidelberg.de

Jain Rajmal
Udaipur Solar Observ
11 Vidya Marg
Udaipur 313 001
India
☎ 91 25 626/27 457

Jakimiec Jerzy
Astronomical Institute
Wroclaw Univ
Ul Kopernika 11
PL 51 622 Wroclaw
Poland
☎ 48 71 372 9373/74
℻ 48 71 372 9378

Jakobsen Peter
ESA/ESTEC
SSD
Box 299
NL 2200 AG Noordwijk
Netherlands
☎ 31 71 565 6555
℻ 31 71 565 4532

Jamar Claude A J
Centre Spatial d Liege
Universite Liege
Av du Pre-Aily
B 4031 Angleur Liege
Belgium
☎ 32 4 367 6668
℻ 32 4 367 5613
✉ cjamar@astro.ulg.ac.be

James John F
Schuster Laboratory
Univ Manchester
Oxford Rd
Manchester M13 9PL
UK
☎ 44 161 275 4224
℻ 44 161 275 4223

James Richard A
Astronomy Dpt
Univ Manchester
Oxford Rd
Manchester M13 9PL
UK
☎ 44 161 275 4224
℻ 44 161 275 4223

Jameson Richard F
Astronomy Dpt
Univ Leicester
University Rd
Leicester LE1 7RH
UK
☎ 44 116 252 2073
℻ 44 113 252 2200

Janes Kenneth A
Astronomy Dpt
Boston Univ
725 Commonwealth Av
Boston MA 02215
USA
☎ 1 617 353 2627
℻ 1 617 353 3200

Janiczek Paul M
USNO
3450 Massachusetts Av NW
Washington DC 20392 5100
USA
☎ 1 202 653 1569
✉ pmj@ceres.usno.navy.mil

Janka Hans Thomas
Astronomy/Astrophys Ctr
Univ of Chicago
5640 S Ellis Av
Chicago IL 60637
USA
☎ 1 312 702 7853
℻ 1 312 702 6645
✉ thomas@granta.
uchicago.edu

Jankovics Istvan
Konkoly Observ
Thege U 13/17
Box 67
H 1525 Budapest
Hungary
☎ 36 1 375 4122
℻ 36 1 275 4668

Jannuzi Buell Tomasson
NOAO/KPNO
Box 26732
950 N Cherry Av
Tucson AZ 85726 6732
USA
☎ 1 520 318 8353
℻ 1 520 318 8360
✉ bjannuzi@noao.edu

Janot-Pacheco Eduardo
IAG
Univ Sao Paulo
CP 9638
01065 970 Sao Paulo SP
Brazil
☎ 55 11 577 8599
℻ 55 11 276 3848

Janssen Michael Allen
CALTECH/JPL
MS 183 301
4800 Oak Grove Dr
Pasadena CA 91109 8099
USA
☎ 1 213 354 7247
℻ 1 818 393 6030

Jardine Moira Mary
Dpt Phys/Astronomy
Univ of St Andrews
St Andrews North Haugh
Fife KY16 9SS
UK
☎ 44 13 34 7661
℻ 44 13 347 4487
✉ mmj@uk.ac.sussex.starlink

Jaroszynski Michal
Astronomical Observ
Warsaw Univ
Al Ujazdowskie 4
PL 00 478 Warsaw
Poland
☎ 48 22 29 4011
℻ 48 22 29 4697

Jarrett Alan H
Boyden Observ
Box 334
9300 Bloemfontein
South Africa
☎ 27 51 37605

Jarzebowski Tadeusz
Ul Jelenia 28/31
PL 54 242 Wroclaw
Poland

Jaschek Carlos O R
39 Teso d l Feria 481
E 37008 Salamanca
Spain
℻ 34 23 21 6884
✉ jaschek@arrakis.es

Jasniewicz Gerard
GRAAL
Univ Montpellier II
Pl E Bataillon
F 34095 Montpellier Cdx 5
France
☎ 33 4 67 14 3567
℻ 33 4 67 14 4535

Jassur Davoud MZ
Dpt Th Phys/Astrophys
Univ of Tabriz
Tabriz 51664
Iran
☎ 98 41 307841
℻ 98 41 341 244
✉ tabriz_u@rose.ipm.
ac.ir

Jastrow Robert
Inst Space Studies
2880 Broadway
New York NY 10025
USA
☎ 1 212 678 5611

Jatenco-Pereira Vera
IAG
Univ Sao Paulo
CP 9638
01065 970 Sao Paulo SP
Brazil
☎ 55 11 577 8599
℻ 55 11 276 3848
✉ jatenco@vax.iagusp.usp.br

Jauncey David L
CSIRO
Div Radiophysics
Box 76
Epping NSW 2121
Australia
☎ 61 2 868 0222
℻ 61 2 868 0310
✉ djauncey@atnf.csiro.au

Jayanthi Udaya B
INPE
Dpt Astronomia
CP 515
12227 010 S Jose dos Campos
Brazil
☎ 55 123 41 8977*392
℻ 55 123 21 8743

Jayarajan A P
IIA
Koramangala
Sarjapur Rd
Bangalore 560 034
India
☎ 91 80 356 6585
℻ 91 80 553 4043

Jedamzik Karsten
MPA
Karl Schwarzschildstr 1
D 85748 Garching
Germany
☎ 49 893 299 3207
℻ 49 893 299 3235
✉ jedamzik@mpa-
garching.mpg.de

Jefferies Stuart
NOAO/NSO
Box 26732
950 N Cherry Av
Tucson AZ 85726 6732
USA
☎ 1 520 323 4182
℻ 1 520 325 9278
✉ stuartj@noao.edu

Jeffers Stanley
Cress Physics Dpt
York Univ
4700 Keele St
Downsview ON M3J 1P3
Canada
☎ 1 416 667 3851

Jeffery Christopher S
Armagh Observ
College Hill
Armagh BT61 9DG
UK
☎ 44 18 61 522 928
℻ 44 18 61 527174
✉ csj@star.arm.ac.uk

Jefferys William H
Astronomy Dpt
Univ of Texas
Rlm 15 308
Austin TX 78712 1083
USA
☎ 1 512 471 4455
℻ 1 512 471 6016
✉ bill@astro.as.utexas.edu

Jenkins Charles R
Royal Greenwich Obs
Madingley Rd
Cambridge CB3 0EZ
UK
☎ 44 12 23 374 000
℻ 44 12 23 374 700
✉ crj@ast.cam.ac.uk

Jenkins Edward B
Princeton Univ Obs
Peyton Hall
Princeton NJ 08544 1001
USA
☎ 1 609 452 3826
℻ 1 609 258 1020
✉ ebj@astro.princeton.edu

Jenkins Louise F
Astronomy Dpt
Yale Univ
Box 208101
New Haven CT 06520 8101
USA
☎ 1 203 432 3000
℻ 1 203 432 5048

Jenkner Helmut
STScI
Homewood Campus
3700 San Martin Dr
Baltimore MD 21218
USA
☎ 1 401 338 4842
℻ 1 401 338 5090
✉ jenkner@stsci.edu

Jenner David C
Astrosft Corporation
3153 Ne 84th St
Seattle WA 98115 4717
USA
☎ 1 425 527 2018
℻ 1 425 527 2019
✉ davej@astrosoft.com

Jennings R E
Dpt Phys/Astronomy
UCLO
Gower St
London WC1E 6BT
UK
☎ 44 171 387 7050
℻ 44 171 380 7145

Jenniskens Petrus Matheus Marie
NASA/ARC
SETI
MS 239 4
Moffett Field CA 94035 1000
USA
☎ 1 650 604 3086
℻ 1 650 604 1088
✉ peter@max.arc.nasa.gov

Jennison Roger C
Electronics Lab
Univ of Kent
Canterbury CT2 7NT
UK
☎ 44 12 27 668 22
℻ 44 12 27 456 084

Jensch A
Pestalozzistr 9
D 07700 Jena
Germany

Jensen Eberhart
Inst Theor Astrophys
Univ of Oslo
Box 1029
N 0315 Blindern Oslo 3
Norway
☎ 47 22 856 501
℻ 47 22 856 505

Jeong Jang Hae
Dpt Astronomy/Space Sci
Chungbuk Ntl Univ
San 48 Gaeshin Dong
Cheongju 360 763
Korea RP
☎ 82 431 61 2313
℻ 82 431 67 4232
✉ jeongjh@astro.
chungbuk.ac.kr

Jerjen Helmut
MSSSO
Weston Creek
Private Bag
Canberra ACT 2611
Australia
☎ 61 6 279 8038
℻ 61 6 249 0233
✉ jerjen@mso.anu.edu.au

Jerzykiewicz Mikolaj
Astronomical Institute
Wroclaw Univ
Ul Kopernika 11
PL 51 622 Wroclaw
Poland
☎ 48 71 372 9373/74
℻ 48 71 372 9378
✉ mjerz@astro.uni.
wroc.pl

Jewell Philip R
NRAO
Campus Bg 65
949 N Cherry Av
Tucson AZ 85721 0655
USA
☎ 1 520 882 8250
℻ 1 520 882 7955

Ji Hongqing
Intl Latitude Observ
Tianjin
China PR

Ji Shuchen
Box 110 Kunming
Yunnan Province
Yunnan
China PR
☎ 86 72 946

Jiang Chongguo
Yunnan Observ
CAS
Kunming 650011
China PR
☎ 86 871 2035
℻ 86 871 717 1845

Jiang Dongrong
Shanghai Observ
CAS
80 Nandan Rd
Shanghai 200030
China PR
☎ 86 21 6438 6191
℻ 86 21 6438 4618

Jiang Shengtao
4 491 West St
Brantford ONT N3T SR7
Canada
✉ sjiang@bfree.on.ca

Jiang Shi-Yang
Beijing Astronomical Obs
CAS
Beijing 100080
China PR
☎ 86 1 28 1698
℻ 86 1 256 10855

Jiang Shuding
Graduate School
Univ Sci/Techn
Box 3908
Beijing 100039
China PR
☎ 86 1 81 7031*253

Jiang Xiaoyuan
Shanghai Observ
CAS
80 Nandan Rd
Shanghai 200030
China PR
☎ 86 21 6438 6191
℻ 86 21 6438 4618
✉ jiangxiaoyuan@
ihw.com.cn

Jiang Yaotiao
Astronomy Dpt
Nanjing Univ
Nanjing 210093
China PR
☎ 86 25 34151
℻ 86 25 330 2728

Jiang Zhaoji
Beijing Astronomical Obs
CAS
Beijing 100080
China PR
☎ 86 1 28 1698
℻ 86 1 256 10855

Jimenez Mancebo A J
IAC
Observ d Teide
via Lactea s/n
E 38200 La Laguna
Spain
☎ 34 2 232 9100
℻ 34 2 232 9117
✉ ajm@ll.iac.es

Jin Biaoren
Inst Geodesy/
Geophysics
Xu Dong Lu
Wuhan 430077
China PR
☎ 86 87 5571

Jin Shenzeng
Beijing Astronomical Obs
CAS
Beijing 100080
China PR
☎ 86 1 28 1698
℻ 86 1 256 10855

Jin Wenjing
Shanghai Observ
CAS
80 Nandan Rd
Shanghai 200030
China PR
☎ 86 21 6438 6191
℻ 86 21 6438 4618
✉ jwj@center.shao.ac.cn

Jockers Klaus
MPI Aeronomie
Max Planck Str 2
Postfach 20
D 37189 Katlenburg Lindau
Germany
☎ 49 555 697 9293
℻ 49 555 697 9240
✉ jockers@linax1.dnet.
gwdg.de

Joench-Soerensen Helge
Astronomical Observ, NBIfAFG
Copenhagen Univ
Juliane Maries Vej 30
DK 2100 Copenhagen
Denmark
☎ 45 35 325 933
℻ 45 35 325 989
✉ helge@astro.ku.dk

Joersaeter Steven
Stockholm Observ
Royal Swedish Acad Sciences
S 133 36 Saltsjoebaden
Sweden
☎ 46 8 716 4463
℻ 46 8 717 4719
✉ steven@astro.su.se

Joeveer Mihkel
Tartu Observ
Estonian Acad Sci
Toravere
EE 2444 Tartumaa
Estonia
☎ 372 7 410 469
℻ 372 7 410 205
✉ mihkel@aai.ee

Jog Chanda J
Physics Dpt
IISc
Bangalore 560 012
India
☎ 91 80 334 4411

Johansen Karen T
Astronomical Observ, NBIfAFG
Copenhagen Univ
Juliane Maries Vej 30
DK 2100 Copenhagen
Denmark
☎ 45 35 32 5928
℻ 45 35 32 5989
✉ karen@astro.ku.dk

Johansson Lars Erik B
Onsala Space Observ
Chalmers Technical Univ
S 439 92 Onsala
Sweden
☎ 46 31 772 5564
℻ 46 31 772 5550
✉ leb@oso.chalmers.se

Johansson Lennart
Astronomical Observ
Box 515
S 755 91 Uppsala
Sweden
☎ 46 18 51 1274
℻ 46 18 52 7583

Johansson Sveneric
Physics Dpt
Univ of Lund
Soelvegatan 14
S 223 62 Lund
Sweden
☎ 46 46 222 6097
℻ 46 46 222 4709
✉ atomsej@seldc52

Johnson Donald R
NBS
Bg 221 Rm A363
Gaithersburg MD 20899
USA
☎ 1 301 921 2828

Johnson Fred M
Physics/Astronomy Dpt
California State Univ
Fullerton CA 92634
USA
☎ 1 714 773 3366
℻ 1 714 449 5810

Johnson Hollis R
Astronomy Dpt
Indiana Univ
Swain W 319
Bloomington IN 47405
USA
☎ 1 812 335 4172
℻ 1 812 855 8725

Johnson Hugh M
1017 Newell Rd
Palo Alto CA 94303
USA
☎ 1 415 326 7223

Johnson Torrence V
CALTECH/JPL
MS 183 601
4800 Oak Grove Dr
Pasadena CA 91109 8099
USA
☎ 1 818 354 2761
℻ 1 818 393 6030

Johnston Kenneth J
USNO
Scientific director
3450 Massachusetts Av NW
Washington DC 20392 5420
USA
☎ 1 202 653 1513
℻ 1 202 653 1497
✉ kjj@astro.usno.navy.mil

Johnstone Roderick
Inst of Astronomy
The Observatories
Madingley Rd
Cambridge CB3 0HA
UK
☎ 44 12 23 337 510
℻ 44 12 23 337 523
✉ rmj@mail.ast.cam.ac.uk

Jokipii J R
Lunar/Planetary Lab
Room 325 Bg 92
Univ of Arizona
Tucson AZ 85721 0092
USA
☎ 1 520 621 4256
✆ 1 520 621 4933

Joly Francois
Univ Bordeaux
Lab Astrophysique
123 r Lamartine
F 33405 Talence
France
☎ 33 5 56 86 4330
✆ 33 5 56 40 4251

Joly Monique
Observ Paris Meudon
DAES
Pl J Janssen
F 92195 Meudon PPL Cdx
France
☎ 33 1 45 07 7441
✆ 33 1 45 07 7971

Jonas Justin Leonard
Dpt Phys/Electronics
Rhodes Univ
Box 94
6140 Grahamstown
South Africa
☎ 27 461 318 454/318 450
✆ 27 461 250 49

Joncas Gilles
Dpt d Physique
Universite Laval
Ste Foy QC G1K 7P4
Canada
☎ 1 418 656 2652
✆ 1 418 656 2040
✉ joncas@phy.ulaval.ca

Jones Albert F
31 Ranui Rd
Stoke
Nelson
New Zealand
☎ 64 54 73 905

Jones Barbara
CASS
UCSD
C 011
La Jolla CA 92093 0216
USA
☎ 1 714 452 4474
✆ 1 619 534 2294

Jones Barrie W
Physics Dpt
The Open University
Walton Hall
Milton Keynes MK7 6AA
UK
☎ 44 190 865 3378
✆ 44 190 865 3744
✉ bw_jones@uk.ac.open.acs.vax

Jones Bernard J T
Warren House
1 East Pilgrim's Way
Otford Kent TN14 5RJ
UK
☎ 44 195 952 5321
✆ 44 195 952 2407

Jones Burton
Lick Observ
Univ of California
Santa Cruz CA 95064
USA
☎ 1 831 459 2384
✆ 1 831 426 3115
✉ jones@ucolick.org

Jones Dayton L
CALTECH/JPL
MS 238 332
4800 Oak Grove Dr
Pasadena CA 91109 8099
USA
☎ 1 818 354 7774
✆ 1 818 393 6030
✉ dj@bllac.jpl.nasa.gov

Jones Derek H P
Royal Greenwich Obs
Madingley Rd
Cambridge CB3 0EZ
UK
☎ 44 12 23 374 000
✆ 44 12 23 374 700
✉ dhpj@ast.cam.ac.uk

Jones Eric M
LANL
MS F665
Box 1663
Los Alamos NM 87545 2345
USA
☎ 1 505 667 6386
✆ 1 505 665 4055

Jones Frank Culver
NASA GSFC
Code 665
Greenbelt MD 20771
USA
☎ 1 301 286 5506
✆ 1 301 286 3391
✉ jones@lheavx.gsfc.nasa.gov

Jones Harrison Price
NOAO/KPNO
Box 26732
950 N Cherry Av
Tucson AZ 85726 6732
USA
☎ 1 520 325 9354
✆ 1 520 325 9360

Jones James
Dpt Phys/Astronomy
Univ W Ontario
London ON N6A 3K7
Canada
☎ 1 519 661 3283*6452
✆ 1 519 661 2033
✉ jimeteor@uwovas.uwo.ca

Jones Janet E
Warren House
1 East Pilgrim's Way
Otford Kent TN14 5RJ
UK
☎ 45 195 952 2407
✆ 45 195 952 5321
✉ janet@astrag.demon.co.uk

Jones Michael
MRAO
Cavendish Laboratory
Madingley Rd
Cambridge CB3 0HE
UK
☎ 44 12 23 337 363
✆ 44 12 23 354 599
✉ mike@mrao.cam.ac.uk

Jones Paul
School of Physics
UNSW
Box 10
Kingswood NSW 2747
Australia
☎ 61 47 360 437
✆ 61 47 360 779
✉ p.jones@st-nepcan.uws.edu.au

Jones Thomas Walter
School Phys/Astronomy
Univ Minnesota
116 Church St SE
Minneapolis MN 55455
USA
☎ 1 612 624 8546
✆ 1 612 626 2029
✉ twj@ast1.spa.umn.edu

Jopek Tadeusz Jan
Astronomical Observ
A Mickiewicz Univ
Ul Sloneczna 36
PL 60 286 Poznan
Poland
☎ 48 61 67 9670
✆ 48 61 53 6536
✉ jopek@phys.amu.edu.pl

Jordan Carole
Dpt Th Physics
Univ of Oxford
1 Keble Rd
Oxford OX1 3NP
UK
☎ 44 186 527 3999
✆ 44 186 527 3947
✉ c.jordan1@physics.oxford.ac.uk

Jordan H L direktor
Inst Plasmaphysik
Kernforschungsanlage
Juelich Gmbh Pf 365
D 5170 Juelich 1
Germany

Jordan Stefan
Inst Theor Phys/
Sternwarte Univ Kiel
Olshausenstr 40
D 24098 Kiel
Germany
☎ 49 431 880 4105
✆ 49 431 880 4100
✉ pas58@rz-uni-kiel.d400.de

Jordan Stuart D
NASA GSFC
Code 682
Greenbelt MD 20771
USA
☎ 1 301 286 7672
✆ 1 301 286 1617

Jorden Paul Richard
Royal Greenwich Obs
Madingley Rd
Cambridge CB3 0EZ
UK
☎ 44 12 23 374 000
✆ 44 12 23 374 700
✉ prj@mail.ast.cam.ac.uk

Jordi Carme
Dpt Astronom Meteo
Univ Barcelona
Avd Diagonal 647
E 08028 Barcelona
Spain
☎ 34 3 402 1126
✆ 34 3 402 1133
✉ carme@mizar.am.ub.es

Jorgensen Henning E
Astronomical Observ, NBIfAFG
Copenhagen Univ
Blegdamsvej 17
DK 2100 Copenhagen
Denmark
☎ 45 35 32 3995
✆ 45 35 32 5989
✉ henning@astro.ku.dk

Jorgensen Inger
McDonald Observ
Univ of Texas
Rlm 15 308
Austin TX 78712 1083
USA
☎ 1 512 471 3464
✆ 1 512 471 6016
✉ inger@astro.as.utexas.edu

Jorgensen Uffe Graae
Astronomical Observ, NBIfAFG
Copenhagen Univ
Juliane Maries Vej 30
DK 2100 Copenhagen
Denmark
☎ 45 35 12 5998
✆ 45 35 32 5989
✉ uffegi@nbivax.nbi.dk

Jorissen Alain
Inst Astrophysique
Univ Bruxelles
Campus Plaine CP 226
B 1050 Brussels
Belgium
☎ 32 2 650 2834
✆ 32 2 650 4226
✉ ajorisse@astro.ulb.ac.be

Joselyn Jo Ann c
NOAA ERL R/E/SE2
Space Environment Lab
325 Broadway
Boulder CO 80303
USA
☎ 1 303 497 5147

Joseph Charles Lynn
Dpt Astrophysical Sci
Princeton Univ
Princeton NJ 08544 1001
USA
☎ 1 609 258 3808
℻ 1 609 258 1020
✉ clj@astro.princeton.edu

Joseph J H
Dpt Geophys/Planet Sci
Tel Aviv Univ
Ramat Aviv
Tel Aviv 69978
Israel
☎ 972 3 642 0633
℻ 972 3 640 9282

Joseph Robert D
Astrophysics Gr
Imperial Coll
Prince Consort Rd
London SW7 2BZ
UK
☎ 44 171 594 7771
℻ 44 171 594 7772

Joshi Mohan N
TIFR/Radio Astronomy Ctr
Box 8
Udhagamandalam 643 001
India
☎ 91 423 2032

Joshi Suresh Chandra
Uttar Pradesh State
Observ
Po Manora Peak 263 129
Nainital 263 129
India
☎ 91 59 422136

Joshi U C
PRL
Navrangpura
Ahmedabad 380 009
India
☎ 91 272 46 2129
℻ 91 272 44 5292
✉ joshi@prl.ernet.in

Joss Paul Christopher
Physics Dpt
MIT Rm 6 203
Box 165
Cambridge MA 02139 4307
USA
☎ 1 617 253 4845
℻ 1 617 253 9798
✉ joss@mitins.mit.edu

Joubert Martine
LAS
Traverse du Siphon
Les Trois Lucs
F 13376 Marseille Cdx 12
France
☎ 33 4 91 05 5900
℻ 33 4 91 66 1855
✉ joubert@frlasm51

Jourdain de Muizon M
ESA
Apt 50727 Villafranca
E 28080 Madrid
Spain
☎ 34 1 813 1100
℻ 34 1 813 1139

Journet Alain
OCA
CERGA
F 06130 Grasse
France
☎ 33 4 93 36 5849
℻ 33 4 93 40 5353

Jovanovic Bozidar
Fac of Agriculture
Inst Waterranging
Veljka Vlahovica 2
21000 Novi Sad
Yugoslavia FR
☎ 381 215366

Joy Marshall J
NASA/MSFC
Space Science Lab
Code ES 65
Huntsville AL 35812
USA
☎ 1 205 544 3423
℻ 1 205 544 7754
✉ joy@ssl.msfc.nasa.gov

Judge Philip
HAO
NCAR
Box 3000
Boulder CO 80307 3000
USA
☎ 1 303 497 1000
℻ 1 303 497 1568

Jugaku Jun
Tokyo Astronomical Obs
NAOJ
Osawa Mitaka
Tokyo 181
Japan
☎ 81 422 32 5111
℻ 81 422 34 3793

Juliusson Einar
Menntaskolinn
IS 840 Laugarvatni
Iceland
☎ 354 86 1133

Jung Jean
22 r Briant
F 92260 Fontenay aux Roses
France

Junkes Norbert
RAIUB
Univ Bonn
auf d Huegel 69
D 53121 Bonn
Germany
☎ 49 228 525 399
℻ 49 228 525 229
✉ njunkes@mpifr-bonn.mpg.de

Junkkarinen Vesa T
CASS
UCSD
C 011
La Jolla CA 92093 0216
USA
☎ 1 619 534 0735
℻ 1 619 534 2294

Jupp Alan H
Dpt Appl Maths/Theor Phys
Univ of Liverpool
Box 147
Liverpool L69 3BX
UK
☎ 44 151 709 6022
℻ 44 151 708 6502

Jura Michael
Dpt Astronomy/Phys
UCLA
Los Angeles CA 90095 1562
USA
☎ 1 213 825 4302
℻ 1 213 206 2096
✉ jura@clotho.astro.ucla.edu

Jurgens Raymond F
CALTECH/JPL
MS 238 420
4800 Oak Grove Dr
Pasadena CA 91109 8099
USA
☎ 1 818 354 4974
℻ 1 818 393 6030

Just Andreas
ARI
Moenchhofstr 12-14
D 69120 Heidelberg
Germany
☎ 49 622 140 5129
℻ 49 622 140 5297
✉ s38@mvs.urz.uni-heidelberg.de

Juszkiewicz Roman
Copernicus Astron Ctr
Polish Acad Sci
Ul Bartycka 18
PL 00 716 Warsaw
Poland
☎ 48 22 41 1086
℻ 48 22 410 046

K

Kaastra Jelle S
SRON
Postbus 800
Sorbonnelaan 2
NL 3584 CA Utrecht
Netherlands
☎ 31 30 253 8570
✆ 31 30 254 0860
✉ J.kaastra@sron.ruu.nl

Kaburaki Osamu
Astron Institute
Tohoku Univ
Aramaki Aoba
Sendai 980
Japan
☎ 81 206 543 2888
✆ 81 222 61 2860
✉ okabu@astroa.astr.tohodu.ac.jp

Kaburaki Osamu
Astronomical Institute
Tohoku Univ
Aramaki Aoba
Sendai 980
Japan
☎ 81 22 222 1800
✆ 81 222 61 2806
✉ okabu@astroa.astr.tohoku.ac.jp

Kadla Zdenka I
Pulkovo Observ
Acad Sciences
10 Kutuzov Quay
196140 St Petersburg
Russia
☎ 7 812 298 2242
✆ 7 812 315 1701
✉ kadla@pulkovo.spb.su

Kadouri Talib Hadi
SARC
Scientific Res Council
Box 2441
Jadiriyah Baghdad
Iraq
☎ 964 1 776 5127

Kaehler Helmuth
Hamburger Sternwarte
Univ Hamburg
Gojensbergsweg 112
D 21029 Hamburg
Germany
☎ 49 407 252 4112
✆ 49 407 252 4198

Kafatos Minas
Physics Dpt
George Mason Univ
4400 University Dr
Fairfax VA 22030
USA
☎ 1 703 993 1997
✆ 1 703 993 1980
✉ mkafatos@compton.gmu.edu

Kafka Peter
MPA
Karl Schwarzschildstr 1
D 85748 Garching
Germany
☎ 49 893 299 00
✆ 49 893 299 3235

Kaftan May A
1432 Corcoran St NW
Apt 2
Washington DC 20009
USA

Kahabka Peter
Astronomical Institute
Univ of Amsterdam
Kruislaan 403
NL 1098 SJ Amsterdam
Netherlands
☎ 31 20 525 7476
✆ 31 20 525 7484
✉ ptk@astro.uva.nl

Kahane Claudine
Observ Grenoble
Lab Astrophysique
BP 53x
F 38041 S Martin Heres Cdx
France
☎ 33 4 76 51 4600
✆ 33 4 76 44 8821
✉ kahane@frgag51

Kahler Stephen W
USAF Phillips Lab/GPSG
Boston College
Hanscom AFB MA 01731 3010
USA
☎ 1 617 377 9665
✆ 1 617 377 3160
✉ kahler@plh.af.mil

Kahlmann Hans Cornelis
Radio Observ Westerbork
Schattenberg 1
NL 9433 TA Zwiggelte
Netherlands
☎ 31 59 39 2421
✆ 31 59 39 2486
✉ kahlmann_hans@nfra.nl

Kahn Franz D
Astronomy Dpt
Univ Manchester
Oxford Rd
Manchester M13 9PL
UK
☎ 44 161 275 4224
✆ 44 161 275 4223

Kaifu Norio
Subaru Telescope
NAOJ
650 N A'Ohoku Place
Hilo HI 96720
USA
☎ 1 808 934 5910
✆ 1 808 934 5984
✉ kaifu@subaru.naoj.org

Kaitchuck Ronald H
Dpt Phys/Astronomy
Ball State Univ
Muncie IN 47306
USA
☎ 1 317 285 8860
✆ 1 317 285 1624

Kajino Toshitaka
Ntl Astronomical Obs
Mitaka
Tokyo 188
Japan
☎ 81 422 34 3740
✆ 81 422 34 3746
✉ kajino@nao.ac.jp

Kakinuma Takakiyo T
Inst Atmospheric Res
Nagoya Univ
3-13 Honohara
Toyokawa Aichi 442
Japan
☎ 81 533 86 3154
✆ 81 533 86 0811

Kakuta Chuichi
105 88 Onigoe
Shirakawa
Fukushima 961
Japan
☎ 81 248 24 3321
✆ 81 248 24 3321

Kalafi Manoucher
Ctr for Astron Res
Univ of Tabriz
Tabriz 51664
Iran
☎ 98 41 32564

Kalandadze N B
Abastumani Astrophysical Observ
Georgian Acad Sci
383762 Abastumani
Georgia
☎ 995 88 32 95 5367
✆ 995 88 32 98 5017

Kalberla Peter
RAIUB
Univ Bonn
auf d Huegel 69
D 53121 Bonn
Germany
☎ 49 228 73 3645
✆ 49 228 73 3672
✉ pkalberl@astro.uni-bonn.de

Kalenichenko Valentin
Astronomical Observ
Taras Shevchenko Univ
3 Observatorna Str
Kyiv 254 053
Ukraine
☎ 380 4421 62762
✆ 380 4421 62630
✉ kalenych@aoku.freenet.kiev.ua

Kaler James B
Dpt Astron/Physics
Univ of Illinois
1002 W Green St
Urbana IL 61801
USA
☎ 1 217 333 9382
✆ 1 217 244 7638

Kalinkov Marin P
Astronomy Dpt
Bulgarian Acad Sci
72 Lenin Blvd
BG 1784 Sofia
Bulgaria
☎ 359 2 75 8927
✆ 359 2 75 5019

Kalkofen Wolfgang
CfA
HCO/SAO
60 Garden St
Cambridge MA 02138 1516
USA
☎ 1 617 495 7285
✆ 1 617 495 7049
✉ wolf@cfa.harvard.edu

Kalloglian Arsen T
Byurakan Astrophys Observ
Armenian Acad Sci
378433 Byurakan
Armenia
☎ 374 88 52 28 3453/4142
✆ 374 88 52 52 3640

Kalman Bela
Debrecen Heliophys Observ
Acad Sciences
Box 30
H 4010 Debrecen
Hungary
☎ 36 5 231 1015

Kalmykov A M
Astronomical Institute
Uzbek Acad Sci
Astronomicheskaya Ul 33
700000 Tashkent
Uzbekistan
☎ 7 3712 35 8102

Kalnajs Agris J
MSSSO
Weston Creek
Private Bag
Canberra ACT 2611
Australia
☎ 61 262 881 111*248
℻ 61 262 490 233
✉ agris@mso.anu.oz

Kaltcheva Nadia
Astronomy Dpt
Univ of Sofia
Anton Ivanov St 5
BG 1126 Sofia
Bulgaria
☎ 359 2 54 4852
℻ 359 2 68 9085

Kaluzny Janusz
Astronomical Observ
Warsaw Univ
Al Ujazdowskie 4
PL 00 478 Warsaw
Poland
☎ 48 22 29 4011
℻ 48 22 29 4697

Kambe Eiji
Dpt Geoscience
Ntl Defence Academy
Yokosuka
Kanagawa 239
Japan
☎ 81 468 41 3810
℻ 81 468 44 5902
✉ kambe@apsgw.aps.
seikei.ac.jp

Kamel Osman M
Astronomy Dpt
Fac of Sciences
Cairo Univ
Geza
Egypt
☎ 20 2 572 7022
℻ 20 2 572 7556

Kameya Osamu
NAOJ
MAO
2-12 Hoshigaoka
Mizusawa Iwate 023
Japan
☎ 81 197 22 7153
℻ 81 197 22 7120
✉ kameya@miz.nao.ac.jp

Kamijo Fumio
Astronomy Dpt
Univ of Tokyo
Bunkyo Ku
Tokyo 113
Japan
☎ 81 381 22 111
℻ 81 338 13 9439

Kaminishi Keisuke
Physics Dpt
Kumamoto Univ
2-39-1 Kurokami
Kumamoto 860
Japan
☎ 81 963 44 2111

Kammeyer Peter C
USNO
3450 Massachusetts Av NW
Washington DC 20392 5100
USA
☎ 1 202 762 1428
℻ 1 202 762 1516

Kamp Lucas Willem
Astronomy Dpt
Boston Univ
725 Commonwealth Av
Boston MA 02215
USA
☎ 1 617 353 2625
℻ 1 617 353 3200

Kamper Karl W
David Dunlap Observ
Univ of Toronto
Box 360
Richmond Hill ON L4C 4Y6
Canada
☎ 1 416 884 9562
℻ 1 416 978 3921

Kanamitsu Osamu
Dpt Earth Sci/Astron
Fukuoka Univ of Edu
729-1 Akama Munakata
Fukuoka 811 41
Japan
☎ 81 940 358 1365
℻ 81 940 33 7730
✉ kanamitu@fukuoka-
edu.ac.jp

Kanayev Ivan I
Pulkovo Observ
Acad Sciences
10 Kutuzov Quay
196140 St Petersburg
Russia
☎ 7 812 123 4401
℻ 7 812 123 1922
✉ ivan@kanaev.spb.su

Kanbach Gottfried
MPE
Postfach 1603
D 85740 Garching
Germany
☎ 49 893 299 3544
℻ 49 893 299 3606
✉ gok@mpe.mpg.de

Kanbur Shashi
Dpt Phys/Astronomy
Univ of Glasgow
Glasgow G12 8QQ
UK
☎ 44 141 339 8855*4268
℻ 44 141 334 9029
✉ shash@astro.gla.ac.uk

Kandalian Rafik A
Byurakan Astrophys Observ
Armenian Acad Sci
378433 Byurakan
Armenia
☎ 374 88 52 28 3453/4142
℻ 374 88 52 52 3640
✉ rkandali@helios.sci.am

Kandel Robert S
LMD
Ecole Polytechnique
F 91128 Palaiseau Cdx
France
☎ 33 1 69 41 8200

Kandemir Guelcin
Istanbul Technical Univ
Fen Fakultesi Fizik B
Maslak
34452 Istanbul
Turkey
☎ 90 212 609 109
℻ 90 212 519 0834

Kandpal Chandra D
Uttar Pradesh State
Observ
Po Manora Peak 263 129
Nainital 263 129
India
☎ 91 59 42 2136/2325

Kandrup Henry Emil
Astronomy Dpt
Univ of Florida
211 SSRB
Gainesville FL 32611
USA
☎ 1 904 392 2681
℻ 1 904 392 9741
✉ kandrup@astro.ufl.edu

Kane Sharad R
Space Sci Laboratory
Univ of California
Grizzly Peak Blvd
Berkeley CA 94720 7950
USA
☎ 1 415 642 1719

Kaneko Noboru
Physics Dpt
Hokkaido Univ
Kita 10 Nishi 8
Sapporo 060
Japan
☎ 81 117 16 2111
℻ 81 222 85 132

Kang Gon Ik
Pyongyang Astron Obs
Acad Sciences DPRK
Taesong district
Pyongyang
Korea DPR
☎ 850 5 3134/5 & 5 3239

Kang Jin Sok
Pyongyang Astron Obs
Acad Sciences DPRK
Taesong district
Pyongyang
Korea DPR
☎ 850 5 3134/5 & 5 3239

Kang Yong Hee
Dpt Earth Science/Ed
Kyungpook Ntl Univ
Taegu
Korea RP
☎ 82 539 50 5916
℻ 82 539 50 5946
✉ yhkang@bh.kyungpook.ac.kr

Kang Young Woon
Dpt Earth Science
King Sejong Univ
89 Koonja-dong
Seoul Sungdong 133 747
Korea RP
☎ 82 467 51 21

Kanyo Sandor
Konkoly Observ
Thege U 13/17
Box 67
H 1525 Budapest
Hungary
☎ 36 1 375 4122
℻ 36 1 275 4668

Kapahi Vijay K
NCRA/TIFR
Pune Univ Campus Pb 3
Ganeshkhind
Pune 411 007
India
☎ 91 212 35 6105
℻ 91 212 35 5149
✉ vijay@gmrt.ernet.in

Kapisinsky Igor
Astronomical Institute
Slovak Acad Sci
Dubravska 9
SK 842 28 Bratislava
Slovak Republic
☎ 421 7 37 5157
℻ 421 7 37 5157

Kaplan George H
USNO
3450 Massachusetts Av NW
Washington DC 20392 5420
USA
☎ 1 202 762 1562
℻ 1 202 762 1612
✉ gkapla@usno.navy.mil

Kaplan J
Dpt Astronomy/Phys
UCLA
Box 951562
Los Angeles CA 90025 1562
USA
☎ 1 310 825 4434
℻ 1 310 206 2096

Kaplan Lewis D
Atmosph/Environmental
Research Inc
840 Memorial Dr
Cambridge MA 02139
USA
☎ 1 617 547 6207

Kapoor Ramesh Chander
IIA
Koramangala
Sarjapur Rd
Bangalore 560 034
India
☎ 91 80 356 6585
℻ 91 80 553 4043

Karaali Salih
Univ Observ
Univ of Istanbul
University 34452
34452 Istanbul
Turkey
☎ 90 212 522 4200*610
℻ 90 212 519 0834

Karachentsev Igor D
SAO
Acad Sciences
Nizhnij Arkhyz
357147 Karachaevo
Russia
☎ 7 878 789 2501
✉ ikar@luna.sao.ru

Karachentseva Valentina
Astronomical Observ
Kyiv State Univ
Observatorna Str 3
254053 Kyiv
Ukraine
☎ 380 4421 61994
✉ aoku@gluk.apc.org

Karas Vladimir
Astronomical Institute
Charles Univ
V Holesovickack 2
CZ 180 00 Praha 8
Czech R
☎ 420 2 2191 2572
℻ 420 2 688 5095
✉ karas@mbox.cesnet.cz

Kardashev Nicolay S
Astro Space Ctr
Lebedev Physical Inst
Profsoyuznaya 84/32
117810 Moscow
Russia
☎ 7 095 333 2378
℻ 7 095 333 2378/310 7023
✉ nkardash@dpc.asc.rssi.ru

Karetnikov Valentin G R
Astronomical Observ
Odessa State Univ
Shevchenko Park
270014 Odessa
Ukraine
☎ 380 4822 20356
℻ 380 4822 28442
✉ baby@paco.net

Karitskaya Eugenia A
SAI
Acad Sciences
Universitetskij Pr 13
119899 Moscow
Russia
☎ 7 095 939 1616
℻ 7 095 230 2081
✉ karitsk@sai.msu.su

Karlicky Marian
Astronomical Institute
Czech Acad Sci
Fricova 1
CZ 251 65 Ondrejov
Czech R
☎ 420 204 85 3356
℻ 420 2 88 1611
✉ karlicky@asu.cas.cz

Karoji Hiroshi
Tokyo Astronomical Obs
NAOJ
Osawa Mitaka
Tokyo 181
Japan
☎ 81 422 41 3643
℻ 81 422 41 3776
✉ karoji@sxt1.mtk.nao.ac.jp

Karovska Margarita
CfA
HCO/SAO
60 Garden St
Cambridge MA 02138 1516
USA
☎ 1 617 495 7347
℻ 1 617 495 7356
✉ karovska@cfa

Karp Alan Hersh
Hewlett Packard Co
Hp Labs 3u 7
1501 Page Mill Rd
Palo Alto CA 94304
USA
☎ 1 415 857 6766
℻ 1 415 857 5172
✉ karp@hpl.hp.com

Karpen Judith T
NRL
Code 4175 K
4555 Overlook Av SW
Washington DC 20375 5000
USA
☎ 1 202 767 3441

Karpinskij Vadim N
Pulkovo Observ
Acad Sciences
10 Kutuzov Quay
196140 St Petersburg
Russia
☎ 7 812 298 2242
℻ 7 812 315 1701

Karttunen Hannu
Turku Univ
Tuorla Observ
Vaeisaelaentie 20
FIN 21500 Piikkio
Finland
☎ 358 2 274 4244
℻ 358 2 243 3767
✉ hannu.karttunen@astro.utu.fi

Karygina Zoya V
Astrophys Inst
Kazakh Acad Sci
480068 Alma Ata
Kazakhstan
☎ 7 62 4040

Kashscheev B L
Inst Radio Astron
Ukrainian Acad Science
4 Chervonopraporna st
310059 Kharkiv
Ukraine

Kasper U
Institut Mathematik
Univ Potsdam
am Neuen Palais 10
D 14469 Potsdam
Germany
✉ ukasper@rz.uni-potsdam.de

Kasturirangan K
ISRO Satellite Ctr
Dpt Space
Vimanapura Post
Bangalore 560 017
India
☎ 91 80 354 779
℻ 91 82 333 2228

Kasuga Takashi
College of Eng
Hosei Univ
Kajinocho Koganei
Tokyo 184
Japan
☎ 81 423 87 6244
℻ 81 423 87 6123

Kasumov Fikret K O
Inst of Physics
Akademgorodoc
Narimanov Ul 33
370122 Baku
Azerbajan
☎ 994 39 6784

Katgert Peter
Leiden Observ
Box 9513
NL 2300 RA Leiden
Netherlands
☎ 31 71 527 5817
℻ 31 71 527 5833
✉ katgert@strw.leidenuniv.nl

Katgert-Merkelijn J K
Rijkuniversiteit te
Huygens Lab
Box 9504
NL 2300 RA Leiden
Netherlands
☎ 31 71 527 5916
℻ 31 71 527 5819
✉ merkelyn@strw.leidenuniv.nl

Kato Ken-ichi
Science Museum of Osaka
4-2-1 Nakanoshima Kita-ku
Osaka 530
Japan
☎ 81 644 45 184
℻ 81 644 45 657
✉ kato@sci-museum.kita.osaka.jp

Kato Mariko
Keio Univ
Hiyoshi 4 1 1
Kouloku-ku
Yokohama 223
Japan
☎ 81 455 63 5753
℻ 81 455 63 1650
✉ mariko@educ.cc.keio.ac.jp

Kato Shoji
Astronomy Dpt
Kyoto Univ
Sakyo Ku
Kyoto 606 01
Japan
☎ 81 757 51 2111
℻ 81 75 753 7010
✉ kato@kusastro.kyoto-u.ac.jp

Kato Takako
Inst Plasma Physics
Nagoya Univ
Furocho Chikusa Ku
Nagoya 464 01
Japan
☎ 81 527 81 5111
℻ 81 527 89 2845

Katz Jonathan I
Physics Dpt
Washington Univ
Campus Box 1105
St Louis MO 63130
USA
☎ 1 314 889 6202
℻ 1 314 935 4083

Katz Joseph
Racah Inst of Phys
Hebrew Univ Jerusalem
Jerusalem 91904
Israel
☎ 972 2 58 4604
℻ 972 2 61 1519

Kaufman Michele
Physics Dpt
Ohio State Univ
174 W 18th Av
Columbus OH 43210 1106
USA
☎ 1 614 422 5713

Kaufmann Jens Peter
Inst Astron/Astrophysik
Technische Uni
Hardenbergstr 36
D 10623 Berlin
Germany
☎ 49 303 145 462
℻ 49 303 142 3018

Kaufmann Pierre
CRAAE INPE
EPUSP/PTR
CP 61548
01065 970 Sao Paulo SP
Brazil
☎ 55 11 815 9322
℻ 55 11 815 4272

Kaul Chaman
Nuclear Res Laboratory
Bhabha Atomic Res Ctr
Mumbai 400 085
India
☎ 91 22 556 4225
℻ 91 22 556 0750
✉ nrl@magnum.barc.ernet.in

Kaula William M
Dpt Earth/Space Sci
UCLA
Box 951567
Los Angeles CA 90024 1567
USA
☎ 1 310 825 3880
℻ 1 310 825 2779

Kawabata Kinaki
Physics Dpt
Nagoya Univ
Furocho Chikusa Ku
Nagoya 464 01
Japan
☎ 81 527 89 2840
℻ 81 527 89 2845

Kawabata Kiyoshi
Physics Dpt/College of Sciences
Science Univ Tokyo
1-3 Kagurazaka Shinjuku
Tokyo 162
Japan
☎ 81 326 04 271
✉ kawabata@rs.kagu.sut.ac.jp

Kawabata Shusaku
Kyoto Gakuen Univ
NANJO
Sogabecho Kameoka
Kyoto 621
Japan
☎ 81 771 22 2001

Kawabe Ryohei
Nobeyama Radio Obs
NAOJ
Minamimaki Mura
Nagano 384 13
Japan
☎ 81 267 63 4385
℻ 81 267 63 4339
✉ kawabe@nro.nao.ac.jp

Kawaguchi Ichiro
Astronomy Dpt
Kyoto Univ
Sakyo Ku
Kyoto 606 01
Japan
☎ 81 757 51 2111
℻ 81 75 753 7010

Kawaguchi Kentarou
Nobeyama Radio Obs
NAOJ
Minamimaki Mura
Nagono 384 13
Japan
☎ 81 267 984 301
℻ 81 267 982 884
✉ nrokent@nro.nao.ac.jp

Kawai Nobuyuki
Cosmic Radiation Lab
RIKEN
2 1 Hirosawa
Wako Saitama 351 01
Japan
☎ 81 484 62 1111*3226
℻ 81 484 62 4640
✉ nkawai@postman.riken.go.jp

Kawaler Steven D
Physics Dpt
Iowa State Univ
Ames IA 50011
USA
☎ 1 515 294 9728
℻ 1 515 294 6027
✉ sdk@iastate.edu

Kawara Kimiaki
Tokyo Astronomical Obs
NAOJ
Osawa Mitaka
Tokyo 181
Japan
☎ 81 422 32 5111
℻ 81 422 34 3793
✉ kkawara@iso.vilspa.esa.es

Kawasaki Masahiro
Inst Cosmic Ray Res
Univ of Tokyo
Midori Cho
Tanashi 188
Japan
☎ 81 424 69 9595
℻ 81 424 62 3096
✉ kawasaki@ctsun1.icrr.u-tokyo.a

Kawata Yoshiyuki
Kanazawa Techn Inst
7 1 Ogigaoka
Nonoichimachi
Ishikawa 921
Japan
☎ 81 762 48 1100

Kayser Rainer
Hamburger Sternwarte
Univ Hamburg
Gojensbergsweg 112
D 21029 Hamburg
Germany
☎ 49 407 252 4126
℻ 49 407 252 4198
✉ st40010@dhhuni4

Kazantzis Panayotis
Mathematics Dpt
Univ of Patras
GR 261 10 Rion
Greece
☎ 30 61 99 7572
℻ 30 61 99 7636

Kazes Ilya
Observ Paris Meudon
ARPEGES
Pl J Janssen
F 92195 Meudon PPL Cdx
France
☎ 33 1 45 07 7606
℻ 33 1 45 07 7971

Keay Colin S l
Physics Dpt
Newcastle Univ
Newcastle NSW 2308
Australia
☎ 61 49 21 5451/5440
℻ 61 49 21 6907
✉ phcslk@cc.newcastle.edu.au

Keel William C
Dpt Phys/Astronomy
Univ of Alabama
Box 870324
Tuscaloosa AL 35487 0324
USA
☎ 1 205 348 5050
℻ 1 205 348 5051
✉ keel@bildad.astr.ua.edu

Keenan Philip C
Perkins Observ
Ohio State Univ
Box 449
Delaware OH 43015
USA
☎ 1 614 363 1257

Keene Jocelyn Betty
CALTECH
MS 320 47
Pasadena CA 91125
USA
☎ 1 818 395 6675
℻ 1 818 796 8806
✉ jbk@tacos.caltech.edu

Kegel Wilhelm H
Inst Theor Physics
Univ Frankfurt
Robert Mayer Str 8-10
D 60054 Frankfurt A M
Germany
☎ 49 69 798 2357
℻ 49 69 798 8350

Keil Klaus
Dpt Geology
Univ New Mexico
800 Yale Blvd Ne
Albuquerque NM 87131
USA
☎ 1 505 277 4204

Keil Stephen L
AFGL
NSO
Sacremento Peak Obs
Sunspot NM 88349
USA
☎ 1 505 434 1390
℻ 1 504 434 7029

Kelemen Janos
Konkoly Observ
Thege U 13/17
Box 67
H 1525 Budapest
Hungary
☎ 36 1 375 4122
℻ 36 1 275 4668
✉ kelemen@ogyalla.konkoly.hu

Keller Charles F
LANL
MS F665
Box 1663
Los Alamos NM 87545 2345
USA
☎ 1 505 667 5648
℻ 1 505 665 4055

Keller Christoph U
NOAO/NSO
Box 26732
950 N Cherry Av
Tucson AZ 85726 6732
USA
☎ 1 520 318 8445
℻ 1 520 318 8278
✉ ckeller@noao.edu

Keller Geoffrey
Astronomy Dpt
Ohio State Univ
174 W 18th Av
Columbus OH 43210 1106
USA
☎ 1 614 422 6279
℻ 1 614 292 2928

Keller Hans Ulrich
Observ/Planetarium
Neckarstr 47
D 70173 Stuttgart
Germany
☎ 49 711 162 920
℻ 49 711 216 3912

Keller Horst Uwe
MPI Aeronomie
Max Planck Str 2
Postfach 20
D 37189 Katlenburg Lindau
Germany
☎ 49 555 979 419
℻ 49 555 979 141
✉ keller@linmpi.mpg.de

Kellermann Kenneth I
NRAO
520 Edgemont Rd
Charlottesville VA 22903
USA
☎ 1 804 296 0240
℻ 1 804 296 0278
✉ kkellerm@nrao.edu

Kellogg Edwin M
CfA
HCO/SAO MS 3
60 Garden St
Cambridge MA 02138 1516
USA
☎ 1 617 495 7156
℻ 1 617 495 7356
✉ emk@cfa.harvard.edu

Kemball Athol
NRAO
Box 0
Socorro NM 87801 0387
USA
☎ 1 505 835 7330
℻ 27 505 835 7027
✉ akemball@nrao.edu

Kembhavi Ajit K
IUCAA
Post Bag 4
Megghnad Saha Rd
Ganeshkind Pune 411 007
India
☎ 91 212 336 415
✆ 91 212 350 760
✉ akk@iucaa.ernet.in

Kenderdine Sidney
MRAO
Cavendish Laboratory
Madingley Rd
Cambridge CB3 0HE
UK
☎ 44 12 23 337 294
✆ 44 12 23 354 599

Kendziorra Eckhard
Astronomisches Institut
Univ Tuebingen
Waldhaeuserstr 64
D 72076 Tuebingen
Germany
☎ 49 707 129 6127
✆ 49 707 129 3458

Kennedy Eugene T
School of Physical Sci
Ntl Inst Higher Education
Glasnevin
Dublin 9
Ireland
☎ 353 1 370 071

Kennedy Hans Daniel
Box 7243
Mail Ctr
Toowoomba QLD 4352
Australia

Kennedy John E
1902 315 5th Av N
Saskatoon SK S7K 5Z8
Canada

Kennicutt Robert C
Dpt Phys/Astronomy
Steward Observ
Univ Arizona
Tucson AZ 85721
USA
☎ 1 520 621 2288
✆ 1 520 621 1532
✉ rkennicutt@as.
arizona.edu

Kenny Harold
Physics Dpt
RMC
Kingston ONT K7K 5L0
Canada
☎ 1 613 541 6000*6042
✆ 1 613 541 6040
✉ kenny-h@rmc.ca

Kent Stephen M
Fermilab
MS 209
Box 500
Batavia IL 60510
USA
☎ 1 708 840 8231
✆ 1 708 840 8231

Kentischer Thomas
Kiepenheuer Institut
f Sonnenphysik
Schoneckstr 6
D 79104 Freiburg Breisgau
Germany
☎ 49 761 319 8158
✆ 49 761 319 8111
✉ tk@kis.uni-freiburg.de

Kenyon Scott J
CfA
HCO/SAO
60 Garden St
Cambridge MA 02138 1516
USA
☎ 1 617 495 7235
✆ 1 617 495 7356

Kepler S O
Instituto Fisica
UFRGS
CP 15051
91501 900 Porto Alegre RS
Brazil
☎ 55 51 316 6556
✆ 55 51 319 1762
✉ kepler@if.ufrgs.br

Keppens Rony
FOM
Box 1207
NL 3430 BE Nieuwegein
Netherlands
☎ 31 30 603 1224
✆ 31 30 6031204
✉ keppens@rijnh.nl

Kerr Frank J
Astronomy Program
Univ of Maryland
College Park MD 20742 2421
USA
☎ 1 301 454 6302
✆ 1 301 314 9067

Kerr Roy P
Dpt Phys/Astronomy
Univ of Canterbury
Private Bag 4800
Christchurch 1
New Zealand
☎ 64 348 2009
✆ 64 3 364 2469
✉ rpk@math.
canterbury.ac.nz

Kerschbaum Franz
Inst Astronomie
Univ Wien
Tuerkenschanzstr 17
A 1180 Wien
Austria
☎ 43 1 470 680056
✆ 43 1 470 6015
✉ kerschbaum@astro1.
ast.univie.ac.at

Keskin Varol
Dpt Astron/Space Sci
Ege Univ
Box 21
35100 Bornova Izmir
Turkey
☎ 90 232 388 0110*1738
✉ efeast05@vm3090.ege.edu.tr

Kessler Martin F
ESA
Apt 50727 Villafranca
E 28080 Madrid
Spain
☎ 34 1 813 1253
✆ 34 1 813 1308
✉ mkessler@iso.vilspa.
esa.es

Kesteven Michael J I
CSIRO
Div Radiophysics
Box 76
Epping NSW 2121
Australia
☎ 61 2 868 0222
✆ 61 2 868 0310
✉ mkevsteve@atnf.
csiro.au

Khachikian E Ye
Byurakan Astrophys Observ
Armenian Acad Sci
378433 Byurakan
Armenia
☎ 374 88 52 28 3453/4142
✆ 374 88 52 52 3640
✉ ekhach@helios.sci.am

Khalesseh Bahram
Physics Dpt
School of Sciences
Univ of Ferdowsi
Mashhad
Iran
☎ 98 51 32021*64
✆ 98 51 87079

Khaliullin Khabibrachman F
SAI
Acad Sciences
Universitetskij Pr 13
119899 Moscow
Russia
☎ 7 095 939 2378
✆ 7 095 939 0126
✉ hfh@sai.msu.su

Khanna Ramon
Landessternwarte
Koenigstuhl
D 69117 Heidelberg
Germany
☎ 49 622 150 9265
✆ 49 622 150 9202
✉ rkhanna@lsw.uni-
heidelberg.de

Kharadze E K
Abastumani Astrophysical
Observ
Georgian Acad Sci
380060 Tbilisi
Georgia
☎ 995 88 32 37 5226
✆ 995 88 32 98 5017
✉ ekhara@dtapha.kheta.
georgie.su

Kharchenko Nina
Main Astronomical Obs
Ukrainian Acad Science
Golosiiv
252650 Kyiv 22
Ukraine
☎ 380 4426 63110
✆ 380 4426 62147
✉ nkhar@mao.kiev.ua

Khare Bishun N
NASA/ARC
MS 239 14
Moffett Field CA 94035 1000
USA
☎ 1 650 604 2465
✆ 1 650 604 1088
✉ bkhare@mail.arc.nasa.gov

Khare Pushpa
Physics Dpt
Utkal Univ
Bhubaneswar 751 004
India
☎ 91 674 481 079
✆ 91 674 481 142
✉ khare@iopb.ernet.in

Kharin Arkadiy S
Main Astronomical Obs
Ukrainian Acad Science
Golosiiv
252650 Kyiv 22
Ukraine
☎ 380 4426 64769/266 1970
✆ 380 4426 62147
✉ kharin@mao.gluk.apc.org

Kharitonov Andrej V
Astrophys Inst
Kazakh Acad Sci
480068 Alma Ata
Kazakhstan
☎ 7 62 4040

Khatisashvili Alfez Sh
Abastumani Astrophysical
Observ
Georgian Acad Sci
383762 Abastumani
Georgia
☎ 995 88 32 95 5367
✆ 995 88 32 98 5017

Khetsuriani Tsiala S
Abastumani Astrophysical
Observ
Georgian Acad Sci
383762 Abastumani
Georgia
☎ 995 88 32 95 5367
✆ 995 88 32 98 5017

Khokhlova Vers L
Inst of Astronomy
Acad Sciences
Pyatnitskaya Ul 48
109017 Moscow
Russia
☎ 7 095 231 5461
✆ 7 095 230 2081

Kholshevnikov Konstatin V
Astronomical Observ
St Petersburg Univ
Bibliotechnaja Pl 2
198904 St Petersburg
Russia
☎ 7 812 257 9488
℻ 7 812 428 4259

Kholtygin Alexander F
Astronomical Institute
St Petersburg Univ
Bibliotechnaja Pl 2
198904 St Petersburg
Russia
☎ 7 812 428 4163
℻ 7 812 428 6649
✉ afk@aispb4.spb.su

Khozov Gennadij V
Astronomical Observ
St Petersburg Univ
Bibliotechnaja Pl 2
198904 St Petersburg
Russia
☎ 7 812 257 9484
℻ 7 812 428 4259

Khromov Gavriil S
Astron Geod Sciety
of Russia
24 Sadovaja Kudrinskaja S
103001 Moscow
Russia
☎ 7 095 291 5896

Kiang Tao
Dunsink Observ
DIAS
Castleknock
Dublin 15
Ireland
☎ 353 1 38 7911
℻ 353 1 38 7090

Kiasatpoor Ahmad
Physics Dpt
Univ of Esfahan
Daneshgah E
Esfahan
Iran
☎ 98 31 44321

Kibblewhite Edward J
Inst of Astronomy
The Observatories
Madingley Rd
Cambridge CB3 0HA
UK
☎ 44 12 23 337 548
℻ 44 12 23 337 523

Kielkopf John F
Physics Dpt
Univ of Louisville
Louisville KY 40292
USA
☎ 1 502 588 6787
℻ 1 502 588 8194

Kiguchi Masayoshi
Res Inst Science/Tech
Kinki Univ
Higashi
Osaka 577
Japan
☎ 81 672 12 332

Kii Tsuneo
ISAS
3 1 1 Yoshinodai
Sagamihara
Kanagawa 229 8510
Japan
☎ 81 427 51 3911*2624
℻ 81 427 59 4253

Kijak Jaroslaw
Astronomy Centre
Lubuska 2
PL 65 001 Zielona Gora
Poland
☎ 48 68 20 2863
℻ 48 68 202 863
✉ jkijak@ca.wsp.zgora.pl

Kikuchi Sadaemon
Astronomical Institute
Tohoku Univ
Sendai Aoba
Miyagi 980
Japan
☎ 81 22 222 1800*3327
℻ 81 22 262 6609

Kiladze R I
Abastumani Astrophysical
Observ
Georgian Acad Sci
383762 Abastumani
Georgia
☎ 995 88 32 95 5367
℻ 995 88 32 98 5017
✉ roki@abao.kheta.
georgia.su

Kilambi G C
Astronomy Dpt
Univ of Osmania
Hyderabad 500 007
India
☎ 91 71 951*247

Kilar Bogdan
Fac of Geodesy
Univ E Kardelj
Jamova 2
Ljubljana
Slovenia

Kilian-Montenbruck Judith
Inst Astron/Astrophysik
Univ Sternwarte
Scheinerstr 1
D 81679 Muenchen
Germany
☎ 49 899 220 9429
℻ 49 899 220 9427
✉ kilian@usm.uni-
muenchen.de

Kilkenny David
SAAO
PO Box 9
7935 Observatory
South Africa
☎ 27 21 47 0025
℻ 27 21 47 3639
✉ dmk@saao.ac.za

Killeen Neil
AAO
ATNF
Box 76
Epping NSW 2121
Australia
☎ 61 2 868 0222
℻ 61 2 868 0400
✉ nikilleen@atnf.
csiro.au

Kilmartin Pamela
Mt John Observ
Box 56
Lake Tekapo 8770
New Zealand
☎ 64 3 680 6817
✉ pokilmartin@csc.
canterbury.ac.nz

Kim Chulhee
Dpt Earth Sci Edu
Chonbuk Ntl Univ
Chonju 560 756
Korea RP

Kim Chun Hwey
Dpt Astronomy/Space Sci
Chungbuk Ntl Univ
San 48 Gaeshin Dong
Cheongju 360 763
Korea RP
☎ 82 431 61 3139
℻ 82 431 61 4232
✉ kimch@astro.
chungbuk.ac.kr

Kim Ho Il
Korea Astronomy Obs/ISSA
36 1 Whaam Dong
Yuseong Gu
Taejon 305 348
Korea RP
☎ 82 428 65 3282
℻ 82 428 65 3282
✉ hikim@hanul.issa.
re.kr

Kim Iraida S
SAI
Acad Sciences
Universitetskij Pr 13
119899 Moscow
Russia
☎ 7 095 939 2245
℻ 7 095 131 1357
✉ kim@sai.msk.su

Kim Jik Su
Pyongyang Astron Obs
Acad Sciences DPRK
Taesong district
Pyongyang
Korea DPR
☎ 850 5 3134/5 & 5 3239

Kim Kap-sung
Dpt Astron/Space Sci
Kyunghee Univ
Yong In Kun
Kyunggee 449 701
Korea RP
☎ 82 331 28 02443
℻ 82 331 28 14964

Kim Kwang-tae
Dpt Astron/Space Res
Chungnam Ntl Univ
Daejoen 304 764
Korea RP
☎ 82 428 21 5463

Kim Tu Hwan
Korea Astronomy Obs/ISSA
36 1 Whaam Dong
Yuseong Gu
Taejon 305 348
Korea RP
☎ 82 428 23 1497
℻ 82 428 65 3282

Kim Yong Hyok
Pyongyang Astron Obs
Acad Sciences DPRK
Taesong district
Pyongyang
Korea DPR
☎ 850 5 3134/5 & 5 3239

Kim Yong Uk
Pyongyang Astron Obs
Acad Sciences DPRK
Taesong district
Pyongyang
Korea DPR
☎ 850 5 3134/5 & 5 3239

Kim Yonggi
Dpt Astronomy/Space Sci
Chungbuk Ntl Univ
Chungbuk
Korea RP
☎ 82 431 61 2312
℻ 82 431 272 0695
✉ ykkim@astro.chungbuk.ac.kr

Kim Yongha
Dpt Astron/Space Sci
Chungnam Ntl Univ
Chungnam
Korea RP
☎ 82 428 21 5461
℻ 82 428 22 8380
✉ ykim@jupiter.chungnam.ac.kr

Kim Yul
Pyongyang Astron Obs
Acad Sciences DPRK
Taesong district
Pyongyang
Korea DPR
☎ 850 5 3134/5 & 5 3239

Kim Zong Dok
Pyongyang Astron Obs
Acad Sciences DPRK
Taesong district
Pyongyang
Korea DPR
☎ 850 5 3134/5 & 5 3239

Kimble Randy A
NASA GSFC
Code 681
Greenbelt MD 20771
USA
☎ 1 301 286 5783
℻ 1 301 286 7642
✉ kimble@stars.gsfc.nasa.gov

Kimura Hiroshi
Purple Mountain Obs
CAS
Nanjing 210008
China PR
☎ 86 25 33921
📠 86 25 301 459

Kimura Toshiya
Ctr Prom Comput Sci
Japan AERI
2 2 54 Nakameguro
Meguro ku Tokyo 153
Japan
✉ kimura@koma.
jaeri.go.jp

King Andrew R
Astronomy Dpt
Univ Leicester
University Rd
Leicester LE1 7RH
UK
☎ 44 116 252 2073
📠 44 113 252 2200

King David Leonard
Royal Greenwich Obs
Madingley Rd
Cambridge CB3 0EZ
UK
☎ 44 12 23 374 000
📠 44 12 23 374 700
✉ king@uk.ac.cam.
ast-star

King David S
Dpt Phys/Astronomy
Univ New Mexico
800 Yale Blvd Ne
Albuquerque NM 87131
USA
☎ 1 505 277 2941

King Henry C
Trillium 206 White Lion Rd
Little Chalfont
Bucks HP7 9NU
UK

King Ivan R
Astronomy Dpt
Univ of California
601 Campbell Hall
Berkeley CA 94720 3411
USA
☎ 1 510 642 2206
📠 1 510 642 3411
✉ iking@astro.berkeley.edu

King-Hele Desmond G
Royal Aircraft Establ
Farnborough Hants
UK
☎ 44 12 522 4461

Kingston Arthur E
Dpt Appl Maths/Theor-Phys
Queen's Univ
Belfast BT7 1NN
UK
☎ 44 12 32 24 5133
📠 44 12 32 23 9182

Kinman Thomas D
NOAO/KPNO
Box 26732
950 N Cherry Av
Tucson AZ 85726 6732
USA
☎ 1 520 327 5511
📠 1 520 325 9360

Kinney Anne L
STScI
Homewood Campus
3700 San Martin Dr
Baltimore MD 21218
USA
☎ 1 301 338 4831
📠 1 301 338 4767

Kinoshita Hiroshi
Tokyo Astronomical Obs
NAOJ
Osawa Mitaka
Tokyo 181
Japan
☎ 81 422 34 3615
📠 81 422 34 3793
✉ kinoshita@nao.ac.jp

Kiplinger Alan L
APAS DPR
Univ of Colorado
Campus Box 389
Boulder CO 80309 0391
USA
☎ 1 303 497 5892

Kippenhahn Rudolf
Rautenbreite 2
D 37083 Goettingen
Germany
☎ 49 551 247 14
📠 49 551 229 02

Kipper Tonu
Tartu Observ
Estonian Acad Sci
Toravere
EE 2444 Tartumaa
Estonia
☎ 372 7 416 625
📠 372 7 410 205

Kiral Adnan
Univ Observ
Univ of Istanbul
University 34452
34452 Istanbul
Turkey
☎ 90 212 522 3597
📠 90 212 519 0834

Kirbiyik Halil
Physics Dpt
Middle East Tech Univ
06531 Ankara
Turkey
☎ 90 41 22 37100/3528

Kirby Kate P
CfA
HCO/SAO
60 Garden St
Cambridge MA 02138 1516
USA
☎ 1 617 495 7237
📠 1 617 495 7356

Kirian Tatiana R
Pulkovo Observ
Acad Sciences
10 Kutuzov Quay
196140 St Petersburg
Russia
☎ 7 812 123 4252
📠 7 812 315 1701
✉ anna@gaoranspb.su

Kirilova Daniela
Astronomy Dpt
Bulgarian Acad Sci
72 Lenin Blvd
BG 1784 Sofia
Bulgaria
☎ 359 2 62 56833
📠 359 2 75 5019
✉ mih@phys.uni-
sofia.bg

Kirk John
MPI Kernphysik
Postfach 103980
D 69029 Heidelberg
Germany
☎ 49 622 151 6248
📠 49 622 151 6482
✉ kirk@kirk0.mpi-
hd.mpg.de

Kirkpatrick Ronald C
LANL
MS 220
Box 1663
Los Alamos NM 87545 2345
USA
☎ 1 505 667 4812
📠 1 505 665 4055

Kirshner Robert Paul
Astronomy Dpt
Harvard Univ MS 46
60 Garden St
Cambridge MA 02138 1516
USA
☎ 1 617 495 7390
📠 1 617 495 7105

Kiselev Nikolai N
Astrophys Inst
Uzbek Acad Sci
Sviridenko Ul 22
734670 Dushanbe
Tajikistan
☎ 7 3770 23 1432
📠 7 3770 27 5483

Kiselman Dan
Stockholm Observ
Royal Swedish Acad Sciences
S 133 36 Saltsjoebaden
Sweden
☎ 46 8 716 4477
📠 46 8 716 4228
✉ dan@astro.su.se

Kiselyov Alexej A
Pulkovo Observ
Acad Sciences
10 Kutuzov Quay
196140 St Petersburg
Russia
☎ 7 812 298 2242
📠 7 812 315 1701

Kislyakov Albert G
Inst Applied Physics
Acad Sciences
Ulyanov Ul 46
603600 N Novgorod
Russia
☎ 7 831 236 7253
📠 7 831 236 2061

Kislyuk Vitalij S
Main Astronomical Obs
Ukrainian Acad Science
Golosiiv
252650 Kyiv 22
Ukraine
☎ 380 4426 63110
📠 380 4426 62147
✉ kislyuk@mao.kiev.ua

Kisseleva Tamara P
Pulkovo Observ
Acad Sciences
10 Kutuzov Quay
196140 St Petersburg
Russia
☎ 7 812 298 2242
📠 7 812 315 1701

Kissell Kenneth E
Physics Dpt
Univ of Maryland
College Park MD 20742 2421
USA
☎ 1 301 314 9531
📠 1 301 314 9531
✉ kkissell@img.umd.edu

Kitai Reizaburo
Hida Observ
Kyoto Univ
Kamitakara
Gifu 506 13
Japan
☎ 81 578 62 311
📠 81 578 62 118
✉ kitai@kusastro.kyoto-u.ac.jp

Kitamoto Shunji
Fac of Sciences
Osaka Univ
Machikaneyama
Toyonaka Osaka 560
Japan
☎ 81 684 41 151

Kitamura M
Tokyo Astronomical Obs
NAOJ
Osawa Mitaka
Tokyo 181
Japan
☎ 81 422 32 5111
📠 81 422 34 3793

Kitchin Christopher R
Hatfield Polytechnic
Observ
Bayfordbury
Hertford Herts SG13 8LD
UK
☎ 44 199 255 8451

Kiziloglu Nilguen
Physics Dpt
Middle East Tech Univ
06531 Ankara
Turkey
☎ 90 41 223 7100*3268
✆ 90 41 286 8638
✉ nlk@trmetu

Kiziloglu Uemit
Physics Dpt
Middle East Tech Univ
06531 Ankara
Turkey
☎ 90 41 22 37100*3275
✆ 90 41 286 8638
✉ umk@trmetu

Kjaergaard Per
Astronomical Observ, NBIfAFG
Copenhagen Univ
Juliane Maries Vej 30
DK 2100 Copenhagen
Denmark
☎ 45 35 32 5999/5987
✆ 45 35 32 5989
✉ per@astro.ku.dk

Kjeldsen Hans
Inst Phys/Astronomy
Univ of Aarhus
Ny Munkegade
DK 8000 Aarhus C
Denmark
☎ 45 89 423 609
✆ 45 86 12 0740
✉ hans@obs.aau.dk

Kjeldseth-Moe Olav
Inst Theor Astrophys
Univ of Oslo
Box 1029
N 0315 Blindern Oslo 3
Norway
☎ 47 22 856 510
✆ 47 22 85 6505

Kjurkchieva Diana
Physics Dpt
Higher Pedagogical Inst
BG 9700 Shoumen
Bulgaria
☎ 359 6 46 3151*289

Klapp Jaime
Dpt Fisica
UAM-I Iztapalapao
Apt 55 534
Mexico DF 09340
Mexico
☎ 52 5 724 4623
✆ 52 5 686 1717

Klare Gerhard
Landessternwarte
Koenigstuhl
D 69117 Heidelberg
Germany
☎ 49 622 110 036
✆ 49 622 150 9202

Klarmann Joseph
Physics Dpt
Washington Univ
Campus Box 1105
St Louis MO 63130
USA
☎ 1 314 889 6299
✆ 1 314 935 4083

Kleczek Josip
Astronomical Institute
Czech Acad Sci
Fricova 1
CZ 251 65 Ondrejov
Czech R
☎ 420 204 85 7157
✆ 420 2 88 1611
✉ kleczek@asu.cas.cz

Klein Karl Ludwig
Observ Paris Meudon
DASOP
Pl J Janssen
F 92195 Meudon PPL Cdx
France
☎ 33 1 45 34 7761
✆ 33 1 45 07 7959

Klein Michael J
CALTECH/JPL
MS 303 401
4800 Oak Grove Dr
Pasadena CA 91109 8099
USA
☎ 1 818 354 7132
✆ 1 818 393 6030

Klein Richard I
LLNL
L 23
Box 808
Livermore CA 94551 9900
USA
☎ 1 415 422 3548
✆ 1 415 423 0238

Klein Ulrich
RAIUB
Univ Bonn
auf d Huegel 69
D 53121 Bonn
Germany
☎ 49 228 73 3644
✆ 49 228 73 3672

Kleinmann Douglas E
Honeywell Electro Optics
Operation
2 Forbes Rd
Lexington MA 02173
USA
☎ 1 617 863 3841

Klemola Arnold R
Lick Observ
Univ of California
Santa Cruz CA 95064
USA
☎ 1 831 459 4049
✆ 1 831 426 3115
✉ klemola@ucolick.org

Klemperer W K
NBS
Electromagnetic Fields D
325 Broadway
Boulder CO 80303
USA
☎ 1 303 497 3757

Klepczynski William J
ISI
1608 Spring Hill Rd
Suite 200
Vienne VA 22182
USA
☎ 1 202 651 7670
✆ 1 202 651 7699
✉ wklepczy@aol.com

Kliem Bernhard
Astrophysik Inst
Potsdam Univ
an d Sternwarte 16
D 14482 Potsdam
Germany
☎ 49 331 749 9208
✆ 49 331 749 200
✉ bkleim@aip.de

Klimchuk James A
Ctr for Space Sci/
Astrophysics
Stanford Univ ERL
Stanford CA 94305 4055
USA
☎ 1 415 723 1765
✆ 1 415 725 2333
✉ klimchuk@flare.
stanford.edu

Klinglesmith Daniel A
NASA GSFC
Code 684
Greenbelt MD 20771
USA
☎ 1 301 286 6541

Klinkhamer Frans
Inst Theor Physik
Univ Karlsruhe
D 76128 Karlsruhe
Germany

Kliore Arvydas Joseph
CALTECH/JPL
4800 Oak Grove Dr
Pasadena CA 91109 8099
USA
☎ 1 818 354 6164
✆ 1 818 393 6030

Klochkova Valentina
SAO
Acad Sciences
Nizhnij Arkhyz
357147 Karachaevo
Russia
☎ 7 878 78 92 501
✆ 7 96 908 2861
✉ valenta@sao.ru

Klock B L
4509 Bayside Dr
Milton FL 32570 8423
USA
☎ 1 904 994 1728
✉ blklock@sprintmail.com

Klocok Lubomir
Astronomical Institute
Slovak Acad Sci
SK 059 60 Tatranska Lomni
Slovak Republic
☎ 421 969 96 7866
✆ 421 969 96 7656

Klokocnik Jaroslav
Astronomical Institute
Czech Acad Sci
Fricova 1
CZ 251 65 Ondrejov
Czech R
☎ 420 204 85 7158
✆ 420 2 88 1611
✉ jklokocn@asu.cas.cz

Klose Sylvio
Thueringer Landessternwarte
Sternwarte 5
D 07778 Tautenburg
Germany
☎ 49 364 278 630
✆ 49 364 278 6329
✉ klose@tls-tautenburg.de

Kluzniak Wlodzimiere
Copernicus Astron Ctr
Polish Acad Sci
Ul Bartycka 18
PL 00 716 Warsaw
Poland
☎ 48 22 41 1086
✆ 48 22 41 0828
✉ wlodek@camk.edu.pl

Klvana Miroslav
Astronomical Institute
Czech Acad Sci
Fricova 1
CZ 251 65 Ondrejov
Czech R
☎ 420 204 85 7221
✆ 420 2 88 1611
✉ mklvana@asu.cas.cz

Klymyshyn I A
Pushkin Ul 96 Apt 66
284000 Ivanofrankovsk
Ukraine

Knacke Roger F
Penn State Erie
The Behrend College
Station Rd
Erie PA 16563 0203
USA
☎ 1 814 898 6105
✆ 1 814 898 6213

Knapen Johan Hendrik
Physical Sci Div
Univ Hertfordshire
College Lane
Hatfield Herts AL10 9AB
UK
☎ 44 170 728 5251
✆ 44 170 728 5279
✉ knapen@star.herts.ac.uk

Knapp Gillian R
Dpt Astrophysical Sci
Princeton Univ
Princeton NJ 08544 1001
USA
☎ 1 609 452 3824
✆ 1 609 258 1020

Knee Lewis
DRAO
Box 248
Penticton BC V2A 6K3
Canada
☎ 1 250 493 2277
℻ 1 250 493 7767
✉ lewis.knee@hia.nrc.ca

Kneer Franz
Univ Sternwarte
Goettingen
Geismarlandstr 11
D 37083 Goettingen
Germany
☎ 49 551 39542
℻ 49 551 395 043

Kneib Jean-Paul
OMP
14 Av E Belin
F 31400 Toulouse Cdx
France
☎ 33 5 61 33 2824
℻ 33 5 61 33 2840
✉ kneib@obs-mip.fr

Knezevic Zoran
Astronomical Observ
Volgina 7
11150 Beograd
Yugoslavia FR
☎ 381 1 419 357/401 320
℻ 381 1 419 553
✉ zoran@aob.aob.bg.
ac.yu

Kniffen Donald A
Physics Dpt
Hampden Sydney Coll
Box 862
Hampden Sydney VA 23943
USA
☎ 1 804 223 6255
℻ 1 804 223 6374
✉ donk@pulsar.hsc.edu

Knoelker Michael
HAO
NCAR
Box 3000
Boulder CO 80307 3000
USA
☎ 1 303 497 1501
℻ 1 303 497 1568
✉ knoe@ncar.edu

Knoska Stefan
Astronomical Institute
Slovak Acad Sciences
SK 059 60 Tatranska Lomni
Slovak Republic
☎ 421 969 96 7866
℻ 421 969 96 7656

Knowles Stephen H
9455 Deramus Farm CT
Vienna VA 22448 5180
USA
☎ 1 202 404 7829
✉ knowles@bdcmail.nrl.
navy.mil

Knude Jens Kirkeskov
Astronomical Observ, NBIfAFG
Copenhagen Univ
Juliane Maries Vej 30
DK 2100 Copenhagen
Denmark
☎ 45 35 32 5986
℻ 45 35 32 5989
✉ indus@astro.ku.dk

Ko Hsien C
Dpt Elect Eng
Ohio State Univ
1958 Neil Av
Columbus OH 43210 1106
USA
☎ 1 614 422 2571

Kobayashi Eisuke
Science Inst of Osaka
13 23 Karita 4 Chome
Sumiyoshi Ku
Osaka 558
Japan
☎ 81 669 21 882

Kobayashi Hideyuki
ISAS
3 1 1 Yoshinodai
Sagamihara
Kanagawa 229 8510
Japan
☎ 81 427 51 3911*2709
℻ 81 427 51 3972
✉ hkobaya@vsop.isas.
ac.jp

Kobayashi Yukisayu
Tokyo Astronomical Obs
NAOJ
Osawa Mitaka
Tokyo 181
Japan
☎ 81 422 32 5111
℻ 81 422 34 3793

Kocer Durcun
Astronomy Dpt
Univ of Istanbul
34452 Istanbul
Turkey
☎ 90 2121 522 3597
℻ 90 212 522 6123

Koch David G
NASA/ARC
MS 245 6
Moffett Field CA 94035 1000
USA
☎ 1 415 604 5528
℻ 1 415 604 6779

Koch Robert H
Dpt Astron/Astrophys
Univ of Pennsylvania
209 S 33rd St
Philadelphia PA 19104
USA
☎ 1 215 898 7882
℻ 1 215 898 9336
✉ rkoch@upenn.sas.edu

Koch-Miramond Lydie
DAPNIA/SAP
CEA Saclay
BP 2
F 91191 Gif s Yvette Cdx
France
☎ 33 1 69 08 4329
℻ 33 1 69 08 6577

Kocharov Grant E
Ioffe Physical Tech Inst
Acad Sciences
Polytechnicheskaya Ul 26
194021 St Petersburg
Russia
☎ 7 812 247 9167
℻ 7 812 247 1017
✉ kocharov@stu.spb.su

Kochhar R K
IIA
Koramangala
Sarjapur Rd
Bangalore 560 034
India
☎ 91 80 356 6585
℻ 91 80 553 4043
✉ rkk@iiap.ernet.in

Kodaira Keiichi
Tokyo Astronomical Obs
NAOJ
Osawa Mitaka
Tokyo 181
Japan
☎ 81 422 32 5111
℻ 81 422 34 3793

Kodama Hideo
Yukawa Istt Theor Physics
Kyoto Univ
Kitashirakawa
Kyoto 606 01
Japan
☎ 81 757 53 7015
℻ 81 75 753 7015
✉ Kodama@yukawa.kyoto-
u.ac.jp

Koeberl Christian
Inst Geochemistry
Univ Wien
Althanstr 14
A 1090 Wien
Austria
☎ 43 1 313 36 1714
℻ 43 1 313 36 781
✉ a8631dab@vm.univie.ac.at

Koechlin Laurent
OMP
14 Av E Belin
F 31400 Toulouse Cdx
France
☎ 33 5 61 33 2887
℻ 33 5 61 33 2840
✉ koechlin@obs-mip.fr

Koehler H
Sauerbruchstr 6
D 7920 Heidenheim
Germany
☎ 49 732 144 560

Koehler James A
Physics Dpt
Univ of Saskatchewan
Saskatoon SK S7N 0W0
Canada
☎ 1 306 966 6442

Koehler Peter
Carl Zeiss Jena Gmbh
Tatzens Promenade 1a
Postfach 125
D 07740 Jena
Germany
☎ 49 364 1640
℻ 49 364 164 2542
✉ koehler@zeiss.de

Koehler-Vogel Suzanne
Hamburger Sternwarte
Univ Hamburg
Gojensbergsweg 112
D 21029 Hamburg
Germany
☎ 49 407 252 4137
℻ 49 407 252 4198
✉ skoehler@hs.uni-hamburg.de

Koempe Carsten
Astrophysik Inst
Univ Sternwarte
Schillergaesschen 2
D 07745 Jena
Germany
☎ 49 364 163 0313
℻ 49 364 163 0417
✉ koempe@betty.astro.uni-
jena.de

Koen Marthinus
SAAO
PO Box 9
7935 Observatory
South Africa
☎ 27 21 47 0025
℻ 27 21 473639
✉ ck@saao.ac.za

Koenigsberger Gloria
Instituto Astronomia
UNAM
Apt 70 264
Mexico DF 04510
Mexico
☎ 52 5 622 3906
℻ 52 5 616 0653

Koeppen Joachim
Inst Theor Phys/
Sternwarte Univ Kiel
Olshausenstr 40
D 24098 Kiel
Germany
☎ 49 431 880 4103
℻ 49 431 880 4432
✉ pas86@rz.uni.kiel.d400.de

Koester Detlev
Inst Theor Phys/
Sternwarte Univ Kiel
Olshausenstr 40
D 24098 Kiel
Germany
☎ 49 431 880 4110
℻ 49 431 880 4432

Kofman Lev
Inst for Astronomy
Univ of Hawaii
2680 Woodlawn Dr
Honolulu HI 96822
USA
☎ 1 808 956 6196
✆ 1 808 956 9590
✉ kofman@ifa.hawaii.edu

Kogoshvili Natela G
Abastumani Astrophysical Observ
Georgian Acad Sci
383762 Abastumani
Georgia
☎ 995 88 32 95 5367
✆ 995 88 32 98 5017

Kogure Tomokazu
Togano-o 1-10
Hashimoto Yawata
Kyoto 614
Japan
☎ 81 759 83 2984

Kohl John L
CfA
HCO/SAO
60 Garden St
Cambridge MA 02138 1516
USA
☎ 1 617 495 7377
✆ 1 617 495 7356

Kohoutek Lubos
Hamburger Sternwarte
Univ Hamburg
Gojensbergsweg 112
D 21029 Hamburg
Germany
☎ 49 407 252 4112
✆ 49 407 252 4198

Koide Shinji
Toyama Univ
3190 Gofuku
Toyama 930
Japan
☎ 81 764 45 6 745
✆ 81 764 456 703
✉ koidesin@ecs.toyama-u.ac.jp

Koike Chiyoe
Kyoto Pharml Univ
Misasagi
Yamashina
Kyoto 607
Japan
☎ 81 75 595 4702
✆ 81 75 595 4793
✉ koike@cr.scphys.kyoto-u.ac.jp

Kojima Masayoshi
Solar Terrestrial Lab
Nagoya Univ
3-13 Honohara Toyokawa
Aichi 442
Japan
☎ 81 533 83 154
✆ 81 533 86 0811
✉ kojima@stelab.nagoya-u.ac.jp

Kojima Yasufumi
Physics Dpt
Hiroshima Univ
Higashi Senda Machi
Hiroshima 739
Japan
☎ 81 824 247 365
✆ 81 824 240 717
✉ kojima@theo.phys.sci.hiroshima-u.ac.jp

Kokkotas Konstantinos
Astronomy Lab
Univ Thessaloniki
GR 540 06 Thessaloniki
Greece
☎ 30 31 99 8185
✆ 30 31 99 5384
✉ kokkotas@astro.auth.gr

Kokott Wolfgang
Observatorium Hoher List
Univ Sternwarte Bonn
D 54550 Daun
Germany
☎ 49 6592 2150
✆ 49 65 92 9851 40
✉ w.kokott@lrz.uni-muenchen.de

Kokurin Yurij L
Lebedev Physical Inst
Acad Sciences
Leninsky Pspt 53
117924 Moscow
Russia
☎ 7 095 135 0360
✆ 7 095 135 7880

Kolaczek Barbara
Planetary Geodesy Dpt
Polish Acad Sci
Ul Bartycka 18
PL 00 716 Warsaw
Poland
☎ 48 22 41 1086
✆ 48 22 41 0828
✉ cbk@camk.edu.pl

Kolb Edward W
Fermilab
MS 209
Box 500
Batavia IL 60510
USA
☎ 1 708 840 8231
✆ 1 708 840 8231
✉ uck@star.le.ac.uk

Kolb Ulrich
Astronomy Group
Univ Leicester
University Rd
Leicester LE1 7RH
UK
☎ 44 116 252 2079
✆ 44 116 252 2070
✉ uck@star.le.ac.uk

Kolchinskij I G
Main Astronomical Obs
Ukrainian Acad Science
Golosiiv
252650 Kyiv 22
Ukraine
☎ 380 4426 63110
✆ 380 4426 62147

Kolesnik Igor G
Main Astronomical Obs
Ukrainian Acad Science
Golosiiv
252650 Kyiv 22
Ukraine
☎ 380 4426 63110
✆ 380 4426 62147

Kolesnik L N
Main Astronomical Obs
Ukrainian Acad Science
Golosiiv
252650 Kyiv 22
Ukraine
☎ 380 4426 60869
✆ 380 4426 62147

Kolesov Alexander K
Astronomical Observ
St Petersburg Univ
Bibliotechnaja Pl 2
199178 St Petersburg
Russia
☎ 7 812 428 7129
✆ 7 812 428 4259

Kolev Dimitar Zdravkov
Ntl Astronomical Obs
Bulgarian Acad Sci
Box 136
BG 4700 Smoljan
Bulgaria
☎ 359 7 341 559

Kolka Indrek
Tartu Observ
Estonian Acad Sci
Toravere
EE 2444 Tartumaa
Estonia
☎ 372 7 410 438
✆ 372 7 410 205
✉ indrek@aai.ee

Kollath Zoltan
Konkoly Observ
Thege U 13/17
Box 67
H 1525 Budapest
Hungary
☎ 36 1 375 4122
✆ 36 1 275 4668

Kollatschny Wolfram
Univ Sternwarte
Goettingen
Geismarlandstr 11
D 37083 Goettingen
Germany
☎ 49 551 395 067
✆ 49 551 395 043

Kollberg Erik L
Dpt Microwave Techn
Chalmers Technical Univ
S 412 96 Goeteborg
Sweden
☎ 46 31 772 1000
✆ 46 31 772 3204

Komarov N S
Astronomical Observ
Odessa State Univ
Shevchenko Park
270014 Odessa
Ukraine
☎ 380 4822 20396
✆ 380 4822 28442

Komberg Boris V
Space Res Inst
Acad Sciences
Profsojuznaya Ul 84/32
117810 Moscow
Russia
☎ 7 095 333 3366
✆ 7 095 333 2378

Komitov Boris
Astronomy Dpt
Bulgarian Acad Sci
72 Lenin Blvd
BG 1784 Sofia
Bulgaria
☎ 359 2 75 8927
✆ 359 2 75 8927
✉ planet@bgearn

Kompaneets Dmitriy A
Astro Space Ctr
Lebedev Physical Inst
Profsoyuznaya 84/32
Moscow 117810
Russia
☎ 7 095 333 3366
✆ 7 095 333 2378
✉ dkompan@dpc.asc.rssi.ru

Kondo Masaaki
Senshu Univ
Higashi-mita Tama-ku
Kawasaki Shi
Kanagawa 214
Japan
☎ 81 449 11 7131
✉ 86123@nacsis.ac.jp

Kondo Masayuki
1 2 25 Osawa
Mitaka
Tokyo 181
Japan
☎ 81 422 32 5111
✆ 81 422 34 3793

Kondo Yoji
NASA GSFC
Code 684
Greenbelt MD 20771
USA
☎ 1 301 286 6247
✆ 1 301 286 1752
✉ kondo@stars.gsfc.nasa.gov

Konigl Arieh
Astronomy/Astrophys Ctr
Univ of Chicago
5640 S Ellis Av
Chicago IL 60637
USA
☎ 1 312 702 7968
✆ 1 312 702 8212

Kononovich Edward V
SAI
Acad Sciences
Universitetskij Pr 13
119899 Moscow
Russia
☎ 7 095 939 2858
✆ 7 095 939 0126

Konopleva Varvara P
Main Astronomical Obs
Ukrainian Acad Science
Golosiiv
252650 Kyiv 22
Ukraine
☎ 380 4426 63110
✆ 380 4426 62147

Konovalenko Olexandr
Inst Radio Astron
Ukrainian Acad Science
4 Chervonopraporna st
310002 Kharkiv
Ukraine
☎ 380 5724 71134
✆ 380 5724 76506
✉ rai@ira.kharkov.ua

Kontizas Evangelos
Astronomical Institute
Ntl Observ Athens
Box 20048
GR 118 10 Athens
Greece
☎ 30 1 346 1191
✆ 30 1 342 1019
✉ ekontiza@astro.noa.gr

Kontizas Mary
Astrophysics Dpt
Ntl Univ Athens
Panepistimiopolis
GR 157 84 Zografos
Greece
☎ 30 1 728 4770
✆ 30 1 723 8413
✉ mkontiza@atlas.uoa.gr

Kontorovich Victor
Inst Radio Astron
Ukrainian Acad Science
4 Chervonopraporna st
310002 Kharkiv
Ukraine
☎ 380 5724 51014
✆ 380 5724 76506
✉ rai@ira.kharkov.ua

Koo Bon Chul
Astronomy Dpt
Seoul Ntl Univ
Kwanak Ku
Seoul 151 742
Korea RP
☎ 82 288 06623
✆ 82 288 71435
✉ koo@astrohi.snu.ac.kr

Koo David C-Y
Lick Observ
Univ of California
Santa Cruz CA 95064
USA
☎ 1 831 459 2130
✆ 1 831 426 3115
✉ koo@ucolick.org

Koornneef Jan
SRON
Univ Groningen
Postbus 800
NL 9700 AV Groningen
Netherlands
☎ 31 50 363 4073
✆ 31 50 363 6100
✉ koornneef@sron.rug.nl

Kopecky Miloslav
Astronomical Institute
Czech Acad Sci
Fricova 1
CZ 251 65 Ondrejov
Czech R
☎ 420 204 85 7118
✆ 420 2 88 1611

Kopp Greg
Meadowlark Optics
7460 Weld County Rd 1
Longmont CO 80504 9470
USA
☎ 1 303 776 4068
✆ 1 303 776 5856
✉ gkopp@meadowlark.com

Kopp Roger A
LANL
MS B259
Box 1663
Los Alamos NM 87545 2345
USA
☎ 1 505 665 3010
✆ 1 505 667 7780
✉ rak@lanl.gov

Kopylov Alexander
SAO
Acad Sciences
Nizhnij Arkhyz
357147 Karachaevo
Russia
☎ 7 878 784 6148
✆ 7 96 908 2861
✉ akop@sao.ru

Kopylov Ivan M
Pulkovo Observ
Acad Sciences
10 Kutuzov Quay
196140 St Petersburg
Russia
☎ 7 812 298 2242
✆ 7 812 315 1701

Korakitis Romylos
National Technical
Univ of Athens
Gen Rogakou str 39
GR 151 25 Maroussi
Greece
☎ 30 772 27 23
✆ 30 772 26 70
✉ romylos@survey.ntua.gr

Koratkar Anuradha P
STScI
Homewood Campus
3700 San Martin Dr
Baltimore MD 21218
USA
☎ 1 301 338 4470
✆ 1 301 338 1592
✉ koratkar@stsci.edu

Korchak Alexander A
ITMIRWP
Acad Sciences
142092 Troitsk
Russia
☎ 7 095 334 0120
✆ 7 095 334 0124

Koribalski Baerbel Silvia
CSIRO
ATNF
Box 76
Epping NSW 2121
Australia
☎ 61 2 9372 4361
✆ 61 2 9372 4310
✉ bkoribal@atnf.csiro.au

Kormendy John
Inst for Astronomy
Univ of Hawaii
2680 Woodlawn Dr
Honolulu HI 96822
USA
☎ 1 808 956 6680
✆ 1 808 988 2790
✉ kormendy@ifa.hawaii.edu

Korovyakovskij Yurij P
SAO
Acad Sciences
Nizhnij Arkhyz
357147 Karachaevo
Russia
☎ 7 878 789 2501

Korsun Alla
Main Astronomical Obs
Ukrainian Acad Science
Golosiiv
252650 Kyiv 22
Ukraine
☎ 380 4426 64759
✆ 380 4426 62147
✉ akorsun@mao.kiev.ua

Korzhavin Anatoly
SAO St Petersburg Br
Acad Sciences
10 Kutuzov Quay
196140 St Petersburg
Russia
☎ 7 812 123 4019
✆ 7 812 123 1922
✉ kor@saoran.spb.su

Kosai Hiroki
Agasaki 2058 1
Tamashima
Kurashiki City 713
Japan
☎ 81 865 26 6233
✆ 81 86 526 6233

Kosek Wieslaw
Space Res Ctr
Polish Acad Sci
Ul Bartycka 18A
PL 00 716 Warsaw
Poland
☎ 48 22 40 3766
✆ 48 22 36 8961
✉ kosek@cbk.waw.pl

Koshiba Masa-Toshi
Tokai Univ
2-28 Tomigaya
Shibuya
Tokyo 151
Japan
☎ 81 334 67 2211*483
✆ 81 334 85 4958

Kosin Gennadij S
Pulkovo Observ
Acad Sciences
10 Kutuzov Quay
196140 St Petersburg
Russia
☎ 7 812 298 2242
✆ 7 812 315 1701

Kosovichev Alexander
Ctr for Space Sci/Astro
Astrophysics
Stanford Univ ERL 328
Stanford CA 94305 4055
USA
☎ 1 415 723 7667
✆ 1 415 725 2333
✉ akosovichev@solar.stanford.edu

Kostik Roman I
Main Astronomical Obs
Ukrainian Acad Science
Golosiiv
252650 Kyiv 22
Ukraine
☎ 380 4426 64762
✆ 380 4426 62147

Kostina Lidija D
Pulkovo Observ
Acad Sciences
10 Kutuzov Quay
196140 St Petersburg
Russia
☎ 7 812 298 2242
✆ 7 812 315 1701

Kostyakova Elena B
SAI
Acad Sciences
Universitetskij Pr 13
119899 Moscow
Russia
☎ 7 095 939 2858
✆ 7 095 939 0126

Kosugi Takeo
Tokyo Astronomical Obs
NAOJ
Osawa Mitaka
Tokyo 181
Japan
☎ 81 422 32 5111
✆ 81 422 34 3793

Kotanyi Christophe
NEPAE
UFSM
Cidado Universitaria
97100 Santa Maria RS
Brazil
☎ 55 552 26 1616

Kotelnikov Vladimir A
Inst Radio/Electron
Acad Sciences
Marx Av 18
103907 Moscow
Russia
☎ 7 095 203 6078

Kotilainen Jari
SISSA
Astronomy Dpt
Via Beirut 2 4
I 34014 Trieste
Italy
☎ 39 40 378 7525
℻ 39 40 378 7528
✉ jkotilai@sissa.it

Kotnik-Karuza Dubravka
Fac of Education
Omladinska 14
HR 51216 Viskovo
HR 51000 Rijeka
Croatia
☎ 385 51 227 474
℻ 385 51 515 142
✉ kotnik@mapef.pefri.hr

Kotov Valery
Crimean Astrophys Obs
Ukrainian Acad Science
Nauchny
334413 Crimea
Ukraine
☎ 380 6554 71161
℻ 380 6554 40704

Kotrc Pavel
Astronomical Institute
Czech Acad Sci
Fricova 1
CZ 251 65 Ondrejov
Czech R
☎ 420 204 85 222
℻ 420 2 88 1611
✉ pkotrc@asu.cas.cz

Koubsky Pavel
Astronomical Institute
Czech Acad Sci
Fricova 1
CZ 251 65 Ondrejov
Czech R
☎ 420 204 85 727
℻ 420 2 88 1611
✉ koubsky@sunstel.asu.cas.cz

Koupelis Theodoros
Marathon Ctr
Univ of Wisconsin
518 7th Av
Wausau WI 54401
USA
☎ 1 715 845 9602
℻ 1 715 848 3568
✉ tkoupeli@uwcmail.uwc.edu

Kourganoff Vladimir
20 Av Paul Apell
F 75014 Paris
France
☎ 33 1 45 40 5053

Koutchmy Serge
IAP
98bis bd Arago
F 75014 Paris
France
☎ 33 1 44 32 8056
℻ 33 1 44 32 8001

Kouveliotou Chryssa
NASA/MSFC
Space Science Lab
Code ES 62
Huntsville AL 35812
USA
☎ 1 205 544 7711
℻ 1 205 544 5800

Kovachev B J
Astronomy Dpt
Bulgarian Acad Sci
72 Lenin Blvd
BG 1784 Sofia
Bulgaria
☎ 359 2 75 8827
℻ 359 2 75 8927

Kovacs Agnes
Debrecen Heliophys Observ
Acad Sciences
Box 30
H 4010 Debrecen
Hungary
☎ 36 5 231 1015

Kovacs Geza
Konkoly Observ
Thege U 13/17
Box 67
H 1525 Budapest
Hungary
☎ 36 1 375 4122
℻ 36 1 275 4668

Koval I K
Main Astronomical Obs
Ukrainian Acad Science
Golosiiv
252650 Kyiv 22
Ukraine
☎ 380 4426 60869
℻ 380 4426 62147

Kovalevsky Jean
OCA
CERGA
F 06130 Grasse
France
☎ 33 4 93 40 5353
℻ 33 4 93 40 5353
✉ kovalevsky@mfg.cnes.fr

Kovar N S
Physics Dpt
Univ of Houston
Houston TX 77004
USA
☎ 1 713 743 3550
℻ 1 713 743 8589

Kovar Robert P
9666 E Orchard Dr
Englewood CO 80111
USA
☎ 1 303 394 4494

Kovetz Attay
Dpt Phys/Astronomy
Tel Aviv Univ
Ramat Aviv
Tel Aviv 69978
Israel
☎ 972 3 420 234
℻ 972 3 640 8179

Kowal Charles Thomas
STScI
Homewood Campus
3700 San Martin Dr
Baltimore MD 21218
USA
☎ 1 301 338 4700
℻ 1 301 338 4767

Koyama Katsuji
Inst Space/Astron Sci
Univ of Tokyo
Meguro Ku
Tokyo 153
Japan
☎ 81 346 71 111
℻ 81 334 85 2904

Koyama Shin
Kagawa Video Study Ctr
Univ of The Air
Saiwai Cho 2 1
Takamatsu 760
Japan

Kozai Yoshihide
Tokyo Astronomical Obs
NAOJ
2-21-1Ohsawa Mitaka
Tokyo 181
Japan
☎ 81 422 34 3650
℻ 81 422 34 3690
✉ kozai@c1.mtk.nao.ac.jp

Kozasa Takashi
Dpt Earth/Planet Sci
Kobe Univ
Nada
Kobe 657 8501
Japan
☎ 81 788 03 0978
℻ 81 788 03 0490
✉ kozasa@kobe-u.ac.jp

Kozlovsky B Z
Dpt Phys/Astronomy
Tel Aviv Univ
Ramat Aviv
Tel Aviv 69978
Israel
☎ 972 3 640 8208
℻ 972 3 640 8179

Kozlowski Maciej
Copernicus Astron Ctr
Polish Acad Sci
Ul Bartycka 18
PL 00 716 Warsaw
Poland
☎ 48 22 41 1086
℻ 48 22 41 0828

Kraan-Korteweg Renee C
Dpt Astronomia
Univ Guanajuato
Apdo 144
Guanajuato GTO 36000
Mexico
☎ 52 4 732 9548
℻ 52 4 732 0253
✉ kraan@astro.ugto.mx

Kraemer Gerhard
Astronomisches Institut
Univ Tuebingen
Waldhaeuserstr 64
D 72076 Tuebingen
Germany
☎ 49 707 129 2486
℻ 49 707 129 34 58

Kraft Robert P
Lick Observ
Univ of California
Santa Cruz CA 95064
USA
℻ 1 831 426 3115
✉ kraft@helios.ucsc.edu

Kraicheva Zdravska
Astronomy Dpt
Bulgarian Acad Sci
7th November St 1
BG 1000 Sofia
Bulgaria
☎ 359 2 80 2831
℻ 359 2 80 2831

Kramer Kh N
4924 SW 59 Av
Portland OR 97221 1163
USA
☎ 1 503 203 8045

Krasinski Andrzej
Copernicus Astron Ctr
Polish Acad Sci
Ul Bartycka 18
PL 00 716 Warsaw
Poland
☎ 48 22 41 1086
℻ 48 22 410 046

Krasinsky George A
Inst Appl Astronomy
Acad Sciences
Zhdanovskaya Ul 8
197042 St Petersburg
Russia
☎ 7 812 230 7414
℻ 7 812 230 7413
✉ kra@isida.ipa.rssi.ru

Kraus John D
Radio Observ
Ohio State Univ
2015 Neil Av
Columbus OH 43210
USA
☎ 1 614 548 7895

Krause Marita
RAIUB
Univ Bonn
auf d Huegel 69
D 53121 Bonn
Germany
☎ 49 228 525 315
℻ 49 228 525 229
✉ mkrause@mpifr-bonn.mpg.de

Kraushaar William L
Physics Dpt
Univ Wisconsin
1150 University Av
Madison WI 53706
USA
☎ 1 608 262 5916
📠 1 608 263 0361

Krautter Joachim
Landessternwarte
Koenigstuhl
D 69117 Heidelberg
Germany
☎ 49 622 150 9209
📠 49 622 150 9202
✉ jkrautte@hp2.lsw.uni-heidelberg.de

Kravchuk Sergei
Main Astronomical Obs
Ukrainian Acad Science
Golosiiv
252650 Kyiv 22
Ukraine
☎ 380 4426 62502
📠 380 4426 62147
✉ kravchuk@mao.kiev.ua

Kreidl Tobias J N
Lowell Observ
1400 W Mars Hill Rd
Box 1149
Flagstaff AZ 86001
USA
☎ 1 602 774 3358
📠 1 520 774 3358

Kreiner Jerzy Marek
Inst of Physics
Pedagogical Univ
Ul Podchorazych 2
PL 30 084 Krakow
Poland
☎ 48 12 37 8286
📠 48 12 372 2243

Kreisel E
Einstein Laboratorium
Telegrafenberg
Rosa Luxemburg Str 17a
D 14482 Potsdam
Germany
☎ 49 762 225

Krelowski Jacek
Inst of Astronomy
N Copernicus Univ
Ul Chopina 12/18
PL 87 100 Torun
Poland
☎ 48 56 26 018
📠 48 56 24 602
✉ jacek@astri.uni.torun.pl

Krempec-Krygier Janina
Inst of Astronomy
N Copernicus Univ
Ul Chopina 12/18
PL 87 100 Torun
Poland
☎ 48 56 26 018
📠 48 56 24 602

Kresakova Margita
Astronomical Institute
Slovak Acad Sci
Dubravska 9
SK 842 28 Bratislava
Slovak Republic
☎ 421 7 37 5157
📠 421 7 37 5157

Kreysa Ernst
RAIUB
Univ Bonn
auf d Huegel 69
D 53121 Bonn
Germany
☎ 49 228 525 269
📠 49 228 525 229

Krieger Allen S
Radiation Science Inc
Box 293
Belmont MA 02178
USA
☎ 1 617 494 0335

Krikorian Ralph
IAP
98bis bd Arago
F 75014 Paris
France
☎ 33 1 44 32 8145
📠 33 1 44 32 8001

Krisciunas Kevin
Astronomy Dpt
Univ of Washington
Box 351580
Seattle WA 98195 1580
USA
☎ 1 425 543 2888
📠 1 425 685 0403
✉ kevin@orca.astro.washington.edu

Krishna Gopal
NCRA/TIFR
Pune Univ Campus Pb 3
Ganeshkhind
Pune 411 007
India
☎ 91 212 35 6105
📠 91 212 33 5760

Krishna Swamy K S
TIFR/Astrophys Gr
Homi Bhabha Rd
Colaba
Bombay 400 005
India
☎ 91 22 219 111
📠 91 22 215 2110

Krishnamohan S
TIFR/Radio Astronomy Ctr
Box 8
Bangalore 560 012
India
☎ 91 80 336 4062

Krishnan Thiruvenkata
Helios Antennas/Electron
234 Avvai Shanmugham Rd
Gopalapuram
Madras 600 086
India
☎ 91 44 827 2680
📠 91 44 825 6510
✉ tkrishna@giasmd01.vsnl.net.in

Kriss Gerard A
Dpt Phys/Astronomy
JHU
Charles/34th St
Baltimore MD 21218
USA
☎ 1 301 338 7679
📠 1 301 546 7279

Kristensen Leif Kahl
Inst Phys/Astronomy
Univ of Aarhus
Ny Munkegade
DK 8000 Aarhus C
Denmark
☎ 45 86 12 8899
📠 45 86 20 2711

Kristenson Henrik
Dorttininggatan 20
S 432 410 Varberg
Sweden

Kristiansson Krister
Physics Dpt
Univ of Lund
Soelvegatan 14
S 223 62 Lund
Sweden
☎ 46 46 222 7726
📠 46 46 222 4709

Krivov Alexander
Astron Inst St Petersburg
State Univ
Bibliotechnaya Pl 2
St Petersburg 198904
Russia
☎ 7 812 428 4163
📠 7 812 428 7129
✉ krivov@aispbu.spb.su

Krivsky Ladislav
Astronomical Institute
Czech Acad Sci
Fricova 1
CZ 251 65 Ondrejov
Czech R
☎ 420 204 85 7225/7111
📠 420 2 88 1611
✉ astsun@csearn

Kriz Svatopluk
SK Press
Masarykovo Nam 35
CZ 25101 Ricany
Czech R
☎ 420 204 2486

Krogdahl W S
Dpt Phys/Astronomy
Univ Kentucky
Lexington KY 40506 0055
USA
☎ 1 606 272 2659
📠 1 606 323 2846

Krolik Julian H
Dpt Phys/Astronomy
JHU
Charles/34th St
Baltimore MD 21218
USA
☎ 1 301 338 7926
📠 1 301 516 7279

Krolikowska-Soltan Malgorzata
Space Res Ctr
Polish Acad Sci
Ul Bartycka 18
PL 00 716 Warsaw
Poland
☎ 48 22 40 3766
📠 48 22 36 8961
✉ mkr@cbk.waw.pl

Kron Richard G
Yerkes Observ
Univ of Chicago
Box 258
Williams Bay WI 53191
USA
☎ 1 312 236 5468
📠 1 414 245 9805

Kronberg Philipp
Astronomy Dpt
Univ of Toronto
60 St George St
Toronto ON M5S 1A1
Canada
☎ 1 416 978 4971
📠 1 416 978 3921

Kroto Harold
School of Chemistry
Univ of Sussex
Falmer
Brighton BN1 9QJ
UK
☎ 44 12 73 67 8329
✉ kafe4@cluster.sussex.ac.uk

Kroupa Pavel
Inst Theor Astrophysik
d Univer
Tiergartenstr 15
D 69121 Heidelberg
Germany
☎ 49 622 154 6710
📠 49 622 154 4221
✉ pavel@wombat.ita.uni-heidelberg.de

Kruchinenko Vitaliy G
Astronomical Observ
Kyiv State Univ
Observatornaya Ul 3
252053 Kyiv
Ukraine
☎ 380 4421 62691

Kruegel Endrik
RAIUB
Univ Bonn
auf d Huegel 69
D 53121 Bonn
Germany
☎ 49 228 525 0
📠 49 228 525 229

Krueger Albrecht
Zntrlinst Astrophysik
Sternwarte Babalsberg
Rosa Luxemburg Str 17a
D 14473 Potsdam
Germany
☎ 49 331 762225

Krumm Nathan Allyn
Physics Dpt
Univ of Cincinnati
210 Braunstein Ml 11
Cincinnati OH 45221 0111
USA
☎ 1 513 475 2232
✆ 1 513 556 3425

Krupp Edwin C
Griffith Observ
2800 East Observ Rd
Los Angeles CA 90027
USA
☎ 1 213 664 1181
✆ 1 213 663 4323

Kruszewski Andrzej
Astronomical Observ
Warsaw Univ
Al Ujazdowskie 4
PL 00 478 Warsaw
Poland
☎ 48 22 29 4011
✆ 48 22 29 4697

Krygier Bernard
Inst Radio Astronomy
N Copernicus Univ
Ul Gagarina 11
PL 87 100 Torun
Poland
☎ 48 56 78 3327
✆ 48 56 11 651

Kryvodubskyj Valery
Astronomical Observ
Kyiv State Univ
Observatorna Str 3
254053 Kyiv
Ukraine
☎ 380 4421 62630
✉ aoku@gluk.apc.org

Krzeminski Wojciech
CIW
Las Campanas Observ
Casilla 601
La Serena
Chile
☎ 56 51 21 3032

Ksanfomaliti Leonid V
Space Res Inst
Acad Sciences
Profsojuznaya Ul 84/32
117810 Moscow
Russia
☎ 7 095 333 2322/3122
✆ 7 095 310 7023

Kubat Jiri
Astronomical Institute
Czech Acad Sci
Fricova 1
CZ 251 65 Ondrejov
Czech R
☎ 420 204 62 0127
✆ 420 2 88 1611
✉ kubat@sunstel.asu.cas.cz

Kubiak Marcin A
Astronomical Observ
Warsaw Univ
Al Ujazdowskie 4
PL 00 478 Warsaw
Poland
☎ 48 22 29 4011
✆ 48 22 29 4697

Kubicela Aleksandar
Astronomical Observ
Volgina 7
11050 Beograd
Yugoslavia FR
☎ 381 1 419 357
✆ 381 1 419 553

Kubo Yoshio
Hydrographic Dpt
Geodesy/Geophys Div
Tsukiji 5 Chuo Ku
Tokyo 104
Japan
☎ 81 354 13 811
✆ 81 3545 2885
✉ kuyoy@ws09.cue.jhd.go.jp

Kubota Jun
Kwasan/Hida Obs
Kyoto Univ
Yamashina
Kyoto 607
Japan
☎ 81 755 81 1235
✆ 81 75 593 9617

Kucera Ales
Astronomical Institute
Slovak Acad Sci
SK 059 60 Tatranska Lomni
Slovak Republic
☎ 421 969 967 866
✆ 421 969 967 656
✉ akucera@auriga.ta3.sk

Kudritzki Rolf-Peter
Inst Astron/Astrophysik
Univ Sternwarte
Scheinerstr 1
D 81679 Muenchen
Germany
☎ 49 899 890 21
✆ 49 899 220 9427

Kuehne Christoph F
Koeflacher Str 36
D 7928 Giengen/brenz
Germany
☎ 49 732 244 48

Kuehr Helmut
MPI Astronomie
Koenigstuhl
D 69117 Heidelberg
Germany
☎ 49 622 152 81
✆ 49 622 152 8246

Kuenzel Horst
Dieselstr 13
D 14482 Potsdam
Germany
☎ 49 331 77318

Kuhi Leonard V
School Phys/Astronomy
Univ Minnesota
116 Church St SE
Minneapolis MN 55455
USA
☎ 1 612 624 0211
✆ 1 612 624 2029

Kuhn Jeffery Richard
Physics/Astronomy Dpt
Michigan State Univ
East Lansing MI 48824
USA
☎ 1 517 353 2986
✆ 1 517 535 4500
✉ kuhn@msupa.edu

Kuijpers H Jan M E
Sterrekundig Inst Utrecht
Box 80000
NL 3508 TA Utrecht
Netherlands
☎ 31 30 253 5209
✆ 31 30 253 1601
✉ j.kuijpers@astro.uu.nl

Kuin Nicolaas Paulus M
NASA/GSFC
Code 633 2
Greenbelt MD 20771
USA
☎ 1 301 286 0677
✆ 1 301 286 1771
✉ kuin@nssdc.gsfc.nasa.gov

Kuiper Thomas B H
CALTECH/JPL
MS 169 5065
4800 Oak Grove Dr
Pasadena CA 91109 8099
USA
☎ 1 818 354 5479
✆ 1 818 393 6030

Kuklin Georgly V
Sibizmir
Acad Sciences
664697 Irkutsk 33
Russia
☎ 7 395 262 9388

Kulcar Ladislav
Pedagogicka Fakulta
Katedra Fyziky
Tajovskeho 40
SK 975 49 Banska Bystrice
Slovak Republic
☎ 421 88 3 4553
✆ 421 88 3 3132

Kulikova Nelli V
INPE
Studgorodok 1
Kaluga Region
Obninck 249020
Russia
☎ 7 843 970 822
✆ 7 095 255 2225
✉ iate@strom.iasnet.com

Kulkarni Prabhakar V
PRL
Navrangpura
Ahmedabad 380 009
India
☎ 91 272 46 2129
✆ 91 272 44 5292

Kulkarni Shrinivas R
CALTECH
Palomar Obs
MS 105 24
Pasadena CA 91125
USA
☎ 1 818 356 4010
✆ 1 818 568 1517

Kulkarni Vasant K
NCRA/TIFR
Pune Univ Campus Pb 3
Ganeshkhind
Pune 411 007
India
☎ 91 212 33 7107
✆ 91 212 33 5760

Kulsrud Russell M
Dpt Astrophys Sci
Princeton Univ
Princeton NJ 08544 1001
USA
☎ 1 609 683 2613
✆ 1 609 258 1020

Kultima Johannes
EISCAT
Geophysical Observ
FIN 99600 Sodankylae
Finland
☎ 358 16 61 9880
✆ 358 16 61 0375

Kumagai Shiomi
Physics Dpt CST
Nihon Univ
Kanda-Surugadai I-8
Chiyoda ku Tokyo 101
Japan
☎ 81 3 3259 0890
✆ 81 3 3259 0890
✉ kumagai@phys.cst.nihon-u.ac.jp

Kumai Yasuki
Dpt Manager
Kumamoto Gakuen Univ
2 5 1 Oe
Kumamoto 862
Japan
☎ 81 963 64 5161*1544
✆ 81 96 372 0702
✉ kumai@kumagaku.ac.jp

Kumajgorodskaya Raisa N
SAO
Acad Sciences
Nizhnij Arkhyz
357147 Karachaevo
Russia
☎ 7 878 789 3515

Kumar C Krishna
Dpt Phys/Astronomy
Howard Unive
Washington DC 20059
USA
☎ 1 202 636 6245

Kumar Shiv S
University Station
Univ of Virginia
Box 3818
Charlottesville VA 22903 0818
USA
☎ 1 804 924 7494
✆ 1 804 924 3104
✉ ssk9u@virginia.edu

Kumkova Irina I
Inst Appl Astronomy
Acad Sciences
Zhdanovskaya Ul 8
197042 St Petersburg
Russia
☎ 7 812 123 4452
✆ 7 812 230 7413
✉ kumkova@ipa.rssi.ru

Kumsiashvily Mzia I
Abastumani Astrophysical
Observ
Georgian Acad Sci
383762 Abastumani
Georgia
☎ 995 88 32 95 5367
✆ 995 88 32 98 5017

Kumsishvili J I
Abastumani Astrophysical
Observ
Georgian Acad Sci
383762 Abastumani
Georgia
☎ 995 88 32 95 5367
✆ 995 88 32 98 5017

Kun Maria
Konkoly Observ
Thege U 13/17
Box 67
H 1525 Budapest
Hungary
☎ 36 1 375 4122
✆ 36 1 275 4668

Kunchev Peter
Astronomy Dpt
Univ of Sofia
Anton Ivanov St 5
BG 1126 Sofia
Bulgaria
☎ 359 2 68 8176
✆ 359 2 68 9085

Kundt Wolfgang
IAEF
Univ Bonn
auf d Huegel 71
D 53121 Bonn
Germany
☎ 49 228 26 7400
✆ 49 228 73 3672

Kundu Mukul R
Astronomy Program
Univ of Maryland
College Park MD 20742 2421
USA
☎ 1 301 454 3005
✆ 1 301 314 9067

Kunieda Hideyo
Astrophysics Dpt
Nagoya Univ
Furocho Chikusa Ku
Nagoya 464 01
Japan
☎ 81 527 81 5111
✆ 81 527 81 3541

Kunitzsch Paul
Davidstr 17
D 81679 Muenchen
Germany

Kunkel William E
CIW
Las Campanas Observ
Casilla 601
La Serena
Chile
☎ 56 51 21 3032
✉ kunkel@charlie.ctio.
noao.edu

Kunth Daniel
IAP
98bis bd Arago
F 75014 Paris
France
☎ 33 1 44 32 8085
✆ 33 1 44 32 8001

Kunze Ruediger
Inst Theor Phys/
Sternwarte Univ Kiel
Olshausenstr 40
D 24098 Kiel
Germany
☎ 49 431 880 1575
✆ 49 431 880 4432
✉ pas29@rz.uni-kiel.
dbp.de

Kuperus Max
Sterrekundig Inst Utrecht
Box 80000
NL 3508 TA Utrecht
Netherlands
☎ 31 30 253 5211
✆ 31 30 253 1601
✉ m.kuperus@fys.ruu.nl

Kupliauskiene Alicija
Inst Theor Physics/
Astronomy
Gostauto 12
Vilnius 2600
Lithuania
☎ 370 2 612 723
✆ 370 2 224 694
✉ kupl@itpa.fi.lt

Kurfess James D
NRL
Code 4150
4555 Overlook Av SW
Washington DC 20375 5000
USA
☎ 1 202 767 3182

Kuril'chik Vladimir N
SAI
Acad Sciences
Universitetskij Pr 13
119899 Moscow
Russia
☎ 7 095 139 1030
✆ 7 095 939 0126

Kurochka L N
Astronomical Observ
Kyiv State Univ
Observatornaya Ul 3
252053 Kyiv
Ukraine
☎ 380 4421 62691

Kurokawa Hiroki
Kwasan/Hida Obs
Kyoto Univ
Yamashina
Kyoto 607
Japan
☎ 81 75 581 1235
✆ 81 75 593 9617

Kurpinska-Winiarska M
Astronomical Observ
Krakow Jagiellonian Univ
Ul Orla 171
PL 30 244 Krakow
Poland
☎ 48 12 25 1294
✆ 48 12 25 1318

Kurt Vitaliy G
Space Res Inst
Acad Sciences
Profsojuznaya Ul 84/32
117810 Moscow
Russia
☎ 7 095 333 3122
✆ 7 095 310 7023

Kurtanidze Omar
Abastumani Astrophysical
Observ
Georgian Acad Sci
383762 Abastumani
Georgia
☎ 995 88 32 95 5367
✆ 995 88 32 98 5017
✉ okur@abao.kheta.ge

Kurtz Donald Wayne
Astronomy Dpt
Univ of Cape Town
Private Bag
7700 Rondebosch
South Africa
☎ 27 21 650 2394
✆ 27 21 650 3342
✉ dkurtz@uctvax.net.ac.za

Kurtz Michael Julian
CfA
HCO/SAO
60 Garden St
Cambridge MA 02138 1516
USA
☎ 1 617 495 7434
✆ 1 617 495 7476
✉ kurtz@cfa.harvard. edu

Kurucz Robert L
CfA
HCO/SAO
60 Garden St
Cambridge MA 02138 1516
USA
☎ 1 617 495 7429
✆ 1 617 495 7356
✉ kurucz@cfa.harvard.edu

Kurzynska Krystyna
Astronomical Observ
A Mickiewicz Univ
Ul Sloneczna 36
PL 60 286 Poznan
Poland
☎ 48 61 67 9670
✆ 48 61 53 6536
✉ kurzastr@plpuamm

Kus Andrzej Jan
Inst Radio Astronomy
N Copernicus Univ
Ul Gagarina 11
PL 87 100 Torun
Poland
☎ 48 56 78 3327
✆ 48 56 11 651

Kushwaha R S
Mathematics Dpt
Univ of Jodhpur
Jodhpur 342 001
India

Kustaanheimo Paul E
Denmarks Tekn Hojskole
Lundtoftevej 7
DK 2800 Lyngby
Denmark
☎ 45 42 88 3022

Kusunose Masaaki
School of Science
Kwansei Gakuin Univ
1 155 Uegahara Ichiban cho
Nishinomiya 662
Japan
☎ 81 798 54 6454
✆ 81 798 51 0914
✉ z96020@kgupyr.kwansei.ac.uk

Kutner Marc Leslie
RPI
Physics Dpt
Troy NY 12180 3590
USA
☎ 1 518 266 6417

Kutter G Siegfried
NASA GSFC
Code 681
Greenbelt MD 20771
USA
☎ 1 301 286 8701

Kutuzov Sergei A
Astronomical Observ
St Petersburg Univ
Bibliotechnaja Pl 2
199164 St Petersburg
Russia
☎ 7 812 428 6649/315 1701
✆ 7 812 428 4259
✉ kvk@ast.lgu.spb.su

Kuyken Koenraad H
Kapteyn Sterrekundig Inst
Univ Groningen
Postbus 800
NL 9700 Av Groningen
Netherlands
☎ 31 50 363 4073
✆ 31 50 363 6100

Kuzmanoski Mike
Astronomy Dpt
Univ of Belgrade
Studentski Trg 16
11000 Beograd
Yugoslavia FR
☎ 381 11 638 715
✆ 381 11 630 151

Kuzmin Andrei V
SAI
Acad Sciences
Universitetskij Pr 13
119899 Moscow
Russia
☎ 7 095 939 1970
✆ 7 095 932 8841
✉ avk@sai.msu.su

Kuzmin Arkadii D
Lebedev Physical Inst
Acad Sciences
Leninsky Pspt 53
117924 Moscow
Russia
☎ 7 095 135 2250
✆ 7 095 135 7880
✉ akuzmin@prao.psn.ru

Kwee K K
Leiden Observ
Box 9513
NL 2300 RA Leiden
Netherlands
☎ 31 71 527 2727
✆ 31 71 527 5819

Kwitter Karen Beth
Astronomy Dpt
Thompson Physics Lab
Williams College
Williamstown MA 01267
USA
☎ 1 413 597 2272
✆ 1 413 597 3200
✉ karen.b.kwitter@
williams.edu

Kwok Sun
Dpt Phys/Astronomy
Univ of Calgary
2500 University Dr NW
Calgary AB T2N 1N4
Canada
☎ 1 403 220 5414
✆ 1 403 289 3331
✉ kwok@iras.ucalgary.ca

Kylafis Nikolaos D
Physics Dpt
Univ of Crete
Box 1527
GR 711 11 Iraklion
Greece
☎ 30 81 23 9757
✆ 30 81 23 9735

L

La Bonte Barry James
Inst for Astronomy
Univ of Hawaii
2680 Woodlawn Dr
Honolulu HI 96822
USA
☎ 1 808 956 6531
✆ 1 808 988 2790

la Dous Constanze A
Sonneberg Observ
Sternwartestr 32
D 96515 Sonneberg
Germany
☎ 49 367 581 210
✆ 49 367 581 219
✉ cld@stw.tu-ilmenau.de

La Padula Cesare
IAS
Area d Ricerca CNR
Via Fosso Cavaliere 100
I 00133 Roma
Italy
☎ 39 6 4993 4473
✆ 39 6 2066 0188

Labay Javier
Dpt Fisica Atmosfera
Univ Barcelona
Avd Diagonal 645
E 08028 Barcelona
Spain
☎ 34 3 330 7311
✆ 34 3 402 1133

Labeyrie Antoine
OHP
F 04870 S Michel Obs
France
☎ 33 4 92 70 6400
✆ 33 4 92 76 6295
✉ labeyrie@obs-hp.fr

Labeyrie Jacques
CEA CEN, CFR
BP 2
F 91191 Gif/Yvette
France
☎ 33 1 69 7828

Labhardt Lukas
Astronomisches Institut
Univ Basel
Venusstr 7
CH 4102 Binningen
Switzerland
☎ 41 61 205 5415
✆ 41 61 205 5455
✉ labhardt@astro.unibas.ch

Labs Dietrich
Landessternwarte
Koenigstuhl
D 69117 Heidelberg
Germany
☎ 49 622 110 036
✆ 49 622 150 9202

Lacey Cedric
Theoretical Astrophysics Centre
Juliane Maries Vej 30
DK 2100 Copenhagen
Denmark
☎ 45 35 32 5902
✆ 45 35 32 5910
✉ lacey@tac.dk

Lachieze-Rey Marc
DAPNIA/SAP
CEA Saclay
BP 2
F 91191 Gif s Yvette Cdx
France
☎ 33 1 69 08 6292
✆ 33 1 69 08 6577

Laclare Francis
OCA
CERGA
F 06130 Grasse
France
☎ 33 4 93 36 5849
✆ 33 4 93 40 5353

Lacy Claud H
Physics Dpt
Univ of Arkansas
104 Physics Bdg
Fayetteville AR 72701
USA
☎ 1 501 575 2506
✆ 1 501 575 4580
✉ clacy@comp.uark.edu

Lacy John H
Astronomy Dpt
Univ of Texas
Rlm 15 308
Austin TX 78712 1083
USA
☎ 1 512 471 1469
✆ 1 512 471 6016

Lada Charles Joseph
CfA
HCO/SAO MS 72
60 Garden St
Cambridge MA 02138 1516
USA
☎ 1 617 495 7017
✆ 1 617 495 7356
✉ clada@cfa.harvard.edu

Laffineur Marius
21 bd Brune
F 75014 Paris
France

Lafon Jean-Pierre J
Observ Paris Meudon
DASGAL
Pl J Janssen
F 92195 Meudon PPL Cdx
France
☎ 33 1 45 07 7858
✆ 33 1 45 07 7971

Lagerkvist Claes-Ingvar
Astronomical Observ
Box 515
S 751 20 Uppsala
Sweden
☎ 46 18 55 3780
✆ 46 18 527 583
✉ classe@astro.uu.se

Lagerqvist Albin
Inst Theoret Phys
Vanadisvaegen 9
S 113 46 Stockholm
Sweden
☎ 46 8 16 4500

Lago Maria Teresa V T
Centro Astrofisica
Univ d Porto
rua d Campo Alegre 823
P 4150 Porto
Portugal
☎ 351 2 600 7081
✆ 351 2 600 7082
✉ mtlago@astro.up.pt

Lagrange Anne-Marie
Observ Grenoble
Lab Astrophysique
BP 53x
F 38041 S Martin Heres Cdx
France
☎ 33 4 76 51 4203
✆ 33 4 76 44 8821
✉ lagrange@

Lahav Ofer
Inst of Astronomy
The Observatories
Madingley Rd
Cambridge CB3 0HA
UK
☎ 44 12 23 337 548
✆ 44 12 23 337 523
✉ ol1@ast-star.cam.ac.uk

Lahulla J Fornies
Observ Astronomico
Nacional
Alfonso XII-3
E 28014 Madrid
Spain
☎ 34 1 227 0107

Lai Sebastiana
Istt Astronomia
Univ d Cagliari
Via Ospedale 72
I 09100 Cagliari
Italy
☎ 39 70 66 3544
✆ 39 70 72 5425

Laing Robert
Royal Greenwich Obs
Madingley Rd
Cambridge CB3 0EZ
UK
☎ 44 12 23 37 4000
✆ 44 12 23 374 700

Laird John B
Dpt Phys/Astronomy
Bowling Green State Univ
Bowling Green OH 43403
USA
☎ 1 419 372 7244
✆ 1 419 372 9938
✉ laird@tycho.bgsu.edu

Lake Kayll William
Physics Dpt
Queen's Univ
Kingston ON K7L 3N6
Canada
☎ 1 613 547 3020
✆ 1 613 545 6463

Lal Devendra
PRL
Navrangpura
Ahmedabad 380 009
India
☎ 91 272 46 2129
✆ 91 272 44 5292

Lala Petr
Office for Outer Space
UN Office in Vienna
Bg F 0818 Box 500
A 1400 Wien
Austria
☎ 43 1 211 31 4952
✆ 43 1 213 45 5830
✉ plala@unov.un.or.at

Lallement Rosine
Service Aeronomie
BP 3
F 91371 Verrieres Buisson
France
☎ 33 1 64 47 4235

Lamb Frederick K
Physics Dpt
Univ of Illinois
1110 W Green St
Urbana IL 61801 3080
USA
☎ 1 217 333 6363
✆ 1 217 333 4990
✉ slamb@rigel.astro.uiuc.edu

Lamb Richard C
Physics Dpt
Iowa State Univ
Ames IA 50011
USA
☎ 1 515 294 3873
✆ 1 515 294 3262

Lamb Susan Ann
Dpt Astron/Physics
Univ of Illinois
1002 W Green St
Urbana IL 61801
USA
☎ 1 217 333 5550
✆ 1 217 244 7638
✉ slamb@rigel.astro.uiuc.edu

Lamb Donald Quincy
Dpt Astron/Astrophys
Univ of Chicago
5801 S Ellis Av
Chicago IL 60637
USA
☎ 1 312 702 7194
✆ 1 312 702 2812
✉ lamb@oddjob.uchicago.edu

Lambeck Kurt
Australian Ntl Univ
Res School Earth Science
Box 4
Canberra ACT 2600
Australia
☎ 61 2 62 49 2487

Lambert David L
Astronomy Dpt
Univ of Texas
Rlm 15 308
Austin TX 78712 1083
USA
☎ 1 512 471 7438
✆ 1 512 471 6016
✉ dll@astro.as.utexas.edu

Lamers Henny J G L M
SRON
Postbus 800
Sorbonnelaan 2
NL 3584 CA Utrecht
Netherlands
☎ 31 30 253 5720
✆ 31 30 254 0860
✉ hennyl@sron.ruu.nl

Lamla Erich E
Bruesselerstr 9
D 5300 Bonn 1
Germany

Lamontagne Robert
Dpt d Physique
Universite Montreal
CP 6128 Succ A
Montreal QC H3C 3J7
Canada
☎ 1 514 342 7273
✆ 1 514 343 2071
✉ 5007@cc.umontreal.ca

Lampens Patricia
Koninklijke Sterrenwacht van Belgie
Ringlaan 3
B 1180 Brussels
Belgium
☎ 32 2 373 0263
✆ 32 2 374 9822
✉ patricia.lampens@oma.be

Lampton Michael
Space Sci Laboratory
Univ of California
Grizzly Peak Blvd
Berkeley CA 94720 7950
USA
☎ 1 510 642 3576
✆ 1 510 643 8303
✉ mlampton@ssl.berkeley.edu

Lamy Philippe
LAS
Traverse du Siphon
Les Trois Lucs
F 13376 Marseille Cdx 12
France
☎ 33 4 91 05 5932
✆ 33 4 91 66 1855
✉ lamy@frlasm51

Lancaster Brown Peter
10a St Peter's Rd
Aldeburgh Suffolk
UK

Lanciano Nicoletta
Dpt Matematica
Univ d Roma La Sapienza
Pl A Moro 2
I 00185 Roma
Italy
☎ 39 6 49 91 3254
✆ 39 6 4470 1007
✉ lanciano@uniroma1.it

Lancon Ariane
Observ Strasbourg
11 r Universite
F 67000 Strasbourg
France
☎ 33 3 88 15 0710
✆ 33 3 88 15 0740
✉ lancon@astro.u-strasbg.fr

Lande Kenneth
Physics Dpt
Univ of Pennsylvania
Philadelphia PA 19104
USA
☎ 1 215 898 8177
✆ 1 215 898 9336

Landecker Peter Bruce
Hughes Space/Comm Co
Bg S50 MS X 375
Box 92919
Los Angeles CA 90009
USA
☎ 1 310 364 8725
✆ 1 310 364 5027
✉ pblandecker@ccgate.hac.com

Landecker Thomas L
DRAO
NRCC/HIA
Box 248
Penticton BC V2A 6K3
Canada
☎ 1 250 493 2277
✆ 1 250 493 7767

Landi Degl-Innocenti E
Dpt Astronomia
Univ d Firenze
Largo E Fermi 5
I 50125 Firenze
Italy
☎ 39 55 27 521
✆ 39 55 22 0039

Landini Massimo
Osserv Astrofis Arcetri
Univ d Firenze
Largo E Fermi 5
I 50125 Firenze
Italy
☎ 39 55 27 52247
✆ 39 55 22 0039

Landman Donald Alan
Mission Res Corp
735 State St
PO Drawer 719
Santa Barbara CA 93102
USA

Landolfi Marco
Osserv Astrofis Arcetri
Univ d Firenze
Largo E Fermi 5
I 50125 Firenze
Italy
☎ 39 55 275 2256
✆ 39 55 22 0039
✉ landolfi@arcetri.astro.it

Landolt Arlo U
Dpt Phys/Astronomy
Louisiana State Univ
Baton Rouge LA 70803 4001
USA
☎ 1 504 388 8276
✆ 1 504 334 1098
✉ landolt@rouge.phys.lsu.edu

Landstreet John D
Dpt Phys/Astronomy
Univ W Ontario
London ON N6A 3K7
Canada
☎ 1 519 679 2111*6707
✆ 1 519 661 2033
✉ jlandstr@uwo.ca

Lane Adair P
CfA
HCO/SAO
60 Garden St
Cambridge MA 02138 1516
USA
☎ 1 617 496 7654
✆ 1 617 495 7356
✉ adain@cfa.harvard.edu

Lane Arthur Lonne
CALTECH/JPL
4800 Oak Grove Dr
Pasadena CA 91109 8099
USA
☎ 1 818 345 2725
✆ 1 818 393 6030

Laney Clifton D
SAAO
PO Box 9
7935 Observatory
South Africa
☎ 27 21 47 0025
✆ 27 21 47 3639

Lang James
Rutherford Appleton Lab
Space/Astrophysics Div
Bg R25/R68
Chilton Didcot OX11 0QX
UK
☎ 44 12 35 821 900
✆ 44 12 35 445 808

Lang Kenneth R
Dpt Phys/Astronomy
Tufts Univ
Robinson Hall
Medford MA 02155
USA
☎ 1 617 381 3390

Lang Mark
Physics Dpt
Univ College
Galway
Ireland
☎ 353 91 524 411*3241
✆ 353 91 525 700
✉ mark.lang@ucg.ie

Langer George Edward
Physics Dpt
Colorado College
Colorado Springs CO 80903
USA
☎ 1 303 473 2233*578

Langer Norbert
Inst Theor Phys/Astroph
Univ Potsdam
am Neuen Palais 10
D 14469 Potsdam
Germany
✉ ntl@astro-physik.uni-potsdam.de

Langer William David
CALTECH/JPL
MS 169 506
4800 Oak Grove Dr
Pasadena CA 91109 8099
USA
☎ 1 818 354 5823
✆ 1 818 393 6030

Langhoff Stephen Robert
NASA/ARC
MS 230 3
Moffett Field CA 94035 1000
USA
☎ 1 415 604 6213
✆ 1 416 604 0350
✉ langhoff@pegasms.arc.nasa.gov

Lannes Andre
OMP
14 Av E Belin
F 31400 Toulouse Cdx
France
☎ 33 5 61 25 2101
✆ 33 5 61 27 3179

Lanning Howard Hugh
STScI/CSC
Homewood Campus
3700 San Martin Dr
Baltimore MD 21218
USA
☎ 1 301 338 4486
✆ 1 301 338 4767
✉ lanning@stsci.edu

Lantos Pierre
Observ Paris Meudon
DASOP
Pl J Janssen
F 92195 Meudon PPL Cdx
France
☎ 33 1 45 07 7767
✆ 33 1 45 07 7971
✉ pierre.lantos@obspm.fr

Lanz Thierry
NASA GSFC
Code 681
Greenbelt MD 20771
USA
☎ 1 301 286 5920
✆ 1 301 286 1753
✉ lanz@stars.gsfc.nasa.gov

Lanza Antonino Francesco
Osserv Astronomico
d Catania
Via A Doria 6
I 95125 Catania
Italy
☎ 39 95 733 2238
✆ 39 95 33 0592
✉ nlanza@alpha4.ct.astro.it

Lapasset Emilio
Observ Astronomico
de Cordoba
Laprida 854
5000 Cordoba
Argentina
☎ 54 51 23 0491
✆ 54 51 21 0613
✉ lapasset@astro.edu.ar

Lapuente Pilar Ruiz
Dpt Astronom Meteo
Univ Barcelona
Avd Diagonal 647
E 08028 Barcelona
Spain
☎ 34 3 402 1121
✆ 34 3 402 1133
✉ pilar@farcm0.ub.es

Lapushka K K
Astronomical Observ
Latvian State Univ
Rainis Bul 19
LV 226098 Riga
Latvia
☎ 371 13 2 223149/611984
✆ 371 782 0180
✉ lapushka@astr2.lu.lv

Laques Pierre
OMP
9 r Pont de La mouette
F 65200 Bagneres Bigorre
France
☎ 33 5 62 95 1969
✆ 33 3 88 25 0160

Lara Martin
ROA
Cecilio Pujazon 22-3 A
E 11110 San Fernando Cadiz
Spain
☎ 34 5 659 9367
✆ 34 5 659 9366
✉ mlara@roast.roa.es

Large Michael I
School of Physics
UNSW
Sydney NSW 2006
Australia
☎ 61 2 692 2222
✆ 61 2 660 2903

Lari Carlo
Istt Radioastronomia
CNR
Via P Gobetti101
I 40129 Bologna
Italy
☎ 39 51 639 9385
✆ 39 51 639 9385

Larionov Mikhael G
SAI
Acad Sciences
Universitetskij Pr 13
119899 Moscow
Russia
☎ 7 095 939 2378
✆ 7 095 932 8841
✉ mgl@sai.msu.su

Larson Harold P
Lunar/Planetary Lab
Room 325 Bg 92
Univ of Arizona
Tucson AZ 85721 0092
USA
☎ 1 520 621 6943
✆ 1 520 621 4933

Larson Richard B
Astronomy Dpt
Yale Univ
Box 208101
New Haven CT 06520 8101
USA
☎ 1 203 432 3015
✆ 1 203 432 5048
✉ larson@astro.yale.edu

Larson Stephen M
Lunar/Planetary Lab
Room 325 Bg 92
Univ of Arizona
Tucson AZ 85721 0092
USA
☎ 1 520 621 4973
✆ 1 520 621 4933

Larsson Stefan
Stockholm Observ
Royal Swedish Acad Sciences
S 133 36 Saltsjoebaden
Sweden
☎ 46 8 716 4464
✆ 46 8 717 4719
✉ larsson@astro.su.se

Larsson-Leander Gunnar
Lund Observ
Box 43
S 221 00 Lund
Sweden
☎ 46 46 222 7000
✆ 46 46 222 4614

Lasala Gerald J
Physics Dpt
Univ Southern Maine
96 Falmouth St
Portland ME 04103
USA
☎ 1 207 780 4557
✆ 1 207 780 4933
✉ lasala@portland

Lasenby Anthony
MRAO
Cavendish Laboratory
Madingley Rd
Cambridge CB3 0HE
UK
☎ 44 12 23 337 274
✆ 44 12 23 354 599

Lasher Gordon Jewett
IBM
T J Watson Res Ctr
Box 218
Yorktown Heights NY 10598
USA

Laskar Jacques
BDL
77 Av Denfert Rochereau
F 75014 Paris
France
☎ 33 1 40 51 2274
✆ 33 1 46 33 2834
✉ laskar@bdl.fr

Laskarides Paul G
Astronomy Dpt
Ntl Univ Athens
Panepistimiopolis
GR 157 84 Zografos
Greece
☎ 30 1 724 3211
✆ 30 1 723 5122
✉ plaskar@atlas.uoa.gr

Lasker Barry M
STScI
Homewood Campus
3700 San Martin Dr
Baltimore MD 21218
USA
☎ 1 301 338 4840
✆ 1 301 338 5075
✉ lasker@stsci.edu

Lasota Jean-Pierre
Observ Paris Meudon
DARC
Pl J Janssen
F 92195 Meudon PPL Cdx
France
☎ 33 1 45 07 7416
✆ 33 1 45 07 7971

Latham David W
CfA
HCO/SAO
60 Garden St
Cambridge MA 02138 1516
USA
☎ 1 617 495 7215
✆ 1 617 495 7467
✉ dlatham@cfa.harvard.edu

Latour Jean J
OMP
14 Av E Belin
F 31400 Toulouse Cdx
France
☎ 33 5 61 25 2101
✆ 33 5 61 27 3179

Lattanzi Mario G
Osserv Astronomico
d Torino
St Osservatorio 20
I 10025 Pino Torinese
Italy
☎ 39 11 461 9000
✆ 39 11 461 9030
✉ lattanzi@to.asto.it

Lattanzio John
Mathematics Dpt
Monash Univ
Wellington Rd
Clayton VIC 3168
Australia
☎ 61 3 9905 4428
✆ 61 3 9905 4930
✉ johnl@flash.maths.monash.edu.au

Latter William B
IPAC/CALTECH
MS 100 22
Pasadena CA 91125
USA
☎ 1 626 397 7356
✆ 1 626 397 9600
✉ latter@ipac.caltech.edu

Lattimer James M
Dpt Earth/Space Sci
Astronomy Program
Suny at Stony Brook
Stony Brook NY 11794 2100
USA
☎ 1 516 632 8227
℻ 1 516 632 8176
✉ lattimer@astro.sunysb.edu

Latypov A A
Astronomical Institute
Uzbek Acad Sci
Astronomicheskaya Ul 33
700052 Tashkent
Uzbekistan
☎ 7 3712 35 8102
✉ admin@astrin.silk.glas.opc.org

Lauberts Andris
FOA 3
Box 1165
S 581 11 Linkoping
Sweden
☎ 46 13 11 8235
℻ 46 13 13 1665

Launay Francoise
Observ Paris Meudon
DASGAL
Pl J Janssen
F 92195 Meudon PPL Cdx
France
☎ 33 1 45 07 7550
℻ 33 1 45 07 7121
✉ francoise.launay@obspm.fr

Launay Jean-Michel
Observ Paris Meudon
Pl J Janssen
F 92195 Meudon PPL Cdx
France
☎ 33 1 45 07 7554
℻ 33 1 45 07 7971

Laureijs Rene J
ESA
Apt 50727 Villafranca
E 28080 Madrid
Spain
☎ 34 1 813 1367
℻ 34 1 813 1308
✉ rlaurey@iso.vilspa.esa.es

Laurent Bertel E
Inst Theoret Phys
Vanadisvaegen 9
S 113 46 Stockholm
Sweden
☎ 46 8 16 4500

Laurent Claudine
Observ Paris
DEMIRM
61 Av Observatoire
F 75014 Paris
France
☎ 33 1 40 51 2221
℻ 33 1 44 54 1804

Laurikainen Eija
Dpt Geosc/Astronomy
Univ of Oulu
Box 333
FIN 90570 Oulu
Finland
☎ 358 81 553 1936
℻ 358 81 553 1934
✉ eija@seita.oulu.fi

Lausberg Andre
Inst Astrophysique
Universite Liege
Av Cointe 5
B 4000 Liege
Belgium
☎ 32 4 254 7510
℻ 32 4 254 7511
✉ lausberg@astro.ulg.ac.be

Lautman D A
CfA
HCO/SAO
60 Garden St
Cambridge MA 02138 1516
USA
☎ 1 617 495 7461
℻ 1 617 495 7356

Laval Annie
Observ Marseille
2 Pl Le Verrier
F 13248 Marseille Cdx 04
France
☎ 33 4 91 95 9088
℻ 33 4 91 62 1190

Lavrov Mikhail I
Engelhardt Astronom Observ
Observatoria Station
422526 Kazan
Russia
☎ 7 32 4827

Lavrukhina Augusta K
Vernadsky Inst Geochem/Analytical Chemistry
Kosygin Str 19
117334 Moscow
Russia
☎ 7 095 137 7538

Lawrence Andrew
Royal Observ
Blackford Hill
Edinburgh EH9 3HJ
UK
☎ 44 131 668 8346
℻ 44 131 668 8356
✉ al@roe.ac.uk

Lawrence Charles R
CALTECH
Palomar Obs
MS 105 24
Pasadena CA 91125
USA
☎ 1 818 356 4976
℻ 1 818 568 1517

Lawrence G M
LASP
Univ of Colorado
Campus Box 392
Boulder CO 80309 0392
USA

Lawrence John Keeler
Dpt Phys/Astronomy
San Fernando Observ
California State Univ.
Northridge CA 91330 8268
USA
☎ 1 818 541 0649
℻ 1 818 541 0663
✉ jlawrence@galileo.csun.edu

Lawrie David G
Aerospace Corp
MS M4 041
Box 92957
Los Angeles CA 90009 2957
USA
☎ 1 310 648 6142
℻ 1 310 336 1636

Lawson Warrick
Senior Lecturer
School of Physics
Australian Defense Force Academy
Canberra ACT 2600
Australia
☎ 61 2 62 688 810
℻ 61 26 268 8786
✉ w-lawson@adfa.oz.au

Layden Andrew Choisy
Astronomy Dpt
Univ of Michigan
Dennison Bldg
Ann Arbor MI 48109 1090
USA
☎ 1 313 763 6318
℻ 1 313 763 6317
✉ layden@astro.lsa.umich.edu

Layzer David
CfA
HCO/SAO MS 31
60 Garden St
Cambridge MA 02138 1516
USA
☎ 1 617 495 7000
℻ 1 617 495 7356

Lazareff Bernard
Observ Grenoble
Lab Astrophysique
BP 53x
F 38041 S Martin Heres Cdx
France
☎ 33 4 76 51 4600
℻ 33 4 76 44 8821

Lazaro Carlos
IAC
Observ d Teide
via Lactea s/n
E 38200 La Laguna
Spain
☎ 34 2 232 9100
℻ 34 2 232 9117

Lazovic Jovan P
Astronomy Dpt
Univ of Belgrade
Studentski Trg 16
11000 Beograd
Yugoslavia FR
☎ 381 11 638 715
℻ 381 11 630 151

Lazzaro Daniela
Observ Nacional
Rua Gl Bruce 586
San Cristovao
20921 030 Rio de Janeiro RJ
Brazil
☎ 55 21 580 7181
℻ 55 21 580 0332
✉ daza@lnccvm

Le Borgne Jean-Francois
OMP
14 Av E Belin
F 31400 Toulouse Cdx
France
☎ 33 5 61 33 2929
℻ 33 5 61 53 6722
✉ leborgne@obs-mip.fr

Le Bourlot Jacques
Observ Paris Meudon
DAEC
Pl J Janssen
F 92195 Meudon PPL Cdx
France
☎ 33 1 45 07 7566
℻ 33 1 45 07 7469
✉ lebourlot@obspm.fr

Le Contel Jean-Michel
OCA
Observ Nice
BP 139
F 06304 Nice Cdx 4
France
☎ 33 4 93 89 0420
℻ 33 4 92 00 3033

Le Dourneuf Maryvonne
Observ Paris Meudon
Pl J Janssen
F 92195 Meudon PPL Cdx
France
☎ 33 1 45 07 7555
℻ 33 1 45 07 7971

Le Fevre Olivier
Observ Paris Meudon
DAEC
Pl J Janssen
F 92195 Meudon PPL Cdx
France
☎ 33 1 45 07 7555
℻ 33 1 45 07 7469

Le Floch Andre
Dpt d Physique
Universite Tours
F 37200 Tours
France
✉ lefloch@univ-tours.fr

Le Poole Rudolf S
Leiden Observ
Box 9513
NL 2300 RA Leiden
Netherlands
☎ 31 71 527 5835
℻ 31 71 527 5819
✉ lepoole@strw.leidenuniv.nl

Le Squeren Anne-Marie
GRAAL
Univ Montpellier II
Pl E Bataillon
F 34095 Montpellier Cdx 5
France
☎ 33 4 67 14 3567
℻ 33 4 67 54 4850

Lea Susan Maureen
Dpt Phys/Astronomy
San Fransisco State Univ
1600 Holloway Av
San Francisco CA 94132
USA
☎ 1 415 338 1655
℻ 1 415 338 2178
✉ lea@stars.sfsu.edu

Leach Sydney
Observ Paris Meudon
DAMAP
Pl J Janssen
F 92195 Meudon PPL Cdx
France
☎ 33 1 45 07 7561
℻ 33 1 45 07 7100
✉ leach@obspm.fr

Leacock Robert Jay
Astronomy Dpt
Univ of Florida
211 SSRB
Gainesville FL 32611
USA
☎ 1 904 392 2052
℻ 1 904 392 5089

Leahy Denis A
Physics Dpt
Univ of Calgary
2500 University Dr NW
Calgary AB T2N 1N4
Canada
☎ 1 403 220 7192
℻ 1 403 289 3331

Leahy J Patrick
NRAL
Univ Manchester
Jodrell Bank
Macclesfield SK11 9DL
UK
☎ 44 14 777 1321
℻ 44 147 757 1618
✉ jpl@uk.ac.man.
jb.star

Leblanc Yolande
Observ Paris Meudon
DESPA
Pl J Janssen
F 92195 Meudon PPL Cdx
France
☎ 33 1 45 07 7759
℻ 33 1 45 07 7971

Lebofsky Larry Allen
Lunar/Planetary Lab
Room 325 Bg 92
Univ of Arizona
Tucson AZ 85721 0092
USA
☎ 1 520 621 6947
℻ 1 520 621 4933

Lebovitz Norman R
Mathematics Dpt
Univ of Chicago
5734 S University Av
Chicago IL 60637
USA
☎ 1 312 702 7329

Lebre Agnes
GRAAL
Univ Montpellier II
Pl E Bataillon
F 34095 Montpellier Cdx 5
France
☎ 33 4 67 14 4735
℻ 33 4 67 14 4535
✉ lebre@graal.univ-
montp2.fr

Lebreton Yveline
Observ Paris Meudon
DASGAL
Pl J Janssen
F 92195 Meudon PPL Cdx
France
☎ 33 1 45 07 7859
℻ 33 1 45 07 7971
✉ lebreton@obspm.fr

Lecar Myron
CfA
HCO/SAO
60 Garden St
Cambridge MA 02138 1516
USA
☎ 1 617 495 7251
℻ 1 617 495 7356

Leckrone David S
NASA GSFC
Code 681
Greenbelt MD 20771
USA
☎ 1 301 286 8904
℻ 1 301 286 8709

Lederle Trudpert
ARI
Moenchhofstr 12-14
D 69120 Heidelberg
Germany
☎ 49 622 140 50
℻ 49 622 140 5297
✉ s00@mvs.urz.uni-
heidelberg.de

Ledlow Michael James
Dpt Phys/Astronomy
Univ New Mexico
800 Yale Blvd NE
Albuquerque NM 87131
USA
☎ 1 505 277 4521
℻ 1 505 277 1520
✉ mledlow@wombat.
phys.unm.edu

Lee Dong Hun
Dpt Astron/Space Sci
Kyunghee Univ
Yong In Kun
Kyunggee 449 701
Korea RP
☎ 82 331 86 131
✉ dhlee@nms.kyunghee.ac.kr

Lee Hyung Mok
Dpt Earth Sciences
Pusan Ntl Univ
Kum Jong Ku
Pusan 609 735
Korea RP
☎ 82 515 10 2702
℻ 82 515 13 7495
✉ hmlee@hyowon.
pusan.ac.kr

Lee Jong Truenliang
Taiwan
China R

Lee Myung Gyoon
Astronomy Dpt
Seoul Ntl Univ
Kwanak Ku
Seoul 151 742
Korea RP
☎ 82 288 06 621
℻ 82 288 71435
✉ mglee@astrog.
snu.ac.kr

Lee Paul D
Dpt Phys/Astronomy
Louisiana State Univ
Baton Rouge LA 70803 4001
USA
☎ 1 504 388 2261
℻ 1 504 388 5855

Lee Sang Gak
Astronomy Dpt
Seoul Ntl Univ
Kwanak Ku
Seoul 151
Korea RP
☎ 82 877 21 31/2139
℻ 82 288 7135

Lee See-woo
Astronomy Dpt
Seoul Ntl Univ
Kwanak Ku
Seoul 151
Korea RP
☎ 82 877 21 31/9*3308
℻ 82 288 71435

Lee Terence J
Royal Observ
Head of Technology
Blackford Hill
Edinburgh EH9 3HJ
UK
☎ 44 131 667 3321
℻ 44 131 668 8356

Lee Thyphoon
Inst Earth Sciences
Academia Sinica
Box 23 59
Taipei 107
China R
☎ 886 2 396 3211

Lee Woo-baik
Korea Astronomy Obs/ISSA
36 1 Whaam Dong
Yuseong Gu
Taejon 305 348
Korea RP
☎ 82 428 61 5611
℻ 82 428 61 5610

Lee Yong-Sam
Dpt Astronomy/Space Sci
Chungbuk Ntl Univ
San 48 Gaeshin Dong
Cheongju 360 763
Korea RP
☎ 82 431 61 2314
℻ 82 431 67 4232
✉ yslee@cbucc.cbnu.ac.kr

Lee Young Wook
Astronomy Dpt
Yonsei Univ
Shinchon 134 Seodaemoon K
Seoul 120 749
Korea RP
☎ 82 236 12 689
℻ 82 231 35 033
✉ ywlee@galaxy.yonsei.ac.kr

Leedjaerv Laurits
Tartu Observ
Estonian Acad Sci
Toravere
EE 2444 Tartumaa
Estonia
☎ 372 7 410 343
℻ 372 7 410 205
✉ leed@aai.ee

Leer Egil
Inst Theor Astrophys
Univ of Oslo
Box 1029
N 0315 Blindern Oslo 3
Norway
☎ 47 22 856 503
℻ 47 22 85 6505

Lefebvre Michel
CNES/GRGS/BGI
18 Av E Belin
F 31055 Toulouse Cdx
France
☎ 33 5 61 27 3131
℻ 33 5 61 27 3179

Lefevre Jean
OCA
Observ Nice
BP 139
F 06304 Nice Cdx 4
France
☎ 33 4 93 89 0420
℻ 33 4 92 00 3033

Leger Alain
IAS
Bt 121
Universite Paris XI
F 91405 Orsay Cdx
France
☎ 33 1 69 85 8580
℻ 33 1 69 85 8675
✉ leger@iaslab.ias.fr

Legg Thomas H
NRCC/HIA
100 Sussex Dr
Ottawa ON K1A 0R6
Canada
☎ 1 613 993 6060
℻ 1 613 952 6602

Lehman Holger
Thueringer Landessternwarte
Sternwarte 5
D 07778 Tautenburg
Germany
☎ 49 364 278 630
℻ 49 364 278 6329
✉ lehm@tls-tautenburg.de

Lehmann Marek
Astronomical Latitude Observ
Polish Acad Sciences
Borowiec
PL 62 035 Kornik
Poland
☎ 48 61 17 0187
℻ 48 61 17 0219
✉ marek@cbk.paznan.pl

Lehnert B P
Dpt Plasma Physics
Royal Inst of Technology
S 100 44 Stockholm 70
Sweden
☎ 46 87 87 7763

Lehto Harry J
Turku Univ
Tuorla Observ
Vaeisaelaentie 20
FIN 21500 Piikkio
Finland
☎ 358 2 274 4263
℻ 358 2 243 3767
✉ hlehto@utu.fi

Leibacher John
NOAO/NSO
Box 26732
950 N Cherry Av
Tucson AZ 85726 6732
USA
☎ 1 520 325 9302
℻ 1 520 325 9305
✉ jleibacher@noao.arizona.edu

Leibowitz Elia M
Dpt Phys/Astronomy
Tel Aviv Univ
Ramat Aviv
Tel Aviv 69978
Israel
☎ 972 3 413 788
℻ 972 3 640 8179

Leibundgut Bruno
ESO
Karl Schwarzschildstr 2
D 85748 Garching
Germany
☎ 49 893 200 6295
℻ 49 893 202 362

Leighly Karen Marie
Astronomy Dpt
Columbia Univ
538 W 120th St
New York NY 10027
USA
☎ 1 212 854 3927
℻ 1 212 854 8121
✉ leighly@postman.riken.go.jp

Leikin Grigerij A
Inst of Astronomy
Acad Sciences
Pyatnitskaya Ul 48
109017 Moscow
Russia
☎ 7 095 231 5461
℻ 7 095 230 2081

Leinert Christoph
MPI Astronomie
Koenigstuhl
D 69117 Heidelberg
Germany
☎ 49 622 152 8264
℻ 49 622 152 8246
✉ leinert@mpia-hd.mpg.de

Leisawitz David
NASA GSFC
Code 631
Greenbelt MD 20771
USA
☎ 1 301 286 0807
℻ 1 301 286 1771
✉ leisawitz@stars.gsfc.nasa.gov

Leister Nelson Vani
IAG
Univ Sao Paulo
CP 9638
01065 970 Sao Paulo SP
Brazil
☎ 55 11 577 8599
℻ 55 11 276 3848
✉ leister@vax.lagusp.usp.br

Leite Scheid Paulo
Observ Nacional
Rua Gl Bruce 586
Sao Cristovao
20921 030 Rio de Janeiro RJ
Brazil
☎ 55 21 580 7313
℻ 55 21 580 0332

Leitherer Claus
STScI
Homewood Campus
3700 San Martin Dr
Baltimore MD 21218
USA
☎ 1 301 338 4425
℻ 1 301 338 4767
✉ leithererstsci.edu

Lekht Evueni
INAOE
Tonantzintlaz
Apdo 51 y 216
Puebla PUE 72000
Mexico
☎ 52 2 247 2011
℻ 52 2 247 2231

Lelievre Gerard
Observ Paris
DASGAL
61 Av Observatoire
F 75014 Paris
France
☎ 33 1 40 51 2255
℻ 33 1 40 51 2232
✉ lelievre@obspm.fr

Lemaire Jean-louis
Observ Paris Meudon
DAMAP
Pl J Janssen
F 92195 Meudon PPL Cdx
France
☎ 33 1 45 07 7563
℻ 33 1 45 07 7469

Lemaire Joseph F
IASB
Ringlaan 3
B 1180 Brussels
Belgium
☎ 32 2 373 0407
℻ 32 2 374 8423
✉ jl@oma.be

Lemaire Philippe
IAS
Bt 121
Universite Paris XI
F 91405 Orsay Cdx
France
☎ 33 1 69 85 8622
℻ 33 1 69 85 8675
✉ lemaire@iaslab.ias.fr

Lemaitre Anne
Dpt Mathematique
FUNDP
Rempart de la Vierge 8
B 5000 Namur
Belgium
☎ 32 81 72 4908
℻ 32 81 72 4914
✉ alemaitre@math.fundp.ac.be

Lemaitre Gerard R
Observ Marseille
2 Pl Le Verrier
F 13248 Marseille Cdx 04
France
☎ 33 4 91 95 9088
℻ 33 4 91 62 1190

Lemke Dietrich
MPI Astronomie
Koenigstuhl
D 69117 Heidelberg
Germany
☎ 49 622 152 8259
℻ 49 622 152 8246
✉ lemke@mpia-hd.mpg.de

Lemke Michael
Dr Remeis Sternwarte
Univ Erlangen-Nuernberg
Sternwartstr 7
D 96049 Bamberg
Germany
☎ 49 951 952 2216
℻ 49 951 952 2222
✉ ai26@a400.sterwarte.uni-erlangen.de

Lena Pierre J
Observ Paris Meudon
DESPA
Pl J Janssen
F 92195 Meudon PPL Cdx
France
☎ 33 1 45 07 7719
℻ 33 1 45 07 7971

Lenhardt Helmut
ARI
Moenchhofstr 12-14
D 69120 Heidelberg
Germany
☎ 49 622 140 5251
℻ 49 622 140 5297
✉ s29@mvs.urz.uni-heidelberg.de

Lenzen Rainer
MPI Astronomie
Koenigstuhl
D 69117 Heidelberg
Germany
☎ 49 622 152 80
℻ 49 622 152 8246
✉ lenzen@mpia-hd.mpg.de

Leonard Peter James T
NASA/GSFC
Code 660 1
Greenbelt MD 20771
USA
☎ 1 301 286 7632
℻ 1 301 286 1681
✉ leonard@cossc.gsfc.nasa.gov

Leone Francesco
Osserv Astronomico
d Catania
Via A Doria 6
I 95125 Catania
Italy
☎ 39 95 733 2229
℻ 39 95 33 0592
✉ fleone@astrct.ct.astro.it

Leorat Jacques
Observ Paris Meudon
DAEC
Pl J Janssen
F 92195 Meudon PPL Cdx
France
☎ 33 1 45 07 7421
℻ 33 1 45 07 7971

Lepine Jacques R D
IAG
Univ Sao Paulo
CP 9638
01065 970 Sao Paulo SP
Brazil
☎ 55 11 275 3720
℻ 55 11 276 3848

Lepp Stephen H
Physics Dpt
Univ of Nevada
4505 S Maryland Parkway
Las Vegas NV 89154
USA
☎ 1 702 739 3653
℻ 1 702 739 0804
✉ lepp@physics.unlv.edu

Lequeux James
DEMIRM
ENS
24 r Lhomond
F 75231 Paris Cdx 05
France
☎ 33 1 43 29 1215
℻ 33 1 40 51 2002
✉ aanda@frmeu51

Leroy Bernard
Observ Paris Meudon
DASOP
Pl J Janssen
F 92195 Meudon PPL Cdx
France
☎ 33 1 45 07 7812
✆ 33 1 45 07 7959

Leroy Jean-Louis
OMP
14 Av E Belin
F 31400 Toulouse Cdx
France
☎ 33 5 61 33 2929
✆ 33 5 61 27 3179
✉ leroy@obs-mip.fr

Lesage Alain
Observ Paris Meudon
DASGAL
Pl J Janssen
F 92195 Meudon PPL Cdx
France
☎ 33 1 45 07 7829
✆ 33 1 45 07 7878

Lesch Harold
RAIUB
Univ Bonn
auf d Huegel 69
D 53121 Bonn
Germany
☎ 49 228 525 1
✆ 49 228 525 229

Leschiutta S
Politecnico d Torino
Univ d Torino
Corso d Azeglio 46
I 10125 Torino
Italy
☎ 39 11 556 7235

Lester Daniel F
Astronomy Dpt
Univ of Texas
Rlm 15 308
Austin TX 78712 1083
USA
☎ 1 512 471 3442
✆ 1 512 471 6016

Lester John B
Astronomy Dpt
Univ of Toronto
Erindale College
Mississauga ON L5l 1C6
Canada
☎ 1 416 828 5356
✆ 1 416 828 5328

Lesteven Soizick
Observ Strasbourg
11 r Universite
F 67000 Strasbourg
France
☎ 33 3 88 15 0747
✆ 33 3 88 15 0740
✉ lesteven@astro.u-strasbg.fr

Lestrade Jean-Francois
Observ Paris Meudon
ARPEGES
Pl J Janssen
F 92195 Meudon PPL Cdx
France
☎ 33 1 45 07 7601
✆ 33 1 45 07 7939
✉ lestrade@obspm.fr

Letfus Vojtech
Astronomical Institute
Czech Acad Sci
Fricova 1
CZ 251 65 Ondrejov
Czech R
☎ 420 204 85 7225
✆ 420 2 88 1611

Leto Giuseppe
Radioastronomy Inst
CNT
Contrada Renna Bassa CP 141
I 96017 Noto
Italy
☎ 39 931 824 111
✆ 39 931 824122
✉ pleto@ira.noto.cnr.it

Leung Chun Ming
Physics Dpt
Rensselaer Polytechn Inst
Troy NY 12180 3590
USA
☎ 1 518 266 6318

Leung Kam Ching
Dpt Phys/Astronomy
Univ Nebraska
Behlen Observatory
Lincoln NE 68588 0111
USA
☎ 1 402 472 2770
✆ 1 402 472 2879

Levasseur-Regourd A.-C.
Service Aeronomie
BP 3
F 91371 Verrieres Buisson
France
☎ 33 1 64 47 4293
✆ 33 1 43 29 8673
✉ chantal.levasseur@aerov.jussieu.fr

Levato Orlando Hugo
Complejo Astronomico
El Leoncito
CC 467
5400 San Juan
Argentina
☎ 54 64 21 3693
✆ 54 64 21 1475
✉ levato@castec.edu.ar

Levine Randolph H
50 Carver Rd
Newton MA 02161
USA
☎ 1 617 965 5953

Levison Harold F
USNO
Flagstaff Station
Box 1149
Flagstaff AZ 86002 1149
USA
☎ 1 602 779 5132
✆ 1 520 774 3626

Levreault Russell M
35 Payson Av
Easthampton MA 01027
USA
☎ 1 413 527 9442

Levy Eugene H
Lunar/Planetary Lab
Room 325 Bg 92
Univ of Arizona
Tucson AZ 85721 0092
USA
☎ 1 520 621 6962
✆ 1 520 621 4933

Levy Jacques R
Observ Paris
61 Av Observatoire
F 75014 Paris
France
☎ 33 1 43 20 1210
✆ 33 1 44 54 1804

Lewin Walter H G
Physics Dpt
MIT Rm 37 627
Box 165
Cambridge MA 02139 4307
USA
☎ 1 617 253 4282
✆ 1 617 253 9798

Lewis Brian Murray
Arecibo Observ
Box 995
Arecibo PR 00612
USA
☎ 1 809 878 2612
✉ blewis@naic.edu

Lewis J S
Lunar/Planetary Lab
Room 325 Bg 92
Univ of Arizona
Tucson AZ 85721 0092
USA
☎ 1 520 621 4972
✆ 1 520 621 4933

Li Chun-Sheng
Astronomy Dpt
Nanjing Univ
Nanjing 210093
China PR
☎ 86 25 34651*2882
✆ 86 25 330 2728

Li Depei
Nanjing Astronomical
Instrument Factory
Box 846
Nanjing
China PR
☎ 86 25 46191
✆ 86 25 71 1256

Li Dongming
Purple Mountain Obs
CAS
Nanjing 210008
China PR
☎ 86 25 46700
✆ 86 25 301 459

Li Gi Man
Pyongyang Astron Obs
Acad Sciences DPRK
Taesong district
Pyongyang
Korea DPR
☎ 850 5 3134/5 & 5 3239

Li Guoping
Ctr for Astronomy
Instrumen Res
CAS
Nanjing 210042
China PR
☎ 86 25 542 1770
✆ 86 25 541 1872

Li Gyong Won
Pyongyang Astron Obs
Acad Sciences DPRK
Taesong district
Pyongyang
Korea DPR
☎ 850 5 3134/5 & 5 3239

Li Hong-Wei
JILA
Univ of Colorado
Campus Box 440
Boulder CO 80309
USA
☎ 1 303 492 7789
✆ 1 303 492 5235

Li Hui
Purple Mountain Obs
CAS
Nanjing 210008
China PR
☎ 86 25 331 7874/443 033
✆ 86 25 330 1459
✉ lihui@public1.ptt.js.cn

Li Hyok Ho
Pyongyang Astron Obs
Acad Sciences DPRK
Taesong district
Pyongyang
Korea DPR
☎ 850 5 3134/5 & 5 3239

Li Jing
Beijing Astronomical Obs
CAS
Beijing 100080
China PR
☎ 86 1 22 040
✆ 86 1 256 10855

Li Jinling
Shanghai Observ
CAS
80 Nandan Rd
Shanghai 200030
China PR
☎ 86 21 6438 6191
✆ 86 21 6438 4618
✉ jll@center.shao.ac.cn

Li Kejun
Yunnan Observ
CAS
Kunming 650011
China PR
☎ 86 871 384 0510
℻ 86 871 717 1845
✉ ynao@public.km.yn.cn

Li Linghuai
Purple Mountain Obs
CAS
Nanjing 210008
China PR
☎ 86 25 330 7609
℻ 86 25 330 1459
✉ linghuai@public1.ptt.js.nj

Li Neng-Yao
Purple Mountain Obs
CAS
Nanjing 210008
China PR
☎ 86 25 37609
℻ 86 25 301 459

Li Qi
Beijing Astronomical Obs
CAS
Beijing 100080
China PR
☎ 86 10 6256 6698
℻ 86 1 256 10855
✉ hyb@bao01.bao.ac.cn

Li Qibin
Beijing Astronomical Obs
CAS
Beijing 100080
China PR
☎ 86 10 6255 1698
℻ 86 1 256 10855
✉ lqb@bao01.bao.ac.cn

Li Sin Hyong
Pyongyang Astron Obs
Acad Sciences DPRK
Taesong district
Pyongyang
Korea DPR
☎ 850 5 3134/5 & 5 3239

Li Son Jae
Pyongyang Astron Obs
Acad Sciences DPRK
Taesong district
Pyongyang
Korea DPR
☎ 850 5 3134/5 & 5 3239

Li Ting
Nanjing Astronomical
Instrument Factory
Box 846
Nanjing
China PR
☎ 86 25 46191
℻ 86 25 71 1256

Li Tipei
Inst High Energy Phys
CAS
Box 918
Beijing
China PR
☎ 86 10 821 3344
℻ 86 10 821 3374

Li Wei
Beijing Astronomical Obs
CAS
Beijing 100080
China PR
☎ 86 10 62 61 2194
℻ 86 1 256 10855
✉ lw@sun10.bao.ac.cn

Li Weibao
Yunnan Observ
CAS
Kunming 650011
China PR
☎ 86 871 2035
℻ 86 871 717 1845

Li Xiaoqing
Purple Mountain Obs
CAS
Nanjing 210008
China PR
☎ 86 25 31096
℻ 86 25 301 459

Li Yan
Yunnan Observ
CAS
Kunming 650011
China PR
☎ 86 871 717 2946
℻ 86 871 717 1845
✉ ynao@public.km.yn.cn

Li Yuanjie
Physics Dpt
Huazhong Normal Univ
Wuhan
China PR
☎ 86 87 0541 7122

Li Zhengxin
Shanghai Observ
CAS
80 Nandan Rd
Shanghai 200030
China PR
☎ 86 21 6438 6191
℻ 86 21 6438 4618
✉ lzx@center.shao.ac.cn

Li Zhengxing
Purple Mountain Obs
CAS
Nanjing 210008
China PR
☎ 86 335 1917
℻ 86 25 330 1459
✉ pmoyl@ba001.bao.ac.cn

Li Zhian
Astronomy Dpt
Beijing Normal Univ
Beijing 100875
China PR
☎ 86 1 201 2255*2618
℻ 86 1 201 3929
✉ lxf@class1.bao.ac.cn

Li Zhifang
Shanghai Observ
CAS
80 Nandan Rd
Shanghai 200030
China PR
☎ 86 21 6438 6191
℻ 86 21 6438 4618

Li Zhigang
Shaanxi Observ
CAS
Box 18
Lintong 710600
China PR
☎ 86 33 2255
℻ 86 9237 3496

Li Zhiping
Beijing Astronomical Obs
CAS
Beijing 100080
China PR
☎ 86 10 6256 7194
℻ 86 1 256 10855
✉ lizhi@class1.bao.ac.cn

Li Zhisen
Beijing Astronomical Obs
CAS

W Suburb
Beijing 100080
China PR
☎ 86 1 28 1698
℻ 86 1 256 10855

Li Zhongyuan
Astrophysics Division
Univ Science/Technology
Hefei 230026
China PR
☎ 86 551 33 1134
℻ 86 551 33 1760
✉ lzy@ms.ess.ustc.cn

Li Zongwei
Astronomy Dpt
Beijing Normal Univ
Beijing 100875
China PR
☎ 86 10 204 2288
℻ 86 1 201 3929

Liang Edison P
Dpt Phys/Astronomy
Rice Univ
Box 1892
Houston TX 77251 1892
USA
☎ 1 713 527 8101*3524
℻ 1 713 285 5143
✉ liang@vega.rice.edu

Liang Shiguang
Shanghai Observ
CAS
80 Nandan Rd
Shanghai 200030
China PR
☎ 86 21 6438 6191
℻ 86 21 6438 4618

Liang Zhonghuan
Shaanxi Observ
CAS
Box 18
Lintong 710600
China PR
☎ 86 33 2255
℻ 86 9237 3496

Liao De-chun
Shanghai Observ
CAS
80 Nandan Rd
Shanghai 200030
China PR
☎ 86 21 6438 6191
℻ 86 21 6438 4618
✉ ldc@center.shao.ac.cn

Liao Xinhao
Astronomy Dpt
Nanjing Univ
Nanjing 210093
China PR
☎ 86 25 63 7651*2884
℻ 86 25 330 2728

Libbrecht Kenneth G
CALTECH
MS 264 33
Big Bear Solar Obs
Pasadena CA 91125
USA
☎ 1 818 356 3722
℻ 1 909 866 4240
✉ kgl@sundog.caltech.edu

Liddell U
NASA Headquarters
Space Sci/Applications
600 Independence Av SW
Washington DC 20546
USA
☎ 1 202 358 1409

Liddle Andrew
Astronomy Centre
Univ of Sussex
Falmer
Brighton BN1 9QH
UK
☎ 44 12 73 606 755 *2933
℻ 44 12 73 678 097
✉ arl@starlink.sussex.ac.uk

Liebert James W
Steward Observ
Univ of Arizona
Tucson AZ 85721
USA
☎ 1 520 621 4513
℻ 1 520 621 1532
✉ liebert@arizrvax

Liebscher Dierck-E
Astrophysik Inst
Potsdam Univ
an d Sternwarte 16
D 14482 Potsdam
Germany
☎ 49 331 749 9231
℻ 49 331 749 9309
✉ deliebscher@aip.de

Lieske Jay H
CALTECH/JPL
MS 301 150
4800 Oak Grove Dr
Pasadena CA 91109 8099
USA
☎ 1 818 354 3642
✆ 1 818 354 3437
✉ jhl@naif.jpl.nasa.gov

Liffman Kurt
Fluid Dynamics Lab
CSIRO/DBCE
Box 56
Highett VIC 3190
Australia

Likkel Lauren Jones
Program in Astronomy
Washington State Univ
Mathematics Dpt
Pullman WA 99165 3113
USA
☎ 1 509 335 3172
✆ 1 509 335 1188
✉ likkel@beta.math.wsu.edu

Lilje Per Vidar Barth
Inst Theor Astrophys
Univ of Oslo
Box 1029
N 0315 Blindern Oslo 3
Norway
☎ 47 22 856 501
✆ 47 22 856 505
✉ per.lilje@astro.uio.no

Liller William
Instituto Isaac Newton
Casilla 8 9
Santiago 9
Chile
☎ 56 2 217 2013
✆ 56 2 217 2352

Lilley Edward A
CfA
HCO/SAO
60 Garden St
Cambridge MA 02138 1516
USA
☎ 1 617 495 3971
✆ 1 617 495 7356

Lillie Charles F
TRW Space/Technology
1 Space Park
Redondo Beach CA 90278
USA
☎ 1 213 812 2248

Lilly Simon J
Astronomy Dpt
Univ of Toronto
60 St George St
Toronto ON M5S 1A1
Canada
☎ 1 416 978 2183
✆ 1 416 971 2026
✉ lilly@astro.utoronto.ca

Lima Botti Luiz Claudio
CRAAE INPE
EPUSP/PTR
CP 61548
01065 970 Sao Paulo SP
Brazil
☎ 55 11 815 5936
✆ 55 11 815 6289
✉ lclbotti@bruspvm

Lin Chia C
Mathematics Dpt
MIT Rm 2 224
Box 165
Cambridge MA 02139
USA
☎ 1 617 253 1796

Lin Douglas N C
Lick Observ
Univ of California
Santa Cruz CA 95064
USA
☎ 1 831 429 2732
✆ 1 831 426 3115

Lin Qinchang
Guangzhou Satellite Station
Acad Sciences
Wushan
Guangzhou 510650
China PR
☎ 86 20 8770 7002*20 8774 3534
✆ 86 20 8772 2081
✉ gzgsos@publicl.guangzhou.gd.cn

Lin Yuanzhang
Beijing Astronomical Obs
CAS
Beijing 100080
China PR
☎ 86 1 28 1698
✆ 86 1 256 10855

Lincoln J Virginia
2005 Alpine Dr
Boulder CO 80304
USA
☎ 1 303 442 6757

Lindblad Bertil A
Lund Observ
Box 43
S 221 00 Lund
Sweden
☎ 46 46 222 7000
✆ 46 46 222 4614
✉ linasu@gemini.ldc.lu.se

Lindblad Per Olof
Stockholm Observ
Royal Swedish Acad Sciences
S 133 36 Saltsjoebaden
Sweden
☎ 46 8 717 0380
✆ 46 8 717 4719
✉ lindblad@astro.su.se

Linde Peter
Lund Observ
Box 43
S 221 00 Lund
Sweden
☎ 46 46 222 4701
✆ 46 46 222 4614
✉ peter@astro.lu.se

Lindegren Lennart
Lund Observ
Box 43
S 221 00 Lund
Sweden
☎ 46 46 222 7309
✆ 46 46 222 4614

Lindgren Harri
Robinson Crusoe 966
Dpt 133
Las Condes
Santiago
Chile
☎ 56 2 201 2492
✉ harri@astro.nastol.lu.se

Lindsey Charles Allan
NOAO/NSO
Box 26732
950 N Cherry Av
Tucson AZ 85726 6732
USA
☎ 1 520 325 9294
✆ 1 520 325 9278

Ling Chih-Bing
Inst of Mathematics
Academia Sinica
Box 143
Taipei
China R

Ling J
Observ Astronomico
Ramon Maria Aller
Avd de Las Ciencias S/n
E Santiago de Compostela
Spain
☎ 34 8 159 2747
✆ 34 8 156 2569

Lingenfelter Richard E
CASS
UCSD
C 011
La Jolla CA 92093 0216
USA
☎ 1 619 452 2464
✆ 1 619 534 2294

Linke Richard Alan
8 Anderson Lane
Princeton NJ 08540
USA

Linnell Albert P
5323 NE 42nd St
Seattle WA 98105 4910
USA
✆ 1 425 522 8319
✉ linnell@msupa.pa.msu.edu

Linsky Jeffrey L
JILA
Univ of Colorado
Campus Box 440
Boulder CO 80309 0440
USA
☎ 1 303 492 7838
✆ 1 303 492 5235
✉ jlinsky@jila.colorado.edu

Linsley John
Dpt Phys/Astronomy
Univ New Mexico
800 Yale Blvd Ne
Albuquerque NM 87131
USA
☎ 1 505 243 1924

Lippincott Sarah Lee
29 Kendal Dr
Kennett Square
Kennet Square PA 19348 2323
USA
☎ 1 610 388 1448

Lipschutz Michael E
Wetherill Chemistry Bldg
Purdue Univ
W Lafayette IN 47907
USA
☎ 1 317 494 5326

Lipunov Vladimir M
SAI
Acad Sciences
Universitetskij Pr 13
119899 Moscow
Russia
☎ 7 095 939 5006
✆ 7 095 939 0126
✉ lipunov@sai.msu.su

Liritzis Ioannis
Res Ctr Astronomy
Acad Athens
14 Anagnostopoulou St
GR 106 73 Athens
Greece
☎ 30 1 361 3589
✆ 30 1 363 1606

Lis Dariusz C
CALTECH
MS 320 47
Downs Laboratory Physics
Pasadena CA 91125
USA
☎ 1 818 395 6617
✆ 1 818 796 8806
✉ dcl@tacos.caltech.edu

Liseau Rene
Stockholm Observ
Royal Swedish Acad Sciences
S 133 36 Saltsjoebaden
Sweden
☎ 46 8 716 4485
✆ 46 8 717 4719
✉ rene@astro.su.se

Lisi Franco
Osserv Astrofis Arcetri
Univ d Firenze
Largo E Fermi 5
I 50125 Firenze
Italy
☎ 39 55 27 52289
✆ 39 55 22 0039

Lissauer Jack J
NASA/ARC
MS 245 3
Moffett Field CA 94035 1000
USA
☎ 1 415 604 2293
📠 1 415 604 6779
✉ lissauer@ringside.arc.nasa.gov

Liszt Harvey Steven
NRAO
520 Edgemomt Rd
Charlottesville VA 22903
USA
☎ 1 804 296 0344

Little Leslie T
Electronics Lab
Univ of Kent
Canterbury CT2 7NT
UK
☎ 44 12 27 668 22
📠 44 12 27 456 084

Little-Marenin Irene R
Whitin Observ
Wellesley Coll
Wellesley MA 02181
USA
☎ 1 617 235 5303
📠 1 617 283 3667
✉ ilittle@lucy.wellesley.edu

Littleton John E
Physics Dpt
WV University
Box 6315
Morgantown WV 26506 6315
USA
☎ 1 304 293 3498
✉ jel@wvnvaxa.wvnet.edu

Litvak Marvin M
Technology Res Ass
1525 Espinosa Circle
Palos Verdes Est CA 90274
USA

Litvinenko Leonid N
Inst Radio Astron
Ukrainian Acad Science
4 Chervonopraporna st
310002 Kharkiv
Ukraine
☎ 380 5724 51009
📠 380 5724 76506

Liu Bao-Lin
Purple Mountain Obs
CAS
Nanjing 210008
China PR
☎ 86 25 42817/46700
📠 86 25 301 459

Liu Caipin
Purple Mountain Obs
CAS
Nanjing 210008
China PR
☎ 86 25 42817
📠 86 25 301 459

Liu Ciyuan
Shaanxi Observ
CAS
Box 18
Lintong 710600
China PR
☎ 86 33 2255
📠 86 9237 3496
✉ pub2@ms.sxso.

Liu Jinming
Shanghai Observ
CAS
80 Nandan Rd
Shanghai 200030
China PR
☎ 86 21 6438 6191
📠 86 21 6438 4618
✉ ljm@center.shao.ac.cn

Liu Liao
Physics Dpt
Beijing Normal Univ
Beijing 100071
China PR
☎ 86 1 201 2255*2618
📠 86 1 201 3929

Liu Lin
Astronomy Dpt
Nanjing Univ
Nanjing 210093
China PR
☎ 86 25 34651*2882
📠 86 25 330 2728

Liu Linzhong
Purple Mountain Obs
CAS
Nanjing 210008
China PR
☎ 86 25 46700
📠 86 25 301 459

Liu Qingyao
Yunnan Observ
CAS
Kunming 650011
China PR
☎ 86 871 2035
📠 86 871 717 1845

Liu Ruliang
Purple Mountain Obs
CAS
Nanjing 210008
China PR
☎ 86 25 330 3921
📠 86 25 330 1459

Liu Sou-Yang
14959 Duflet Dr
Gaithersburg MD 20878
USA

Liu Xiaowei
Dpt Phys/Astronomy
UCLO
Gower St
London WC1E 6BT
UK
☎ 44 171 387 7050
📠 44 171 380 7145
✉ xwl@star.ucl.ac.uk

Liu Xinping
Inst of Mechanics
CAS
Beijing 100080
China PR
☎ 86 1 28 4185
📠 86 1 25 61284

Liu Xuefu
Astronomy Dpt
Beijing Normal Univ
Beijing 100875
China PR
☎ 86 1 65 6531*6285
📠 86 1 201 3929

Liu Yongzhen
Graduate School
Univ Sci/Techn
Box 3908
Beijing 100039
China PR
☎ 86 1 81 7031*713

Liu Yuying
Beijing Astronomical Obs
CAS
Beijing 100080
China PR
☎ 86 10 6271 2537
📠 86 1 256 10855
✉ qjfu@cenpok.net

Liu Zongli
Beijing Astronomical Obs
CAS
Beijing 100080
China PR
☎ 86 1 28 1698
📠 86 1 256 10855

Livingston William C
NOAO/NSO
Box 26732
950 N Cherry Av
Tucson AZ 85726 6732
USA
☎ 1 520 327 5511
📠 1 520 325 9278
✉ wdl@noao.edu

Livio Mario
Physics Dpt
IIT
Technion City
Haifa 32000
Israel
☎ 972 4 293 549
✉ phr81ml@technion

Livshits Mikhail A
ITMIRWP
Acad Sciences
142092 Troitsk
Russia
☎ 7 095 334 0120
📠 7 095 334 0124

Lizano-Soberon Susana
Instituto Astronomia
UNAM
Apt 70 264
Mexico DF 04510
Mexico
☎ 52 5 622 3906
📠 52 5 548 3712
✉ lizano@alfa.astroscu.unam.mx

Lloyd Huw
Astrophysics Group
Liverpool J M Univ
Byrom St
Liverpool L3 3AF
UK
☎ 44 151 231 4103
📠 44 151 231 2337
✉ hml@staru1.livjm.ac.uk

Lloyd Evans Thomas Harry
SAAO
PO Box 9
7935 Observatory
South Africa
☎ 27 21 47 3639
📠 27 21 47 3639
✉ tle@saao.ac.za

Lo Kwok-Yung
Dpt Astron/Physics
Univ of Illinois
1002 W Green St
Urbana IL 61801
USA
☎ 1 217 333 9381
📠 1 217 244 7638
✉ kyl@sgr.astro.uiuc.edu

Locanthi Dorothy Davis
Carnegie Observatories
813 Santa Barbara St
Pasadena CA 91101 1292
USA

Lochman Jan
Astronomical Institute
Czech Acad Sci
Dvorakova 298
CZ 511 01 Turnov
Czech R
☎ 420 43 62 2913
📠 420 43 62 2913

Lochner James Charles
NASA/GSFC
Code 660 2
Greenbelt MD 20771
USA
☎ 1 301 286 9711
📠 1 301 286 1684
✉ lochner@xeric.gsfc.nasa.gov

Locke Jack L
NRCC
Ottawa ON K1H 7J5
Canada
☎ 1 613 523 0812

Lockman Felix J
NRAO
Box 2
Green Bank Wv 24944
USA
☎ 1 304 456 2011
📠 1 304 456 2271

Lockwood G Wesley
Lowell Observ
1400 W Mars Hill Rd
Box 1149
Flagstaff AZ 86001
USA
☎ 1 607 774 3358
📠 1 520 774 3358
✉ gwl@lowell.edu

Loden Kerstin R
Stockholm Observ
Royal Swedish Acad Sciences
S 133 36 Saltsjoebaden
Sweden
☎ 46 8 717 0195
✆ 46 8 717 4719
✉ loden_k@astro.su.se

Loden Lars Olof
Stockholm Observ
Royal Swedish Acad Sciences
S 133 36 Saltsjoebaden
Sweden
☎ 46 8 717 2095
✆ 46 8 717 4719
✉ lodenl@astro.su.se

Loiseau Nora
ESA
Apt 50727 Villafranca
E 28080 Madrid
Spain
☎ 34 1 813 1356
✆ 34 1 813 1139
✉ nll@vilsp.esa.es

Lomb Nicholas Ralph
Sydney Observ
Museum Appl Arts/Sci
Box K346
Haymarket NSW 2000
Australia

Lominadze Jumber
Abastumani Astrophysical
Observ
Georgian Acad Sci
380060 Tbilisi
Georgia
☎ 995 88 32 37 5226
✆ 995 88 32 98 5017
✉ Jlomin@parliament.ge

Long Knox S
STScI
Homewood Campus
3700 San Martin Dr
Baltimore MD 21218
USA
☎ 1 301 338 4862
✆ 1 301 338 5090
✉ long@stsci.edu

Longair Malcolm S
MRAO
Cavendish Laboratory
Madingley Rd
Cambridge CB3 0HE
UK
☎ 44 12 23 337 083
✆ 44 12 23 354 599
✉ msl@uk.ac.cam.phy.ras

Longbottom Aaron
Mathematics Dpt
Univ of St Andrews
St Andrews North Haugh
Fife KY16 9SS
UK
☎ 44 13 34 463735
✆ 44 13 344 63748
✉ aaron@dcs.st-and.
ac.uk

Longmore Andrew J
Joint Astronomy Ctr
660 N A'ohoku Pl
University Park
Hilo HI 96720
USA
☎ 1 808 961 3756
✆ 1 808 961 3756

Lonsdale Carol J
CALTECH
MS 100 22
IPAC
Pasadena CA 91125
USA
☎ 1 818 397 9529
✆ 1 818 397 9600
✉ cjl@ipac.caltech.edu

Lopes Dalton De faria
Observ Nacional
Rua Gl Bruce 586
Sao Cristovao
20921 030 Rio de Janeiro RJ
Brazil
☎ 55 21 580 781
✆ 55 21 580 0332

Lopes-Gautier Rosaly
CALTECH/JPL
MS 183 601
4800 Oak Grove Dr
Pasadena CA 91109 8099
USA
☎ 1 818 393 4584
✆ 1 818 354 0966

Lopez Carlos
Observ Astronomico
Felix Aguilar
Av Benavidez 8175 Oeste
5407 Marquesado (SJ)
Argentina
☎ 54 64 23 1467
✆ 54 64 23 4980/23 8494
✉ celopez@unsjf.edu.ar

Lopez De Coca M D P
Inst Astrofisica
Andalucia Apt 3004
Prof Albareda 1
E 18080 Granada
Spain
☎ 34 5 812 1311
✆ 34 5 881 4530
✉ pilar@iaa.es

Lopez Garcia Zulema I
Observ Astronomico
Felix Aguilar
Av Benavidez 8175 Oeste
5407 Marquesado (SJ)
Argentina
☎ 54 64 23 1494

Lopez Jose A
Observ Astronomico
Felix Aguilar
Av Benavidez 8175 Oeste
5407 Marquesado (SJ)
Argentina
☎ 54 64 23 1494

Lopez Jose Alberto
Observ Astronomico Nacional
UNAM
Apt 877
Ensenada BC 22800
Mexico
☎ 52 6 174 4593
✆ 52 6 174 4607
✉ jal@bufadora.
astrosen.unam.mx

Lopez Rosario
Dpt Fisica Atmosfera
Univ Barcelona
Avd Diagonal 647
E 08028 Barcelona
Spain
☎ 34 3 330 7311
✆ 34 3 402 1133

Lopez-Arroyo M
Observ Astronomico
Nacional
Alfonso XII-5
E 28014 Madrid
Spain
☎ 34 1 227 0107

Lopez-Cruz Omar
INAOE
Tonantzintlaz
Apdo 51 y 216
Puebla PUE 72000
Mexico
☎ 52 2 247 2011
✆ 52 2 247 2231
✉ omar@inaoep.mx

Lopez-Garcia Francisco
Observ Astronomico
Felix Aguilar
Av Benavidez 8175 Oeste
5407 Marquesado (SJ)
Argentina
☎ 54 64 23 1494

Lopez-Gonzalez Maria J
Inst Astrofisica
Andalucia Apt 3004
Prof Albareda 1
E 18080 Granada
Spain
☎ 34 5 812 1311
✆ 34 5 881 4530

Lopez-Moreno Jose Juan
Inst Astrofisica
Andalucia Apt 3004
Prof Albareda 1
E 18080 Granada
Spain
☎ 34 5 812 1311
✆ 34 5 881 4530
✉ lopez@iaa.es

Lopez-Puertas Manuel
Inst Astrofisica
Andalucia Apt 3004
Prof Albareda 1
E 18080 Granada
Spain
☎ 34 5 812 1300
✆ 34 5 881 4530

Lopez-Valverde M A
Inst Astrofisica
Andalucia Apt 3004
Prof Albareda 1
E 18080 Granada
Spain
☎ 34 5 812 1311
✆ 34 5 881 4530

Lord Steven Donald
IPAC/CALTECH
MS 100 22
Pasadena CA 91125
USA
☎ 1 818 397 9502
✆ 1 818 397 9600
✉ lord@gal.arc.nasa.gov

Loren Robert Bruce
Box 2915
Silver City NM 88062
USA

Lorenz Hilmar
Zntrlinst Astrophysik
Sternwarte Babalsberg
Rosa Luxemburg Str 17a
D 14473 Potsdam
Germany
☎ 49 331 762225

Lortet Marie-Claire
Observ Paris Meudon
DAPHE
Pl J Janssen
F 92195 Meudon PPL Cdx
France
☎ 33 1 45 07 7426
✆ 33 1 45 07 7100
✉ Lortet@obspm.fr

Losco Lucette
Fac des Sciences
F 25030 Besancon Cdx
France

Lotova Natalia A
ITMIRWP
Acad Sciences
142092 Troitsk
Russia
☎ 7 095 334 0120
✆ 7 095 334 0124

Loucif Mohammed Lakhdar
IAP
98bis bd Arago
F 75014 Paris
France
☎ 33 1 44 32 8049
✆ 33 1 44 32 8001
✉ mohal@iap.fr

Louise Raymond
Fac des Sciences
Dpt Physique
33 r St-leu
F 80039 Amiens
France

Loulergue Michelle
Observ Paris Meudon
ARPEGES
Pl J Janssen
F 92195 Meudon PPL Cdx
France
☎ 33 1 45 07 7455
✆ 33 1 45 07 7971

Loup Cecile
IAP
98bis bd Arago
F 75014 Paris
France
☎ 33 1 44 32 8054
℻ 33 1 44 32 8001
✉ loup@iap.fr

Lovas Francis John
NBS
Div 545
Molecular Spectroscopic
Washington DC 20234
USA
☎ 1 301 921 2023

Lovas Miklos
Konkoly Observ
Thege U 13/17
Box 67
H 1525 Budapest
Hungary
☎ 36 1 375 4122
℻ 36 1 275 4668

Lovelace Richard V E
Applied Phys Dpt
Cornell Univ
Clark Hall
Ithaca NY 14853 6801
USA
☎ 1 607 255 3968
℻ 1 607 255 7658
✉ rvll@cornell.edu

Lovell Sir Bernard
NRAL
Univ Manchester
Jodrell Bank
Macclesfield SK11 9DL
UK
☎ 44 14 777 1321
℻ 44 147 757 1618

Low Boon Chye
HAO
NCAR
Box 3000
Boulder CO 80307 3000
USA
☎ 1 303 497 1553
℻ 1 303 497 1568

Low Frank J
4940 Calle Barril
Tucson AZ 85718
USA
☎ 1 520 621 2779

Lowe Robert P
Dpt Phys/Astronomy
Univ W Ontario
London ON N6A 3K7
Canada
☎ 1 519 661 3929
℻ 1 519 661 3129
✉ lowe@canlon.physics.uwo.ca

Loyola Patricio
Dpt Astronomia
Univ Chile
Casilla 36 D
Santiago
Chile
☎ 56 2 229 4101
℻ 56 2 229 4002

Lozinskaya Tat-yana A
SAI
Acad Sciences
Universitetskij Pr 13
119899 Moscow
Russia
☎ 7 095 139 1030
℻ 7 095 939 0126

Lozinskij Alexander M
Inst of Astronomy
Acad Sciences
Pyatnitskaya Ul 48
109017 Moscow
Russia
☎ 7 095 231 5461
℻ 7 095 230 2081

Lozitskij Vsevolod
Astronomical Observ
Kyiv State Univ
Observatorna Str 3
254053 Kyiv
Ukraine
☎ 380 4421 63910
✉ aoku@gluk.apc.org

Lu Benkui
Purple Mountain Obs
CAS
Nanjing 210008
China PR
☎ 86 25 32893
℻ 86 25 301 459

Lu Chunlin
Purple Mountain Obs
CAS
Nanjing 210008
China PR
☎ 86 25 42700
℻ 86 25 301 459
✉ luchln@public1.
ptt.js.cn

Lu Jufu
Astrophysics Division
Univ Science/Technology
Hefei 230026
China PR
☎ 86 551 33 1134
℻ 86 551 33 1760

Lu Limin
CALTECH
Palomar Obs
MS 105 24
Pasadena CA 91125
USA
☎ 1 818 395 4421
℻ 1 818 568 1517
✉ ll@astro.caltech.edu

Lu Phillip K
Dpt Phys/Astronomy
W Connecticut State Univ
181 White St
Danbury CT 06810 7859
USA
☎ 1 203 837 8422
℻ 1 203 837 8673
✉ lu@wcsu.
ctstateu.edu

Lu Ruwei
Yunnan Observ
CAS
Kunming 650011
China PR
☎ 86 871 2035
℻ 86 871 717 1845

Lu Tan
Astronomy Dpt
Nanjing Univ
Nanjing 210093
China PR
☎ 86 25 663 7551
℻ 86 25 330 7965

Lu Yang
Astronomy Dpt
Nanjing Univ
Nanjing 210093
China PR
☎ 86 25 34651*2882
℻ 86 25 330 2728

Lub Jan
Rijkuniversiteit te
Huygens Lab
Box 9504
NL 2314 EX Leiden
Netherlands
☎ 31 71 527 5840/5835
℻ 31 71 527 5819
✉ lub@strw.leidenuniv.nl

Lubowich Donald A
American Inst Phys
500 Sunnyside Blvd
Woodbury NY 11797
USA
☎ 1 516 576 2468
℻ 1 516 921 4320
✉ dal@aip.org

Lucas Robert
IRAM
300 r La Piscine
F 38406 S Martin Heres Cdx
France
☎ 33 4 76 82 4942
℻ 33 4 76 51 5938
✉ lucas@iram.fr

Lucchin Francesco
Dpt Fisica G Galilei
Univ d Padova
Via Marzolo 8
I 35131 Padova
Italy
☎ 39 49 84 4333
℻ 39 49 84 4245

Lucey John
Physics Dpt
Univ of Durham
South Rd
Durham DH1 3LE
UK
☎ 44 191 374 2000
℻ 44 191 374 3749

Luck John M
Orroral Geodetic Obs
AUSLIG
Box 2
Belconnen ACT 2616
Australia
☎ 61 62 525 172

Luck R Earle
Physics Dpt
CWRU
Rock Bdg
Cleveland OH 44106
USA
☎ 1 216 368 6697

Lucke Peter B
Dpt Phys/Astronomy
Mount Union College
Alliance OH 44601
USA
☎ 1 216 821 5320

Lucy Leon B
ESO
Karl Schwarzschildstr 2
D 85748 Garching
Germany
☎ 49 893 2006249
℻ 49 893 202 362

Ludmany Andras
Debrecen Heliophys Observ
Acad Sciences
Box 30
H 4010 Debrecen
Hungary
☎ 36 5 231 1015
✉ ludmany@tigris.klte.hu

Luest Reimar
MPI Meteorology
Grenzstr 4
D 20146 Hamburg
Germany
☎ 49 404 117 3300
℻ 49 404 117 3390
✉ luest@dkrz.d400.de

Lugger Phyllis M
Astronomy Dpt
Indiana Univ
Swain W 319
Bloomington IN 47405
USA
☎ 1 812 335 6929
℻ 1 812 855 8725

Lukacevic Ilija S
Astronomy Dpt
Univ of Belgrade
Studentski Trg 16
11000 Beograd
Yugoslavia FR
☎ 381 11 638 715
℻ 381 11 630 151

Lukash Vladimir N
Astro Space Ctr
Lebedev Physical Inst
Profsoyuznaya 84/32
Moscow 117810
Russia
☎ 7 095 333 3366
℻ 7 095 333 2378
✉ lukash@dpc.asc.
rssi.ru

Luks Thomas
Astronomisches Institut
Ruhr Univ Bochum
Postfach 102148
D 44780 Bochum
Germany
☎ 49 234 700 5802/6660
℻ 49 234 709 4169
✉ thomas.luks@astro.
ruhr-uni-bochum.de

Luminet Jean-Pierre
Observ Paris Meudon
DARC
Pl J Janssen
F 92195 Meudon PPL Cdx
France
☎ 33 1 45 07 7423
℻ 33 1 45 07 7971

Lumme Kari A
Helsinky Univ
Observ
Box 14
FIN 00014 Helsinki
Finland
☎ 358 19 12 2910
℻ 358 19 12 2952

Luna Homero G
IAR
CC 5
1894 Villa Elisa (Bs As)
Argentina
☎ 54 21 4 3793
℻ 54 21 211 761

Lund Niels
Danish Space Res Inst
Lundtoftevej 7
DK 2800 Lyngby
Denmark
☎ 45 42 88 2277

Lundquist Charles A
Res Institute
Univ of Alabama
Box 209
Huntsville AL 35899
USA
☎ 1 205 895 6100

Lundqvist Peter
Stockholm Observ
Royal Swedish Acad Sciences
S 133 36 Saltsjoebaden
Sweden
☎ 46 8 716 4489
℻ 46 8 717 4719
✉ peter@astro.su.se

Lundstedt Henrik
Lund Observ
Box 43
S 221 00 Lund
Sweden
☎ 46 46 222 7294
℻ 46 46 222 4614
✉ henrik@astro.lu.se

Lundstrom Ingemar
Lund Observ
Box 43
S 221 00 Lund
Sweden
☎ 46 46 222 7300
℻ 46 46 222 4614

Lunel Madeleine
Observ Lyon
Av Ch Andre
F 69561 S Genis Laval cdx
France
☎ 33 4 78 56 0705
℻ 33 4 72 39 9791

Lungu Nicolaie
Inst Politechnic
Catedra Matematica
Ul Emil Isac 15
RO 3400 Cluj Napoca
Rumania
☎ 40 64 117 229
℻ 40 64 19 2820

Luo Baorong
Yunnan Observ
CAS
Kunming 650011
China PR
☎ 86 871 2035
℻ 86 871 717 1845

Luo Dingchang
Beijing Astronomical Obs
CAS
Beijing 100080
China PR
☎ 86 1 27 5580
℻ 86 1 256 10855

Luo Dingjiang
Beijing Astronomical Obs
CAS
Beijing 100080
China PR
☎ 86 1 28 1698
℻ 86 1 256 10855

Luo Guoquan
Yunnan Observ
CAS
Kunming 650011
China PR
☎ 86 871 384 0501
℻ 86 871 717 1845
✉ luogq@public.km.
yn.cn

Luo Shaoguang
Dpt Geophysics
Beijing Univ
Beijing 100871
China PR
☎ 86 10 6275 1141
℻ 86 10 6256 4095

Luo Shi-Fang
Shanghai Observ
CAS
80 Nandan Rd
Shanghai 200030
China PR
☎ 86 21 6438 6191
℻ 86 21 6438 4618

Luo Xianhan
Dpt Geophysics
Beijing Univ
Beijing 100871
China PR
☎ 86 1 22 239

Lupishko Dmitrij F
Astronomical Observ
Kharkiv State Univ
Sumskaja Ul. 35
310002 Kharkiv
Ukraine
☎ 380 5724 32428

Luri Xavier
Dpt Astronom Meteo
Univ Barcelona
Avd Diagonal 647
E 08028 Barcelona
Spain
☎ 34 3 402 1126
℻ 34 3 402 1133
✉ xluri@mizar.am.
ub.es

Lustig Guenter
Inst Astronomie
Karl-Franzens-Univ
Universitaetsplatz 5
A 8010 Graz
Austria
☎ 43 316 380 5272
℻ 43 316 384 091
✉ lustig@bkug.
kfunigraz.ac.at

Luttermoser Donald
Physics Dpt
E Tennessee State Univ
Box 70652
Johnson City TN37614
USA
☎ 1 423 439 4231
℻ 1 423 439 6905

Lutz Barry L
Dpt Phys/Astronomy
N Arizona Univ
Box 6010
Flagstaff AZ 86011 6010
USA
☎ 1 520 523 2661
℻ 1 520 523 2626

Lutz Julie H
Program in Astronomy
Washington State Univ
Pullman WA 99164 2930
USA
☎ 1 509 335 3136

Lynas-Gray Anthony E
Dpt Phys/Astronomy
UCLO
Gower St
London WC1E 6BT
UK
☎ 44 171 387 7050
℻ 44 171 380 7145

Lynch David K
Aerospace Corp
MS M2 226
Box 92957
Los Angeles CA 90009 2957
USA
☎ 1 310 648 6686
℻ 1 310 336 1636

Lynden-Bell Donald
Inst of Astronomy
The Observatories
Madingley Rd
Cambridge CB3 0HA
UK
☎ 44 12 23 337 548
℻ 44 12 23 337 523

Lynds Beverly T
3244 6th St
Boulder CO 80304
USA
✉ blynds@unidata.ucar.edu

Lynds Roger C
NOAO/KPNO
Box 26732
950 N Cherry Av
Tucson AZ 85726 6732
USA
☎ 1 520 327 5511
℻ 1 520 325 9360

Lyne Andrew G
NRAL
Univ Manchester
Jodrell Bank
Macclesfield SK11 9DL
UK
☎ 44 14 777 1321
℻ 44 147 757 1618

Lyubimkov Leonid S
Crimean Astrophys Obs
Ukrainian Acad Science
Nauchny
334413 Crimea
Ukraine
☎ 380 6554 71161
℻ 380 6554 40704

Lyuty Victor M
Crimean Station of
Sternberg Inst
Nauchny
334413 Crimea
Ukraine
☎ 380 6554 71161
℻ 380 6554 40704

M

Ma Chopo
NASA/GSFC
Code 926
Greenbelt MD 20771
USA
☎ 1 301 286 3992
✆ 1 301 286 0213
✉ cma@virgo.gsfc.nasa.gov

Ma Chun-yu
Dpt Math
Memorial Univ of Newfoundland
St John's NF A1C 5S7
Canada
☎ 1 709 737 4358
✆ 1 709 737 3010
✉ cyma@math.mun.ca

Ma Er
Beijing Astronomical Obs
CAS
Beijing 100080
China PR
☎ 86 1 28 1698
✆ 86 1 256 10855

Ma Xingyuan
Dpt Geography
Capital Normal Univ
Beijing 1000037
China PR

Ma Yuehua
Purple Mountain Obs
CAS
Nanjing 210008
China PR
☎ 86 25 330 3583
✆ 86 25 330 1459
✉ pmoyl@pub.nj-online.nj.js.cn

Ma YuQian
Inst High Energy Phys
CAS
Box 918
Beijing
China PR
☎ 86 10 821 3344

Mac Low Mordecai-Mark
MPI Astronomie
Koenigstuhl
D 69117 Heidelberg
Germany
☎ 49 622 152 8224
✆ 49 622 152 8246
✉ mordecai@mpia-hd.mpg.de

Macalpine Gordon M
Astronomy Dpt
Univ of Michigan
Dennison Bldg
Ann Arbor MI 48109 1090
USA
☎ 1 313 764 3433
✆ 1 313 764 2211

Maccacaro Tommaso
Osserv Astronomico
d Brera
Via Brera 28
I 20121 Milano
Italy
☎ 39 2 720 23751
✆ 39 2 720 01600
✉ tommaso@bach.mi.astro.it

Maccagni Dario
IFCTR CNR
Univ d Milano
Via E Bassini 15
I 20133 Milano
Italy
☎ 39 2 236 3542
✆ 39 2 266 5753
✉ dario@ifctr.mi.cnr.it

MacCallum Malcolm A H
School Mathemat Sc
QMWC
Mile End Rd
London E1 4NS
UK
☎ 44 171 980 4811
✆ 44 181 975 5500

Macchetto Ferdinando
STScI
Homewood Campus
3700 San Martin Dr
Baltimore MD 21218
USA
☎ 1 301 338 4790
✆ 1 301 338 2617
✉ macchetto@stsci.edu

MacConnell Darrell J
STScI/CSC
Homewood Campus
3700 San Martin Dr
Baltimore MD 21218
USA
☎ 1 301 338 4800
✆ 1 301 338 4767
✉ macconnell@stsci.edu

MacDonald Geoffrey H
Electronics Lab
Univ of Kent
Canterbury
Kent CT2 7NT
UK
☎ 44 1227 823 707
✆ 44 1227 456 084
✉ ghm@star.ukc.ac.uk

MacDonald James
Physics Dpt
Univ of Delaware
Newark DE 19716
USA
☎ 1 302 831 6855
✆ 1 302 831 1637

Maceroni Carla
Osserv Astronomico
d Roma
Via d osservatorio 5
I 00040 Monteporzio
Italy
☎ 39 6 944 8028
✆ 39 6 944 7243
✉ maceroni@astrmp.astro.it

MacGillivray Harvey T
Royal Observ
Blackford Hill
Edinburgh EH9 3HJ
UK
☎ 44 131 667 3321
✆ 44 131 668 8356

Machado Luiz E da silva
Univ Fed Rio d Janeiro
Av Sernambetiba 3300 Bl7
22 630 Barra Da Tijuca
20080 Rio de Janeiro RJ
Brazil
☎ 55 213 99 2589

Machado Marcos
CNIE
Avenida Mitre 3100
1663 San Miguel (Bs As)
Argentina
☎ 54 2 664 8371

Machalski Jerzy
Astronomical Observ
Jagiellonian Univ
Ul Mazowiecka 36/33
PL 30 019 Krakow
Poland

Maciejewski Andrzej J
Inst of Astronomy
N Copernicus Univ
Ul Chopina 12/18
PL 87 100 Torun
Poland
☎ 48 56 26 018*53
✆ 48 56 24 602
✉ maciejka@pltumk11

Maciel Walter J
IAG
Univ Sao Paulo
CP 9638
01065 970 Sao Paulo SP
Brazil
☎ 55 11 577 8599
✆ 55 11 276 3848
✉ marciel@orion.iagusp.usp.br

Mack Peter
Arizona Guest Observ
4414 W Plantation St
Tucson AZ 85741 4034
USA
☎ 1 520 579 0698

MacKay Craig D
Inst of Astronomy
The Observatories
Madingley Rd
Cambridge CB3 0HA
UK
☎ 44 12 23 337 548
✆ 44 12 23 337 523

MacKinnon Alexander L
Astronomy Dpt
Univ of Glasgow
Glasgow G12 8OW
UK
☎ 44 141 339 8855
✆ 44 141 334 9029

MacLeod John M
NRCC/HIA
100 Sussex Dr
Ottawa ON K1A 0R6
Canada
☎ 1 613 993 6060
✆ 1 613 952 6602

MacQueen Robert M
Physics Dpt
Rhodes College
2000 N Parkway
Memphis TN 38112
USA
☎ 1 901 726 3000

Macrae Donald A
David Dunlap Observ
Univ of Toronto
Box 360
Richmond Hill ON L4C 4y6
Canada
☎ 1 416 884 9562
✆ 1 416 978 3921
✉ macrae@vela.astro.utoronto.ca

Macy William Wray
151 Melville Av
Palo Alto CA 94304
USA

Madau Piero
STScI
Homewood Campus
3700 San Martin Dr
Baltimore MD 21218
USA
☎ 1 301 338 2622
✆ 1 301 338 5085
✉ madau@stsci.edu

Madden Suzanne
Service Astrophysique
CEA Saclay
Orme d Merisiers Bt 709
F 91190 Gif s Yvette Cdx
France
☎ 33 1 63 08 9276
✆ 33 1 69 08 6577
✉ madden@discovery.saclay.cea.fr

Maddison Ronald Ch
Physics Dpt
Univ of Keele
Keele ST5 5BG
UK
☎ 44 178 262 1111
✆ 44 178 271 1093

Maddox Stephen
Royal Greenwich Obs
Madingley Rd
Cambridge CB3 0EZ
UK
☎ 44 12 23 37 4000
✆ 44 12 23 374 700
✉ sjm@ast.cam.ac.uk

Madej Jerzy
Astronomical Observ
Warsaw Univ
Al Ujazdowskie 4
PL 00 478 Warsaw
Poland
☎ 48 22 29 4011
✆ 48 22 29 4967
✉ jm@alkor.astrouw.edu.pl

Madore Barry Francis
CALTECH
NASA/IPAC
MS 100-22
Pasadena CA 91125
USA
☎ 1 818 397 9512
✆ 1 626 397 9600
✉ barry@ipac.caltech.edu

Madsen Jes
Inst Phys/Astronomy
Univ of Aarhus
Ny Munkegade
DK 8000 Aarhus C
Denmark
☎ 45 86 12 8899
✆ 45 86 20 2711

Maeda Kei-ichi
Physics Dpt
Waseda Univ
Okubo 3-4-1 Shinjuku-ku
Tokyo 160
Japan
☎ 81 320 34 141

Maeda Koitiro
Physics Dpt
College Medicine
Nishinomiya
Hyogo 663
Japan
☎ 81 798 45 6111

Maeder Andre
Observ Geneve
Chemin d Maillettes 51
CH 1290 Sauverny
Switzerland
☎ 41 22 755 2611
✆ 41 22 755 3983
✉ andre.maeder@obs.unige.ch

Maehara Hideo
Tokyo Astronomical Obs
NAOJ
Osawa Mitaka
Tokyo 181
Japan
☎ 81 865 44 2155
✆ 81 865 44 2360
✉ maehara@oao.nao.ac.jp

Maetzler Christian
Physikalisches Institut
Univ Bern
Sidlerstr 5
CH 3012 Bern
Switzerland
☎ 41 31 65 4589

Maffei Paolo
Cattedra Astrofisca
Univ d Perugia
Via d elce d Sotto
I 06100 Perugia
Italy
☎ 39 75 45 647

Magain Pierre
Inst Astrophysique
Universite Liege
Av Cointe 5
B 4000 Liege
Belgium
☎ 32 4 254 7510
✆ 32 4 254 7511
✉ magain@astro.ulg.ac.be

Magakian Tigran Y
Byurakan Astrophys Observ
Armenian Acad Sci
378433 Byurakan
Armenia
☎ 374 88 52 28 3453/4142
✆ 374 88 52 52 3640

Magalhaes Antonio A S
Observ Astronomico
Univ d Porto
Monte d Virgem
P 4400 Vila Nova Gaia
Portugal
☎ 351 2 782 0404
✆ 351 2 782 7253

Magalhaes Antonio Mario
IAG
Univ Sao Paulo
CP 9638
01065 970 Sao Paulo SP
Brazil
☎ 55 11 577 8599
✆ 55 11 576 3848
✉ magalhaes@vax.iagusp.usp.br

Magazzu Antonio
Osserv Astronomico
d Catania
Via A Doria 6
I 95125 Catania
Italy
☎ 39 95 733 2242
✆ 39 95 33 0592
✉ antonio@ct.astro.it

Maggio Antonio
Osserv Astronomico
Univ d Palerno
Piazza Parlemento 1
I 90134 Palermo
Italy
☎ 39 91 651 8132
✆ 39 91 651 7292
✉ maggio@oapa.astropa.it

Magnan Christian
GRAAL
Univ Montpellier II
Pl E Bataillon
F 34095 Montpellier Cdx 5
France
☎ 33 4 67 14 3902
✆ 33 4 67 54 4850
✉ magnan@graal.univ-montp2.fr

Magnani Loris Alberto
Dpt Phys/Astronomy
Univ of Georgia
Athens GA 30602 2451
USA
☎ 1 706 542 2876
✆ 1 706 542 2492
✉ loris@jove.physast.uga.edu

Magnaradze Nina G
Abastumani Astrophysical Observ
Georgian Acad Sci
380060 Tbilisi
Georgia
☎ 995 88 32 37 5226
✆ 995 88 32 98 5017

Magni Gianfranco
IAS
Rpto Planetologia
Via d universita 11
I 00133 Roma
Italy
☎ 39 6 4993 4473
✆ 39 6 2066 0188

Magnusson Per
Swedish Ntl Space Board
Box 4006
S 171 04 Solna
Sweden
☎ 46 8 627 6492
✆ 46 8 627 5014
✉ per.magnusson@astro.uu.se

Magris C Gladis
CIDA
Box 264
Merida 5101 A
Venezuela
☎ 598 7 471 2780/3883
✆ 598 7 471 2459
✉ magris@cida.ve

Magun Andreas
Physikalisches Institut
Univ Bern
Sidlerstr 5
CH 3012 Bern
Switzerland
☎ 41 31 65 8914

Mahat Rosli H
Astronomy Dpt
Univ of Malaya
59100 Kuala Lumpur
Malaysia

Maheswaran Murugesapillai
Inst Fundament Studies
380/72 Bauddhaloka
Mawatha
Colombo 7
Sri Lanka
☎ 94 1 597538

Mahmoud Farouk M A B
NAIGR
Helwan Observ
Cairo 11421
Egypt
☎ 20 78 0645/2683
✆ 20 62 21 405 297

Mahra H S
Uttar Pradesh State Observ
Po Manora Peak 263 129
Nainital 263 129
India
☎ 91 59 42 2136/2583

Mahtessian Abraham P
Byurakan Astrophys Observ
Armenian Acad Sci
378433 Byurakan
Armenia
☎ 374 88 52 28 3453/4142
✆ 374 88 52 52 3640

Maia Marcio A G
Observ Nacional
Rua Gl Bruce 586
Sao Cristovao
20921 030 Rio de Janeiro RJ
Brazil
☎ 55 21 589 6504
✆ 55 21 580 0332
✉ maia@on.br

Maihara Toshinori
Physics Dpt
Kyoto Univ
Sakyo Ku
Kyoto 606 01
Japan
☎ 81 757 51 2111
✆ 81 75 753 7010

Maillard Jean-Pierre
IAP
98bis bd Arago
F 75014 Paris
France
☎ 33 1 44 32 8139
✆ 33 1 44 32 8001
✉ maillard@iap.fr

Maitzen Hans M
Inst Astronomie
Univ Wien
Tuerkenschanzstr 17
A 1180 Wien
Austria
☎ 43 1 345 36094
✆ 43 1 470 6015
✉ maitzen@astro.ast.univie.ac.at

Majid Abdul Bin A H
Wisma Mahyuddin Dan Siew
79 Jalan 1/91 Taman Shamelin Perkasa
Batu 31/2 Jalan Cheras
55100 Kuala Lumpur
Malaysia
☎ 60 03 9831 988
✆ 60 03 9832 000
✉ msurver@tm.net.my

Major John
Physics Dpt
Univ of Durham
South Rd
Durham DH1 3LE
UK
☎ 44 191 374 2111
✆ 44 191 374 3749

Makarenko Ekaterina N
Astronomical Observ
Odessa State Univ
Shevchenko Park
270014 Odessa
Ukraine
☎ 380 4822 28442
✆ 380 4822 28442

Makarov Valentine I
Kislovodsk Station of the Pulkovo Observ
357741 Kislovodsk
Russia
☎ 7 865 373 3088
✆ 7 812 123 1922

Makarov Valeri
Astronomical Observ, NBIfAFG
Copenhagen Univ
Juliane Maries Vej 30
DK 2100 Copenhagen
Denmark
☎ 45 35 32 5966
✆ 45 35 32 5989
✉ makarov@astro.ku.dk

Makarova Elena A
SAI
Acad Sciences
Universitetskij Pr 13
119899 Moscow
Russia
☎ 7 095 139 1973
✆ 7 095 939 0126

Makhlouf Amar
DREAN
14 r Zighoud Youcef
Altarf 36
Algeria
☎ 213 8 68 0792

Makino Fumiyoshi
ISAS
3 1 1 Yoshinodai
Sagamihara
Kanagawa 229 8510
Japan
☎ 81 814 27 513911
✆ 81 814 275 94253

Makino Junichiro
Dpt Info Sci/gra
Univ of Tokyo
Meguro Ku
Tokyo 153
Japan
☎ 81 334 65 3925
✆ 81 334 65 3925
✉ makino@kyohou.c.u-tokyo.ac.jp

Makishima Kazuo
Inst Space/Astron Sci
Univ of Tokyo
Meguro Ku
Tokyo 153
Japan
☎ 81 346 71 111*303
✆ 81 334 85 2904

Makita Mitsugu
Osaka Gakuin Jn College
2 37 1 Kishibe Minami
Suita
Osaka 564
Japan

Malacara Daniel
CTIO
Apdo Postal 948
Leon GTO 37000
Mexico
☎ 52 4 718 4425
✆ 52 4 717 5000

Malagnini Maria Lucia
OAT
Box Succ Trieste 5
Via Tiepolo 11
I 34131 Trieste
Italy
☎ 39 40 31 99255
✆ 39 40 30 9418

Malaise Daniel J
DMO Consulting SA
r Varin 141 A
B 4000 Liege
Belgium
☎ 32 4 254 0034

Malakpur Iradj
Inst of Geophysics
Univ of Tehran
Kargar Shomali
Tehran 14394
Iran
☎ 98 21 631081/3

Malaroda Stella M
Complejo Astronomico
El Leoncito
CC 467
5400 San Juan
Argentina
☎ 54 64 22 5718

Malasan Hakim Luthfi
Bosscha Observ
ITB Lembang
Bandung 40391
Indonesia
☎ 62 22 278 6001
✆ 62 22 250 9170
✉ hakim@sirius.as.itb.ac.id

Malawi Abdulrahman
Astronomy Dpt
Box 9028
KAAU
Jeddah 21413
Saudi Arabia
☎ 966 2 687 0571
✆ 966 2 687 0571
✉ scf3010@sakaau03

Malherbe Jean-Marie
Observ Paris Meudon
DASOP
Pl J Janssen
F 92195 Meudon PPL Cdx
France
☎ 33 1 45 07 7629
✆ 33 1 45 07 7634
✉ malherrbe@obspm.fr

Malin David F
AAO
Box 296
Epping NSW 2121
Australia
☎ 61 2 9372 4800
✆ 61 2 9372 4860
✉ dfm@aaoepp.aao.gov.au

Malina Roger Frank
Ctr for EUV Astrophys
Univ of California
2150 Kittredge St
Berkeley CA 94720 5030
USA
☎ 1 510 643 5636
✆ 1 415 643 5660
✉ rmalina@ssl.berkeley.edu

Malitson Harriet H
13315 Magellan Av
Rockville MD 20853
USA
☎ 1 301 946 0496

Malkamaeki Lauri J
Elson Res Inc
Box 6356
Ingwood TX 77325 6356
USA

Malkan Matthew Arnold
Dpt Astronomy
UCLA
Box 951562
Los Angeles CA 90025 1562
USA
☎ 1 310 825 3404
✆ 1 310 206 2096
✉ malkan@bonnie.astro.ucla.edu

Malkin Zinovy M
Inst Appl Astronomy
Zhdanovskaya 8
St Petersburg 197110
Russia
☎ 7 812 235 3216
✆ 7 812 230 7413
✉ malkin@ipa.rssi.ru

Malkov Oleg Yu
Inst of Astronomy
Acad Sciences
Pyatnitskaya st 48
109017 Moscow
Russia
☎ 7 095 953 1702
✆ 7 095 230 2081
✉ malkov@inasan.rssi.ru

Mallia Edward A
Astrophysics Dpt
Univ of Oxford
Keble Rd
Oxford OX1 3RQ
UK
☎ 44 186 527 3999
✆ 44 186 527 3947

Mallik D C V
IIA
Koramangala
Sarjapur Rd
Bangalore 560 034
India
☎ 91 80 356 6585/6497
✆ 91 80 553 4043

Malofeev Valerij M
Lebedev Physical Inst
Leninsky Pspt 53
Moscow 117924
Russia
☎ 7 096 7734 187
✆ 7 095 135 7880
✉ malofeev@rasfian.sezpukhov.su

Maltby Per
Inst Theor Astrophys
Univ of Oslo
Box 1029
N 0315 Blindern Oslo 3
Norway
☎ 47 22 856 509
✆ 47 22 85 6505

Malumian Vigen
Byurakan Astrophys Observ
Armenian Acad Sci
378433 Byurakan
Armenia
☎ 374 88 52 28 3453/4142
℻ 374 88 52 52 3640
✉ malumian@
byruakan.sci.am

Malville J Mckim
APAS
Univ of Colorado
Campus Box 391
Boulder CO 80309 0391
USA
☎ 1 303 492 8788
✉ malville@spot.
colorado.edu

Malyuto Valeri
Tartu Observ
Estonian Acad Sci
Toravere
EE 2444 Tartumaa
Estonia
☎ 372 7 410 305
℻ 372 7 410 205
✉ valeri@aai.ee

Mammano Augusto
Osserv Astrofisico
Univ d Padova
Via d Osservatorio 8
I 36012 Asiago
Italy
☎ 39 424 462 665
℻ 39 424 462 884

Mampaso Antonio
IAC
Observ d Teide
via Lactea s/n
E 38200 La Laguna
Spain
☎ 34 2 232 9100
℻ 34 2 232 9117
✉ amr@iac.es

Manabe Seiji
Intl Latitude Observ
NAOJ
Hoshigaoka Mizusawa Shi
Iwate 023
Japan
☎ 81 197 24 7111
℻ 81 197 22 7120

Manara Alessandro A
Osserv Astronomico
d Brera
Via Brera 28
I 20121 Milano
Italy
☎ 39 2 720 23751
℻ 39 2 720 01600

Manchado Arturo
IAC
Observ d Teide
via Lactea s/n
E 38200 La Laguna
Spain
☎ 34 2 232 9100
℻ 34 2 232 9117

Manchanda R K
TIFR
Homi Bhabha Rd
Colaba
Bombay 400 005
India
☎ 91 22 219 111*336
℻ 91 22 495 2110

Manchester Richard N
CSIRO
Div Radiophysics
Box 76
Epping NSW 2121
Australia
☎ 61 2 868 0225
℻ 61 2 868 0310
✉ rmanches@atnf.
csiro.au

Mancuso Santi
Dpt Scienze Fisiche
Univ d Napoli
Mostra D Oltremare Pad 19
I 80125 Napoli
Italy
☎ 39 81 725 3447

Mandel Holger
Landessternwarte
Koenigstuhl
D 69117 Heidelberg
Germany
☎ 49 622 150 9234
℻ 49 622 150 9202
✉ hmandel@lsw.uni-
heidelberg.de

Mandolesi Nazzareno
TeSRE
CNR
Via P Gobetti101
I 40129 Bologna
Italy
☎ 39 51 639 8665
℻ 39 51 639 9385

Mandrini Cristina Hemilse
IAFE
CC 67 Suc 28
1428 Buenos Aires
Argentina
☎ 54 1 781 6755
℻ 54 1 786 8114
✉ mandrini@iafe.uba.ar

Mandzhos Andrej V
SAO
Acad Sciences
Nizhnij Arkhyz
357147 Karachaevo
Russia
☎ 7 865 789 2501
✉ amand@sao.
stavropol.su

Manfroid Jean
Inst Astrophysique
Universite Liege
Av Cointe 5
B 4000 Liege
Belgium
☎ 32 4 254 7510
℻ 32 4 254 7511
✉ manfroid@astro.
ulg.ac.be

Mangeney Andre
Observ Paris Meudon
DESPA
Pl J Janssen
F 92195 Meudon PPL Cdx
France
☎ 33 1 45 07 7661
℻ 33 1 45 07 7971

Mangum Jeffrey Gary
NRAO
Campus Bg 65
949 North Cherry Av
Tucson AZ 85721 0655
USA
☎ 1 520 621 5685
℻ 1 520 621 5554
✉ jmangum@nrao.edu

Mann Gottfried
Zntrlinst Astrophysik
Sternwarte Babelsberg
Rosa Luxemburg Str 17a
D 14473 Potsdam
Germany
☎ 49 331 762225

Mann Ingrid
MPI Aeronomie
Max Planck Str 2
Postfach 20
D 37189 Kaltenburg Lindau
Germany
☎ 49 555 697 9291
℻ 49 555 697 9240
✉ mann@linax1.dnet.
gwdg.de

Mann Patrick J
Dpt Phys/Astronomy
Univ W Ontario
London ON N6A 3K7
Canada
☎ 1 519 661 3183
℻ 1 519 661 3486
✉ 2014-562@uwovax

Mannheim Karl
Univ Sternwarte
Goettingen
Geismarlandstr 11
D 37083 Goettingen
Germany
☎ 49 551 395 050
℻ 49 551 395 043
✉ kmannhe@uni-sw.
gwdg.de

Mannino Giuseppe
Istt Matematico
Via Campi 181
I 41100 Modena
Italy

Mannucci Filippo
Osserv Astrofis Arcetri
Univ d Firenze
Largo E Fermi 5
I 50125 Firenze
Italy
☎ 39 55 27 52230
℻ 39 55 22 0039
✉ filippo@arcetri.
astro.it

Manoharan P K
Radio Astronomy Centre
Tata Inst Fundamental Res
PO Box 8
Udhagamandalam (Ooty) 643 001
India
☎ 91 423 42 032
℻ 91 423 42 588
✉ mano@racooty.ernet.in

Manrique Walter T
Observ Astronomico
Felix Aguilar
Av Benavidez 8175 Oeste
5407 Marquesado (SJ)
Argentina
☎ 54 64 23 1494

Mansfield Victor N
Dpt Phys/Astronomy
Colgate Univ
Hamilton NY 13346
USA
☎ 1 315 824 1000

Mansouri Reza
Physics Dpt
Sharif Univ
Box 11365 9161
Tehran 11365
Iran
☎ 98 21 918 2619
℻ 98 21 601 2983
✉ mansouri@netware2.imp.ac.ir

Mantegazza Luciano
Osserv Astronomico
d Milano
Via E Bianchi 46
I 22055 Merate
Italy
☎ 39 990 6412
℻ 39 990 8492

Mao Wei
Yunnan Observ
CAS
Kunming 650011
China PR
☎ 86 871 2035
℻ 86 871 717 1845

Marabini Rodolfo Jose
Observ Astronomico
Paseo d Bosque S/n
1900 La Plata (Bs As)
Argentina
☎ 54 21 217 308
℻ 54 21 211 761

Maran Stephen P
NASA GSFC
Code 600
Greenbelt MD 20771
USA
☎ 1 301 286 5154
℻ 1 301 286 1772
✉ hrsmaran@eclair.gsfc.nasa.gov

Marano Bruno
Dpt Astronomia
Univ d Bologna
Via Zamboni 33
I 40126 Bologna
Italy
☎ 39 51 259 301
℻ 39 51 259 407

Marar T M k
ISRO Satellite Ctr
Airport Rd
Vimanapura Post
Bangalore 560 017
India
☎ 91 80 566 251

Maraschi Laura
IFCTR CNR
Univ d Milano
Via E Bassini 15
I 20133 Milano
Italy
☎ 39 2 236 3542
✆ 39 2 266 5753

Marcaide Juan-Maria
Dpt Astron/Astrofisica
Univ Valencia
Dr Moliner 50
E 46100 Burjassot
Spain
☎ 34 6 398 3079
✆ 34 6 398 3084
✉ jmm@vlbi.uv.es

Marcelin Michel
Observ Marseille
2 Pl Le Verrier
F 13248 Marseille Cdx 04
France
☎ 33 4 91 95 9088
✆ 33 4 91 62 1190

Marchal Christian
Dpt Etudes Synthese
ONERA
F 92320 Chatillon ss Bagneux
France
☎ 33 1 46 57 1160

Marcialis Robert
Lunar/Planetary Lab
Room 325 Bg 92
Univ of Arizona
Tucson AZ 85721 0092
USA
☎ 1 520 327 4827
✆ 1 520 621 4933
✉ umpire@lpl.
arizona.edu

Mardirossian Fabio
Dpt Astronomia
Univ d Trieste
Via Tiepolo 11
I 34131 Trieste
Italy
☎ 39 40 79 3921*221
✆ 39 40 30 9418

Mardling Rosemary
Mathematics Dpt
Monash Univ
Clayton 3168
Australia
☎ 61 3 9905 4506
✆ 61 3 9905 3867
✉ r.mardling@2maths.
monash.edu.au

Marek John
44 Percy Rd
Wrexham Clwyd
UK

Margon Bruce H
Astronomy Dpt
Univ of Washington
Box 351580
Seattle WA 98195 1580
USA
☎ 1 425 543 0089
✆ 1 425 685 0403
✉ margon@janus.
astro.washington.

Margoni Rino
Osserv Astrofisico
Univ d Padova
Via d Osservatorio 8
I 36012 Asiago
Italy
☎ 39 424 462 665
✆ 39 424 462 884

Margrave Thomas Ewing
400 Johnson St
Vienna VA 22180
USA

Marie M A
Astronomy Dpt
Fac of Sciences
Cairo Univ
Geza
Egypt
☎ 20 2 572 7022
✆ 20 2 572 7556

Marik Miklos
Astronomy Dpt
Eotvos Univ
Ludovika ter 2
H 1083 Budapest
Hungary
☎ 36 1 114 1019
✆ 36 1 210 1089
✉ marik@innin.elte.hu

Marilli Ettore
Osserv Astronomico
d Catania
Via A Doria 6
I 95125 Catania
Italy
☎ 39 95 733 2246
✆ 39 95 33 0592
✉ emarilli@astrct.
ct.astro.it

Marino Brian F
156 Queen St
Northcote
Auckland 9
New Zealand

Mariotti Jean-Marie
Observ Paris Meudon
Pl J Janssen
F 92195 Meudon PPL Cdx
France
☎ 33 1 45 07 7570
✆ 33 1 45 07 7971

Maris Georgeta
Astronomical Institute
Romanian Acad Sciences
Cutitul de Argint 5
RO 75212 Bucharest
Rumania
☎ 40 1 641 3686
✆ 40 1 312 3391
✉ gmaris@imar.ro

Mariska John Thomas
NRL
Code 7673
4555 Overlook Av SW
Washington DC 20375 5000
USA
☎ 1 202 767 2605
✆ 1 202 404 7997
✉ jmariska@solar.
stanford.edu

Mark James Wai-Kee
4510 Fox Run Dr
Plainsboro NJ 08536
USA

Markellos Vassilis V
Dpt Eng Science
Univ of Patras
GR 261 10 Rion
Greece
☎ 30 61 99 7572
✆ 30 61 99 7636

Markkanen Tapio
Helsinky Univ
Observ
Box 14
FIN 00014 Helsinki
Finland
☎ 358 19 12 2335
✆ 358 19 12 2952

Markova Nevjana
Ntl Astronomical Obs
Bulgarian Acad Sci
Box 136
BG 4700 Smoljan
Bulgaria
☎ 359 3 01 2890
✆ 359 3 02 1356

Markowitz William
651 SW 6th St
Ct 1012
Pompano Beach Fl 33060
USA
☎ 1 305 941 0083

Markworth Norman Lee
Dpt Phys/Astronomy
Stephen F Austin State Un
Box 13044
Nacogdoches TX 75962
USA
☎ 1 409 468 3001
✆ 1 409 468 1226
✉ f_markwort@titan.
sfasu.edu

Marlborough J M
Dpt Phys/Astronomy
Univ W Ontario
London ON N6A 3K7
Canada
☎ 1 519 679 3184
✆ 1 519 661 3486

Marmolino Ciro
Dpt Fisica
Univ d Napoli
Mostra D Oltremare Pad 19
I 80125 Napoli
Italy
☎ 39 81 725 3428

Marochnik L S
Computor Sciences Corp
System Sciences div
10000 A Areospace Rd
Lauham Seabrook Ma 20706
USA
☎ 1 301 794 1483
✆ 1 301 459 4482
✉ leonid@cexsels.gsfc.nasa.gov

Marov Mikhail Ya
Keldysh Inst Applied Maths
Acad Sciences
Miusskaja Sq 4
125047 Moscow
Russia
☎ 7 095 250 0485
✆ 7 095 972 0737
✉ marov@applmat.msk.su

Marques Dos Santos P
IAG
Univ Sao Paulo
CP 9638
01065 970 Sao Paulo SP
Brazil
☎ 55 11 276 3941
✆ 55 11 276 3848

Marques Manuel N
OAL
Tapada d Ajuda
P 1300 Lisboa 3
Portugal
☎ 351 1 363 7351
✆ 351 1 362 1722

Marraco Hugo G
Observ Astronomico
Paseo d Bosque S/n
1900 La Plata (Bs As)
Argentina
☎ 54 21 21 7308
✆ 54 21 211 761

Marschall Laurence A
Physics Dpt
Gettysburg College
Gettysburg PA 17325
USA
☎ 1 717 337 6062
✆ 1 717 337 6666
✉ marshall@gettysburg.edu

Marscher Alan Patrick
Astronomy Dpt
Boston Univ
725 Commonwealth Av
Boston MA 02215
USA
☎ 1 617 353 5029
✆ 1 617 353 5704

Marsden Brian G
CfA
HCO/SAO
60 Garden St
Cambridge MA 02138 1516
USA
☎ 1 617 495 7244
✆ 1 617 495 7231
✉ marsden@cfa.harvard.edu

Marsh Julian C D
Hatfield Polytechnic
Observ
Bayfordbury
Hertford Herts SG13 8LD
UK
☎ 44 199 255 8451

Marsh Thomas
Astrophysics Dpt
Univ of Oxford
Keble Rd
Oxford OX1 3RH
UK
☎ 44 186 527 3303
℻ 44 186 527 3418

Marshall Herman Lee
Ctr for EUV Astrophys
Univ of California
2150 Kittredge St
Berkeley CA 94720 5030
USA
☎ 1 415 643 5671
℻ 1 415 643 5660
✉ hermanm@ssl.
bereley.edu

Marshall Kevin P
Medellin Planetarium
Carrera 52 No 71 177
Medellin
Columbia
℻ 57 233 21 59
✉ planetar@educame.
gov.co

Marsoglu A
Univ Observ
Univ of Istanbul
University 34452
34452 Istanbul
Turkey
☎ 90 212 522 3597
℻ 90 212 519 0834

Marston Anthony Philip
Dpt Phys/Astronomy
Drake Univ
Des Moines Ia 50311
USA
☎ 1 515 271 3034
℻ 1 515 271 3977
✉ tm9991r@acad.drake.edu

Martens Petrus C
ESA/ESTEC
SSD
Box 299
NL 2200 AG Noordwijk
Netherlands
☎ 31 71 565 6555
℻ 31 71 565 4690
✉ pmartens@soho.
esa.estec.nl

Martin Anthony R
UK Culham Laboratory
Rm F4/135
Abingdon OX14 3DB
UK
☎ 44 12 35 21840

Martin Derek H
Astrophysics Group
QMWC
Mile End Rd
London E1 4NS
UK
☎ 44 171 975 5555
℻ 44 181 975 5500

Martin Eduardo
IAC
Observ d Teide
via Lactea s/n
E 38200 La Laguna
Spain
☎ 34 2 232 9100
℻ 34 2 232 9117
✉ ege@iac.es

Martin Francois
Dpt Astrophysique
Universite Nice
Parc Valrose
F 06034 Nice Cdx
France
☎ 33 4 93 51 9100
℻ 33 4 93 52 9806

Martin Inacio Malmonge
Univ Estadual Campinas
Instituto Fisica
CP 6165
13083 970 Campinas SP
Brazil
☎ 55 192 398 112
℻ 55 192 393 127
✉ martin@ifi.unicamp.br

Martin Jean-Michel P
Observ Paris Meudon
ARPEGES
Pl J Janssen
F 92195 Meudon PPL Cdx
France
☎ 33 1 45 07 7608
℻ 33 1 45 07 7971
✉ jmmartin@obspm.fr

Martin Maria Cristina
IAR
CC 5
1894 Villa Elisa (Bs As)
Argentina
☎ 54 21 4 3793
℻ 54 21 211 761

Martin Peter G
Cita Mclennan Labs
Univ of Toronto
60 St George St
Toronto ON M5A 1A1
Canada
☎ 1 416 978 6840
℻ 1 416 978 3921
✉ pgmartin@cita.
utoronto.ca

Martin Robert N
Steward Observ
Univ of Arizona
Tucson AZ 85721
USA
☎ 1 520 621 1539
℻ 1 520 621 1532

Martin William C
NBS
Physics Bg A167
Gaithersburg MD 20899
USA
☎ 1 301 921 2011

Martin William L
Royal Greenwich Obs
Madingley Rd
Cambridge CB3 0EZ
UK
☎ 44 12 23 374 000
℻ 44 12 23 374 700

Martin-Diaz Carlos
IAC
Observ d Teide
via Lactea s/n
E 38200 La Laguna
Spain
☎ 34 2 232 9100
℻ 34 2 232 9117
✉ cmd@iac.es

Martin-Loron M
Hermanos Miralles 14
E Madrid 1
Spain

Martin-Pintado Jesus
Ctr Astron d Yebes
OAN
Apt 148
E 19080 Guadalajara
Spain
☎ 34 1 122 3358
℻ 34 1 129 0063

Martinet Louis
Observ Geneve
Chemin d Maillettes 51
CH 1290 Sauverny
Switzerland
☎ 41 22 755 2611
℻ 41 22 755 3983
✉ louis.martinet@obs.
unige.ch

Martinez Mario
Dpt Geofisica
CIESE
Apt 2732
Ensenada BC 22860
Mexico

Martinez Peter
SAAO
PO Box 9
7935 Observatory
South Africa
☎ 27 21 47 0025
℻ 27 21 47 3639
✉ peter@saao.ac.za

Martinez Pillet Valentin
IAC
Observ d Teide
via Lactea s/n
E 38200 La Laguna
Spain
☎ 34 2 232 9100
℻ 34 2 232 9117
✉ vmp@iac.es

Martinez Roger Carlos
IAC
Observ d Teide
via Lactea s/n
E 38200 La Laguna
Spain
☎ 34 2 232 9100
℻ 34 2 232 9117

Martinez-Gonzalez E
Istt Fisica d Cantabria
C/o Faculdad Ciencias
Avda Los Castros s/n
E 39005 Santander
Spain
☎ 34 4 220 1468
℻ 34 4 220 1402

Martini Aldo
IAS
Area d Ricerca CNR
Via Fosso Cavaliere 100
I 00133 Roma
Italy
☎ 39 6 4993 4473
℻ 39 6 2066 0188

Martins Donald Henry
Dpt Phys/Astronomy
Univ of Alaska
3221 Uaa Dr
Anchorage AK 99508
USA
☎ 1 907 786 1238

Martres Marie-Josephe
Observ Paris Meudon
F 92195 Meudon PPL Cdx
France

Marvin Ursula B
CfA
HCO/SAO
60 Garden St
Cambridge MA 02138 1516
USA
☎ 1 617 495 7270
℻ 1 617 495 7356

Marx Gyorgy
Dpt Atomic Physics
Eotvos Univ
Pushkin U 5 7
H 1088 Budapest
Hungary
☎ 36 1 187 902
℻ 36 1 118 0206
✉ kuerti@awiraf

Marzari Francesco
Dpt Fisica G Galilei
Univ d Padova
Via Marzolo 8
I 35131 Padova
Italy
☎ 39 49 827 7190
℻ 39 49 827 7102
✉ mazari@pd.infn.it

Masai Kuniaki
Physics Dpt
Tokyo Metropol Univ
1-1 Minami Ohsawa
Hachioji Tokyo 192 03
Japan
☎ 81 426 77 2516
℻ 81 426 77 2483
✉ masai@phys.metro-u.ac.jp

Masani A
Osserv Astronomico
d Brera
Via Brera 28
I 20100 Milano
Italy
☎ 39 2 874 444
✆ 39 2 720 01600

Masegosa Gallego J
Inst Astrofisica
Andalucia Apt 3004
Prof Albareda 1
E 18080 Granada
Spain
☎ 34 5 812 1311
✆ 34 5 881 4530

Masheder Michael
Physics Dpt
Univ of Bristol
Tyndall Av
Bristol BS8 1TL
UK
☎ 44 11 79 288 716
✆ 44 11 79 255 624
✉ mike.masheder@
bristol.ac.uk

Mashonkina Lyudmila
Astronomy Dpt
Kazan State Univ
Kremlevskaya str 18
Kazan 420008
Russia
☎ 7 843 264 3092
✉ ml@astro.ksu.ras.ru

Maslennikov Kirill L
Pulkovo Observ
Acad Sciences
10 Kutuzov Quay
196140 St Petersburg
Russia
☎ 7 812 123 4493
✆ 7 812 123 3360
✉ cyrill@pulkovo.
spb.su

Maslowski Jozef
Astronomical Observ
Krakow Jagiellonian Univ
Ul Orla 171
PL 30 244 Krakow
Poland
☎ 48 12 25 1294
✆ 48 12 25 1318

Masnou Francoise
28 All Gambauberie
F 91190 Gif s Yvette
France

Masnou Jean-Louis
Observ Paris Meudon
DARC
Pl J Janssen
F 92195 Meudon PPL Cdx
France
☎ 33 1 45 07 7427
✆ 33 1 45 07 7971

Mason Glenn M
Physics Dpt
Univ of Maryland
College Park MD 20742 2421
USA
☎ 1 301 405 6203
✆ 1 301 314 9547
✉ mason@sampx3.
umd.edu

Mason Helen E
Dpt Appl Maths/Theor Phys
Silver Street
Cambridge CB3 9EW
UK
☎ 44 12 23 337 98
✆ 44 12 23 337 918

Mason John William
51 Orchard Way West Barnham
Bognor Regis
West Sussex PO22 0HX
UK
☎ 44 124 35 53244
✆ 44 124 355 4272

Mason Keith Owen
Mullard Space Science Lab
Univ College London
Holmbury St Mary
Dorking Surrey RH5 6NT
UK
☎ 44 13 06 702 92
✆ 44 14 83 278 312

Massa Derck Louis
Hughes STX Corp
7701 Greenbelt Rd
Suite 400
Greenbelt MD 20770
USA
☎ 1 301 286 5767
✆ 1 301 286 1771
✉ derck.massa@gsfc.
nasa.gov

Massaglia Silvano
Istt Fisica
Univ d Torino
Corso d Azeglio 46
I 10125 Torino
Italy
☎ 39 11 657 694

Massaguer Josep
Dpt Fisica Aplicada
U P Cataluna
Campus Nord Ed C3
E 08034 Barcelona
Spain
☎ 34 3 401 6827
✆ 34 3 401 6090
✉ massaguer@fa.upc.es

Massevich Alla G
Inst of Astronomy
Acad Sciences
Pyatnitskaya Ul 48
109017 Moscow
Russia
☎ 7 095 231 5461
✆ 7 095 230 2081

Massey Philip L
NOAO/KPNO
Box 26732
950 N Cherry Av
Tucson AZ 85726 6732
USA
☎ 1 520 327 5511
✆ 1 520 325 9360

Masson Colin R
CfA
HCO/SAO
60 Garden St
Cambridge MA 02138 1516
USA
☎ 1 617 495 7000
✆ 1 617 495 7356

Masuda Satoshi
STEL Nagoya Univ
3-13 Honohara Toyokawa
Aich
Nagoya 442
Japan
☎ 81 53 389 5186
✆ 81 53 389 5090
✉ masuda@stelab.
nagoya-u.ac.jp

Matas Vladimir R
ARI
Moenchhofstr 12-14
D 69120 Heidelberg
Germany
☎ 49 622 140 50
✆ 49 622 140 5297

Materne Juergen
Aretinstr 27
D 81545 Muenchen
Germany

Matese John J
Physics Dpt
USL
Box 44210
Lafayette LA 70504 4210
USA
☎ 1 318 482 6697
✆ 1 318 482 6699
✉ matese@usl.edu

Mather John Cromwell
NASA GSFC
Code 685
Greenbelt MD 20771
USA
☎ 1 301 286 8720
✆ 1 301 286 1617
✉ mather@stars.gsfc.
nasa.gov

Matheson David Nicholas
Rutherford Appleton Lab
Space/Astrophysics Div
Bg R25/R68
Chilton Didcot OX11 0QX
UK
☎ 44 12 35 821 900
✆ 44 12 35 821 900

Mathews William G
Lick Observ
Univ of California
Santa Cruz CA 95064
USA
☎ 1 831 429 2074
✆ 1 831 426 3115

Mathewson Donald S
MSSSO
Weston Creek
Private Bag
Canberra ACT 2611
Australia
☎ 61 262 881 111
✆ 61 262 490 233
✉ dsm@mso.anu.edu.au

Mathez Guy
OMP
14 Av E Belin
F 31400 Toulouse Cdx
France
☎ 33 5 61 25 2101
✆ 33 5 61 27 3179

Mathieu Robert D
Astronomy Dpt
Univ Wisconsin
475 N Charter St
Madison WI 53706 1582
USA
☎ 1 608 262 5679
✆ 1 608 263 0361
✉ mathieu@madraf.astro.
wisc.edu

Mathioudakis Mihalis
Astrophysics Dpt
Ntl Univ Athens
Panepistimiopolis
GR 157 84 Zografos
Greece
☎ 30 1 724 3211
✆ 30 1 723 8413
✉ mm@rigel.da.uoa.gr

Mathis John S
Astronomy Dpt
Univ Wisconsin
475 N Charter St
Madison WI 53706 1582
USA
☎ 1 608 262 5994
✆ 1 608 263 0361
✉ mathis@wiscmacc

Mathur B S
Ntl Physical Laboratory
Time/Frequency Section
Hillside Rd
New Delhi 110 012
India
☎ 91 11 586 168

Mathys Gautier
ESO
Santiago Office
Casilla 19001
Santiago 19
Chile
☎ 56 2 698 8757
✆ 56 2 695 42 63
✉ gmathys@eso.org

Matsakis Demetrios N
USNO
3450 Massachusetts Av NW
Washington DC 20392 5100
USA
☎ 1 202 653 1823
✉ dnm@orion.usno.navy.mil

Matson Dennis L
CALTECH/JPL
MS 183 501
4800 Oak Grove Dr
Pasadena CA 91109 8099
USA
☎ 1 213 354 2984
℻ 1 818 393 6030

Matsuda Takuya
Dpt Earth/Planet Sci
Kobe Univ
Nada
Kobe 657 8501
Japan
☎ 81 788 81 1212*4421
℻ 81 788 82 1549
✉ tmatsuda@icluna.kobe-u.ac.jp

Matsuhara Hideo
Astrophysics Dpt
Nagoya Univ
Furocho Chikusa Ku
Nagoya 464 01
Japan
☎ 81 527 89 2560
℻ 81 527 89 2919
✉ maruma@toyo.phys.nagoya-u.ac.jp

Matsui Takafumi
Dpt Earth/Planet Phy
Univ of Tokyo
Bunkyo Ku
Tokyo 113
Japan
☎ 81 338 12 2111*4305
℻ 81 338 18 3247

Matsumoto Masamichi
Fukui Univ of Technology
3 6 1 Gakuen
Fukui 910 8505
Japan
✉ masm@ccmails.fukui-ut.ac.jp

Matsumoto Ryoji
Dpt Phys/Fac of Sci
Chiba Univ
1-33 Yayoicho Inage Ku
Chiba 263
Japan
☎ 81 43 290 3724
℻ 81 43 290 3720
✉ matumoto@c.chiba-u.ac.jp

Matsumoto Toshio
ISAS
3 1 1 Yoshinodai
Sagamihara
Kanagawa 229 8510
Japan
☎ 81 427 51 3911*2638
℻ 81 427 59 4253
✉ matsumo@koala.astro.isas.ac.jp

Matsumura Masafumi
Fac of Education
Kagawa Univ
Saiwai Cho Takamatsushi
Kagawa 760
Japan
☎ 81 878 61 4141*400
℻ 81 878 34 7144

Matsuo Hiroshi
Nobeyama Radio Obs
NAOJ
Minamimaki Mura
Nagano 384 13
Japan
☎ 81 267 984 333
℻ 81 267 982 884
✉ matsuo@nro.nao.ac.jp

Matsuoka Masaru
RIKEN
2-1 Hirosawa
Wako-shi
Saitama 351 01
Japan
☎ 81 484 62 1111*3221
℻ 81 484 62 4640
✉ matsuoka@postman.riken.go.jp

Matsuura Oscar T
IAG
Univ Sao Paulo
CP 9638
01065 970 Sao Paulo SP
Brazil
☎ 55 11 275 3720
℻ 55 11 276 3848

Mattei Janet Akyuz
AAVSO
25 Birch St
Cambridge MA 02138 1205
USA
☎ 1 617 354 0484
℻ 1 617 354 0665
✉ jmattei@aavso.org

Matteucci Francesca
IAS
Area d Ricerca CNR
Via Fosso Cavaliere 100
I 00133 Roma
Italy
☎ 39 6 4993 4473
℻ 39 6 2066 0188

Matthews Clifford
Dpt Chemistry M/C111
Univ of Illinois
Box 4338
Chicago IL 60680
USA
☎ 1 312 996 3161
℻ 1 312 996 0431

Matthews Henry E
Joint Astronomy Ctr
660 N A'ohoku Pl
University Park
Hilo HI 96720
USA
☎ 1 808 961 3756
℻ 1 808 961 6516

Matthews Jaymie
Dpt Geophys/Astronomy
UBC
2075 Wesbrook Pl
Vancouver BC V6T 1Z4
Canada
☎ 1 250 822 2696/2267
℻ 1 250 822 6047
✉ matthews@astro.ubc.ca

Matthews Thomas A
Astronomy Program
Univ of Maryland
College Park MD 20742 2421
USA
☎ 1 301 454 6650
℻ 1 301 314 9067

Mattig W
Kiepenheuer Inst
Sonnenphysik
Schoeneckstr 6
D 79104 Freiburg Breisgau
Germany
☎ 49 761 328 64
℻ 49 761 319 8111

Mattila Kalevi
Helsinky Univ
Observ
Box 14
FIN 00014 Helsinki
Finland
☎ 358 19 12 2947
℻ 358 19 12 2952
✉ kalevi.mattila@helsinki.fi

Mattox John
Astronomy Dpt
Boston Univ
725 Commonwealth Av
Boston MA 02215
USA
☎ 1 617 353 5354
℻ 1 617 353 5704
✉ mattox@bu.edu

Matveyenko Leonid I
Space Res Inst
Acad Sciences
Profsojuznaya Ul 84/32
117810 Moscow
Russia
☎ 7 095 333 3122
℻ 7 095 310 7023

Matz Steven Micheal
Dearborn Observ
NW Univ
2131 Sheridan Rd
Evanston IL 60208 3112
USA
☎ 1 847 491 8643
℻ 1 847 491 3135
✉ matz@ossenu.astro.nwu.edu

Matzner Richard A
Astronomy Dpt
Univ of Texas
Rlm 15 308
Austin TX 78712 1083
USA
☎ 1 512 471 5062
℻ 1 512 471 6016

Mauas Pablo
IAFE
CC 67 Suc 28
1428 Buenos Aires
Argentina
☎ 54 1 781 6755
℻ 54 1 786 8114
✉ pablo@iafe.uba.ar

Mauche Christopher W
LLNL
L 41
Box 808
Livermore CA 94551 9900
USA
☎ 1 415 422 7017
℻ 1 415 423 7228
✉ mauche@llnl.gov

Maucherat J
LAS
Traverse du Siphon
Les Trois Lucs
F 13376 Marseille Cdx 12
France
☎ 33 4 91 05 5900
℻ 33 4 91 66 1855

Mauder Horst
Astronomisches Institut
Univ Tuebingen
Waldhaeuserstr 64
D 72076 Tuebingen
Germany
☎ 49 707 129 2486
℻ 49 707 129 3458

Mauersberger Rainer
Steward Observ
Univ of Arizona
Tucson AZ 85721
USA
☎ 1 520 621 5751
℻ 1 520 621 1532
✉ mauers@as.arizona.edu

Maurice Eric N
Observ Marseille
2 Pl Le Verrier
F 13248 Marseille Cdx 04
France
☎ 33 4 91 95 9088
℻ 33 4 91 62 1190
✉ maurice@obmara.cnrs-mrs.fr

Maurogordato Sophie
Observ Paris Meudon
Pl J Janssen
F 92195 Meudon PPL Cdx
France
☎ 33 1 45 07 7407
℻ 33 1 45 07 7469
✉ maurogordato@obspm.fr

Mauron Nicolas
GRAAL
Univ Montpellier II
Pl E Bataillon
F 34095 Montpellier Cdx 5
France
☎ 33 4 67 14 3567
℻ 33 4 67 14 4535

Mavraganis A G
Dpt Eng/Sect Mech
Ntl Techn Univ/5 Heroes
Panepistimiopolis
GR 157 84 Zografos
Greece
☎ 30 1 643 3170
℻ 30 1 723 5122

Mavrides Stamatia
Observ Paris Meudon
DESPA
Pl J Janssen
F 92195 Meudon PPL Cdx
France
☎ 33 1 45 07 7597
✆ 33 1 45 07 7469

Mavridis L N
Dpt Geodesy/Astron
Univ Thessaloniki
GR 540 06 Thessaloniki
Greece
☎ 30 31 99 2693

Mavromichalaki Helen
Dpt Physics/Univ
Nuclear Physics Section
104 Solonos St
GR 106 80 Athens
Greece
☎ 30 1 363 9439

Max Claire E
LLNL
L 413
Box 808
Livermore CA 94551 9900
USA
☎ 1 415 422 5442
✆ 1 415 423 0238
✉ max1@llnl.gov

Maxwell Alan
CfA
HCO/SAO
60 Garden St
Cambridge MA 02138 1516
USA
☎ 1 617 495 9059
✆ 1 617 495 7356

May J
Observ Radioastr d Maipu
Univ Chile
Casilla 68
Santiago
Chile
☎ 56 2 229 4101

Mayer Cornell H
1209 Villamay Blvd
Alexandria VA 22307
USA

Mayer Pavel
Astronomical Institute
Charles Univ
V Holesovickack 2
CZ 180 00 Praha 8
Czech R
☎ 420 2 2191 2572
✆ 420 2 688 5095
✉ mayer@mbox.cesnet.cz

Mayfield Earle B
Californian Polytechnic
State Univ
1427 Bayview Heights Dr
Los Osos CA 93403
USA
☎ 1 213 528 5231

Mayor Michel
Observ Geneve
Chemin d Maillettes 51
CH 1290 Sauverny
Switzerland
☎ 41 22 755 2611
✆ 41 22 755 3983
✉ michel.mayor@obs.unige.ch

Mayya Divakara
INAOE
Tonantzintlaz
Apdo 51 y 216
Puebla PUE 72000
Mexico
☎ 52 2 247 2011
✆ 52 2 247 2231
✉ ydm@inaoep.mx

Maza Jose
Dpt Astronomia
Univ Chile
Casilla 36 D
Santiago
Chile
☎ 56 2 229 4101
✆ 56 2 229 4002
✉ masa@uchcecvm

Mazeh Tsevi
Wise Observ
Tel Aviv Univ
Ramat Aviv
Tel Aviv 69978
Israel
☎ 972 3 640 8729
✆ 972 3 640 8149
✉ mazeh@wise7.tau.ac.il

Mazure Alain
LAS
Traverse du Siphon
Les Trois Lucs
F 13376 Marseille Cdx 12
France
☎ 33 4 91 05 5902
✆ 33 4 91 66 1855
✉ amazure@astrsp-mrs.fr

Mazurek Thaddeus John
6920 Av Rotella
San Jose CA 95139
USA

Mazzitelli Italo
IAS
Area d Ricerca CNR
Via Fosso Cavaliere 100
I 00133 Roma
Italy
☎ 39 6 4993 4473
✆ 39 6 2066 0188

Mazzoni Massimo
Dpt Astronomia
Univ d Firenze
Largo E Fermi 5
I 50125 Firenze
Italy
☎ 39 55 27 521
✆ 39 55 22 0039

Mazzucconi Fabrizio
Osserv Astrofis Arcetri
Univ d Firenze
Largo E Fermi 5
I 50125 Firenze
Italy
☎ 39 55 27 52250
✆ 39 55 22 0039

McAdam W Bruce
School of Physics
UNSW
Sydney NSW 2006
Australia
☎ 61 2 692 2222
✆ 61 2 660 2903

McAlister Harold A
CHARA
Georgia State Univ
Atlanta GA 30303 3083
USA
☎ 1 404 658 2932
✆ 1 404 651 1389
✉ hal@chara.gsu.edu

McBreen Brian Philip
Physics Dpt
Univ College
Belfield
Dublin 4
Ireland
☎ 353 1 693 244
✆ 353 1 671 1759

McCabe Marie K
1617 S Beretania St 801
Honolulu HI 96826
USA
☎ 1 808 956 0923
✆ 1 808 988 2790

McCall Marshall Lester
Physics Dpt
York Univ
4700 Keele St
North York ON M3J 1P3
Canada
☎ 1 416 736 2100
✆ 1 416 736 5386

McCammon Dan
Physics Dpt
Univ Wisconsin
1150 University Av
Madison WI 53706
USA
☎ 1 608 262 5916
✆ 1 608 262 3077

McCarroll Ronald
Univ Bordeaux
Lab Astrophysique
123 r Lamartine
F 33405 Talence
France
☎ 33 5 56 86 4330
✆ 33 5 56 40 4251

McCarthy Dennis D
USNO
3450 Massachusetts Av NW
Washington DC 20392 5100
USA
☎ 1 202 762 1627
✆ 1 202 652 0587
✉ dmc@maia.usno.navy.mil

McCarthy Martin F
Specola Vaticana
I 00120 Citta del Vaticano
Vatican City State
☎ 39 6 698 3411/5266
✆ 39 6 698 84671
✉ mmcarthy@as.arizona.edu

McClain Edward F
4133 Maple Rd
Morningside MD 20746
USA
☎ 1 301 736 8933

McClintock Jeffrey E
CfA
HCO/SAO
60 Garden St
Cambridge MA 02138 1516
USA
☎ 1 617 495 7136
✆ 1 617 495 7356

McClure Robert D
NRCC/HIA
DAO
5071 W Saanich Rd
Victoria BC V8X 4M6
Canada
☎ 1 250 388 0230
✆ 1 250 363 0045
✉ robert.mcclure@hia.nrc.ca

McCluskey George E
Astronomy Div/Maths Dpt
Leigh Univ
14 East Packer Av
Bethlehem PA 18015
USA
☎ 1 215 861 3721
✉ cgm0@lehigh.edu

McConnell David
CSIRO
ATNF
Locked Bag 194
Narrabri NSW 2390
Australia
☎ 61 67 904 000
✆ 61 67 904 090
✉ dmcconne@atnf.csiro.au

McCord Thomas B
Inst of Geophysics
Univ of Hawaii
2525 Correa Rd
Honolulu HI 96822
USA
☎ 1 808 956 6488
✆ 1 808 988 2790

McCray Richard
JILA
Univ of Colorado
Campus Box 440
Boulder CO 80309 0440
USA
☎ 1 303 492 7835
✆ 1 303 492 5235
✉ dick@jila.colorado.edu

McCrea J Dermott
Physics Dpt
Univ College
Belfield
Dublin 4
Ireland
☎ 353 1 693 244
📠 353 1 671 1759

McCrosky Richard E
CfA
HCO/SAO
60 Garden St
Cambridge MA 02138 1516
USA
☎ 1 617 495 7212
📠 1 617 495 7356

McCulloch Peter M
Physics Dpt
Univ Tasmania
GPO Box 252c
Hobart TAS 7001
Australia
☎ 61 2 202 420
📠 61 2 202 410

McCutcheon William H
Physics Dpt
UBC
2075 Wesbrook Pl
Vancouver BC V6T 2A6
Canada
☎ 1 604 228 3853
📠 1 604 228 5324

McDonald Frank B
Inst Physical Sci/Tech
Univ of Maryland
College Park MD 20742 2421
USA
☎ 1 301 405 4874
📠 1 301 314 9363

McDonald J K Petrie
768 Richmond Av
Victoria BC V8S 3Z1
Canada
☎ 1 250 592 6880

McDonnell J A M
Unit for Space Sci
Univ of Kent
Canterbury CT2 7NR
UK
☎ 44 12 27 459 616
📠 44 12 27 762 616

McDonough Thomas R
CALTECH
500 S Oak Knoll No 46
Pasadena CA 91101
USA
☎ 1 818 795 0147

McElroy M B
Dpt Earth/Planet Sci
Harvard Univ MS 46
60 Garden St
Cambridge MA 02138 1516
USA
☎ 1 617 495 7100
📠 1 617 495 7356

McFadden Lucy Ann
Astronomy Program
Univ of Maryland
College Park MD 20742 2421
USA
☎ 1 301 454 6650
📠 1 301 314 9067

McGaugh Stacy Sutton
Dpt Terrestr Magnetism
Carnegie Inst Washington
5241 Bd Branch Rd NW
Washington DC 20015 1305
USA
☎ 1 202 686 4370*4399
📠 1 202 364 8726
✉ ssm@dtm.ciw.edu

McGee Richard X
CSIRO
Div Radiophysics
Box 76
Epping NSW 2121
Australia
☎ 61 2 868 0222
📠 61 2 868 0310

McGimsey Ben Q
Dpt Phys/Astronomy
Georgia State Univ
Atlanta GA 30303 3083
USA
☎ 1 404 658 2279
📠 1 404 542 2492

McGrath Melissa Ann
STScI
Homewood Campus
3700 San Martin Dr
Baltimore MD 21218
USA
☎ 1 301 338 4545
📠 1 301 338 4767
✉ mcgrath@stsci.edu

McGraw John T
Steward Observ
Univ of Arizona
Tucson AZ 85721
USA
☎ 1 520 621 5381
📠 1 520 621 1532

McGregor Peter John
MSSSO
Weston Creek
Private Bag
Canberra ACT 2611
Australia
☎ 61 262 79 80 33
📠 61 262 49 02 33
✉ peter@mso.anu.edu.au

McHardy Ian Michael
Physics Dpt
Southampton Univ
Astro/Space Physics Gp
Southampton SO9 5NH
UK
☎ 44 170 359 2079
📠 44 170 358 5813

McIntosh Bruce A
1007 655 Windemere Rd
London ON N5X 2W8
Canada
✉ mcintosh@sympatico.ca

McIntosh Patrick S
NOAA ERL R/E/SE3
Space Environment Lab
325 Broadway
Boulder CO 80303
USA
☎ 1 303 497 3795

McKee Christopher F
Physics Dpt
Univ of California
366 LeConte Hall
Berkeley CA 94720
USA
☎ 1 415 642 0805
📠 1 510 642 3411

McKeith Niall Enda
Physics Dpt
St Patrick's College
Maynooth
Maynooth Co Kildare
Ireland
☎ 353 1 285 222

McKenna Lawlor Susan
Physics Dpt
St Patrick's College
Maynooth
Maynooth Co Kildare
Ireland
☎ 353 1 285 222

McKinnon William Beall
Dpt Earth/Planet Sci
Washington Univ
Campus Box 1105
St Louis MO 63130
USA
☎ 1 314 935 5604
📠 1 314 935 7361
✉ mckinnon@wunder.
wustl.edu

McLaren Robert A
Inst for Astronomy
Univ of Hawaii
2680 Woodlawn Dr
Honolulu HI 96822
USA
☎ 1 808 956 8768
📠 1 808 946 3467
✉ mclaren@ifa.hawaii.edu

McLean Brian J
STScI
Homewood Campus
3700 San Martin Dr
Baltimore MD 21218
USA
☎ 1 301 338 4900
📠 1 301 338 4767
✉ mclean@stsci.edu

McLean Donald J
CSIRO
Div Radiophysics
Box 76
Epping NSW 2121
Australia
☎ 61 2 868 0222
📠 61 2 868 0310
✉ dmclean@rp.csiro.au

McLean Ian S
Dpt Astronomy
UCLA
Box 951562
Los Angeles CA 90025 1562
USA
☎ 1 310 825 1140
📠 1 310 206 2096
✉ mclean@bonnie.astro.ucla.edu

McMahan Robert Kenneth
Dpt Phys/Astronomy
Univ North Carolina
204 Phillips Hall 039a
Chapel Hill NC 27599 3255
USA
☎ 1 919 962 7168
📠 1 919 962 0480

McMahon Richard
Inst of Astronomy
The Observatories
Madingley Rd
Cambridge CB3 0HA
UK
☎ 44 12 23 337 548
📠 44 12 23 337 523
✉ rgm@mail.ast.cam.ac.uk

McMillan Robert S
Lunar/Planetary Lab
Room 325 Bg 92
Univ of Arizona
Tucson AZ 85721 0092
USA
☎ 1 520 621 6968
📠 1 520 621 1940
✉ bob@lpl.arizona.edu

McMullan Dennis
MP Group
Cavendish Laboratory
Madingley Rd
Cambridge CB3 0HE
UK
☎ 44 12 23 337 274
📠 44 12 23 363 263
✉ dennis@mcmullan.demon.
co.uk

McNally Derek
UCLO
ULO
Mill Hill Park
London NW7 2QS
UK
☎ 44 181 959 0421
📠 44 181 906 4161
✉ dmn@starlink.ucl.ac.uk

Mcnamara Delbert H
Dpt Phys/Astronomy
Brigham Young Univ
Provo Ut 84602
USA
☎ 1 801 378 2298
✉ mcnamara@astro.byu.edu

McNaught Robert H
AAO
Box 650
Coonabarabran NSW 2357
Australia
☎ 61 68 426 269
📠 61 68 842 298
✉ rmn@aaocbn2.aao.
gov.au

McWhirter R W Peter
Rutherford Appleton Lab
Space/Astrophysics Div
Bg R25/R68
Chilton Didcot OX11 0QX
UK
☎ 44 12 35 446 424
℻ 44 12 35 44 5808

Meaburn John
Astronomy Dpt
Univ Manchester
Oxford Rd
Manchester M13 9PL
UK
☎ 44 161 275 4224
℻ 44 161 275 4223

Mead Jaylee Montague
2700 Virginia Av NW
Apt 701
Washington DC 20037
USA
☎ 1 202 338 0208
℻ 1 202 338 4407
✉ jmead@blackhole.aas.org

Meadows A Jack
Astronomy Dpt
Univ Leicester
University Rd
Leicester LE1 7RH
UK
☎ 44 116 252 2073
℻ 44 113 252 2200

Meatheringham Stephen
39 / 60 Henty Street
Braddon ACT 2612
Australia
☎ 61 2 62 68 8142
℻ 61 2 62 68 8150
✉ stephen.meatheringham@adfa.edu.au

Mebold Ulrich
RAIUB
Univ Bonn
auf d Huegel 71
D 53121 Bonn
Germany
☎ 49 228 73 3658
℻ 49 228 73 3672

Mediavilla Evencio
IAC
Observ d Teide
via Lactea s/n
E 38200 La Laguna
Spain
☎ 34 2 232 9100
℻ 34 2 232 9117

Medina Jose
Dpt Fisica
Univ Alcala
Apt 20
E 28800 Alcala d Henares
Spain
☎ 34 1 885 4940
℻ 34 1 885 4953

Medvedev Yuri A
Astronomical Observ
Odessa State Univ
Shevchenko Park
270014 Odessa
Ukraine
☎ 380 4822 28442
℻ 380 4822 28442

Medvedev Yuriy D
Inst Theoret Astron
Acad Sciences
10 Nab Kutuzova
191187 St Petersburg
Russia
☎ 7 812 275 1064
℻ 7 812 272 7968
✉ medvedev@ita.spb.su

Meech Karen
Inst for Astronomy
Univ of Hawaii
2680 Woodlawn Dr
Honolulu HI 96822
USA
☎ 1 808 956 6828
℻ 1 808 988 2790
✉ meech@ifa.hawaii.edu

Meeks M Littleton
Meeks Associates Inc
Box 643
Lincoln MA 01773
USA
☎ 1 617 259 0093

Meerson Baruch
Racah Inst of Phys
Hebrew Univ Jerusalem
Jerusalem 91904
Israel
☎ 972 2 584 470
℻ 972 2 61 1519
✉ meerson@hujivms

Megessier Claude
Observ Paris Meudon
DASGAL
Pl J Janssen
F 92195 Meudon PPL Cdx
France
☎ 33 1 45 07 7862
℻ 33 1 45 07 7971

Megevand Denis
Observ Geneve
Chemin d Maillettes 51
CH 1290 Sauverny
Switzerland
☎ 41 22 755 2611
℻ 41 22 755 3983
✉ denis.megevand@obs.unige.ch

Mehringer David Michael
CALTECH
MC 320 47
Pasadena CA 91125
USA
☎ 1 818 395 6610
℻ 1 818 796 8806
✉ dmehring@socrates.caltech.edu

Meidav Meir
School of Education
Tel Aviv Univ
Tel Aviv 69978
Israel
☎ 972 3 545 0840
℻ 972 3 64 13944

Meier David L
CALTECH/JPL
238 332 JPL
4800 Oak Grove Dr
Pasadena CA 91109 8099
USA
☎ 1 818 354 5062
℻ 1 818 393 6890
✉ dlm@cena.jpl.nasa.gov

Meier Robert R
NRL
Code 7640
4555 Overlook Av SW
Washington DC 20375 5000
USA
☎ 1 202 767 2773

Meikle William P S
Astrophysics Gr
Imperial Coll
Prince Consort Rd
London SW7 2BZ
UK
☎ 44 171 594 7771
℻ 44 171 594 7772

Meiksin Avery Abraham
Royal Observ
Blackford Hill
Edinburgh EH9 3HJ
UK
☎ 44 131 668 8355
℻ 44 131 668 8416
✉ A.Meiksin@roe.ac.uk

Mein Nicole
Observ Paris Meudon
DASOP
Pl J Janssen
F 92195 Meudon PPL Cdx
France
☎ 33 1 45 07 7801
℻ 33 1 45 07 7959

Mein Pierre
Observ Paris Meudon
DASOP
Pl J Janssen
F 92195 Meudon PPL Cdx
France
☎ 33 1 45 07 7801
℻ 33 1 45 07 7971
✉ meinp@obspm.fr

Meinig Manfred
Bund Kartograp/Geodesy
Aussenstelle Potsdam
Michendorfer Chaussee 23
D 14473 Potsdam
Germany
☎ 49 331 316 609
℻ 49 331 316 602
✉ mg@potsdam.ifag.de

Meire Raphael
Wang Europe
Weidestr 11
B 9950 Evergem
Belgium
☎ 32 9 253 8755
℻ 32 2 714 2177
✉ rmeire@be.wang.com

Meisel David D
Dpt Phys/Astronomy
State Univ College
SUNY
Geneseo NY 14454
USA
☎ 1 716 245 5284
✉ meisel@uno.cc.geneseo.edu

Meisenheimer Klaus
MPI Astronomie
Koenigstuhl
D 69117 Heidelberg
Germany
☎ 49 622 152 8206
℻ 49 622 152 8246
✉ meise@dhdmpi5v

Meister Claudia Veronika
Inst Theor Phys/Astroph
Univ Potsdam
am Neuen Palais 10
D 14469 Potsdam
Germany
☎ 49 331 7499 327
℻ 49 331 7499 309

Mekarnia Djamel
OCA
Observ Nice
BP 139
F 06304 Nice Cdx 4
France
☎ 33 4 92 00 3161
℻ 33 4 92 00 3033
✉ mekarnia@obs-nice.fr

Mekler Yuri
Dpt Geophys/Planet Sci
Tel Aviv Univ
Ramat Aviv
Tel Aviv 69978
Israel
☎ 972 3 641 3505
℻ 972 3 640 9282

Melbourne William G
CALTECH/JPL
MS 238 540
4800 Oak Grove Dr
Pasadena CA 91109 8099
USA
☎ 1 818 354 5071
℻ 1 818 393 6030
✉ ren@logos.jpl.nasa.gov

Melchior Paul J
Koninklijke Sterrenwacht
van Belgie
Ringlaan 3
B 1180 Brussels
Belgium
☎ 32 2 373 0267
℻ 32 2 374 9822
✉ paul.Melchoir@oma.be

Melia Fulvio
Dpt Phys/Astronomy
Steward Observ
Univ of Arizona
Tucson AZ 85721
USA
☎ 1 520 621 9651
✆ 1 520 621 5698
✉ melia@spacetime.physics.arizona.edu

Melik-Alaverdian Yu
Byurakan Astrophys Observ
Armenian Acad Sci
378433 Byurakan
Armenia
☎ 374 88 52 28 3453/4142
✆ 374 88 52 52 3640

Melikian Norair D
Byurakan Astrophys Observ
Armenian Acad Sci
378433 Byurakan
Armenia
☎ 374 88 52 28 3453/4142
✆ 374 88 52 52 3640
✉ byurakan@adonis.ias.msk.su

Mellema Garrelt
Stockholm Observ
Royal Swedish Acad Sciences
S 133 36 Saltsjoebaden
Sweden
☎ 46 8 716 4462
✆ 46 8 717 4719
✉ garrelt@astro.su.se

Mellier Yannick
IAP
98bis bd Arago
F 75014 Paris
France
☎ 33 1 44 32 8140
✆ 33 1 44 32 8001
✉ mellier@iap.fr

Melnick Gary J
CfA
HCO/SAO
60 Garden St
Cambridge MA 02138 1516
USA
☎ 1 617 495 7388
✆ 1 617 495 7356
✉ melnick@cfa.harvard.edu

Melnick Jorge
Dpt Astronomia
Univ Chile
Casilla 36 D
Santiago
Chile
☎ 56 2 229 4101
✆ 56 2 229 4002

Melott Adrian L
Dpt Phys/Astronomy
Univ of Kansas
Lawrence KS 66045
USA
☎ 1 913 864 4626
✆ 1 913 864 5262

Melrose Donald B
Dpt Th Physics
Univ of Sydney
Sydney NSW 2006
Australia
☎ 61 2 692 2222
✆ 61 2 660 2903

Men A V
Inst Radio Astron
Ukrainian Acad Science
4 Chervonopraporna st
310085 Kharkiv
Ukraine

Mendes Da Costa Aracy
CRAAE INPE
EPUSP/PTR
CP 61548
01065 970 Sao Paulo SP
Brazil
☎ 55 11 815 5936
✆ 55 11 815 6289

Mendes de Oliveira Claudia
IAG/USP
Miguel Stefano 4200
04301 904 Sao Paulo SP
Brazil
☎ 55 11 577 8599
✆ 55 11 577 0270
✉ oliveira@andromeda.iagusp.usp.br

Mendez Manuel
Instituto Astronomia
UNAM
Apt 70 264
Mexico DF 04510
Mexico
☎ 52 5 622 3906
✆ 52 5 616 0653

Mendez Mariano
Observ Astronomico
Paseo d Bosque S/n
1900 La Plata (Bs As)
Argentina
☎ 54 21 216 357
✆ 54 21 211 761
✉ mmendez@fcaglp.edu.ar

Mendez Roberto H
Inst Astron/Astrophysik
Univ Sternwarte
Scheinerstr 1
D 81679 Muenchen
Germany
☎ 49 899 220 9442
✆ 49 899 220 9427
✉ mendez@usm.uni-muenchen.de

Mendis Devamitta Asoka
EECS
UCSD
La Jolla CA 92093 0216
USA
☎ 1 619 452 2719

Mendoza Claudio
IBM Venezuela Scient Ctr
Box 388
Caracas 1010 A
Venezuela
☎ 58 2 908 8697

Mendoza V Eugenio E
Instituto Astronomia
UNAM
Apt 70 264
Mexico DF 04510
Mexico
☎ 52 5 622 3906
✆ 52 5 616 0653

Mendoza-Torres Jose-Eduardo
INAOE
Tonantzintlaz
Apdo 51 y 216
Puebla PUE 72000
Mexico
☎ 52 2 247 2011
✆ 52 2 247 2231

Meneguzzi Maurice M
DAPNIA/SAP
CEA Saclay
BP 2
F 91191 Gif s Yvette Cdx
France
☎ 33 1 69 08 4438
✆ 33 1 69 08 6577

Meng Xinmin
Yunnan Observ
CAS
Kunming 650011
China PR
☎ 86 871 2035
✆ 86 871 717 1845

Mennella Vito
Osserv d Capodimonte
Via Moiariello 16
I 80131Napoli
Italy
☎ 39 81 298384
✆ 39 81 45 6710
✉ mennella@astrna.na.astro.it

Mennessier Marie-Odile
GRAAL
Univ Montpellier II
Pl E Bataillon
F 34095 Montpellier Cdx 5
France
☎ 33 4 67 52 3548
✆ 33 4 67 14 4534
✉ memes@graal.univ-monrp2.fr

Menon T K
Dpt Geophys/Astronomy
UBC
2219 Main Mall
Vancouver BC V6T 1Z4
Canada
☎ 1 604 822 2267
✆ 1 604 822 6047

Mentese Huseyin
Univ Observ
Univ of Istanbul
University 34452
34452 Istanbul
Turkey
☎ 90 212 522 3597
✆ 90 212 519 0834

Menzies John W
SAAO
PO Box 9
7935 Observatory
South Africa
☎ 27 21 47 0025
✆ 27 21 47 3639
✉ jwm@saao.ac.za

Merat Parviz
IAP
98bis bd Arago
F 75014 Paris
France
☎ 33 1 44 32 8108
✆ 33 1 44 32 8001

Mercier Claude
Observ Paris Meudon
DASOP
Pl J Janssen
F 92195 Meudon PPL Cdx
France
☎ 33 1 45 07 7815
✆ 33 1 45 07 7959

Merighi Roberto
Osserv Astronom Bologna
Univ d Bologna
Via Zamboni 33
I 40126 Bologna
Italy
☎ 39 51 259 401
✆ 39 51 259 407

Merkle Fritz
Carl Zeiss Jena Gmbh
Tatzens Promenade 1a
Postfach 125
D 07740 Jena
Germany
☎ 49 364 1640
✆ 49 364 164 2542

Merman G A
Inst Theoret Astron
Acad Sciences
N Kutuzova 10
192187 St Petersburg
Russia
☎ 7 812 278 8898
✆ 7 812 272 7968

Merman Natalia V
Pulkovo Observ
Acad Sciences
10 Kutuzov Quay
196140 St Petersburg
Russia
☎ 7 812 298 2242
✆ 7 812 315 1701

Mermilliod Jean-Claude
Institut Astronomie
Univ Lausanne
CH 1290 Chavannes d Bois
Switzerland
☎ 41 22 755 2611
✆ 41 22 755 3983
✉ jean-claude.mermilliod@obs.unige.ch

Merriam James B
Dpt Geological Sci
Univ of Saskatchewan
Saskatoon SK S7N 0W0
Canada
☎ 1 306 966 5716
✆ 1 306 966 8593
✉ merriam@geoid.usask.ca

Mertz Lawrence N
287 Fairfield Ct
Palo Alto CA 94306 4619
USA
☎ 1 650 494 8578
✉ digiphase@worldnet.att.net

Merzanides Constantinos
Zalogou 15
GR 65403 Kavala
Greece
☎ 30 51 22 840

Message Philip J
Dpt Appl Maths/Theor Phys
Univ of Liverpool
Box 147
Liverpool L69 3BX
UK
☎ 44 151 709 6022
✆ 44 151 708 6502

Messerotti Mauro
Astronomical Observ
OAT
Basovizza 302
I 34012 Trieste
Italy
☎ 39 40 22 6176
✆ 39 40 226761
✉ messerotti@oat.ts.astro.it

Messina Antonio
Dpt Astronomia
Univ d Bologna
Via Zamboni 33
I 40126 Bologna
Italy
☎ 39 51 259 301
✆ 39 51 259 407

Mestel Leon
Astronomy Centre
Univ of Sussex
Falmer
Brighton BN1 9QH
UK
☎ 44 12 73 60 6755
✆ 44 12 73 67 8097

Meszaros Attila
Astronomical Institute
Charles Univ
V Holesovickack 2
CZ 180 00 Praha 8
Czech R
☎ 420 2 2191 2572
✆ 420 2 688 5095
✉ mezsaros@mbox.cesnet.cz

Meszaros Peter
Astronomy Dpt
Pennsylvania State Univ
525 Davey Lab
University Park PA 16802
USA
☎ 1 814 865 0418
✆ 1 814 863 3399
✉ pmeszaros@astro.psu.edu

Metaxa Margarita
Inst Arsakeio High Sch
63 Eth Andistaseos Str
GR 152 31 Athens
Greece
☎ 30 1 674 2825
✆ 30 1 675 6968
✉ mmetaxa@compulink.gr

Metcalfe Leo
ESA/ESTEC
SSD
Box 299
NL 2200 AG Noordwijk
Netherlands
☎ 31 71 565 6555
✆ 31 71 565 4532

Metz Klaus
Novalis-Haus Zimmer 217
Ghersburgstr 19
D 83043 Bad Aibling
Germany

Meurs Evert
Dunsink Observ
DIAS
Castleknock
Dublin 15
Ireland
☎ 353 1 38 7911
✆ 353 1 38 7090

Meusinger Helmut
Thueringer Landessternwarte
Sternwarte 5
D 07778 Tautenburg
Germany
☎ 49 364 278 6362
✆ 49 364 278 6329
✉ meus@tls-tautenburg.de

Mewe R
SRON
Postbus 800
Sorbonnelaan 2
NL 3584 CA Utrecht
Netherlands
☎ 31 30 253 5600
✆ 31 30 254 0860

Meyer Claude
OCA
CERGA
F 06130 Grasse
France
☎ 33 4 93 40 5379
✆ 33 4 93 40 5353
✉ meyer@ocar01.obs-azur.fr

Meyer David M
Dearborn Observ
NW Univ
2131 Sheridan Rd
Evanston IL 60208 3112
USA
☎ 1 847 491 4516
✆ 1 847 491 3135

Meyer Friedrich
MPA
Karl Schwarzschildstr 1
D 85748 Garching
Germany
☎ 49 893 299 00
✆ 49 893 299 3235

Meyer Jean-Paul
DAPNIA/SAP
CEA Saclay
BP 2
F 91191 Gif s Yvette Cdx
France
☎ 33 1 69 08 5025
✆ 33 1 69 08 6577

Meyer-Hofmeister Eva
MPA
Karl Schwarzschildstr 1
D 85748 Garching
Germany
☎ 49 893 299 00
✆ 49 893 299 3235

Meyers Karie Ann
NOAO
Box 26732
950 N Cherry Av
Tucson AZ 85726 6732
USA
☎ 1 520 325 9202
✆ 1 520 325 9360
✉ kmeyers@noao.edu

Meylan Georges
ESO
Karl Schwarzschildstr 2
D 85748 Garching
Germany
☎ 49 893 200 6293
✆ 49 893 202 362
✉ gmeylan@eso.org

Meynet Georges
Observ Geneve
Chemin d Maillettes 51
CH 1290 Sauverny
Switzerland
☎ 41 22 755 2611
✆ 41 22 755 3983
✉ georges.meynet@obs.unige.ch

Mezger Peter G
RAIUB
Univ Bonn
auf d Huegel 69
D 53121 Bonn
Germany
☎ 49 228 525 297
✆ 49 228 525 229

Mezzetti Marino
SISSA
Astronomy Dpt
c/o SISSA Via Beirut 4
I 34014 Trieste
Italy
☎ 39 40 378 7478
✆ 39 40 378 7528
✉ mezzetti@gandalf.sissa.it

Mianes Pierre
OMP
14 Av E Belin
F 31400 Toulouse Cdx
France
☎ 33 5 61 25 2101
✆ 33 5 61 27 3179

Miao Yongkuan
Astronomy Dpt
Nanjing Univ
Nanjing 210093
China PR
☎ 86 25 663 7551*2882
✆ 86 25 330 2728

Miao Yongrui
Shanghai Observ
CAS
80 Nandan Rd
Shanghai 200030
China PR
☎ 86 21 6438 6191
✆ 86 21 6438 4618

Micela Giuseppina
Osserv Astronomico
Univ d Palerno
Piazza Parlemento 1
I 90134 Palermo
Italy
☎ 39 91 657 0451
✆ 39 91 48 8900

Michalec Adam
Astronomical Observ
Krakow Jagiellonian Univ
Ul Orla 171
PL 30 244 Krakow
Poland
☎ 48 12 25 1294
✆ 48 12 25 1318

Michalowski Tadeusz
Astronomical Observ
A Mickiewicz Univ
Ul Sloneczna 36
PL 60 286 Poznan
Poland
☎ 48 61 67 9670
✆ 48 61 53 6536
✉ tmich@phys.amu.edu.pl

Michard Raymond
OCA
Observ Nice
BP 139
F 06304 Nice Cdx 4
France
☎ 33 4 93 89 0420
✆ 33 4 92 00 3033

Michaud Georges J
250 Du Finistere
St Lambert J4S 1P5
Canada
☎ 1 514 343 6672
✉ michaudg@cerca.
umontreal.ca

Michel Eric
Observ Paris Meudon
Pl J Janssen
F 92195 Meudon PPL Cdx
France
☎ 33 1 45 07 7872
✆ 33 1 45 07 7872
✉ Eric.Michel@obspm.fr

Michel F Curtis
Dpt Phys/Astronomy
Rice Univ
Box 1892
Houston TX 77251 1892
USA
☎ 1 713 527 4925
✆ 1 713 285 5143

Mickaelian Areg Martin
Byurakan Astrophys Observ
Armenian Acad Sci
378433 Byurakan
Armenia
☎ 374 88 52 28 3453/4142
✆ 374 88 52 52 3640
✉ aregmick@bao.sci.am
byurakan@pnas.sci.am

Mickelson Michael E
Dpt Phys/Astronomy
Denison Univ
Granville OH 43023
USA
☎ 1 614 587 6467
✆ 1 614 587 6240
✉ mickelson@denison.edu

Mietelski Jan S
Astronomical Observ
Krakow Jagiellonian Univ
Ul Orla 171
PL 30 244 Krakow
Poland
☎ 48 12 25 1294
✆ 48 12 25 1318

Migenes Victor
Dpt Astronomia
Univ Guanajuato
Apdo 144
Guanajuato GTO 36000
Mexico
☎ 52 4 732 9548
✆ 52 4 732 0253

Mighell Kenneth John
NOAO/KPNO
Box 26732
950 N Cherry Av
Tucson AZ 85726 6732
USA
☎ 1 520 318 8391
✆ 1 520 318 8360
✉ mighell@noao.edu

Mignard Francois
OCA
CERGA
F 06130 Grasse
France
☎ 33 4 93 40 5382
✆ 33 4 93 40 5353
✉ miguard

Mihaila Ieronim
Bucharest Univ
Str Academiei 14
RO 70109 Bucharest
Rumania
☎ 40 1 623 0819

Mihalas Barbara R Weibel
NCSA
Beckman Inst Draw 25
405 Mathews Av
Urbana IL 61801
USA

Mihalas Dimitri
Dpt Astron/Physics
Univ of Illinois
1011 W Springfield Av
Urbana IL 61801
USA
☎ 1 217 333 3090
✆ 1 217 244 7638

Mikami Takao
Osaka Gakuin Univ
2-36-1 Kishibe Minami
Suita Shi
Osaka 564
Japan
☎ 81 638 18 434

Mikesell Alfred H
8316 Waldnut Rd NE
Olympia WA 98506 69550
USA
☎ 1 206 493 1457

Mikhail Fahmy I
Ain Shams Univ
Fac Sciences
Cairo Univ
Cairo
Egypt
☎ 20 2 257 5887

Mikhail Joseph Sidky
NAIGR
Helwan Observ
Cairo 11421
Egypt
☎ 20 78 0645/2683
✆ 20 62 21 405 297
✉ astro.frcu.eun.eg

Mikhelson Nikolaj N
Pulkovo Observ
Acad Sciences
10 Kutuzov Quay
196140 St Petersburg
Russia
☎ 7 812 297 9465
✆ 7 812 315 1701

Mikkola Seppo
Turku Univ
Tuorla Observ
Vaeisaelaentie 20
FIN 21500 Piikkio
Finland
☎ 358 2 274 4256
✆ 358 2 243 3767

Mikolajewska Joanna
Copernicus Astron Ctr
Polish Acad Sci
Ul Bartycka 18
PL 00 716 Warsaw
Poland
☎ 48 22 41 1086
✆ 48 22 41 0828
✉ mikolaj@camk.
edu.pl

Mikulasek Zdenek
N Copernicus Observ
and Planetarium
Kravi Hora 2
CZ 616 00 Brno 16
Czech R
☎ 420 5 4132 1287
✉ mikulas@dior.ics.
sci.muni.cz

Milani Andrea
Dpt Matematica
Univ d Pisa
Via Buonarroti 2
I 56127 Pisa
Italy

Milano Leopoldo
Dpt Scienze Fisiche
Univ d Napoli
Mostra D Oltremare Pad 19
I 80125 Napoli
Italy
☎ 39 81 725 3447

Miles Howard G
Lane Park Ptiyne
St Minver
Wadebridge PL27 6PN
UK
☎ 44 120 88 63153

Milet Bernard L
OCA
Observ Nice
BP 139
F 06304 Nice Cdx 4
France
☎ 33 4 93 89 0420
✆ 33 4 92 00 3033

Miley George K
Leiden Observ
Box 9513
NL 2333 RA Leiden
Netherlands
☎ 31 71 527 5849
✆ 31 71 527 5819

Milkey Robert W
AAS
2000 Florida Av Nw
Suite 400
Washington DC 20009 1231
USA
☎ 1 202 328 2010
✆ 1 202 234 2560
✉ milkey@aas.org

Millar Thomas J
Mathematics Dpt
UMIST
Box 88
Manchester M60 1QD
UK
☎ 44 161 200 3677
✆ 44 161 200 3669

Miller Freeman D
Astronomy Dpt
Univ of Michigan
Dennison Bldg
Ann Arbor MI 48109 1090
USA
☎ 1 313 764 3447
✆ 1 313 764 2211

Miller Guy Scott
Dearborn Observ
NW Univ
2131 Sheridan Rd
Evanston IL 60208 3112
USA
☎ 1 847 491 8647
✆ 1 847 491 3135
✉ gsmiller@ossenu.
astro.nwu.edu

Miller Hugh R
Dpt Phys/Astronomy
Georgia State Univ
Atlanta GA 30303 3083
USA
☎ 1 404 658 2279
✆ 1 404 542 2492

Miller John C
Astrophysics Dpt
Univ of Oxford
Keble Rd
Oxford OX1 3RQ
UK
☎ 44 186 551 1336
✆ 44 186 527 3390

Miller Joseph S
Lick Observ
Univ of California
Santa Cruz CA 95064
USA
☎ 1 831 429 2135
✆ 1 831 426 3115

Miller Richard H
Astronomy/Astrophys Ctr
Univ of Chicago
5640 S Ellis Av
Chicago IL 60637
USA
☎ 1 773 702 8201
✆ 1 312 702 8212
✉ rhm@oddjob.uchicago.edu

Millet Jean
LAS
Traverse du Siphon
Les Trois Lucs
F 13376 Marseille Cdx 12
France
☎ 33 4 91 05 5900
✆ 33 4 91 66 1855

Milliard Bruno
LAS
Traverse du Siphon
Les Trois Lucs
F 13376 Marseille Cdx 12
France
☎ 33 4 91 05 5900
✆ 33 4 91 66 1855
✉ milliard@frlasm51

Milligan J E
NASA GSFC
IR Astrophys Br
Greenbelt MD 20771
USA

Millikan Allan G
7061 Boughton Hill Rd
Victor NY 14564
USA
☎ 1 716 924 9802

Millis Robert L
Lowell Observ
1400 W Mars Hill Rd
Box 1149
Flagstaff AZ 86001
USA
☎ 1 602 774 3358
✆ 1 520 774 3358

Mills Allan A
Astronomy Dpt
Univ Leicester
University Rd
Leicester LE1 7RH
UK
☎ 44 116 252 2073
✆ 44 113 252 2200

Mills Bernard Y
School of Physics
UNSW
Sydney NSW 2006
Australia
☎ 61 2 692 2544
✆ 61 2 660 2903

Milne Douglas K
CSIRO
Div Radiophysics
Box 76
Epping NSW 2121
Australia
☎ 61 2 868 0222
✆ 61 2 868 0310
✉ dmilne@atnf.csiro.au

Milogradov-Turin Jelena
Astronomy Dpt
Univ of Belgrade
Studentski Trg 16
11000 Beograd
Yugoslavia FR
☎ 381 11 638 715
✆ 381 11 630 151
✉ epmfm21@yubgss21.bg.ac.yu

Milone Eugene F
Dpt Phys/Astronomy
Univ of Calgary
2500 University Dr NW
Calgary AB T2N 1N4
Canada
☎ 1 403 220 5412
✆ 1 403 289 3331
✉ milone@acs.
ucalgary.ca

Milone Luis A
Observ Astronomico
de Cordoba
Laprida 854
5000 Cordoba
Argentina
☎ 54 51 23 0491
✆ 54 51 21 0613
✉ milone@astro.
edu.ar

Minarovjech Milan
Astronomical Institute
Slovak Acad Sci
SK 059 60 Tatranska Lomni
Slovak Republic
☎ 421 969 96 7866
✆ 421 969 96 7656

Mineshige Shin
Astronomy Dpt
Kyoto Univ
Sakyo Ku
Kyoto 606 01
Japan
☎ 81 757 53 7008
✆ 81 75 753 7010
✉ minesige@kusastro.
kyoto-u.ac.j

Mineva Veneta
Astronomy Dpt
Bulgarian Acad Sci
72 Lenin Blvd
BG 1784 Sofia
Bulgaria
☎ 359 2 75 8827
✆ 359 2 75 5019

Mingaliev Marat G
SAO
Acad Sciences
Nizhnij Arkhyz
357147 Karachaevo
Russia
☎ 7 878 789 2501
✆ 7 871 140 6337
✉ marat@sao.stavropol.su

Minh Young Chol
Korea Astronomy Obs/ISSA
36 1 Whaam Dong
Yuseong Gu
Taejon 305 348
Korea RP
☎ 82 428 65 3282
✆ 82 428 65 3282
✉ minh@hanul.issa.re.kr

Minikulov Nasridin K
Astrophys Inst
Uzbek Acad Sci
Sviridenko Ul 22
734670 Dushanbe
Tajikistan
☎ 7 3772 27 4351
✆ 7 3770 27 5483

Minin Igor N
Astronomical Observ
St Petersburg Univ
Bibliotechnaja Pl 2
198904 St Petersburg
Russia
☎ 7 812 257 9489
✆ 7 812 428 4259

Minn Young Key
Dpt Astron/Space Sci
Kyunghee Univ
Yong In Kun
Kyunggee 449 701
Korea RP
☎ 82 276 46 131

Minnet Harry C
CSIRO
Div Radiophysics
Box 76
Epping NSW 2121
Australia
☎ 61 2 868 0222
✆ 61 2 868 0310

Minniti Dante
ESO
Karl Schwarzschildstr 2
D 85748 Garching
Germany
☎ 49 893 200 6532
✆ 49 893 202 362
✉ dante@eso.org

Mintz Blanco Betty
NOAO
CTIO
Casilla 603
La Serena
Chile
☎ 56 51 22 5415
✆ 56 51 20 5342
✉ bblanco@noao.edu

Mioc Vasile
Astronomical Institute of the
Romanian Academy
Romanian Acad Sciences
Cutitul de Argint 5
RO 75212 Bucharest
Rumania
☎ 40 1 335 3692
✆ 40 1 337 3389
✉ vmioc@roastro.
astro.ro

Mirabel Igor Felix
DAPNIA/SAP
CEA Saclay
BP 2
F 91191 Gif s Yvette Cdx
France
☎ 33 1 69 08 3492
✆ 33 1 69 08 6577

Mironov Nikolay T
Main Astronomical Obs
Ukrainian Acad Science
Golosiiv
252650 Kyiv 22
Ukraine
☎ 380 4426 64759
✆ 380 4426 62147

Mirzoyan Ludwik V
Byurakan Astrophys Observ
Armenian Acad Sci
378433 Byurakan
Armenia
☎ 374 88 52 28 3453/4142
✆ 374 88 52 52 3640
✉ gaabao96@pnas.sci.am

Misconi Nebil Yousif
Dpt Phys/Space Sci
Florida Inst Techn
150 W University Blvd
Melbourne FL 32901
USA
☎ 1 407 768 8000
✆ 1 407 984 8461

Mishenina Tamara
Astronomical Observ
Odessa State Univ
Shevchenko Park
Odessa 270014
Ukraine
☎ 380 4822 20396
✆ 380 4822 28442
✉ root@astro.odessa.ua

Misner Charles W
Astronomy Program
Univ of Maryland
College Park MD 20742 2421
USA
☎ 1 301 454 6650
✆ 1 301 314 9067

Missana Marco
Osserv Astronomico
d Brera
Via Cremagnani 13/11
I 20059 Vimercate
Italy
☎ 39 2 723 20302
✉ missana@brera.mi.astro.it

Mitalas Romas Assoc
Dpt Phys/Astronomy
Univ W Ontario
London ON N6A 3K7
Canada
☎ 1 519 679 3184
✆ 1 519 661 3486

Mitchell George F
Astronomy Dpt
St Mary's Univ
Halifax NS B3H 3C3
Canada
☎ 1 902 429 9780
✆ 1 902 420 5561

Mitchell Kenneth J
General Sciences Corp
6100 Chevy Chase Dr
Laurel MD 20707
USA
☎ 1 301 513 7815
✉ mitchell@aruba.gsfc.nasa.gov

Mitchell Peter
School of Physics
UNSW
Sydney NSW 2052
Australia
☎ 61 2 9385 5168
✆ 61 2 660 2903

Mitic Ljubisa A
Astronomical Observ
Volgina 7
11050 Beograd
Yugoslavia FR
☎ 381 1 419 357/421 875
✆ 381 1 419 553

Mitra A P
Ntl Physical Laboratory
Hillside Rd
New Delhi 110 012
India
☎ 91 11 585 298/440

Mitrofanova Lyudmila A
Pulkovo Observ
Acad Sciences
10 Kutuzov Quay
196140 St Petersburg
Russia
☎ 7 812 298 2242
✆ 7 812 315 1701

Mitsuda Kazuhisa
ISAS
3 1 1 Yoshinodai
Sagamihara
Kanagawa 229 8510
Japan
☎ 81 427 51 3911
✆ 81 427 59 4253

Mitton Jacqueline
8a Canterbury Close
Cambridge CB4 3QQ
UK
☎ 44 12 23 355 24

Mitton Simon
CUP
Shaftsbury Rd
Cambridge CB2 2RU
UK
☎ 44 12 23 31 2393
✉ sam11@phx.cam.ac.uk

Miyaji Shigeki
Physics Dpt
Chiba Univ
1-33 Yayoicho Inage ku
Chiba 263
Japan
☎ 81 43 290 3719
✆ 81 43 290 3720
✉ miyaji@c.chiba-u.ac.jp

Miyama Syoken
Physics Dpt
Kyoto Univ
Sakyo Ku
Kyoto 606 01
Japan
☎ 81 757 51 2111
✆ 81 75 753 7010

Miyamoto Masanori
Tokyo Astronomical Obs
NAOJ
Osawa Mitaka
Tokyo 181
Japan
☎ 81 422 32 5111
✆ 81 422 32 1924
✉ ma@uranus.mtk.mao.jp.psomaxx@c1.mtk.nao.ac.jp

Miyamoto Sigenori
Physics Dpt/Fac Sciences
Osaka Univ
Machikaneyama
Toyonaka Osaka 560
Japan
☎ 81 684 41 151

Miyoshi Makoto
National Astronomical Obs
2 10 Hoshigaoka
Mizusawa Shi
Iwate 023
Japan
☎ 81 197 22 7138
✆ 81 197 222 715
✉ miyoshi@miz.nao.ac.jp

Miyoshi Shigeru
Physics Dpt
Kyoto Sangyo Univ
Kamigamo
Kyoto 603
Japan
☎ 81 757 05 1609
✆ 81 75 705 1640
✉ miyoshi@asca.kyoto-su.ac.jp

Mizumoto Yoshihiko
Ntl Astronomical Obs
Osawa 2-21-1
Mitaka
Tokyo 181
Japan
☎ 81 422 34 3702
✆ 81 422 34 3708
✉ mizumoto@optik.mtk.nao.ac.jp

Mizuno Akira
Astrophysics Dpt
Nagoya Univ
Furocho Chikusa Ku
Nagoya 464 01
Japan
☎ 81 527 81 6769
✆ 81 527 82 0647
✉ mizuno@a.phys.nagoya-u.ac.jp

Mizuno Shun
Kanazawa Techn Inst
7 1 Ogigaoka
Nonoichimachi
Ishikawa 921
Japan
☎ 81 762 48 1100

Mizuno Takao
Tokyo Gakugei Univ
Koganei City
Tokyo 184
Japan
☎ 81 423 25 2111
✆ 81 423 24 9832
✉ mizuno@yamabuki.u-gakugei.ac.jp

Mizutani Kohei
Physics Dpt
Saitama Univ
Z55 Shimo-Okubo
Urawa 338
Japan
☎ 81 488 58 3376
✆ 81 488 58 3698
✉ mizutani@crspark.cr.phy.saitama-u.ac.jp

Mnatsakanian Mamikon A
Byurakan Astrophys Observ
Armenian Acad Sci
378433 Byurakan
Armenia
☎ 374 88 52 28 3453/4142
✆ 374 88 52 52 3640

Mo Houjun
MPA
Karl Schwarzschildstr 1
D 85748 Garching
Germany
☎ 49 893 299 3246
✆ 49 893 299 3235
✉ hom@mpa-garching.mpg.de

Mo Jing-er
Purple Mountain Obs
CAS
Nanjing 210008
China PR
☎ 86 25 36967
✆ 86 25 301 459

Mochkovitch Robert
IAP
98bis bd Arago
F 75014 Paris
France
☎ 33 1 44 32 8187
✆ 33 1 44 32 8001

Mochnacki Stephan W
David Dunlap Observ
Univ of Toronto
60 St George St
Toronto ON M5s 1A1
Canada
☎ 1 416 978 2016
✆ 1 416 978 3921

Modali Sarma B
SM Systems/Res Co
8401 Corporate Dr
Suite 450
Landover MD 20785
USA
☎ 1 301 459 3322

Modisette Jerry L
18323 Hereford Ln
Houston TX 77058
USA

Moehler Sabine
Dr Remeis Sternwarte
Univ Erlangen-Nuernberg
Sternwartstr 7
D 96049 Bamberg
Germany
☎ 49 951 952 2217
✆ 49 951 952 2222
✉ moehler@a400.sternwarte.uni-erlangen.de

Moehlmann Diedrich
Inst Raumsimulation
Postfach 90 60 58
D 51140 Koeln
Germany
☎ 49 220 360 13205
✆ 49 220 360 12094/2352
✉ dirk.moehlmann@europa.rs.kp.dl

Moellenhoff Claus
Landessternwarte
Koenigstuhl
D 69117 Heidelberg
Germany
☎ 49 622 150 9210
✆ 49 622 150 9202
✉ cmoellen@hp2.lsw.uni-heidelberg.de

Moeller Palle
ESO
Karl Schwarzschildstr 2
D 85748 Garching
Germany
☎ 49 893 20 060
✆ 49 893 202 362

Moerdijk Willy G
Sterrekundig Observ
Universiteit Gent
Krijgslaan 281 S9
B 9000 Gent
Belgium
☎ 32 9 264 4763
✆ 32 92 64 4989

Moesgaard Kristian P
History of Sci Inst
Univ of Aarhus
Bygadan 1/Torrild
DK 8300 Odder
Denmark
☎ 45 86 53 1004

Moffat Anthony F J
Dpt d Physique
Universite Montreal
CP 6128 Succ A
Montreal QC H3C 3J7
Canada
☎ 1 514 343 6682
✆ 1 514 343 2071

Moffat John W
Physics Dpt
Univ of Toronto
60 St George St
Toronto ON M5S 1A1
Canada
☎ 1 416 978 2949
✆ 1 416 978 3921

Moffatt Henry Keith
Isaac Newton Inst for Mathematical Sci
20 Clarkson Rd
Cambridge CB3 9EH
UK
☎ 44 12 23 335 80
✆ 44 12 23 330 508
✉ hkm2@newton.cam.ac.uk

Moffett Thomas J
Physics Dpt
Purdue Univ
West Lafayette IN 47907
USA
☎ 1 317 494 5508
✆ 1 317 494 0706
✉ moffett@physics.purdue.edu

Mogilevskij Eh I
ITMIRWP
Acad Sciences
142092 Troitsk
Russia
☎ 7 095 232 1931
℻ 7 095 334 0124

Mohan Chander
Mathematics Dpt
Univ of Roorkee
Roorkee 247 667
India

Mohan Vijay
Uttar Pradesh State
Observ
Po Manora Peak 263 129
Nainital 263 129
India
☎ 91 59 42 2136

Mohd Zambri Zainuddin
Ctr for Foundation
Univ of Malaysia
59100 Kuala Lumpur
Malaysia
☎ 60 3 755 2744
℻ 60 3 757 3661

Moiseev I G
Crimean Astrophys Obs
Ukrainian Acad Science
Nauchny
334413 Crimea
Ukraine
☎ 380 6554 71161
℻ 380 6554 40704

Molaro Paolo
OAT
Box Succ Trieste 5
Via Tiepolo 11
I 34131 Trieste
Italy
☎ 39 40 31 99255
℻ 39 40 30 9418

Molchanov Andrea P
Astronomical Observ
St Petersburg Univ
Bibliotechnaja Pl 2
199178 St Petersburg
Russia
☎ 7 812 428 7129
℻ 7 812 428 4259

Moles Mariano J
Inst Astrofisica
Andalucia Apt 3004
Prof Albareda 1
E 18080 Granada
Spain
☎ 34 5 812 1311
℻ 34 5 881 4530

Molina Antonio
Inst Astrofisica
Andalucia Apt 3004
Prof Albareda 1
E 18080 Granada
Spain
☎ 34 5 812 1300
℻ 34 5 881 4530

Molinari Emilio
Osserv Astronomico
d Milano
Via E Bianchi 46
I 22055 Merate
Italy
☎ 39 990 6412
℻ 39 990 8492
✉ molinari@merate.
mi.astro.it

Molnar Michael R
Molnar Technologies
3 Stoningham Dr
Warren NJ 07060
USA
☎ 1 201 580 1404

Molotaj Olexandr
Astronomical Observ
Observatorna st 3
Kyiv 254 053
Ukraine
☎ 380 4421 62391
℻ 380 4426 64507
✉ mol@aoku.freenet.
kiev.ua

Momchev Gospodin
Astronomical Observ
Box 7
BG 8800 Sliven
Bulgaria
☎ 359 4 42 8353

Monaghan Joseph J
Mathematics Dpt
Monash Univ
Wellington Rd
Clayton VIC 3168
Australia
☎ 61 3 541 2563

Monet Alice K B
USNO
Flagstaff Station
Box 1149
Flagstaff AZ 86002 1149
USA
☎ 1 602 779 5132
℻ 1 602 774 3626
✉ alice@nofs.navy.mil

Monet David G
USNO
Flagstaff Station
Box 1149
Flagstaff AZ 86002 1149
USA
☎ 1 520 779 5132
℻ 1 520 774 3626
✉ dgm@nofs.navy.mil

Monfils Andre G
NEOC SA
Chemin Macors 17
B 4052 Beaufays
Belgium
☎ 32 4 368 7929
℻ 32 4 368 7929

Monier Richard
Observ Strasbourg
11 r Universite
F 67000 Strasbourg
France
☎ 33 3 88 15 0753
℻ 33 3 88 37 1408
✉ rmonier@cdsxba.u-
strasbg.fr

Monin Jean-Louis
Observ Grenoble
Lab Astrophysique
BP 53x
F 38041 S Martin Heres Cdx
France
☎ 33 4 76 51 4786
℻ 33 4 76 44 8821

Monnet Guy J
Observ Lyon
Av Ch Andre
F 69561 S Genis Laval Cdx
France
☎ 33 4 78 56 0705
℻ 33 4 72 39 9791

Monteiro Mario J P F G
Centro Astrofisica
Univ d Porto
rua d Campo Alegre 823
P 4150 Porto
Portugal
☎ 351 2 600 7081
℻ 351 2 600 7082
✉ mjm@astro.up.pt

Monteiro Tania S
Mathematics Dpt
Royal Holloway College
Univ of London
Egham Surrey TW20 0EX
UK
☎ 44 17 843 4455*3106
✉ asl0703@ulcc

Montes Carlos
OCA
Observ Nice
BP 139
F 06304 Nice Cdx 4
France
☎ 33 4 93 89 0420
℻ 33 4 92 00 3033

Montmerle Thierry
DAPNIA/SAP
CEA Saclay
BP 2
F 91191 Gif s Yvette Cdx
France
☎ 33 1 69 08 4722
℻ 33 1 69 08 9266
✉ montmerle@sapvxg.
saclay.cea.fr

Moody Joseph Ward
Dpt Phys/Astronomy
263 FB
BYU
Provo UT 84602
USA
☎ 1 801 378 4347
℻ 1 801 378 2265
✉ jmoody@astro.byu.edu

Mook Delo E
Dpt Phys/Astronomy
Dartmouth College
6127 Wilder Lab
Hanover NH 03755
USA
☎ 1 603 646 2972
℻ 1 603 646 1446

Moon Shin Haeng
Korea Astronomy Obs/ISSA
36 1 Whaam Dong
Yuseong Gu
Taejon 305 348
Korea RP
☎ 82 428 61 1497
℻ 82 428 81 5610

Moons Michele B M M
Dpt Mathematique
FUNDP
Rempart de la Vierge 8
B 5000 Namur
Belgium
☎ 32 81 72 4940
℻ 32 81 72 4914
✉ mmoons@math.fundp.ac.be

Moore Daniel R
Mathematics Dpt
Imperial Coll
Prince Consort Rd
London SW7 2BZ
UK
☎ 44 171 594 7771
℻ 44 171 594 7772

Moore Elliott P
Physics Dpt
Campus Station
Socorro NM 87801
USA
☎ 1 505 835 5431
℻ 1 505 835 5707
✉ moore@arctic.nmt.edu

Moore Patrick
Farthings
39 West St
Selsey Sussex
UK
☎ 44 124 360 3668

Moore Ronald L
NASA/MSFC
Space Science Lab
Code ES 52
Huntsville AL 35812
USA
☎ 1 205 453 0118
℻ 1 205 544 7754

Moorhead James M
Dpt Phys/Astronomy
Univ W Ontario
London ON N6A 3K7
Canada
☎ 1 519 679 2111*6712
℻ 1 519 661 2009
✉ moorhead@phobos.
astro.uwo.ca

Moorwood Alan F M
ESO
Karl Schwarzschildstr 2
D 85748 Garching
Germany
☎ 49 893 200 6294
✆ 49 893 202 362

Moos Henry Warren
Dpt Phys/Astronomy
JHU
Charles/34th St
Baltimore MD 21218
USA
☎ 1 301 516 7337
✆ 1 301 516 5494
✉ hwm@pha.jhu.edu

Morales-Duran Carmen
ESA
Apt 50727 Villafranca
E 28080 Madrid
Spain
☎ 34 1 675 0700
✆ 34 1 813 1139

Moran James M
CfA
HCO/SAO MS 42
60 Garden St
Cambridge MA 02138 1516
USA
☎ 1 617 495 7477
✆ 1 617 495 7345
✉ moran@cfa.harvard.edu

Morbey Christopher L
NRCC/HIA
DAO
5071 W Saanich Rd
Victoria BC V8X 4M6
Canada
☎ 1 250 388 0220
✆ 1 250 363 0045
✉ chris.morbey@hia.
nrc.ca

Morbidelli Alessandro
OCA
Observ Nice
BP 139
F 06304 Nice Cdx 4
France
☎ 33 4 92 00 3126
✆ 33 4 92 00 3033
✉ morby@obs-nice.fr

Morbidelli Lorenzo
Osserv Astrofis Arcetri
Univ d Firenze
Largo E Fermi 5
I 50125 Firenze
Italy
☎ 39 55 27 521
✆ 39 55 22 0039
✉ lorenzo@arcetri.astro.it

Morbidelli Roberto
Osserv Astronomico
d Torino
St Osservatorio 20
I 10025 Pino Torinese
Italy
☎ 39 11 461 9000
✆ 39 11 461 9030
✉ morbidelli@to.astro.it

Morcos Abd El Fady B
NAIGR
Helwan Observ
Cairo 11421
Egypt
☎ 20 78 0645/2683
✆ 20 62 21 405 297
✉ astro@frcu.eun.eg

Moreau Olivier
Forum des Sciences
Planetarium
1 Pl Hotel de ville
F 59650 Villeneuve d'Ascq
France
☎ 33 3 20 19 3600
✆ 33 3 20 19 3601
✉ moreau@astro.ulg.
ac.be

Moreels Guy
Observ Besancon
41bis Av Observatoire
BP 1615
F 25000 Besancon Cdx
France
☎ 33 3 81 50 2266
✆ 33 3 81 66 6944

Morel Pierre-Jacques
OCA
Observ Nice
BP 139
F 06304 Nice Cdx 4
France
☎ 33 4 93 89 0420
✆ 33 4 92 00 3033

Moreno Corral Marco A
Observ Astronomico Nacional
UNAM
Apt 877
Ensenada BC 22800
Mexico
☎ 52 6 174 4580
✆ 52 6 174 4607

Moreno Edmundo
Instituto Astronomia
UNAM
Apt 70 264
Mexico DF 04510
Mexico
☎ 52 5 622 3906
✆ 52 5 548 3712

Moreno Fernando
Inst Astrofisica
Andalucia Apt 3004
Prof Albareda 1
E 18080 Granada
Spain
☎ 34 5 812 1311
✆ 34 5 881 4530

Moreno Hugo
Dpt Astronomia
Univ Chile
Casilla 36 D
Santiago
Chile
☎ 56 2 229 4101/4002
✆ 56 2 229 4002
✉ hmoreno@das.
uchile.cl

Moreno-Insertis Fernando
IAC
Observ d Teide
via Lactea s/n
E 38200 La Laguna
Spain
☎ 34 2 232 9100
✆ 34 2 232 9117
✉ fmi@iac.es

Moreton G E
15 5 The Esplanade
Balmoral Beach NSW 2088
Australia

Morgan Brian Lealan
Astrophysics Gr
Imperial Coll
Prince Consort Rd
London SW7 2BZ
UK
☎ 44 171 594 7771
✆ 44 171 594 7772
✉ b.morgan@ic.ac.uk

Morgan David H
Royal Observ
Blackford Hill
Edinburgh EH9 3HJ
UK
☎ 44 131 667 3321
✆ 44 131 662 1668

Morgan John Adrian
Aerospace Corp
MS M4 041
Box 92957
Los Angeles CA 90009 2957
USA
☎ 1 310 336 5000
✆ 1 310 336 1636

Morgan Peter
Canberra Coll Adv Educ
School Applied Science
Box 1
Belconnen ACT 2616
Australia
☎ 61 62 522 557

Morgan Thomas H
Southwest Res Inst
Po Drawer 28510
San Antonio TX 78228 0510
USA
☎ 1 210 522 3985
✆ 1 210 647 4325
✉ tmorgan@swri.dnet.
nasa.gov

Morganti Raffaella
Istt Radioastronomia
CNR
Via P Gobetti101
I 40129 Bologna
Italy
☎ 39 51 639 9385
✆ 39 51 639 9431
✉ morganti@astbo1.bo.cnr.it

Moriarty-Schieven G H
Joint Astronomy Ctr
660 N A'ohoku Pl
University Park
Hilo HI 96720
USA
☎ 1 808 969 6531
✆ 1 808 961 6516
✉ gms@jach.hawaii.edu

Morimoto Masaki
Tokyo Astronomical Obs
NAOJ
Osawa Mitaka
Tokyo 181
Japan
☎ 81 992 85 8960
✆ 81 992 58 4866

Morison Ian
NRAL
Univ Manchester
Jodrell Bank
Macclesfield SK11 9DL
UK
☎ 44 14 777 1321
✆ 44 147 757 1618

Morita Kazuhiko
Physics Dpt
Hokkaido Univ
Kita 10 Nishi 8
Sapporo 060
Japan
☎ 81 117 11 2111
✆ 81 222 85 132

Morita Koh-ichiro
Nobeyama Radio Obs
NAOJ
Minamimaki Mura
Nagano 384 13
Japan
☎ 81 267 63 4331
✆ 81 267 63 4339
✉ morita@nro.nao.ac.jp

Moriyama Fumio
3 6 23 Nogami
Takarazuka
Hyogo 665
Japan
☎ 81 638 18 434

Morossi Carlo
OAT
Box Succ Trieste 5
Via Tiepolo 11
I 34131 Trieste
Italy
☎ 39 40 31 99255
✆ 39 40 30 9418

Moroz Vasilis I
Space Res Inst
Acad Sciences
Profsojuznaya Ul 84/32
117810 Moscow
Russia
☎ 7 095 333 3122
✆ 7 095 310 7023

Morozhenko A V
Main Astronomical Obs
Ukrainian Acad Science
Golosiiv
252650 Kyiv 22
Ukraine
☎ 380 4426 63110
℻ 380 4426 62147

Morozhenko N N
Main Astronomical Obs
Ukrainian Acad Science
Golosiiv
252650 Kyiv 22
Ukraine
☎ 380 4426 63110
℻ 380 4426 62147

Morras Ricardo
IAR
CC 5
1894 Villa Elisa (Bs As)
Argentina
☎ 54 21 25 4909
℻ 54 21 254 909
✉ r morras@irma.edu.ar

Morrell Nidia
Observ Astronomico
Paseo d Bosque S/n
1900 La Plata (Bs As)
Argentina
☎ 54 21 217 308
℻ 54 21 211 761
✉ nidia@fcaglp.edu.ar

Morris Charles S
CALTECH/JPL
MS 300 319
4800 Oak Grove Dr
Pasadena CA 91109 8099
USA
☎ 1 818 354 8074
℻ 1 818 393 6030

Morris David
IRAM
300 r La Piscine
F 38406 S Martin Heres Cdx
France
☎ 33 4 76 82 4930
℻ 33 4 76 51 5938
✉ morris@iram.fr

Morris Mark Root
Dpt Astronomy/Phys
UCLA
Box 951562
Los Angeles CA 90025 1562
USA
☎ 1 310 825 3320
℻ 1 310 206 2096
✉ morris@astro.ucla.edu

Morris Michael C
Royal Greenwich Obs
Madingley Rd
Cambridge CB3 0EZ
UK
☎ 44 12 23 374 000
℻ 44 12 23 374 700

Morris Simon
NRCC/HIA
DAO
5071 W Saanich Rd
Victoria BC V8X 4M6
Canada
☎ 1 250 363 0062
℻ 1 250 363 0045
✉ simon@dao.nrc.ca

Morris Stephen C
NRCC/HIA
DAO
5071 W Saanich Rd
Victoria BC V8X 4M6
Canada
☎ 1 250 363 0023
℻ 1 250 363 0045
✉ morris@dao.nrc.ca

Morris Steven
Apt 2
2860 W 235th St
Torrance CA 90505
USA
☎ 1 213 530 8708

Morrison David
NASA/ARC
MS 200 7
Moffett Field CA 94035 1000
USA
☎ 1 415 604 5028
℻ 1 415 604 6779
✉ david_morrison@
qmgate.arc.nasa.gov

Morrison Leslie V
Royal Greenwich Obs
Madingley Rd
Cambridge CB3 0EZ
UK
☎ 44 1223 374 771
℻ 44 12 23 374 700
✉ lvm@ast.cam.ac.uk

Morrison Nancy Dunlap
Dpt Phys/Astronomy
Univ of Toledo
2801 W Bancroft St
Toledo OK 43606
USA
☎ 1 419 537 2659
℻ 1 419 530 2723

Morrison Philip
Physics Dpt
MIT Rm 6 205
Box 165
Cambridge MA 02139 4307
USA
☎ 1 617 253 5086
℻ 1 617 253 9798

Morton Donald C
NRCC/HIA
DAO
5071 W Saanich Rd
Victoria BC V8X 4M6
Canada
☎ 1 250 363 0040
℻ 1 250 363 8483
✉ don.morton@hia.
nrc.ca

Morton G A
1122 Skycrest Dr
APT 6
Walnut Creek CA 94595
USA
☎ 1 415 933 3802

Moscardini Lauro
Dpt Astronomia
Univ d Padova
Vic d Osservatorio 5
I 35122 Padova
Italy
☎ 39 49 829 3474
℻ 39 49 875 9840
✉ moscardini@astrpd.
pd.astro.it

Moskalik Pawel
Copernicus Astron Ctr
Polish Acad Sci
Ul Bartycka 18
PL 00 716 Warsaw
Poland
☎ 48 22 41 1086
℻ 48 22 410 046

Moss Christopher
Inst of Astronomy
The Observatories
Madingley Rd
Cambridge CB3 0HA
UK
☎ 44 12 23 337 548
℻ 44 12 23 337 523

Moss David L
Mathematics Dpt
Univ Manchester
Oxford Rd
Manchester M13 9PL
UK
☎ 44 161 275 4224
℻ 44 161 275 4223

Mosser Benoit
IAP
98bis bd Arago
F 75014 Paris
France
☎ 33 1 44 32 8149
℻ 33 1 44 32 8001
✉ mosser@iap.fr

Motta Santo
Dpt Matematica
Citta Universitaria
Via A Doria 6
I 95125 Catania
Italy
☎ 39 95 733 0533*668
℻ 39 95 33 0592

Motz Lloyd
Astronomy Dpt
Columbia Univ
538 W 120th St
New York NY 10027
USA
☎ 1 212 280 3279
℻ 1 212 316 9504

Mouchet Martine
Observ Paris Meudon
DAEC
Pl J Janssen
F 92195 Meudon PPL Cdx
France
☎ 33 1 45 07 7522
℻ 33 1 45 07 7469
✉ mouchet@obspm.fr

Mould Jeremy R
MSSSO
Weston Creek
Private Bag
Canberra ACT 2611
Australia
☎ 61 262 490 266
℻ 61 262 490 233
✉ jrm@mso.anu.edu.au

Mouradian Zadig M
Observ Paris Meudon
DASOP
Pl J Janssen
F 92195 Meudon PPL Cdx
France
☎ 33 1 45 07 7800
℻ 33 1 45 07 7959

Mourard Denis
OCA
Observ Calern/Caussols
2130 r Observatoire
F 06460 S Vallier Thiey
France
☎ 33 4 93 40 5492
℻ 33 4 93 40 4431
✉ mourard@obs-nice.fr

Mouschovias Telemachos Ch
Dpt Astron/Physics
Univ of Illinois
1011 W Springfield Av
Urbana IL 61801
USA
☎ 1 217 333 3090
℻ 1 217 244 7638

Moussas Xenophon
Astrophysics Dpt
Ntl Univ Athens
Panepistimiopolis
GR 157 84 Zografos
Greece
☎ 30 1 723 5122
℻ 30 1 723 5122
✉ moussas@grathon1.earn

Mozurkewich David
USNO
AD5 Bg 22 Rm 220
3450 Massachusetts Av NW
Washington DC 20392 5100
USA
☎ 1 202 762 1520

Muecket Jan P
Astrophysik Inst
Potsdam Univ
an d Sternwarte 16
D 14482 Potsdam
Germany
☎ 49 331 749 90
℻ 49 331 749 9200
✉ jpmuecket@aip.de

Mueller Ewald
MPA
Karl Schwarzschildstr 1
D 85748 Garching
Germany
☎ 49 893 299 00
✆ 49 893 299 3235

Mueller Ivan I
Geodetic Sci/Surveying
Ohio State Univ
1958 Neil Av
Columbus OH 43210 1247
USA
☎ 1 614 422 2269
🖂 mueller@mps.ohio-state.edu

Mueller Volker
Astrophysik Inst
Potsdam Univ
an d Sternwarte 16
D 14482 Potsdam
Germany
☎ 49 331 749 90
✆ 49 331 749 9200
🖂 vmueller@aip.de

Muench Guido
Casa de Manana
849 Coast Blvd
La Jolla CA 92037
USA
☎ 1 619 454 2151
✆ 1 619 454 7537

Muerset Urs
Inst Astronomie
ETH Zentrum
CH 8092 Zuerich
Switzerland
☎ 41 1 632 3633
✆ 41 1 632 1205
🖂 muerset@astro.phys.ethz.ch

Mufson Stuart Lee
Astronomy Dpt
Indiana Univ
Swain W 319
Bloomington IN 47405
USA
☎ 1 812 335 6927
✆ 1 812 855 8725

Muinonen Karri
Helsinky Univ
Observ
Box 14
FIN 00014 Helsinki
Finland
☎ 358 19 12 2941
✆ 358 19 12 2952
🖂 karri.Muinonen@Helsinki.fi

Muinos Jose L
ROA
Cecilio Pujazon 22-3 A
E 11110 San Fernando
Spain
☎ 34 5 688 3548
✆ 34 5 688 1732
🖂 ppmu@roa.cica.es

Mukai Koji
NASA/GSFC
Code 660 2
Greenbelt MD 20771
USA
☎ 1 301 286 9447
✆ 1 301 286 1684
🖂 mukai@lheaux.gsfc.nasa.gov

Mukai Sonoyo
Dpt Industrial Eng
Kinki Univ
Kowakae 3 4 1
Osaka 577
Japan
☎ 81 6 721 2332*4007
✆ 81 6 730 1320
🖂 mukai@im.kindai.ac.jp

Mukai Tadashi
Dpt Earth/Planet Sci
Kobe Univ
Nada
Kobe 657 8501
Japan
☎ 81 788 03 0574
✆ 81 788 03 0490
🖂 mukai@kobe-u.ac.jp

Mukherjee Krishna
Dpt Phys/Astronomy
Slippery Rock Univ Pennsylvania
327 Vincent Science
Slippery Rock PA 16057
USA
☎ 1 414 738 2858

Mulholland J Derral
Inst Azur Espace
4 r de la Fontaine
F 06620 Le Bar s Loup
France
☎ 33 4 93 42 9385
✆ 33 4 93 42 9385
🖂 100073.3636@compuserve.com

Mullaly Richard F
31 Eden Av
Turramurra NSW 2074
Australia

Mullan Dermott J
Bartol Res Inst
Univ of Delaware
Newark DE 19716
USA
☎ 1 302 831 2170
✆ 1 302 831 1843
🖂 mullan@bartol.udel.edu

Muller Andre B
Thopmaslaan 40
NL 5631 GM Eindhoven
Netherlands
☎ 31 40 43 0322

Muller Paul
3 r Chauvain
F 06000 Nice
France

Muller Richard
OMP
14 Av E Belin
F 31400 Toulouse Cdx
France
☎ 33 5 62 95 0069
✆ 33 5 61 27 3179

Muller Richard A
Lawrence Berkeley Lab
Univ of California
Bg 50 Rm 238
Berkeley CA 94720
USA
☎ 1 415 486 5235
✆ 1 510 486 6738

Muller C A
Odinksveld 8
NL 7491 HD Delden
Netherlands
☎ 31 54 07 2428

Mumford George S
Dpt Phys/Astronomy
Tufts Univ
Robinson Hall
Medford MA 02155
USA
☎ 1 617 653 8923
🖂 gmumford@tufts.edu

Mumma Michael Jon
NASA GSFC
Code 693
Greenbelt MD 20771
USA
☎ 1 301 286 6994
🖂 mmumma@lepvax.dnet.nasa.gov

Mundt Reinhard
MPI Astronomie
Koenigstuhl
D 69117 Heidelberg
Germany
☎ 49 622 152 8227
✆ 49 622 152 8246

Mundy Lee G
Astronomy Program
Univ of Maryland
College Park MD 20742 2421
USA
☎ 1 301 454 6650
✆ 1 301 314 9067

Munoz-Tunon Casiana
IAC
Observ d Teide
via Lactea s/n
E 38200 La Laguna
Spain
☎ 34 2 232 9100
✆ 34 2 232 9117

Munro Richard H
Ball Aerospace Systems Division
Box 1062
Boulder CO 80306
USA
☎ 1 303 939 4591
✆ 1 303 939 4656
🖂 rmunro@ball.com

Murakami Hiroshi
ISAS
3 1 1 Yoshinodai
Sagamihara
Kanagawa 229 8510
Japan
☎ 81 427 51 3911 *2603
✆ 81 427 59 4253
🖂 hmurakam@astro.isas.ac.jp

Murakami Izumi
Ntl Inst Fusion Sci
Data/Planning Ctr
Oroshi-cho
Toki 509 5292
Japan
☎ 81 572 58 2264
✆ 81 572 58 2628
🖂 mizumi@nifs.ac.jp

Murakami Toshio
ISAS
3 1 1 Yoshinodai
Sagamihara
Kanagawa 229 8510
Japan
☎ 81 427 51 3911
✆ 81 427 59 4253
🖂 murakami@astro.isas.ac.jp

Muratorio Gerard
Observ Marseille
2 Pl Le Verrier
F 13248 Marseille Cdx 04
France
☎ 33 4 95 04 4146
✆ 33 4 91 62 1190
🖂 muratorio@observatoire.cnrs-mrs.fr

Murdin Paul G
Part Phys/Astron Res co
Polaris House
North Star Av
Swindon SN22 1SZ
UK
☎ 44 17 934 42075
✆ 44 17 934 42036
🖂 murdinp@pparc.ac.uk

Murdoch Hugh S
Astrophysics Dpt
Univ of Sydney
Sydney NSW 2006
Australia
☎ 61 2 692 2222
✆ 61 2 660 2903

Murdock Thomas Lee
General Res Corp
Dpt Technology
5 Cherry Hill
Danvers MA 01923
USA
☎ 1 617 777 6323

Mureddu Leonardo
Stazione Astronomica
Univ d Cagliari
Via Ospedale 72
I 09100 Cagliari
Italy
☎ 39 70 66 3544
✆ 39 70 65 7657

Muriel Hernan
Observ Astronomico
de Cordoba
Laprida 854
5000 Cordoba
Argentina
☎ 54 51 214 059
✆ 54 51 21 0613
✉ muriel@uncbob.edu

Murphy Brian William
Dpt Phys/Astronomy
Butler Univ
4600 Sunset Av
Indianapolis IN 46208
USA
☎ 1 317 283 9282
✆ 1 317 283 9950
✉ murphy@astrosun.tn.cornell.edu

Murphy Robert E
NASA Headquarters
Code Y
600 Independence Av SW
Washington DC 20546
USA
☎ 1 202 453 1720

Murray C Andrew
Derwent Cott 12 Derwent Rd
Meads
Eastbourne BN20 7PH
UK
☎ 44 132 364 7600
✆ 44 132 364 7600

Murray Carl D
Astronomy Unit
QMWC
Mile End Rd
London E1 4NS
UK
☎ 44 171 975 5456
✆ 44 181 981 9587
✉ C.D.Murray@qmw.ac.uk

Murray John B
Dpt Earth Sci
The Open University
Walton Hall
Milton Keynes MK7 6AA
UK
☎ 44 190 865 2118
✆ 44 190 865 5151
✉ j.b.murray@open.ac.uk

Murray Stephen David
LLNL
L 58
Box 808
Livermore CA 94551 9900
USA
☎ 1 415 423 9382
✆ 1 415 422 5102
✉ murray@astron.berkeley.edu

Murray Stephen S
CfA
HCO/SAO
60 Garden St
Cambridge MA 02138 1516
USA
☎ 1 617 495 7205
✆ 1 617 495 7356

Murtagh Fionn
Observ Strasbourg
11 r Universite
F 67000 Strasbourg
France
☎ 33 3 88 15 0710
✆ 33 3 88 15 0740
✉ fmurtagh@cdsxb6.u-strasbg.fr

Murthy Jayant
Dpt Phys/Astronomy
JHU
Charles/34th St
Baltimore MD 21218
USA
☎ 1 301 516 7027
✆ 1 301 516 4109
✉ murthy@pha.jhu.edu

Musatenko Sergij
Astronomical Observ
Observatorna st 3
Kyiv 254 053
Ukraine
☎ 380 4426 64457
✆ 380 4426 64507
✉ musat@aoku.freenet.kiev.ua

Musen Peter
8804 Orbit Lane
Lanham MD 20801
USA
☎ 1 301 552 3848

Mushotzky Richard
NASA/GSFC
Code 662
Greenbelt MD 20771
USA
☎ 1 301 286 7579
✆ 1 301 286 1684
✉ mushotzky@lheaux.gsfc.nasa.gov

Musielak Zdzislaw E
NASA/MSFC
Space Science Lab
Code ES 52
Huntsville AL 35812
USA
☎ 1 205 544 7619
✆ 1 205 547 7754

Musman Steven
NOAA
Gioscience Lab
1305 East West Highway
Silver Spring MD 20910
USA

Mutel Robert Lucien
Dpt Phys/Astronomy
Univ of Iowa
605 Brookland Park Dr
Iowa City IA 52242 1479
USA
☎ 1 319 335 1950
✆ 1 319 335 1753

Mutschlecner J Paul
Astronomy Dpt
Indiana Univ
Swain W 319
Bloomington IN 47405
USA
☎ 1 812 855 6911
✆ 1 812 855 8725

Muxlow Thomas
NRAL
Univ Manchester
Jodrell Bank
Macclesfield SK11 9DL
UK
☎ 44 14 777 1321
✆ 44 147 757 1618

Muzzio Juan C
Observ Astronomico
Paseo d Bosque S/n
1900 La Plata (Bs As)
Argentina
☎ 54 21 217 308
✆ 54 21 211 761/25 5004
✉ jcmuzzio@fcaglp.fcaglp.unlp.edu.ar

Myachin Vladimir F
Inst Theoret Astron
Acad Sciences
N Kutuzova 10
192187 St Petersburg
Russia
☎ 7 812 278 8898
✆ 7 812 272 7968

Myers Philip C
CfA
HCO/SAO MS 42
60 Garden St
Cambridge MA 02138 1516
USA
☎ 1 617 495 7295
✆ 1 617 495 7345
✉ myers@cfa.harvard.edu

N

Nacozy Paul E
Federal Space Systems
Box 26712
Austin TX 78755
USA
☎ 1 512 467 6659

Nadal Robert
OMP
14 Av E Belin
F 31400 Toulouse Cdx
France
☎ 33 5 61 25 2101
✆ 33 5 61 27 3179
✉ nadal@astro.obs-mip.fr

Nadeau Daniel
Dpt d Physique
Universite Montreal
CP 6128 Succ A
Montreal QC H3C 3J7
Canada
☎ 1 514 343 6676
✆ 1 514 343 2071

Nadyozhin Dmittris K
Inst Theor/Exper Phys
Cheremushkinskaya Ul 25
117259 Moscow
Russia
☎ 7 095 123 0292

Nagase Fumiaki
ISAS
3 1 1 Yoshinodai
Sagamihara
Kanagawa 229 8510
Japan
☎ 81 427 51 3911
✆ 81 427 59 4253
✉ nagase@astro.isas.ac.jp

Nagata Tetsuya
Physics Dpt
Nagoya Univ
Furo-Cho Chikusa-Ku
Nagoya 464 81
Japan
☎ 81 527 89 2926
✆ 81 527 89 2922
✉ nagata@zlab.phys.nagoya-u.ac.jp

Nagendra K N
IIA
Koramangala
Sarjapur Rd
Bangalore 560 034
India
☎ 91 80 553 0672
✆ 91 80 553 4043
✉ knn@iiap.ernet.in

Nagirner Dmitrij I
Astronomical Observ
St Petersburg Univ
Bibliotechnaja Pl 2
198904 St Petersburg
Russia
☎ 7 812 257 9489
✆ 7 812 428 4259

Nagnibeda Valery G
Astronomical Observ
St Petersburg Univ
Bibliotechnaja Pl 2
198904 St Petersburg
Russia
☎ 7 812 257 9491
✆ 7 812 428 4259

Nagovitsyn Yuri A
Pulkovo Observ
Acad Sciences
10 Kutuzov Quay
196140 St Petersburg
Russia
☎ 7 812 123 1997
✆ 7 812 315 1701
✉ sol@gaoran.spb.su

Nahon Fernand
25 Av de L'Europe
F 92310 Sevres
France
☎ 33 1 45 34 1805

Naidenov Victor O
Ioffe Physical Tech Inst
Acad Sciences
Polytechnicheskaya Ul 26
194021 St Petersburg
Russia
☎ 7 812 247 9167
✆ 7 812 247 1017

Nair Sunita
NRAL
Univ Manchester
Jodrell Bank
Macclesfield SK11 9DL
UK
☎ 44 14 777 1321
✆ 44 147 757 1618
✉ sunita@jb.man.ac.uk

Nakada Yoshikazu
Kiso Observ
Inst of Astronomy
Mitake Kiso
Nagano 397 01
Japan
☎ 81 264 52 3360
✆ 81 264 52 3361
✉ lnakada@c1.mtk.nao.ac.jp

Nakagawa Naoya
Univ Electro-communicatio
Chofu-shi
Tokyo 182
Japan
☎ 81 424 83 2161

Nakagawa Takao
ISAS
3 1 1 Yoshinodai
Sagamihara
Kanagawa 229 8510
Japan
☎ 81 427 51 3911
✆ 81 427 59 4253
✉ nakagawa@astro.isas.ac.jp

Nakagawa Yoshinari
Chiba Inst Technology
Narashino 275
Japan
☎ 81 474 75 2111

Nakagawa Yoshitsugu
Dpt Earth/Planet Sci
Kobe Univ
Nada
Kobe 657 8501
Japan
☎ 81 788 03 0563
✆ 81 788 82 1549
✉ yoshi@saturn.phys.kobe-u.ac.jp

Nakai Naomasa
Nobeyama Radio Obs
NAOJ
Minamimaki Mura
Nagano 384 13
Japan
☎ 81 267 63 4367
✆ 81 267 63 4387
✉ nronaka@jpnnro

Nakai Yoshihiro
Kwasan/Hida Obs
Kyoto Univ
Yamashina
Kyoto 607
Japan
☎ 81 755 81 1235
✆ 81 755 93 9617

Nakajima Hiroshi
Nobeyama Radio Obs
NAOJ
Minamimaki Mura
Nagano 384 13
Japan
☎ 81 267 98 2034
✆ 81 267 98 2884

Nakajima Koichi
Lab of Astron/Geophys
Hitotsubashi Univ
Naka 2-1 Kunitachi
Tokyo 186
Japan
☎ 81 435 72 1101
✉ nakajmki@cc.nao.ac.jp

Nakajima Tadashi
NAOJ
Osawa 2- 21-2
Mitaka 181 0015
Japan
☎ 81 422 34 3643
✆ 81 422 34 3776
✉ tadashi@dodgers.mtk.nao.ac.jp

Nakamoto Taishi
Ctr for Comput Physics
Univ of Tsukuba
Tennoudai 1 1 1
Tsukuba 305 Ibaraki
Japan
☎ 81 298 53 6490
✆ 81 298 53 6406
✉ nakamoto@rccp.tsukuba.ac.jp

Nakamura Akiko M
ISAS
3 1 1 Yoshinodai
Sagamihara
Kanagawa 229 8510
Japan
☎ 81 427 51 3911
✆ 81 427 59 4237
✉ akiko@planeta.sci.isas.ac.jp

Nakamura Takashi
Physics Dpt
Kyoto Univ
Sakyo Ku
Kyoto 606 01
Japan
☎ 81 757 53 7008
✆ 81 75 753 7010

Nakamura Tsuko
Tokyo Astronomical Obs
NAOJ
Osawa Mitaka
Tokyo 181
Japan
☎ 81 381 22 111
✆ 81 422 34 3698
✉ nakamura@c1.mtk.nao.ac.jp

Nakamura Yasuhisa
Dpt Sci Education
Fukushima Univ
Matsukawa Machi
Fukushima 960 12
Japan
☎ 81 245 48 8200
✆ 81 245 48 3181
✉ nakamura@educ.
fukushima-u.ac.jp

Nakano Makoto
Fac of Education
Oita Univ
700 Dannoharu
Oita 870 11
Japan
☎ 81 975 69 3311*360
✆ 81 975 68 8319
✉ mnakano@cc.oita-
u.ac.jp

Nakano Syuichi
3 19 1 Chomo
Takenokuchi Sumoto
Hyogo Ken 656
Japan

Nakano Takenori
Nobeyama Radio Obs
NAOJ
Minamimaki Mura
Nagano 384 13
Japan
☎ 81 267 98 4361
✆ 81 267 98 2884

Nakariakov Valery
School Maths/Comp Sci
Univ of St Andrews
North Haugh
St Andrews Fife KY16 9SS
UK
☎ 44 1334 76161
✆ 44 1334 74487
✉ valery@dcs.st-
andrews.ac.uk

Nakayama Kunji
Fac of Education
Kochi Univ
Kochi 780
Japan
☎ 81 888 44 8417
✆ 81 888 44 8453
✉ knakayam@cc.kochi-
u.ac.jp

Nakayama Shigeru
3 7 11 Chuou
Nakano Ku
Tokyo 164
Japan

Nakazawa Kiyoshi
Tokyo Inst of Tech
Ohokayama 2-12-1
Meguroku Tokyo 152
Japan

Namba Osamu
Marco Pololaan 319
NL 3526 GE Utrecht
Netherlands

Namboodiri P M S
IIA
Koramangala
Sarjapur Rd
Bangalore 560 034
India
☎ 91 80 356 6585
✆ 91 80 553 4043
✉ pmsn@iiap.ernet.in

Nambu Yasusada
Physics Dpt
Nagoya Univ
Chikusa
Nagoya 464 01
Japan
☎ 81 527 89 2913
✆ 81 527 89 2916
✉ nambu@allegro.phys.
nagoya-u.ac.jp

Nan Rendong
Beijing Astronomical Obs
CAS
Beijing 100080
China PR
☎ 86 1 28 1698
✆ 86 1 256 10855

Nandy Kashinath
Royal Observ
Blackford Hill
Edinburgh EH9 3HJ
UK
☎ 44 131 667 3321
✆ 44 131 662 1668

Napier William M
Armagh Observ
College Hill
Armagh BT61 9DG
UK
☎ 44 18 61 522 928
✆ 44 18 61 527 174
✉ wmn@star.arm.ac.uk

Napiwotzki Ralf
Dr Remeis Sternwarte
Univ Erlangen-Nuernberg
Sternwartstr 7
D 96049 Bamberg
Germany
☎ 49 951 952 2217
✆ 49 951 952 2222
✉ napiwotzki@sternwarte.
uni-erlangen.de

Naqvi S I H
Physics Dpt
Univ of Regina
Regina
Saskatchewan S4S OA2
Canada
☎ 1 306 585 4258
✆ 1 306 585 4894
✉ naqvi@max.cc.
uregina.ca

Narain Udit
Astrophys Res Gp
Meeut College
Meerut 250 001
India

Naranan S
TIFR
Homi Bhabha Rd
Colaba
Bombay 400 005
India
☎ 91 22 219 111
✆ 91 22 215 2110

Narasimha Delampady
TIFR
Homi Bhabha Rd
Colaba
Bombay 400 005
India
☎ 91 22 495 2311
✆ 91 22 495 2110

Narayan Ramesh
CfA
HCO/SAO MS 51
60 Garden St
Cambridge MA 02138 1516
USA
☎ 1 617 495 7000
✆ 1 617 495 7356

Narayana J V
Regional Meteorological
Office
4 College Rd
Madras 600 006
India

Narita Shinji
Doshisha Univ
Kyoto 602
Japan

Narlikar Jayant V
IUCAA
PO Box 4
Ganeshkhind
Pune 411 007
India
☎ 91 212 351 414/5
✆ 91 212 350 760
✉ jvn@iucaa.ernet.in

Natali Giuliano
IAS
Area d Ricerca CNR
Via Fosso Cavaliere 100
I 00133 Roma
Italy
☎ 39 6 4993 4473
✆ 39 6 2066 0188
✉ somp@saturn.rm.
fra.cnr.it

Nath Mishra Kameshwar
Bhilai Inst of Tech
Durg 491 001 MP
India
☎ 91 788 323 997
✆ 91 788 358 361

Nather R Edward
Astronomy Dpt
Univ of Texas
Rlm 15 308
Austin TX 78712 1083
USA
☎ 1 512 471 4461
✆ 1 512 471 6016

Natta Antonella
Dpt Astronomia
Univ d Firenze
Largo E Fermi 5
I 50125 Firenze
Italy
☎ 39 55 27 52239
✆ 39 55 22 0039

Naumov Vitalij A
Pulkovo Observ
Acad Sciences
10 Kutuzov Quay
196140 St Petersburg
Russia
☎ 7 812 298 2242
✆ 7 812 315 1701

Navarro Julio Fernando
Steward Observ
Univ of Arizona
Tucson AZ 85721
USA
☎ 1 520 621 5950
✆ 1 520 621 1532
✉ jnavarro@as.arizona.edu

Nawar Samir
NAIGR
Helwan Observ
Cairo 11421
Egypt
☎ 20 78 0645/2683
✆ 20 62 21 405 297
✉ altadross@fruc.eun.eg

Neckel Heinz
Hamburger Sternwarte
Univ Hamburg
Gojensbergsweg 112
D 21029 Hamburg
Germany
☎ 49 407 252 4130
✆ 49 407 252 4198

Neckel Th
MPI Astronomie
Koenigstuhl
D 69117 Heidelberg
Germany
☎ 49 622 152 8288
✆ 49 622 152 8246

Nee Tsu-Wei
Physics Dpt
Ntl Central Univ
Chung Li Taiwan 32054
China R
☎ 886 3 462 2302
✆ 886 3 426 2304

Neeman Yuval
Dpt Phys/Astronomy
Tel Aviv Univ
Ramat Aviv
Tel Aviv 69978
Israel
☎ 972 3 640 9279
✆ 972 3 640 8179

Nefedeva Antonina I
Engelhardt Astronom
Observ
Observatoria Station
422526 Kazan
Russia
☎ 7 32 4827

Neff James Edward
Dpt Phys/Astronomy
Coll of Charleston
Charleston SC 29424
USA
☎ 1 803 953 5593
✆ 1 803 953 4824
✉ neffj@cofc.edu

Neff John S
Dpt Phys/Astronomy
Univ of Iowa
605 Brookland Park Dr
Iowa City IA 52242 1479
USA
☎ 1 319 353 4340
✆ 1 319 335 1753

Neff Susan Gale
NASA GSFC
Code 684 1
Greenbelt MD 20771
USA
☎ 1 301 286 5137
✆ 1 301 286 8709
✉ neff@cobblr.gsfc.
nasa.gov

Neidig Donald F
AFGL
NSO
Sunspot NM 88349
USA
☎ 1 505 434 7000
✆ 1 504 434 7029

Neizvestny Sergei
SAO
Acad Sciences
Nizhnij Arkhyz
357147 Karachaevo
Russia
☎ 7 878 784 6278
✆ 7 96 908 2861
✉ nws@sao.ru

Nelson Alistair H
Physics Dpt
Univ Wales College
Box 913
Cardiff CF1 3TH
UK
☎ 44 12 22 874 785
✆ 44 12 22 371 921
✉ nelsona@uk.
ac.cardiff

Nelson Burt
Astronomy Dpt
San Diego State Univ
San Diego CA 92182
USA
☎ 1 619 265 6175
✆ 1 619 594 1413
✉ nelson@minaka.
sdsu.edu

Nelson George Driver
Astronomy Dpt
Univ of Washington
Box 351580
Seattle WA 98195 1580
USA
☎ 1 425 543 6616
✆ 1 425 685 3218
✉ pnelson@cosmos.astro.
washinton

Nelson Graham John
CSIRO
Div Radiophysics
Box 76
Epping NSW 2121
Australia
☎ 61 2 868 0222
✆ 61 2 868 0310
✉ gnelson@atnf.
csiro.au

Nelson Jerry E
Lick Observ
Univ of California
Santa Cruz CA 95064
USA
☎ 1 831 459 2513
✆ 1 831 426 3115

Nelson Robert M
CALTECH/JPL
MS 183 501
4800 Oak Grove Dr
Pasadena CA 91109 8099
USA
☎ 1 213 354 6893
✆ 1 818 393 6030

Nemec James
Intl Statistics &
Research Corp
Box 496
Brentwood Bay BC V8M 1R3
Canada
✉ nemec@dao.nrc.ca

Nemiroff Robert
Michigan Technical Univ
Physics Dpt
1400 Townsend Dr
Houghton MI 49931
USA
☎ 1 906 487 2198
✉ nemiroff@mtu.edu

Nesci Roberto
Istt Astronomico
Univ d Roma La Sapienza
Via G M Lancisi 29
I 00161 Roma
Italy
☎ 39 6 44 03734
✆ 39 6 44 03673
✉ nesci@astrm2.rm.
astro.it

Nesis Anastasios
Kiepenheuer Inst
Sonnenphysik
Schoeneckstr 6
D 79104 Freiburg Breisgau
Germany
☎ 49 761 382 067
✆ 49 761 32280

Ness Norman F
Bartol Res Inst
Univ of Delaware
Newark DE 19716
USA
☎ 1 302 451 8116
✆ 1 302 510 6665

Nesterov Nikolai S
RT 22
Katsiveli
334247 Crimea
Ukraine
☎ 380 6547 27952
✆ 380 6557 27961/6554 40704
✉ nesterov@rt22.
crimea.ua

Netzer Hagai
Dpt Phys/Astronomy
Tel Aviv Univ
Ramat Aviv
Tel Aviv 69978
Israel
☎ 972 3 545 0208
✆ 972 3 640 8179

Neugebauer Gerry
CALTECH
MS 320 47
Downes Lab of Physics
Pasadena CA 91125
USA
☎ 1 818 356 4284
✆ 1 818 796 8806

Neukirch Thomas
Mathematics Dpt
Univ of St Andrews
St Andrews North Haugh
Fife KY16 9SS
UK
☎ 44 13 34 7661
✆ 44 13 347 4487
✉ thomas@dcs.st-
andrews.ac.uk

Neukum G
DLR
Inst f Planetenerkundg
Rudower Chaussee 5
D 12489 Berlin
Germany
☎ 49 306 954 5300
✆ 49 306 954 5303

Neupert Werner M
817 E Moorhead Circle
Boulder CO 80303
USA
☎ 1 301 286 8169

Neuzil Ludek
Astronomical Institute
Czech Acad Sci
Fricova 1
CZ 251 65 Ondrejov
Czech R
☎ 420 204 85 7331
✆ 420 2 88 1611
✉ astsun@csearn

Neves de Araujo Jose Carlos
IAG/USP
Miguel Stefano 4200
04301 904 Sao Paulo SP
Brazil
☎ 55 11 577 8599*235
✆ 55 11 577 8599
✉ jcarlos@andromeda.
iagusp.usp.br

New Roger
Sheffield Hallam Univ
City Campus
Pond St
Sheffield S1 1WB
UK
☎ 44 114 253 3056
✆ 44 114 253 3085
✉ r.new@shu.ac.uk

Newburn Ray L
CALTECH/JPL
MS 169 237
4800 Oak Grove Dr
Pasadena CA 91109 8099
USA
☎ 1 818 354 2319
✆ 1 818 393 4619
✉ ray@sch5.jpl.nasa.gov

Newell Edward B
MSSSO
Weston Creek
Private Bag
Canberra ACT 2611
Australia
☎ 61 262 881 111
✆ 61 262 490 233

Newhall X X
25913 Carillo Dr
Valencia CA 91355 2147
USA
☎ 1 805 259 9999
✆ 1 805 254 4444
✉ xxn@logos.jpl.nasa.gov

Newman Michael John
LANL
MS B220 X 2
Box 1663
Los Alamos NM 87545 2345
USA
☎ 1 505 667 7698
✆ 1 505 665 4055

Newsom Gerald H
Astronomy Dpt
Ohio State Univ
174 W 18th Av
Columbus OH 43210 1106
USA
☎ 1 614 422 7082
✆ 1 614 292 2928
✉ gnewsom@astronomy.ohio-
state.edu

Newton Gavin
Dpt Phys/Astronomy
Univ of Glasgow
Glasgow G12 8QQ
UK
☎ 44 141 339 8855*4196
✆ 44 1413 349 029

Newton Robert R
APL
JHU
Johns Hopkins Rd
Laurel MD 20723 6099
USA
☎ 1 301 953 7100
✆ 1 301 953 1093

Nezlin Mikhail
RRC Kurchatov Inst
Kurchatov Sq 1
123182 Moscow
Russia
☎ 7 095 196 7976
✆ 7 095 943 0073
✉ nezlin@wowa.net.
kiae.su

Ng Kin-Wang
Inst of Physics
Academia Sinica
Taipei 11529
China R
☎ 886 2 782 3075
✆ 886 2 783 4187
✉ phkwng@twnas886

Nguyen Mau Tung
Committee for Space Res
201 K16 Bach Khoa
Box 429 Bo Ho
Ha Noi 10 000
Vietnam
☎ 84 422 58 333
✆ 84 42 524 83

Nguyen-Quang Rieu
Observ Paris
DEMIRM
61 Av Observatoire
F 75014 Paris
France
☎ 33 1 40 51 2103
✆ 33 1 40 51 2002

Nha Il-Seong
Yonsei Univ Obs
134 Sinchon-dong
Seodaemum ku
Seoul 120 449
Korea RP
☎ 82 392 20 131
✆ 82 236 50 937

Niarchos Panayiotis
Astronomy Dpt
Ntl Univ Athens
Panepistimiopolis
GR 157 84 Zografos
Greece
☎ 30 1 724 3414
✆ 30 1 723 5122

Niazy Adnan Mohammad
KACST
Box 6086
Riyadh 11442
Saudi Arabia
☎ 966 1 488 3751
✆ 966 1 488 3756

Nicastro Luciano
TeSRE
CNR
Via P Gobetti101
I 40129 Bologna
Italy
☎ 39 51 639 8667
✆ 39 51 639 8723
✉ nicastro@tesre.bo.cnr.it

Nicholls Jennifer
School of Physics
UNSW
RCFTA
Sydney NSW 2006
Australia
☎ 61 2 9351 2621
✆ 61 2 660 2903
✉ jan@physics.su.
oz.au

Nicholls Ralph W
Dpt Phys/Astronomy
York Univ
4700 Keele St
North York ON M3J 1P3
Canada
☎ 1 416 736 5247
✆ 1 416 736 5626
✉ fs300003@yusol

Nichols-Bohlin Joy
NASA GSFC
Code 684 9
Greenbelt MD 20771
USA
☎ 1 301 794 1410
✆ 1 301 459 4482
✉ nichols@iue.gsfc.
nasa.gov

Nicolaidis Efthymios
NRF
48 Vas Constantinou Av
GR 116 35 Athens
Greece
☎ 30 1 721 0554
✆ 30 1 724 6212
✉ efnicol@eie.gr

Nicolas Kenneth Robert
NRL
Code 4163
4555 Overlook Av SW
Washington DC 20375 5000
USA
☎ 1 202 767 2517

Nicolet Bernard
Observ Geneve
Chemin d Maillettes 51
CH 1290 Sauverny
Switzerland
☎ 41 22 755 2611
✆ 41 22 755 3983
✉ bernard.nicolet@obs.
unige.ch

Nicoll Jeffrey Fancher
Inst Defense Analyses
1801 N Beauregard St
Alexandria VA 22311
USA
☎ 1 703 578 2987
✆ 1 703 578 2877
✉ jnicoll@ida.org

Nicollier Claude
NASA/JSC
Code CB
Houston TX 77058
USA
☎ 1 713 244 8888
✆ 1 713 244 8873
✉ claude.nicollier1@jsc.
nasa.gov

Nicolov Nikolai S
Astronomy Dpt
Univ of Sofia
Anton Ivanov St 5
BG 1126 Sofia
Bulgaria
☎ 359 2 68 8176
✆ 359 2 68 9085

Nicolson George D
HartRAO
FRD
PO Box 443
1740 Krugersdorp
South Africa
☎ 27 11 642 4692
✆ 27 11 642 2446

Nicolson Iain
Univ of Hertfordshire
Bayfordbury
Lower Hatfield Rd
Hertford Herts SG13 8LD
UK
☎ 44 170 728 5560
✆ 44 170 728 5562

Niedner Malcolm B
NASA GSFC
Code 680
Greenbelt MD 20771
USA
☎ 1 301 286 2000

Niedzielski Andrzej
Inst of Astronomy
N Copernicus Univ
Ul Chopina 12/18
PL 87 100 Torun
Poland
☎ 48 56 26 018
✆ 48 56 24 602
✉ aniedzi@astri.uni.
torun.pl

Niell Arthur E
Haystack Observ
Westford MA 01886
USA
☎ 1 617 692 4764
✆ 1 617 981 0590

Niemela Virpi S
Observ Astronomico
Paseo d Bosque s/n
1900 La Plata (Bs As)
Argentina
☎ 54 21 217 308
✆ 54 21 211 761
✉ virpi@colihue.fcaglp.
unlp.edu.ar

Niemi Aimo
Turku Univ
Tuorla Observ
Vaeisaelaentie 20
FIN 21500 Piikkio
Finland
☎ 358 2 274 4268
✆ 358 2 243 3767

Nieuwenhuijzen Hans
Sterrekundig Inst Utrecht
Box 80000
NL 3508 TA Utrecht
Netherlands
☎ 31 30 253 5237
✆ 31 30 253 1601

Niimi Yukio
Tokyo Astronomical Obs
NAOJ
Osawa Mitaka
Tokyo 181
Japan
☎ 81 422 32 5111
✆ 81 422 34 3793
✉ time@spacetime.mtk.nao.jp

Nikitin Alexsey A
Astronomical Observ
St Petersburg Univ
Bibliotechnaja Pl 2
198904 St Petersburg
Russia
☎ 7 812 293 2262
✆ 7 812 428 4259

Nikoghossian Arthur G
Byurakan Astrophys Observ
Armenian Acad Sci
378433 Byurakan
Armenia
☎ 374 88 52 28 3453/4142
✆ 374 88 52 52 3640

Nikoloff Ivan
Perth Obşerv
Walnut Rd
Bickley WA 6076
Australia
☎ 61 8 92 93 1865
✆ 61 8 92 93 8138

Nikolov Andrej
Astronomy Dpt
Univ of Sofia
Anton Ivanov St 5
BG 1126 Sofia
Bulgaria
☎ 359 2 68 8176
✆ 359 2 68 9085

Nilson Peter
Astronomical Observ
Box 515
S 751 20 Uppsala
Sweden
☎ 46 18 53 0265
✆ 46 18 52 7583

Nilsson Carl
CfA
HCO/SAO
60 Garden St
Cambridge MA 02138 1516
USA
☎ 1 617 495 7461
✆ 1 617 495 7356

Ninkov Zoran
Ctr for Imaging Sci
Rochester Inst of
Technology
Rochester NY 14623 5604
USA
☎ 1 716 475 7195
✆ 1 716 475 5988
✉ zxnpci@borg.cis.rit.edu

Ninkovic Slobodan
Astronomical Observ
Volgina 7
11050 Beograd
Yugoslavia FR
☎ 381 1 419 357/421 875
✆ 381 1 419 553

Nishi Keizo
Tokyo Astronomical Obs
NAOJ
Osawa Mitaka
Tokyo 181
Japan
☎ 81 422 32 5111
✆ 81 422 34 3793

Nishi Ryoichi
Physics Dpt
Kyoto Univ
Sakyo-Ku
Kyoto 606 01
Japan
☎ 81 757 53 3880
✆ 81 75 753 3886
✉ nishi@tap.scphys.kyoto-u.ac.jp

Nishida Minoru
Physics Dpt
Kyoto Univ
Sakyo Ku
Kyoto 606 01
Japan
☎ 81 757 53 7008
✆ 81 75 753 7010

Nishida Mitsugu
Dpt Literature
Kobe Women-s Univ
Suma Ku
Kobe 654
Japan
☎ 81 787 31 4416

Nishikawa Jun
Ntl Astronomical Obs
Osawa
Mitaka
Tokyo 181
Japan
☎ 81 422 34 3643
✆ 81 422 34 3608
✉ nisikawa@optik.mtk.nao.ac.jp

Nishimura Jun
Inst Space/Aeron Sci
Univ of Tokyo
Meguro Ku
Tokyo 153
Japan
☎ 81 346 71 111*388
✆ 81 334 85 2904

Nishimura Masaki
Physics Dpt
Hokkaido Univ
Kita 10 Nishi 8
Sapporo 060
Japan
☎ 81 117 11 2111
✆ 81 222 85 132

Nishimura Shiro Emer
Ntl Astronomical Obs
NAOJ
Osawa Mitaka
Tokyo 181
Japan
☎ 81 422 34 3761
✆ 81 422 34 3793
✉ bnishim@c1.mtk.nao.ac.jp

Nishimura Tetsuo
Steward Observ
Univ of Arizona
Tucson AZ 85721
USA
☎ 1 520 621 2054
✆ 1 520 621 1532

Nishio Masanori
Nobeyama Radio Obs
NAOJ
Minamimaki Mura
Nagano 384 13
Japan
☎ 81 267 63 4381
✆ 81 267 98 2506

Nissen Poul E
Inst Phys/Astronomy
Univ of Aarhus
Ny Munkegade
DK 8000 Aarhus C
Denmark
☎ 45 86 12 8899
✆ 45 86 12 0740
✉ pen@obs.aau.dk

Nittmann Johann
Digital Equipment Cbt
Favoritenstr 7
A 1040 Wien
Austria
☎ 43 1 505 48 7012
✆ 43 1 505 48 7022
✉ johannnittmann@cbt.mts.dec.com

Nityananda R
RRI
Sadashivanagar
CV Raman Av
Bangalore 560 080
India
☎ 91 80 336 0122
✆ 91 80 334 0492
✉ najaram@rri.ernet.in

Nobili Anna M
Dpt Matematica
Univ d Pisa
Via Buonarroti 2
I 56127 Pisa
Italy

Nobili L
Dpt Fisica G Galilei
Univ d Padova
Via Marzolo 8
I 35131 Padova
Italy
☎ 39 49 84 4205*111
✆ 39 49 84 4245

Nocera Luigi
Ist Atomica/Molecolare
Univ d Pisa
Via Giardino 7
I 56127 Pisa
Italy
☎ 39 50 501 384
✆ 39 50 251 75
✉ bistab @ icnucevm

Noci Giancarlo
Dpt Astronomia
Univ d Firenze
Largo E Fermi 5
I 50125 Firenze
Italy
☎ 39 55 27 521
✆ 39 55 22 0039

Noel Fernando
Dpt Astronomia
Univ Chile
Casilla 36 D
Santiago
Chile
☎ 56 2 229 4101
✆ 56 2 229 4002

Noels Arlette
Inst Astrophysique
Universite Liege
Av Cointe 5
B 4000 Liege
Belgium
☎ 32 4 254 7510
✆ 32 4 254 7511
✉ noels@astro.ulg.ac.be

Noens Jacques-Clair
OMP
9 r Pont de La mouette
F 65200 Bagneres Bigorre
France
☎ 33 5 62 95 1969
✆ 33 5 62 95 1070

Noerdlinger Peter D
NASA/GSFC
Code 902
Greenbelt MD 20771
USA
☎ 1 301 614 5475
✆ 1 301 614 5268
✉ pnoerdli@eos.hitc.com

Noguchi Kunio
Astrophysics Dpt
Nagoyo Univ
Furocho Chikusa Ku
Nagoyo 464 01
Japan
☎ 81 527 89 2926
✆ 81 527 89 2922

Noguchi Masafumi
Astron Institute
Tohoku Univ
Aoba Ku
Sendai 980
Japan
☎ 81 206 543 2888

Noh Hyerim
Senior Res Scientist
Korea Astronomy Obs
San 36 1 Hwaam-dong
Yusung gu Taejeon
Korea RP
☎ 82 428 65 3332
✆ 82 563 36 9450
✉ hr@hanul.issa.re.kr

Noll Keith Stephen
STScI
Homewood Campus
3700 San Martin Dr
Baltimore MD 21218
USA
☎ 1 301 338 5080
✆ 1 301 338 4767
✉ noll@snoqualm.stsci.edu

Nollez Gerard
IAP
98bis bd Arago
F 75014 Paris
France
☎ 33 1 44 32 8142
✆ 33 1 44 32 8001

Nomoto Ken'ichi
Astronomy Dpt
Univ of Tokyo
Bunkyo Ku
Tokyo 113
Japan
☎ 81 358 00 6882
✆ 81 338 13 9439
✉ nomoto@astron.s.u-tokyo.ac.jp

Noonan Thomas W
2133 Lynbridge Dr
Charlotte NC 28270
USA

Norci Laura
Dunsink Observ
DIAS
Castleknock
Dublin 15
Ireland
☎ 353 1 38 7911
✆ 353 1 38 7090
✉ ln@dunsink.dias.ie

Nordh Lennart H
Stockholm Observ
Royal Swedish Acad Sciences
S 133 36 Saltsjoebaden
Sweden
☎ 46 8 717 0195
✆ 46 8 717 4719
✉ nordh@astro.su.se

Nordlund Aake
Astronomical Observ, NBIfAFG
Copenhagen Univ
Juliane Maries Vej 30
DK 2100 Copenhagen
Denmark
☎ 45 35 32 5968
✆ 45 35 32 5989

Nordstroem Birgitta
Astronomical Observ, NBIfAFG
Copenhagen Univ
Juliane Maries Vej 30
DK 2100 Copenhagen
Denmark
☎ 45 35 32 5950
✆ 45 35 32 5989
✉ birgitta@astro.ku.dk

Norgaard-Nielsen Hans U
Danish Space Res Inst
Juliane Maries Vej 30
DK 2100 Copenhagen
Denmark
☎ 45 35 32 5728
✆ 45 35 36 2475

Noriega-Crespo Alberto
Astronomy Dpt
Univ of Washington
Box 351580
Seattle WA 98195 1580
USA
☎ 1 425 685 2155
✆ 1 425 685 0403
✉ aipsac@phast.phys.
washinton.ed

Norman Colin A
CAS
JHU
Charles/34th St
Baltimore MD 21218 2695
USA
☎ 1 301 516 7329
✆ 1 301 338 5090
✉ norman@stsci.edu

Norris John
MSSSO
Weston Creek
Private Bag
Canberra ACT 2611
Australia
☎ 61 262 881 111
✆ 61 262 490 233
✉ jen@mso.anu.edu.au

Norris Raymond Paul
CSIRO
Div Radiophysics
Box 76
Epping NSW 2121
Australia
☎ 61 2 9372 4416
✆ 61 2 9372 4400
✉ rnorris@rp.csiro.au

North John David
Filosofisch Inst
Rijksuniversiteit
A Weg 30
NL 9718 CW Groningen
Netherlands
☎ 31 59 07 1846
✆ 31 50 363 6160

North Pierre
Institut Astronomie
Univ Lausanne
CH 1290 Chavannes d Bois
Switzerland
☎ 41 22 755 2611
✆ 41 22 755 3983
✉ Pierre.north@obs.unige.ch

Noskov Boris N
SAI
Acad Sciences
Universitetskij Pr 13
119899 Moscow
Russia
☎ 7 095 939 2858
✆ 7 095 939 0126

Notni P
Astrophysik Inst
Potsdam Univ
an d Sternwarte 16
D 14482 Potsdam
Germany
☎ 49 331 749 90
✉ pnotni@aip.de

Nottale Laurent
Observ Paris Meudon
DAF
Pl J Janssen
F 92195 Meudon PPL Cdx
France
☎ 33 1 45 07 7403
✆ 33 1 45 07 7469

Noumaru Junichi
Subaru Telescope
NAOJ
650 N A'Ohoku Place
Hilo HI 96720
USA
☎ 1 808 934 5088
✆ 1 808 934 5984
✉ noumaru@naoj.org

Novello Mario
Ctr Bras Pesquisas Fisic
Rua Dr Xavier Sigaud
150 Urca
22290 Rio de Janeiro RJ
Brazil
☎ 55 21 541 0337

Novick Robert
Physics Dpt
Columbia Univ
538 W 120th St
New York NY 10027
USA
☎ 1 212 280 3293
✉ rn@cuphy3.phys.
columbia.edu

Novikov Igor D
Theoretical Astrophysics Centre
Juliane Maries Vej 30
DK 2100 Copenhagen
Denmark
☎ 45 35 32 5901
✆ 45 35 32 5910
✉ novikov@nordita.dk

Novikov Sergej B
SAI
Acad Sciences
Universitetskij Pr 13
119899 Moscow
Russia
☎ 7 095 939 2858
✆ 7 095 939 0126

Novoselov Victor S
Astronomical Observ
St Petersburg Univ
Bibliotechnaja Pl 2
198904 St Petersburg
Russia
☎ 7 812 257 9491
✆ 7 812 428 4259

Novosyadlyj Bohdan
Astronomical Observ
Lviv State Univ
Kyryla i Mephodia st 8
290005 Lviv
Ukraine
☎ 380 3227 29088
✆ 380 3227 28088
✉ novos@astro.lviv.ua

Novotny Jan
Dpt General Physics
Masaryk Univ
Kotlarska 2
CZ 611 37 Brno
Czech R
☎ 420 5 4112 9466
✆ 420 5 4121 1214
✉ novotny@elanor.sci.
muni.cs

Noyes Robert W
CfA
HCO/SAO
60 Garden St
Cambridge MA 02138 1516
USA
☎ 1 617 495 7424
✆ 1 617 495 7356
✉ noyes@cfa.
harvard.edu

Nugis Tiit
Tartu Observ
Estonian Acad Sci
Toravere
EE 2444 Tartumaa
Estonia
☎ 372 7 410 265
✆ 372 7 410 205

Nulsen Paul
Physics Dpt
Univ of Wollongong
Northfields Av
Wollongong NSW 2522
Australia
☎ 61 42 21 3517
✆ 61 42 21 3151
✉ p.nulsen@now.edu.au

Nunes Rogerio S de sousa
Gr Matem Aplicada
Univ Porto
Rua Das Taipas 135
P 4000 Porto
Portugal
☎ 351 3 803 13/769
✆ 351 2 782 0404

Nunez Jorge
Dpt Astronom Meteo
Univ Barcelona
Avd Diagonal 647
E 08028 Barcelona
Spain
☎ 34 3 247 5736
✆ 34 3 402 1133
✉ jorge@fajnml.am.ub.es

Nunez Josue Arturo
Observ Astronomico
Paseo d Bosque S/n
1900 La Plata (Bs As)
Argentina
☎ 54 21 217 308
✆ 54 21 258 985
✉ jan@fcaglp.edu.ar

Nuritdinov Salakhutdin
Astronomical Institute
Uzbek Acad Sci
Astronomicheskaya Ul 33
700052 Tashkent
Uzbekistan
☎ 7 3712 35 8102
✆ 7 37 11 44 7728
✉ snur@tsu.silk.org

Nussbaumer Harry
Inst Astronomie
ETH Zentrum
CH 8092 Zuerich
Switzerland
☎ 41 1 632 3631
✆ 41 1 632 1205

Nuth Joseph A III
NASA GSFC
Code 691
Greenbelt MD 20771
USA
☎ 1 301 286 9467

Nyman Lars-Aake
Onsala Space Observ
Chalmers Technical Univ
S 439 92 Onsala
Sweden
☎ 46 31 772 5500
✆ 46 31 772 5550

O

O'Brien Paul Thomas
Dpt Phys/X-Ray Astron
Univ Leicester
University Rd
Leicester LE1 7RH
UK
☎ 44 116 252 2073
✆ 44 116 250 182
✉ pto@star.le.ac.uk

O'Brien Tim
Astrophysics Group
Liverpool J M Univ
Byrom St
Liverpool L3 3AF
UK
☎ 44 151 231 2337
✆ 44 151 231 2337
✉ tob@staru1.livjm.ac.uk

O'Byrne John
Astronomy Dpt
School of Physics
Univ of Sydney
Sydney NSW 2006
Australia
☎ 61 2 935 13184
✆ 61 2 9351 7726
✉ j.obyrne@physics.usyd.edu.au

O'Connell Robert F
Dpt Phys/Astronomy
Louisiana State Univ
Baton Rouge LA 70803 4001
USA
☎ 1 504 388 6848
✆ 1 504 388 5855

O'Connell Robert West
University Station
Univ of Virginia
Box 3818
Charlottesville VA 22903 0818
USA
☎ 1 804 924 7494
✆ 1 804 924 3104
✉ rwo@virginia.edu

O'Connor Seamus L
Physics Dpt
Univ College
Belfield
Dublin 4
Ireland
☎ 353 1 693 244
✆ 353 1 671 1759

O'Dea Christopher P
STScI
Homewood Campus
3700 San Martin Dr
Baltimore MD 21218
USA
☎ 1 301 338 2590
✆ 1 301 338 5085

O'Dell Charles R
Dpt Phys/Astronomy
Rice Univ
Box 1892
Houston TX 77251 1892
USA
☎ 1 713 527 8101
✆ 1 713 285 5143

O'Dell Stephen L
NASA/MSFC
Space Science Lab
Code ES 65
Huntsville AL 35812
USA
☎ 1 205 544 7708
✆ 1 205 547 7754

O'Donoghue Darragh
Astronomy Dpt
Univ of Cape Town
Private Bag
7700 Rondebosch
South Africa
☎ 27 21 650 2391
✆ 27 21 54 3726

O'Handley Douglas A
NASA/ARC
Ctr Mars Expl
Moffett Field CA 94035 1000
USA
☎ 1 415 604 5028
✆ 1 415 604 6779
✉ doug_ohandley@qmgate.arc.nasa.gov

O'Keefe John A
NASA GSFC
Code 681
Greenbelt MD 20771
USA
☎ 1 301 286 8445

O'Leary Brian T
Future Focas
5136 E Karen Dr
Scottsdale AZ 85254
USA

O'Mara Bernard J
Physics Dpt
Univ of Queensland
St Lucia
Brisbane QLD 4067
Australia
☎ 61 7 377 3429
✆ 61 7 371 5896

O'Mongain Eon
Physics Dpt
Univ College
Belfield
Dublin 4
Ireland
☎ 353 1 693 244
✆ 353 1 671 1759

O'Sullivan Creidhe
Physics Dpt
Univ College
Galway
Ireland
☎ 353 91 524 411*3329
✆ 353 91 525 700
✉ creidhe@physics.ucg.ie

O'Sullivan Denis F
DIAS
School Cosmic Phys
5 Merrion Sq
Dublin 2
Ireland
☎ 353 1 662 1333
✆ 353 1 662 1477

O'Sullivan John David
CSIRO
Div Radiophysics
Box 76
Epping NSW 2121
Australia
☎ 61 2 868 0222
✆ 61 2 868 0310

Obashev Saken O
Astrophys Inst
Kazakh Acad Sci
480068 Alma Ata
Kazakhstan
☎ 7 62 4040

Obi Shinya
Univ of the Air
2-11 Wakaba
Mihama Ku
Chiba 261 8586
Japan
☎ 81 43 276 5111
✆ 81 43 276 6130

Oblak Edouard
Observ Besancon
41bis Av Observatoire
BP 1615
F 25000 Besancon Cdx
France
☎ 33 3 81 66 6933
✆ 33 3 81 66 6944
✉ oblak@obs-besancon.fr

Obregon Diaz Octavio J
Dpt Fisica
UAM-I Iztapalapao
Apt 55 534
Mexico DF 09340
Mexico
☎ 52 5 724 4623
✆ 52 5 686 1717

Obridko Vladimir N
ITMIRWP
Acad Sciences
142092 Troitsk
Russia
☎ 7 095 232 1921
✆ 7 095 334 0124
✉ obridko@lars.izmiran.troitsk.s

Occhionero Franco
OAR
Via d Parco Mellini 84
I 00136 Roma
Italy
☎ 39 6 34 7056
✆ 39 6 349 8236

Ochsenbein Francois
Observ Strasbourg
11 r Universite
F 67000 Strasbourg
France
☎ 33 3 88 15 0755
✆ 33 3 88 25 0160
✉ francois@simbad.u-strasbg.fr

Oda Minoru
Wako Shi
Saitama 351 01
Japan
☎ 81 484 62 1111
✆ 81 484 67 5942

Oda Naoki
NEC Corporation
Material Devel Ctr/Ist Dev
Miyazaki 4 1 1 Miyamae-ku
Kawasaki Kanagawa 216
Japan
☎ 81 448 56 2291
✆ 81 448 56 2244
✉ oda@mdc.cl.nec.co.jp

Odell Andrew P
Dpt Phys/Astronomy
N Arizona Univ
Box 6010
Flagstaff AZ 86011 6010
USA
☎ 1 520 523 2661
✆ 1 520 523 2626
✉ Andy.Odell@nau.edu

Odenwald Sten F
NRL
Code 4138 0
4555 Overlook Av SW
Washington DC 20375 5000
USA
☎ 1 202 767 3010

Odgers Graham J
3924 Scolton Rd
Victoria BC V8N 4E2
Canada

Odstrcil Dusan
Astronomical Institute
Czech Acad Sci
Fricova 1
CZ 251 65 Ondrejov
Czech R
☎ 420 204 85 7126
℻ 420 2 88 1611
✉ adstrcil@asu.cas.cz

Oegelman Hakki B
Physics Dpt
Univ Wisconsin
1150 University Av
Madison WI 53706
USA
☎ 1 608 265 2052
℻ 1 608 262 3077
✉ ogelman@astrog.
physics.wisc.edu

Oegerle William R
Dpt Phys/Astronomy
JHU
Charles/34th St
Baltimore MD 21218
USA
☎ 1 301 516 6732
℻ 1 301 516 5494
✉ oegerle@pha.jhu.edu

Oehman Yngve
Thulelem 53
S 223 67 Lund
Sweden
☎ 46 46 14 3362

Oekten Adnan
Univ Observ
Univ of Istanbul
University 34452
34452 Istanbul
Turkey
☎ 90 212 522 3597
℻ 90 212 522 6123
✉ jk017@triuvm11

Oemler Augustus
Astronomy Dpt
Yale Univ
Box 208101
New Haven CT 06520 8101
USA
☎ 1 203 436 3460
℻ 1 203 432 5048

Oertel Goetz K
AURA Inc
Suite 350
1200 New York Av NW
Washington DC 20005
USA
☎ 1 202 483 2101
℻ 1 202 483 2106
✉ goertel@stsci.edu

Oestgaard Erlend
Physics Dpt
Univ of Trondheim
AVH
N 7055 Dragvollm
Norway
☎ 47 7 92 0411*117

Oestreicher Roland
Landessternwarte
Koenigstuhl
D 69117 Heidelberg
Germany
☎ 49 622 110 036
℻ 49 622 152 8246

Oetken L
Zntrlinst Astrophysik
Sternwarte Babelsberg
Rosa Luxemburg Str 17a
D 14482 Potsdam
Germany
℻ 49 331 749 9309

Oezel Mehmet Emin
Space Sci Dpt
Marmara Res Centre
PK 21
41470 Gebze Kocaeli
Turkey
☎ 90 262 641 2300
℻ 90 262 641 2309
✉ ozel@trmbeam.tr

Oezkan Mustafa Tuerker
Univ Observ
Univ of Istanbul
University 34452
34452 Istanbul
Turkey
☎ 90 212 528 3847
℻ 90 212 522 6123
✉ jk017@triuvm11

Ofman Leon
NASA/GSFC
Code 662 1
Greenbelt MD 20771
USA
☎ 1 301 286 9913
℻ 1 301 286 1617
✉ leon.ofman@gsfc.
nasa.gov

Ogawa Hideo
Nagoya Univ
Chikusa-ku Nagoya
Aichi 464 01
Japan
☎ 81 527 89 2840
℻ 81 52 789 2845
✉ ogawa@a.phys.
nagoya-u.ac.jp

Ogawara Yoshiaki
Inst Space/Astron Sci
Univ of Tokyo
3-1-1 Yoshinodia
Sagamihara Kanagawa 229
Japan
☎ 81 427 51 3911
℻ 81 427 59 4253
✉ ogawara@astro.
isas.ac.jp

Ogelman Hakki B
Physics Dpt
Univ Wisconsin
1150 University Av
Madison WI 53706
USA
☎ 1 608 265 2052
℻ 1 608 263 0800
✉ ogelman@astrog.physics.
wisc.edu

Ogura Katsuo
College of Literature
Kokugakuin Univ
Higashi 4-10-28
Shibuyaku Tokyo 150
Japan
☎ 81 298 42 6913

Oh Kap-Soo
Dpt Astronomy/Space Sci
Chungnam Ntl Univ
Taejon 305 764
Korea RP
☎ 82 428 21 5464
℻ 82 428 22 8380
✉ ksoh@astrol.
chungnam.ac.kr

Oh Kyu Dong
Dpt Earth Science
Chonnam Ntl
University
Kwangju Chonnan
Korea RP
✉ ohkd@chonnam.
chonnam.ac.kr

Ohashi Nagayoshi
CfA
HCO/SAO MS 78
60 Garden St
Cambridge MA 02138 1516
USA
☎ 1 617 495 7003
℻ 1 617 496 7554
✉ nohashi@cfa.
harvard.edu

Ohashi Takaya
Physics Dpt
Tokyo Metropol Univ
1-1 Minami Ohsawa
Hachioji Tokyo 192 03
Japan
☎ 81 426 77 1111 *3245
℻ 81 426 77 2483
✉ ohashi@phys.metro-
u.ac.jp

Ohashi Yukio
3 5 26 Hiroo
Shibuya-ku
Tokyo
Japan
☎ 81 334 00 7438

Ohishi Masatoshi
Nobeyama Radio Obs
NAOJ
Minamimaki Mura
Nagano 384 13
Japan
☎ 81 267 98 4433
℻ 81 267 98 2884
✉ ohishi@nro.noa.
ac.jp

Ohki Kenichiro
Tokyo Astronomical Obs
NAOJ
Osawa Mitaka
Tokyo 181
Japan
☎ 81 422 32 5111
℻ 81 422 34 3793

Ohnishi Kouji
Nagano Ntl College of Techn
716 Tokuma
Nagaro 381
Japan
☎ 81 262 957 027
℻ 81 262 957 027
✉ ohnishi@ge.nagano-nct.ac.jp

Ohring George
Dpt Geophys/Planet Sci
Tel Aviv Univ
Ramat Aviv
Tel Aviv 69978
Israel
☎ 972 3 640 8633
℻ 972 3 640 9282

Ohta Kouji
Astronomy Dpt
Kyoto Univ
Sakyo Ku
Kyoto 606 01
Japan
☎ 81 757 53 3896
℻ 81 75 753 3897
✉ ohta@kusastro.kyoto-u.ac.jp

Ohtani Hiroshi
Astronomy Dpt
Kyoto Univ
Sakyo Ku
Kyoto 606 01
Japan
☎ 81 757 51 2111
℻ 81 75 753 7010
✉ ohtani@kusastro.kyoto-u.ac.jp

Ohtsubo Junji
Fac of Eng
Shizuoka Univ
3 Chome Jyohoku
Hamamatsu 432
Japan
☎ 81 534 71 1171*585
℻ 81 534 75 1764

Ohyama Noboru
Fac of Eng
Shizuoka Univ
3 Chome Jyohoku
Hamamatsu 432
Japan
☎ 81 534 71 1171

Oja Heikki
Helsinky Univ
Observ
Box 14
FIN 00014 Helsinki
Finland
☎ 358 19 12 2942
℻ 358 19 12 2952

Oja Tarmo
Kvistaberg Observ
S 197 00 Bro
Sweden
☎ 46 85 824 0157
℻ 46 18 527 583
✉ oja@astro.uu.se

Oka Takeshi
Chemistry Dpt
Univ of Chicago
5735 S Ellis Av
Chicago IL 60637
USA
☎ 1 312 962 7070
✆ 1 708 571 8813

Okamoto Isao
Intl Latitude Observ
NAOJ
Hoshigaoka Mizusawa Shi
Iwate 023
Japan
☎ 81 197 24 7111
✆ 81 197 22 7120

Okamura Sadanori
Astronomy Dpt
Univ of Tokyo
Bunkyo Ku
Tokyo 113
Japan
☎ 81 358 00 6880
✆ 81 338 13 9439
✉ okamura@apsunl.astron.s.u-tokyo.jp

Okazaki Akira
Dpt Sci Education
Gunma Univ
Maebashi
Gunma 371
Japan
☎ 81 272 32 1611
✆ 81 272 33 9231

Okazaki Atsuo T
Coll General Education
Hokkai Gakuen Univ
Toyohira-ku
Sapporo 062
Japan
☎ 81 118 41 1161*284
✆ 81 118 24 3141
✉ okazaki@phys.hokusai.ac.jp

Okazaki Seichi
2 4 4 Osawa
Mitaka
Tokyo 181
Japan
☎ 81 422 31 6770
✆ 81 422 34 3793

Oke J Beverley
DAO
NRCC/HIA
5071 West Saanich Rd
Victoria BC V8X 4M6
Canada
☎ 1 205 363 0005
✆ 1 250 363 0045
✉ oke@dao.nrc.ca

Okeke Pius N
Dpt Phys/Astronomy
Univ of Nigeria
Nsukka
Nigeria
☎ 234 42 77 1911
✆ 234 42 77 0644

Okoye Samuel E
Dpt Phys/Astronomy
Univ of Nigeria
Nsukka
Nigeria
☎ 234 42 77 0752
✆ 234 42 77 0644

Okuda Haruyuki
Inst Space/Astron Sci
Univ of Tokyo
Meguro Ku
Tokyo 153
Japan
☎ 81 346 71 111
✆ 81 334 85 2904
✉ kuda@astro.isas.ac.jp

Okuda Toru
Inst Earth Science
Hokkaido Univ Educat
1-2 Hachiman Cho
Hakodate 040
Japan
☎ 81 138 41 1121

Okumura Sachiko
Nobeyama Radio Obs
NAOJ
Minamimaki Mura
Nagano 384 13
Japan
☎ 81 267 63 4366
✆ 81 267 63 4339
✉ sokumura@nro.nao.ac.jp

Olah Katalin
Konkoly Observ
Thege U 13/17
Box 67
H 1525 Budapest
Hungary
☎ 36 1 375 4122
✆ 36 1 275 4668
✉ olah@buda.konkoly.hu

Olano Carlos Alberto
IAR
CC 5
1894 Villa Elisa (Bs As)
Argentina
☎ 54 21 4 3793
✆ 54 21 211 761

Olberg Michael
Onsala Space Observ
Chalmers Technical Univ
S 439 92 Onsala
Sweden
☎ 46 31 772 5500
✆ 46 31 772 5550
✉ olberg@oso.chalmers.se

Oleak H
Zntrlinst Astrophysik
Sternwarte Babelsberg
Rosa Luxemburg Str 17a
D 14473 Potsdam
Germany
☎ 49 331 762225

Oliva Ernesto
Osserv Astrofis Arcetri
Univ d Firenze
Largo E Fermi 5
I 50125 Firenze
Italy
☎ 39 55 27 52310
✆ 39 55 22 0039

Oliveira Grijo A K
Observ Nacional
Rua Gl Bruce 586
San Cristovao
20921 030 Rio de Janeiro RJ
Brazil
☎ 55 21 580 7181
✆ 55 21 580 0332

Oliver John Parker
Astronomy Dpt
Univ of Florida
211 SSRB
Gainesville FL 32611
USA
☎ 1 904 392 2052
✆ 1 904 392 5089

Olivier Scot Stewart
LLNL
L 413
Box 808
Livermore CA 94551 9900
USA
☎ 1 415 423 8129
✆ 1 415 423 0238
✉ olivier@sunlight.llnl.gov

Ollongren A
Leiden Observ
Dpt Maths/Computer Sci
Box 9512
NL 2300 RA Leiden
Netherlands
☎ 31 71 527 2727*5006
✆ 31 71 527 6985

Olmi Luca
Arecibo Observ
Box 995
Arecibo PR 00613
USA
☎ 1 809 878 2612
✆ 1 809 878 1861
✉ olmi@naic.edu

Olnon Friso
NFRA
Postbus 2
NL 7990 AA Dwingeloo
Netherlands
☎ 31 521 59 5100
✆ 31 52 159 7332

Olofsson Goeran S
Stockholm Observ
Royal Swedish Acad Sciences
S 133 36 Saltsjoebaden
Sweden
☎ 46 8 717 2639
✆ 46 8 717 4719
✉ olofsson@astro.su.se

Olofsson Hans
Stockholm Observ
Royal Swedish Acad Sciences
S 133 36 Saltsjoebaden
Sweden
☎ 46 8 716 4448
✆ 46 8 717 4719
✉ hans@astro.su.se

Olofsson Kjell
Astronomical Observ
Box 515
S 751 20 Uppsala
Sweden
☎ 46 18 51 0736
✆ 46 18 52 7583
✉ kjell.olofsson@astro.uu.se

Olowin Ronald Paul
Dpt Phys/Astronomy
Saint Mary's College
207 D Galileo Hall
Moraga CA 94575
USA
☎ 1 510 631 4428
✆ 1 510 376 4027
✉ rpolowin@galileo.stmarys-ca.ed

Olsen Erik H
Astronomical Observ, NBIfAFG
Copenhagen Univ
Juliane Maries Vej 30
DK 2100 Copenhagen
Denmark
☎ 45 35 32 5929
✆ 45 35 32 5989
✉ eho@astro.ku.dk

Olsen Fogh H J
Astronomical Observ, NBIfAFG
Copenhagen Univ
Juliane Maries Vej 30
DK 2100 Copenhagen
Denmark
☎ 45 35 32 5932
✆ 45 35 32 5989
✉ fogh@astro.ku.dk

Olsen Kenneth H
1029 187th Pl SW
Lynnwood WA 98036
USA
☎ 1 206 776 7007
✉ kolsen@geophys.washington.edu

Olson Edward C
Dpt Astron/Physics
Univ of Illinois
1002 W Green St
Urbana IL 61801
USA
☎ 1 217 333 5531
✆ 1 217 244 7638
✉ olsomed@uxh.cso.uiuc.edu

Olthof Hindericus
ESA/ESTEC
SSD
Box 299
NL 2200 AG Noordwijk
Netherlands
☎ 31 71 565 6555
✆ 31 71 565 4532

Omarov Tuken B
Astrophys Inst
Kazakh Acad Sci
480068 Alma Ata
Kazakhstan
☎ 7 64 4040

Omnes Roland
LPTHE
Bt 211
Universite Paris XI
F 91405 Orsay Cdx
France
☎ 33 1 69 41 7744

Omont Alain
IAP
98bis bd Arago
F 75014 Paris
France
☎ 33 1 44 32 8071
📠 33 1 44 32 8001

Onaka Takashi
Astronomy Dpt
Univ of Tokyo
Bunkyo Ku
Tokyo 113
Japan
☎ 81 381 22 111
📠 81 338 13 9439

Onegina A B
Main Astronomical Obs
Ukrainian Acad Science
Golosiiv
252650 Kyiv 22
Ukraine
☎ 380 4426 63744
📠 380 4426 62147

Onello Joseph S
Physics Dpt
NY State Univ
Bowers Hall 133
Cortland NY 13045
USA
☎ 1 607 753 2915
📠 1 607 753 2927
✉ onello@snycorva.
cortland.edu

Ono Yoro
Physics Dpt
Hokkaido Univ
Kita 10 Nishi 8
Sapporo 063
Japan
☎ 81 117 11 2111
📠 81 222 85 132

Onuora Lesley Irene
Physics/Astronomy Div
Univ of Sussex
Falmer
Brighton BN1 PQH
UK
☎ 44 12 73 678 971
✉ lonuora@astr.cpes.
susx.ac.uk

Ooe Masatsugu
Intl Latitude Observ
NAOJ
Hoshigaoka Mizusawa Shi
Iwate 023
Japan
☎ 81 197 24 7111
📠 81 197 22 7120

Oohara Ken-ichi
Physics Dpt
Niigata Univ
Niigata 950 21
Japan
☎ 81 252 64 2281
📠 81 252 64 2282
✉ oohara@astro.sc.
niigata-u.ac.jp

Oosterloo Thomas
CSIRO
ATNF
Box 76
Epping NSW 2121
Australia
☎ 61 2 9372 4451
📠 61 2 9372 4400
✉ toosterl@atnf.csiro.au

Opendak Michael
Dpt Astron/Astrophys
Univ of Pennsylvania
209 South 33 St
Philadelphia PA 19104
USA
☎ 1 215 898 5066
📠 1 215 898 9336
✉ mopendak@mail.sas.
upenn.edu

Opher Reuven
IAG
Univ Sao Paulo
CP 9638
01065 970 Sao Paulo SP
Brazil
☎ 55 11 275 3720
📠 55 11 276 3848

Opolski Antoni
Astronomical Institute
Wroclaw Univ
Ul Kopernika 11
PL 51 622 Wroclaw
Poland
☎ 48 71 372 9373/74
📠 48 71 372 9378

Oprescu Gabriela
Astronomical Institute
Romanian Acad Sciences
Cutitul de Argint 5
RO 75212 Bucharest
Rumania
☎ 40 1 641 3686
📠 40 1 312 3391
✉ goprescu@imar.ro

Oproiu Tiberiu
Astronomical Observ
Romanian Acad Sciences
Ul Ciresilor 19
RO 3400 Cluj Napoca
Rumania
☎ 40 64 194 592
📠 40 64 194 592

Orchiston Wayne
Carter Observ
Ntl Observ New Zealand
Box 2909
Wellington 1
New Zealand
☎ 64 4 472 8167
📠 64 4 472 8320
✉ wayne.cartdir@
vuw.ac.nz

Orellana Rosa Beatriz
Observ Astronomico
Paseo d Bosque S/n
1900 La Plata (Bs As)
Argentina
☎ 54 21 217 308
📠 54 21 211 761
✉ rorellana@fcaglp.
edu.ar.

Orlin Hyman
Ntl Acad Sciences
NRC
2101 Constitution Av NW
Washington DC 20418
USA
☎ 1 202 334 3520
📠 1 202 334 2791

Orlov Mikhail
Main Astronomical Obs
Ukrainian Acad Science
Golosiiv
252650 Kyiv 22
Ukraine
☎ 380 4426 63110
📠 380 4426 62147

Ormes Jonathan F
NASA GSFC
Code 660
Greenbelt MD 20771
USA
☎ 1 301 286 8801

Orsatti Ana M
Observ Astronomico
Paseo d Bosque S/n
1900 La Plata (Bs As)
Argentina
☎ 54 21 217 308
📠 54 21 258 985
✉ amo@fcaglp.edu.ar

Orte Alberto
ROA
Cecilio Pujazon 22-3 A
E 11100 San Fernando
Spain
☎ 34 5 689 5441
📠 34 5 659 9366

Ortolani Sergio
Osserv Astronom Padova
Univ d Padova
Vic d Osservatorio 5
I 35122 Padova
Italy
☎ 39 49 829 3411
📠 39 49 875 9840

Orton Glenn S
CALTECH/JPL
MS 183 301
4800 Oak Grove Dr
Pasadena CA 91109 8099
USA
☎ 1 818 354 4321
📠 1 818 393 6030

Orus Juan J
C/Josep Tarradellas
134 Atc 3
E 08028 Barcelona
Spain
☎ 34 3 405 2888
✉ jorge@fajnm1.am.ub.es

Osaki Toru
Ryukoku Univ
Fukakusa Tsukamoto
Fushimi Ku
Kyoto 612
Japan
☎ 81 756 42 1111

Osaki Yoji
Astronomy Dpt
Univ of Tokyo
Bunkyo Ku
Tokyo 113
Japan
☎ 81 381 22 111
📠 81 338 13 9439
✉ osaki@dept.astron.s.u-
tokyo.ac.jp

Osborn Wayne
Physics Dpt
Central Michigan Univ
Mt Pleasant MI 48859
USA
☎ 1 517 774 3321
📠 1 517 774 2697
✉ osborn@dune.phy.cmich.edu

Osborne John L
Physics Dpt
Univ of Durham
South Rd
Durham DH1 3LE
UK
☎ 44 191 374 2178
📠 44 191 374 3749
✉ j.l.osborne@durham.ac.uk

Osborne Julian P
Dpt Phys/X-Ray Astron
Univ Leicester
University Rd
Leicester LE1 7RH
UK
☎ 44 116 252 2073
📠 44 116 250 182
✉ julo@star.le.ac.uk

Osman Anas Mohamed
NAIGR
Helwan Observ
Cairo 11421
Egypt
☎ 20 78 0645/2683
📠 20 6221 405 297

Osmer Patrick S
Astronomy Dpt
Ohio State Univ
174 W 18th Av
Columbus OH 43210 1106
USA
☎ 1 614 292 6789
℻ 1 614 292 2928

Osorio Isabel Maria T V P
Observ Astronomico
Univ Porto
Monte d Virgem
P 4400 Vila Nova de gaia
Portugal
☎ 351 2 782 0404
℻ 351 2 782 7253
✉ iposorio@oa.fc.up.pt

Osorio Jose J S P
Observ Astronomico
Univ d Porto
Monte d Virgem
P 4400 Vila Nova de gaia
Portugal
☎ 351 2 782 0404
℻ 351 2 782 7253
✉ posorio@oa.fc.up.pt

Oster Ludwig F
NSF
Div Astron Sci
4201 Wilson Blvd
Arlington VA 22230
USA
☎ 1 202 357 9857
℻ 1 703 306 0525

Osterbrock Donald E
Lick Observ
Univ of California
Santa Cruz CA 95064
USA
☎ 1 831 429 2605
℻ 1 831 426 3115
✉ don@ucolick.org

Ostriker Jeremiah P
Princeton Univ Obs
Peyton Hall
Princeton NJ 08544 1001
USA
☎ 1 609 258 3800
℻ 1 609 258 1020
✉ jpo@astro.princeton.edu

Ostro Steven J
CALTECH/JPL
MS 300 233
4800 Oak Grove
Pasadena CA 91109 8099
USA
☎ 1 818 354 3173
℻ 1 818 354 9476
✉ ostro@echo.jpl.nasa.gov

Ostrowski Michal
Astronomical Observ
Krakow Jagiellonian Univ
Ul Orla 171
PL 30 244 Krakow
Poland
☎ 48 12 25 1294
℻ 48 12 25 1318
✉ mio@oa.uj.edu.pl

Oswalt Terry D
Dpt Phys/Space Sci
Florida Inst Techn
150 W University Blvd
Melbourne FL 32901
USA
☎ 1 407 768 8000*7325
℻ 1 407 984 8461
✉ oswawlt@tycho.pss.fit.edu

Oterma Liisi
Sirkkalankatu 31
FIN 20700 Turku
Finland
☎ 358 2 133 2081

Othman Mazlan
Planetarium Division
Prime Minister's Dpt
2731/5 Jalanjohor Selatan
50480 Kuala Lumpur
Malaysia
☎ 60 3 282 4463
℻ 60 3 282 4507

Otmianowska-Mazur Katarzyna
Astronomical Observ
Krakow Jagiellonian Univ
Ul Orla 171
PL 30 244 Krakow
Poland
☎ 48 12 25 1294
℻ 48 12 25 1318
✉ otmian@oa.uj.edu.pl

Ott Heinz-Albert
Astronomisches Institut
Univ Muenster
Wilhelm Klemm Str 10
D 48149 Muenster
Germany
☎ 49 251 833 561
℻ 49 251 833 669
✉ ott@cygnus.uni-muenster.de

Ottelet I J
Inst Astrophysique
Universite Liege
Av Cointe 5
B 4000 Liege
Belgium
☎ 32 4 254 7510
℻ 32 4 254 7511

Ouhrabka Miroslav
College Education In
Hradec Kralove
Nam Svobody 301
CZ 501 91 Hradec Kralove
Czech R
☎ 420 49 25226
℻ 420 49 25785

Oukbir Jamila
Observ Strasbourg
11 r Universite
F 67000 Strasbourg
France
☎ 33 3 88 15 0710
℻ 33 3 8815 0740
✉ oukbir@wirtz.u-strasbg.fr

Overbeek Michiel Daniel
Box 212
1610 Edenvale 1610
South Africa
☎ 27 11 453 6918
✉ danieo@global.co.za

Owaki Naoaki
Dpt Astron/Earth Sci
Tokyo Gakugei Univ
Koganei
Tokyo 184
Japan
☎ 81 423 25 2111
℻ 81 423 25 4219

Owen Frazer Nelson
NRAO
VLA
Box 0
Socorro NM 87801 0387
USA
☎ 1 505 772 4011
℻ 1 505 835 7027

Owen Tobias C
Inst for Astronomy
Univ of Hawaii
2680 Woodlawn Dr
Honolulu HI 96822
USA
☎ 1 808 956 8007
℻ 1 808 988 2790
✉ owen@hubble.ifa.hawaii.edu

Owen William Mann
CALTECH/JPL
MS 301 150
4800 Oak Grove Dr
Pasadena CA 91109 8099
USA
☎ 1 818 354 2505
℻ 1 818 393 6388
✉ wmo@jpl.nasa.gov

Owocki Stanley Peter
Bartol Res Inst
Univ of Delaware
Newark DE 19716
USA
☎ 1 302 451 8357
℻ 1 302 451 1843
✉ owocki@bartol.udel.edu

Oxenius Joachim
Dpt d Physique
Univ Bruxelles
Campus Plaine CP 226
B 1050 Brussels
Belgium
☎ 32 2 650 5818
℻ 32 2 650 5045

Ozeki Hiroyuki
Inst Molecular Science
Myodaiji
Okazaki 444
Japan
☎ 81 564 55 7322
℻ 81 564 55 4639
✉ ozeki@ims.ac.jp

Ozernoy Leonid M
NASA GSFC
Code 685
Greenbelt MD 20771
USA
☎ 1 301 286 8801
✉ ozernoy@lheavx.gsfc.nasa.gov

Ozguc Atila
Kandilli Observ
Bugazici Univ
Cengelkoy
81220 Istanbul
Turkey
☎ 90 216 308 0514
℻ 90 216 332 1711
✉ ozguc@boun.edu.tr

Ozsvath I
Univ of Texas
Programs in Mathemat Sci
Box 830688
Richardson TX 75083 0688
USA
☎ 1 214 690 2174

P

Pacharin-Tanakun P
Physics Dpt
Chulalongkorn Univ
Bangkok 10330
Thailand
☎ 66 2 251 4902
℻ 66 2 253 1150

Pachner Jaroslav
2101
465 Richmond Rd
Ottawa ON K2A 1Z1
Canada
☎ 1 613 728 0740
℻ 1 613 728 0740

Pacholczyk Andrzej G
Steward Observ
Univ of Arizona
Tucson AZ 85721
USA
☎ 1 520 621 6928
℻ 1 520 621 1532

Paciesas William S
Physics Dpt
Univ of Alabama
Huntsville AL 35899
USA
☎ 1 205 544 7712
℻ 1 205 544 5800
✉ paciesas@ssl.msfc.nasa.gov

Pacini Franco
Dpt Astronomia
Univ d Firenze
Largo E Fermi 5
I 50125 Firenze
Italy
☎ 39 55 27 52232
℻ 39 55 22 0039
✉ pacini@arcetri.astro.it

Paczynski Bohdan
Copernicus Astron Ctr
Polish Acad Sci
Ul Bartycka 18
PL 00 716 Warsaw
Poland
☎ 48 22 41 1086
℻ 48 22 41 0828

Padalia T D
Uttar Pradesh State Observ
Po Manora Peak 263 129
Nainital 263 129
India
☎ 91 59 42 2136

Padman Rachael
MRAO
Cavendish Laboratory
Madingley Rd
Cambridge CB3 0HE
UK
☎ 44 12 23 337 274
℻ 44 12 23 354 599
✉ padman@mrao.cam.ac.uk

Padmanabhan Janardhan
PRL
Navrangpura
Ahmedabad 380 009
India
☎ 91 272 46 2129
℻ 91 272 44 5292

Padmanabhan T
IUCAA
PO Box 4
Ganeshkhind
Pune 411 007
India
☎ 91 212 351 414/5
℻ 91 212 350 760
✉ paddy@iucaa.ernet.in

Padovani Paolo
STScI
Homewood Campus
3700 San Martin Dr
Baltimore MD 21218
USA
☎ 1 301 338 4742
℻ 1 301 338 5075
✉ padovani@stsci.edu

Padrielli Lucia
Istt Radioastronomia
CNR
Via P Gobetti101
I 40129 Bologna
Italy
☎ 39 51 639 0384
℻ 39 51 639 9431
✉ padrielli@astbo1.bo.cnr.it

Paerels Frederik B S
Astronomy Dpt
Columbia Univ
538 W 120th St
New York NY 10027
USA
☎ 1 212 854 8125
℻ 1 212 854 8121
✉ frits@naima.phys.columbia.edu

Pagano Isabella
Osserv Astronomico d Catania
Via A Doria 6
I 95125 Catania
Italy
☎ 39 95 733 2243
℻ 39 95 33 0592
✉ ipagano@astrct.ct.astro.it

Page Arthur
Physics Dpt
Univ of Queensland
St Lucia
Brisbane QLD 4072
Australia
☎ 61 7 365 2422
℻ 61 7 365 1199

Page Clive G
Dpt Phys/X-Ray Astron
Univ Leicester
University Rd
Leicester LE1 7RH
UK
☎ 44 116 252 2073
℻ 44 116 250 182

Page Dany
Instituto Astronomia
UNAM
Apt 70 264
Mexico DF 04510
Mexico
☎ 52 5 622 3910
℻ 52 5 616 0653
✉ page@astroscu.unam.mx

Page Don Nelson
Physics Dpt
Univ of Alberta
Edmonton AB T6G 2J1
Canada
☎ 1 403 492 4129
℻ 1 403 492 0714
✉ don@phys.ualberta.ca

Pagel Bernard E J
Groombridge Lewes Rd.
Ringmer
E Sussex BN8 5ER
UK
☎ 44 31 42 1616
℻ 44 31 38 9157
✉ bejp@star.cpes.susx.ac.uk

Pakvor Ivan
Astronomical Observ
Volgina 7
11050 Beograd
Yugoslavia FR
☎ 381 1 419 357/421 875
℻ 381 1 419 553

Pal Arpad
Fac of Mathematics
Univ of Cluj Napoca
Str Kogalniceanu 1
RO 3400 Cluj Napoca
Rumania
☎ 40 64 116 101
℻ 40 64 194 592

Palagi Francesco
Osserv Astrofis Arcetri
Univ d Firenze
Largo E Fermi 5
I 50125 Firenze
Italy
☎ 39 55 27 521
℻ 39 55 22 0039

Paletou Frederic
OCA
Observ Nice
BP 139
F 06304 Nice Cdx 4
France
☎ 33 4 92 00 30 50
℻ 33 4 92 00 3121
✉ paletou@obs-nice.fr

Palla Francesco
Osserv Astrofis Arcetri
Univ d Firenze
Largo E Fermi 5
I 50125 Firenze
Italy
☎ 39 55 27 52249
℻ 39 55 22 0039
✉ palla@arcetri.astro.it

Pallavicini Roberto
Osserv Astronomico
Univ d Palerno
Piazza Parlemento 1
I 90134 Palermo
Italy
☎ 39 91 233 251
℻ 39 91 233 444
✉ pallavic@oapa.astropa.unipa.it

Palle Pere-Lluis
IAC
Observ d Teide
via Lactea s/n
E 38200 La Laguna
Spain
☎ 34 2 232 9100
℻ 34 2 232 9117

Palmeira Ricardo A R
INPE
Dpt Astronomia
CP 515
12227 010 S Jose dos Campos
Brazil
☎ 55 123 22 9977
℻ 55 123 21 8743

Palmer Patrick E
Astronomy/Astrophys Ctr
Univ of Chicago
5640 S Ellis Av
Chicago IL 60637
USA
☎ 1 312 962 7972
✆ 1 312 702 8212
✉ ppalmer@oskar.
uchicago.edu

Palmer Philip
Astronomy Unit
QMWC
Mile End Rd
London E1 4NS
UK
☎ 44 171 975 5462
✆ 44 181 981 9587
✉ philip @ qmc.maths.

Palous Jan
Astronomical Institute
Czech Acad Sci
Bocni Ii 1401
CZ 141 31 Praha 4
Czech R
☎ 420 2 67 10 3065
✆ 420 2 76 90 23
✉ palous@ig.cas.cz

Paltani Stephane
INTEGRAL
Science Data Ctr
16 Chemin Ecogia
CH 1290 Versoix
Switzerland
☎ 41 22 950 9141
✆ 41 22 950 9133
✉ stephane.paltani@
obs.unige.ch

Palumbo Giorgio G C
Dpt Astronomia
Univ d Bologna
Via Zamboni 33
I 40126 Bologna
Italy
☎ 39 51 259 424
✆ 39 51 259 407
✉ ggcpalumbo@
alma02.cineca.it

Palus Pavel
Dpt Astron/Astrophys
Comenius Univ
Mlynska Dolina
SK 842 15 Bratislava
Slovak Republic
☎ 421 7 72 400
✆ 421 7 32 5882

Pamyatnikh Alexsey A
Inst of Astronomy
Acad Sciences
Pyatnitskaya Ul 48
109017 Moscow
Russia
☎ 7 095 231 5461
✆ 7 095 230 2081
✉ alosza@camk.edu.pl

Pan Junhua
Nanjing Astronomical
Instrument Factory
Box 846
Nanjing
China PR
☎ 86 25 46191
✆ 86 25 71 1256

Pan Liande
Shaanxi Observ
CAS
Box 18
Lintong 710600
China PR
☎ 86 33 2255
✆ 86 9237 3496

Pan Ning-Bao
Beijing Astronomical Obs
CAS
Beijing 100080
China PR
☎ 86 1 28 1698
✆ 86 1 256 10855

Pan Rong-Shi
Shanghai Observ
CAS
80 Nandan Rd
Shanghai 200030
China PR
☎ 86 21 6438 6191
✆ 86 21 6438 4618

Pan Xiao-Pei
CALTECH
Palomar Obs
MS 105 24
Pasadena CA 91125
USA
☎ 1 818 356 4015
✆ 1 818 568 1517
✉ xpp@deimos.
caltech.edu

Panagia Nino
STScI
Homewood Campus
3700 San Martin Dr
Baltimore MD 21218
USA
☎ 1 301 338 4916
✆ 1 301 338 5090
✉ panagia@stsci.edu

Panchuk Vladimir E
SAO
Acad Sciences
Nizhnij Arkhyz
357147 Karachaevo
Russia
☎ 7 865 789 3527
✉ panchuk@sao.
stavropol.su

Pande Girish Chandra
126 Aryanagar
Lucknow 226 004
India

Pande Mahesh Chandra
Uttar Pradesh State
Observ
Po Manora Peak 263 129
Nainital 263 129
India
☎ 91 59 42 2136

Pandey A K
Uttar Pradesh State
Observ
Po Manora Peak 263 129
Nainital 263 129
India
☎ 91 59 42 2136

Pandey S K
Physics Dpt
Univ of Ravishankar
Raipur 492 010
India
☎ 91 27 064

Pandey Uma Shankar
Physics Dpt
Univ of Gorakhpur
Gorakhpur 273 009
India
☎ 91 551 336 601

Panek Robert J
Astronomy Dpt
Pennsylvania State Univ
525 Davey Lab
University Park PA 16802
USA
☎ 1 814 865 3631
✆ 1 814 863 7114

Pang Kevin
CALTECH/JPL
MS T11 823
4800 Oak Grove Dr
Pasadena CA 91109 8099
USA
☎ 1 818 354 5392
✆ 1 818 393 6030

Pankonin Vernon Lee
NSF
Div Astron Sci
4201 Wilson Blvd
Arlington VA 22230
USA
☎ 1 703 306 1820
✆ 1 703 306 0525
✉ vpankoni@nsf.gov

Pannunzio Renato
Osserv Astronomico
d Torino
St Osservatorio 20
I 10025 Pino Torinese
Italy
☎ 39 11 461 9034
✆ 39 11 461 9030
✉ pannunzio@to.astro.it

Panov Kiril
Astronomy Dpt
Bulgarian Acad Sci
7th November St 1
BG 1000 Sofia
Bulgaria
☎ 359 2 80 2831
✆ 359 2 80 2831

Paolicchi Paolo
Dpt Fisica
Univ d Pisa
Piazza Torricelli 2
I 56100 Pisa
Italy
☎ 39 50 43343

Pap Judit
CALTECH/JPL
MS 171 400
4800 Oak Grove Dr
Pasadena CA 91109 8099
USA
☎ 1 818 354 2662
✆ 1 818 354 4707
✉ jpap@solar.stanford.edu

Papaelias Philip
Astrophysics Dpt
Ntl Univ Athens
Panepistimiopolis
GR 157 84 Zografos
Greece
☎ 30 1 723 3122
✆ 30 1 723 5122

Papagiannis Michael D
Astronomy Dpt
Boston Univ
725 Commonwealth Av
Boston MA 02215
USA
☎ 1 617 353 5705
✆ 1 617 353 5704

Papaliolios Costas
CfA
HCO/SAO
60 Garden St
Cambridge MA 02138 1516
USA
☎ 1 617 495 7123
✆ 1 617 495 7356

Papaloizou John C B
School Mathemat Sc
QMWC
Mile End Rd
London E1 4NS
UK
☎ 44 171 980 4811
✆ 44 181 975 5500

Paparo Margit
Konkoly Observ
Thege U 13/17
Box 67
H 1525 Budapest
Hungary
☎ 36 1 375 4122
✆ 36 1 275 4668
✉ paparo@ogyalla.konkoly.hu

Papathanasoglou D
Astronomy Dpt
Ntl Univ Athens
Panepistimiopolis
GR 157 84 Zografos
Greece
☎ 30 1 724 3414
✆ 30 1 723 5122

Papayannopoulos Th
Astrophysics Dpt
Ntl Univ Athens
Panepistimiopolis
GR 157 84 Zografos
Greece
☎ 30 1 723 7924
✆ 30 1 723 8413
✉ tpapagia@atlas.uoa.
ariadna-t.gr

Papousek Jiri
Mlynska 2
CZ 602 00 Brno
Czech R
☎ 420 5 332 517

Paquet Paul Eg
Koninklijke Sterrenwacht
van Belgie
Ringlaan 3
B 1180 Brussels
Belgium
☎ 32 2 373 0249
✆ 32 2 374 9822
✉ paul.paquet@oma.be

Parcelier Pierre
Observ Paris
61 Av Observatoire
F 75014 Paris
France
☎ 33 1 43 20 1210
✆ 33 1 44 54 1804

Paredes Jose Maria
Dpt Astronom Meteo
Univ Barcelona
Avd Diagonal 647
E 08028 Barcelona
Spain
☎ 34 3 402 1130
✆ 34 3 402 1133
✉ josepmp@mizar.am.ub.es

Paresce Francesco
ESO
Karl Schwarzschildstr 2
D 85748 Garching
Germany
☎ 49 893 200 6297
✆ 49 893 200 6480
✉ fparesce@eso.org

Parfinenko Leonid D
Pulkovo Observ
Acad Sciences
10 Kutuzov Quay
196140 St Petersburg
Russia
☎ 7 812 123 4400
✆ 7 812 123 1922
✉ makarov@gaosun.spb.su

Parijskij Yuri N
Pulkovo Observ
Acad Sciences
10 Kutuzov Quay
196140 St Petersburg
Russia
☎ 7 812 297 9452
✆ 7 812 315 1701

Parise Ronald A
NASA GSFC
Code 684 9
Greenbelt MD 20771
USA
☎ 1 301 286 3896

Parisot Jean-Paul
Observ Bordeaux
BP 89
F 33270 Floirac
France
☎ 33 5 56 86 4330
✆ 33 5 56 40 4251

Park Changbom
Astronomy Dpt
Seoul Ntl Univ
Kwanak Ku
Seoul 151 742
Korea RP
☎ 82 288 06 769
✆ 82 288 71435
✉ cbp@astrogate.snu.ac.kr

Park Hong Suh
Dpt Earth Sci
Korea Ntl Univ of Educat
Cheongwon
Choongbook 373 791
Korea RP
✉ hspark@knuecc-sun.knue.ac.kr

Park Myeong-gu
Dpt Astron/Meteorology
College of Ntl Sciences
Kyungpook Ntl Univ
Taegu 702 701
Korea RP
☎ 82 539 50 6364
✆ 82 539 57 0431

Park Seok Jae
Korea Astronomy Obs/ISSA
36 1 Whaam Dong
Yuseong Gu
Taejon 305 348
Korea RP
☎ 82 428 65 3266
✆ 82 428 61 5610
✉ sjpark@apissa.issa.re.kr

Park Young-Deuk
Korea Astron Observ
Jachun Box 1
Whabook Youngchun
Kyungbook
Korea RP
☎ 82 563 301 015
✆ 82 563 369 450
✉ ydpark@flare.boao.re.kr

Parker Edward A
Electronics Lab
Univ of Kent
Canterbury CT2 7NT
UK
☎ 44 12 27 668 22
✆ 44 12 27 456 084

Parker Eugene N
Astrophys/Space Res Lab
Univ of Chicago
933 E 56th St
Chicago IL 60637
USA
☎ 1 312 702 7847
✆ 1 312 702 6647

Parker Neil
Royal Greenwich Obs
Madingley Rd
Cambridge CB3 0EZ
UK
☎ 44 12 23 374 000
✆ 44 12 23 374 700
✉ nmp@ast.cam.ac.uk

Parker Quentin
Royal Observ
Blackford Hill
Edinburgh EH9 3HJ
UK
☎ 44 131 668 8379
✆ 44 131 668 8356

Parker Robert A R
NASA/JSC
Code CB
Houston TX 77058
USA
☎ 1 713 483 2221
✆ 1 281 483 5347

Parkinson John H
School of Science/Maths
City Campus
Pond St
Sheffield S1 1WB
UK
☎ 44 114 253 3086
✆ 44 114 253 3066

Parkinson Truman
NOAO/KPNO
Box 26732
950 N Cherry Av
Tucson AZ 85726 6732
USA
☎ 1 520 327 5511
✆ 1 520 325 9360

Parkinson William H
CfA
HCO/SAO MS 50
60 Garden St
Cambridge MA 02138 1516
USA
☎ 1 617 495 4865
✆ 1 617 495 7455
✉ parkinson@cfa.harvard.edu

Parma Paola
Istt Radioastronomia
CNR
Via P Gobetti101
I 40129 Bologna
Italy
☎ 39 51 639 9380
✆ 39 51 639 9431
✉ parma@astbo1.bo.cnr.it

Parmar Arvind Nicholas
ESA/ESTEC
SSD
Box 299
NL 2200 AG Noordwijk
Netherlands
☎ 31 71 565 4532
✆ 31 71 565 4690
✉ aparmar@estsa2.estec.esa.nl

Parnovsky Sergei
Astronomical Observ
Observatorna St 3
Kyiv 254 053
Ukraine
☎ 380 4429 54515
✉ par@aoku.freenet.ua

Parravano Antonio
Ctro Astrofis Teorica
Univ Los Andes
Apdo 26
Merida 5251 A
Venezuela
☎ 58 7 440 1330/1331
✆ 58 7 440 1286
✉ parravan@ciens.ula.ve

Parrish Allan
Astronomy Dpt
Univ Massachusetts
GRC 6191
Amherst MA 01003
USA

Parsamyan Elma S
Byurakan Astrophys Observ
Armenian Acad Sci
378433 Byurakan
Armenia
☎ 374 88 52 28 3453/4142
✆ 374 88 52 52 3640

Parsons Sidney B
STScI
Homewood Campus
3700 San Martin Dr
Baltimore MD 21218
USA
☎ 1 301 338 4807
✆ 1 301 338 4767

Parthasarathy M
IIA
Koramangala
Sarjapur Rd
Bangalore 560 034
India
☎ 91 80 356 6585/6497
✆ 91 80 553 4043

Partridge Robert B
Haverford College
Haverford PA 19041
USA
☎ 1 215 896 1145
✆ 1 215 896 1224
✉ bpartrid@haverford.edu

Parv Bazil
Fac of Mathematics
Univ of Cluj Napoca
Ul Kogalniceanu 1
RO 3400 Cluj Napoca
Rumania
☎ 40 64 193 415
✆ 40 64 191 901
✉ bparv@cs.ubbcluj.ro

Pasachoff Jay M
Hopkins Observ
Williams College
Williamstown MA 01267
USA
☎ 1 413 597 2105
✆ 1 413 597 3200
✉ jay.m.pasachoff@williams.edu

Pascoal Antonio J B
Observ Astronomico
Univ d Porto
Monte d Virgem
P 4400 Vila Nova de gaia
Portugal
☎ 351 2 782 0404
✆ 351 2 782 7253

Pascu Dan
USNO
3450 Massachusetts Av NW
Washington DC 20392 5100
USA
☎ 1 202 653 1178
✉ pas@clem.usno.navy.mil

Pasian Fabio
OAT
Box Succ Trieste 5
Via Tiepolo 11
I 34131 Trieste
Italy
☎ 39 40 31 99221
Fax 39 40 30 9418
✉ pasian@ts.astro.it

Pasinetti Laura E
Dpt Fisica
Univ d Milano
Via Celoria 16
I 20133 Milano
Italy
☎ 39 2 239 2272
Fax 39 2 706 38413
✉ pasinetti@milano.infn.it

Pastori Livio
Osserv Astronomico d Milano
Via E Bianchi 46
I 22055 Merate
Italy
☎ 39 990 6412
Fax 39 990 8492

Pastoriza Miriani G
Instituto Fisica
UFRGS
CP 15051
91501 900 Porto Alegre RS
Brazil
☎ 55 512 36 4677

Paterno Lucio
Istt Astronomia
Citta Universitaria
Via A Doria 6
I 95125 Catania
Italy
☎ 39 95 733 2235
Fax 39 95 33 0592
✉ lpaterno@astrct.ct.astro.it

Pathria Raj K
Physics Dpt
Univ Waterloo
Waterloo ON N2L 3G1
Canada
☎ 1 519 885 1211
Fax 1 519 746 8115

Pati A K
IIA
Koramangala
Sarjapur Rd
Bangalore 560 034
India
☎ 91 80 553 4043
Fax 91 80 553 0672
✉ pati@iiap.ernet.in

Patkos Laszlo
Konkoly Observ
Thege U 13/17
Box 67
H 1525 Budapest
Hungary
☎ 36 1 375 4122
Fax 36 1 275 4668

Patnaik Alok Ranjan
RAIUB
Univ Bonn
auf d Huegel 69
D 53121 Bonn
Germany
☎ 49 228 525 264
Fax 49 228 525 229
✉ apatnaik@mpifr-bonn.mpg.de

Patriarchi Patrizio
Osserv Astrofis Arcetri
Univ d Firenze
Largo E Fermi 5
I 50125 Firenze
Italy
☎ 39 55 27 52282
Fax 39 55 22 0039

Paturel Georges
Observ Lyon
Av Ch Andre
F 69561 S Genis Laval Cdx
France
☎ 33 4 78 56 0705
Fax 33 4 72 39 9791
✉ patu@image.univ-lyon1.fr

Paul Jacques
DAPNIA/SAP
CEA Saclay
BP 2
F 91191 Gif s Yvette Cdx
France
☎ 33 1 69 08 4462
Fax 33 1 69 08 6577

Pauldrach Adalbert W A
Inst Astron/Astrophysik
Univ Sternwarte
Scheinerstr 1
D 81679 Muenchen
Germany
☎ 49 899 220 9436
Fax 49 899 220 9427

Pauliny Toth Ivan K K
RAIUB
Univ Bonn
auf d Huegel 69
D 53121 Bonn
Germany
☎ 49 228 525 243
Fax 49 228 864 40
✉ p033ptt@mpirbn.mpifr-bonn.mpg.de

Pauls Thomas Albert
NRL
Code 7213
4555 Overlook Av SW
Washington DC 20375 5351
USA
☎ 1 202 767 0171
Fax 1 202 404 8894
✉ pauls@atlas.nrl.navy.mil

Pauwels Thierry
Koninklijke Sterrenwacht van Belgie
Ringlaan 3
B 1180 Brussels
Belgium
☎ 32 2 373 0225
Fax 32 2 374 9822
✉ thierry.pauwels@oma.be

Pavlenko Elena
Crimean Astrophys Obs
Ukrainian Acad Science
Nauchny
334413 Crimea
Ukraine
☎ 380 6554 71124
Fax 380 6554 40704
✉ pavlenko@crao.crimea.ua

Pavlenko Yakov V
Main Astronomical Obs
Ukrainian Acad Science
Golosiiv
252650 Kyiv 22
Ukraine
☎ 380 4426 64771
Fax 380 4426 62147
✉ maouas@gluk.apc.org

Pavlinsky Mikhail A
Space Res Inst
Acad Sciences
Profsojuznaya Ul 84/32
117810 Moscow
Russia
☎ 7 095 333 2366
Fax 7 095 333 5377
✉ mykle@rea.iki.rssi.ru

Pavlovski Kresimir
Hvar Observ
Fac Geodesy
Kaciceva 26
HR 10000 Zagreb
Croatia
☎ 385 1 442 600
Fax 385 1 445 410

Paxton Harold J B R
Royal Greenwich Obs
Madingley Rd
Cambridge CB3 0EZ
UK
☎ 44 12 23 374 000
Fax 44 12 23 374 700

Payne David G
CALTECH/JPL
MS 264 748
4800 Oak Grove Dr
Pasadena CA 91109 8099
USA
☎ 1 213 351 1321
Fax 1 818 393 6030

Peach Gillian
Dpt Phys/Astronomy
UCLO
Gower St
London WC1E 6BT
UK
☎ 44 171 387 7050
Fax 44 171 380 7145
✉ ucap22g@ucl.ac.uk

Peach John V
Astrophysics Dpt
Univ of Oxford
Keble Rd
Oxford OX1 3RQ
UK
☎ 44 186 551 1336
Fax 44 186 527 3947

Peacock Anthony
ESA/ESTEC
SSD
Box 299
NL 2200 AG Noordwijk
Netherlands
☎ 31 71 565 6555
Fax 31 71 565 4690
✉ astro@astro.estec.esa.nl

Peacock John Andrew
Royal Observ
Blackford Hill
Edinburgh EH9 3HJ
UK
☎ 44 131 667 3321
Fax 44 131 668 8356

Peale Stanton J
Physics Dpt
Univ of California
Santa Barbara CA 93106
USA
☎ 1 805 961 2977

Pearce Gillian
Astrophysics Dpt
Univ of Oxford
Keble Rd
Oxford OX1 3RH
UK
☎ 44 186 527 3297
Fax 44 186 527 3947

Pearson Timothy J
CALTECH
Palomar Obs
MS 105 24
Pasadena CA 91125
USA
☎ 1 818 356 4980
Fax 1 818 568 9352

Pecina Petr
Astronomical Institute
Czech Acad Sci
Fricova 1
CZ 251 65 Ondrejov
Czech R
☎ 420 204 85 729
Fax 420 2 88 1611
✉ ppecina@asu.cas.cz

Pecker Jean-Claude
College d France
3 r Ulm
F 75331 Paris Cdx 05
France
☎ 33 1 44 27 1695
Fax 33 1 44 27 1185
✉ j.c.pecker@wanadoo.fr

Pedersen Bent M
Observ Paris Meudon
ARPEGES
Pl J Janssen
F 92195 Meudon PPL Cdx
France
☎ 33 1 45 07 7809
℻ 33 1 45 07 7971

Pedersen Holger
Astronomical Observ, NBIfAFG
Copenhagen Univ
Juliane Maries Vej 30
DK 2100 Copenhagen
Denmark
☎ 45 35 32 5980
℻ 45 35 32 5989

Pedersen Olaf
Astronomical Observ, NBIfAFG
Copenhagen Univ
Juliane Maries Vej 30
DK 2100 Copenhagen
Denmark
☎ 45 35 32 5934
℻ 45 35 32 5989
✉ holger@astro.ku.dk

Pedlar Alan
NRAL
Univ Manchester
Jodrell Bank
Macclesfield SK11 9DL
UK
☎ 44 14 777 1321
℻ 44 147 757 1618

Pedoussaut Andre
OMP
14 Av E Belin
F 31400 Toulouse Cdx
France
☎ 33 5 61 25 2101
℻ 33 5 61 53 6722
✉ pedouss@obs-mip.fr

Pedreros Mario
Dpt Fisica
Univ Tarapaca
Casilla 7 D
Arica
Chile
✉ mpedrero@vitor.faci.uta.cl

Peebles P James E
Physics Dpt
Princeton Univ
Jadwin Hall
Princeton NJ 08544 1001
USA
☎ 1 609 452 4386
℻ 1 609 258 6853

Peery Benjamin F
Dpt Phys/Astronomy
Howard Unive
Washington DC 20059
USA
☎ 1 202 636 6267

Pei Chunchuan
Purple Mountain Obs
CAS
Nanjing 210008
China PR
☎ 86 25 330 3738
℻ 86 25 330 1459
✉ ccpei@nj.col.com.cn

Peimbert Manuel
Instituto Astronomia
UNAM
Apt 70 264
Mexico DF 04510
Mexico
☎ 52 5 622 3906
℻ 52 5 616 0653
✉ peimbert@astroscu.unam.mx

Pekeris Chaim Leib
Applied Maths Dpt
Weizmann Inst of Sci
Box 26
Rehovot 76100
Israel
☎ 972 8 483 292
℻ 972 8 34 4106

Pekuenlue E Rennan
Dpt Astron/Space Sci
Ege Univ
Box 21
35100 Bornova Izmir
Turkey
☎ 90 232 388 0110*2322

Pel Jan Willem
Kapteyn Sterrekundig Inst
Univ Groningen
Postbus 800
NL 9700 AV Groningen
Netherlands
☎ 31 50 363 4082
℻ 31 50 363 4033
✉ j.w.pel@astro.rug.nl

Peletier Reynier Frans
Kapteyn Sterrekundig Inst
Univ Groningen
Postbus 800
NL 9700 AV Groningen
Netherlands
☎ 31 50 363 4073
℻ 31 50 363 6100

Pellas Paul
Laboratoire Mineralogie
61 r Buffon
F 75005 Paris
France
☎ 33 1 47 07 2824

Pellegrini Paulo S S
Observ Nacional
Rua Gl Bruce 586
Sao Cristovao
20921 030 Rio de Janeiro RJ
Brazil
☎ 55 21 580 3683
℻ 55 21 580 0332
✉ pssp@incc

Pellerin Charles J
NASA Headquarters
Code EZ
600 Independence Av SW
Washington DC 20546
USA
☎ 1 202 453 1437

Pellet Andre
Observ Marseille
2 Pl Le Verrier
F 13248 Marseille Cdx 04
France
☎ 33 4 91 95 9088
℻ 33 4 91 62 1190

Pelletier Guy
Observ Grenoble
Lab Astrophysique
BP 53x
F 38041 S Martin Heres Cdx
France
☎ 33 4 76 51 4570
℻ 33 4 76 44 8821

Pello Roser Descayre
OMP
14 Av E Belin
F 31400 Toulouse Cdx
France
☎ 33 5 61 33 2812
℻ 33 5 61 53 6722
✉ roser@obs-mip.fr

Pelt Jaan
Tartu Observ
Estonian Acad Sci
Toravere
EE 2444 Tartumaa
Estonia
☎ 372 7 410 242
℻ 372 7 410 205
✉ pelt@aai.ee

Pena Jose
Instituto Astronomia
UNAM
Apt 70 264
Mexico DF 04510
Mexico
☎ 52 5 622 3906
℻ 52 5 548 0653
✉ penas@astroscu.unam.mx

Pena Miriam
Instituto Astronomia
UNAM
Apt 70 264
Mexico DF 04510
Mexico
☎ 52 5 622 3908
℻ 52 5 616 0653
✉ miriam@astroscu.unam.mx

Pendleton Yvonne Jean
NASA/ARC
MS 245 3
Moffett Field CA 94035 1000
USA
☎ 1 415 604 4391
℻ 1 415 604 6779
✉ pendleton@galileo.arc.nasa.gov

Peng Bo
Beijing Astronomical Obs
CAS
Beijing 100080
China PR
☎ 86 10 6262 5498
℻ 86 1 256 10855
✉ pb@bao01.bao.ac.cn

Peng Qiuhe
Astronomy Dpt
Nanjing Univ
Nanjing 210093
China PR
☎ 86 25 34651*2882
℻ 86 25 330 7965

Peng Yunlou
Astronomy Dpt
Nanjing Univ
Nanjing 210093
China PR
☎ 86 25 37551*2882
℻ 86 25 330 2728

Peniche Rosario
Instituto Astronomia
UNAM
Apt 70 264
Mexico DF 04510
Mexico
☎ 52 5 622 3906
℻ 52 5 548 3712
✉ penas @astroscu.unam.mx

Penny Alan John
Rutherford Appleton Lab
Space/Astrophysics Div
Bg R25/R68
Chilton Didcot OX11 0QX
UK
☎ 44 12 35 445 675
℻ 44 12 35 44 6667
✉ a.j.penny@rl.ac.uk

Pensado Jose
Observ Astronomico Nacional
Alfonso XII-5
E 28014 Madrid
Spain
☎ 34 1 227 0107

Penston Margaret
Royal Greenwich Obs
Madingley Rd
Cambridge CB3 0EZ
UK
☎ 44 12 23 374 000
℻ 44 12 23 374 700

Penzias Arno A
Bell Labs
Rm 6A409
700 Mountain Av
Murray Hill NJ 07974
USA
☎ 1 908 582 3361
℻ 1 908 582 4702

Pequignot Daniel
Observ Paris Meudon
DAF
Pl J Janssen
F 92195 Meudon PPL Cdx
France
☎ 33 1 45 07 7438
℻ 33 1 45 07 7469

Peraiah Annamaneni
IIA
Koramangala
Sarjapur Rd
Bangalore 560 034
India
☎ 91 80 356 6585/6497
✆ 91 80 553 4043

Perault Michel
DEMIRM
ENS
24 r Lhomond
F 75231 Paris Cdx 05
France
☎ 33 1 43 29 1225
✆ 33 1 45 87 3489

Percy John R
Science Div
Univ of Toronto
Erindale College
Mississauga ON L5l 1C6
Canada
☎ 1 905 828 5351
✆ 1 905 828 5328
✉ jpercy@credit.erin.utoronto.ca

Perdang Jean M
Inst Astrophysique
Universite Liege
Av Cointe 5
B 4000 Liege
Belgium
☎ 32 4 254 7510
✆ 32 4 254 7511
✉ perdang@astro.ulg.ac.be

Perdomo Raul
Observ Astronomico
Paseo d Bosque S/n
1900 La Plata (Bs As)
Argentina
☎ 54 21 217 308
✆ 54 21 211 761
✉ perdomo@fcagcp.edu.ar

Perdrix John
Geochemistry Res Ctr
2/2/11 Brodie Hall Dr
Bentley WA 6102
Australia
☎ 61 9 361 4410
✆ 61 9 361 4418
✉ geochem@techpkwa.curtin.edu.au

Perea-Duarte Jaime D
Inst Astrofisica
Andalucia Apt 3004
Prof Albareda 1
E 18080 Granada
Spain
☎ 34 5 812 1311
✆ 34 5 881 4530
✉ jaime@iaa.s

Perek Lubos
Astronomical Institute
Czech Acad Sci
Bocni Ii 1401
CZ 141 31 Praha 4
Czech R
☎ 420 2 67 10 3068
✆ 420 2 76 90 23
✉ perek@ig.cas.cz

Peres Giovanni
Osserv Astronomico
Univ d Palerno
Piazza Parlemento 1
I 90134 Palermo
Italy
☎ 39 91 651 7998
✆ 39 91 651 7292
✉ peres@oapa.astrpa.unipa.it

Perez Enrique
IAC
Observ d Teide
via Lactea s/n
E 38200 La Laguna
Spain
☎ 34 2 232 9100
✆ 34 2 232 9117
✉ epj@iac.es

Perez Fournon Ismael
IAC
Observ d Teide
via Lactea s/n
E 38200 La Laguna
Spain
☎ 34 2 232 9100
✆ 34 2 232 9117

Perez de Tejada H A
Observ Astronomico Nacional
UNAM
Apt 877
Ensenada BC 22800
Mexico
☎ 52 6 174 4580
✆ 52 6 174 4607

Perez-Peraza Jorge A
Instituto Geofisica
UNAM
Delegation Coyoacan
Mexico DF 04510
Mexico
☎ 52 5 548 1375
✆ 52 5 550 2486
✉ japerez@astroscu.unam.mx

Perinotto Mario
Dpt Astronomia
Univ d Firenze
Largo E Fermi 5
I 50125 Firenze
Italy
☎ 39 55 27 521
✆ 39 55 22 0039

Perkins Francis W
Plasma Physics Lab
Princeton Univ
Box 451
Princeton NJ 08544 1001
USA
☎ 1 609 683 2603
✆ 1 609 243 2751

Perley Richard Alan
NRAO
Box 0
Socorro NM 87801 0387
USA
☎ 1 505 772 4011
✆ 1 505 835 7027

Perola Giuseppe C
Istt Astronomico
Univ d Roma La Sapienza
Via G M Lancisi 29
I 00161 Roma
Italy
☎ 39 6 44 03734
✆ 39 6 44 03673

Perrier Christian
Observ Grenoble
Lab Astrophysique
BP 53x
F 38041 S Martin Heres Cdx
France
☎ 33 4 76 51 4788
✆ 33 4 76 44 8821

Perrin Jean-Marie
LAS
Traverse du Siphon
Les Trois Lucs
F 13376 Marseille Cdx 12
France
☎ 33 4 91 05 5900
✆ 33 4 91 66 1855
✉ perrin@frlasm51

Perrin Marie-Noel
Observ Paris
DASGAL
61 Av Observatoire
F 75014 Paris
France
☎ 33 1 40 51 2245
✆ 33 1 44 54 1804

Perry Charles L
Dpt Phys/Astronomy
Louisiana State Univ
Baton Rouge LA 70803 4001
USA
☎ 1 504 388 8287
✆ 1 504 388 5855
✉ perry@lsumvs

Perry Judith J
Inst of Astronomy
The Observatories
Madingley Rd
Cambridge CB3 0HA
UK
☎ 44 12 23 337 548
✆ 44 12 23 337 523

Perry Peter M
915 Harwood Rd
Harwood MD 20706
USA
☎ 1 301 261 7527
✆ 1 301 261 4279
✉ pmperry@erols.com

Perryman Michael A C
ESA/ESTEC
SSD
Box 299
NL 2200 AG Noordwijk
Netherlands
☎ 31 71 565 3615
✆ 31 71 565 4690
✉ mperryma@estsa2.estec.esa.nl

Persi Paolo
IAS
Area d Ricerca CNR
Via Fosso Cavaliere 100
I 00133 Roma
Italy
☎ 39 6 4993 4473
✆ 39 6 2066 0188
✉ persi@ifsi.rm.fra.cnr.it

Persides Sotirios C
Astronomy Lab
Univ Thessaloniki
GR 540 06 Thessaloniki
Greece
☎ 30 31 99 1357

Pesch Peter
Physics Dpt
CWRU
Rock Bdg
Cleveland OH 44106
USA

Pesek Ivan
Astronomical Observ
Technical Univ
Thakurova 7
CZ 166 29 Praha 6
Czech R
✉ pesek@fsv.cvut.cz

Peters Geraldine Joan
Space Sci Ctr
USC
SHS MC 1341
Los Angeles CA 90089 1341
USA
☎ 1 213 740 6336
✆ 1 213 740 6342
✉ gjpeters@mucen.usc.edu

Peters William L III
Steward Observ
Univ of Arizona
Tucson AZ 85721
USA
☎ 1 520 621 5380
✆ 1 520 621 1532
✉ peters@as.arizona.edu

Petersen J Otzen
Astronomical Observ, NBIfAFG
Copenhagen Univ
Juliane Maries Vej 30
DK 2100 Copenhagen
Denmark
☎ 45 35 32 5994
✆ 45 35 32 5989

Peterson Bradley Michael
Astronomy Dpt
Ohio State Univ
174 W 18th Av
Columbus OH 43210 1106
USA
☎ 1 614 292 7886
℻ 1 614 292 2928
✉ peterson@payne.mps.
ohio-state.edu

Peterson Bruce A
MSSSO
Weston Creek
Private Bag
Canberra ACT 2611
Australia
☎ 61 262 881 111
℻ 61 262 490 233

Peterson Charles John
Dpt Phys/Astronomy
Univ of Missouri
223 Physics Bldg
Columbia MO 65211
USA
☎ 1 314 882 3217

Peterson Deane M
Dpt Earth/Space Sci
Astronomy Program
Suny at Stony Brook
Stony Brook NY 11794 2100
USA
☎ 1 516 632 8223
℻ 1 516 632 8240
✉ dpeterson@astro.
sunysb.edu

Peterson Laurence E
CASS
UCSD
C 011
La Jolla CA 92093 0216
USA
☎ 1 619 534 3461
℻ 1 619 534 2294
✉ lepeterson@ucsd.edu

Peterson Ruth Carol
607 Marion Pl
Palo Alto CA 94301
USA
☎ 1 408 459 3559
℻ 1 408 426 3115
✉ peterson@lick.
ucsc.edu

Petford A David
Astrophysics Dpt
Univ of Oxford
South Parks Rd
Oxford OX1 3RQ
UK
☎ 44 186 551 1336
℻ 44 186 527 3947

Pethick Christopher J
NORDITA
Blegdamsvej 17
DK 2100 Copenhagen
Denmark
☎ 45 35 32 5500
℻ 45 31 38 9157
✉ pethick@nbivax.nbi.dk

Petit Gerard
BIPM
Pavillon de Breteuil
F 92312 Sevres Cdx
France
☎ 33 1 45 07 7067
℻ 33 1 45 34 2021
✉ gpetit@bipm.fr

Petit Jean-Marc
OCA
Observ Nice
BP 139
F 06304 Nice Cdx 4
France
☎ 33 4 92 00 3089
℻ 33 4 92 00 3033
✉ petit@obs-nice.fr

Petitjean Patrick
IAP
98bis bd Arago
F 75014 Paris
France
☎ 33 1 44 32 8150
℻ 33 1 44 32 8001
✉ petitjean@iap.fr

Peton Alain
Observ Marseille
2 Pl Le Verrier
F 13248 Marseille Cdx 04
France
☎ 33 4 91 95 9088
℻ 33 4 91 62 1190

Petre Robert
NASA GSFC
Code 666
Greenbelt MD 20771
USA
☎ 1 301 286 3844
℻ 1 301 286 1684
✉ petre@lheavx.gsfc.
nasa.gov

Petri Winfried
Unterleiten 2
Postfach 106
D 83722 Schliersee
Germany
☎ 49 802 664 28

Petrini Daniel
OCA
Observ Nice
BP 139
F 06304 Nice Cdx 4
France
☎ 33 4 93 89 0420
℻ 33 4 92 00 3033

Petro Larry David
STScI
Homewood Campus
3700 San Martin Dr
Baltimore MD 21218
USA
☎ 1 301 338 4501
℻ 1 301 338 1592
✉ petro@stsci.edu

Petropoulos Basil Ch
Res Ctr Astronomy
Acad Athens
14 Anagnostopoulou St
GR 106 73 Athens
Greece
☎ 30 1 361 3589
℻ 30 1 363 1606

Petrosian Artashes R
Byurakan Astrophys Observ
Armenian Acad Sci
378433 Byurakan
Armenia
☎ 374 88 52 28 3453/4142
℻ 374 88 52 52 3640
✉ vahe@ipia.sci.am

Petrosian Vahe
Ctr for Space Sci/
Astrophysics
Stanford Univ ERL 304
Stanford CA 94305 4055
USA
☎ 1 415 723 1436
℻ 1 415 723 4840
✉ vahe@bigbang.
stanford.edu

Petrov G M
Nikolaev Observ
Observatorna 1
327030 Nikolaev
Ukraine
☎ 380 5123 75206
℻ 380 5123 62420
✉ root@mao.nikolaev.ua

Petrov Gennadij M
Inst Radio/Electron
Acad Sciences
Marx Av 18
103907 Moscow
Russia

Petrov Georgy Trendafilov
Astronomy Dpt
Bulgarian Acad Sci
72 Lenin Blvd.
BG 1784 Sofia
Bulgaria
☎ 359 2 75 8927
℻ 359 2 75 5019

Petrov Peter P
Crimean Astrophys Obs
Ukrainian Acad Science
Nauchny
334413 Crimea
Ukraine
☎ 380 6554 71161
℻ 380 6554 40704

Petrovay Kristof
Astronomy Dpt
Eotvos Univ
Ludovika ter 2
H 1083 Budapest
Hungary
☎ 36 1 114 1019
℻ 36 1 210 1089
✉ kris@innin.elte.hu

Petrovskaya Irina
Astron Inst St Petersburg
State Univ
Bibliotechnaya Pl 2
St Petersburg 198904
Russia
☎ 7 812 428 4268
℻ 7 812 428 4677/7129
✉ ivp@aispbu.spb.su

Petrovskaya Margarita S
Inst Theoret Astron
Acad Sciences
N Kutuzova 10
191187 St Petersburg
Russia
☎ 7 812 278 8898
℻ 7 812 272 7968

Pettengill Gordon H
Ctr for Space Res
MIT Rm 37 641
Box 165
Cambridge MA 02139 4307
USA
☎ 1 617 253 4281
℻ 1 617 253 0861
✉ ghp@space.mit.edu

Pettersen Bjoern Ragnvald
Geodetic Inst
Norw Mapping Authority
N 3500 Honefoss
Norway
☎ 47 32 11 8100
℻ 47 32 11 8101
✉ bjornrp@gdiv.statkart.no

Pettini Marco
Osserv Astrofis Arcetri
Univ d Firenze
Largo E Fermi 5
I 50125 Firenze
Italy
☎ 39 55 27 52282
℻ 39 55 22 0039

Pettini Max
Royal Greenwich Obs
Madingley Rd
Cambridge CB3 0EZ
UK
☎ 44 12 23 374 000
℻ 44 12 23 374 700

Petuchowski Samuel J
Bromberg/Sunstein LLP
125 Summer St
Boston MA 02146
USA
☎ 1 617 443 9292
℻ 1 617 353 3200
✉ spetuchowski@bromsun.com

Pevtsov Alexei A
Physics Dpt
Montana State Univ
Bozeman MT 59717 0350
USA
☎ 1 406 994 7839
℻ 1 406 994 4452
✉ pevtsov@physics.montana.edu

Peyturaux Roger H
IAP
98bis bd Arago
F 75014 Paris
France
☎ 33 1 44 32 8007
✆ 33 1 44 32 8001

Pfau Werner
Astrophysik Inst
Univ Sternwarte
Schillergaesschen 2
D 07745 Jena
Germany
☎ 49 364 163 0320
✆ 49 364 163 0417
🖂 pfau@astro.uni-jena.de

Pfeiffer Raymond J
8 Barbara Lane
Titusville NJ 08560
USA
☎ 1 609 883 4612

Pfennig Hans H
Birnbaumweg 1
D 29223 Celle
Germany

Pfenniger Daniel
Observ Geneve
Chemin d Maillettes 51
CH 1290 Sauverny
Switzerland
☎ 41 22 755 2611
✆ 41 22 755 3983
🖂 Daniel.pfenniger@obs.unige.ch

Pfleiderer Jorg
Inst Astronomie
Technikerstr 15
A 6020 Innsbruck
Austria
☎ 43 512 507 6030
✆ 43 512 507 2923
🖂 chef@ast1.uibk.ac.at

Pflug Klaus
Zntrlinst Astrophysik
Sternwarte Sonnenberg
Rosa Luxemburg Str 17a
D 14473 Potsdam
Germany
☎ 49 331 275 3731

Pham-Van Jacqueline
OCA
CERGA
F 06130 Grasse
France
☎ 33 4 93 36 5849
✆ 33 4 93 40 5353

Phelps Randy L
Carnegie Observatories
813 Santat Barbara St
Pasadena CA 91101 1292
USA
☎ 1 818 304 0247
✆ 1 818 795 8136
🖂 phelps@ociw.edu

Philip A G Davis
1125 Oxford Pl
Schenectady NY 12308
USA
☎ 1 518 374 5636
✆ 1 518 346 5781
🖂 agdp@union.edu

Phillips John G
Astronomy Dpt
Univ of California
601 Campbell Hall
Berkeley CA 94720 3411
USA
☎ 1 415 642 5275
✆ 1 510 642 3411

Phillips John Peter
Astrophysics Group
QMWC
Mile End Rd
London E1 4NS
UK
☎ 44 171 975 5555
✆ 44 181 975 5500

Phillips Kenneth J H
Rutherford Appleton Lab
Space/Astrophysics Div
Bg R25/R68
Chilton Didcot OX11 0QX
UK
☎ 44 12 35 446 424
✆ 44 12 35 446 667
🖂 phillips@solg2.bnsc.rl.ac.uk

Phillips Mark M
NOAO
CTIO
Casilla 603
La Serena
Chile
☎ 56 51 22 5415
✆ 56 51 20 5342

Phillips Thomas Gould
CALTECH
MS 320 47
Pasadena CA 91125
USA
☎ 1 818 356 4278

Piacentini Ruben
Observ Astronomico
de Rosario
Cc 606
2000 Rosario
Argentina
☎ 54 41 63 084
✆ 54 41 63 084
🖂 ruben@ uunet.uu.net

Piatti Andres Eduardo
Observ Astronomico
de Cordoba
Laprida 854
5000 Cordoba
Argentina
☎ 54 51 331 064
✆ 54 51 331 063
🖂 andres@oac.uncor.edu

Piazza Liliana Rizzo
CRAAE INPE
EPUSP/PTR
CP 61548
01065 970 Sao Paulo SP
Brazil
☎ 55 11 815 9322*3620
✆ 55 11 815 6289

Picat Jean-Pierre
OMP
14 Av E Belin
F 31400 Toulouse Cdx
France
☎ 33 5 61 25 2101
✆ 33 5 61 27 3179

Picca Domenico
Dpt Fisica
Univ d Bari
I 70123 Bari
Italy
☎ 39 80 24 3215
✆ 39 80 24 2434

Piccioni Adalberto
Osserv Astronom Bologna
Univ d Bologna
Via Zamboni 33
I 40126 Bologna
Italy
☎ 39 51 259 401
✆ 39 51 259 407

Pick Monique
Observ Paris Meudon
DASOP
Pl J Janssen
F 92195 Meudon PPL Cdx
France
☎ 33 1 45 07 7811
✆ 33 1 45 07 7959

Pickles Andrew John
Inst for Astronomy
Univ of Hawaii
2680 Woodlawn Dr
Honolulu HI 96822
USA
☎ 1 808 956 6756
✆ 1 808 988 2790
🖂 pickles@ifa.hawaii.edu

Piddington Jack H Res Fel
CSIRO NmML
Lindfield
Sydney NSW 2070
Australia
☎ 61 2 9467 6211
✆ 61 2 660 2903

Pier Jeffrey R
USNO
Flagstaff Station
Box 1149
Flagstaff AZ 86002 1149
USA
☎ 1 602 779 5132
✆ 1 520 774 3626

Pierce A Keith
NOAO/NSO
Box 26732
950 N Cherry Av
Tucson AZ 85726 6732
USA
☎ 1 520 327 5511
✆ 1 520 325 9360

Pierce David Allen
7706 Wastlawn Av
Los Angeles CA 90045
USA

Pierre Marguerite
DAPNIA/SAP
CEA Saclay
BP 2
F 91191 Gif s Yvette Cdx
France
☎ 33 1 69 08 3492
✆ 33 1 69 08 6577
🖂 mpierre@cea.fr

Pigatto Luisa
Osserv Astronom Padova
Univ d Padova
Vic d Osservatorio 5
I 35122 Padova
Italy
☎ 39 49 829 3411
✆ 39 49 875 9840

Pigulski Andrzej
Astronomical Institute
Wroclaw Univ
Ul Kopernika 11
PL 51 622 Wroclaw
Poland
☎ 48 71 372 9373/74
✆ 48 71 372 9378
🖂 pigulski@astro.uni.wroc.pl

Piirola Vilppu E
Turku Univ
Tuorla Observ
Vaeisaelaentie 20
FIN 21500 Piikkio
Finland
☎ 358 2 274 4274
✆ 358 2 243 3767
🖂 vilppu.piirola@astro.utu.fi

Pijpers Frank Peter
Inst Phys/Astronomy
Univ of Aarhus
Ny Munkegade
DK 8000 Aarhus C
Denmark
☎ 45 89 423 714
✆ 45 86 12 0740
🖂 fpp@aauobs.obs.aau.dk

Pike Christopher David
Rutherford Appleton Lab
Space/Astrophysics Div
Bg R25/R68
Chilton Didcot OX11 0QX
UK
☎ 44 12 35 821 900
✆ 44 12 35 44 5808

Pikichian Hovhannes
Byurakan Astrophys Observ
Armenian Acad Sci
378433 Byurakan
Armenia
☎ 374 88 52 28 3453/4142
✆ 374 88 52 52 3640

Pilachowski Catherine
NOAO/KPNO
Box 26732
950 N Cherry Av
Tucson AZ 85726 6732
USA
☎ 1 520 327 5511
✆ 1 520 325 9360
✉ catyp@noao.edu

Pilcher Carl Bernard
233 Kent Oaks Way
Gaithersburg MD 20878
USA

Pilkington John D H
22 Hawthylands Cres
Hailsham
East Sussex BN27 1 HG
UK
✉ jdhp@mail.ast.cam.
ac.uk

Pillinger Colin
Dpt Earth Sci
The Open University
Walton Hall
Milton Keynes MK7 6AA
UK
☎ 44 190 865 2119
✆ 44 190 865 5910

Pilyugin Leonid
Main Astronomical Obs
Ukrainian Acad Science
Golosiiv
252650 Kyiv 22
Ukraine
☎ 380 4426 63110
✆ 380 4426 62147

Pineau des Forets Guillaume
Observ Paris Meudon
DAEC
Pl J Janssen
F 92195 Meudon PPL Cdx
France
☎ 33 1 45 07 7454
✆ 33 1 45 07 7469
✉ forets@obspm.fr

Pineault Serge
Dpt d Physique
Universite Laval
Sainte Foy QC G1K 7P4
Canada
☎ 1 416 656 3901
✆ 1 418 656 2040
✉ pineault@phy.
ulaval.ca

Pineda de Carias Maria Cristina
CAAOS
UNAH
Apdo 4432
Tegucigalpa MDC
Honduras
☎ 504 32 21 10*230
✆ 504 31 95 08
✉ mpineda@ns.
hondunet.net

Pines David
Physics Dpt
Univ of Illinois
1110 W Green St
Urbana IL 61801 3080
USA
☎ 1 217 333 0115
✆ 1 217 244 7638

Pingree David
Brown Univ
Box 1900
Providence RI 02912
USA
☎ 1 401 863 2101

Pinigin Gennadij I
Nikolaev Observ
Observatorna 1
327030 Nikolaev
Ukraine
☎ 380 5123 75206
✆ 380 5123 62420
✉ root@mao.nikolaev.ua

Pinkau K
MPI Plasmaphysik
Postfach 1603
D 85740 Garching
Germany
☎ 49 893 299 342
✆ 49 893 299 3235

Pinotsis Antonis D
Astronomy Dpt
Ntl Univ Athens
Panepistimiopolis
GR 157 84 Zografos
Greece
☎ 30 1 724 3414
✆ 30 1 723 5122

Pinsonneault Marc Howard
Astronomy Dpt
Ohio State Univ
174 W 18th Av
Columbus OH 43210 1106
USA
☎ 1 614 292 5346
✆ 1 614 292 2928
✉ pinsono@payne.mps.
ohio.state.edu

Pintado Olga Ines
Instituto Fisica
Univ Nacional Tucuman
Congreso 330 7B
San Miguel de Tucuman
Argentina
☎ 54 81 229073
✆ 54 81 228255
✉ opintado@tucbbs.
com.ar

Pinto Girolamo
Osserv Astronom Padova
Univ d Padova
Vic d Osservatorio 5
I 35122 Padova
Italy
☎ 39 49 829 3411
✆ 39 49 875 9840

Pinto Philip Alfred
Steward Observ
Univ of Arizona
Tucson AZ 85721
USA
☎ 1 520 621 8678
✆ 1 520 621 1532
✉ ppinto@as.arizona.edu

Piotto Giampaolo
Dpt Astronomia
Univ d Padova
Vic d Osservatorio 5
I 35122 Padova
Italy
☎ 39 49 829 3435
✆ 39 49 875 9840
✉ piotto@astrpd.pd.astro.it

Pipher Judith L
Dpt Phys/Astronomy
Univ of Rochester
Rochester NY 14627 0171
USA
☎ 1 716 275 4402
✆ 1 716 275 8527
✉ jlpipher@sherman.pas.
rochester.edu

Piro Luigi
IAS
Area d Ricerca CNR
Via Fosso Cavaliere 100
I 00133 Roma
Italy
☎ 39 6 4993 4473
✆ 39 6 2066 0188
✉ persi@alpha1.rm.
fra.cnr.it

Pirronello Valerio
Osserv Astronomico
d Catania
Via A Doria 6
I 95125 Catania
Italy
☎ 39 95 733 21111
✆ 39 95 33 0592

Piskunov Anatoly E
Inst of Astronomy
Acad Sciences
Pyatnitskaya Ul 48
109017 Moscow
Russia
☎ 7 095 231 5461
✆ 7 095 230 2081
✉ piskunov@inasan.
rssi.ru

Piskunov Nikolai E
Astronomical Observ
Box 515
S 751 20 Uppsala
Sweden
☎ 46 18 51 4490
✆ 46 18 52 7583
✉ piskunov@astro.uu.se

Pismis de Recillas Paris
Instituto Astronomia
UNAM
Apt 70 264
Mexico DF 04510
Mexico
☎ 52 04510
✆ 52 5 616 0653

Pittich Eduard M
Astronomical Institute
Slovak Acad Sci
Dubravska 9
SK 842 28 Bratislava
Slovak Republic
☎ 421 7 37 5157
✆ 421 7 37 5157
✉ pittich@savba.sk

Pitz Eckhart
MPI Astronomie
Koenigstuhl
D 69117 Heidelberg
Germany
☎ 49 622 152 81
✆ 49 622 152 8246

Pizzella G
Dpt Fisica
Univ d Roma
Pl A Moro 2
I 00185 Roma
Italy
☎ 39 6 49 40156
✆ 39 6 202 3507

Pizzichini Graziella
TeSRE
CNR
Via P Gobetti101
I 40129 Bologna
Italy
☎ 39 51 639 8694
✆ 39 51 639 8724
✉ graziella@astbo1.bo.cnr.it

Planesas Pere
Observ Astronomico
Nacional
Apt 1143
E 28800 Alcala d Henares
Spain
☎ 34 1 885 5060
✆ 34 1 885 5062
✉ planesas@oan.es

Plassard J
Ksara Observ
Ksara
Lebanon

Plastino Angel Ricardo
Observ Astronomico
Paseo d Bosque S/n
CC 727
1900 La Plata (Bs As)
Argentina
☎ 54 21 523 995
✆ 54 21 258 138
✉ plastino@fcaglp.fcaglp.unlp.
edu.ar

Platais Imants K
Astronomy Dpt
Yale Univ
Box 208101
New Haven CT 06520 8101
USA
☎ 1 203 432 3021
✆ 1 203 432 5048
✉ imants@astro.yale.edu

Plavec Mirek J
Dpt Astronomy/Phys
UCLA
Box 951562
Los Angeles CA 90025 1562
USA
☎ 1 310 825 1672
✆ 1 310 206 2096

Plavec Zdenka
Dpt Astronomy/Phys
UCLA
Box 951562
Los Angeles CA 90025 1562
USA
☎ 1 310 206 8596
✆ 1 310 206 2096

Plez Bertrand
Atomspektroskopi
Fysiska Inst
Box 118
S 221 00 Lund
Sweden
☎ 46 46 12 6097
✆ 46 46 10 4709
✉ plez@ferrum.fysik.lu.se

Plionis Manolis
Astronomical Institute
Ntl Observ Athens
Box 20048
GR 118 10 Athens
Greece
☎ 30 1 346 1191
✆ 30 1 346 4566
✉ plionis@sapfo.astro.noa.gr

Pneuman Gerald W
550 Roe Rd
Paradise CA 95969
USA

Podsiadlowski Philipp
Oxford Univ
Nuclear/Astrophysics Lab
Keble Rd
Oxford OX1 3RH
UK
☎ 44 186 527 3343
✆ 44 186 527 3390
✉ podso@astro.ox.ac.uk

Poeckert Roland H
Defense Res
Establishment Pacific
Fmo Cfb Esquimalt
Victoria BC V0S 1B0
Canada

Poedts Stefaan
Ctr Plasma Astrophys
Katholicke Univ Leuven
Celestijnenlaan 200B
B 3001 Heverlee
Belgium
☎ 32 16 327 023
✆ 32 16 32 7999
✉ stefaan.poedts@wis.kuleuven.ac.be

Poeppel Wolfgang G I
IAR
CC 5
1894 Villa Elisa (Bs As)
Argentina
☎ 54 21 254 909/21 87 0230
✆ 54 21 254 909
✉ wpoppel@irma.edu.ar

Pogge Richard William
Astronomy Dpt
Ohio State Univ
174 W 18th Av
Columbus OH 43210 1106
USA
☎ 1 614 292 0279
✆ 1 614 292 2928
✉ pogge@payne.mps.ohio-state.edu

Pogodin Mikhail A
Pulkovo Observ
Acad Sciences
10 Kutuzov Quay
196140 St Petersburg
Russia
☎ 7 812 123 4400
✆ 7 812 314 3360
✉ pogodin@pulkovo.spb.su

Pojmanski Grzegorz
Astrononomical Observ
Warsaw Univ
Al Ujazdowskie 4
PL 00 478 Warsaw
Poland
☎ 48 22 29 4011
✆ 48 22 629 4967
✉ gp@sirius.astronw.edu.pl

Pokorny Zdenek
N Copernicus Observ
& Planetarium
Kravi Hora
CZ 616 00 Brno 16
Czech R
☎ 420 5 4132 1287
✉ zpokorny@sci.muni.cz

Pokrzywka Bartlomiej
Mt Suhora Astronomical Obs
Krakow Pedagocial Univ
Ul Podchorazych 2
PL 30 084 Krakow
Poland
☎ 48 12 374 777*280
✆ 48 12 372 243
✉ sfpokrzy@cyf-kr.edu.pl

Poland Arthur I
NASA GSFC
Code 682
Greenbelt MD 20771
USA
☎ 1 301 286 7334

Polcaro V F
IAS
Area d Ricerca CNR
Via Fosso Cavaliere 100
I 00133 Roma
Italy
☎ 39 6 4993 4473
✆ 39 6 2066 0188

Polechova Pavla
Obs/Planetarium
Petrin 205
CZ 118 46 Praha 1
Czech R
☎ 420 2 53 53513
✆ 420 2 258 940
✉ Pavla.Polechov@ecn.cz

Poletto Giannina
Osserv Astrofis Arcetri
Univ d Firenze
Largo E Fermi 5
I 50125 Firenze
Italy
☎ 39 55 27 52252
✆ 39 55 22 0039

Poliakov Eugene V
Pulkovo Observ
Acad Sciences
10 Kutuzov Quay
196140 St Petersburg
Russia
☎ 7 812 123 4555
✆ 7 812 315 1701
✉ poliakov@gao.spb.su

Polidan Ronald S
NASA GSFC
Code 681 0
Greenbelt MD 20771
USA
☎ 1 301 286 5039
✆ 1 301 286 8709

Pollacco Don
Astrophysics Group
Liverpool J M Univ
Byrom St
Liverpool L3 3AF
UK
☎ 44 151 231 2337
✆ 44 151 231 2337
✉ dlp@staru1.livjm.ac.uk

Pollard Karen
Dpt Phys/Astronomy
Univ of Canterbury
Private Bag 4800
Christchurch 1
New Zealand
☎ 64 3 364 2987*7579
✆ 64 3 364 2469
✉ k.pollard@phys.canterbury.ac.nz

Pollas Christian
OCA
Observ Calern/Caussols
2130 r Observatoire
F 06460 S Vallier Thiey
France
☎ 33 4 93 40 5437
✆ 33 4 93 40 5433
✉ pollas@ocar01.obs.azur.fr

Polnitzky Gerhard
Inst Astronomie
Univ Wien
Tuerkenschanzstr 17
A 1180 Wien
Austria
☎ 43 1 345 36090
✆ 43 1 470 6015

Polosukhina-Chuvaeva Nina
Crimean Astrophys Obs
Ukrainian Acad Science
Nauchny
334413 Crimea
Ukraine
☎ 380 6554 71161
✆ 380 6554 40704
✉ polo@crao.crimea.ua.

Polozhentsev Dimitrij D
Pulkovo Observ
Acad Sciences
10 Kutuzov Quay
196140 St Petersburg
Russia
☎ 7 812 123 4453
✆ 7 812 123 1922
✉ ddp@mahis.spb.su

Polyachenko Valerij L
Inst of Astronomy
Acad Sciences
Pyatnitskaya Ul 48
109017 Moscow
Russia
☎ 7 095 231 3980
✆ 7 095 230 2081
✉ iaas@node.ias.msk.su

Polymilis Chronis
Astronomy Dpt
Ntl Univ Athens
Panepistimiopolis
GR 157 84 Zografos
Greece
☎ 30 1 724 3414
✆ 30 1 723 5122
✉ spmso@grathun1

Poma Angelo
Osserv Astronomico
d Cagliari
Poggio d Pini 54
I 09012 Capoterra
Italy
☎ 39 70 725 246
✆ 39 70 72 5425
✉ poma@ca.astro.it

Ponce G A
Apdo P 3023
Tegucigalpa MDC
Honduras
✆ 504 327 196
✉ gponce@ns.hondunet.net

Pongracic Helen
School of Physics
UNSW
RCFTA
Sydney NSW 2006
Australia
☎ 61 2 9351 2546
✆ 61 2 660 2903
✉ helenp@physics.su.oz.au

Ponman Trevor
Dpt Space Res
Univ Birmingham
Box 363
Birmingham B15 2TT
UK
☎ 44 12 14 72 1301

Ponsonby John E B
NRAL
Univ Manchester
Jodrell Bank
Macclesfield SK11 9DL
UK
☎ 44 14 777 1321
✆ 44 147 757 1618

Poole Graham
Dpt Phys/Electronics
Rhodes Univ
Box 94
6140 Grahamstown
South Africa
☎ 27 461 318 454/318450
✆ 27 461 250 49
✉ phgp@ruchem.
ru.ac.za

Pooley Guy
MRAO
Cavendish Laboratory
Madingley Rd
Cambridge CB3 0HE
UK
☎ 44 12 23 337 294
✆ 44 12 23 354 599

Pop Vasile
Fac of Mathematics
Univ of Cluj Napoca
Str Kogalniceanu 1
RO 3400 Cluj Napoca
Rumania
☎ 40 64 116 101
✆ 40 64 194 592

Popelar Josef
Geodetic Survey Division
Canada Ctr for Survey
615 Booth St
Ottawa ON K1A 0E9
Canada

Popescu Cristina Carmen
Astronomical Institute
Romanian Acad Sciences
Cutitul de Argint 5
RO 75212 Bucharest
Rumania
☎ 40 1 623 6892
✆ 40 1 312 3391
✉ popescu@levi.mpi-hd.
mpg.de

Popescu Petre
Astronomical Institute
Romanian Acad Sciences
Cutitul de Argint 5
RO 75212 Bucharest
Rumania
☎ 40 1 641 3686
✆ 40 1 312 3391
✉ ppopescu@imar.ro

Popov Michkail V
Space Res Inst
Acad Sciences
Profsojuznaya Ul 84/32
117810 Moscow
Russia
☎ 7 095 333 2512
✆ 7 095 333 2378
✉ pian@sovam.com

Popov Vasil Nikolov
Astronomy Dpt
Bulgarian Acad Sci
72 Lenin Blvd
BG 1784 Sofia
Bulgaria
☎ 359 2 75 8927
✆ 359 2 75 5019

Popov Victor S
Pulkovo Observ
Acad Sciences
10 Kutuzov Quay
196140 St Petersburg
Russia
☎ 7 812 298 2242
✆ 7 812 315 1701
✉ gnedin@pulkovo.
spb.su

Popovic Georgije
Astronomical Observ
Volgina 7
11050 Beograd
Yugoslavia FR
☎ 381 1 419 357
✆ 381 1 419 553

Popper Daniel M
Dpt Astronomy/Phys
UCLA
Box 951562
Los Angeles CA 90025 1562
USA
☎ 1 310 825 3622
✆ 1 310 206 2096

Poquerusse Michel
Observ Paris Meudon
DESPA
Pl J Janssen
F 92195 Meudon PPL Cdx
France
☎ 33 1 45 07 7530
✆ 33 1 45 07 2806

Porcas Richard
RAIUB
Univ Bonn
auf d Huegel 69
D 53121 Bonn
Germany
☎ 49 228 525 282
✆ 49 228 525 229
✉ porcas@mpifr-bonn.
mpg.de

Porceddu Ignazio E P
Osserv Astronomico
d Cagliari
Poggio d Pini 54
I 09012 Capoterra
Italy
☎ 39 70 72 5246
✆ 39 70 72 5425
✉ iporcedd@ca.astro.it

Poretti Ennio
Osserv Astronomico
d Milano
Via E Bianchi 46
I 22055 Merate
Italy
☎ 39 990 6412
✆ 39 990 8492
✉ poretti@astmib.infn.it

Porfir'ev Vladimir V
Pedagogic Inst
Ministry of Education
107846 Moscow
Russia

Porter Jason G
NASA/MSFC
Space Science Lab
Code ES 52
Huntsville AL 35812
USA
☎ 1 205 544 7607
✆ 1 205 544 5862

Porter Neil A
Physics Dpt
Univ College
Belfield
Dublin 4
Ireland
☎ 353 1 693 244*211
✆ 353 1 671 1759

Porubcan Vladimir
Astronomical Institute
Slovak Acad Sci
Dubravska 9
SK 842 28 Bratislava
Slovak Republic
☎ 421 7 37 5157
✆ 421 7 37 5157
✉ astropor@savba.sk

Postnov Konstantin A
SAI
Acad Sciences
Universitetskij Pr 13
119899 Moscow
Russia
☎ 7 095 939 5006
✆ 7 095 932 8841
✉ pk@sai.msu.su

Pottasch Stuart R
Kapteyn Sterrekundig Inst
Univ Groningen
Postbus 800
NL 9700 AV Groningen
Netherlands
☎ 31 50 363 4073
✆ 31 50 363 6100

Potter Andrew E
NASA/JSC
Code SN3
Houston TX 77058
USA
☎ 1 713 483 5276
✆ 1 713 483 5347

Potter Heino I
Pulkovo Observ
Acad Sciences
10 Kutuzov Quay
196140 St Petersburg
Russia
☎ 7 812 123 4412
✆ 7 812 123 1922
✉ hip@aee.spb.su

Poulakos Constantine
Res Ctr Astronomy
Acad Athens
14 Anagnostopoulou St
GR 106 73 Athens
Greece
☎ 30 1 361 3589
✆ 30 1 363 1606

Poulle Emmanuel
Ecole Ntle des Chartes
19 r de La Sorbonne
F 75005 Paris
France
☎ 33 1 45 89 4857

Pounds Kenneth A
Dpt Phys/X-Ray Astron
Univ Leicester
University Rd
Leicester LE1 7RH
UK
☎ 44 116 252 2073
✆ 44 116 250 182

Pouquet Annick
OCA
Observ Nice
BP 139
F 06304 Nice Cdx 4
France
☎ 33 4 92 00 3057
✆ 33 4 92 00 3033
✉ pouquet@rameau.obs-nice.fr

Poutanen Juri
Astronomical Observ
Box 515
S 751 20 Uppsala
Sweden
☎ 46 18 50 8331
✆ 46 18 52 7583
✉ juri@astro.uu.se

Poveda Arcadio
Instituto Astronomia
UNAM
Apt 70 264
Mexico DF 04510
Mexico
☎ 52 5 622 3906
✆ 52 5 616 0653

Povel Hanspeter
Inst Astronomie
ETH Zentrum
CH 8092 Zuerich
Switzerland
☎ 41 1 632 4222
✆ 41 1 632 1205
✉ povel@astro.phys.ethz.ch

Poyet Jean-Pierre
OMP
14 Av E Belin
F 31400 Toulouse Cdx
France
☎ 33 5 61 25 2101
✆ 33 5 61 27 3179

Prabhakaran Nayar S R
Physics Dpt
Univ of Kerala
Kariyavattom
Trivandrum 695 581
India
☎ 91 471 41 8920

Prabhu Tushar P
IIA
Koramangala
Sarjapur Rd
Bangalore 560 034
India
☎ 91 80 553 0672
✆ 91 80 553 4043
✉ tpp@iiap.ernet.in

Praderie Francoise
Observ Paris
DEMIRM
61 Av Observatoire
F 75014 Paris
France
☎ 33 1 40 51 2116
✆ 33 1 40 51 2002

Pradhan Anil
Astronomy Dpt
Ohio State Univ
174 W 18th Av
Columbus OH 43210 1106
USA
☎ 1 614 292 5850
✆ 1 614 292 2928
✉ pradhan@ohstpy

Prantzos Nikos
IAP
98bis bd Arago
F 75014 Paris
France
☎ 33 1 44 32 8188
✆ 33 1 44 32 8001
✉ nikos@friap51

Prasad Sheo S
Lockheed Palo Alto Res Lb
Dpt 91 20 Bg 255
3251 Hanover St
Palo Alto CA 94304 1191
USA
☎ 1 415 424 2659
✆ 1 415 424 3333

Prasanna A R
PRL
Navrangpura
Ahmedabad 380 009
India
☎ 91 272 46 2129
✆ 91 272 44 5292

Pratap R
Inst Applied Sciences
Cochin 682 317
India

Pravdo Steven H
CALTECH/JPL
MS 168 222
4800 Oak Grove Dr
Pasadena CA 91109 8099
USA
☎ 1 818 354 4134
✆ 1 818 393 6030

Predeanu Irina
Astronomical Institute
Romanian Acad Sciences
Cutitul de Argint 5
RO 75212 Bucharest
Rumania
☎ 40 1 641 3686
✆ 40 1 312 3391
✉ ipredeanu@imar.ro

Preite Martinez Andrea
IAS
Area d Ricerca CNR
Via Fosso Cavaliere 100
I 00133 Roma
Italy
☎ 39 6 4993 4473
✆ 39 6 2066 0188
✉ andrea@saturn.ias.
fra.cnr.it

Preka-Papadema P
Astrophysics Dpt
Ntl Univ Athens
Panepistimiopolis
GR 157 84 Zografos
Greece
☎ 30 1 723 5122
✆ 30 1 723 5122
✉ spm75@grathun1

Prendergast Kevin H
Astronomy Dpt
Columbia Univ
538 W 120th St
New York NY 10027
USA
☎ 1 212 280 3280
✆ 1 212 316 9504

Prentice Andrew J R
Mathematics Dpt
Monash Univ
Wellington Rd
Clayton VIC 3168
Australia
☎ 61 3 541 0811

Press William H
CfA
HCO/SAO
60 Garden St
Cambridge MA 02138 1516
USA
☎ 1 617 495 4908
✆ 1 617 495 7356

Preston George W
Carnegie Observatories
813 Santa Barbara St
Pasadena CA 91101 1292
USA
☎ 1 818 577 1122
✆ 1 818 795 8136
✉ gwp@ociw.edu

Preston Robert Arthur
CALTECH/JPL
MS 138 307
4800 Oak Grove Dr
Pasadena CA 91109 8099
USA
☎ 1 818 354 6895
✆ 1 818 393 4965
✉ rap@sgra.jpl.nasa.gov

Preuss Eugen
RAIUB
Univ Bonn
auf d Huegel 69
D 53121 Bonn
Germany
☎ 49 228 525 1
✆ 49 228 525 229
✉ epreuss@mpifr-
bonn.mpg.de

Prevot Louis
Observ Marseille
2 Pl Le Verrier
F 13248 Marseille Cdx 04
France
☎ 33 4 91 95 9088
✆ 33 4 91 62 1190
✉ lprevot@obsmara.
cnrs-mrs.fr

Prevot-Burnichon Marie-Louise
Observ Marseille
2 Pl Le Verrier
F 13248 Marseille Cdx 04
France
☎ 33 4 91 95 9088
✆ 33 4 91 62 1190

Prialnik-Kovetz Dina
Dpt Geophys/Planet Sci
Tel Aviv Univ
Ramat Aviv
Tel Aviv 69978
Israel
☎ 972 3 545 0633
✆ 972 3 640 9282

Price Michael J
Science Applications
5151 E Broadway
Suite 1100
Tucson AZ 85711
USA
☎ 1 520 748 7400

Price R Marcus
Dpt Phys/Astronomy
Univ New Mexico
800 Yale Blvd Ne
Albuquerque NM 87131
USA
☎ 1 505 277 2616

Price Stephan Donald
USAF Phillips Lab/GPSG
Boston College
Hanscom AFB MA 01731 3010
USA
☎ 1 617 377 4552
✆ 1 617 377 3138
✉ price@plh.af.mil

Priest Eric R
Applied Maths Dpt
Univ of St Andrews
North Haugh
St Andrews Fife KY16 9SS
UK
☎ 44 1334 76161*815
✆ 44 1334 74487
✉ eric@cs.st-andrews.ac.uk

Priester Wolfgang
IAEF
Univ Bonn
auf d Huegel 71
D 53121 Bonn
Germany
☎ 49 228 73 3671
✆ 49 228 73 3672

Prieto Cristina
Univ d Vigo
Dpt Matemat Aplicada
EUITI
E Vigo
Spain
✉ oacris@usc.es

Prieto Mercedes
IAC
Observ d Teide
via Lactea s/n
E 38200 La Laguna
Spain
☎ 34 2 232 9100
✆ 34 2 232 9117

Prieur Jean-Louis
OMP
14 Av E Belin
F 31400 Toulouse Cdx
France
☎ 33 5 61 33 2929
✆ 33 5 61 53 6722

Prince Helen Dodson
4800 Fillmore Av
Alexandria VA 22311
USA
☎ 1 703 578 1000

Pringle James E
Inst of Astronomy
The Observatories
Madingley Rd
Cambridge CB3 0HA
UK
☎ 44 12 23 337 548
✆ 44 12 23 337 523

Prinja Raman
Dpt Phys/Astronomy
UCLO
Gower St
London WC1E 6BT
UK
☎ 44 171 387 7050
✆ 44 171 380 7145

Pritchet Christopher J
Physics Dpt
Univ of Victoria
Box 1700
Victoria BC V8W 2Y2
Canada
☎ 1 250 721 7704
✆ 1 250 721 7715

Probstein R F
Dpt Mechan Eng
MIT
Box 165
Cambridge MA 02139
USA
☎ 1 617 253 2240

Prochazka Franz V
Inst Interdisziplinare
Forschung/Fortbildung
Sterneckstr 15
A 9010 Klagenfurt
Austria
☎ 43 463 2700 754
✆ 43 463 2700 759
✉ franz.prochazka@
uni-Klu.ac.at

Prodan Yuri I
SAI
Acad Sciences
Universitetskij Pr 13
119899 Moscow
Russia
☎ 7 095 139 5543
✆ 7 095 939 0126

Proffitt Charles R
NASA GSFC
Code 684 9
Greenbelt MD 20771
USA
☎ 1 301 286 3608
℻ 1 301 286 7642
✉ proffitt@iuesoc.gsfc.nasa.gov

Proisy Paul E
Observ Lyon
Av Ch Andre
F 69561 S Genis Laval cdx
France
☎ 33 4 78 56 0705
℻ 33 4 72 39 9791

Prokakis Theodore J
Astronomical Institute
Ntl Observ Athens
Box 20048
GR 118 10 Athens
Greece
☎ 30 1 346 1191/1 804 0619
℻ 30 1 346 4566

Prokof'eva Valentina V
Crimean Astrophys Obs
Ukrainian Acad Science
Nauchny
334413 Crimea
Ukraine
☎ 380 6554 71161
℻ 380 6554 40704

Prokof-Eva Irina A
Pulkovo Observ
Acad Sciences
10 Kutuzov Quay
196140 St Petersburg
Russia
☎ 7 812 298 2242
℻ 7 812 315 1701

Pronik I I
Crimean Astrophys Obs
Ukrainian Acad Science
Nauchny
334413 Crimea
Ukraine
☎ 380 6554 71161
℻ 380 6554 40704

Pronik V I
Crimean Astrophys Obs
Ukrainian Acad Science
Nauchny
334413 Crimea
Ukraine
☎ 380 6554 71161
℻ 380 6554 40704

Prosser Charles Franklin
NOAO/KPNO
Box 26732
950 N Cherry Av
Tucson AZ 85726 6732
USA
☎ 1 520 318 8411
℻ 1 520 318 8360
✉ cprosser@noao.edu

Proszynski Mieczyslaw
Copernicus Astron Ctr
Polish Acad Sci
Ul Bartycka 18
PL 00 716 Warsaw
Poland
☎ 48 22 41 1086
℻ 48 22 121 273

Protheroe Raymond J
Physics Dpt
Univ of Adelaide
Box 498
Adelaide SA 5001
Australia
☎ 61 8 228 5996

Protheroe William M
Astronomy Dpt
Ohio State Univ
174 W 18th Av
Columbus OH 43210 1106
USA
☎ 1 614 422 7891
℻ 1 614 292 2928

Protich Milorad B
Astronomical Observ
Volgina 7
11050 Beograd
Yugoslavia FR
☎ 381 1 402 365
℻ 381 1 419 553

Proust Dominique
Observ Paris Meudon
DAEC
Pl J Janssen
F 92195 Meudon PPL Cdx
France
☎ 33 1 45 07 7411
℻ 33 1 45 07 7469
✉ proust@obspm.fr

Proverbio Edoardo
Osserv Astronomico
d Brera
Via Brera 28
I 20121 Milano
Italy
☎ 39 2 723 201
℻ 39 2 720 01600
✉ proverbio@astrco.astro.it

Provost Janine
OCA
Observ Nice
BP 139
F 06304 Nice Cdx 4
France
☎ 33 4 93 89 0420
℻ 33 4 92 00 3033

Prusti Timo
ESA
Apt 50727 Villafranca
E 28080 Madrid
Spain
☎ 34 1 813 1248
℻ 34 1 813 1308
✉ tprusti@iso.vilspa.esa.es

Pryce Maurice H l
Physics Dpt
UBC
2075 Wesbrook Pl
Vancouver BC V6T 1W5
Canada
☎ 1 604 228 6417
℻ 1 604 228 5324

Pryor Carlton Philip
Dpt Phys/Astronomy
Rutgers Univ
Box 849
Piscataway NJ 08854 0849
USA
☎ 1 908 932 5462
℻ 1 908 932 4343
✉ pryor@pryor.rutgers.edu

Pskovskij Juri P
SAI
Acad Sciences
Universitetskij Pr 13
119899 Moscow
Russia
☎ 7 095 139 3721
℻ 7 095 939 0126

Puche Daniel
CfA
HCO/SAO
60 Garden St
Cambridge MA 02138 1516
USA
☎ 1 617 495 7344
℻ 1 617 495 7014
✉ dpuche@cfa.harvard.edu

Pucillo Mauro
OAT
Box Succ Trieste 5
Via Tiepolo 11
I 34131 Trieste
Italy
☎ 39 40 31 99255
℻ 39 40 30 9418
✉ pucillo@oat.ts.astro.it

Puerari Ivanio
INAOE
Tonantzintlaz
Apdo 51 y 216
Puebla PUE 72000
Mexico
☎ 52 2 247 2011
℻ 52 2 247 2231
✉ puerari@inaoep.mx

Puetter Richard C
CASS
UCSD
C 011
La Jolla CA 92093 0216
USA
☎ 1 619 534 4995
℻ 1 619 534 2294

Pugach Alexander F
Main Astronomical Obs
Ukrainian Acad Science
Golosiiv
252650 Kyiv 22
Ukraine
☎ 380 4426 64771
℻ 380 4426 62147

Puget Jean-Loup
IAS
Bt 121
Universite Paris XI
F 91405 Orsay Cdx
France
☎ 33 1 69 85 8500
℻ 33 1 69 85 8675

Pugliano Antonio
Ist Universitario Navale
Via Acton 38
I 80133 Napoli
Italy
☎ 39 81 551 2330
℻ 39 81 552 1485
✉ pugliano@naval.uniav.it

Puls Joahim
Inst Astron/Astrophysik
Univ Sternwarte
Scheinerstr 1
D 81679 Muenchen
Germany
☎ 49 899 220 9436
℻ 49 899 220 9427

Punetha Lalit Mohan
Uttar Pradesh State
Observ
Po Manora Peak 263 129
Nainital 263 129
India
☎ 91 59 42 2136

Purton Christopher R
DRAO
NRCC/HIA
Box 248
Penticton BC V2A 6K3
Canada
☎ 1 250 493 2277
℻ 1 250 493 7767

Puschell Jeffery John
Lockheed Martin A&NS
103 Chesapeake Park Plaza
Baltimore MD 21220 4295
USA

Pushkin Sergey B
Time/Frequency Service
Gosstandard Ussr
117049 Moscow
Russia

Pustylnik Izold B
Tartu Observ
Estonian Acad Sci
Toravere
EE 2444 Tartumaa
Estonia
☎ 372 7 410 465
℻ 372 7 410 205
✉ izold@aai.ee

Pyatunina Tamara B
Inst Appl Astronomy
Acad Sciences
Zhdanovskaya St 8
197042 St Petersburg
Russia
☎ 7 812 230 7414
℻ 7 812 230 7413
✉ ptb@ipa.rssi.ru

Pye John P
Dpt Phys/X-Ray Astron
Univ Leicester
University Rd
Leicester LE1 7RH
UK
☎ 44 116 252 2073
✆ 44 116 250 182

Pyper Smith Diane M
Physics Dpt
Univ of Nevada
4505 S Maryland Parkway
Las Vegas NV 89154
USA
☎ 1 702 739 3653
✆ 1 702 739 0804

Q

Qi Guanrong
Shaanxi Observ
CAS
Box 18
Lintong 710600
China PR
☎ 86 33 2255
℻ 86 9237 3496

Qian Bochen
Shanghai Observ
CAS
80 Nandan Rd
Shanghai 200030
China PR
☎ 86 21 6438 6191
℻ 86 21 6438 4618

Qian Shanjie
Beijing Astronomical Obs
CAS
W Suburb
Beijing 100080
China PR
☎ 86 1 28 2194
℻ 86 1 256 10855

Qian Zhihan
Shanghai Observ
CAS
80 Nandan Rd
Shanghai 200030
China PR
☎ 86 21 6438 6191
℻ 86 21 6438 4618

Qiao Guojun
Dpt Geophysics
Beijing Univ
Beijing 100871
China PR
☎ 86 1 28 2471*3888
℻ 86 1 25 64095

Qin Dao
Purple Mountain Obs
CAS
Nanjing 210008
China PR
☎ 86 25 46700
℻ 86 25 301 459

Qin Songnian
Yunnan Observ
CAS
Kunming 650011
China PR
☎ 86 871 2035
℻ 86 871 717 1845

Qin Zhihai
Astronomy Dpt
Nanjing Univ
Nanjing 210093
China PR
☎ 86 25 34651*2882
℻ 86 25 330 2728

Qiu Puzhang
Yunnan Observ
CAS
Kunming 650011
China PR
☎ 86 871 2035
℻ 86 871 717 1845

Qiu Yaohui
Yunnan Observ
CAS
Kunming 650011
China PR
☎ 86 871 335 7813
℻ 86 871 391 1845
✉ gfb@public.km.yn.cn

Qiu Yuhai
Beijing Astronomical Obs
CAS
Beijing 100080
China PR
☎ 86 1 28 1698
℻ 86 1 256 1085

Qu Qinyue
Astronomy Dpt
Nanjing Univ
22 Hankou Lu
Nanjing 210093
China PR
☎ 86 25 637 551*3186
℻ 86 25 302 728

Quamar Jawaid
Univ of Karachi
C 6 Staff Town
Karachi 75270
Pakistan
☎ 92 46 54 91
✉ bartel@sgl.ists.ca

Quan Hejun
Shanghai Observ
CAS
80 Nandan Rd
Shanghai 200030
China PR
☎ 86 21 6438 6191
℻ 86 21 6438 4618

Quarta Maria Lucia
IAG
Univ Sao Paulo
CP 9638
01065 970 Sao Paulo SP
Brazil
☎ 55 11 577 8599
℻ 55 11 276 3848

Quast Germano Rodrigo
Observ Nacional
Rua Colonel Renno 07
CP 21
37500 Itajuba Mg
Brazil
☎ 55 356 22 0788

Queloz Didier
Observ Geneve
Chemin d Maillettes 51
CH 1290 Sauverny
Switzerland
☎ 41 22 755 2611
℻ 41 22 755 3983
✉ didier.queloz@obs.unige.ch

Quemerais Eric
Service Aeronomie
BP 3
F 91371 Verrieres Buisson
France
☎ 33 1 64 47 4317
℻ 33 1 69 20 2999
✉ eric.quemerais@aerov.jussieu.fr

Quenby John J
Astrophysics Gr
Imperial Coll
Prince Consort Rd
London SW7 2BZ
UK
☎ 44 171 594 7771
℻ 44 171 594 7772

Querci Francois R
OMP
14 Av E Belin
F 31400 Toulouse Cdx
France
☎ 33 5 61 33 2879
℻ 33 5 61 53 2840
✉ querci@astro.obs-mip.fr

Querci Monique
OMP
14 Av E Belin
F 31400 Toulouse Cdx
France
☎ 33 5 61 33 2879
℻ 33 5 61 53 2840
✉ querci@astro.obs-mip.fr

Quesada Vinicio
Stazione Astronomicia
Univ d Cagliari
Via Ospedale 72
I 09124 Cagliari
Italy
☎ 39 70 66 3544
℻ 39 70 65 7657

Quinn Peter
MSSSO
Weston Creek
Private Bag
Canberra ACT 2611
Australia
☎ 61 262 490 272
℻ 61 262 490 233
✉ pjq@minuet.anu.oz.au

Quintana Hernan
Gr Astrofisica
Univ Catolica
Casilla 104
Santiago 22
Chile
☎ 56 2 686 4938
℻ 56 2 686 4948
✉ hquintana@astrouc.puc.cl

Quintana Jose M
Inst Astrofisica
Andalucia Apt 3004
Prof Albareda 1
E 18080 Granada
Spain
☎ 34 5 812 1300
℻ 34 5 881 4530

Quirk William J
LLNL
L 389
Box 808
Livermore CA 94551 9900
USA
☎ 1 415 422 1852
℻ 1 415 422 24563
✉ quirk@pop.net

R

Raadu Michael A
Dpt Plasma Physics
Royal Inst of Technology
S 100 44 Stockholm 70
Sweden
☎ 46 87 90 6000
✉ raadu@plasma.kth.se

Rabbia Yves
OCA
CERGA
F 06130 Grasse
France
☎ 33 4 93 36 5849
✆ 33 4 93 40 5353

Rabin Douglas Mark
NOAO/NSO
Box 26732
950 N Cherry Av
Tucson AZ 85726 6732
USA
☎ 1 520 325 9331
✆ 1 520 325 9360
✉ rabin@noao.edu

Rabolli Monica
Observ Astronomico
Paseo d Bosque S/n
1900 La Plata (Bs As)
Argentina
☎ 54 21 217 308
✆ 54 21 211 761

Rachkovsky D N
Crimean Astrophys Obs
Ukrainian Acad Science
Nauchny
334413 Crimea
Ukraine
☎ 380 6554 71161
✆ 380 6554 40704

Racine Rene
Dpt d Physique
Universite Montreal
CP 6128 Succ A
Montreal QC H3C 3J7
Canada
☎ 1 514 343 6718
✆ 1 514 343 2071

Radford Simon John E
NRAO
Campus Bg 65
949 N Cherry Av
Tucson AZ 85721 0655
USA
☎ 1 520 882 8250*125
✆ 1 520 882 7955
✉ sradford@nrao.edu

Radhakrishnan V
Astronomical Institute
Univ of Amsterdam
Kruislaan 403
NL 1098 SJ Amsterdam
Netherlands
☎ 31 20 525 7491/2
✆ 31 20 525 7484
✉ rad@astro.uva.nl

Radick Richard R
AFGL
NSO
Sunspot NM 88349
USA
☎ 1 505 434 1390
✆ 1 504 434 7029

Radiman Iratius
Bosscha Observ
ITB Lembang
Lembang 40391
Indonesia
☎ 62 22 229 6001
✆ 62 22 278 7289

Radoski Henry R
AFOSR/NP
NSO
Bolling Air Force Base
Washington DC 20332
USA
☎ 1 202 767 4906

Radu Eugenia
Astronomical Observ
Romanian Acad Sciences
Ul Ciresilor 19
RO 3400 Cluj Napoca
Rumania
☎ 40 64 194 592
✆ 40 64 194 592

Raedler K H
Zntrlinst Astrophysik
Sternwarte Babalsberg
Rosa Luxemburg Str 17a
D 14473 Potsdam
Germany
☎ 49 331 275 3731

Rafanelli Piero
Osserv Astronom Padova
Univ d Padova
Vic d Osservatorio 5
I 35122 Padova
Italy
☎ 39 49 829 3411
✆ 39 49 875 9840

Rafert James Bruce
Dpt Phys/Space Sci
Florida Inst Techn
150 W University Blvd
Melbourne FL 32901
USA
☎ 1 407 768 8000
✆ 1 407 984 8461

Rafferty Theodore J
USNO
Astrometry Dpt
3450 Massachusetts Av NW
Washington DC 20392 5400
USA
☎ 1 202 762 1471
✆ 1 202 653 1497
✉ tjr@sicon.usno.
navy.mil

Raghavan Nirupama
IIT
6 West Av
New Delhi 110 016
India

Raharto Moedji
Astronomy Dpt
Bandung Instte Techn
Jl Ganesha 10
Bandung 40132
Indonesia
☎ 62 22 244 0252
✆ 62 22 2505442

Rahunen Timo
Tampereen Saerkaenniemi Oy
Saerkaenniemi
FIN 33230 Tampere
Finland
☎ 358 3 131 333

Raikova Donka
Astronomy Dpt
Bulgarian Acad Sci
7th November St 1
BG 1000 Sofia
Bulgaria
☎ 359 2 80 2831
✆ 359 2 80 2831

Raimond Ernst
NFRA
Postbus 2
NL 7990 AA Dwingeloo
Netherlands
☎ 31 521 59 5100
✆ 31 52 159 7332
✉ exr@nfra.nl

Raine Derek J
Astronomy Dpt
Univ Leicester
University Rd
Leicester LE1 7RH
UK
☎ 44 116 252 2073
✆ 44 113 252 2200

Raitala Jouko T
Astronomy Dpt
Univ of Oulu
Box 333
FIN 90570 Oulu
Finland
☎ 358 81 553 1945
✆ 358 81 553 1934
✉ jouko.raitala@oulu.fi

Rajamohan R
IIA
Koramangala
Sarjapur Rd
Bangalore 560 034
India
☎ 91 80 356 6497/6585
✆ 91 80 553 4043

Rajchl Jaroslav
Astronomical Institute
Czech Acad Sci
Fricova 1
CZ 251 65 Ondrejov
Czech R
☎ 420 204 85 745
✆ 420 2 88 1611
✉ astmph@csearn

Raju P K
IIA
Koramangala
Sarjapur Rd
Bangalore 560 034
India
☎ 91 80 356 6585
✆ 91 80 553 4043

Raju Vasundhara
IIA
Koramangala
Sarjapur Rd
Bangalore 560 034
India
☎ 91 80 553 0672 *76
✆ 91 80 553 4043
✉ rvas@iiap.ernet.in

Rakavy Gideon
Einstein Inst of Phys
Hebrew Univ Jerusalem
Jerusalem 91904
Israel

Rakos Karl D
Inst Astronomie
Univ Wien
Tuerkenschanzstr 17
A 1180 Wien
Austria
☎ 43 1 345 36095
✆ 43 1 470 6015

Rakshit H
Bengal Engineerg College
Sibpore
Hewrah
India

Ram Sagar
Uttar Pradesh State
Observ
Monora Peak
Nainital 263 129
India
☎ 91 5942 35136/35583
✆ 91 5942 35583/36281
✉ sagar@upso.
ernet.in

Ramadurai Souriraja
TIFR
Homi Bhabha Rd
Colaba
Bombay 400 005
India
☎ 91 22 495 2311
✆ 91 22 495 2110

Ramamurthy Swaminathan
CASA
Univ of Osmania
Hyderabad 500 007
India
☎ 91 868951*247

Ramana Murthy P V
TIFR
Homi Bhabha Rd
Colaba
Bombay 400 005
India
☎ 91 22 495 2979
✆ 91 22 495 2110
✉ ra1mana@tifrvax

Ramaty Reuven
NASA GSFC
Code 665
Greenbelt MD 20771
USA
☎ 1 301 286 8715

Ramella Massimo
OAT
Box Succ Trieste 5
Via Tiepolo 11
I 34131 Trieste
Italy
☎ 39 40 31 99255
✆ 39 40 30 9418

Rampazzo Roberto
Osserv Astronomico
d Brera
Via Brera 28
I 20121 Milano
Italy
☎ 39 2 723 20319
✆ 39 2 720 01600
✉ rampazzo@brere.mi.astro.it

Ramsey Lawrence W
Astronomy Dpt
Pennsylvania State Univ
525 Davey Lab
University Park PA 16802
USA
☎ 1 814 865 0418
✆ 1 814 863 7114

Randic Leo
Geodetical Faculty
Univ of Zagreb
Horvatovac BB
HR 10000 Zagreb
Croatia
☎ 385 1 420 222
✆ 385 1 432 462

Rangarajan K E
IIA
Koramangala
Sarjapur Rd
Bangalore 560 034
India
☎ 91 80 553 0672
✆ 91 80 553 4043
✉ rangaraj@iiap.
ernet.in

Ranieri Marcello
IAS
Area d Ricerca CNR
Via Fosso Cavaliere 100
I 00133 Roma
Italy
☎ 39 6 4993 4473
✆ 39 6 2066 0188

Rank David M
Lick Observ
Univ of California
Santa Cruz CA 95064
USA
☎ 1 831 429 2277
✆ 1 831 426 3115

Rankin Joanna M
Physics Dpt
Univ of Vermont
A405 Cook Building
Burlington VT 05405
USA
☎ 1 802 656 2644
✉ rankin@merlin.uvm-
gen.uvm.edu

Rao A Pramesh
NCRA/TIFR
Pune Univ Campus Pb 3
Ganeshkhind
Pune 411 007
India
☎ 91 212 33 7107
✆ 91 212 33 5760

Rao Arikkala Raghurama
TIFR/Space Phys Gr
Homi Bhabha Rd
Colaba
Bombay 400 005
India
☎ 91 22 215 2971
✆ 91 22 215 2110
✉ arrao@tifrvax.tifr.res.in

Rao D Mohan
IIA
Koramangala
Sarjapur Rd
Bangalore 560 034
India
☎ 91 80 553 0672
✆ 91 80 553 4043
✉ mohan@iiap.
ernet.in

Rao K Narahari
Physics Dpt
Ohio State Univ
174 W 18th Av
Columbus OH 43210 1106
USA
☎ 1 614 422 6505

Rao K Ramanuja
C/o Dr K Surendra
Rua Cel Joao Cursino 210
Apt 92 Vila Adyana
12200 S Jose dos Campos
Brazil

Rao M N
PRL
Navrangpura
Ahmedabad 380 009
India
☎ 91 272 46 2129
✆ 91 272 44 5292

Rao N Kameswara
IIA
Koramangala
Sarjapur Rd
Bangalore 560 034
India
☎ 91 80 356 6585
✆ 91 80 553 4043

Rao P Vivekananda
Astronomy Dpt
Univ of Osmania
Hyderabad 500 007
India
☎ 91 71 951

Rao Ramachandra V
ISRO Satellite Ctr
Peenya
Vimanapura Post
Bangalore 560 058
India
☎ 91 80 526 6251
✆ 91 80 526 5407

Raoult Antoinette
Observ Paris Meudon
DASOP
Pl J Janssen
F 92195 Meudon PPL Cdx
France
☎ 33 1 45 07 7766
✆ 33 1 45 07 7959

Rapaport Michel
Observ Bordeaux
BP 89
F 33270 Floirac
France
☎ 33 5 56 86 4330
✆ 33 5 56 40 4251

Rapley Christopher G
Mullard Space Science Lab
Univ College London
Holmbury St Mary
Dorking Surrey RH5 6NT
UK
☎ 44 13 06 702 92
✆ 44 14 83 278 312

Rasio Frederic A
Physics Dpt
MIT
Box 165
Cambridge MA 02139 4307
USA
☎ 1 617 253 5084
✆ 1 617 253 9798
✉ rasio@mit.edu

Rastorguev Alexey S
SAI
Acad Sciences
Universitetskij Pr 13
119899 Moscow
Russia
☎ 7 095 939 1622
✆ 7 095 932 8841
✉ ras@sai.msu.su

Ratag Mezak Arnold
LAPAN
Ntl Inst Aeron/Space
Ul Junjunan 133
Bandung 40173
Indonesia
✉ mezakr@indosat.net.id

Ratnatunga Kavan U
Physics Dpt
Carnegie Mellon Univ
5000 Forbes Av
Pittsburgh PA 15213 3890
USA
☎ 1 412 268 1888
✆ 1 412 681 0648
✉ kavan@astro.phys.cmu.edu

Raubenheimer Barend C
Cosmic Ray Res Unit
Potchefstroom Univ
for CHE
2520 Potchefstroom
South Africa
☎ 27 148 299 2423
✆ 27 148 299 2421

Rauch Thomas
Astronomisches Institut
Univ Tuebingen
Waldhaeuserstr 64
D 72076 Tuebingen
Germany
☎ 49 70 71 297 8614
✆ 49 707 129 3458
✉ rauch@astro.uni-tuebingen.de

Rautela B S
Uttar Pradesh State
Observ
Po Manora Peak 263 129
Nainital 263 129
India
☎ 91 59 42 2136

Raveendran A V
IIA
Koramangala
Sarjapur Rd
Bangalore 560 034
India
☎ 91 80 553 0672
✆ 91 80 553 4043
✉ avr@iiap.ernet.in

Rawlings Jonathan
Astrophysics Dpt
Univ of Oxford
Keble Rd
Oxford OX1 3RH
UK
☎ 44 186 527 3303
✆ 44 186 527 3418
✉ jr@uk.ac.ox.astro

Rawlings Steven
Astrophysics Dpt
Univ of Oxford
Keble Rd
Oxford OX1 3RH
UK
☎ 44 186 527 3303
✆ 44 186 527 3418
✉ sr@uk.ac.ox.astro

Ray Alak
TIFR
Homi Bhabha Rd
Colaba
Bombay 400 005
India
☎ 91 22 215 2971
✆ 91 22 215 2110
✉ akr@tifrvax

Ray James R
USNO
EO Dpt
3450 Massachusetts Av NW
Washington DC 20392 5420
USA
✉ jimr@maia.usno.navy.mil

Ray Thomas P
DIAS
School Cosmic Phys
5 Merrion Sq
Dublin 2
Ireland
☎ 353 1 662 1333
✆ 353 1 662 1477
✉ tr@cp.dias.ie

Raychaudhuri Amalkumar
Presidency College
College St
Calcutta 73
India

Raychaudhury Somak
IUCAA
PO Box 4
Ganeshkhind
Pune 411 007
India
☎ 91 212 351 414/5
✆ 91 212 350 760
✉ somak@iucaa.ernet.in

Rayet Marc
Inst Astrophysique
Univ Bruxelles
Campus Plaine CP 226
B 1050 Brussels
Belgium
☎ 32 2 650 3572
✆ 32 2 650 4226
✉ mrayet@astro.ulb.ac.be

Raymond John Charles
CfA
HCO/SAO
60 Garden St
Cambridge MA 02138 1516
USA
☎ 1 617 495 7000
✆ 1 617 495 7356
✉ jraymond@cfa.harvard.edu

Rayrole Jean R
Observ Paris Meudon
DASOP
Pl J Janssen
F 92195 Meudon PPL Cdx
France
☎ 33 1 45 07 7789
✆ 33 1 45 07 7971

Razdan Hiralal
Technical Coordination Gp
Bhabha Atomic Res Ctr
7F Cntrl Complex Trombay
Bombay 400 085
India
☎ 91 22 551 1859
✆ 91 22 495 2110

Razin Vladimir A
Radiophysical Res Inst
Lyadov Ul 25/14
603600 N Novgorod
Russia
☎ 7 831 236 7294
✆ 7 831 236 9902

Reach William
CALTECH
MS 100 22
Pasadena CA 91001
USA
☎ 1 626 397 7057
✆ 1 626 397 9600
✉ reach@ipac.caltech.edu

Readhead Anthony C S
CALTECH
Robinson Bldg
Pasadena CA 91125
USA
☎ 1 213 356 4972

Reames Donald V
NASA GSFC
Code 661
Greenbelt MD 20771
USA
☎ 1 301 286 6454
✆ 1 301 286 1682
✉ reames@lheavx.gsfc.nasa.gov

Reasenberg Robert D
CfA
HCO/SAO Rm B 217
60 Garden St
Cambridge MA 02138 1516
USA
☎ 1 617 495 7108
✆ 1 617 495 7356
✉ reasenberg@cfa.harvard.edu

Reaves Gibson
Dpt Phys/Astronomy
USC
Los Angeles CA 90089 1342
USA
☎ 1 213 740 6330
✆ 1 213 740 7254
✉ reaves@mizar.usc.edu

Reay Newrick K
Astrophysics Gr
Imperial Coll
Prince Consort Rd
London SW7 2BZ
UK
☎ 44 171 594 7771
✆ 44 171 594 7772

Rebeirot Edith
Observ Marseille
2 Pl Le Verrier
F 13248 Marseille Cdx 04
France
☎ 33 4 91 95 9088
✆ 33 4 91 62 1190

Reber Grote
C/o Post office
Bothwell TAS 7030
Australia
☎ 61 2 23 7371

Rebolo Rafael
IAC
Observ d Teide
via Lactea s/n
E 38200 La Laguna
Spain
☎ 34 2 232 9100
✆ 34 2 232 9117

Recillas-Cruz Elsa
Instituto Astronomia
UNAM
Apt 70 264
Mexico DF 04510
Mexico
☎ 52 5 622 3906
✆ 52 5 616 0653

Redfern Michael R
Physics Dpt
Univ College
Galway
Ireland
☎ 353 91 524 411
✆ 353 91 525 700

Reed B Cameron
Physics Dpt
Alma College
Alma MI 48801 1599
USA
☎ 1 517 463 7266
✆ 1 517 463 7277
✉ reed@alma.edu

Rees David Elwyn
CSIRO
Div Radiophysics
Box 76
Epping NSW 2121
Australia
☎ 61 2 868 0493
✆ 61 2 868 0411
✉ drees@rp.csiro.au

Rees Martin J
Inst of Astronomy
The Observatories
Madingley Rd
Cambridge CB3 0HA
UK
☎ 44 12 23 337 548
✆ 44 12 23 337 523
✉ jm@ast-star.cam.ac.uk

Reeves Edmond M
NASA Headquarters
Off Life/Microgravity
600 Independence Av SW
Washington DC 20546
USA
☎ 1 202 358 2560
✆ 1 202 358 4166
✉ ereeves@hq.nasa.gov

Reeves Hubert
DAPNIA/SAP
CEA Saclay
BP 2
F 91191 Gif s Yvette Cdx
France
☎ 33 1 69 08 5159
✆ 33 1 69 08 6577

Refsdal Sjur
Hamburger Sternwarte
Univ Hamburg
Gojensbergsweg 112
D 21029 Hamburg
Germany
☎ 49 407 252 4124
✆ 49 407 252 4198

Regev Oded
Physics Dpt
IIT
Technion City
Haifa 32000
Israel
☎ 972 4 293 992
✆ 972 4 221 514
✉ phr91or@technion

Reglero-Velasco Victor
Dpt Mat y Astron
Univ Valencia
Dr Moliner 50
E 46100 Burjassot
Spain
☎ 34 6 386 4326
✆ 34 6 386 4302
✉ reglero@evalvx.ific.uv.es

Rego Fernandez M
Dpt Astrofisica
Fac Fisica
Univ Complutense
E 28040 Madrid
Spain
☎ 34 1 449 5316
✆ 34 1 394 5195

Regoes Enikoe
Inst of Astronomy
The Observatories
Madingley Rd
Cambridge CB3 0HA
UK
☎ 44 12 23 330 894
✆ 44 12 23 337 523
✉ eniko@mail.ast.cam.ac.uk

Regulo Clara
IAC
Observ d Teide
via Lactea s/n
E 38200 La Laguna
Spain
☎ 34 2 232 9100
✆ 34 2 232 9117

Reich Wolfgang
RAIUB
Univ Bonn
auf d Huegel 69
D 53121 Bonn
Germany
☎ 49 228 525 0
✆ 49 228 525 229

Reichert Gail Anne
NASA GSFC
Code 631
Greenbelt MD 20771
USA
☎ 1 301 286 0615
✆ 1 301 286 1771
✉ gail.reichert.2@gsfc.nasa.gov

Reid Mark Jonathan
CfA
HCO/SAO
60 Garden St
Cambridge MA 02138 1516
USA
☎ 1 617 495 7470
✆ 1 617 495 7356

Reid Neill
CALTECH
Palomar Observ MS 105 24
Pasadena CA 91125
USA
☎ 1 818 356 6586
✆ 1 815 568 1517

Reif Klaus
RAIUB
Univ Bonn
auf d Huegel 71
D 53121 Bonn
Germany
☎ 49 228 73 3657
✆ 49 228 73 3672
✉ reif@astro.uni-bonn.de

Reig Pablo
Dpt Phys/Astronomy
Southampton SO17 1BJ
UK
☎ 44 170 359 2079
✆ 44 170 359 3910
✉ pablo@astro.soton.ac.uk

Reimers Dieter
Hamburger Sternwarte
Univ Hamburg
Gojensbergsweg 112
D 21029 Hamburg
Germany
☎ 49 407 252 4112
✆ 49 407 252 4198

Reinisch Gilbert
OCA
Observ Nice
BP 139
F 06304 Nice Cdx 4
France
☎ 33 4 93 89 0420
✆ 33 4 92 00 3033

Reinsch Klaus
Univ Sternwarte
Goettingen
Geismarlandstr 11
D 37083 Goettingen
Germany
☎ 49 551 394 037
✆ 49 551 395 043
✉ reinsch@uni-sw.gwdg.de

Reipurth Bo
CASA
Univ of Colorado
Campus Box 389
Boulder CO 80309 0389
USA
☎ 1 303 492 4050
✆ 1 303 492 7178
✉ reipurth@casa.olorado.edu

Reitsema Harold J
Ball Aerospace Systems
Division
Box 1062
Boulder CO 80306
USA
☎ 1 303 939 5026
✆ 1 303 939 6602
✉ hreitsem@ball.com

Reiz Anders
Maltgatan 2
S 24014 Veberöd
Sweden

Remy Battiau Liliane G A
Conseil de la Recherche
Universite Liege
7 Pl du XX Aout
B 4000 Liege
Belgium
☎ 32 4 242 0080
✆ 32 4 366 5558
✉ l.remy@ulg.ac.be

Rengarajan T N
TIFR/IR Astronomy
Homi Bhabha Rd
Colaba
Bombay 400 005
India
☎ 91 22 219 111
✆ 91 22 215 2110
✉ renga@tifrc3.tifr.res.in

Rense William A
204 Birch Dr
Lafayette LA 70506
USA
☎ 1 318 981 0769

Renson P F M
Inst Astrophysique
Universite Liege
Av Cointe 5
B 4000 Liege
Belgium
☎ 32 4 254 7510
✆ 32 4 254 7511
✉ renson@astro.ulg.ac.be

Renzini Alvio
Dpt Astronomia
Univ d Bologna
Via Zamboni 33
I 40126 Bologna
Italy
☎ 39 51 259 301
✆ 39 51 259 407

Rephaeli Yoel
Dpt Phys/Astronomy
Tel Aviv Univ
Ramat Aviv
Tel Aviv 69978
Israel
☎ 972 3 640 8208
✆ 972 3 640 8179

Requieme Yves
Observ Bordeaux
BP 89
F 33270 Floirac
France
☎ 33 5 56 86 4330
✆ 33 5 56 40 4251
✉ requieme@obsboa.observ.u-bordeaux.fr

Revelle Douglas Orson
LANL
MS F665
Box 1663
Los Alamos NM 87545 2345
USA
☎ 1 505 667 2897
✆ 1 505 665 4055
✉ dor@eddie.lanl.gov

Reyes Francisco
Astronomy Dpt
Univ of Florida
211 SSRB
Gainesville FL 32611
USA
☎ 1 904 392 2049
✆ 1 904 392 5089
✉ freyes@ufpine

Reynolds John
AAO
ATNF
Box 76
Epping NSW 2121
Australia
☎ 61 2 868 0222
✆ 61 2 868 0400
✉ jreynold@atnf.csiro.au

Reynolds John H
Physics Dpt
Univ of California
366 LeConte Hall
Berkeley CA 94720
USA
☎ 1 415 642 4863
✆ 1 510 642 3411

Reynolds Ronald J
Physics Dpt
Univ Wisconsin
1150 University Av
Madison WI 53706
USA
☎ 1 608 262 5916
✆ 1 608 262 3077

Reynolds Stephen P
Physics Dpt
N Carolina State Univ
Box 8202
Raleigh NC 27695 8202
USA
☎ 1 919 515 7751
✆ 1 919 515 1971
✉ stephen_reynolds@ncsu.edu

Rhodes Edward J
11801 Killimore Av
Northridge CA 91326
USA

Riazi Nematollah
Physics Dpt
Shiraz Univ
Shiraz 71454
Iran
☎ 98 71 24609
✆ 98 71 20 027
✉ riazi@sun01.sci.shirazu.ac.ir

Ribes Jean-Claude
INSU
3 r Michel Ange
F 75794 Paris Cedex 16
France
☎ 33 1 44 96 5000
✆ 33 1 44 96 4387

Rice John B
Dpt Phys/Astronomy
Univ of Brandon
270 18th Street
Brandon MB R7A 6A9
Canada
☎ 1 204 727 9693
✆ 1 204 728 7346
✉ rice@brandonu.ca

Rich Robert Michael
Astronomy Dpt
Columbia Univ
538 W 120th St
New York NY 10027
USA
☎ 1 212 854 6837
✆ 1 212 854 8121
✉ rmr@figaro.phys.columbia.edu

Richards Mercedes T
University Station
Univ of Virginia
Box 3818
Charlottesville VA 22903 0818
USA
☎ 1 804 924 4895
✆ 1 804 924 3104
✉ mtr8r@virginia.edu

Richardson E Harvey
1871 Elmhurst Pl
Victoria BC V8N 1R1
Canada

Richardson Kevin J
Astrophysics Group
QMWC
Mile End Rd
London E1 4NS
UK
☎ 44 171 975 5555
✆ 44 181 975 5500

Richardson Lorna Logan
Dpt Phys/Astronomy
Univ of Glasgow
Glasgow G12 8QQ
UK
☎ 44 141 339 8855*4153
✆ 44 141 334 9029
✉ lorna@astro.gla.ac.uk

Richardson R S
Griffith Observ
Box 27787
Los Felix Station
Los Angeles CA 90027
USA
☎ 1 213 664 1181

Richer Harvey B
Dpt Geophys/Astronomy
UBC
2075 Wesbrook Pl
Vancouver BC V6T 1W5
Canada
☎ 1 604 228 4134
✆ 1 604 228 6047

Richer John
MRAO
Cavendish Laboratory
Madingley Rd
Cambridge CB3 0HE
UK
☎ 44 12 23 339 992
✆ 44 12 23 354 599
✉ jsr@mrao.cam.ac.uk

Richichi Andrea
Osserv Astrofis Arcetri
Univ d Firenze
Largo E Fermi 5
I 50125 Firenze
Italy
☎ 39 55 27 52230
✆ 39 55 220 039
✉ richichi@arcetri.astro.it

Richstone Douglas O
Astronomy Dpt
Univ of Michigan
Dennison Bldg
Ann Arbor MI 48109 1090
USA
☎ 1 313 764 3441
✆ 1 313 764 2211
✉ d_richstone@ub.cc.umich.edu

Richter G A
Zntrlinst Astrophysik
Sternwarte Babelsberg
an d Sternwarte 16
D 96575 Sonneberg
Germany
☎ 49 96 74 2287

Richter Gotthard
Astrophysik Inst
Potsdam Univ
an d Sternwarte 16
D 14482 Potsdam
Germany
☎ 49 331 749 9301
✆ 49 331 749 9309
✉ gmrichter@aip.de

Richter Johannes
Inst Theor Phys/
Sternwarte Univ Kiel
Olshausenstr 40
D 24098 Kiel
Germany
☎ 49 431 880 3835
✆ 49 431 880 4432

Richtler Tom
Univ Sternwarte
Univ Bonn
auf d Huegel 71
D 53121 Bonn
Germany
☎ 49 228 73 3669
✆ 49 228 73 3672

Rickard James Joseph
Box 777
Borrego Springs CO 92004
USA
☎ 1 714 767 5462

Rickard Lee J
NRL
Code 4138 Rrd
4555 Overlook Av SW
Washington DC 20375 5000
USA
☎ 1 202 767 2495

Ricker George R
Ctr for Space Res
MIT Rm 37 527
Box 165
Cambridge MA 02139 4307
USA
☎ 1 617 253 7532
✆ 1 617 253 0861

Rickett Barnaby James
EECS
UCSD
La Jolla CA 92093 0216
USA
☎ 1 619 452 2731

Rickman Hans
Astronomical Observ
Box 515
S 751 20 Uppsala
Sweden
☎ 46 18 51 35 22
✆ 46 18 52 7583
✉ hans.rickman@astro.uu.se

Ricort Gilbert
Dpt Astrophysique
Universite Nice
Parc Valrose
F 06034 Nice Cdx
France
☎ 33 4 93 51 9100
✆ 33 4 93 52 9806

Riddle Anthony C
700 Grant Pl
Boulder CO 80302
USA
☎ 1 303 447 8127

Riegel Kurt W
Dpt Navy
3601 5th St
S #201
Arlington VA 22204 1600
USA
☎ 1 703 695 3363
✉ riegel.kurt@hq.navy.mil

Rieger Erich
MPE
Postfach 1603
D 85740 Garching
Germany
☎ 49 893 299 3511
✆ 49 893 299 3569
✉ eir@dgaippls

Riegler Guenter R
NASA Headquarters
Code SZ
600 Independence Av SW
Washington DC 20546
USA
☎ 1 202 358 0370
✆ 1 202 358 3096
✉ griegler@gm.ossa.hq.nasa.gov

Rieutord Michel
OMP
14 Av E Belin
F 31400 Toulouse Cdx
France
☎ 33 5 61 33 2949
✆ 33 5 61 53 6722
✉ rieutord@obs-mip.fr

Righini Alberto
Dpt Astronomia
Univ d Firenze
Largo E Fermi 5
I 50125 Firenze
Italy
☎ 39 55 27 521
✆ 39 55 22 0039

Riihimaa Jorma J
Astronomy Dpt
Univ of Oulu
Box 333
FIN 90570 Oulu
Finland
☎ 358 81 555 2022
✆ 358 81 553 1934

Rijnbeek Richard
Space/Plasma Phys
School Mathes/Phys Sci
Univ of Sussex
Brighton BN1 9QH
UK
✆ 44 12 73 678097
✉ richardr@central.sussex.ac.uk

Rijsdijk Case
SAAO
PO Box 9
7935 Observatory
South Africa
☎ 27 21 47 0025
✆ 27 21 47 3639
✉ case@saao.ac.za

Riley Julia M
MRAO
Cavendish Laboratory
Madingley Rd
Cambridge CB3 0HE
UK
☎ 44 12 23 337 274
✆ 44 12 23 354 599

Rindler Wolfgang
Univ of Texas
UTD
Box 830688
Richardson TX 75083 0688
USA
☎ 1 214 690 2885

Ring James
Astrophysics Gr
Imperial Coll
Prince Consort Rd
London SW7 2BZ
UK
☎ 44 171 594 7771
✆ 44 171 594 7772

Ringnes Truls S
Inst Theor Astrophys
Univ of Oslo
Box 1029
N 0315 Blindern Oslo 3
Norway
☎ 47 22 856 503
✆ 47 22 85 6505

Ringuelet Adela E
Observ Astronomico
Paseo d Bosque S/n
1900 La Plata (Bs As)
Argentina
☎ 54 21 217 308
✆ 54 21 211 761

Ringwald Frederick Arthur
Dpt Astron/Astrophys
Pennsylvania State Univ
525 Davey Lab
University Park PA 16802 6305
USA
☎ 1 814 863 1756
℻ 1 814 863 3399
✉ ringwald@astro.
psu.edu

Ripken Hartmut W
Dara German
Space Agency
Koenigwintererstr 522-524
D 53227 Bonn
Germany
☎ 49 228 447 203
℻ 49 228 447 700
✉ ripken@t-online.de

Ritter Hans
MPA
Karl Schwarzschildstr 1
D 85748 Garching
Germany
☎ 49 893 299 880
℻ 49 893 299 3235

Rivolo Arthur Rex
Dpt Astron/Astrophys
Univ of Pennsylvania
Philadelphia PA 19104
USA
☎ 1 215 898 6250
℻ 1 215 898 9336

Rizvanov Naufal G
Engelhardt Astronom
Observ
Observatoria Station
422526 Tatarstan
Russia
☎ 7 32 4827
✉ rng@astro.ksu.ras.ru

Rizzo Jose Ricardo
IAR
CC 5
1894 Villa Elisa (Bs As)
Argentina
☎ 54 21 254 909
℻ 54 21 254 909
✉ rizzo@irma.ar
jrizzo@volta.ing.unlp.edu.ar

Roark Terry P
Dpt Phys/Astronomy
Univ of Wyoming
Box 3905
Laramie WY 82071
USA
☎ 1 307 766 6513
℻ 1 307 766 2652
✉ troark@uwyo.edu

Robb Russell M
Dpt Phys/Astronomy
Univ of Victoria
Box 3055
Victoria BC V8W 3P6
Canada
☎ 1 250 721 7750
℻ 1 250 721 7715
✉ robb@uvic.ca

Robbins R Robert
Astronomy Dpt
Univ of Texas
Rlm 15 308
Austin TX 78712 1083
USA
☎ 1 512 471 7312
℻ 1 512 471 6016

Robe H A G
Inst Astrophysique
Universite Liege
Av Cointe 5
B 4000 Liege
Belgium
☎ 32 4 254 7510
℻ 32 4 254 7511

Roberge Wayne G
Physics Dpt
Rensselaer Polytechn Inst
Troy NY 12180 3590
USA
☎ 1 518 276 6454
℻ 1 518 276 6680
✉ roberge@orion.phys.
rpi.edu

Roberti Giuseppe
Dpt Fisica
Univ d Napoli
Mostra D Oltremare Pad 19
I 80125 Napoli
Italy
☎ 39 81 725 3428

Roberts David Hall
Physics Dpt
Brandeis Univ
Waltham MA 02254
USA
☎ 1 617 647 2846
℻ 1 781 736 2915

Roberts Morton S
NRAO
520 Edgemont Rd
Charlottesville VA 22903
USA
☎ 1 804 296 0233
℻ 1 804 296 0278
✉ mroberts@nrao.edu

Roberts William W
Applied Maths Dpt
Box 3818
Charlottesville VA 22903
USA
☎ 1 804 924 1038

Robertson Douglas S
C/o CIRES
Univ of Colorado
Campus Box 216
Boulder CO 80309 0216
USA

Robertson James Gordon
School of Physics
UNSW
Sydney NSW 2006
Australia
☎ 61 2 935 1825
℻ 61 2 9351 7726
✉ jgr@physics.usyd.
edu.au

Robertson John Alistair
Dpt Applied Maths
Univ of St Andrews
North Haugh
St Andrews Fife KY16 9SS
UK
☎ 44 1334 76161
℻ 44 1334 74487

Robertson Norna
Dpt Phys/Astronomy
Univ of Glasgow
Glasgow G12 8QQ
UK
☎ 44 141 339 8855
℻ 44 1413 349 029
✉ gw05@uk.ac.gla.ph.il

Robillot Jean-Maurice
Observ Bordeaux
BP 89
F 33270 Floirac
France
☎ 33 5 56 86 4330
℻ 33 5 56 40 4251

Robin Annie C
Observ Besancon
41bis Av Observatoire
BP 1615
F 25010 Besancon Cdx
France
☎ 33 3 81 66 6941
℻ 33 3 81 66 6944
✉ robin@obs-besancon.fr

Robinson Andrew
Physical Sci Div
Univ Hertfordshire
College Lane
Hatfield Herts AL10 9AB
UK
☎ 44 170 728 5253
℻ 44 170 728 5179
✉ ar@star.herts.ac.uk

Robinson Brian J
Box 256
Milsons Point NSW 2061
Australia

Robinson Edward Lewis
Astronomy Dpt
Univ of Texas
Rlm 15 308
Austin TX 78712 1083
USA
☎ 1 512 471 3401
℻ 1 512 471 6016
✉ elr@astron.as.
utexas.edu

Robinson Garry
Physics Dpt Univ College
UNSW
Northcott Dr
Campbell ACT 2600
Australia
☎ 61 2 62 68 8800
℻ 61 26 268 8786
✉ garry@phadfa.ph.
adfa.oz.au

Robinson I
Univ of Texas
MS Be 32
Box 830688
Richardson TX 75080 0688
USA
☎ 1 214 690 2176

Robinson Leif J
Sky/Telescope
49 Bay State Rd
Cambridge MA 02238
USA
☎ 1 617 864 7360
℻ 1 617 253 9798

Robinson Lloyd B
Lick Observ
Univ of California
Santa Cruz CA 95064
USA
☎ 1 831 429 2437
℻ 1 831 426 3115

Robinson William J
18 Hollingwood Rise
Ilkley
W Yorkshire LS29 9PW
UK

Robinson Richard D
NASA GSFC
Code 681
Greenbelt MD 20771
USA
☎ 1 301 286 8701

Robley Robert
9 All Verdier
F 31000 Toulouse
France
☎ 33 5 61 52 2273

Robson Ian E
Joint Astronomy Ctr
660 N A'ohoku Pl
University Park
Hilo HI 96720
USA
☎ 1 808 961 3756
℻ 1 808 961 6516

Roca Cortes Teodoro
IAC
Observ d Teide
via Lactea s/n
E 38200 La Laguna
Spain
☎ 34 2 232 9100
℻ 34 2 232 9117

Rocca-Volmerange Brigitte
IAP
98bis bd Arago
F 75014 Paris
France
☎ 33 1 44 32 8091
℻ 33 1 44 32 8001

Roche Patrick F
Astrophysics Dpt
Univ of Oxford
Keble Rd
Oxford OX1 3RH
UK
☎ 44 186 527 3338
℻ 44 186 527 3418
✉ pfr@uk.ac.oxford.astrphysics

Rochester Michael G
Dpt Earth Sciences
Memorial Univ of
Newfoundland
St Johns NF A1B 3X57
Canada
☎ 1 709 737 7565
℻ 1 709 737 2589
✉ mrochest@kean.
ucs.mun.ca

Roddier Claude
NOAO/ADP
Box 26732
950 N Cherry Av
Tucson AZ 85726 6732
USA
☎ 1 520 325 9220
℻ 1 510 318 8360

Roddier Francois
NOAO/ADP
Box 26732
950 N Cherry Av
Tucson AZ 85726 6732
USA
☎ 1 520 325 9220
℻ 1 510 318 8360

Rodgers Alex W
MSSSO
Weston Creek
Private Bag
Canberra ACT 2611
Australia
☎ 61 262 881 111
℻ 61 262 490 233

Rodionova Janna F
SAI
Acad Sciences
Universitetskij Pr 13
119899 Moscow
Russia
☎ 7 095 939 1649
℻ 7 095 932 8841
✉ jeanna@sai.msu.su

Rodman Richard B
65 Locust Av
Lexington MA 02173
USA
☎ 1 617 861 8149

Rodono Marcello
Istt Astronomia
Citta Universitaria
Via A Doria 6
I 95125 Catania
Italy
☎ 39 95 733 21111
℻ 39 95 33 0592
✉ mrodono@alpha4.
ct.astro.it

Rodrigo Rafael
Inst Astrofisica
Andalucia Apt 3004
Prof Albareda 1
E 18080 Granada
Spain
☎ 34 5 812 1300
℻ 34 5 881 4530

Rodriguez Eloy
Inst Astrofisica
Andalucia Apt 3004
Prof Albareda 1
E 18080 Granada
Spain
☎ 34 5 812 1311
℻ 34 5 881 4530

Rodriguez Hildago Ines I
IAC
Observ d Teide
via Lactea s/n
E 38200 La Laguna
Spain
☎ 34 2 232 9100
℻ 34 2 232 9117
✉ irh@iac.es

Rodriguez Luis F
Instituto Astronomia
UNAM
Apt 70 264
Mexico DF 04510
Mexico
☎ 52 5 622 3906
℻ 52 5 616 0653

Rodriguez-Eillamil R
ROA
Cecilio Pujazon 22-3 A
E 11110 San Fernando
Spain
☎ 34 5 688 3548
℻ 34 5 689 9302

Rodriguez-Espinosa Jose M
IAC
Observ d Teide
via Lactea s/n
E 38200 La Laguna
Spain
☎ 34 2 232 9100
℻ 34 2 232 9117
✉ jre@ll.iac.es

Rodriguez-Franco Arturo
Observ Astronomico
Nacional
Apt 1143
E 28800 Alcala d Henares
Spain
☎ 34 1 885 5060
℻ 34 1 885 5062
✉ arturo@oan.es

Roelfsema Peter
Kapteyn Sterrekundig Inst
Univ Groningen
Postbus 800
NL 9700 AV Groningen
Netherlands
☎ 31 50 363 4073
℻ 31 50 363 6100
✉ pjutr@hgrrug

Roemer Elizabeth
Lunar/Planetary Lab
Room 325 Bg 92
Univ of Arizona
Tucson AZ 85721 0092
USA
☎ 1 520 621 2897
℻ 1 520 621 4933
✉ eroemer@pirl.lpl.arizona.edu

Roemer Max
IAEF
Univ Bonn
auf d Huegel 71
D 53121 Bonn
Germany
☎ 49 228 73 3670
℻ 49 228 73 3672

Roennaeng Bernt O
Onsala Space Observ
Chalmers Technical Univ
S 439 92 Onsala
Sweden
☎ 46 31 772 5500
℻ 46 31 772 5550

Roeser Hans-peter
DLR
Inst f Planetenerkundg
Rudower Chaussee 5
D 12489 Berlin
Germany
☎ 49 306 954 5500
℻ 49 306 954 5502
✉ hans-peter.roeser@
dlr.de

Roeser Hermann-Josef
MPI Astronomie
Koenigstuhl
D 69117 Heidelberg
Germany
☎ 49 622 152 8206
℻ 49 622 152 8246

Roeser Siegfried
ARI
Moenchhofstr 12-14
D 69120 Heidelberg
Germany
☎ 49 622 140 5158
℻ 49 622 140 5297
✉ s19@ix.urz.uni-
heidelberg.de

Roessiger Siegfried
Zntrlinst Astrophysik
Sternwarte Sonneberg
an d Sternwarte 16
D 96575 Sonneberg
Germany
☎ 49 96 74 2287
℻ 49 96 74 2836

Roger Robert S
DRAO
NRCC/HIA
Box 248
Penticton BC V2A 6K3
Canada
☎ 1 250 493 2277
℻ 1 250 493 7767

Rogers Alan E E
Haystack Observ
Westford MA 01886
USA
☎ 1 617 692 4764
℻ 1 617 981 0590

Rogers Christopher
DRAO
NRCC/HIA
Box 248
Penticton BC V2A 6K3
Canada
☎ 1 250 493 2277
℻ 1 250 493 7767
✉ crogers@drao.nrc.ca

Rogers Forrest J
LLNL
Box 808
Livermore CA 94551 9900
USA
☎ 1 415 422 7351
℻ 1 925 423 7228
✉ rogers4@llnl.gov

Rogerson John B
Dpt Astrophysical Sci
Princeton Univ
Peyton Hall
Princeton NJ 08544 1001
USA
☎ 1 609 452 3806
℻ 1 609 258 1020

Rogstad David H
CALTECH/JPL
MS 11/116
4800 Oak Grove Dr
Pasadena CA 91109 8099
USA
☎ 1 213 351 1321
℻ 1 818 393 6030

Rohlfs Kristen
Inst Theor Physik
Ruhr Univ Bochum
Postfach 102148
D 44780 Bochum
Germany
☎ 49 234 700 5802
℻ 49 234 709 4169

Roland Ginette
Inst Astrophysique
Universite Liege
Av Cointe 5
B 4000 Liege
Belgium
☎ 32 4 254 7510
℻ 32 4 254 7511
✉ roland@astro.ulg.ac.be

Rolland Angel
Inst Astrofisica
Andalucia Apt 3004
Prof Albareda 1
E 18080 Granada
Spain
☎ 34 5 812 1300
℻ 34 5 881 4530

Roman Nancy Grace
4260 N Park Av
Apt 306w
Chevy Chase MD 20815
USA
☎ 1 301 656 6092
℻ 1 301 286 1771
✉ nancy.g.roman@gsfc.nasa.gov

Romanchuk Pavel R
Astronomical Observ
Kyiv State Univ
Observatornaya Ul 3
252053 Kyiv
Ukraine
☎ 380 4421 62691

Romani Roger William
Physics Dpt
Stanford Univ
Stanford CA 94305
USA
☎ 1 415 725 7595
✆ 1 415 723 9389
✉ rwr@geminga.
stanford.edu

Romanishin William
Dpt Phys/Astronomy
Univ of Oklahoma
Norman OK 73019
USA
☎ 1 405 325 3961
✆ 1 405 325 7557

Romano Giuliano
V S Antonio d Padova 7
I 31100 Treviso
Italy

Romanov Yuri S
Astronomical Observ
Odessa State Univ
Shevchenko Park
270014 Odessa
Ukraine
☎ 380 4822 20396
✆ 380 4822 28442
✉ root@astro.odessa.ua

Romero Gustavo Esteban
IAR
CC 5
1894 Villa Elisa (Bs As)
Argentina
☎ 54 21 254 909
✆ 54 21 254 909
✉ romero@irma.iar.
unlp.edu.ar

Romero Perez M Pilar
Inst Astron/Geodesia
Fac Ciencias Matemat
Univ Complutense
E 28040 Madrid
Spain
☎ 34 1 244 2501
✆ 34 1 394 5195

Romney Jonathan D
NRAO
Box 0
Socorro NM 87801 0387
USA
☎ 1 505 772 4011
✆ 1 505 835 7027

Rompolt Bogdan
Astronomical Institute
Wroclaw Univ
Ul Kopernika 11
PL 51 622 Wroclaw
Poland
☎ 48 71 372 9373/74
✆ 48 71 372 9378

Ron Cyril
Astronomical Institute
Czech Acad Sci
Bocni Ii 1401
CZ 141 31 Praha 4
Czech R
☎ 420 2 67 10 3030
✆ 420 2 76 90 23
✉ ron@ig.cas.cz

Ronan Colin A
Flat 6 Bourne Ct The Bourne
Hastings
East Sussex TN34 3UZ
UK
☎ 44 142 444 6362

Roncin Jean-Yves
Lab Traitement du Signal
et Instrumentation
23 r Dr P Michelon
F 42023 Saint Etienne
France
☎ 33 4 77 48 5168
✆ 33 4 77 48 5120
✉ roncin@univ-
st-etienne.fr

Rong Jianxiang
Astronomy Dpt
Nanjing Univ
Nanjing 210093
China PR
☎ 86 25 34651*2882
✆ 86 25 330 2728

Rood Herbert J
Inst Advanced Study
School Natural Science
Princeton NJ 08540
USA
☎ 1 609 734 8055

Rood Robert T
University Station
Univ of Virginia
Box 3818
Charlottesville VA 22903 0818
USA
☎ 1 804 924 4904
✉ rtr@virginia.edu

Roos Matts
Helsinky Univ
Observ
Box 14
FIN 00014 Helsinki
Finland
☎ 358 19 12 2940
✆ 358 191 8366
✉ mroos@phcu.
helsinki.fi

Roos Nicolaas
Leiden Observ
Box 9513
NL 2300 RA Leiden
Netherlands
☎ 31 71 527 5864
✆ 31 71 527 5819
✉ roos@strw.Leidenuniv.nl

Roosen Robert G
6208 Univ Av
Suite A
San Diego CA 92115
USA
☎ 1 619 229 1449
✉ roosen@ax.com

Roques Francoise
Observ Paris Meudon
DESPA
Pl J Janssen
F 92195 Meudon PPL Cdx
France
☎ 33 1 45 07 7409
✆ 33 1 45 07 7469
✉ roques@obspm.fr

Roques Sylvie
OMP
14 Av E Belin
F 31400 Toulouse Cdx
France
☎ 33 5 61 25 2101
✆ 33 5 61 27 3179

Ros Rosa M
Dpt Mat Aplicada
U P Cataluna
Victor Balaguer Sn
E 08800 Vilanova
Spain
☎ 34 3 896 7720
✆ 34 3 896 7700
✉ ros@mat.upc.es

Rosa Dorothea
Heidelbergerstr 31 B
D 69151 Neckargemuend
Germany
✉ mrosa@eso.org

Rosa Dragan
Astronomical Observ
Opaticka 22
pp 943
HR 10000 Zagreb
Croatia
☎ 385 1 442 271
✆ 385 1 422 271
✉ dragen.rosa@public.
srce.hr

Rosa Michael Richard
ESO
ST/ECF
Karl Schwarzschildstr 2
D 85748 Garching
Germany
☎ 49 893 200 60
✆ 49 893 202 362

Rosado Margarita
Instituto Astronomia
UNAM
Apt 70 264
Mexico DF 04510
Mexico
☎ 52 5 622 3906
✆ 52 5 616 0653

Rosch Jean
OMP
9 r Pont de La mouette
F 65204 Bagneres Bigorre
France
☎ 33 5 62 95 1969
✆ 33 3 88 25 0160

Rose James Anthony
Dpt Phys/Astronomy
Univ North Carolina
204 Phillips Hall 039a
Chapel Hill NC 27514
USA
☎ 1 919 962 7170
✆ 1 919 962 0480

Rose William K
Astronomy Program
Univ of Maryland
College Park MD 20742 2421
USA
☎ 1 301 299 2777
✆ 1 301 314 9067

Rosen Edward
Dpt History
City College of NY
138 St Convene Av
New York NY 10031
USA
☎ 1 212 690 6823

Rosendhal Jeffrey D
NASA Headquarters
Code SZ
600 Independence Av SW
Washington DC 20546
USA
☎ 1 202 358 0738
✆ 1 202 358 3096
✉ jrosendhal@smtpgmgw.ossa.
hq.nasa.gov

Roslund Curt
Dpt Astron/Astrophys
Chalmers Technical Univ
S 412 96 Goeteborg
Sweden
☎ 46 31 772 1000
✆ 46 31 772 3204

Rosner Robert
Astronomy/Astrophys Ctr
Univ of Chicago
5640 S Ellis Av
Chicago IL 60637
USA
☎ 1 312 702 0560
✆ 1 312 702 8212
✉ rreosner@oddjob.uchicago.edu

Rosquist Kjell
Inst Theoret Phys
Vanadisvaegen 9
S 113 46 Stockholm
Sweden
☎ 46 82 28 160*225

Ross Dennis K
Physics Dpt
Iowa State Univ
Ames IA 50011
USA
☎ 1 515 294 6010
✆ 1 515 294 3262

Ross John E R
Physics Dpt
Univ of Queensland
St Lucia
Brisbane QLD 4072
Australia
☎ 61 7 3365 3429
✆ 61 7 371 5896
✉ ross@physics.uq.oz.au

Rossello Gaspar
Dpt Fisica Atmosfera
Univ Barcelona
Avd Diagonal 645
E 08028 Barcelona
Spain
☎ 34 3 402 1125
✆ 34 3 402 1133

Rossi Corinne
Istt Astronomico
Univ d Roma La Sapienza
Via G M Lancisi 29
I 00161 Roma
Italy
☎ 39 6 44 03734
✆ 39 6 44 03673

Rossi Lucio
IAS
Area d Ricerca CNR
Via Fosso Cavaliere 100
I 00133 Roma
Italy
☎ 39 6 4993 4473
✆ 39 6 2066 0188

Rostas Francois
Observ Paris Meudon
DAMAP
Pl J Janssen
F 92195 Meudon PPL Cdx
France
☎ 33 1 45 07 7565
✆ 33 1 4507 7469
✉ rostas@obspm.fr

Roth Miguel R
CIW
Las Campanas Observ
Casilla 601
La Serena
Chile
☎ 56 51 21 3032
✉ mroth@uchcevm

Roth-Hoppner Maria Luise
Hamburger Sternwarte
Univ Hamburg
Gojensbergsweg 112
D 21029 Hamburg
Germany
☎ 49 407 252 4112
✆ 49 407 252 4198

Rothenflug Robert
DAPNIA/SAP
CEA Saclay
BP 2
F 91191 Gif s Yvette Cdx
France
☎ 33 1 69 08 4327
✆ 33 1 69 08 6577
✉ rothenflug@32779.
decnet.cern

Rots Arnold H
CfA
HCO/SAO MS 81
60 Garden St
Cambridge MA 02138 1516
USA
☎ 1 617 495 7701
✆ 1 617 495 7356
✉ arots@head-cfa.harvard.edu

Rouan Daniel
Observ Paris Meudon
DESPA
Pl J Janssen
F 92195 Meudon PPL Cdx
France
☎ 33 1 45 07 7715
✆ 33 1 45 07 2806
✉ rouan@obspm.fr

Roudier Thierry
OMP
14 Av E Belin
F 31400 Toulouse Cdx
France
☎ 33 5 61 25 2101
✆ 33 5 61 27 3179

Roueff Evelyne M A
Observ Paris Meudon
DAF
Pl J Janssen
F 92195 Meudon PPL Cdx
France
☎ 33 1 45 07 7435
✆ 33 1 45 07 7435

Rountree Janet
Box 65285
Tucson AZ 85728
USA
✉ rountree@nssdca.
gsfc.nasa.gov

Rouse Carl A
627 15 Th Str
Del Mar CA 92014
USA
☎ 1 619 455 4015

Rousseau Jean-Michel
Observ Bordeaux
BP 89
F 33270 Floirac
France
☎ 33 5 56 86 4330
✆ 33 5 56 40 4251

Rousseau Jeanine
Observ Lyon
Av Ch Andre
F 69561 S Genis Laval cdx
France
☎ 33 4 78 56 0705
✆ 33 4 72 39 9791

Rousselot Philippe
Observ Besancon
41bis Av Observatoire
BP 1615
F 25010 Besancon Cdx
France
☎ 33 3 81 66 6900
✆ 33 3 81 66 6944
✉ philippe@obs-
besancon.fr

Routledge David
Dpt Electric Engin
Univ of Alberta
Edmonton AB T6G 2J1
Canada
☎ 1 403 432 5668

Routly Paul M
USNO
3450 Massachusetts Av NW
Washington DC 20392 5100
USA
☎ 1 202 762 1434
✆ 1 202 762 1516

Rovero Adrian Carlos
IAFE
CC 67 Suc 28
1428 Buenos Aires
Argentina
☎ 54 1 781 6755
✆ 54 1 786 8114
✉ rovero@iafe.uba.ar

Rovira Marta Graciela
IAFE
CC 67 Suc 28
1428 Buenos Aires
Argentina
☎ 54 1 781 6755
✆ 54 1 786 8114

Rovithis Peter
Astronomical Institute
Ntl Observ Athens
Box 20048
GR 118 10 Athens
Greece
☎ 30 1 346 3803
✆ 30 1 346 4566

Rovithis-Livaniou Helen
Astrophysics Dpt
Ntl Univ Athens
Panepistimiopolis
GR 157 84 Zografos
Greece
☎ 30 1 724 3414
✆ 30 1 723 5122

Rowan-Robinson Michael
Astrophysics Gr
Imperial Coll
Prince Consort Rd
London SW7 2BZ
UK
☎ 44 171 594 7771
✆ 44 171 594 7772

Rowson Barrie
21 Buttermere Rd
Gatley SK8 4RQ
UK
☎ 44 16 14 284306
✉ barrie_Rowson@
Compuserve.com

Roxburgh Ian W
School Mathemat Sc
QMWC
Mile End Rd
London E1 4NS
UK
☎ 44 171 980 4811
✆ 44 181 975 5500

Roy Archie E
Astronomy Dpt
Univ of Glasgow
Glasgow G12 8QQ
UK
☎ 44 141 339 8855*502
✆ 44 141 334 9029

Roy Jean-Rene
Dpt d Physique
Universite Laval
Fac Sciences/Genie
Laval QC G1K 7P4
Canada
☎ 1 418 656 5816
✆ 1 418 656 2040
✉ jrroy@phy.ulaval.ca

Rozas Maite
IAC
Observ d Teide
via Lactea s/n
E 38200 La Laguna
Spain
☎ 34 2 232 9100
✆ 34 2 232 9117
✉ mrozas@ll.iac.es

Rozelot Jean-Pierre
OCA
CERGA
F 06130 Grasse
France
☎ 33 4 93 36 5849
✆ 33 4 93 40 5353

Rozhkovskij Dimitrij A
Astrophys Inst
Kazakh Acad Sci
480068 Alma Ata
Kazakhstan
☎ 7 62 4040

Rozyczka Michal
Copernicus Astron Ctr
Polish Acad Sci
Ul Bartycka 18
PL 00 716 Warsaw
Poland
☎ 48 22 41 1086
✆ 48 22 41 0828
✉ mnr@camk.edu.pl

Ruben G
Sternwarte Babelsberg
Kantstr 6
D 14473 Potsdam
Germany

Rubin Robert Howard
NASA/ARC
MS 245 6
Moffett Field CA 94035 1000
USA
☎ 1 415 604 6450
✆ 1 415 604 6779
✉ rubin@cygnus.arc.nasa.gov

Rubin Vera C
Dpt Terrestr Magnetism
Carnegie Inst Washington
5241 Bd Branch Rd NW
Washington DC 20015 1305
USA
☎ 1 202 966 0863
✆ 1 202 364 8726
✉ rubin@gal.ciw.edu

Rubio Monica
Dpt Astronomia
Univ Chile
Casilla 36 D
Santiago
Chile
☎ 56 2 229 4101
✆ 56 2 229 4002

Rucinski Daniel
Space Res Ctr
Polish Acad Sci
Ul Bartycka 18
PL 00 716 Warsaw
Poland
☎ 48 22 40 3766
✆ 48 22 36 8961
✉ rucinski@cbk.
waw.pl

Rucinski Slavek M
CFHT Corp
Box 1597
Kamuela HI 96743
USA
☎ 1 808 885 3161
✆ 1 808 885 7288

Rudak Bronislaw
NCAC
Polish Acad Sciences
Ul Rabianska 8
PL 87 100 Torun
Poland
☎ 48 56 19 319
✆ 48 56 19 381
✉ bronek@ncac.
torun.pl

Ruder Hanns
Lehrstuhl Theoret Astro
Physik Univ Tuebingen
Auf d Morgenstelle 12 C
D 72076 Tuebingen
Germany
☎ 49 707 129 2487
✆ 49 7071 29 7575

Ruderman Malvin A
Physics Dpt
Columbia Univ
538 W 120th St
New York NY 10027
USA
☎ 1 212 280 3317

Rudkjobing Mogens
Inst Phys/Astronomy
Univ of Aarhus
Ny Munkegade
DK 8000 Aarhus C
Denmark
☎ 45 86 12 8899
✆ 45 86 20 2711

Rudnick Lawrence
School Phys/Astronomy
Univ Minnesota
116 Church St SE
Minneapolis MN 55455
USA
☎ 1 612 624 3396
✆ 1 612 624 2029
✉ larry@astro.spa.
umn.edu

Rudnicki Konrad
Astronomical Observ
Krakow Jagiellonian Univ
Ul Orla 171
PL 30 244 Krakow
Poland
☎ 48 12 25 1294
✆ 48 12 25 1318

Rudnitskij Georgij M
SAI
Acad Sciences
Universitetskij Pr 13
119899 Moscow
Russia
☎ 7 095 939 1030
✆ 7 095 939 1661
✉ gmr@sai.msk.su

Rudzikas Zenonas B
Inst Theor Physics/
Astronomy
Gostauto 12
Vilnius 2600
Lithuania
☎ 370 2 620 939
✆ 370 2 224 694
✉ astro@itpa.fi.lt

Ruediger Guenther
Astrophysik Inst
Potsdam Univ
an d Sternwarte 16
D 14482 Potsdam
Germany
☎ 49 331 749 9512
✉ gruediger@aip.de

Ruelas-Mayorga R A
Instituto Astronomia
UNAM
Apt 70 264
Mexico DF 04510
Mexico
☎ 52 5 622 3906
✆ 52 5 616 0653
✉ rarm@astroscu.
unam.mx

Rufener Fredy G
Observ Geneve
Chemin d Maillettes 51
CH 1290 Sauverny
Switzerland
☎ 41 22 755 2611
✆ 41 22 755 3983

Ruffert Maximilian
Dpt Maths/Statistics
Univ of Edinburgh
Edinburgh EH9 3JZ
UK
☎ 44 131 650 5039
✆ 44 131 650 6553
✉ max@maths.ed.
ac.uk

Ruffini Remo
Dpt Fisica
Univer d Roma
Pl A Moro 2
I 00185 Roma
Italy
☎ 39 6 49 76304
✆ 39 6 202 3507

Rugge Hugo R
Aerospace Corp
MS M2 226
Box 92957
Los Angeles CA 90009 2957
USA
☎ 1 310 648 7086
✆ 1 310 336 1636

Ruiz Maria Teresa
Dpt Astronomia
Univ Chile
Casilla 36 D
Santiago
Chile
☎ 56 2 229 4101
✆ 56 2 229 4002

Rule Bruce H
Hale Observatories
2205 Monte Vista St
Pasadena CA 91107
USA
☎ 1 818 794 6593

Rumsey Norman J
21 Malone Rd
Lower Hutt
New Zealand
☎ 64 4 69 6787

Rupprecht Gero
ESO
Karl Schwarzschildstr 2
D 85748 Garching
Germany
☎ 49 893 200 6355
✆ 49 893 202 362
✉ grupprec@eso.org

Ruprecht Jaroslav
Astronomical Institute
Czech Acad Sci
Bocni Ii 1401
CZ 141 31 Praha 4
Czech R
☎ 420 2 67 10 3041
✆ 420 2 76 90 23
✉ astdss@csearn

Rusconi Luigia
Dpt Astronomia
Univ d Trieste
Via Tiepolo 11
I 34131 Trieste
Italy
☎ 39 40 79 4863
✆ 39 40 30 9418

Rusin Vojtech
Astronomical Institute
Slovak Acad Sci
SK 059 60 Tatranska Lomni
Slovak Republic
☎ 421 969 96 7866
✆ 421 969 96 7656

Ruskol Eugenia L
Inst Phys of the Earth
Acad Sciences
Gruzinskaya 10
123342 Moscow
Russia
☎ 7 095 252 0726

Russell Christopher T
Inst Geophys/Planets
UCLA
Los Angeles CA 90095 1567
USA
☎ 1 310 825 3188
✆ 1 310 206 3051
✉ ctrussell@igpp.ucla.edu

Russell Jane L
NSF
Div Astron Sci
4201 Wilson Blvd
Arlington VA 22230
USA
☎ 1 703 306 1827
✆ 1 703 306 0525
✉ jrussell@rosserv.gsfc.nasa.gov

Russell John A
Dpt Phys/Astronomy
USC
Los Angeles CA 90089 1342
USA
☎ 1 213 743 0231
✆ 1 213 740 7254

Russell Kenneth S
AAO
UK Schmidt Telescope
Private Bag
Coonabarabran NSW 2357
Australia
☎ 61 68 426 311
✆ 61 68 846 298

Russell Stephen
DIAS
School Cosmic Phys
5 Merrion Sq
Dublin 2
Ireland
☎ 353 1 662 1333
✆ 353 1 682 003
✉ sr@dias.ie

Russev Ruscho
Astronomy Dpt
Univ of Sofia
Anton Ivanov St 5
BG 1126 Sofia
Bulgaria
☎ 359 2 68 8176
✆ 359 2 68 9085

Russeva Tatjana
Astronomy Dpt
Bulgarian Acad Sci
72 Lenin Blvd
BG 1784 Sofia
Bulgaria
☎ 359 2 75 8927
✆ 359 2 75 5019

Russo Guido
Dpt Scienze Fisiche
Univ d Napoli
Mostra D Oltremare Pad 19
I 80125 Napoli
Italy
☎ 39 81 725 3447
✉ russo@napoli.infn.it

Rust David M
APL
JHU
Johns Hopkins Rd
Laurel MD 20723 6099
USA
☎ 1 301 953 5000
✆ 1 301 953 1093
✉ David.rust@jhuapl.edu

Rusu I
Astronomical Institute
Romanian Acad Sciences
Cutitul de Argint 5
RO 75212 Bucharest
Rumania
☎ 40 1 335 6892
✆ 40 1 312 3391

Rusu L
Str Mitropolitul Iosif 47
RO 75217 Bucharest
Rumania

Rutten Renee G M
Royal Greenwich Obs
Apt 321
Santa Cruz de La Palma
E 38780 Santa Cruz
Spain
☎ 34 2 240 5500
✆ 34 2 240 5646/405 501

Rutten Robert J
Sterrekundig Inst Utrecht
Box 80000
NL 3508 TA Utrecht
Netherlands
☎ 31 30 253 5200
✆ 31 30 253 1601
✉ R.J.Rutten@astro.uu.nl

Ruzdjak Vladimir
Inst of Physics
Univ of Zagreb
Box 304
HR 10000 Zagreb
Croatia
☎ 385 1 271 211

Ruzickova-Topolova B
Astronomical Institute
Czech Acad Sci
Fricova 1
CZ 251 65 Ondrejov
Czech R
☎ 420 204 85 201
✆ 420 2 88 1611

Ryabchikova Tanya
Inst of Astronomy
Acad Sciences
Pyatnitskaya Ul 48
109017 Moscow
Russia
☎ 7 095 231 5461
✆ 7 095 230 2081

Ryabov Yu A
Mathematics Dpt
MADI
Leningradsky Pr 64
125319 Moscow
Russia
☎ 7 095 155 0326

Ryan Sean Gerard
AAO
Box 296
Epping NSW 2121
Australia
☎ 61 2 9372 4843
✆ 61 2 9372 4880
✉ sgr@aaoepp.aao.gov.au

Rybansky Milan
Astronomical Institute
Slovak Acad Sci
SK 059 60 Tatranska Lomni
Slovak Republic
☎ 421 969 96 7866
✆ 421 969 96 7656

Rybicki George B
CfA
HCO/SAO
60 Garden St
Cambridge MA 02138 1516
USA
☎ 1 617 495 7452
✆ 1 617 495 7093
✉ grybicki@cfa.harvard.edu

Rydbeck Gustaf H B
Onsala Space Observ
Chalmers Technical Univ
S 439 92 Onsala
Sweden
☎ 46 31 772 5500
✆ 46 31 772 5550

Rydbeck Olof E H
Onsala Space Observ
Chalmers Technical Univ
S 439 92 Onsala
Sweden
☎ 46 31 772 5500
✆ 46 31 772 5550

Ryder Stuart
Joint Astronomy Ctr
660 N A'ohoku Pl
University Park
Hilo HI 96720
USA
☎ 1 808 961 3756
✆ 1 808 961 6516
✉ sryder@jach.hawaii.edu

Rydgren Alfred Eric
Boeing Aerospace Co
MS 87 08
Box 3999
Seattle WA 98124 2499
USA
☎ 1 425 773 2155

Rykhlova Lidija V
Inst of Astronomy
Acad Sciences
Pyatnitskaya Ul 48
109017 Moscow
Russia
☎ 7 095 231 5461
✆ 7 095 230 2081

Rys Stanislaw
Astronomical Observ
Krakow Jagiellonian Univ
Ul Orla 171
PL 30 244 Krakow
Poland
☎ 48 12 25 1294
✆ 48 12 25 1318
✉ strys@oa.uj.edu.pl

Ryu Dongsu
Dpt Astronomy/Space Sci
Chungnam Ntl Univ
Daejeon
Korea RP
☎ 82 428 21 5466
✆ 82 428 22 8380
✉ ryu@sirius.chungnam.ac.kr

Ryutova Margarita P
Inst Theor Astronomy
Acad Sciences
Siberian Div
630090 Novosibirsk
Russia
☎ 7 383 235 9943
✆ 7 383 235 2163
✉ ryutova@vxinpb.inp.nsk.su

Rzhiga Oleg N
Inst Radio/Electron
Acad Sciences
Marx Av 18
103907 Moscow
Russia

S

Saar Enn
Tartu Observ
Estonian Acad Sci
Toravere
EE 2444 Tartumaa
Estonia
☎ 372 7 416 625
✆ 372 7 410 205
✉ saar@aai.tartu.ew.su

Sabano Yutaka
Astronomical Institute
Tohoku Univ
Sendai Aoba
Miyagi 980
Japan
☎ 81 222 22 1800
✆ 81 22 262 6609

Sabau-Graziati Lola
Inst Nacional Tec
Aeroespacial
Ctr Aljarvir km 4 500
E 28850 Torrejon d Ardoz
Spain
☎ 34 1 520 1681
✆ 34 1 520 1945
✉ sabaumd@inta.es

Sabbadin Franco
Osserv Astrofisico
Univ d Padova
Via d Osservatorio 8
I 36012 Asiago
Italy
☎ 39 424 462 665
✆ 39 424 462 884

Sack Noam
Dpt Theor Phys
Hebrew Univ Jerusalem
Jerusalem 91904
Israel
☎ 972 2 658 4605
✆ 972 2 61 1519

Sackett Penny
Kapteyn Sterrekundig Inst
Univ Groningen
Postbus 800
NL 9700 AV Groningen
Netherlands
☎ 31 50 363 4073
✆ 31 50 363 6100
✉ psackett@astro.rug.nl

Sackmann Ingrid Juliana
CALTECH
Kellogg Radiation Lab
Pasadena CA 91125
USA
☎ 1 818 356 4256
✆ 1 818 564 8708

Sadakane Kozo
Astronomical Institute
Osaka Kyoiku Univ
4 88 Minamikawahoricho
Osaka 543
Japan
☎ 81 729 76 3211 *3124
✆ 81 729 76 3269

Sadat Rachida
Observ Strasbourg
11 r Universite
F 67000 Strasbourg
France
☎ 33 3 88 15 0757
✆ 33 3 88 25 0160
✉ sadat@astro.u-strasbg.fr

Sadeh D
Dpt Phys/Astronomy
Tel Aviv Univ
Ramat Aviv
Tel Aviv 69978
Israel
☎ 972 3 420 553
✆ 972 3 640 8179

Sadik Aziz R
SARC
Scientific Res Council
Box 2441
Jadiriyah Baghdad
Iraq
☎ 964 1 776 5127

Sadler Elaine Margaret
School of Physics
UNSW
Sydney NSW 2006
Australia
☎ 61 2 935 12622
✆ 61 2 9351 7726
✉ ems@physics.usyd.edu.au

Sadollah Nasiri Gheidari
Physics Dpt
Univ of Zanjan
Km6 Tabriz Rd 45195 313
Zanjan
Iran
☎ 98 2821 27001 4

Sadun Alberto Carlo
Univ of Colorado
Campus Box 157
Box 173364
Denver CO 80217 3364
USA
☎ 1 303 556 8344
✆ 1 303 556 6257
✉ asadun@castle.cudenver.edu

Sadzakov Sofija
Astronomical Observ
Volgina 7
11050 Beograd
Yugoslavia FR
☎ 381 1 419 357/421 875
✆ 381 1 419 553

Saemundson Thorsteinn
Raunvisindastofnun
Haskolans
Dunhaga 3
IS 107 Reykjavik
Iceland
☎ 354 525 4800
✆ 354 552 8911
✉ halo@ravnis.hi.is

Safko John L
Dpt Phys/Astronomy
Univ S Carolina
Columbia SC 29208
USA
☎ 1 803 777 6466

Safronov Victor S
Inst Phys of the Earth
Acad Sciences
Gruzinskaya 10
123242 Moscow
Russia
☎ 7 095 252 0726

Sagdeev Roald Z
Space Res Inst
Acad Sciences
Profsojuznaya Ul 84/32
117810 Moscow
Russia
☎ 7 095 333 14 66
✆ 7 095 310 7023

Sage Leslie John
Nature
968 Ntl Press Bldg
529 14th St NW
Washington DC 20045 1938
USA
☎ 1 202 737 2355
✆ 1 202 628 1609
✉ nature@naturedc.com

Saggion Antonio
Dpt Fisica G Galilei
Univ d Padova
Via Marzolo 8
I 35131 Padova
Italy
☎ 39 49 84 4254
✆ 39 49 84 4245

Saglia Roberto Philip
Inst Astron/Astrophysik
Univ Sternwarte
Scheinerstr 1
D 81679 Muenchen
Germany
☎ 49 899 220 9425
✆ 49 899 220 9427
✉ saglia@hal1.usm.uni-muenchen.de

Saha Abhijit
NOAO/KPNO
Box 26732
950 N Cherry Av
Tucson AZ 85726 6732
USA
☎ 1 520 318 8288
✆ 1 520 318 8360
✉ saha@noao.edu

Saha Swapan Kumar
IIA
Koramangala
Sarjapur Rd
Bangalore 560 034
India
☎ 91 80 356 9902
✆ 91 80 553 4043

Sahade Jorge
Observ Astronomico
Paseo d Bosque S/n
CC 677
1900 La Plata (Bs As)
Argentina
☎ 54 21 217 308
✆ 54 21 211 761/25 8985
✉ sahade@fcaglp.fcaglp.unlp.edu.ar

Sahai Raghvendra
CALTECH/JPL
MS 169 506
4800 Oak Grove Dr
Pasadena CA 91109 8099
USA
☎ 1 818 354 0452
✆ 1 818 354 8895
✉ sahai@jplsp.jpl.nasa.gov

Sahal-Brechot Sylvie
Observ Paris Meudon
DAMAP
Pl J Janssen
F 92195 Meudon PPL Cdx
France
☎ 33 1 45 07 7442
✆ 33 1 45 07 7469
✉ sahal@obspm.fr

Sahibov Firuz H
Astrophys Inst
Uzbek Acad Sci
Sviridenko Ul 22
734670 Dushanbe
Tajikistan
☎ 7 3770 23 1432
✆ 7 3770 27 5483

Sahni Varun
IUCAA
PO Box 4
Ganeshkhind
Pune 411 007
India
☎ 91 212 351 414/5
✆ 91 212 350 760
✉ varun@iucaa.
ernet.in

Sahu Kailash C
STScI
Homewood Campus
3700 San Martin Dr
Baltimore MD 21218
USA
☎ 1 301 338 4930
✆ 1 301 338 4796
✉ ksahu@stsci.edu

Saijo Keiichi
Dpt Physical Sci
National Science Museum
7-20 Ueno Park Taito Ku
Tokyo 110
Japan
☎ 81 382 20 111

Saikia Dhruba Jyoti
NCRA/TIFR
Pune Univ Campus Pb 3
Ganeshkhind
Pune 411 007
India
☎ 91 212 33 7107
✆ 91 212 33 5760

Saio Hideyuki
Astronomical Institute
Tohoku Univ
Sendai Aoba
Miyagi 980
Japan
☎ 81 22 222 1800*3327
✆ 81 22 262 6609
✉ saio@astroa.astr.
tohoku.ac.jp

Saissac Joseph
7 r Auguste Limouzi
F 11100 Narbonne
France

Saito Kuniji
Tokyo Astronomical Obs
NAOJ
Osawa Mitaka
Tokyo 181
Japan
☎ 81 422 32 5111
✆ 81 422 34 3793

Saito Mamoru
Astronomy Dpt
Kyoto Univ
Sakyo Ku
Kyoto 606 01
Japan
☎ 81 757 51 2111*3904
✆ 81 75 753 7010
✉ saitom@kusastro.
kyoto-u.ac.jp

Saito Sumisaburo
Kwasan/Hida Obs
Kyoto Univ
Yamashina
Kyoto 607
Japan
☎ 81 755 81 1235
✆ 81 755 93 9617

Saito Takao
Taihaku 3 6 29
Sendai 980
Japan
☎ 81 22 245 8437
✆ 81 222 45 8437

Saito Takao
Geophys Inst
Tohoku Univ
Aoba Aoba-ku
Sendai 980
Japan
☎ 81 22 268 4508
✆ 81 222 68 4508

Sakai Junichi
Fac of Eng
Toyama Univ
Toyama 930
Japan
☎ 81 764 41 1271

Sakai Shoko
NOAO/KPNO
Box 26732
950 N Cherry Av
Tucson AZ 85726 6732
USA
☎ 1 520 318 8255
✆ 1 520 318 8360
✉ shoko@noao.edu

Sakamoto Seiichi
Nobeyama Radio Obs
NAOJ
Minamimaki Mura
Nagano 384 13
Japan
☎ 81 267 984 396
✆ 81 267 984 339
✉ seiichi@nro.nao.ac.jp

Sakao Taro
Ntl Astronomical Obs
Osawa 2-21-1
Mitaka
Tokyo 181
Japan
☎ 81 422 34 3715
✆ 81 422 34 3700
✉ sakao@solar.mtk.nao.ac.jp

Sakashita Shiro
Physics Dpt
Hokkaido Univ
Kita 10 Nishi 8
Sapporo 060
Japan
☎ 81 117 16 2111
✆ 81 222 85 132

Sakhibov Firouz
Gartenfeldstr 41
D 61350 Bad Homburg
Germany
☎ 49 6172 93 7106
✆ 49 617 293 7106
✉ fsakhibov@metronet.de

Sakhibullin Nail A
Astronomy Dpt
Kazan State Univ
Lenin Ul 18
420008 Kazan
Russia
☎ 7 32 3641

Sakurai Kunitomo
Physics Dpt
Kanagawa Univ
Kanagawaku
Yokohama 221
Japan
☎ 81 454 81 5661

Sakurai Takashi
Tokyo Astronomical Obs
NAOJ
Osawa Mitaka
Tokyo 181
Japan
☎ 81 422 32 5111
✆ 81 422 34 3793

Sakurai Takeo T
1 10 2 Takadai
Nagaokakyo shi 617
Japan

Sala Ferran
Dpt Fisica Atmosfera
Univ Barcelona
Avd Diagonal 647
E 08028 Barcelona
Spain
☎ 34 3 330 7311
✆ 34 3 402 1133

Salas Luis
Observ Astronomico
Nacional
Box 439027
San Diego CA 92143 9027
USA
☎ 1 526 174 4580
✆ 1 52 617 44607
✉ salas@bufadora.
astrosen.unam.mx

Salazar Antonio
ROA
Cecilio Pujazon 22-3 A
E 11110 San Fernando
Spain
☎ 34 5 688 3548
✆ 34 5 689 9302
✉ ccgeneral@czv1.uca.es

Saletic Dusan
Astronomical Observ
Volgina 7
11050 Beograd
Yugoslavia FR
☎ 381 1 157 022
✆ 381 1 419 553

Salinari Piero
Osserv Astrofis Arcetri
Univ d Firenze
Largo E Fermi 5
I 50125 Firenze
Italy
☎ 39 55 27 52231
✆ 39 55 22 0039

Salisbury J W
US Geological Survey
927 National Ctr
Reston VA 22092
USA
☎ 1 703 860 6668

Salo Heikki
Astronomy Dpt
Univ of Oulu
Box 333
FIN 90570 Oulu
Finland
☎ 358 81 553 1931
✆ 358 81 553 1934
✉ heikki@seita.oulu.fi

Salpeter Edwin E
Astronomy Dpt
Cornell Univ
512 Space Sc Bldg
Ithaca NY 14853 6801
USA
☎ 1 607 255 3302
✆ 1 607 255 5907
✉ hann@astrosun.tn.cornell.edu

Salter Christopher John
Arecibo Observ
Box 995
Arecibo PR 00613
USA
☎ 1 809 878 2612
✆ 1 809 878 1861
✉ csalter@naic.edu

Salukvadze G N
Abastumani Astrophysical
Observ
Georgian Acad Sci
383762 Abastumani
Georgia
☎ 995 88 32 95 5367
✆ 995 88 32 98 5017

Salvador-Sole Eduardo
Dpt Fisica Atmosfera
Univ Barcelona
Avd Diagonal 645
E 08028 Barcelona
Spain
☎ 34 3 330 7311
✆ 34 3 402 1133

Salvati Marco
Osserv Astrofis Arcetri
Univ d Firenze
Largo E Fermi 5
I 50125 Firenze
Italy
☎ 39 55 27 52258
℻ 39 55 22 0039
✉ salvati.astro.it

Salzer John Joseph
Astronomy Dpt
Van Vleck Observ
Wesleyan Univ
Middletown CT 06457
USA
☎ 1 203 347 9411*2827
℻ 1 203 344 7981
✉ slaz@parcha.astro.wesleyan.edu

Samain Denys
IAS
Bt 121
Universite Paris XI
F 91405 Orsay Cdx
France
☎ 33 1 69 85 8578
℻ 33 1 69 85 8675

Samec Ronald G
Physics Dpt
Bob Jones Univ
1700 Wade Hampton Blvd
Greenville SC 36014
USA
☎ 1 864 370 1800 *2223
℻ 1 864 242 5100 *2223
✉ rsamec@wpo.bju.edu

Sampson Douglas H
Astronomy Dpt
Pennsylvania State Univ
525 Davey Lab
University Park PA 16802
USA
☎ 1 814 865 0261
℻ 1 814 863 7114

Sams Bruce Jones III
MPE
Postfach 1603
D 85740 Garching
Germany
☎ 49 893 299 3587
℻ 49 893 299 3569
✉ sams@mpa-garching.mpg.de

Samus Nikolai N
Inst of Astronomy
Acad Sciences
Pyatnitskaya Ul 48
109017 Moscow
Russia
☎ 7 095 939 3318
℻ 7 095 230 2081
✉ samus@sai.msu.su

Sanahuja Blas
Dpt Fisica Atmosfera
Univ Barcelona
Avd Diagonal 645
E 08028 Barcelona
Spain
☎ 34 3 330 7311*298
℻ 34 3 402 1133

Sanamian V A
Byurakan Astrophys Observ
Armenian Acad Sci
378433 Byurakan
Armenia
☎ 374 88 52 28 3453/4142
℻ 374 88 52 52 3640

Sanchez Almeida Jorge
IAC
Observ d Teide
via Lactea s/n
E 38200 La Laguna
Spain
☎ 34 2 232 9100
℻ 34 2 232 9117
✉ jos@ias.es

Sanchez Filomeno
Dpt Astron/Astrofisica
Univ Valencia
Dr Moliner 50
E 46100 Burjassot
Spain
☎ 34 6 386 4755
℻ 34 6 398 3084
✉ filomeno@evalvx.ific.uv.es

Sanchez Francisco
IAC
Observ d Teide
via Lactea s/n
E 38200 La Laguna
Spain
☎ 34 2 232 9100
℻ 34 2 232 9117

Sanchez Manuel
ROA
Cecilio Pujazon 22-3 A
E 11110 San Fernando
Spain
☎ 34 5 688 3548
℻ 34 5 659 9366
✉ sanchez@roa.es

Sanchez-Lavega Agustin
Dpt Fisica Aplicada
Ets Ingenieros Ind y Tel
Avda Urquijo S/n
E 18013 Bilbao
Spain
☎ 34 4 441 6400*353
℻ 34 4 441 4041

Sanchez-Saavedra M Luisa
Fac Ciencias
Univ Granada
E 18080 Granada
Spain
☎ 34 5 820 2212

Sancisi Renzo
Kapteyn Sterrekundig Inst
Univ Groningen
Postbus 800
NL 9700 AV Groningen
Netherlands
☎ 31 50 363 4073
℻ 31 50 363 6100

Sandage Allan
Carnegie Observatories
813 Santa Barbara St
Pasadena CA 91101 1292
USA
☎ 1 818 577 1122
℻ 1 818 795 8136

Sandell Goran Hans l
Joint Astronomy Ctr
660 N A'ohoku Pl
University Park
Hilo HI 96720
USA

Sanders David B
Inst for Astronomy
Univ of Hawaii
2680 Woodlawn Dr
Honolulu HI 96822
USA
☎ 1 808 956 5055
℻ 1 808 956 9580
✉ sanders@ifa.hawaii.edu

Sanders Robert
Kapteyn Sterrekundig Inst
Univ Groningen
Postbus 800
NL 9700 AV Groningen
Netherlands
☎ 31 50 363 4073
℻ 31 50 363 6100

Sanders Walt L
2395 Delaware Av 68
Santa Cruz CA 95060
USA
✉ wsanders@nmsu.edu

Sanders Wilton Turner III
Physics Dpt
Univ Wisconsin
1150 University Av
Madison WI 53706
USA
☎ 1 608 262 5916
℻ 1 608 263 6738
✉ sanders@dxs.ssec.wisc.edu

Sandford Maxwell T II
LANL
Box 1663
Los Alamos NM 87545 2345
USA
☎ 1 505 667 6384
℻ 1 505 665 4055

Sandford Scott Alan
NASA/ARC
MS 245 6
Moffett Field CA 94035 1000
USA
☎ 1 415 604 6849
℻ 1 415 604 6779
✉ sandford@ssa1.arc.nasa.gov

Sandmann William Henry
Physics Dpt
Harvey Mudd College
Claremont CA 91711
USA
☎ 1 714 621 8024

Sandqvist Aage
Stockholm Observ
Royal Swedish Acad Sciences
S 133 36 Saltsjoebaden
Sweden
☎ 46 8 717 0380
℻ 46 8 717 4719
✉ sandqvis@astro.su.se

Sanford Peter William
Dpt Phys/Astronomy
UCLO
Gower St
London WC1E 6BT
UK
☎ 44 171 387 7050
℻ 44 171 380 7145

Saniga Metod
Astronomical Institute
Slovak Acad Sci
SK 059 60 Tatranska Lomni
Slovak Republic
☎ 421 969 967 866
℻ 421 969 967 656
✉ msaniga@auriga.ta3.sk

Sanroma Manuel
Osserv d Ebro
URL
E 43520Tarragona
Spain
☎ 34 7 750 0511
℻ 34 7 750 4660
✉ msanroma@etse.urv.es

Sansaturio Maria E
ETS
Ingen Industriales
Paseo d Cauce S/n
E 47011 Valladolid
Spain
☎ 34 8 330 4899
℻ 34 8 339 2026

Santamaria Raffaele
Ist Universitario Navale
Via Acton 38
I 80133 Napoli
Italy
☎ 39 81 547 5135
℻ 39 81 552 1485

Santin Paolo
OAT
Box Succ Trieste 5
Via Tiepolo 11
I 34131 Trieste
Italy
☎ 39 40 31 99255
℻ 39 40 30 9418

Santos Filipe D
Centro Fisica Nuclear
Univ Lisboa
Av Prof Gamma Pinto N2
P 1699 Lisboa Codex
Portugal
☎ 351 1 795 0790
℻ 351 1 765 622
✉ santos@ptifm

Santos Nilton Oscar
Observ Nacional
Rua Gl Bruce 586
Sao Cristovao
20921 030 Rio de Janeiro RJ
Brazil
☎ 55 21 580 0235
✆ 55 21 580 0332

Sanwal Basant Ballabh
Uttar Pradesh State
Observ
Po Manora Peak 263 129
Nainital 263 129
India
☎ 91 59 42 2136/2583
✉ sanwal@bison.obs-besancon.fr

Sanwal N B
Astronomy Dpt
Univ of Osmania
Hyderabad 500 007
India
☎ 91 71 951*247
✉ nbs@ouastr.ernet.in

Sanyal Ashit
7505 Ridgewell Ct
Beltsville MD 20705
USA

Sanz I Subirana Jaume
Dpt Mat Aplicada
U P Cataluna
Box 30002
E 08080 Barcelona
Spain
☎ 34 3 401 6799
✆ 34 3 401 6801
✉ matjss@mat.upc.es

Sanz Jose L
Istt Fisica d Cantabria
C/o Faculdad Ciencias
Avda Los Castros s/n
E 39005 Santander
Spain
☎ 34 4 220 1452
✆ 34 4 220 1402

Sapar Arved
Tartu Observ
Estonian Acad Sci
Toravere
EE 2444 Tartumaa
Estonia
☎ 372 7 416 625
✆ 372 7 410 205

Sapre A K
Physics Dpt
Univeristy of Ravishankar
Raipur 429 010
India
☎ 91 27 064

Sarajedini Ata
San Francisco State Univ
Dpt Phys/Astronomy
1600 Holloway Av
San Francisco CA 94132
USA
☎ 1 408 459 3771
✆ 1 408 426 3115
✉ ata@stars.sfsu.edu

Sarasso Maria
Osserv Astronomico
d Torino
St Osservatorio 20
I 10025 Pino Torinese
Italy
☎ 39 11 8101 924
✆ 39 11 8101 930
✉ sarasso@to.astro.it

Sarazin Craig L
University Station
Univ of Virginia
Box 3818
Charlottesville VA 22903 0818
USA
☎ 1 804 924 4903
✆ 1 804 924 3104
✉ cls7i@virginia.edu

Sareyan Jean-Pierre
OCA
Observ Nice
BP 139
F 06304 Nice Cdx 4
France
☎ 33 4 93 89 0420
✆ 33 4 92 00 3033

Sargent Anneila I
CALTECH
MS 320 47
Downes Lab of Physics
Pasadena CA 91125
USA
☎ 1 818 356 6622
✆ 1 818 568 9352
✉ afs@astro.caltech.edu

Sargent Wallace L W
CALTECH
Palomar Obs
MS 105 24
Pasadena CA 91125
USA
☎ 1 818 356 4055
✆ 1 818 568 1517

Sarma M B K
Astronomy Dpt
Univ of Osmania
Hyderabad 500 007
India
☎ 91 65 228
✉ mbks@ouastr.ernet.in

Sarma N V G
RRI
Sadashivanagar
CV Raman Av
Bangalore 560 080
India
☎ 91 80 336 0122
✆ 91 80 334 0492
✉ sarma@rri.ernet.in

Sarmiento-Galan A F
Instituto Astronomia
UNAM
Apt 70 264
Mexico DF 04510
Mexico
☎ 52 5 622 3906
✆ 52 5 616 0653
✉ ansar@astroscu.unam.mx

Sarna Marek Jacek
Copernicus Astron Ctr
Polish Acad Sci
Ul Bartycka 18
PL 00 716 Warsaw
Poland
☎ 48 22 403 766
✆ 48 22 41 0828

Sarris Eleftherios
Astronomical Institute
Ntl Observ Athens
Box 20048
GR 118 10 Athens
Greece
☎ 30 1 346 1191
✆ 30 1 346 4566

Sarris Emmanuel T
Dpt Elect Eng
Democritos Univ of Thrace
GR 671 00 Xanthi
Greece
☎ 30 54 12 6948

Sartori Leo
Dpt Phys/Astronomy
Univ Nebraska
Behlen Observatory
Lincoln NE 68588 0111
USA
☎ 1 402 472 2770
✆ 1 402 472 2879

Sasaki Minoru
Shimonoseki City Univ
2-1 Daigaku-cho
Shimonoseki 751
Japan
☎ 81 832 52 0288
✆ 81 832 54 8691
✉ sasaki-m@shimonoeski-cu.ac.jp

Sasaki Misao
Physics Dpt
Kyoto Univ
Sakyo Ku
Kyoto 606 01
Japan
☎ 81 757 51 3883
✆ 81 75 753 3886
✉ misao@jpnyitp

Sasaki Shin
Physics Dpt
Tokyo Metropol Univ
1-1 Minami Ohsawa
Hachioji Tokyo 192 03
Japan
☎ 81 426 77 111*3346
✆ 81 426 77 2483
✉ sasaki@phys.metro-u.ac.jp

Sasaki Toshiyuki
Ntl Astronomical Obs
2 21 1 Osawa
Mitaka
Tokyo 181
Japan
☎ 81 422 34 3075
✆ 81 422 34 3793
✉ sasaki@oao.nao.ac.jp

Sasao Tetsuo
Intl Latitude Observ
NAOJ
Hoshigaoka Mizusawa Shi
Iwate 023
Japan
☎ 81 197 24 7111
✆ 81 197 22 7120

Saslaw William C
University Station
Univ of Virginia
Box 3818
Charlottesville VA 22903 0818
USA
☎ 1 804 924 4892

Sasselov Dimitar D
CfA
HCO/SAO
60 Garden St
Cambridge MA 02138 1516
USA
☎ 1 617 495 7451
✆ 1 617 495 7049
✉ sasselov@cfa.harvard.edu

Sastri Hanumath J
IIA
Koramangala
Sarjapur Rd
Bangalore 560 034
India
☎ 91 80 356 6585
✆ 91 80 553 4043

Sastry Ch V
IIA
Koramangala
Sarjapur Rd
Bangalore 560 034
India
☎ 91 80 356 6585
✆ 91 80 553 4043

Sastry Shankara K
Astronomy Dpt
Univ of Osmania
Hyderabad 500 007
India
☎ 91 71 951

Sato Fumio
Dpt Astron/Earth Sci
Tokyo Gakugei Univ
Koganei
Tokyo 184
Japan
☎ 81 423 25 2111
✆ 81 423 25 4219

Sato Humitaka
Physics Dpt
Kyoto Univ
Sakyo Ku
Kyoto 606 01
Japan
☎ 81 757 53 7008
✆ 81 75 753 7010

Sato Katsuhiko
Physics Dpt
Univ of Tokyo
Bunkyo Ku
Tokyo 113
Japan
☎ 81 381 22 111*4207
✆ 81 356 89 0465
✉ sato@phys.s.u-tokyo.ac.jp

Sato Koichi
Intl Latitude Observ
NAOJ
Hoshigaoka Mizusawa Shi
Iwate 023
Japan
☎ 81 197 24 7111
✆ 81 197 22 7120

Sato Massae
IAG
Univ Sao Paulo
CP 9638
01065 970 Sao Paulo SP
Brazil
☎ 55 11 577 8599
✆ 55 11 276 3848

Sato Naonobu
Akita Univ
1-1 Tegata Gakuencho
Akita 010
Japan
☎ 81 188 33 5261

Sato Shinji
Astrophysics Dpt
Nagoya Univ
Furocho Chikusa Ku
Nagoya 464 01
Japan
☎ 81 527 89 2953
✆ 81 527 89 2919
✉ sato@toyo.phys.nagoya-u.ac.jp

Sato Shuji
Tokyo Astronomical Obs
NAOJ
Osawa Mitaka
Tokyo 181
Japan
☎ 81 422 41 3643
✆ 81 422 41 3776

Sato Yuzo
4 8 19 Osawa
Mitaka
Tokyo 181
Japan
☎ 81 422 32 5111
✆ 81 422 34 3793

Sault Robert
AAO
ATNF
Box 76
Epping NSW 2121
Australia
☎ 61 2 868 0222
✆ 61 28 680 310
✉ rsault@atnf.csiro.auu

Saunders Richard D E
MRAO
Cavendish Laboratory
Madingley Rd
Cambridge CB3 0HE
UK
☎ 44 12 23 337 274
✆ 44 12 23 354 599

Sauval A Jacques
Koninklijke Sterrenwacht
van Belgie
Ringlaan 3
B 1180 Brussels
Belgium
☎ 32 2 373 0203
✆ 32 2 374 9822
✉ Jacques.Sauval@oma.be

Savage Ann
AAO
UK Schmidt Telescope
Private Bag
Coonabarabran NSW 2357
Australia
☎ 61 68 426 311
✆ 61 68 846 298

Savage Blair D
Astronomy Dpt
Univ Wisconsin
475 N Charter St
Madison WI 53706 1582
USA
☎ 1 608 262 3072
✆ 1 608 263 0361
✉ savage@madraf.astro.wisc.edu

Savanov Igor S
Crimean Astrophys Obs
Ukrainian Acad Science
Nauchny
334413 Crimea
Ukraine
☎ 380 6554 71113
✆ 380 6554 40704
✉ savanov@crao.crimea.ua

Savedoff Malcolm P
Dpt Phys/Astronomy
Univ of Rochester
Bausch & Lomb Bldg
Rochester NY 14627 0171
USA
☎ 1 716 275 3089
✆ 1 716 275 4351
✉ mpsa@pas.rochester.edu

Saverio Delli Santi
Dpt Astronomia
Univ d Bologna
Via Zamboni 33
I 40126 Bologna
Italy
☎ 39 51 259 334
✆ 39 51 259 407
✉ dellisanti@astbo3.bo.astro.it

Savonije Gerrit Jan
Astronomical Institute
Univ of Amsterdam
Kruislaan 403
NL 1098 SJ Amsterdam
Netherlands
☎ 31 20 525 7497/7491
✆ 31 20 525 7484
✉ gertjan@astro.uva.nl

Sawa Takeyasu
Dpt Phys/Astronomy
Aichi Univ of Edu
Kariya
Aichi 448
Japan
☎ 81 566 36 3111
✆ 81 566 36 4337
✉ sawa@auephyas.aichi-edu.ac.jp

Sawant Hanumant S
INPE
Dpt Astronomia
CP 515
12227 010 S Jose dos Campos
Brazil
☎ 55 123 22 9977
✆ 55 123 21 8743

Sawyer Constance B
850 20th St 705
Boulder CO 80302
USA
✉ sawyer@stripe.colorado.edu

Saxena A K
IIA
Koramangala
Sarjapur Rd
Bangalore 560 034
India
☎ 91 80 356 6585/6497
✆ 91 80 553 4043

Saxena P P
Dpt Maths/Astronomy
Univ of Lucknow
Lucknow
India
✆ 91 522 330 065
✉ lkuniv@sirnetd.ernet.in

Saygac A Talat
Univ Observ
Univ of Istanbul
University 34452
34452 Istanbul
Turkey
☎ 90 212 522 3597
✆ 90 212 519 0834
✉ IK131@triuvm11.iap.fr

Sazhin Michail
SAI
Acad Sciences
Universitetskij Pr 13
119899 Moscow
Russia
☎ 7 095 939 5006
✆ 7 095 939 0126
✉ snn@sai.msk.su

Sbirkova-Natcheva T
Ntl Astronomical Obs
Astronomical Observ
Box 136
BG 4700 Smoljan
Bulgaria
☎ 359 3 02 2953

Scalise Eugenio
INPE
Dpt Astronomia
CP 515
12227 010 S Jose dos Campos
Brazil
☎ 55 123 45 6814
✆ 55 12 345 6822
✉ scalise@das.inpe.br

Scalo John Michael
Astronomy Dpt
Univ of Texas
Rlm 15 308
Austin TX 78712 1083
USA
☎ 1 512 471 4461
✆ 1 512 471 6016

Scaltriti Franco
Osserv Astronomico
d Torino
St Osservatorio 20
I 10025 Pino Torinese
Italy
☎ 39 11 461 9000
✆ 39 11 461 9030

Scappini Flavio
Istt Spectroscop Mol
CNR
Via P Gobetti101
I 40129 Bologna
Italy
☎ 39 51 639 8510
✆ 39 51 639 8540
✉ scappini@astbo1.bo.cnr.it

Scardia Marco
Osserv Astronomico
d Milano
Via E Bianchi 46
I 22055 Merate
Italy
☎ 39 990 6412
✆ 39 990 8492

Scarfe Colin D
Dpt Phys/Astronomy
Univ of Victoria
Box 3055
Victoria BC V8W 3P6
Canada
☎ 1 250 721 7740
✆ 1 250 721 7715
✉ scarfe@uvphys.phys.uvic.ca

Scargle Jeffrey D
NASA/ARC
MS 245 3
Moffett Field CA 94035 1000
USA
☎ 1 415 694 6330
✆ 1 415 604 6779

Scarrott Stanley M
Physics Dpt
Univ of Durham
South Rd
Durham DH1 3LE
UK
☎ 44 191 374 2000
✆ 44 191 374 3749

Schaal Ricardo E
CRAAE INPE
EPUSP/PTR
CP 61548
01065 970 Sao Paulo SP
Brazil
☎ 55 11 815 6289
✆ 55 11 815 4272
✉ ecorreia@brusp.ansp.br

Schadee Aert
Sterrekundig Inst Utrecht
Box 80000
NL 3508 TA Utrecht
Netherlands
☎ 31 30 253 5208
✆ 31 30 253 1601
✉ a.schadee@astro.uu.nl

Schaefer Gerhard
Max Planck Gesellscaft
AG Gravitationstheorie
Max Wien Platz 1
D 07745 Jena
Germany
☎ 49 364 163 6643
✆ 49 364 163 6728

Schaerer Daniel
OMP
14 Av E.Berlin
F 31400 Toulouse Cdx
France
☎ 33 5 61 33 2929
✆ 33 5 61 33 2840
✉ schaerer@srvdec.obs-mip.fr

Schaifers Karl
Steinbachweg 37
D 69000 Heidelberg
Germany
☎ 49 622 180 1511

Schaller Gerard
Universite Geneve
Universite Dufour
Service Informatiques
CH 1211 Geneve 4
Switzerland
☎ 41 22 705 7187
✆ 41 22 705 7986
✉ Gerard.schaller@seinf.unige.ch

Schanda Erwin
Buendackerstr 108
CH 3047 Bremgarten BE
Switzerland
☎ 41 31 65 8910

Scharmer Goeran Bjarne
Stockholm Observ
Royal Swedish Acad Sciences
S 133 36 Saltsjoebaden
Sweden
☎ 46 8 717 0195
✆ 46 8 717 4719
✉ scharmer@astro.su.se

Schatten Kenneth H
NASA/GSFC
Code 925
Greenbelt MD 20771
USA
☎ 1 301 286 3831
✆ 64 301 286 1757
✉ schatten@gsfc.nasa.gov

Schatzman Evry
Observ Paris Meudon
DASGAL
Pl J Janssen
F 92195 Meudon PPL Cdx
France
☎ 33 1 45 07 7873
✆ 33 1 45 07 7878

Schechter Paul L
Physics Dpt
MIT Rm 6 206
Box 165
Cambridge MA 02139 4307
USA
☎ 1 617 253 0690
✆ 1 617 253 9798
✉ schech@achernar.mit.edu

Scheepmaker Anton
Cosmic Ray WG
Huygens Lab
Wassenaarseweg 78
NL 2300 RA Leiden
Netherlands

Scheffler Helmut
Carl-orff-weg 16
D 6906 Leimen 3
Germany

Scheidecker Jean-Paul
OCA
Observ Nice
BP 139
F 06304 Nice Cdx 4
France
☎ 33 4 93 89 0420
✆ 33 4 92 00 3033

Scherb Frank
Physics Dpt
Univ Wisconsin
1150 University Av
Madison WI 53706
USA
☎ 1 608 262 6879
✆ 1 608 262 3077

Scherrer Philip H
HEPL
Stanford Univ
Annex B211
Stanford CA 94305 4085
USA
✉ pscherrer@solar.stanford.edu

Scheuer Peter A G
MRAO
Cavendish Laboratory
Madingley Rd
Cambridge CB3 0HE
UK
☎ 44 12 23 337 307
✆ 44 12 23 354 599
✉ pags@mrao.cam.ac.uk

Schilbach Elena
Astrophysik Inst
Potsdam Univ
an d Sternwarte 16
D 14482 Potsdam
Germany
☎ 49 331 749 9335
✆ 49 331 749 9309
✉ eschilbach@aip.de

Schild Hansruedi
Inst Astronomie
ETH Zentrum
CH 8092 Zuerich
Switzerland
☎ 41 1 632 3633
✆ 41 1 632 1205
✉ hschild@astro.phys.ethz.ch

Schild Rudolph E
CfA
HCO/SAO
60 Garden St
Cambridge MA 02138 1516
USA
☎ 1 617 495 7426
✆ 1 617 495 7356

Schildknecht Thomas
Astronomical Institute
Univ Bern
Sidlerstr 5
CH 3012 Bern
Switzerland
☎ 41 31 631 8591
✆ 41 31 631 3869
✉ schild@aiub.unibe.ch

Schilizzi Richard T
NFRA
Postbus 2
NL 7990 AA Dwingeloo
Netherlands
☎ 31 521 59 5100
✆ 31 52 159 7332
✉ rts@nfra.nl

Schiller Stephen
Physics Dpt
South Dakota State Univ
Box 2219 Room 310b
Brookings SD 57007
USA
☎ 1 605 688 4293

Schindler Karl
Inst Theor Physik
Ruhr Univ Bochum
Postfach 102148
D 44780 Bochum
Germany
☎ 49 234 700 3454
✆ 49 234 709 4169

Schindler Sabine
MPE
Postfach 1603
D 85740 Garching
Germany
☎ 49 893 299 3219
✆ 49 893 299 3235
✉ sas@mpa-garching.mpg.de

Schlegel Eric Matthew
NASA GSFC
Code 668
Greenbelt MD 20771
USA
☎ 1 301 286 6636
✆ 1 301 286 3391
✉ eric@heasfs.gsfc.nasa.gov

Schleicher David G
Lowell Observ
1400 W Mars Hill Rd
Box 1149
Flagstaff AZ 86001
USA
☎ 1 602 774 3358
✆ 1 520 774 3358

Schleicher Helmold
Kiepenheuer Institut
f Sonnenphysik
Schoneckstr 6
D 79104 Freiburg Breisgau
Germany
☎ 49 761 319 8226
✆ 49 761 319 8111
✉ schleicher@kis.uni-freiburg.de

Schlesinger Barry M
Hughes STX Corp
7701 Greenbelt Rd
Suite 400
Greenbelt MD 20770
USA
☎ 1 301 286 1069
✉ bschlesinger@nssdca.gsfc.nasa.gov

Schlickeiser Reinhard
RAIUB
Univ Bonn
auf d Huegel 69
D 53121 Bonn
Germany
☎ 49 228 525 1
✆ 49 228 525 229

Schloerb F Peter
Dpt Phys/Astronomy
Univ Massachusetts
GRC
Amherst MA 01003
USA
☎ 1 413 545 4303

Schlosser Wolfhard
Astronomisches Institut
Ruhr Univ Bochum
Postfach 102148
D 44780 Bochum
Germany
☎ 49 234 700 3454
✆ 49 234 709 4169

Schlueter A
MPI Plasmaphysik
Postfach 1603
D 85740 Garching
Germany
☎ 49 893 299 347
✆ 49 893 299 3235

Schlueter Dieter
Inst Theor Phys/
Sternwarte Univ Kiel
Olshausenstr 40
D 24098 Kiel
Germany
☎ 49 431 880 4109
✆ 49 431 880 4432

Schmadel Lutz D
ARI
Moenchhofstr 12-14
D 69120 Heidelberg
Germany
☎ 49 622 140 50
✆ 49 622 140 5297
✉ s31@mvs.urz.uni-heidelberg.de

Schmahl Edward J
Astronomy Program
Univ of Maryland
College Park MD 20742 2421
USA
☎ 1 301 454 6074
✆ 1 301 314 9067

Schmalberger Donald C
The Albany Academy
Academy Rd
Albany NY 12208
USA
☎ 1 518 465 1461

Schmeidler F
Mauerkircherstr 17
D 8000 Muenchen 80
Germany

Schmelz Joan T
NASA GSFC
Smm Xrp
Greenbelt MD 20771
USA
☎ 1 301 220 4164
✆ 1 301 220 4171

Schmid Hans Martin
Landessternwarte
Koenigstuhl
D 69117 Heidelberg
Germany
☎ 49 622 150 9222
✆ 49 622 150 9202
✉ hschmid@lsw.uni-heideberg.de

Schmid-Burgk J
RAIUB
Univ Bonn
auf d Huegel 69
D 53121 Bonn
Germany
☎ 49 228 525 271
✆ 49 228 525 229

Schmidt Edward G
Dpt Phys/Astronomy
Univ Nebraska
Behlen Observatory
Lincoln NE 68588 0111
USA
☎ 1 402 472 2788
✆ 1 402 472 2879
✉ eschmidt@unlinfo.unl.edu

Schmidt H U
MPA
Karl Schwarzschildstr 1
D 85748 Garching
Germany
☎ 49 893 299 9413/4
✆ 49 893 299 3235

Schmidt Hans
Wachsbleiche 5
D 53111 Bonn
Germany

Schmidt K H
Zntrlinst Astrophysik
Sternwarte Babelsberg
Rosa Luxemburg Str 17a
D 14473 Potsdam
Germany
☎ 49 331 762225
✉ khschmidt@aip.de

Schmidt Maarten
CALTECH
Palomar Obs
MS 105 24
Pasadena CA 91125
USA
☎ 1 818 356 4204
✆ 1 818 568 9352

Schmidt Thomas
Rudolf Steiner Schule
an d Probstei 23
D 48000 Bielefeld 1
Germany
☎ 49 521 880 407

Schmidt Wolfgang
Kiepenheuer Inst
Sonnenphysik
Schoeneckstr 6
D 79104 Freiburg Breisgau
Germany
☎ 49 761 319 8162
✆ 49 761 3198 111
✉ wolfgang@kis.uni-freiburg.de

Schmidt-Kaler Theodor
Georg Buechner Str 37
D 97176 Margetschoechheim
Germany

Schmidtke Paul C
Physics Dpt
Arizona State Univ
Astronomy Program
Tempe AZ 85287 1504
USA
☎ 1 602 965 2918
✆ 1 605 965 7954

Schmieder Brigitte
Observ Paris Meudon
DASOP
Pl J Janssen
F 92195 Meudon PPL Cdx
France
☎ 33 1 45 07 7817
✆ 33 1 45 07 7959
✉ schmieder@obspm.fr

Schmitt Dieter
Univ Sternwarte
Goettingen
Geismarlandstr 11
D 37083 Goettingen
Germany
☎ 49 551 395 046
✆ 49 551 395 043

Schmitter Edward F
Physics Dpt
Univ of Lagos
Akoka
Lagos PMB 1004
Nigeria
☎ 234 1 83 7864

Schmutz Werner
Inst Astronomie
ETH Zentrum
CH 8092 Zuerich
Switzerland
☎ 41 1 632 3806
✆ 41 1 632 1205

Schneider Donald P
Astronomy Dpt
Pennsylvania State Univ
525 Davey Lab
University Park PA 16802
USA
☎ 1 814 863 9554
✆ 1 814 863 3399
✉ dps@astro.psu.edu

Schneider Glenn H
Steward Observ
Univ of Arizona
Tucson AZ 85721
USA
☎ 1 520 621 5865
✆ 1 520 621 1891
✉ gschneider@as.arizona.edu

Schneider Jean
Observ Paris Meudon
DARC
Pl J Janssen
F 92195 Meudon PPL Cdx
France
☎ 33 1 45 07 7430
✆ 33 1 45 07 7971
✉ schneider@obspm.fr

Schneider Nicholas M
LASP
Univ of Colorado
Campus Box 392
Boulder CO 80309 0392
USA
☎ 1 303 492 7672
✆ 1 303 492 6946
✉ nick@pele.colorado.edu

Schneider Peter
MPA
Karl Schwarzschildstr 1
D 85748 Garching
Germany
☎ 49 893 299 00
✆ 49 893 299 3235
✉ peter@mpa-garching.mpg.de

Schnell Anneliese
Inst Astronomie
Univ Wien
Tuerkenschanzstr 17
A 1180 Wien
Austria
☎ 43 1 345 36093
✆ 43 1 470 6015

Schneps Matthew H
CfA
HCO/SAO
60 Garden St
Cambridge MA 02138 1516
USA
☎ 1 617 495 7472
✆ 1 617 495 7356

Schnopper Herbert W
CfA
HCO/SAO
60 Garden St
Cambridge MA 02138 1516
USA
☎ 1 617 496 7763
✆ 1 617 496 7577
✉ hscnopper@cfa.harvard.edu

Schober Hans J
Inst Astronomie
Karl-Franzens-Univ
Universitaetsplatz 5
A 8010 Graz
Austria
☎ 43 316 380 5273
✆ 43 316 384 091

Schoeffel Eberhard F
Merianerstr 42
D 8600 Bamberg
Germany

Schoenberner Detlef
Astrophysik Inst
Potsdam Univ
an d Sternwarte 16
D 14482 Potsdam
Germany
☎ 49 331 749 9395
✆ 49 331 749 9526
✉ DeSchoenberner@aip.de

Schoeneich W
Zntrlinst Astrophysik
Sternwarte Babelsberg
Rosa Luxemburg Str 17a
D 14473 Potsdam
Germany
☎ 49 331 762225

Schoenfelder Volker
MPE
Postfach 1603
D 85740 Garching
Germany
☎ 49 893 299 578
✆ 49 893 299 3569

Scholl Hans
OCA
Observ Nice
BP 139
F 06304 Nice Cdx 4
France
☎ 33 4 93 89 0420
℻ 33 4 92 00 3033

Scholz Gerhard
AIP
SOE
Telegrafenberg
D 14473 Potsdam
Germany
☎ 49 331 288 2309
℻ 49 331 288 2310
✉ gscholz@aip.fr

Scholz M
Inst Theor Astrophysik
d Univer
Tiergartenstr 15
D 69121 Heidelberg
Germany
☎ 49 622 154 4837
℻ 49 622 154 4221
✉ b15@rw.iwe.uni-heidelberg.de

Scholz Ralf Dieter
Astrophysik Inst
Potsdam Univ
an d Sternwarte 16
D 14482 Potsdam
Germany
☎ 49 331 749 9336
℻ 49 331 749 9309
✉ rdscholz@aip.de

Schoolman Stephen A
Lockheed Palo Alto Res Lb
Dpt 91 20 Bg 255
3251 Hanover St
Palo Alto CA 94304 1191
USA
☎ 1 415 858 4074
℻ 1 415 424 3994

Schramm K Jochen
Megaphot Ev
Univ Hamburg
Gojenbergsweg 112
D 21029 Hamburg
Germany
☎ 49 407 252 4121
℻ 49 407 252 4198
✉ jschramm@hs.uni-hamburg.de

Schramm Thomas
Hamburger Sternwarte
Univ Hamburg
Gojensbergsweg 112
D 21029 Hamburg
Germany
☎ 49 407 252 4126
℻ 49 407 252 4198
✉ tschramm@hs.uni-hamburg.de

Schreiber Roman
NCAC
Polish Acad Sciences
Ul Rabianska 8
PL 87 100 Torun
Poland
☎ 48 56 19 319
℻ 48 56 19 381
✉ schreibe@ncac.torun.pl

Schreier Ethan J
STScI
Homewood Campus
3700 San Martin Dr
Baltimore MD 21218
USA
☎ 1 301 338 4740
℻ 1 301 338 2519
✉ schreier@stsci.edu

Schrijver C J
Sterrekundig Inst Utrecht
Box 80000
NL 3508 TA Utrecht
Netherlands
☎ 31 30 253 5224
℻ 31 30 253 1601
✉ kschrijver@solar

Schrijver Johannes
SRON
Postbus 800
Sorbonnelaan 2
NL 3584 CA Utrecht
Netherlands
☎ 31 30 253 5728
℻ 31 30 254 0860
✉ hanss@sron.ruu.nl

Schroeder Daniel J
Dpt Phys/Astronomy
Beloit College
Beloit WI 53511
USA
☎ 1 608 365 3391

Schroeder Klaus Peter
Hamburger Sternwarte
Univ Hamburg
Gojensbergsweg 112
D 21029 Hamburg
Germany
☎ 49 407 252 4141
℻ 49 407 252 4198

Schroeder Rolf
DESY Deutsches Elek
Synchrotron
Moeoerkenweg 37
D 21029 Hamburg
Germany
☎ 49 407 244 650
℻ 49 40 8998 4303
✉ r.schroeder@bergedorf.de
schroeder@desy.de

Schroeter Egon H
Kiepenheuer Inst
Sonnenphysik
Schoeneckstr 6
D 79104 Freiburg Breisgau
Germany
☎ 49 761 328 64
℻ 49 761 319 8111

Schroll Alfred
Sonnenobservatorium
Kanzelhoehe
A 9521 Treffen
Austria
☎ 43 424 82717

Schruefer Eberhard
IAEF
Univ Bonn
auf d Huegel 71
D 53121 Bonn
Germany
☎ 49 228 73 3390
℻ 49 228 73 3672

Schubart Joachim
ARI
Moenchhofstr 12-14
D 69120 Heidelberg
Germany
☎ 49 622 140 5153
℻ 49 622 140 5297
✉ s24@aixterm1.urz.uni-heidelberg.de

Schuch Nelson Jorge
Observ Nacional
UFSM/ctro Tecnologia
Cidade Universitaria
97100 Santa Maria RS
Brazil
☎ 55 552 26 1616

Schuecker Peter
Astronomisches Institut
Univ Muenster
Wilhelm Klemm Str 10
D 48149 Muenster
Germany
☎ 49 251 83 9128
℻ 49 251 833 669
✉ peter@cygnus.uni-muenster.de

Schuecking E L
Physics Dpt
New York Univ
4 Washington
New York NY 10003
USA
☎ 1 212 998 770
℻ 1 212 995 4016

Schuessler Manfred
Kiepenheuer Inst
Sonnenphysik
Schoeneckstr 6
D 79104 Freiburg Breisgau
Germany
☎ 49 761 328 64
℻ 49 761 319 8111

Schuler Walter
Sonnenrain 15
CH 4533 Riedholz
Switzerland
☎ 41 65 23 2055

Schulte D H
ITEK Corp
10 Maguire Rd
Lexington MA 02173
USA
☎ 1 617 276 2532

Schulte-Ladbeck Regina E
Dpt Phys/Astronomy
Univ of Pittsburgh
3941 O-hara St
Pittsburgh PA 15260
USA
☎ 1 412 624 9013
℻ 1 412 624 1833
✉ rsl@vms.cis.pitt.edu

Schultz Alfred Bernard
STScI/CSC
Homewood Campus
3700 San Martin Dr
Baltimore MD 21218
USA
☎ 1 301 338 5044
℻ 1 301 338 4767
✉ schultz@stsci.edu

Schultz G V
RAIUB
Univ Bonn
auf d Huegel 69
D 53121 Bonn
Germany
☎ 49 228 525 291
℻ 49 228 525 229

Schulz Hartmut
Astronomisches Institut
Ruhr Univ Bochum
Postfach 102148
D 44780 Bochum
Germany
☎ 49 234 700 3454
℻ 49 234 709 4169

Schulz Rolf Andreas
RAIUB
Univ Bonn
auf d Huegel 69
D 53121 Bonn
Germany
☎ 49 228 525 232
℻ 49 228 525 229

Schumacher Gerard
OCA
Observ Calern/Caussols
2130 r Observatoire
F 06460 S Vallier Thiey
France
☎ 33 4 93 42 6270

Schumann Joerg Dieter
Observatorium Hoher List
Univ Sternwarte Bonn
D 54550 Daun
Germany
☎ 49 65 92 2937
℻ 49 65 92 9851 40

Schuster William John
Observ Astronomico Nacional
UNAM
Apt 877
Ensenada BC 22800
Mexico
☎ 52 6 174 4580
℻ 52 6 174 4607
✉ schuster@bufadora.astrsen.unam.mx

Schutte Willem Albert
Leiden Observ
Box 9513
NL 2300 RA Leiden
Netherlands
☎ 31 71 527 5944
✆ 31 71 5275 819
🖃 schutte@strwchem.
leidenuniv.nl

Schutz Bernard F
MPI Gravitational Phys
Haus d Wirtschaft
Schlaatzweg 1
D 14473 Potsdam
Germany

Schutz Bob Ewald
CSR
Univ of Texas
Austin TX 78759
USA
☎ 1 512 471 1356
✆ 1 512 471 3570

Schwan Heiner
ARI
Moenchhofstr 12-14
D 69120 Heidelberg
Germany
☎ 49 622 140 50
✆ 49 622 140 5297
🖃 s25@ix.urz.uni-
heidelberg.de

Schwartz Daniel A
CfA
HCO/SAO
60 Garden St
Cambridge MA 02138 1516
USA
☎ 1 617 495 7232
✆ 1 617 495 7356
🖃 das@cfa.harvard.edu

Schwartz Philip R
NRL
Code 4138
4555 Overlook Av SW
Washington DC 20375 5000
USA
☎ 1 202 767 3391

Schwartz Richard D
Physics Dpt
Univ of Missouri
8001 Natural Bridge Rd
St Louis MO 63121
USA
☎ 1 314 553 5025
✆ 1 314 516 6152

Schwartz Rolf
RAIUB
Univ Bonn
auf d Huegel 69
D 53121 Bonn
Germany
☎ 49 228 525 303
✆ 49 228 525 229

Schwartz Steven Jay
Astronomy Unit
QMWC
Mile End Rd
London E1 4NS
UK
☎ 44 171 975 5454
✆ 44 181 981 9587

Schwarz Hugo E
Nordic Optical Telescope
Apt 474
E 38700 Santa Cruz
Spain
☎ 34 2 240 5500
✆ 34 2 240 5646/405 501
🖃 hschwarz@not.iac.es

Schwarz Ulrich J
Kapteyn Sterrekundig Inst
Univ Groningen
Postbus 800
NL 9700 AV Groningen
Netherlands
☎ 31 50 363 4073
✆ 31 50 363 6100

Schwarzenberg-Czerny A
Astronomical Observ
Warsaw Univ
Al Ujazdowskie 4
PL 00 478 Warsaw
Poland
☎ 48 22 29 4011
✆ 48 22 29 4967
🖃 czerny@camk.edu.pl

Schwehm Gerhard
ESA/ESTEC
SSD
Box 299
NL 2200 AG Noordwijk
Netherlands
☎ 31 71 565 6555
✆ 31 71 565 4690

Schweizer Francois
Dpt Terrestr Magnetism
Carnegie Inst Washington
5241 Bd Branch Rd NW
Washington DC 20015 1305
USA
☎ 1 202 686 4370*4401
🖃 schweizer@bmrt.
ciw.edu

Schwekendiek Peter
ARI
Moenchhofstr 12-14
D 69120 Heidelberg
Germany
☎ 49 622 140 5128
✆ 49 622 140 5297
🖃 peter@auriga.ari.uni-
heidelberg.de

Schwope Axel
Astrophysik Inst
Potsdam Univ
an d Sternwarte 16
D 14482 Potsdam
Germany
☎ 49 331 749 9232
✆ 49 331 749 9309
🖃 aschwope@aip.de

Sciama Dennis W
ICTP
St Costiera 11
Miramare
I 34014 Trieste
Italy
☎ 39 40 22 4118
✆ 39 40 22 4163

Sciortino Salvatore
Osserv Astronomico
Univ d Palerno
Piazza Parlemento 1
I 90134 Palermo
Italy
☎ 39 91 657 0451
✆ 39 91 48 8900

Sconzo Pasquale
29 Old Mystic St
Arlington MA 02174
USA
☎ 1 617 646 9315

Scorza de Appl Cecilia
Landessternwarte
Koenigstuhl
D 69117 Heidelberg
Germany
☎ 49 622 150 9214
✆ 49 622 150 9202
🖃 cscorza@hp2.lsw.uni-
heidelberg.de

Scott Douglas
Dpt Phys/Astronomy
UBC
2219 Main Mall
Vancouver BC V6T 1Z4
Canada
☎ 1 604 822 2802
✆ 1 604 822 6047
🖃 dscott@astro.ubc.ca

Scott Eugene Howard
NASA GSFC
Code 684 9
Greenbelt MD 20771
USA
☎ 1 301 286 8746

Scott John S
Steward Observ
Univ of Arizona
Tucson AZ 85721
USA
☎ 1 520 621 2288
✆ 1 520 621 1532

Scott Paul F
MRAO
Cavendish Laboratory
Madingley Rd
Cambridge CB3 0HE
UK
☎ 44 12 23 337 294
✆ 44 12 23 354 599

Scoville Nicholas Z
CALTECH
Palomar Obs
MS 105 24
Pasadena CA 91125
USA
☎ 1 818 356 4979
✆ 1 818 568 9352

Scrimger J Norman
Astronomy Dpt
St Mary's Univ
Halifax NS B3H 3C3
Canada
☎ 1 902 420 5633
✆ 1 902 420 5561

Scuflaire Richard
Inst Astrophysique
Universite Liege
Av Cointe 5
B 4000 Liege
Belgium
☎ 32 4 254 7530
✆ 32 4 254 7511
🖃 R.Scuflaire@ulg.ac.be

Seaquist Ernest R
Astronomy Dpt
Univ of Toronto
60 St George St
Toronto ON M5S 1A1
Canada
☎ 1 416 978 3146
✆ 1 416 978 3921
🖃 seaquist@astro.
utoronto.ca

Searle Leonard
Carnegie Observatories
813 Santa Barbara St
Pasadena CA 91101 1292
USA
☎ 1 818 577 1122
✆ 1 818 795 8136

Sears Richard Langley
Astronomy Dpt
Univ of Michigan
Dennison Bldg
Ann Arbor MI 48109 1090
USA
☎ 1 313 763 3295
✆ 1 313 764 2211

Seaton Michael J
Dpt Phys/Astronomy
UCLO
Gower St
London WC1E 6BT
UK
☎ 44 171 387 7050
✆ 44 171 380 7145

Secco Luigi
Dpt Astronomia
Univ d Padova
Vic d Osservatorio 5
I 35122 Padova
Italy
☎ 39 49 829 3411
✆ 39 49 875 9840

Sedlmayer Erwin
Inst Astron/Astrophysik
Technische Uni
Hardenbergstr 36
D 10623 Berlin
Germany
☎ 49 303 142 3783
✆ 49 303 142 3018

Sedmak Giorgio
Dpt Astronomia
Univ d Trieste
Via Tiepolo 11
I 34131 Trieste
Italy
☎ 39 40 79 4863
✆ 39 40 30 9418

Sedrakian David
Ntl Acad Sci
Bagramian 24
375049 Yerevan
Armenia
☎ 374 88 52 52 7051

Seeds Michael August
Astronomy Program
Franklin/Marshall College
Lancaser PA 17604 3003
USA
☎ 1 717 291 3800
✆ 1 717 291 4143
✉ m_seeds@fandm

Seeger Charles Louis III
San Francisco State Univ
473 James Rd
Palo Alto CA 94306
USA
☎ 1 415 493 6005

Seeger Philip A
LANL
MS H805
Box 1663
Los Alamos NM 87545 2345
USA
☎ 1 505 667 8843
✆ 1 505 665 4055

Segal Irving E
Mathematics Dpt
MIT Rm 2 224
Box 165
Cambridge MA 02139
USA
☎ 1 617 253 4985

Segaluvitz Alexander
Box 659
Kefar Sava
Israel

Segan Stevo
Astronomy Dpt
Univ of Belgrade
Studentski Trg 16
11000 Beograd
Yugoslavia FR
☎ 381 11 638 715
✆ 381 11 630 151

Seggewiss Wilhelm
Observatorium Hoher List
Univ Sternwarte Bonn
D 54550 Daun
Germany
☎ 49 65 92 2150
✆ 49 65 92 9851 40
✉ seggewis@astro.uni-bonn.de

Sehnal Ladislav
Astronomical Institute
Czech Acad Sci
Fricova 1
CZ 251 65 Ondrejov
Czech R
☎ 420 204 85 7113
✆ 420 2 88 1611
✉ isehnal@asu.cas.cz

Seidelmann P Kenneth
USNO
3450 Massachusetts Av NW
Washington DC 20392 5100
USA
☎ 1 202 762 1441
✆ 1 202 762 1516
✉ pks@spica.usno.navy.mil

Seiden Philip E
IBM
T J Watson Res Ctr
Box 218
Yorktown Heights NY 10598
USA
☎ 1 914 945 1424

Seidov Zakir F
Shemakha Astrophysical Observ
Azerbajan Acad Sci
373243 Shemakha
Azerbajan
☎ 994 89 22 39 8248

Seielstad George A
UND Aeospace
Box 9007
Grand Forks ND 58202
USA
✉ landei@aero.und.nodak.edu

Seifert Walter
Landessternwarte
Koenigstuhl
D 69117 Heidelberg
Germany
☎ 49 622 150 9232
✆ 49 622 150 9202
✉ wseifert@lsw.uni-heidelberg.de

Seimenis John
Mathematics Dpt
Univ of The Aegean
GR 83200 Samos
Greece
☎ 30 94 39 2999
✆ 30 1 268 1990
✉ isei@iu.aegean.gr

Sein-Echaluce M Luisa
Dpt Mat Aplicada
Univ Zaragoza
E 50009 Zaragoza
Spain
☎ 34 7 676 1000
✆ 34 7 676 1140

Seiradakis John Hugh
Astronomy Lab
Univ Thessaloniki
GR 540 06 Thessaloniki
Greece
☎ 30 31 99 8173
✆ 30 31 99 5384
✉ jhs@astro.auth.gr

Seitter Waltraut C
Astronomisches Institut
Univ Muenster
Wilhelm Klemm Str 10
D 48149 Muenster
Germany
☎ 49 251 833 561
✆ 49 251 833 669
✉ waltraut@cygnus.uni-muenster.de

Sekanina Zdenek
CALTECH/JPL
MS 169 237
4800 Oak Grove Dr
Pasadena CA 91109 8099
USA
☎ 1 818 354 7589
✆ 1 818 393 6030
✉ zs@sek.jpl.nasa.gov

Seki Munezo
Astron Institute
Tohoku Univ
Kawauchi
Sendai 980 77
Japan
☎ 81 222 217 7757
✆ 81 222 263 9279
✉ seki@astroa.astr.tohku.ac.jp

Sekiguchi Kazuhiro
Joint Astronomy Ctr
660 N A'ohoku Pl
University Park
Hilo HI 96720
USA
☎ 1 808 934 5905
✆ 1 808 934 5984
✉ kaz@subaru.naoj.org

Sekiguchi Maki
Ntl Astronomical Obs
2 21 1 Osawa
Mitaka
Tokyo 181
Japan
☎ 81 422 34 3643
✆ 81 422 34 3776
✉ osekigu@cl.mtk.nao.ac.jp

Sekiguchi Naosuke
Musashidai 3-16-8
FUCHU
Tokyo 183
Japan

Sekimoto Yutaro
Physics Dpt
Univ of Tokyo
7-3-1 Hongo Bunkyo-ku
Tokyo 113
Japan
☎ 81 338 12 2111*4217
✆ 81 356 84 5291
✉ sekimoto@phys.s.u-tokyo.ac.jp

Sellgren Kristen
Astronomy Dpt
Ohio State Univ
Rm 5036 Smith Lab
Columbus OH 43210
USA
☎ 1 614 282 1898
✆ 1 614 292 2928
✉ sellgren@payne.mps.ohio-state.edu

Sellwood Jerry A
Dpt Phys/Astronomy
Rutgers Univ
Box 849
Piscataway NJ 08854 0849
USA
☎ 1 908 445 2501
✆ 1 908 932 4343

Selvelli Pierluigi
OAT
Box Succ Trieste 5
Via Tiepolo 11
I 34131 Trieste
Italy
☎ 39 40 31 99269
✆ 39 40 30 9418

Semel Meir
Observ Paris Meudon
DASOP
Pl J Janssen
F 92195 Meudon PPL Cdx
France
☎ 33 1 45 07 7790
✆ 33 1 45 07 7971

Semeniuk Irena
Astronomical Observ
Warsaw Univ
Al Ujazdowskie 4
PL 00 478 Warsaw
Poland
☎ 48 22 29 4011
✆ 48 22 29 4697

Semenzato Roberto
Dpt Fisica G Galilei
Univ d Padova
Via Marzolo 8
I 35131 Padova
Italy
☎ 39 49 84 4247
✆ 39 49 84 4245

Semkov Evgeni
Inst of Astronomy
Bulgarian Acad Sci
72 Lenin Blvd
BG 1784 Sofia
Bulgaria
☎ 359 2 743 1739
✆ 359 2 975 3201
✉ evgeni@wfpa.acad.bg

Sengbusch Kurt V
MPA
Karl Schwarzschildstr 1
D 85748 Garching
Germany
☎ 49 893 299 00
✆ 49 893 299 3235

Sequeiros Juan
Dpt Fisica
Univ Alcala
Apt 20
E 28800 Alcala d Henares
Spain
☎ 34 1 889 4940
✆ 34 1 889 4953

Serabyn Eugene
CALTECH
MS 320 47
Pasadena CA 91125
USA
☎ 1 818 395 6664
✆ 1 818 796 8806
✉ serabyn@tacos.caltech.edu

Serafin Richard A
Uhlandstr 46
D 46047 Oberhausen
Germany
☎ 49 208 889 102

Serio Salvatore
Osserv Astronomico
Univ d Palerno
Piazza Parlemento 1
I 90134 Palermo
Italy
☎ 39 91 592 451
✆ 39 91 48 8900

Serrano Alfonso
Instituto Astronomia
UNAM
Apt 70 264
Mexico DF 04510
Mexico
☎ 52 5 622 3906
✆ 52 5 616 0653
✉ ping@inaoep.mx

Servan Bernard
Observ Paris
DASGAL
61 Av Observatoire
F 75014 Paris
France
☎ 33 1 40 51 2236
✆ 33 1 44 54 1804

Seshadri Sridhar
IUCAA
PO Box 4
Ganeshkhind
Pune 411 007
India
☎ 91 212 351 414/5
✆ 91 212 350 760
✉ sridhar@iucaa.ernet.in

Sessin Wagner
INPE
Dpt Astronomia
CP 515
12227 010 S Jose dos Campos
Brazil
☎ 55 123 22 9088

Setti Giancarlo
Istt Radioastronomia
CNR
Via P Gobetti101
I 40129 Bologna
Italy
☎ 39 51 639 93 65
✆ 39 51 639 94 31

Sevarlic Branislav M
Astronomical Observ
Volgina 7
11050 Beograd
Yugoslavia FR
☎ 381 1 419 357
✆ 381 1 419 553

Severino Giuseppe
Osserv d Capodimonte
Via Moiariello 16
I 80131 Napoli
Italy
☎ 39 81 44 0101
✆ 39 81 45 6710

Sevilla Miguel J
Inst Astron/Geodesia
Fac Ciencias Matemat
Univ Complutense
E 28040 Madrid
Spain
☎ 34 1 244 2501
✆ 34 1 394 5195

Seward Frederick D
CfA
HCO/SAO
60 Garden St
Cambridge MA 02138 1516
USA
☎ 1 617 495 7282
✆ 1 617 495 7356
✉ fds@cfa.harvard.edu

Seymour P A H
112 Powisland Dr
Plymouth PL6 6AF
UK

Sezer Cengiz
Dpt Astron/Space Sci
Ege Univ
Box 21
35100 Bornova Izmir
Turkey
☎ 90 232 388 0110*2322

Shaffer David B
1742 Saddleback Ct
Henderson Nv 89014
USA
☎ 1 702 451 5562
✆ 1 702 451 5562
✉ dbs@bootes.gsfc.nasa.gov

Shafter Allen W
Astronomy Dpt
San Diego State Univ
San Diego CA 92182
USA
☎ 1 617 594 6170
✆ 1 619 594 1413
✉ shafter@proteus.sdsu.edu

Shaham Jacob
Physics Dpt
Columbia Univ
538 W 120th St
New York NY 10027
USA
☎ 1 212 280 3349

Shahul Hameed Mohin
IIA
Koramangala
Sarjapur Rd
Bangalore 560 034
India
☎ 91 80 553 0672*76
✆ 91 80 553 4043
✉ mohin@iiap.ernet.in

Shakeshaft John R
MRAO
Cavendish Laboratory
Madingley Rd
Cambridge CB3 0HE
UK
☎ 44 12 23 337 294
✆ 44 12 23 354 599
✉ jrs@phy-ravx.cam.ac.uk

Shakhbazian Romelia K
Byurakan Astrophys Observ
Armenian Acad Sci
378433 Byurakan
Armenia
☎ 374 88 52 28 3453/4142
✆ 374 88 52 52 3640

Shakhbazyan Yurij L
Byurakan Astrophys Observ
Armenian Acad Sci
378433 Byurakan
Armenia
☎ 374 88 52 28 3453/4142
✆ 374 88 52 52 3640

Shakhovskaya Nadejda I
Crimean Astrophys Obs
Ukrainian Acad Science
Nauchny
334413 Crimea
Ukraine
☎ 380 6554 71164
✆ 380 6554 40704
✉ nish@crao.crimea.ua

Shakhovskoj Nikolay M
Crimean Astrophys Obs
Ukrainian Acad Science
Nauchny
334413 Crimea
Ukraine
☎ 380 6554 71161
✆ 380 6554 40704
✉ shakh@crao.crimea.ua

Shakura Nicholaj I
SAI
Acad Sciences
Universitetskij Pr 13
119899 Moscow
Russia
☎ 7 095 939 2858
✆ 7 095 939 0126

Shalabiea Osama M A
NAIGR
Helwan Observ
Cairo 11421
Egypt
☎ 20 78 0645/2683
✆ 20 62 21 405 297
✉ omarsh@frcu.eun.eg

Shallis Michael J
Astrophysics Dpt
Univ of Oxford
South Parks Rd
Oxford OX1 3RQ
UK
☎ 44 186 527 3999
✆ 44 186 527 3947

Shaltout Mosalam A M
NAIGR
Helwan Observ
Cairo 11421
Egypt
☎ 20 78 0645/2683
✆ 20 62 21 405 297

Shandarin Sergei F
Inst Physics Problems
Kosygin Str 2
117334 Moscow
Russia
☎ 7 095 137 3248

Shane William W
9095 Coker Rd
Prunedale CA 93907
USA

Shanks Thomas
Physics Dpt
Univ of Durham
South Rd
Durham DH1 3LE
UK
☎ 44 191 374 2171
✆ 44 191 374 3749
✉ tom.shanks@durham.ac.uk

Shao Cheng-yuan
CfA
HCO/SAO
60 Garden St
Cambridge MA 02138 1516
USA
☎ 1 617 495 7212
✆ 1 617 495 7356

Shao Zhengyi
Shanghai Observ
CAS
80 Nandan Rd
Shanghai 200030
China PR
☎ 86 21 6438 6191
✆ 86 21 6438 4618
✉ zyshao@center.shao.ac.cn

Shapero Donald C
Ntl Acad Sciences
NRC
2101 Constitution Av NW
Washington DC 20418
USA
☎ 1 202 334 3520
✆ 1 202 334 2791
✉ dshapero@nas.edu

Shapiro Irwin I
CfA
HCO/SAO Rm P 209
60 Garden St
Cambridge MA 02138 1516
USA
☎ 1 617 495 7100
℻ 1 617 495 7356
✉ shapiro@cfa.
harvard.edu

Shapiro Maurice M
Univ of Maryland
Suite 1514
205 Yoakum Parkway
Alexandria VA 22304
USA
☎ 1 703 370 1985

Shapiro Stuart L
CRSR
Cornell Univ
306 Space Sc Bldg
Ithaca NY 14853 6801
USA
☎ 1 607 256 4936
℻ 1 607 255 9888

Shapley Alan H
NOAA ERL R/E/SE
Space Environment Lab
325 Broadway
Boulder CO 80303
USA
☎ 1 303 497 3978

Shaposhnikov Vladimir E
Instte Appl Physics RAS
46 Ulyanov St
Nizhny Novgorod 603600
Russia
☎ 7 831 238 4341
℻ 7 831 236 2061
✉ sh130@appl.sci-
nnov.ru

Shara Michael
STScI
Homewood Campus
3700 San Martin Dr
Baltimore MD 21218
USA
☎ 1 301 338 4743
℻ 1 301 338 4767

Sharaf Mohamed Adel
Astronomy Dpt
Fac of Sciences
Cairo Univ
Geza
Egypt
☎ 20 2 572 7022
℻ 20 2 572 7556

Sharaf Sh G
Inst Theoret Astron
Acad Sciences
N Kutuzova 10
192187 St Petersburg
Russia
☎ 7 812 278 8898
℻ 7 812 272 7968

Sharma A Surjalal
Astronomy Dpt
Univ of Maryland
College Park MD 20742 2421
USA
☎ 1 301 405 1528
℻ 1 301 405 9966
✉ ssh@astro.umd.edu

Sharma Dharma Pal
Box 5
New Town TAS 7008
Australia
✉ sharma@physvax.phys.
utas.edu.au

Sharov A S
SAI
Acad Sciences
Universitetskij Pr 13
119899 Moscow
Russia
☎ 7 095 139 2657
℻ 7 095 939 0126

Sharp Christopher
Service Astrophysique
CEA Saclay
Orme d Merisiers Bt 709
F 91191 Gif s Yvette Cdx
France
☎ 33 1 69 08 9276
℻ 33 1 69 08 6577

Sharples Ray
Physics Dpt
Univ of Durham
South Rd
Durham DH1 3LE
UK
☎ 44 191 374 2000
℻ 44 191 374 3749

Sharpless Stewart
Dpt Phys/Astronomy
Univ of Rochester
Rochester NY 14627 0171
USA
☎ 1 716 275 4389
℻ 1 716 275 4351

Shastri Prajval
IIA
Koramangala
Sarjapur Rd
Bangalore 560 034
India
☎ 91 80 553 4043
℻ 91 80 553 0672
✉ pshastri@iiap.ernet.in

Shaver Peter A
ESO
Karl Schwarzschildstr 2
D 85748 Garching
Germany
☎ 49 893 200 6233
℻ 49 893 200 6480
✉ pshaver@eso.org

Shaviv Giora
Physics Dpt
IIT
Technion City
Haifa 32000
Israel
☎ 972 4 293 020
℻ 972 22 1680

Shaw James Scott
Dpt Phys/Astronomy
Univ of Georgia
Athens GA 30602 2451
USA
☎ 1 706 542 2485
℻ 1 706 542 2492

Shaw John H
Astronomy Dpt
Ohio State Univ
174 W 18th Av
Columbus OH 43210 1106
USA
☎ 1 614 422 7968
℻ 1 614 292 2928

Shawl Stephen J
Dpt Phys/Astronomy
Univ of Kansas
Lawrence KS 66045
USA
☎ 1 913 864 4626
℻ 1 913 864 5262
✉ Shawl@kuphsx.phsx.
ukans.edu

Shaya Edward J
Astronomy Program
Univ of Maryland
College Park MD 20742 2421
USA
☎ 1 301 454 6650
℻ 1 301 314 9067

Shcheglov P V
SAI
Acad Sciences
Universitetskij Pr 13
119899 Moscow
Russia
☎ 7 095 139 1973
℻ 7 095 939 0126

Shcherbakov Alexander
Crimean Astrophys Obs
Ukrainian Acad Science
Nauchny
334413 Crimea
Ukraine
☎ 380 6554 71194
℻ 380 6554 40704
✉ sherb@crao.crimea.ua

Shcherbina-Samojlova I
Inst Science/Tech
125219 Moscow
Russia
☎ 7 095 155 4237

Shchukina Nataliya
Main Astronomical Obs
Ukrainian Acad Science
Golosiiv
252650 Kyiv 22
Ukraine
☎ 380 4426 64762
℻ 380 4426 62147
✉ shchukin@mao.kiev.ua

Shea Margaret A
AFPL
Space Physics Div
Hanscom AFB
Bedford MA 01731 3010
USA

Shearer Andrew
Physics Dpt
Univ College
Galway
Ireland
☎ 353 91 524 411*3114
℻ 353 91 525 700
✉ shearer@epona.physics.ucg.ie

Sheeley Neil R
NRL
Code 4172
4555 Overlook Av SW
Washington DC 20375 5000
USA
☎ 1 202 767 2777

Sheffer Eugene K
SAI
Acad Sciences
Universitetskij Pr 13
119899 Moscow
Russia
☎ 7 095 939 2046
℻ 7 095 939 0126

Sheffield Charles
Earth Satellite Corp
6011 Executive Blvd
Suite 400
Rockville MD 20852 3804
USA
☎ 1 301 231 0660
℻ 1 301 231 5020
✉ csheffie@earthsat.com

Shefov Nicolai N
Inst Phys of Atmosph
Acad Sciences
Pyzhevsky 3
109017 Moscow
Russia

Shelus Peter J
Astronomy Dpt
Univ of Texas
Rlm 15 316
Austin TX 78712 1083
USA
☎ 1 512 471 3339
℻ 1 512 471 6016

Shematovich Valery I
Inst Theoret Astron
Acad Sciences
Pyatnitskaya 48
109017 Moscow
Russia
☎ 7 095 231 7375
℻ 7 095 230 2081
✉ shematov@inasan.rssi.ru

Shen Benjamin S P
Dpt Phys/Astronomy
Univ of Pennsylvania
Philadelphia PA 19104 6396
USA
☎ 1 215 898 8141
℻ 1 215 898 9336

Shen Changjun
Purple Mountain Obs
CAS
Nanjing 210008
China PR
☎ 86 25 46700
℻ 86 25 301 459

Shen Chun-Shan
Astron Sty China
Ntl Tsing Hua Univ
Hsin Chu 300043
China R
☎ 886 35 71 9039

Shen Kaixian
Shaanxi Observ
CAS
Box 18
Lintong 710600
China PR
☎ 86 33 2255
✆ 86 9237 3496

Shen Liangzhao
Beijing Astronomical Obs
CAS
Beijing 100080
China PR
☎ 86 1 28 1698
✆ 86 1 256 10855

Shen Longxiang
Beijing Astronomical Obs
CAS

W Suburb
Beijing 100080
China PR
☎ 86 1 28 1698
✆ 86 1 256 10855

Shen Parnan
Nanjing Astronomical
Instrument Factory
Box 846
Nanjing
China PR
☎ 86 25 46191
✆ 86 25 71 1256

Sher David
Box 9624
Cincinnati OH 452098
USA
☎ 1 513 871 8850

Sheridan K V
17b/23 Thornton St
Darling Point NSW 2027
Australia

Sherwood William A
RAIUB
Univ Bonn
auf d Huegel 69
D 53121 Bonn
Germany
☎ 49 228 525 362
✆ 49 228 525 229

Shevchenko Vladislav V
SAI
Acad Sciences
Universitetskij Pr 13
119899 Moscow
Russia
☎ 7 095 939 2858
✆ 7 095 939 0126

Shevgaonkar R K
Dpt Electric Engin
IIT
POWAI
Bombay
India
☎ 91 22 578 2545*2440
✆ 91 22 578 3480

Shi Guang-Chen
Purple Mountain Obs
CAS
Nanjing 210008
China PR
☎ 86 25 33921
✆ 86 25 301 459

Shi Shengcai
Purple Mountain Obs
CAS
Nanjing 210008
China PR
☎ 86 25 330 3738
✆ 86 25 330 1459
🖂 shencai@nro.nao.ac.jp

Shi Zhongxian
Beijing Astronomical Obs
CAS
Beijing 100080
China PR
☎ 86 1 28 1698
✆ 86 1 256 10855

Shibahashi Hiromoto
Astronomy Dpt
Univ of Tokyo
Bunkyo Ku
Tokyo 113
Japan
☎ 81 338 12 2111*4274
✆ 81 338 13 9439
🖂 shibahashi@astron.s.u-tokyo.ac.jp

Shibai Hiroshi
Nagoya Univ
Furo cho
Chikusa ku
Nagoya 464 8602
Japan
☎ 81 527 89 2452
✆ 81 527 89 2919
🖂 shibai@u.phys.nagoya-u.ac.jp

Shibasaki Kiyoto
Nobeyama Radio Obs
NAOJ
Minamimaki Mura
Nagano 384 13
Japan
☎ 81 267 63 4300
✆ 81 267 98 2884

Shibata Katsunori M
Ntl Astronomical Obs
Osawa 2-21-1
Mitaka
Tokyo 181
Japan
☎ 81 422 34 3750
✆ 81 422 34 3649
🖂 shibata@hotaka.mtk.nao.ac.jp

Shibata Kazunari
Ntl Astronomical Obs
Solar Physics div
Mitaka
Tokyo 181
Japan
☎ 81 422 34 3712
✆ 81 422 34 3700
🖂 shibata@spot.mtk.nao.ac.jp

Shibata Masaru
Dpt Earth/Space Sci
Osaka Univ
Toyonaka
Osaka 560
Japan
☎ 81 685 05 483
✆ 81 685 05 504
🖂 shibata@vega.ess.sci.osaka-u.ac.jp

Shibata Shinpei
Physics Dpt
Yamagata Univ
Kojirakawa
Yamagata 990
Japan
☎ 81 236 31 1421

Shibata Yukio
Res Inst Scientific Meast
Tohoku Univ
Aramaki
Sendai 980
Japan
☎ 81 206 543 2888

Shibazaki Noriaki
Physics Dpt
Rikkyo Univ
Nishi-Ikebukuro
Tokyo 171
Japan
☎ 81 339 85 2389
✆ 81 359 92 3434
🖂 shibazak@rikkyo.ac.jp

Shields Gregory A
Astronomy Dpt
Univ of Texas
Rlm 15 212
Austin TX 78712 1083
USA
☎ 1 512 471 4461
✆ 1 512 471 6016
🖂 shields@astro.as.texas.edu

Shields Joseph C
Dpt Phys/Astronomy
Ohio Univ
Clippinger Res Labs
Athens OH 45701 2979
USA
☎ 1 614 593 0336
✆ 1 614 593 0433
🖂 shields@helios.phy.ohiou.edu

Shigeyama Toshikazu
Astronomy Dpt
Univ of Tokyo
Bunkyo Ku
Tokyo 113
Japan
☎ 81 338 12 9224
✆ 81 338 13 9439
🖂 shigeyama@astron.s.u-tokyo.ac.jp

Shimasaku Kazuhiro
Astronomy Dpt
Univ of Tokyo
2-11-16 Yayoi Bunkyo-ku
Tokyo 113
Japan
☎ 81 356 84 0516
✆ 81 338 13 9439
🖂 shimasaku@astron.s.u-tokyo.ac.jp

Shimizu Mikio
ISAS
3 1 1 Yoshinodai
Sagamihara
Kanagawa 229 8510
Japan
☎ 81 427 51 3911
✆ 81 427 59 4253

Shimizu Tsutomu Emer
Terada Ootanti 26-16
Joyo Shi
Kyoto 610 01
Japan

Shimmins Albert John
5/36 Philipson St
Albert Park
Melbourne VIC 3206
Australia
☎ 61 3 9690 3803

Shine Richard A
Lockheed Palo Alto Res Lb
Dpt 91 30 Bg 256
3251 Hanover St
Palo Alto CA 94304 1211
USA
☎ 1 415 858 4135
✆ 1 415 424 3994

Shipman Harry L
Physics Dpt
Univ of Delaware
Newark DE 19716
USA
☎ 1 302 451 2986
✆ 1 302 831 1637

Shiryaev Alexander A
ASRI Techsat Project
IIT
Technion City
Haifa 32000
Israel
☎ 972 4 292 398
✆ 972 4 230 956
🖂 astchsat@vmsa.technion.ac.il

Shishov Vladimir I
Lebedev Physical Inst
Acad Sciences
Leninsky Pspt 53
117924 Moscow
Russia
☎ 7 095 135 2250
℻ 7 095 135 7880

Shitov Yuri P
Lebedev Physical Inst
Leninsky Pspt 53
Moscow 117924
Russia
☎ 7 096 773 2649
℻ 7 095 135 7880
✉ shitov@rasfian.serpukhov.su

Shivanandan Kandiah
NRL
Code 4138 S
4555 Overlook Av SW
Washington DC 20375 5000
USA
☎ 1 202 767 2749
℻ 1 202 767 6473

Shkodrov V G
Astronomy Dpt
Bulgarian Acad Sci
72 Lenin Blvd
BG 1784 Sofia
Bulgaria
☎ 359 2 75 8927
℻ 359 2 75 5019

Shkuratov Yurii
Astronomical Observ
Kharkiv State Univ
Sumskaja Ul. 35
310002 Kharkiv
Ukraine
☎ 380 5724 32428
✉ shkuratov@astron.kharkov.ua

Shlosman Isaac
Dpt Phys/Astronomy
Univ Kentucky
Lexington KY 40506 0055
USA
☎ 1 606 257 3461
℻ 1 606 323 2846
✉ shlosman@asta.pa.uky.edu

Shmeld Ivar
Radioastrophys Observ
Latvian Acad Sci
Akademijas Sq 1
LV 1527 Riga
Latvia
☎ 371 722 8321
℻ 371 782 1153
✉ shmeld@acad.latnet.lv

Shobbrook Robert R
Astronomy Dpt
Univ of Sydney
Sydney NSW 2006
Australia
☎ 61 2 692 3604
℻ 61 2 660 2903

Sholomitsky G B
Space Res Inst
Acad Sciences
Profsojuznaya Ul 84/32
117810 Moscow
Russia
☎ 7 095 333 3122
℻ 7 095 310 7023

Shone David
NRAL
Univ Manchester
Jodrell Bank
Macclesfield SK11 9DL
UK
☎ 44 14 777 1321
℻ 44 147 757 1618
✉ dls@jb.man.ac.uk

Shor Viktor A
Inst Theoret Astron
Acad Sciences
N Kutuzova 10
191187 St Petersburg
Russia
☎ 7 812 278 8809
℻ 7 812 272 7968

Shore Bruce W
LLNL
Box 808
Livermore CA 94551 9900
USA
☎ 1 415 422 6204
℻ 1 415 423 0238
✉ shore2@llnl.gov

Shore Steven N
Dpt Phys/Astronomy
Indiana Univ
1700 Mishawaka Av
South Bend IN 46634 7111
USA
☎ 1 219 237 4401

Shortridge Keith
AAO
Box 296
Epping NSW 2121
Australia
☎ 61 2 9372 4822
℻ 61 2 9372 4880
✉ ks@aaoepp.aao.gov.au

Shostak G Seth
1372 Cuernavaca Circ
Mountain View CA 94040
USA
☎ 1 415 967 8193
✉ Seth_Shostak@setigate.seti-inst.edu

Shu Chenggang
Shanghai Observ
CAS
80 Nandan Rd
Shanghai 200030
China PR
☎ 86 21 6438 6191
℻ 86 21 6438 4618

Shu Frank H
Astronomy Dpt
Univ of California
601 Campbell Hall
Berkeley CA 94720 3411
USA
☎ 1 415 642 2529
℻ 1 510 642 3411

Shukla K
Dpt Maths/Astronomy
Univ of Lucknow
Lucknow
India

Shukre C S
RRI
Sadashivanagar
CV Raman Av
Bangalore 560 080
India
☎ 91 80 336 0122
℻ 91 80 334 0492
✉ shukre@rri.ernet.in

Shul-man L M
Main Astronomical Obs
Ukrainian Acad Science
Golosiiv
252650 Kyiv 22
Ukraine
☎ 380 4426 63110
℻ 380 4426 62147

Shulga Valery
Inst Radio Astron
Ukrainian Acad Science
4 Chervonopraporna st
310002 Kharkiv
Ukraine
☎ 380 5724 48591
℻ 380 5724 76506
✉ shulga@rian.kharkov.ua

Shull John Michael
JILA
Univ of Colorado
Campus Box 440
Boulder CO 80309 0440
USA
☎ 1 303 492 7827
℻ 1 303 492 5235

Shull Peter Otto
Physics Dpt
Oklahoma State Univ
Stillwater OK 74078 3072
USA
☎ 1 405 744 5785
℻ 1 405 744 6811
✉ pos@okstate.edu

Shulov Oleg S
Astronomical Observ
St Petersburg Univ
Bibliotechnaja Pl 2
199178 St Petersburg
Russia
☎ 7 812 428 7129
℻ 7 812 428 4259

Shustov Boris M
Inst of Astronomy
Acad Sciences
Pyatnitskaya Ul 48
109017 Moscow
Russia
☎ 7 095 231 5461
℻ 7 095 230 2081
✉ bshustov@inasan.rssi.ru

Sibille Francois
Observ Lyon
Av Ch Andre
F 69561 S Genis Laval cdx
France
☎ 33 4 78 56 0705
℻ 33 4 72 39 9791

Sicardy Bruno
Observ Paris Meudon
EUROPA
Pl J Janssen
F 92195 Meudon PPL Cdx
France
☎ 33 1 45 07 7409
℻ 33 1 45 07 7469
✉ sicardy@obspm.fr

Sidlichovsky Milos
Astronomical Institute
Czech Acad Sci
Bocni Ii 1401
CZ 141 31 Praha 4
Czech R
☎ 420 2 67 10 3078
℻ 420 2 76 90 23
✉ sidlich@csearn

Sidorenkov Nikolay S
Hydrometeorologic Res Ctr
Bolshoi Predtechensky Per. 9-11
123342 Moscow
Russia
☎ 7 095 255 5026
℻ 7 095 255 1582

Sieber Wolfgang
FN Niederrhein
Postfach 2850
D 47728 Krefeld
Germany
☎ 49 215 182 2317
℻ 49 215 182 2317
✉ sieber@kr.fh-niederrhein.de

Sienkiewicz Ryszard
Copernicus Astron Ctr
Polish Acad Sci
Ul Bartycka 18
PL 00 716 Warsaw
Poland
☎ 48 22 41 1086
℻ 48 41 0828

Signore Monique
DEMIRM
ENS
24 r Lhomond
F 75231 Paris Cdx 05
France
☎ 33 1 44 32 3989
℻ 33 1 44 32 3992
✉ signore@physique.ens.fr

Sigurdsson Steinn
Dpt Astron/Astrophys
Pennsylvania State Univ
525 Davey Laboratory
University Park PA 16802
USA
☎ 1 814 865 3631
✆ 1 814 863 7114
✉ steinn@ast.cam.ac.uk

Sikora Marek
Copernicus Astron Ctr
Polish Acad Sci
Ul Bartycka 18
PL 00 716 Warsaw
Poland
☎ 48 22 41 1086
✆ 48 41 0828

Sikorski Jerzy
Inst Theor Physics
Univ of Gdansk
Ul Wita Stwosza 57
PL 80 952 Gdansk
Poland
☎ 48 41 5241*188
✉ fizjks@halina.univ.gda.pl

Sil'chenko Olga K
SAI
Acad Sciences
Universitetskij Pr 13
119899 Moscow
Russia
☎ 7 095 939 2657
✆ 7 095 932 8841
✉ olga@sai.msu.su

Silant-ev Nikolai
Pulkovo Observ
Acad Sciences
10 Kutuzov Quay
196140 St Petersburg
Russia
☎ 7 812 123 4090
✆ 7 812 315 1701
✉ alexeeva@gaoran.spb.su

Silberberg Rein
NRL
Code 4154
4555 Overlook Av SW
Washington DC 20375 5000
USA
☎ 1 202 767 2803

Silich Sergey
Main Astronomical Obs
Ukrainian Acad Science
Golosiiv
252650 Kyiv 22
Ukraine
☎ 380 4426 64771
✆ 380 4426 62147
✉ maouas@gluk.apc.org

Silk Joseph I
Astronomy Dpt
Univ of California
601 Campbell Hall
Berkeley CA 94720 3411
USA
☎ 1 415 642 2113
✆ 1 510 642 3411

Sillanpaa Aimo Kalevi
Turku Univ
Tuorla Observ
Vaeisaelaentie 20
FIN 21500 Piikkio
Finland
☎ 358 2 274 4262
✆ 358 2 243 3767
✉ aimosill@kontu.utu.fi

Silverberg Eric C
McDonald Observ
Univ of Texas
Box 1337
Fort Davis TX 79734 1337
USA

Silvestro Giovanni
Istt Fisica
Univ d Torino
Corso d Azeglio 46
I 10125 Torino
Italy
☎ 39 11 658 623

Sim Mary E
Royal Observ
Blackford Hill
Edinburgh EH9 3HJ
UK
☎ 44 131 667 3321
✆ 44 131 668 8356

Sima Zdislav
Astronomical Institute
Czech Acad Sci
Bocni Ii 1401
CZ 141 31 Praha 4
Czech R
☎ 420 2 67 10 3042
✆ 420 2 76 90 23
✉ sima@ig.cas.cz

Simek Milos
Astronomical Institute
Czech Acad Sci
Fricova 1
CZ 251 65 Ondrejov
Czech R
☎ 420 204 85 7252
✆ 420 2 88 1611
✉ semik@asu.cas.cz

Simien Francois
Observ Lyon
Av Ch Andre
F 69561 S Genis Laval cdx
France
☎ 33 4 78 56 0705
✆ 33 4 72 39 9791

Simkin Susan M
Physics/Astronomy Dpt
Michigan State Univ
East Lansing MI 48824
USA
☎ 1 517 353 4540
✆ 1 517 353 4500

Simmons John Francis l
31 Havelock St
Glasgow G11 5HA
UK

Simnett George M
Dpt Space Res
Univ Birmingham
Box 363
Birmingham B15 2TT
UK
☎ 44 12 14 72 1301

Simo Charles
Fac Matematicas
Univ Barcelona
Av Jose Antonio 585
E 08028 Barcelona
Spain

Simoda Mahiro
1 362 6
Suzuki Kodaira
Tokyo 187
Japan

Simon George W
AFGL
NSO
Sunspot NM 88349
USA
☎ 1 505 434 1390
✆ 1 504 434 7029

Simon Guy
Observ Paris Meudon
DASGAL
Pl J Janssen
F 92195 Meudon PPL Cdx
France
☎ 33 1 45 07 7787
✆ 33 1 45 07 7971

Simon Jean-Louis
BDL
77 Av Denfert Rochereau
F 75014 Paris
France
☎ 33 1 43 20 1210
✆ 33 1 46 33 2834
✉ jean-louis.simon@bdl.fr

Simon Klaus Peter
Inst Astron/Astrophysik
Univ Sternwarte
Scheinerstr 1
D 81679 Muenchen
Germany
☎ 49 899 890 21
✆ 49 899 220 9427

Simon Michal
Dpt Earth/Space Sci
Astronomy Program
Suny at Stony Brook
Stony Brook NY 11794 2100
USA
☎ 1 516 246 7672
✆ 1 515 632 8240

Simon Norman R
Dpt Phys/Astronomy
Univ Nebraska
Behlen Observatory
Lincoln NE 68588 0111
USA
☎ 1 402 472 2788
✆ 1 402 472 2879

Simon Paul C
IASB
Ringlaan 3
B 1180 Brussels
Belgium
☎ 32 2 373 0400
✆ 32 2 375 1579
✉ paul.simon@oma.be

Simon Rene L E
Inst Astrophysique
Universite Liege
Av Cointe 5
B 4000 Liege
Belgium
☎ 32 4 254 7510
✆ 32 4 254 7511

Simon Theodore
Inst for Astronomy
Univ of Hawaii
2680 Woodlawn Dr
Honolulu HI 96822
USA
☎ 1 808 956 6317
✆ 1 808 988 2790

Simonneau Eduardo
IAP
98bis bd Arago
F 75014 Paris
France
☎ 33 1 44 32 8143
✆ 33 1 44 32 8001

Simons Stuart
School Mathemat Sc
QMWC
Mile End Rd
London E1 4NS
UK
☎ 44 171 980 4811
✆ 44 181 975 5500

Simonson S Christian
1061 Russell Av
Los Altos CA 94022
USA
☎ 1 415 968 0473

Simovljevitch Jovan L
Astronomy Dpt
Univ of Belgrade
Studentski Trg 16
11000 Beograd
Yugoslavia FR
☎ 381 11 638 715
✆ 381 11 630 151

Sims Kenneth P
Sydney Observ
Observ Park
Sydney NSW 2000
Australia
☎ 61 2 217 0485
✆ 61 2 217 0489.

Sinachopoulos D
Koninklijke Sterrenwacht
van Belgie
Ringlaan 3
B 1180 Brussels
Belgium
☎ 32 2 373 0291
✆ 32 2 374 9822
✉ dimitris@oma.be

Sinclair Andrew T
Royal Greenwich Obs
Madingley Rd
Cambridge CB3 0EZ
UK
☎ 44 12 23 374 000
✆ 44 12 23 374 700

Singh H P
Dpt Phys/Delhi Univ
Sri Venkateswara College
Dhaula Kuan
New Delhi 110 021
India
☎ 91 11 729 1309

Singh Jagdev
IIA
Koramangala
Sarjapur Rd
Bangalore 560 034
India
☎ 91 80 356 6585/6497
✆ 91 80 553 4043

Singh Kulinder Pal
TIFR
Homi Bhabha Rd
Colaba
Bombay 400 005
India
☎ 91 22 215 2971*2376
✆ 91 22 495 2110
✉ singh@tifrvax.tifr.res.in

Singh Patan Deen
IAG
Univ Sao Paulo
CP 9638
01065 970 Sao Paulo SP
Brazil
☎ 55 11 275 3720
✆ 55 11 276 3848
✉ iagusp@brfapesp

Sinha K
Indira Gandhi Planet
9 Nabiullah Rd
Surajkund Park
Lucknow 226 018
India
☎ 91 522 229 176
✆ 91 522 229 176/211 793

Sinha Rameshwar P
8700 Brickyard Rd
Potomac MD 20854
USA
☎ 1 301 983 3899
✉ rsinha@nesdis.noaa.gov

Sinnerstad Ulf E
Stockholm Observ
Royal Swedish Acad Sciences
S 133 36 Saltsjoebaden
Sweden
☎ 46 8 717 0195
✆ 46 8 717 4719

Sinton William M
850 E David Dr
Flagstaff AZ 86001
USA
☎ 1 602 774 8308

Sinvhal Shambhu Dayal
175DH Scheme 74C
Vijalnagar
Indore 452 010
India

Sinzi Akira M
Hydrographic Dpt
Geodesy/Geophys Div
Tsukiji 5 Chuo Ku
Tokyo 104
Japan
☎ 81 335 41 3816
✆ 81 3545 2885

Sion Edward Michael
Astronomy Dpt
Villanova Univ
Villanova PA 19085
USA
☎ 1 215 645 4822
✉ scion@scivax.stsci.edu

Siregar Suryadi
Astronomy Dpt
Bandung Instte Techn
Jl Ganesha 10
Bandung 40132
Indonesia
☎ 62 22 244 0252
✆ 62 22 2505442

Siroky Jaromir
Palacky Univ
Dpt Physics/Astronomy
Lenin St 26
CZ 771 46 Olomouc
Czech R
☎ 420 68 22451

Sironi Giorgio
Dpt Fisica
Univ d Milano
Via Celoria 16
I 20133 Milano
Italy
☎ 39 2 239 2272
✆ 39 2 706 38413
✉ giorgio.sironi@mi.infn.it

Sirousse Zia Haydeh
Gravit/Cosmologie
Relativi UPMC Tour 22
4 Pl Jussieu
F 75252 Paris Cedex 05
France
☎ 33 1 44 27 7292
✆ 33 1 44 27 7287
✉ strousse@ccr.jussieu.fr

Siry Joseph W
4438 42nd St Nw
Washington DC 20016
USA

Sisson George M
Planetrees
Wall
Hexham NE46 4EQ
UK
☎ 44 14 348 1434

Sistero Roberto F
Observ Astronomico
de Cordoba
Laprida 854
5000 Cordoba
Argentina
☎ 54 51 23 0491
✆ 54 51 21 0613
✉ sistero@astro.edu.ar

Sitarski Grzegorz
Space Res Ctr
Polish Acad Sci
Ul Ordona 21
PL 01 237 Warsaw
Poland
☎ 48 22 40 3766
✆ 48 22 36 8961

Sitko Michael L
Physics Dpt
Univ of Cincinnati
210 Braunstein Ml 11
Cincinnati OH 45221 0011
USA
☎ 1 513 556 0501
✆ 1 513 556 3425
✉ sitko@dusty.phy.uc.edu

Sitnik G F
SAI
Acad Sciences
Universitetskij Pr 13
119899 Moscow
Russia
☎ 7 095 139 1973
✆ 7 095 939 0126

Sivan Jean-Pierre
OHP
F 04870 St Michel Obs
France
☎ 33 4 92 70 6400
✆ 33 4 92 76 6295
✉ sivan@obs-hp.fr

Sivaram C
IIA
Koramangala
Sarjapur Rd
Bangalore 560 034
India
☎ 91 80 356 6585/6497
✆ 91 80 553 4043

Sivaraman K R
IIA
Koramangala
Sarjapur Rd
Bangalore 560 034
India
☎ 91 80 356 6585
✆ 91 80 553 4043

Sjogren William L
CALTECH/JPL
MS 264 664
4800 Oak Grove Dr
Pasadena CA 91109 8099
USA
☎ 1 818 354 4868
✆ 1 818 393 6030

Skalafuris Angelo J
NRL
Code 5307
4555 Overlook Av SW
Washington DC 20375 5000
USA
☎ 1 302 767 3227

Skillen Ian
Inst of Astronomy
The Observatories
Madingley Rd
Cambridge CB3 0HA
UK
☎ 44 12 23 337 548
✆ 44 12 23 337 523

Skilling John
Dpt Appl Maths/Theor Phys
Silver Street
Cambridge CB3 9EW
UK
☎ 44 12 23 337 87
✆ 44 12 23 337 918

Skillman Evan D
School Phys/Astronomy
Univ Minnesota
116 Church St SE
Minneapolis MN 55455
USA
☎ 1 612 624 4523
✆ 1 612 626 2029
✉ skillman@ast1.spa.umn.edu

Skinner Gerald
School Physics/Res
Univ Birmingham
Box 363
Birmingham B15 2TT
UK
☎ 44 12 14 14 6450

Skopal Augustin
Astronomical Institute
Slovak Acad Sci
SK 059 60 Tatranska Lomni
Slovak Republic
☎ 421 969 967 866
✆ 421 969 967 656
✉ astrskop@ta3.sk

Skripnichenko Vladimir
Inst Appl Astronomy
Acad Sciences
Zhdanovskaya Ul 8
197042 St Petersburg
Russia
☎ 7 812 230 7414
✆ 7 812 230 7413

Skulachov Dmitry
Space Res Inst
Acad Sciences
Profsojuznaya Ul 84/32
117810 Moscow
Russia
☎ 7 095 333 2588
✆ 7 095 310 7023

Skulskyj Mychajlo Y
Physics Dpt
State Univ Lvivska Polite
S Bandera Str
290646 Lviv
Ukraine
☎ 380 3227 44300
✆ 380 3227 44300
✉ msky@astro.lviv.ua

Skumanich Andre
HAO
NCAR
Box 3000
Boulder CO 80307 3000
USA
☎ 1 303 497 1528
✆ 1 303 497 1568

Slade Martin A III
CALTECH/JPL
MS 238 420
4800 Oak Grove Dr
Pasadena CA 91109 8099
USA
☎ 1 818 354 6538
✆ 1 818 393 6030

Sleath John
6 The Quadrant
Little Earling Lane
London W5 4EE
UK
☎ 44 181 579 0222

Slee O B
CSIRO
Div Radiophysics
Box 76
Epping NSW 2121
Australia
☎ 61 2 868 0222
✆ 61 2 868 0310

Slettebak Arne
Astronomy Dpt
Ohio State Univ
174 W 18th Av
Columbus OH 43210 1106
USA
☎ 1 614 292 7861
✆ 1 614 292 2928
✉ slettebk@ohstpy.mps.ohio-state

Slezak Eric
OCA
Observ Nice
BP 139
F 06304 Nice Cdx 4
France
☎ 33 4 92 00 3124
✆ 33 4 92 00 3033
✉ slezak@obs-nice.fr

Sloan Gregory Clayton
NASA/ARC
MS 245 6
Moffett Field CA 94035 1000
USA
☎ 1 415 604 5495
✆ 1 415 604 6779
✉ sloan@ssa1.arc.nasa.gov

Slonim E M
Astronomical Institute
Uzbek Acad Sci
Astronomicheskaya Ul 33
700000 Tashkent
Uzbekistan
☎ 7 3712 35 8102

Slovak Mark Haines
Po Box 751044
Memphis TN 38175
USA

Slysh Viacheslav I
Astro Space Ctr
Lebedev Physical Inst
Profsoyuznaya 84/32
117810 Moscow
Russia
☎ 7 095 333 2167
✆ 7 095 333 2378
✉ vslysh@dpc.asc.rssi.ru

Smak Joseph I
Copernicus Astron Ctr
Polish Acad Sci
Ul Bartycka 18
PL 00 716 Warsaw
Poland
☎ 48 22 41 1086
✆ 48 22 41 0828
✉ jis@alfa.camk.edu.pl

Smaldone Luigi Antonio
Dpt Fisica
Univ d Napoli
Mostra D Oltremare Pad 19
I 80125 Napoli
Italy
☎ 39 81 725 3428

Smale Alan Peter
NASA GSFC
Code 668
Greenbelt MD 20771
USA
☎ 1 301 286 7063
✆ 1 301 286 3391
✉ smale@lheaux.gsfc.nasa.gov

Smalley Barry
Physics Dpt
Univ of Keele
Keele ST5 5BG
UK
☎ 44 178 262 1111
✆ 44 178 271 1093
✉ bs@astro.keele.ac.uk

Smecker-Hane Tammy A
Dpt Phys/Astronomy
UCI
Irvine CA 92697 4575
USA
☎ 1 714 824 7773
✆ 1 714 824 2174
✉ tsmecker@uci.edu

Smette Alain
NASA GSFC
Code 681
Greenbelt MD 20771
USA
☎ 1 301 286 8619
✆ 1 301 286 1752
✉ asmette@band3.gsfc.nasa.gov

Smeyers Paul
Inst v Sterrenkunde
Katholicke Univ Leuven
Celestijnenlaan 200B
B 3001 Heverlee
Belgium
☎ 32 16 32 7033
✆ 32 16 32 7999

Smirnov Michael
Inst of Astronomy
Acad Sciences
Pyatnitskaya Ul 48
109017 Moscow
Russia
☎ 7 095 233 1624
✆ 7 095 230 2081
✉ rykhlova@inasan.rssi.ru

Smit J A
Sterrekundig Inst Utrecht
Box 80000
NL 3508 TA Utrecht
Netherlands
☎ 31 30 253 5200
✆ 31 30 253 1601

Smith Alex G
Astronomy Dpt
Univ of Florida
211 SSRB
Gainesville FL 32611
USA
☎ 1 904 392 6135
✆ 1 904 392 5089

Smith Barham W
LANL
MS D436
Box 1663
Los Alamos NM 87545 2345
USA
☎ 1 505 667 1585
✆ 1 505 665 4055
✉ bwsmith@lanl.gov

Smith Bradford A
Inst f Astronomy
Univ of Hawaii
82-6012 Pu-uhonua Rd
Napo'opo'o HI 96740 8226
USA
☎ 1 808 328 2568
✆ 1 808 328 9472
✉ brad@mahina.ifa.hawaii.edu

Smith Bruce F
NASA/ARC
MS 245 3
Moffett Field CA 94035 1000
USA
☎ 1 415 694 5515
✆ 1 415 604 6779

Smith Craig H
Physics Dpt Univ College
UNSW
Northcott Dr
Campbell ACT 2600
Australia
☎ 61 2 62 68 8790
✆ 61 26 268 8786
✉ craig@phadfa.ph.adfa.oz.au

Smith Dean F
Berkeley Res Ass
290 Green Rock Dr
Boulder CO 80302
USA
☎ 1 303 444 1922

Smith Eric Philip
NASA GSFC
Code 681
Greenbelt MD 20771
USA
☎ 1 301 286 8549
✆ 1 301 286 1753
✉ eric.p.smith@gsfc.nasa.gov

Smith Francis Graham
NRAL
Univ Manchester
Jodrell Bank
Macclesfield SK11 9DL
UK
☎ 44 14 777 1321
✆ 44 147 757 1618

Smith Geoffrey
Astrophysics Dpt
Univ of Oxford
Keble Rd
Oxford OX1 3RH
UK
☎ 44 186 527 3304
✆ 44 186 527 3947
✉ gs@astro.ox.ac.uk

Smith Graeme H
Lick Observ
Univ of California
Santa Cruz CA 95064
USA
☎ 1 831 459 2513
✆ 1 831 426 3115

Smith Haywood C
Astronomy Dpt
Univ of Florida
Box 112055
Gainesville FL 32611
USA
☎ 1 904 392 7744
✆ 1 904 392 5089
✉ hsmith@astro.ufl.edu

Smith Humphry M
23 Normandale
Bexhill on Sea TN39 3LU
UK
☎ 44 14 24 21 4288

Smith Keith Colin
Dpt Phys/Astronomy
UCLO
Gower St
London WC1E 6BT
UK
☎ 44 171 387 7050*3410
✆ 44 171 380 7145
✉ kcs@star.ucl.ac.uk

Smith Linda J
Dpt Phys/Astronomy
UCLO
Gower St
London WC1E 6BT
UK
☎ 44 171 387 7050*788
✆ 44 171 380 7145

Smith Malcolm G
NOAO
CTIO
Casilla 603
La Serena
Chile
☎ 56 51 22 5415
✆ 56 51 20 5342
✉ msmith@noao.edu

Smith Michael
Astronomisches Institut
Univ Wuerzburg
am Hubland
D 97074 Wuerzburg
Germany
☎ 49 931 888 5038
✆ 49 931 888 4603
✉ smith@astro.uni-wuerzburg.de

Smith Myron A
STScI
Homewood Campus
3700 San Martin Dr
Baltimore MD 21218
USA
☎ 1 301 338 5036
✆ 1 301 338 4767
✉ msmith@stsci.edu

Smith Niall
Applied Physics/Instrum Dpt
Rossa Av
Bishopstown
Cork
Ireland
☎ 353 2 132 6369
✆ 353 2 134 5191
✉ nsmith@rtc-cork.ie

Smith Peter L
CfA
HCO/SAO MS 50
60 Garden St
Cambridge MA 02138 1516
USA
☎ 1 617 495 4984
✆ 1 617 495 7455
✉ plsmith@cfa.harvard.edu

Smith Robert Connon
Astronomy Centre
Univ of Sussex
Falmer
Brighton BN1 9QH
UK
☎ 44 12 73 678 974
✆ 44 12 73 678 097
✉ r.c.smith@sussex.ac.uk

Smith Robert G
Physics Dpt Univ College
UNSW
Northcott Dr
Campbell ACT 2600
Australia
☎ 61 2 62 68 8746
✆ 61 26 268 8786
✉ rgs@phadfa.ph.adfa.oz.au

Smith Rodney M
Dpt Phys/Astronomy
Univ of Wales
Box 913
Cardiff CF2 3YB
UK
☎ 44 12 22 874 000*5282
✆ 44 12 22 874 056
✉ r.smith@astro.cf.ac.uk

Smith Verne V
Astronomy Dpt
Univ of Texas
Rlm 15 308
Austin TX 78712 1083
USA
☎ 1 512 471 3351
✆ 1 512 471 6016

Smith W Hayden
Physics Dpt
Washington Univ
Mcdonnel Ctr Space Sci
St Louis MO 63130
USA
☎ 1 314 889 6574
✆ 1 314 935 4083

Smith Harding E
CASS
UCSD
C 011
La Jolla CA 92093 0216
USA
☎ 1 419 534 4558
✆ 1 419 534 2294

Smits Derck P
HartRAO
FRD
PO Box 443
1740 Krugersdorp
South Africa
☎ 27 11 642 4692
✆ 27 11 642 2424
✉ derck@bootes.hartrao.ac.za

Smol-Kov Gennadij Ya
Sibizmir
Acad Sciences
664697 Irkutsk 33
Russia
☎ 7 395 262 9388

Smoot III George F
Lawrence Berkeley Lab
Univ of California
Bg 50 230
Berkeley CA 94720
USA
☎ 1 415 486 5237
✆ 1 510 486 6738

Smriglio Filippo
Istt Astronomico
Univ d Roma La Sapienza
Via G M Lancisi 29
I 00161 Roma
Italy
☎ 39 6 44 03734
✆ 39 6 44 03673

Smylie Douglas E
Dpt Phys/Earth/Atm Sci
York Univ
4700 Keele St
Downsview ON M3J 1P3
Canada
☎ 1 416 736 5245

Smyth Michael J
Royal Observ
Blackford Hill
Edinburgh EH9 3HJ
UK
☎ 44 131 667 3321
✆ 44 131 668 8356
✉ mjs@roe.ac.uk

Sneden Christopher A
Astronomy Dpt
Univ of Texas
Rlm 16.324
Austin TX 78712 1083
USA
☎ 1 512 471 1349
✆ 1 512 471 6016
✉ chris@verdi.as.texas.edu

Snell Ronald L
Five College
RAO
B619 Lederle Grad Res Twr
Amherst MA 01003
USA
☎ 1 413 545 1949

Snezhko Leonid I
SAO
Acad Sciences
Nizhnij Arkhyz
357147 Karachaevo
Russia
☎ 7 878 789 3513

Snijders Mattheus A J
IRAM
300 r La Piscine
F 38406 S Martin Heres Cdx
France
☎ 33 4 76 42 3383
✆ 33 4 76 51 5938

Snow Theodore P
CASA
Univ of Colorado
Campus Box 389
Boulder CO 80309 0389
USA
☎ 1 303 492 4050
✆ 1 303 492 7178
✉ tsnow@casa.colorado.edu

Snyder Lewis E
Dpt Astron/Physics
Univ of Illinois
1011 W Springfiels Av
Urbana IL 61801
USA
☎ 1 217 333 5530
✆ 1 217 244 7638

Soares Domingos S L
Physics Dpt
UFMG
CP 702
30161 970 Belo Horizonte
Brazil
☎ 55 31 499 5649
✆ 55 31 499 5600
✉ dsoares@fisica.ufmg.br

Soberman Robert K
Dpt Astron/Astrophys
Univ of Pennsylvania
David Rittenhouse Lab
Philadelphia PA 19104
USA
☎ 1 215 898 8176
✆ 1 215 898 9336

Sobieski Stanley
NASA GSFC
Code 673
Greenbelt MD 20771
USA

Sobolev Andrej M
Ural State Univ
Astronomical Observ
Lenin Av 51
Ekaterinburgh 620083
Russia
☎ 7 343 261 5431
✆ 7 343 255 5964
✉ Andrej.sobolev@usu.ru

Sobolev V V
Astronomical Observ
St Petersburg Univ
Bibliotechnaja Pl 2
199178 St Petersburg
Russia
☎ 7 812 428 7129
✆ 7 812 428 4259

Soboleva N S
Pulkovo Observ
Acad Sciences
10 Kutuzov Quay
196140 St Petersburg
Russia
☎ 7 812 298 2242
✆ 7 812 315 1701

Sobotka Michal
Astronomical Institute
Czech Acad Sci
Fricova 1
CZ 251 65 Ondrejov
Czech R
☎ 420 204 85 201
✆ 420 2 88 1611
✉ msobotka@asu.cas.cz

Sobouti Yousef
Inst Advanced Studies
in Basic Sciences
Gaveh Zang
Zanjan
Iran
☎ 98 241 449 0212
✆ 98 241 449 023
✉ sobouti@rose.ipm.ac.ir

Sochilina Alla S
Inst Theoret Astron
Acad Sciences
N Kutuzova 10
191187 St Petersburg
Russia
☎ 7 812 278 8898
✆ 7 812 272 7968
✉ 1099@ita.spb.su

Sodemann M
Inst Phys/Astronomy
Univ of Aarhus
Ny Munkegade
DK 8000 Aarhus C
Denmark
☎ 45 86 12 8899
✆ 45 86 20 2711

Soderblom David R
STScI
Homewood Campus
3700 San Martin Dr
Baltimore MD 21218
USA
☎ 1 301 338 4543
✆ 1 301 338 5090
✉ soderblom@stsci.edu

Soderblom Larry
USGS
Br of Astrogeology
2255 N Gemini Dr
Flagstaff AZ 86001
USA

Soderhjelm Staffan
Lund Observ
Box 43
S 221 00 Lund
Sweden
☎ 46 46 222 7303
✆ 46 46 222 4614
✉ staffan@astro.lu.se

Sodin Leonid
Inst Radio Astron
Ukrainian Acad Science
4 Chervonopraporna st
310002 Kharkiv
Ukraine
☎ 380 5724 71134
✆ 380 5724 76506
✉ rai@ira.kharkov.ua

Sodre Laerte
IAG
Univ Sao Paulo
CP 9638
01065 970 Sao Paulo SP
Brazil
☎ 55 11 5778 599
✆ 55 11 2763 848
✉ laerte@astro1.iagusp.
usp.br

Soffel Michael
Inst Planet Geodaesie
Technische Univ Dresden
Mommsenstr 13
D 01062 Dresden
Germany
☎ 49 351 463 4200
✆ 49 351 463 7019
✉ soffel@rcs.urz.tu-dresden.de

Sofia Sabatino
Astronomy Dpt
Yale Univ
Box 208101
New Haven CT 06520 8101
USA
☎ 1 203 432 3011
✆ 1 203 432 5048
✉ sofia@astro.yale.edu

Sofue Yoshiaki
Inst of Astronomy
Univ of Tokyo
Osawa Mitaka
Tokyo 181
Japan
☎ 81 422 41 3734
✆ 81 422 41 3749
✉ y.sifyi@tansei.cc.
u-tokyo.ac.

Soifer Baruch T
CALTECH
MS 320 47
Downes Lab of Physics
Pasadena CA 91125
USA
☎ 1 818 356 6626
✆ 1 818 796 8806

Sokolov Konstantin
Inst Radio Astron
Ukrainian Acad Science
4 Chervonopraporna st
310002 Kharkiv
Ukraine
☎ 380 5724 51014
✆ 380 5724 76506
✉ rai@ira.kharkov.ua

Sokolov Viktor G
Inst Theoret Astron
Acad Sciences
10 Nab Kutuzova
191187 St Petersburg
Russia
☎ 7 812 275 1090
✆ 7 812 272 7968
✉ 1095@ita.spb.su

Sokolowski Lech
Astronomical Observ
Krakow Jagiellonian Univ
Ul Orla 171
PL 30 244 Krakow
Poland
☎ 48 12 25 1294
✆ 48 12 25 1318

Sokolsky Andrej G
Inst Theoret Astron
Acad Sciences
N Kutuzova 10
191187 St Petersburg
Russia
☎ 7 812 279 0667
✆ 7 812 272 7968
✉ sokolsky@iiii.spb.su

Sol Helene
Observ Paris Meudon
DAEC
Pl J Janssen
F 92195 Meudon PPL Cdx
France
☎ 33 1 45 07 7428
✆ 33 1 45 07 7469

Solanes Majua Jose M
Astronomy Dpt
Cornell Univ
512 Space Sc Bldg
Ithaca NY 14853 6801
USA
☎ 1 607 255 6915
✆ 1 607 255 8803
✉ solanes@astrosun.tn.
cornell.edu

Solanki Sami K
Inst Astronomie
ETH Zentrum
CH 8092 Zuerich
Switzerland
☎ 41 1 632 3810
✆ 41 1 632 1205
✉ solanki@astro.phys.
ethz.ch

Solaric Nikola
Hvar Observ
Fac Geodesy
Kaciceva 26
HR 10000 Zagreb
Croatia
☎ 385 1 521 548
✆ 385 1 445 410

Solc Martin
Astronomical Institute
Charles Univ
V Holesovickack 2
CZ 180 00 Praha 8
Czech R
☎ 420 2 2191 2572
✆ 420 2 2191 1292
✉ solc@mbox.troja.
mff.cuni.cz

Solf Josef
Thueringer Landessternwarte
Sternwarte 5
D 07778 Tautenburg
Germany
☎ 49 364 278 630
✆ 49 364 278 6329

Solheim Jan Erik
Physics Dpt
Univ of Tromso
N 9037 Tromso
Norway
☎ 47 77 64 5191
✆ 47 77 64 5595
✉ janerik@phys.uit.no

Soliman Mohamed Ahmed
NAIGR
Helwan Observ
Cairo 11421
Egypt
☎ 20 78 0645/2683
✆ 20 62 21 405 297

Solivella Gladys Rebecca
Observ Astronomico
Paseo d Bosque S/n
1900 La Plata (Bs As)
Argentina
☎ 54 21 217 308
✆ 54 21 38 810
✉ gladys@fcaglp.edu.ar

Sollazzo Claudio
ESOC
Robert-Bosch Str 5
D 64293 Darmstadt
Germany
☎ 49 615 188 61

Solomon Philip M
Dpt Earth/Space Sci
Astronomy Program
Suny at Stony Brook
Stony Brook NY 11794 2100
USA
☎ 1 516 246 8383
✆ 1 515 632 8240

Solovaya Nina A
SAI
Acad Sciences
Universitetskij Pr 13
119899 Moscow
Russia
☎ 7 095 939 3764
✆ 7 095 939 0126
✉ ursa@sai.msk.su

Soltan Andrzej Maria
Copernicus Astron Ctr
Polish Acad Sci
Ul Bartycka 18
PL 00 716 Warsaw
Poland
☎ 48 22 41 1086
✆ 48 22 41 0046

Soltau Dirk
Kiepenheuer Institut
f Sonnenphysik
Schoneckstr 6
D 79104 Freiburg Breisgau
Germany
☎ 49 761 319 8154
✆ 49 761 319 8111
✉ soltau@kis.uni-freiburg.de

Soltynski Maciej
ASSA
P O Box 9
7935 Observatory
South Africa
☎ 27 21 918 4152
✆ 27 21 918 4146
✉ mgs@maties.sun.ac.za

Soma Mitsuru
Ntl Astron Observ Japan
Osawa
Mitaka
Tokyo 181 8588
Japan
☎ 81 422 34 3788
✆ 81 422 32 1924
✉ somamt@cc.nao.ac.jp

Somerville William B
Dpt Phys/Astronomy
UCLO
Gower St
London WC1E 6BT
UK
☎ 44 171 382 7050
✆ 44 171 380 7145

Sommer-Larsen Jesper
Theoretical Astrophysics Centre
Juliane Maries Vej 30
DK 2100 Copenhagen
Denmark
☎ 45 35 32 5909
℻ 45 35 32 5910

Somov Boris V
SAI
Acad Sciences
Universitetskij Pr 13
119899 Moscow
Russia
☎ 7 095 939 1644
℻ 7 095 939 01 26
✉ snn@sai.msk.su

Sonett Charles P
Lunar/Planetary Lab
Room 325 Bg 92
Univ of Arizona
Tucson AZ 85721 0092
USA
☎ 1 520 621 6935
℻ 1 520 621 4933

Song Doo Jong
Korea Astronomy Obs/ISSA
36 1 Whaam Dong
Yuseong Gu
Taejon 305 348
Korea RP
☎ 82 428 61 1502
℻ 82 428 61 5610
✉ djsong@apiss.issa.re.kr@garam

Song Guoxuan
Shanghai Observ
CAS
80 Nandan Rd
Shanghai 200030
China PR
☎ 86 21 6438 6191
℻ 86 21 6438 4618

Song Jin'an
Shaanxi Observ
CAS
Box 18
Lintong 710600
China PR
☎ 86 33 2255
℻ 86 9237 3496

Song Mutao
Purple Mountain Obs
CAS
Nanjing 210008
China PR
☎ 86 25 46700
℻ 86 25 301 459

Songsathaporn Ruangsak
Physics Dpt
Chiang Mai Univ
Faculty of Sciences
Chiang Mai 50002
Thailand
☎ 66 53 22 1934*135

Sonneborn George
NASA GSFC
Code 681
Greenbelt MD 20771
USA
☎ 1 301 286 3665
℻ 1 301 286 1753
✉ sonneborn@stars.gsfc.nasa.gov

Sonti Sreedhar Rao
Astronomy Dpt
Univ of Osmania
Hyderabad 500 007
India
☎ 91 868951*247

Sood Ravi
Physics Dpt
Univ College Adfa
UNSW
Canberra ACT 2600
Australia
☎ 61 2 62 688 765
℻ 61 26 2688 786
✉ r-sood@adfa.edu.au

Soon Willie H
CfA
HCO/SAO MS 16
60 Garden St
Cambridge MA 02138 1516
USA
☎ 1 617 495 7488
℻ 1 617 495 7049
✉ wsoon@cfassp34.harvard.edu

Soonthornthum Boonrucksar
Physics Dpt
Fac Sciences
Chiang Mai Univ
Chiang Mai 50200
Thailand
☎ 66 53 22 1699*3367
℻ 66 53 22 2268
✉ boonraks@cmu.chianmai.ac.th

Sorensen Gunnar
Inst Phys/Astronomy
Univ of Aarhus
Ny Munkegade
DK 8000 Aarhus C
Denmark
☎ 45 86 12 8899
℻ 45 86 20 2711

Sorensen Soren-Aksel
Dpt Computer Science
UCLO
Gower St
London WC1E 6BT
UK
☎ 44 171 387 7050
℻ 44 171 380 7145

Sorochenko R L
Lebedev Physical Inst
Acad Sciences
Leninsky Pspt 53
117924 Moscow
Russia
☎ 7 095 135 0171
℻ 7 095 135 7880

Sorokin Nikolai A
Inst Theoret Astron
Acad Sciences
Pyatnitskaya 48
109017 Moscow
Russia
☎ 7 095 231 2923
℻ 7 095 230 2081
✉ nsorokin@inasan.rssi.ru

Soru-Escaut Irina
Observ Paris Meudon
Pl J Janssen
F 92195 Meudon PPL Cdx
France
☎ 33 1 45 34 7530
℻ 33 1 45 07 7971

Sotirovski Pascal
Observ Paris Meudon
DASOP
Pl J Janssen
F 92195 Meudon PPL Cdx
France
☎ 33 1 45 07 7802
℻ 33 1 45 07 7971

Soubiran Caroline
Observ Bordeaux
BP 89
F 33270 Floirac
France
☎ 33 5 56 86 4330
℻ 33 5 56 40 4251
✉ soubiran@observ.u-bordeaux.fr

Soucail Genevieve
OMP
14 Av E Belin
F 31400 Toulouse Cdx
France
☎ 33 5 61 33 2819
℻ 33 5 61 53 6722
✉ soucail@obs-mip.fr

Souchay Jean
Observ Paris
DANOF
61 Av Observatoire
F 75014 Paris
France
☎ 33 1 40 51 2322
℻ 33 1 40 51 2291
✉ souchay@obspm.fr

Souffrin Pierre B
OCA
Observ Nice
BP 139
F 06304 Nice Cdx 4
France
☎ 33 4 93 89 0420
℻ 33 4 92 00 3033

Soulie Guy
Observ Bordeaux
BP 89
F 33270 Floirac
France
☎ 33 5 56 86 4330
℻ 33 5 56 40 4251

Souriau Jean-Marie
10 r Mazarine
F 13100 Aix en Provence
France
☎ 33 4 42 26 2580
✉ Jean-Marie.Souriau@gyptis.univ-mrs.fr

Sowell James Robert
Georgia Inst
of Technology
School of Physics
Atlanta GA 30332
USA
☎ 1 404 894 3628
✉ js58@hydra.gatech.edu

Spada Gianfranco
TeSRE
CNR
Via P Gobetti101
I 40129 Bologna
Italy
☎ 39 51 639 8665
℻ 39 51 639 8724

Spadaro Daniele
Osserv Astronomico
d Catania
Via A Doria 6
I 95125 Catania
Italy
☎ 39 95 733 21111
℻ 39 95 33 0592

Spaenhauer Andreas Martin
Astronomisches Institut
Univ Basel
Venusstr 7
CH 4102 Binningen
Switzerland
☎ 41 61 271 7711/12
℻ 41 61 205 5455

Spagna Alessandro
Osserv Astronomico
d Torino
St Osservatorio 20
I 10025 Pino Torinese
Italy
☎ 39 11 461 9034
℻ 39 11 461 9030
✉ spagna@to.astro.it

Sparke Linda
Washburn Observ
Univ Wisconsin
475 N Charter St
Madison WI 53706
USA
☎ 1 608 262 3071
℻ 1 608 263 0361
✉ sparke@wiscmac3

Sparks Warren M
LANL
MS F669
Box 1663
Los Alamos NM 87545 2345
USA
☎ 1 505 667 4922
℻ 1 505 665 4055

Sparks William Brian
STScI
Homewood Campus
3700 San Martin Dr
Baltimore MD 21218
USA
☎ 1 301 338 4843
✆ 1 301 338 4767
✉ sparks@stsci.edu

Sparrow James G
Aeronautical Res
Laboratories
4331
Melbourne VIC 3001
Australia
☎ 61 36 477 623

Spasova Nedka Marinova
Astronomy Dpt
Bulgarian Acad Sci
72 Lenin Blvd
BG 1784 Sofia
Bulgaria
☎ 359 2 75 8927
✆ 359 2 75 8927

Speer R J
Physics Dpt
Imperial Coll
Prince Consort Rd
London SW7 2BZ
UK
☎ 44 171 594 7771
✆ 44 171 594 7772

Spencer John Howard
NRL
Code 7214
4555 Overlook Av SW
Washington DC 20375 5351
USA
☎ 1 202 767 3050
✉ spencer@proteus.nrl.
navy.mil

Spencer Ralph E
NRAL
Univ Manchester
Jodrell Bank
Macclesfield SK11 9DL
UK
☎ 44 14 777 1321
✆ 44 147 757 1618

Sperauskas Julius
Astronomical Observ
Ciurlionio 29
Vilnius 2009
Lithuania
☎ 370 2 633 343
✆ 370 2 223 563

Spergel David N
Princeton Univ Obs
Peyton Hall
Princeton NJ 08544 1001
USA
☎ 1 609 258 3589
✆ 1 609 258 1020
✉ dns@astro.princeton.edu

Spicer Daniel Shields
NASA GSFC
Code 682
Greenbelt MD 20771
USA
☎ 1 301 286 7334

Spiegel E
Astronomy Dpt
Columbia Univ
538 W 120th St
New York NY 10027
USA
☎ 1 212 854 3278
✆ 1 212 316 9504

Spielfiedel Annie
Observ Paris Meudon
DAMAP
Pl J Janssen
F 92195 Meudon PPL Cdx
France
☎ 33 1 45 07 7453
✆ 33 1 45 07 7469
✉ spilfild@obspm.fr

Spinrad Hyron
Astronomy Dpt
Univ of California
601 Campbell Hall
Berkeley CA 94720 3411
USA
☎ 1 415 642 2078
✆ 1 510 642 3411

Spite Francois M
Observ Paris Meudon
DASGAL
Pl J Janssen
F 92195 Meudon PPL Cdx
France
☎ 33 1 45 07 7840
✆ 33 1 45 07 7878
✉ francois.spite@
obspm.fr

Spite Monique
Observ Paris Meudon
DASGAL
Pl J Janssen
F 92195 Meudon PPL Cdx
France
☎ 33 1 45 07 7839
✆ 33 1 45 07 7971
✉ monique.spite@
obspm.fr

Spithas Elefterios N
Astronomy Dpt
Ntl Univ Athens
Panepistimiopolis
GR 157 84 Zografos
Greece
☎ 30 1 724 3414
✆ 30 1 723 5122

Spoelstra T A Th
NFRA
Postbus 2
NL 7991 Pd Dwingeloo
Netherlands
☎ 31 521 59 5100
✆ 31 52 159 7332

Sprague Ann Louise
Lunar/Planetary Lab
Room 325 Bg 92
Univ of Arizona
Tucson AZ 85721 0092
USA
☎ 1 520 621 2282
✆ 1 520 621 4933
✉ sprague@titan.lpl.arizona.edu

Spruit Henk C
MPA
Karl Schwarzschildstr 1
D 85748 Garching
Germany
☎ 49 893 299 00
✆ 49 893 299 3235

Spurny Pavel
Astronomical Institute
Czech Acad Sci
Fricova 1
CZ 251 65 Ondrejov
Czech R
☎ 420 204 85 7153
✆ 420 2 88 1611
✉ spurny@asu.cas.cz

Spurzem Rainer
ARI
Moenchhofstr 12-14
D 69120 Heidelberg
Germany
☎ 49 622 140 5230
✆ 49 622 140 5297
✉ spurzem@relay.ari.uni-
heidelberg.de

Spyrou Nicolaos
Astronomy Lab
Univ Thessaloniki
GR 540 06 Thessaloniki
Greece
☎ 30 31 99 2658

Sramek Richard A
NRAO
Box 0
Socorro NM 87801 0387
USA
☎ 1 505 835 7394
✆ 1 505 835 7027

Sreekantan B V
TIFR
Homi Bhabha Rd
Colaba
Bombay 400 005
India
☎ 91 22 219 111
✆ 91 22 215 2110

Sreenivasan S Ranga
Physics Dpt
Univ of Calgary
2500 University Dr NW
Calgary AB T2N 1N4
Canada
☎ 1 403 284 5385
✆ 1 403 289 3331

Srinivasan G
RRI
Sadashivanagar
CV Raman Av
Bangalore 560 080
India
☎ 91 80 334 0122
✆ 91 80 334 0492
✉ srini@rri.ernet.in

Srivastava Dhruwa
Physics Dpt
Univ of Gorakhpur
Gorakhpur 273 009
India
☎ 91 551 332 398
✆ 91 551 340 459
✉ dcs@gkpu.ernet.in

Srivastava J B
Uttar Pradesh State
Observ
Po Manora Peak 263 129
Nainital 263 129
India
☎ 91 59 42 2136
✉ srini@rri.ernet.in

Srivastava Ram Kumar
Uttar Pradesh State
Observ
Po Manora Peak 263 129
Nainital 263 129
India
☎ 91 59 42 2136

St-Louis Nicole
Dpt d Physique
Universite Montreal
CP 6128 Succ A
Montreal QC H3C 3J7
Canada
☎ 1 514 343 6932
✆ 1 514 343 2071
✉ stlouis@astro.umontreal.ca

Stabell Rolf
Inst Theor Astrophys
Univ of Oslo
Box 1029
N 0315 Blindern Oslo 3
Norway
☎ 47 22 856 530
✆ 47 22 85 6505
✉ rolf.stabell@astro.uio.no

Stacey Gordon J
Astronomy Dpt
Cornell Univ
512 Space Sc Bldg
Ithaca NY 14853 6801
USA
☎ 1 607 255 5900
✆ 1 607 255 5875
✉ stacey@astrosun.tn.edu

Stachnik Robert V
NASA Headquarters
Astrophysics Div
Code SZ
Washington DC 20546
USA
☎ 1 202 358 0351
✆ 1 202 358 3096
✉ rstachnik@gm.ossa.hq.
nasa.gov

Stagg Christopher
Dpt Phys/Astronomy
Univ of Calgary
2500 University Dr NW
Calgary AB T2N 1N4
Canada
☎ 1 403 220 7423
✆ 1 403 289 3331
✉ azpiazu@wesson.phys.
ucalgary.ca

Stagni Ruggero
Osserv Astrofisico
Univ d Padova
Via d Osservatorio 8
I 36012 Asiago
Italy
☎ 39 424 462 665
✆ 39 424 462 884

Stahl Otmar Richard
Landessternwarte
Koenigstuhl
D 69117 Heidelberg
Germany
☎ 49 622 150 9231
✆ 49 622 150 9202
✉ o.stahl@lsw.uni-heidelberg.de

Stahler Steven W
Physics Dpt
MIT
Box 165
Cambridge MA 02139 4307
USA
☎ 1 617 253 0905
✆ 1 617 253 9798

Stahr-Carpenter M
1101 Hill Top Rd
Charlottesville VA 22903
USA
☎ 1 804 293 7063

Stalio Roberto
Dpt Astronomia
Univ d Trieste
Via Tiepolo 11
I 34131 Trieste
Italy
☎ 39 40 79 3921*221
✆ 39 40 30 9418

Standish E Myles
CALTECH/JPL
MS 301 150
4800 Oak Grove Dr
Pasadena CA 91109 8099
USA
☎ 1 818 354 3959
✆ 1 818 393 6388
✉ ems@smyles.jpl.nasa.gov

Stanford Spencer A
LLNL
L 413
Box 808
Livermore CA 94551 9900
USA
☎ 1 415 423 6013
✆ 1 415 423 0238
✉ adam@igpp.llnl.gov

Stanga Ruggero
Dpt Astronomia
Univ d Firenze
Largo E Fermi 5
I 50125 Firenze
Italy
☎ 39 55 27 521
✆ 39 55 22 0039

Stange Lothar
Technical Univ
Dresden
Mommsenstr 13
D 8027 Dresden
Germany
☎ 49 51 463 4652

Stanghellini Letizia
Osserv Astronom Bologna
Univ d Bologna
Via Zamboni 33
I 40126 Bologna
Italy
☎ 39 51 259 333
✆ 39 51 259 407
✉ stanghellini@astb03.bo.astro.it

Stanila George
Mitropolit Grigore 28
Box 28
RO 75215 Bucharest
Rumania
☎ 40 1 623 3426

Stankevich Kazimir S
Radiophysical Res
Inst
Lyadov Ul 25/14
603600 N Novgorod
Russia
☎ 7 832 338 9091
✆ 7 831 236 9902

Stanley G J
Box 1348
Carmel Valley CA 93924
USA
☎ 1 408 659 2940

Stannard David
NRAL
Univ Manchester
Jodrell Bank
Macclesfield SK11 9DL
UK
☎ 44 14 777 1321
✆ 44 147 757 1618

Stark Antony A
CfA
HCO/SAO MA 78
60 Garden St
Cambridge MA 02138 1516
USA
☎ 1 617 496 7648
✆ 1 617 496 7554
✉ aas@cfa.harvard.edu

Stark Glen
Whitin Observ
Wellesley Coll
Wellesley MA 02181
USA
☎ 1 617 235 0320
✆ 1 617 283 3667
✉ gstark@lucy.wellesley.edu

Starrfield Sumner
Phys/Astronomy Dpt
Arizona State Univ
Astronomy Program
Tempe AZ 85287 1504
USA
☎ 1 602 965 7569
✆ 1 602 965 7954
✉ sumner.starrfield@asu.edu

Stasinska Grazyna
Observ Paris Meudon
DAEC
Pl J Janssen
F 92195 Meudon PPL Cdx
France
☎ 33 1 45 07 7422
✆ 33 1 45 07 7971
✉ grazyna@obspm.fr

Stathopoulou Maria
Astrophysics Dpt
Ntl Univ Athens
Panepistimiopolis
GR 157 84 Zografos
Greece
☎ 30 1 724 3414
✆ 30 1 722 8981
✉ marstath@grathun1

Staubert Ruediger Prof
Astronomisches Institut
Univ Tuebingen
Waldhaeuserstr 64
D 72076 Tuebingen
Germany
☎ 49 707 129 4980
✆ 49 707 129 3458
✉ staubert@ait.physik.uni.tuebingen.de

Staude Hans Jakob
MPI Astronomie
Koenigstuhl
D 69117 Heidelberg
Germany
☎ 49 622 152 8229
✆ 49 622 152 8246

Staude Juergen
AIP
SOE
Telegrafenberg
D 14473 Potsdam
Germany
☎ 49 331 288 2300
✆ 49 331 288 2310
✉ jstaude@aip.de

Stauffer John Richard
117 Sylvester Av
Winchester MA 01890
USA
☎ 1 617 495 7024
✆ 1 617 495 7490
✉ stauffer@cfa.harvard.edu

Staveley-Smith Lister
CSIRO
Div Radiophysics
Box 76
Epping NSW 2121
Australia
☎ 61 2 868 0222
✆ 61 2 868 0310
✉ lstavele@atnf.csiro.au

Stavinschi Magdalena
Astronomical Institute of the
Romanian Academy
Romanian Acad Sciences
Cutitul de Argint 5
RO 75212 Bucharest
Rumania
☎ 40 1 641 3686
✆ 40 1 312 3391
✉ magda@roastro.astro.ro

Stawikowski Antoni
Inst of Astronomy
N Copernicus Univ
Ul Chopina 12/18
PL 87 100 Torun
Poland
☎ 48 56 26 018
✆ 48 56 24 602

Stebbins Robin
JILA
Univ of Colorado
Campus Box 440
Boulder CO 80309 0440
USA
☎ 1 303 492 6073
✆ 1 303 492 5235

Stecher Theodore P
NASA GSFC
Code 680
Greenbelt MD 20771
USA
☎ 1 301 286 8718
✆ 1 301 286 1753
✉ stecher@uit.gsfc.nasa.gov

Stecker Floyd W
NASA GSFC
Code 660
Greenbelt MD 20771
USA
☎ 1 301 286 6057

Stecklum Bringfried
Thueringer Landessternwarte
Sternwarte 5
D 07778 Tautenburg
Germany
☎ 49 364 278 6354
✆ 49 364 278 6329
✉ stecklum@tls-tautenburg.de

Steel Duncan I
Spaceguard Australia P/L
Box 3303
Rundle Mall
Adelaide SA 5000
Australia
☎ 61 8 8232 6133
✆ 61 8 8232 6103
✉ dis@a011.aone.net.au-

Steele Colin D C
Mathematics Dpt
UMIST
Box 88
Manchester M60 1QD
UK
☎ 44 161 200 3632
✆ 44 161 200 3669
✉ cds@ast.ma.umist.ac.uk

Steenman-Clark Lois
OCA
Observ Nice
BP 139
F 06304 Nice Cdx 4
France
☎ 33 4 93 89 0420
✆ 33 4 92 00 3033

Stefanik Robert
Oak Ridge Observ
Harvard Smithsonian Ctr
Pinnacle Rd
Harvard MA 01451
USA
☎ 1 617 495 7070
✉ stefanik@cfa.harvard.edu

Stefano Andreon
Osserv d Capodimonte
Via Moiariello 16
I 80131 Naples
Italy
☎ 39 81 557 5546
✆ 39 81 45 6710
✉ andreon@na.astro.it

Steffen Matthias
Inst Theor Phys/
Sternwarte Univ Kiel
Olshausenstr 40
D 24098 Kiel
Germany
☎ 49 431 880 4101
✆ 49 431 880 4432

Stefl Stanislav
Astronomical Institute
Czech Acad Sci
Fricova 1
CZ 251 65 Ondrejov
Czech R
☎ 420 204 85 7143
✆ 420 2 88 1611
✉ sstefl@sunstel.asu.cas.cz

Stefl Vladimir
Dpt Th Phys/Astrophys
Masaryk Univ
Kotlarska 2
CZ 611 37 Brno
Czech R
☎ 420 5 4112 9482
✆ 420 5 4121 1214
✉ stehl@astro.sci.muni.cz

Stehle Chantal
Observ Paris Meudon
DARC
Pl J Janssen
F 92195 Meudon PPL Cdx
France
☎ 33 1 45 07 7453
✆ 33 1 45 07 7971
✉ chantal.stehle@obspm.fr

Steiger W R
30 Kiele Pl
Hilo HI 96720
USA
☎ 1 808 961 4980
✆ 1 808 961 6516

Steigman Gary
Physics Dpt
Ohio State Univ
174 W 18th Av
Columbus OH 43210 1106
USA
☎ 1 614 292 1999

Steiman-Cameron Thomas
NASA/ARC
Theor Studies Br
MS 245 3
Moffett Field CA 94035 1000
USA
☎ 1 415 694 3120
✆ 1 415 604 6779

Stein John William
LaRoche College
Dpt Maths/Natural Sci
9000 Babcock Blvd
Pittsburgh PA 15237
USA
☎ 1 412 367 9300
✆ 1 412 367 9277
✉ jstein@unix.cis.pitt.edu

Stein Robert F
Physics/Astronomy Dpt
Michigan State Univ
East Lansing MI 48824
USA
☎ 1 517 353 8661
✆ 1 517 353 4500

Stein Wayne A
School Phys/Astronomy
Univ Minnesota
116 Church St SE
Minneapolis MN 55455
USA
☎ 1 612 373 9963
✆ 1 612 624 2029

Steiner Joao E
INPE
Dpt Astronomia
CP 515
12227 010 S Jose dos Campos
Brazil
☎ 55 123 22 9977
✆ 55 123 21 8743

Steiner Oskar
Kiepenheuer Institut
f Sonnenphysik
Schoneckstr 6
D 79104 Freiburg Breisgau
Germany
☎ 49 761 319 8222
✆ 49 761 31 98 111
✉ steiner@kis.uni-freiburg.de

Steinert Klaus Guenter
Lorhmann Observ
Dresden Univ
Mommsenstr 13
D 01062 Dresden
Germany
☎ 49 351 463 4097
✆ 49 351 463 7019
✉ lormobs@rcs.urz.tu-dresden.de

Steinitz Raphael
Physics Dpt
Ben Gurion Univ
Box 653
Beersheva 84105
Israel
☎ 972 57 70 985

Steinle Helmut
MPE
Postfach 1603
D 85740 Garching
Germany
☎ 49 893 299 3374
✆ 49 893 299 3569
✉ hcs@mpe-garching.mpg.de

Steinlin Uli
Schulgasse 7
CH 4105 Biel Benken
Switzerland

Steinolfson Richard S
Southwest Res Inst
6220 Culebra Rd
San Antonio TX 78228 2510
USA
☎ 1 512 522 2822
✆ 1 512 647 4325

Stellingwerf Robert F
LANL
MS F645 X DIV
Box 1663
Los Alamos NM 87545 2345
USA
☎ 1 505 667 4370
✆ 1 505 665 4055

Stellmacher Goetz
IAP
98bis bd Arago
F 75014 Paris
France
☎ 33 1 44 32 8057
✆ 33 1 44 32 8001

Stellmacher Irene
BDL
77 Av Denfert Rochereau
F 75014 Paris
France
☎ 33 1 43 20 1210
✆ 33 1 46 33 2834

Stencel Robert Edward
Physics Dpt
Univ of Denver
2112 E Wesley Av
Denver CO 80208
USA
☎ 1 303 871 2135
✆ 1 303 871 4405
✉ rstencel@diana.cair.du.edu

Stenflo Jan O
Inst Astronomie
ETH Zentrum
CH 8092 Zuerich
Switzerland
☎ 41 1 632 3804
✆ 41 1 632 1205
✉ stenflo@astro.phys.ethz.ch

Stenholm Bjoern
Lund Observ
Box 43
S 221 00 Lund
Sweden
☎ 46 46 222 7306
✆ 46 46 222 4614
✉ bjorn@astro.lu.se

Stenholm Lars
Astronomical Observ
Box 515
S 751 20 Uppsala
Sweden
☎ 46 18 11 2488
✆ 46 18 52 7583
✉ stenholm@laban.uv.se

Stepanian A A
Crimean Astrophys Obs
Ukrainian Acad Science
Nauchny
334413 Crimea
Ukraine
☎ 380 6554 71179
✆ 380 6554 40704

Stepanian Jivan A
Byurakan Astrophys Observ
Armenian Acad Sci
378433 Byurakan
Armenia
☎ 374 88 52 28 3453/4142
✆ 374 88 52 52 3640

Stepanian N N
Crimean Astrophys Obs
Ukrainian Acad Science
Nauchny
334413 Crimea
Ukraine
☎ 380 6554 71161
✆ 380 6554 40704

Stepanov Alexander V
Pulkovo Observ
Acad Sciences
10 Kutuzov Quay
196140 St Petersburg
Russia
☎ 7 812 298 2242
✆ 7 812 315 1701

Stephens S A
TIFR
Homi Bhabha Rd
Colaba
Bombay 400 005
India
☎ 91 22 219 111
✆ 91 22 495 2110

Stephenson C Bruce
Physics Dpt
CWRU
Rock Bdg
Cleveland OH 44106
USA
☎ 1 216 368 6699
✉ cbs3@po.cwru.edu

Stephenson F Richard
Physics Dpt
Univ of Durham
South Rd
Durham DH1 3LE
UK
☎ 44 191 374 2153
✆ 44 191 374 3749

Stepien Kazimierz
Astronomical Observ
Warsaw Univ
Al Ujazdowskie 4
PL 00 478 Warsaw
Poland
☎ 48 22 29 4011
✆ 48 22 29 4967
✉ kst@taurus.astrouw.edu.pl

Stepinski Tomasz
LPI
3600 Bay Area Blvd
Houston TX 77058
USA
☎ 1 713 486 2170

Steppe Hans
IRAM
Avd Divina Pastora 7
Bloque 6/2b
E 18012 Granada
Spain
☎ 34 5 827 9508
✆ 34 5 820 7662

Sterken Christiaan Leo
Astrofysisch Inst
Vrije Univ Brussel
Pleinlaan 2
B 1050 Brussels
Belgium
☎ 32 2 629 3469
✆ 32 9 362 3976
✉ csterken@vub.ac.be

Stern Robert Allan
Lockheed Palo Alto Res Lb
Dpt 91 30 Bg 252
3251 Hanover St
Palo Alto CA 94304 1191
USA
☎ 1 416 424 3272
✆ 1 415 424 3994
✉ stern@sag.space.lockheed.com

Stern S Alan
Southwest Res Inst
6220 Culebra Rd
San Antonio TX 78238 5166
USA
☎ 1 303 546 9670
✆ 1 303 546 9687
✉ alan@swri.space.swri.edu

Steshenko N V
Crimean Astrophys Obs
Ukrainian Acad Science
Nauchny
334413 Crimea
Ukraine
☎ 380 6554 71161
✆ 380 6554 40704

Stetson Peter B
NRCC/HIA
DAO
5071 W Saanich Rd
Victoria BC V8X 4M6
Canada
☎ 1 250 363 0029
✆ 1 250 363 0045
✉ stetson@dao.nrc.ca

Stevens Gerard A
SRON
Postbus 800
Sorbonnelaan 2
NL 3584 CA Utrecht
Netherlands
☎ 31 30 253 5600
✆ 31 30 254 0860

Stevens Ian
School of Physics
Univ Birmingham
Box 363
Birmingham B15 2TT
UK
☎ 44 12 14 14 3565
✆ 44 12 14 14 3722
✉ irs@star.sr.bham.ac.uk

Steves Bonita Alice
Mathematics Dpt
Glasgow Caledonia Univ
Cowcaddens Rd
Glasgow G4 0BA
UK
☎ 44 141 331 3619
✆ 44 141 331 3608
✉ bst@gcal.ac.uk

Stewart John Malcolm
Dpt Appl Maths/Theor Phys
Silver Street
Cambridge CB3 9EW
UK
☎ 44 12 23 337 00
✆ 44 12 23 337 918

Stewart Paul
Mathematics Dpt
Univ Manchester
Oxford Rd
Manchester M13 9PL
UK
☎ 44 161 275 4224
✆ 44 161 275 4223

Stewart Ronald T
CSIRO
Div Radiophysics
Box 76
Epping NSW 2121
Australia
☎ 61 2 868 0222
✆ 61 2 868 0310
✉ rstewart@atnf.csiro.au

Steyaert Herman
Toegepaste Wiskunde Inf
Universiteit Gent
Krijgslaan 281 S9
B 9000 Gent
Belgium
☎ 32 9 264 4805
✆ 32 9 264 4995

Stiavelli Massimo
Via F Corridoni 25
I 56100 Pisa
Italy
☎ 39 50 48806

Stibbs Douglas W N
MSSSO
Weston Creek
Private Bag
Canberra ACT 2611
Australia
☎ 61 262 881 111
✆ 61 262 490 233
✉ dwns@mso.anu.edu.au

Stickland David J
Rutherford Appleton Lab
Space/Astrophysics Div
Bg R25/R68
Chilton Didcot OX11 0QX
UK
☎ 44 12 35 446 523
✆ 44 12 35 44 5808

Stier Mark T
Perkin Elmer Corp
MS 842
100 Wooster Heights Rd
Danbury CT 06810 7859
USA
☎ 1 203 797 5708
✆ 1 203 797 6259
✉ mstier@west.raytheon.com

Stift Martin Johannes
Inst Astronomie
Univ Wien
Tuerkenschanzstr 17
A 1180 Wien
Austria
☎ 43 1 345 36096
✆ 43 1 470 6015

Stinebring Daniel R
Physics Dpt
Oberlin College
Oberlin OH 44074 1088
USA

Stirpe Giovanna M
Osserv Astronom Bologna
Univ d Bologna
Via Zamboni 33
I 40126 Bologna
Italy
☎ 39 51 25 9301
✆ 39 51 25 9407
✉ stirpe@astbo3.cineca.it

Stix Michael
Kiepenheuer Inst
Sonnenphysik
Schoeneckstr 6
D 79104 Freiburg Breisgau
Germany
☎ 49 761 328 64
✆ 49 761 319 8111

Stobie Robert S
SAAO
PO Box 9
7935 Observatory
South Africa
☎ 27 21 47 0025
✆ 27 21 47 3639
✉ rss@saao.ac.za

Stock Jurgen D
CIDA
Box 264
Merida 5101 A
Venezuela
☎ 58 7 471 2780/3883
✆ 58 7 471 2459
✉ stock@cida.ve/stock@milkyway.cc.fc.ul.pt

Stocke John T
CASA
Univ of Colorado
Campus Box 389
Boulder CO 80309 0389
USA
☎ 1 303 492 1521
✆ 1 303 492 7178

Stockman Hervey S
STScI
Homewood Campus
3700 San Martin Dr
Baltimore MD 21218
USA
☎ 1 301 338 4730
✆ 1 301 338 2519

Stockton Alan N
Inst for Astronomy
Univ of Hawaii
2680 Woodlawn Dr
Honolulu HI 96822
USA
☎ 1 808 956 8566
✆ 1 808 988 2790

Stoeger William R
Specola Vaticana
I 00120 Citta del Vaticano
Vatican City State
☎ 39 6 698 3411/5266
✆ 39 6 698 84671

Stoev Alexei
People's Astronom Obs
& Planetaria in Bulgaria
Yuri Gagarin
BG 6000 Stara Zagora
Bulgaria
☎ 359 4 24 3183

Stoker Pieter H
Cosmic Ray Res Unit
Potchefstroom Univ
for CHE
2520 Potchefstroom
South Africa
☎ 27 148 299 2423
✆ 27 148 299 2421

Stone Edward C
CALTECH/JPL
MS 180 904
4800 Oak Grove Dr
Pasadena CA 91109 8099
USA
☎ 1 818 354 3405
✆ 1 818 393 4218

Stone James McLellan
Astronomy Dpt
Univ of Maryland
College Park MD 20742 2421
USA
☎ 1 301 405 2103
✆ 1 301 314 9067
✉ jstone@astro.umd.edu

Stone R G
NASA GSFC
Code 690
Greenbelt MD 20771
USA
☎ 1 301 286 8631
✆ 1 301 286 1683

Stone Remington P S
Lick Observ
Mount Hamilton CA 95140
USA
☎ 1 408 274 1809
✉ rem@lick.ucsc.edu

Stone Ronald Cecil
USNO
Flagstaff Station
Box 1149
Flagstaff AZ 86002 1149
USA
☎ 1 520 779 5132
✆ 1 520 774 3626
✉ rcs@nofs.navy.mil

Storchi-Bergman Thaisa
Instituto Fisica
UFRGS
CP 15051
91501 900 Porto Alegre RS
Brazil
☎ 55 512 36 4677

Storey John W V
School of Physics
UNSW
BOX 1
Kensington NSW 2033
Australia
☎ 61 26 974 591
✆ 61 6 268 8786

Storey Michelle
Dpt Th Physics
Univ of Sydney
Sydney NSW 2006
Australia
☎ 61 2 692 2538
✆ 61 2 660 2903

Storm Jesper
ESO
Santiago Office
Casilla 19001
Santiago 19
Chile
☎ 56 2 228 5006
✆ 56 2 228 5132

Stotskii Alexander A
Inst Appl Astronomy
Zhdanovskaya 8
St Petersburg 197110
Russia
☎ 7 812 235 3316*118
✆ 7 812 230 7413
✉ stotskii@isida.ipa.rssi.ru

Strachan Leonard Jr
CfA
HCO/SAO MS 50
60 Garden St
Cambridge MA 02138 1516
USA
☎ 1 617 496 7569
✆ 1 617 495 7455
✉ lstrachan@cfa.harvard.edu

Strafella Francesco
Dpt Fisica
Univ d Lecce
Via Per Arnesano
I 73100 Lecce
Italy
☎ 39 83 26 27/247

Straizys V
Inst Theor Physics/
Astronomy
Gostauto 12
Vilnius 2600
Lithuania
☎ 370 2 613 440
✆ 370 2 818 464
✉ straizys@itpa.fi.lt
straizys@itpa.elnet.lt

Strand Kaj Aa
3200 Rowland Pl Nw
Washington DC 20008
USA
☎ 1 202 966 0495

Strassmeier Klaus G
Inst Astronomie
Univ Wien
Tuerkenschanzstr 17
A 1180 Wien
Austria
☎ 43 1 340 31695
✆ 43 1 470 6015
✉ strassmeier@avia.
una.ac.at

Straumann Norbert
Inst Theor Physics
Univ Zuerich
Winterthurerstr 190
CH 8057 Zuerich
Switzerland
☎ 41 1 257 5815
✆ 41 1 257 5704

Strazzulla Giovanni
Osserv Astronomico
d Catania
Via A Doria 6
I 95125 Catania
Italy
☎ 39 95 733 21111
✆ 39 95 33 0592

Strelnitski Vladimir
Maria Mitchell Observ
3 Vestal St
Nantucket MA 02554
USA
✉ vladimir@mmo.org

Strigatchev Anton
Astronomy Dpt
Bulgarian Acad Sci
72 Lenin Blvd
BG 1784 Sofia
Bulgaria
☎ 359 2 75 8927
✆ 359 2 75 8927
✉ antonst@phys.acad.bg

Stringfellow Guy Scott
CASA
Univ of Colorado
Campus Box 389
Boulder CO 80309 0389
USA
☎ 1 303 492 6056
✆ 1 303 492 4052
✉ guy@casa.
colorado.edu

Strittmatter Peter A
Steward Observ
Univ of Arizona
Tucson AZ 85721
USA
☎ 1 520 621 6532
✆ 1 520 621 1532

Strobel Andrzej
Inst of Astronomy
N Copernicus Univ
Ul Chopina 12/18
PL 87 100 Torun
Poland
☎ 48 56 26 018
✆ 48 56 24 602

Strobel Darrell F
Dpt Earth/Planet Sci
JHU
Charles/34th St
Baltimore MD 21218
USA
☎ 1 301 338 7034
✆ 1 301 338 7933

Strohmeier Wolfgang
Volkfeldstr 5
D 8600 Bamberg
Germany
☎ 49 951 553 94

Strom Karen M
Astronomy Dpt
Univ Massachusetts
GRC 518 B 6732
Amherst MA 01002
USA
☎ 1 413 545 2290

Strom Richard G
NFRA
Postbus 2
NL 7990 AA Dwingeloo
Netherlands
☎ 31 521 59 5100
✆ 31 52 159 7332

Strom Robert G
Lunar/Planetary Lab
Room 325 Bg 92
Univ of Arizona
Tucson AZ 85721 0092
USA
☎ 1 520 621 2720
✆ 1 520 621 4933

Strom Stephen E
Dpt Phys/Astronomy
Univ Massachusetts
GRC 518 B
Amherst MA 01003
USA
☎ 1 413 545 2290
✆ 1 413 545 4223
✉ sstrom@donald.phast.
umass.edu

Strong Ian B
229 Rio Brave Dr
White Rock NM 87544
USA

Strong Keith T
Lockheed Palo Alto Res Lb
Dpt 91 30 Bg 252
3251 Hanover St
Palo Alto CA 94304 1191
USA
☎ 1 415 354 5136
✆ 1 415 424 3994
✉ strong@sag.space.
lockheed.com

Struble Mitchell F
Dpt Astron/Astrophys
Univ of Pennsylvania
David Rittenhouse Lab
Philadelphia PA 19104
USA
☎ 1 215 243 8176
✆ 1 215 898 9336

Struck-Marcell Curtis J
Physics Dpt
Iowa State Univ
Ames IA 50011
USA
☎ 1 515 294 5440
✆ 1 515 294 3262

Strukov Igor A
Space Res Inst
Acad Sciences
Profsojuznaya Ul 84/32
117810 Moscow
Russia
☎ 7 095 333 1466
✆ 7 095 310 7023

Stryker Linda L
Arizona State Univ West
Box 37100
Phoenix AZ 85069 7100
USA
☎ 1 602 543 6000
✆ 1 602 543 6004
✉ stryker@phyast.la.asu.edu

Stumpff Peter
RAIUB
Univ Bonn
auf d Huegel 69
D 53121 Bonn
Germany
☎ 49 228 525 360
✆ 49 228 525 229

Sturch Conrad R
STScI/CSC
Homewood Campus
3700 San Martin Dr
Baltimore MD 21218
USA
☎ 1 301 338 4856
✆ 1 301 338 4767

Sturrock Peter A
Ctr for Space Sci/
Astrophysics
Stanford Univ ERL 306
Stanford CA 94305 4055
USA
☎ 1 415 723 1438
✆ 1 415 725 2333
✉ sturrock@flare.staford.edu

Stutzi Juergen
Erstes Physikal Institut
Univ Koeln
Zuelpicherstr 77
D 50937 Koeln
Germany
☎ 49 221 470 3494/3567
℻ 49 221 470 5162
✉ stutzki@ph1.uni-koeln.de

Su Dingqiang
Ctr for Astronomy
Instrumen Res
CAS
Nanjing 210042
China PR
☎ 86 25 5411 776
℻ 86 25 5411 872
✉ dqsu@public1.ptt.js.cn

Su Hongjun
Beijing Astronomical Obs
CAS
Beijing 100080
China PR
☎ 86 10 6254 5179
℻ 86 1 256 10855

Subrahmanya C R
NCRA/TIFR
Pune Univ Campus Pb 3
Ganeshkhind
Pune 411 007
India
☎ 91 212 33 7107
℻ 91 212 33 5760

Subrahmanyam P V
Astronomy Dpt
Univ of Osmania
Hyderabad 500 007
India
☎ 91 71 951*247

Subramanian K R
IIA
Koramangala
Sarjapur Rd
Bangalore 560 034
India
☎ 91 80 553 0672 76
℻ 91 80 553 4043
✉ subra@iiap.ernet.in

Subramanian Kandaswamy
NCRA/TIFR
Pune Univ Campus Pb 3
Ganeshkhind
Pune 411 007
India
☎ 91 212 33 7107
℻ 91 212 33 5760

Suda Kazuo
Astronomical Institute
Tohoku Univ
Sendai Aoba
Miyagi 980
Japan
☎ 81 222 22 1800
℻ 81 22 262 6609

Sudzius Jokubas
Astronomical Observ
Ciurlionio 29
Vilnius 2009
Lithuania
☎ 370 2 633 343
℻ 370 2 635 648/223 563
✉ jokubas.sudzius@ff.vu.lt

Suematsu Yoshinori
Tokyo Astronomical Obs
NAOJ
Osawa Mitaka
Tokyo 181
Japan
☎ 81 422 41 3705
℻ 81 422 41 3700
✉ suematsu@spot.mtk.nao.ac.jp

Suess Steven T
NASA/MSFC
Space Science Lab
Code ES 52
Huntsville AL 35812
USA
☎ 1 205 453 2824
℻ 1 205 544 7754

Sugai Hajime
Astronomy Dpt
Kyoto Univ
Sakyo-ku
Kyoto 606 01
Japan
☎ 81 75 753 3898
℻ 81 75 753 3897
✉ sugai@kusastro.kyoto-u.ac.jp

Sugawa Chikara
Hananoi 1586 25
Kashiwa Shi
Chiba Ken 277
Japan
☎ 81 471 33 3825

Sugimoto Daiichiro
Univ of the Air
2-11 Wakaba
Mihama-ku
Chiba 261 8586
Japan
☎ 81 43 298 4180
℻ 81 43 298 4180
✉ sugimoto@u-air.ac.jp

Suginohara Tatsushi
Physics Dpt
Univ of Tokyo
7-3-1 Hongo Bunkyo-ku
Tokyo 113
Japan
☎ 81 338 12 2111*4177
℻ 81 356 84 9642
✉ tatsushi@phys.s.u-tokyo.ac.jp

Sugitani Koji
College of General Education
Nagoya City Univ
Mizuho Ku
Nagoya 467
Japan
☎ 81 528 72 5846
℻ 81 528 82 3075
✉ d43000@jpnkudpc

Sugiyama Naoshi
Physics Dpt
Univ of Tokyo
Hongo 7-3-1 Bunkyo-ku
Kyoto 113
Japan
☎ 81 338 12 2111*4191
℻ 81 356 89 0465

Suh Kyung-Won
Dpt Astronomy/Space Sci
Chungbuk Ntl Univ
San 48 Gaeshin Dong
Cheongju 360 763
Korea RP
☎ 82 431 61 2315
℻ 82 431 67 4232
✉ kwsuh@cbucc.cbnu.ac.kr

Sukartadiredja Darsa
Planetarium Dki
Jl Cikini Raya 73
Jakarta 10330
Indonesia
☎ 62 21 377 530

Sukumar Sundarajan
DRAO
NRCC/HIA
Box 248
Penticton BC V2A 6K3
Canada
☎ 1 250 493 2277
℻ 1 250 493 7767

Sulentic Jack W
Dpt Phys/Astronomy
Univ of Alabama
Box 870324
Tuscaloosa AL 35487 0324
USA
☎ 1 205 348 5050
℻ 1 205 348 5051
✉ jsulentic@ua1vm

Sullivan Denis John
Physics Dpt
Victoria Univ
Private Bag
Wellington
New Zealand
☎ 64 721 000
✉ sullivan@matai.vuw.ac.nz

Sullivan Woodruff T
Astronomy Dpt
Univ of Washington
Box 351580
Seattle WA 98195 1580
USA
☎ 1 425 543 7773
℻ 1 425 685 0403
✉ woody@astro.washington.edu

Sultanov G F
Shemakha Astrophysical Observ
Azerbajan Acad Sci
373243 Shemakha
Azerbajan
☎ 994 89 22 39 8248

Summers Hugh P
UK Culham Lab
Jet Joint Undertaking
Abingdon OX14 3EA
UK
☎ 44 12 35 28822

Sun Jin
Astronomy Dpt
Beijing Normal Univ
Beijing 100875
China PR
☎ 86 1 65 6531*6285
℻ 86 1 201 3929

Sun Kai
Dpt Geophysics
Beijing Univ
Beijing 100871
China PR
☎ 86 1 28 2471*3888

Sun Wei-Hsin
Inst Phys/Astronomy
Ntl Central Univ
Chung Li Taiwan 32054
China R
☎ 886 3 425 4960
℻ 886 3 426 2304
✉ sun@angel.phy.ncu.edu.tw

Sun Yisui
Astronomy Dpt
Nanjing Univ
Nanjing 210093
China PR
☎ 86 25 37551
℻ 86 25 330 2728

Sun Yongxiang
Inst Geodesy/Geophysics
Xu Dong Lu
Wuchang 430077
China PR

Sunada Kazuyoshi
Nobeyama Radio Obs
NAOJ
Minamimaki Mura
Nagano 384 13
Japan
☎ 81 267 984 384
℻ 81 267 982 927
✉ sunada@nro.nao.ac.jp

Sundelius Bjoern
Mechanics Div
Chalmers Technical Univ
S 412 96 Goeteborg
Sweden
☎ 46 31 722 1517
℻ 46 31 772 3477
✉ bjsu@mec.chalmers.se

Sundman Anita
Stockholm Observ
Royal Swedish Acad Sciences
S 133 36 Saltsjoebaden
Sweden
☎ 46 8 717 0634
℻ 46 8 717 4719
✉ sundman@astro.su.se

Suntzeff Nicholas B
NOAO
CTIO
Casilla 603
La Serena
Chile
☎ 56 51 22 5415
📠 56 51 20 5342
✉ nsuntzeff@noao.edu

Sunyaev Rashid A
Space Res Inst
Acad Sciences
Profsojuznaya Ul 84/32
117810 Moscow
Russia
☎ 7 095 333 3373
📠 7 095 310 7023
✉ rsunyaev@esoci

Suran Marian Doru
Astronomical Institute
Romanian Acad Sciences
Cutitul de Argint 5
RO 75212 Bucharest
Rumania
☎ 40 1 641 3686
📠 40 1 312 3391
✉ suran@roastro.
astro.ro

Surdej Jean M G
STScI
Homewood Campus
3700 San Martin Dr
Baltimore MD 21218
USA
☎ 1 301 338 4984
📠 1 301 338 4767
✉ surdej@stsci.edu

Surdin Vladimir G
SAI
Acad Sciences
Universitetskij Pr 13
119899 Moscow
Russia
☎ 7 095 939 1616
📠 7 095 932 8841
✉ surdin@sai.msu.su

Sutantyo Winardi
Bosscha Observ
ITB Lembang
Lembang 40391
Indonesia
☎ 62 22 229 6001
📠 62 22 278 7289

Sutherland Peter G
Physics Dpt
Mcmaster Univ
Hamilton ON L8S 4M1
Canada
☎ 1 416 525 9140
📠 1 416 546 1252

Sutherland William
Astrophysics Dpt
Univ of Oxford
Keble Rd
Oxford OX1 3RH
UK
☎ 44 186 527 3310
📠 44 186 527 3418
✉ wjs@uk.ac.ox.astro

Suto Yasushi
Physics Dpt
Univ of Tokyo
Bunkyo Ku
Tokyo 113
Japan
☎ 81 338 12 3111*4195
📠 81 356 84 9642

Sutton Edmund Charles
Dpt Astron/Physics
Univ of Illinois
1002 W Green St
Urbana IL 61801
USA
☎ 1 217 333 9339
📠 1 217 244 7638
✉ sutton@astro.uiuc.edu

Suzuki Hideyuki
Kek Ntl Laboratory For
High Energy Physics
1 1 Oho
Tsukuba Ibaraki 305
Japan
☎ 81 298 64 5400
📠 81 298 64 2580
✉ suzuki@kethl.kek.jp

Suzuki Yoshimasa
23-1 Nakajima
Hironomachi
Uji 611
Japan

Svalgaard Leif
Hertogenlaan 31
B 3202 Lubbeek
Belgium

Svechnikova Maria A
Astronomy Dpt
Uralskij State Univ
Lenin Pr 51
620083 Sverdlovsk
Russia
☎ 7 343 222 0729
📠 7 343 222 3386

Svensson Roland
Stockholm Observ
Royal Swedish Acad Sciences
S 133 36 Saltsjoebaden
Sweden
☎ 46 8 716 4472
📠 46 8 717 4719
✉ svensson@astro.su.se

Sveshnikov Mikhail
Inst Theoret Astron
Acad Sciences
N Kutuzova 10
191187 St Petersburg
Russia
☎ 7 812 275 0360
📠 7 812 272 7968
✉ 1088@iipah.spb.su

Svestka Jiri
Obs/Planetarium
Petrin 205
CZ 118 46 Praha 1
Czech R
☎ 420 2 53 53513
📠 420 2 258 940

Svestka Zdenek
SRON
Postbus 800
Sorbonnelaan 2
NL 3584 CA Utrecht
Netherlands
☎ 31 30 253 5600
📠 31 30 254 0860

Svolopoulos Sotirios
Astrophysics Dpt
Ntl Univ Athens
Panepistimiopolis
GR 157 84 Zografos
Greece
☎ 30 1 724 3414
📠 30 1 723 5122

Svoren Jan
Astronomical Institute
Slovak Acad Sci
SK 059 60 Tatranska Lomni
Slovak Republic
☎ 421 969 96 7866
📠 421 969 96 7656

Swade Daryl Allen
STScI
Homewood Campus
3700 San Martin Dr
Baltimore MD 21218
USA
☎ 1 301 338 4480
📠 1 301 338 4767
✉ swade@stsci.edu

Swanenburg B N
SRON
Postbus 800
Sorbonnelaan 2
NL 3584 CA Utrecht
Netherlands
☎ 31 30 253 5600
📠 31 30 254 0860

Swank Jean Hebb
NASA GSFC
Code 666
Greenbelt MD 20771
USA
☎ 1 301 286 9167
📠 1 301 286 3391

Swarup Govind
NCRA/TIFR
Pune Univ Campus Pb 3
Ganeshkhind
Pune 411 007
India
☎ 91 212 33 6111
📠 91 212 33 5760

Sweet Peter A
Astronomy Dpt
Univ of Glasgow
Glasgow G12 8QQ
UK
☎ 44 141 339 8855
📠 44 141 334 9029

Sweigart Allen V
NASA GSFC
Code 681
Greenbelt MD 20771
USA
☎ 1 301 286 6274

Sweitzer James Stuart
Rose Ctr Earth/Space
Americ Museum Ntrl History
Central Park West at 79th Street
New York NY 10024
USA
☎ 1 212 769 5808
📠 1 212 496 3555
✉ sweitzer@amnh.org

Swenson George W
Dpt Elect/Computer Eng
Univ of Illinois
1406 W Green St
Urbana IL 61801
USA
☎ 1 217 333 4498

Swensson John W
Physics Dpt
Univ of Lund
Soelvegatan 14 A
S 223 62 Lund
Sweden
☎ 46 46 222 7000
📠 46 46 222 4709

Swerdlow Noel
Astronomy/Astrophys Ctr
Univ of Chicago
5640 S Ellis Av
Chicago IL 60637
USA
☎ 1 312 962 7969
📠 1 312 702 8212

Swihart Thomas L
448 W Hardy Rd
Tucson AZ 85737
USA

Swings Jean-Pierre
Inst Astrophysique
Universite Liege
Av Cointe 5
B 4000 Liege
Belgium
☎ 32 4 254 7510
📠 32 4 254 7511
✉ swings@astro.ulg.ac.be

Syer David
MPA
Karl Schwarzschildstr 1
D 85748 Garching
Germany
☎ 49 893 299 3216
📠 49 893 299 3235
✉ syer@mpa-garching.mpg.de

Sygnet Jean-Francois
IAP
98bis bd Arago
F 75014 Paris
France
☎ 33 1 44 32 8094
📠 33 1 44 32 8001
✉ sygnet@friap51

Sykes Mark Vincent
Steward Observ
Univ of Arizona
Tucson AZ 85721
USA
☎ 1 520 621 2054
📠 1 520 621 1532
✉ msykes@as.arizona.edu

Sykes-Hart Avril B
Astrophysics Dpt
Univ of Oxford
South Parks Rd
Oxford OX1 3RQ
UK
☎ 44 186 527 3999
✆ 44 186 527 3947

Sykora Julius
Astronomical Institute
Slovak Acad Sci
SK 059 60 Tatranska Lomni
Slovak Republic
☎ 421 969 96 7866
✆ 421 969 96 7656

Sylvester Roger
Dpt Phys/Astronomy
UCLO
Gower St
London WC1E 6BT
UK
☎ 44 171 387 7050
✆ 44 171 380 7145
✉ rjs@star.ucl.ac.uk

Sylwester Barbara
Space Res Ctr
Acad Sciences
Ul Kopernika 11
PL 51 622 Wroclaw
Poland
☎ 48 71 372 9373/74
✆ 48 71 372 9378
✉ bs@cbk.pan.wroc.pl

Sylwester Janusz
Space Res Ctr
Acad Sciences
Ul Kopernika 11
PL 51 622 Wroclaw
Poland
☎ 48 71 372 9373/74
✆ 48 71 372 9378
✉ js@cbk.pan.wroc.pl

Synnott Stephen P
CALTECH/JPL
MS 264 686
4800 Oak Grove Dr
Pasadena CA 91109 8099
USA
☎ 1 818 354 6933
✆ 1 818 393 6030

Szabados Laszlo
Konkoly Observ
Thege U 13/17
Box 67
H 1525 Budapest
Hungary
☎ 36 1 375 4122
✆ 36 1 275 4668

Szafraniec Rozalia
Ul Kopernika 27
PL 31 501 Krakow
Poland

Szalay Alex
Dpt Phys/Astronomy
JHU
Charles/34th St
Baltimore MD 21218
USA
☎ 1 301 516 7217
✆ 1 301 516 5096
✉ szalai@pha.jhu.edu

Szatmary Karoly
Physics Dpt
Jate Univ
Dom ter 9
H 6720 Szeged
Hungary
☎ 36 6 231 1622
✆ 36 3 631 1154
✉ k.szatmary@physx.u-szeged.hu

Szczerba Ryszard
NCAC
Polish Acad Sciences
Ul Rabianska 8
PL 87 100 Torun
Poland
☎ 48 56 19 249
✆ 48 56 19 381
✉ szczerba@ncac.torun.pl

Szecsenyi-Nagy Gabor
Astronomy Dpt
Eotvos Univ
Ludovika ter 2
H 1083 Budapest
Hungary
☎ 36 1 114 1019
✆ 36 1 210 1089
✉ szena@ludens.elte.hu

Szego Karoly
Central Res Inst
for Physics
Box 49
H 1525 Budapest
Hungary
☎ 36 1 551 682
✆ 36 1 696 567

Szeidl Bela
Konkoly Observ
Thege U 13/17
Box 67
H 1525 Budapest
Hungary
☎ 36 1 375 4122
✆ 36 1 275 4668

Szkody Paula
Astronomy Dpt
Univ of Washington
Box 351580
Seattle WA 98195 1580
USA
☎ 1 425 543 1988
✆ 1 425 685 0403
✉ szkody@astro.washington.edu

Szostak Roland
Inst Didaktik Physik
Westfalische Wilhelms Uni
Wilhelm Klemm Str 10
D 48149 Muenster
Germany
☎ 49 251 839 386/7
✆ 49 251 832 090

Szymanski Michal
Astrononomical Observ
Warsaw Univ
Al Ujazdowskie 4
PL 00 478 Warsaw
Poland
☎ 48 22 29 4011
✆ 48 22 629 4967
✉ msz@sirius.astrouw.edu.pl

T

Taam Ronald Everett
Dearborn Observ
NW Univ
2131 Sheridan Rd
Evanston IL 60208 3112
USA
☎ 1 847 491 7528
✆ 1 847 491 3135

Tabara Hiroto
Fac of Education
Utsunomiya Univ
Minemachi
Utsunomiya 321
Japan
☎ 81 286 36 1515

Taboada Ramon Rodriguez
Inst Geophys/Astron
C 212 N 2906/29 Y 31
Lisa
La Habana
Cuba
☎ 53 21 8435

Taborda Jose Rosa
Fac of Sciences
Astronomical Observ
R Esc Politecnica 58
P 1200 Lisboa
Portugal

Tacconi Linda J
MPE
Postfach 1603
D 85740 Garching
Germany
☎ 49 893 299 3873
✆ 49 893 299 3569
✉ linda@hethp.mpe-garching.mpg.de

Tacconi-Garman Lowell E
MPE
Postfach 1603
D 85740 Garching
Germany
☎ 49 893 299 3288
✆ 49 893 299 3569
✉ lowell@mpa-garching.mpg.de

Tademaru Eugene
Astronomy Dpt
Univ Massachusetts
Amherst MA 01002
USA
☎ 1 413 545 4301

Taff Laurence G
5208 Springlake Way
Baltimore MD 21212 3240
USA
☎ 1 301 338 4799
✆ 1 301 338 4796

Taffara Salvatore
Via Calza 5bis
I 35128 Padova
Italy
☎ 39 49 807 1624

Tagger Michel
Service Astrophysique
CEA Saclay
Orme d Merisiers Bt 709
F 91190 Gif s Yvette Cdx
France
☎ 33 1 69 08 5607
✆ 33 1 69 08 9266
✉ tagger@cea.fr

Tagliaferri Gianpiero
Osserv Astronomico
d Milano
Via E Bianchi 46
I 22055 Merate
Italy
☎ 39 990 6412
✆ 39 990 8492
✉ gtagliaf@merate.mi.astro.it

Tagliaferri Giuseppe
Osserv Astrofis Arcetri
Univ d Firenze
Largo E Fermi 5
I 50125 Firenze
Italy
☎ 39 55 27 521
✆ 39 55 22 0039

Tago Erik
Tartu Observ
Estonian Acad Sci
Toravere
EE 2444 Tartumaa
Estonia
☎ 372 7 410 434
✆ 372 7 410 205
✉ erik@jupiter.aai.ee

Takaba Hiroshi
Fac of Eng
Gifu Univ
1 1 Yanagito
Gifu 501 1193
Japan
☎ 81 582 93 3054
✆ 81 81 293 3057
✉ takaba@cc.gifu-u.ac.jp

Takada-Hidai Masahide
Res Inst of Civilization
Tokai Univ
1117 Kitakaname Hiratsuka
Kanagawa 259 12
Japan
☎ 81 463 58 1211*4813
✆ 81 463 29 4047
✉ hidai@keyaki.cc.u-tokai.ac.jp

Takagi Kojiro
Physics Dpt
Toyama Univ
3190 Gofuku
Toyama 930
Japan
☎ 81 764 23 4716
✆ 81 764 45 6549

Takagi Shigetsugu
Dpt Fisica/CCE/UFRN
Univ Federal
Do Rio Grande d Norte
59072 970 Natal RN
Brazil
☎ 55 84 231 9586
✆ 55 84 231 9749

Takahara Fumio
Fac of Sciences
Osaka Univ
Machinkaneyama
Toyonaka Osaka 560 0043
Japan
☎ 81 426 77 1111*3348
✆ 81 426 77 2483
✉ takahara@vega.ess.sci.osaka-u.ac.jp

Takahara Mariko
Doshisha Col Liberal Arts
Imadegawa-dori nishiiru
Teramachi-kamigyo-ku
Kyoto 602
Japan
☎ 81 75 251 4211
✆ 81 75 251 4289
✉ e50334@sakura.kudpc.kyoto-u.ac.jp

Takahashi Koji
ISAS
3 1 1 Yoshinodai
Sagamihara
Kanagawa 229 8510
Japan
☎ 81 427 51 3911
✆ 81 427 59 4253
✉ takahasi@astro.isas.ac.jp

Takahashi Masaaki
Dpt Phys/Astronomy
Aichi Univ of Edu
Kariya
Aichi 448
Japan
☎ 81 566 36 3111*538
✆ 81 566 36 4337
✉ takahasi@auephyas.aichi-edu.ac.jp

Takahashi Tadayuki
ISAS
3 1 1 Yoshinodai
Sagamihara
Kanagawa 229 8510
Japan
☎ 81 427 51 3911
✆ 81 427 59 4253
✉ takahasi@astro.isas.ac.jp

Takakubo Keiya
Mukai Yama 1 3 20
Taihaku
Sendai 982
Japan

Takakura Tatsuo Emer
Astronomy Dpt
Univ of Tokyo
Bunkyo Ku
Tokyo 113
Japan
☎ 81 381 22 111
✆ 81 338 13 9439

Takalo Leo O
Turku Univ
Tuorla Observ
Vaeisaelaentie 20
FIN 21500 Piikkio
Finland
☎ 358 2 274 4258
✆ 358 2 243 3767
✉ takalo@kontu.utu.fi

Takami Hideki
Ntl Astronomical Obs
2 21 1 Osawa
Mitaka
Tokyo 181
Japan
☎ 81 422 34 3865
✆ 81 422 34 3864
✉ takami@optik.mtk.nao.ac.jp

Takano Toshiaki
Nobeyama Radio Obs
NAOJ
Minamimaki Mura
Nagano 384 13
Japan
☎ 81 267 63 4487
✆ 81 267 63 4444
✉ takano@nro.nao.ac.jp

Takarada Katsuo
Kyoto Inst Techn
Matsugasaki
Sakyo Ku
Kyoto 606 01
Japan
☎ 81 757 91 3211
✆ 81 75 753 7010

Takase Bunshiro
Tokyo Astronomical Obs
NAOJ
Osawa Mitaka
Tokyo 181
Japan
☎ 81 422 32 5111
✆ 81 422 34 3793

Takato Naruhisa
Ntl Astronomical Obs
Osawa 2-21-1
Mitaka
Tokyo 181
Japan
☎ 81 422 343 909
✆ 81 422 34 3864
✉ takato@optik.mtk.nao.ac.jp

Takayanagi Kazuo
Oji Honcho 1 3 10
Kita Ku
Tokyo 114
Japan
☎ 81 339 09 8591
✆ 81 339 09 8594
✉ takakazu@se.shibaura-it.ac.jp

Takeda Hidenori
Dpt Aeronautic Engineer
Kyoto Univ
Sakyo Ku
Kyoto 606 01
Japan
☎ 81 757 53 7008
✆ 81 75 753 7010

Takeda Yoichi
Inst of Astronomy
Univ of Tokyo
Osawa Mitaka
Tokyo 181
Japan
☎ 81 422 41 3739
✆ 81 422 41 3749

Takenouchi Tadao
1 28 30 Kichijyoji Kita Machi
Musashino Shi
Tokyo 180
Japan

Takens Roelf Jan
Astronomical Institute
Univ of Amsterdam
Kruislaan 403
NL 1098 SJ Amsterdam
Netherlands
☎ 31 20 525 7491
✆ 31 20 525 7484
✉ roelf@astro.uva.nl

Takeuti Mine
Astron Institute
Tohoku Univ
Aoba ku
Sendai 980 77
Japan
☎ 81 22 217 6512
✆ 81 222 17 6513
✉ takeuti@astroa.astr.tohoku.ac.j
p

Talavera A
ESA
Apt 50727 Villafranca
E 28080 Madrid
Spain
☎ 34 1 813 1100
✆ 34 1 813 1139

Talbot Raymond J
The Aerospace Corporation
1927 Curtis Av
Redondo Beach CA 90278
USA
☎ 1 213 379 9927

Talon Raoul
CESR
BP 4346
F 31029 Toulouse Cdx
France
☎ 33 5 61 55 6666
✆ 33 5 61 55 6701

Talwar Satya P
Dpt Phys/Astrophys
Univ of Delhi
New Delhi 110 007
India
☎ 91 11 291 8993

Tamenaga Tatsuo
Fac of Education
Mie Univ
Tsu-shi
Mie 514
Japan

Tammann Gustav Andreas
Astronomisches Institut
Univ Basel
Venusstr 7
CH 4102 Binningen
Switzerland
☎ 41 61 271 7711/12
✆ 41 61 271 7810
✉ tammann@ubaclu.unibas.ch

Tamura Motohide
Ntl Astronomical Obs
2 21 1 Osawa
Mitaka
Tokyo 181
Japan
☎ 81 422 34 3705
✆ 81 422 34 3608
✉ otamura@c1.mtk.nao.ac.jp

Tamura Shin'ichi
Astronomy Dpt
Tohoku Univ
Aramaki
Sendai 980
Japan
☎ 81 222 22 1800
✉ tamura@astroa.astr.
tohoku.ac.jp

Tan Detong
Shanghai Observ
CAS
80 Nandan Rd
Shanghai 200030
China PR
☎ 86 21 6438 6191
✆ 86 21 6438 4618

Tan Huisong
Yunnan Observ
CAS
Kunming 650011
China PR
☎ 86 871 2035
✆ 86 871 717 1845

Tanabe Hiroyoshi
Tokyo Astronomical Obs
NAOJ
Osawa Mitaka
Tokyo 181
Japan
☎ 81 422 32 5111
✆ 81 422 34 3793

Tanabe Kenji
Okayama Univ of Science
1-1 Ridai-cho
Okayama 700
Japan
☎ 81 862 52 3161
✆ 81 862 55 3847

Tanabe Toshihiko
Tokyo Astronomical Obs
NAOJ
Osawa Mitaka
Tokyo 181
Japan
☎ 81 422 32 5111
✆ 81 422 34 3793

Tanaka Masuo
Inst of Astronomy
Univ of Tokyo
Osawa Mitaka
Tokyo 181
Japan
☎ 81 422 41 3743
✆ 81 422 41 3749

Tanaka Riichiro
Fac of Education
Niigata Univ
8050 Ikarashi 2
Niigata 950 21
Japan
☎ 81 252 62 7269

Tanaka Wataru
Opt/IR Astro Div
NAOJ
Osawa Mitaka
Tokyo 181
Japan
☎ 81 422 34 3603
✆ 81 422 34 3608
✉ otanak2@c1.mtk.nao.ac.jp

Tanaka Yasuo
ISAS
3 1 1 Yoshinodai
Sagamihara
Kanagawa 229 8510
Japan
☎ 81 427 51 3911
✆ 81 427 59 4253
✉ tanaka@

Tanaka Yasuo
Physics Dpt
Ibaraki Univ
Bunkyo
Mito 310
Japan
☎ 81 292 26 1621*372

Tanaka Yutaka D
Kobe Yamate Women's
Junior College
Suwayama Chuo-ku
Kobe 650
Japan
☎ 81 783 41 6060

Tancredi Gonzalo
Dpt Astronomia
Fac Humani/Ciencia
Tristan Naraja 1674
Montevideo 11200
Uruguay
☎ 598 2 418 004
✆ 598 2 421 957
✉ gonzalo@fisica.edu.uy

Tandberg-Hanssen Einar A
NASA/MSFC
Space Science Lab
Code ES 01
Huntsville AL 35812
USA
☎ 1 205 544 7578
✆ 1 205 544 7754
✉ tandberg@msfc.nasa.gov

Tandon Jagdish Narain
Dpt Phys/Astrophys
Univ of Delhi
New Delhi 110 007
India
☎ 91 11 252 1521

Tandon S N
IUCAA
PO Box 4
Ganeshkhind
Pune 411 007
India
☎ 91 212 351 414/5
✆ 91 212 350 760
✉ sntandon@iucaa.ernet.in

Tang Yuhua
Astronomy Dpt
Nanjing Univ
Nanjing 210093
China PR
☎ 86 25 37651
✆ 86 25 330 2728

Tango William J
School of Physics
UNSW
Sydney NSW 2006
Australia
☎ 61 2 692 3953
✆ 61 2 660 2903
✉ tango@physics.su.oz.au

Taniguchi Yoshiaki
Astron Institute
Tohoku Univ
Aramaki Aoba
Sendai 980
Japan
☎ 81 222 22 1800 *3319
✆ 81 222 61 2806
✉ tani@astroa.astr.tohoku.ac.jp

Tanikawa Kiyotaka
Ntl Astronomical Obs
2 21 1 Osawa
Mitaka
Tokyo 181
Japan
☎ 81 442 34 3634
✆ 81 442 34 3746
✉ tanikawa@ferio.mtk.nao.ac.jp

Tanzella-Nitti Giuseppe
Roman Athenaeum of The
Holy Cross
S Girolamo d Carita 64
I 00186 Roma
Italy

Tanzi Enrico G
IFCTR CNR
Univ d Milano
Via E Bassini 15
I 20133 Milano
Italy
☎ 39 2 236 3542
✆ 39 2 266 5753

Tapde Suresh Chandra
NCRA/TIFR
Pune Univ Campus Pb 3
Ganeshkhind
Pune 411 007
India
☎ 91 212 351 414/5
✆ 91 212 350 760
✉ sct@gmrt.ernet.in

Tapia Mauricio
Observ Astronomico Nacional
UNAM
Apt 877
Ensenada BC 22800
Mexico
☎ 52 6 174 4580
✆ 52 6 174 4607

Tapia-Perez Santiago
Phillips Lab
LIMA
Kirtland Afb NM 87117
USA
☎ 1 505 846 5045
✉ tapia@plk.af.mil

Tapley Byron D
Aerospace Eng Dpt
Univ of Texas
WRW 402
Austin TX 78712
USA
☎ 1 512 471 1356

Tapping Kenneth F
NRCC/HIA
100 Sussex Dr
Ottawa ON K1A 0R6
Canada
☎ 1 613 991 5842
✆ 1 613 952 6602

Tarady Vladimir K
Main Astronomical Obs
Ukrainian Acad Science
Golosiiv
252650 Kyiv 22
Ukraine
☎ 380 4426 62286
✆ 380 4426 62147
✉ tarady@mao.apc.org

Tarenghi Massimo
ESO
Karl Schwarzschildstr 2
D 85748 Garching
Germany
☎ 49 893 200 6236
✆ 49 893 202 362

Tarnstrom Guy
MIT Lincoln Laboratory
MS S4 241
Box 73
Lexington MA 02173
USA
☎ 1 617 863 5500

Tarrab Irene
221 r La Fayette
F 75010 Paris
France

Tarter C Bruce
LLNL
L 295
Box 808
Livermore CA 94551 9900
USA
☎ 1 415 422 4169
✆ 1 415 423 0238

Tarter Jill C
SETI Institute
Project Phoenix Office
2035 Landings Dr
Mountain View CA 94043
USA
☎ 1 415 960 4555
✆ 1 415 968 5830
✉ tarter@vger.seti-inst.edu

Tashiro Makoto
Physics Dpt
Univ of Tokyo
7-3-1 Hongo Bunkyo-ku
Tokyo 113
Japan
☎ 81 338 12 2111*4171
✆ 81 338 12 6938
✉ tashiro@phys.s.u-tokyo.ac.jp

Tassoul Jean-Louis
6000 Ch Deacon App J2
Montreal QC H3S 2T9
Canada
✉ tassoulj@magellan.
umontreal.ca

Tassoul Monique
6000 Ch deacon App J2
Montreal QC H3S 2T9
Canada
✉ tassoul@magellan.
umontreal.ca

Tatematsu Ken-ichi
Nobeyama Radio Obs
NAOJ
Minamimaki Mura
Nagano 384 13
Japan
☎ 81 267 98 4376
✆ 81 267 98 4339
✉ tatmatsu@nro.nao.ac.jp

Tatevyan S K
Inst of Astronomy
Acad Sciences
Pyatnitskaya Ul 48
109017 Moscow
Russia
☎ 7 095 231 5461
✆ 7 095 230 2081

Tateyama Claudio Eiichi
CRAAE INPE
EPUSP/PTR
CP 61548
01065 970 Sao Paulo SP
Brazil
☎ 55 11 815 5936
✆ 55 11 815 6289

Taton Rene
Centre Alexandre Koyre
57 r Cuvier
Pavillon Cheveul
F 75231 Paris Cdx 05
France

Tatum Jeremy B
Climenhoga Observ
Univ of Victoria
Box 1700
Victoria BC V8W 2Y2
Canada
☎ 1 250 721 7750
✆ 1 250 721 7715

Tauber Jan
ESA/ESTEC
Astrophysics Division
Box 299
NL 2201 AZ Noordwijk
Netherlands
☎ 31 71 565 5342
✆ 31 71 565 4690
✉ jtauber@estsa2.estec.esa.nl

Tautvaisiene Grazina
Inst Theor Physics/
Astronomy
Gostauto 12
Vilnius 2600
Lithuania
☎ 370 2 621 117
✆ 370 2 224 694
✉ taut@itpa.fi.lt

Tavakol Reza
School Mathemat Sc
QMWC
Mile End Rd
London E1 4NS
UK
☎ 44 171 980 4811
✆ 44 181 975 5500

Tavares J T l
Av dias Da Silva
173 R/c Esq
P 3000 Coimbra
Portugal

Tawadros Maher Jacoub
NAIGR
Helwan Observ
Cairo 11421
Egypt
☎ 20 78 0645/2683
✆ 20 62 21 405 297

Tawara Yuzuru
Astrophysics Dpt
Nagoya Univ
Furocho Chikusa Ku
Nagoya 464 01
Japan
☎ 81 527 81 5111
✆ 81 527 81 3541

Taylor A R
Physics Dpt
Univ of Calgary
2500 University Dr NW
Calgary AB T2N 1N4
Canada
☎ 1 403 220 5385
✆ 1 403 289 3331

Taylor Andrew
Dpt Physics/
Math Physics
Univ Adelaide
Adelaide SA 5005
Australia
☎ 61 8 303 5313
✆ 61 8 303 4380
✉ ataylor@physics.
adelaide.edu.au

Taylor Donald Boggia
Royal Greenwich Obs
Madingley Rd
Cambridge CB3 0EZ
UK
☎ 44 12 23 374 739
✆ 44 12 23 374 700
✉ dbt@uk.ac.ro-
greenwich.starlin

Taylor Donald J
Dpt Phys/Astronomy
Univ Nebraska
Behlen Observatory
Lincoln NE 68588 0111
USA
☎ 1 402 472 3686
✆ 1 402 472 2879

Taylor Gregory Benjamin
CALTECH
Palomar Obs
MS 105 24
Pasadena CA 91125
USA
☎ 1 818 395 4024
✆ 1 818 568 9352
✉ gbt@astro.caltech.edu

Taylor Joseph H
Physics Dpt
Princeton Univ
Jadwin Hall
Princeton NJ 08544 1001
USA
☎ 1 609 452 4368
✆ 1 609 258 6853

Taylor Keith
AAO
Box 296
Epping NSW 2121
Australia
☎ 61 2 868 0222
✆ 61 2 868 0310

Taylor Kenneth N R
105a Copeland Rd
Beecroft NSW 2119
Australia

Tchang-Brillet Lydia
Observ Paris Meudon
DAMAP
Pl J Janssen
F 92195 Meudon PPL Cdx
France
☎ 33 1 45 07 7576
℻ 33 1 45 07 7469
✉ brillet@fobspm.fr

Tchouikova Nadejda A
SAI
Acad Sciences
Universitetskij Pr 13
119899 Moscow
Russia
☎ 7 095 939 5024
℻ 7 095 939 0126
✉ chujkova@sai.msu.su

te Lintel Hekkert Peter
Parkes Observ
Box 276
Parkes NSW 2870
Australia
☎ 61 68 611 726
℻ 61 68 611 730
✉ plintel@atnf.csiro.au

Teays Terry J
NASA GSFC
Code 684 9
Greenbelt MD 20771
USA
☎ 1 301 286 5749
℻ 1 301 286 7642

Tedesco Edward F
TerraSystems Inc
Space Sci Res Div
59 Wednesday Hill Rd
Lee NH 03824 6537
USA
✉ etedesco@terrasys.com

Teerikorpi Veli Pekka
Turku Univ
Tuorla Observ
Vaeisaelaentie 20
FIN 21500 Piikkio
Finland
☎ 358 2 143 5822
℻ 358 2 143 3767

Teherany D
83 Av Rey
Tehran
Iran

Teixeira Ramachrisna
IAG
Univ Sao Paulo
CP 9638
01065 970 Sao Paulo SP
Brazil
☎ 55 11 577 8599
℻ 55 11 576 3848
✉ teixeira@iag.usp.ansp.br

Tejfel Viktor G
Astrophys Inst
Kazakh Acad Sci
480068 Alma Ata
Kazakhstan
☎ 7 68 3053
✉ tejf@afi.academ.alma-ata.su

Tektunali H Gokmen
Univ Observ
Univ of Istanbul
University 34452
34452 Istanbul
Turkey
☎ 90 212 522 3597
℻ 90 212 519 0834

Telesco Charles M
Astronomy Dpt
Univ of Florida
211 SSRB
Gainesville FL 32611
USA
☎ 1 904 392 2052
℻ 1 904 392 5089

Telles Eduardo
Observ Nacional
Rua Gl Bruce 586
Sao Cristovao
20921 030 Rio de Janeiro RJ
Brazil
☎ 55 21 580 781
℻ 55 21 580 0332
✉ etelles@on.br

Telnyuk-Adamchuk V
Astronomical Observ
Kyiv State Univ
Observatorna Str 3
254053 Kyiv
Ukraine
☎ 380 4421 62691
℻ 380 4422 46387
✉ aoku@gluk.apc.org

Tempesti Piero
Istt Astronomico
Univ d Roma La Sapienza
Via G M Lancisi 29
I 00161 Roma
Italy
☎ 39 6 44 03734
℻ 39 6 44 03673

Tenorio-Tagle Guillermo
IAC
Observ d Teide
via Lactea s/n
E 38200 La Laguna
Spain
☎ 34 2 232 9100
℻ 34 2 232 9117
✉ gtt@ast.cam.ac.uk

Teplitskaya R B
Sibizmir
Acad Sciences
664697 Irkutsk 33
Russia
☎ 7 395 262 9388

Ter Haar Dirk
Box 10
349 Middle St
Petworth GU28 0RY
UK

Terashita Yoichi
Kanazawa Techn Inst
7 1 Ogigaoka
Nonoichimachi
Ishikawa 921
Japan
☎ 81 762 48 1100

Terebizh Valery Yu
Crimean Astrophys Obs
Ukrainian Acad Science
Nauchny
334413 Crimea
Ukraine
☎ 380 6554 71161
℻ 380 6554 40704

Terekhov Oleg V
Astro Space Ctr
Lebedev Physical Inst
Profsoyuznaya 84/32
Moscow 117810
Russia
☎ 7 095 333 5300
℻ 7 095 333 5377
✉ terekhov@hea.iki.rssi.ru

Terentjeva Alexandra K
Inst of Astronomy
Acad Sciences
Pyatnitskaya Ul 48
109017 Moscow
Russia
☎ 7 095 231 5461
℻ 7 095 230 2081
✉ ater@airas.msk.su

Terlevich Elena
Inst of Astronomy
The Observatories
Madingley Rd
Cambridge CB3 0HA
UK
☎ 44 12 23 337 548
℻ 44 12 23 337 523
✉ et@ast.cam.ac.uk

Terlevich Roberto Juan
Royal Greenwich Obs
Madingley Rd
Cambridge CB3 0EZ
UK
☎ 44 12 23 374 000
℻ 44 12 23 374 700

Ternullo Maurizio
Osserv Astronomico
d Catania
Via A Doria 6
I 95125 Catania
Italy
☎ 39 95 733 21111
℻ 39 95 33 0592

Terranegra Luciano
Osserv d Capodimonte
Via Moiariello 16
I 80131 Naples
Italy
☎ 39 81 557 5111
℻ 39 81 45 6710
✉ terranegra@astrna.na.astro.it

Terrell Nelson James
LANL
MS D 436 NIS 2
Box 1663
Los Alamos NM 87545 2345
USA
☎ 1 505 667 2044
℻ 1 505 665 4414
✉ jterell@lanl.gov

Terrile Richard John
CALTECH/JPL
MS 183 501
4800 Oak Grove Dr
Pasadena CA 91109 8099
USA
☎ 1 213 351 1321
℻ 1 818 393 6030

Terzan Agop
Observ Lyon
Av Ch Andre
F 69561 S Genis Laval Cdx
France
☎ 33 4 78 56 0705
℻ 33 4 72 39 9791

Terzian Yervant
Astronomy Dpt
Cornell Univ
512 Space Sc Bldg
Ithaca NY 14853 6801
USA
☎ 1 607 255 4935
℻ 1 607 255 9817
✉ terzian@astrosun.tn.cornell.edu

Terzides Charalambos
Astronomy Lab
Univ Thessaloniki
GR 540 06 Thessaloniki
Greece
☎ 30 31 99 1357

Teske Richard G
Astronomy Dpt
Univ of Michigan
Dennison Bldg
Ann Arbor MI 48109 1090
USA
☎ 1 313 764 3398
℻ 1 313 764 2211

Teuben Peter J
Astronomy Program
Univ of Maryland
College Park MD 20742 2421
USA
☎ 1 301 405 1540
℻ 1 301 314 9067
✉ teuben@astro.umd.edu

Thaddeus Patrick
CfA
HCO/SAO
60 Garden St
Cambridge MA 02138 1516
USA
☎ 1 617 495 7340
℻ 1 617 495 7356

Thakur Ratna Kumar
Physics Dpt
Univ of Ravishankar
Raipur 492 010
India
☎ 91 27 064

The Pik-Sin
Astronomical Institute
Univ of Amsterdam
Kruislaan 403
NL 1098 SJ Amsterdam
Netherlands
☎ 31 20 525 7491
✆ 31 20 525 7484
✉ psthe@astro.uva.nl

Theis Christian
Inst Theor Phys/
Sternwarte Univ Kiel
Olshausenstr 40
D 24098 Kiel
Germany
☎ 49 431 880 4110
✆ 49 431 880 4100
✉ theis@astrophysik.
uni-kiel.de

Thejll Peter Andreas
Danish Meteorol Inst
Solar/Terrestrial Physics Div
Lyngbyvej 100
DK 2100 Copenhagen
Denmark
☎ 45 39 15 7477
✆ 45 39 15 7460
✉ thejll@dmi.dk

Theodossiou Efstratios
Astrophysics Dpt
Ntl Univ Athens
Panepistimiopolis
GR 157 84 Zografos
Greece
☎ 30 1 724 3414
✆ 30 1 723 5122

Thevenin Frederic
OCA
Observ Nice
BP 139
F 06304 Nice Cdx 4
France
☎ 33 4 92 00 3011
✆ 33 4 92 00 3033
✉ thevenin@froni51

Thielemann Friedrich-Karl
Physics Dpt
Univ Basel
Klingelbergst 82
CH 4056 Basel
Switzerland
☎ 41 61 267 3748
✆ 41 61 267 3784
✉ fkt@quasar.physik.
unibas.ch

Thielheim Klaus O
Abt Mathemati/physik
Univ Kiel
Postfach 5151
D 24063 Kiel
Germany
☎ 49 431 880 3216
✉ thielheim@email.
uni-kiel.de

Thiry Yves R
Universite Paris VI
4 Pl Jussieu Tour 66
F 75230 Paris Cdx 05
France
☎ 33 1 43 36 2525

Thoburn Christine
Royal Greenwich Obs
Madingley Rd
Cambridge CB3 0EZ
UK
☎ 44 12 23 374 000
✆ 44 12 23 374 700

Tholen David J
Inst for Astronomy
Univ of Hawaii
2680 Woodlawn Dr
Honolulu HI 96822
USA
☎ 1 808 956 6930
✆ 1 808 988 2790
✉ tholen@ifa.hawaii.edu

Thomas Claudine
BIPM
Pavillon de Breteuil
F 92312 Sevres Cdx
France
☎ 33 1 45 07 7073
✆ 33 1 45 34 2021
✉ cthomas@bipm.fr

Thomas David V
6 Livonia Rd
Sidmouth
Devon EX10 9JB
UK
☎ 44 13 95 515 156
✉ thomas@mail.ph.ed.ac.uk

Thomas Hans -Christoph
MPA
Karl Schwarzschildstr 1
D 85748 Garching
Germany
☎ 49 893 299 00
✆ 49 893 299 3235

Thomas John H
Dpt Phys/Astronomy
Univ of Rochester
Rochester NY 14627 0171
USA
☎ 1 716 275 4083
✆ 1 716 273 1049
✉ thomas@astro.me.
rochester.edu

Thomas Peter A
Astronomy Centre
Univ of Sussex
Falmer
Brighton BN1 9QH
UK
☎ 44 12 73 60 6755*3099
✆ 44 12 73 67 8097
✉ petert@syma.sussex.ac.uk

Thomas Roger J
NASA GSFC
Code 682
Greenbelt MD 20771
USA
☎ 1 301 286 7921
✆ 1 301 286 1617
✉ thomas@jet.gsfc.nasa.gov

Thomasson Magnus
Onsala Space Observ
Chalmers Technical Univ
S 439 92 Onsala
Sweden
☎ 46 31 772 5534
✆ 46 31 772 5590
✉ magnus@oso.chalmers.se

Thomasson Peter
NRAL
Univ Manchester
Jodrell Bank
Macclesfield SK11 9DL
UK
☎ 44 14 777 1321
✆ 44 147 757 1618

Thompson A Richard
NRAO
2015 Ivy Rd
Charlottesville VA 22903
USA
☎ 1 804 296 0211
✆ 1 804 296 0278

Thomsen Bjarne B Lect
Inst Phys/Astronomy
Univ of Aarhus
Ny Munkegade
DK 8000 Aarhus C
Denmark
☎ 45 86 12 8899
✆ 45 86 20 2711

Thomson Robert
9 Bellevue Pl
Edinburgh EH7 4BS
UK
☎ 44 131 556 0998
✉ rct@mail.ast.cam.ac.uk

Thonnard Norbert
Atom Sciences
114 Ridgeway Ctr
Oak Ridge TN 37830
USA
☎ 1 615 483 1113

Thorne Kip S
CALTECH
MS 130 33
Pasadena CA 91125
USA
☎ 1 818 395 4598
✆ 1 818 796 5675
✉ kip@tapir.caltech.edu

Thorsett Stephen Erik
Physics Dpt
Princeton Univ
Jadwin Hall
Princeton NJ 08544 1001
USA
☎ 1 609 258 1245
✆ 1 609 258 6853
✉ steve@pulsar.princeton.edu

Thorstensen John R
Dpt Phys/Astronomy
Dartmouth College
6127 Wilder Lab
Hanover NH 03755
USA
☎ 1 603 646 2869
✆ 1 603 646 1446

Thronson Harley Andrew
Dpt Phys/Astronomy
Univ of Wyoming
Box 3905
Laramie WY 82071
USA
☎ 1 307 766 6150
✆ 1 307 766 2652
✉ hthronso@hq.nasa.gov

Thuan Trinh Xuan
Astronomy Dpt
Univ of Virginia
Box 3818
Charlottesville VA 22903 0818
USA
☎ 1 804 924 4894
✆ 1 804 924 3104
✉ txt@starburst.astro.virinia.edu

Thuillot William
BDL
77 Av Denfert Rochereau
F 75014 Paris
France
☎ 33 1 40 51 2262
✆ 33 1 46 33 2834
✉ thuillot@bdl.fr

Thum Clemens
IRAM
Avd Divina Pastora 7
Bloque 6/2b
E 18012 Granada
Spain
☎ 34 5 827 9508
✆ 34 5 820 7662

Tian Jing
Beijing Astronomical Obs
CAS
Beijing 100080
China PR
☎ 86 10 6254 5179
✆ 86 1 256 10855

Tiersch Heinz
Zntrlinst Astrophysik
Sternwarte Babelsberg
Rosa Luxemburg Str 17a
D 14482 Potsdam
Germany
☎ 49 331 749 9206
✆ 49 331 749 9309
✉ htiersch@aip.de

Tifft William G
Steward Observ
Univ of Arizona
Tucson AZ 85721
USA
☎ 1 520 621 6532
✆ 1 520 621 1532

Tifrea Emilia
Astronomical Institute
Romanian Acad Sciences
Cutitul de Argint 5
RO 75212 Bucharest
Rumania
☎ 40 1 641 3686
✆ 40 1 312 3391

Timothy J Gethyn
Ctr for Space Sci/
Astrophysics
Stanford Univ ERL 314
Stanford CA 94305 4055
USA
☎ 1 415 497 0059
℻ 1 415 725 2333

Tinbergen Jaap
ASTROM
Postbus 2
NL 7990 AA Dwingeloo
Netherlands
☎ 31 52 159 5100
℻ 31 52 159 7332
✉ tinbergen@astro.rug.nl

Ting Yeou-Tswen
Astronomy Section
Central Weather Bureau
64 Kung Yuen Rd
Taipei 100
China R
☎ 886 2 371 3181*281

Tinney Christopher
AAO
Box 296
Epping NSW 2121
Australia
☎ 61 2 372 4849
℻ 61 2 372 4880
✉ cgt@aaoepp.aao.gov.au

Tipler Frank Jennings
Physics Dpt
Tulane Univ
New Orleans LA 70118
USA
✉ frank.tipler@tulane.edu

Title Alan Morton
Advance Techn Ctr
Org H1 A2 Bg 252
3251 Hanover St
Palo Alto CA 94304
USA
☎ 1 415 424 4034
℻ 1 415 424 3994
✉ title@space.lockheed.com

Tiuri Martti
Helsinki Univ Technology
Radio Laboratory
Otakaari 5 A
FIN 02150 Espoo 15
Finland
☎ 358 9 451 2545
℻ 358 9 451 2378

Tjin-a-Djie Herman R E
Koekoelaan 106
NL 1403 EJ Bussum
Netherlands
☎ 31 21 591 7076

Tlamicha Antonin
Astronomical Institute
Czech Acad Sci
Fricova 1
CZ 251 65 Ondrejov
Czech R
☎ 420 204 85 7324
℻ 420 2 88 1611
✉ tlamicha@asu.cas.cz

Tobin William
Dpt Phys/Astronomy
Univ of Canterbury
Private Bag 4800
Christchurch 1
New Zealand
☎ 64 3 364 2531
℻ 64 3 364 2469
✉ w.tobin@csc.canterbury.
ac.nz

Todoran Ioan
Astronomical Observ
Romanian Acad Sciences
Ul Ciresilor 19
RO 3400 Cluj Napoca
Rumania
☎ 40 64 194 592
℻ 40 64 19 2820

Tofani Gianni
Osserv Astrofis Arcetri
Univ d Firenze
Largo E Fermi 5
I 50125 Firenze
Italy
☎ 39 55 27 52217
℻ 39 55 22 0039

Tohline Joel Edward
Dpt Phys/Astronomy
Louisiana State Univ
Baton Rouge LA 70803 4001
USA
☎ 1 504 388 6851
℻ 1 504 388 5855

Tokarev Yurij V
Radiophysical Res
Inst
Lyadov Ul 25/14
603600 N Novgorod
Russia
☎ 7 831 236 0188
℻ 7 831 236 9902

Tokovinin Andrej A
SAI
Acad Sciences
Universitetskij Pr 13
119899 Moscow
Russia
☎ 7 095 939 3318
℻ 7 095 939 1661
✉ toko@sai.msu.su

Tokunaga Alan Takashi
Inst for Astronomy
Univ of Hawaii
2680 Woodlawn Dr
Honolulu HI 96822
USA
☎ 1 808 956 6691
℻ 1 808 988 2790
✉ tokunaga@galileo.ifa.
hawaii.edU

Tolbert Charles R
University Station
Univ of Virginia
Box 3818
Charlottesville VA 22903 0818
USA
☎ 1 804 924 7494
✉ crt@virginia.edu

Toller Gary N
9364 Dewlit Way
Columbia MD 21045 5118
USA
✉ toller@stars.gsfc.nasa.gov

Tomasi Paolo
Istt Radioastronomia
CNR
Via P Gobetti101
I 40129 Bologna
Italy
☎ 39 51 639 9385
℻ 39 51 639 9385

Tomasko Martin G
Lunar/Planetary Lab
Room 325 Bg 92
Univ of Arizona
Tucson AZ 85721 0092
USA
☎ 1 520 621 6969
℻ 1 520 621 4933

Tomczak Michal
Astronomical Institute
Univ of Wroclaw
Ul Kopernika 11
PL 51 622 Wroclaw
Poland
☎ 48 71 372 9373/74
℻ 48 71 372 9378
✉ tomczak@astro.uni.wroc.pl

Tomimatsu Akira
Physics Dpt
Nagoya Univ
Furocho Chikusa Ku
Nagoya 464 01
Japan
☎ 81 527 89 2840
℻ 81 527 89 2845

Tomisaka Kohji
Fac of Education
Niigata Univ
8050 Ikarashi 2
Niigata 950 21
Japan
☎ 81 252 62 7269
℻ 81 252 64 2026
✉ tomisaka@ed.niigata-u.ac.jp

Tomita Kenji
Yukawa Istt Theor Physics
Kyoto Univ
Kitashirakawa
Kyoto 606 01
Japan
☎ 81 75 753 7017
℻ 81 75 753 7017
✉ tomita@yukawa.kyoto-u.ac.jp

Tomita Koichiro
4-11-20 Yoga
Setagayaku
Tokyo 158
Japan
☎ 81 370 00 066

Tomov Nikolai
Ntl Astronomical Obs
Bulgarian Acad Sci
Box 136
BG 4700 Smoljan
Bulgaria
☎ 359 3 02 1357*283
℻ 359 3 02 1356

Tomov Toma V
Ntl Astronomical Obs
Bulgarian Acad Sci
Box 136
BG 4700 Smoljan
Bulgaria
☎ 359 7 341 599

Toner Clifford George
NOAO/NSO
Box 26732
950 N Cherry Av
Tucson AZ 85726 6732
USA
☎ 1 520 323 4111
℻ 1 520 325 9278
✉ toner@noao.edu

Tong Fu
Purple Mountain Obs
CAS
Nanjing 210008
China PR
☎ 86 25 33921
℻ 86 25 301 459

Tong Yi
Astronomy Dpt
Beijing Normal Univ
Beijing 100875
China PR
☎ 86 1 65 6531*6285
℻ 86 1 201 3929

Tonry John
Mathematics Dpt
MIT Rm 6 204a
Box 165
Cambridge MA 02139
USA
☎ 1 617 253 7528
✉ jt@anatares.mit.edu

Tonwar Suresh C
TIFR
Homi Bhabha Rd
Colaba
Bombay 400 005
India
☎ 91 22 495 2311
℻ 91 22 495 2110

Toomre Alar
Mathematics Dpt
MIT Rm 2 372
Box 165
Cambridge MA 02139
USA
☎ 1 617 253 4326

Toomre Juri
JILA/DAG
Univ of Colorado
Campus Box 400
Boulder CO 80309 0440
USA
☎ 1 303 492 7854
℻ 1 303 492 5235

Topaktas Latif A
KACST
Box 2455
Riyadh 11453
Saudi Arabia
☎ 966 1 467 6324
℻ 966 1 467 4253

Torao Masahisa
Senpuku Ga Oka 2-11-9
Susono City 410 11
Japan

Torelli M
OAR
Via d Parco Mellini 84
I 00136 Roma
Italy
☎ 39 6 34 7056
℻ 39 6 349 8236

Tornambe Amedeo
IAS
Area d Ricerca CNR
Via Fosso Cavaliere 100
I 00133 Roma
Italy
☎ 39 6 4993 4473
℻ 39 6 2066 0188

Toro Tibor
Inst of Theoretical
Physics
Univ Timisoara
RO 1900 Timisoara
Rumania
☎ 40 56 130 823

Toroshlidze Teimuraz I
Abastumani Astrophysical
Observ
Georgian Acad Sci
383762 Abastumani
Georgia
☎ 995 88 32 95 5367
℻ 995 88 32 98 5017

Torra Jordi
Dpt Fisica Atmosfera
Univ Barcelona
Avd Diagonal 647
E 08028 Barcelona
Spain
☎ 34 3 330 7311
℻ 34 3 402 1133

Torrejon Jose Miguel
Dto Fisica e Ing de
Esc Politecnica Superior
Apt 99 Univ Alicante
E 03080 Alicante
Spain
☎ 34 6 590 3682
✉ jmt@castor.daa.uv.es

Torrelles Jose M
Inst Astrofisica
Andalucia Apt 3004
Prof Albareda 1
E 18080 Granada
Spain
☎ 34 5 812 1311
℻ 34 5 881 4530
✉ torrelles@iaa.es

Torres Carlos
Dpt Astronomia
Univ Chile
Casilla 36 D
Santiago
Chile
☎ 56 2 229 4101
℻ 56 2 229 4002

Torres Carlos Alberto
Observ Nacional
Rua Coronel Renno 07
CP 21
37500 Itajuba Mg
Brazil
☎ 55 356 22 0788

Torres Guillermo
CfA
HCO/SAO MS 20
60 Garden St
Cambridge MA 02138 1516
USA
☎ 1 617 495 7335
℻ 1 617 495 7467
✉ gtorres@cfa.harvard.edu

Torres Dodgen Ana V
15735 Forest Hill Dr
Boulder Creek CA 95006
USA
☎ 1 408 338 0158

Torres-Peimbert Silvia
Instituto Astronomia
UNAM
Apt 70 264
Mexico DF 04510
Mexico
☎ 52 5 622 3906
℻ 52 5 616 0653
✉ silvia@astroscu.unam.mx

Torricelli Guidetta
Osserv Astrofis Arcetri
Univ d Firenze
Largo E Fermi 5
I 50125 Firenze
Italy
☎ 39 55 27 52260
℻ 39 55 22 0039

Torroja J
Dpt Astrofisica
Fac Fisica
Univ Complutense
E 28040 Madrid
Spain
☎ 34 2 244 2501
℻ 34 2 394 5195

Tosa Makoto
Astron Institute
Tohoku Univ
Sendai Aoba
Miyagi 980
Japan
☎ 81 22 217 6501
℻ 81 22 217 6513
✉ tosa@astroa.astr.tohoku.
ac.jp

Tosi Monica
Osserv Astronom Bologna
Univ d Bologna
Via Zamboni 33
I 40126 Bologna
Italy
☎ 39 51 259 401
℻ 39 51 259 407

Toth Imre
Konkoly Observ
Thege U 13/17
Box 67
H 1525 Budapest
Hungary
☎ 36 1 375 4122
℻ 36 1 275 4668
✉ h697kno@ogyalla.
konkoly.hu

Totsuka Yoji
Inst Cosmic Ray Res
Univ of Tokyo
Midoricho Tanashi
Tokyo 188
Japan
☎ 81 424 61 4131
℻ 81 424 68 1438
✉ totsuka@jpnutins

Touma Hamid
CNCPRST
NRC
BP 8027
Rabat Agdal
Morocco
☎ 212 7 72 803/7 74 215
℻ 212 7 77 1288
✉ touma@cnr.ac.ma

Tout Christopher
Inst of Astronomy
The Observatories
Madingley Rd
Cambridge CB3 0HA
UK
☎ 44 12 23 337 548
℻ 44 12 23 337 523
✉ ct23@phoenix.cambridge.
ac.uk

Tovmassian Gaghik
Observ Astronomico Nacional
UNAM
Apt 877
Ensenada BC 22800
Mexico
☎ 52 6 174 4580
℻ 52 6 174 4607
✉ gag@bufadora.astrosen.
unam.mx

Tovmassian Hrant M
INAOE
Tonantzintlaz
Apdo 51 y 216
Puebla PUE 72000
Mexico
☎ 52 2 247 2011
℻ 52 2 247 2231
✉ hrant@tonali.inaop.mx

Townes Charles Hard
Physics Dpt
Univ of California
366 LeConte Hall
Berkeley CA 94720
USA
☎ 1 510 642 1128
℻ 1 510 643 8497
✉ cht@sunspot.ssl.berkeley.
edu

Toyama Kiyotaka
Hokkaido Inform Univ
Nishinopporo 59-2
Ebetsu 069
Japan
☎ 81 113 85 4411
℻ 81 113 84 0134

Tozer David C
School of Physics
Univ of Newcastle
Newcastle/Tyne NE1 7RU
UK
☎ 44 191 222 7411
℻ 44 191 261 1182

Tozzi Gian Paolo
Osserv Astrofis Arcetri
Univ d Firenze
Largo E Fermi 5
I 50125 Firenze
Italy
☎ 39 55 27 52250
℻ 39 55 22 0039

Traat Peeter
Tartu Observ
Estonian Acad Sci
Toravere
EE 2444 Tartumaa
Estonia
☎ 372 7 434 932
℻ 372 7 410 205
✉ traat@obs.ee

Trafton Laurence M
Astronomy Dpt
Univ of Texas
Rlm 15 308
Austin TX 78712 1083
USA
☎ 1 512 471 1476
℻ 1 512 471 6016

Tran Minh Nguyet
Observ Paris Meudon
DAMAP
Pl J Janssen
F 92195 Meudon PPL Cdx
France
☎ 33 1 45 07 7447
℻ 33 1 45 07 7511

Tran-Minh Francoise
Observ Paris Meudon
DASGAL
Pl J Janssen
F 92195 Meudon PPL Cdx
France
☎ 33 1 45 07 7553
℻ 33 1 45 07 7469

Traub Wesley Arthur
CfA
HCO/SAO
60 Garden St
Cambridge MA 02138 1516
USA
☎ 1 617 495 7406
℻ 1 617 495 7356
✉ traub@cfa.harvard.edu

Traving Gerhard
Inst Theor Astrophysik
d Univer
Im Neuenheimer Feld 561
D 69120 Heidelberg
Germany
☎ 49 622 156 2815
✆ 49 622 154 4221

Treder H J
Zntrlinst Astrophysik
Sternwarte Babalsberg
Rosa Luxemburg Str 17a
D 14473 Potsdam
Germany
☎ 49 331 275 3731

Treffers Richard R
Astronomy Dpt
Univ of California
601 Campbell Hall
Berkeley CA 94720 3411
USA
☎ 1 415 642 4223
✆ 1 510 642 3411

Trefftz Eleonore E
Musenbergstr 28b
D 81929 Muenchen
Germany

Trefzger Charles F
Astronomisches Institut
Univ Bern
Venusstr 7
CH 4102 Binningen
Switzerland
☎ 41 61 271 7711/12
✆ 41 61 205 5455

Trehan Surindar K
Mathematics Dpt
Univ of Panjab
Chandigarh 160 014
India
☎ 91 29 938

Trellis Michel
OCA
Observ Nice
BP 139
F 06304 Nice Cdx 4
France
☎ 33 4 93 89 0420
✆ 33 4 92 00 3033

Tremaine Scott Duncan
Dpt Astrophys Sci
Princeton Univ
Peyton Hall
Princeton NJ 08544 1001
USA
☎ 1 609 258 3810
✆ 1 609 258 1020
✉ tremaine@astro.priceton.edu

Tremko Jozef
Astronomical Institute
Slovak Acad Sci
SK 059 60 Tatranska Lomni
Slovak Republic
☎ 421 969 96 7866
✆ 421 969 96 7656

Treumann Rudolf A
MPE
Postfach 1603
D 85740 Garching
Germany
☎ 49 893 299 831
✆ 49 893 299 3569

Trevese Dario
Istt Astronomico
Univ d Roma La Sapienza
Via G M Lancisi 29
I 00161 Roma
Italy
☎ 39 6 44 03734
✆ 39 6 44 03673

Trexler James H
1921 So Abrego Dr
Greenvalley AZ 85614 1403
USA

Trimble Virginia L
Dpt Phys/Astronomy
UCI
Irvine CA 92697 4575
USA
☎ 1 714 824 6948/301 405 5822
✆ 1 714 824 2174/301 314 9067
✉ vtrimble@uci.edu
astro.umd.edu

Trinchieri Ginevra
Osserv Astronomico
d Brera
Via Brera 28
I 20121 Milano
Italy
☎ 39 2 723 20327
✆ 39 2 720 01600
✉ ginevra@brera.mi.astro.it

Tripathi B M
Uttar Pradesh State
Observ
Po Manora Peak 263 129
Nainital 263 129
India
☎ 91 59 42 2136

Tripicco Michael J
NASA GSFC
Code 664
Greenbelt MD 20771
USA
✉ miket@ros10.gsfc.nasa.gov

Tritakis Basil P
Res Ctr Astronomy
Acad Athens
14 Anagnostopoulou St
GR 106 73 Athens
Greece
☎ 30 1 361 3589
✆ 30 1 363 1606

Tritton Keith P
Magpie Cottage
Fox St
Great Gransden
Sandy SG19 3AA
UK
☎ 44 176 767 7219
✆ 44 1767 677 219
✉ kpt2@tutor.open.ac.uk

Tritton Susan Barbara
Royal Observ
Blackford Hill
Edinburgh EH9 3HJ
UK
☎ 44 131 667 3321
✆ 44 131 662 1668
✉ sbt@roe.ac.uk

Troche-Boggino A E
Fac Cienc Exactas/Natural
Univ Nacional Asuncion
Suc de Correos 19
Agen Pos Campus Una Km10
Paraguay
☎ 595 92 501 517
✆ 595 21 502 239
✉ atroche@pol.com.py

Troland Thomas Hugh
Dpt Phys/Astronomy
Univ Kentucky
Lexington KY 40506 0055
USA
☎ 1 606 257 8620
✆ 1 606 323 2846

Trottet Gerard
Observ Paris Meudon
DASOP
Pl J Janssen
F 92195 Meudon PPL Cdx
France
☎ 33 1 45 07 7808
✆ 33 1 45 07 7959

Truemper Joachim
MPE
Postfach 1603
D 85740 Garching
Germany
☎ 49 893 299 3559
✆ 49 893 299 3569

Trujillo Bueno Javier
IAC
Observ d Teide
via Lactea s/n
E 38200 La Laguna
Spain
☎ 34 2 232 9100
✆ 34 2 232 9117

Trullols I Farreny Enric
Dpt Mat Aplicada
U P Cataluna
Victor Balaguer Sn
E 08800 Vilanova
Spain
☎ 34 3 896 7738
✆ 34 3 896 7700
✉ enric@mat.upc.es

Trulsen Jan K
Inst Theor Astrophys
Univ of Oslo
Box 1029
N 0315 Blindern Oslo 3
Norway
☎ 47 22 856 501
✆ 47 22 85 6505

Truong Bach
Observ Paris
DEMIRM
61 Av Observatoire
F 75014 Paris
France
☎ 33 1 40 51 2221
✆ 33 1 40 51 2002

Truran James W
Dpt Astron/Astrophys
Univ of Chicago
5640 S Ellis Av
Chicago IL 60637
USA
☎ 1 312 702 9584
✆ 1 312 702 6645
✉ truran@nova.uchicago.edu

Trushkin Sergei Anatol'evich
SAO
Acad Sciences
Nizhnij Arkhyz
357147 Karachaevo
Russia
☎ 7 878 784 6103
✆ 7 96 908 2861
✉ satr@sao.ru

Trussoni Edoardo
Istt Cosmo Geofisica
CNR
Corso Fiume 4
I 10133 Torino
Italy
☎ 39 11 657 694/8979

Trutse Yu L
Inst Phys of Atmosph
Acad Sciences
Pyzhevsky 3
109017 Moscow
Russia

Tsai Chang-Hsien
Taipei Astronomical
Museum
N 363 Kee-Ho Rd
Taipei
China R
☎ 886 2 594 7432

Tsamparlis Michael
Astrophysics Dpt
Ntl Univ Athens
Panepistimiopolis
GR 157 84 Zografos
Greece
☎ 30 1 724 3414*211
✆ 30 1 962 4430
✉ rich@grathun1

Tsao Mo
No 47 Sec 3
Hsin-i Rd
Taipei 106
China R
☎ 886 2 704 7795

Tsap T T
Crimean Astrophys Obs
Ukrainian Acad Science
Nauchny
334413 Crimea
Ukraine
☎ 380 6554 71161
✆ 380 6554 40704

Tsarevsky Gregory
Astro Space Ctr
Lebedev Physical Inst
Profsoyuznaya 84/32
117810 Moscow
Russia
☎ 7 095 333 2133
℻ 7 095 333 2378
✉ gtsarevs@atnf.csiro.au

Tsay Wean-Shun
Inst Phys/Astronomy
Ntl Central Univ
Chung Li Taiwan 32054
China R
☎ 886 3 422 7151*5335/5300
℻ 886 3 425 1175
✉ tsay@phyast.dnet.ncu.edu.tw

Tscharnuter Werner M
Inst Theor Astrophysik
d Univer
Tiergartenstr 15
D 69121 Heidelberg
Germany
☎ 49 622 154 4837
℻ 49 622 154 4221
✉ wmt@ita.uni-heidelberg.de

Tseytlin Naum M
Radiophysical Res
Inst
Lyadov Ul 25/14
603600 N Novgorod
Russia
☎ 7 831 236 0129
℻ 7 831 236 9902

Tsikoudi Vassiliki
Physics Dpt
Univ of Ioannina
GR 451 10 Ioannina
Greece
☎ 30 651 98 481
℻ 30 651 45 697
✉ vtikoud@cc.uoi.gr

Tsinganos Kanaris
Physics Dpt
Univ of Crete
Box 1527
GR 711 11 Heraklion
Greece
☎ 30 81 23 9757*154
℻ 30 81 23 9735
✉ tsingan@iesl.forth.gr

Tsioumis Alexandros
Dpt Geodesy/Astron
Univ Thessaloniki
GR 540 06 Thessaloniki
Greece
☎ 30 31 99 2693

Tsiropoula Georgia
Astronomical Institute
Ntl Observ Athens
Box 20048
GR 118 10 Athens
Greece
☎ 30 1 346 1191
℻ 30 1 346 4566

Tsubaki Tokio
Dpt Earth Science
Shiga Univ
2-5-1 Hiratsu
Ohtsu 520
Japan
☎ 81 775 37 081

Tsuboi Masato
Nobeyama Radio Obs
NAOJ
Minamimaki Mura
Nagano 384 13
Japan
☎ 81 267 63 4314
℻ 81 267 98 2927
✉ nrrotsub@jpnnro

Tsubota Yukimasa
Krio Senior High School
4-1-2 Hiyoshi
Kouhoku-ku
Yokohama 223
Japan
☎ 81 455 63 1111
℻ 81 455 62 8291
✉ tsubota@hc.keio.ac.jp

Tsuchida Masayoshi
IAG
Univ Sao Paulo
CP 9638
01065 970 Sao Paulo SP
Brazil
☎ 55 11 577 8599
℻ 55 11 276 3848
✉ masa@brfapesp

Tsuchiya Atsushi
Omachi 4 2 18
Kamakura 248
Japan

Tsuchiya Toshio
Astronomy Dpt
Kyoto Univ
Sakyo-ku
Kyoto 606 01
Japan
☎ 81 75 753 4288
℻ 81 75 753 3897
✉ tsuchiya@kusastro.kyoto-u.ac.jp

Tsuji Takashi
Inst of Astronomy
Univ of Tokyo
Osawa Mitaka
Tokyo 181
Japan
☎ 81 422 32 5111
℻ 81 422 34 3793

Tsujimoto Takuji
Ntl Astronomical Obs
Osawa 2-21-1
Mitaka
Tokyo 181
Japan
☎ 81 422 34 3617
℻ 81 422 32 1924
✉ tsujmttk@cc.nao.ac.jp

Tsunemi Hiroshi
Fac of Sciences
Osaka Univ
Machikaneyama
Toyonaka Osaka 560
Japan
☎ 81 684 41 151
✉ tsunemi@jpnoskfm

Tsuneta Saku
Inst of Astronomy
Univ of Tokyo
Osawa Mitaka
Tokyo 181
Japan
☎ 81 422 32 4710
℻ 81 422 34 3793

Tsuru Takeshi
Physics Dpt
Kyoto Univ
Sakyo-ku
Kyoto 606 01
Japan
☎ 81 75 753 3843
℻ 81 75 753 5377
✉ tsuru@cr.scphys.kyoto-u.ac.jp

Tsuruta Sachiko
Physics Dpt
Montana State Univ
Bozeman MT 59717 0350
USA
☎ 1 406 994 6779
℻ 1 406 994 4452
✉ uphst@gemini.oscs.montana.edu

Tsutsumi Takahiro
Nobeyama Radio Obs
NAOJ
Minamimaki Mura
Nagano 384 13
Japan
☎ 81 267 984 371
℻ 81 267 982 884
✉ tsutsumi@nro.nao.ac.jp

Tsvetanov Zlatan I
Dpt Phys/Astronomy
JHU
Charles/34th St
Baltimore MD 21218
USA
☎ 1 301 516 8585
℻ 1 301 516 8260
✉ ztsvetanov@pha.jhu.edu

Tsvetkov Milcho K
Inst of Astronomy
Bulgarian Acad Sci
72 Lenin Blvd
BG 1784 Sofia
Bulgaria
☎ 359 2 743 1414
℻ 359 2 740 831
✉ tsvetkov@wfpa.bas.bg

Tsvetkov Tsvetan
Astronomy Dpt
Univ of Sofia
Anton Ivanov St 5
BG 1126 Sofia
Bulgaria
☎ 359 2 68 8176
℻ 359 2 68 9085

Tsvetkova Katya
Astronomy Dpt
Bulgarian Acad Sci
72 Lenin Blvd
BG 1784 Sofia
Bulgaria
☎ 359 2 743 1739
℻ 359 2 975 3201
✉ katya@wfpa.acad.bg

Tsygan Anatolii I
Ioffe Physical Tech Inst
Acad Sciences
Polytechnicheskaya Ul 26
194021 St Petersburg
Russia
☎ 7 812 247 9326
℻ 7 812 247 1963
✉ varsh@eo.pti.spb.su

Tuchman Ytzhak
Racah Inst of Phys
Hebrew Univ Jerusalem
Jerusalem 91904
Israel
☎ 972 2 584 417
℻ 972 2 61 1519

Tucholke Hans-Joachim
Univ Sternwarte
Univ Bonn
auf d Huegel 71
D 53121 Bonn
Germany
☎ 49 228 733 649
℻ 49 228 733 672
✉ tucholke@astro.uni-bonn.de

Tucker Wallace H
Box 266
Bonsall CA 92003
USA
☎ 1 619 728 7103

Tueg Helmut
Alfred Wegener Institut
f Polarforschung
Columbus Ctr
D 2850 Bremerhaven
Germany

Tufekcioglu Zeki
Astronomy Dpt
Univ of Ankara
Fen Fakultesi
06100 Besevler
Turkey
☎ 90 41 212 6720
℻ 90 312 223 2395

Tull Robert G
Astronomy Dpt
Univ of Texas
Rlm 15 308
Austin TX 78712 1083
USA
☎ 1 512 471 3337
℻ 1 512 471 6016

Tully John A
OCA
Observ Nice
BP 139
F 06304 Nice Cdx 4
France
☎ 33 4 93 89 0420
℻ 33 4 92 00 3033

Tully Richard Brent
Inst for Astronomy
Univ of Hawaii
2680 Woodlawn Dr
Honolulu HI 96822
USA
☎ 1 808 956 8606
℻ 1 808 988 2790

Tunca Zeynel
Dpt Astron/Space Sci
Ege Univ
Box 21
35100 Bornova Izmir
Turkey
☎ 90 232 388 0110*2322

Tuohy Ian R
British Aerosp Australia
14 Park Way
Technology Park
Salisbury 5095
Australia
☎ 61 83 432 2111
℻ 61 8 349 6629

Tuominen Ilkka V
Astronomy Dpt
Univ of Oulu
Box 333
FIN 90570 Oulu
Finland
☎ 358 81 553 1930
℻ 358 81 553 1934
✉ ilkka.tuominen@oulu.fi

Turatto Massimo
Osserv Astronom Padova
Univ d Padova
Vic d Osservatorio 5
I 35122 Padova
Italy
☎ 39 49 829 3411
℻ 39 49 875 9840

Turck-Chieze Sylvaine
Service Astrophysique
CEA Saclay
Orme d Merisiers Bt 709
F 91190 Gif s Yvette Cdx
France
☎ 33 1 69 08 4387
℻ 33 1 69 08 6577
✉ turck@discovery.saclay.cea.fr

Turlo Zygmunt
Inst of Astronomy
N Copernicus Univ
Ul Chopina 12/18
PL 87 100 Torun
Poland
☎ 48 56 26 018
℻ 48 56 24 602

Turner Barry E
NRAO
520 Edgemont Rd
Charlottesville VA 22903
USA
☎ 1 804 296 0337

Turner David G
Astronomy Dpt
St Mary's Univ
Halifax NS B3H 3C3
Canada
☎ 1 902 429 9780*2254
℻ 1 902 420 5561

Turner Edwin L
Princeton Univ Obs
Peyton Hall
Princeton NJ 08544 1001
USA
☎ 1 609 258 3577
℻ 1 608 258 1020
✉ zlt@astrouvav.princeton.edu

Turner Jean L
Dpt Astronomy/Phys
UCLA
Los Angeles CA 90095 1562
USA
☎ 1 213 825 4305
℻ 1 213 206 2096
✉ turner@astro.ucla.edu

Turner Kenneth C
NSF
Div Astron Sci
4201 Wilson Blvd
Arlington VA 22230
USA
☎ 1 703 306 1820
℻ 1 703 306 0525
✉ kturner@note.nsf.gov

Turner Martin J l
Dpt Phys/X-Ray Astron
Univ Leicester
University Rd
Leicester LE1 7RH
UK
☎ 44 116 252 2073
℻ 44 116 250 182

Turner Michael S
Astronomy/Astrophys Ctr
Univ of Chicago
5640 S Ellis Av
Chicago IL 60637
USA
☎ 1 312 962 7974
℻ 1 312 702 8212

Turnshek David A
Dpt Phys/Astronomy
Univ of Pittsburgh
Pittsburgh PA 15260
USA
☎ 1 412 624 9015
℻ 1 412 624 1833
✉ turnshek@cis.pitt.edu

Turolla Roberto
Dpt Fisica G Galilei
Univ d Padova
Via Marzolo 8
I 35131 Padova
Italy
☎ 39 49 827 7139
℻ 39 49 827 7102
✉ turolla@astaxp.pd.infn.it

Turon Catherine
Observ Paris Meudon
DASGAL
Pl J Janssen
F 92195 Meudon PPL Cdx
France
☎ 33 1 45 07 7837
℻ 33 1 45 07 7878
✉ Catherine.turon@obspm.fr

Turtle A J
Physics Dpt
Univ of Sydney
Sydney NSW 2006
Australia
☎ 61 2 692 2222
℻ 61 2 660 2903

Tutukov A V
Inst of Astronomy
Acad Sciences
Pyatnitskaya Ul 48
109017 Moscow
Russia
☎ 7 095 231 5461
℻ 7 095 230 2081

Twarog Bruce A
Dpt Phys/Astronomy
Univ of Kansas
Lawrence KS 66045
USA
☎ 1 913 864 5163
℻ 1 913 864 5262

Twiss R Q
C/o A R Boschi
96a Holland Rdoschi
London W14 8BD
UK

Tworkowski Andrzej S
School Mathemat Sc
QMWC
Mile End Rd
London E1 4NS
UK
☎ 44 171 980 4822
℻ 44 181 975 5500

Tylenda Romuald
Inst of Astronomy
N Copernicus Univ
Ul Chopina 12/18
PL 87 100 Torun
Poland
☎ 48 56 26 018*10
℻ 48 56 24 602

Tyler G Leonard
Radar Astronomy Inst
Stanford Univ
Stanford CA 94305 4035
USA
☎ 1 415 497 3535
℻ 1 415 723 4840

Tylka Allan J
NRL
Code 4154
4555 Overlook Av SW
Washington DC 20375 5000
USA
☎ 1 202 767 2200
℻ 1 202 767 6473

Tyson John Anthony
Bell Labs
Rm ID432
700 Mountain Av
Murray Hill NJ 07974
USA
☎ 1 908 582 6028
℻ 1 908 582 4702
✉ tyson@lucent.com

Tytler David
CASS
UCSD
C 011
La Jolla CA 92093 0216
USA
☎ 1 619 534 3460
℻ 1 619 534 2294

Tzioumis Anastasios
CSIRO
ATNF
Box 76
Epping NSW 2121
Australia
☎ 61 2 9372 4350
℻ 61 2 9372 4400
✉ atzioumi@atnf.csiro.au

U

Ubertini Pietro
IAS
Area d Ricerca CNR
Via Fosso Cavaliere 100
I 00133 Roma
Italy
☎ 39 6 4993 4473
✆ 39 6 2066 0188
✉ ubertini@alpha1.rm.fra.cnr.it

Uchida Juichi
Tohoku Gakuen Univ
Tagayo Univ
Miyagi 985
Japan

Uchida Yutaka
Astronomy Dpt
Univ of Tokyo
Bunkyo Ku
Tokyo 113
Japan
☎ 81 757 53 3890
✆ 81 757 53 3897

Udal'tsov V A
Lebedev Physical Inst
Acad Sciences
Leninsky Pspt 53
117924 Moscow
Russia
☎ 7 095 135 8560
✆ 7 095 135 7880

Udalski Andrzej
Astronomical Observ
Warsaw Univ
Al Ujazdowskie 4
PL 00 478 Warsaw
Poland
☎ 48 22 29 4011
✆ 48 22 29 4697
✉ vdalski@plwavw61

Udaya Shankar N
RRI
Sadashivanagar
CV Raman Av
Bangalore 560 080
India
☎ 91 80 336 0122
✆ 91 80 334 0492
✉ uday@rri.ernet.in

Udry Stephane
Observ Geneve
Chemin d Maillettes 51
CH 1290 Sauverny
Switzerland
☎ 41 22 755 2611
✆ 41 22 755 3983
✉ stephane.udry@obs.unige.ch

Ueno Munetaka
Dpt Earth/Astronomy
Univ of Tokyo
3-8-1 Komaba Meguro
Tokyo 153
Japan
☎ 81 354 54 6616
✆ 81 334 65 3925
✉ ueno@chianti.c.u-tokyo.ac.jp

Ueno Sueo
Kanazawa Techn Inst
7 1 Ogigaoka
Nonoichimachi
Ishikawa 921
Japan
☎ 81 762 48 1100

Uesugi Akira
Astronomy Dpt
Kyoto Univ
Sakyo Ku
Kyoto 606 01
Japan
☎ 81 757 51 2111
✆ 81 75 753 7010

Ukita Nobuharu
Nobeyama Radio Obs
NAOJ
Minamimaki Mura
Nagano 384 13
Japan
☎ 81 267 98 4397
✆ 81 267 98 4387
✉ ukita@nro.nao.ac.jp

Ulfbeck Ole
Niels Bohr Inst
Copenhagen Univ
Blegdamsvej 17
DK 2100 Copenhagen
Denmark
☎ 45 31 42 1616
✆ 45 31 38 9157

Ulich Bobby Lee
Kaman Aerospace Corp
Electro Optics Dvpt Ct
3480 E Britannia Dr
Tucson AZ 85711
USA
☎ 1 520 295 2104
✆ 1 520 889 0211

Ulmer Melville P
Dearborn Observ
NW Univ
2131 Sheridan Rd
Evanston IL 60208 3112
USA
☎ 1 847 491 5633
✆ 1 847 491 3135

Ulmschneider Peter
Inst Theor Astrophysik
d Univer
Tiergartenstr 15
D 69121 Heidelberg
Germany
☎ 49 622 156 2837
✆ 49 622 154 4221
✉ i98@dhdurz1

Ulrich Bruce T
Hiltenspergerstr 93
D 8000 Muenchen 40
Germany

Ulrich Marie-Helene D
ESO
Karl Schwarzschildstr 2
D 85748 Garching
Germany
☎ 49 893 200 6229
✆ 49 893 202 362

Ulrich Roger K
Dpt Astronomy/Phys
UCLA
Box 951562
Los Angeles CA 90025 1562
USA
☎ 1 310 825 4270
✆ 1 310 206 2096
✉ ulrich@bonnie.astro.ucla.edu

Ulvestad James Scott
NRAO
Box 0
Socorro NM 87801 0387
USA
☎ 1 505 835 7298
✆ 1 505 835 7027
✉ julvesta@aoc.nrao.edu

Umemura Masayuki
Tokyo Astronomical Obs
NAOJ
Osawa Mitaka
Tokyo 181
Japan
☎ 81 422 41 3731
✆ 81 422 41 3746
✉ umemura@ume.mtk.nao.ac.jp

Umlenski Vasil
Astronomy Dpt
Bulgarian Acad Sci
72 Lenin Blvd
BG 1784 Sofia
Bulgaria
☎ 359 2 75 8927
✆ 359 2 75 5019

Underhill Anne B
4696 West 10th Av 301
Vancouver BC V6R 2J5
Canada
☎ 1 250 224 3552

Underwood James H
Lawrence Berkeley Lab
Univ of California
X Ray Optics Lab 80 101
Berkeley CA 94720
USA
☎ 1 415 486 4958
✆ 1 510 486 6738

Unger Stephen
Royal Greenwich Obs
Madingley Rd
Cambridge CB3 0EZ
UK
☎ 44 12 23 374 000
✆ 44 12 23 374 700

Unno Wasaburo
Res Inst Science/Tech
Kinki Univ
Higashi
Osaka 577
Japan
☎ 81 672 12 332

Unwin Stephen C
CALTECH/JPL
MS 301 370
4800 Oak Grove Dr
Pasadena CA 91109 8099
USA
☎ 1 818 354 5066
✆ 1 818 393 5239
✉ unwin@huey.jpl.nasa.gov

Uomoto Alan K
Dpt Phys/Astronomy
JHU
Charles/34th St
Baltimore MD 21218
USA
☎ 1 301 338 8594
✆ 1 301 516 7279
✉ au@pha.jhu.edu

Upgren Arthur R
349 Sciences Ctr
Wesleyan Univ
Middletown CT 06459
USA
☎ 1 860 685 3678
✆ 1 860 685 2131
✉ aupgren@wesleyan.edu

Upson Walter L II
Princeton Univ Obs
Peyton Hall
Princeton NJ 08544 1001
USA
☎ 1 609 258 3801
✆ 1 609 258 1020

Upton E K l
Dpt Astronomy/Phys
UCLA
Box 951562
Los Angeles CA 90025 1562
USA
☎ 1 310 825 4434
✆ 1 310 206 2096

Uras Silvano
Istt Astronomia
Univ d Cagliari
Via Ospedale 72
I 09100 Cagliari
Italy
☎ 39 70 71 1246
✆ 39 70 72 5425

Urasin Lirik A
Engelhardt Astronom
Observ
Observatoria Station
422526 Kazan
Russia
☎ 7 32 4827

Urban Sean Eugene
USNO
3450 Massachusetts Av NW
Washington DC 20392 5420
USA
☎ 1 202 762 1445
✆ 1 202 762 1516
✉ seu@pyxis.usno.navy.mil

Urbanik Marek
Astronomical Observ
Krakow Jagiellonian Univ
Ul Orla 171
PL 30 244 Krakow
Poland
☎ 48 12 25 1294
✆ 48 12 25 1318

Ureche Vasile
Fac of Mathematics
Univ of Cluj Napoca
Str M Kogalniceanu 1
RO 3400 Cluj Napoca
Rumania
☎ 40 64 116 101
✆ 40 64 194 592

Urpo Seppo I
Helsinki Univ Technology
Radio Laboratory
Otakaari 5 A
FIN 02150 Espoo 15
Finland
☎ 358 9 451 2548
✆ 358 9 451 2378

Urry Claudia Megan
STScI
Homewood Campus
3700 San Martin Dr
Baltimore MD 21218
USA
☎ 1 301 338 4593
✆ 1 301 338 4767
✉ cmu@stsci.edu

Usher Peter D
Astronomy Dpt
Pennsylvania State Univ
507 Davey Lab
University Park PA 16802
USA
☎ 1 814 865 3509
✆ 1 814 863 7114

Uson Juan M
NRAO
VLA
Box 0
Socorro NM 87801 0387
USA
☎ 1 505 835 7237
✆ 1 505 835 7027

Usowics Jerzy Bogdan
Inst Radio Astronomy
N Copernicus Univ
Ul Gagarina 11
PL 87 100 Torun
Poland
☎ 48 56 78 3327
✆ 48 56 11 651

Utsumi Kazuhiko
Astronomy Dpt
Hiroshima Univ
Higashi Senda Machi
Hiroshima 730
Japan
☎ 81 822 41 1221

Uus Undo
Tartu Observ
Estonian Acad Sci
Toravere
EE 2444 Tartumaa
Estonia
☎ 372 7 416 625
✆ 372 7 410 205

V

Vager Zeev
Physics Dpt
Weizmann Inst of Sci
Box 26
Rehovot 76100
Israel
☎ 972 8 34 3835
✆ 972 8 34 4106

Vaghi Sergio
ESA/ESTEC
PLP
Box 299
NL 2200 AG Noordwijk
Netherlands
☎ 31 71 565 3453
✆ 31 71 565 4894
✉ svaghi@estec.esa.nl

Vagnetti Fausto
Dpt Fisica/Astrofisica
Univ d Roma Tor Vergata
Via Ric Scientifica
I 00133 Roma
Italy
☎ 39 6 7259 4426
✆ 39 6 202 3507
✉ vagnetti@itovf2.roma2.infn.it

Vahia Mayank N
TIFR
Homi Bhabha Rd
Colaba
Bombay 400 005
India
☎ 91 22 215 2971
✆ 91 22 215 2110
✉ vahia@158.144.1.11

Vaidya P C
34 Sharda Naga
PALDI
Ahmedabad 380 007
India
☎ 91 272 41 3322

Vainstein L A
Lebedev Physical Inst
Acad Sciences
Leninsky Pspt 53
117924 Moscow
Russia
☎ 7 095 135 2250
✆ 7 095 135 7880

Vakili Farrokh
OCA
Observ Calern/Caussols
2130 r Observatoire
F 06460 S Vallier Thiey
France
☎ 33 4 93 42 6270
✆ 33 4 93 09 2613
✉ vakili@froni51

Val'tts Irina E
Astro Space Ctr
Lebedev Physical Inst
Profsoyuznaya 84/32
Moscow 117810
Russia
☎ 7 095 333 2167
✆ 7 095 333 2378
✉ ivaltts@dpc.asc.rssi.ru

Valbousquet Armand
Observ Strasbourg
11 r Universite
F 67000 Strasbourg
France
☎ 33 3 88 35 8200
✆ 33 3 88 25 0160

Valentijn Edwin A
Kapteyn Sterrekundig Inst
Univ Groningen
Postbus 800
NL 9700 AV Groningen
Netherlands
☎ 31 50 363 4073
✆ 31 50 363 6100

Valiron Pierre
Observ Grenoble
Lab Astrophysique
BP 53x
F 38041 S Martin Heres Cdx
France
☎ 33 4 76 51 4787
✆ 33 4 76 44 8821

Vallee Jacques P
NRCC/HIA
5071 W Saanich Rd
Victoria BC V8X 4M6
Canada
☎ 1 250 363 6952
✆ 1 250 363 8483
✉ jacques.vallee@hia.nrc.ca

Vallejo Miguel
ROA
Cecilio Pujazon 22-3 A
E 11110 San Fernando Cadiz
Spain
☎ 34 5 659 9000
✆ 34 5 659 9366
✉ vallejo@roa.cica.es

Valls-Gabaud David
Observ Strasbourg
11 r Universite
F 67000 Strasbourg
France
☎ 33 3 88 15 0758
✆ 33 3 88 15 0760
✉ dvg@astro.u-strasbg.fr

Valnicek Boris
Astronomical Institute
Czech Acad Sci
Fricova 1
CZ 251 65 Ondrejov
Czech R
☎ 420 204 85 7324
✆ 420 2 88 1611
✉ astsun@csearn

Valsecchi Giovanni B
IAS
Rep Planetologia
Via d universita 11
I 00133 Roma
Italy
☎ 39 6 4993 4473
✆ 39 6 2066 0188
✉ giovanni@ias.rm.cnr.it

Valtaoja Esko
Turku Univ
Tuorla Observ
Vaeisaelaentie 20
FIN 21500 Piikkio
Finland
☎ 358 2 274 4251
✆ 358 2 243 3767

Valtaoja Leena
Turku Univ
Tuorla Observ
Vaeisaelaentie 20
FIN 21500 Piikkio
Finland
☎ 358 2 143 5822
✆ 358 2 243 3767
✉ lvaltaoja@kontu.utu.fi

Valtier Jean-Claude
OCA
Observ Nice
BP 139
F 06304 Nice Cdx 4
France
☎ 33 4 93 89 0420
✆ 33 4 92 00 3033

Valtonen Mauri J
Turku Univ
Tuorla Observ
Vaeisaelaentie 20
FIN 21500 Piikkio
Finland
☎ 358 2 274 4245
✆ 358 2 243 3767

Valyaev Valery
10 Nab Kutusova
St Petersburg
Russia
☎ 7 812 275 1006
✆ 7 812 272 7968
✉ sokolsky@iipah.spb.su

van Albada tjeerd S
Kapteyn Sterrekundig Inst
Univ Groningen
Postbus 800
NL 9700 AV Groningen
Netherlands
☎ 31 50 363 4073
✆ 31 50 363 6100

van Allen James A
Dpt Phys/Astronomy
Univ of Iowa
605 Brookland Park Dr
Iowa City IA 52242 1479
USA
☎ 1 319 353 4531
✆ 1 319 335 1753

van Altena William F
Astronomy Dpt
Yale Univ
Box 208101
New Haven CT 06520 8101
USA
☎ 1 203 436 8318
✆ 1 203 432 5048
✉ vanalten@yale.astro.edu

van Beek Frank
Dpt Mechan Eng
Techn Univ Delft
Mekelweg 2
NL 2628 CD Delft
Netherlands
☎ 31 15 278 5396
✆ 31 15 278 3444

van Blerkom David J
Astronomy Dpt
Univ Massachusetts
Amherst MA 01002
USA
☎ 1 413 545 2290

van Breda Ian G
Dunsink Observ
DIAS
Castleknock
Dublin 15
Ireland
☎ 353 1 38 7911
✆ 353 1 38 7090

van Breugel Wil
Radio Astronomy Lab
Univ of California
601 Campbell Hall
Berkeley CA 94720
USA
☎ 1 415 642 5275
✆ 1 415 642 3411

van Bueren Hendrik G
Meidoornlaan 13
NL 3461 ES Linschoten
Netherlands
☎ 31 34 801 5406

van Citters Gordon W
NSF
Div Astron Sci
4201 Wilson Blvd
Arlington VA 22230
USA
☎ 1 703 306 1820
℻ 1 703 306 0525
✉ gvancitt@nsf.gov

van de Hulst H C
Leiden Observ
Box 9513
NL 2300 RA Leiden
Netherlands
☎ 31 71 527 2727
℻ 31 71 527 5819

van de Stadt Herman
SRON
Univ Groningen
Postbus 800
NL 9700 AV Groningen
Netherlands
☎ 31 50 363 4073
℻ 31 50 363 6100

van den Bergh Sidney
NRCC/HIA
DAO
5071 W Saanich Rd
Victoria BC V8X 4M6
Canada
☎ 1 250 363 0006
℻ 1 250 363 0045
✉ sidney.vandenbergh@nrc.ca

van den Heuvel Edward P J
Astronomical Institute
Univ of Amsterdam
Kruislaan 403
NL 1098 SJ Amsterdam
Netherlands
☎ 31 20 525 7491
℻ 31 20 525 7484
✉ edvdh@astro.uva.nl

van den Oord Bert H J
Sterrekundig Inst Utrecht
Box 80000
NL 3508 TA Utrecht
Netherlands
☎ 31 30 253 5200
℻ 31 30 253 1601
✉ oord@astro.uu.nl

van der Borght Rene
31 The Promenade
Isle of Capri
Surfers Paradise 4217
Australia
☎ 61 38 5712

van der Hucht Karel A
SRON
Postbus 800
Sorbonnelaan 2
NL 3584 CA Utrecht
Netherlands
☎ 31 30 253 5600
℻ 31 30 254 0860
✉ K.vanderHucht@sron.ruu.nl

van der Hulst Jan M
Kapteyn Sterrekundig Inst
Univ Groningen
Postbus 800
NL 9700 AV Groningen
Netherlands
☎ 31 50 363 4054
℻ 31 50 363 4033
✉ j.m.van.der.hulst@astro.rug.nl

van der Klis Michiel
Astronomical Institute
Univ of Amsterdam
Kruislaan 403
NL 1098 SJ Amsterdam
Netherlands
☎ 31 20 525 7498/7491/7492
℻ 31 20 525 7484
✉ michiel@astro.uva.nl

van der Kruit Pieter C
Kapteyn Sterrekundig Inst
Univ Groningen
Postbus 800
NL 9700 AV Groningen
Netherlands
☎ 31 50 363 4073
℻ 31 50 363 6100
✉ vdkruit@astro.rug.nl

van der Laan Harry
Sterrekundig Inst Utrecht
Box 80000
NL 3508 TA Utrecht
Netherlands
☎ 31 30 253 5200
℻ 31 30 253 1601
✉ vdlaan@astro.uu.nl

van der Linden Ronald
Ctr Plasma Astrophys
Katholieke Univ Leuven
Celestijnenlaan 200B
B 3001 Heverlee
Belgium
☎ 32 16 327 003
℻ 32 16 32 7998
✉ ronald.vanderlinden@wis.kuleuven.ac.be

van der Marel Roeland P
STScI/RPO
Homewood Campus
3700 San Martin Dr
Baltimore MD 21218
USA
☎ 1 301 338 4931
℻ 1 301 338 4596
✉ marel@stsci.edu

van der Raay Herman B
Physics Dpt
Univ Birmingham
Box 363
Birmingham B15 2TT
UK
☎ 44 12 14 72 1301

van der Veen Wilhelmus EC
Astronomy Dpt
Columbia Univ
538 W 120th St
New York NY 10027
USA
☎ 1 212 854 6831
℻ 1 212 854 9504
✉ wecj@carmen.phys.columbia.edu

van der Walt D J
Physics Dpt
Potchefstroom Univ
for CHE
2520 Potchefstroom
South Africa
☎ 27 148 299 2408
℻ 27 148 299 2421
✉ fskdjvdw@pukvml.puk.ac.za

van der Werf Paul P
Leiden Observ
Box 9513
NL 2300 RA Leiden
Netherlands
☎ 31 71 527 5861
℻ 31 71 527 5819
✉ pvdwerf@strw.leidenuniv.nl

van Dessel Edwin Ludo
Koninklijke Sterrenwacht
van Belgie
Ringlaan 3
B 1180 Brussels
Belgium
☎ 32 2 373 0240
℻ 32 2 374 9822
✉ edwin.vandessel@oma.be

van Diggelen J
Observ Utrecht
Aetsveldselaan 12
NL 1381 EA Weesp
Netherlands

van Dishoeck Ewine F
Leiden Observ
Box 9513
NL 2300 RA Leiden
Netherlands
☎ 31 71 527 5814/5833
℻ 31 71 527 5819
✉ ewine@strw.leidenuniv.nl

van Dorn Bradt Hale
Ctr for Space Res
MIT Rm 37 587
Box 165
Cambridge MA 02139 4307
USA
☎ 1 617 253 7550
℻ 1 617 253 0861
✉ hale@space.mit.edu

van Driel Willem
Observ Paris Meudon
USN
Pl J Janssen
F 92195 Meudon PPL Cdx
France
☎ 33 1 45 07 7609
℻ 33 1 45 07 7879
✉ wim.van.driel@obspm.fr

van Driel Gesztzlyi L
Observ Paris Meudon
DASOP
Pl J Janssen
F 92195 Meudon PPL Cdx
France
☎ 33 1 45 07 7530
℻ 33 1 45 07 7959

van Duinen R J
Chairman N W O
Laan van Nw Oost Indie 131
Box 93138
NL 2509 AC The Hague
Netherlands

van Dyk Schuyler
Astronomy Dpt
Univ of California
601 Campbell Hall
Berkeley CA 94720 3411
USA
☎ 1 510 643 8143
℻ 1 510 642 3411
✉ vandyk@popsicle.berkeley.edu

Van Flandern Tom
Meta Res
6327 Western Av NW
Washington DC 20015
USA
☎ 1 202 362 9176
℻ 1 202 362 8279
✉ metares@well.sf.ca.us

van Genderen Arnoud M
Leiden Observ
Box 9513
NL 2300 RA Leiden
Netherlands
☎ 31 71 527 5856
℻ 31 71 527 5819
✉ genderen@strw.leiden-univ.nl

van Gorkom Jacqueline H
NRAO
Box 0
Socorro NM 87801 0387
USA
☎ 1 505 772 4302
℻ 1 505 835 7027

van Groningen Ernst
Astronomical Observ
Box 515
S 751 20 Uppsala
Sweden
☎ 46 18 53 0265
℻ 46 18 52 7583

van Haarlem Michiel
NFRA
Postbus 2
NL 7990 AA Dwingeloo
Netherlands
☎ 31 521 59 5100
℻ 31 52 159 7332
✉ haarlem@nfra.nl

van Hamme Walter
Dpt Phys/Astronomy
Florida Itl Univ
University Park
Miami FL 33199
USA

van Herk Gijsbert
Leiden Observ
Box 9513
NL 2300 RA Leiden
Netherlands
☎ 31 71 527 2727
✆ 31 71 527 5819

van Hoolst Tim
Koninklijke Sterrenwacht
van Belgie
Ringlaan 3
B 1180 Brussels
Belgium
☎ 32 2 373 0668
✆ 32 2 374 9822
✉ timvh@oma.be

van Horn Hugh M
NSF
Div Astron Sci
4201 Wilson Blvd
Arlington VA 22230
USA
☎ 1 703 306 1820
✆ 1 703 306 0525
✉ hvanhorn@nsf.gov

van Houten C J
Leiden Observ
Box 9513
NL 2300 RA Leiden
Netherlands
☎ 31 71 527 2727
✆ 31 71 527 5819

van Houten-Groeneveld I
Leiden Observ
Box 9513
NL 2300 RA Leiden
Netherlands
☎ 31 71 527 2727
✆ 31 71 527 5819

van Hoven Gerard
Univ of California
158 Sunny Cove
Santa Cruz CA 95062
USA

van Langevelde Huib Jan
JIVE
Postbus 2
NL 7990 AA Dwingeloo
Netherlands
☎ 31 521 59 5220
✆ 31 52 159 7332
✉ huib@jive.nfra.nl

van Leeuwen Floor
Royal Greenwich Obs
Madingley Rd
Cambridge CB3 0EZ
UK
☎ 44 12 23 374 765
✆ 44 12 23 374 700

van Moorsel Gustaaf
NRAO
Box 0
Socorro NM 87801 0387
USA
☎ 1 505 835 7396
✆ 1 505 835 7027
✉ gvanmoor@nrao.edu

van Nieuwkoop JIR
Prinsesselaan 12
NL 7316 CN Apeldoorn
Netherlands

van Paradijs Johannes
Astronomical Institute
Univ of Amsterdam
Kruislaan 403
NL 1098 SJ Amsterdam
Netherlands
☎ 31 20 525 7491
✆ 31 20 525 7484
✉ jvp@astro.uva.nl

van Regemorter Henri
Observ Paris Meudon
DAMAP
Pl J Janssen
F 92195 Meudon PPL Cdx
France
☎ 33 1 45 07 7444
✆ 33 1 45 07 7971

van Rensbergen Walter
Astrofysisch Inst
Vrije Univ Brussel
Pleinlaan 2
B 1050 Brussels
Belgium
☎ 32 2 629 3497
✆ 32 2 629 3424
✉ wvanrens@vnet3.vub.ac.be

van Riper Kenneth A
Box 4729
Los Alamos NM 87544
USA
☎ 1 505 672 1105
✉ kvr@rt66.com

van Santvoort Jacques
Dpt Astrophysique
Univ Mons Hainaut
Pl du Parc 20
B 7000 Mons
Belgium
☎ 32 65 373 727
✆ 32 65 373 054
✉ jvs@umh.ac.be

van Speybroeck Leon P
CfA
HCO/SAO
60 Garden St
Cambridge MA 02138 1516
USA
☎ 1 617 495 7233
✆ 1 617 495 7356
✉ lvs@cfa.harvard.edu

van Winckel Hans
Inst v Sterrenkunde
Katholicke Univ Leuven
Celestijnenlaan 200B
B 3001 Heverlee
Belgium
☎ 32 16 327 039
✆ 32 16 32 7999
✉ hans.vanwinckel@ster.kuleuven.ac.be

van Woerden Hugo
Kapteyn Sterrekundig Inst
Univ Groningen
Postbus 800
NL 9700 AV Groningen
Netherlands
☎ 31 50 363 4073/4066
✆ 31 50 363 6100
✉ secr@astro.rug.nl

van't Veer-Menneret Claude
Observ Paris
DASGAL
61 Av Observatoire
F 75014 Paris
France
☎ 33 1 40 51 2249
✆ 33 1 45 07 7878

van't-Veer Frans
IAP
98bis bd Arago
F 75014 Paris
France
☎ 33 1 44 32 8048
✆ 33 1 44 32 8001

Vandas Marek
Astronomical Institute
Czech Acad Sci
Bocni Ii 1401
CZ 141 31 Praha 4
Czech R
☎ 420 2 67 10 3061
✆ 420 2 76 90 23
✉ vandas@ig.cas.cz

Vanden Bout Paul A
NRAO
520 Edgemont Rd
Charlottesville VA 22903
USA
☎ 1 804 296 0241
✆ 1 804 296 0385
✉ pvandenb@nrao.edu

Vandenberg Don
Physics Dpt
Univ of Victoria
Box 1700
Victoria BC V8W 2Y2
Canada
☎ 1 250 721 7739
✆ 1 250 721 7715
✉ davb@uvvm.uvic.ca

Vandervoort Peter O
Astronomy/Astrophys Ctr
Univ of Chicago
5640 S Ellis Av
Chicago IL 60637
USA
☎ 1 312 962 8209
✆ 1 312 702 8212

Vapillon Loic J
Observ Paris Meudon
DESPA
Pl J Janssen
F 92195 Meudon PPL Cdx
France
☎ 33 1 45 07 7623
✆ 33 1 45 07 7971

Vardanian R A
Byurakan Astrophys Observ
Armenian Acad Sci
378433 Byurakan
Armenia
☎ 374 88 52 28 3453/4142
✆ 374 88 52 52 3640

Vardavas Ilias Mihail
Physics Dpt
Univ of Crete
Box 1527
GR 711 11 Iraklion
Greece
☎ 30 81 23 6589
✆ 30 81 23 9735

Vardya M S
Seven Bunglows 502 Vigyanshila
Juhu-Versova Link Rd
Mumbai 400 061
India

Varma Ram Kumar
PRL
Navrangpura
Ahmedabad 380 009
India
☎ 91 272 46 2129
✆ 91 272 44 5292

Varshalovich Dimitrij
Ioffe Physical Tech Inst
Acad Sciences
Polytechnicheskaya Ul 26
194021 St Petersburg
Russia
☎ 7 812 247 2255
✆ 7 812 247 1017

Varvoglis H
Astronomy Lab
Univ Thessaloniki
GR 540 06 Thessaloniki
Greece
☎ 30 31 99 8024
✆ 30 31 99 5384
✉ varvogli@astro.auth.gr

Vashkov'Yak Sof-Ya N
SAI
Acad Sciences
Universitetskij Pr 13
119899 Moscow
Russia
☎ 7 095 139 3764
✆ 7 095 939 0126

Vasileva Galina J
Pulkovo Observ
Acad Sciences
10 Kutuzov Quay
196140 St Petersburg
Russia
☎ 7 812 298 2242
✆ 7 812 315 1701

Vass Gheorghe
Astronomical Institute
Romanian Acad Sciences
Cutitul de Argint 5
RO 75212 Bucharest
Rumania
☎ 40 1 641 3686
✆ 40 1 312 3391
✉ ghevass@roean

Vassiliev Nikolay N
Inst Theoret Astron
Acad Sciences
10 Nab Kutuzova
191187 St Petersburg
Russia
☎ 7 812 275 4413
✆ 7 812 272 7968
✉ vasiliev@ita.spb.su

Vasu-Mallik Sushma
IIA
Koramangala
Sarjapur Rd
Bangalore 560 034
India
☎ 91 80 356 9179/9180
✆ 91 80 553 4043

Vats Hari OM
PRL
Navrangpuva
Ahmedabad 380 009
India
☎ 91 272 46 2129
✆ 91 272 44 5292
✉ ips@prl.ernet.in

Vauclair Gerard P
OMP
14 Av E Belin
F 31400 Toulouse Cdx
France
☎ 33 5 61 25 2101
✆ 33 5 61 27 3179

Vauclair Sylvie D
OMP
14 Av E Belin
F 31400 Toulouse Cdx
France
☎ 33 5 61 25 2101
✆ 33 5 61 27 3179

Vaughan Alan
School Maths/Physics
Computing/ Electron
Macquarie Univ
Macquarie 2109
Australia
☎ 61 28 058 904
✆ 61 2 805 8983
✉ alanv@macastro.mpcs.mq.oz.au

Vaughan Arthur H
Perkin Elmer Corp
7421 Orangewood Av
Garden Grove CA 92641
USA
☎ 1 714 895 1667

Vauglin Isabelle
Observ Lyon
Av Ch Andre
F 69561 S Genis Laval cdx
France
☎ 33 4 72 39 9098
✆ 33 4 72 39 9791
✉ vauglin@castor.univ-lyon1.fr

Vavrova Zdenka
Na Kopecku 346 II
CZ 379 01 Trebon
Czech R
☎ 420 333 2779

Vaz Luiz Paulo Ribeiro
Physics Dpt
UFMG
CP 702
30161 Belo Horizonte MG
Brazil
☎ 55 314 41 2541
✆ 55 314 49 5649
✉ lpv@fisica.ufmg.br
luiz@astro.ku.dk

Vazquez Manuel
IAC
Observ d Teide
via Lactea s/n
E 38200 La Laguna
Spain
☎ 34 2 232 9100
✆ 34 2 232 9117

Vazquez Ruben Angel
Observ Astronomico
Paseo d Bosque S/n
1900 La Plata (Bs As)
Argentina
☎ 54 21 217 308
✆ 54 21 211 761
✉ rvazquez@fcaglp.edu.ar

Vazquez-Semadeni Enrique
Instituto Astronomia
UNAM
Apt 70 264
Mexico DF 04510
Mexico
☎ 52 5 622 3906
✆ 52 5 616 0653
✉ enro@astroscu.unam.mx

Veck Nicholas
Marconi Res Centre
West Hanningfield Rd
Gt Baddow
Chelmsford Essex CM2 8HN
UK
☎ 44 12 45 733 31
✆ 44 12 457 5244

Vedel Henrik
Theoretical Astrophysics Centre
Juliane Maries Vej 30
DK 2100 Copenhagen
Denmark
☎ 45 35 32 5903
✆ 45 35 32 5910
✉ vedel@tac.dk

Veeder Glenn J
CALTECH/JPL
MS 183 501
4800 Oak Grove Dr
Pasadena CA 91109 8099
USA
☎ 1 213 354 7388
✆ 1 818 393 6030

Vega E Irene
Observ Astronomico
Paseo d Bosque S/n
1900 La Plata (Bs As)
Argentina
☎ 54 21 217 308
✆ 54 21 211 761

Veiga Carlos Henrique
CNPq
r Gl Jose Cristino 77
21041 000 Rio de Janeiro
Brazil
☎ 55 21 589 2955
✆ 55 21 589 8972
✉ cave@on.br

Veillet Christian
CFHT Corp
Box 1597
Kamuela HI 96743
USA
☎ 1 808 885 3161
✆ 1 808 885 7288
✉ veillet@cfht.hawaii.edu

Veilleux Sylvain
Astronomy Dpt
Univ of Maryland
College Park MD 20742 2421
USA
☎ 1 301 405 0282
✆ 1 301 314 9067
✉ veilleux@astro.umd.edu

Veis George
Geodesy Lab
Ntl Techn Univ
Panepistimiopolis
GR 157 84 Zografos
Greece
☎ 30 1 724 3414
✆ 30 1 723 5122

Veismann Uno
Tartu Observ
Estonian Acad Sci
Toravere
EE 2444 Tartumaa
Estonia
☎ 372 7 416 625
✆ 372 7 410 205

Vekstein Gregory
Physics Dpt
UMIST
Box 88
Manchester M60 1QD
UK
☎ 44 161 200 3913
✆ 44 161 200 3669
✉ grigory.vekstein@umist.ac.uk

Velkov Kiril
Astronomy Dpt
Bulgarian Acad Sci
72 Lenin Blvd
BG 1784 Sofia
Bulgaria
☎ 359 2 75 8927
✆ 359 2 75 5019

Velli Marco
Dpt Astronomia
Univ d Firenze
Largo E Fermi 5
I 50125 Firenze
Italy
☎ 39 55 27 521
✆ 39 55 22 0039

Velli Marco
Observ Paris Meudon
DESPA
Pl J Janssen
F 92195 Meudon PPL Cdx
France
☎ 33 1 45 07 7659
✆ 33 1 45 07 2806
✉ velli@obspm.fr

Velusamy T
CALTECH/JPL
MS 169 506
4800 Oak Grove Dr
Pasadena CA 91109 8099
USA
☎ 1 818 354 6112
✆ 1 818 354 8895
✉ velu@kuiper.jpl.nasa.gov

Venkatakrishnan P
IIA
Koramangala
Sarjapur Rd
Bangalore 560 034
India
☎ 91 80 553 0672*263
✆ 91 80 553 4043
✉ pvk@iiap.ernet.in

Venkatesan Doraswamy
Physics Dpt
Univ of Calgary
2500 University Dr NW
Calgary AB T2N 1N4
Canada
☎ 1 403 2205389/3689
✆ 1 403 289 3331

Vennik Jaan
Tartu Observ
Estonian Acad Sci
Toravere
EE 2444 Tartumaa
Estonia
☎ 372 7 410 274
✆ 372 7 410 205
✉ vennik@aai.ee

Ventura Joseph
Physics Dpt
Univ of Crete
Box 2208
GR 710 03 Heraklion
Greece
☎ 30 81 394 216
✆ 30 81 394 201
✉ ventura@physics.uch.gr

Ventura Rita
Osserv Astronomico
d Catania
Via A Doria 6
I 95125 Catania
Italy
☎ 39 95 733 2258
✆ 39 95 33 0592
✉ rventura@astrct.ct.astro.it

Venturi Tiziana
Istt Radioastronomia
CNR
Via P Gobetti101
I 40129 Bologna
Italy
☎ 39 51 639 9385
✆ 39 51 639 9431
✉ tventuri@astbo1.bo.cnr.it

Venugopal V R
TIFR/Radio Astronomy Ctr
Box 8
Udhagamandalam 643 001
India
☎ 91 423 2651/2032

Verbeek Paul
Toegepaste Wiskunde Inf
Universiteit Gent
Krijgslaan 281 S9
B 9000 Gent
Belgium
☎ 32 9 264 4763
✆ 32 9 264 4995

Verbunt Franciscus
Sterrekundig Inst Utrecht
Box 80000
NL 3508 TA Utrecht
Netherlands
☎ 31 30 253 5207
✆ 31 30 253 1601
✉ f.w.m.verbunt@astro.uu.nl

Verdes-Montenegro Lourdes
Inst Astrofisica
Andalucia Apt 3004
Prof Albareda 1
E 18080 Granada
Spain
☎ 34 5 812 1311
✆ 34 5 881 4530
✉ lourdes@iaa.es

Verdet Jean-Pierre
Observ Paris
DANOF
61 Av Observatoire
F 75014 Paris
France
☎ 33 1 40 51 2206
✆ 33 1 44 54 1804

Veres Ferenc
Konkoly Observ
Thege U 13/17
Box 67
H 1525 Budapest
Hungary
☎ 36 1 375 4122
✆ 36 1 275 4668

Vereshchagin Sergei V
Inst of Astronomy
Acad Sciences
Pyatnitskaya Ul 48
109017 Moscow
Russia
☎ 7 095 592 1207
✆ 7 095 230 2081

Vergez Madeleine
OMP
9 r Pont de La mouette
F 65200 Bagneres Bigorre
France
☎ 33 5 62 95 1969
✆ 33 3 88 25 0160

Vergnano A
Osserv Astronomico
d Torino
St Osservatorio 20
I 10025 Pino Torinese
Italy
☎ 39 11 461 9000
✆ 39 11 461 9030

Vergne Maria Marcela
Observ Astronomico
Paseo d Bosque S/n
1900 La Plata (Bs As)
Argentina
☎ 54 21 217 308
✆ 54 21 258 985
✉ mvergne@fcaglp.edu.ar

Verheest Frank
Sterrekundig Observ
Universiteit Gent
Krijgslaan 281 S9
B 9000 Gent
Belgium
☎ 32 9 264 4799
✆ 32 9 264 4989
✉ frank.verheest@rug.ac.be

Verkhodanov Oleg
SAO
Acad Sciences
Nizhnij Arkhyz
357147 Karachaevo
Russia
☎ 7 878 784 4749
✆ 7 96 908 2861
✉ vo@sao.ru

Verma R P
TIFR
Homi Bhabha Rd
Colaba
Bombay 400 005
India
☎ 91 22 219 111
✆ 91 22 495 2110
✉ vermarp@tifrvax.tifr.res.in

Verma Satya Dev
Dpt Physics/Space Sci
Univ School of Sci
Gujarat Univ
Ahmedabad 380 009
India
☎ 91 272 44 0920

Verma V K
Uttar Pradesh State
Observ
Po Manora Peak 263 129
Nainital 263 129
India
☎ 91 59 42 2136

Vermeulen Rene Cornelis
NFRA
Postbus 2
NL 7990 AA Dwingeloo
Netherlands
☎ 31 521 59 5262
✆ 31 52 159 7332
✉ rcv@nfra.nl

Verniani Franco
Istt d Fisica
Univ d Bologna
Via Irnerio 46
I 40126 Bologna
Italy
☎ 39 51 26 0991

Veron Marie-Paule
OHP
F 04870 S Michel Obs
France
☎ 33 4 92 76 6368
✆ 33 4 92 76 6295

Veron Philippe
OHP
F 04870 S Michel Obs
France
☎ 33 4 92 76 6368
✆ 33 4 92 76 6295

Verschueren Werner
Astrophysics Res Gp
Univ Ctr Antwerpen
Groenenborgerlaan 171
B 2020 Antwerpen
Belgium
☎ 32 3 218 0356
✆ 32 3 218 0204
✉ verschue@ruca.ua.ac.be

Verschuur Gerrit L
4125 Yellow Cedar Cove
Lakeland TN 38002
USA
☎ 1 901 372 5932
✉ verschuur@aol.com

Verter Frances
NASA GSFC
Code 685
Greenbelt MD 20771
USA
☎ 1 301 286 7860

Vesecky J F
Radar Astronomy Inst
Stanford Univ
233 Durand
Stanford CA 94305 4035
USA
☎ 1 415 723 1435
✆ 1 415 723 4840

Vetesnik Miroslav
Dpt Th Phys/Astrophys
Masaryk Univ
Kotlarska 2
CZ 611 37 Brno
Czech R
☎ 420 5 4112 9481
✆ 420 5 4121 1214
✉ vetesnik@astro.sci.muni.cz

Vettolani Giampaolo
Istt Radioastronomia
CNR
Via P Gobetti101
I 40129 Bologna
Italy
☎ 39 51 639 9385
✆ 39 51 639 9385

Veverka Joseph
Astronomy Dpt
Cornell Univ
512 Space Sc Bldg
Ithaca NY 14853 6801
USA
☎ 1 607 256 3507
✆ 1 607 255 1767

Vial Jean-Claude
IAS
Bt 121
Universite Paris XI
F 91405 Orsay Cdx
France
☎ 33 1 69 85 8631
✆ 33 1 69 85 8675
✉ vial@iaslab.ias.fr

Viala Yves
Observ Paris
DEMIRM
61 Av Observatoire
F 75014 Paris
France
☎ 33 1 40 51 2116
✆ 33 1 40 51 2002

Viallefond Francois
Observ Paris
DEMIRM
61 Av Observatoire
F 75014 Paris
France
☎ 33 1 40 51 2116
✆ 33 1 40 51 2002

Vicente Raimundo O
R Mestre Aviv 30 R/C
P 1495 Lisboa
Portugal
☎ 351 1 211 2666

Vidal Jean-Louis
GRAAL
Univ Montpellier II
Pl E Bataillon
F 34095 Montpellier Cdx 5
France
☎ 33 4 67 14 3901
✆ 33 4 67 14 4535
✉ vidal@graal.univ-montp2.fr

Vidal Nissim V
Inst Sci/Techn
92 Bayit Vegan St
Jerusalem 96427
Israel
☎ 972 2 413 411
✆ 972 2 430040

Vidal-Madjar Alfred
IAP
98bis bd Arago
F 75014 Paris
France
☎ 33 1 44 32 8073
✆ 33 1 44 32 8001

Viegas Aldrovandi S M
IAG
Univ Sao Paulo
CP 9638
01065 970 Sao Paulo SP
Brazil
☎ 55 11 577 8599
✆ 55 11 276 3848
✉ viegas@vax.iagusp.usp.br

Vieira Martins Roberto
Observ Nacional
Rua Gl Bruce 586
Sao Cristovao
20921 030 Rio de Janeiro RJ
Brazil
☎ 55 21 580 7313
✆ 55 21 580 0332

Vienne Alain
Laboratoire Astronomie
Universite Lille
1 Impasse Observatoire
F 59000 Lille
France
☎ 33 3 20 52 4424
✆ 33 3 20 58 0328
✉ vienne@gat.uni-lille1.fr

Vietri Mario
Dpt Fisica E Amaldi
Universita d Roma 3
via Vasca Navale 84
I 00147 Roma
Italy
☎ 39 6 551 77025
✆ 39 6 557 9303
✉ vietri@corelli.fis.uniroma3.it

Vigier Jean-Pierre
Institut H Poincare
11 r P/M curie
F 75005 Paris
France

Vigotti Mario
Istt Radioastronomia
CNR
Via P Gobetti101
I 40129 Bologna
Italy
☎ 39 51 639 9385
✆ 39 51 639 9385

Vigroux Laurent
Service Astrophysique
CEA Saclay
Orme d Merisiers Bt 709
F 91190 Gif s Yvette Cdx
France
☎ 33 1 69 08 3912
✆ 33 1 69 08 6577
✉ lvigroux@cea.fr

Viik Tonu
Tartu Observ
Estonian Acad Sci
Toravere
EE 2444 Tartumaa
Estonia
☎ 372 3 410 265
✆ 372 7 410 205
✉ viik@jupiter.aai.ee

Vila Samuel C
Dr Ulles 184
E 03224 Tarrasa (brna)
Spain
☎ 34 3 788 0310

Vilas Faith
NASA/JSC
Code SN3
Houston TX 77058
USA
☎ 1 713 483 5056
✆ 1 713 483 5347

Vilas-Boas Jose W
CRAAE INPE
EPUSP/PTR
CP 61548
01065 970 Sao Paulo SP
Brazil
☎ 55 11 815 5936
✆ 55 11 815 6289
✉ jwdsvboa@brusp

Vilchez Medina Jose M
IAC
Observ d Teide
via Lactea s/n
E 38200 La Laguna
Spain
☎ 34 2 232 9100
✆ 34 2 232 9117

Vilhena De Moraes R
INPE
Dpt Astronomia
CP 515
12227 010 S Jose dos Campos
Brazil
☎ 55 123 22 9088
✆ 55 123 21 8743

Vilhu Osmi
Helsinky Univ
Observ
Box 14
FIN 00014 Helsinki
Finland
☎ 358 19 12 2801
✆ 358 19 12 2952
✉ osmi.vilhu@hesinki.fr

Vilkki Erkki U
4 Parc La Londe
F 76130 Mont S Aignan
France
☎ 33 2 35 71 3562

Vilkoviskij Emmanuil Y
Astrophys Inst
Kazakh Acad Sci
480068 Alma Ata
Kazakhstan
☎ 7 62 4040

Villada Monica Maria
Observ Astronomico
de Cordoba
Laprida 854
5000 Cordoba
Argentina
☎ 54 51 230 491
✆ 54 51 21 0613
✉ villada@uncbob.edu.ar

Villela Thyrso Neto
INPE
Dpt Astronomia
CP 515
12227 010 S Jose dos Campos
Brazil
☎ 55 123 41 8977*278
✆ 55 123 21 8743

Vilmer Nicole
Observ Paris Meudon
DASOP
Pl J Janssen
F 92195 Meudon PPL Cdx
France
☎ 33 1 45 07 7806
✆ 33 1 45 07 7959

Vince Istvan
Astronomical Observ
Volgina 7
11050 Beograd
Yugoslavia FR
☎ 381 1 419 357/421 875
✆ 381 1 419 553
✉ ivince@aob.aob.bg.ac.yu

Vinko Jozsef
Optics Dpt
Jate Univ
Dom ter 9
H 6720 Szeged
Hungary
☎ 36 6 231 1622
✆ 36 3 631 1154

Vinluan Renato
Univ of Southern
Philippines
Obrero Davao City 9501
Philippines

Vinod S Krishan
IIA
Koramangala
Sarjapur Rd
Bangalore 560 034
India
☎ 91 80 356 6585/6497
✆ 91 80 553 4043

Viotti Roberto
IAS
Area d Ricerca CNR
Via Fosso Cavaliere 100
I 00133 Roma
Italy
☎ 39 6 4993 4473
✆ 39 6 2066 0188
✉ viotti@irmias

Virgopia Nicola
Dpt Matematica
Univ d Roma La Sapienza
Citta Universitaria
I 00185 Roma
Italy
☎ 39 6 44 03734

Vishniac Ethan T
Astronomy Dpt
Univ of Texas
Rlm 15 308
Austin TX 78712 1083
USA
☎ 1 512 471 1429
✆ 1 512 471 6016

Vishveshwara C V
RRI
Sadashivanagar
CV Raman Av
Bangalore 560 080
India
☎ 91 80 336 0122
✆ 91 80 334 0492

Visvanathan Natarajan
MSSSO
Weston Creek
Private Bag
Canberra ACT 2611
Australia
☎ 61 262 881 111
✆ 61 262 490 233
✉ vis@mso.anu.edu.au

Vitinskij Yurij I
Pulkovo Observ
Acad Sciences
10 Kutuzov Quay
196140 St Petersburg
Russia
☎ 7 812 298 2242
✆ 7 812 315 1701

Viton Maurice
LAS
Traverse du Siphon
Les Trois Lucs
F 13376 Marseille Cdx 12
France
☎ 33 4 91 05 5900
✆ 33 4 91 66 1855
✉ viton@frlasm51

Vittone Alberto Angelo
Osserv d Capodimonte
Via Moiariello 16
I 80131 Napoli
Italy
☎ 39 81 44 0101
✆ 39 81 45 6710

Vittorio Nicola
Istt Astronomico
Univ d Roma La Sapienza
Via G M Lancisi 29
I 00161 Roma
Italy
☎ 39 6 44 03734
✆ 39 6 44 03673

Vityazev Andrei
Inst Dynamics Geosp
Acad Sciences
Leninsky Pr 38 bld 6
117979 Moscow
Russia
☎ 7 095 939 7516
✆ 7 095 137 6511
✉ avit@orig.ipg.msk.su

Vityazev Veneamin V
Astronomical Observ
St Petersburg Univ
Bibliotechnaja Pl 2
198904 St Petersburg
Russia
☎ 7 812 428 7129
✆ 7 812 428 4259

Vivekanand M
RRI
Sadashivanagar
CV Raman Av
Bangalore 560 080
India
☎ 91 80 336 0122
✆ 91 80 334 0492

Vivekananda Rao
CASA
Univ of Osmania
Hyderabad 500 007
India
☎ 91 868951*247

Vives Teodoro Jose
Ctr Astron Hispano Aleman
Reina 66 9b
Apt C 511
E 04002 Almeria
Spain
☎ 34 5 023 0988

Vlachos Demetrius G
Dpt Geodesy/Astron
Univ Thessaloniki
GR 540 06 Thessaloniki
Greece
☎ 30 31 99 1520

Vladilo Giovanni
OAT
Box Succ Trieste 5
Via Tiepolo 11
I 34131 Trieste
Italy
☎ 39 40 31 99255
✆ 39 40 30 9418

Vladimirov Simeon
Astronomical Observ
Bulgarian Acad Sci
Box 15
BG 1309 Sofia
Bulgaria
☎ 359 2 75 8927

Vlahos Loukas
Astronomy Lab
Univ Thessaloniki
GR 540 06 Thessaloniki
Greece
☎ 30 31 99 1357

Voelk Heinrich J
MPI Kernphysik
Postfach 103 980
D 69029 Heidelberg
Germany
☎ 49 622 151 6295
✆ 49 622 151 5640

Vogel Manfred
Inst Astronomie
ETH Zentrum
CH 8092 Zuerich
Switzerland
☎ 41 1 632 3806
✆ 41 1 262 0003
✉ vogel@czhethsa

Vogel Stuart Newcombe
Astronomy Program
Univ of Maryland
College Park MD 20742 2421
USA
☎ 1 301 405 1543
✆ 1 301 314 9067

Voglis Nikos
Astrophysics Dpt
Ntl Univ Athens
Panepistimiopolis
GR 157 84 Zografos
Greece
☎ 30 1 724 3414
✆ 30 1 723 5122

Vogt Nikolaus
Luckengasse 26
D 86720 Noedlingen
Germany
✉ nikolaus.vogt@t-online.de

Vogt Steven Scott
Lick Observ
Univ of California
Santa Cruz CA 95064
USA
☎ 1 831 429 2844
✆ 1 831 426 3115

Voigt Hans H
Charlottenburger Str 19
D 37070 Goettingen
Germany

Vokrouhlicky David
Astronomical Institute
Charles Univ
V Holesovickack 2
CZ 180 00 Praha 8
Czech R
☎ 420 2 2191 2572
✆ 420 2 688 5095
✉ vokrouhl@mbox.cesnet.cz

Volk Kevin
Dpt Phys/Astronomy
Univ of Calgary
2500 University Dr NW
Calgary AB T2N 1N4
Canada
☎ 1 403 931 2366
✆ 1 403 289 3331

Volland H
Astronomisches Institut
Univ Bonn
auf d Huegel 71
D 53121 Bonn
Germany
☎ 49 228 73 3674
✆ 49 228 73 3672

Volonte Sergio
ESA
8 10 r Mario Nikis
F 75738 Paris Cdx 15
France
☎ 33 1 53 69 7654
✆ 33 1 53 69 7236

Volyanskaya Margarita Yu
Astronomy Dpt
Odessa State Univ
Schechenko Park
270014 Odessa
Ukraine
☎ 380 4822 28442
✆ 380 4822 28442
✉ root@astro.odessa.ua

von Appen-Schnur Gerhard F O
Astronomisches Institut
Ruhr Univ Bochum
Postfach 102148
D 44780 Bochum
Germany
☎ 49 234 700 4584
✆ 49 234 709 4169
✉ gfo.von-appen-schnur@astro.ruhr-uni-bochum.de

von Borzeszkowski H H
Einstein Laboratorium
Telegrafenberg
Rosa Luxemburg Str 17a
D 14473 Potsdam
Germany
☎ 49 331 275 3731

von der Luehe Oskar
Kiepenheuer Institut
f Sonnenphysik
Schoneckstr 6
D 79104 Freiburg Breisgau
Germany
☎ 49 761 319 80
✆ 49 761 319 8111
✉ ovdluhe@kis.uni-freiburg.de

von Hippel Theodore A
NOAO
Box 26732
950 N Cherry Av
Tucson AZ 85726 6732
USA
☎ 1 510 318 8142
✆ 1 510 318 8360
✉ ted@noao.edu

von Hoerner Sebastian
Krummenackerstr 186
D 73000 Esslingen
Germany

von Montigny Corinna
Landessternwarte
Koenigstuhl
D 69117 Heidelberg
Germany
☎ 49 622 150 9223
✆ 49 622 150 9202
✉ cvmontig@lsw.uni-heidelberg.de

von Steiger Rudolf
ISSI
Hallerstr 6
CH 3012 Bern
Switzerland
☎ 41 31 631 4890
✆ 41 31 631 4897
✉ vsteiger@phim.unibe.ch

von Weizsaecker C F
Maximillianstr 15
D 8130 Starnberg
Germany

Vondrak Jan
Astronomical Institute
Czech Acad Sci
Bocni II 1401
CZ 141 31 Praha 4
Czech R
☎ 420 2 67 10 3043
✆ 420 2 76 90 23
✉ vondrak@ig.cas.cz

Voroshilov V I
Main Astronomical Obs
Ukrainian Acad Science
Golosiiv
252650 Kyiv 22
Ukraine
☎ 380 4426 63110
✆ 380 4426 62147

Vorpahl Joan A
748 23rd St
Santa Monica CA 90402
USA

Voshchinnikov Nicolai
Astronomical Observ
St Petersburg Univ
Bibliotechnaja Pl 2
198904 St Petersburg
Russia
☎ 7 812 428 4162
✆ 7 812 428 4259

Vrba Frederick J
USNO
Flagstaff Station
Box 1149
Flagstaff AZ 86002 1149
USA
☎ 1 602 779 5132
✆ 1 520 774 3626
✉ fjv@nofs.navy.mil

Vreux Jean Marie
Inst Astrophysique
Universite Liege
Av Cointe 5
B 4000 Liege
Belgium
☎ 32 4 254 7510
✆ 32 4 254 7511
✉ vreux@astro.ulg.ac.be

Vrsnak Bojan
Hvar Observ
Fac Geodesy
Kaciceva 26
HR 10000 Zagreb
Croatia
☎ 385 1 442 600*335
✆ 385 1 445 410
✉ bojan.vrsnak@uni-fg.ac.mail.yu

Vrtilek Jan M
CfA
HCO/SAO MS 3
60 Garden St
Cambridge MA 02138 1516
USA
☎ 1 617 495 7127
✆ 1 617 495 7356
✉ jvrtilek@cfa.harvard.edu

Vrtilek Saeqa Dil
CfA
HCO/SAO MS 83
60 Garden St
Cambridge MA 02138 1516
USA
☎ 1 617 495 7094
✆ 1 617 496 7577
✉ svrtilek@cfa.harvard.edu

Vu Duong Tuyen
BDL
77 Av Denfert Rochereau
F 75014 Paris
France
☎ 33 1 45 07 2262
✆ 33 1 46 33 2834

Vucetich Hector
Univ Nacional La Plata
Dpt Fisica
CC 67
1900 La Plata (Bs As)
Argentina
☎ 54 21 217 308
✆ 54 21 252 006

Vuillemin Andre
LAS
Traverse du Siphon
Les Trois Lucs
F 13376 Marseille Cdx 12
France
☎ 33 4 91 05 5900
✆ 33 4 91 66 1855
✉ andre@frlasm51

Vujnovic Vladis
Inst of Physics
Univ of Zagreb
Box 304
HR 10000 Zagreb
Croatia
☎ 385 1 271 211

Vukicevic K M
Astronomy Dpt
Univ of Belgrade
Studentski Trg 16
11000 Beograd
Yugoslavia FR
☎ 381 11 638 715
✆ 381 11 630 151

Vyalshin Gennadij F
Pulkovo Observ
Acad Sciences
10 Kutuzov Quay
196140 St Petersburg
Russia
☎ 7 812 298 2242
✆ 7 812 315 1701

Vykutilova Marie
Marsovice 47
CZ 592 31 Nove Mesto N.M.
Czech R
☎ 420 616 916 304

W

Wachlin Felipe Carlos
Observ Astronomico
Paseo d Bosque s/n
1900 La PLata (Bs As)
Argentina
☎ 54 21 258 985
📠 54 21 258 985
✉ fcw@fcaglp.fcaglp.unlp.edu.ar

Wada Keiichi
NAOJ
Osawa 2- 21-2
Mitaka 181 0015
Japan
☎ 81 422 34 3632
📠 81 422 34 3746
✉ wada@th.nao.ac.jp

Waddington C Jake
School Phys/Astronomy
Univ Minnesota
116 Church St SE
Minneapolis MN 55455
USA
☎ 1 612 624 2566
📠 1 612 624 2029
✉ wadd@physics.spa.umn.edu

Wade Richard Alan
Astronomy Dpt
Pennsylvania State Univ
525 Davey Lab
University Park PA 16802
USA
☎ 1 814 865 3631
📠 1 814 863 7114
✉ wade@astro.psu.edu

Waelkens Christoffel
Inst v Sterrenkunde
Katholieke Univ Leuven
Celestijnenlaan 200B
B 3001 Heverlee
Belgium
☎ 32 16 32 7036
📠 32 16 32 7999
✉ christoffel@ster.kuleuven.ac.be

Wagner Raymond L
Rockwell Intl
Box 3105
MS 031 BA06
Anaheim CA 92803 3105
USA
☎ 1 714 762 1754
📠 1 714 762 2007
✉ raymond_L._wagner@ccmail.anatcp.rockwell.com

Wagner Robert M
Lowell Observ
1400 W Mars Hill Rd
Box 1149
Flagstaff AZ 86001
USA
☎ 1 602 779 0106
📠 1 520 774 3358

Wagner Stefan
Landessternwarte
Koenigstuhl
D 69117 Heidelberg
Germany
☎ 49 622 150 9212
📠 49 622 150 9202
✉ swagner@mail.lsw.uni-heidelberg.de

Wagner William J
NASA Headquarters
Code SS
Space Physics Div
Washington DC 20546
USA

Wagoner Robert V
Stanford Univ
Varian Physics Bldg
Stanford CA 94305
USA
☎ 1 415 723 4561

Wahlgren Glenn Michael
Physics Dpt
Univ of Lund
Soelvegatan 14
S 223 62 Lund
Sweden
☎ 46 46 222 4874
📠 46 46 222 4709
✉ spek_glenn@garbo.lucas.lu.se

Wainwright John
Applied Maths Dpt
Univ Waterloo
Waterloo ON N2L 3G1
Canada
☎ 1 519 885 1211
📠 1 519 746 6530
✉ jwainwright@math.uwaterloo.ca

Wakamatsu Ken-ichi
Fac of Eng
Gifu Univ
1 1 Yanagito
Gifu 501 1193
Japan
☎ 81 582 30 1111

Wakker Bastiaan Pieter
Astronomy Dpt
Univ Wisconsin
475 N Charter St
Madison WI 53706 1582
USA
☎ 1 605 263 6589
📠 1 605 263 0361
✉ wakker@astro.wisc.edu

Wako Kojiro
Intl Latitude Observ
NAOJ
Hoshigaoka Mizusawa Shi
Iwate 023
Japan
☎ 81 197 24 7111
📠 81 197 22 7120

Walborn Nolan R
STScI
Homewood Campus
3700 San Martin Dr
Baltimore MD 21218
USA
☎ 1 301 338 4915
📠 1 301 338 4767
✉ walborn@stsci.edu

Walch Jean-Jacques
OCA
CERGA
F 06130 Grasse
France
☎ 33 4 93 36 5849
📠 33 4 93 40 5353

Walder Rolf
Inst Astronomie
ETH Zentrum
CH 8092 Zuerich
Switzerland
☎ 41 1 632 4217
📠 41 1 632 4105
✉ walder@astro.phys.ethz.ch

Waldhausen Silvia
Observ Astronomico
Paseo d Bosque S/n
1900 La PLata (Bs As)
Argentina
☎ 54 21 217 308
📠 54 21 258 985
✉ silvia@fcaglp.fcaglp.unlp.edu.ar

Waldmeier Max
Swiss Federal Observ
Wirzenweid 15
CH 8053 Zuerich
Switzerland
☎ 41 1 381 6242

Walker Alistair Robin
NOAO
CTIO
Casilla 603
La Serena
Chile
☎ 56 51 22 5415
📠 56 51 20 5342
✉ awalker@noao.edu

Walker Alta Sharon
Box 1101
McLean VA 22101
USA
☎ 1 703 428 7236
📠 1 703 428 6070
✉ flexus@erols.com

Walker David Douglas
Dpt Phys/Astronomy
UCLO
Gower St
London WC1E 6BT
UK
☎ 44 171 387 7050*3510
📠 44 171 380 7145
✉ ddw@uk.ac.ulc.starlink

Walker Edward N
Deudreys Cott Old Rd
Hailsham
East Sussex BN27 1PU
UK

Walker Gordon A H
Dpt Geophys/Astronomy
UBC
2075 Wesbrook Pl
Vancouver BC V6T 1W5
Canada
☎ 1 604 228 4133
📠 1 604 228 6047
✉ walker@astro.ubc.ca

Walker Helen J
Rutherford Appleton Lab
Space/Astrophysics Div
Bg R25/R68
Chilton Didcot OX11 0QX
UK
☎ 44 12 35 821 900
📠 44 12 35 445 808

Walker Ian Walter
Astronomy Dpt
Univ of Glasgow
Glasgow G12 8QQ
UK
☎ 44 141 339 8855
📠 44 141 334 9029

Walker Merle F
Lick Observ
Univ of California
Santa Cruz CA 95064
USA
☎ 1 831 429 2526
✆ 1 831 426 3115

Walker Richard L
USNO
Flagstaff Station
Box 1149
Flagstaff AZ 86002 1149
USA
☎ 1 502 779 5132
✆ 1 520 774 3626
✉ wak@nofs.navy.mil

Walker Robert C
NRAO
Box 0
Socorro NM 87801 0387
USA
☎ 1 505 835 7247
✆ 1 505 835 7027

Walker Robert M A
Physics Dpt
Washington Univ
Campus Box 1105
St Louis MO 63130
USA
☎ 1 314 889 6225
✆ 1 314 935 4083

Walker William S G
Box 173
Awanui 0552
New Zealand
✉ astronman@voyager.co.nz

Walker Arthur B C
Ctr for Space Sci/
Astrophysics
Stanford Univ ERL 310
Stanford CA 94305 4055
USA
☎ 1 415 497 1486
✆ 1 415 725 2333

Wall J W
Royal Greenwich Obs
Madingley Rd
Cambridge CB3 0EZ
UK
☎ 44 12 23 374 000
✆ 44 12 23 374 700

Wall Jasper V
Royal Greenwich Obs
Madingley Rd
Cambridge CB3 0EZ
UK
☎ 44 1223 374 000
✆ 44 12 23 374 700

Wallace Lloyd V
NOAO/KPNO
Box 26732
950 N Cherry Av
Tucson AZ 85726 6732
USA
☎ 1 520 327 5511
✆ 1 520 325 9360

Wallace Patrick T
Rutherford Appleton Lab
Space/Astrophysics Div
Bg R25/R68
Chilton Didcot OX11 0QX
UK
☎ 44 12 35 445 472
✆ 44 12 35 44 5808
✉ ptw@star.rl.ac.uk

Wallace Richard K
LANL
MS B257 X 7
Box 1663
Los Alamos NM 87545 2345
USA
☎ 1 505 667 5000
✆ 1 505 665 4055

Waller William H
NASA GSFC
Code 681
Greenbelt MD 20771
USA
☎ 1 301 286 5351
✆ 1 301 286 1753
✉ waller@stars.gsfc.nasa.gov

Wallerstein George
Astronomy Dpt
Univ of Washington
FM 20
Seattle WA 98195
USA
☎ 1 206 543 2888
✆ 1 206 685 0403
✉ wall@astro.washington.edu

Wallin John Frederick
CSI MS 5C3
George Mason Univ
4400 University Dr
Fairfax VA 22030
USA
☎ 1 703 993 3617
✆ 1 703 993 1980
✉ jwallin@gmu.edu

Wallinder Frederick
Univ of Oerebro
Inst Techn/Science
S 701 82 Oerebro
Sweden
✉ fredrik.wallinder@hoe.se

Wallis Max K
Dpt Applied Maths/Astron
Univ College
Box 78
Cardiff CF1 1XL
UK
☎ 44 12 22 442 11

Walmsley C Malcolm
RAIUB
Univ Bonn
auf d Huegel 69
D 53121 Bonn
Germany
☎ 49 228 525 305
✆ 49 228 525 229

Walraven Th
Box 98
9850 Orange Freestate
South Africa

Walsh Dennis
NRAL
Univ Manchester
Jodrell Bank
Macclesfield SK11 9DL
UK
☎ 44 14 777 1321
✆ 44 147 757 1618
✉ dw@jb.man.ac.uk

Walsh Robert
School of Mathematics
Univ of St Andrews
North Haugh
St Andrews Fife KY16 9SS
UK
☎ 44 1334 76161
✆ 44 1334 74487
✉ robert@dcs.st.and.ac.uk

Walter Frederick M
Dpt Phys/Astronomy
Astronomy Program
Suny at Stony Brook
Stony Brook NY 11794 3800
USA
☎ 1 516 632 8232
✆ 1 513 632 8176
✉ fwalter@astro.sunysb.edu

Walter Hans G
ARI
Moenchhofstr 12-14
D 69120 Heidelberg
Germany
☎ 49 622 140 5134
✆ 49 622 140 5297
✉ walter@relay.ari.uni-heidelberg.de

Walter Roland
Observ Geneve
Chemin d Maillettes 51
CH 1290 Sauverny
Switzerland
☎ 41 22 755 2611
✆ 41 22 755 3983
✉ roland.walter@obs.unige.ch

Walterbos Rene A M
Astronomy Dpt
NMSU
Box 30001 Dpt 4500
Las Cruces NM 88003
USA
☎ 1 505 646 6522
✆ 1 505 646 1602
✉ rwalterb@nmsu.edu

Walton Nicholas A
Royal Greenwich Obs
Apt 321
Santa Cruz de La Palma
E 38780 Santa Cruz
Spain
☎ 34 2 240 5500
✆ 34 2 240 5646/405 501
✉ naw@ing.iac.es

Wambsganss Joachim
Astrophysik Inst
Potsdam Univ
an d Sternwarte 16
D 14482 Potsdam
Germany
☎ 49 331 749 9316
✆ 49 331 749 9267
✉ jwambsganss@aip.de

Wampler E Joseph
418 Walnut Av
Santa Cruz CA 95060
USA
✉ jwampler @eso.org

Wamsteker Willem
ESA
Apt 50727 Villafranca
E 28080 Madrid
Spain
☎ 34 1 813 1100
✆ 34 1 813 1139
✉ ww@vilspa.esa.es

Wan Fook Sun
Mathematics Dpt
Ntl Univ Singapore
Kent Ridge
Singapore 0511
Singapore
☎ 65 772 2742

Wan Lai
Shanghai Observ
CAS
80 Nandan Rd
Shanghai 200030
China PR
☎ 86 21 6438 6191
✆ 86 21 6438 4618
✉ xytan@fudan.ihep.ac.cn

Wan Tongshan
Shanghai Observ
CAS
80 Nandan Rd
Shanghai 200030
China PR
☎ 86 21 6438 6191
✆ 86 21 6438 4618

Wanas Mamdouh Ishaac
Astronomy Dpt
Fac of Sciences
Cairo Univ
Geza
Egypt
☎ 20 2 572 7022
✆ 20 2 572 7556
✉ wanas@frcu.eun.eg

Wang Chuanjin
Purple Mountain Obs
CAS
Nanjing 210008
China PR
☎ 86 25 46700
✆ 86 25 301 459

Wang dechang
Purple Mountain Obs
CAS
Nanjing 210008
China PR
☎ 86 25 64 6700/4205
✆ 86 25 301 459

Wang Deyu
Purple Mountain Obs
CAS
Nanjing 210008
China PR
☎ 86 25 42817/46700
✆ 86 25 301 459

Wang Gang
Beijing Astronomical Obs
CAS
Beijing 100080
China PR
☎ 86 1 256 9840
✆ 86 1 256 10855

Wang Haimin
Physics Dpt
NJ Inst of Technology
323 High St
Newark NJ 07102
USA
☎ 1 201 596 3562

Wang Jia-Long
Beijing Astronomical Obs
CAS
Beijing 100080
China PR
☎ 86 1 28 1698
✆ 86 1 256 10855

Wang Jiaji
Shanghai Observ
CAS
80 Nandan Rd
Shanghai 200030
China PR
☎ 86 21 6438 6191
✆ 86 21 6438 4618
✉ wangjj@center.shao.ac.cn

Wang Jingsheng
Purple Mountain Obs
CAS
Nanjing 210008
China PR
☎ 86 25 31 3738
✆ 86 25 30 1459

Wang Jingxiu
Beijing Astronomical Obs
CAS
Beijing 100080
China PR
☎ 86 1 28 1698
✆ 86 1 256 10855

Wang Junjie
Beijing Astronomical Obs
CAS
Beijing 100080
China PR
☎ 86 10 6257 8265
✆ 86 1 256 10855
✉ wangjj@bao01.bao.ac.cn

Wang Kemin
Beijing Astronomical Obs
CAS
Beijing 100080
China PR
☎ 86 10 6256 6698
✆ 86 1 256 10855

Wang Lan-Juan
Shanghai Observ
CAS
80 Nandan Rd
Shanghai 200030
China PR
☎ 86 21 6438 6191
✆ 86 21 6438 4618

Wang Qingde Daniel
Dearborn Observ
NW Univ
2131 Sheridan Rd
Evanston IL 60208 3112
USA
☎ 1 847 491 6446
✆ 1 847 491 3135
✉ wqd@nwu.edu

Wang Renchuan
Astrophysics Division
Univ Science/Technology
Hefei 230026
China PR
☎ 86 551 33 1134
✆ 86 551 33 1760

Wang Shouguan
Beijing Astronomical Obs
CAS
Beijing 100080
China PR
☎ 86 10 255 1968
✆ 86 1 256 10855

Wang Shui
Astrophysics Division
Univ Science/Technology
Hefei 230026
China PR
☎ 86 551 33 1134*209
✆ 86 551 33 1760

Wang Shunde
Beijing Astronomical Obs
CAS
Beijing 100080
China PR
☎ 86 1 256 1264
✆ 86 1 256 10855

Wang Sichao
Purple Mountain Obs
CAS
Nanjing 210008
China PR
☎ 86 25 44205
✆ 86 25 301 459

Wang Tinggui
Astrophysics Division
Univ Science/Technology
Hefei 230026
China PR
☎ 86 551 33 1134
✆ 86 551 331 760

Wang Ya'nan
Ctr for Astronomy
Instrumen Res
CAS
Nanjing 210042
China PR
☎ 86 25 5411 776
✆ 86 25 5411 872

Wang Yi-ming
NRL
Code 4172 W
4555 Overlook Av SW
Washington DC 20375 5000
USA
☎ 1 202 404 8460
✆ 1 202 404 7997

Wang Yiming
Yunnan Observ
CAS
Kunming 650011
China PR
☎ 86 871 2035
✆ 86 871 717 1845

Wang Yong
Box 846
Nanjing 210042
China PR
☎ 86 25 550 7485
✆ 86 25 550 7872
✉ ylab@bao01.bao.ac.cn

Wang Zhengming
Shaanxi Observ
CAS
Box 18
Lintong 710600
China PR
☎ 86 33 2255
✆ 86 9237 3496

Wang Zhenru
Astronomy Dpt
Nanjing Univ
Nanjing 210093
China PR
☎ 86 25 37551*2685
✆ 86 25 330 2728
✉ zrwang@nju.edu.cn

Wang Zhenyi
Purple Mountain Obs
CAS
Nanjing 210008
China PR
☎ 86 25 46700
✆ 86 25 301 459

Wang Zhong
CfA
HCO/SAO MS 66
60 Garden St
Cambridge MA 02138 1516
USA
☎ 1 617 496 7632
✆ 1 617 495 7490
✉ zwang@cfa.harvard.edu

Wannier Peter Gregory
CALTECH/JPL
MS 169 506
4800 Oak Grove Dr
Pasadena CA 91109 8099
USA
☎ 1 818 354 3347
✆ 1 818 393 6030

Ward Henry
Dpt Phys/Astronomy
Univ of Glasgow
Glasgow G12 8QQ
UK
☎ 44 141 339 8855*4705
✆ 44 141 334 9029
✉ gw10@uk.ac.gla.ph.il

Ward Martin John
Inst of Astronomy
The Observatories
Madingley Rd
Cambridge CB3 0HA
UK
☎ 44 12 23 337 548
✆ 44 12 23 337 523

Ward Richard A
LLNL
L 13
Box 808
Livermore CA 94551 9900
USA
☎ 1 415 423 2679
✆ 1 415 422 4643
✉ raward@llnl.gov

Ward William R
CALTECH/JPL
MS 183 501
4800 Oak Grove Dr
Pasadena CA 91109 8099
USA
☎ 1 213 351 1321
✆ 1 818 393 6030

Wardle John F C
Physics Dpt
Brandeis Univ
Waltham MA 02154
USA
☎ 1 617 647 2889
✆ 1 781 736 2915

Warman Josef
Instituto Astronomia
UNAM
Apt 70 264
Mexico DF 04510
Mexico
☎ 52 5 622 3906
✆ 52 5 616 0653

Warmels Rein Herm
ESO
Karl Schwarzschildstr 2
D 85748 Garching
Germany
☎ 49 893 200 6292
✆ 49 893 202 362
✉ rwarmels@eso.org

Warner Brian
Astronomy Dpt
Univ of Cape Town
Pricate Bag
7700 Rondebosch
South Africa
☎ 27 21 650 2391
✉ warner@physci.uct.ac.za

Warner John W
Perkin Elmer Corp
MS 892
100 Wooster Heights Rd
Danbury CT 06810 7859
USA
☎ 1 203 796 7919

Warner Peter J
MRAO
Cavendish Laboratory
Madingley Rd
Cambridge CB3 0HE
UK
☎ 44 12 23 337 274
✆ 44 12 23 354 599

Warren Wayne H
NASA GSFC
Code 681
Greenbelt MD 20771
USA
☎ 1 301 286 8701*5419
✉ wayne.h.warren.1@gsfc.nasa.gov

Warwick James W
Radiophysics Corp
5475 Western Av
Boulder CO 80301
USA
☎ 1 303 447 9524

Warwick Robert S
Dpt Phys/X-Ray Astron
Univ Leicester
University Rd
Leicester LE1 7RH
UK
☎ 44 116 252 2073
℻ 44 116 250 182

Washimi Haruichi
Inst Atmospheric Res
Nagoya Univ
3-13 Honohara
Toyokawa Aichi 442
Japan
☎ 81 533 86 3154
℻ 81 533 86 0811

Wasserman Lawrence H
Lowell Observ
1400 W Mars Hill Rd
Box 1149
Flagstaff AZ 86001
USA
☎ 1 520 774 3358
℻ 1 520 774 3358
✉ lhw@lowell.edu

Wasson John T
Inst Geophys/Planets
UCLA
Los Angeles CA 90095 1567
USA
☎ 1 310 825 1986
℻ 1 310 206 3051

Watanabe Jun-ichi
National Astronomical Obs
NAOJ
Osawa Mitaka
Tokyo 181
Japan
☎ 81 422 34 3644
℻ 81 422 34 3810
✉ owatana@cl.mtk.nao.ac.jp

Watanabe Noriaki
Chiba Univ Commerce
1 3 1 Konodai
Ichikawa
Ishikawa 272
Japan
☎ 81 473 72 4111
℻ 81 473 75 1105
✉ watanabe@cuc.ac.jp

Watanabe Takashi
Inst Atmospheric Res
Nagoya Univ
3-13 Honohara
Toyokawa 442
Japan
☎ 81 533 86 3154
℻ 81 533 86 0811

Watanabe Tetsuya
Ntl Astronomical Obs
2 21 1 Osawa
Mitaka
Tokyo 181
Japan
☎ 81 422 34 3714
℻ 81 422 34 3700
✉ watanabe@uvlab.mtk.nao.ac.jp

Waters Laurens B F M
SRON
Univ Groningen
Postbus 800
NL 9700 AV Groningen
Netherlands
☎ 31 50 363 4090
℻ 31 50 363 4033
✉ rensw@rug.nl

Waterworth Michael
School of Physics
QLD Univ of Techn
Gpo Box 2434
Brisbane QLD 4001
Australia

Watson Frederick Garnett
AAO
Private Bag
Coonabarabran NSW 2357
Australia
☎ 61 68 421 622
℻ 61 68 422 288
✉ FGW@aaocbn1.aao.gov.au

Watson Micheal G
Dpt Phys/X-Ray Astron
Univ Leicester
University Rd
Leicester LE1 7RH
UK
☎ 44 116 252 3491
℻ 44 116 250 182
✉ ngw@uk.ac.le.star

Watson Robert
IAC
Observ d Teide
via Lactea s/n
E 38200 La Laguna
Spain
☎ 34 2 260 5276
℻ 34 2 240 5646/405 501
✉ raw@ll.iac.es

Watson Robert
Physics Dpt
Univ Tasmania
GPO Box 252c
Hobart TAS 7001
Australia
☎ 61 2 202 415
℻ 61 2 202 410

Watson William D
Physics Dpt
Univ of Illinois
1110 W Green St
Urbana IL 61801 3080
USA
☎ 1 217 333 7240
℻ 1 217 333 9819

Watt Graeme David
Joint Astronomy Ctr
660 N A'ohoku Pl
University Park
Hilo HI 96720
USA
☎ 1 808 961 3756
℻ 1 808 961 6516
✉ gdw@jach.hawaii.edu

Wayman Patrick A
Glebe Cottage
Glebe Av
Wicklow
Ireland
☎ 353 4 046 9695
℻ 353 1 387 090
✉ pawayman@ccvax.ucd.ie

Wdowiak Thomas J
Physics Dpt
Univ of Alabama
Birmingham AL 35294
USA
☎ 1 205 934 4736

Weaver Harold F
STScI
Homewood Campus
3700 San Martin Dr
Baltimore MD 21218
USA
☎ 1 301 338 5004
℻ 1 301 338 4767
✉ weaver@stsci.edu

Weaver Thomas A
884 Holly Hill Dr
Walnut Creek CA 94596
USA
☎ 1 415 423 1850
✉ weaver@kepler.llnl.gov

Weaver William Bruce
Monterey Inst Res
in Astronomy
200 Eighth St
Marina CA 93933
USA
☎ 1 408 883 1000
℻ 1 408 883 1031
✉ lc@mira.org

Webb David F
USAF Phillips Lab/GPSG
Boston College
Hanscom AFB MA 01731 3010
USA
☎ 1 617 377 3970
℻ 1 617 377 3160

Webb John
School of Physics
UNSW
BOX 1
Kensington NSW 2033
Australia
☎ 61 6 268 8801
℻ 61 2 663 3420
✉ jkw@edwin.phys.unisw.edu.au

Webber John C
NRAO
520 Edgemont Rd
Charlottesville VA 22903
USA
☎ 1 804 296 0287
✉ jwebber@nrao.edu

Webbink Ronald F
Dpt Astron/Physics
Univ of Illinois
1011 W Springfield Av
Urbana IL 61801
USA
☎ 1 217 333 9582
℻ 1 217 244 7638

Weber Joseph
Physics Dpt
Univ of Maryland
College Park MD 20742 2421
USA
☎ 1 301 405 6081
℻ 1 301 314 9525
✉ jw116@umail.umd.edu

Weber Stephen Vance
LLNL
L 477
Box 808
Livermore CA 94551 9900
USA
☎ 1 415 422 5433
℻ 1 415 423 0238
✉ svweber@llnl.gov

Webrova Ludmila
Na Malem Klinu 16
CZ 182 00 Praha 8
Czech R
☎ 420 2 841 301

Webster Adrian S
Joint Astronomy Ctr
660 N A'ohoku Pl
University Park
Hilo HI 96720
USA
☎ 1 808 961 3756
℻ 1 808 961 6516

Webster Rachel
School of Physics
Univ of Melbourne
Parkville VIC 3052
Australia
☎ 61 33 475 450
℻ 61 3 347 4783
✉ webster@tauon.ph.unimelb.edu.a

Weedman Daniel W
Astronomy Dpt
Pennsylvania State Univ
525 Davey Lab
University Park PA 16802
USA
☎ 1 814 865 0418
✆ 1 814 863 7114

Weekes Trevor C
FLWO
CfA
Box 97
Amado AZ 85645 0097
USA
☎ 1 617 495 7461
✆ 1 617 495 7326

Wegner Gary Alan
Dpt Phys/Astronomy
Dartmouth College
6127 Wilder Lab
Hanover NH 03755
USA
☎ 1 603 646 2359
✆ 1 603 646 1446
✉ gary.wegner@dartmouth.edu

Wehinger Peter A
Phys/Astronomy Dpt
Arizona State Univ
Astronomy Program
Tempe AZ 85287 1504
USA
☎ 1 602 965 4063
✆ 1 602 727 6019
✉ peter.wehinger@asu.edu

Wehlau Amelia
Dpt Phys/Astronomy
Univ W Ontario
London ON N6A 3K7
Canada
☎ 1 519 679 3186
✆ 1 519 661 3486
✉ afwehlau@nve.uwo.ca

Wehrle Ann Elizabeth
CALTECH
MS 100 22
IPAC
Pasadena CA 91125
USA
☎ 1 818 397 9588
✆ 1 818 397 9600
✉ aew@ipac.caltech.edu

Wehrse Rainer
Inst Theor Astrophysik
d Univer
Im Neuenheimer Feld 561
D 69120 Heidelberg
Germany
☎ 49 622 156 2837
✆ 49 622 154 4221
✉ b28@rw.iwe.uni-heidelberg.de

Wei Mingzhi
Lick Observ
Univ of California
Santa Cruz CA 95064
USA
☎ 1 831 459 4911
✆ 1 831 426 3115
✉ wmz@ucscloa.ucsc.edu

Weidemann Volker
Inst Theor Phys/
Sternwarte Univ Kiel
Olshausenstr 40
D 24098 Kiel
Germany
☎ 49 431 880 4110
✆ 49 431 880 4432
✉ supas058@astrophysik.
uni-kiel.d400.de

Weidenschilling S J
Planetary Science Inst
620 N Sixth Av
Tucson AZ 85705 8331
USA
☎ 1 520 662 6300
✆ 1 520 622 8060
✉ sjw@psi.edu

Weigelt Gerd
RAIUB
Univ Bonn
auf d Huegel 69
D 53121 Bonn
Germany
☎ 49 228 525 243
✆ 49 228 525 229
✉ p561gwe@mpifr-
bonn.mpg.de

Weiler Edward J
NASA Headquarters
Code SZ
600 Independence Av SW
Washington DC 20546
USA
☎ 1 202 358 0370
✆ 1 202 358 3096
✉ eweiler@gm.ossa.hq.
nasa.gov

Weiler Kurt W
NRL
Code 4131
4555 Overlook Av SW
Washington DC 20375 5000
USA
☎ 1 202 767 0292
✆ 1 202 404 8894
✉ kweiler@shimmer.nrl.
navy.mil

Weill Gilbert M
Spot Image Corp
1897 Preston White Dr
Reston VA 22091 4326
USA
☎ 1 703 620 2200

Weimer Theophile P F
Observ Paris
61 Av Observatoire
F 75014 Paris
France
☎ 33 1 43 20 1210
✆ 33 1 44 54 1804

Weinberg J L
MK Industries Inc
2137 E Flintsone Dr
Tucker GA 30084
USA
☎ 1 770 491 8700
✆ 1 770 491 9109

Weinberg Steven
Astronomy Dpt
Univ of Texas
Rlm 15 308
Austin TX 78712 1083
USA
☎ 1 512 471 4394
✆ 1 512 471 6016

Weinberger Ronald
Inst Astronomie
Technikerstr 15
A 6020 Innsbruck
Austria
☎ 43 512 507 6035
✆ 43 512 507 2923
✉ Ronald.Weinberger@uibk.
ac.at

Weis Edward W
Astronomy Dpt
Van Vleck Observ
Wesleyan Univ
Middletown CT 06457
USA
☎ 1 203 347 9411
✆ 1 860 865 2131

Weisberg Joel Mark
Dpt Phys/Astronomy
Carleton College
Northfield MN 55057
USA
☎ 1 507 663 4367

Weisheit Jon C
Dpt Phys/Astronomy
Rice Univ
Box 1892
Houston TX 77251 1892
USA
☎ 1 713 527 4654
✆ 1 713 285 5143
✉ jonw@zeno.rice.edu

Weiss Achim
MPA
Karl Schwarzschildstr 1
D 85748 Garching
Germany
☎ 49 893 299 00
✆ 49 893 299 3235
✉ acw@dgaippls

Weiss Nigel O
Dpt Appl Maths/Theor Phys
Silver Street
Cambridge CB3 9EW
UK
☎ 44 12 23 351 45
✆ 44 12 23 337 918

Weiss Werner W
Inst Astronomie
Univ Wien
Tuerkenschanzstr 17
A 1180 Wien
Austria
☎ 43 1 470 6800/7*6683
✆ 43 1 470 6015
✉ weiss@galileo.ast.univie.ac.at

Weisskopf Martin Ch
NASA/MSFC
Space Science Lab
Code ES 01
Huntsville AL 35812
USA
☎ 1 205 544 7740
✆ 1 205 547 7754
✉ weisskopf@sslmsfc.
nasa.gov

Weissman Paul Robert
CALTECH/JPL
MS 183 601
4800 Oak Grove Dr
Pasadena CA 91109 8099
USA
☎ 1 818 354 2636
✆ 1 818 393 6030

Weistrop Donna
Physics Dpt
Univ of Nevada
4505 S Maryland Parkway
Las Vegas NV 89154
USA
☎ 1 702 739 3653
✆ 1 702 739 0804
✉ weistrop@nevada.edu

Welch Douglas L
Dpt Phys/Astronomy
McMaster Univ
Hamilton ON L8S 4M1
Canada
☎ 1 905 525 9140*23186
✆ 1 905 546 1252
✉ welch@physics.mcmaster.ca

Welch Gary A
Astronomy Dpt
St Mary's Univ
Halifax NS B3H 3C3
Canada
☎ 1 902 429 9780
✆ 1 902 420 5561

Welch William J
Radio Astronomy Lab
Univ of California
601 Campbell Hall
Berkeley CA 94720
USA
☎ 1 415 642 6679
✆ 1 415 642 6424

Weller Charles S
David Taylor Res Ctr
Code 7200
9500 MacArthur Blvd
W Bethesda MD 20817 5700
USA
✉ weller@oasys.dt.navy.mil

Wellgate G Bernard
Caneheath House
Arlington
Polegate BN26 6SJ
UK

Wellington Kelvin
CSIRO
Div Radiophysics
Box 76
Epping NSW 2121
Australia
☎ 61 2 9372 4375
✆ 61 2 9372 4430
✉ kwelling@rp.csiro.au

Wellmann Peter
Inst Astron/Astrophysik
Univ Sternwarte
Scheinerstr 1
D 81679 Muenchen
Germany
☎ 49 899 890 21
✆ 49 899 220 9427

Wells Donald C
NRAO
520 Edgemont Rd
Charlottesville VA 22903
USA
☎ 1 804 296 0277
✆ 1 804 296 0278
✉ dwells@nrao.edu

Wells Eddie Neil
STScI
Homewood Campus
3700 San Martin Dr
Baltimore MD 21218
USA
☎ 1 301 338 4788
✆ 1 301 338 4767
✉ wells@stsci.edu

Wendker Heinrich J
Hamburger Sternwarte
Univ Hamburg
Gojensbergsweg 112
D 21029 Hamburg
Germany
☎ 49 407 252 4112
✆ 49 407 252 4198

Wenger Marc
Observ Strasbourg
11 r Universite
F 67000 Strasbourg
France
☎ 33 3 88 15 0761
✆ 33 3 88 25 0160
✉ wenger@aastro.u-strasbg.fr

Weniger Schame
23bis r R Schuman
F 94270 Kremlin Bicetre
France

Wentzel Donat G
Astronomy Program
Univ of Maryland
College Park MD 20742 2421
USA
☎ 1 301 405 1518
✆ 1 301 314 9067
✉ wentzel@astro.umd.edu

Wenzel W
Zntrlinst Astrophysik
Sternwarte Sonneberg
an d Sternwarte 16
D 96575 Sonneberg
Germany
☎ 49 96 74 2287

Werner Klaus
Inst Theor Phys/
Sternwarte Univ Kiel
Olshausenstr 40
D 24098 Kiel
Germany
☎ 49 431 880 4106
✆ 49 431 880 4100
✉ supas075@astrophysik.uni-kiel.d400.de

Wesemael Francois
Dpt d Physique
Universite Montreal
CP 6128 Succ A
Montreal QC H3C 3J7
Canada
☎ 1 514 343 7355
✆ 1 514 343 2071

Wesselius Paul R
SRON
Univ Groningen
Postbus 800
NL 9700 AV Groningen
Netherlands
☎ 31 50 363 4074
✆ 31 50 363 4033
✉ paul@guspace.rug.nl

Wesson Paul S
Physics Dpt
Univ Waterloo
Waterloo ON N2L 3G1
Canada
☎ 1 519 885 1211
✆ 1 519 746 8115

West Michael
Dpt Astron/Physics
Saint Mary's Univ
Halifax
Nova Scotia B3H 3C3
Canada
☎ 1 902 496 8175
✆ 1 902 4220 5141
✉ west@sisyphus.stmarys.ca

West Richard M
ESO
Karl Schwarzschildstr 2
D 85748 Garching
Germany
☎ 49 893 200 6276
✆ 49 89 320 2362
✉ rwest@eso.org

Westergaard Niels J
Danish Space Res Inst
Juliane Maries Vej 30
DK 2100 Copenhagen
Denmark
☎ 45 35 32 5705
✆ 45 35 36 2475
✉ njw@dsri.dk

Westerhout Gart
811 West 38th Str
Baltimore MD 21211 2203
USA
☎ 1 301 235 5834

Westerlund Bengt E
Astronomical Observ
Box 515
S 751 20 Uppsala
Sweden
☎ 46 18 13 5157
✆ 46 18 52 7583

Westfold Kevin C
7 Beamsley St
Malvern
Victoria 3144
Australia

Westphal James A
CALTECH
MS 170 25
Pasadena CA 91125
USA
☎ 1 818 395 4900
✆ 1 818 585 1917
✉ jaw@caltech.edu

Wetherill George W
Dpt Terrestr Magnetism
Carnegie Inst Washington
5241 Bd Branch Rd NW
Washington DC 20015 1305
USA
☎ 1 202 686 4370 *4375
✆ 1 202 364 8726
✉ wetherill@eros.ciw.edu

Weymann Ray J
Carnegie Observatories
813 Santa Barbara St
Pasadena CA 91101 1292
USA
☎ 1 818 577 1122
✆ 1 818 795 8136

Wheeler J Craig
Astronomy Dpt
Univ of Texas
Rlm 15 308
Austin TX 78712 1083
USA
☎ 1 512 471 6407
✆ 1 512 471 6016
✉ wheel@astro.as.texas.edu

Wheeler John A
Physics Dpt
Princeton Univ
Jadwin Hall
Princeton NJ 08544 1001
USA
☎ 1 609 258 4400
✆ 1 609 258 1124
✉ dwns@mso.anu.oz.au

Whipple Arthur L
McDonald Observ
Univ of Texas
Rlm 15 308
Austin TX 78712 1083
USA
☎ 1 512 471 6332
✆ 1 512 471 6016

Whipple Fred L
CfA
HCO/SAO
60 Garden St
Cambridge MA 02138 1516
USA
☎ 1 617 495 7200
✆ 1 617 495 7356

Whitaker Ewen A
Lunar/Planetary Lab
Room 325 Bg 92
Univ of Arizona
Tucson AZ 85721 0092
USA
☎ 1 520 621 2888
✆ 1 520 621 4933

White Glenn J
Astrophysics Group
QMWC
Mile End Rd
London E1 4NS
UK
☎ 44 171 975 5555
✆ 44 181 975 5500

White Graeme Lindsay
Fac Science/Tech
UWS
Box 10
Kingswood NSW 2747
Australia
☎ 61 47 360 835
✆ 61 47 3360 779
✉ gwhite@st.nepean.uws.edu.au

White Nathaniel M
Lowell Observ
1400 W Mars Hill Rd
Box 1149
Flagstaff AZ 86001
USA
☎ 1 602 774 3358
✆ 1 520 774 3358
✉ nmw@lowell.edu

White Nicholas Ernest
NASA/GSFC
Code 662
Greenbelt MD 20771
USA
☎ 1 301 286 8443
✆ 1 301 286 1684
✉ white@adhoc.gsfc.nasa.gov

White Oran R
HAO
NCAR
Box 3000
Boulder CO 80307 3000
USA
☎ 1 303 497 1000
✆ 1 303 497 1568
✉ white@hao.ucar.edu

White R Stephen
IGPP
Univ of California
Riverside CA 92521
USA
☎ 1 714 787 4503

White Raymond E
Steward Observ
Univ of Arizona
Tucson AZ 85721
USA
☎ 1 520 621 6528
✆ 1 520 621 1532
✉ rwhite@as.arizona.edu

White Raymond Edwin III
Dpt Phys/Astronomy
Univ of Alabama
Box 870324
Tuscaloosa AL 35487 0324
USA
☎ 1 205 348 1640
℻ 1 520 621 1532
✉ white@merkin.astr.ua.edu

White Richard Allan
NASA/GSFC
Code 631
Greenbelt MD 20771
USA
☎ 1 301 286 7802
℻ 1 301 286 1771
✉ richard.a.white.l@gsfc.nasa.go v

White Richard E
Astronomy Dpt
Smith College
Clark Science Ctr
Northampton MA 01063
USA
☎ 1 413 584 2700

White Richard L
STScI
Homewood Campus
3700 San Martin Dr
Baltimore MD 21218
USA
☎ 1 301 338 4797
℻ 1 301 338 4767

White Simon David Manion
Inst of Astronomy
The Observatories
Madingley Rd
Cambridge CB3 0HA
UK
☎ 44 12 23 337 548
℻ 44 12 23 337 523
✉ swhite@ast.star.camm.ac.uk

White Stephen Mark
Astronomy Dpt
Univ of Maryland
College Park MD 20742 2421
USA
☎ 1 301 405 1547
℻ 1 301 314 9067
✉ white@astro.umd.edu

Whitelock Patricia Ann
SAAO
PO Box 9
7935 Observatory
South Africa
☎ 27 21 47 0025
℻ 27 21 47 3639
✉ paw@saao.ac.za

Whiteoak John B
CSIRO
ATNF
Box 76
Epping 2121 NSW
Australia
☎ 61 2 9372 4110
℻ 61 2 9372 4310
✉ jwhiteoa@atnf.csiro.au

Whitford Albert E
Lick Observ
Univ of California
Santa Cruz CA 95064
USA
☎ 1 831 429 2149
℻ 1 831 426 3115

Whitmore Bradley C
STScI
Homewood Campus
3700 San Martin Dr
Baltimore MD 21218
USA
☎ 1 301 338 4713
℻ 1 301 338 4767

Whitney Charles A
CfA
HCO/SAO
60 Garden St
Cambridge MA 02138 1516
USA
☎ 1 617 496 5405
℻ 1 617 495 7356
✉ cwhitney@cfa.harvard.edu

Whitrow Gerald James
41 Home Park Rd
Wimbledon
London SW19 7HS
UK
☎ 44 181 947 343 467

Whittet Douglas C B
Physics Dpt
Rensselaer Polytechn Inst
Troy NY 12180 3590
USA
☎ 1 518 276 6310
℻ 1 518 276 6680

Whittle D Mark
University Station
Univ of Virginia
Box 3818
Charlottesville VA 22903 0818
USA
☎ 1 864 924 4900
✉ dmw8f@virginia.edu

Whitworth Anthony Peter
Dpt Applied Maths/Astron
Univ College
Box 78
Cardiff CF1 1XL
UK
☎ 44 12 22 442 11

Wickramasinghe D T
Australian Ntl Univ
Applied Maths Dpt
Box 4
Canberra Act 2600
Australia

Wickramasinghe N C
Dpt Applied Maths/Astron
Univ College
Box 78
Cardiff CF1 1XL
UK
☎ 44 12 22 442 11

Widemann Thomas
Service Aeronomie
BP 3
F 91371 Verrieres Buisson
France
☎ 33 1 64 47 5256
℻ 33 1 69 20 2999
✉ widemann@aerov.jussieu.fr

Widing Kenneth G
NRL
Code 7144
4555 Overlook Av SW
Washington DC 20375 5000
USA
☎ 1 202 767 2605

Wiedling Tor
Ostra Villavagen 15
S 611 36 Nykoping
Sweden

Wiehr Eberhard
Univ Sternwarte
Goettingen
Geismarlandstr 11
D 37083 Goettingen
Germany
☎ 49 551 395 053
℻ 49 551 395 043

Wielebinski Richard
RAIUB
Univ Bonn
auf d Huegel 69
D 53121 Bonn
Germany
☎ 49 228 525 300
℻ 49 228 525 436

Wielen Roland
ARI
Moenchhofstr 12-14
D 69120 Heidelberg
Germany
☎ 49 622 140 5122
℻ 49 622 140 5297
✉ wielen@ari.uni-heidelberg.de

Wiese Wolfgang L
NIST
Physics Division
270/Quince Orchard Rd
Gaithersburg MD 20899
USA
☎ 1 301 975 3201
℻ 1 301 975 3038
✉ wolfgang.wiese@nist.gov

Wiita Paul Joseph
Dpt Phys/Astronomy
Georgia State Univ
Atlanta GA 30303 3083
USA
☎ 1 404 651 1367
℻ 1 404 651 1389
✉ wiita@chara.gsu.edu

Wijnbergen Jan
Kapteyn Sterrekundig Inst
Univ Groningen
Postbus 800
NL 9700 AV Groningen
Netherlands
☎ 31 50 363 4073
℻ 31 50 363 6100

Wiklind Tommy
Onsala Space Observ
Chalmers Technical Univ
S 439 92 Onsala
Sweden
☎ 46 31 772 5537
℻ 46 31 772 5590
✉ tommy@oso.chalmers.se

Wild John Paul
CSIRO
Limestone Av
Box 225
Dickson ACT 2602
Australia
☎ 61 62 484 595

Wild Paul
Astronomisches Institut
Univ Bern
Sidlerstr 5
CH 3012 Bern
Switzerland
☎ 41 31 631 3892
℻ 41 31 631 3869

Wilkening Laurel L
Univ of Arizona
Admin Bg 601
Tucson AZ 85721
USA
☎ 1 520 626 3513

Wilkes Belinda J
CfA
HCO/SAO
60 Garden St
Cambridge MA 02138 1516
USA
☎ 1 617 495 7268
℻ 1 617 495 7356

Wilkins George A
Windward Higher Brook Meadow
Sidford
Sidmouth Devon EX10 9SS
UK
☎ 44 139 557 9641
✉ g.a.wilkins@exeter.ac.uk

Wilkinson Althea
Astronomy Dpt
Univ Manchester
Oxford Rd
Manchester M13 9PL
UK
☎ 44 161 275 4224
℻ 44 161 275 4223

Wilkinson David T
Physics Dpt
Princeton Univ
Jadwin Hall
Princeton NJ 08544 1001
USA
☎ 1 609 452 4406
℻ 1 609 258 6853

Wilkinson Peter N
NRAL
Univ Manchester
Jodrell Bank
Macclesfield SK11 9DL
UK
☎ 44 14 777 1321
℻ 44 147 757 1618

Will Clifford M
Physics Dpt
Washington Univ
Campus Box 1105
St Louis MO 63130
USA
☎ 1 314 935 6244
✆ 1 314 935 6219
🖂 cmw@wuphys.wustl.edu

Williamon Richard M
Fernbank Sci Ctr
156 Heaton Park Dr
156 Heaton Park Dr
Atlanta GA 30307
USA
☎ 1 404 378 4313

Williams Barbara A
Physics Dpt
Univ of Delaware
Newark DE 19716
USA
☎ 1 302 451 6526
✆ 1 302 831 1637

Williams Carol A
Mathematics Dpt
Univ of S Florida
4202 E Fowler Av
Tamp FL 33620 5700
USA
☎ 1 813 974 2643
✆ 1 813 974 2700
🖂 cw@kepler.math.usf.edu

Williams David A
Dpt Phys/Astronomy
UCLO
Gower St
London WC1E 6BT
UK
☎ 44 171 387 7050
✆ 44 171 380 7145
🖂 daw@star.ucl.ac.uk

Williams Glen A
Physics Dpt
Central Michigan Univ
Mt Pleasant MI 48858
USA
☎ 1 517 774 3365
✆ 1 517 774 2697
🖂 32nsqsv@cmuvm

Williams Iwan P
Astronomy Unit
QMWC
Mile End Rd
London E1 4NS
UK
☎ 44 171 975 5452
✆ 44 181 983 3522
🖂 i.p.williams@qmw.ac.uk

Williams James G
CALTECH/JPL
MS 264 700
4800 Oak Grove Dr
Pasadena CA 91109 8099
USA
☎ 1 818 354 6466
✆ 1 818 393 6030
🖂 jgw@logos.jpl.nasa.gov

Williams John A
Physics Dpt
Albion College
Albion MI 49224
USA
☎ 1 517 629 5511

Williams Peredur M
Royal Observ
Blackford Hill
Edinburgh EH9 3HJ
UK
☎ 44 131 667 3321
✆ 44 131 668 8356

Williams Robert E
STScI
Homewood Campus
3700 San Martin Dr
Baltimore MD 21218
USA
☎ 1 301 338 4710
✆ 1 301 338 2519
🖂 wms@stsci.edu

Williams Robin
Physics Dpt
Univ of Leeds
Leeds LS2 9JT
UK
☎ 44 113 233 3860
✆ 44 113 233 3846
🖂 rjrw@ast.leeds.ac.uk

Williams Theodore B
Dpt Phys/Astronomy
Rutgers Univ
Box 849
Piscataway NJ 08854 0849
USA
☎ 1 201 932 2516
✆ 1 908 445 4343

Willis Allan J
Dpt Phys/Astronomy
UCLO
Gower St
London WC1E 6BT
UK
☎ 44 171 387 7050
✆ 44 171 380 7145
🖂 ajw@uk.ac.ucl.star

Willis Anthony Gordon
DRAO
NRCC/HIA
Box 248
Penticton BC V2A 6K3
Canada
☎ 1 250 493 2277
✆ 1 250 493 7767

Willmer Christopher N A
Observ Nacional
Rua Gl Bruce 586
Sao Cristovao
20921 030 Rio de Janeiro RJ
Brazil
☎ 55 21 589 6504
✆ 55 21 580 0332
🖂 chaw@on.br

Willmore A Peter
Dpt Space Res
Univ Birmingham
Box 363
Birmingham B15 2TT
UK
☎ 44 12 14 72 1301

Willner Steven Paul
CfA
HCO/SAO
60 Garden St
Cambridge MA 02138 1516
USA
☎ 1 617 495 7123
✆ 1 617 495 7490
🖂 swilner@cfa.harvard.edu

Wills Beverley J
Astronomy Dpt
Univ of Texas
Rlm 15 308
Austin TX 78712 1083
USA
☎ 1 512 471 3424
✆ 1 512 471 6016

Wills Derek
Astronomy Dpt
Univ of Texas
Rlm 15 308
Austin TX 78712 1083
USA
☎ 1 512 471 4461
✆ 1 512 471 6016

Willson Lee Anne
Physics Dpt
Iowa State Univ
Ames IA 50011
USA
☎ 1 515 294 6765
✆ 1 515 294 3262

Willson Robert Frederick
Dpt Phys/Astronomy
Tufts Univ
Robinson Hall
Medford MA 02155
USA
☎ 1 617 628 5000

Willstrop Roderick V
Inst of Astronomy
The Observatories
Madingley Rd
Cambridge CB3 0HA
UK
☎ 44 12 23 337 548
✆ 44 12 23 337 523
🖂 rvw@ast.cam.ac.uk

Wilner David J
CfA
HCO/SAO
60 Garden St
Cambridge MA 02138 1516
USA
☎ 1 617 496 7623
✆ 1 617 495 7345
🖂 dwilner@cfa.harvard.edu

Wilson Albert G
Box 1871
Sebastopol CA 95473
USA

Wilson Andrew S
Astronomy Program
Univ of Maryland
College Park MD 20742 2421
USA
☎ 1 301 405 1519
✆ 1 301 314 9067

Wilson Brian G
Univ of Queensland
55 Walcott St
St Lucia QLD 4067
Australia
☎ 61 73 772 200

Wilson Christine
Dpt Astron/Physics
McMaster Univ
Hamilton
Ontario L8S 4MI
Canada
☎ 1 905 525 9140*27483
✆ 1 905 546 1252
🖂 wilson@physics.mcmaster.ca

Wilson Curtis A
St John's College
Box 2880
Annapolis MD 21404
USA
☎ 1 301 263 2371

Wilson James R
737 South M
Livermore CA 94550
USA
☎ 1 415 422 1659
✆ 1 412 423 0238

Wilson Lionel
Env Science Dpt
Lancaster Univ
Lancaster LA1 4YQ
UK
☎ 44 15 246 5201*4075

Wilson Michael John
Dpt Appl Maths
Univ of Leeds
Leeds L52 9JT
UK
☎ 44 113 233 5110
✆ 44 113 242 9525

Wilson P
Geoforschungs Zentrum
Telegrafenberg A17
D 14407 Potsdam
Germany

Wilson Peter R
Applied Maths Dpt
Univ of Sydney
Sydney NSW 2006
Australia
☎ 61 2 692 2222
✆ 61 2 660 2903

Wilson Richard
CfA
HCO/SAO MS 42
60 Garden St
Cambridge MA 02138 1516
USA
☎ 1 617 495 7000

Wilson Robert
Dpt Phys/Astronomy
UCLO
Gower St
London WC1E 6BT
UK
☎ 44 171 380 7154
📠 44 171 380 7145
✉ rw@uk.ac.ucl.star

Wilson Robert E
Astronomy Dpt
Univ of Florida
211 SSRB
Gainesville FL 32611
USA
☎ 1 904 392 1182
📠 1 904 392 5089

Wilson Robert W
CfA
HCO/SAO MS 42
60 Garden St
Cambridge MA 02138 1516
USA
☎ 1 617 495 7000

Wilson S J
Mathematics Dpt
Ntl Univ Singapore
Kent Ridge
Singapore 0511
Singapore
☎ 65 775 6666

Wilson Thomas L
RAIUB
Univ Bonn
auf d Huegel 69
D 53121 Bonn
Germany
☎ 49 228 525 378
📠 49 228 525 229
✉ p073twi@mpifr-bonn.mpg.ge

Wilson William J
CALTECH/JPL
MS 168 327
4800 Oak Grove Dr
Pasadena CA 91109 8099
USA
☎ 1 818 354 5699
📠 1 818 393 6030

Winckler John R
School Phys/Astronomy
Univ Minnesota
116 Church St SE
Minneapolis MN 55455
USA
☎ 1 612 373 4688
📠 1 612 624 2029

Windhorst Rogier A
Physics Dpt
Arizona State Univ
Astronomy Program
Tempe AZ 85287 1504
USA
☎ 1 602 965 7143
📠 1 605 965 7954

Wing Robert F
Astronomy Dpt
Ohio State Univ
174 W 18th Av
Columbus OH 43210 1106
USA
☎ 1 614 422 7876
📠 1 614 292 2928
✉ ts4718@ohstmvsa

Winget Donald E
Astronomy Dpt
Univ of Texas
Rlm 15 308
Austin TX 78712 1083
USA
☎ 1 512 471 4461
📠 1 512 471 6016

Winiarski Maciej
Astronomical Observ
Krakow Jagiellonian Univ
Ul Orla 171
PL 30 244 Krakow
Poland
☎ 48 12 25 1294
📠 48 12 25 1318

Wink Joern Erhard
IRAM
300 r La Piscine
F 38406 S Martin Heres Cdx
France
☎ 33 4 76 42 3383
📠 33 4 76 51 5938

Winkler Christoph
ESA/ESTEC
SSD
Box 299
NL 2200 AG Noordwijk
Netherlands
☎ 31 71 565 3591
📠 31 71 565 4690
✉ cwinkler@estsa2.estec.
esa.nl

Winkler Gernot M R
USNO
Time Service Dpt
3450 Massachusetts Av NW
Washington DC 20392 5100
USA
☎ 1 202 762 1450
📠 1 202 653 1507
✉ gw@cassini.usno.navy.mil

Winkler Karl-Heinz A
LANL
MS B260
Box 1663
Los Alamos NM 87545 2345
USA
☎ 1 505 667 2897
📠 1 505 665 4055
✉ khw@lanl.gov

Winkler Paul Frank
Physics Dpt
Middlebury Coll
Middlebury VT 05753
USA
☎ 1 802 388 3711
📠 1 802 388 0739

Winnberg Anders
Onsala Space Observ
Chalmers Technical Univ
S 439 92 Onsala
Sweden
☎ 46 31 772 5527
📠 46 31 772 5590
✉ anders@oso.chalmers.se `

Winnewisser Gisbert
Erstes Physikal Institut
Univ Koeln
Zuelpicherstr 77
D 50937 Koeln
Germany
☎ 49 211 470 3567
📠 49 221 470 5162

Winter Othon Cabo
DMA UNESP
CP 205
12500 000 Guaratingueta
Brazil
☎ 55 12 525 2800*105
📠 55 12 525 2466
✉ ocwinter@feg.unesp.br

Wiramihardja Suhardja
Bosscha Observ
ITB Lembang
Lembang 40391
Indonesia
☎ 62 22 229 6001
📠 62 22 278 7289

Wisotzki Lutz
Hamburger Sternwarte
Univ Hamburg
Gojensbergsweg 112
D 21029 Hamburg
Germany
☎ 49 407 252 4139
📠 49 407 252 4198
✉ lwisotzki@hs.uni-hamburg.de

Withbroe George L
CfA
HCO/SAO
60 Garden St
Cambridge MA 02138 1516
USA
☎ 1 617 495 7438
📠 1 617 495 7356

Witt Adolf N
Dpt Phys/Astronomy
Univ of Toledo
2801 W Bancroft St
Toledo OH 43606
USA
☎ 1 419 530 2709
📠 1 419 530 2723
✉ awitt@anwsun.astro.utoledo.ed
u

Witten Louis
Physics Dpt
Univ of Cincinnati
210 Braunstein Ml 11
Cincinnati OH 45221 0111
USA
☎ 1 513 475 6492
📠 1 513 556 3425

Wittmann Axel D
Univ Sternwarte
Goettingen
Geismarlandstr 11
D 37083 Goettingen
Germany
☎ 49 551 395 042
📠 49 551 395 043

Witzel Arno
RAIUB
Univ Bonn
auf d Huegel 69
D 53121 Bonn
Germany
☎ 49 228 525 211
📠 49 228 525 229

Wiyanto Paulus
Astronomy Dpt
Bandung Instte Techn
Jl Ganesha 10
Bandung 40132
Indonesia
☎ 62 22 244 0252
📠 62 22 2505442

Wlerick Gerard
Observ Paris Meudon
DASGAL
Pl J Janssen
F 92195 Meudon PPL Cdx
France
☎ 33 1 45 07 2240
📠 33 1 45 07 7971

Wnuk Edwin
Astronomical Observ
A Mickiewicz Univ
Ul Sloneczna 36
PL 60 286 Poznan
Poland
☎ 48 61 67 9670
📠 48 61 53 6536

Woan Graham
Dpt Phys/Astronomy
Univ of Glasgow
Glasgow G12 8QQ
UK
☎ 44 141 339 8855
📠 44 141 330 5183
✉ graham@astro.gla.ac.uk

Woehl Hubertus
Kiepenheuer Inst
Sonnenphysik
Schoeneckstr 6
D 79104 Freiburg Breisgau
Germany
☎ 49 761 319 8258
📠 49 761 3198 111
✉ hw@kis.uni-freiburg.de

Wolf Bernhard
Landessternwarte
Koenigstuhl
D 69117 Heidelberg
Germany
☎ 49 622 150 9213
📠 49 622 150 9202
✉ bwolf@hp2.uni-
heidelberg.de

Wolf Marek
Astronomical Institute
Charles Univ
V Holesovickack 2
CZ 180 00 Praha 8
Czech R
☎ 420 2 2191 2572
✆ 420 2 688 5095
✉ wolf@mbox.cesnet.cz

Wolf Rainer E A
MPI Astronomie
Koenigstuhl
D 69117 Heidelberg
Germany
☎ 49 622 152 81
✆ 49 622 152 8246

Wolfe Arthur M
CASS
UCSD
C 011
La Jolla CA 92093 0216
USA
☎ 1 619 534 7435
✆ 1 619 534 6316
✉ awolfe@ucsd

Wolfendale Arnold W
Physics Dpt
Univ of Durham
South Rd
Durham DH1 3LE
UK
☎ 44 191 3742 160
✆ 44 191 374 3749
✉ a.w.wolfendale@durham.ac.uk

Wolff Sidney C
NOAO/KPNO
Box 26732
950 N Cherry Av
Tucson AZ 85726 6732
USA
☎ 1 520 327 5511
✆ 1 520 325 9360

Wolfire Mark Guy
2205 Sulgrave Av
Baltimore MD 21209
USA
✉ mworlfire@nasm.edu

Wolfson C Jacob
Lockheed Palo Alto Res Lb
Dpt 91 30 Bg 256
3251 Hanover St
Palo Alto CA 94304 1191
USA
☎ 1 415 424 2855
✆ 1 415 424 3994

Wolfson Richard
Physics Dpt
Middlebury Coll
Middlebury VT 05753
USA
☎ 1 802 388 3711
✆ 1 802 388 0739
✉ wolfson@midd

Wolstencroft Ramon D
Royal Observ
Blackford Hill
Edinburgh EH9 3HJ
UK
☎ 44 131 668 9307
✆ 44 131 662 1668
✉ rdw@roe.ac.uk

Wolszczan Alexander
Arecibo Observ
Box 995
Arecibo PR 00613
USA
☎ 1 809 878 2612

Wolter Anna
Osserv Astronomico
d Brera
Via Brera 28
I 20121 Milano
Italy
☎ 39 2 723 20321
✆ 39 2 720 01600
✉ anna@brera.mi.astro.it

Woltjer Lodewijk
OHP
F 04870 S Michel Obs
France
☎ 33 4 92 76 6662
✆ 33 4 92 76 6295
41 22 788 3551

Woo Jong Ok
Korean Ntl Univ
of Education
Chungwon-gun
Chungbuk 320 23
Korea RP
☎ 82 431 60 3712

Wood Douglas O S
55 Whitney Ridge Terrace
North Haven CT 06473
USA
✉ dougwood@kodak.com

Wood H J
Astronomy Dpt
Indiana Univ
Swain W 319
Bloomington IN 47405
USA
☎ 1 812 855 6911
✆ 1 812 855 8725

Wood Janet H
Physics Dpt
Univ of Keele
Keele ST5 5BG
UK
☎ 44 178 262 1111
✆ 44 178 271 1093
✉ jhw@uk.ac.keele.
ph.starlink

Wood John A
CfA
HCO/SAO
60 Garden St
Cambridge MA 02138 1516
USA
☎ 1 617 495 7278
✆ 1 617 495 7356
✉ jwood@cfa.harvard.edu

Wood Matthew Alan
Dpt Phys/Space Sci
Florida Inst Techn
150 W University Blvd
Melbourne FL 32901
USA
☎ 1 407 768 8000*7207
✆ 1 407 984 8461
✉ wood@kepler.pss.f
it.edu

Wood Peter R
MSSSO
Weston Creek
Private Bag
Canberra ACT 2611
Australia
☎ 61 262 881 111
✆ 61 262 490 233

Wood Roger
Satellite Laser Ranger Gp
Herstmonceux Castle
Hailsham
East Sussex BN27 1RP
UK
☎ 44 132 383 3171*3391

Woodsworth Andrew W
NRCC/HIA
DAO
5071 W Saanich Rd
Victoria BC V8X 4M6
Canada
☎ 1 250 388 0024
✆ 1 250 363 0045
✉ wdswrth@dao.nrc.ca

Woodward Paul R
School Phys/Astronomy
Univ Minnesota
116 Church St SE
Minneapolis MN 55455
USA
☎ 1 612 624 0211
✆ 1 612 624 2029

Woolf Neville J
Steward Observ
Univ of Arizona
Tucson AZ 85721
USA
☎ 1 520 621 2288
✆ 1 520 621 1532

Woolfson Michael M
Physics Dpt
Univ of York
Heslington York YO1 5DD
UK
☎ 44 19 045 9861

Woolsey E G
1909 Lauder Dr
Ottawa ON K2A 1A9
Canada

Woosley Stanley E
Lick Observ
Univ of California
Santa Cruz CA 95064
USA
☎ 1 831 429 2976
✆ 1 831 426 3115
✉ woosley@helios.ucsc.edu

Wootten Henry Alwyn
NRAO
520 Edgemont Rd
Charlottesville VA 22903
USA
☎ 1 804 296 0329
✆ 1 804 296 0278
✉ awootten@nrao.edu

Worden Simon P
6757 N 27th St
Arlington VA 22213
USA

Worrall Diana Mary
Physics Dpt
Univ of Bristol
Tyndall Av
Bristol BS8 1TL
UK
☎ 44 11 79 288 787
✆ 44 11 79 255 624
✉ d.worrall@bristol.ac.uk

Worrall Gordon
Birdswood
Eardisley
Herefds
UK

Worswick Susan
Royal Greenwich Obs
Madingley Rd
Cambridge CB3 0EZ
UK
☎ 44 12 23 374 000
✆ 44 12 23 374 700

Woszczyk Andrzej
Inst of Astronomy
N Copernicus Univ
Ul Chopina 12/18
PL 87 100 Torun
Poland
☎ 48 56 26 018
✆ 48 56 24 602

Woszczyna Andrzej
Astronomical Observ
Krakow Jagiellonian Univ
Ul Orla 171
PL 30 244 Krakow
Poland
☎ 48 12 25 1294
✆ 48 12 25 1318
✉ woszcz@oa.uj.edu.pl

Wouterloot Jan Gerard A
Erstes Physikal Institut
Univ Koeln
Zuelpicherstr 77
D 50937 Koeln
Germany
☎ 49 221 470 4528
✆ 49 221 470 5162
✉ wauterloot@ph1.uni-
koeln.de

Wozniak Herve
Observ Marseille
2 Pl Le Verrier
F 13248 Marseille Cdx 04
France
☎ 33 4 91 95 9088
✆ 33 4 91 62 1190
✉ wozniak@observatoire.
cnrs-mrs.fr

Wramdemark Stig S
Lund Observ
Box 43
S 221 00 Lund
Sweden
☎ 46 46 222 7303
✆ 46 46 222 4614
✉ stig@astro.lu.se

Wray James D
21200 Todd Valley Rd
Trlr 54
Foresthill CA 95631 9511
USA

Wright Alan E
Parkes Observ
ATNF
Private Bag 276
Parkes NSW 2870
Australia
☎ 61 68 61 1732
✆ 61 68 61 1730
✉ awright@atnf.csiro.au

Wright Andrew
Mathematics Dpt
Univ of St Andrews
St Andrews North Haugh
Fife KY16 9SS
UK
☎ 44 13 34 7661
✆ 44 13 347 4487

Wright Edward L
Dpt Astronomy/Phys
UCLA
Box 951562
Los Angeles CA 90025 1562
USA
☎ 1 310 825 5755
✆ 1 310 206 2096

Wright Helen Greuter
Thomas House Apt 517
1330 Massachusetts Av
Washington DC 20005
USA

Wright James P
NSF
Div Astron Sci
4201 Wilson Blvd
Arlington VA 22230
USA
☎ 1 703 306 1819
✆ 1 703 306 0525
✉ jwright@nsf.gov

Wright Melvyn C H
Radio Astronomy Lab
Univ of California
601 Campbell Hall
Berkeley CA 94720
USA
☎ 1 415 642 0420
✆ 1 415 642 6424

Wrixon Gerard T
Ntl Microelectron Res Ctr
Univ College Cork
Cork
Ireland
☎ 353 1 215 08375

Wrobel Joan Marie
NRAO
Box 0
Socorro NM 87801 0387
USA
☎ 1 505 835 7000
✆ 1 505 835 7027
✉ jwrobel@nrao.edu

Wroblewski Herbert
Dpt Astronomia
Univ Chile
Casilla 36 D
Santiago
Chile
☎ 56 2 229 4101
✆ 56 2 229 4002

Wu Bin
Inst Geodesy/
Geophysics
Xu Dong Lu
Wuchang 430077
China PR
☎ 86 27 681 3855
✆ 86 27 781 1242

Wu Chi Chao
STScI/CSC
Homewood Campus
3700 San Martin Dr
Baltimore MD 21218
USA
☎ 1 301 338 4770
✆ 1 301 459 4482

Wu Guichen
Shaanxi Observ
CAS
Box 18
Lintong 710600
China PR
☎ 86 29 323 2255
✆ 86 9237 313 496

Wu Hong'ao
Purple Mountain Obs
CAS
Nanjing 210008
China PR
☎ 86 44 3281 7278
✆ 86 25 301 459

Wu Hsin-Heng
Physics Dpt
Ntl Central Univ
Chung Li Taiwan 32054
China R
☎ 886 3 462 2302
✆ 886 3 426 2304

Wu Huai-Wei
Shanghai Observ
CAS
80 Nandan Rd
Shanghai 200030
China PR
☎ 86 21 6438 6191
✆ 86 21 6438 4618

Wu Lianda
Purple Mountain Obs
CAS
Nanjing 210008
China PR
☎ 86 25 32893
✆ 86 25 301 459

Wu Linxiang
Dpt Geophysics
Beijing Univ
Beijing 100871
China PR
☎ 86 1 28 2471*3888

Wu Mingchan
Yunnan Observ
CAS
Kunming 650011
China PR
☎ 86 871 2035
✆ 86 871 717 1845

Wu Nailong
STScI
Homewood Campus
3700 San Martin Dr
Baltimore MD 21218
USA
☎ 1 301 516 6864
✆ 1 301 516 8720
✉ nailong@stsci.edu

Wu Shengyin
Beijing Astronomical Obs
CAS
Beijing 100080
China PR
☎ 86 1 28 1698
✆ 86 1 256 10855

Wu Shi Tsan
School Engin
Univ of Alabama
Huntsville AL 35899
USA
☎ 1 205 895 6413

Wu Shouxian
Shaanxi Observ
CAS
Box 18
Lintong 710600
China PR
☎ 86 33 55951
✆ 86 9237 3496

Wu Xiangping
Beijing Astronomical Obs
CAS
Beijing 100080
China PR
☎ 86 1 254 6089
✆ 86 1 256 10855
✉ wxp@bao01.bao.ac.cn

Wu Xinji
Dpt Geophysics
Beijing Univ
Beijing 100871
China PR
☎ 86 1 28 2471*3929

Wu Xuejun
Physics Dpt
Nanjing Normal Univ
Nanjing 210097
China PR
☎ 86 25 371 6641
✆ 86 25 372 3207
✉ xjwu@pine.njnu.edu.cn

Wu Yuefang
Dpt Geophysics
Beijing Univ
Beijing 100871
China PR
☎ 86 1 255 9461 *2963
✆ 86 1 256 1085

Wu Zhiren
Shanghai Scientifical/
Tech Edu Publishing House
393 Guan Shen Yaun Rd
Shanghai 200233
China PR
☎ 86 21 36 5791

Wuelser Jean-Pierre
Lockheed Palo Alto Res Lb
Dpt 91 30 Bg 252
3251 Hanover St
Palo Alto CA 94304 1191
USA
☎ 1 415 424 3289
✆ 1 415 424 3994
✉ wuelser@sag.space.
lockheed.com

Wuensch Johann Jakob
Stifterstr 8
D 86672 Thierhaupten
Germany
✉ wuen@gfz-potsdam.de

Wunner Guenter
Inst Theor Physik
Ruhr Univ Bochum
Postfach 102148
D 44780 Bochum
Germany
☎ 49 234 700 3454
✆ 49 234 709 4169

Wyckoff Susan
Physics Dpt
Arizona State Univ
Astronomy Program
Tempe AZ 85287 1504
USA
☎ 1 602 965 3561
✆ 1 605 965 7954
✉ wyckoff@asycos

Wyller Arne A
Box 5501
Santa Fe NM 87502 5501
USA
☎ 1 505 473 5849
✆ 1 505 473 5849

Wynn-Williams C G
Inst for Astronomy
Univ of Hawaii
2680 Woodlawn Dr
Honolulu HI 96822
USA
☎ 1 808 956 8807
✆ 1 808 988 2790

Wyse Rosemary F
Dpt Phys/Astronomy
JHU
Charles/34th St
Baltimore MD 21218
USA
☎ 1 301 516 5392
✆ 1 301 516 8260

M257

Wytrzyszczak Iwona
Astronomical Observ
A Mickiewicz Univ
Ul Sloneczna 36
PL 60 286 Poznan
Poland
☎ 48 61 67 9670
✆ 48 61 53 6536
✉ iwona@phys.amu.edu.pl

X

Xi Zezong
Inst History Ntl Science
1 Gong Yuan West Rd
Beijing
China PR
☎ 86 1 55 7180

Xia Xiaoyang
Physics Dpt
Tianjin Normal Univ
Tianjin 300074
China PR
☎ 86 22 71 6989

Xia Yifei
Astronomy Dpt
Nanjing Univ
Nanjing 210093
China PR
☎ 86 25 34651*2882
℻ 86 25 330 2728

Xia Zhiguo
Yunnan Observ
CAS
Kunming 650011
China PR
☎ 86 871 2035
℻ 86 871 717 1845

Xian Ding-Zhang
Purple Mountain Obs
CAS
Nanjing 210008
China PR
☎ 86 25 37609
℻ 86 25 301 459

Xiang Delin
Purple Mountain Obs
CAS
Nanjing 210008
China PR
☎ 86 25 33738
℻ 86 25 301 459

Xiang Shouping
Astrophysics Division
Univ Science/Technology
Hefei 230026
China PR
☎ 86 551 33 1134
℻ 86 551 33 1760

Xiao Naiyuan
Astronomy Dpt
Nanjing Univ
Nanjing 210093
China PR
☎ 86 25 34651*2882
℻ 86 25 330 2728

Xie Guangzhong
Yunnan Observ
CAS
Kunming 650011
China PR
☎ 86 871 2035
℻ 86 871 717 1845

Xie Liangyun
Inst Geodesy/
Geophysics
Xu Dong Lu
Wuchang 430077
China PR

Xiong Darun
Purple Mountain Obs
CAS
Nanjing 210008
China PR
☎ 86 25 42817
℻ 86 25 301 459

Xiradaki Evangelia
Astrophysics Dpt
Ntl Univ Athens
Panepistimiopolis
GR 157 84 Zografos
Greece
☎ 30 1 723 5122/9628 306
℻ 30 1 723 5122

Xu Aoao
Astronomy Dpt
Nanjing Univ
Nanjing 210093
China PR
☎ 86 25 663 7551*2882
℻ 86 25 330 2728

Xu Bang-Xin
Astronomy Dpt
Nanjing Univ
Nanjing 210093
China PR
☎ 86 25 663 7551*2882
℻ 86 25 330 2728

Xu Chongming
Physics Dpt
Nanjing Normal Univ
Nanjing 210097
China PR
☎ 86 25 371 6641
℻ 86 25 372 3207
✉ cmxu@pine.njnu.edu.cn

Xu Jiayan
Shaanxi Observ
CAS
Box 18
Lintong 710600
China PR
☎ 86 33 2255
℻ 86 9237 3496
✉ tt.ww@company.bjbta.chinamail.sprint.com

Xu Jihong
URUMQI Astronomical Stat
CAS
40 South Beijing Rd
Xinjiang 830011
China PR
☎ 86 33 5757

Xu Peiyuan
Inst Electronic Physics
SUST
Jia ding
Shanghai
China PR
☎ 86 21 95 1602

Xu Pinxin
Purple Mountain Obs
CAS
Nanjing 210008
China PR
☎ 86 25 32893
℻ 86 25 301 459

Xu Tong-Qi
Shanghai Observ
CAS
80 Nandan Rd
Shanghai 200030
China PR
☎ 86 21 6438 6191
℻ 86 21 6438 4618
✉ lpz@center.shao.ac.cn

Xu Wenli
Ctr for Astronomy
Instrumen Res
CAS
Nanjing 210042
China PR
☎ 86 25 541 1776
℻ 86 25 541 1872

Xu Zhentao
Purple Mountain Obs
CAS
Nanjing 210008
China PR
☎ 86 25 31096
℻ 86 25 301 459

Xu Zhicai
Purple Mountain Obs
CAS
Nanjing 210008
China PR
☎ 86 25 46700
℻ 86 25 301 459

Y

Yabushita Shin A
Dpt Appl Maths/Phys
Kyoto Univ
Sakyo Ku
Kyoto 606 01
Japan
☎ 81 757 51 2111
📠 81 75 753 7010

Yabuuti Kiyoshi
20 Tanaka HigaskI
Hinokuch Mach
Kyoto 606
Japan

Yahil Amos
Dpt Earth/Space Sci
Astronomy Program
Suny at Stony Brook
Stony Brook NY 11794 2100
USA
☎ 1 516 632 8224
📠 1 516 632 8240
✉ Amos.Yahil@sunysb.edu

Yakovlev Dmitry
Ioffe Physical Tech Inst
Acad Sciences
Polytechnicheskaya Ul 26
194021 St Petersburg
Russia
☎ 7 812 247 9368
📠 7 812 247 1017
✉ yak%astro.pti.spb.su@main.ioffe.rssi.ru

Yallop Bernard D
Royal Greenwich Obs
Madingley Rd
Cambridge CB3 0EZ
UK
☎ 44 1223 374 735
📠 44 12 23 374 700
✉ bdy@mail.ast.cam.ac.uk

Yamada Shoichi
MPE
Postfach 1603
D 85740 Garching
Germany
☎ 49 893 299 3247
📠 49 893 299 3235
✉ shoichi@mpa-garching.mpg.de

Yamada Yoshiyaki
Physics Dpt
Kyoto Univ
Sakyo Ku
Kyoto 606 01
Japan
☎ 81 757 53 3844
📠 81 75 753 3886
✉ yamada@scphys.kyoto-u.ac.jp

Yamagata Tomohiko
Element/Second Edu Br
Minstry Edu Sci/Cult
Kasumigaseki Chiyoda
Tokyo 100
Japan
☎ 81 335 81 4211*2403
📠 81 424 40 8460
✉ yamagata@omega.mtk.ioa.s.u-tokyo.ac.jp

Yamaguchi Shichiro
Fac of Eng
Gifu Univ
1 1 Yanagito
Gifu 501 1193
Japan
☎ 81 582 30 1111

Yamamoto Satoshi
Physics Dpt
Univ of Tokyo
7-3-1 Hongo Bunkyo-ku
Tokyo 113
Japan
☎ 81 338 12 2111
📠 81 358 02 3325
✉ yamamoto@phys.s.u-tokyo.ac.jp

Yamamoto Tetsuo
ISAS
3 1 1 Yoshinodai
Sagamihara
Kanagawa 229 8510
Japan
☎ 81 427 51 3911
📠 81 427 59 4253

Yamamoto Yoshiaki
Physics Dpt
Konan Univ
Okamoto 8 Higashinada
Kobe 658
Japan
☎ 81 784 35 2485
📠 81 784 52 9502
✉ yamamoto@apsun2.hep.konan-u.ac.jp

Yamaoka Hitoshi
Dpt Physics/Faculty Sci
Kyushu Univ
4-2-1 Ropponmatsu Chuo-ku
Fukuoka 810
Japan
☎ 81 92 726 4739
📠 81 92 726 4841
✉ yamaoka@rc.kyushu-u.ac.jp

Yamasaki Atsuma
Dpt Geoscience
National defense Academy
Hashirimizu
Yokosuka 239
Japan
☎ 81 468 41 3810
📠 81 468 44 5902

Yamashita Kojun
ISAS
3 1 1 Yoshinodai
Sagamihara
Kanagawa 229 8510
Japan
☎ 81 427 51 3911
📠 81 427 59 4253

Yamashita Takuya
Ntl Astronomical Obs
2 21 1 Osawa
Mitaka
Tokyo 181
Japan
☎ 81 422 34 3710
📠 81 422 34 3608
✉ oyamash@cl.mtk.nao.ac.jp

Yamashita Yasumasa
Tokyo Astronomical Obs
NAOJ
Osawa Mitaka
Tokyo 181
Japan
☎ 81 422 41 3740
📠 81 422 34 3793

Yamauchi Makoto
Fac of Eng
Miyazaki Univ
Gakuen Kibanadai
Miyazaki 889 21
Japan
☎ 81 985 58 2811*4401
📠 81 985 58 0913
✉ yamauchi@astro.miyazaki-u.ac.jp

Yamauchi Shigeo
College Human Social Sc
Iwate Univ
Ueda 3
Morioka 020
Japan
☎ 81 196 21 6817
📠 81 196 21 6715
✉ yamauchi@hiryu.hss.iwate-u.ac.jp

Yamazaki Akira
Hydrographic Dpt
Geodesy/Geophys Div
Tsukiji 5 Chuo Ku
Tokyo 104
Japan
☎ 81 354 13 811
📠 81 3545 2885

Yan Haojian
Shanghai Observ
CAS
80 Nandan Rd
Shanghai 200030
China PR
☎ 86 21 6438 6191
📠 86 21 6438 4618

Yan Jun
Purple Mountain Obs
CAS
Nanjing 210008
China PR
☎ 86 25 443 2817
📠 86 25 330 1459

Yan Lin-shan
Shanghai Observ
CAS
80 Nandan Rd
Shanghai 200030
China PR
☎ 86 21 6438 6191
📠 86 21 6438 4618

Yang Fumin
Shanghai Observ
CAS
80 Nandan Rd
Shanghai 200030
China PR
☎ 86 21 6438 6191
📠 86 21 6438 4618

Yang Ji
Purple Mountain Obs
CAS
Nanjing 210008
China PR
☎ 86 25 443 2817
📠 86 25 330 1459

Yang Jian
Purple Mountain Obs
CAS
Nanjing 210008
China PR
☎ 86 25 46700
📠 86 25 301 459

Yang Ke-jun
Shaanxi Observ
CAS
Box 18
Lintong 710600
China PR
☎ 86 332 255
℻ 86 923 73 496

Yang Lantian
Physics Dpt
Huazhong Normal Univ
Wuhan
China PR
☎ 86 75 601300

Yang Pibo
Central China Normal Univ
Physics Dpt
Wuhan 430079
China PR
☎ 86 27 787 8444*2889
℻ 86 27 787 6070
✉ pbyang@ccnu.edu.cn

Yang Shijie
Purple Mountain Obs
CAS
Nanjing 210008
China PR
☎ 86 25 46700
℻ 86 25 301 459

Yang Stephenson L S
Dpt Phys/Astronomy
Univ of Victoria
Victoria BC V8W 3P6
Canada
☎ 1 250 721 8655
℻ 1 250 721 7715
✉ yang@beluga.phys.univ.ca

Yang Tinggao
Shaanxi Observ
CAS
Box 18
Lintong 710600
China PR
☎ 86 33 2255
℻ 86 29 389 0196
✉ csau@ms.sxso.ac.cn

Yang Zhigen
Shanghai Observ
CAS
80 Nandan Rd
Shanghai 200030
China PR
☎ 86 21 6487 6589
℻ 86 21 6438 4618
✉ yangz@center.shao.ac.cn

Yankulova Ivanka
Astronomy Dpt
Univ of Sofia
Anton Ivanov St 5
BG 1126 Sofia
Bulgaria
☎ 359 2 68 8176
℻ 359 2 68 9085

Yano Hajime
NASA/JSC
Code SN2
Houston TX 77058
USA
☎ 1 281 244 5018
℻ 1 281 483 5347
✉ hajime.yano1@.jsc.nasa.gov

Yanovitskij Edgard G
Main Astronomical Obs
Ukrainian Acad Science
Golosiiv
252650 Kyiv 22
Ukraine
☎ 380 4426 63110
℻ 380 4426 62147

Yao Baoan
Shanghai Observ
CAS
80 Nandan Rd
Shanghai 200030
China PR
☎ 86 21 6438 6191
℻ 86 21 6438 4618

Yao Jinxing
Purple Mountain Obs
CAS
Nanjing 210008
China PR
☎ 86 25 46700
℻ 86 25 301 459

Yao Zhengqiu
NAIRC
Box 864
210042 Nanjing
China PR
☎ 86 255 40 5562
℻ 86 255 40 5562
✉ njalmost@public1.ptt.js.cn

Yaplee B S
8 Crest View Ct
Rockville MD 20854
USA
☎ 1 301 762 0935

Yarov-Yarovoj M S
Mathematics Dpt
MVTU
Vtoraya Baumanskaya 5
107005 Moscow
Russia
☎ 7 095 267 0392

Yasuda Haruo
Nishihara Greenhaitsu 6-405
4-5-37 Nishihara-machi
Tanashi-shi
Tokyo 188
Japan
✉ ya@po.teleway.ne.jp

Yatskiv Ya S
Main Astronomical Obs
Ukrainian Acad Science
Golosiiv
252650 Kyiv 22
Ukraine
☎ 380 4426 63110
℻ 380 4422 46387
✉ yatskiv@mao.kiev.ua

Yau Kevin K C
CALTECH/JPL
MS 230 101
4800 Oak Grove Dr
Pasadena CA 91109 8099
USA
☎ 1 818 393 5880
℻ 1 818 393 1105
✉ kevin.k.yau@jpl.nasa.gov

Yavnel Alexander A
Meteorite Committee
Acad Sciences
Ulianovoj M Ul 3 K 1
117313 Moscow
Russia
☎ 7 095 137 7538

Ye Binxun
Beijing Astronomical Obs
CAS
Beijing 100080
China PR
☎ 86 1 28 1203
℻ 86 1 256 10855

Ye Shi-hui
Purple Mountain Obs
CAS
Nanjing 210008
China PR
☎ 86 25 46700
℻ 86 25 301 459

Ye Shuhua
Shanghai Observ
CAS
80 Nandan Rd
Shanghai 200030
China PR
☎ 86 21 6438 6191
℻ 86 21 6438 4618
✉ ysh@center.shao.ac.cn

Ye Wenwei
Inst Seismology
State Seismo Bureau
Xiao Hong Shan Wuhan
Hubei 230026 Anhui
China PR
☎ 86 81 3401

Yee Howard K C
Astronomy Dpt
Univ of Toronto
60 St George St
Toronto ON M5S 1A1
Canada
☎ 1 416 978 4833
℻ 1 416 978 3921
✉ hyee@utorphys

Yeh Tyan
HAO
NCAR
Box 3000
Boulder CO 80307 3000
USA
☎ 1 303 497 5401
℻ 1 303 497 1568

Yeivin Y
Dpt Phys/Astronomy
Tel Aviv Univ
Ramat Aviv
Tel Aviv 69978
Israel
☎ 972 3 640 8208
℻ 972 3 640 8179

Yengibarian Norair
Byurakan Astrophys Observ
Armenian Acad Sci
378433 Byurakan
Armenia
☎ 374 88 52 28 3453/4142
℻ 374 88 52 52 3640
✉ yengib@bao.sci/bagrat@ffpmc.am

Yeomans Donald K
CALTECH/JPL
MS 301 150 G
4800 Oak Grove Dr
Pasadena CA 91109 8099
USA
☎ 1 818 354 2127
℻ 1 818 393 1159
✉ donald.k.yeomans@jpl.nasa.gov

Yershov Vladimir N
Pulkovo Observ
Acad Sciences
10 Kutuzov Quay
196140 St Petersburg
Russia
☎ 7 812 123 4494
℻ 7 812 315 1701
✉ yersh@gao.spb.su

Yi Meiliang
Ctr for Astronomy
Instrumen Res
CAS
Nanjing 210042
China PR
☎ 86 25 541 1776
℻ 86 25 541 1872

Yi Zhaohua
Astronomy Dpt
Nanjing Univ
Nanjing 210093
China PR
☎ 86 25 46700
℻ 86 25 330 2728

Yilmaz Fatma
Univ Observ
Univ of Istanbul
University 34452
34452 Istanbul
Turkey
☎ 90 212 522 3597
℻ 90 212 519 0834

Yilmaz Nihal
Astronomy Dpt
Univ of Ankara
Fen Fakultesi
06100 Besevler
Turkey
☎ 90 41 23 6550
℻ 90 312 223 2395

Yin Jisheng
Beijing Astronomical Obs
CAS
Beijing 100080
China PR
☎ 86 1 28 1203
✆ 86 1 256 10855

Yin Qi-Feng
NRAO
520 Edgemont Rd
Charlottesville VA 22903
USA
☎ 1 804 296 0267
✉ qyun@nrao.edu

Yin Xinhui
Purple Mountain Obs
CAS
Nanjing 210008
China PR
☎ 86 25 330 3738
✆ 86 25 330 1459
✉ yangguo@public1.ptt.js.cn

Yock Philip
Physics Dpt
Univ of Auckland
Private Bag
Auckland
New Zealand
☎ 64 9 373 7999* 6838
✆ 64 9 308 2377
✉ p.yock@auckland.ac.nz

Yoder Charles F
CALTECH/JPL
MS 183 150
4800 Oak Grove Dr
Pasadena CA 91109 8099
USA
☎ 1 818 354 2444
✆ 1 818 393 6030

Yokosawa Masayoshi
Physics Dpt
Ibaraki Univ
2-1-1 Bunkyo Mito
Ibaraki 310
Japan
☎ 81 292 26 1621

Yokoyama Jun-ichi
YITP
Kyoto Univ
Gokanosho
Uji 611
Japan
☎ 81 774 31 7431
✆ 81 774 33 6226
✉ yokoyama@yisun1.
yukawa.kyoto-u.ac.jp

Yokoyama Koichi
Intl Latitude Observ
NAOJ
Hoshigaoka Mizusawa Shi
Iwate 023
Japan
☎ 81 197 24 7111
✆ 81 197 22 7120

Yokoyama Tadashi
Univ Estadual Paulista
Instituto Fisica
CP 178
13500 Rio Claro
Brazil
☎ 55 195 34 0122

Yoneyama Tadaoki
2-1-16 Hibarigaoka-kita
Tokyo 202
Japan

Yoon Tae
Dpt Astron/Atmosph Sci
Kyungpook Ntl Univ
Taegu 702 701
Korea RP
☎ 82 53 950 6359
✆ 82 539 50 6360
✉ yoonts@bh.kyungpook.
ac.kr

York Donald G
Astronomy/Astrophys Ctr
Univ of Chicago
5640 S Ellis Av
Chicago IL 60637
USA
☎ 1 312 962 8930
✆ 1 312 702 8212

Yorke Harold W
Astronomisches Institut
Univ Wuerzburg
am Hubland
D 97074 Wuerzburg
Germany
☎ 49 931 888 5301
✆ 49 931 888 4603

Yoshida Atsumasa
Inst Physical/Chem Res
RIKEN
2 1 Hirosawa
Wako Saitama 351 01
Japan
☎ 81 484 62 1111 *3226
✆ 81 484 62 4640
✉ ayoshida@postman.
riken.go.jp

Yoshida Haruo
Tokyo Astronomical Obs
NAOJ
Osawa Mitaka
Tokyo 181
Japan
☎ 81 422 41 3614
✆ 81 442 41 3793
✉ yoshida@c1.mtk.
nao.ac.jp

Yoshida Junzo
Physics Dpt
Kyoto Sangyo Univ
Kamigamo
Kyoto 603
Japan
☎ 81 757 01 2151
✆ 81 75 705 1640

Yoshida Michitoshi
Okayama Astrophys Obs
Kamogata-cho
Asakuchi-gun
Okayama 719 02
Japan
☎ 81 865 44 2155
✆ 81 865 44 2360
✉ yoshida@oao.nao.ac.jp

Yoshida Shigeomi
Kiso Observ
Inst of Astronomy
Mitake Kiso
Nagano 397 01
Japan
☎ 81 264 52 3360
✆ 81 264 52 3361
✉ yoshida@kiso.iao.s.u-
tokyo.ac.jp

Yoshii Yuzuru
Tokyo Astronomical Obs
NAOJ
Osawa Mitaka
Tokyo 181
Japan
☎ 81 422 32 5111
✆ 81 422 34 3793
✉ yoshii@top.mtk.nao.ac.jp

Yoshikawa Makoto
ISAS
3 1 1 Yoshinodai
Sagamihara
Kanagawa 229 8510
Japan
☎ 81 427 59 8341
✆ 81 427 59 8341
✉ makoto@pub.isas.ac.jp

Yoshimura Hirokazu
Astronomy Dpt
Univ of Tokyo
Bunkyo Ku
Tokyo 113
Japan
☎ 81 381 22 111
✆ 81 338 13 9439

Yoshimura Motohiko
Physics Dpt
Tohoku Univ
Aramaki
Sendai 980
Japan
☎ 81 222 22 1800
✆ 81 222 25 1891

Yoshino Kouichi
CfA
HCO/SAO MS 50
60 Garden St
Cambridge MA 02138 1516
USA
☎ 1 617 495 2796
✆ 1 617 495 7455
✉ yoshino@cfa.harvard.edu

Yoshioka Kazuo
Univ of the Air
Gumma Study Ctr
1-13-2 Wakamiya-cho
Maebashi City Gunma 371
Japan
☎ 81 27 230 1134
✆ 81 27 230 1094
✉ yoshioka@u-air.ac.jp

Yoshizawa Masanori
Tokyo Astronomical Obs
NAOJ
Osawa Mitaka
Tokyo 181
Japan
☎ 81 422 32 5111
✆ 81 422 32 1924

Yoss Kenneth M
Dpt Astron/Physics
Univ of Illinois
1011 W Springfield Av
Urbana IL 61801
USA
☎ 1 217 333 3295
✆ 1 217 244 7638

You Jianqi
Purple Mountain Obs
CAS
Nanjing 210008
China PR
☎ 86 25 46700
✆ 86 25 301 459

You Junhan
Astrophysics Division
Univ Science/Technology
Hefei 230026
China PR
☎ 86 551 33 1134
✆ 86 551 33 1760

Young Andrew T
Astronomy Dpt
San Diego State Univ
San Diego CA 92182
USA
☎ 1 619 594 5817
✆ 1 619 594 1413
✉ aty@mintaka.sdsu.edu

Young Arthur
Astronomy Dpt
San Diego State Univ
San Diego CA 92182
USA
☎ 1 619 265 6167
✆ 1 619 594 1413
✉ young@minaka.
sdsu.edu

Young Judith Sharn
Five College
RAO
B619 Lederle Grad Res Twr
Amherst MA 01003
USA
☎ 1 413 545 0789

Young Louise Gray
Astronomy Dpt
San Diego State Univ
San Diego CA 92182
USA
☎ 1 619 287 8890
✆ 1 619 594 1413

Younis Saad M
SARC
Scientific Res Council
Box 2441
Jadiriyah Baghdad
Iraq
☎ 964 1 776 5127

Yousef Shahinaz M
Astronomy Dpt
Fac of Sciences
Cairo Univ
Geza
Egypt
☎ 20 2 572 7022
✆ 20 2 572 7556

Youssef Nahed H
Astronomy Dpt
Fac of Sciences
Cairo Univ
Geza
Egypt
☎ 20 2 572 7022
✆ 20 2 572 7556

Yu Kyung-Loh
Astronomy Dpt 55
Seoul Ntl Univ
Kwanak-ku
Seoul 151 742
Korea RP
☎ 82 288 06621
✆ 82 288 71435

Yu Xin Alfred
Dpt Appl Maths
Hong Kong Polytechnic
Hung Hom
Kowloon
China PR
☎ 852 376 66951
✆ 852 336 29 045

Yu Yan
Dpt Radiation Oncology
Univ of Rochester
601 Elmwood Av Box 647
Rochester NY 14642
USA
✉ yu@deneb.mps.ohio-state.edu

Yu Zhiyao
Shanghai Observ
CAS
80 Nandan Rd
Shanghai 200030
China PR
☎ 86 21 6438 6191
✆ 86 21 6438 4618
✉ zyyu@center.shao.ac.cn

Yuan Chi
Physics Dpt
City College of NY
138 St Convene Av
New York NY 10031
USA
☎ 1 212 690 6823

Yuan Kuo-Chuan
Taipei Astronomical
Museum
N 363 Kee-Ho Rd
Taipei
China R
☎ 886 2 594 9516
✆ 886 2 591 0763

Yuasa Manabu
Dpt Math/Physics
Kinki Univ
Higashi
Osaka 577
Japan

Yudin Boris F
SAI
Acad Sciences
Universitetskij Pr 13
119899 Moscow
Russia
☎ 7 095 939 1661
✆ 7 095 939 1661
✉ yudin@sai.msu.su

Yue Zengyuan
Dpt Geophysics
Beijing Univ
Beijing 100871
China PR
☎ 86 1 28 2471*3888

Yuldashbaev Taimas S
Astronomical Institute
Uzbek Acad Sci
Astronomicheskaya Ul 33
700052 Tashkent
Uzbekistan
☎ 7 3712 35 8102

Yumi Shigeru
Keyakidai 1-12-2 Kiyamacho
Miyakigun
Saga 841 02
Japan

Yun Hong-Sik
Astronomy Dpt
Seoul Ntl Univ
Kwanak Ku
Seoul 151
Korea RP
☎ 82 288 06 621/22
✆ 82 288 71435

Yungelson Lev R
Inst of Astronomy
Acad Sciences
Pyatnitskaya Ul 48
109017 Moscow
Russia
☎ 7 095 231 5461
✆ 7 095 230 2081

Z

Zabolotny Vladimir F
Astro Space Ctr
Lebedev Physical Inst
Profsoyuznaya 84/32
Moscow 117924
Russia
☎ 7 095 333 1555
℻ 7 095 333 2378

Zabriskie F R
RD 1
Alexandria PA 16611
USA
☎ 1 814 669 4483

Zachariadis Theodosios
Res Ctr Astronomy
Acad Athens
14 Anagnostopoulou St
GR 106 73 Athens
Greece
☎ 30 1 361 3589
℻ 30 1 363 1606
✉ exaka20@grathun1

Zacharias Norbert
USNO/CTIO
950 N Cherry Av.
Tuscon AZ 85719
USA
☎ 1 520 318 470
✉ nz@ctioj8.ctio.noao.edu

Zacharov Igor
Astronomical Institute
Czech Acad Sci
Fricova 1
CZ 251 65 Ondrejov
Czech R
☎ 420 204 85 7239
℻ 420 2 88 1611
✉ astsun@csearn

Zachilas Loukas
Dpt Civil Eng
Univ of Thessalia
Pedio Areos
GR 383 34 Volos
Greece
☎ 30 42 16 9781*122
℻ 30 42 162660
✉ zachilas@rigas.uth.gr

Zadunaisky Pedro E
Univers Buenos Aires
Fac Cienc Exactas Math
Ciud Univer Pabl
1428 Buenos Aires
Argentina

Zafiropoulos Basil
Physics Dpt
Univ of Patras
GR 261 10 Rion
Greece
☎ 30 61 99 7572
℻ 30 61 99 7636

Zagretdinov Renat
Astronomy Dpt
Kazan State Univ
Kremlevskaya Str 18
Kazan 420008
Russia
☎ 7 843 238 7695
✉ Renat.Zagretdinov@ksu.ru

Zahn Jean-Paul
Observ Paris Meudon
DASGAL
Pl J Janssen
F 92195 Meudon PPL Cdx
France
☎ 33 1 45 07 7804
℻ 33 1 45 07 7878
✉ zahn@obspm.fr

Zaitsev Valerii V
Inst Applied Physics
Acad Sciences
Ulyanov Ul 46
603600 N Novgorod
Russia
☎ 7 831 236 7253
℻ 7 831 236 2061

Zakharova Polina E
Ural State Univ
Astronomical Observ
Lenin Av 51
Ekaterinburgh 620083
Russia
☎ 7 343 261 5431
℻ 7 343 255 5964
✉ Polina.Zakharova@usu.ru

Zakhozhaj Volodimir
Timurovtsev St 70 Apt 58
310204 Kharkiv
Ukraine
☎ 380 5726 34630
✉ zkh@astron.kharkov.ua

Zambon Giulio
21 Res Sequoia
F 91400 Orsay
France

Zambrano Alejandro
ROA
Cecilio Pujazon 22-3 A
E 11110 San Fernando
Spain
☎ 34 5 688 3548
℻ 34 5 689 9302
✉ ccgeneral@czv1.uca.es

Zamorani Giovanni
Istt Radioastronomia
CNR
Via P Gobetti101
I 40129 Bologna
Italy
☎ 39 51 639 9385
℻ 39 51 639 9385

Zamorano Jaime
Dpt Astrofisica
Fac Fisica
Univ Complutense
E 28040 Madrid
Spain
☎ 34 1 394 5195
℻ 34 1 394 4590
✉ jaz@ucmast.fis.ucm.es

Zampieri Luca
Physics Dpt
Univ of Illinois
1110 W Green St
Urbana IL 61801 3080
USA
☎ 1 217 333 2807
℻ 1 217 333 9819
✉ zampieri@donald.physics.uiuc.edu

Zander Rodolphe
Inst Astrophysique
Universite Liege
Av Cointe 5
B 4000 Liege
Belgium
☎ 32 4 254 7510
℻ 32 4 254 7511
✉ zander@astro.ulg.ac.be

Zaninetti Lorenzo
Istt Fisica
Univ d Torino
Corso d Azeglio 46
I 10125 Torino
Italy
☎ 39 11 657 694

Zappala Rosario Aldo
Istt Astronomia
Citta Universitaria
Via A Doria 6
I 95125 Catania
Italy
☎ 39 95 733 *493
℻ 39 95 33 0592

Zappala Vincenzo
Osserv Astronomico
d Torino
St Osservatorio 20
I 10025 Pino Torinese
Italy
☎ 39 11 461 9035
℻ 39 11 461 9030
✉ zappala@to.astro.it

Zare Khalil
Aerospace Eng Dpt
Univ of Texas
WRW 414
Austin TX 78712
USA
☎ 1 512 471 4234
℻ 1 512 471 3788
✉ aozy757@emx.cc.utexas.edu

Zarnecki Jan Charles
Unit for Space Sci
Univ of Kent
Canterbury CT2 7NR
UK
☎ 44 12 27 764 000
℻ 44 12 27 762 616
✉ jcz@ukc.ac.uk

Zarro Dominic M
NASA GSFC
Code 682
Greenbelt MD 20771
USA
☎ 1 301 286 4689
℻ 1 301 286 1617

Zasov Anatole V
SAI
Acad Sciences
Universitetskij Pr 13
119899 Moscow
Russia
☎ 7 095 939 2858
℻ 7 095 939 0126

Zavagno Annie
Observ Marseille
2 Pl Le Verrier
F 13248 Marseille Cdx 04
France
☎ 33 4 91 95 9088
℻ 33 4 91 62 1190
✉ zavagno@obmara.cnrs-mrs.fr

Zavatti Franco
Dpt Astronomia
Univ d Bologna
Via Zamboni 33
I 40126 Bologna
Italy
☎ 39 51 259 301
℻ 39 51 259 407
✉ zavatti@astbo4.bo.astro.it

Zayer Igor
Lockheed Palo Alto Res Lb
Dpt 91 30 Bg 252
3251 Hanover St
Palo Alto CA 94304 1191
USA
☎ 1 415 424 3545
℻ 1 415 424 3994
✉ izayer@solar.standford.edu

Zdanavicius Kazimeras
Inst Theor Physics/
Astronomy
Gostauto 12
Vilnius 2600
Lithuania
☎ 370 2 612 898
℻ 370 2 224 696
✉ astro@itpa.ft.ly

Zdunik Julian
Copernicus Astron Ctr
Polish Acad Sci
Ul Bartycka 18
PL 00 716 Warsaw
Poland
☎ 48 22 41 1086
℻ 48 22 41 0828
✉ jlz@camk.edu.pl

Zdziarski Andrzej
Copernicus Astron Ctr
Polish Acad Sci
Ul Bartycka 18
PL 00 716 Warsaw
Poland
☎ 48 22 41 1086
℻ 48 22 41 0828
✉ aaz@camk.edu.pl

Zealey William J
Dpt Eng Physics
Univ of Wollongong
Northfields Av
Wollongong NSW 2522
Australia
☎ 61 242 213 522
℻ 61 242 215 944
✉ b.zealey@uow.edu.au

Zeilik Michael Ii
Dpt Phys/Astronomy
Univ New Mexico
800 Yale Blvd Ne
Albuquerque NM 87131
USA
☎ 1 505 277 4442

Zeilinger Werner W
Inst Astronomie
Univ Wien
Tuerkenschanzstr 17
A 1180 Wien
Austria
☎ 43 1 470 6800*175
℻ 43 1 470 6015
✉ wzeil@doradus.ast
univie.ac.at

Zeippen Claude
Observ Paris Meudon
DAEC
Pl J Janssen
F 92195 Meudon PPL Cdx
France
☎ 33 1 45 07 7443
℻ 33 1 45 07 7971

Zekl Hans Wilhelm
Odenwaldstr 7
D 64683 Einhausen
Germany
☎ 49 625 158 8650
℻ 49 625 170 0140
✉ zeklh@rhein-neckar.
netsurf.de

Zelenka Antoine
Dachslenbergstr 56
CH 8180 Buelach
Switzerland
☎ 41 1 860 5156

Zellner Benjamin H
Physics Dpt
Georgia Southern Univ
Landrum Box 8031
Statesboro GA 30460
USA
☎ 1 912 681 0080
℻ 1 912 681 0471
✉ zellner@gsvms2.cc.
gasou.edu

Zeng Qin
Purple Mountain Obs
CAS
Nanjing 210008
China PR
☎ 86 25 30 8516
℻ 86 25 301 459
✉ qinzeng@public1.ptt.
js.cn

Zeng Zhifang
Beijing Astronomical Obs
CAS
Beijing 100080
China PR
☎ 86 10 6254 5179
℻ 86 1 256 10855
✉ zfzeng@bao01.bao.
ac.cn

Zensus J-Anton
RAIUB
Univ Bonn
auf d Huegel 96
D 53121 Bonn
Germany
☎ 1 228 525 378
℻ 1 228 525 435
✉ azensus@mpifr-bonn.
mpg.de

Zepf Stephen Edward
Astronomy Dpt
Univ of California
601 Campbell Hall
Berkeley CA 94720 3411
USA
☎ 1 510 642 4075
℻ 1 510 642 3411
✉ zepf@astron.berkeley.edu

Zerull Reiner H
Bereich Extraterr Physik
Ruhr Univ Bochum
Postfach 102148
D 46047 Bochum
Germany
☎ 49 234 700 4576
℻ 49 234 709 4169

Zhagar Youri H
Astronomical Observ
Latvian State Univ
Rainis Bul 19
LV 226098 Riga
Latvia
☎ 371 13 2 223149
℻ 371 782 0180

Zhai Disheng
Beijing Astronomical Obs
CAS

W Suburb
Beijing 100080
China PR
☎ 86 1 28 1698
℻ 86 1 256 10855

Zhai Zaocheng
Shanghai Observ
CAS
80 Nandan Rd
Shanghai 200030
China PR
☎ 86 21 6438 6191
℻ 86 21 6438 4618

Zhanf Shouzhong
414 West 120 St
Apt 401
New York NY 10027
USA
☎ 1 212 666 4689

Zhang Bairong
Yunnan Observ
CAS
Kunming 650011
China PR
☎ 86 871 2035
℻ 86 871 717 1845

Zhang Bin
Dpt Geophysics
Beijing Univ
Beijing 100871
China PR
☎ 86 1 28 2471*3888

Zhang Cheng-Yue
STScI
Homewood Campus
3700 San Martin Dr
Baltimore MD 21218
USA
☎ 1 301 516 6864
℻ 1 301 516 8981
✉ zhang@stsci.edu

Zhang Er-Ho
Astronomy Dpt
Univ of Texas
Rlm 15 220
Austin TX 78712 1083
USA
☎ 1 512 471 4462
℻ 1 512 471 6016

Zhang Fujun
Shanghai Observ
CAS
80 Nandan Rd
Shanghai 200030
China PR
☎ 86 21 6438 6191
℻ 86 21 6438 4618

Zhang Guo-Dong
Beijing Astronomical Obs
CAS
Beijing 100080
China PR
☎ 86 1 28 1698
℻ 86 1 256 10855

Zhang Heqi
Purple Mountain Obs
CAS
Nanjing 210008
China PR
☎ 86 25 46700
℻ 86 25 301 459

Zhang Hui
Shaanxi Observ
CAS
Box 18
Lintong 710600
China PR
☎ 86 33 2255
℻ 86 9237 3496

Zhang Jialu
Astrophysics Division
Univ Science/Technology
Hefei 230026
China PR
☎ 86 551 33 1134
℻ 86 551 33 1760

Zhang Jian
Astrophysics Div
Dpt Geophysics
Beijing Univ
Beijing 100871
China PR
☎ 86 10 6275 3146
℻ 86 10 6256 4095
✉ jianzhan@pku.edu.cn

Zhang Jiaxiang
Purple Mountain Obs
CAS
Nanjing 210008
China PR
☎ 86 25 46700
℻ 86 25 301 459

Zhang Jin
URUMQI Astronomical Stat
CAS
40 South Beijing Rd
Xinjiang 830011
China PR
☎ 86 99 138 39979
℻ 86 0991 3838628
✉ uao@rose.cnc.ac.cn

Zhang Jintong
Inst Geodesy/
Geophysics
Xu Dong Lu
Wuchang 430077
China PR

Zhang Peiyu
Purple Mountain Obs
CAS
Nanjing 210008
China PR
☎ 86 25 37521
℻ 86 25 301 459

Zhang Xiaolei
CfA
HCO/SAO MS 78
60 Garden St
Cambridge MA 02138 1516
USA
☎ 1 617 495 7020
℻ 1 617 496 7554
✉ xzhang@cfa.harvard.edu

Zhang Xiuzhong
Shanghai Observ
CAS
80 Nandan Rd
Shanghai 200030
China PR
☎ 86 21 6438 6191
✆ 86 21 6438 4618

Zhang Xizhen
Beijing Astronomical Obs
CAS
Beijing 100080
China PR
☎ 86 10 6262 5498
✆ 86 1 256 10855
✉ zxz@class1.bao.ac.cn

Zhang Yang
Astrophysics Division
Univ Science/Technology
Hefei 230026
China PR
☎ 86 551 360 3801
✆ 86 551 33 1760
✉ yz@cfasun.cfa.ustc.edu.cn

Zhang Youyi
Purple Mountain Obs
CAS
Nanjing 210008
China PR
☎ 86 25 46700
✆ 86 25 301 459

Zhang Zhen-Jiu
Physics Dpt
Huazhong Normal Univ
Wuhan
China PR
☎ 86 75 601 6908

Zhang Zhenda
Astronomy Dpt
Nanjing Univ
Nanjing 210093
China PR
☎ 86 25 34651*2882
✆ 86 25 330 2728

Zhang Zheng-Pan
CRESS
York Univ
4700 Keele St
North York ON M3J 1P3
Canada
☎ 1 416 512 1400
✆ 1 416 512 0791
✉ sheng@stpl.ists.ca

Zhao Changyin
Purple Mountain Obs
CAS
Nanjing 210008
China PR
☎ 86 25 331 1632
✆ 86 25 331 1632
✉ wlda@public2.bta.net.cn

Zhao Gang
Beijing Astronomical Obs
CAS
Beijing 100080
China PR
☎ 86 1 257 8296
✆ 86 1 256 10855

Zhao Gang
Shanghai Observ
CAS
80 Nandan Rd
Shanghai 200030
China PR
☎ 86 21 6438 6191
✆ 86 21 6438 4618
✉ gzhao@bao01.bao.ac.cn

Zhao Jun Hui
CfA
HCO/SAO MS 78
60 Garden St
Cambridge MA 02138 1516
USA
☎ 1 617 495 7294
✆ 1 617 496 7554
✉ jzhao@cfa.harvard.edu

Zhao Junliang
Shanghai Observ
CAS
80 Nandan Rd
Shanghai 200030
China PR
☎ 86 21 6438 6191
✆ 86 21 6438 4618

Zhao Ming
Shanghai Observ
CAS
80 Nandan Rd
Shanghai 200030
China PR
☎ 86 21 6438 6191
✆ 86 21 6438 4618

Zhao Renyang
Beijing Astronomical Obs
CAS
Beijing 100080
China PR
☎ 86 1 28 1698
✆ 86 1 256 10855

Zhao Yongheng
Beijing Astronomical Obs
CAS
Beijing 100080
China PR
☎ 86 1 257 8296
✆ 86 1 256 10855
✉ yac@bao01.bao.ac.cn

Zhao Zhaowang
Beijing Astronomical Obs
CAS
Beijing 100080
China PR
☎ 86 10 6256 9840
✆ 86 1 256 10855
✉ zzwang@bao01.bao.ac.cn

Zharkov Vladimir N
Inst Phys of the Earth
Acad Sciences
Gruzinskaya 10
123342 Moscow
Russia
☎ 7 095 254 5251

Zharkova Valentina
Dpt Phys/Astronomy
Univ of Glasgow
Glasgow G12 8QQ
UK
☎ 44 141 339 8855*4153
✆ 44 141 334 9029
✉ vzharkova@solar.stanford.edu

Zhdanov Valery
Astronomical Observ
Kyiv State Univ
Observatorna Str 3
254053 Kyiv
Ukraine
☎ 380 4451 31492/216 2691
✉ aoku@gluk.apc.org

Zhekov Svetozar A
Space Res Inst
Bulgarian Acad Sci
Moskovska Str 6
BG 1000 Sofia
Bulgaria
☎ 359 2 88 3503
✆ 359 2 80 1347
✉ szhekov@bgearn.acad.bg

Zheleznìakov Vladimir V
Inst Applied Physics
Acad Sciences
Ulyanov Ul 46
603600 N Novgorod
Russia
☎ 7 831 236 7253
✆ 7 831 236 2061

Zhelyazkov Ivan
Fac of Physics
Univ of Sofia
Anton Ivanov St 5
BG 1164 Sofia
Bulgaria
☎ 359 2 62 56641
✆ 359 2 96 25 276
✉ izh@phys.uni-sofia.bg

Zheng Dawei
Shanghai Observ
CAS
80 Nandan Rd
Shanghai 200030
China PR
☎ 86 21 6438 6191
✆ 86 21 6438 4618

Zheng Jia-Qing
Turku Univ
Tuorla Observ
Vaeisaelaentie 20
FIN 21500 Piikkio
Finland
☎ 358 2 274 4254
✆ 358 2 243 3767
✉ zheng@sara.utu.fi

Zheng Wei
CAS
JHU
Charles/34th St
Baltimore MD 21218 2695
USA
☎ 1 301 516 5459
✆ 1 410 516 8260
✉ zheng@pha.jhu.edu

Zheng Xinwu
Astronomy Dpt
Nanjing Univ
Nanjing 210093
China PR
☎ 86 25 3593445
✆ 86 25 330 2728
✉ xwzheng@nju.edu.cn

Zheng Xuetang
Applied Phys Dpt
E China Inst of Technology
Nanjing 210014
China PR

Zheng Yijia
Beijing Astronomical Obs
CAS
Beijing 100080
China PR
☎ 86 1 28 1698
✆ 86 1 256 10855

Zheng Ying
Purple Mountain Obs
CAS
Nanjing 210008
China PR
☎ 86 25 46700
✆ 86 25 301 459

Zhevakin S A
Radiophysical Res Inst
Lyadov Ul 25/14
603600 N Novgorod
Russia
☎ 7 831 236 6751
✆ 7 831 236 9902

Zhilyaev Boris
Main Astronomical Obs
Ukrainian Acad Science
Golosiiv
252650 Kyiv 22
Ukraine
☎ 380 4426 64769
✆ 380 4426 62147
✉ maouas@gluk.apc.org

Zhong Hongqi
Beijing Astronomical Obs
CAS
Beijing 100080
China PR
☎ 86 10 6261 2194
✆ 86 1 256 10855
✉ zhg@sun10.bao.ac.cn

Zhong Min
Inst Geodesy/Geophysics
Xu Dong Rd
Wuhan 430077
China PR
☎ 86 27 681 3841
✆ 86 27 681 3841
✉ lab@asch.whigg.ac.cn

Zhou Aihua
Purple Mountain Obs
CAS
Nanjing 210008
China PR
☎ 86 25 330 5218/335 0752
✆ 86 25 330 1459
✉ pmoyl@ba001.bao.ac.cn

Zhou Daoqi
Dpt Geophysics
Beijing Univ
Beijing 100871
China PR
☎ 86 1 28 2471*3888

Zhou Hongnan
Physics Dpt
Nanjing Normal Univ
Nanjing 210097
China PR
☎ 86 25 371 6641
℻ 86 25 330 2728

Zhou Ti-jian
Dpt Geophysics
Beijing Univ
Beijing 100871
China PR
☎ 86 1 28 2471*3888

Zhou Xu
Beijing Astronomical Obs
CAS
Beijing 100080
China PR
☎ 86 10 6253 6437
℻ 86 1 256 10855
✉ zhouxu@qso.bao.ac.cn

Zhou Youyuan
Astrophysics Division
Univ Science/Technology
Hefei 230026
China PR
☎ 86 551 33 1134
℻ 86 551 33 1760

Zhou ZhenpPu
Purple Mountain Obs
CAS
Nanjing 210008
China PR
☎ 86 25 33738
℻ 86 25 301 459

Zhu Cisheng
Astronomy Dpt
Nanjing Univ
Nanjing 210093
China PR
☎ 86 25 37551*2882
℻ 86 25 330 2728

Zhu Jin
Beijing Astronomical Obs
CAS
Beijing 100080
China PR
☎ 86 10 6253 6437
℻ 86 1 256 10855
✉ jinzhu@sun.ihep.ac.cn

Zhu Nenghong
Shanghai Observ
CAS
80 Nandan Rd
Shanghai 200030
China PR
☎ 86 21 6438 6191
℻ 86 21 6438 4618

Zhu Shichang
Physics Dpt
Shanghai Teachers Univ
10 Gillin Rd
Shanghai
China PR
☎ 86 21 384 301

Zhu Wenyao
Shanghai Observ
CAS
80 Nandan Rd
Shanghai 200030
China PR
☎ 86 21 6438 6191
℻ 86 21 6438 4618

Zhu Xingfeng
Astrophysics Division
Univ Science/Technology
Hefei 230026
China PR
☎ 86 551 33 1134
℻ 86 551 33 1760

Zhu Yaozhong
Inst Geodesy/
Geophysics
Xu Dong Lu
Wuchang 430077
China PR
☎ 86 551 81 3401

Zhu Yonghe
Beijing Astronomical Obs
CAS
Beijing 100080
China PR
☎ 86 1 28 1698
℻ 86 1 256 10855

Zhu Yongtian
Ctr for Astronomy
Instrumen Res
CAS
Nanjing 210042
China PR
☎ 86 25 541 1776
℻ 86 25 540 5562

Zhu Zi
Shaanxi Observ
CAS
Box 18
Lintong 710600
China PR
☎ 86 29 389 0425
℻ 86 29 3890196
✉ pub2@ms.sxso.ac.cn

Zhuang Qixiang
Bondar Clegg/Co Ltd
NRCC
5420 Canotec Rd
Ottawa ON K1J 8X5
Canada
☎ 1 613 749 2220

Zhuang Weifeng
Library
Shantou Univ
Shantou
China PR

Zhugzhda Yuzef D
ITMIRWP
Acad Sciences
142092 Troitsk
Russia
☎ 7 095 334 0120
℻ 7 095 334 0124

Zickgraf Franz Josef
Landessternwarte
Koenigstuhl
D 69117 Heidelberg
Germany
☎ 49 622 150 9223
℻ 49 622 150 9202
✉ fzickgr@mail.lsw.uni-heidelberg.de

Zieba Stanislaw
Astronomical Observ
Krakow Jagiellonian Univ
Ul Orla 171
PL 30 244 Krakow
Poland
☎ 48 12 25 1294
℻ 48 12 25 1318

Ziegler Harald
HRZ
Univ Dortmund
Postfach 1621
D 44221 Dortmund
Germany
☎ 49 231 755 2039
℻ 49 231 755 2731
✉ h.ziegler@hrz.uni-dortmund.de

Zijlstra Albert
Physics Dpt
UMIST
Box 88
Manchester M60 1QD
UK
☎ 44 161 200 3925
℻ 44 161 200 3669
✉ aaz@iapetus.phy.umist.ac.uk

Zikides Michael C
Astronomy Dpt
Ntl Univ Athens
Panepistimiopolis
GR 157 84 Zografos
Greece
☎ 30 1 724 3414
℻ 30 1 723 5122

Zimmermann Helmut
Astrophysik Inst
Univ Sternwarte
Schillergaesschen 2
D 07745 Jena
Germany
☎ 49 364 163 0323
℻ 49 364 163 0417

Zinn Robert J
Astronomy Dpt
Yale Univ
Box 208101
New Haven CT 06520 8101
USA
☎ 1 203 436 3460
℻ 1 203 432 5048

Zinnecker Hans
Astrophysik Inst
Potsdam Univ
an d Sternwarte 16
D 14482 Potsdam
Germany
☎ 49 331 749 9347
℻ 49 331 749 9267
✉ hzinnecker@aip.de

Ziolkowski Janusz
Copernicus Astron Ctr
Polish Acad Sci
Ul Bartycka 18
PL 00 716 Warsaw
Poland
☎ 48 22 41 1086
℻ 48 22 41 0828

Ziolkowski Krzysztof
Space Res Ctr
Polish Acad Sci
Ul Ordona 21
PL 01 237 Warsaw
Poland
☎ 48 22 40 3766
℻ 48 22 36 8961

Zirin Harold
CALTECH
MS 264 33
Pasadena CA 91125
USA
☎ 1 818 356 3857
℻ 1 818 395 3814

Zirker Jack B
AFGL
NSO
Sunspot NM 88349
USA
☎ 1 505 434 1390
℻ 1 504 434 7029

Zitelli Valentina
Dpt Astronomia
Univ d Bologna
Via Zamboni 33
I 40126 Bologna
Italy
☎ 39 51 259 301
℻ 39 51 259 407

Ziurys Lucy Marie
Chemistry Dpt
Arizona State Univ
Tempe AZ 85287 1604
USA
☎ 1 602 965 7278
℻ 1 602 965 2747
✉ ziurys@asuchm.la.asu.edu

Ziznovsky Jozef
Astronomical Institute
Slovak Acad Sci
SK 059 60 Tatranska Lomni
Slovak Republic
☎ 421 969 96 7866
℻ 421 969 96 7656
✉ ziga@ta3.sk

Zlobec Paolo
OAT
Box Succ Trieste 5
Via Tiepolo 11
I 34131 Trieste
Italy
☎ 39 40 31 99255
✆ 39 40 30 9418

Zlotnik Elena Ya
Inst Applied Physics
Acad Sciences
Ulyanov Ul 46
603600 N Novgorod
Russia
☎ 7 831 236 3519
✆ 7 831 236 2081

Zola Stanislaw
Astronomical Observ
Krakow Jagiellonian Univ
Ul Orla 171
PL 30 244 Krakow
Poland
☎ 48 12 25 1294
✆ 48 12 25 1318
✉ zola@oa.uj.edu.pl

Zombeck Martin V
CfA
HCO/SAO
60 Garden St
Cambridge MA 02138 1516
USA
☎ 1 617 495 7227
✆ 1 617 495 7700
✉ mvz@cfa.harvard.edu

Zorec Juan
IAP
98bis bd Arago
F 75014 Paris
France
☎ 33 1 44 32 8121
✆ 33 1 44 32 8001

Zosimovich Irina D
Inst of History
Ukrainian Acad Science
Kirov Ul 4
252001 Kyiv
Ukraine
☎ 380 4429 0272

Zou Huicheng
Shanghai Observ
CAS
80 Nandan Rd
Shanghai 200030
China PR
☎ 86 21 6438 6191
✆ 86 21 6438 4618

Zou Yixin
Beijing Astronomical Obs
CAS
Beijing 100080
China PR
☎ 86 1 28 1261
✆ 86 1 256 10855

Zou Zhenlong
Beijing Astronomical Obs
CAS
Beijing 100080
China PR
☎ 86 1 28 1698
✆ 86 1 256 10855

Zsoldos Endre
Konkoly Observ
Thege U 13/17
Box 67
H 1525 Budapest
Hungary
☎ 36 1 375 4122
✆ 36 1 275 4668
✉ zsoldos@ogyalla.konkoly.hu

Zuccarello Francesca
Istt Astronomia
Citta Universitaria
Via A Doria 6
I 95125 Catania
Italy
☎ 39 95 733 21111
✆ 39 95 33 0592

Zuckerman Ben M
Dpt Astronomy/Phys
UCLA
Box 951562
Los Angeles CA 90025 1562
USA
☎ 1 310 825 9338
✆ 1 310 206 2096

Zuiderwijk Edwardus J
Royal Greenwich Obs
Madingley Rd
Cambridge CB3 0EZ
UK
☎ 44 12 23 374 868
✆ 44 12 23 374 700
✉ ejz@ast.cam.ac.uk

Zverko Juraj
Astronomical Institute
Slovak Acad Sci
SK 059 60 Tatranska Lomni
Slovak Republic
☎ 421 969 96 7866
✆ 421 969 96 7656
✉ zverko@ta3.sk

Zvolankova Judita
Astronomical Institute
Slovak Acad Sci
Dubravska 9
SK 842 28 Bratislava
Slovak Republic
☎ 421 7 37 5157
✆ 421 7 37 5157

Zwaan Cornelis
Sterrekundig Inst Utrecht
Box 80000
NL 3508 TA Utrecht
Netherlands
☎ 31 30 253 5223
✆ 31 30 253 1601
✉ zwaan@astro.uu.nl

Zwitter Tomaz
Astron Observ
Univ E Kardelj
Jadranska 19
Ljubljana
Slovenia
☎ 386 61 265 061
✆ 386 61 217 281
✉ zwitter@itssissa

Zylka Robert
RAIUB
Univ Bonn
auf d Huegel 69
D 53121 Bonn
Germany
☎ 49 228 525 376
✆ 49 228 525 229
✉ rzylka@mpifr-bonn.mpg.de

Batch number: 10615115

Printed by Printforce, the Netherlands